"Clay's is an essential part of the environmental health profession and this update is essential for an everchanging subject presenting new challenges and new legislation. Its contribution informs the profession and its practitioners and helps to raise the standards of health across the world."

– *Tim Deveaux, EH Consultant, UK*

Clay's Handbook of Environmental Health

Since its first publication in 1933, *Clay's Handbook of Environmental Health* (under its different names) has provided a definitive guide for the environmental health practitioner (EHP), and an essential reference for the consultant and student. This 22nd edition continues with its more recent successful structure, reviewing the core principles, techniques, competencies and skills required of an EHP, and then outlining the specialist subjects without getting too bogged down in a legalistic approach, seeking to broaden the content for a more global audience.

This new edition seeks to educate the EHP on the public health impacts of global heating and the climate emergency and also reflects the COVID-19 pandemic, as might be expected. Although seeking to have global appeal, the impact of the UK leaving the EU is also addressed. The book examines environmental health in different settings, including in the military, working in both conflict and natural disaster settings, and environmental health at sea and in airports. In line with previous editions, case studies are used to illustrate how EH problems have been resolved. This new edition includes information on key issues in public and environmental health including air pollution, contaminated land, housing and health, noise, water, food safety, pests and vector control, chemicals in the environment and radiation, as well as sustainability and public health and humanitarian crises.

This handbook aims to give a basic understanding of the philosophical basis of environmental health, as well as the required technical aspects and an understanding of environmental health in different settings. All chapters have sections on further reading and sources of information. *Clay's Handbook* is essential reading for all practitioners, students and researchers in environmental and public health wherever they are working.

Stephen Battersby is a chartered environmental health practitioner and has been an independent environmental health and housing consultant for 32 years. His first degree was in chemistry and applied zoology, before he trained as a public health inspector, qualifying in 1974. He was awarded his PhD from the University of Surrey in 2002. After working for a number of local authorities in England, he was employed by the environmental health professional body (now the CIEH) from 1980 to 1988. Since then, he has been associated with both Warwick and Surrey Universities. As a freelance consultant, he has also been an expert witness in housing litigation. At Warwick University, among other research projects he was part of the team that developed the Housing Health and Safety Rating System, the basis for assessing housing conditions under the Housing Act 2004. He has undertaken research (including on the public health implications of urban rat infestations) and lectured on environmental health at the University of Surrey. He has also been a visiting lecturer at King's College London and Kingston University. He also helped establish RHE Global as a resource for those working in housing. He remains a visiting senior research fellow in the Centre for Environment and Sustainability at the University of Surrey. He chaired the CIEH Council in 2005 and was President from 2008 to 2011 and is currently Vice President. He was awarded an MBE for services to environmental health in the 2014 Queen's Birthday Honours List. He is the series editor of Routledge's Focus on Environmental Health series and has also contributed to publications in the series. He has written and spoken extensively on environmental health and housing matters.

Clay's Handbook of Environmental Health

Twenty-Second Edition

Edited by
Stephen Battersby

Routledge
Taylor & Francis Group

LONDON AND NEW YORK

Twenty-second edition published 2023
by Routledge
4 Park Square, Milton Park, Abingdon, Oxon, OX14 4RN

and by Routledge
605 Third Avenue, New York, NY 10158

Routledge is an imprint of the Taylor & Francis Group, an informa business

First edition published by HK Lewis 1933
Twenty-first edition published by Routledge 2016

British Library Cataloguing-in-Publication Data
A catalogue record for this book is available from the British Library

Library of Congress Cataloging-in-Publication Data
Names: Clay, Henry Hurrell, 1882– author. | Battersby, Stephen, editor.
Title: Clay's handbook of environmental health / edited Stephen Battersby.
Other titles: Handbook of environmental health | Handbook of environmental health
Description: Twenty-second edition. | Milton Park, Abingdon, Oxon ; New York, NY :
    Routledge, 2022. | Includes bibliographical references and index.
Identifiers: LCCN 2021060875 (print) | LCCN 2021060876 (ebook) |
    ISBN 9780367476502 (hardback) | ISBN 9781032280738 (paperback) |
    ISBN 9781003035640 (ebook)
Subjects: LCSH: Environmental health—Great Britain—Handbooks, manuals, etc. |
    Environmental health—Great Britain—Administration—Handbooks, manuals, etc.
Classification: LCC RA566.5.G7 C53 2022 (print) | LCC RA566.5.G7 (ebook) |
    DDC 362.10941—dc23/eng/20211214
LC record available at https://lccn.loc.gov/2021060875
LC ebook record available at https://lccn.loc.gov/2021060876

ISBN: 978-0-367-47650-2 (hbk)
ISBN: 978-1-032-28073-8 (pbk)
ISBN: 978-1-003-03564-0 (ebk)

DOI: 10.1201/9781003035640

Typeset in Bembo and Stone Sans
by Apex CoVantage, LLC
Printed and bound by CPI Group (UK) Ltd, Croydon, CR0 4YY

Contents

Contents

Contributors

Darren Addison: Electromagnetic Fields Group, UK, Health Security Agency

Mutahir Ahmad: Operational Protection Department, UK Health Security Agency

John Ambrose: Retired, CIEH Port Health Ambassador and Port Health Trainer with 40 years' experience working in port health at major seaports

Dr David J. Baker: Former Consultant Medical Toxicologist, Health Protection Agency (Retired)

Dr Caroline Barratt: Senior Lecturer and Deputy Director of Education, School of Health and Human Sciences, University of Essex and co-founder of the UK Environmental Health Research Network

Dr Stephen Battersby MBE: Environmental Health and Housing Consultant (retired) and Visiting Senior Research Fellow, Centre for Environment & Sustainability, University of Surrey

Dr Tom Bond: Senior Lecturer, Department of Civil and Environmental Engineering, University of Surrey

Dr Naima Bradley: Honorary Associate Professor Centre for Environmental Health and Sustainability, University of Leicester

Roger Braithwaite: Consultant and trainer specialising in environmental protection and contaminated land. Past winner of the CIEH President's Award in recognition of his contribution to the advancement of environmental health

John Bryson: Environmental Health and Urban Renewal Consultant (retired)

Sian Buckley: Senior Lecturer at the University of the West of England, Bristol

Paul Charlson: Head of Planning and Regulatory Services at West Lancashire Borough Council

Amanda Cleary: Barrister and Lecturer in Law, School of Law, University of Surrey

Andrew Colthurst: Chartered Environmental Health Practitioner and noise consultant (retired)

Jeff Cooper: Editor of the ICE's *Waste and Resource Management* journal, past-President of CIWM and of the International Solid Waste Association (previously worked for the Environment Agency)

Dr Rob Couch: Senior Public Health Officer (Health Protection – COVID-19), Department of Public Health, Bedford Borough, Central Bedfordshire and Milton Keynes Councils and co-founder of the UK Environmental Health Research Network

Caryn L. Cox: Consultant in Health Protection, UK Health Security Agency and Public Health Wales NHS Trust

Sarah Daniels: Director and founder of the Red Cat Partnership

Dr Surindar Dhesi: Senior Lecturer in Environmental Health and Risk Management, School of Geography, Earth and Environmental Sciences, University of Birmingham and co-founder of the UK Environmental Health Research Network

Alec Dobney: Environmental Hazards and Emergencies Department, Centre for Radiation, Chemical and Environmental Hazards UKHSA

Prof. Raquel Duarte-Davidson: Head of the Chemicals and Environmental Effects Department, UK Health Security Agency

Dr Véronique Ezratty MD: Service des Etudes Médicales d'EDF, Paris, France

Lt Colonel James Fawcett MBE: Head of Army Environmental Health, Headquarters Army, Marlborough Lines

Steve Fisher: Technical Director WSP Acoustics

Prof. Timothy W Gant: Head of the Toxicology Department, UK Health Security Agency, Visiting Professor, Imperial College, London

Dr Emma Gillingham: Climate Change and Health, UK Health Security Agency, Chilton Oxford, UK

Tracy Gooding: Principal Radiation Protection Scientist, Radon Group Leader, UK Health Security Agency

Angela Hands: Chartered Environmental Health Practitioner and Public Health Practitioner

Jonathan Hayes: Risk Management Director, Moto Hospitality Ltd

Tony Lewis: Senior Lecturer at Middlesex University, London, and Training Programme Development Manager at RHE Global

Zena Lynch: Senior Lecturer in Environmental Health, School of Geography, Earth and Environmental Sciences, University of Birmingham

Dr Helen L. Macintyre: Climate Change and Health, UK Health Security Agency, Chilton, Oxford, UK

Prof. Rosalind Malcolm: Professor of Law and Director of the Environmental Regulatory Research Group, School of Law, University of Surrey

Dr Emma L. Marczylo: Centre for Radiation, Chemical and Environmental Hazards, UK Health Security Agency

Dr Andrew Mathieson: Senior Lecturer, Medical School, Australian National University Canberra, and Chair International

Environmental Health Faculty Forum and Honorary Editor IFEH

Wing Commander Gary Moyes: Officer Commanding Environmental and Occupational Health Wing, Royal Air Force Centre of Aviation Medicine

Prof. Virginia Murray: Head of Global Disaster Risk Reduction, UK Health Security Agency

Prof. John O'Hagan: Laser and Optical Radiation Dosimetry Group, UK Health Security Agency

Prof. David Ormandy: School of Law, University of Warwick

Dr Alan Page: Head of Department, Natural Sciences, and Associate Professor of Environmental and Public Health, School of Science and Technology, Middlesex University and co-founder of the UK Environmental Health Research Network

Roger Pearce: Senior Director, HSE and Security for global AST at STERIS Corporation

Dr Revati Phalkey: Climate Change and Health, UK Health Security Agency, Chilton, Oxford, UK

Iain Pocknell: Chartered Environmental Health Practitioner with over twenty-one years' experience in port health at airports and imported food controls

John Pointing: Barrister, specialising in environmental health law, and the Legal Partner of Statutory Nuisance Solutions

Dr Kathy Pond: Senior Lecturer, Department of Civil and Environmental Engineering, University of Surrey

Dr Jill Stewart: Senior Lecturer in Environmental Health and Housing, School of Science and Technology, Middlesex University and co-founder of the UK Environmental Health Research Network

Levente Szentkirályi: Assistant Teaching Professor with the Social Responsibility and Sustainability Division, in the Leeds School of Business at the University of Colorado at Boulder, USA

Dr Harriet Whiley: Associate Professor in Environmental Health, College of Science and Engineering, Flinders University, Adelaide, Australia

Stuart Wiggans: Senior Director, Operational Compliance, Asda Stores Ltd

Foreword

It is a considerable honour to be asked to write a foreword to a publication, and particularly one as esteemed as *Clay's Handbook of Environmental Health*. While considering what to say in my foreword to this, the 22nd edition, I looked back to see what the author of the foreword of the previous 21st edition had said. In his foreword, Tim Everett commented on emerging challenges, such as diabetes and climate change whilst noting that the old scourges of public health still held sway in many parts of the globe. He noted almost in passing the threat that new pathogenic strains of viruses might pose and commented that inequalities in health remained an issue that had to be addressed.

In the seven years since his foreword and this one, much of what he said has been proved to be correct. There can be no doubt, despite what those who protest would say, that climate change is real, and its catastrophic effects are being felt across the globe. Temperature records are broken on a regular basis. Wildfires, floods and droughts are a common occurrence. The damage that humans have done and continue to do to the planet is clear, and there is a very real concern that we may have left it too late to reverse it. It is indicative of the lack of environmental justice that while the developed world scrambles to reverse the damage that industrialisation and our modern lifestyle has done to the planet, populations in large parts of the world lack the basic amenities for life and the possibility of them levelling up seems as remote as it ever did. The 'Think Global, Act Local' mantra is one that environmental health practitioners know well. Our day-to-day activities contribute to helping to reduce the impact of climate change on individuals and communities, our lobbying and research work helps to inform the national and global debate and to drive change at government and inter-government levels.

Despite our access to modern technology, health surveillance and tracking systems and sophisticated modern medicine, the COVID-19 pandemic that started in early 2020 has demonstrated clearly that public and environmental health are as relevant and essential as they ever were. We have long argued that prevention is better than cure, but the truth of this was starkly demonstrated at the beginning of the pandemic when no effective vaccines were available and preventing the spread of the virus was the only defence we had at our disposal. It is sobering that the first line of defence against a disease that has killed over 150,000 people in the UK was, and remains, proper and effective handwashing, wearing a mask to prevent inhalation of viruses, social distancing and effective ventilation. All of these most basic of public health interventions were all we had to deal with a previously unknown virus that decimated populations. These are interventions that doubtless featured in the first edition of *Clay's Handbook of Environmental Health* over 80 years ago but remain as relevant today.

The new challenges of addressing climate change and the COVID-19 pandemic focused environmental health practitioners' attention but the traditional challenges have not gone away. Poor housing conditions continue to detrimentally impact on the health of their occupants, with the poorest and least able to advocate for themselves being most affected. Food safety and integrity remains a huge issue, with investigations showing food fraud becoming the crime of choice for professional criminals and being a multi-million-pound global business. The issue of allergens as a public health issue sits alongside food contamination as a potent risk to those vulnerable individuals exposed to them. Many of the traditional heavy industries have declined, but the move to a gig economy presents new health and safety challenges and the growth of remote and home working has highlighted mental health and well-being as a new health and safety challenge. Pest control remains as important as ever, as the species of pest change due to the warming climate, and the spread of traditional pests is facilitated by international travel and trade. The COVID-19 pandemic proves we lose sight of traditional solutions to public health issues at our peril, and the new edition of *Clay's* recognises this. The new sits alongside the old; both are of equal value to practitioners.

When I started out on my environmental health career 40 years ago, *Clay's Handbook of Environmental Health* was the go-to source of reference. As students, we referred to it on an almost daily basis, and it informed and underpinned much of my early practice. Our research methods were largely paper based, with further reliance on microfiche indexes, and were slow and very often frustrating to pursue. The rise of the internet where a huge

amount of information is just a click away has changed the way we search for information. Everything you need to know about whatever you are researching is accessible immediately. The internet, however, for all its convenience and almost limitless information, needs to be treated with caution. In addition to good, authoritative information, it harbours baseless ideas presented as fact and conspiracy theories dressed up as authoritative sources. Mark Twain is reputed to have said that a lie can be halfway round the world and back again while the truth is lacing up its boots. It is certainly true that a conspiracy theory promoted on the internet can undermine an evidence-based public health message, can make baseless or dangerous solutions and misguided or deluded individuals appear credible and can be very difficult to counter. In addition to our practical problem-solving skills, advocacy has always been one of the tools in the environmental health practitioners' toolbox; new ways of promoting public health messages though social media is a necessary refinement of that skill. In an age where faceless communication is becoming the norm our practice must embrace the change to ensure that our messages are clear and unequivocal and are heard and understood. The need to be the calm voice of reason and persuasion and where necessary of authority has never been greater.

The practice of environmental health has many new challenges in addition to the old ones. We need to address them all. The way in which we tackle them has new tools in addition to the old ones. They all have their place. This edition of *Clay's Handbook of Environmental Health* follows the long tradition of its predecessors in being a comprehensive and reliable exposition of environmental health theory and practice, and it continues to be of value to practitioners for that reason.

Julie Barratt
President, Chartered Institute of
Environmental Health
BSc LLB CEnvH Barrister at Law

Preface

This, the 22nd edition of *Clay's Handbook*, is the third I have edited and is also the last that I will be editing. There comes a time when it is right to step aside and let someone else take on the challenge of preparing this pillar of the environmental health world. Indeed, the time has perhaps come to consider the format so as to make the work more widely accessible.

Preparing this edition has been particularly difficult having been delayed by the 2020–21 COVID-19 pandemic. Indeed, since early 2020 many of the contributors have been working tirelessly in their various organisations, trying to protect public health during the pandemic and also supporting businesses to operate safely – there should not be any dichotomy between the economy and public health; one cannot thrive without the other. These colleagues continue to do so, having also, in the case of colleagues in Public Health England, had to endure (my word) reorganisation into the UK Health Security Agency (UKHSA) or the Office for Health Improvement and Disparities (OHID) (it seems "inequalities" is not politically acceptable). I am grateful to all contributors for their efforts.

Clay's Handbook of Environmental Health reflects the diversity of environmental health work not only in the British Isles but around the world. I have continued to try and make this textbook relevant to colleagues wherever they practise and most importantly to those embarking on their careers in environmental health (and related professions). My attempts to include an increasing number of contributors from countries beyond the British Isles has not been as successful as I hoped originally and I trust that future editors will be more successful because as we are seeing, so many environmental health issues are global. The COVID-19 pandemic has shown, and is illustrated in the book on the subject in the companion Focus on Environmental Health series, that EHPs around the world are playing an important role in protecting and improving public health whether they are working for government or commercial enterprises. It is just unfortunate, to put it mildly, that EHPs, certainly in England, were not involved in tackling the pandemic from the outset, but then as has been concluded by House of Commons Committees, a number of unnecessary mistakes were made.

As ever there are some new contributors to this edition and some who have contributed previously. Some chapters have been completely re-written while some are very much updates of what was written six years ago and there are some wholly new chapters given the UK's exit from the EU and the climate emergency. The approach is in part the consequence of time constraints caused by the pressures of the pandemic, but I hope that should not detract from the value of the work. Anyway, is some continuity not a good thing?

There are a number of chapters that have been written in sections, in some cases each section has different authors. Particularly in larger chapters and where there are different authors, the notes (indicated by a superscript number) are at the end of the relevant section in the hope that this will be more helpful to readers and easier to find. In most cases the list of references are included at the end of the chapter (except again in the largest or multi-author chapters). I hope readers find the organisation of the chapters helpful but any feedback to the publishers will, I am sure, help the next editor.

This book is intended as a first point of reference – a starting point. It is impossible in one volume to cover every aspect of environmental health in the minutest detail. Quite obviously the world has changed enormously since the first edition, when our predecessors' work was narrower in scope and there were very few other sources of information. This is one of the reasons for creating the "Focus on Environmental Health" series.

As is apparent from early in the book and becomes more evident looking at the later chapters, as a concept environmental health is both simple and complex. The chapters on the climate emergency and air pollution highlight this, but also the inter-connectedness of all the areas of environmental health work. The work of EHPs is crucial in facing the challenge of global heating effectively. We can only hope that the COP 26, held as this book is being completed, is successful. If not, while the planet itself might survive in some form or other, the human race might not.

Environmental health practitioners or environmental health officers are very much part of the public health workforce undertaking work from the prosaic to the exciting and ground-breaking, so as to tackle

threats to health. Yet too often EHPs have allowed themselves to be put into to silos (and in England subsumed into "regulatory services", and important as effective regulation is, this can also be very limiting). Such an approach will not help the profession nor more importantly the public. Firstly, regulations can often reflect political priorities which can be short-term, rather than matching public health needs in the longer term. Secondly this approach can then be conveniently reduced to a "tick box approach" that can be easily measured but requires little training or education. It certainly reduces the need for professional judgement which in turn may lead to greater problems or risks being overlooked or ignored with efforts (and resources) misdirected. Whatever else this book does, it provides information on which rational professional judgement can be made. It also provides information that can be used to help educate the lay person (including politicians) who again have failed too often in recent years to understand the value of properly educated and trained EHPs.

Of the three issues that occur throughout the book – Global Heating (the Climate Emergency), the COVID-19 pandemic and Brexit – two are global so it is hoped that this book has global relevance. However, it has to be acknowledged that the primary audience has always been in the UK, and so the impact of the UK leaving the European Union (and particularly leaving the Single Market and Customs Union) cannot be ignored. It has major implications for public health and as the justifiable furore over the discharge of raw sewage into rivers and coastal waters showed in October 2021, there is a risk of the UK becoming once more the "dirty man of Europe", with standards to protect public health reduced. Worse, the prevailing attitude at government level seems to be that anything "European" is to be avoided and the UK's closest neighbours antagonised – this cannot help in tackling the climate crisis nor future pandemics, and the SARS-CoV-2 pandemic will not be the last and note too that most emerging diseases are zoonotic.

What editing this book has shown me is that it is time for EHPs to break free from the shackles imposed by others and to assert themselves and advocate better not in their own cause but to demonstrate just how they can make even greater contributions to protecting and improving public health, and address health inequalities within and between countries (or disparities if one wants to use the term). It is telling that in the UK, EHPs would be equally at home in the UKHSA or OHID. The CIEH has argued for a Chief Environmental Health Officer at national level, and with the right person in post and with the right backing, that would make a lot of sense in either or both organisations. When one looks at the content of this book and the important areas it covers, one might ask why no such a post exists?

Stephen Battersby MBE
November 2021

1

Historical context, philosophy and principles of environmental health

Stephen Battersby

SECTION 1: A HISTORICAL PERSPECTIVE

Introduction

The history and development in Britain and Ireland of what we now call environmental health is long and complex, reflecting the changing nature of societies and differences between countries, from the localised challenges faced in the 19th century to those of global heating and the COVID-19 pandemic of 2020–21. Some of these challenges are new and some persist and in one form or another have always existed.

Even the term 'environmental health' is interpreted differently in different parts of the world and those professionals called 'environmental health officers' (EHOs) or 'environmental health practitioners' (EHPs) carry out different functions reflecting this. The term EHO and EHP are used interchangeably in this book, recognising that not all members of the profession are officers of a local authority or municipality.

As so-called Western or now high-income societies developed, so the stressors on human health increased despite economic growth and increasing levels of income. There were both environmental and health impacts from the moment that fire was discovered, and wherever on this planet humans gathered. Historically many of the concerns were with the effects of waste on health and as Mellanby [1] said, quoting archaeologist Jacquetta Hawkes,

> waste disposal by human beings may be said to have begun when hunters of the earliest stone age tossed their gnawed bones over their shoulders' to such an extent that caves became so reduced in size that they could no longer be occupied, and were certainly not "cave proud".

Until the 19th century systems of waste disposal, the sewerage and sewage treatment in the cities of Britain were very ineffective although through the centuries the great houses and monasteries may have had better organised (and healthier) systems for dealing with waste. In many towns refuse was allowed to rot in the streets, and there was no concern with contamination of water supplies. In some low-income countries municipal solid waste management remains a problem. According to the World Bank 90% of waste is openly dumped or burned in low-income countries, and it is the poor and most vulnerable who are disproportionately affected [2].

DOI: 10.1201/9781003035640-1

The history of what we in this book consider to be 'environmental health' has been closely allied to the social history of the UK since the 18th century. In the British Isles there were great economic social and environmental changes in the 18th and 19th centuries, not least of which was the first industrial revolution that started in Britain and spread to the rest of the world. Urbanisation and industrialisation brought millions of people together in crowded, sprawling and insanitary settlements. In response, society underwent a period of rapid change particularly motivated when the effects of the insanitary conditions spread to the middle and upper classes and the causes of cholera and typhoid were identified.

New forms of public services were introduced aimed at protecting public health and maintaining both a healthier workforce and reducing the strain on the Poor Law Commission (and later Board) and parishes, which are discussed later. The municipal corporations of the 1830s and the new Boards of Health were largely conceived to deal with outbreaks of communicable disease and to afford basic measures of health protection. Subsequently, a new concept of health arose: the idea of public health. New philosophies for the delivery of education and other public services began to take shape, and a new form of democracy evolved – extended suffrage, which only became universal in the early 20th century. The nature of a country's governance has an impact on public health.

In 1780 the electorate in England and Wales consisted of just 214,000 people – less than 3% of the total population. Change came gradually and often only as a response to campaigns for extended suffrage which included threats of violence and even revolution such as that which had occurred in France. The second Reform Act (1867) described by Lord Derby as *a leap in the dark* still only meant two in every five Englishmen had the vote in 1870. The third Reform Act enfranchised all male house owners in both urban and rural areas but only added 6 million people to the voting

registers and Britain was probably by then less democratic than many other countries in Europe. For many people, 19th-century parliamentary reform was a disappointment because political power was still left in the hands of the aristocracy and the middle classes. Universal suffrage, with voting rights for women (though not for those under 30), did not arrive in Britain until 1918.[1] Only in 1928 were women given the vote on the same terms as men when all adults over 21 could then vote. In a democracy the extent of the vote reflects political decisions and how threats to public health will be tackled. Why does this matter?

It can be argued that the wider the franchise the greater the chance that those elected to govern will respond positively to calls for measures to improve, or at least protect public health. It could be argued that even those in the UK who voted to leave the EU in the 2016 Referendum did so thinking it was a way to improve the health of communities. As Cicero wrote as a principle in *De Legibus*, *Salus populi suprema lex esto* (the welfare of the people shall be the supreme law). John Locke (1632–1704) the English philosopher and physician used it as an epigraph in his treatises[2] in which he said this was a fundamental rule for government.

Gostin [3] argued,

> while countries without democracy are capable of securing some level of public health, the level of commitment to public health goals is dictated by the government (which may place a greater value on remaining in power) instead of being determined by the people. The collective efforts of the body politic to protect and promote the population's health represent a central theoretical tenet of what we call public health ethics.

Considering the COVID-19 pandemic it is worth pointing out that Alan Maryon-Davis[3] [4] some years prior had said *denial and lack of information about Ebola caused hundreds of avoidable deaths in west Africa. From Spanish flu to AIDS to SARS, cover-ups and misinformation have fuelled epidemics.* A number of commentators have argued that in the UK mixed and inconsistent messaging (or attempts to transfer responsibility to individuals) and indecision have been the reasons why there have been so many COVID-19 cases in the UK (England in particular). Furthermore, some suggest that reliance on medical rather than public health expertise has led to an approach that is only concerned with preventing hospitals and medical practitioners from being overwhelmed rather than preventing illness in the population in general.[4]

Moving along from the influence of societal/political organisation; environmental health is about focussing on threats to public health including public mental health, from different stressors rather than concern for the environment alone. This is particularly true if we consider the 'environment' to be limited to what are interpreted as 'green' or 'conservation' issues (though these too can have a public health dimension). We should recognise that such distinctions mislead. As Einstein said, *the environment is everything that is not me.* The concern for environmental health practitioners is with the impact of stressors (that are the consequence of human activity) on human health rather than what might be called 'behavioural' or 'medical' aspects of public health. What this means is that EHOs or EHPs (whatever term is used) are very much part of the public health workforce in any country and should be seen as such.

Impacts are not always the same as issues, and too often whether an adverse health impact becomes an issue depends on the public response and perception. This is often linked to how and where it is reported in the media. Whether something is perceived as a public health issue can be a reflection of public or political pressure.

Although this book is not law-led (the law can often lag behind public opinion or understanding of risks) the development of environmental health as considered here can

be demonstrated by the development of the legal framework for 'environmental health'. However environmental health is more than just a subject that relies on legislation and the implementation of those laws for its existence. Environmental health could exist without any legislation, which provides only one means of addressing problems – law is only a tool and should not constrain or limit the work of environmental health practitioners.

That said, if EHOs (EHPs) are part of the public health workforce then environmental health law is an element of public health law and Gostin [3] suggested five essential characteristics of the latter. At the government level public health is a special responsibility in collaboration with partners in the community, business, the media and academia. Then at the population level public health focusses on the health of populations rather than the individual. There is then the relationships between the state and the population (or between the state and individuals who place themselves or the community at risk) that public health addresses. Public health then deals with the provision of population-based services that are underpinned on the scientific methodologies of public health such as epidemiology. Finally, public health authorities possess the power to regulate individuals and businesses to secure safeguarding or protection of the community – active intervention rather than reliance on 'voluntarism' or ethical voluntary action.

From the outset with 'inspectors of nuisances' environmental health was concerned with those factors easily identifiable as affecting human health, but as the interaction of the environment with human health became better understood the role of environmental health expanded. In some cases, this was also accidental as environmental health included a cadre of people able to deal with issues (and legislation) for which there was no obvious body to implement control measures – for example licensing of animal boarding establishments or other licensable activities involving animals. Earlier editions of this book sought to divide the evolution of environmental health into eras. The historical perspective follows from the 21st edition and provides some pointers and key issues in this evolution that have helped to define environmental health.

This assumes that development continues in one direction only – continuously positive – but there is always the possibility that the mistakes and tragedies of the past will be repeated, and not necessarily as farce, to misquote Karl Marx. For example, should the UK move to become a low regulation society while at the same time the economic impact of leaving the EU is negative, then there could be serious ramifications for environmental health.

The agricultural revolution

The period from 1700 to the early 19th century led to movement from the country into towns as improvements in husbandry and arable farming reduced the need for labour. The Enclosure Acts allowed those with the wealth to purchase public fields and push out small-scale farmers, causing a migration of men looking for wage labour in cities and urban areas.

Changes in agriculture did mean that although workers' diets improved this gave rise to only a slow decline in the rural death rate. Those who remained as farm workers had a low living standard often in damp and crowded housing.

Social changes and growth of towns made it increasingly difficult to provide a consistent and satisfactory food supply for those on low incomes who had moved into the urban areas. The increased productivity of the rural economy did not lead to improved food quality; indeed, the conditions were ripe for fraudulent substitution, and for food to be adulterated.

The industrial revolution

The industrial revolution did not follow directly on the agricultural revolution; they overlapped considerably, for example Abram

Darby first smelted local iron ore with coke made from Coalbrookdale coal in 1709, and in the coming decades Shropshire became a centre for industry due to the low price of fuel from local mines. The Iron Bridge over the River Severn used a major transport system, opened in 1781.

Waterpower was essential for increasing mechanisation. The Derwent Valley, upstream from Derby on the southern edge of the Pennines, contains a series of 18th and 19th century cotton mills and an industrial landscape of high historical and technological significance. It began with the construction of the Silk Mill in Derby in 1721. The scale, output and numbers of workers employed were without precedent. However, it was not until Richard Arkwright constructed a water-powered cotton spinning mill at Cromford in 1771, and a second, larger mill in 1776–77 that the 'Arkwright System' was established. In terms of industrial buildings, the Derwent Valley mills can be considered to be the first of what was to become the model for factories throughout the world in subsequent centuries.

Changes to the rural economy and the development of industry with more opportunity for paid work (often in small towns and villages in relatively rural areas) contributed further to the movement of people away from rural areas. The pace of industrialisation quickened with the development of the steam engine, which enabled more factories, and especially textile mills, to be set up in favourable locations, where there were ample supplies of coal and water. These developments were accompanied by adverse environmental impacts as the amount of waste increased, the environment was seen as a 'free good' and waste also became increasingly complex.

Machinery enabled production to be vastly increased with advantage taken of economies of scale. Increasingly large factories meant a need for more workers. These had to be housed nearby because, in the absence of public transport, they had to be able to walk to work. The need for houses near to the factories resulted in street after street of small, poorly constructed houses. There was little provision for drainage or refuse disposal; water supplies were inadequate and usually grossly polluted. To add to these unfavourable conditions, ever larger factories overshadowed the houses where the factory workers lived. They were polluted by smoke, grit and dust from both the factory furnaces and domestic fires emitting smoke at low levels, as chimney heights were only sufficient for combustion.

For many workers, factory practices involved working with dangerous machinery, and fatal or disabling accidents to workers were common. If the adult workers suffered from bad working conditions, children had it worse. Child labour was often considered an essential factor in production. Children started to work at an early age and, in the textile industry particularly, they formed part of a team cleaning under looms and spinning machines for example. Adults who could not take a child with them might find work hard to get and, where the shortage of children was acute, mill owners imported children from poor-law institutions as apprentices. The general social and economic conditions in the early 19th century have to be viewed against a background of, initially, an unreformed Parliament, an inadequate local government system and a political philosophy that permitted almost anything in the name of individual liberty.

While this was true of the industrial towns and cities, Liverpool was an exception as it was a mercantile city − the port from which many of the manufactured goods were exported or raw material such as cotton and sugar were imported. It was the rapid growth of the city with the construction by manual labour of the docks[5] that led to the squalid living conditions for the relatively unskilled labour that either built the docks or loaded and unloaded the ships.[6] This is discussed later with the sanitary movement.

The political and economic views current at the time meant there was no political will to ameliorate the conditions endured by most urban workers and their families. The French Revolution and the end of the Napoleonic Wars however did bring pressure for political change, although this was rarely successful. The social changes brought about by the agricultural and industrial revolutions did bring working people together, and along with their recognition of their exploitation by the political, factory- and farm-owning classes, led to growing pressure for change in employment practices and everyday life.

Not all those in the upper classes were immune to the plight of the working classes. Some politicians and philanthropists adopted causes with which they sympathised and fought for changes using their social and political connections. In the more urban areas, groups of people emerged who became interested in the welfare of their fellow city dwellers, and supported inquiries and investigations. Physicians, such as Thomas Perceval of Manchester and James Currie of Liverpool, visited slum areas and reported what they had found. Environmental disadvantage was not uniformly spread. Some districts were comparatively pleasant, even though few adequate sanitary amenities were available. Some more prosperous boroughs spent money to improve parts of their areas. In many instances these improvements were carried out by improvement commissioners appointed to implement environmental improvements in particular localities and specifically authorised by a private act of Parliament. The cost of the works was then being recouped by a 'rate' levied on the householders in that area. In 1795, a contagious disease swept through children working at a mill near Manchester, and the local justices of the peace took what action they could to prevent a recurrence. Early in the 19th century there was the first move towards improving the conditions of some juvenile workers. Sir Robert Peel, himself a mill owner, introduced a bill that became the Health and Morals of Apprentices Act 1802.

With industrialisation came increasing air pollution. Although the workers in those factories were earning a living they were also living in close proximity and were the worst affected by the pollution that grounded close by. Eventually the Alkali Inspectorate was established in January 1864 following Parliament's approval of the Alkali Act in 1863.[7] The Inspectorate's role was to control the release of damaging acid gas (hydrogen chloride) from alkali works. The legislation marked a change from the existing laissez faire approach to industry. By the 1850s, it was estimated that some 250,000 tons of salt were being decomposed annually in alkali works in the UK, resulting in the release of about 115,000 tons of acid gas. Legal proceedings against these works had been unsuccessful because it had been difficult to attribute damage from the gas to a particular work and to conclusively attribute any damage to acid gas. Serious lobbying began only when the landed gentry and wealthy landowners experienced reductions in their land values and extensive damage to their woodlands resulting from this air pollution and any action taken. It had little to do with the health effects on those living close to the factories. Nevertheless, the main approach was to have tall chimneys so that the emissions were dispersed and diluted.

Chartism

Chartism was a working-class national protest movement for political reform in Britain active from 1838 to 1858. It took its name from the People's Charter of 1838 with particular strongholds of support in the North of England, the east Midlands, the Potteries, the Black Country and South Wales. Support for the movement was greatest in 1839, 1842 and 1848 when petitions signed by millions of working people were presented to the House of Commons. The strategy adopted was to use the massive scale of support to put pressure on politicians to concede manhood suffrage.

The People's Charter called for six reforms to make the political system more democratic:

- A vote for every man 21 years of age;
- The Secret Ballot – to protect the elector in the exercise of his vote;

- No property qualification for Members of Parliament – thus enabling the constituencies to return the man of their choice, *be he rich or poor*;
- Payment of members, which would enable a trades person or others to serve a constituency and not lose out financially by having to give up from his business;
- Equal constituencies, securing the same amount of representation for the same number of electors, instead of allowing small constituencies to swamp the votes of large ones;
- Annual Parliament elections, as a means of stopping bribery and intimidation (a constituency might be bought once in seven years, but no one could buy a constituency under a system of universal suffrage every year). A Member elected annually would be less able betray their constituents.

Chartism can be interpreted as a continuation of the 18th-century fight against corruption and for democracy in an industrial society but attracted considerably more support than some more radical groups because of wage cuts and unemployment.

Trades unions and the Tolpuddle martyrs

Skilled workers in Britain had begun organising themselves into trade unions in the 17th century following on the model of the medieval guilds. During the 18th century, when the industrial revolution prompted a wave of new trade disputes, the government introduced measures to prevent collective action on the part of workers. The Combination Acts, passed in 1799 and 1800, during the Napoleonic wars, made any sort of strike action illegal – and workmen could receive up to three months' imprisonment or two months' hard labour if they broke these new laws. Unions in Britain were subject to severe repression until 1824, but despite this were widespread in cities such as London. After violent Luddite protests in 1811 and 1812,

Parliament repealed the Combination Acts in 1824 and 1825. Trade unions were legalised in 1824, where growing numbers of factory workers joined these associations in their efforts to achieve better wages and working conditions.

In the 1830s life in villages like Tolpuddle in Dorset was hard and deteriorating. Farm workers could not bear yet more cuts to their pay. Some fought back by smashing the new threshing machines, but this brought harsh punishments. In 1834, farm workers in west Dorset formed a trade union so that by combining together they could be more effective at pressurising employers to act reasonably. Unions were actually lawful and growing fast, but six leaders of the union were arrested and sentenced to seven years' transportation (to Australia) for taking an oath of secrecy. Their real crime in the eyes of the establishment was to have formed a trade union to protest about their meagre pay of six shillings (30p) a week.

The French Revolution and the Swing Rebellion were fresh in the minds of the British establishment. The latter started in 1830, in East Kent, when angry labourers destroyed a threshing machine. This was the start of a revolt that spread across the south. Incidents in Dorset villages probably involved farm workers who were later to join the Tolpuddle union, so landowners were determined to stamp out any form of organised protests.

Poor laws

From its beginnings in the 14th century, up to the inauguration of the National Health Service in 1948, England's poor laws evolved in such a way as to make significant impacts on the health of the population. It also is important in the context of the development of local government, as local authorities eventually replaced the 'Vestrys', the governing body of the parishes. The parish had been the basic unit of local government since at least the 14th century, although Parliament imposed few if any civic functions on parishes before the 16th century.

The history of the poor laws has two elements: the 'Old Poor Law' primarily in the 1601 Act for the Relief of the Poor, and the New Poor Law brought in by the Poor Law Amendment Act of 1834.

The Old Poor Law is characterised as being parish-centred, haphazardly implemented, locally enforced, and with, one of its more notorious aspects, the operation of workhouses being completely voluntary. The New Poor Law, based on a new administrative unit of the Poor Law Union, aimed to introduce a rigorously implemented, centrally enforced, standard system that was to be imposed on all and was centred on the workhouse. The underpinning idea was that life in the workhouse should be so bad that it would provide an incentive to do anything to avoid it.

The new system introduced in 1834 was very similar to that before except for the way in which poor law relief was administered. The grouping of parishes into unions, the deterrent workhouse and the workhouse test had all existed under the Old Poor Law. The New Poor Law saw a fundamental change in the way that the poor were viewed by many of their 'betters'. The traditional attitude had been one of poverty being inevitable (exemplified by the oft-quoted biblical text *For the poor always ye have with you*), the poor essentially were victims of their situation, and their relief a Christian duty. The 1834 Act was guided by a growing view that the poor were largely responsible for their own situation and which they could change if they chose to do so. The concept of the deserving and undeserving poor now existed (and still survives in some quarters).

The 1834 Act was welcomed by some thinking it would reduce the cost of looking after the poor, take beggars off the streets and encourage poor people to work hard to support themselves. The new Poor Law ensured that the poor were housed in workhouses, clothed and fed. Children who entered the workhouse would receive some schooling. In return for this care, all workhouse residents would have to work for several hours each day.

Not all Victorians supported this approach, and some spoke out against the new Poor Law, calling the workhouses 'Prisons for the Poor'. The poor themselves hated and feared the threat of the workhouse so much that there were riots in northern towns.

One important and complex piece of poor law legislation which had originated in 1662, and which did not finally disappear until 1948, was the Settlement Act (otherwise the Poor Relief Act of 1662). It gave a newcomer to a parish the right to a 'settlement' and the right to poor relief in any place where they had lived unchallenged for 40 days. During this period, following complaint to the churchwardens, that person could be ordered back to his place of last settlement, unless he was renting a property worth £10 a year. It may be difficult to believe, but parts of the 1601 Poor Law Act were not finally repealed until 1967. However, some of the same attitudes can be seen in more recent approaches by governments.

Under the Poor Laws the poor were divided into three groups: able-bodied adults, children and the old or non-able-bodied (impotent). Overseers were instructed to put the able-bodied to work, to give apprenticeships to poor children and to provide 'competent sums of money' to relieve the 'impotent'.

The Poor Law Commissioners

The increase in spending on poor relief in the late 18th and early 19th centuries, combined with the attacks on the Poor Laws by Thomas Malthus and other political economists as well as riots, led the government in 1832 to appoint the Royal Commission to investigate the Poor Laws. The Commission published its report, written by Nassau Senior and Edwin Chadwick, in March 1834. The report, described by the historian R. H. Tawney [5] as *brilliant, influential and wildly unhistorical*, called for sweeping reforms of the Poor Law, including the grouping of parishes into Poor Law Unions, the abolition of outdoor relief for the able-bodied and their families, and the appointment of a centralised Poor Law Commission to direct the administration of poor relief. It was soon after the report was published that Parliament adopted the Poor Law Amendment Act of 1834. The Act implemented some of the report's recommendations and left others, like the regulation of outdoor relief, to the three newly appointed Poor Law Commissioners. Edwin Chadwick had hoped to be one of them but was appointed Secretary to the Commission.

Relief expenditure was financed by a tax levied on all parishioners whose property value exceeded a minimum level. A rural parish's taxpayers could be divided into two groups, the labour-hiring farmers and the non-labour-hiring taxpayers such as family farmers (landed gentry), shopkeepers and artisans. In grain-producing areas, where there were large seasonal variations in the demand for labour, labour-hiring farmers were able to reduce costs by laying off unneeded workers during quiet seasons (during the winter and between sowing and harvest) and having them collect poor relief. Large farmers used their political power to tailor the administration of poor relief so as to lower their labour costs. In the early 19th century poor relief represented a subsidy to labour-hiring farmers rather than a transfer from farmers and other taxpayers to agricultural labourers. It seems the increase in relief spending in the late 18th and early 19th centuries was partly a result of politically dominant farmers taking advantage of the poor relief system to shift some of their labour costs onto other taxpayers [6]. Most rural parish vestries were dominated by labour-hiring farmers as a result of *the principle of weighting the right to vote according to the amount of property occupied.*

In livestock farming areas, where demand for labour was constant, it had not been in farmers' interests to shed labour during the winter, and the number of able-bodied labourers receiving casual relief was smaller. The Poor Law Amendment Act of 1834 reduced the political power of labour-hiring farmers.

Even today the social security and tax credit system in Britain could be seen as subsidising employers who pay low wages or flexible (zero-hours) contracts. As figures from the Department of Work and Pensions have shown[8] in 2017/18, 57% of all working-age adults on low income were in working families, and as of April 2019 HMRC figures showed that 2.26 million in-work families were claiming tax credits. Cash benefits have been shown to have the largest effect on reducing income inequality.[9]

The Economic History Association (USA)[10] has provided an insight in a paper by George Boyer of Cornell University [6] where he points out that what was interesting was the increase in the number of able-bodied males on relief during the late 17th and 18th centuries. In the second half of the 18th century, a large share of rural households in southern England suffered significant declines in real income. Farm-level data has shown real wages in the south-east declined by 13% from 1770–79 to 1800–09, and remained low until the 1820s [7]. Boyer suggests the effect of late 18th-century enclosures on agricultural labourers' living standards might have been overstated, although those labourers who had common rights must have been hurt by enclosures.

In some parts of the South and East, women and children were employed in wool spinning, lace making, straw plaiting and other cottage industries. Employment opportunities in wool spinning, the largest cottage industry, declined in the late 18th century, and employment in the other cottage industries declined in the early 19th century [6, 8–9]. The decline of cottage industry reduced the ability of women and children to contribute to household income. This, in combination with the decline in agricultural labourers' wage rates and, in some villages, the loss of common rights, caused many rural households' incomes in southern England to fall dangerously close to subsistence by 1795.

The situation was different in the north and midlands. The real wages of day labourers in agriculture remained roughly constant from 1770 to 1810, and then increased sharply, so that by the 1820s wages were about 50% higher than they were in 1770 [7]. Moreover, while some parts of the north and midlands experienced a decline in cottage industry, in Lancashire and the West Riding of Yorkshire the concentration of textile production and the industrial revolution led to increased employment opportunities for women and children.

A comparison of English poor relief with poor relief on the European continent reveals that from 1795 to 1834 relief expenditures per capita, and expenditure as a share of national product, were significantly higher in England than on the continent. However, differences in spending between England and the continent were relatively small before 1795 and after 1834 [10]. It has been argued that simple economic explanations cannot account for the different patterns of English and continental relief.

The Poor Law Commissioners were not popular with the parishes. The Commission had the power to issue directives but there was no way to make parishes do what the Commission wanted them to do. It did have powers over workhouse diets and could veto appointments to the Boards of Guardians. Chadwick had wanted the 1834 Act to be implemented in the North of England first where, at that time, there were fewer economic problems, there was high employment and food was plentiful.

It is interesting to read quotes from the Poor Law Commissioners Report of 1834.

> The Out-door Relief of the Able-bodied, when given in kind, consists rarely of food, rather less unfrequently of fuel, and still less unfrequently of clothes, particularly shoes; but its most usual form is that of relieving the applicants, either wholly or partially, from the expense of obtaining house-room.
>
> In-doors Relief, that which is given within the walls of the Poor-house, or

as it is usually, but very seldom, properly denominated the Workhouse, is also subject to great mal-administration.

But it will be seen that the process of dispauperizing the able-bodied is in its ultimate effects a process which elevates the condition of the great mass of society.

The principle adopted in the parish of Cookham, Berks, is thus stated:

As regards the able-bodied labourers who apply for relief, giving them hard work at low wages by the piece, and exacting more work at a lower price than is paid for any other labour in the parish. In short, to adopt the maxim of Mr. Whately, to let the labourer find that the parish is the hardest taskmaster and the worst paymaster he can find, and thus induce him to make his application to the parish his last and not his first resource.

By 1839 the vast majority of rural parishes had been grouped into Poor Law Unions, and most of these had built or were building workhouses. On the other hand, the Commission met with strong opposition when it attempted in 1837 to set up Poor Law Unions in the industrial north, and the implementation of the New Poor Law was delayed in several industrial cities. In an attempt to regulate the granting of relief to able-bodied males, the Commission, and its replacement in 1847, the Poor Law Board, issued several orders to selected Poor Law Unions. The Outdoor Labour Test Order of 1842, sent to unions without workhouses or where the workhouse test was deemed unenforceable, said that able-bodied males could be given outdoor relief only if they were set to work by the Union. The Outdoor Relief Prohibitory Order of 1844 prohibited outdoor relief for both able-bodied males and females except on account of sickness or 'sudden and urgent necessity'. The Outdoor Relief Regulation Order of 1852 extended the labour test for those receiving relief outside of workhouses.

The share of the population on relief fell sharply from 1871 to 1876, and then continued to decline, until 1914. Real per capita relief expenditure increased from 1876 to 1914, largely because the Poor Law provided increasing amounts of medical care for the poor. The role played by the Poor Law declined due to an increase in the availability of alternative sources of assistance. There was a sharp increase in the second half of the 19th century in philanthropy, increased membership of friendly societies (mutual help associations), the co-operative movement,[11] and of trade unions providing mutual insurance policies. The benefits provided workers and their families with some protection against income loss, and few who belonged to friendly societies or unions providing 'friendly' benefits ever needed to apply to the Poor Law for assistance.

Between 1906 and 1911 Parliament passed several pieces of social welfare legislation collectively known as the Liberal welfare reforms. These laws provided free meals and medical inspections (later treatment) for needy school children (1906, 1907, 1912) and weekly pensions for poor persons over age 70 (1908) and established national sickness and unemployment insurance (1911). The Liberal reforms purposely reduced the role played by poor relief and paved the way for the abolition of the Poor Law.

Sanitary movement

Towards the end of the 18th century the phrase *cleanliness is next to Godliness* came into common use and can be found in a sermon by John Wesley. Into the 19th century that concept seems to have been taken literally by Christian activists and the more socially progressive or liberal Christians seem to have made bathing one of their causes and allied themselves with the new soap companies. It was probably seen as a part of the Victorian idea of self-improvement and the notion of the deserving and undeserving poor but bathing and keeping clean was not easy for those in poverty and towards the bottom of society.

Parkinson [11] refers to the period that began in 1834 with the Poor Law Commission's Report and ended with the passing of the 1848 Public Health Act as being the period of 'the Public Health Agitation'. It was the start of the era of 'sanitary reform' that shaped the genesis and evolution of the EHO/EHP. The 'agitators' were influential individuals and voluntary movements that helped to change the prevailing attitude and approach to public health from 'laissez faire' to 'interventionist', and furthered the social and political acceptance of the institution and ideology of 'inspection'. Edwin Chadwick was the most celebrated individual within the movement (he was not knighted until 1889). Chadwick's best-known work, *The Sanitary Condition of the Labouring Population of Great Britain*, published in 1842, followed the failure to reform the Poor Laws. Due to political pressures, he had been instructed to drop his work. With a change of political control, he was required to complete it but again there was political interference and, on the grounds that it would cause offence, the government refused to publish it as a government report. Chadwick was allowed to print it under his own name as his personal view and at his own expense.

Chadwick had looked at the causes of poverty, and from his investigations into the living conditions of the working population in England and Wales he along with others such as Duncan in Liverpool, recognised the connection between poverty and ill health and an insanitary environment.

Chadwick was also a follower of Jeremy Bentham (he was his secretary for a time from 1830), the man who founded the theory of utilitarianism and leader of a group known as the Philosophical Radicals. The basis of this philosophy was that actions were right if they tended to produce *the greatest happiness for the greatest number of people*. Bentham argued that the idea of rights was 'nonsense', and the idea of natural rights was 'nonsense on stilts'.

Bentham wanted to abolish all legislation that did not bring about the *greatest happiness for the greatest number of people*. Bentham's ideas were applied to the 1834 Act but at this time 'people' were considered to be only those who had the vote. To give the *greatest happiness to the greatest number of people* meant cutting the poor rates. Consequently, the poor actually suffered, and his ideas of utilitarianism were called 'Brutilitarianism' by the working classes. Bentham was also a law reformer and his followers in stages brought about penal reform, extension of suffrage and an improvement in the sanitary conditions of the poor.

The most celebrated voluntary organisation was the Health of Towns Association. Indeed, Parkinson [11] has argued that *after 1844, agitation was mainly carried on through the medium of the Health of Towns Association* and the inaugural meeting of the Health of Towns Association (HTA) organised by Dr Thomas Southwood Smith, a friend and colleague of Chadwick's, took place in December 1844 at Exeter Hall, in the Strand, London. This venue was strongly associated with reform movements such as the Anti-Slavery Society, the Reformation Society, the Royal Humane Society and the Anti-Corn Law League, which all met there. This was an age of radicals, evangelists and reform. For more on the Health of Towns Association see [11].

Their efforts produced an improvement in the public health. One of the recommendations in *The Sanitary Condition of the Labouring Population of Great Britain* was that for the prevention of disease it would be 'good economy' to appoint a district medical officer of health independent of private practice. Chadwick was also of the view that it made more economic sense to have a healthier workforce that would be more productive. His thinking was not based on altruism or human rights.

As can be the case, and arguably a ruse to delay action, a Royal Commission was established to examine the problem. The Royal Commission on the Health of Towns was appointed in 1843 to investigate the sanitary arrangements of 50 English towns. Its findings confirmed the findings of Chadwick and it recognised and insisted that the state must take responsibility for public health and so resulted in the Public Health Act in 1848.

To consider the development of the sanitary movement in practice, it is useful to consider what happened in Liverpool was then reflected across the country. Liverpool as a mercantile port grew rapidly in the early part of the 18th century and into the 19th century. The growth of global trade involved construction of many new docks and buildings and the city population increased with unskilled labour as well as more prosperous business and merchant (including shipowning) classes. For example, many of the best-known British insurance companies were founded in the city and unlike many European countries that went the route of mutual insurers, the expansion of the UK insurance sector was privately funded, usually by local merchants and manufacturers. Unlike other industrial towns there was little or no skilled working class and no control over construction until 1842. Liverpool, with a large proportion of casual labourers, suffered even more from a low standard of living and insanitary housing than other major cities.

Unusually even for then, Liverpool had a large proportion of people living in cellars. Until the Liverpool Improvement Act 1842 there was no control over these cellar dwellings. The local Council however was active and wanted to improve conditions further. In 1845 the Mayor convened a meeting to establish a Health of Towns Association, and under the Liverpool Sanitary Act 1846 Dr William Duncan was appointed Liverpool's (and Britain's) first Medical Officer of Health (MOH). The Health of Towns Committee appointed Thomas Fresh as the Inspector of Nuisances in September 1844 [12] before Duncan's appointment. In 1847 he was appointed the Inspector of Nuisances under the 1846 Liverpool Act. Of all the officers in charge of separate departments it is said that the one most closely in touch with Dr Duncan was the Inspector of Nuisances [12].

Fresh not only acted under the Liverpool Sanitary Act 1847, but subsequently the Nuisances Removal and Diseases Prevention Act 1848 and the Nuisances Removal and Diseases Prevention Amendment Act 1849. Frazer [13] also records how there was daily communication and co-operation among the Town Clerk, MOH, Borough Engineer, Building Surveyor, Water Engineer, the Head Constable and the Inspector of Nuisances. The aim was to promote *unity of action in all departments under the Health Committee.*

The office of 'inspector of nuisances' and then 'sanitary inspector' featured nationally for the first time in the Town Improvement Clauses Act of 1847 and was subsequently confirmed by the more comprehensive, though still permissive, Public Health Act 1848 which also established the General Board of Health. The office was made compulsory, as was that of the MOH by the Public Health Act 1872 [14].

Hamlin [14] reports that by the end of the 1880s the inspector of nuisances was a fixture in every town, a member of a subordinate profession whose domain, skills and responsibilities were exactly demarcated. Towards the end of the 19th century inspectors were identifying and abating hundreds, even thousands of nuisances per year even in towns of modest size. There are however a number of parallels with the present situation. Crook [15] has said *inspection is widely recognised as a defining feature of the modern British state and the 19th Century is really where it started with the establishment of a range of inspectorates.*

Crook has also argued that

> it is surprising, in fact, that sanitary inspectors should have received relatively scant attention. Not only were they more numerous than central inspectors, they were also more engaged with members of the public, and at a point of intense sensitivity: private property.

By the start of the 20th century hundreds of thousands of sanitary inspections were taking place each year in Britain's towns and cities. Furthermore, it is recorded that other contemporaries were impressed by the powers they possessed.

Hamlin [14] also argues that in the domain of nuisances, inspection is important in a number of contexts. Firstly, it is the history of urban public health. Then, however the inspector of nuisances was the bottom layer of local public health administration. Too often historians encounter their work only through the filter of the MOH who digests mundane detection and remediation into columns in an annual report. The 1872 Public Health Act made inspectors underlings to medical officers, not least because the notion of health was based on the 'medical model'. While sanitary inspectors exercised professional discretion, they were not part of a professional class. In letters to complainants or offenders, inspectors did not discuss standards, agendas or the greater good. They merely set out what they had found, dictated remedies and set deadlines. By the 1970s it was argued that Public Health Inspectors should be known as environmental health officers and should not be subservient to MOHs, as they were a profession (by education, role and to some extent the ability to use discretion) and members of such a profession should not play second fiddle to medically qualified practitioners. This view took a lead in the World Health Organization's definition of health in the preamble to its constitution. Here health is *a state of complete physical, mental and social well-being and not merely the absence of disease or infirmity* – a much broader definition than the *medical model.*

Inspectors of nuisances also reflected the growth of a comprehensive and sometimes intrusive state in response to the growth of nuisances – or did the growth of the inspectorate increase the number of identified nuisances? Hamlin [14] argues that while town dwellers might wish to avoid the police, they did not avoid the inspector. By the 1880s many towns had instituted systematic inspections of dwellings. Nuisance inspection may

be seen as, in part, the creating and maintaining of bourgeois communities and providing a means of mediating local conflict. It was (and is now) a hybrid inspectorate belonging both to the local sphere and to the central state.

The sanitary movement, as has been pointed out [14], also brought 'nuisances' into the realm of policing and transferred it from an allegation that could be contested, and subject to civil action, to summary judgement and the criminal. It also represented the ascendancy of public nuisance, that is those affecting the population in general over private. Public nuisance as a Common Law tort and related to property rights was largely replaced by Statutory Nuisances as set out in Acts of Parliament and addressed defined circumstances.[12]

Crook [15] has also said that sanitary inspection was a means of surveillance and a means of intervention. That said, sanitary inspection is better understood as a mode of liberal surveillance and as part of a liberal culture of governance. That is, it is a broad culture of governance committed to both actively securing a moralised, pluralistic and healthy society and respecting society's capacity to govern itself, free of intervention. Liberal governance must also have inherent tensions and antagonisms, such as those between state and society, local and central. In the modern context, this can be seen in the tension between having control of certain activities and also a desire to remove 'red-tape' (deregulatory environment). This has perhaps led to the so-called 'better regulation' agenda in England. It has been made apparent during the COVID-19 pandemic that many (usually on the political extremes) are opposed to state intervention even when it comes to protecting public health, whether 'lockdowns', the mandatory wearing of masks or vaccination programmes. Crook suggests that these antagonisms are internal to the practice of liberal governance, which does not resolve or overcome these tensions and antagonisms, but rather keeps them in play, as a recurrent form of political struggle and critique. Public diplomacy was crucial to the work of the inspectors of nuisances and sanitary inspectors. The public had to be worked with (the educational role) and through, sometimes making the job relatively easy, at other times slow and confrontational. This public dimension it is argued made sanitary inspection a political and an ethical art [15].

This is again illustrated when coming to the present time and considering the COVID-19 pandemic, when in the UK after years of criticism of the 'nanny state' and policies of deregulation, there had to be a concerted educational campaign on how to wash hands correctly. Reluctance to follow government advice during the pandemic may reflect cynicism of government at all levels as the result of previous experience of the state which has increased inequality and exacerbated disadvantage or used language that increases social tensions and divisions. Why should one follow government advice when that government has previously made one's life harder? Requiring action to prevent the spread of the SARS-CoV-2 virus such as mandating a 'COVID Vaccination Pass' or ('Pass Sanitaire') in closed spaces is also seen by some as an infringement of personal liberty. Yet this fails to recognise that others equally have a right to protection from infection (or at least risks reduced).

In the 19th century public health as a matter of intervention, was a reflection of an interventionist state even when economically and politically it had a *laissez faire* approach. That state (at the local and national level) was also publicly accountable, and the tension meant the job of the inspector of nuisances or sanitary inspector could be fraught when employed by a local authority, the elected members of which could have vested interests.

A good example of inspector-as-social-worker was the issue of 'overcrowding' – or too many people living in cramped and crowded conditions, as the term 'overcrowding' implies that some 'crowding' is acceptable. Originally

it had been a matter for medieval courts, more for economic than health-related reasons, as refugees and beggars would lower local wages and demand alms. Lodging-house inspection had been an important part of the early sanitary law. Although ostensibly the prime concern was disease, there was also political motivation (Chartist subversives could be found) and moral (against promiscuous bed-sharing). The 1855 Nuisances Removal Act allowed action against overcrowding (and gave rise to much discussion of the minimum volume for healthy living). Hamlin [14] reports that the figure settled on in the Sanitary Act of 1866 was 300 cubic feet per person, with children counting as half persons and cows as five person-equivalents. Interestingly, Dr Duncan had referred to the Inspectors of Prisons who had recommended not less that 1000 cubic feet for every prisoner, that is 28.3 m^3. For comparison a room of 6.5 m^2 with an average ceiling height of 2.4 m would have 15.6 m^3 (551 cubic feet). It was 'overcrowding' that led to night inspections, which caused considerable problems. Many also saw concentration of bodies and breath as the most proximate cause of typhus. Crowding remains an issue and again the COVID-19 pandemic highlights this, but before the pandemic the WHO[13] had reported that studies were consistent in showing that crowding is associated with increased risks of TB and it seems highly plausible that this will also be true for COVID-19, and indeed there is some evidence to support this[14] (see also Chapter 12).

There were moral concerns in the Victorian age, but many insisted that morals were not the business of sanitary inspection. Inspectors faced overcrowding because of public outrage at this common problem, but it was not an issue they could solve at the individual level. For some in authority when individual cases were used to highlight the wider issue, the view was taken that it was failure of the Nuisance Inspector to exercise their duty properly. Reference has been made [14] to how an inspector of nuisance might often identify overcrowding but rarely acted. It was obvious to them that chasing persons into the street did not increase housing stock or augment the ability to afford more space. Those evicted from one place would then crowd another, so it was a continuing dilemma. The same dilemma exists today. While it might be difficult to argue for a better solution at the local level (rather than treating a symptom of the underlying problem) one justification for any professional association to exist is that by using evidence from the local inspector (practising EHPs) it is possible to formulate arguments for change at the national level.

As well as and overlapping with complaint-based regulation enforcement by inspectors, they also had to inspect problematic industries such as public slaughtering and other offensive trades, bakeries and food markets. They had to enforce other legislative initiatives, like child labour laws, wharfs and canal boat-dwellers, which occupied other inspectors. Such programmes of inspection involved their own paperwork, *by the end of century one senses that, for many inspectors, form, particularly the filling out of forms, has overtaken substance* [14].

When it came to professional development, Inspectors of Nuisances (and Sanitary Inspectors) were often drawn from trades (which in itself could cause conflict) and indeed some of the elected members of the local authorities could have vested interests as they would be drawn from both business and landlords. The pecuniary interests of elected councillors could be at odds with public health. Councils and the Local Government Board could limit the actions of inspectors by requiring enforcement of by-laws and statutes, which then limited discretionary freedom or judgement (professionalism) and could overwhelm them with work. Endless filling up of returns drove out flexibility and bureaucratic transparency displaced discretion [14].

More subtle, but perhaps more profound according to Hamlin [14], was the introduction of form (standard) letters for the conduct of routine business. By the 1870s, Knight and Company, publishers for the Local Government Board, had cornered the market both for commentaries on local government law with towns buying multiple copies to educate their staff and form letters for particular purposes. These included for all types of nuisances and might be customised but often cited the act or by-law authorising the inspector's involvement. One needed only to fill in name, address, nuisance and deadline. Earlier, enforcement had often been seen as a delicate dance, involving power, deference, cajoling and negotiation. It could be said inspectors 'wait on' landowners whose consent they seek. Form letters (and busy work) made officials officious. Interestingly in modern times officers often want standard forms or *pro-formas* even when there are no 'prescribed forms'.

Hamlin [14] refers to the professionalisation of sanitary inspectors but still with subordination to medical officers. This expansion in the number of sanitary inspectors was accompanied by a growing sense of professional identity and collective interest. The first national gathering of sanitary inspectors was held in 1876 as part of the inaugural congress of the Sanitary Institute of Great Britain. As career paths became national rather than local, inspectors began their own professional organisation, the Association of Public Sanitary Inspectors in 1883.[15] Hamlin argues the enterprise flourished only through acknowledgement of its subordinate status to the newly professionalising medical officers of the British Institute of Public Health (eligible only to those with both medical and public health qualifications) who had overall control. Yet they also encouraged and supported the inspectors too.

Crook [15] has suggested that sanitary inspection was only in part about the bureaucratic application of scientific reasoning. In practice, the situation was more nuanced and complex and the inspectors' worked taking account of science, elements of probability, interpersonal relationships and civility (in effect what seemed decent). As such, sanitary inspections turned upon a rationality that was flexible and could vary with particular circumstances. Its nature was partly a product of the constraints intrinsic to practice, but it was also a product of the fact that sanitary inspection was based on the notion of what he calls the *liberal culture of governance*. Central to this was the maintenance of a civilised, respectable public sphere, free of sources of disgust and disquiet and committed to the integrity and enjoyment of property – in truth a reflection of the value system and norms of the time. It is suggested that in the role of an inspector, while science played a part, the task involved a commitment to empiricism. This is why it can be argued that sanitary inspection was both a political and ethical task and the work of the EHP remains so. Interestingly he suggests that sanitary inspectors were encouraged even then to develop a 'habit of close observation' in the manner of the natural scientist. The Medical Officer of Health for Birkenhead [16] urged inspectors to simulate a meticulous attention to detail, the sort of which had informed the work of Lyell, Darwin and Pasteur. Any given inspection involved a thoroughgoing examination of any potential 'nuisance'. In the case of private houses, it was said that every *facet of design should be carefully scrutinized, including pipes, joints, traps and chimneys, and even features like wallpaper.* This empiricism required the use of all the senses in carrying out an inspection, and this remains true today. Empirical observation and publication of findings can be the basis of research, something that is intrinsic to a 'profession', and yet something that even today, with inspectors who are ever better educated can be problematic to achieve. It can be difficult to get practitioners to publish papers or reports on their work and findings from empirical observation. This is evidenced by the apparent reluctance of EHPs to contribute to the Routledge Focus on Environmental Health series,[16] a vehicle that exists, at least in part, to promote their work.

The main function of the meetings of the Association of Public Sanitary Inspectors appears to have been to reflect on precisely the ways in which this independence was thwarted in practice. Although other matters such as pay and pensions, to qualifications and tenure were part of the deliberations. Inspectors had to negotiate a number of tensions. On the one hand, they were empowered in the name of the public: they were public servants charged with promoting public well-being. On the other hand, they were accountable to the public and its representatives, both locally and nationally. In essence inspectors inspected the public and in turn the inspectors were accountable to the public and its representatives on the Council and in Parliament. As now, there were also complaints about rulings in the Courts and how the interpretation of the law by lawyers which could limit the inspectors' ability to deal with problems that may have been highlighted by the public.

The birth of epidemiology and a more scientific approach

Cholera first reached Britain from continental Europe in October 1831 and during the subsequent year resulted in over 30,000 deaths. This had a major impact and caused concerns to society not least because it could affect those other than the poor. It caused considerable consternation to the authorities as to what were the causes. It is often thought Dr John Snow solved the problem and frequent mention is made of his removal of the Broad Street pump handle. Snow's 1855 book, *On The Mode of Communication of Cholera*, is held up as a demonstration of 'the epidemiological imagination'. Snow's contribution, and the context in which he worked is covered in detail in a paper by Smith [17]. As is usually the case with any discovery the person making the discovery has built on the work of others and also to counter alternative theories. In this case it was the 'miasma theory', which held that transmission was airborne and associated with bad smells. As Smith [17] points out, Snow's work *appeared amidst a veritable spate of speculation, experiment, investigation and recommendations regarding* cholera, and some of these less celebrated (at least now) contributions remain instructive. Dr John Sutherland's report for the General Board of Health on the 1848–1849 British cholera epidemic included his investigation of the effect of water source on cholera risk in Salford, Manchester. Smith [17] says that Sutherland found a strong association between household water supply and the occurrence of cholera. Diarrhoea showed a similar, but less marked, association with water supply. Sutherland concluded that with respect to water obtained from wells into which the contents of sewers or privies had escaped, the predisposition occasioned by the continued use of such water is perhaps the most fatal of all. Duncan in Liverpool had also taken an epidemiological approach in that city that had some of the worst conditions.

In his work Snow demonstrated an attack rate many times higher in those receiving water from the Southwark and Vauxhall company than those receiving it from the Lambeth Company. Others pointed out that those parts of London which had high overall mortality in the years before the cholera epidemic tended to have high death rates from cholera during the epidemic and attributed this to environmental factors which increased the risk of both cholera and other causes of death. By tracking the cases Snow was also able to identify a focus of infection.

When considering the cholera outbreak in Golden Square, Broad Street, London 1854, the Reverend Henry Whitehead who tracked the cases over time, formed the view that when Snow removed the pump handle the epidemic was already waning and he argued that appeared to have had no effect, although he did concede that the closure of the pump may have prevented recurrence of the epidemic. Snow accepted that the mortality rate was diminished. He had not been able to identify the causative organism via microscopic examination or any chemical pollutant so was not able to prove the water's danger conclusively. However, he indicated that the reduction in cases might have been

because much of the population moved away, which started soon after the outbreak. He conceded that it was impossible to decide whether the well still contained the cholera poison in an active state, or whether for whatever reason the water had become free from the cause.

Something that helped Snow and his investigation were several unexplained deaths from cholera that did not at first appear to be linked to the Broad Street pump water. In particular a widow living in Hampstead had died of cholera in early September, and her niece, who lived in Islington, had succumbed with the same symptoms the following day. Neither of these women had been near Broad Street for a long time. Snow was intrigued so visited and interviewed the widow's son. He discovered from him that she had once lived in Broad Street and liked the taste of the water from the pump in Broad Street so much that she sent her servant down every day to bring back a large bottle of it for her by cart. The last bottle of water, which her niece had also drunk from, had been brought from Broad Street in late August at the very start of the epidemic.

It was discovered later that the public well serving the pump had been dug only 0.9 m from an old cesspit that had begun to leak faecal bacteria. A baby who had contracted cholera from another source had its nappies washed in this cesspit, the opening of which was under a nearby house.[17]

Smith [17] refers also to Snow's criticism of Chadwick's policy of flushing the sewers and draining cesspools. Snow maintained that this increased the contamination of drinking water. Chadwick defended his action, on the basis that a small increase in the pollution of the Thames was better than allowing the sewers to continue giving off the pestilent exhalations into the atmosphere as he was a supporter of the miasma theory.

It was also common to blame not only poverty but drunkenness for cholera. Snow attributed the lack of cholera in brewery workers near Broad Street to the fact that they drank only beer. The brewery actually had a different water source.

The work of Duncan in Liverpool should not be overlooked when considering the development of the epidemiology. This is discussed in the first publication of the Routledge Focus on Environmental Health series (see Note 5).

In mid-19th century Britain, cholera was seen to be a disease of the poor. This attitude was shared by those who thought cholera caused by a specific agent, and those who saw it as a consequence of more general environmental causes. Smith [17] says that it was also common to think that that cholera in poor communities represented a threat to the health of the wealthier classes. The thought was that the disease could spread from epicentres in poor areas to the better-off areas and led to a fear of the poor, and to calls that something must be done to improve their circumstances.

Interestingly Smith [17] refers to these concerns having some similarities with current thinking and the role of income inequality in health. He highlights the work of Wilkinson [18] who has argued that higher levels of inequality within societies are not just associated with worse health amongst the poor, but with overall worse health, affecting the poor and rich alike in highly unequal societies.

In this instance it was argued that where the needs of the poor were ignored this translated into worse health for the rich for psychosocial reasons. In the Victorian era there were more material reasons for these concerns. Yet the descriptions of miasmas from the 19th century share some similarities with modern accounts of adverse psychosocial environments. Smith [17] gives an example of how unfairness to one's fellows could be rewarded by ill health reported by Snow. He documented the case of a landlord whose tenants complained that drainage from cesspools was entering their water supply. The landlord's agent maintained there was no problem, but the tenants still complained. So, the landlord went himself, and said he could not see anything wrong and drank a glass of

it. This was a Wednesday, on returning home he was taken ill with cholera and died the following Saturday. There was no cholera in the vicinity of where the landlord lived.

Crook [15] has also argued that the greater significance assumed by epidemiological science and detailed legal norms led in time to the formal provision of vocational qualifications for sanitary inspectors. Initially drawn from the ranks of tradesmen, police officers and even soldiers, by the late Victorian period there was a growing consensus that inspectors required technical instruction. The first exams were carried out in 1877 under the auspices of the Sanitary Institute of Great Britain. Only those appointed in London were required to have any certification. In 1899 a Local Government Board approved the formation of the Sanitary Inspectors' Examination Board to regulate the provision of training. In order for a candidate to pass they were required to have a basic grasp of the various facets of 'sanitary science'. Most exams covered four topics: elementary physics and chemistry; elementary statistical methods; practices of municipal hygiene [15].

Sanitary inspection was seen as an observational, fact-based exercise incorporating a scientific ethos. Its relation to the science of epidemiology and evolving theories of disease transmission was more complex. However, there were many professional journals. Crook [15] refers to *The Sanitary Record*, *Public Health* and the *Journal of State Medicine* to ensure that sanitary inspectors could be informed of all the latest innovations in the theory and practice of disease prevention. Then as now, it was seen as important for practitioners to be well-informed and up to date. This is what became known as continuing professional development, and indeed any professional including EHPs should undertake sufficient research or study to ensure they are properly informed so as to be able to make valid judgements and assessments. Something that for example was referred to in the original 2006 Housing Health and Safety Rating System Operating Guidance.[18]

As we will see in the next chapter when considering the attributes of an environmental health practitioner or officer (and subsequent chapters), it is apparent that they are part of the public health workforce. It is generally accepted that the three pillars of public health are:

- Protection;
- Prevention of disease;
- Promotion of well-being;

and through this book the role of EHPs in each pillar will become apparent.

Public health actions under these pillars should be based upon evidence-based knowledge and enabled by good governance, advocacy and the capacity to ensure fair, secure and sustainable health and well-being for all and public health is the art and science of organising collective efforts to achieve this [19].

This section, while taking a historical perspective has to some extent also conjured up different images of 'public health' (some not necessarily coinciding with the modern image of environmental health practitioners) and this will be of further relevance in the remainder of the chapter (and this publication). These images are well summarised by Geoff Rayner and Tim Lang [20] as:

1 People constructing knowledge-based interventions
2 Public health as state intervention – uniformed officers
3 Public health as responding to pandemic disease
4 Public health as rescue, saving individual lives
5 Focussing on individuals shaped by wider determinants
6 Health as the operation of systems
7 Public health as interactions and transitions
8 Public health as social movements and causes
9 Public health as nanny or Big Brother

In addition, it can be argued that environmental health is about preventative (upstream) interventions to reduce the need for greater medical action downstream whether to treat an individual or population.

The fact that EHPs are part of the public health workforce can be easily forgotten, despite the history. This is particularly true when public health responsibilities are split between a range of agencies at different levels of government. This has been further illustrated in the UK during the COVID-19 pandemic when it took some time for EHOs to be deployed effectively and best use made of their skills and knowledge.[19]

Public and environmental health has changed many times since the Second World War and although the many Public Health Inspectors and environmental health officers in the UK were glad when no longer directly under the control of the Medical Officer of Health, one must wonder whether the changes since 1974 and into the 21st century have been wholly beneficial. As Gorsky [21] said back in 2007 the NHS is about curative rather than preventive medicine and the *economic constraints on local authority health service expansion limited their room for manoeuvre* while the monitoring of local population health remained essential. The question is, have the structures and funding in England and other countries allowed this? Although outside the scope of this book, it is also right to question continually whether in any country the legal and administrative structures and funding are appropriate for protecting public and environmental health and preventing ill health and unintentional injuries.

Legislative history of environmental health

One way to consider or to visualise the historical development of environmental health is to consider key pieces of legislation to address some of the social ills that arose in the 18th and 19th centuries and continued from then. See Table 1.1. Table 1.1 sets out

Table 1.1 A legislative history of environmental health in England and Wales

Year	Legislation	Purpose
1846	Nuisances Removal and Diseases Prevention Act	Temporary legislation to help address the spread of cholera, setting out procedures for the removal of 'nuisances'. The Privy Council given power to make regulations for the prevention of infectious disease
1848	Public Health Act	Provisions of a sanitary code and administrative machine to give it effect, the General Board of Health, plus a permissive power to establish local boards of health (local boards) who would have responsibility for all sewers
1848	Nuisances Removal and Diseases Prevention Act	A further response to the threat of cholera and providing for the speedier removal of certain nuisances and the 'revention of contagious and epidemic diseases'
1851	Common Lodging Houses Act	Control over common lodging houses used by the poor where many families shared
1851	Labouring Classes Lodging Act	Gave local boards (local authorities) powers to establish lodging houses for the homeless
1855	Nuisances Removal Act	For the first time included 'statutory nuisances' such as 'premises in such a state as to be a nuisance or injurious to health' and a number of others still existing.
1858	Local Government Act	Made changes to the procedure for constituting a local board and gave them some additional powers. There was also a change in nomenclature: the authorities created by the 1858 Act were simply titled 'Local Boards' and their areas as 'Local Government Districts'
1860	Adulteration of Food Act	Limited responses to the increasing problem of food adulteration; allowed for the appointment of public analysts but no sampling powers
1863	Alkali Act	Created the Alkali Inspectorate to control emissions to air (noxious vapours) from certain processes
1866	Sanitary Act	Local authorities to supply water and sewerage system and also to control smoke
1866	Labouring Classes Dwellings Act	Empowered Public Works Loan Commissioners to lend money to councils for the provision of housing
1867	Factory Act	This brought any workplace employing more than 50 persons in manufacture under the factory acts
1867	Workshop Regulation Act	Workplaces with less than 50 persons were prevented from employing children under 8 years old. Children aged 8 to 13 were restricted to half time working
1868	Artisans and Labourers' Dwellings Improvement Act (Torrens Act)	Provided for the demolition or improvement of unsanitary dwellings
1871	Factories Act	A reaction to the failure of local authorities to enforce the law and creating duties to enforce on the inspectors of factories
1872	Public Health Act	Local government board made a series of orders prescribing qualifications, appointment, salaries, and tenure of office of medical officers of health and inspectors of nuisances and also created sanitary authorities. Local government board could approve the establishment of Port Sanitary Authorities

Table 1.1 (continued)

Year	Legislation	Purpose
1872	Adulteration of Food, Drink, and Drugs Act	Made it an offence to sell food, drink or drugs that were not of the 'nature, substance or quality' demanded by the purchaser
1875	Sale of Food and Drugs Act	Extended and amended the 1872 Act
1875	Public Health Act	Gave wide ranging powers to local authorities (created urban and rural sanitary authorities), from dealing with unfit food, to sewers, from statutory nuisances, to building by-laws
1875	Artisans and Labourers' Dwellings Improvement Act (Cross Act)	Allowed councils to deal with unhealthy houses by buying the land and buildings for the purpose of improvement. Councils were allowed to build houses. Named after Richard Cross, Home Secretary
1879	Artisans and Labourers Dwellings Improvement Act	Changes that enabled local authorities to deal with individual insanitary houses. Useful powers, but they did not permit authorities to deal with areas of bad housing
1882	Artisans Dwellings Act	Extended provisions in earlier Acts to all urban sanitary districts and insofar as it allowed demolition of obstructive buildings when dealing with unhealthy or unfit housing
1882	Municipal Corporations Act	Created a modernised form of municipal borough
1885	Housing of the Working Classes Act	Required local authorities to use the powers they already had with regard to insanitary housing and amended the Artisans and Labourers Dwellings Improvement Acts and Artisans Dwellings Acts
1886	Shop Hours Act	Limited the working hours of young people to 74 per week including meal times.
1888	Local Government Act	Established county councils and county boroughs and also created the London County Council and metropolitan boroughs replaced a mixture of local authorities
1889	Infectious Diseases Notification Act	Allowed local authorities to adopt powers to require notification
1890	Housing of the Working Classes Act	A consolidation Act but which was also the first that based rehabilitation legislation with provisions for gradual renewal and environmental improvements
1891	Factory Act	Local authorities made responsible for sanitary conveniences in factories, and for cleanliness, ventilation, overcrowding and lime washing in workshops
1894	Local Government Act	Urban District and Rural District Councils replaced urban and rural sanitary districts
1892	Shops Act	To control the hours, young persons can be employed in shops and warehouse
1899	Infectious Diseases Act (Extension) Act	Required all sanitary authorities to adopt the 1889 Act (it had been a power not a duty)
1901	Factory and Workshops Act	Consolidation and extension of previous provisions
1909	Housing and Town Planning Act	Provided permissive powers to control land use
1912	Shops Act	Extension of protection to shop workers

(Continued on next page)

Table 1.1 (continued)

Year	Legislation	Purpose
1919	Housing and Town Planning Act (Addison Act)	Followed the 1917 Tudor Walters Committee Report and stimulated house building with subsidies to local authorities to build
1921	Public Health (Officers) Act	Set conditions of office and tenure of medical officers and sanitary inspectors and required that their dismissal could only be with the consent of the minister, and it also required that the term 'inspector of nuisances' be replaced with the designation 'sanitary inspector'
1924	Housing Act (Wheatley Act)	Consolidation of Acts to stimulate house building and to deal with repair, maintenance and sanitary condition of housing, including subsidy for the clearance of unhealthy housing
1925	Public Health Act	First powers to deal with food premises and extended to water and washing facilities and cleanliness
1926	Housing (Rural Workers) Act	The first national legislation to provide discretionary improvement grants in rural areas (maximum £100) and half the grant provided by the Exchequer
1926	Public Health (Smoke Abatement) Act	Was the first attempt to deal with visible air pollution, after many years of campaigning
1928	Food and Drugs (Adulteration) Act 1928	Requiring that foods and drugs sold shall be of the nature, substance, and quality demanded. It was an offence under that Act to sell articles which do not conform to standards
1930	Housing Act (Greenwood Act)	Reformed 19th century legislation and provision for the clearance of insanitary housing (slum clearance) with compulsory purchase and clearance area procedures with subsidies for re-housing displaced families. There was little scope for authorities to ignore the provisions as five-year programmes were required
1934	Shops Act	Controlled hours of work for those under 18 years of age and introduced health and welfare provisions for shop and office workers
1935	Housing Act	Amended procedures for slum clearance and introduced new powers to deal with overcrowding including a statutory definition of overcrowding (still in force in 2015)
1936	Public Health Act	Consolidated previous provisions, extended statutory nuisance powers and required compulsory notification of specified communicable diseases and was the basis of many environmental health interventions until replaced by specific later legislation
1937	Public Health (Drainage of Trade Premises) Act	Sought to control non-domestic discharges to public sewers
1938	Food and Drugs Act	Provision on protection of food supplies, control of certain foods, on inspection of food and control of some food premises, also included provisions on infectious diseases
1949	Housing Act	Extended the grant provisions in the 1926 Act to urban areas particularly for the installation of the basic amenities. Local authorities could acquire housing for improvement
1949	Prevention of Damage by Pests Act	Provisions on the control of rats and mice primarily to protect food from damage

Table 1.1 (continued)

Year	Legislation	Purpose
1950	Shops Act	A consolidating measure
1951	Factory Act	A consolidating measure
1954	Housing Rents and Repairs Act	Required local authorities to survey all houses in their district to determine which were unfit (for the first time a fitness standard was defined); included provision for the clearance and redevelopment of areas of unfit housing, for securing or promoting the reconditioning and maintenance of houses via improvement grants, and amended the law relating to housing and the exercise of certain powers relating to land, and rent control
1955	Sanitary Inspectors (Change of Designation) Act	Sanitary Inspectors were designated Public Health Inspectors
1955	Food and Drugs Act	Updating of controls on food and food premises including regulation making powers, such as food hygiene regulations
1956	Agriculture (Safety, Health and Welfare Provisions) Act	Enforcement was entrusted mainly to the Agricultural Inspectorate and not to local authorities, whose role was equated with their, then, limited duties under factory legislation
1956	Clean Air Act	Following the killer London smogs and the Beaver Committee, led to action to reduce domestic smoke emissions via smoke control areas and increased monitoring and action to reduce smoke emissions
1957	Housing Act	Fundamental restatement of housing law with changes to the law on repair and clearance
1959	House Purchase and Housing Act	Introduced the first mandatory grant aid (standard grant) for owners of eligible housing (including landlords) and relaxed discretionary grant conditions
1961	Clean Air Act	Extended to provisions on smoke control including controls on chimney heights
1963	Offices Shops and Railway Premises Act	Applied occupational health and safety matters to largely non-industrial workers such as those in offices in wholesale and retail premises. Under this Act, regulations were made prescribing health and safety standards. Most enforcement duties were assigned to local authorities and whose role was entrusted to EHPs. Inspectors of Factories had oversight over local authority work
1964	Housing Act	Dealt with grant aid for water supply and provision of the standard amenities as well as area improvement
1969	Housing Act	Followed the Dennington Committee Report (Older Homes – a call for action) and introduced provisions to support housing renewal, taking account of bad external arrangement
1972	Local Government Act	Reformed local government structures in England and Wales, created two-tier structure. The 1300+ councils in England and Wales were reduced to about 400 shire districts. (Scotland had its own Act in 1973 to restructure local government)

(*Continued on next page*)

Table 1.1 (continued)

Year	Legislation	Purpose
1973	Water Act	Created ten Regional Water Authorities that took over water functions from local authorities, with regulation based on river basin management
1974	Health and Safety at Work etc. Act	Following on from the Robens Committee it provided new duties on employers and employees to protect health and safety at work. Created Health and Safety Executive and new powers for local authorities, enforcement provisions same for local authorities and HSE but applied to different classes of different premises. Powers to make Regulations and for HSE to issue Codes of Practice
1974	Housing Act	Introduced Housing Action Areas and changes to grant provisions with a new range of grants to secure a 30-year life
1974	Control of Pollution Act	Included provisions on smoke pollution, noise
1976	Food and Drugs (Amendment) Act	Introduced powers to close insanitary food premises
1984	Food Act	Consolidation Act
1984	Public Health (Control of Disease) Act	Established Port Health Authorities, made provisions on control of disease and to regulate common lodging houses and canal boats
1985	Housing Act	Consolidation Act
1989	Water Act	Privatised the water industry; until 1974 public water supplies had been a local authority function then passed to Regional Water Authorities. Also created National Rivers Authority to deal with water pollution
1991	Water Resources Act	Introduced new provisions to control discharges to water and also allowed for water quality standards
1991	Water Industry Act	Among other things sought to update regulation of industrial discharges to sewers previously covered by the 1937 Act
1989	Local Government and Housing Act	Introduced first update of the Fitness Standard since 1957 (included the basic amenities for the first time) and mandatory grants for owner occupiers of unfit houses and no upper limit to grant but subject to means testing, Group Repair, Minor Works Assistance and Disabled Facilities Grants introduced
1990	Food Safety Act	Updating of the law and increasing powers to deal with unsafe food and unhygienic premises
1990	Environmental Protection Act	Introduced new powers to deal with industrial emissions, not just visible pollutants giving local authorities powers similar to that of the central inspectorate, new regulation of waste, and updated statutory nuisance provisions and later included provisions on contaminated land
1995	Environment Act	Created the Environment Agency which became the central pollution control agency among other functions, dealing with waste, water and air emissions as well as flood control and environmental protection. Revised provisions in the 1990 Act including on contaminated land
1996	Housing Grants, Construction and Regeneration Act	Limits to renovation grants for unfit houses

Table 1.1 (continued)

Year	Legislation	Purpose
1996	Housing Act	Introduced new provisions on registration schemes for HMOs
2000	Local Government Act	Power of local authorities to promote well-being (as defined in the Act)
2000	Regulation of Investigatory Powers Act	Regulates the powers of public bodies to carry out surveillance and investigation, covering the interception of communications.
2003	Water Act	Included provisions on contaminated land and water pollution and on changes to drainage law
2004	Housing Act	Introduced powers to deal with hazards (the Housing Health and Safety Rating System introduced by Regulations made under the Act) plus HMO and Selective licensing introduced
2005	Clean Neighbourhoods and Environment Act	Added to the list of statutory nuisances, and provisions on litter, noise, waste management and control of dogs
2006	Wales Act	Relating to devolution of powers to the Welsh Assembly
2007	Local Government and Public Involvement in Health Act	Introduced local area agreements and community strategies
2007	Sustainable Communities Act (Amended 2010)	Provisions to promote the sustainability of local communities
2010	Flood and Water Management Act	Includes provisions on sustainable drainage
2011	Localism Act	Introduced changes to planning, repealed requirement for regional strategies and some changes to housing law
2015	Deregulation Act	Repealed a number of provisions, such as duty to prepare sustainable community strategies and local area agreements; decriminalised some offences (e.g. on household waste); made changes to health and safety at work legislation and also changes to licensing law
2015	Modern Slavery Act	Consolidated the existing slavery and human trafficking offences and introduction of two new civil orders to enable the courts to place restrictions on those convicted of modern slavery offences
2016	Housing and Planning Act	Included provisions about housing supply, estate agents, rent charges, planning and compulsory purchase and provisions on banning orders for 'rogue landlords'
2016	Investigatory Powers Act	Legislation on interception of communications, equipment interference and the acquisition and retention of communications data, bulk personal datasets and other information; and the treatment of material held as a result of such interception, or acquisition or retention
2016	Immigration Act	Part 2 related to landlords in England knowingly leasing premises to a person who is disqualified by reason of their immigration status
2017	Homelessness Reduction Act	Duty on local housing authority to provide advice to prevent homelessness
2017	Neighbourhood Planning Act	Revised provisions on land-use planning and compulsory purchase by amending previous legislation

(*Continued on next page*)

Table 1.1 (continued)

Year	Legislation	Purpose
2017	Wales Act	Amending previous devolution legislation
2018	Homes (Fitness for Human Habitation) Act	Introduced Housing Health and Safety Rating System hazards into fitness criteria for the Landlord and Tenant Act 1985 and landlord's obligation for rented property to be 'fit' (tenants able to take own legal action where dwelling not suitable for occupation)
2019	Wild Animals in Circuses Act	Prohibiting the use of wild animals in travelling circuses in England
2019	Tenant Fees Act	Act to prevent landlords and letting agents in England from charging certain fees
2020	European Union (Withdrawal Agreement) Act	Act to set out terms of UK's withdrawal from the EU
2020	Coronavirus Act	Emergency measures to deal with the COVID-19 pandemic including measures on the food supply chain
2021	Fire Safety Act	Amending application of the Regulatory Reform (Fire Safety) Order 2005 where a building contains two or more sets of domestic premises
2021	Environment Act	Requires statements and reports about environmental protection and establishes the Office for Environmental Protection; Requires Secretary of State to set long-term legally binding targets on air quality, biodiversity, water, resource efficiency and waste reduction. Duty on government to reduce discharges from storm overflows. Includes provisions on the regulation of chemicals and on biodiversity

a chronology of some key provisions since the 1848 Public Health Act relevant to England and Wales. Other countries, particularly those in the former Commonwealth may have started from a similar point but will no doubt have diverged as different problems and issues have emerged.

Table 1.1 sets out Acts only and is included for illustrative purposes only, setting out primary legislation relevant to environmental health practitioners particularly in the 20th century. Relevant law is often contained in secondary legislation made under the primary legislation. Legislation is often a reaction to campaigns by individuals or organisations or particular events and may lag behind public concerns. Indeed, legislation may be seen as the minimum action Parliament (and the government as the Executive) think they can do in the circumstances and is often the result of negotiation and trade-offs so that legislative standards tend to be the minimum not the optimum for public health.

Section 1 notes

1 http://www.nationalarchives.gov.uk/pathways/citizenship/struggle_democracy/getting_vote.htm

2 Locke J. 1690 *Two Treatise of Government* available via The Project Gutenberg. https://english.hku.hk/staff/kjohnson/PDF/LockeJohnSECONDTREATISE1690.pdf

3 Honorary professor of public health at Kings College London and former director of public health for the London Borough of Southwark.

4 See for example Dr Gabriel Scally in the *Observer* newspaper 12 September 2021, The winter Covid plan will be marked by delay, confusion and ignorance. Sound familiar? https://www.theguardian.com/commentisfree/2021/sep/12/the-winter-covid-plan-will-be-marked-by-delay-confusion-and-ignorance-sound-familiar

5 Liverpool's first dock was the world's first enclosed commercial dock, built in 1715 and at one time Liverpool's docks stretched along 7.5 miles of the banks of the River Mersey.

6 See Battersby S. (2017) Duncan of Liverpool – The first medical officer of health. In *Pioneers in Public Health – Lessons from History*, Stewart J (Ed.), Routledge, Abingdon.

7 See 150th anniversary of the establishment of the Alkali Inspectorate at https://www.envchemgroup.com/150th-anniversary-of-the-establishment-of-the-alkali-inspectorate.html

8 https://assets.publishing.service.gov.uk/government/uploads/system/uploads/attachment_data/file/789997/households-below-average-income-1994-1995-2017-2018.pdf

9 https://assets.publishing.service.gov.uk/government/uploads/system/uploads/attachment_data/file/821693/CWTC_Commentary_April_2019.pdf

10 See http://eh.net/

11 In 1844 the Rochdale Pioneers founded the modern Co-operative Movement, to provide an affordable alternative to poor-quality and adulterated food and provisions, using any surplus to benefit the community.

12 Interestingly, criminal public nuisance is an either-way offence, and as a common law and not a statutory offence there is no limit to the sentence that can be imposed provided it is not 'inordinate'. Police therefore have the power to arrest for the crime of public nuisance using their general arrest powers in s24(1)(b) Police and Criminal Evidence Act 1984. Conspiracy to commit a public nuisance is also an offence contrary to section 1(1) of the Criminal Law Act 1977.

13 https://www.ncbi.nlm.nih.gov/books/NBK535293/pdf/Bookshelf_NBK535293.pdf

14 Rader B, Nande A, et al. (2020) *Crowding and the Epidemic Intensity of COVID-19 Transmission*, a pre-print paper. https://www.medrxiv.org/content/10.1101/2020.04.15.20064980v2; See also https://www.bmj.com/content/370/bmj.m3181 where a study is reported as saying contact within households is thought to be responsible for roughly 70% of SARS-CoV-2 transmission when widespread community control measures are in place.

15 For one review of the historical development of environmental health in the UK, and the professional body see Johnson R. (1983) *A Century of Progress – History of the Institution of Environmental Health Officers, 1883–1983*, CIEH, London.

16 https://www.routledge.com/Routledge-Focus-on-Environmental-Health/book-series/CONENHE

17 In light of the COVID-19 pandemic the issue of testing, contact-tracing and isolation, basic and traditional public health response dating from the 19th century has become a real debate. Those countries that took this approach appear to have dealt with the disease better. Infectious disease contact tracing and investigation is something that EHOs have been trained to do in the past. See https://www.ecdc.europa.eu/sites/default/files/documents/Contact-tracing-Public-health-management-persons-including-healthcare-workers-having-had-contact-with-COVID-19-cases-in-the-European-Union%E2%80%93second-update_0.pdf and https://apps.who.int/iris/bitstream/handle/10665/185258/WHO_EVD_Guidance_Contact_15.1_eng.pdf;jsessionid=768915DE811CCC1D7883827BB073107C?sequence=1

18 See https://assets.publishing.service.gov.uk/government/uploads/system/uploads/attachment_data/file/15810/142631.pdf

19 See for example Day C. (2021) *COVID-19: The Global Environmental Health Experience*, Routledge Focus on Environmental Health Series, Routledge, Abingdon, Oxon. https://www.routledge.com/COVID-19-The-Global-Environmental-Health-Experience/Day/p/book/9780367743161

Section 1 References

[1] Mellanby K. (1992) *Waste and Pollution – the Problem for Britain*, Harper Collins, London.

[2] World Bank. (2018) *What a Waste: An Updated Look into the Future of Solid Waste Management*, September. https://www.worldbank.org/en/news/immersive-story/2018/09/20/what-a-waste-an-updated-look-into-the-future-of-solid-waste-management (Accessed 1 June 2021).

[3] Gostin LO. (2004) Health of the people: The highest law? *The Journal of Law, Medicine & Ethics*, 32: 509–515. https://scholarship.law.georgetown.edu/facpub/1806.

[4] Maryon-Davis A. (2015) Outbreaks under wraps: How denials and cover-ups spread Ebola, SARS and AIDS. *Index on Censorship, Sage*, 44(1): 72–75. Article first published online 12 March, issue published 1 March. https://doi.org/10.1177/0306422015570534.

[5] Tawney RH. (1926) *Religion and the Rise of Capitalism: A Historical Study*, J. Murray, London.

[6] Boyer GR. (2002) English poor laws. In *EH. Net Encyclopaedia*, Whaples R. (Ed.). http://eh.net/encyclopedia/english-poor-laws/ and http://eh.net/?s=Poor+Laws.

[7] Clark G. (2001) Farm wages and living standards in the industrial revolution: England, 1670–1869. *Economic History Review*, 2nd series, 54: 477–505.

[8] Pinchbeck I. (1930) *Women Workers and the Industrial Revolution, 1750–1850*, Routledge, London.

[9] Boyer GR. (1990) *An Economic History of the English Poor Law, 1750–1850*, Cambridge University Press, Cambridge.

[10] Lindert PH. (1998) Poor relief before the welfare state: Britain versus the continent, 1780–1880. *European Review of Economic History*, 2: 101–140.

[11] Parkinson N. (2014) The health of towns association and the genesis of the environmental health practitioner. *Journal of Environmental Health Research*, 14(1): 5–16.

[12] Parkinson N. (2013) Thomas Fresh (1803–1861), inspector of nuisances, Liverpool's first public health officer. *Journal of Medical Biography*, 4: 238–249, 21 November. https://doi.org/10.1177/0967772013479277. Epub 26 July 2013.

[13] Frazer WM. (1997) *Duncan of Liverpool*, Carnegie Publishing, Preston, Lancs.

[14] Hamlin C. (2013) Nuisances and community in mid-Victorian England: The attractions of inspection. *Social History*, 38(3): 346–379. http://doi.org/10.1080/03071022.2013.81706.

[15] Crook T. (2007) Sanitary inspection and the public sphere in late Victorian and Edwardian Britain: A case study in liberal governance. *Social History*, 32(4), November.

[16] Vacher F. (1894) Address: Conference of sanitary inspectors. *Journal of the Sanitary Institute*, XV: 410–411.

[17] Smith GD. (2002) Commentary: Behind the broad street pump: Aetiology, epidemiology and prevention of cholera in mid 19th century Britain. *International Journal of Epidemiology*, 31: 920–932.

[18] Wilkinson R. (2002) Commentary: Liberty, fraternity, equality. *International Journal of Epidemiology*, 31: 538–543.

[19] The Commonwealth. (2017) *Health Protection Policy Toolkit – Health and an Essential Component of Global Security*, Commonwealth Secretariat, London, 2nd ed. https://www.thecommonwealth-healthhub.net/wp-content/uploads/2017/05/HPToolkitwordversionEd2-CHMM-2017.pdf.

[20] Rayner G, Lang T. (2012) *Ecological Public Health – Reshaping Conditions for Good Health*, Earthscan, Routledge, Abingdon, Oxon.

[21] Gorsky M. (2007) Local leadership in public health: The role of the medical officer of health in Britain 1872–1974. *Journal of Epidemiology and Community Health*, 61: 468–472. https://doi.org/10.1136/jech.2006.046326.

SECTION 2: THE 21ST CENTURY; THE PRINCIPLES AND PHILOSOPHY OF ENVIRONMENTAL HEALTH

Introduction and context

Having looked at some of the history – where environmental health came from with some reference to current issues and practice – we now turn to the second decade of the 21st century. Many challenges facing environmental health and societies generally around the globe are not always so different in nature from those in the past. These challenges are both economic and environmental (and there remain parallels with the 19th century given the widespread acceptance of neoliberal economics at least until the COVID-19 pandemic). It can be argued that unless solutions to economic problems take account fully of the environmental imperative presented by the climate crisis and global heating (and the

fundamental risks to public health), then the responses will be inadequate. It might be thought that economic policies and structures have nothing to do with environmental health or EHOs. However, looking back at the history of environmental health it is apparent that many of the problems that EHOs have had to face have been a consequence of society's approach to economics and indeed the economic and political philosophies of governments. As Kate Raworth has said [1], *economics is the mother tongue of public policy, the language of public life, and the mindset that shapes society*.

The COVID-19 pandemic has shown the fallacy of the belief that economic progress and unprecedented prosperity around the world means humans are not part of the natural world. For too many there has been the view that somehow humans have created an autonomous ecosystem. The events of 2020–21 have shown this to be both dangerous and wrong.

It has been shown that more unequal societies are also unhealthier, and in any pandemic and in extremes of weather as the result of global heating, everyone will be affected but the poorest will suffer most. The issue of climate change should actually dictate how economic problems are resolved. Such challenges are fundamental to the philosophy underpinning environmental health as also is the notion of equity or fairness, and in particular equity in health.

Michael Marmot said in 2010 [2] and reiterated in 2020 [3], *Health inequalities are not inevitable and can be significantly reduced . . . avoidable health inequalities are unfair and putting them right is a matter of social justice.* In answer to the frequent response that the recommendations of the 2010 report could not be afforded because of the economic climate, he said *it is inaction that cannot be afforded, for the human and economic costs are too high.*

Equity in health means that there should be no disparities in health as the result of social systems between social groups that have an underlying social advantage or disadvantage (as the result of economic wealth or power). Equally there should be no disparities as the result of the major social determinants of health. Inequities in health come about when society puts groups of people who are already disadvantaged, as the result of poverty, gender, race or sexual orientation or are otherwise disenfranchised, at further disadvantage with respect to their health. Equity is an ethical principle; it also reflects and is closely related to human rights principles. So it should be recognised that global heating, pandemics and economic crises could increase inequity, and that the further down the social ladder one is (that is the less wealth one has), the greater will be the adverse effects, and this applies as much between countries as within countries.

One problem is however that the main indicator of a country's economic performance remains the Gross Domestic Product (GDP) and this is used to measure 'progress'. This is a crude measure of total economic activity, and as such is a poor reflection of health and well-being. So that for example expenditure on tobacco appears in GDP, as would money spent on food that has little nutritional benefit, never mind in the UK and some other countries the sale of illegal drugs and prostitution. One might ask for example in such an economic situation where should the effort be put to achieve more in environmental health terms; more food safety inspections or actions to achieve healthier eating and better diets for those on lower incomes which could also reduce obesity. Or, should we spend money on drugs and treatment for those living in damp and cold homes or in improving those unhealthy and unsafe homes? From a public health rather than merely consumer protection or 'service' perspective, in both cases the upstream intervention (prevention) would be the obvious course.

The reality is that the effects of economic growth have made us worse off in terms of our health. Using GDP as a measure takes no account of increasing inequality pollution or damage to people's health and the environment. Indeed, it also treats crime, divorce and other elements of social breakdown as economic gains. This model will always undermine the notion of sustainable development (considered in the Annex to this chapter), and as such it cheats other countries, future generations and ourselves. We do need to redefine progress, and replace GDP with new indicators of progress; however it will be brave politicians that take this on as there are so many vested interests. There are two problems. First more government finance departments (Treasuries) take a short-term approach and don't want to lead. Secondly despite the science, there are too many 'climate change deniers' (or sceptics) in positions of power who it seems are linked to vested interests in the form of business who do not have a long-term vision. It seems change and a new approach will come about only when catastrophe as the result of global heating is at the door and the general population are directly and adversely affected. There is maybe a hope that the COVID-19 pandemic will lead to a change in attitudes and a recognition that the economic paradigms are not the way to achieve healthy societies.

Jackson [4] considered what prosperity should look like in a finite world with limited resources and asked whether prosperity is possible without 'growth'. He undermines the notion that prosperity is the same as conventional economic growth as measured by GDP. Prosperity *isn't obviously synonymous with income or wealth* and *rising prosperity isn't self-evidently the same thing as economic growth.* An alternative such as an Index of Sustainable Economic Welfare (ISEW) takes account of damage to the environment and public health and would be a better way of reflecting economic performance, but that is not for this book. However, this does put into context the issues facing environmental health and why inequality is such an important issue in environmental health.

The emergence of the concept of Doughnut Economics [1] should therefore be of interest to EHOs/EHPs. Kate Raworth argues that we should move away from the use of GDP as a measure, as she points out as an example of how we have got it wrong the G20 leaders in 2014 pledged to increase the GDP of their countries by 2.1% just days after the Intergovernmental Panel on Climate Change (IPCC) warned of the severe and irreversible damage from rising greenhouse gas emissions. The argument is that GDP growth should be put aside, and we should start afresh, asking *what enables human beings to thrive.* The concept can be demonstrated by the diagram of the ring doughnut (see Figure 1.1). This is a concept that has already been adopted by one city, Amsterdam, to plot how it will rebuild after COVID-19.

Figure 1.1 is the 'doughnut' which has been described as a 21st-century compass,

and between its social foundation of human well-being and ecological ceiling of planetary pressure lies the safe and just place for humanity. The inner ring of her doughnut sets out the minimum needed to 'lead a good life', derived from the UN's Sustainable Development Goals (discussed later). It ranges from food and clean water to a certain level of housing, sanitation, energy, education, healthcare, gender equality, income and political voice. Anyone not attaining such minimum standards is living in the doughnut's hole. The outer ring of the doughnut, where you might find the sprinkles, represents the ecological ceiling drawn up by earth-system scientists. It highlights the boundaries across which humankind should not go to avoid damaging the climate, soils, oceans, the ozone layer, freshwater and abundant biodiversity.

As the earlier part of this chapter has shown, environmental health (even if we did not always call it that) has seen similar momentous changes and crucial moments in history to those we now face. In the first quarter of the 21st century, we are at a comparable watershed. For many countries, particularly in the Northern Hemisphere (or 'the West'), urbanisation and industrialisation are largely behind us; we live in post-industrial economies but that does not protect us from catastrophes be it the result of global heating or pandemics.

Despite the UK leaving the EU the horizons, including those of EHOs, are increasingly global, not merely national. Manufacturing industries employ fewer people, and much of the economy is driven by ideas and as we have seen financial markets can impact on a country's economic health almost by the touch of a computer key. It is apparent that engineered or technological solutions in themselves can no longer be the answer to the earth's problems and where they have a role to play they need to be managed and those affected need to be involved in the solutions. The problems themselves have assumed less tractable, more qualitative dimensions to do with 'lifestyles', quality of

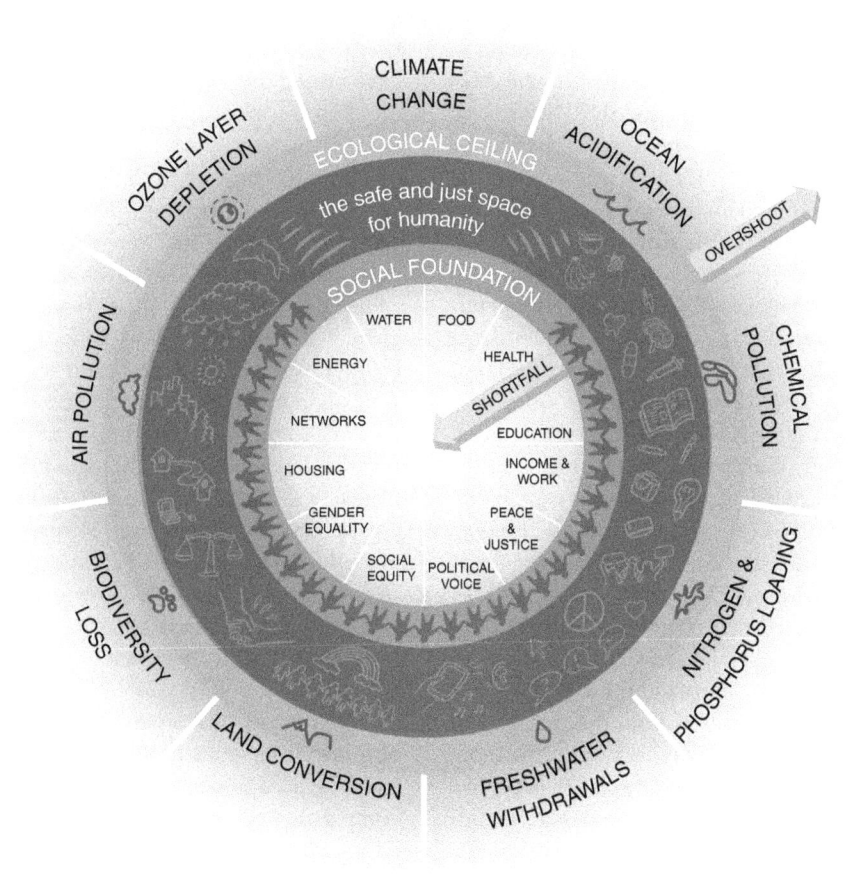

Figure 1.1 The principle of doughnut economics – a fun version developed by DEAL[20] – was created in conjunction with an animation that also shows the significance of Universal Basic Income in doughnut economics

Source: This was presented at the 2021 Basic Income Earth Network Congress in 2021. This tool is licensed under a Creative Commons BY SA 4.0 license

life and psychological (mental) health and well-being. More importantly, perhaps, the manageable, predominantly urban concept of public health that was developed in the 19th century has been replaced by a much bigger concept, that of the global biosphere. As a result, we are grappling with the much bigger issue of the future of life and health on earth, a question that rarely occurred to the Victorians.

As has been shown earlier environmental health came about as part of the sanitary movement in the 19th century but needs always to be reinvented or at least to be reassessed. The work by the WHO Commission on Social Determinants of Health in its final report *Closing the Gap in a Generation: Health Equity through Action on the Social Determinants of Health* [5] highlighted the inequalities or inequity that exists within and between countries and indeed between generations (an issue that was in part recognised in the 18th century by some more enlightened thinkers). These inequities in health, avoidable health inequalities, arise because of the circumstances in which people grow, live, work and age, and the systems put in place to deal with illness. The issue of inequities in health were further highlighted in the *Marmot Review* undertaken for the Department of Health by the Commission chaired by Michael Marmot [2]. This has not led to the changes required for a healthier society in the UK (or at least in England).

Marmot's ten-year review of the 2010 report [3] showed that government spending as a percentage of gross domestic product (GDP) declined by seven percentage points between 2009/10 and 2018/19, from 42% to 35% and the health spending was reduced accordingly. Cuts to local authorities over the decade have *been hugely significant; local government allocations from the Ministry of Housing, Communities and Local Government declined by 77 percent between 2009/10 and 2018/19.*

As the 2020 report highlights it is not just the impact of overall cuts but where they have fallen which has impacted most on inequalities. Reporting analysis from the Institute for Fiscal Studies, the report shows that the more deprived areas, those with greatest need, had the greatest reductions in per person spending. Yet, while there has been limited action on health inequalities nationally, many local authorities have taken forward the recommendations and approaches outlined in the 2010 *Marmot Review*.

A survey by the King's Fund conducted in 2011 found that over 75% of local authorities had incorporated the approach directly into their health and well-being strategies and local government continues to see this as important,[21] using Marmot principles to tackle health inequalities. The social determinants are highly relevant to local authorities, particularly given the strong focus on place, well-being and cross-sectoral working by local governments, which social determinants approaches require, and which local government is well set up to deliver. Community empowerment is central to efforts to reduce health inequalities and was one of the key features of the original *Marmot Review*. What Marmot has also shown is that the problems

associated with pandemics, the climate emergency and other 'environmental health' issues have caused those lower down the social gradient to be more likely to suffer the worst consequences.

Though many of the new health threats are global, the problems and the responses will often be regional or local and reflect cultural differences. They may also be experienced at the individual level. Without change and political leadership there will be increased inequity in health, continuing environmental degradation, a decaying family and community fabric, increasing stress and the deterioration of the state into a kind of reactive, coping mechanism. Against such a background, the emphasis within environmental health has largely become curative (a sticking plaster approach) or enforcement of Regulations to address an issue rather than prevention, hard technological solutions rather than a 'softer' people-based response. There is a risk of too much measuring of outputs rather than health outcomes of the work. This approach leads to professional compartmentalisation and marginalisation rather than integration, since the former may appear to offer more certain career gains to individuals. This is reflected in the purely mechanical regulatory non-strategic approach to soften the edges of adverse health impacts from activities and policies. Yet advocates of environmental health emphasise the preventive approach as fundamental to the environmental health work, and that environmental health practitioners have a strong overview of the risks to public health as the result of their breadth of knowledge (if not always depth).

Many diseases of the modern way of life, such as cardiovascular disease, obesity and cancer, will take much longer to conquer if prevention, or upstream intervention, remains a low priority and integrated planning and delivery of services are compromised by demands for quick fixes. For example, the research needed to understand them will be slower to materialise and this provides an opportunity for powerful rear-guard actions by well-organised vested interests including industrial lobbies. There is, however, a further complexity to consider; when prevention does become a higher priority, it is often in the context of finite or limited resources. This is where some hard decisions may have to be made about the shifting of resources from treatment and care to prevention and all the ethical issues that this entails. A review of National Health Service expenditure in 2002 [6] highlighted the economic benefits of upstream investment and the potential contribution of public health measures in reducing the burden of disease. However, there has often been a lack of solid evidence on which measures to use and indeed on outcomes of those measures. Environmental health practitioners who are the most likely readership of this book have not been good at providing this evidence and reporting on issues of concern, unlike other professions of publishing research and findings.

Success requires 'full engagement', that is, increased, or better directed, spending and high public engagement to achieve improved health status and increased healthy life expectancy across the whole of society. In recent years, while those lower on the social scale have barely seen an increase in healthy life expectancy, at the top there has been a marked increase. This also has economic implications with an increase in the age for eligibility for state pension, if those on lower incomes are unfit to work from their early fifties. Disability-Free Life Expectancy (DFLE) was found to be about 15 years between those in the most deprived neighbourhoods and those in the least deprived [2] who on average have a DFLE of 70 years (see also Annex 1.1).

The social determinants of health are the conditions in which people are born, grow, live, work and age, including the health system. These circumstances are shaped by the distribution of money, power and resources at global, national and local levels, which are themselves influenced by policy choices. The social determinants of health are mostly responsible for health inequities – the unfair and avoidable differences in health status seen within and between countries.

The environmental impact on health is an inevitable by-product of human activity, and it is therefore the nature of that activity, and the attitudes that go with it, that hold the key. It will be to the soft technologies – the technologies of mind, reason and social organisation – that we should look for solutions. Hard 'engineered' technologies are at best a partial answer, at worst a diversion. The answers offered are much more difficult. They require vision, ambition and leadership; they also require debate, consensus and agreement. In the worst-case scenarios described, these solutions will be more, not less, difficult to realise and they will need a new kind of environmental health practitioner to tackle them. The World Health Organization (WHO) has contextualised environmental health into human rights and human rights cannot be secured in a degraded or polluted environment. The fundamental right to life is threatened by soil degradation and deforestation and by exposures to toxic chemicals, hazardous wastes and contaminated drinking water.

> This is because social and environmental conditions determine the extent to which people enjoy their basic rights to life, health, adequate food and housing, and traditional livelihood and culture. It is time to recognize that those who pollute or destroy the natural environment (and indeed those who provide financial backing for such activities whether as investors or customers) are not just committing a crime against nature but are violating human rights as well.

So said Klaus Toepfer, Executive Director of the United Nations Environment Programme at the 57th Session of the Commission on Human Rights, Geneva, 2001.

To fulfil their future role, professional environmental health practitioners should have an understanding of the following:

- The *definitions and principles* of environmental health;
- The *agenda* with which they need to be engaged;
- The *skills and expertise* required of their professional practice;
- The *objective* of their environmental health activities.

The matters of skills, expertise and competencies are addressed in detail in Chapter 2, but as we will see this points to the environmental health practitioner being more than a 'sanitary police officer'. However, some of the problems facing the modern EHP are a consequence of the difficulties in securing recognition and how the profession is perceived by administrators or legislators is considered later in this chapter with a contribution from Australia. This issue also overlaps with the next section on seeking to define environmental health.

What is environmental health?

In 1991 Dahlgren and Whitehead set out a famous and influential model on the social determinants of health [5] (see Figure 1.2). It can be seen that what we consider to be environmental health and areas of environmental health practice, feature strongly. Environmental health practitioners are therefore part of the public health workforce. From this perspective then, 'environmental health' is also a component of the public health movement and this becomes more apparent on the following pages.

Figure 1.3 reflects the wider concerns with global and economic issues affecting health and demonstrates even more fully how environmental health fits within society.

The term environmental health in many respects is a catch-all that creates unease and misunderstanding. By separating the two dimensions of *environment* and *health* we can illustrate the all-encompassing nature of the combined term. Albert Einstein noted that environment was *everything that's not me*, and WHO in its constitution considers health to

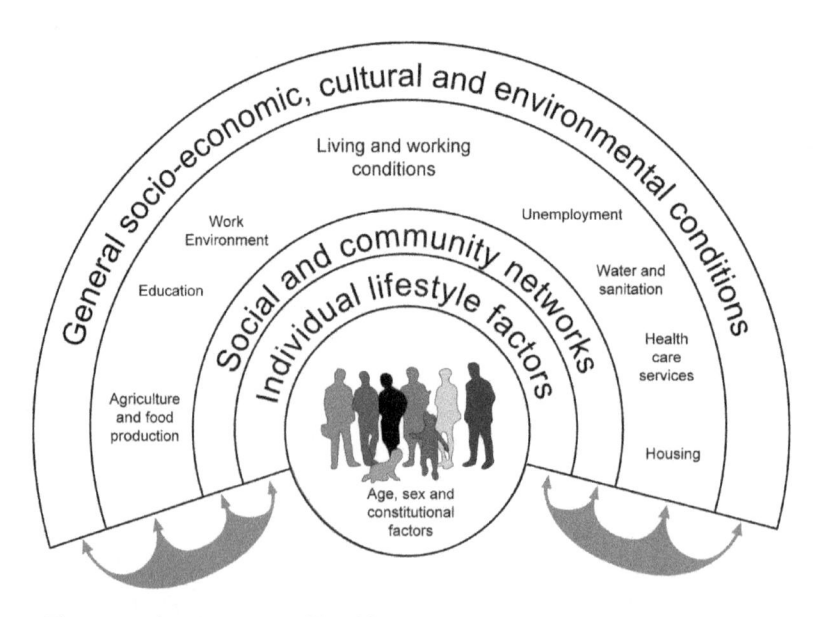

Figure 1.2 The main determinants of health
Source: Dahlgren & Whitehead[22] [7]

be a state of complete physical, mental and social well-being.

Despite the broad challenge of these concepts, several definitions exist for environmental health. In 1989,[23] the WHO defined it *as comprising of those aspects of human health and disease that are determined by factors in the environment. It also refers to the theory and practice of assessing and controlling factors in the environment that can potentially affect health* [9]. A further attempt at defining the term emerged from a meeting of WHO European Member States in 1993 in Sofia. The proposed definition was:

> Environmental health comprises of those aspects of human health, including quality of life, that are determined by physical, biological, social and psycho-social factors in the environment. It also refers to the theory and practice of assessing, correcting and preventing those factors in the environment that can potentially affect adversely the health of present and future generations.
>
> *[10]*

This definition concludes that environmental health is in fact two things: first, certain aspects of human health; and second, a means by which to address these issues. This was finalised by a conference in Vilnius, Lithuania in 1994.

The Sixth Ministerial Conference on Environment and Health was held in Ostrava,

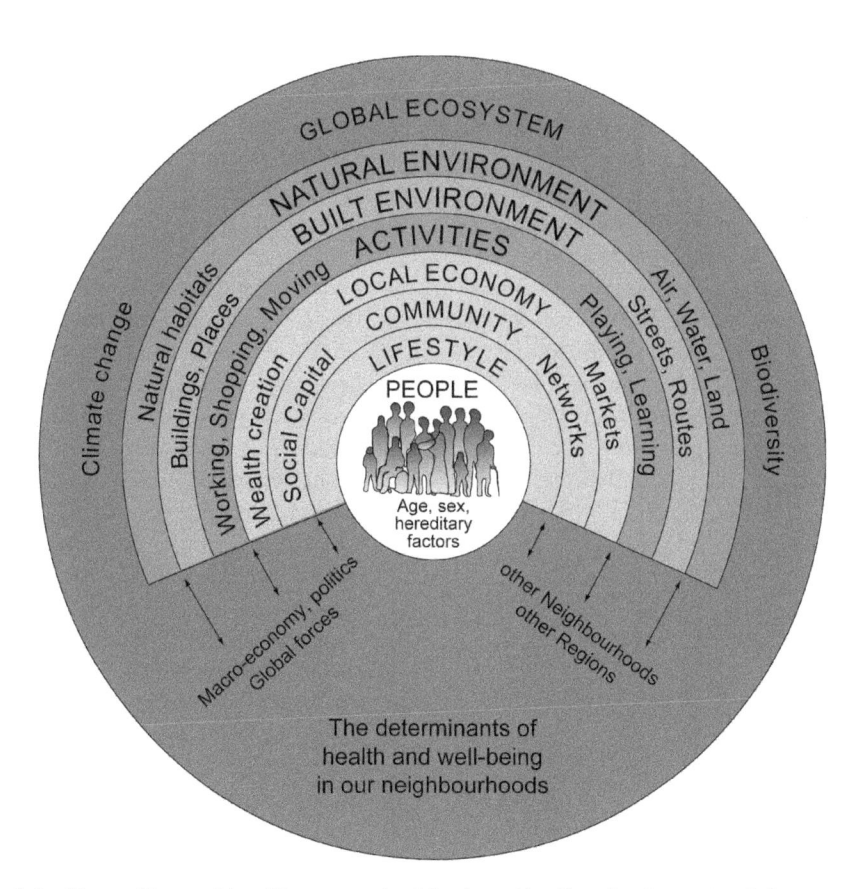

Figure 1.3 The settlement health map: a tool for investigating the impacts of the natural and built environment on public health

Source: Barton & Grant (2006) [8]. Based on the public health concept by Dahlgren & Whitehead [7]

Czech Republic in 2017[24] to review developments since the previous ministerial conference, held in Parma, Italy in 2010. The declaration from that conference included *inter alia* a call to strengthen the public health functions of the health systems, and expand capacities across all sectors, levels of government and stakeholders to reduce environment-related health risks.

The WHO has also made clear *healthier environments could prevent almost one quarter of the global burden of disease*. It has said

> the COVID-19 pandemic is a further reminder of the delicate relationship between people and our planet. Clean air, stable climate, adequate water, sanitation and hygiene, safe use of chemicals, protection from radiation, healthy and safe workplaces, sound agricultural practices, health-supportive cities and built environments, and a preserved nature are all prerequisites for good health.[25]

Figure 1.4 demonstrates the interface of environmental health. It shows that it is a discipline concerned primarily with human health, noting that humans operate in different environments such as the living environment, the home environment, the work environment and the recreational environment. It is from this perspective that a holistic view can be taken of human health and

the environments in which people live. As a concept and as a means of delivering practical solutions, environmental health provides a strong basis upon which decision-makers can work towards sustainable development – if attention is paid to the experiences of EHOs/EHPs and this requires those professionals to be more effective advocates for change. That said, it has also been clearly recognised that environmental health is a wide-ranging discipline that relies upon inter-sectoral co-operation and action. It is therefore essential that all the potential contributors to the development and implementation of environmental health programmes can recognise their role. EHPs/EHOs and other 'environmental health' professionals cannot deal with all aspects of environmental health. As a concept it is far larger than the roles traditionally played by EHOs employed in local authorities.

More recently in the USA in the *Environmental Health Playbook*, the National Environmental Health Partnership Council[26] has said Environmental Health [11] is the

> branch of public health that focuses on the interrelationships between people and their environment, promotes human health and well-being, and fosters healthy and safe communities. As a fundamental component of a comprehensive public health system, environmental health works to advance policies and programs to reduce chemical and other environmental exposures in air, water, soil, and food to protect residents and provide communities with healthier environments. Environmental health protects the public by tracking environmental exposures in communities across the United States and potential links with disease outcomes.

The Playbook says that to achieve a healthy community, homes should be safe, affordable and healthy places for families to gather. Workplaces, schools and child-care centres

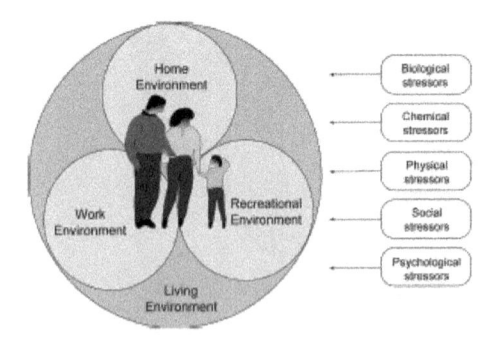

Figure 1.4 The interface of environment health

Source: MacArthur & Bonnefoy (1998) [12]

should be free of exposures that negatively impact the health of workers or children. Nutritious, affordable foods should be safe for all community members. Access to safe and affordable multimodal transportation options, including biking and public transit, improves the environment and drives down obesity and other chronic illnesses. Outdoor and indoor air quality in all communities should be healthy and safe to breathe for everyone. Children and adults alike should have access to safe and clean public spaces, such as parks. In the case of a disaster or some traumatic event, a community needs to be prepared; it should have the tools and resources to be resilient against physical (infrastructure and human) and emotional damage. All these activities require the participation of federal, state, local and tribal governments, the Partnership Council has argued.

This publication makes it clear that many practitioners in the public and private sectors make invaluable contributions towards environmental health and protection or improvement of public health; they may not perceive themselves as environmental health professionals, but they nevertheless perform tasks that contribute to the whole. It is wholly consistent with this approach that socially responsible commercial enterprises might want to employ their own cadre of EHPs. There are many people, for example public health specialists, health visitors, practice nurses, general practitioners, occupational health and safety staff and technical specialists, who work within the broader public health movement and who can, and do, work with EHOs in a way that maximises the human resources available (a point made clear by the WHO in the context of the profile of the EHO). Professionals, whose work is on the periphery of traditional environmental health activities, for example urban planners, would also greatly benefit the system if they also understood the role of environmental health and could relate its principles to their work. It is important that they understand the principles of environmental health and how

their diagnosis and subsequent action can assist in developing effective environmental health interventions. Furthermore, building capacities in environmental health should not be restricted to the public sector workforce. There is an ever-increasing number of practitioners employed in the private sector, not just those in mainstream environmental management positions or where public services have been 'contracted out'. These practitioners recognise and build the preventative approach of environmental health management into everyday business practices. In the same way EHPs can help build capacity within community groups.

The principles of environmental health

The development of the environmental health approach has grown organically in response to societal changes and pressures rather than by design. However, it has demonstrated that it can bring rhetoric to life and adds considerable value to the process of improving human health and quality of life.

In a world that is subject to constant and turbulent change, it is important to retain some sense of core values or principles as touchstones for environmental health work. Environmental health and the mechanisms to deliver it are founded upon such fundamental principles. They do not apply simply to environmental health at the community level, where in the UK the main focus has been on local government. These principles apply to all levels of government and all sectors that contribute to environmental health. They relate to many government issues and depend upon the way in which governments at all levels relate and interact among themselves and with communities. This is a major challenge as many governments still fail to understand the true nature of environmental health, and therefore the significance and value of its approach.

Environmental health is relevant in three different time phases. It must work to repair historical damage – damage that has already occurred – to identify and manage current risks and finally to prevent future problems (prevention being better – and usually cheaper – than cure). The emphasis given to each phase is determined by a complex formula of factors, depending largely on an assessment of the risks and resources available. It is of course important and correct to address the most pressing issues relating to environmental health urgently, but emphasis should also be given to addressing, and so avoiding, future problems. That is, prevention is better and usually cheaper than cure and this is the basis of the precautionary principle, which has become widely accepted in many policies and programmes but has been threatened by 'neoliberal' economic thinking and the pressure to 'deregulate' and cut 'red tape'. Adhering to the principle should however ensure that environmental health action remains at the leading edge of improving the quality of life. The boxed text (overpage) sets out the six principles.

In the context of Principle 3, the Nuffield Council on Bioethics[28] (and in the context of COVID-19) has considered the issues of how the government deals with both the pandemic and how to move from the lockdown. There are ethical questions about how we balance different interests (e.g. individual and collective; economic and social) and different risks (e.g. of COVID infection, and of poor health associated with poverty and isolation). Furthermore, what and who should be prioritised when it comes to the hard decisions (e.g. COVID-19 over other health needs; the young, the elderly or key workers)? Who bears responsibilities for supporting those in need and about whether a country has not only national but also international responsibilities? There is the issue about how privacy will be protected with contact-tracing apps and about the implications of mass testing for disease or immunity. There are issues around the validity of tests; who gets an 'immunity certificate', and where does that leave the rest of us? These are seen as critically important issues for the Council and ones that affect

everybody so there is a need for openness and public discussions.

Yet, in England at least, the Westminster Government does not seem to want to engage or take on board other views on any of these issues; nor is it evident that they are thinking about them. 'We are following the science' was the supposedly reassuring message. However, following the science is not politically or morally neutral. Every scientist knows (or should) that science does not provide certainty and is usually subject to debate or argument; it does not deliver policy answers which involve values and judgements for which people are responsible and should be scrutinised, and accountable, thus requiring transparency. This requires that the public know which values are in play and what judgements are being made, by whom and on what advice.

Decisions are being made that go to the very heart of what governments are there to do. However, they must do so openly, transparently and accountably, especially where those decisions impinge on precisely that freedom or aspects of well-being.

Democratic governments must be subjected to public debate and challenge. The existence of an emergency or crisis makes things difficult but is no justification for closing down on public discourse. On the contrary, if the whole of society is at risk, all society needs to know and to have a voice according to the Nuffield Council on Bioethics.

Reports on these issues and environmental health

There have been a number of reports relevant to the development of environmental health and its principles, philosophical basis and the environmental health profession in the 21st century.

Environmental Health 2012 – In 2002 it was felt that for a variety of reasons, environmental health had become disconnected from the public health agenda and other

Principle 1

The maintenance and improvement of the human condition, and prevention of ill health is at the centre of all environmental health action.

A similar principle was contained in *Agenda 21* [13] and in 2012 the UN Conference on Sustainable Development (Rio+20) reaffirmed the commitment to Agenda 21 in 'The Future We Want'.[27] The principle is recognition that the main target of environmental action is the well-being of the human race and those factors in the environment, however wide, that may affect it. This also implies that prevention (upstream interventions) is better than cure (treatment). The failure to prevent avoidable ill health leads to greater pressures on systems for the treatment of ill health. So that prevention is often cheaper than cure.

Principle 2

The groups within society whose health is most at risk are often those who have least control over their own lives and have little economic power, living in the worst housing with poor environmental conditions, working in the most dangerous or insecure occupations, and with limited access to a wholesome and varied food supply.

Those lower down the social scale, and those with less economic power bear a disproportionate share of the global burden of ill health and suffering. They often live in unsafe and overcrowded housing, in under-served rural areas or peri-urban slums. They are more likely than the well-off to be exposed to pollution and other health risks at home, at work and in their communities. They are also more likely to consume insufficient food, and food of poor quality, to smoke tobacco, and to be exposed to other risks harmful to health. This undermines their ability to lead socially and economically productive lives [14]. Those who are disadvantaged do not form a single homogeneous group: different people are at a disadvantage in different contexts. For example, low-income households in northern European countries may be at risk of poor health because of damp and cold housing conditions, fuel poverty and/or inadequate nutrition; in parts of Africa it may be a lack of safe water and sanitation, or indoor pollution because of the fuel used for cooking and heating in the home.

This phenomenon has been recognised by the WHO, which acknowledges that access to the appropriate medical technology cannot in itself offset the adverse effects of environmental derogation and that good health will remain unobtainable unless the environments in which people live are health promoting. A reduction in inequality or inequity requires equal access to environmental health services and an uptake of services that relates to need. The provision of high technology services should not be restricted to certain sections of the population because of social or economic disadvantage in the others, and services should be sensitive to the needs of minority groups. Equity is a core and primary element that underpins any action on environmental health. To achieve this, the more disadvantaged within the population will require particular assistance and attention to redress the imbalance.

Marmot's conclusions on the social gradient in health [2–3] – the lower a person's social position, the worse his or her health – are relevant. Action should focus on reducing the gradient in health. Health inequalities (or health inequity) result from social inequalities. Action has to be universal to reduce the steepness of the social gradient in health but with a scale and intensity that is proportionate to the level of disadvantage. This has been described as 'proportionate universalism'.

As Wilkinson and Picket [15] have pointed out also the health of a society (and almost everything else) is not affected by how wealthy that society is, but how equal. The bigger the gap between rich and poor the worse it is for everyone in that society including the well-off.

Principle 3

A range of governance issues that can be described as the conditions for civic engagement must be in place.

As seen in the first section of this chapter adoption of the democratic principles of government is the cornerstone to the effective management of environmental health. For example, the European Charter of Environment and Health [9] sets out the basic entitlements of individuals, including the rights of full information, active consultation and genuine participation in environmental health decisions. The traditional approach is no longer appropriate and times are changing and demands have increased for greater public participation in all aspects of society.

Democratic principles also require a two-way exchange and the involvement of non-governmental organisations and an informed public in the decision-making process is both necessary and practicable. Recent developments in information technology including social networking make this easier. It is therefore important that decision-makers are not only held accountable, but that they owe their accountability to the public. Modernisation of local services requires that people are not merely represented but actively participate in the development and delivery of local policies, services and projects and in doing so they can ensure that these meet their priorities and needs. This power-sharing agenda can feel challenging and there are a variety of skills that people working in environmental health need to develop to ensure that a broad cross-section of the community is engaged and not only the 'usual suspects'. It is therefore important to join up and to communicate with others within a locality and ensure that partner agencies work together to plan for and focus on community engagement.

Many of the environmental and public health problems facing society can only be solved by communities acting together in concert, rather than as individuals. This is one of the key areas to challenge the traditional approach to solving environmental health problems. Environmental health practitioners do need to consider how well equipped they are for helping with the development of communities and indeed developing an advocacy role.

Over-centralisation and reduction of local services can lead to slow or inappropriate responses as was illustrated by the different responses around the world to COVID-19 with inadequate use of local level government and the workforce as part of a testing and tracing strategy.

Principle 4

Co-operation and partnership.

Isolated decisions and actions cannot normally solve problems in environmental health: an inter-sectoral approach is needed. The practice of co-operation and partnership in pursuit of improvements in environmental health, not only between the health and environment sectors, but also with economic sectors and with all social partners, is a crucial element. This is the principle that lies at the heart of effective environmental health management.

If properly interpreted and applied, inter-sectoral co-operation and co-ordination means that:

- The problems tackled are common ones in which all participants have a stake;
- All the public and private sector organisations and interests active in the sector are involved, not just government;
- Policy makers, technical and service staff and volunteers at both national and local levels have actual or potential functions to perform;
- Various participants may play leading and supporting roles with respect to specific issues
- Co-operation consists not only of ratifying proposals, but also of participation in defining issues, prioritising needs, collecting and interpreting information, shaping and evaluating alternatives and building the capabilities necessary for implementation;
- Stable co-operative mechanisms are established, nurtured and revised according to experience.

Principle 5

Sustainable development or sustainability (see also Annex 1.1).

In a similar way to the term environmental health, this concept does not just encompass certain issues, but also requires particular ways of managing them. In the policy-making process relating to environmental health three particularly important threads serve to confirm the overlapping nature of environmental health and sustainable development. This obviously links with Principle 1 and prevention being better than cure.

Principle 6

Environmental health issues are truly international in their character.

The ease of international communication and travel is making the world seem an ever-smaller place. Environmental health professionals have long recognised the fragility of the planet, and that the contaminants in our environment do not respect national boundaries. The world of environmental health is also small. The worldwide community of professionals who dedicate their working lives to improving and protecting the places we live in to safeguard health are tiny compared with those who work to exploit and deplete the world's resources in pursuit of 'wealth creation' but which might be both short term and short sighted. However, the diminutive character of the world's environmental health community brings advantages. Communication at the international level is crucial. Although languages may be different and the heritage and culture places EHPs in different systems, environmental health problems and approaches are mutual. The common wealth of knowledge can provide an irresistible resource for solving many of the perplexing dichotomies facing practitioners in their daily work. International co-operation and collaboration, such as through the International Federation of Environmental Health, is therefore a key principle for environmental health and is one that should not be overlooked, despite the distraction of our immediate surroundings and problems.

public health organisations. The CIEH and then Health Education Authority[29] looked at this, recognising that there had to be multidisciplinary delivery of public health of which environmental health is a part. The Environmental Health 2012 vision [16] captured the challenges, constraints and ideas from environmental health, public health and health improvement professionals from the various sectors and it incorporated the advice of key practitioners, professionals and academics from environmental health, local authorities, health services, voluntary and community groups.

Participants in that project reported that the mainstream practice of environmental health had become fixed on the delivery of a narrow agenda (and that may well have gotten worse as the result of financial cuts to local government). A number of factors were identified as preventing it from achieving its traditional involvement in addressing the wider determinants of health. They expressed concern about the fragmentation of environmental health services (more marked since the report was published), and lack of clarity on the nature of future environmental health roles and their contribution to health improvement and tackling health inequalities.

A previous edition of the book included that this group had said

> there was a reported lack of available resources to deliver the new approaches and initiatives called for by the modern public health agenda, or to participate fully in the new organizational structures for public health. In particular, the culture of performance management of specific enforcement targets has resulted in environmental health officers having to take on predominantly technical and enforcement roles.

This report is considered further in the next chapter.

This trend has continued and at the expense of the effective practice of the wider principles of environmental health

protection, and has had, and continues to have, the effect of deskilling many in the profession, leading to dissatisfaction among many environmental health practitioners. This has perhaps also been reflected in the difficulties in securing placements for environmental health students and as an example, the problem of both lack of skills and absence from the 'top tables' of policy and strategy development was highlighted in the report of the CIEH Commission on Housing Renewal and Public Health [17]. The Environmental Health 2012 project was carried out in England, but the findings have been relevant to all the countries within the United Kingdom and may have wider international application because similar issues have also been reported in other parts of the world, from America [18] to Australia [19] (also see later in this chapter).

An understanding of the role of environmental health in public health is important. If public health is the *science and art of preventing disease, prolonging health through organised efforts and informed choices of society, organisations, public and private, communities and individuals*[30] then environmental health is a major part of public health and environmental health practitioners are part of the wider public health workforce. Addressing the wider determinants of health, including food, housing standards, occupational health and safety, air quality, noise and environmental issues generally (upstream interventions), should mean that environmental health makes a fundamental contribution to the maintenance and improvement of public health, improving the quality of life and well-being of individuals and communities.

In order to demonstrate the importance of these contributions it was necessary to describe the sphere of environmental health. Figure 1.5 takes as its starting point the interface of environmental health as shown in Figure 1.4 [12]. The original interface has been developed and extended so as to include the main areas of activity of environmental health in food, water, air, land and buildings. The

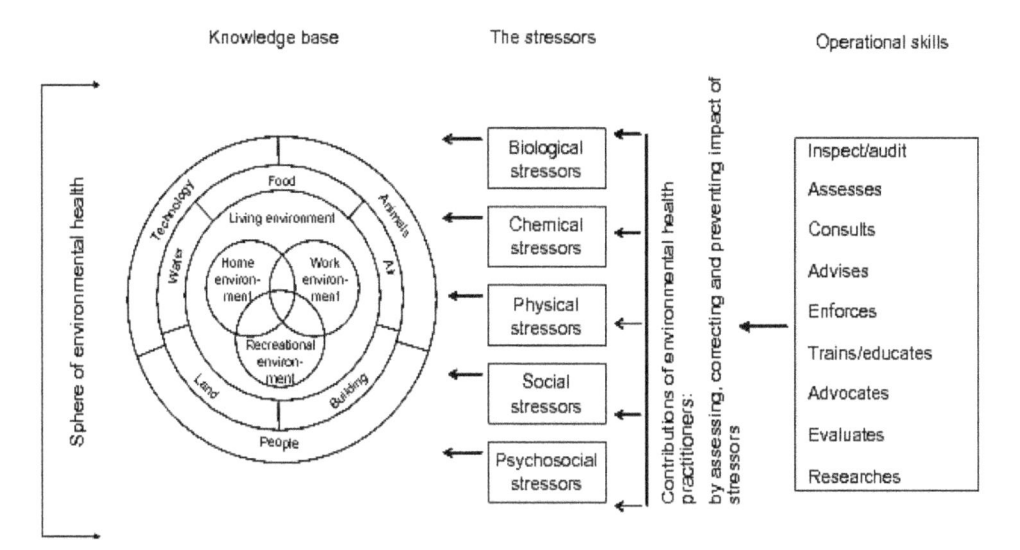

Figure 1.5 The 2012 vision model of environmental health
Source: Developed from Burke et al. [16]

Environmental health stressors

Environmental health stressors are features of the environment that may induce harm in or damaging responses to a living system or organism.

Biological stressors

Those biological elements of the natural and human-made world which present a direct risk to human health through ingestion, inhalation, inoculation or physical contact, and those miscellaneous elements that may influence biological systems to the detriment of humans and their environments.

Chemical stressors

Those chemical entities (or their intermediates) which, through their presence in a particular environment, expose the human to risk through ingestion, inhalation, inoculation or absorption and/or which interfere in biological systems to the detriment of humans and their environments.

Physical stressors

Those measurable physical manifestations induced naturally or through human activity that may impact unfavourably on human health through their damaging effects on cells, tissues, organs and homeostatic systems, as well as their impact on mental and social well-being.

Social stressors

Those behaviours associated with human life that are a consequence of settlement in communities and habitation and which have impacts on health and well-being.

Psychosocial stressors

Those attitudes of mind and mental processes that may have an adverse impact on the health of a person or community.

stressors have also been extended and specified (see boxed text on stressors).

A further explanation of the environmental stressors is given in the boxed text.

It is interesting to note that in the context of housing (the home environment) and the Housing Health and Safety Rating System (HHSRS) that the hazards in the home environment have been grouped into Physiological Requirements, Psychological Requirements, Protection against Infection and Protection against Accidents, reflecting work of the American Public Health Association from before the Second World War.[31]

The contributions of the environmental health practitioner to public health have been defined as assessing, correcting and preventing the impact of the stressors on the living environment. The operational skills required by the environmental health practitioner have been identified as assessing, consulting, advising, enforcing, training/educating, advocating, evaluating and researching and these are considered in more detail in Chapter 2.

Agendas for Change [20] had preceded the 2012 Vision [16] and was a further influence on the subject of environmental health and was carried out by the UK Commission on Environmental Health. The Commission was established by the Chartered Institute of Environmental Health (CIEH) in 1996 and recognised that the famous names of public health had all reacted to the conditions of the industrial revolution, and the accompanying physical, economic and social changes, as well as increased pressures on the environment,

but times had changed. While society still bore some physical hallmarks of the Victorian age it operated in a completely different, and more complex manner and one that changed more rapidly than any before it.

The Commission's thinking recognised that many of the environmental health problems that face society today are created by society and can only be tackled by the whole of society taking action. Figure 1.6 is indicative of this thinking (although coming from another source). It is relatively easy to get individuals to take action when it is a matter of individual health and self-interest. However, the new challenge is how to get whole communities to act, not for individual benefit but for the advantage of society as a whole or even a community. The example of road transport and air quality illustrates this point well, as it is only when a complete modal shift

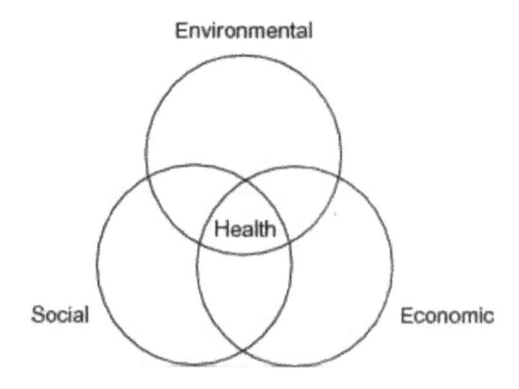

Figure 1.6 The social, economic and environmental axis

Source: Reproduced from WHO Europe [14]

in transport patterns occurs that air quality has a chance of improving. Individuals acting in isolation and not using their cars will not make a significant difference to air quality and will ultimately simply mean greater inconvenience for those people.

The Commission recognised, in particular, that there was a strong need for the integration of public and environmental health. Although since that time 'public health' has returned to local government in England, whether this integration has been achieved is open to question and is certainly variable. Firstly, in two-tier areas public health is an upper-tier function while environmental health is a lower-tier function. Furthermore, an independent assessment [21] published in 2020 conclude that the reforms, *coincided with austerity where local government funding, in the specific public health grant and more widely, was not prioritised by the government compared to NHS funding, receiving real-terms cuts from central government.*

The report also concluded that there is a need for clear local accountability in service delivery, and sustainable development processes provide many of the elements of good environmental health and these elements need to be nurtured to ensure sustainable change.

The CIEH Commission [20] identified five major principles as key components of environmental health, and these informed the Commission's approach:

- Precautionary approach (already touched upon);
- Inter-sectoral collaboration;
- Addressing inequalities and inequities;
- Community participation;
- Sustainable development.

It was recognised that equality and efficiency go together. Even if lip service is being paid to inequalities being a problem for all,[32] an unequal and divided society is a malfunctioning society, and malfunction carries a heavy and measurable penalty, in ill health, in deteriorating environment and in the resulting social and financial burden. In this context the reader is referred to Wilkinson and Pickett [15] where an epidemiological approach has been taken and similar conclusions reached.

There has to be a question whether we can rely on the assumption that the inclusion of health and the 'green' dimension in our planning, as well as managing our lifestyles better, will be effective in securing the changes required.

The fundamental message of the Commission's report was that as human beings we can only be truly healthy in a healthy environment [20]. It recognised that we cannot insulate ourselves from our surroundings – the air we breathe, the water we drink, the food we eat, the buildings and the landscapes we inhabit. Directly or indirectly, they will affect our health and well-being. These may seem to be unexceptional statements, yet they carry implications for the development of policy at the national and international level that are far from understood. They mean, for example, that we need a far more integrated and comprehensive approach to policy formulation, one that embraces health impacts even where they may not be obvious. The problem – and one of the reasons why this integrated approach does not yet exist – is that the target seems to be constantly shifting.

A new terminology evolved, speaking of interdependence and interconnectedness, of the importance of a holistic perspective and has been recognised for some time and as the LGMB set out [22] in the context of sustainable development and Local Agenda 21.[33] From professionals, this demands a flexibility of response that may run counter to specialisms. For the public, it can mean that patience with established procedures is lower and expectations much higher. It was argued that people not only expect their environment not to damage their health, but expect that it should promote their health and well-being, and that remains the case more than ever.

A number of studies have also shown that people feel better – less stressed, more content, less aggressive – in the presence of natural landscape features: greenery, trees, water (not all issues that traditionally EHOs address). Their mental performance improves, they feel refreshed, relaxed and reinvigorated, and with these feelings comes improved physical health. These elusive, qualitative issues probably underpin the movement of people from the cities in search of better health and quality of life. Of course, many more people do not have such choices.

In this context the environmental health practitioner should also consider the work of the Department of Health on public mental health; see for example *Confident Communities Brighter Futures – A Framework for Developing Well-Being* [23], which built on the evidence base for public mental health. It is no coincidence either that in the Housing Act 2004 health in the context of housing hazards and the Housing Health and Safety Rating System, health is defined as to include mental health.

Although often referred to as inequality, to a large extent what we are talking about is inequity, which is 'a want of (social) justice' and the fact or quality of being unfair. So unlike the term inequality it is clear that differences in health outcomes are not wholly a matter of some immutable law of nature. The equality issue – the gap between the haves, the have-less and the have-nots – emerged as one of the dominant themes of the 1990s. This is a gap that matters in human terms, and many examples exist of real health differentials[34] and this is highlighted by the work of Michael Marmot and the Institute of Health Equity. By and large, those on lowest incomes and the most vulnerable live in the unhealthiest housing, suffer the most degraded living environments, work in the worst jobs, have the lowest level of educational attainment and eat the least wholesome food (and pay proportionally more for it). Those higher up the social gradient will also have worse health than those further up the social gradient from themselves as Marmot has demonstrated. To say that poor environments and unhealthy lifestyles go together is not to say

that they always go together and that individuals, through effort or initiative, cannot transcend them; but people born into such circumstances have the dice loaded against them. Growing social and environmental inequality exact a great toll on health and life prospects. Resolving these issues requires political commitment.

The Commission considered lifestyle and the two main components to the lifestyle agenda, both of which depended on the idea that there is such a thing as an identifiable Western lifestyle and that this carries implications for the health of individuals, society and the global environment. The first is that such lifestyles are unhealthy, and that they tend to reinforce social and environment inequality. The second is that they are unsustainable – that the Western oil-based lifestyle cannot be transferred wholesale to the rest of the world without enormous damage to the biosphere.

Yet, healthy and sustainable lifestyles, like green consumerism, require knowledge and empowerment that comes with financial security as well as an understanding of the consequences of choices made. It is not just a matter of being 'energy efficient' but overall of using and consuming less by those higher up the social scale. However, in an highly unequal society it is also very difficult for individuals to 'make healthy choices' when the economic pressures mean that keeping a cold house warm could increase personal indebtedness.

The matter of democracy has already been considered. Several factors prompted the search for a new democratic model, including the loss of local control to the new global economy and global trade with global businesses being richer than many countries; the difficulty of reconciling long-term issues of sustainable development with short-term political preoccupations; and the development of local action around specifically local environmental or health issues [24–25].

A widespread cynicism also attaches to politics and politicians. Local government has been emasculated: its powers and competencies have been undermined and funding reduced. At the same time there is an impatience with established procedures. Over the past three decades there has been a rapid expansion of civil society – the network of charities, voluntary organisations and campaigning and pressure groups through which much political activity is focussed and directed. Cumulatively, such pressures point towards a new model for devolving power and involving people, perhaps a new model of community ownership. Whether that comes about remains to be seen, again the influence of the COVID-19 pandemic might bring about a 'new normal'.

When looking at the history of environmental or public health, in addition to the images of public health set out in the first section referring to Rayner and Lang [26] there are probably four eras reflected in this chapter. The first could be typified as the 'miasma period' when disease was thought to be caused by malodours and air pollution (which did at least lead to improved sanitation). Then the period when 'germ theory' held sway – the era of infectious disease and individual action including vaccination as well as improved sanitation and living conditions. The third and more recent period can be typified as concerned with chronic disease so that poor health was linked to individual behaviour and choices and was considered as the 'black box' era.[35] The final and more current era is the era of 'ecological public health', when public health is seen as the result of complex interactions of social, behavioural and physical factors. Health services have tended to be 'medical services' treating the effects of disease and ill health rather than trying to prevent ill health in the first place.

Figure 1.7 sets out what Rayner and Lang [26] have called the complex model of

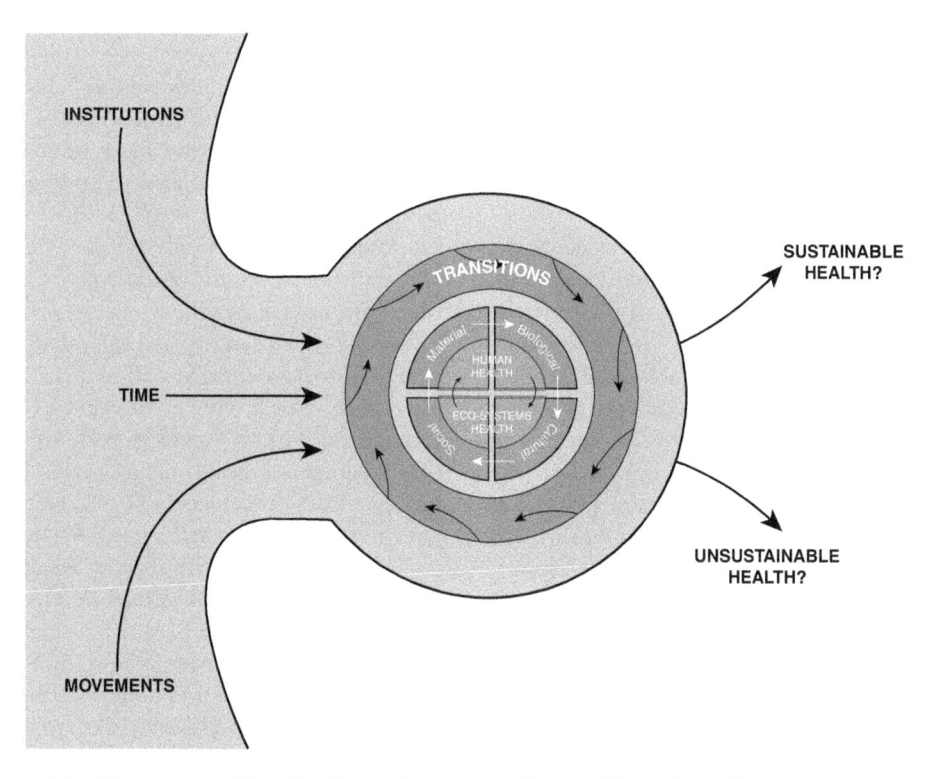

Figure 1.7   The model within the flow of events leading to different health outcomes
Source: Rayner & Lang [26], at page 326 and reproduced by kind permission of Taylor and Francis

ecological public health and in this chapter an attempt has been made to show how EHPs can contribute to ecological public health (and be equipped to do so). The figure represents how the flow of events can contribute to sustainable or unsustainable health outcomes. In that sense anything that does not address, and redress health inequalities must be an unsustainable health outcome. Environmental health practitioners are indeed part of a movement that seeks to improve public health by addressing those stressors that adversely impact on public health.

Readers can themselves consider the areas of work that can contribute to addressing these issues. This handbook provides pointers on how those contributions can be made. A wish to make a positive difference in people's lives, to improve their health and to prevent exploitation by those with economic or other power that prejudices that, is a common motive of environmental health practitioners. The authors hope that readers will accord with that view and rise to the challenges that this presents.

An objective for environmental health

The aim of this first chapter is to consider the history, philosophy, purpose and practice of environmental health. It is clearly the case that environmental health, whilst firmly rooted in the history of public health, needs to adapt and modernise in order to meet the demands and expectations of changing societies and new generations. It is more than the application of technical expertise whether as a generalist or specialist. Fundamentally and regardless of governments, or government policies, the modern practice of environmental health must be embedded within sustainable development in order to deliver the long-term and long-lasting improvements in both our public health and our environment.

The WHO's Global Strategy on Health Environment and Climate Change [27] aims to provide a vision and way forward on how the world and its health community *need to respond to environmental health risks and challenges until 2030, and to ensure safe, enabling and equitable environments for health by transforming our way of living, working, producing, consuming and governing.* In this context, environmental risks to health are defined as all the environmental physical, chemical, biological and work-related factors external to a person, and all related behaviours. The strategy focusses especially on the part of the environment that can reasonably be modified. Avoidable environmental risks are said to cause about one quarter of all deaths and disease burden worldwide, amounting to some 13 million deaths each year. A healthy environment is vital for human health and development. The sustainability of health systems is put at risk if the upstream determinants of disease and ill health are not seriously tackled. The WHO argues that *approaches that focus on treatment of individual diseases rather than reducing the adverse impact of determinants of health will be insufficient to tackle modern environmental health challenges.* In order to address the challenges in health, environment and climate change, governments and society, including professions and individuals, will all need to continue to rethink the way we live, work, produce, consume and govern. There needs to be effective and equitable action on the drivers of environmental risks to health.

The acceptance of this and the incorporation of previous definitions offers a challenging objective for future environmental health practice. For example, until recently the role of environmental health in improving public mental health had hardly been considered but the report and underlying evidence base published by the Department of Health in 2010 demonstrates that mental health and well-being is as much a part of environmental health as accident prevention [23]. Social and health inequalities can both result in and be caused by mental ill health. One in six of the adult population in the UK experiences mental ill health at any one time, often as the result of, or exacerbated by, those factors and social determinants where the EHP can intervene.

Sometime EHPs fail to see their role in improving mental well-being which can be defined as

> a dynamic state, in which the individual is able to develop their potential, work productively and creatively, build strong and positive relationships with others, and contribute to their community. It is enhanced when an individual is able to fulfil their personal and social goals and achieve a sense of purpose in society.
>
> [28]

Taking account of what has been written thus far and also such matters as the WHO Ostrava and Helsinki Declarations,[36] and starting from what has been written in previous editions of this book the objective for environmental health can be set out as to:

> Achieve strategies and equitable action to intervene between people and the environments in which they live, work and take leisure while at the same time reconciling personal choice and social responsibilities in health so as to prevent premature death, reduce disease and ill health, disability, disfunction, harm, discomfort, stigma, disadvantage and prejudice and to create supportive and sustainable environments that allow and help people to live healthy lives.

Remembering, as has already been said, prevention is better (and usually cheaper) than cure, and as the NEHPC [11] has argued, *for every $1 spent on national and state-level programs, $71 in asthma-related expenditure is saved.*

In England outside the clinical arena the key responsibility for improving the health of local populations, including reducing health inequalities, now rests with upper (county) tier and unitary local authorities. In this context it is unfortunate that environmental health is a lower-tier function in two-tier authorities. The Health and Social Care Act 2012 gave each unitary and upper-tier local authority in England the duty to *take such steps as it considers appropriate for improving the health of the people in its area.* Local authorities set up statutory health and well-being boards to drive local commissioning and integration of all health services, based upon local needs, giving new opportunities to improve the health and well-being of local communities right across the life course. This arrangement provides an opportunity for EHPs to contribute ever more to improving public health, given their skills, knowledge and expertise. Public Health Outcomes Framework consists of two overarching outcomes that set the vision for the whole public health system. These outcomes are increased healthy life expectancy, which takes account of the health quality as well as the length of life and reduced differences in life expectancy and healthy life expectancy between communities (through greater improvements in more disadvantaged communities). In essence the Public Health Outcomes Framework sets out a vision for public health, to improve and protect the nation's health, and improve the health of the poorest fastest.

Table 1.2 Setting out the main headings of the Public Health Outcomes Framework and it highlights some areas as examples where it is envisaged EHPs can contribute

---

1 Improving the wider determinants of health
  Objective: Improvements against wider factors which affect health and well-being and health inequalities
  The percentage of the population affected by transport noise;
  Fuel poverty.

(Continued on next page)

Table 1.2 (continued)

2 Health improvement

Objective: People are helped to live healthy lifestyles, make healthy choices and reduce health inequalities

The percentage of those 18+ who are overweight or obese;
Diet;
Smoking prevalence – adults (over 15);
Falls and injuries in people aged 65 and over.

3 Health protection

Objective: The population's health is protected from major incidents and other threats whilst reducing health inequalities

Fraction of mortality attributable to particulate air pollution;
Population vaccination coverage;
NHS organisations with board-approved sustainable development management plan;
Comprehensive, agreed inter-agency plans for responding to health protection incidents and emergencies;
Anti-microbial resistance.

4 Healthcare public health and preventing premature mortality

Objective: Reduced numbers of people living with preventable ill health and people dying prematurely, while reducing the gap between communities.

Infant mortality;
Mortality rate from causes considered preventable;
Mortality from communicable diseases;
Hip fractures in people aged 65 and over;
Excess winter deaths.

Source: Public Health Outcomes Framework at https://assets.publishing.service.gov.uk/government/uploads/system/uploads/attachment_data/file/862264/At_a_glance_document2.pdf (Accessed 26 July 2021)

Overarching indicators

Wider determinants	Health improvement	Health protection	Healthcare and premature mortality
Rate of complaints about noise	Smoking prevalence (a number of specific indicators)	Fraction of mortality attributable to particulate air pollution	Infant mortality rate
Percentage of the population exposed to road, rail and air transport noise of 65dB(A) or more during the daytime	Self-reported well-being (a number of specific indicators)	TB incidence	Mortality rate from causes considered preventable
Percentage of the population exposed to road, rail and air transport noise of 55dB(A) or more during the night-time	Emergency hospital admissions due to falls in people aged 65 and over (for EHPs the home is the area of concern)	NHS organisations with a board-approved sustainable development plan	Under 75 mortality rate from respiratory disease considered preventable
Fuel poverty			Excess winter deaths
			Hip fractures in older people (a number of age-related indicators)

Source: Derived from https://fingertips.phe.org.uk/profile/public-health-outcomes-framework

The framework focusses on the two high level outcomes we want to achieve across the public health system and beyond:

1 Increased healthy life expectancy
2 Reduced differences in life expectancy and healthy life expectancy between communities

Section 2 References

[1] Raworth K. (2017) *Doughnut Economics – Seven Ways to Think Like a 21st-Century Economist*, Random House Business, London.

[2] Marmot M, Allen J, Goldblatt P, Boyce T, McNeish D, Grady M, Geddes I. (2010) *Fair Society, Healthy Lives: The Marmot Review – Strategic Review of Health Inequalities in England Post-2010*, UCL, London. http://www.instituteofhealthequity.org/resources-reports/fair-society-healthy-lives-the-marmot-review/fair-society-healthy-lives-full-report-pdf.pdf.

[3] Marmot M, Allen, J, Boyce T, Goldblatt P, Morrison J. (2020) *Health Equity in England: The Marmot Review 10 Years on*, Institute of Health Equity, UCL, London.

[4] Jackson T. (2011) *Prosperity Without Growth – Economics for a Finite Planet*, Earthscan, London.

[5] WHO. (2008) *Closing the Gap in a Generation Health Equity Through Action on the Social Determinants of Health*, Final Report of the Commission on Social Determinants of Health, WHO, Geneva.

[6] HM Treasury. (2002) *Securing Our Future Health: Taking a Long-Term View – the Wanless Report*, HM Treasury, London.

[7] Dahlgren G, Whitehead M. (2007) *Policies and Strategies to Promote Social Equity in Health*, Institute for Futures Studies, Background Document to WHO – Strategy Paper for Europe, Stockholm, This Working Paper Was Originally Published in Print Form in September 1991. The Figure "The Main Determinants of Health" Was Revised in This Version. https://www.iffs.se/media/1326/20080109110739filmz8uvqv2wqfshmrf6cut.pdf.

[8] Barton H, Grant M. (2006) A health map for the local human habitat. *Journal for the Royal Society for the Promotion of Health*, 126(6): 252–261.

[9] WHO. (1989) *Environment and Health: A European Charter and Commentary*, WHO, Copenhagen.

[10] WHO. (1994) *Action Plan for Environmental Health Services in Central and Eastern Europe and the Newly Independent States*, Report on a WHO consultation, Sofia, Bulgaria, 19–22 October 1993. WHO Regional Office for Europe Copenhagen (document EUR/ICP/CEH123).

[11] National Environmental Health Partnership Council. (2017) *Environmental Health Playbook: Investing in a Robust Environmental Health System*, APHA. https://apha.org/-/media/files/pdf/topics/environment/eh_playbook.ashx?la=en&hash=FBAE72B837D58A3C3145602B1043250DE7BD41BD.

[12] MacArthur ID, Bonnefoy X. (1998) *Environmental Health Services in Europe – Policy Options*, WHO, Copenhagen.

[13] UN. (1992) *Earth Summit. Agenda 21: The United Nations Programme of Action from Rio*, United Nations Department of Information, New York.

[14] WHO. (1998) *Health for All in the Twenty First Century*, WHO, Geneva.

[15] Wilkinson R, Pickett K. (2009) *The Spirit Level – Why Equality Is Better for Everyone*, Allen Lane, London.

[16] Burke S, Gray I, Paterson K, Meyrick J. (2002) *Environmental Health 2012 – A Key Partner in Delivering the Public Health Agenda*, Health Development Agency, London.

[17] CIEH. (2007) *Commission on Housing Renewal and Public Health – Final Report*, CIEH, London. https://www.cieh.org/media/1246/commission-on-housing-renewal-and-public-health-final-report.pdf.

[18] CDC. (2002) *A Strategy to Revitalize Environmental Health Services in the United States*, Centres for Disease Control and Prevention, Department of Health and Human Services, Atlanta Georgia, Unpublished Working Draft.

[19] Bell S. (2002) Environmental health: Victorian anachronism or dynamic discipline? *Journal of the Australian Institute of Environmental Health*, 2(4).

[20] CIEH. (1997) *Agendas for Change – Report of the Environmental Health Commission*, Chartered Institute of Environmental Health, London.

[21] Buck D. (2020) *The English Local Government Public Health Reforms: An Independent Assessment*, The Kings Fund, London.

[22] LGMB. (1996) *Health and Sustainable Development – Local Agenda 21 – Roundtable Guidance*, Local Government Management Board, Luton.

[23] Department of Health. (2010) *New Horizons, Confident Communities, Brighter Futures, A Framework for Developing Well-Being*, DoH, London.

[24] WHO. (1997) *Health and Environment in Sustainable Development – Five Years After the Earth Summit*, WHO, Geneva.

[25] WHO, EURO. (1997) *Sustainable Development and Health: Concepts, Principles and Framework for Action for European Cities and Towns*, World Health Organization Regional Office for Europe, Copenhagen.

[26] Rayner G, Lang T. (2012) *Ecological Public Health – Reshaping Conditions for Good Health*, Earthscan, Routledge, Abingdon, Oxon.

[27] WHO. (2020) *Global Strategy on Health, Environment and Climate Change: The Transformation Needed to Improve Lives and Well-Being Sustainably Through Healthy Environments*, World Health Organization, Geneva. Licence: CC BY-NC-SA 3.0 IGO.

[28] Foresight Mental Capital and Wellbeing Project. (2008) *Mental Capital and Wellbeing – Making the Most of Ourselves in the 21st Century*. https://www.gov.uk/government/uploads/system/uploads/attachment_data/file/292450/mental-capital-wellbeing-report.pdf.

Other useful reading and sources of information

Department of Health (1998) *Our Healthier Nation: A Contract for Health*. Cm.3852, The Stationery Office, London. https://assets.publishing.service.gov.uk/government/uploads/system/uploads/attachment_data/file/265721/title.pdf (Accessed September 2021).

Department of Health (2009) *New Horizons – A shared vision for mental health, CoI. London* https://www.nhs.uk/NHSEngland/NSF/Documents/NewHorizonsConsultation_ACC.pdf (Accessed September 2021).

Faculty of Public Health. (2016) *Good Public Health Practice Framework*. https://www.fph.org.uk/media/1304/good-public-health-practice-framework_-2016_final.pdf (Accessed September 2021) and (2018) *Embedding Sustainable Development in UK Public Health Training*. https://www.fph.org.uk/media/2268/sdn-report-final-2017-2018-nov1-1.pdf.

Public Health Research and Practice (formerly *NSW Public Health Bulletin*), New South Wales Ministry of Health. https://www.phrp.com.au/.

Stewart, J, Cornish, Y, Eds. (2009) *Professional Practice in Public Health*, Reflect Press, Exeter, Devon, UK

WHO Europe http://www.euro.who.int/en/home.

WHO (1998) *Environmental Health Services in Europe – An Overview of Practice in the 1990s*. London World Health Organization, Stationery Office. https://www.euro.who.int/__data/assets/pdf_file/0003/109875/E66792.pdf

Section 2 notes

20 See https://doughnuteconomics.org/tools-and-stories/111- Doughnut Economics Action Lab (DEAL) is part of the emerging global movement of new economic thinking and doing that is rising to this challenge. Our aim is to help create 21st-century economies that are regenerative and distributive by design, so that they can meet the needs of all people within the means of the living planet.

21 See for example https://www.local.gov.uk/using-marmot-principles-tackle-health-inequalities-and-covid-19-23-june-2020

22 For a discussion of the model and its relevance 30 years on see Dahlgren G, Whitehead M. (2021) The Dahlgren-Whitehead model of health determinants: 30 years on and still chasing rainbows. *Public Health*, 199: 20–24.

23 The first European Conference on Environment and Health was held in Frankfurt-am-Main, Germany and adopted the European Charter on Environment and Health available at https://www.euro.who.int/__data/assets/pdf_file/0019/114085/ICP_RUD_113.pdf

24 Declaration of the Sixth Ministerial Conference on Environment and Health, Ostrava, Czech Republic June 2017. https://www.euro.who.int/en/media-centre/events/events/2017/06/sixth-ministerial-conference-on-environment-and-health/documentation#336668

25 https://www.who.int/health-topics/environmental-health#tab=tab_1; Environment and Health for European Cities in the 21st Century, WHO 2017. https://www.euro.who.int/__data/assets/pdf_file/0020/341615/bookletdef.pdf

26 APHA is a champion and founding partner of the National Environmental Health Partnership Council. The Council strives to support healthy people by working for healthier environments. And the Council brings together diverse stakeholders to help expand and sustain awareness, education, policies and practices related to environmental health, available at https://apha.org/topics-and-issues/environmental-health/partners/national-environmental-health-partnership-council

27 http://www.un.org/disabilities/documents/rio20_outcome_document_complete.pdf

28 https://www.nuffieldbioethics.org/news/ten-questions-on-the-next-phase-of-the-uks-covid-19-response; Statement of 25 April 2020 at https://www.nuffieldbioethics.org/news/statement-covid-19-and-the-basics-of-democratic-governance

29 The Health Education Authority, established in 1987 as a special health authority to encourage

health education and health promotion, and largely funded by the UK government's Department of Health. It was subsequently replaced by the Health Development Agency, which became part of the National Institute for Health and Clinical Excellence (NICE) which in turn became the National Institute for Health and Care Excellence.

30 As in Acheson D, (1988) *Public Health in England: Report to the Committee of Inquiry into the Future of the Public Health Function*, HMSO, London, and amended in Wanless D. (2004) *Securing Good Health for the Whole Population: Final Report*, DoH, London.

31 This can be seen in the Basic Principles of Healthful Housing. New York: American Public Health Association, 1938.

32 See for example http://www.oecd.org/health/inequalities-in-health.htm

33 Local Agenda 21 is conceptualised in chapter 28 of Agenda 21, which was adopted by 178 governments at the 1992 Rio Conference. Agenda 21 recognised that many environmental problems can be traced back to local communities and that local governments have an important role to play in implementing environmental programmes and gathering community support.

34 For example, see the Acheson Report (*An Independent Inquiry into Inequalities in Health in England*) https://www.gov.uk/government/uploads/system/uploads/attachment_data/file/265503/ih.pdf

35 See for example Weed DL. (1998) Beyond black box epidemiology. *Am J Public Health*, 88(1): 12–14.

36 Declaration on Action for Environment and Health in Europe, Second European Conference on Environment and Health, Helsinki, Finland, June 1994.

SECTION 3: FURTHER THOUGHTS ON ENVIRONMENTAL HEALTH

This section includes two essays that consider different aspects of the issues facing environmental health practitioners. The first looks at environmental health in Australia and how EHPs are viewed. The second essay looks at the precautionary principle and ethics looking particularly at e-cigarettes and 'vaping'.

Environmental health in Australia

Harriet Whiley

Environmental health

The Australia Department of Health subscribes to the World Health Organization definition of Environmental Health (considered earlier in this chapter) and goes on to state:

Environmental health involves those aspects of public health concerned with factors, circumstances, and conditions in the environment or surroundings of humans that can exert an influence on health and well-being.

Environmental health provides the basis of public health. Improvements in sanitation, drinking water quality, food safety, disease control, and housing conditions have been central to the significant improvement in quality of life and longevity experienced over the last hundred years.

Environmental health practice addresses emerging health risk arising from the pressures that human development places on the environment.

[1–2]

This definition of environmental health represents a broader, more holistic, approach and understanding. It acknowledges that environmental health covers a range of complex and multidisciplinary issues and moves away from the traditional perspective that environmental health practice is primarily focussed on regulation and enforcement.

In Australia, the peak environmental health advisory group is the National Environmental

53

Health Standing Committee (enHealth), which is part of the Australian Health Protection Principal Committee under the Australian Health Ministers' Advisory Council. This enHealth is responsible for providing nationally agreed policy advice, co-ordination and support needed to respond to environmental health issues. Presently, enHealth consists of representatives from Commonwealth, state and territory health departments; the New Zealand Ministry of Health; and the National Health and Medical Research Council (NHMRC) and provides advice to the Australian Health Protection Principal Committee.

Environmental health practice

Australia is a federation of six states and two territories. There is both Commonwealth legislation that covers the whole federation, and state legislation created by each state or territory. The federal government can only pass laws under the powers granted to it under the Constitution of Australia. If there is a discrepancy between two pieces of legislation, then federal law overrides state and territory law [3]. Each state and territory create local governments/councils and these take primary responsibility for delivering environmental health services.

In Australia the term Environmental Health Officer (EHO) refers to an Environmental Health Practitioner (EHP) with an appropriate tertiary degree level qualification recognised by the relevant state or territory authority. EHOs work in federal, state and local government, in the defence forces and within private practice; however, the majority are employed by local governments (sometimes referred to as municipal governments or councils) [4–5]. The term EHP refers to any individual employed in the environmental health space and encompasses Indigenous Environmental Health Workers (IEHW) and can refer to any person with or without environmental health qualifications including Environmental Health Coordinators,

Environmental Health Supervisors, Environmental Health Technicians, Healthy Housing Workers and Animal Welfare Workers [6–7].

In Australia, EHOs are authorised officers responsible for the regulation of environment-related activities that can impact human health. This requires them to undertake investigations, assess health hazards and risks, determine and direct remedial action, enforce legal requirements, prosecute legal cases, educate and develop public health and emergency management policy and plans. The specific functional areas EHOs are involved in include food safety, infection control, water management, environmental protection, planning development, buildings and accommodation, tobacco and control of drugs and poisons [2]. One interesting difference between Australian EHOs and UK EHPs is that Australian EHOs have limited involvement in housing. Australian EHOs deal with environmental health issues related to supported residential facilities and different types of accommodation. However, they would only get involved with government or private housing if the risks are perceived to be a public health threat that affects neighbouring environments (e.g. rats or mosquitoes that are entering a neighbouring property or garden). Australian EHOs do not regulate housing conditions that only affect the health and safety of the occupier, such as mould, lack of personal hygiene facilities, fire hazards or indoor pollutants. An exception to this is methamphetamine contamination. If the police alert local government that a house has been used to manufacture illicit drugs, EHOs will place a notice on the house that it should not be inhabited until assessment and remediation has been conducted. This notice will serve to warn any potential buyers if the house is sold before remediation is completed.

EHOs in Australia overlooked and underrated

In Australia, environmental health is under-recognised and undervalued. It is the invisible profession, and no one knows about the essential work of EHOs. In part this is due to the preventative nature of environmental health which means that it only receives attention when things go wrong and all the barriers and defences that have been put in place fail. Environmental health is also misunderstood by the general public. This was illustrated by a survey of university students who were given a series of roles and activities and asked to indicate which were the responsibility of an EHO. The students answered wrong more times than if they had just guessed the answer for every single question [8].

Furthermore, EHOs are invisible within their own workplaces. Municipal councils have a key role in planning and delivering environmental health services and enHealth stipulates that a risk-based approach is needed to achieve this.

> A risk based approach underpins management of health protection in Australia. This means a standardised approach to the classification, identification and management of environmental health risks, with clear delineation of risk mitigation responsibilities.
>
> *[2]*

However, due to the increasing pressure placed on local government to increase productivity and efficiency with diminishing resources (a consequence of capping of local government rates) [9] the description of EHO roles and responsibilities have been condensed to focus on regulatory responsibilities. For example, current key performance indicators for EHOs include the total number of inspections rather than the quality of the preventive measures put in place. This is further evidenced by a survey of councils' annual reports, and websites, which found the environmental health profession was considered to have a specific focus on securing minimum levels of compliance required by legislation. The skills and knowledge of the profession and their capability utilise a risk-based approach to ensure public health protection was not conveyed in these documents

[10]. This issue has been further compounded by the increasing scope and breadth of environmental health due to emerging issues. As a consequence, EHOs have increasingly limited time to become involved in important but non-regulatory responsibilities, such as planning and emergency management (Psarras, pers comm. 2018 in [8]).

The lack of recognition by their own organisations often results in experienced EHOs leaving the profession. A workforce survey conducted in South Australia in 2010 found that 36.5% of EHOs were looking to move organisations within the next year or two and 16.2% were looking to leave soon [6]. This supported the findings of a previous workforce review conducted in Victoria that found one in three EHOs were looking to leave their job in the next five years. In the Victorian workforce review, lack of support was identified as a factor that really impacted on job satisfaction, with a number of EHOs voicing their frustration about their capacity to effectively fulfil their role due to an overwhelming workload [7].

Limited opportunities for career progression is another issue that impacts on job satisfaction, with just under a third of Victorian EHOs identifying career opportunities as an issue of high or very high importance and only 13% identifying that they were satisfied with current prospects [7]. One possible reason for the lack of progression within an organisation is the absence of a formalised national professional development scheme. This differs from the current Chartered Institute of Environmental Health (CIEH) accreditation scheme in the UK. Another reason might be the invisibility of the profession and the lack of advocacy at the higher levels within organisations. In both the UK and in Australia, this often results in EHOs/EHPs being managed by non-environmental health professionals who do not appreciate the importance, responsibilities and complexities involved in environmental health practice. This has a snowball effect; EHOs/EHPs are overworked and too time poor to make advocacy a priority and as a consequence are under-resourced and overworked [4, 6–7].

EHOs and the COVID-19 pandemic

The COVID-19 pandemic has shown the world that environmental health is one of the most important professions for public health protection [11–12]. It has also highlighted the importance of ensuring that the environments we live in are safe. In Australia, EHOs were assigned with a range of COVID-19 tasks that varied between states and locations. Responsibilities included, but were not limited to, isolation and quarantine compliance checks, contract tracing, new food safety inspections for take-away options, and advising social distancing restrictions for public spaces [13] (under review). Despite this increase in workload, the role of Australian EHOs in the pandemic response has been successful in raising the profile [14]. This was particularly evident within some municipal councils, where EHOs were considered the 'go-to people' for infection control advice [15].

The future

The role of the environmental health profession in today's society is complex and constantly evolving to address emerging health risks. These new risks are arising from the pressures that human development places on the environment, climate change and the increasing vulnerabilities of local communities. Currently the environmental health workforce in Australia is limited. It is insufficient to cover current environmental health problems and therefore has limited capability to tackle emerging issues. It is essential that the profile of environmental health is raised and the future workforce needs are secured to protect public health now and into the future [8]. To achieve this, environmental health in Australia, and perhaps globally, needs to proactively engage with the media and develop a marketing and communication strategy to raise the profile of the profession [16].

Essay references

[1] Australia Department of Health. (2014) *Overview of Environmental Health.* https://www1. health.gov.au/internet/main/publishing.nsf/ Content/health-pubhlth-strateg-envhlth-index.htm (Accessed 12 October 2020).

[2] Enhealth. (2016) *Preventing Disease and Injury Through Healthy Environments: Environmental Health Standing Committee (enHealth) Strategic Plan 2016 to 2020,* Environmental Health Standing Committee (Enhealth) and Endorsed by the Australian Health Protection Principal Committee, Canberra, Australia. https:// www1.health.gov.au/internet/main/publishing.nsf/content/A12B57E41EC9F326CA257 BF0001F9E7D/$File/Standing-Committee-Strategic-Plan-2016-2020.pdf (Accessed 12 October 2020).

[3] Parliament of Australia. (2018) *Infosheet 20 – The Australian System of Government,* Canberra, Australia. https://www.aph.gov.au/About_ Parliament/House_of_Representatives/Powers_practice_and_procedure/00_-_Infosheets/ Infosheet_20_-_The_Australian_system_of_ government (Accessed 20 October 2020).

[4] Australian Institute of Environmental Health. (2007) *National Local Government Environmental Health Workforce Summit: The Changing Landscape of the Environmental Health Workforce in Local Government: Turning Threats into Opportunities,* Brisbane, Australia. https://www. eh.org.au/documents/item/358 (Accessed 10 May 2018).

[5] Parliament of Australia. (2018). *Infosheet 20 – The Australian System of Government.* https://www.aph.gov.au/About_Parliament/ House_of_Representatives/Powers_practice_ and_procedure/00_-_Infosheets/Infosheet_20_- _The_Australian_system_of_government (Accessed 20 October 2020).

[6] Bartosak C. (2012) *Environmental Health Workforce Attraction and Retention – Research Paper,* Environmental Health Australia, Adelaide, South Australia. https://www.eh.org.au/documents/item/247 (Accessed 10 May 2018).

[7] Windsor and Associates. (2005) *Environmental Health Officer Workforce Review,* Victorian Department of Human Services, Victoria, Australia. https://www2.health.vic.gov.au/ about/publications/policiesandguidelines/ Environmental-Health-Officer-Workforce-Review (Accessed 12 October 2020).

[8] Whiley H, Willis E, Smith J, Ross K. (2019) Environmental health in Australia: Overlooked and underrated. *Journal of Public Health*, 41: 470–475.

[9] Productivity Commission. (2017) *Shifting the Dial: 5 Year Productivity Review*, Supporting Paper No. 16, Local Government, Canberra, ustralia.

[10] Smith J, Whiley H, Ross K. (under review) *The New Environmental Health in Australia: Failure to Launch*.

[11] Morse T, Chidziwisano K, Musoke D, Beattie TK, Mudaly S. (2020) Environmental health practitioners: A key cadre in the control of COVID-19 in sub-Saharan Africa. *BMJ Global Health*, 5: e003314.

[12] US National Environmental Health Association. (2020) *2020 World Environmental Health Day Declaration*, NEHA, Denver. https://www.neha.org/sites/default/files/news-events/press-releases/2020-World-EH-Day-Declaration.pdf (Accessed 27 October 2020).

[13] Rodrigues MA, Silva MV, Errett NA, Davis G, Lynch Z, Dhesi S, Hannelly T, Mitchell G, Dyjack D, and Ross KE. (2021) How can Environmental Health Practitioners contribute to ensure population safety and health during the COVID-19 pandemic?. *Safety Science*, 136, p. 105–136.

[14] Porter M. (2020) Meet the public health unit's environmental health workers keeping us safe from COVID-19. *The Leader*, NSW Australia. https://www.theleader.com.au/story/6930340/meet-the-public-health-workers-keeping-us-safe/ (Accessed 12 October 2020).

[15] Moore N. (2020) *EHO COVID-19 Snapshot*, COVID-19 Seminar: Environmental Health Australia, South Australia.

[16] Houghton F. (2019) Re: Environmental health in Australia: Overlooked and underrated. *Journal of Public Health*, 42: 435–436.

MORAL FOUNDATIONS OF THE PRECAUTIONARY PRINCIPLE: JUUL AND THE HAZARDS OF VAPING

Levente Szentkirályi

Consider the ongoing prescription opioid crisis, involving companies like Purdue Pharma, which claims the lives of thousands of Americans each year[37] [1–3], or the preventable incidences of various cancers over several decades that have been attributed to Johnson and Johnson's 'asbestos-tainted' talc body products [4], or the disquieting cases of non-Hodgkin's lymphoma associated with exposure to Monsanto's *Roundup* herbicide products [5–7] or the growing rates of nicotine addiction and lung-related illness among our youth caused by e-cigarettes, such as JUUL and its iconic, sleek vaping devices. American environmental history is replete with examples of corporations deliberately falsifying or withholding scientific data that verifies risks to human health, misleading the public with embellished or false public-relations campaigns about the actual health threats their business practices

create, manufacturing products and emitting substances without adequately testing their potentially harmful health effects, delaying the implementation of policies that could prevent harm to public health and using their political clout to influence business-friendly environmental policies.[38] The case of JUUL and vaping, which this brief discussion highlights, is no exception, and illustrates the sort of corporate irresponsibility and lack of regulatory oversight we should strive to avoid.

The challenge, however, with such cases that involve novel technologies or substances, like aerosol nicotine delivery mechanisms, is that the potential adverse health effects are often uncertain or scientifically unverified. Also, within the dominant framework for environmental risk management in the United States – grounded in the cold, consequentialist calculus of quantitative risk assessment and risk-benefit balancing – until a preponderance of scientific evidence corroborates actual risks of harm, environmental health issues are not ripe for resolution. In other words, prevailing uncertainty is alleged to undermine the basis for preventative measures that may mitigate the possibility for harm. Yet, paralleling this approach to managing environmental health risk, where the onus is squarely on public health scientists and regulatory agencies to prove the existence of unreasonable risks to the public, are policies that protect the intellectual property rights of industry, such as patents and trade secrets. These protections create incentives for industry to withhold proprietary data about the actual health risks that their products and business practices may entail, which undermine *both* the ability of regulatory agencies to effectively monitor industry and hold corporations accountable, as well as the ability of legislative bodies to implement laws and regulations that can effectively safeguard our communities from potential health hazards.

Indeed, these dynamics defined our high-profile experiences, with tetraethyl lead (fuel additive) in tailpipe emissions in the 1930s, DDT in the 1950s, DES (synthetic estrogen administered to pregnant women) in the 1970s, traditional tobacco products in the 1990s, and they currently define the myriad current challenges we face with poorly understood substances like e-cigarette aerosols – which I have defined elsewhere as *uncertain environmental threats* [8]. By creating the *de facto* implication that under conditions of uncertainty – when we neither know the outcome of exposure, nor the probability that exposure will beget some harm – an abundant sum of actual cases of illness or death is a necessary condition for justifying health regulations, this risk management framework allows industry to gamble with public welfare by exposing our communities to uncertain threats that may well prove to be harmful or fatal. Expressed differently, under conditions of uncertainty, the standard that this framework endorses would have us accept that before we can be protected from environmental harms, we must first become victims

of the very harm that environmental safeguards aim to prevent.

As I have argued in detail in *The Ethics of Precaution*,[39] such a standard is entirely untenable, and is founded on the faulty assumption – both among policy analysts and ethicists – that when we lack the evidence to ascertain credible risks to human health, industry has no obligation to take reasonable strides to nevertheless protect the public against uncertain possibilities of harm. Uncertainty does *not* give industry license to act in ways that may prove injurious to others, and it does not attenuate industry's culpability for failing to exercise precaution or engendering actual harm. In briefly noting some of the questionable actions that JUUL Labs, Inc. took in helping to bring about what the Food and Drug Administration has termed an *epidemic* of youth vaping [11], this discussion cursorily introduces the Kantian-inspired ethics of precaution that I have argued should guide environmental health regulations whenever uncertainty prevails – which I hope will help analysts and practitioners who utilise this edited volume hold industry accountable to the public as socially responsible members of our communities.

As others have written in great detail about the public health controversy that JUUL Labs, Inc. has created [12] in utilising a *proprietary nicotine salt-based e-liquid formula* [13], JUUL framed its new aerosol nicotine delivery system as a safe alternative to traditional, *combustible* tobacco products, with the potential of mitigating the lung-related illnesses and death associated with smoking [14]. Yet, while little was known about the short- and long-term health effects of inhaling nicotine aerosols, JUUL's vaping products were rushed to mass production without substantive toxicological testing to ascertain its health safety [15]. Even now, more than a decade after e-cigarettes were introduced on the market, the Johns Hopkins Ciccarone Center for the Prevention of Heart Disease admits that vaping exposes users to various chemicals *that we don't yet understand and that are probably not safe* [16] – aerosols that the Surgeon General underscores may

> contain harmful and potentially harmful chemicals, including nicotine; ultrafine particles that can be inhaled deep into the lungs; flavoring such diacetyl, a chemical linked to a serious lung disease; volatile organic compounds such as benzene, which is found in car exhaust; and heavy metals, such as nickel, tin, and lead.[40]
>
> [17–21]

What *is* well understood, however, is the addictive nature of nicotine, which poses unique and lasting developmental problems in adolescents (which also makes them more vulnerable to other forms of substance abuse) [22–23], and which has compounded growing concerns over lung-related illnesses [24–25]. And instead of heeding the unique health risks that adolescents face, JUUL targeted this vulnerable group with its innovative marketing campaigns [26] and product designs, not the least of which included an assortment of flavoured aerosols (including Mango, Creme, Fruit, and Cucumber) [27–28] that entice our youth [29], and a sleek, concealable design and unobtrusive smell that *has led to students vaping in places where cigarette smoking would normally be prohibited, including classrooms and school bathrooms* [30].

Indeed, JUUL quickly proved to dominate the e-cigarette market, accounting for nearly three-fourths of all sales by 2019 [31] (falling since to controlling 40% of the market [32]) during which time the rate of teen vaping has ballooned. For instance, according to the National Institutes of Health (NIH), the average population of 8th graders who reported to have vaped nicotine in the past month increased from 3.5% in 2017 to 10.5% in 2020; this figure more than doubled among 10th graders from 8.2% in 2017 to 19.3% in 2020, and that while only one in ten 12th graders reported to have vaped nicotine in the past month in 2017, by 2020 this figure

had increased 25% [33]. Moreover, the percent of adolescents in this NIH study who self-reported vaping nicotine at some point in their lives is also on the rise: as of 2020, 22.7% of 8th graders, 38.7% of 10th graders and 44.3% of 12th graders have experimented with nicotine e-cigarettes – with 16.9% of 8th graders, 30.7% of 10th graders, and 36.2% of 12th graders reporting to have used JUUL products at some point in their young lives [34]. These trends parallel the FDA's recent findings that

> almost 40 percent of high school students using e-cigarettes were using them on 20 or more days out of the month and almost a quarter of them used e-cigarettes every single day, indicating a strong dependence on nicotine among youth [35].[41]

Beyond intentionally targeting adolescents [36], taking advantage of the Food and Drug Administration's limited purview over the manufacture and sale of e-cigarettes until 2016 – when the regulatory authority of the FDA finally extended to the vaping industry [37] – JUUL has also withheld proprietary toxicological data on likely health effects of its products, undermining *the public's right to know* ([38] at p. 122), and it has tried to influence the scientific community and the discourse over the potential dangers of vaping by publishing privately funded studies that offer favourable analyses of the public health benefit of its products (as offering an alternative to traditional tobacco or combustible cigarettes).

Coupled to these considerations are ostensible conflicts of interest between JUUL's business practices and public health priorities, not the least of which include the following:

- Hiring as its new CEO K.C. Crosthwaite, an executive from leading tobacco corporation, Altria (maker of Marlboro) – shortly after Altria had invested $12.8 billion in JUUL and assumed 35% ownership in the company [39–41];

- Suspending the manufacture of its flavoured vaping products (in October 2019) only after the Trump Administration had announced its intentions (in September 2019) to prohibit the sale of flavoured e-cigarettes [42–43];
- Hurrying the $40 million settlement in the high-profile lawsuit with North Carolina in late June 2020, in order *to avoid courtroom testimony from parents and teenagers while the [Food and Drug Administration] is reviewing its vaping products* [44]; and
- Neglecting to release to the public its colossal 125,000-page application to the FDA, which aims to sway the regulatory agency to allow the company to continue the sale of its electronic nicotine delivery system (ENDS) products – by demonstrating the public benefit of ENDS alternatives to traditional tobacco cigarettes [45].

As *The Ethics of Precaution* argues, industry cannot rationalise the uncertain public health threats it creates by appealing to the inherent limitations of toxicological sciences and complexities of quantitative risk assessment, cannot excuse these uncertain threats by appealing to arguments from moral luck and claiming ignorance of the potential for harm, cannot validate these uncertain threats with good intentions and foreseeable benefits, and cannot dismiss substantive obligations to exercise due care under conditions of uncertainty by appealing to the well-worn (and nevertheless misplaced) objections against the precautionary principle. By failing to take reasonable measures to preventatively safeguard the public against uncorroborated possibilities of harm, industry wrongfully gambles with the welfare of our community members and violates their equal moral standing, by projecting its own risk perceptions and tolerances on others and ignoring the rights and interests of others not to be put in potential harm's way.

In this spirit, before and as JUUL began to manufacture and sell e-cigarettes, and before we began to better understand the actual adverse health effects of exposure to vaping aerosols – which followed actual illnesses, addictions and deaths – on the industrial standard of due care that I defend in *The Ethics of Precaution*, JUUL at least had the following moral obligations:

1 To perform more rigorous toxicological tests to try to ascertain the actual effects of exposure to its aerosols, the costs of which are nominal relative to the profits JUUL customarily enjoys;

2 To fully and truthfully disclose its safety assessment data to regulatory agencies so they may more effectively mitigate the possibility of harm associated with vaping nicotine;

3 To proactively provide details to the public on the potentially adverse health effects, even if they remain uncertain, in easily accessible and understandable ways for individuals to be able to make more informed choices and to better insulate themselves from potential harm;

4 To strive to find feasible substitute substances whose health effects are known and which do not entail unreasonable risks of harm; and

5 To ensure that vulnerable populations (such as infants and children, the elderly, pregnant women, racial and ethnic minorities, the poor, and so forth) are neither targeted nor unintentionally affected by its business practices.

And in neglecting to heed such duties of due care, beyond its liability for the actual injuries that its ENDS products created, JUUL is also culpable to blame and punishment for the uncertain threats it failed to preventatively strive to mitigate.

Uncertain threats of environmental harm are pervasive and pose a unique and exigent moral problem in large measure because of the complex nature of environmental harms. The effects of long-term low-dose exposure are difficult to measure, in part because epidemiological and toxicity testing capabilities are limited and extrapolation from controlled animal studies are problematic. Diverse intervening causal factors, cumulative exposures, deferred effects, different pathways and timings and lengths of exposure, and varying physiological dispositions (vulnerabilities) to being harmed upon exposure, all serve to alter the kind, magnitude, consistency and observability of demonstrable harm. Yet, these challenges and prevailing uncertainty should not give industry license to gamble with human health. It is both naïve to ground regulatory decisions about environmental health exclusively or even chiefly in the imperfect toxicological sciences when uncertainty prevails, and it is also immoral for us to tolerate a system of environmental risk management that would have us delay regulation under uncertainty until the imperfect toxicological sciences can yield preponderances of evidence to verify actual risks to public health. As others have aptly insisted, *justice delayed is justice denied* [46], and it is imperative that we hold industry to higher standards of corporate social responsibility. We must acknowledge that what is at stake is not the deterrence of statistical illnesses and deaths, but rather the safeguard of the welfare and equality of particular people, and that what should shape decisions about whether to regulate uncertain threats (as well as quantifiable environmental risks) is what we deserve and owe each other as moral equals – obligations of mutual respect that are not diminished by prevailing uncertainty and economic and political costs.

'Moral foundations' notes

37 According to the National Institutes of Health (2021), more than 14,000 deaths were associated with prescription opioids in 2019.

38 See [46] Shrader-Frechette (2007), Chapters 1–5.

39 See Szentkirályi (2019) [8], Chapters 2–5; see also Szentkirályi (2018) [9] and Szentkirályi (2020) [10].

40 And this says nothing of the vitamin E acetate in THC vaping products, which have caused dozens of deaths and thousands of hospitalisations. See Blount (2020) and Centers for Disease Control and Prevention (3 August 2021).

41 The FDA notes that "While we are seeing some progress in youth prevalence rates, the fact that there are still 3.6 million youth e-cigarettes users in 2020 is deeply concerning" (FDA, 23 June 2021).

Essay references

[1] Steenhuysen J, Trotta D. (2021) U.S. drug overdose deaths rise 30% to record during pandemic. *Reuters*, 14 July. https://reuters.com/world/us/us-drug-overdose-deaths-rise-30-record-during-pandemic-2021-07-14/.

[2] U.S. Department of Health and Human Services. (2021) *Opioid Crisis Statistics*, 12 February. https://www.hhs.gov/opioids/about-the-epidemic/opioid-crisis-statistics/index.html (Accessed 1 September 2021).

[3] National Institutes of Health, National Institute on Drug Abuse. (2021) *Overdose Death Rates*, 9 January. https://drugabuse.gov/drug-topics/trends-statistics/overdose-death-rates.

[4] Girion L. (2018) Johnson and Johnson knew for decades that asbestos lurked in its baby powder. *Reuters*, 14 December. https://www.reuters.com/investigates/special-report/johnsonandjohnson-cancer/ (Accessed 1 September 2021).

[5] Zhang L, et al. (2019) Exposure to glyphosate-based herbicides and risk for non-Hodgkin lymphoma. *Mutation Research/Reviews in Mutation Research*, 781: 186–206.

[6] Cohen P. (2020) Roundup maker to pay $10 billion to settle cancer suits. *New York Times*, 24 June. https://nytimes.com/2020/06/24/business/roundup-settlement-lawsuits.html.

[7] Hals T. (2021) Bayer to rethink roundup in U.S. residential market after judge nixes $2 billion settlement. *Reuters*, 27 May. https://reuters.com/business/healthcare-pharmaceuticals/us-judge-rejects-bayers-2-bln-deal-resolve-future-roundup-lawsuits-2021-05-26/.

[8] Szentkirályi L. (2019) *The Ethics of Precaution: Uncertain Environmental Health Threats and Duties of Due Care*, Routledge, New York.

[9] Szentkirályi L. (2018) A Rights-Based Conception of the Precautionary Principle. In *Handbook of Philosophy and Public Policy*, Boonin D (Ed.), Palgrave Macmillan, New York, Chapter 56. *Policy, and Environment*, 23: 261–280.

[10] Szentkirályi L. (2020) Luck has nothing to do with it: Prevailing uncertainty and responsibilities of due care. *Ethics, Policy, and Environment*, 23: 261–280.

[11] Food and Drug Administration. (2019) *The Federal Response to the Epidemic of E-Cigarette Use, Especially Among Children, and the FDA's*

Compliance Policy, 4 December. https://fda.gov/news-events/congressional-testimony/federal-response-epidemic-e-cigarette-use-especially-among-children-and-food-and-drug.

[12] Ducharme J. (2021) *Big Vape: The Incendiary Rise of JUUL*, Henry Holt and Company, New York.

[13] JUUL Labs, Inc. (2019) *Defining E-Cigarettes and Why JUUL Is the Market's Best Alternative to Cigarettes*, 2 July. https://juul.com/resources/Defining-E-cigarettes-and-Why-JUUL-is-the-Market-Best-Alternative-to-Cigarettes.

[14] JUUL Labs, Inc. *About Webpage*. https://juul.com/about-juul (Accessed 19 September 2021).

[15] Ducharme J. (2021) *Big Vape: The Incendiary Rise of JUUL*, Henry Holt and Company, New York.

[16] Johns Hopkins Medicine. *Five Vaping Facts You Need to Know*. https://hopkinsmedicine.org/health/wellness-and-prevention/5-truths-you-need-to-know-about-vaping (Accessed 19 September 2021).

[17] U.S. Surgeon General. *Know the Risks: E-Cigarettes and Young People*. https://e-cigarettes.surgeongeneral.gov/knowtherisks.html (Accessed 12 September 2021).

[18] Centers for Disease Control and Prevention. (2021) *Quick Facts on the Risks of E-Cigarettes for kids, Teens, and Young Adults*, 25 August. https://cdc.gov/tobacco/basic_information/e-cigarettes/Quick-Facts-on-the-Risks-of-E-cigarettes-for-Kids-Teens-and-Young-Adults.html?s_cid=OSH_emg_GL0001 (Accessed 12 September 2021).

[19] U.S. Surgeon General. (2016) *E-Cigarette Use Among Youth and Young Adults: A Report of the Surgeon General*, Department of Health and Human Services, Rockville.

[20] Blount B, et al. (2020) Vitamin E acetate in bronchoalveolar-lavage fluid associated with EVALI. *New England Journal of Medicine*, 382: 697–705.

[21] Centers for Disease Control and Prevention. (2021) *Outbreak of Lung Injury Associated with the Use of E-cigarette, or Vaping, Products*, 3 August. https://cdc.gov/tobacco/basic_information/e-cigarettes/severe-lung-disease.html (Accessed 14 September 2021).

[22] U.S. Surgeon General. (2016) *E-Cigarette Use Among Youth and Young Adults: A Report of the Surgeon General*, Department of Health and Human Services, Rockville.

[23] Balingit M. (2019) In the 'JUUL room': E-cigarettes spawn a form of teen addiction that worries doctors, parents and schools. *Washington Post*, 26 July. https://www.washingtonpost.com/local/education/helpless-to-the-draw-of-nicotine-doctors-parents-and-schools-grapple-with-teens-addicted-to-e-cigarettes/2019/07/25/e1e8ac9c-830a-11e9-933d-7501070ee669_story.html.

[24] Centers for Disease Control and Prevention. (2021) *Outbreak of Lung Injury Associated with the Use of E-cigarette, or Vaping, Products*, 3 August. https://cdc.gov/tobacco/basic_information/e-cigarettes/severe-lung-disease.html (Accessed 14 September 2021).

[25] Johns Hopkins Medicine. (2020) *'Vaping' Increases Odds of Asthma and COPD*, 1 January. https://hopkinsmedicine.org/news/newsroom/news-releases/vaping-increases-odds-of-asthma-and-copd.

[26] Huang J, et al. (2019) Vaping versus JUULing: How the extraordinary growth and marketing of JUUL transformed the U.S. retail e-cigarette market. *Tobacco Control*, 28: 146–151.

[27] U.S. Congress Committee on Energy and Commerce Subcommittee on Oversight and Investigations. (2020) *Hearing on Vaping in America: E-Cigarette Manufacturers' Impact on Public Health*, 5 February. https://congress.gov/116/meeting/house/110462/witnesses/HHRG-116-IF02-Wstate-CrosthwaiteK-20200205-SD002.pdf.

[28] Spindle T, Eissenberg T. (2018) Pod mod electronic cigarettes: An emerging threat to public health. *JAMA Network Open*, 1: e183518.

[29] Ambrose B, et al. (2015) Flavored tobacco product use among U.S. youth aged 12–17 years, 2013–2014. *JAMA*, 314: 1871–1873.

[30] Ramamurthi D, et al. (2019) JUUL and other stealth vaporisers: Hiding the habit from parents and teachers. *Tobacco Control*, 28: 610–616.

[31] Cox D. (2019) Nicotine sickness: The latest vaping scare. *The Guardian*, 30 November. https://theguardian.com/society/2019/nov/30/nicotine-sickness-the-latest-vaping-scare.

[32] Sullivan B. (2021) The FDA postpones a long-awaited decision on JUUL's vaping products. *National Public Radio*, 10 September. https://npr.org/2021/09/09/1035288252/fda-vaping-ecigarettes-smoking.

[33] National Institutes of Health, National Institute on Drug Abuse. (2020) *Monitoring the Future Study: Trends in Prevalence of Various Drugs*, 17 December. https://drugabuse.gov/drug-topics/trends-statistics/monitoring-future/monitoring-future-study-trends-in-prevalence-various-drugs (Accessed 14 September 2020).

[34] National Institutes of Health, National Institute on Drug Abuse. (2020) *Monitoring the Future Study: Trends in Prevalence of Various*

Drugs, 17 December. https://drugabuse.gov/drug-topics/trends-statistics/monitoring-future/monitoring-future-study-trends-in-prevalence-various-drugs (Accessed 14 September 2020).

[35] Food and Drug Administration. (2021) *An Epidemic Continues: Youth Vaping in America*, 23 June. https://fda.gov/news-events/congressional-testimony/epidemic-continues-youth-vaping-america-06232021.

[36] Food and Drug Administration. (2019) *FDA Warns JUUL Labs for Marketing Unauthorized Modified Risk Tobacco Products, Including in Outreach to Youth*, 9 September. https://fda.gov/news-events/press-announcements/fda-warns-juul-labs-marketing-unauthorized-modified-risk-tobacco-products-including-outreach-youth.

37 Centers for Disease Control and Prevention. (2021) *Quick Facts on the Risks of E-Cigarettes for Kids, Teens, and Young Adults*, 25 August. https://cdc.gov/tobacco/basic_information/e-cigarettes/Quick-Facts-on-the-Risks-of-E-cigarettes-for-Kids-Teens-and-Young-Adults.html?s_cid=OSH_emg_GL0001 (Accessed 12 September 2021).

[38] Shrader-Frechette K. (2007) *Taking Action, Saving Lives: Our Duties to Protect Environmental and Public Health*, Oxford University Press, Oxford, 122.

[39] Kaplan S. (2021) JUUL is fighting to keep its e-cigarettes on the U.S. market. *New York Times*, 5 July. https://nytimes.com/2021/07/05/health/juul-vaping-fda.html.

[40] Kaplan S, et al. (2019) JUUL replaces its CEO with a tobacco executive. *New York Times*, 25 September. https://nytimes.com/2019/09/25/health/juul-vaping.html.

[41] LaVito A. (2018) Tobacco giant Altria takes 35% stake in JUUL, valuing e-cigarette company at $38 billion. *CNBC News*, 20 December. https://cnbc.com/2018/12/20/altria-takes-stake-in-juul-a-pivotal-moment-for-the-e-cigarette-maker.html.

[42] Bettelheim A, et al. (2019) Trump moves to ban flavored e-cigarettes. *Politico*, 11 September. https://politico.com/story/2019/09/11/trump-weighs-ban-on-flavored-e-cigarettes-1489507.

[43] U.S. Congress Committee on Energy and Commerce Subcommittee on Oversight and Investigations. (2020) *Hearing on Vaping in America: E-Cigarette Manufacturers' Impact on Public Health*, 5 February. https://congress.gov/116/meeting/house/110462/witnesses/HHRG-116-IF02-Wstate-CrosthwaiteK-20200205-SD002.pdf.

[44] Kaplan S. (2021) JUUL is fighting to keep its e-cigarettes on the U.S. market. *New York Times*, 5 July. https://nytimes.com/2021/07/05/health/juul-vaping-fda.html.

[45] Kaplan S. (2021) JUUL is fighting to keep its e-cigarettes on the U.S. market. *New York Times*, 5 July. https://nytimes.com/2021/07/05/health/juul-vaping-fda.html.

[46] Shrader-Frechette K. (2007) *Taking Action, Saving Lives: Our Duties to Protect Environmental and Public Health*, Oxford University Press, Oxford, 97.

ANNEX 1.1 KEY CONCEPTS AND TERMS

Sustainable development and environmental health

The World Commission on Environment and Development (WCED) (the Brundtland Commission) was convened by the United Nations in 1983 to address growing concern *about the accelerating deterioration of the human environment and natural resources and the consequences of that deterioration for economic and social development*. In its 1987 Report (*Our Common Future*) it defined Sustainable Development (SD) as *development that meets the needs of the present without compromising the ability of future generations to meet their own needs*. It introduced the notion of generational equity as well as equity within society being an issue. This also recognises that economic growth and consumption cannot continue at the present rate as planet earth has finite resources.

There has been slow progress since then. In June 1992, at the Earth Summit in Rio de Janeiro, more than 178 countries adopted Agenda 21, a comprehensive plan of action to build a global partnership for sustainable development to improve human lives and protect the environment. Member States of the UN unanimously adopted the Millennium Declaration at the Summit in September 2000. The Summit led to the elaboration of eight Millennium Development Goals to reduce extreme poverty by 2015. The Johannesburg Declaration on Sustainable Development and the Plan of Implementation, adopted at the Summit on Sustainable

Development in South Africa in 2002, reaffirmed the global community's commitments to poverty eradication and the environment, and built on Agenda 21 and the Millennium Declaration by including more emphasis on multilateral partnerships.

At the UN Conference Rio+20[42] in June 2012, Member States adopted the outcome document 'The Future We Want' in which they decided to launch a process to develop a set of SDGs to build upon the MDGs and to establish the UN High Level Political Forum. The Rio+20 outcome also contained other measures for implementing sustainable development, including mandates for future programmes of work in development financing, small island developing states and more. In 2013, the General Assembly set up a working group to develop a proposal on the SDGs. In January 2015, the General Assembly began the negotiation process on the post-2015 agenda.

The 2030 Agenda for Sustainable Development was adopted by all United Nations Member States in 2015, and was intended to provide a shared blueprint for people and the planet. At its heart are the 17 Sustainable Development Goals (SDGs), which are a call for action by all countries in a global partnership. They recognise that ending poverty and other deprivations must go hand in hand with strategies that improve health and education, reduce inequality and spur economic growth – all while tackling climate change and working to preserve our oceans and forests.

The 17 SDGs[43] are:

1 No poverty – end poverty in all its forms everywhere
2 Zero hunger – end hunger, achieve food security and improved nutrition and promote sustainable agriculture
3 Good health and well-being – ensure healthy lives and promote well-being for all at all ages
4 Quality education – ensure inclusive and equitable quality education and promote life-long learning opportunities for all
5 Gender equality – achieve gender equality and empower all women and girls
6 Clean water and sanitation – ensure availability and sustainable management of water and sanitation for all
7 Affordable and clean energy – ensure access to affordable, reliable, sustainable and modern energy for all
8 Decent work and economic growth – promote sustained inclusive and sustainable economic growth, full and productive employment and decent work for all
9 Industry innovation and infrastructure – build resilient infrastructure, promote inclusive and sustainable industrialisation and foster innovation
10 Reduced inequalities – reduce inequalities within and among countries
11 Sustainable cities and communities – make cities and human settlements, inclusive, safe resilient and sustainable
12 Responsible consumption and production – ensure sustainable consumption and production patterns
13 Climate action – take urgent action to combat climate change and its impacts
14 Life below water – conserve and sustainably use the oceans, seas and marine resources for sustainable development
15 Life on land – protect, restore and promote sustainable use of terrestrial ecosystems, sustainably manage forests, combat desertification and halt and reverse land degradation
16 Peace, justice and strong institutions – promote peaceful and inclusive societies for sustainable development, provide access to justice for all and build effective, accountable and inclusive institutions at all levels
17 Partnerships for the goals – strengthen the means of implementation and revitalise the global partnership for sustainable development

It can be seen from this list that the work of EHPs contributes to a number of the SDGs as also illustrated by Figure 1.8 giving

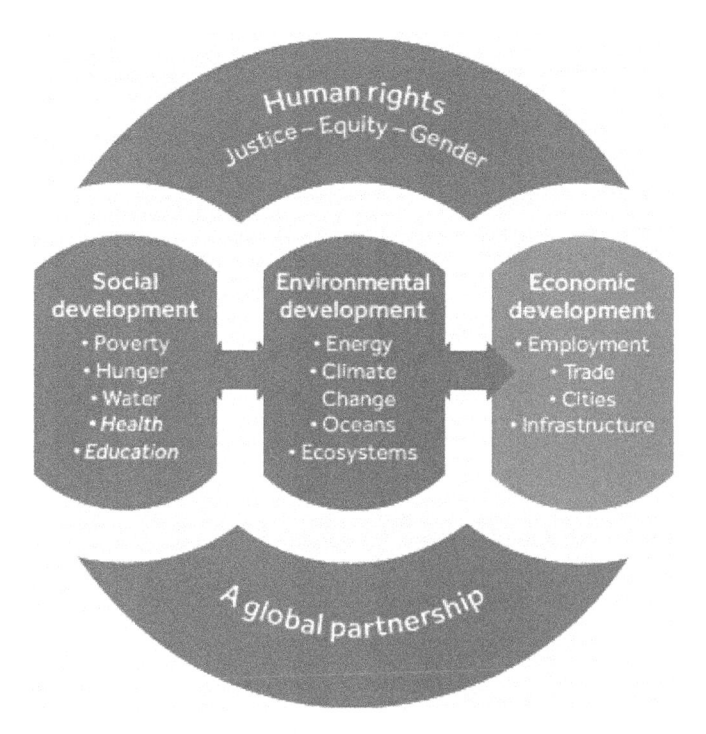

Figure 1.8   Overview of the Sustainable Development Goals 2015–2030

From Commonwealth Secretariat (2016) A Systems Framework for Healthy Policy – Advancing Global Health Security and Sustainable Well-being for All – Implementation Tool for the 'Global Charter for the Public's Health'. Commonwealth. Secretariat, London, UK, A systems framework for Healthy Policy

an overview of the SDGs, and Figure 1.9 is another idea of the components of sustainable development.

The UK Commission on Sustainable Development (which had its funding withdrawn by the Coalition Government after 2010) said that climate change (global heating) resulting from carbon and greenhouse gas emissions posed (and poses) potentially catastrophic risks to human health and threatened to widen health inequalities between rich and poor populations in the UK. It would impact on insurance and other material resources such infrastructure and the ability to cope with the effects of climate change. Thus, sustainable development, health equity and the climate emergency are all interconnected.

Matters that contribute to SD such as active travel, promoting green spaces and healthy eating (including consumption of less, but better-quality meat), will yield co-benefits for both health and carbon emissions given that many crops are grown merely to feed animals.[44] The provision of opportunities for healthy, low-carbon living distributed in ways that favour people with low incomes and so helps to reduce their vulnerability to ill health is therefore clearly part of sustainable development and is inherent to the idea of sustainable communities.

In common with other notions of 'development', sustainable development is concerned with the improvement of the human condition, but the difference is that the emphasis is less on economic growth or production; rather it considers both environmental and social capital. This reflects the thinking behind 'doughnut economics' mentioned previously. Thus sustainable development is means that whatever

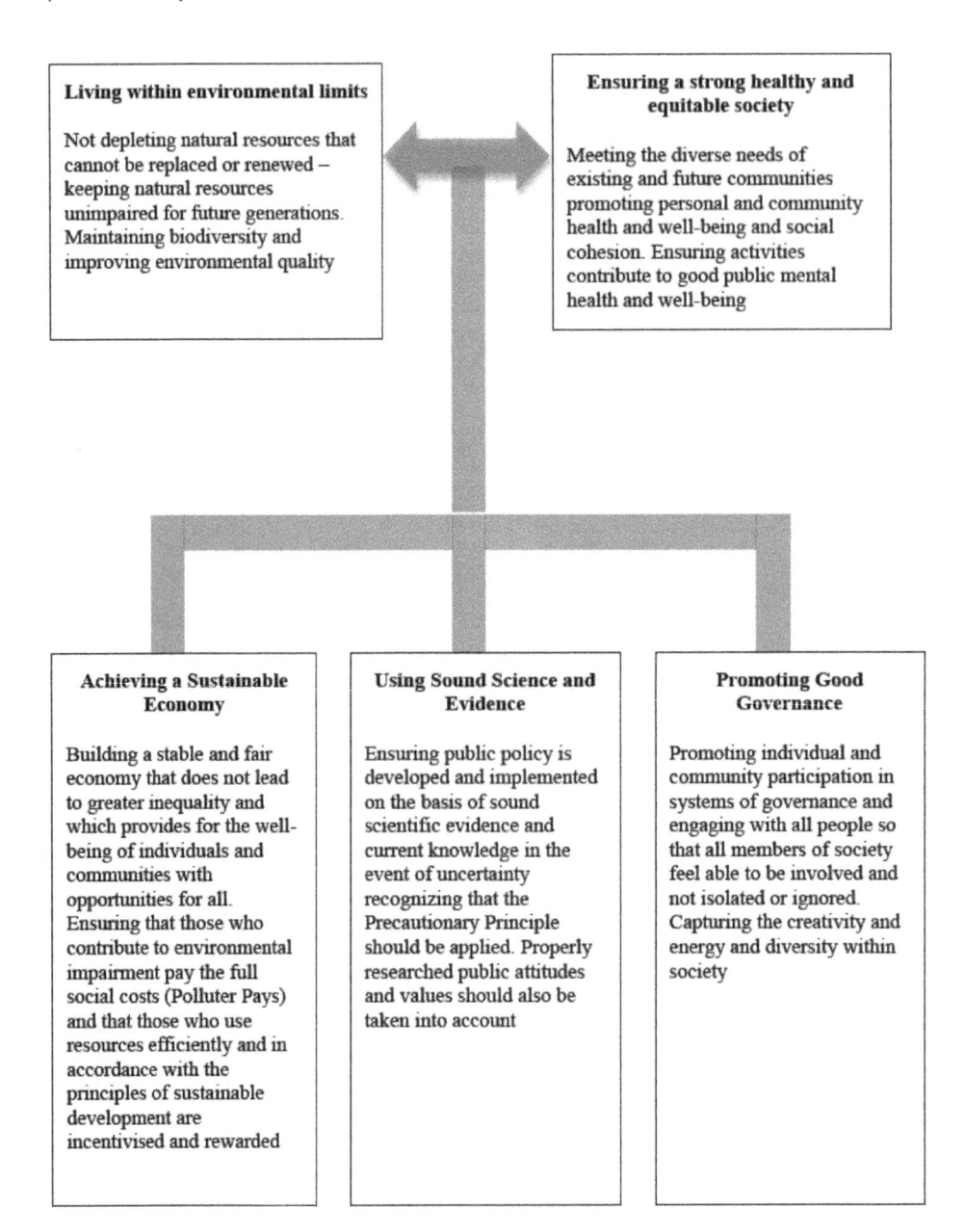

Figure 1.9 One idea of the components of sustainable development

is done to improve the quality of life of people should not lead to a degradation of the environment. The environment itself should be interpreted in both the bio-physical sense and the socio-economic sense (see Bell and Morse 2003[45] for more information).

The notion of sustainable development relies on no longer treating the environment as a 'free good'. A proper value has to be placed on the 'services' provided by the natural environment. There has also to be economic transformation and sustainable development provides an opportunity to lift people out of poverty, advance social justice and protect the natural environment. There is a considerable amount of information on the related websites.[46] For public health it is worth noting that deforestation to permit the growing of foodstuffs for livestock also increases the risks of pandemics as destroying habitats makes viruses and other pathogens more likely to infect humans according to a paper in 2020 in the *Scientific American*.[47]

The 2020 annual UN report on progress on sustainable development made for disappointing reading, perhaps not surprising with the pandemic as the website says, *Now, in only a short period of time, the COVID-19 pandemic has unleashed an unprecedented crisis, causing further disruption to SDG progress, with the world's poorest and most vulnerable affected the most.*

An estimated[48] 71 million people are expected to be pushed back into extreme poverty in 2020, the first rise in global poverty since 1998. Lost incomes, limited social protection and rising prices mean even those who were previously secure could find themselves at risk of poverty and hunger. Underemployment and unemployment due to the crisis mean some 1.6 billion already vulnerable workers in the informal economy – half the global workforce – may be significantly affected, with their incomes estimated to have fallen by 60% in the first month of the crisis.

More than 1 billion slum dwellers worldwide are acutely at risk from the effects of COVID-19, suffering from a lack of adequate housing, no running water at home, shared toilets, little or no waste management systems, overcrowded public transport and limited access to formal healthcare facilities. Women and children are also among those bearing the heaviest brunt of the pandemic's effects. Disruption to health and vaccination services and limited access to diet and nutrition services have the potential to cause hundreds of thousands of additional under-5 deaths and tens of thousands of additional maternal deaths in 2020. Many countries have seen a surge in reports of domestic violence against women and children.

If education is one means of lifting children out of poverty and reducing health inequity, then school closures as the result of the pandemic have kept 90% of students worldwide (1.57 billion) out of school and caused over 370 million children to miss out on school meals they depend on. Lack of access to computers and the internet at home means remote learning is out of reach of many. About 70 countries reported moderate to severe disruptions or a total suspension of childhood vaccination services during March and April of 2020.

The UN report suggests that as more families fall into extreme poverty, children in poor and disadvantaged communities are at much greater risk of child labour, child marriage and child trafficking. In fact, the global gains in reducing child labour are likely to be reversed for the first time in 20 years. The report also shows that climate change is still occurring much faster than anticipated.

It is worth noting that the Commonwealth Secretariat also produced a framework (The Systems Framework for Healthy Policy)[49] to advance global health security and sustainable well-being for all, recognising that sustainable development and health go hand in hand (see Figure 1.10).

The objectives of the framework are

- To strengthen health systems' ability to implement universal health coverage (UHC) by facilitating global health security and sustainable, fair health outcomes;
- To support wider economic growth and the post-2015 Sustainable Development Goals (SDGs);
- To provide a flexible policy framework for tools that can be applied to different countries and settings to strengthen health systems, including for assessment, planning, training, evaluation and accreditation;
- To strengthen leadership and governance, scale up workforce planning and public heath capacity-building, standardise assessment and improve the quality of health systems.

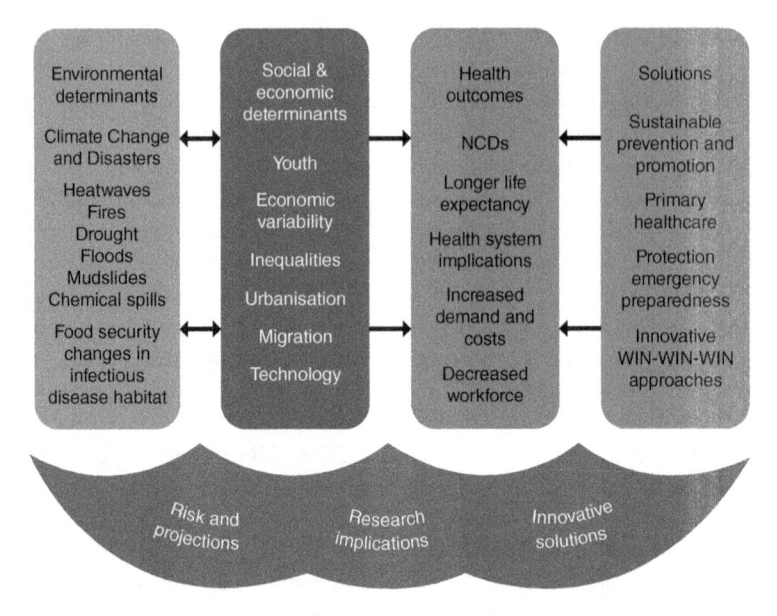

Figure 1.10 Summary of public health challenges and solutions for 2050

Adopted from WHO EURO (2012),'Gastein presentation', available at http://www.euro.who.int/en/
what-we-do/health-topics/noncommunicable-diseases) in Commonwealth Secretariat (2016) A Systems
Framework for Healthy Policy – Advancing Global Health Security and Sustainable Well-being for All
Commonwealth. Secretariat London, UK

Health systems are already struggling to meet current demands and the challenges facing society over the next 40 years will only add further stress. As a result, solutions may require more innovative approaches that address the wider determinants of health and require multidisciplinary partnerships, which are traditionally outside the remit of health policy makers.

The policy framework under the health protection component has included environmental health (along with communicable disease control and emergency preparedness). The environmental health component is seen as assessing, monitoring, planning and delivering action to mitigate harm and to increase the benefits to health from environmental determinants including:

- Air, soil, food, water and sanitation, for which safety, security and quality are concerns;

- Built environment, including housing, urban and rural planning, noise and transport;
- Occupational health;
- Port health;
- Chemical, biological, physical and radiation hazards;
- A 'One Health' approach, including human, animal and plant health as part of a planetary health approach.

Underlying all of the protection components is climate change and sustainability – adaption and mitigation, which make clear that health policy should assess the health impacts and provide advocacy and policy advice on risks to health and strengthen relevant public health functions to support:

- Adaptation planning and strengthening of health resilience;
- Cross–sector sustainability and mitigation plans that benefit health, the economy and

the environment, including safe roads and green spaces that promote active transport, insulation of buildings, reduction of unhealthy food, clean cookstoves;

- Link environmental health determinants and benefits with health promotion.

Risk, Risk Assessment and Risk Management

Much in this book in later chapters will consider the idea of risk and risk management in the different contexts. Here is merely an introduction. No human activity is risk free, and hazards are all around whether at home or at work or travelling between the two.

Risk – is a combination of the probability or frequency of an occurrence of a defined hazard and the magnitude of the consequences of the occurrence. Thus risk can depend upon:

- The frequency of exposure;
- Precautions taken;
- Other relevant factors that affect the magnitude of the consequences.

The risk may depend upon the number potentially affected, or the severity of the harm. Thus the harm may be relatively minor but could potentially affect a large number of people. On the other hand the potential harm could be very severe such as death, but potentially there may only be a very small number of people affected.

Risk assessment – in the workplace (and in general) risk assessment is nothing more than a careful examination of what, in the particular location or occupation, could cause harm to you or others and is a means of helping you to weigh up whether adequate precautions are in place or should more be done to prevent harm. An accident is an unplanned and uncontrolled event leading to injury damage or loss or error. Accidents do not always result in injury or damage; there can be near misses or dangerous occurrences. If

these are reported then steps can be taken to prevent those near misses becoming real accidents, and the information provided will inform future risk assessments. At its simplest a five-step process is involved.

1. Look for the hazards (a situation that could occur or some aspect of an operation that has the potential for harm, that is, human injury, damage to property or the environment)
2. Decide who might be harmed and how
3. Evaluate the risks and decide whether the existing precautions are adequate or whether more should be done (which is merging into risk management)
4. Record the findings
5. Review the assessment and revise it if necessary, in the light of experience or more information

The assessment can produce a rating of the severity of risk which can be represented as

Risk rating = likelihood × severity of harm

This formula assumes that the person undertaking the assessment is competent to do so. Indeed an assessor if not competent is unlikely to produce an assessment that is suitable and sufficient. This means that the assessor should be familiar with the environment being assessed and the operations there.

It is possible to assign a number from 1–3 to indicate the severity of harm or consequences of the hazard (this could also be a letter from A–C). A number from 1–3 can also be assigned to indicate the likelihood of the event causing the hazard to actually occur.

Using the following chart it is possible to produce a score and link it to action criteria. Thus

Severity of hazard outcome or consequence

1 = death, major injury, major damage or loss to equipment or property

2 = >3-day injury, damage to property or equipment

3 = Minor injury, minor damage to property or equipment

Likelihood of the event occurring

1 = extremely likely to occur

2 = Frequent, often or likely to occur

3 = Slight chance of occurrence

The allocation of likelihood and severity will then produce a score using the formula. See the following chart and Figure 1.11.

	Severity of harm reducing →		
Risk – likelihood of event	1	2	3
1	1	2	3
2	2	4	6
3	3	6	9

Implications

Score

1 Unacceptable: must receive immediate attention to remove or reduce risk

2 Urgent: must receive attention as soon as possible to reduce likelihood or hazard

3 Must receive attention to verify how hazard or likelihood can be reduced. Are satisfactory systems in place?

4 Should be given attention to ascertain if likelihood or hazard can be reduced and to check if adequate procedures are in place

6 Low priority

9 Low priority

Figure 1.11 Risk rating chart

A simpler but similar method is shown in Figure 1.12.

Where risk here is used to mean chance or probability.

Implications: A1 Unacceptable; A2/B1 Urgent; A3/C1 Must receive attention; B2 Should receive attention; B3/C2 Low priority; C3 Very low priority

Risk management – although the whole process including the assessment can be seen as risk management, it is also the arrangements, works or precautions or other steps taken to reduce the likelihood of an occurrence that could cause harm. In some instances, the management process might be designed to minimise the harm outcomes as well as prevent the occurrence. There may already some precautions or arrangement in place, but the assessment may have deemed them inadequate and so the precautions may be enhanced on the basis of that evaluation. These arrangements or precautions can be preventative measures, such as the use of mechanical lifting rather than manual handling of materials or they can be protective measures, so that any harm is minimised, for example the use of smoke detection and alarm systems and protected escape routes or the use of domestic sprinkler systems in multi-occupied housing where the likelihood of a fire might be high (and remain high) but the harm can be minimised. This in the context of environmental health will often reflect how well the business is run. A well-run business will have identified the risks, and then assessed and managed them.

	Hazard/severity		
Risk	A	B	C
1	A1	B2	C3
2	A2	B4	C6
3	A3	B6	C9

Where risk here is used to mean chance or probability

Figure 1.12 Risk assessment matrix

Joint Strategic Needs Assessments (JSNA)

Unlike the other concepts that are generic and largely applicable globally, Joints Strategic Needs Assessments are more specific to Britain. It does represent an important concept as despite the substantial changes to public health structures. The Health and Social Care Act of 2012 placed a statutory duty on both upper-tier local authorities and Clinical Commissioning Groups (CCGs) in England and Wales to prepare a JSNA together; to commission with regard to the JSNA; and refer to the JSNA in the development of the local Joint Health and Wellbeing Strategy. The purpose of JSNAs and Joint Health and Wellbeing Strategies (JHWSs) is to improve the health and well-being results of the local community and reduce inequalities for all ages. Local authorities and CCGs have a joint and equal statutory responsibility. In 2013 the government issued statutory guidance on these.[50]

In Scotland the integration of health and social care is a key Scottish government initiative that will bring together the planning of adult social care services, NHS community services and some NHS hospital-based services under a single body known as an 'integration authority'. The legislation relating to the integration of health and social care is set out in the Public Bodies (Joint Working) (Scotland) Act 2014 and a set of linked regulations. A key requirement of the legislation is that each integration authority must produce a strategic plan that:

- Divides the local authority area for which the integration authority is responsible into at least two localities;
- Sets out how the functions and services that the integration authority is responsible for will be delivered and how the related budget will be used;
- Explains how the integration authority intends to achieve a set of outcomes known as the national health and well-being outcome.

A Joint Strategic Needs Assessment (JSNA) is a key element of the process of preparing a strategic plan, providing an assessment and forecast of needs to enable investment to be linked to all agreed desired outcomes, considering options, planning the nature, range and quality of future services and working in partnership to put these in place. The JSNA is part of a cycle, which will inform strategic planning, which in turn will be used to develop a monitoring and performance framework.

Thus, a JSNA is a tool intended to enable local partners to co-ordinate in understanding the future health, care and well-being needs of the local community. Environmental health practitioners should seek to be involved (if not already) as much of their work can make a significant contribution to health improvement and they should also have relevant data. An understanding of JSNAs is therefore essential to EHOs and it is apparent that until recently knowledge and involvement has been highly variable.

Health and well-being boards reconstituted as a partnership forum rather than an executive decision-making body were established under the Health and Social Care Act 2012 to act as a forum in which key leaders from the local health and care systems could work together to improve the health and well-being of their local population. They became fully operational on 1 April 2013 in all 152 local authorities with adult social care and public health responsibilities.

The 'responsible local authority' and the CCGs in England Wales have to prepare the JSNAs and JHWSs and these functions have to be exercised by the health and well-being board established by the local authority. The responsibility falls on the health and well-being board[51] as a whole to ensure its quality and effectiveness. This will in turn depend upon all members working together throughout the process. Two or more health and well-being boards could choose to work together to produce JSNAs and JHWSs, covering their combined geographical area. Some health and well-being boards may collaborate with neighbouring areas where they share common problems as this can prove to be more cost effective than working in isolation. Local authorities and health and well-being boards can also decide to include additional members on the board beyond the core members. These additional members could be other service providers (NHS, private or voluntary and the community sector), health and care professionals, representatives of criminal justice agencies, fire and rescue services, local voluntary and community sector organisations, universities, or representatives of military populations, who can bring expertise and knowledge of the local community to enhance JSNAs and JHWSs.

The policy intention is for health and well-being boards to also consider wider factors that impact on their communities' health and well-being, and local assets that can help to improve outcomes and reduce inequalities. Local areas are free to undertake JSNAs in a way best suited to their local circumstances – there is no template or format that must be used and no mandatory data set to be included.

A range of quantitative and qualitative evidence should be used in JSNAs. There is a range of data sources and tools that health and well-being boards may find useful for obtaining quantitative data. Qualitative information can be gained via a number of avenues, including but not limited to views collected by the local Healthwatch organisation or by local voluntary sector organisations, feedback given to local providers by service users; and views fed in as part of community participation within the JSNA and JHWS process. JSNAs can also be informed by more detailed local needs assessments such as at a district or ward level; looking at specific groups (such as those likely to have poor health outcomes); or on wider issues that affect health such as employment, crime, community safety, transport, planning or housing. Boards will need to ensure that the staff who are supporting JSNAs have easy access to the evidence they need to undertake any analysis they needed to support the board's decisions.

Public Health England (PHE) was expected to support local authorities to deliver locally appropriate interventions and services. It was intended that PHE would provide data, interpretation and evidence to enable local public health teams in England to improve the public's health. It should be noted that the regional Public Health Observatories (PHOs) transferred along with the specialist observatories and the National Cancer Intelligence Network into PHE in 2013 and so at the time of writing, with the demise of PHE, it is not known how JSNAs will be supported but it is anticipated that this will fall within the remit of the Office for Health Improvement and Disparities.

An early study found that NHS managers were more aware of the health implications of local authorities' work than local authority top managers were themselves. This seemed to indicate that managers within local authorities were not highlighting impacts of their work on the health of the local community as well as they could.

As JSNAs set out the health and well-being needs of the local population, they should identify the groups whose needs are not being met whether by their own means or the statutory and voluntary agencies. This will enable these needs to be considered when setting priorities and commissioning services. For example, this will include the housing needs of people living in poor-condition private sector housing, whether owner occupiers or renters on low incomes whose needs have often been overlooked. Environmental health practitioners will hold information on housing conditions and actions, and also on other environmental health matters such as poor air quality and other environmental factors. Data should be included for existing needs and projected needs for up to ten years ahead, recognising the impact of changing demographics with an increase in older people and the effects of climate change.

JSNAs should be seen as providing a framework within which to examine all factors that impact on health and well-being of local communities, including employment (and issues on occupational health and safety), education, housing and other environmental factors. Evidence on local housing conditions and their effect on health also need to be included.

The JSNA should also involve a mapping exercise to map local services relating to health and well-being and how they are used. This will facilitate a gap analysis for service and identify barriers to people accessing them to be highlighted and considered in commissioning decisions.

Local people and agencies including advice agencies and those who work with them can also provide information to support the quantitative data in the JSNA and they should be consulted throughout the process on the findings. Consultees need to include service users, carers, voluntary organisations and service providers, which can include private sector providers, and in the housing context this would include landlords. The following questions are examples that could be used in the consultation. What health inequalities or inequities need to be addressed? What client groups have unmet needs? What services need to be changed?

The final output of the JSNA is a high-level strategic assessment of qualitative and quantitative data on the major health and well-being needs of an area. It is good practice for the findings to be made available in an accessible format that can be readily understood by the community affected and should include a list of key priorities or recommendations.

Early in the existence of JSNAs, a survey by a Local Government Association agency, the IDeA, to assess the progress with JSNAs found that the process and the output varied considerably. Some two-tier authorities were found to suffer from arguments about relative need in their areas, which hampered their development. It remains a concern that the lower-tier authorities (or at least their EHPs) are not adequately engaged.

The LGA produced further guidance,[52] remarking that the emphasis in JSNAs initially had been largely about producing a statement of need which could drive commissioning across local government, the NHS and other local partners. National guidance had recommended that JSNAs needed to acknowledge local people as an important source of information; in practice they tended to be more quantifiable than qualitative. In truth, community engagement in JSNAs had been variable. 'A Glass Half Full' (guidance prepared by IDeA in 2010)[53] had outlined the case for balancing the needs/'deficits'-based approach with an approach based on community assets/strengths and detailed some techniques for discovering and mobilising community assets. The argument being that when practitioners begin with what communities have – their assets – as opposed to what they don't have – their needs – a community's ability to address those needs increases and the LGA's later guidance built on that experience in one local authority.

Quantifying the burden of disease from mortality and morbidity

There are a number of ways of quantifying the burden of disease or metrics used in health economics two commonly seen are:

> Disability-Adjusted Life Year (DALY) – The DALY was developed originally as an input into the World Bank's "World Development Report 1993: Investing in Health" and has subsequently been adopted by the WHO as a measure of the "burden of disease".[54]

The WHO says one DALY can be thought of as one lost year of 'healthy' life. The sum of these DALYs across the population, or the burden of disease, can be thought of as a measurement of the gap between current health status and an ideal health situation where the entire population lives to an advanced age, free of disease and disability.

DALYs for a disease or health condition are calculated as the sum of the Years of Life Lost (YLL) due to premature mortality in the population and the Years Lost due to Disability (YLD) for people living with the health condition or its consequences:

This is calculated by:

$$DALY = YLL + YLD$$

The YLL basically correspond to the number of deaths multiplied by the standard life expectancy at the age at which death occurs. The basic formula for YLL (without yet including other social preferences considered), is the following for a given cause, age and sex:

$$YLL = N \times L$$

where:
N = number of deaths
L = standard life expectancy at age of death in years

YLL measures the incident stream of lost years of life due to deaths; an incidence perspective has also been taken for the calculation of YLD in the original Global Burden of Disease Study for year 1990 and in subsequent WHO updates for years 2000 to 2004.[55]

To estimate YLD for a particular cause in a particular time period, the number of incident cases in that period is multiplied by the average duration of the disease and a weight factor that reflects the severity of the disease on a scale from 0 (perfect health) to 1 (dead). The basic formula for YLD is the following (again, without applying social preferences):

$$YLD = I \times DW \times L$$

where:
I = number of incident cases
DW = disability weight
L = average duration of the case until remission or death (years)

This approach is not without its critics however[56] as it attempts both to measure the global burden of disease and to guide the allocation of resources. The DALY also fails at its other goal: that of measurement of disease burden, for both statistical and ethical reasons.

The Global Burden of Disease analysis for WHO is intended to provide a comprehensive and comparable assessment of mortality and loss of health due to diseases, injuries and risk factors for all regions of the world. The overall burden of disease is assessed using the DALY which is a time-based measure that combines years of life lost due to premature mortality and years of life lost due to time lived in states of less than full health. One DALY can be thought of as one lost year of 'healthy' life. The sum of the DALYs across the population, or the burden of disease, can be thought of as a measurement of the gap between current health status and an ideal health situation where the entire population lives to an advanced age, free of disease and disability. The DALY measures the gap between the actual health status of a population and some 'ideal' or reference status, using time as the measure.

The WHO has said that 'egalitarian principles' are explicit in the DALY metric, and the global burden of disease studies apply these to all regions of the world. This is because the studies use the same 'ideal' life expectancy for all population subgroups and exclude all non-health characteristics (such as race, socio-economic status or occupation) apart from age and sex from consideration in calculating lost years of healthy life. They use the same 'disability weight' for everyone living a year in a specified health state. The disability weight is a weighting factor that reflects the severity of the disease on a scale from 0 (perfect health) to 1 (equivalent to death). Years Lost due to Disability (YLD) are calculated by multiplying the incident cases by duration and disability weight for the condition. The disability weights used for the WHO Global Burden of Disease are on the WHO website.[57] It is possible to use the same approach for example to assess the burden of disease attributable to poor housing conditions.

Quality-Adjusted Life Year – The Quality-Adjusted Life Year (QALY) has been an approach used to measure the health outcomes associated with different healthcare interventions. It is a relatively crude measure to provide an indication of the benefits gained from a variety of medical procedures in terms of quality and life and survival for the patient. It is obviously relevant in making decisions on the allocation of resources. The assessment of outcomes from treatment and other influences on health have two essential elements: the quantity (time) and quality of life. Life expectancy is a traditional measure with few problems of comparison – people are either alive or dead so this approach takes account of the quality of life.

The QALY is therefore a measure of the state of health of a person or group in which the benefits, in terms of length of life, are adjusted to reflect the quality of life. One QALY is equal to one year of life in perfect health.

QALYs are calculated by estimating the years of life remaining for a patient following a particular treatment or intervention and weighting each year with a quality-of-life score (on a zero to 1 scale). It is often measured in terms of the person's ability to perform the activities of daily life, freedom from pain and mental disturbance.

Disability-Free Life Expectancy (DFLE) – DFLE is the average number of years an individual is expected to live free of disability if current patterns of mortality and disability continue to apply. This indicator has been developed in a number of OECD countries since the 1970s.

It is the average number of years a person aged 'x' would live disability-free (no limiting long-term illness) if he or she experienced the particular area's age-specific mortality and health rates for 2001 throughout their life. They are estimates and are calculated by combining age and sex-specific mortality rates, with age and sex-specific rates on

general health and limiting long-term illness from census data.

Healthy Life Expectancy – The Office for National Statistics (ONS) in the UK uses 'healthy life expectancy' (HLE) which estimates a lifetime spent in 'very good' or 'good' health and is based on how individuals perceive their general health. So it is given as the average number of years a person aged 'x' would live in good/fairly good health. ONS also uses Disability-Free Life Expectancy (DFLE) estimates for lifetime free from a limiting persistent illness or disability; this is based upon a self-rated assessment of how health conditions and illnesses reduce an individual's ability to carry out day-to-day activities.

Health expectancies add a quality-of-life dimension to estimates of life expectancy (LE) by dividing expected lifespan into time spent in different states of health. The ONS routinely publish two types of health expectancies. The first is the HLE and DFLE.

Both health expectancies are summary measures of population health and important indicators of the well-being of society.

In 2019 it was reported by ONS[58] that for the UK healthy life expectancy in 2016 to 2018 was 63.1 years but disability-free life expectancy was 62.6 years. This did, however, vary by regions so that for North-East England the figures were 59.4 and 58.3 years respectively. While for males in London the healthy life expectancy was 64.2 years and disability-free life expectancy was 65.0 years and the figures for Wales were 61.4 years and 59.9 and for Scotland 61.9 and 61.2.

For women, healthy life expectancy in the UK was 63.6 and a disability-free life expectancy was 61.6 years. For Wales the figures were 62.0 and 59.3 and in Scotland 62.2 and 60.7.

The ONS reported that changes in life expectancy and healthy life expectancy, the proportion of life spent in good health in the UK has decreased, from 79.9% to 79.5% for males, and from 77.4% to 76.7% for females. This is because actual life expectancy has increased slightly but the size of these increases was substantially smaller than those observed during the first decade of the 21st century. Changes to healthy life expectancy at birth were smaller than life expectancy in the UK between 2009 to 2011 and 2016 to 2018, causing the years lived in poorer health to increase more than the years lived in good health.

An English male could expect to live 79.6 years in 2016–18, but the average healthy life expectancy was only 63.4 years meaning 16.2 of those years (20%) would be spent in 'not good' health. In 2016–18 an English woman could expect to live 83.2 years, of which 19.3 years (23%) would have been spent in 'not good' health. Although females live an average of 3.6 years longer than males, much of that time is spent in poor health.

As a comparison in Australia a boy born between 2014 and 2016 has a life of 80.4 years and female life expectancy is 84.6 years, according to the Australian Institute of Health and Welfare.[59] Australia also publishes a burden of disease study and the latest reported 4.8 million years of healthy life was lost in 2015 equivalent to 199 DALY per 1,000 population and 38% of the burden of disease was preventable.[60]

Notes

42 UN. The Future We Want at https://sustainabledevelopment.un.org/content/documents/733FutureWeWant.pdf

43 Progress on the goals is reported at https://sustainabledevelopment.un.org/?menu=1300

44 26% of the planet's ice-free land is used for livestock grazing and 33% of croplands are used for livestock feed production. Livestock contributes to 7% of the total greenhouse gas emissions through enteric fermentation and manure according to the Food and Agriculture Organization of the UN – http://www.fao.org/nr/sustainability/sustainability-and-livestock; The UK National Food Strategy Independent Review – The Plan published in 2021 says more than 75% of the world's farmland and 85% of the farmland in the UK and abroad is used to graze animals or to produce crops to feed animals; see https://www.nationalfoodstrategy.org

45 Bell S, Morse S. (2003) *Measuring Sustainability – Learning from Doing*, Earthscan, London.

46 http://www.un.org/en/sustainablefuture/; http://www.uncsd2012.org/content/documents/814UNCSD%20REPORT%20final%20revs.pdf; and https://sustainabledevelopment.un.org/

47 https://www.scientificamerican.com/article/stopping-deforestation-can-prevent-pandemics1/

48 https://www.un.org/sustainabledevelopment/progress-report

49 See https://drive.google.com/file/d/0B8wr6920su0aeXNVR01IeHdTYmc/view

50 See https://www.gov.uk/government/uploads/system/uploads/attachment_data/file/223842/Statutory-Guidance-on-Joint-Strategic-Needs-Assessments-and-Joint-Health-and-Wellbeing-Strategies-March-2013.pdf

51 For an explanation of these see https://www.kingsfund.org.uk/publications/health-wellbeing-boards-explained

52 https://www.local.gov.uk/sites/default/files/documents/developing-rich-and-vibra-308.pdf

53 http://www.assetbasedconsulting.net/uploads/publications/A%20glass%20half%20full.pdf

54 See http://www-wds.worldbank.org/external/default/WDSContentServer/WDSP/IB/1996/07/01/000009265_3970311114344/Rendered/PDF/multi0page.pdf

55 https://www.who.int/healthinfo/global_burden_disease/metrics_daly/en/

56 See Anand S, Hanson K. (1997) Disability-adjusted life years: A critical review. *Journal of Health Economics*, 16: 685–702, for example.

57 https://www.who.int/healthinfo/global_burden_disease/daly_disability_weight/en/

58 See https://www.ons.gov.uk/peoplepopulationandcommunity/healthandsocialcare/healthandlifeexpectancies/bulletins/healthstatelifeexpectanciesuk/2016to2018#healthy-and-disability-free-life-expectancy-in-the-uk

59 https://www.aihw.gov.au/reports/life-expectancy-death/deaths-in-australia/contents/life-expectancy

60 https://www.aihw.gov.au/getmedia/08eb5dd0-a7c0-429a-b35f-c8275e7a1dbf/aihw-bod-21.pdf.aspx?inline=true

2

The environmental health practitioner

Stephen Battersby

Introduction

One of the purposes of this book is to provide the basic information that an environmental health practitioner (EHP) or officer (EHO)[1] anywhere in the world can use to develop their competencies. Chapter 1 included a section that set out what is understood by the term 'environmental health' and also what can be considered 'an objective of environmental health'. As hopefully was made apparent, 'environmental health' is something more than that which an EHP 'does' as is well illustrated by World Health Organization (WHO) documents on the topic.[2]

In this chapter the aim is to be a little more focussed on what makes a competent environmental health practitioner whether they work for governmental agencies (central or local government) or in private practice. Reference is made to a number of documents that have been published over the years, but readers will be able to discern some common themes.

The personal skills and requirements needed for a competent EHP, taking account of that set out for Australia[3] can be summarised by the following:

- Have good communication and interview skills;
- Have good negotiation skills;
- Be tactful and courteous;
- Be able to be firm and impartial when making decisions;
- Be resourceful;
- Be able to take initiative;
- Have good problem-solving skills;
- Be able to work independently or as a part of a team;
- Be able to investigate, observe and inspect (including of high-rise buildings), then identify problems and record findings accurately;
- Be able to identify risks objectively, analyse the risk and manage risk;
- Be able to solve a problem or establish the most appropriate response.

Given this list, it can be seen why EHPs should be involved in tracing and advising contacts of COVID-19 cases and these skills have been utilised by many countries when tackling the COVID-19 pandemic. After all, throughout the centuries one of the fundamental tasks of the EHO/EHP has been sampling and tracing contacts of those suspected

DOI: 10.1201/9781003035640-2

of or confirmed as having infectious diseases and ensuring the necessary precautions are taken.

As was set out in the 21st edition [1] environmental health is a

> discipline that recognises the need for its practitioners to see things 'in the round', and where it is no longer acceptable to work or act without regard to the impact on all who might be affected by their actions or inactions.

This means that while seeking to protect and improve the health and well-being of others, practitioners need to seek a point of common understanding with those that may be responsible for compromising the health of others. So *there can be little doubt that this raises questions about the skills-set demanded of an EHP performing regulatory work today, even more so the practitioner who will be working in this field in five- or ten-years' time.* As the President of NEHA said in 2016, *being an environmental health professional is a difficult and complicated career that has its rewards and satisfactions* [2].

This touches on the ethical dimension of environmental health practice and it cannot be ignored that any competent EHP must have regard to professional ethics, and this includes human rights.

The rest of the chapter looks at these matters in a little more detail, but it will be seen that ethics arise throughout, as that dictates how the skills, knowledge and competencies are used.

Ethics and environmental health

Ethics can be defined as the study of what is morally right and wrong or a set of beliefs about what is morally right and wrong. In the context of this book ethics is more the principle of moral conduct that makes a distinction between good and bad/evil, right and wrong, virtue and non-virtue. The word

ethics is derived from a Greek word 'ethos' meaning character. So as a general point it is a branch of knowledge that governs right and wrong conduct and behaviours of an individual, profession, group or organisation.

Environmental health practitioners, unlike medical practitioners with the Hippocratic Oath, do not have a universal code; however their professional bodies will or should have such codes (for example that of the CIEH Code of Ethics for members and fitness to practise[4] and NEHA Code of Ethics for NEHA credentialed professionals[5]), yet their work has an ethical dimension, but practitioners are not always required to be a member of such professional bodies. Environmental health practitioners, whether members of professional bodies or not, do however have to make ethical decisions. For example, what should they do should Parliament pass a law that they are required to enforce but which will actually harm public health – say that smoking bans were reversed, or steps to improve air quality were reversed, or the government (local or national) has policies that condemn people to live in housing that harms their health? What if a policy to improve human health harms or depletes the natural environment with habitat or species loss? Korthals [3] has highlighted how many think that fast food outlets on the streets of cities and the advertising of and sponsorship by these chains contributes to increasing obesity and other diseases such as type two diabetes, cardiovascular disease and cancer of the intestines. *However, one of the most powerful food companies, thinks other-wise and acts accordingly.* As he says, *environmental risks to health are for the most part not equally distributed, so that rich people tend to live in healthier environments than do poor people.* So how should EHPs respond? Should they focus only on food hygiene, their traditional area of work, in such premises?

David E Riggs [2] has suggested that ethics is a set of values that guide the decisions of EHPs, influence actions and give purpose to the lives of EHPs. Professional and personal ethics in environmental health practice are the cornerstones that help good decisions to be made on a daily basis which protect public health and promote the public's trust in our profession. Ethics not only guide *our decisions, but they also help us navigate complex situations that have no easy or clear answers.* So it is that the ethical professional should recognise that every decision or action might have significant consequences further down the line. In an *ever-changing profession, along with technical expertise, ethical behaviour is arguably the foundation of success personally and professional.* Professional bodies should not only provide an ethical code (which should be applied as part of regulating membership) but also provide a means whereby members of the profession can raise issues particularly where there is concern that there is a conflict between an employer's or client's demands and their professional ethics.

The CIEH Code says it is based on four ethical principles, which constitute the main domains of responsibility within which ethical issues are considered; these are integrity, competence, responsibility and respect. As the Faculty of Public Health has said, *Public health is grounded in values and moral norms that guide decisions, behaviours, policy and practice. Ethics and law are accordingly core public health competencies for professional practice.*[6] The Health Professions Council of South Africa[7] says that the fundamental duty of EHPs is to serve humankind, to safeguard public health and to protect them against diseases.

> This invariably affects the rights and interests of all the citizens. As a result, this compels them to conform to the basic principles of ethics such as honesty, trust, courtesy, fairness, transparency, promise keeping, respect for others and maintaining high integrity which serves as a moral compass.

This is also reflected in the 'objective' set out in the previous chapter.

It is perhaps also appropriate at this stage to consider some parts of the Universal Declaration of Human Rights as this provides a starting point for any EHPs, particularly those working for the state (nationally or locally) in a regulatory capacity.

The preamble says that *recognition of the inherent dignity and of the equal and inalienable rights of all members of the human family is the foundation of freedom, justice and peace in the world.* The preamble also says that it is essential, *if man is not to be compelled to have recourse, as a last resort, to rebellion against tyranny and oppression, that human rights should be protected by the rule of law.* Some relevant articles out of the 30 are set out next.

Article 1 says that *all human beings are born free and equal in dignity and rights. They are endowed with reason and conscience and should act towards one another in a spirit of brotherhood.*

Under Article 2 everyone is

> entitled to all the rights and freedoms set forth in this Declaration, without distinction of any kind, such as race, colour, sex, language, religion, political or other opinion, national or social origin, property, birth or other status. Furthermore, no distinction shall be made on the basis of the political, jurisdictional or international status of the country or territory to which a person belongs, whether it be independent, trust, non-self-governing or under any other limitation of sovereignty.

Under Article 3 everyone has the right to life, liberty and the security of person, while Article 4 bans slavery. Under Article 5 no one should be subjected to torture or to cruel, inhuman or degrading treatment or punishment. Article 7 says that all are equal before the law and are entitled without any discrimination to equal protection of the law. All are entitled to equal protection against any discrimination in violation of this Declaration and against any incitement to such discrimination.

Article 10 requires that everyone is entitled *in full equality to a fair and public hearing by an independent and impartial tribunal, in the determination of his rights and obligations and of any criminal charge against him.* While Article 11 says that *everyone charged with an offence has the right to be presumed innocent until proved guilty according to law in a public trial at which he has had all the guarantees necessary for his defence.* Nor should anyone be held guilty of an offence on account of any act or omission which was not an offence when it was committed.

Under Article 12 no one shall be subjected to *arbitrary interference with their privacy, family, home or correspondence, nor to attacks upon their honour and reputation* and there is a right to the protection of the law against such interference or attacks.

Everyone has the right to freedom of movement and residence within the borders of each state as well as the right to leave any country including their own, and to return to their country under Article 13. While under Article 14 everyone has the right to seek and to enjoy in other countries asylum from persecution. Everyone has the right to a nationality, and no one shall be arbitrarily deprived of their nationality nor denied the right to change their nationality under Article 15.

Among other things Article 16 says the family is the natural and fundamental group unit of society and is entitled to protection by society and the state and in Article 17 along with the right to own property, no one shall be arbitrarily deprived of their property. Article 18 is about the right to freedom of thought, conscience and religion. While under Article 19 everyone has the right to freedom of opinion and expression.

Article 21 provides also that everyone has the right to equal access to public service in their country. Article 23 provides that everyone has the right to work, to free choice of employment, to just and favourable conditions of work and to protection against unemployment, to equal pay for equal work.

Article 25 provides that everyone has the right to a standard of living adequate for the health and well-being of themselves and of their family, including food, clothing, housing and medical care and necessary social services, and the right to security in the event of unemployment, sickness, disability, widowhood, old age or other lack of livelihood in circumstances beyond their control. While Article 26 provides that everyone has a right to education.

Resnick [4] has suggested not unreasonably that environmental health ethics issues range from local concerns, such as waste management and urban development and housing, to national ones, such as air and water quality, to global concerns, such as climate change and pandemics [4]. So the moral dilemmas related to environmental health ethics go beyond conflicts involving public health and protection of species, habitats or ecosystems. Furthermore, other values are in play, including respect for autonomy and human rights, social justice, economic development, animal welfare, sustainability and obligations to future generations.

This book aims to provide information relevant to the technical competence of EHPs as well as highlight policy and ethical issues. The rest of the chapter focusses on what various sources suggest makes a competent EHP.

The requirements for the environmental health practitioner

Although written some years ago the CIEH and then Health Development Agency prepared a strategic vision for environmental health in 2002 [5] and identified environmental health officers (practitioners) as having their prime focus on improving and maintaining health rather than curing illness, and as having a unique contribution to make to public health. This remains true and there are certain operational skills required and this book provides some pointers.

The use of problem-solving skills, supported by legal powers, to intervene in the causes of ill health in the home, workplace and community placed them in this unique role. There will inevitably be trade-offs between different risks and EHPs will have to reconcile different demands. The actions of EHPs can directly influence health determinants and maintain healthy environments for the benefit of both individuals and communities, while also extending to the protection of the environment for future generations. The functions and activities of a professional cadre of environmental health practitioners should not be dependent on or be dictated to by the legislation. Rather the starting point should be predominantly about identifying problems at the local level and using their powers available (both persuasive and legislative) to secure remedies and resolve issues. Although written in 2002 the vision for 2012 remains relevant in that EHPs should *also contribute to tackling public health issues at regional, national and international levels.*

Given the pressures on societies from infectious diseases, global heating and other risks to public health, the aim must be for environmental health practitioners to:

- Play lead roles in local government community health and well-being strategies and actively contribute to the public health agenda;
- Be key partners in protecting and improving the health and quality of life of individuals and communities and reducing health inequalities;
- Tackle the wider determinants of population health by identifying, controlling and preventing current and future risks from environmental stressors.

The personal characteristics of a competent EHP are summarised in Figure 2.1.

The introduction and Figure 2.1 set out the personal skills required of EHPs (EHOs). In the boxed text are the necessary operational skills.

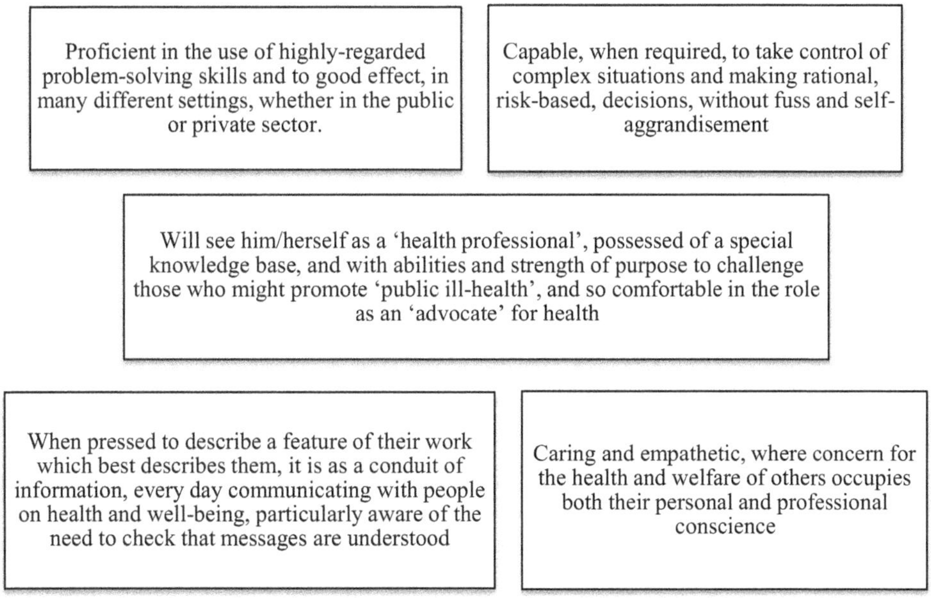

Proficient in the use of highly-regarded problem-solving skills and to good effect, in many different settings, whether in the public or private sector.

Capable, when required, to take control of complex situations and making rational, risk-based, decisions, without fuss and self-aggrandisement

Will see him/herself as a 'health professional', possessed of a special knowledge base, and with abilities and strength of purpose to challenge those who might promote 'public ill-health', and so comfortable in the role as an 'advocate' for health

When pressed to describe a feature of their work which best describes them, it is as a conduit of information, every day communicating with people on health and well-being, particularly aware of the need to check that messages are understood

Caring and empathetic, where concern for the health and welfare of others occupies both their personal and professional conscience

Figure 2.1 Personal characteristics of an EHP

The operational skills of environmental health practitioners can be set out as an ability to:

Assess – be aware of, and able to practise, those analytical skills that form the basis of professional judgement based on available evidence.

Consult – be aware of, and able to practise, the full range of techniques for giving and receiving information.

Advise – be able to communicate technically correct information for the purpose of informing colleagues, clients, communities and others of the most appropriate course of action to be taken in a wide range of circumstances.

Enforce effectively (be 'smart'[8] about enforcement) – be aware of, and be able to use, the full range of mechanisms available for securing compliance with legislative provisions, statutory requirements and standards and good practice, commensurate with the perceived level of risk – that include sorting the well-intentioned but ignorant non-compliant operation from the wilfully (whether knowledgeable or ignorant) non-compliant individual or organisation and be less concerned with the knowledgeable compliant.

Train/educate – be aware of and able to use a range of practical skills associated with education in an environmental health context for the purpose of:

Acquiring knowledge;
Raising public awareness (including with employers); and
Modifying behaviour to meet changing public health needs (including for example understanding and utilising the concept of social marketing).

Advocate – be able to support, promote and campaign on a range of issues to highlight environmental health issues and to address inequity in health.

Evaluate – be able to consider all aspects of an environmental health issue and be able to apportion values which can be supported and defended.

Research – be able to:

Discover, identify and use appropriate sources of information;
Critically assess options in day-to-day practice; and
Undertake a research exercise from planning to report stages.

Report – be willing to and able to write up and publish reports and reviewed papers on experiences, successes and failures from which other environmental health practitioners can learn.

The competent EHO can use the operational skills and competencies to assess and monitor environments; to identify and quantify risks; to determine the most appropriate intervention to be exercised in order to secure improvements in public health and address issues of health equity. As has been made clear enforcement of legislation is but one way of identifying and dealing with issues of public health. The law can also underpin the power and authority of the environmental health practitioner, but it has to be understood and interpreted correctly in the light of reputable advice and cases. In order to undertake these tasks, the practitioner operates within a framework of professional attributes, ethics and an evidence base. The practitioner should also engage in 'dynamic updating' by taking account of and working within relevant policies, current strategies and standards and be aware of current topics or issues and new health risks. Legislation can change as the result of political thinking and changes in political priorities, and indeed environmental stressors can change and additional stressors identified yet the need for good public health will remain. This requires the professional to read relevant journals and papers.

Competencies

As has been indicated already, practitioners contributing to environmental health can include more than those qualified as environmental practitioners. A report to the International Federation of Environmental Health Council, September 2010 [6] set out details of more broadly drawn baseline competencies to be displayed and maintained by those working within environmental health practice, irrespective of their professional background. It argued that the issue was *competence not qualifications is what matters and competencies are the small building blocks of professional practice that contribute to the development of the capability in a practitioner to become competent.*

It endeavoured to reconcile the requirements for those with environmental health practitioner qualifications and those who are qualified in related fields but operate in the broader spectrum of environmental health practice. It also provided a set of higher-level discretionary competencies which could, additionally, be displayed as part of the environmental health professional profile. These remain valid.

Baseline competencies

Be able to identify and articulate the range of actual and potential biological, chemical, physical, social and psychosocial stressors that may adversely impact upon human health and the environment.

Be able to identify and articulate the points at which practitioners may intervene to prevent, control or mitigate the impact by applying the principles of risk management.

Be able to identify and articulate the most appropriate interventions, having regard to the evidence available, and taking into account such other political, financial and technological factors that are likely to influence the decision-making process.

Be able to communicate and engage with partner organisations, agencies, fellow professionals and others with whom appropriate intervention strategies might need to be formulated.

Be able to design and manage the implementation of appropriate interventions.

Be able to monitor, review and evaluate the effectiveness of the intervention strategy; altering or adapting it, where necessary, according to the actual or predicted outcome.

Be able to display transferable skills such as personal and organisational management, communication skills, project management and critical thinking.

Discretionary competencies

Design, develop and implement continuing education sessions for peers and other stakeholders on the success and challenges in delivering environmental health-related programmes.

Design, implement and evaluate quality assurances processes for all programmes, policies and best practice.

Advocate and implement practice policies, legislation and standards designed to improve current levels of protection to human health and the environment.

Be able to apply regulatory tools and enforcement options when non-compliance is observed.

Gather evidence for publication of novel research, and disseminate good practice.

Contribute to the development and actions of integrated multidisciplinary teams in pursuits of environmental health goals.

Support the integration of environmental health practice goals within other areas of related professional practice.

Monitor and review the effectiveness of the intervention strategy; altering or adapting it, where necessary, according to the actual or predicted outcome.

Be able to advise on the strategic consequences of the interventions that are proposed.

Provide direction and leadership to support individuals and teams to achieve their goals and work collaboratively.

All professionals working within environmental health practice should exhibit basic competencies. Discretionary competencies are more likely to be demonstrated by those who are working at a high level within the environmental health field. They are not essential to be recognised as working with an environmental health profile. Additional competencies can be included which relate to specific local, regional or national circumstances and these may be very different in differing countries and continents (e.g. interventions dealing with impact of extreme heat or cold). These are summarised in the following boxed text.

At the risk of further repetition, it should be recognised that EHPs/EHOs are part of the public health workforce. This was highlighted in England in what Public Health England set out in the Public Health Skills and Knowledge Framework[9] and [7]. The framework provides an architecture to describe generic activities undertaken by the public health workforce and is presented as a hierarchy of functions. Public health functions are described in areas of activities: A – Technical; B – Contextual; and C – Delivery. The Function level of the framework describes a group of skills, for example C1 provides leadership to drive improvement in health outcomes and the reduction of health inequalities and then has sub-functions carried out across the public health workforce. The sub-functions describe the activity that is attributable to an individual in their role (e.g. A2.5 – design and/or implement sustainable and multi-faceted programmes interventions or services to address complex problems; or B3.3 – commission and/or provide services and interventions in ways that involve end users and support community interests to achieve equitable person-centred delivery). Those sub-functions relevant to EHOs are set out in Figure 2.2.

That EHPs or EHOs are part of the public health workforce is also exemplified by the WHO-ASPHER Competency Framework for the Public Health Workforce in the European Region. The WHO-ASPHER Competency Framework is intended to support the implementation of the European Programme of Work 2020–2025 (EPW) – United action for better health in Europe and the pursuit of Sustainable Development Goals, in particular Sustainable Development Goal Three on

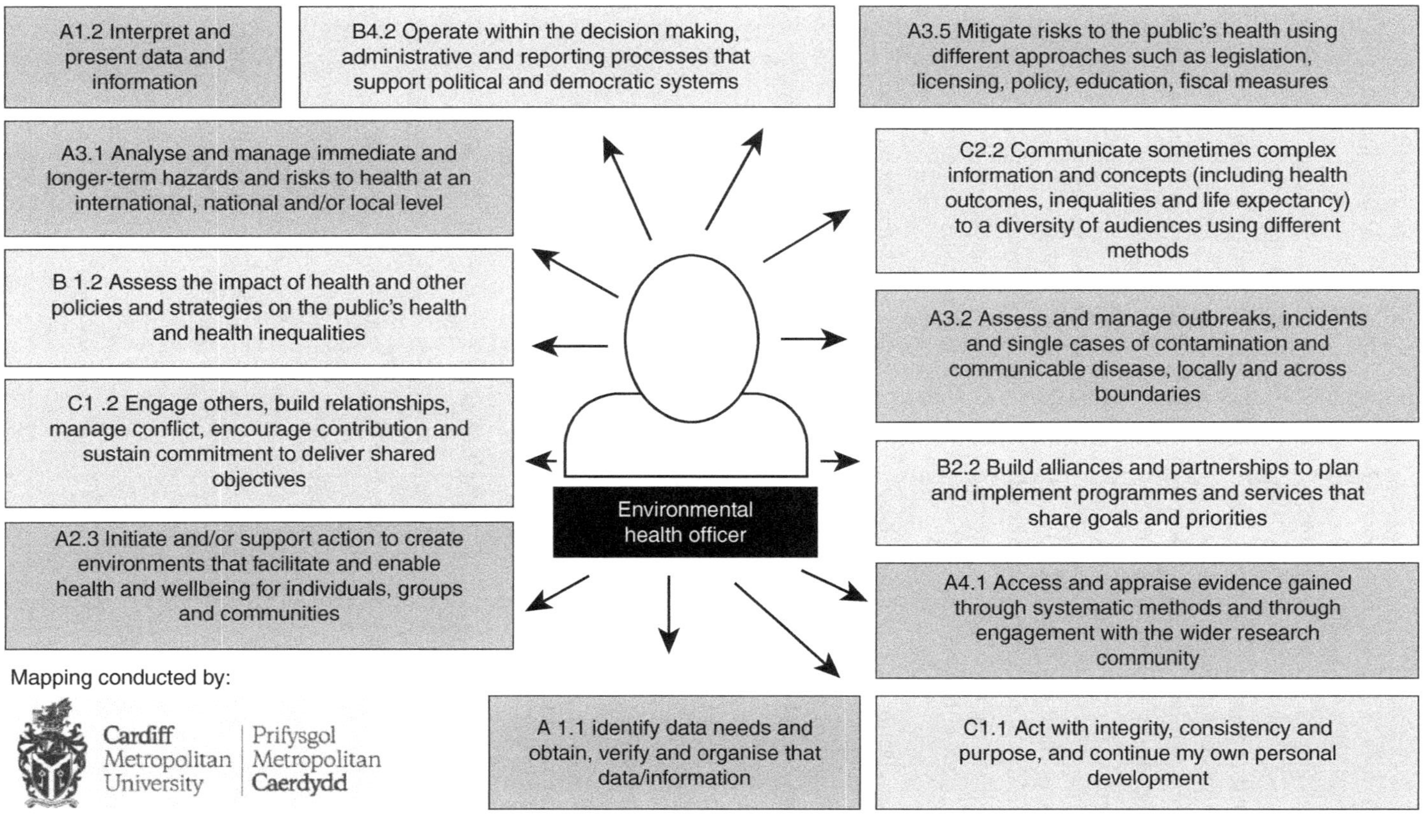

Figure 2.2 The requirements of the EHO as set out in the Good Public Health Practice Framework Use Guide [7]

Mapping conducted by:

Cardiff Prifysgol
Metropolitan Metropolitan
University Caerdydd

A 1.1 Identify data needs and obtain, verify and organise that data/information

C1.1 Act with integrity, consistency and purpose, and continue my own personal development

Environmental health officer

A2.3 Initiate and/or support action to create environments that facilitate and enable health and wellbeing for individuals, groups and communities

A4.1 Access and appraise evidence gained through systematic methods and through engagement with the wider research community

C1.2 Engage others, build relationships, manage conflict, encourage contribution and sustain commitment to deliver shared objectives

B2.2 Build alliances and partnerships to plan and implement programmes and services that share goals and priorities

B 1.2 Assess the impact of health and other policies and strategies on the public's health and health inequalities

A3.2 Assess and manage outbreaks, incidents and single cases of contamination and communicable disease, locally and across boundaries

A3.1 Analyse and manage immediate and longer-term hazards and risks to health at an international, national and/or local level

C2.2 Communicate sometimes complex information and concepts (including health outcomes, inequalities and life expectancy) to a diversity of audiences using different methods

A1.2 Interpret and present data and information

B4.2 Operate within the decision making, administrative and reporting processes that support political and democratic systems

A3.5 Mitigate risks to the public's health using different approaches such as legislation, licensing, policy, education, fiscal measures

universal health coverage [8]. Interestingly this framework

> responds to a need to move beyond professional silos and adopt a broader understanding of public health professionals based on inclusive knowledge and shared experience. The health needs of the population, not the interests of the profession, form the point of departure for all the efforts to professionalize public health.

It refers also to the Essential Public Health Operations (EPHOs) to deliver public health services [8–9] (see Figure 2.3).

The EPHOs are as follows:

EPHO 1 Surveillance of population health and well-being

EPHO 2 Monitoring and response to health hazards and emergencies

EPHO 3 Health protection including environmental occupation, food safety and others

EPHO 4 Health promotion including action to address social determinants and health inequity

EPHO 5 Disease prevention including early detection of illness

EPHO 6 Assuring governance for health and well-being

EPHO 7 Assuring a sufficient and competent public health workforce

EPHO 8 Assuring sustainable organisational structures and financing

EPHO 9 Advocacy, communication and social mobilisation for health

EPHO 10 Advancing public health research to inform policy and practice

It is also worth reiterating what was quoted in the 21st edition from Professor George Morris and Ruth Robertson who said in their report commissioned by the Royal

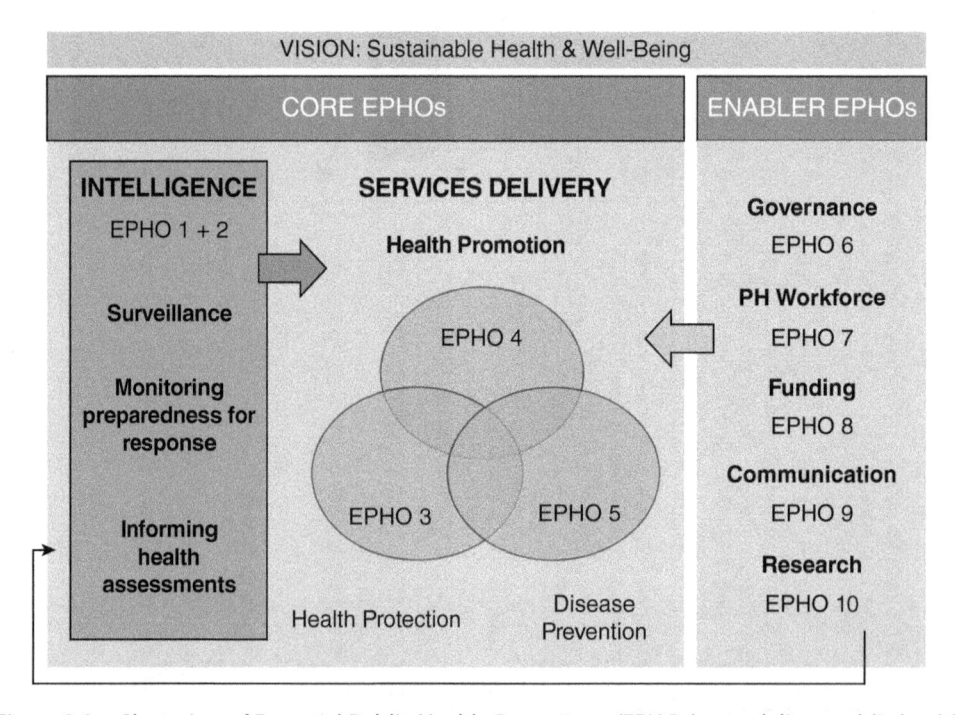

Figure 2.3 Clustering of Essential Public Health Operations (EPHOs) – to deliver public health services [9]

Environmental Health Institute of Scotland as long ago as November 2003:

> the profession undoubtedly faces its own challenge, which can be seen as having two primary elements. One is about the need to re-energize a beleaguered profession, to propel it from its current dependent, reactive state to rival the most effective of modern professions. The second, and inter-related element is about rediscovering effectiveness as a force in public health. Much of the profession's task is about seeing the world in a quite different way. There is a need to create a clear statement of identity and to nurture the skills and maturity to work effectively in the new networks forming at local level to address the Health Improvement Challenge.
>
> *[10]*

These challenges remain as can be illustrated by the varied deployment and work of EHPs in the COVID-19 pandemic around the world.

The challenges of both resources and identity are forcing both academic organisations and professional bodies alike to fundamentally rethink how to plan to produce the next generation of practitioners and support the development of the existing cadre. However, it has to be said that EHPs, as individuals, have to demonstrate their essential contribution to both safeguarding and improving public health. It may be that they act at the individual case level, such as an individual business, or dwelling but collectively they can contribute effectively when forming alliances and networks. As a joint report from a number of agencies said in 2020 in the context of the COVID-19 pandemic a good Local Outbreak Plan and contact tracing system will be led by public health, working as a *system within the local system* and in particular *the local environmental health function will be an equally crucial part of the public health core capability in the application of their capabilities and expertise* [11].

In Australia in 2009 the final report of enHealth, 'Environmental Health Officer Skills, Knowledge and Experience Workforce Project' [12], was published and shows many similarities with what has already been considered earlier.[10] This project was established and managed by the Victorian Department of Human Services (DHS) on behalf of the Environmental Health Committee and already introduced to the reader in the essay at the end of Chapter 1. Its purpose was to develop a skills and knowledge matrix that could underpin a national approach to managing environmental health workforce issues. Identifying the skills and knowledge to undertake environmental health roles was seen as a first step towards delivering a sustainable level of environmental workforce capability. It includes a description of the whole EHO role, not a specific EHO job; a statement of common EHO skills and knowledge; a statement of a minimum level of EHO skills and knowledge to perform the role competently; a statement of skills and knowledge required regardless of whether this is attained through formal education, training and/or experience; a description of skills and knowledge directly related to the environmental health role; and support for emerging roles based on reported developments in the field (anticipating developments in the role in the short to medium term). The document sets out in some detail the necessary underpinning skills and knowledge required of an EHP (EHO). A separate part then sets out in further detail the applied skills and knowledge required of EHPs (EHOs). Readers interested in the full document should use the web address given. As an example, Figure 2.4 sets out the applied skills and knowledge in the context of prevention and control of notifiable and communicable conditions. The document recognises that requirements can vary slightly between different states and in this example the optional skills relate to Queensland (Qld) and the ability to conduct facility audits.

Communication	Risk management	Administering and reporting

• Provide information and advice to businesses and the public on public and environmental health issues, preventative strategies, risk management and outbreak management
• Communicate information on nature and level of public and environmental health risk
• Apply interpersonal skills to interview people about issues which may be personal and sensitive
• Develop and implement health promotion campaigns
• Provide information, advice and training to a range of stakeholder groups on the prevention and control of communicable diseases
• Apply mediation skills to facilitate agreements and resolve conflict (e.g. with neighbouring properties)
	• Apply research skills (e.g. to assess public and environmental health status, identify health trends and establish priorities)
	• Design studies to investigate public and environmental health risk levels and conduct investigations associated with notifiable and communicable conditions
	• Identify common types, symptoms and control measures for diseases that are:
	  – Waterborne
	  – Blood borne
	  – Zoonotic
	  – Vectors
	  – Other infectious diseases
	• Apply understanding of typical facilities, equipment and processes or procedures to identify public and environmental health risks (e.g. associated with cooling towers, personal service businesses, brothels, cemeteries, crematoriums and mortuaries, human parasites, pests, animals and birds)
	• Apply general principles of disinfection, sterilisation, sanitation and infection control to assess or determine action required to eliminate, reduce or control public health risks related to communicable disease
	• Assess options and adequacy of mitigation measures
	• Design and implement studies, policies and programs to minimise risk (e.g. immunisation programmes)
	• Identify, assess and manage personal OHS and safety of others
		• Contribute to/lead planning processes (e.g. disease control, environmental health plans, environmental impact plans)
		• Develop quality assurance plans to ensure delivery of quality service (e.g. to manage quality assurance related to vaccination programmes)
		• Maintain and evaluate public and environmental health records and databases
		• Administer issuing of licences, legal notices, orders and fines
		• Liaise with and report to partner agencies/ departments involved in communicable disease control and management to improve and protect public and environmental health
		• Manage/participate in investigations and related processes associated with notifiable diseases (e.g. contact tracing)

Optional

Specific to EHOs in some jurisdictions:
• Conduct facility audits (specific to Qld EHOs)

Figure 2.4 Applied skills and knowledge: prevention and control of notifiable and communicable conditions [12]

A profile of the environmental health officer as set out by the World Health Organization

We have considered the skills and competencies necessary and later will discuss some routes to qualification, but here the profile of an environmental health practitioner is considered, setting out what can be expected of a competent environmental health officer wherever in the world they are working. Some years back the CIEH designed a profile in order to produce practitioners capable of applying their core knowledge, skills and competencies; although this has been supplanted by a new approach (the professional standards framework) this profile of the EHP as set out in Figure 2.5 has some relevance, and it is useful to compare this with what was produced previously by the WHO.

A consultation by the WHO on the Role, Functions and Training Requirements of Environmental Health Officers (Sanitarians) in Europe in 1978, concluded:

> Experience of those countries in the Region which have officers with specialized training in environmental health who are recognized as constituting a specific profession clearly demonstrates their value. Thus, it would be to the advantage of all Member states to introduce into their environmental health service staff of this kind whom the consultation called in English, environmental health officers.

That conclusion was followed by a further series of considerations by the WHO, one of which looked at the development of environmental health 'manpower' [13], and as well as addressing governments, it contained a professional profile of the environmental health officer and given the nature of environmental health still has relevance in the 21st century.

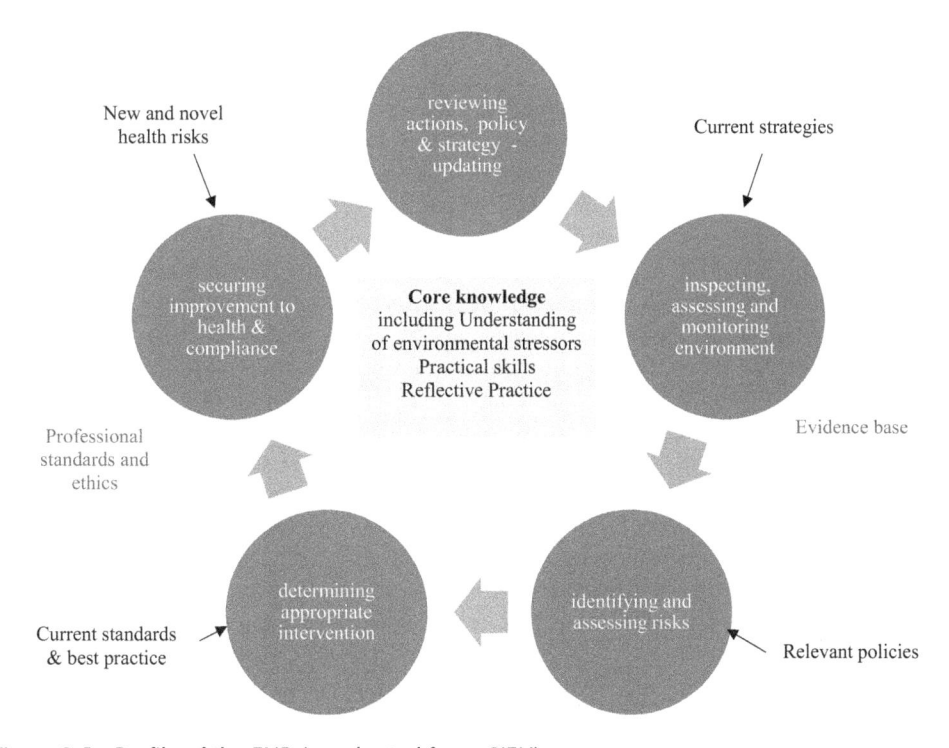

Figure 2.5 Profile of the EHP (as adapted from CIEH)

1   The EHO is concerned with administration, inspection, education and regulations in respect of environmental health (as defined in Chapter 1).

2   The numbers of EHOs should be sufficient to exercise adequate surveillance over health-related environmental conditions and monitoring activities. They must have a close association with the people in their area and be accessible to them. They should be members of multidisciplinary primary healthcare teams delivering comprehensive healthcare at community level.

3   EHOs act as public arbiters of environmental health standards, maintaining close contact with the community. They must at all times be aware of the general environmental circumstances in the areas or districts in which they work, and must know what industrial hazards to health may arise there and what resources are available in the event of an emergency.

4   These are professional officers capable of developing professional standards and applying them to their own work in relation to that of non-professionals involved in environmental health linking with a range of other professionals such as physicians, veterinarians, toxicologists, engineers, nurses and other specialists.

5   A vital function is to maintain effective liaisons with other professional officers who have a contribution to make in the promotion of environmental health in the different settings.

6   EHOs carry out the well-established duties of sanitarians/sanitary inspectors, including inspection of housing and food hygiene, and monitoring and controlling new hazards which could harm the health of a community.

7   Greater emphasis should be placed on the preventive role of EHOs in relation to environmental hazards to health.

8   While not having expertise of other specialists such as veterinarians, toxicologists and microbiologists they will have sufficient background and practical knowledge to understand the principles underlying the work of these others and may develop some specialist expertise. They will be able to work easily with the other professionals. Their wider experience should enable them to formulate an approach on a broader base, to contribute to the decisions to be made, or to make these decisions alone in cases where they have the necessary authority.

9   They should be able to plan and co-ordinate activities between different professional disciplines, official agencies and authorities. They need to have continuing links with other professionals involved in environmental and health-related work. The other professionals with whom liaisons will be appropriate include physicians, physicists, microbiologists, chemists, civil/building/sanitary engineers, veterinarians and lawyers.

10   In some countries and situations, the environmental health officers will initiate the collaboration; in others, they will provide the information and advice that are sought. This liaison role will extend beyond the other professionals to technicians and a range of other specialists, including those concerned with the public health laboratory services. While being able to act independently in both advisory and enforcement capacities, exercising self-reliance and initiative, they should also be able to function as members of a team with other professionals in implementing environmental health programmes.

11   In industry and commerce, environmental control specialists interpret legislation, promote and maintain standards, and solve the problems that may come to light through,

for example, a system of internal control or 'self-inspection' (e.g. via self-regulation or earned autonomy).

12 An important part of their functions must be to acquaint themselves with actual or potential environmental hazards and to ensure that appropriate action is taken to deal with them, for example, to safeguard the public from the hazards associated with microbiological contamination of food and with chemical residues in food substances, and to monitor and control potential and existing environmental hazards, with the backing of strong legislation.

13 A combination of training in public health and toxicology should enable them to cope with such problems as soil pollution due to degradation-resistant agricultural pesticides, leachates from industrial wastes, fallout from the plumes of chemical works, liquid radioactive wastes from industry and research; chemical pollution of the work environment from solvents and from dust arising from processes using silica, asbestos and lead; pollution of the home environment due to such products as cosmetics, detergents, paints, pesticides and gas used as fuel; heavy contamination of water resources by mercury, antimony, barium, cobalt and other metals due to industrial wastes, pesticides used in agriculture, etc.; and new problems in food safety such as the irradiation of food.

14 Environmental health officers in the public service should have the following basic functions:

a Improving human health and protecting it from environmental hazards
b Enforcing environmental legislation
c Developing liaisons between the inhabitants and the local authority, and between the local and higher levels of administration
d Acting independently to provide advice on environmental matters
e Initiating and implementing health education programmes to promote an understanding of environmental principles

15 Because of the range of functions, EHOs may operate in a managerial capacity and in collaboration with other environmental agencies and services. Increasing complexities of environmental health problems require a continuing development of expertise and an updating of knowledge. EHOs must be in a position to respond to challenges presented by new hazards in the environment, and exercise influence in promoting and regulating environmental health activities. Their training should equip them with the necessary expertise to act at any level – national, provincial (or intermediate) or local – or within any sector, private or public.

EHOs/EHPs are usually trained as generalists across the range of basic environmental health activities. This means they should be able to see problems in context and utilise that breadth of knowledge to resolve the problems identified. There will be occasions when they will need to consult other professional officers, and to make use of laboratory and other expert scientific services. A vital part of their functions will, therefore, be to maintain effective liaisons with other relevant professional officers who have a contribution to make in the promotion of environmental health. Environmental health is very much a team concept, and this must be recognised in any organisational arrangement.

The professional profile of the environmental health officer in the above boxed text summarises what was in that WHO document [13] and also addresses issues that governments should take into account.

Point 2 in the boxed text on the numbers of EHOs does not define what would be sufficient. It is hard to find any recommendation on the number per 1000 population but the World Bank reported in 2019 that the UK had 7.86 nurses per 1000 population, Ireland had 11.61, and Australia 11.64.[11] WHO reported that in 1997 the UK had 14,439 Environmental and Occupational Health and Hygiene Personnel (not just EHOs) – for a population of 58.24 million a ratio of 24.8 per 100,000 population.[12]

In a rare example of the availability of data in 2020, in South Australia with a population of 1.677 million according to Data SA, local councils had 97 full-time and 85 part-time EHO posts (although some might be working for more than one council so could be double counted). This equates to more than ten EHOs per 100,000 population. This perhaps reflects the geography of the state but is probably a greater ratio than in England where the workforce survey of 2014/15 [14] indicated a ratio of about eight per 100,000. The most recent CIEH survey covering England [15] indicates that at 2020 the ratio was six fully qualified EHO/EHP employed for every 100,000 population[13] (assuming the upper figure for the estimated FTEs) and four out of five local authorities are using agency staff to cover environmental health work. These figures are indicative only as the two reports used different methods. It does also have to be recognised that there are now a substantial number of environmental health practitioners in practise outside local government and who are working to protect public health.

It is not only a matter of numbers but how competent EHPs and their skills are utilised. In the context of the COVID-19 pandemic this was considered by Day [16]. It has also been reported that in England some local authorities had few if any qualified EHPs undertaking housing inspections, what should be one of the core areas of work of the EHP [17]. As the CIEH reports also [15] around eight out of ten local authority EHPs were redeployed in response to the pandemic.

This issue of the professional profile was further explored in a subsequent WHO document in the Environmental Health Services in Europe series [18]. This publication also points out that in a survey of professions involved in 'environmental health services' they had found 31 categories of professionals involved in environmental health-related areas from a sample of 26 countries, although not all categories were found in every state. At the same time, it is said that many of the professionals listed might not consider themselves as environmental health professionals nor would they describe the organisations in which they work as 'environmental health services'. This does show the inter-sectoral and interdisciplinary nature of environmental health and just why environmental health practitioners at whom this book is aimed need to be able to work with other professionals (as pointed out in 8 in the boxed text earlier).

Qualifying and appointment as an EHO or EHP

Around the world the route to qualification varies although there are great similarities too and no doubt changes will continue in response to the needs of different societies. We look briefly at the approaches to qualifying in a number of different countries. For more detail readers should consult the relevant professional bodies.

England, Wales and Northern Ireland

The Environmental Health Registration Board (EHRB) issued Certificates of Registration to those persons who had successfully completed an approved course of study in the subject of environmental health that includes an accredited academic course of study, work-based experiential learning and professional examinations. The EHRB and the CIEH agreed on changes to the way that qualification as an Environmental Health Practitioner (EHP) will be recognised in future (and the following paragraphs reflect that) and it was proposed to close the EHRB. However, in view of the general response to this proposal further changes are likely, and this closure was put on hold pending an announcement in autumn 2021. The main argument with the Chartered Programme (the initial alternative to EHRB registration) was that it was too long at two years and that people could choose to demonstrate their competence in the workplace in a minimum of two subdisciplines, rather than all five of them. There was also an argument that there needed to be some independent registration system outside the membership structures of the CIEH as the EHRB had originally included other 'non-institution' stakeholders.

As a result, the stated aim of the CIEH has been firstly to produce an accessible learning Portfolio that can be completed in an environmental health workplace context in 6–12 months. So that completing an accredited environmental health degree, the EHP Portfolio and the Professional Discussion (a face-to-face oral assessment carried out by two assessors via teleconference) will lead to Registration as an Environmental Health Practitioner with CIEH.

It has been announced[14] that a candidate who completes this pathway will become a Registered EHP with CIEH. This means a person would have to complete an accredited BSc or MSc in Environmental Health (including the practical food ID exam), demonstrate their knowledge in practice through an approved assessment such as the EHP Portfolio and Professional Discussion and commit to maintaining their competence through CPD and adhering to the CIEH Code of Ethics and fitness to practise rules. A criticism of the EHRB was the lack of a mechanism

to remove a practitioner from the EHRB register for any reason, including serious misconduct even if they could be removed from CIEH membership for breaching the Code of Ethics.

Continuing professional registration under the new system will be tied to CIEH membership. From September 2021 there will be three registers on the CIEH website:

- A Register of Chartered Environmental Health Practitioners;
- A Register of Environmental Health Practitioners;
- A Register of Food Safety Practitioners.

These registers will be linked to the CIEH membership database that can be updated automatically. Practitioners will be added to the registers on qualification and removed as appropriate. When the closure of EHRB has been completed, the CIEH will be publishing its registers on their website as a permanent record. These Registers will be open to the public, updated in real time, and therefore dynamic, as status changes, determined by completion of CPD and adherence to the Code of Ethics. This is a departure from (improvement upon) the EHRB Register which was an unchanging 'permanent record', and therefore not a reliable way of verifying someone's currency and suitability to practise as an EHP. Existing EHRB registrants will not be required to join the CIEH registers, but there will be a simple process for this should they wish to do so.

At the time of writing the different membership grades (see later) have different requirements, and the recent focus has been on the Chartered Environmental Health Practitioner Programme, a two-year, practice-based route to Chartered Status built on the CIEH Professional Standards. This may well be changing with the move towards registration and the Chartered Programme may well be reviewed.

The CIEH Professional Standards which define what is expected of environmental health practitioners who exist *to improve and protect people's health, safety and well-being by securing safer, cleaner and healthier environments (or aspire to do so)*. The Standards apply to EHPs, Chartered EHPs and others and include commitments, behaviours, skills and technical expertise. The Standards reflect the CIEHs' desire to capture the common attributes of a diverse group of professionals, working in different disciplines, roles and sectors, showcasing what it means to be an environmental health professional (as opposed to the traditional EHO-only organisation). The Standards facilitate professional development, learning and progression and ultimately uphold and improve standards within environmental health for public protection.

Thus, the aspiring Chartered practitioner has to hold either the Member (MCIEH) or Fellowship (FCEIH) grade; hold a CIEH accredited qualification in environmental health at a minimum of level 6; be practising in an environmental health setting; and have a statement of employer support when applying to join the two-year programme. For those environmental health practitioners who have held the Environmental Health Registration Board (EHRB) Certification of Registration there is a one-year conversion programme enabling practitioners to achieve Chartered Status through the completion of specific elements of the full programme. Applicants have to pay a fee to join either programme.

All members have to meet the requirements on Continuous Professional Development (CPD). There are four grades of membership: Affiliate, Associate (who has Level 3 [or higher] qualification related to environmental health; or a minimum of two years' relevant practice in an associated role), Member (who has a Level 4 qualification and at least five years' experience in environmental health; or a Level 6 qualification and is currently working in environmental health; or ten years' relevant practice in any area of environmental health) and Fellow (who has at least five years' practice in the Member grade plus a CIEH accredited degree [or the diploma which preceded it]; or at least ten years' practice in the Member grade plus a relevant Level 6 qualification). At least 20 hours a year of CPD is required for all grades apart from Associates where that is ten hours (there is no CPD requirement for Affiliates). However, for 'Chartered' practitioners there is a requirement for 30 hours CPD to be completed annually.

The CIEH Professional Standards define what is expected of EHPs who improve and protect people's health, safety and well-being by securing safer, cleaner and healthier environments (or aspire to do so). See Figure 2.6 [19] for how the CIEH saw the requirements for an EHP.[15] It can be seen how it reflects some of the matters discussed previously.

The CIEH has also supported a group of employers who with some academics have developed an apprenticeship standard as a route to qualification.[16] This is being actively promoted. An apprenticeship in environmental health is seen as an alternative route towards becoming a competent and qualified Environmental Health Practitioner (EHP). Unlike a traditional degree course, an apprenticeship works by combining study and work to provide both practical experience and academic knowledge of the subject. The Environmental Health Practitioners BSc Apprenticeship (currently only available in England) lasts for four years, and on completion the apprentice will gain a degree-level qualification and the Environmental Health apprenticeship certificate from the Institute for Apprenticeships and Technical Education.[17] For apprentices this provides a more accessible pathway into the profession and ideal for those who would not be able to afford university fees. The academic fees are covered by government funding, while apprentices are also paid a salary by their employer for the duration of their apprenticeship.

Scotland

Here the Royal Environmental Health Institute of Scotland (REHIS) has a system more

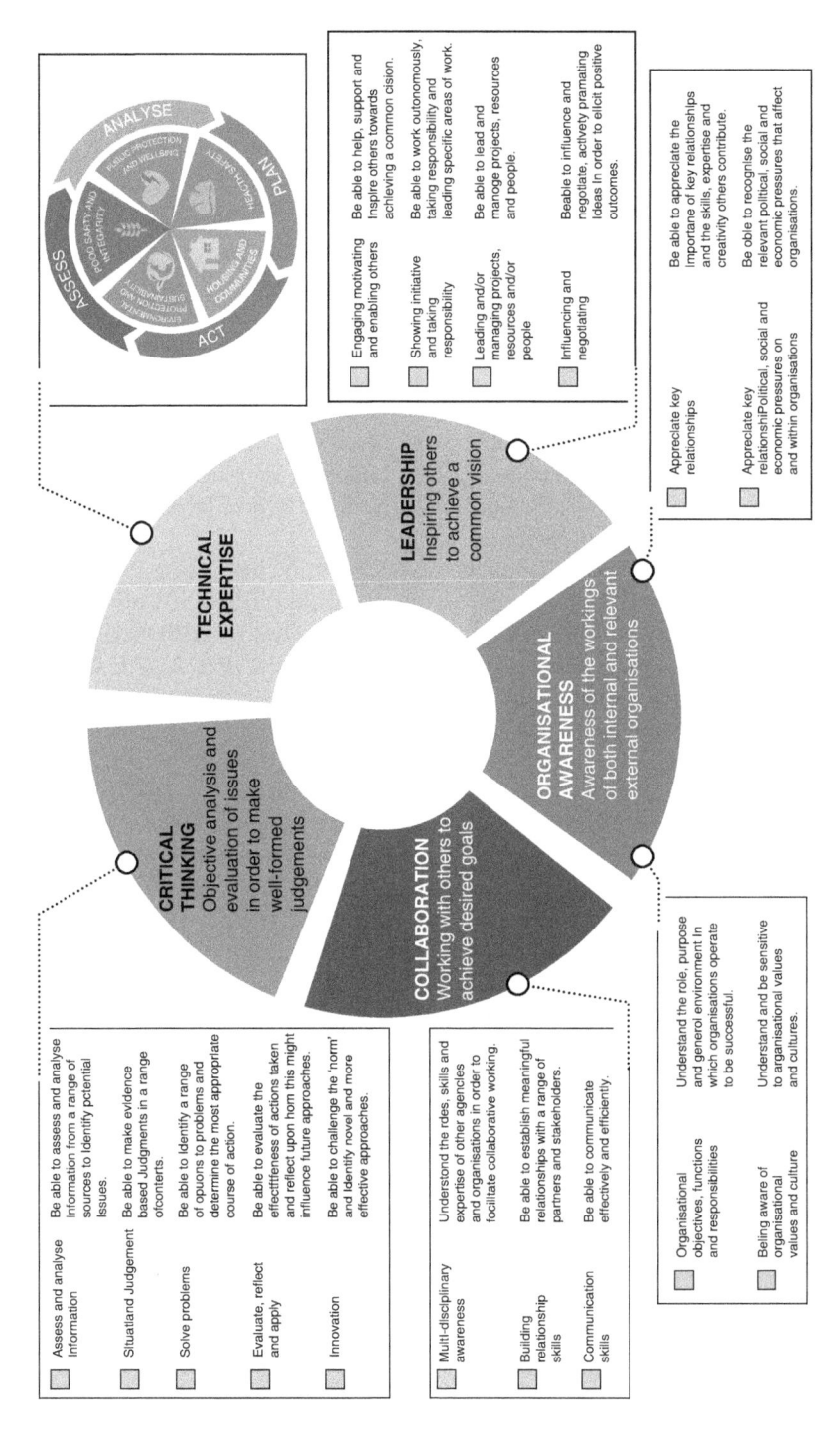

Figure 2.6 Professional commitment and professional behaviours (CIEH) [19]

like it used to be previously in other parts of the UK (and is not related to membership status). It says that qualifying as an EHO in Scotland requires academic study and structured professional practice training. The academic study involves acceptance onto a REHIS-accredited MSc or BSc (Hons) Environmental Health degree course and satisfactory completion. Structured professional practice training must also be undertaken. The REHIS Scheme of Professional Practice requires the completion of a Portfolio through learning in a relevant work environment. This can be undertaken through a formal training placement with a local authority or by gaining experience learning from a number of relevant organisations. It may be possible to carry out the training in vacation periods throughout the course of academic study (integrated training) or defer all training until the completion of academic study (end-on training).

Candidates must register with REHIS at the commencement of the professional practice and a Portfolio has to be developed in accordance with the requirements set out in the Scheme of Professional Practice. The Portfolio will be assessed prior to the Professional Examination. After obtaining the degree in environmental health and successfully completing the Portfolio candidates are required to pass the Professional Examination which assesses professional competence in each of the programme areas. On passing the Professional Examination, the REHIS Diploma in Environmental Health, the qualification required to become an Environmental Health Officer, will be awarded.

REHIS says it is the only organisation in the world which can confer the title 'Chartered Environmental Health Officer'. Chartered Status is the highest possible professional status and recognises an individual's professional qualifications, training, experience within, and commitment to, environmental health practice. For comparison members of the CIEH can attain *Chartered Status and use the post-nominals of CEnvH as they will not necessarily be 'officers'*.

All members of REHIS can achieve and maintain Chartered Status subject to complying with the Institute's Scheme of CPD for a minimum continuous three-year period. In addition, members must also comply with the Institute's regulations governing professional conduct and discipline.

All applications from EHOs who qualified outside Scotland will be assessed on an individual basis. The Institute provides advice for EHOs who qualified:

- In England, Wales and Northern Ireland;
- In the European Union or in Norway, Iceland, Liechtenstein and Switzerland;
- Overseas (other than in the European Union or in Norway, Iceland, Liechtenstein and Switzerland).

The Institute has had a reciprocal agreement with the Chartered Institute of Environmental Health (CIEH), and any EHO who holds the CIEH/EHRB Certificate of Registration may automatically and without assessment work in that capacity in Scotland. The Institute currently also has mutual recognition with the Republic of Ireland and recognises the Irish Environmental Health Officer qualification. EHOs from countries outside the UK, and this includes the Republic of Ireland, are required to have their qualifications assessed by this Institute before they can work as an EHO in Scotland. However, as Tom Bell, then Secretary of REHIS wrote in the Spring 2020 edition of *Environmental Health Scotland* (the REHIS Journal)[18] the system now in England, Wales and Northern Ireland is for aspiring environmental health practitioners to meet the membership-based criteria of the CIEH's Chartered Environmental Health Practitioner Programme/Professional Standards Framework (as set out previously). He wrote the *Environmental Health Registration Board's (EHRB's) Certificate of Registration and the Institute's Diploma in Environmental Health have long been the point at which broad equivalence and reciprocity have been established*. The difficulty for REHIS was that the proposal to abolish the EHRB by the CIEH meant an 'on qualification' reciprocity could no longer be established. At the time of writing the CIEH has announced that its Board took the decision in September 2020 to delay the closure of EHRB, pending a review. The CIEH set up a project which will see the introduction of *a new modernised registration process replacing the existing EHRB registration scheme and underpinning Chartered Status as the ultimate symbol of professionalism.* The project announced a new system from September 2021 (see above).

REHIS is a UK Competent Authority under Regulation 4 of the European Communities (Recognition of Professional Qualifications) Regulations 2007 for the professional titles 'Environmental Health Officer' and 'Chartered Environmental Health Officer'. The situation after the end of the transition period when the UK has left the EU is unclear. At the time of writing anyone wishing to have their qualifications assessed with the intention of working in Scotland will have to submit the following documentation to allow an evaluation of the qualification to be undertaken.

- A copy of the applicant's birth certificate;
- A copy of any degree (or equivalent) certificate from the relevant educational institute along with full details of the syllabus of the course of education and training which has been undertaken;
- A copy of any professional qualification certificate and/or any documentation which indicates the qualification entitles the applicant to practise as an environmental health officer in the country where they qualified including details of the type of training practice undertaken;
- Details of which organisation, if any, provided and supervised any training practice;
- Evidence of post-qualification professional experience;
- Documentation which provides evidence of nationality/citizenship.

Ireland

Ireland is somewhat different in that an Environmental Health Officer is a regulated profession, one where access to or practice of a profession is restricted to those who meet the professional qualifications required by law and the Department of Health is the Competent Authority for the validation of qualifications for EHOs. The minimum approved academic qualifications necessary for appointment as an Environmental Health Officer in the publicly funded health sector are a BSc. (Environmental Health) Level 8 – Dublin Institute of Technology; or a BSc. degree Level 8 – University of Dublin and a Diploma in Environmental Health – Dublin Institute of Technology; or a Diploma in Health Inspection awarded prior to 10th December 1982 and recognised by the Minister for Health or a non-Irish professional qualification in Environmental Health, recognised by the Minister for Health. Becoming qualified as an Environmental Health Officer in Ireland involves both the academic study and practical training.

Healthcare professionals including EHOs with foreign qualifications who have not completed the validation process may not work in the Irish publicly funded health sector. If a person has EHO qualification from outside the Republic of Ireland, it must be recognised under Directive 2005/36/EC for eligibility to be employed as an Environmental Health Officer in Ireland. Directive 2005/36/EC applies to European Economic Area (EEA) nationals with EEA qualifications who wish to practise a regulated profession in an EEA member state other than that in which they obtained their professional qualifications, on either a self-employed or employed basis.

Its intention is to make it easier for qualified professionals to practise their professions in European countries other than their own. Public health and safety and consumer protection are seen as being safeguarded through the qualification recognition process.

Canada

In Canada there are certified public health inspectors. To become a certified public health inspector in Canada a prescribed course of studies in environmental and public health fields has to be completed at one of the seven educational institutes in Canada.[19] After graduation, the candidate has to complete a three-month filed practicum with a local public health unit and follow by a certification exam by the Canadian Institute of Public Health Inspectors (CIPHI).

All candidates intending to obtain the CPHI(C) designation must satisfactorily complete a minimum of 12 weeks of practical experience with a public health unit. The practicum provides candidates with exposure to various inspection programmes and is supervised by an individual at the health unit who holds a CPHI(C) and is involved in a supervisory position.

The Certificate in Public Health Inspection, or CPHI(C), is granted by Canadian Institute of Public Health Inspectors and is recognised throughout Canada by boards of health and other local, provincial and federal agencies. It demonstrates that an individual in possession of this certification has received proper education and training in the field of environmental public health. Qualified candidates apply to sit the Canadian Institute of Public Health Inspectors Board of Certification (BOC) exam, which consists of an oral exam in addition to two written reports based on their 12 weeks practicum.

Australia

Australian Environmental Health Degree and Graduate Diploma programmes are accredited in accordance with the Environmental Health Australia Accreditation Policy to ensure course content meets nationally consistent requirements for practice as an EHO anywhere in Australia. As from 1 July 2009 there have been EHA-accredited universities in every state and the Northern Territory.

Environmental Health Australia (EHA) is the main environmental health professional organisation in Australia which advocates environmental health issues and represents the professional interests of all environmental health practitioners. The membership of EHA is made up of the individual state associations (there is a service agreement with the individual state associations) to which practitioners belong and is responsible for both national and international representation of its membership. Different states have their own legislation covering the qualifications necessary for appointment as an EHO, and EHOs are statutory appointments in that they will normally be appointed under a state's public health legislation.

New Zealand

New Zealand has specific legislation relating to the appointment of EHOs, the Environmental Health Officers Qualifications Regulations 1993 which state that no person shall be appointed as an Environmental Health Officer unless they are qualified, that is, hold the National Diploma in Environmental Health Science or a qualification accepted by the Director-General of Health as at least equivalent to it. The Ministry of Health lists New Zealand and overseas qualifications accepted as being qualifications equivalent to the National Diploma in Environmental Health Science; as required by the Regulations, this list includes those from specific institutions including in the UK and Ireland. This means that a person holding a recognised qualification will be able to apply directly to a potential employer without the need for a NZQA or Ministry of Health assessment.[20]

Sri Lanka

Sri Lanka still has Public Health Inspectors (PHIs) and these are appointed by the Department of Health. They must first pass the Public Health Inspectors Examination conducted by the Department.

Singapore

This is another state where the profession is regulated by legislation. The Environmental Public Health (Qualifications of Environmental Control Officers) Notification was made under the Environmental Public Health Act 1999. For a person to be employed as an Environmental Control Officer they have to have successfully completed a training course conducted by the Ministry of the Environment and have obtained a degree or diploma as specified in the legislation and be registered under other legal provisions.

South Africa[21]

In South Africa the Health Professions Council of South Africa is a statutory body, established by The Health Professions Act and has the role of protecting the public and guiding the health professions. Regulations made under The Health Professions Act 1974 require *environmental health practitioners with the following professions to be registered under the auspices of the Professional Board for Environmental Health Practitioners, namely, EHPs, student EHPs, Food Inspectors and Environmental Health Assistants.*

Conclusion

This chapter has examined the skills and competencies required of an EHP/EHO, including routes to qualification. It is apparent that in some countries EHOs (EHPs) are a regulated profession and the state itself has set out regulations governing qualification. It is somewhat ironic that in England where the profession came into being there are no similar controls and indeed that there is no statutory requirement for local council to employ any EHOs. The notion of deregulation was perhaps highlighted by the original proposal by the CIEH to abolish the Environmental Health Registration Board but it is now known that the outcome of the review means that there will be a new system of professional registration for practising EHPs but that remains a long way from EHPs being a statutory appointment.

Chapter notes

1 As pointed out in Chapter 1 in some jurisdictions the term EHO is used (particularly for those employed in government) and in others the term used is EHP.

2 See https://www.who.int/health-topics/environmental-health#tab=tab_1

3 See The Good Universities Guide at https://www.gooduniversitiesguide.com.au/careers-guide/environmental-health-officer

4 https://www.cieh.org/media/3973/code-of-ethics-for-members-and-ftp.pdf

5 https://www.neha.org/professional-development/credentials/code-ethics

6 https://www.fph.org.uk/public-health-ethics-and-law/

7 https://www.hpcsa-blogs.co.za/ethical-conduct-of-environmental-health-practitioners-a-constitutional-imperative/

8 Smart in this context is not wholly or directly related to the management term which is a mnemonic for "Specific, Measurable, Achievable, Relevant, and Time-based" but enforcement has to be clearly done correctly and efficiently (it can be frustrating for a 'guilty party' to avoid conviction as the result of a failure to follow the correct procedure). This is effective for protecting health and safety, and proportionate as well as timely.

9 See https://www.gov.uk/government/publications/public-health-skills-and-knowledge-framework-phskf/public-health-skills-and-knowledge-framework-august-2019-update and https://assets.publishing.service.gov.uk/government/uploads/system/uploads/attachment_data/file/584408/public_health_skills_and_knowledge_framework.pdf

10 See https://www.lga.sa.gov.au/webdata/resources/files/enHealth%20Skills%20and%20Knowledge%20Matrix%20final%20for%20web.pdf

11 See https://www.nuffieldtrust.org.uk/chart/number-of-nurses-per-1-000-population-for-oecd-countries-1.

12 See https://www.who.int/data/gho/data/indicators/indicator-details/GHO/environmental-and-occupational-health-and-hygiene-professionals-(number)

13 Taking the population of England as 56 million and the workforce survey estimated between 3240 and 3360 full-time equivalent fully qualified EHOs working for local authorities.

14 https://www.cieh.org/news/blog/2021/registration-registers-and-registrants/

15 https://www.cieh.org/media/2225/cieh-professional-standards-framework-guide.pdf

16 See https://www.cieh.org/professional-development/apprenticeships/

17 https://www.instituteforapprenticeships.org

18 https://www.rehis.com/sites/default/files/Spring%202020%20Journal_0.pdf

19 https://www.ciphi.on.ca/career

20 https://www.health.govt.nz/our-work/environmental-health/environmental-health-officers-qualifications-regulations-1993

21 See https://www.hpcsa.co.za/?contentId=0&menuSubId=46&actionName=For%20Professionals for more information

Chapter references

[1] Day C. (2016) Environmental health – a changing practice. In *Clay's Handbook of Environmental Health*, Battersby S (Ed.), Routledge, Abingdon, Oxon, 21st ed., 60–89.

[2] Riggs DE. (2016) The environmental health profession. *Journal of Environmental Health*, 6–7. National Environmental Health Association. www.neha.org.

[3] Korthals M. (2011) Ethics of environmental health. *The Sage Handbook of Health Care Ethics: Core and Emerging Issues*, 413–426. https://doi.org/10.4135/9781446200971.N34.

[4] Resnik DB. (2019) An overview of ethics and environmental health. In *The Oxford Handbook of Public Health Ethics*, Mastroianni AC, Kahn JP, Kass NE (Eds.), Oxford University Press, Oxford.

[5] Burke S, Gray I, Paterson K, Meyrick J. (2002) *Environmental Health 2012 – A Key Partner in Delivering the Public Health Agenda*, Health Development Agency, London.

[6] IFEH (2010) *Developing an International Competence-Based Curriculum for Environmental Health*, Revised Report on Behalf of the International Faculty Forum of Environmental Health, Vancouver, Canada. https://www.ifeh.org/docs/meetings/ifeh_council/Item%2015a-%20International%20competency%20framework%20for%20Environmental%20Health%20-Report%20to%20IFEH%20Council%20,4-5%20th%20September%202010.pdf.

[7] Public Health England. (2016) *User Guide [Version 1] for the PHSKF*, PHE, London. https://www.gov.uk/government/publications/public-health-skills-and-knowledge-framework-phskf/public-health-skills-and-knowledge-framework-august-2019-update.

[8] WHO. (2020) *WHO-ASPER Competency Framework for the Public Health Workforce in the European Region*, WHO Regional Office for Europe, Copenhagen. https://www.euro.who.int/__data/assets/pdf_file/0003/444576/WHO-ASPHER-Public-Health-Workforce-Europe-eng.pdf.

[9] WHO. (2012) *Review of Public Health Capacities and Services in the European Region*, WHO Regional Office for Europe, Copenhagen. https://www.euro.who.int/__data/assets/pdf_file/0010/172729/Review-of-public-health-capacities-and-services-in-the-European-Region.pdf.

[10] Morris G, Robertson R. (2003) *Environmental Health in Scotland and the Health Improvement Challenge*, A Report Commissioned by the Royal Environmental Health Institute of Scotland, Edinburgh. http://www.rehis.com/sites/default/files/Morris-Robertson%20Report.pdf.

[11] Association of Directors of Public Health (ADPH). (2020) *Public Health Leadership, Multi-Agency Capability: Guiding Principles for Effective Management of COVID-19 at a Local Level*, London. https://www.adph.org.uk/wp-content/uploads/2020/06/Guiding-Principles-for-Making-Outbreak-Management-Work-Final.pdf.

[12] Environmental Health Committee (enHealth). (2009) *enHealth Environmental Health Officer Skills and Knowledge Matrix*, Department of Health and Ageing, Canberra, Australia.

[13] WHO. (1987) *Development of Environmental Health Manpower, Environmental Health Series No. 18*, World Health Organization Regional Office for Europe, Copenhagen.

[14] CIEH. (2015) *Environmental Health Workforce Survey 2014/15 – Phase 1 and 2 Summary Report*, CIEH, London.

[15] CIEH. (2021) *Environmental Health Workforce Survey Report: Local Authorities in England*, CIEH, London.

[16] Day C. (2021) *COVID-19 – The Global Environmental Health Experience*, Routledge Focus on Environmental Health Series, Routledge, Abingdon, Oxon.

[17] Battersby SA. (2018) *Private Rented Sector Inspections and Local Housing Authority Staffing*, Supplementary Report for Karen Buck MP. http://www.sabattersby.co.uk/documents/Final_Staffing_Report_Master.pdf (Accessed 23 February 2021).

[18] Fitzpatrick M, Bonnefoy X. (1998) *Environmental Health Services in Europe 3 – Professional*

Profiles, WHO Regional Publications. European Series No. 82, WHO, Copenhagen.

[19] CIEH. (2018) *CIEH Professional Standards Framework, Professionalism in Practice: Guide Version 1.0*, CIEH, London.

Other useful reading and sources of information

Drew CH, van Duivenboden J, Bonnefoy X. (2000) *Guidelines for Evaluating Environmental Health Services*, Environmental Health Services in Europe WHO Regional Publications, European Series No. 90, WHO, Copenhagen. https://apps.who.int/iris/bitstream/handle/10665/272701/9289013575-eng.pdf.

Stewart, J, Cornish, Y, Eds. (2009) *Professional Practice in Public Health*, Reflect Press, Exeter, Devon, UK.

WHO (1998) Environmental Health Services in Europe – An Overview of Practice in the 1990s: v. 1: Environmental Health Services in Europe (WHO Regional Publications, European S.) MacArthur I and Bonnefoy X.

3

Communication in environmental health

Sarah Daniels

Effective communication is 20% what you know and 80% how you feel about what you know.

– Jim Rohn

Introduction

Communication is an act of interacting with people and sharing information with them. It involves the transmission of verbal (e.g., words and sentences) and non-verbal messages (e.g., body postures, facial gestures, tone of voice).

It consists of a sender, a receiver, and a channel of communication. Sometimes during the process of communication, the clarity and accuracy may be altered. Hence the need for effective communication, and 'training' in the art and science of communicating effectively.

Communication combines various skills, such as active listening, the ability to deal with the stress or pressure of the present situation, plus the capability to deal with your own emotions and those of whom you are communicating with.

The word 'communication' originates from the Latin word 'communicare' which means to share. Information can be conveyed and understood through the process of communication.

Communication is a fundamental skill, one that is often taken for granted, but one that environmental health practitioners (EHPs) in both the public and private sectors need to excel at in order to 'get the message across.' Good intentions and personal experience are rarely sufficient to rise to the challenges inherent in the work of EHPs whatever aspect of environmental health in which they may be involved.

The nature of communication can be theorised in many different ways, for example, as ritual, storytelling, deliberation, rational argument, dissemination, articulation, translation, and even failure [1]. However, it is theorised, EHPs (as is made clear in a number of chapters of the book) need to ensure that any message they send is understood in the way it was meant in the different contexts in which they operate.

Communication is (or should be) a two-way process. We are giving or exchanging information in a number of formats, but it must be received by another or others for that true exchange to have taken place. That is a criticism of the Shannon Weaver model (Figure 3.1) in that one direction dominates – it is based on a sender and receiver with the latter being passive. Yet whether the recipient understands the message depends on factors such as language skills, existing knowledge,

DOI: 10.1201/9781003035640-3

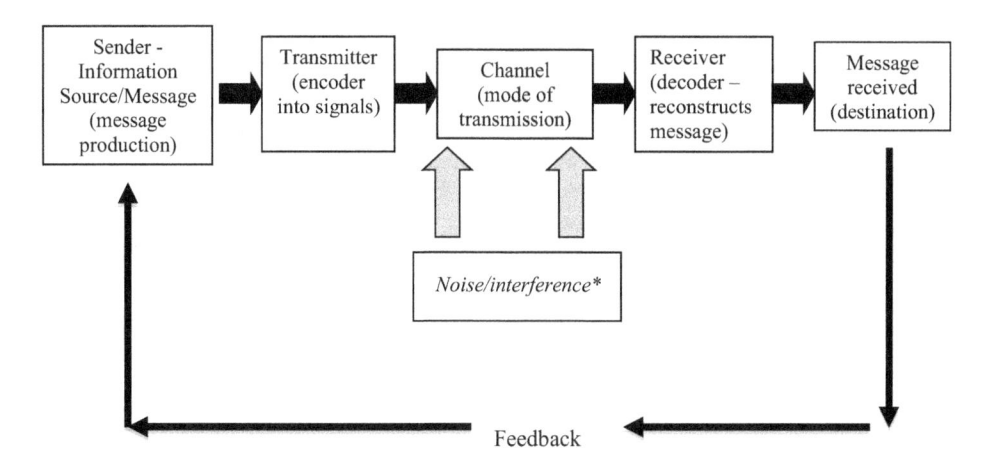

Figure 3.1 The Shannon Weaver Model of Communication

* Noise is any interference that affects the information flow between sender and receiver/recipient

Source: Based on Shannon CE. (1948) A mathematical theory of communication. *Bell System Technical Journal*, 27(3): 379–423 and reproduced in Shannon CE and Weaver W. (1963) *A Mathematical Theory of Communication*, University of Illinois Press.[1] See also other sources of information at the end of the chapter.

and understanding and their aptitude in written communication (reading skills) and there is an interaction between sender and receiver. In the context of environmental health work, that may mean literacy levels. People with whom EHPs come into contact, and with whom they have communicate, are likely to cover the whole gambit. EHPs are required to communicate with their work colleagues, employers and clients, the government, business, and members of the public in their home and at work, to name but a few. EHPs will also be communicating as enforcement officers in a regulatory capacity, as consultants, trainers, academics, and as government officials (as an arm of the state). So, the style and content may vary but the skill in effective communication remains.

Communication will need to be in a number of different formats, for example the spoken word, in writing, or via social media platforms. Whichever platform is used, communication still needs to be effective. The needs of the receiver need to be considered and an absolute focus on the desired outcome kept in mind.

How EHPs present their message is critical to a successful outcome. It will depend on how the message is presented as to how successful the exchange is.

EHPs as communicators need to be mindful of

- How much the receiver of the message already understands;
- The sensitivity of the listener;
- The enthusiasm and/or passion needed to get the message effectively across.

In this chapter the author will look at communication in accordance with Rudyard Kipling's six honest serving wo(men)[2] or EHPs, namely, What, and Why and When and How and Where and Who.

Key matters in communication

The what

This could also be formally called an agenda. All EHPs have their own agendas in their heads and in their daily lives. EHPs need to have a clear idea of what our agenda is during any communication, what is the desired outcome or consequence of the communication.

Are we a Food Safety Officer that needs to seek improvements? A Trainer who needs to change the behaviour of the delegates? A board/government member who wishes to seek influence/direction? Are we coaching or disciplining a colleague, or are we being interviewed?

EHPs must always consider what they want the outcome of their communication to be. We should never assume that because we have told someone something that they have heard it or indeed understood it. The message may need to be repeated, in another format, or a summary given.

EHPs will spend much of their working lives communicating to others; and must develop the skills to provide information, advice, which will motivate and influence behaviour.

The why

Why do we need to communicate the environmental health message to all that will listen? Simply put, it is about achieving better public health and wellbeing. EHPs have some important messages to get out there to protect and improve public health. No more so than during the recent pandemic. EHPs in their chosen area of work must ensure that they understand all the relevant information that they need to communicate, and that they are indeed competent to do so. Communication is a key element of the practitioner's skill set. If the message is not completely understood by the giver, then the audience may be unable to move forward with the message's contents.

It may be as basic as providing simple communications or information over the telephone to an enquiry or providing more detailed reports concerning a complex issue to the media, elected members or a board.

It is essential that EHPs practise their communication skills. Effective communication cannot be left to chance. There must be

planning; there must be a honing of presentation skills; an understanding of the receiver(s) of the communication; and use made for feedback mechanisms to enable improvements to be made.

Communication may not come easy to all. For most people it is a learnt skill. Some communications will be brief, others more detailed, but in all cases, we need to remain focussed on the needs of those with whom we are communicating, or trying to get the message out to, so that the message is received and understood. It is essential to check in with the receiver that the message has been received.

The when

When we communicate is a key element to how the message is received. If it is not the right time to have the conversation, then the effectiveness will be lessened. When the receiver is unfocussed the message will be lost. The communicator needs to be mindful of this and back up a verbal communication with a written narrative. A summary of the key points is useful in many situations.

This is the same with training skills. A Theorist/Reflector such as identified by Honey and Mumford [2] (see Figure 3.2) will need time to take on knowledge/skills. The learning styles identified include:

- Activists – who are 'hands-on' learners and prefer to have a go and learn through trial and error;
- Reflectors – who are 'tell me' learners and prefer to be thoroughly briefed before proceeding;
- Theorists – who are 'convince me' learners and want reassurance that a project makes sense;
- Pragmatists – who are 'show me' learners and want a demonstration from an acknowledged expert.

The communicator will need to be intuitive to the needs of the listener, and if

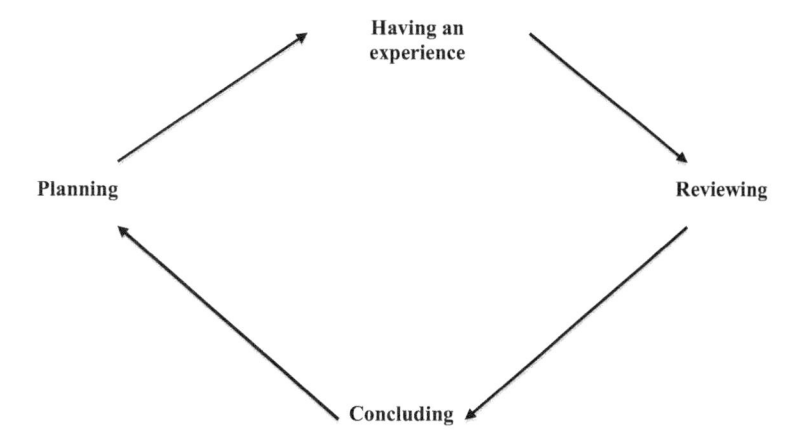

Figure 3.2 Learning Styles as identified by Honey and Mumford [2]

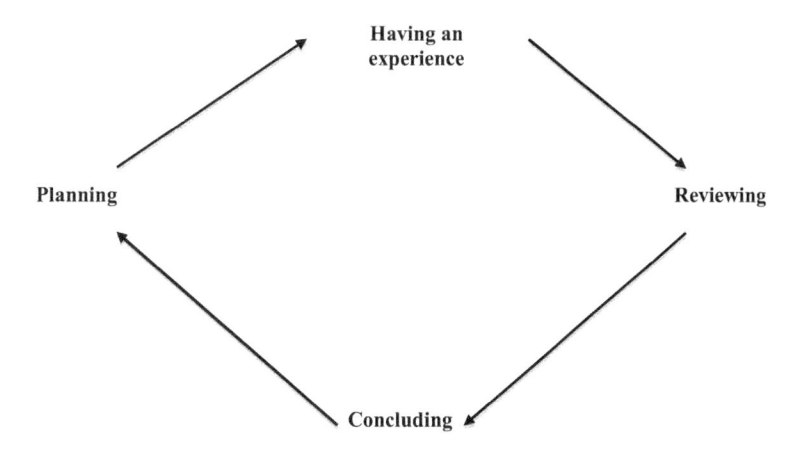

Figure 3.3 The learning cycle based on Mumford (1997) [3]

face-to-face then body language can be read. Alternatively, if on the telephone, listen out for key pointers, as not all message receivers/recipients will be able to communicate their needs.

The learning cycle as in Figure 3.3 underlines why the communicator needs to understand and be prepared for the different learning styles.

Written communications can be accessed more readily at a time that suits and sufficient time can be set aside. However, if reading detailed reports five minutes before a key meeting will not normally enable the true

facts to be assimilated, it is one skill that Barristers do seem to have acquired.

The how

Communication can be non-verbal and verbal, formal and informal. Speech is used in verbal communications, which should be in such a format that it is understood by all. Non-verbal communication can include vocal cues such as tone, pitch, volume, intonation, speed of the words being spoken, and even silence, but also facial expressions, body movements, and gestures.

The message needs to have a logical arrangement; in its simplest terms a message has a beginning, middle, and end. It should include a 'call to action' that is something that the receiver/listener can do with the information received. A 'call to action' is a marketing plan for the message. Why should anyone take on board the contents of the message?

EHPs may need to plan some communications more thoroughly than others. Answering a phone call enquiry often requires the answer to be given there and then. However, planning a public health campaign or a training session you have much longer, and will potentially involve a team of people all with their own agendas and communication styles.

How EHPs present or deliver their message can be spilt into four key groups.

1   Body language
2   Intonation
3   Use of words
4   Ambience

For *body language*, there have been numerous studies on this form of non-verbal communication,[3] which includes facial expressions, body posture, positioning, and gestures. Body language can account for 57% of our communication, and it is the message we convey without speaking. A good simple example is if someone wishes to disengage with the conversation they could look away, cross their arms, or sigh.

Some examples of body language:

- If we tower over someone when talking to them, most people will find this intimidating and threatening. And any message being given would certainly have a negative association with it. Towering over someone or getting too close to a person will ensure the person receiving the message feels subservient and uncomfortable.
- Crossing our arms can mean that we feel we are being attacked and are being defensive. This portrays a negative feeling; or it could mean that the recipient

is feeling cold, or that it is a position of comfort.

- Facial expressions often tell us what a person is feeling. If you look at the pupils of someone who is pleased to see you, they will be dilated. It has been thought that movement of the eyes can be a sure sign that either someone is lying or telling the truth; dependant on which way they look. This is fraught with the danger of misinterpretation.
- Most of us however should be able to recognise when someone is showing stronger emotions such as boredom, anger, or indeed stress.

For *intonation*, around one-third of verbal communication is about intonation, which can be broken down into speed, volume, and emphasis. If we speak quickly due to urgency, irritation, excitement, or uncertainty, this will be picked up by the receiver or recipient. If we slow our speech the recipient will notice this calmer voice and is more likely to relax and subsequently respond more calmly. We often mimic what we hear and see. To reach a calm conclusion we need to communicate in a calm way; speaking steadily will help get our message across in the way we intend, and will ensure the other person can listen. Shouting on the other hand begets shouting! The volume aspect. Therefore, by keeping the volume to a sensible level so as not to be irritating is a sure-fire way of getting the message across.

Emphasis is an intriguing element; it is easy to do in the spoken word but not so in written messages (unless we embolden words or use italics); however, emphasis can be added using capitals, emboldened words, etc.

Emphasis in sentences is key to a successful meaning:

I think you should do X – the emphasis on the I could mean that it is only me who thinks this.

I *think* you should do X – the emphasis is now on the think, which means I am in doubt.

I think *you* should do X – means you are the best person responsible to do this X.

I think you should do *X* – means X needs to be done.

Raising your voice whilst saying these sentences will also change the meaning. Raising your voice at the end turns the sentence into a question.

With written texts we need to consider how we communicate non-verbally through areas such as the font we use. Writing a technical report in, say, Comic Sans will not convey the serious of the contents. Nor would be using emoticons ☺. However, both may be appropriate for some public health campaigns, or indeed informal communications between colleagues.

The *use of words* is important too and when communicating we need to choose our words wisely. Words can evoke both logical and emotional responses. All the words we use have an emotional aspect to them and will subsequently cause the receiver to react. We need to consider using words that do not evoke a non-desired response or that will cause confusion.

Typical words – would, be, could, and should. When you are trying to find an amicable solution, using could gives the receiver of the message options and creates more positivity. The same can be said of the use of want and would like. The first is a command – I *want* you to improve X situation and the second a suggestion, I *would* like you to investigate ways of improving X situation. Very few people like being told what to do by someone else, and using want is like towering over someone. It implies power and authority.

However, when you need to be assertive, 'I want' is appropriate. Particularly when it is a matter of a statement of fact.

Ambience – EHPs need to be able to read or understand the ambience of a situation (a feel for the 'atmosphere'). Some people are unable to mask how they are feeling. We all want to be happy, relaxed, and pleasant folks, but for many there are other things going on in their worlds, which may not mean they are receptive to our communications. There may be places where people are not so receptive. For example, scolding a manager, during an inspection or audit, in front of his or her subordinates will achieve a different result from sitting privately discussing the issues.

EHPs therefore need to consider their body language, intonation, use of words, and ambience to ensure the effectiveness of their communications.

Written communications can take many forms, such as reports, letters, emails, posters, and social media. The same rules apply to all. There are many guides on how to write a good report – most universities have these on their websites.[4] Local authorities and larger businesses will have guidelines on how to write letters and indeed emails. These often must fit a template for consistency and corporate branding.

The one issue that appears consistently is having a logical approach (although the logic might vary). Indeed, what is logical to the speaker (particularly a trained technical person) may not of course be logical to the listener. So, there will be views about what constitutes 'logical,' but there are two good, very simple structures that seem to work well for most people. This is first simply to have a:

- Beginning/introduction;
- Middle/content;
- End/summary.

This is sometimes set out as *Tell them what you are going to tell them, then tell them, then tell them what you have told them*). This works well as a check that the communication is introduced, sets the communication in context for the listener or reader, does not start with too much information, and them leaves them with a repetition of key points to reinforce and aid memory.

The second system has been called 4MAT:[5]

- Rationale or *Why* – the attention grabber; why listener or reader should be interested;

- Content – *What* the core information is – the overview;
- Context – *How* the information applies to the listener, the detail;
- Call to action – *What* the listener or reader (recipient) should do with the information and what the result or outcome will look like.

One issue that is arising is the use of mobile communications methods, which have simplified the speed of communications, but have also resulted in a much less formal style to communications. For professional communications, the writer needs to be wary of slipping into 'text speak' or missing grammar and punctuation.

Verbal communications can take place in a number of situations:

1 One to ones
2 Small groups
3 Meetings – both formal and informal
4 Training sessions – accredited, bespoke, or bite-sized

Each will require some planning into ensuring that they are effective:

- An EHP may not have time to plan a phone call. But the option is to always ask about calling the person back.
- When dealing with someone in the regulatory situation, including advising on compliance (or on non-compliance), on-site communication will be oral. To ensure that what has been heard has been understood in the way it was intended (the message received was the same as that was sent), it is wise (and good practice) to confirm what has been said and agreed in writing, so written communication follows the oral. And in this example, it is essential that an accurate record is maintained on what has been communicated, and if appropriate, an agreed action plan.
- Group communications will need some managing to ensure that all parties have an appropriate time to speak.

- Training should include planning to consider a syllabus, aims, and objectives but also the learning styles of the delegates (see further reading).

For communication to be effective, it should conclude; this can be a simple summing up of the points discussed if a verbal communication, or some key points if a letter/training course. But letting someone know that you are grateful by saying thank you is not just polite, it acknowledges that you respect the time taken in communicating.

The where

There is a vast array of means of communication available to EHPs and where these conversations take place will have an impact on how effectively the message is received.

The written word allows the reader to take the time to digest the message. The written word has progressed at a rapid pace in recent years from the stalwarts of books and newspapers. There are many guides to *good writing style* – see further reading section.

Not so long ago, the only means of verbal communication was face to face; here we can fully utilise all our skills/senses including our body language to ensure our message is understood. Then the telephone arrived, initially one in a fixed location at a premises; now almost everyone has a mobile phone, which means verbal communications can be had anytime and anywhere. A reminder of the modern world of communications is people can make phone calls and pick up emails on top of remote Scottish Munros. Would anyone wish to go back to the days of teams of people sharing one phone?

As with all behaviours there is a need for boundaries, although these may reflect cultural norms. Using a mobile device in a theatre is still largely frowned upon. At the same time theatres are using mobile communications to entice a younger audience to share the theatre experience.

The use of social media platforms has made communications both to singular recipients and a mass audience more rapid and it could be said that this has changed. Indeed, while social media can help get a message or information out to many people quickly, it should not be forgotten that not everyone has access to social media, and careful thought must be given to ensure that the people for whom the message is intended are not excluded.

There are a number of social media platforms that if used wisely are appropriate for EHPs to communicate their message. Their use of course depends on having internet access and following some rules! Next are 12 ways of communicating across social media platforms.

1   *Video calling sites such as Teams/Zoom/Go-to-webinar*

The global pandemic of 2020/21 saw a huge rise in the use of video-calling platforms, and the unprecedented rise in people working form home. The pandemic changed the way many of us communicated both professionally and socially on a day-to-day basis. However, this did give rise to 'zoom fatigue' in some, as it was often the only means of effective communication, and many experienced back-to-back work calls, some dealing with a multitude of faces on a screen. These calls were often interrupted with poor internet/Wi-Fi issues and family matters, including home schooling woes. Then individuals were only able to 'see' and communicate with their family using the same media. Often from an uncomfortable position.

The CIEH used regular online communications to stay connected with members during the pandemic, including hosting regular chat sessions to keep the community feel, and to knowledge share.

The platforms enabled members who would otherwise not be able to attend sessions due to geography or cost to be able to attend either live or at a time that suited on replay.

2   *Websites and apps*

Websites and apps are an excellent communication resource tool. If well designed they ensure that information can be communicated 24/7; services and products can be listed and described, including 'how to' guides and frequently asked questions, especially using a chatbot. They do need to be kept up-to-date and designed appropriately for all to use effectively.

The one drawback is that the message is only received if the person visits the website or app and finds the information. That can be overcome in part by having a system of 'alerts' and 'push notifications,' although these have to be signed-up for.

Many businesses are using WhatsApp as a communication tool; it enables quick communications, including file sharing to individuals or to groups. It can show receipt of the message. And for customers it enables a quick route into a company. WhatsApp is owned by Facebook, which has resulted in some concerns regarding the privacy of data.

3   *Intranets*

An intranet is a great way to share information internally such as policies and procedures; it is good for communicating quickly and efficiently across organisations; but again, a similar drawback to an intranet is that people need to actively seek out the information. It can also result in colleagues communicating online to everyone rather than being specific and targeted. It has the same appeal to some as a static notice board. As an example, the CIEH has myCIEH which is a member's intranet, although within the membership there appear to be differences of opinion as to its value and the extent to which it is used.

4   *Online chat forums*

These can be used professionally to discuss issues and concerns; one of the best utilised is the Knowledge Hub which covers Food Safety. The Institute of Food Safety Integrity and Protection uses expert-led forums.

5 *LinkedIn*[6]

LinkedIn is considered the professional's media platform. Each person using LinkedIn has a profile that can include his or her contact details, qualifications, and previous roles. A person's profile can be used to find an expert, or new employee/employer.

A useful feature are the groups/forums which enable professional debate and discussion.

The CIEH has a LinkedIn presence in the form of a company profile, including groups covering Education and Research and Commercial and Independent Special Interest groups. The International Federation of Environmental Health and the Royal Society for Public Health also have LinkedIn groups.

LinkedIn can also be used to post articles, which are visible to your connections, thus encouraging communications through comment/debate.

6 *Facebook*

Facebook, for those who do not know, started out as a College Yearbook, but has grown massively – it is a way of sharing news and views, largely personal. But there remains a growth in business/organisational use. Organisations via their pages can informally share news and views to connections/friends/likers. The CIEH has encouraged regions and branches to have pages to communicate with the broader membership. The Food Standards Agency has an active Facebook presence, sharing campaigns and updates on a regular basis.

7 *Twitter*

Twitter members can communicate easily and freely; who can forget the communication style of Donald Trump, when he was President of the USA? There is software to shorten posts, or to create

links to direct readers to a web page or blog. Twitter is used well by many organisations to share news instantly, but also to offer a route to asking questions and advice.

There are many organisations that purely push out marketing messages via Twitter. There are also many personal postings, but as with all social media it is not necessary to read everything that is posted, and you can use Twitter Lists to group people and organisations that you wish to listen to (or follow).

Twitter can be used to live stream videos so the platform can be effective for campaigns or showing how things should be done.

The CIEH remains active on Twitter.

8 *Instagram*

Instagram uses pictures to communicate, and is used mainly for personal purposes, and by the creative industries. The CIEH uses Instagram, as photos and videos are a good way of getting a public health message to a wider audience, and especially for campaigns highlighting good and bad practice.

9 *Pinterest*

Pinterest is again a visual platform, and relies on users creating boards which group together interests; it can be used effectively for posting infographics and for sharing pictures with like-minded people.

It has a mixed use by organisations, and can be used to promote such areas as Corporate Social Responsibility, and by training organisations to share tips and knowledge.

10 *YouTube*

YouTube is a series of video clips of varying length and quality. It is a great way for organisations to share knowledge in a short, snappy visual way. The CIEH has a YouTube channel.[7]

11 *Webinars*

A webinar is a short web-based presentation. It is a great professional communications tool. The audience does not need to be in the room with you and the scope is vast. Webinars can be free to attend, although generally require pre-registration. They can be static presentations, for example, talking about slides, or they can be truly interactive. Many of the apps allow polls and chat. They have been used extensively during the global pandemic and are likely to remain a vital tool in professional communications. One particular bonus is the sustainability element, as this media reduces the need to travel.

12 *Emails*

Today's students and younger EHPs will not have lived in a world without emails and may not consider email to be a social media platform. Emails allow quick communication, but the expectation is a quick reply! You can access an organisation directly but also include many recipients and attachments to an email. Thus, if used wisely email can be an effective way to communicate. It does however lack the dimension of body language – hence emojis and emoticons have been developed which in reality should be used sparingly in professional emails 😊.

The etiquette for email compilation should follow that of a letter; and many organisations will have strict guidance on use.

Emails can also be used with templates to create newsletters; subscribers can then be communicated to on a regular basis.

Social media in some respects has made communications more effective or at least quicker; but EHPs do still need to ensure that they are at least communicating with the correct audience. It can be difficult to ascertain whether the message has been received and understood as intended.

The who

When communicating we must be mindful of who we are communicating to, as this can affect how we package up our message.

Over the timespan of our careers, we will need to communicate to a number of different people. These people will have their own agendas, but whoever the EHP is communicating with it is imperative that the communication is received and understood.

It is also important to ensure that the EHP is talking to the right or most appropriate person/audience. In their work EHPs must be always professional; nevertheless, it is possible to have informal communications and still be professional. We can be assertive rather than aggressive.

Barriers to effective communication

This chapter has so far focussed on effective communication, but we must appreciate that poor communication will influence how the intended message is conveyed and therefore received. There is a need to appreciate potential barriers to, or interference with, good communications. This can assist in modifying or improving communications.

The most common barriers are

- Language;
- Physical (visual and auditory limitations and difficulties);
- Resistance to the message;
- Resistance to the sender/transmitter/platform being used;
- Perceived intellect, status, and education;
- Motivation.

Looking at the barriers to communication in more detail:

- *Language* – we should use plain language that the audience will understand (so an appreciation of who comprises the audience is essential). Do not use jargon or acronyms, or at least without some explanation if they are unavoidable. The message should be broken down into three key points as this will be easy to remember. Ambiguous terms or complex legal words should be avoided (unless the audience are lawyers!).

- *Physical challenges including disabilities (both seen and unseen)* – be aware whether receivers or the audience are standing or sitting, including at desks. Some people prefer meetings face-to-face. This was historically difficult if people were in different buildings, or different parts of the country or world. Equally outdated technology can impede communication. There is a recent suggested update to Maslow's Hierarchy of Needs that includes Wi-Fi and batteries – reflecting the reliance on mobile technology [4].[8]
- *Competing and conflicting messages* – be clear of the message/agenda. Be aware of other communications that may have been received from another interested party. Be cognisant of a conflicting need of the receiver; if I do X I can no longer do Y.
- *State of mind of the communicator and the receiver* – further reading here would be on Transactional Analysis (Child/Adult/ Parent)[9] or Neuro-linguistic Programming (how we can reframe our thoughts). There is a whole movement dedicated to how humans interact with each other, including aspects such as the state of the individuals, the purpose of communications, and how they make people feel. The different psychological state we are in when we communicate, for example such as a parent, adult, or child, is an important consideration. A recipient of a message who is in a child state (made to feel so by the transmitter) may well try to please the transmitter (who may be in a parent state) so that could indicate understanding even when the recipient does not understand. This leads to poor and ineffective communication.
- *Learning and communication difficulties* – know your audience. The language used must be appropriate and use unambiguous words or phrases. Words can have different meanings, and the communicator must ensure that the message is received as intended.

Using appropriate language is particularly important with written communications, especially emails where meaning can be misunderstood as there is no body language to convey the true meaning.

- *Resistance to the message* – there is a resistance to, say, hearing an enforcement officer or a public health campaign on stopping smoking when the recipient has no desire to make changes, or thinks the changes are too difficult to make; hence the message should be on reducing rather than stopping an activity completely.
- *Resistance to the transmitter (sender)* – this could be because of physiological barriers – say a person's discomfort, caused by ill health, poor eyesight or hearing difficulties.
- *Motivation* – the recipient does need to want to hear the message, or social marketing will need to be exceptional to ensure the message is received.
- *Intellect and education* – this can be exacerbated by the use of jargon, or difficult or inappropriate words. Or poorly explaining the message can result in confusion.
- *System design* – for example, if a business or organisation structure is unclear, persons may not know who they can communicate with. It can be equally unclear if roles are muddled or confused. Or if there is a poor information system. Poor management could also result in lack of consultation with employees or the wider population.

Public speaking

> *According to most studies, people's number one fear is public speaking. Number two is death. Death is number two. Does that sound right? This means to the average person, if you go to a funeral, you're better off in the casket than doing the eulogy.*
>
> – Jerry Seinfeld

> *There are only two types of speakers in the world. 1. The nervous and 2. Liars.*
>
> – Mark Twain

Something that can come as a surprise to observers, given the training and role of EHPs, is their frequent inability or at least reluctance to speak in public, whether that be to their peers or to a community group in a public meeting. If that is the case, then it will be more difficult to give evidence well in court or other hearings. Indeed, experience of training courses for EHPs has revealed a surprising reluctance to participate in role-play exercises. Inevitably there will be nerves; indeed nervousness is almost a prerequisite for a good 'performance,' but nerves have to be managed. There are many people who have set a number of rules for good public speaking. The following is a distillation and some pointers rather than rules:

- Prepare – know your subject, be it your evidence or the contents of a report, then rehearse if possible. It has been said that there are always three speeches for every one speech you have actually given – the one you practised, the one you gave, and the one you wish you had given.
- Do not read your presentation, unless it is a formal lecture, the contents of which may be published, but even then, make use of asides or ad libs to add 'colour' – use notes or key pointers/headings on cards – Abraham Lincoln said *extemporaneous speaking should be practised and cultivated. It is the lawyer's avenue to the public.*
- Make the audience the centre of your world – focus on them, make eye contact with different members, look around the room from front to back and side to side (the lighthouse effect). Think how you can establish a relationship with the audience – what are the points of contact?
- Give your purpose the most attention – do not confuse topic and purpose (aim). Think about how you would respond if someone asked you the 'purpose' of your speech and you started to tell them about what you intended to say – you are responding by talking about the topic, not the actual purpose of your speech. As an example, when

speaking to a community group about food safety in the home, your purpose is to reduce gastro-intestinal disease in the home and to help them achieve that, but your topic is safe food preparation. Thinking about your purpose (or motivation) can enthuse you, and if you are enthusiastic there is more chance your audience will be too.

- Use your body – but keep your hands out of your pockets if you can (make sure your pockets are empty of loose change if you like to walk about and cannot help putting your hands in your pockets occasionally to appear relaxed).
- Speak up – do not mumble, project your voice, but do not shout. Speak clearly. The way you breathe affects the way your voice comes out so breathing exercises before you even start can help. Air flowing over your vocal cords is the reason you have a voice at all. Breathing shallowly means you will soon run out of air and your voice becomes strained. Take your time to fill your lungs (pauses are an important part of delivery too; well-timed silence can be eloquent so rehearse these in the preparation) and breathe from your diaphragm. Your voice will be supported by the air and your throat muscles will relax, and your voice will carry better.
- Finally remember to say what you have to say and finish, leaving time for questions. The question-and-answer session can be the most informative for the audience. No one ever complains about a speech being too short.

> Make sure you have finished speaking before your audience has finished listening.– Dorothy Sarnoff (singer, actor, and self-help guru)

To draw this chapter to its conclusion, and to bring out some of the important points, the eight principles of communication can be summarised as follows.

1 *Clarity* – the message conveyed should aim to be clear, meaningful, simple, easy, and

unambiguous. It should not be too general or vague. The purpose should be clear.

2 *Complete and timeliness* – communications should include all the facts required by the receiving audience and received to avoid a delay in action. Communications should be delivered at the right time for effective action.

3 *Concise* – the length of the communication should convey the necessary message, without waffle. It should be as succinct as it can be.

4 *Consideration* – to be effective, the giver of the message should step into the shoes of the receiver. How would the receiver be most responsive?

5 *Courtesy* – the sender of the message should be polite and sincere, and be mindful of being inclusive to the needs of the receiver/audience.

6 *Correct* – the sender should strive to be grammatically correct. There are online tools to ensure that written communications comply with the rules of the language being used.

7 *Consistency* – for example, policy and objectives across an organisation should be consistent. U-turns should be left for others, unless specifically stated due to the receipt of further information!

8 *Comment/Feedback* – this enables an effective two-way communication. The individual communicating can then assess their strengths and weaknesses.

Summary of key points from the chapter for all EHPs

- Effective communication is critical to all EHP functions. We must all take responsibility for communicating effectively;
- Active listening is integral to be an effective communicator. We should be engaged and aware of the responses given;
- The skill of communication is no different from other core EHP skills and will take time and effort to develop and maintain. EHPs must not take this skill for granted;

- There are many techniques that you can develop to improve your communication skills – see the further reading section;
- To communicate effectively the message needs to be received by the right person or audience, at the right time, place, and in a way that they can receive it;
- How the message is presented will affect how it is received and understood; think body language, words, tone, and emphasis – all are critical to how the message is received;
- Have a clear outcome/agenda to communications. Know what you are communicating – be competent – what is the purpose and what is the topic, and what is the consequence of the message being sent;
- Always be prepared to recap or summarise;
- Use technology effectively whether in one-to-one communications or broadcasting messages to a larger audience.

Notes

1 The Shannon Weaver model has some criticisms, such as it is more effective in person-to-person communication than in a group or with a mass audience; the sender plays the primary role and receiver plays the secondary role and communication is not a one-way process. However, understanding 'noise' will help to solve problems in communications.

2 I keep six honest serving-men
(They taught me all I knew).
Their names are What and Why and When
And How and Where and Who.
I send them over land and sea,
I send them east and west.
But after they have worked for me,
I give them all a rest.

from The Elephant's child

3 See for example http://www.businessballs.com/body-language.htm#body_language_information_books

4 See for example http://web.mit.edu/msrp/myMSRP/docs/Tips%20for%20writing%20a%20good%20report.pdf (Massachusetts Institute of technology) or https://www.reading.ac.uk/internal/studyadvice/StudyResources/Essays/sta-featuresreports.aspx (University of Reading) or https://www.reading.ac.uk/internal/studyadvice/StudyResources/Essays/sta-featuresreports.aspx

5 http://www.janesunley.com/4mat

6 https://www.linkedin.com/

7 See https://www.youtube.com/user/TheCIEH

8 Changes to the original five-stage model of Maslow's hierarchy are the following 5, 6, and 8 points, developed during the 1960s and 1970s [4]:
1 Biological and physiological needs – air, food, drink, shelter, warmth, sex, sleep, etc.
2 Safety needs – protection from elements, security, order, law, stability, etc.
3 Love and belongingness needs – friendship, intimacy, affection, and love – from work group, family, friends, romantic relationships.
4 Esteem needs – self-esteem, achievement, proficiency, independence, status, dominance, prestige, managerial responsibility, etc.
5 Cognitive needs – knowledge, meaning, etc.
6 Aesthetic needs – appreciation and search for beauty, balance, form, etc.
7 Self-actualisation needs – realising personal potential, self-fulfilment, seeking personal growth and peak experiences.
8 Transcendence needs – helping others to achieve self-actualisation.

9 Transactional Analysis was a model of people and relationships that was developed during the 1960s by Dr Eric Berne.

References

[1] Klyukanov I. (2012) Communication – that which befalls us. *Empedocles: European Journal for the Philosophy of Communication*, 4(1): 15–27.
[2] Honey P, Mumford A. (1992) *Manual of Learning Styles*, Peter Honey, Maidenhead Berks, 3rd ed.
[3] Mumford A. (1997) *Action Learning at Work*, Mumford A (Ed.), Gower, Aldershot, Hants.
[4] McLeod SA. (2014). *Maslow's Hierarchy of Needs*. www.simplypsychology.org/maslow.html.

Further reading and sources of information

1. Claude Shannon and Warren Weaver, *The mathematical theory of communication*. University of Illinois Press. This is the first major model for communication and was devised for Bell Laboratories in 1949. It is related to the functioning of radio and telephone technologies. The simple model is often referred to the transmission model. And is based on the following elements:

    An information source – which produces a message;

    A transmitter – which encodes the message into signals;

    A channel – to which signals are adapted for transmission;

    A receiver – which decodes/reconstructs the message from the signal;

    A destination – where the message arrives.

    It also details three levels of problems for communication:

    The technical problem – how accurately the message can be transmitted;

    The semantic problem – how precisely the meaning is conveyed;

    The effectiveness problem – how effectively the received meaning affects behaviour.

2. Transactional Analysis: https://www.uka4ta.co.uk/what-is-transactional-analysis/. This looks at how we interact with others, in adult, parent, or child mode.

3. Training materials: the CIEH offers two training skills courses; the Level 3 Award in Training Skills and Practice and the Level 3 Award in Education and Training. This will enable delegates to deliver great training presentations.

4. Body Language http://www.businessballs.com/body-language.htm

5. Kate Burton Romilla, *Neuro-linguistic Programming (for Dummies)*. Ready NLP use of positive language and expectations, to reframe how we speak.

6. Honey and Mumford learning styles: https://www.talentlens.com/uk/career-development/honey-and-mumford-learning-styles-questionnaire.html?gclid=Cj0KCQiAmpyRBhC-ARIsABs2EAqwnBqUIGRe5bRsGvF8sJDY-V90OA_q8I7UzoilJois6aBaURHSJp-QUaAmENEALw_wcB

7. Principles of good writing: https://ekmillerdesign.wordpress.com/2015/04/10/principles-of-good-writing-allan-little/

8. How to write a good letter: http://www.goodletterwriting.com/

4

Research and evidence for environmental health policy and practice

Rob Couch, Surindar Dhesi, Jill Stewart, Alan Page and Zena Lynch

We also acknowledge Dr Caroline Barratt, for her contribution to this chapter in the previous edition.

Introduction

Environmental health (EH) professionals have often spoken of the need to become more research active [1–2] and make their work more evidence based, but to date little has been written about how to achieve this in practice. It can also be argued that having a research base is needed in order to advocate effectively for action to improve environmental health. This chapter is therefore written as an introductory guide to research for EH professionals, students and policymakers. By developing knowledge, it is hoped the practitioner will feel more confident navigating the world of research, motivated towards making their own work more evidence based and enthused about contributing to the evidence base from which others can learn. This chapter is not a research methods textbook, a step-by-step guide to research or evidence-based environmental health, nor does it seek to make definitive statements about these complex areas. However, it highlights the most important issues regarding research in environmental

health, considers the importance of research to the environmental health profession and provides useful signposts towards further resources.

The chapter is divided into three sections. The first defines evidence-based environmental health and why it remains a priority for EH professionals. The second section explores the key stages of environmental health research and provides guidance on the development of your reading skills. The final section suggests ways to become more research active and evidence based, acknowledging the many challenges EH professionals face and concluding with a vision for evidence-based environmental health. The chapter ends with an annex including a glossary of environmental health research terms, a list of references and suggested further reading.

Since the previous edition of *Clay's* where this research chapter first appeared there have been many advances in environmental health research and more evidence-based practice by EH professionals. These include more practitioners completing postgraduate research degrees, making time to write books [3–4] and papers and utilising social media (e.g. Twitter, podcasts) to communicate their knowledge and be

DOI: 10.1201/9781003035640-4

a good example to others. The COVID-19 pandemic has also brought the vital role of research to inform environmental health policy and practice to the fore and has seen EH professionals across the world [5–7] documenting their critical pandemic work and sharing this critical knowledge. But this chapter is really about culture change, challenging EH professionals to realise the potential for research to improve and defend what they do and, perhaps, by doing so to help make them more visible? With this in mind it's encouraging to see in a recent membership survey EH professionals identifying 'accessible research' amongst their top five areas in which they thought the Chartered Institute of Environmental Health could be doing better [8]. This chapter therefore seeks to provide you with some of the skills and approaches towards becoming an environmental health researcher-practitioner.

SECTION 1: INTRODUCING EVIDENCE-BASED ENVIRONMENTAL HEALTH

What is evidence-based environmental health?

Environmental health is a relatively new term [9] and does not have a simple definition [10–11]. A definition is problematic for many reasons, not least because 'environment' and 'health' are themselves difficult to define and then combine [12]. However, definitions can provide a useful starting point, and this chapter is based around the wording developed during a series of World Health Organization (WHO) conferences:

> Environmental health comprises those aspects of human health, including quality of life, that are determined by physical, chemical, biological, social and psychosocial factors in the environment. It also refers to the theory and practice of assessing, correcting, controlling and preventing those factors in the environment that can potentially affect adversely the health of present and future generations.
>
> *([13] – page unknown)*

Reflecting on this definition, perhaps the greatest challenge facing potential researchers is grappling with the interdisciplinary nature of environmental health. The WHO definition and the historical development of the role of EH professionals suggests that environment-health relations and their management have always been shaped by many disciplines including biology, chemistry, physics, psychology, law, politics, philosophy, economics, sociology and history. It can therefore be useful to view environmental health as a complex subject shaped by many disciplines. Indeed, the interdisciplinary skills of EH professionals are, arguably, one of their greatest strengths, but simultaneously this lack of a single disciplinary 'home' or body of knowledge presents many challenges. Recognising and managing this complexity is critical for all EH professionals, students and policymakers, and highlights how much could be gained by collaborating with researchers and others with greater knowledge of the many disciplines underpinning environmental health.

Organisations and individuals have long sought to influence environmental health policy and practice using evidence. In his 1842 *Report on the Sanitary Condition of the Labouring Population of Great Britain* Edwin Chadwick described environmental health conditions and inequalities across industrialising Britain and argued that these could be addressed by his 'great preventives' (e.g. household water supplies, toilets and sewerage) delivered by a cadre of EH professionals [14].

Many factors continue to influence the use of evidence in policy and practice including greater pressures towards productivity and competitiveness, an increasingly knowledgeable and well-informed public, declining trust in the expertise of professionals, and greater scrutiny and accountability of governments [15]. As in Chadwick's day ideology remains a powerful driver of environmental health policy [16–17] and politicians can always be heard jousting over 'the evidence' that supports their arguments or undermines their opponents.

An exploration of the origins and development of the 'evidence-based' movement in clinical medicine is beyond the scope of this chapter (see [18]), but its main principles have been incorporated into our definition of 'evidence-based environmental health' as:

> *environmental health policy and practice supported by the best available evidence, taking into account the preferences of citizens and the wider public and our own professional judgment.* ([19] – page 6)

Implicit in this definition is the ability of EH professionals to provide a clear and up-to-date rationale for their work that goes beyond default responses such as 'it's what the law says' or 'that's how we've always done it here' and enables them to challenge engrained attitudes. Policy is included because in the form of legislation and guidance it remains an important driver of environmental health practice. This definition also recognises that evidence is often uncertain, changing, vulnerable to politics and can be difficult to access (hence 'best available'), but EH professionals should have the confidence to embrace its uncertainties and use them to improve public health.

However, the application of evidence works alongside professional judgement because of the limits of the available evidence and the unique and complex nature of environmental health cases. Critically, judgements should also consider the preferences of citizens and the wider public influenced by environmental health activities. The term 'citizens' is used to include all those EH professionals encounter during their daily work (e.g. business owners/operators, employees, the public)

and recognises their legal rights and responsibilities. The terms 'client' and 'customer' are avoided because of their associations with market-derived neoliberal ideologies that remain powerful but can exacerbate health inequalities causing avoidable morbidity and mortality [20]. This can be exemplified by EHPs themselves being uncertain as to who would be the client and customer in the context of a tenant complaining about their housing conditions and a notice served on the landlord, the owner of the property.

A word of warning is also needed here. The term 'evidence-based' has become increasingly politicised and is often used to support the dominant opinion or those with the most powerful voices. In response there has been something of a backlash towards the term, but its use is recommended provided you take the following course of action:

- Tune your 'warning antennae' to provide an alert every time terms like evidence or evidence-based are seen in a publication or someone describes their work in this way;
- Examine the references or challenge the speaker about what they mean by evidence (e.g. what evidence has been used?);
- If there are no references, or the publication is poorly referenced, or the references are based on single studies or personal experiences or have been carried out by those with vested interests they might not have declared – treat the 'evidence' with extreme caution.

Why is evidence-based environmental health needed?

Before grappling with this question, it is important to consider the potential of environmental health research to contribute to a better understanding of some of the greatest and most persistent challenges faced by societies today such as poverty, inequality, the climate emergency, urbanisation and the need for more sustainable economies. Research into the complex relationships between human health and the environment has a long history and is constantly being re-focused, for example in the UK there is now a greater emphasis on the impacts of environmental health on mental health and wellbeing [21–22]. But for EH professionals probably the most powerful argument for research is its potential for better understanding how environment-health relationships are managed, particularly the effectiveness of environmental health interventions. Academics have been researching the work of EH professionals for years (e.g. [23–26]) but research by EH professionals themselves remains rare, as is their engagement with academic research.

Returning to why evidence-based environmental health is needed, Greenhalgh [27] provides warnings of the alternative drivers of decision-making by health professionals:

- Decision-making by anecdote – where decisions are based solely on personal experience;
- Decision-making by press cutting – where decisions are based on single published studies without consideration given to the methods used or the results of alternative studies;
- Decision-making by GOBSAT (Good Old Boys Sat Around Tables) – the product of biased, 'expert opinion' that in reality could simply consist of the bad habits and personal experiences of ageing professionals; and
- Decision-making by cost minimisation – where the cheapest option is followed, regardless of its effectiveness.

In 2015, UK government economic policies could be characterised by austerity that has brought the cost minimisation driver to the fore. For example, policy recommending the re-organisation of local government environmental health services has been advocated by powerful organisations like the Auditor General for Wales [28] but supported by little or no evidence that the recommended models of collaboration and outsourcing will 'improve efficiency' and 'maintain performance'. Until EH professionals can better demonstrate the effectiveness of their work to those making

funding decisions, they are likely to remain a vulnerable and unstable workforce [29] as confirmed by the CIEH's latest England local authority study [30]. In early 2021 this described a national workforce characterised by a lack of resources, a reliance on agency staff, shortages of qualified and experienced staff and limited or no budgets for training the next generations of EH professionals.

Another justification for evidence-based environmental health is to question the models that continue to dominate policy and practice. For example, in the UK the development of food hygiene risk rating systems and their powers to predict the epidemiological risk associated with food premises provides another example of how questionable science can become standard practice. Day [31] describes how the original systems on which scores were based were only designed to provide a 'quick and dirty' means of prioritising inspection resources. Two case control studies [32–33] question the effectiveness of food hygiene risk rating systems and the EH professionals applying them, but these systems remain largely unchanged and have been expanded by initiatives like Scores on the Doors and in England and Wales. Further, the questions raised by the research of EH professionals Fairman and Yapp [34–35] into risk rating systems, as summarised in Box 4.1, remain as relevant as ever and continue to make for uncomfortable reading for all EH professionals.

EH professionals should be utilising evidence which is based on research and evaluation in their policy and practice. Aveyard and Sharp ([36] – pages 40–45) usefully categorise this evidence to include, with a few examples from our favourite research publications:

- Evidence for effectiveness [37];
- Direct evidence – from studies that relate directly to practice [38];
- Indirect evidence – from studies relevant but not directly related to practice [39];
- Evidence deduced from scientific knowledge – where scientific principles are applied to practice to explain how things work [40].

Box 4.1 Enforced self-regulation and SMEs: the Fairman and Yapp debate

Fairman and Yapp [34] evaluated the impact of enforcement approaches and local deprivation on compliance levels and decision-making in UK food sector small and medium-sized enterprises (SMEs) using mixed methods including interviews, focus groups, inspections and document analysis. They found most SMEs were poor at identifying legal non-compliance and perceived compliance as the norm in their businesses until told otherwise by EH professionals. Further, the most successful enforcement approaches by local authorities were high levels of education (costly) and the breaking down of self-regulatory requirements like hazard analysis into prescriptive ('just tell me what to do') instructions that pragmatic EH professionals were willing to do. For similar findings in a health and safety context see Fairman and Yapp [35].

This evidence might be based on ideas constructed to explain phenomena, so-called theoretical research, or on empirical research founded on observation or experience. In the category 'evidence for effectiveness', concerns about the effectiveness of medical interventions has led to the development of 'hierarchies of evidence' where some evidence is considered more trustworthy than others [27]. For example, reviews of well-designed research evidence (e.g. systematic reviews) are generally considered the most trusted forms of evidence at the top of the hierarchy. Some environmental health-related systematic reviews already exist and we discuss these in Section 2, but their development and communication remains a long-standing but overlooked priority for EH professionals. Next is experimental research (i.e. randomised controlled trials) followed by observational research including cohort studies, case control studies and cross-sectional surveys. Case reports are listed as the least trustworthy [27].

Qualitative research is not included in the hierarchy because it cannot be directly compared with quantitative research design. It is not that it is better or worse, but it answers different types of research questions. It is therefore unfair to directly compare them. The quality of research methods needs to be assessed in relation to each piece of research – even systematic reviews and randomised trials that are ranked highly in the hierarchy can be done badly. These terms are discussed in more detail later and defined at the end of this chapter, but at this early stage EH professionals need to be utilising all the research tools available towards a more evidence-based environmental health. Further, EH professionals should be aware of how evidence plays out in 'real world' practice that will be familiar to many. For example public health policy advisor Dr Adrian Davis has created the wonderful 'policymakers hierarchy of evidence' where expert advice is the most powerful, followed by ideological evidence (e.g. party manifestos), evidence from professional associations and that from other sources (e.g. media, lobbyists) with research appearing towards the bottom and thus far less likely to influence policy in practice [41]. This chapter now explores a cycle model of environmental health research.

SECTION 2: THE ENVIRONMENTAL HEALTH RESEARCH CYCLE

Introduction

There are many models describing the research process, but here research is viewed as a cycle of seven linked and frequently overlapping stages as summarised in Figure 4.1. This model is based on the work of Sumner

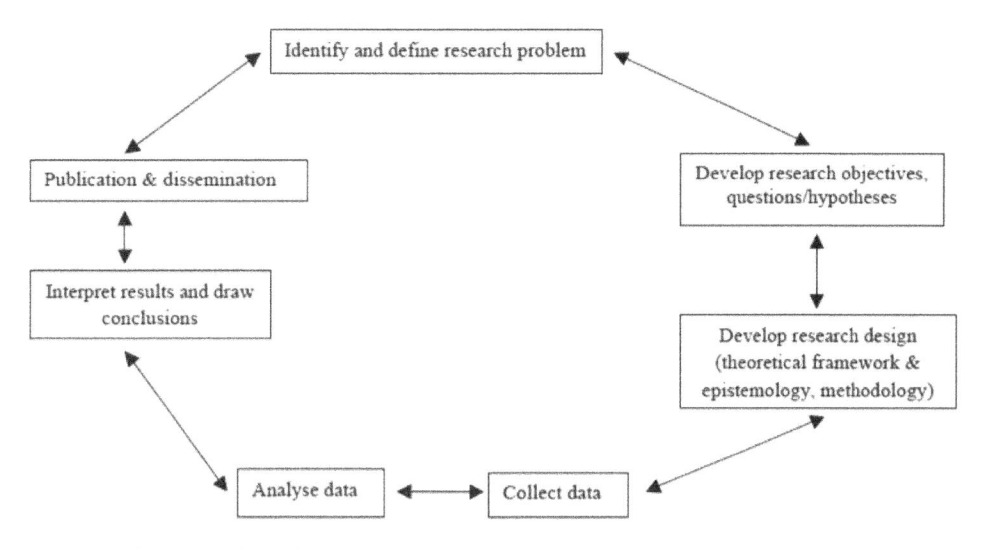

Figure 4.1   The research cycle
Source: Adapted from Couch et al. ([19] – page 27)

and Tribe ([42] – page 102) and is a gross simplification but is useful for exploring the research process. The cycle is used because research is not a linear process. Instead, the stages constantly inform one another and are frequently revisited and improved during the research process, where developing the confidence to move between these stages is part of being a good researcher. In this section we introduce you to each stage and important things to think about towards producing high-quality research. To end the section, we consider research ethics and your responsibilities as an EH researcher.

Identifying and defining the research problem

Well-planned research seeks to build on what is already known and address gaps in our current knowledge about a particular issue. Choosing a topic can be challenging, as can refining an idea into a piece of research that can be carried out using appropriate methodology (see Annex 4.1) within the time and resources available. The

best advice is to choose a topic of genuine interest, as this will sustain you through the process. Inspiration may be found from articles, papers, seminars or colleagues; or there may be issues encountered by the practitioner in the field (see Box 4.2) or wider policies that could be investigated more deeply. A good supervisor or editor can really help here.

Two recent publications brought many of these aspects together. First, one of us (Jill Stewart) led a team of interested EH professionals, many of whom were first-time writers, to convert their passion and curiosity into a book on the early pioneers of environmental health [4]. Second, in 2019 the 'Environmental Health across the world' group formed to share knowledge of different environmental health work across the world. This recently resulted in their first publication documenting the role of EH professionals during the COVID-19 pandemic [7].

Reviewing the literature – Having identified a research topic you will need to carry out a literature review. The purpose of the review is to understand the knowledge that exists,

---

Box 4.2 The art of communication: a landscape shared between regulator and ethnic employer [43]

Empirical research identifies problems in communications between UK local government EH professionals and ethnic minority food businesses [25, 44]. Building on this and his own experiences, Northern Irish EH professional Eamonn Toner of Derry City Council conducted a literature review exploring the influence of culture and communications between the Chinese and EH professionals. This informed his research design, where a sample of 56 EH professionals and 91 Chinese caterers from five local authorities in Northern Ireland were investigated using both qualitative and quantitative methods (focus groups and questionnaires) to explore their views and experiences of shared communications.

Toner concluded that relationships between EH professionals and Chinese businesses were driven largely by the former with little consideration of the unique and complex needs of the latter, particularly the importance of non-verbal communications and the limited effectiveness of simply translating information. Evidence of innovative working by individual EHPs and their departments was also uncovered and there was much goodwill between both parties. The work has informed evidence-based practice and ongoing outcomes including improved awareness and understanding and much closer working relationships between EH professionals, Chinese employers and the wider Chinese community. Non-verbal communications training courses have been developed for EH professionals and the work has been presented at conferences in Ireland and the UK and influenced guidance documents including the UK Food Standards Agency's Resource Handbook on *Working Effectively with Minority Ethnic Food Businesses* [45].

Further, in 2013 Toner was awarded a Fellowship of the Chartered Institute of Environmental Health in recognition of this work and he remains a passionate advocate of the need to improve understandings between regulators and ethnic minority businesses and to support practitioners towards publication.

---

and to identify a 'gap' which your work will help to fill. Here the following quote from Sir Isaac Newton is useful:

> If I have seen further it is by standing on the shoulders of giants.
>
> *[46]*

At these early stages it is tempting to follow initial ideas and to start developing questionnaires, interview schedules, etc. immediately. But following Newton's advice it is important to channel enthusiasm into a review of the knowledge that engages with the existing work of the 'giants' in the area of interest. This will help to identify research trends, gaps in existing knowledge and ensure that the research has not been done before. Exploring

this work can also assist in focusing the topic further, identifying suitable research methods, highlighting potential challenges of researching the topic that has been chosen and informing the theoretical framework. In recent months for example research output has been significantly skewed by COVID-19 with a significant focus on the impact and efficacy of interventions and increasingly the wider impacts of the pandemic, particularly in areas like health inequalities and mental health, all of which is relevant to the work of EH professionals (see [5]). The literature review therefore informs all stages of the research cycle; even when discussing findings reference will be made to work of others to put the results into context. Given the significance of reviewing previous literature in this process we now turn briefly to provide additional guidance on critical reading for research.

EH professionals need to read around their subject as part of their everyday role as recognised in official guidance for EH professionals like that for the Housing Health and Safety Rating System [47]. But Horder's research [48] exploring the lack of reading in the practice of social workers is relevant here. He identified themes that are also common to environmental health including an oral working culture, where knowledge is often passed down from more experienced staff within a working context focused on 'getting things done', and limited access to reading materials in the workplace and the time to read them during the working day. These barriers remain, but by encouraging more research by EH professionals we hope the value of reading for research and evidence-based practice will become recognised as an essential part of professional life.

Read and read again; as an EH professional find out everything about your area of interest and consider reading an active and critical process. That said, how relevant information can be found depends on access to academic literature resources. In academic research, peer-reviewed papers in academic journals are typically the preferred sources of information because the peer-review process is one of the most accepted quality controls available. Ideally, systematic reviews of peer-reviewed research are the best place to start because they identify, collate, appraise, analyse and summarise good quality research around precisely defined research questions. Some already exist for environmental health (e.g. [49–52]) and organisations like the Cochrane Collaboration's Public Health Group[1] are constantly publishing open access systematic reviews of the best available environmental health evidence. However, systematic reviews of environmental health knowledge and its dissemination to those working in policy and practice remains a priority for EH professionals. One major problem is that access to

academic journals and books can be expensive for those not attached to an academic institution, though this might be slowly changing as the shift towards open access gains momentum. If access to an academic library is not possible, EHPs may be entitled to library access as a graduate of an institution where they have previously studied.

Alternatively, access to professional libraries may be possible with relevant collections or via the inter-library loans schemes operated by public libraries in countries like the UK.

Beyond systematic reviews and peer-reviewed papers and books, organisations such as governments, charities and think-tanks all provide important resources for researchers. These resources include technical reports and policy statements and are sometimes referred to as 'grey literature'. They can be of the highest academic quality (e.g. [53–54]) but are so called because they have not been subject to formal publication. However, returning to the previous warnings about evidence, many organisations present their evidence to support and promote their causes and therefore – as with all reading – requires a critical mind.

To find relevant resources, electronic databases enable searches across a variety of literature by subject, author, key words, etc. Web-based databases relevant to environmental health include Assia, Embase, Web of Science and Medline and practitioners are encouraged to visit them and follow their instructions to maximise the effectiveness of searches. Google Scholar is also a very powerful general database, whilst databases like the UK's Fingertips Public Health Profiles can provide very powerful national, regional and local public health knowledge. Although it is not always possible to access journal papers free of charge, an increasing number (especially those relating to COVID-19) are becoming open access and even if the whole article is protected it is often possible to see the abstract summary for free. It might also be worth contacting your old university to explore whether access options for alumni exist.

In the experience of the authors, 'environmental health' and its core areas (e.g. food safety, health and safety, housing, environmental protection, etc.) on their own are not very useful search terms, not least because these areas are so broad themselves and sometimes have limited recognition outside the UK and Commonwealth countries. Instead, widening your searches towards the many disciplines that underpin environmental health could be more productive. For example, using key words often listed below abstracts at the start of journal papers and/or focusing on key authors could yield better search results. The authors themselves are often happy to hear from those interested in their work and they could help guide you further.

Critical evaluation is an essential part of a literature review and many checklists have been developed to help readers identify and interpret the best available evidence. In the UK the health research checklists developed by the Critical Appraisal Skills Programme (CASP) are available for free[2] and are particularly useful, but most checklists are based around the following critical questions (based on [27]):

- Is the research question clear?
- Can the methods answer the question, for example:

  - How was the sample chosen?
  - How was the data collected?
  - How was the data analysed?

- Has the researcher's perspective been discussed?
- Are the results credible?
- Are the conclusions justified?

Lastly, amidst information overload it's easy to lose track of what has been read, where you found it and what the key points were. There are now many ways to manage what is read and to make sure that the correct references are ready for when needed. Electronic reference management software (e.g. *End-Note*, *Mendeley*, *RefWorks*, *Zotero* etc.) are very useful for anyone who is reading for research and have many features to make it quick and easy to look up references and insert them into the work. Basic formats of these systems can often be downloaded for free, whilst more sophisticated versions can be purchased.

Developing research aims and objectives, questions and hypotheses

The literature review will help refine the particular research problem to be focused on, and the best research has very clearly defined aims and objectives, research questions and/or hypotheses to provide a solid foundation from which the research has developed. Brainstorming potential research questions (e.g. who, what, when where, why, how, etc.) can also help to develop and focus ideas. Writing purpose statements is another useful exercise and the following examples in Table 4.1 draw on the authors' PhD experiences.

The objectives of the research study are informed by the research problem and should be clear and realistic, not least because of the inevitable limitations all researchers face, primarily funding and time. It is easy to be overambitious and become overwhelmed at these early stages and therefore it is recommended that the project is as focused as possible. Developing research questions/ hypotheses takes time and many attempts, but this stage is very important because it establishes the basis for the whole study. It is recommended that the practitioner work closely on the research design with a supervisor or more experienced mentor, and also with their employer or university as there may be local guidelines or criteria that need to be considered. It is also recommended that a research diary is maintained from an early stage, recording developments in the research and also recording thoughts, ideas and learning.

Table 4.1 Examples of purpose statements to develop environmental health research

Rob's PhD	
Problem	Large and persistent environmental health inequalities in South African cities
Topic	Environmental health regulation in one city by local government EHPs
Purpose	To describe and explain the factors influencing the decision-making of local government EHPs and their implications for urban environmental health

Surindar's PhD	
Problem	Persistent health inequalities between people in different socio-economic groups
Topic	Public health policymakers approaches to tackling health inequalities
Purpose	To understand public health policymaking in relation to health inequalities and environmental health in England

In the last few years a preference for hypothesis-testing has been observed amongst UK EH professionals. It is suspected that this might be a legacy of university research methods modules taught largely by non-environmental health academics with a preference for quantitative research methods, but these methods are much broader than just hypothesis-testing and what really matters is evaluating the strength or quality of the evidence presented instead of whether hypotheses are proved/disproved [27]. Research questions instead can provide broader and more flexible methods of inquiry and can be:

- Descriptive or exploratory (e.g. how does x vary with y?);
- Explanatory (e.g. which x causes y?);
- Interpretative (e.g. what is x?);
- Driven by the type of study envisaged (e.g. action research – see later)
- (based on Mikkelsen in [42] – page 103).

Developing your research design: theoretical frameworks and epistemology

Having clearly defined what the research aims to achieve, it is time to develop the research design. In summary environmental health researchers must be clear about their theoretical and epistemological assumptions because they have such an influence on the whole research process. A theoretical framework and discussion of epistemology are central to the development of the research design but can be unfamiliar to EH professionals who may not have encountered them at university and who are accustomed to practical approaches and 'solving problems' in their daily work. Here we describe what theoretical frameworks are and why they are important before providing three examples in Annex 4.2 at the end of this chapter. We then describe why epistemological considerations are so important to research.

Theoretical frameworks – In scientific research the term 'theory' is used to refer to a system of ideas constructed to explain phenomena. Theories help us to understand why and how things happen and to make predictions and they are built up over time, not simply from one piece of research. The theoretical framework of a piece of research explains the theories that are relevant to a practitioner's research and how they interact:

> A theoretical framework consists of concepts, together with their definitions, and existing theory/theories that are used for your particular study. The theoretical framework must demonstrate an understanding of theories and concepts that are relevant to the topic of your research paper and that will relate it to the broader fields of knowledge in the class you are taking.
>
> *[55]*

Using a theoretical framework helps to embed the research within previous knowledge and enables the researcher to make it clear what their contribution to knowledge will be, for example:

- Are you hoping to test a current theory?
- Are you providing evidence in support of another theory?
- Are you trying to generate a new theory because existing theories fail?

Less experienced researchers can become quite distressed at the idea of a theoretical framework, but it is not the terrifying proposition one may think. If the researcher has a research question and has done some reading around it, the chances are that they already have one but maybe just don't know it! Think about the definitions, terms and concepts that are regularly used in the research – where did they come from? To construct a theoretical framework, it will have been necessary to read around the subject to be researched, to have identified the previous theories that have been developed or used in relation to the research problem and to describe how the particular question relates to those previous theories.

Theory can also help to shape the methods chosen and help the researcher to interpret the results and draw conclusions. Therefore, becoming more aware of the role of theory in shaping the research will enable the EHP to make much more informed decisions at all stages and to write in more powerful ways. To illustrate their potential the following three theoretical frameworks were chosen for their relevance for informing questions about why EH professionals do what they do. They are summarised in Annex 4.2 at the end of this chapter:

- *Street level bureaucracy* – by Michael Lipsky [26];
- *Why EH regulators generally consider prosecution as the last resort?* – by Steve Tombs and Dave Whyte [56];
- *Environmental health regulation as modern state power?* – by Tom Crook [23].

Epistemology – Epistemology is concerned with the theory of knowledge and asks what the researcher considers 'knowledge' to be, or when can we say that we *know* something about the world. The researcher's epistemological position has important implications for the methods used, what the research is trying to achieve and the nature of the relationship between the researcher and the researched.

The origins of epistemological thinking date back to the Ancient Greeks, but since this period Western philosophers have identified and debated different ways of knowing the world and the work of EH professionals has long been informed by many different epistemologies. For example, those with positivist views of the world argue that there is one, observable and measurable reality and that the researcher can remain objective and independent of the researched [42]. For EH professionals, epidemiology for example utilises inherently positivist positions associated with quantitative research methods like randomised control trials (see later and Annex 4.1).

Alternatively, relativists argue that there are multiple realities in the world that can be experienced. In this case the researcher is subjective and not independent of the researched [29]. For EH professionals immersed in the messy realities of the streets every day, the relativist tools of the social sciences like qualitative methods (see later and glossary) could be more suitable. For example in her research on compliance and environmental health, Lange [57] uses relativist arguments to reject over-simplistic and legalistic descriptions of offences and instead argues that compliance is constructed in the field from the relationships between rules and social practices.

Critiques of the positivist-relativist debate provide alternative assumptions about knowledge and reality that the practitioner should be aware of and could help to shape the research. For example, feminist approaches broadly question the relationships between knowledge and power, particularly how 'knowledge' is not objective and typically reflects a male world view [58]. The aim of feminist inquiry is to facilitate female emancipation and greater understanding of female world views [59]. Alternatively, those following participatory approaches consider research as a cooperative and emancipatory activity; participatory action research for example considers the aim of knowledge inquiry to be liberation and empowerment of the community [58]. It is hoped that participatory action research could empower EH professionals as researchers and they are urged to explore how other professions like nursing have made progress here (e.g. [60–61]).

Developing your research design: how is the research going to be carried out?

Building on the previous stages and other factors (e.g. personal limitations) the choice of methodology is now considered, along with methods and analytical techniques. The term methodology refers to the overall research strategy followed to answer the questions/test the hypotheses and includes the researcher's theoretical position and the methods used to collect, analyse and report the data [42]. In contrast, methods are the detailed tools and techniques used to collect and analyse primary or secondary data (see Annex 4.1). The choice of methodology will be shaped by the epistemology, theoretical framework, the discipline(s) underpinning the study and any limitations faced by the researcher. Another important factor is choosing between quantitative or qualitative methodologies or a combination of both?

Quantitative and qualitative research

Quantitative research generally refers to studies that collect and analyse numerical data and often includes high numbers of participants but little or no direct involvement between the researcher and participant [36].

A wide range of quantitative research designs and methods are available and include randomised controlled trials, cohort studies and case control studies (see glossary). Sampling is often random and data analysis structured around tests of statistical significance. Both these aspects will require careful planning long before any data is collected.

Qualitative studies tend to use data derived from language (written and oral) [58], not numbers, to *explore the meaning and develop in-depth understanding of the research topic as experienced by the participants of the research* and the researcher is often more closely involved with the participant who may play a role in shaping data collection and analysis [36]. Sampling here tends to be more focused on which participants are related to the area of interest, with data analysis based on the coding of data and the development of themes [36]. Commonly used qualitative research methods include interviews, observations, focus groups and questionnaires. Examples of qualitative approaches include:

- *Grounded theory* – data is collected and analysed to generate theory (e.g. explanations of social phenomena) [59];
- *Ethnography* – a community is observed in real time to answer questions about how the community behaves [59];
- *Action research* – practitioners and researchers work together to address everyday issues about practice and develop a systematic approach to implement and evaluate change [36].

The flexibility of the research is another important consideration and will depend on the methodology chosen. A fixed design is integral to most quantitative research designs (e.g. randomised controlled trials) and sets out very specific requirements for the research process, particularly in defining sample size and how data are to be collected and analysed. But for other approaches, especially for some qualitative designs, a flexible design may be more appropriate because the stages of the research process can overlap and inform each another. A flexible design is also important if dynamic workplaces are being researched, like those of many EH professionals where interviews, for example, could be cancelled at the last minute due to reactive workloads.

Environmental health researchers commonly draw on both quantitative and qualitative approaches. These so-called 'mixed' or 'multi-methods' approaches are the subject of some debate, but examples include Hutter's [25] classic study of the work of UK local government EH professionals. This was based on qualitative data (via interview, observation and document analysis) but supported by quantitative data (e.g. workplace performance and law enforcement data). Similarly, Fairman and Yapp [34–35] use mixed methods to investigate compliance with environmental health law in small and medium-sized food businesses in the UK.

Therefore, the authors agree with Baum [63] and reject the argument that randomised trials (towards the top of hierarchies of evidence) are the 'gold standard' and priority for all public health research; instead, EH professionals must utilise all the research tools available to develop an understanding of environmental health and select the most appropriate tool(s) for the problem being investigated.

Data collection and analysis

Data collection sees the research design come to life, but before embarking on the main study it can be very useful to pilot test the design on a small sample and refine it if necessary. For example, even pilot testing and analysing draft questionnaires with family/friends or even on the researcher themselves can provide invaluable information about the process, not least the (considerable) time it can take to collect and analyse data!

A detailed examination of the many data collection methods available is beyond the scope of this chapter, but the reading and references at the end of the chapter will be useful. Things can and do go wrong during data collection, but the response to these difficulties is important. Responsibilities as an EH researcher (see later) will also come to the fore at this stage and the practitioner should be careful to ensure that ethics are not compromised in the pursuit of interesting research.

Depending on the methodology, data analysis might take place during or after data collection and typically starts with organising and then processing the data. For those using quantitative methodologies this stage could involve the careful entry of numerical data into computer software packages (e.g. *SPSS*, *Minitab*) for further statistical analysis. For qualitative methodologies, language data are likely to require transcription and then coding and comparison around themes in accordance with the chosen approach (e.g. grounded theory, action research, etc). Computer software packages can assist the transcription (e.g. *Express Scribe*) and coding process (e.g. *Nvivo*), but Greenhalgh [27] warns about the rule of GIGO (garbage in, garbage out) and that other older techniques like VLDRT (very large dining room table) can also provide excellent qualitative analysis.

Interpretation of results and conclusions

The interpretation of data and the drawing of conclusions is perhaps the most difficult stage of the research process and will be framed by the previous stages and factors, particularly:

- The research problem;
- The aims, objectives, questions/hypotheses;
- The research design – especially your theoretical framework and methodology;
- The values, responsibilities and ethics.

Fundamentally, the researcher's conclusions must be justified by the results and this requires consideration of the credibility of the results and the interpretations of them. For quantitative studies, determining the credibility of the results might involve consideration of the precision of measuring devices or error in tests of significance (e.g. confidence intervals). For qualitative studies, credibility could include results supported by verbatim quotes that can be traced back to the original source [27] and a full description of the position of the researcher (see reflexivity in the glossary). For quantitative studies distinguishing between the results obtained and the interpretation of those results is fairly straightforward, but for qualitative studies this is more difficult because the results are themselves an interpretation of the data [27]. However, Mays and Pope [64] suggest three questions for determining whether the conclusions of a qualitative study are valid:

- How well does this analysis explain why people behave in the way they do?
- How comprehensible would this explanation be to a thoughtful participant in the setting?
- How well does the explanation cohere with what we already know?

Publication and dissemination

The publication and sharing of results should be an integral part of the research process, but this stage is particularly important because it has been observed that EH professionals often do not publish their work. There are many formats available for publication. Alongside more traditional peer-reviewed journals, books, newspapers and professional magazines are an increasingly wide range of more accessible social media formats like websites, blogs, Twitter and podcasts. Ideally EH professionals should be aiming to publish in peer-reviewed journals, preferably those with policies of open access to ensure all can read them.

Publishing in peer-reviewed journals makes the researcher's work available in the databases mentioned earlier and available to be cited in other research as valid, high-quality work. As well as traditional research reports some journals offer the opportunity for new authors to publish and others accept

short opinion pieces (e.g. 500–1000 words) that might not be so daunting. Even experienced authors sometimes find an opinion piece a relatively quick way of disseminating their work in a good journal, but the practitioner needs to study the guidelines for authors before deciding which are likely to accept the topic/argument. The EH professional must also ensure that it is their work to publish in terms of the ownership/permissions/acknowledgements and that the work is original, high quality and ethically sound (see later). The work must not have been published before and should only be submitted to one journal at a time.

Other options for publication include books/book chapters, newspapers or writing for environmental health-related professional magazines. Conferences, seminars and workshops are a great way for an EH professional to disseminate their work and to test the research findings with interested people. The EH professional may be invited or choose to submit an abstract for consideration. Sometimes, a guaranteed publication follows or it may be possible to convert the presentation into an article for publication further widening its impact. Much will also be learned during the process, particularly when questioned by peers. As the UK Environmental Health Research Network, the authors have developed their own social media[3] which are free and took minutes to set up. They are still very new and as yet, it is not possible to say how effective they have been, but it is encouraging that in the UK there is an emerging research debate amongst EH professionals.

Having now explored the main stages of the research cycle introduced at the start of this section, two important themes that cut across all the stages are now explored – ensuring high-quality and ethical research.

What is high-quality environmental health research?

Establishing the quality of research is the subject of much debate and has historically been influenced by what Becker et al. describe as four 'traditional criteria' derived from quantitative research:

- Validity – the extent to which there is a correspondence between the data and the 'truth';
- Reliability – the extent to which observations are consistent when the study is repeated;
- Replicability – the extent to which it is possible to reproduce an investigation;
- Generalisability – the extent to which it is possible to generalise findings to similar cases which have not been studied ([65] – page 7).

However, most of the 250 social policy researchers and research users in this study only considered validity and reliability appropriate quality measures for qualitative research, whilst replicability and generalisability were considered much less crucial. The same study therefore revisited four alternative criteria originally developed by Lincoln and Guba [66] for qualitative research:

- *Credibility* – the extent to which a set of findings are trusted;
- *Transferability* – the extent to which a set of findings are relevant to settings other than the one or ones from which they are derived;
- *Dependability* – the extent to which a set of findings are likely to be relevant to a different time than the one in which it was conducted;
- *Confirmability* – the extent to which the researcher has not allowed personal values to intrude to an excessive degree ([65] – page 8).

In Becker et al.'s study the majority of social policy researchers considered credibility and confirmability the most important quality measures for qualitative research, with dependability and transferability much less important. Several researchers

also considered reflexivity (see later) an important quality criterion for qualitative research.

However, establishing the quality of environmental health research is further complicated by two factors. First, the use of mixed methods is not uncommon in environmental health research. Becker et al. found most researchers suggested a combination of traditional and alternative quality criteria should be used, but with different criteria for the quantitative and qualitative components. Other quality criteria for mixed methods include a clear rationale for using mixed methods and transparency in their use [65]. Second, different perceptions exist between the disciplines about what constitutes high-quality research and therefore the interdisciplinary nature of environmental health has the potential to create further complications. Thus, more work is needed towards establishing what constitutes high-quality environmental health research, but engaging with debates about the quality of the data and how it relates to other populations and settings is critical. For example, assuming that it's possible to describe what the 'best practice' of EH professionals could look like, how can it be known that the 'best practice' within one sample population would be as effective in another?

Lastly, the concerns of environmental health with policy and practice raise important issues about subjectivity and bias throughout the research process. Arguably all research is biased to a degree. This could be unintentional (e.g. personal values) or deliberate (e.g. not declaring the vested interests of your funders), but what's important is recognising and controlling acceptable bias throughout the research process whilst avoiding unacceptable bias (e.g. rejecting data that contradicts your position) [42]. There are no easy ways around these issues and their consideration is another research priority for EH professionals, but research will be compromised if bias is ignored.

Values, responsibilities and ethics of EH researchers

EH professionals are accountable in many ways, not least to their employers and the ethical codes of conduct of the professional organisations to which they belong. By its nature environmental health work is shaped by moral and ethical issues, for example balancing the tensions between economic growth, environmental degradation and the public's health. Conducting research is no different and requires EH professionals to engage with the values, responsibilities and ethics of their study which will now be explored.

All researchers must conform to established standards of ethical practice; aside from being a moral imperative it is also a standard condition for many publishers. In practice, this could involve written approval by university ethics committees, or from a similar body within the workplace, or perhaps from the professional organisation. Green and Thorogood [58] identify three typologies to help understand the relationship between researchers and the wider society in which they operate:

- *The neutral outsider* – Researchers should strive to be disinterested in political and social values, as their role is to produce knowledge for its own sake. This approach implies that researchers should not be concerned about the impacts of their research on individuals or society.
- *The liberal relativist* – Researchers should follow their own (professional) conscience, because ethical standards are not uniform and are differently constructed in different settings.
- *The radical* – Researchers should be openly partisan about their work, striving to redress inequalities and increase social justice through their practice. This is because we do not exist in isolation and the proper role of research is to improve society ([58] – page 55).

Arguably, environmental health research leans towards the radical typology, but Green and Thorogood also acknowledge that a researcher's position might not fit solely with one typology and could change. The principle of informed consent is a cornerstone of ethical practice, and it must be ensured that individuals participating in the study have given their informed consent. This means that people cannot be forced to participate, must be aware of their participation and must understand the consequences of their participation [58]. To illustrate, researchers observing the practice of local government EH professionals might also need to inform all those using their offices (e.g. administrators, non-EH staff, cleaners), the regulated themselves (e.g. business owners) and the wider public (e.g. complainants) about their study.

Maintaining data confidentiality is vital and covers issues such as data security and protecting the identities of individuals and fieldwork sites [58]. These safeguards need to be considered at the earliest planning stages, particularly for those researching potentially vulnerable groups (e.g. children, harassed tenants) or unique cases where reassurances of anonymity and confidentiality could be unrealistic [58]. This is important because the removal of names alone maybe insufficient to prevent identification.

Researchers also have a responsibility to consider how power and values can shape the research process. For example, when interviewing it is important to ensure that individuals are respected and not reduced to mere carriers of 'good data' [58]. There are other issues which can cause power imbalances between the researcher and interviewee which should be taken into account, for example the location of data collection and consideration of whether others are present, and their relationships. During research one of us found that several interviewees chose to speak in cafes and locations away from their places of work, whereas others invited their colleagues to take part in interviews held in their offices. Also important is consideration of whether the researcher is a professional 'equal' or has something in common with those being researched such as occupation or connections which can lead to a greater intimacy and candidness [67]. Chew-Graham et al. [68] also found that people were also willing to express some vulnerability where there are shared backgrounds, but because of this, important issues could remain unchallenged. Conversely, another of us (Caroline Barratt) found that her position as an 'outsider' meant that some participants were more willing to share issues and opinions that may have been considered controversial within the community.

Factors such as experience and background are particularly relevant, where an experienced EH professional might have a different perspective to an inexperienced one or someone from a different public health background, which in turn could influence how research is carried out and understood. These are examples of reflexivity, that is, the reflections of the practitioner as a researcher upon their actions and values during the research process and the effects they might have [59].

Personal safety is also critical, and it is important to properly risk-assess planned actions. Those researching areas like outbreaks of infectious disease or poor housing might find it emotionally difficult and require additional support. Checking that there is full insurance coverage is also critical, particularly if the research is not part of the day job or involves lone working or brings the researcher into contact with areas affected by infectious disease or war for example.

In summary, engaging with and balancing the values, responsibilities and ethics surrounding the study requires careful and ongoing consideration throughout the research process. Sound ethical practice is essential if environmental health researchers are to realise the potential of the research to contribute to the development of the EH profession and to improve the public's health.

SECTION 3: HOW CAN I BECOME MORE RESEARCH ACTIVE AND MAKE WORK MORE EVIDENCE BASED?

This final section explores some of the challenges EH professionals are likely to face in trying to become more research active and evidence based and how they might be overcome. A series of questions raised repeatedly in EHRNet workshops and by the recommended Aveyard and Sharp [36] is used to structure this final section, but there are no easy answers here for any EH professional.

Are EH professionals researching already?

In one word, yes, but the problem is more that EH professionals often don't see their work as research nor perhaps more importantly realise its potential for improving environmental health policy and practice. Returning to the discussion in Section 1, the need for EH professionals to engage with the best available evidence during their day jobs remains a priority. But returning to the work in Couch et al. [19] it is argued here that EH professionals could find they already have many transferrable skills for research.

Evaluating evidence and using it to piece together a 'picture', interviewing people, being able to communicate effectively at all levels, carrying out critical analysis of documents, skills of observation and an ability to make accurate notes, and being well organised and tenacious (and sceptical) are common attributes of both good EH professionals and researchers [67]. By viewing their daily work as a research cycle it is suggested here that EH professionals could become better at maintaining and improving the public's health.

Dr Richard North's *Some observations on food hygiene* [69] continues to be essential reading for all EH professionals, whichever area they work in, because of his application of critical research eyes to the inspection process. North's work is summarised in Table 4.2 to argue that EH professionals are researching already because in the first column the basic stages of the research process (from Figure 4.1 – research cycle model) closely mirror those of North's own inspection stages. Additional comments and advice have been added to further illustrate research as inspection or vice versa.

The authors agree with North that inspection, like research, should be viewed as a cycle, part of a continuous process of maintenance and/or improvement. For example, North argues that post-inspection discussions, revisits and additional works identified in future inspections are part of a programme of continuous development and not, as some argue, due to the inadequacies (e.g. inconsistencies) of previous inspections. Further, North has much to say about cross-cutting themes like the values, responsibilities and ethics of EH professionals and what could characterise high-quality research/inspection as summarised at the bottom of Table 4.2.

The following quote is drawn from the analysis and interpretation stage in Table 4.2

Table 4.2 The inspection process as a research process?

Research cycle stages	North's inspection stages and further comments/advice to illustrate inspection as research
Identify and define research problem	Prevention of food poisoning
Develop research questions/hypothesis	Developing generic model • Using standards prescribed by law, codes of practice, principles of hygiene, etc.
Develop research design	Developing sector- and site-specific models • To reflect sector (e.g. butchers/caterer) and unique circumstances of each premises.
Collect data	The conduct of inspection, inspection techniques and data recording • Utilising observation, interview, document analysis methods in busy kitchens; • Observation notes require great discipline, where plans and photographs can be invaluable.
Analyse data Interpret results and draw conclusions	Analysis and interpretation • Reviewing findings to identify patterns. • Not all findings are easy to interpret; • Relate findings to site-specific model and wider context (e.g. other kitchens in hotel group) and benchmark for future inspections.
Publish research findings	Framing the report • Does the report clearly indicate the risks in the operation and set out recommendations in a way that, if followed, would adequately control/remove/contain all risks identified? • Are requirements framed to enable understanding and implementation without specialist advice?
Your values, responsibilities and ethics	• Recognising your outsider status and how your presence influences observed events (see also [30]); working with key insiders (e.g. chefs) to advise on, explain and interpret inspection data.
High-quality EH research	• Establishing reliability of data by comparing observation results to interview responses of staff and key documents.

Source: Adapted from ([19] – page 74)

to bring the 'inspection as research' argument to life for busy EH professionals. Here North [69] explores why inspection findings must be seen in context:

> Analysis [of inspection findings] requires data and the more data available, the more accurate it can be. In particular, the inspector must be aware that visual observation of conditions may not always provide sufficient evidence on which to base judgements. Therefore, the fact that data are to be analysed itself provides the incentive for a more thorough inspection. Where cleanliness in a kitchen was observed to be substandard, one might expect any harassed manager confronted by an inspector to claim that any drop in standards was temporary – the result of meeting unusual pressures. The inspector will have to determine whether that claim is true. For the skilled inspector, this is not too difficult. In the same way that there is a contrast between soiling levels where there has been a rapid clean-up, there is usually a distinct difference between recent and long-standing accumulations of dirt. Again, the difference will be at its most pronounced in the contrast between visible and less visible areas. But the difference will be that visible soiling may be present, but less obvious areas will show signs of good maintenance, i.e. absence of long-term soiling. Only then can it be assumed that the overall standard reflects short term neglect. If, however, in addition to visible soiling, long term soiling is present in less obvious areas, claims that the standard overall is simply a short-term problem may be less credible.
>
> *([69] – page 90)*

More work is needed on the development of higher-quality models and methods of inspection (e.g. better observation, interview techniques), data analysis and report writing across the breadth of environmental health

practice but we agree with North's comment that EH professionals could 'make the difference' they aspire to

> not by retreating into the bunker and issuing forth a stream of edicts couched in a language which has been inelegantly but accurately called 'corporate-speak drivel', but by getting back to its roots and exploiting the skills for which the profession has in the past been justifiably proud.
>
> *([69] – page 127)*

How can I fit research-related activities into my day job?

Finding the time and resources to conduct research and review the best available evidence is hard enough at the best of times, but one way is to think of all research activities as an investment in personal and professional development. One of the lesser publicised benefits of research is in the creation of space in an otherwise hectic day for thinking and reflection. The creative power of having time to think like when travelling to and from your workplace should not be underestimated. Further, publications should go straight on the CV and are recognised as continuous professional development.

Team meetings are an obvious opportunity to discuss research and evidence but there may be a need to change the culture of meetings, where the usual day-to-day operational issues are covered alongside more in-depth reviews of recent cases or the exploration of what works based on the latest research evidence. When an EHP attends a conference, any notes can be written up and circulated to colleagues, presented to them and discussed at team meetings. Similarly in the housing context, when considering the most appropriate course of action following an HHSRS inspection under Part 1 of the Housing Act 2004,[4] a 'case conference' approach might be better to discuss the issues and evidence rather than leaving the practitioner to make the decision alone. Such meetings should be written up anyway for use in any future appeal.

Do not underestimate the power of environmental health students. Could they help identify the best evidence available or discuss current thinking about a topic from their university studies? The access students have to academic resources could prove particularly useful, but at the same time EH professionals must not delegate their research responsibilities to inexperienced and under-resourced students. Further, can the practitioner become more involved in existing professional environmental health networks or those at your local university? They might welcome a presentation or debate about research and evidence from someone in the field. For example, during the Selly Oak project at the University of Birmingham, students were recruited and trained to complete semi-structured interviews and thus received a small income as well as practical research experience. Similarly, at Middlesex University students participated in research funded by the Association of London Environmental Health Managers exploring private-sector housing support to older people, with an emphasis on dementia [70]. This included valuable experience in research ethics, interviewing skills, data transcription and writing up for peer-reviewed publication.

Am I a good enough role model for evidence-based environmental health?

One of the greatest influences on practice for students and practitioners comes from the role modelling of other EH professionals. Indeed, in her classic study of local government EH professionals, Hutter [25] found the influence of colleagues to be a powerful determinant of law enforcement decision-making but individuals countering the predominant enforcement cultures in offices risked being ostracised by their colleagues. Given this, and a mixed response to

constructive challenge seen amongst EH professionals, it will take time for evidence-based environmental health to become embedded in daily practice. However, Aveyard and Sharp ([36] – page 139) have some good collaborative suggestions that could make the EHP a more effective role model which has been adapted here:

- Ask colleagues for a rationale for their decision-making and judge what they provide, particularly whether the same decision would be made based on the evidence available. If such a rationale cannot be provided, suggest ways in which it would be possible to work on this together.
- Consider reactions to having practice challenged (personal and that of colleagues) – are such challenges seen as personal criticism or an opportunity for professional development? Could more be done to encourage challenge in one's own practice?
- Could links be established with more involvement with the public health programmes at the local university in an attempt to bridge the gaps between research, policy and practice?

How do I challenge the practice of others?

Challenging the work of any professionals must be approached carefully and constructively. The following quote about UK EH professionals from a trade official suggests that inexperience, lack of confidence and a macho culture could explain why challenge is not always welcome amongst EH professionals:

Newly qualified officers in particular tend to be very officious, arrogant, defensive, prickly, unwilling to listen, unreasonable in their demands (everything is black and white), more likely to serve improvement notices and prosecute, and tend to exaggerate the seriousness of the situation and use threatening language e.g. 'you realise that I could close you down' when there is no justification for such a statement. Much of this behaviour I believe is borne out of a lack of experience and maturity. They are not used to inspecting, are unfamiliar with the industry, are unsure where to set the standard so go for perfection, feel their professional competency is threatened if any of their views, statements, are questioned and feel they have to prove themselves to their superiors.

(Bushell in North [69] – pages 107–108)

It is also likely that EH professionals sometimes lack confidence and become defensive because of a lack of a research and knowledge culture that invites criticism and debate and embraces uncertainty.

This is understandable where there can be an overemphasis on education or enforcement and intervention options are limited. Instead of being motivated by not knowing all the answers, this is seen as a threat and practitioners retreat back into their legal and technical comfort zones. Developing more of a research culture will take many years and until this time it is important that any challenge to the practice of others is approached with care. With this in mind, Aveyard and Sharp provided more suggestions that can be adapted for EH professionals:

- Plan and discuss with colleagues/academics/students what to do if practice is observed that conflicts with the best available evidence.

Before challenging the practice of others:

- Consider whether that practice is inappropriate or unsafe and your responsibility as an EH professional to advocate for those whose environmental health is adversely affected;
- Consider what you don't know (e.g. vital information you are not yet aware of) and why your evidence suggests a different course of action;

- Unless immediate action is necessary, avoid challenging others in public;
- Compile your evidence and be prepared to hand it over for review;
- Be ready to present your evidence in the form of questions, not accusations, and invite the perspectives of others on this issue (adapted from [36] – page 140).

Why doesn't evidence influence environmental health policy and practice more?

The work of the Research and Policy in Development (RAPID) group of the Overseas Development Institute is useful here for exploring research and policy relationships, particularly why some research findings influence policy and others don't and how to promote more research informed policy-making [71]. Their framework rejects simplistic 'research produces policy' relationships in favour of more complex and dynamic relationships shaped by the relationships between evidence, its political context and those who bring research to life like EH professionals. Further, they found these relationships are also shaped by wider economic and cultural factors. Research exploring these relationships and how to make environmental health more evidence based is another priority for EH professionals, but research and experience suggests the following could be important:

- Many EH professionals do not know what is known about environmental health, hence the need for systematic reviews and other initiatives to provide a foundation for evidence-based environmental health. Towards this end one of us [72] worked with more than 20 other EH professionals to compile case studies of environmental health interventions and strategies in UK private-sector housing.
- A professional culture where 'solving problems' predominates and EH professionals are not encouraged to engage with

the more complex reasons and theories (see Annex 4.2) about why environmental health problems persist [25];

- The reluctance of some EH professionals to think critically and get political. The historian Dr Tom Crook [23] identifies such attitudes dating back to late Victorian times and associates them with the justifications of 'science', the gradual professionalisation of environmental health characterised by self-proclaimed values like independence and impartiality and a reluctance of EH professionals (as public servants) to openly criticise the institutions upon which their status depends;

- The evidence doesn't sit comfortably with how EH professionals like to see themselves. For example, in her study of UK local government EH professionals, sociologist Professor Bridget Hutter [25] found that being 'reasonable' was the hallmark of their work and she concluded that EH professionals considered their moral mandate at least as important as their legal mandate, or even more so when the law conflicts with popular or individual morality. It is likely that some EH professionals will be uncomfortable with evidence describing their decision-making influenced as much by stereotypes and personal beliefs as law and science.

- EH professionals in many areas lack strong networks, particularly between those researching, teaching and practising environmental health. Other organisations and networks (e.g. policymakers, professional organisations, think-tanks, charities) could provide much-needed expertise to help EH professionals communicate better. Investment is needed to build stronger networks and one open access journal paper provides some useful advice here from EH professionals in South West England [29]. Work is also needed to explore further why some evidence-policy-practice initiatives in environmental health are more influential than others [17].

Further, in a chapter titled 'Evidence isn't enough' UK EH professional Peter Wright [73] provides six tips for more effective policymaking as summarised next that we highly recommend:

1 *Generate viscerality* – get a strong sense of purpose for our evidence. Viscerality is crucial to getting important things done. If your evidence is going to be important in getting vital public policy made, you need to recognise that there is a lot more to this than publishing or presenting a paper. You're going to need to adapt your strategy to counter (many problems) . . . and this needs viscerality, a fire in the belly, more than it needs you to make any intellectual or style changes. Of course, if your evidence isn't important you won't need to do this.

2 *You need a plan* – It should be clear that a one-off presentation of your evidence isn't going to be enough to counter some of the obstacles I've described here. You need to plan how to gain power for your evidence from a very early stage. You'll need to identify advocates, mechanisms for how to drip feed your knowledge at the right times and in the right places.

3 *Research your policymakers* – WIIFM is an important thing to consider in any approach to a policymaker. It stands for 'what's in it for me?' and it is worth trying to find what makes them tick, whether they have any known prejudices and whether there could be any negative consequences to them in turning your evidence into policy.

4 *Become an expert in agnotology (the study of wilful acts to spread confusion and deceit) and the techniques of Denialism* – Remember that there may be people whose job it is to sabotage the adoption of your evidence into policy. Learn the techniques they'll use and learn how to work against them. Become confident to publicly laugh at nonsense when the need arises.

5 *Deal with regulatory capture* – This can be potentially impossible to get past or

work around. On an interpersonal level, it's embarrassing to admit that you've been cleverly brainwashed, and you certainly won't gain influence by pointing this out to someone who you feel has been influenced by their involvement with industry representatives. If a public-sector organisation has taken the decision to work closely with business representatives, it's normally necessary to provide evidence to the top people that regulatory capture has occurred and is causing difficulty. Leadership and management action will be needed if the organisation is to reassert the necessary independent thought for it to properly represent the needs of stakeholders.

6 *If your evidence is important, follow the samurai code* – Protect the weak; be equal to the strong; crush the wicked who would harm our people. Memorise it and turn it into one of your very own cognitive biases.

Lastly, becoming more research active and evidence based could help make environmental health more visible. Rayner and Lang's *Ecological public health* [74] is highly relevant here, particularly its first chapter exploring why public health suffers from 'cultural invisibility'. They recognise that such invisibility has always dogged its history and how the case for public health *always has to be built, argued and won. And, once won, it continues to need to be argued for* ([74] – page 6). The many (centuries old) arguments against public health persist across the world, but they believe that public health is deeply ingrained in the structures of all our societies; what's needed are *stronger and more daring combinations of interdisciplinary work, movements and professions locally, nationally and globally* ([74] – page i).

The EHRNet vision for evidence-based environmental health

This chapter ends with a vision for a more research active and evidence-based environmental health, but the question remains whether there is a critical mass of EH professionals with the will to become more research active and evidence based? Returning to Day's [75] use of Schön's metaphor [76], are EH professionals prepared to descend into the swamp of complex problems that defy ready solution, or will they remain largely in the hills managing tasks and solving relatively unimportant problems using traditional methods? Having got this far the reader is probably a swamp convert, but it is believed the following vision is not as daunting as it sounds and could be achieved by building on what EH professionals already have.

EHRNet dream of a time when environmental health evidence:

- Is accessible to all EH professionals and those affected by their decisions;
- Informs debate about EH policy and practice in the classroom, offices and streets;
- Shapes EH policy and practice at all levels and alongside professional judgement and the preferences of citizens and the public.

EHRNet also dream of a time when organisations and individual EH professionals:

- Understand, value and support evidence and research activity;
- Read beyond traditional media (e.g. law and guidance documents) to encompass wider reflection on research from other disciplines;
- Learn how other professions have become more research active and made their policy and practice more evidence based;
- Welcome criticism, debate and challenge as opportunities to improve EH policy and practice;
- Organise to support individuals and organisations with research and direct their research activities towards known gaps and priorities;
- Move outside their comfort zones and build stronger links with other public health professionals, researchers and wider society for the benefit of all;

- Follow Peter Wright's tips for better public health policymaking.

The authors hope this chapter encourages and supports colleagues to embark on their own swampy research journey towards a better environmental health for all.

Chapter notes

1 See http://ph.cochrane.org/
2 Available via www.casp-uk.net
3 Access it at http://ukehrnet@wordpress.com
4 See also for example ODPM (2006) HHSRS Enforcement Guidance (under review) https://assets.publishing.service.gov.uk/government/uploads/system/uploads/attachment_data/file/7853/safetyratingsystem.pdf; Battersby SA and Pointing J, (2019) *Statutory Nuisance and Residential Property*, Routledge Focus on Environmental Health Series, Routledge, Abingdon Oxon; and Carr H, Cottle S, and Ormandy D. (2008) *Using the Housing Act 2004*, Jordan Publishing Ltd, Bristol, on the options for action.

References

[1] Burke S, Gray I, Paterson K, Meyrick J. (2002) *Environmental Health 2012: A Key Partner in Delivering the Public Health Agenda*, Health Development Agency, London.

[2] McCarthy A. (1996) Protecting the public health – the role of environmental health. *Public Health*, 110: 77–80. https://doi.org/10.1016/S0033-3506(96)80050-3.

[3] Oatt P. (2020) *Selective Licensing: The Basis for a Collective Approach to Addressing Health Inequalities*, Routledge, London. ISBN:9780367429195.

[4] Stewart J. (Ed.) (2017) *Pioneers in Public Health: Lessons from History*, Routledge, London, ISBN:9781032179155.

[5] Day C, Couch R, Dhesi S. (2021) *COVID-19: The Global Environmental Health Experience*, Routledge, London, https://doi.org/10.1201/9781003157229.

[6] Morse T, Chidziwisano K, Musoke D, Beattie T, Mudaly S. (2020) Environmental health practitioners: A key cadre in the control of COVID-19 in sub-Saharan Africa. *BMJ Global Health*. http://doi.org/10.1136/bmjgh-2020-003314.

[7] Rodrigues MA, Silva M, Errett N, Davis G, Lynch Z, Dhesi S, Hannelly T, Mitchell G, Dyjack D, Ross K. (2021) How can environmental health practitioners contribute to ensure population safety and health during the COVID-19 pandemic? *Safety Science*, 136. ISSN:0925-7535. https://doi.org/10.1016/j.ssci.2020.105136.

[8] Chartered Institute of Environmental Health. (2021) CIEH is listening to you. *Environmental Health News*, July–August.

[9] MacArthur I, Bonnefoy X. (1997) *Environmental Health Services in Europe 1: An Overview Of Practice in the 1990s*, WHO Regional Publications No. 76, WHO Regional Office for Europe, Copenhagen, Denmark. PMID:9557583.

[10] CEH Commission on Environmental Health. (1997) *Agendas for Change*, Chadwick House Group Ltd., London.

[11] Smith K, Corvalan C, Kjellstrom T. (1999) How much global ill health is attributable to environmental factors? *Epidemiology*, 10(5): 573–584, September. PMID: 10468437.

[12] Eyles J. (1997) Environmental health research: Setting an agenda by spinning our wheels or climbing a mountain? *Health and Place*, 3(1): 1–13. https://doi.org/10.1016/S1353-8292(96)00031-7.

[13] WHO, Europe – World Health Organization Regional Office for Europe. (1994) *Action Plan for Environmental Health Services in Central and Eastern Europe and the Newly Independent States*, Report on a WHO Consultation, Sofia, Bulgaria, 19–22 October 1993 (Document EUR/ICP/CEH 123) WHO Regional Office for Europe, Copenhagen, Denmark.

[14] Chadwick E. (1842/1965) *Report on the Sanitary Condition of the Labouring Population of Great Britain 1842*, Flinn MW (Ed.), Edinburgh University Press, Edinburgh.

[15] Davies H, Nutley S, Smith P. (2000) Chapter 1: Introducing evidence based policy and practice in public services. In *What Works? Evidence-Based Policy and Practice in Public Services*, Davies H, Nutley S, Smith P (Eds.), The Policy Press, Bristol.

[16] Stewart J. (2005) A review of UK housing policy: Ideology and public health. *Public Health*, 119(6): 525–534. https://doi.org/10.1016/j.puhe.2004.07.006.

[17] Vickers I. (2008) Better regulation and enterprise: The case of environmental health risk. Regulation in Britain. *Policy Studies*, 29(2): 215–232. https://doi.org/10.1080/01442870802033514.

[18] Sackett D, Rosenberg W, Muir Gray J, Haynes R, Richardson W. (1996) Evidence based medicine: What it is and what it isn't. *British Medical Journal*, 312: 71–72. https://doi.org/10.1136/bmj.312.7023.71.

[19] Couch R, Stewart J, Barratt C, Dhesi S, Page A. (2012) *Evidence, Research and Publication: A Guide for Environmental Health Professionals*, Lulu Publications, eBook. https://www.researchgate.net/publication/309672772_Evidence_research_and_publication_a_guide_for_environmental_health_professionals.

[20] Scott-Samuel A, Bambra C, Collins C, Hunter DJ, McCartney G, Smith K. (2014) The impact of Thatcherism on health and well-being in Britain. *International Journal of Health Services*, 44(1): 53–71. https://doi.org/10.2190/HS.44.1.d.

[21] Page A. (2002) Poor housing and mental health in the United Kingdom: Changing the focus for intervention. *Journal of Environmental Health Research*, 1(1).

[22] Barratt C, Kitcher C, Stewart J. (2012) Beyond safety to wellbeing: How local authorities can mitigate the mental health risks of living in houses in multiple occupation. *Journal of Environmental Health Research*, 12(1): 39–51.

[23] Crook T. (2007) Sanitary inspection and the public sphere in late Victorian and Edwardian Britain: A case study in liberal governance. *Social History*, 32(4): 369–393. https://doi.org/10.1080/03071020701616654.

[24] Crook R, Ayee J. (2006) Urban service partnerships, 'street-level bureaucrats' and environmental sanitation in Kumasi and Accra, Ghana: Coping with organisational change in the public bureaucracy. *Development Policy Review*, 24(1): 51–73. https://doi.org/10.1111/j.1467-7679.2006.00313.x.

[25] Hutter B. (1988) *The Reasonable Arm of the Law? The Law Enforcement Procedures of Environmental Health Officers*. Clarendon Press, Oxford. ISBN:978-0198255949.

[26] Lipsky M. (1980) *Street Level Bureaucracy: Dilemmas of the Individual in Public Services*, Russell Sage Foundation, New York. https://doi.org/10.1177/003232928001000113.

27. Greenhalgh T. (2019) *How to Read a Paper: The Basics of Evidence-Based Medicine and Healthcare*, Blackwell Publishing Ltd., Oxford, 6th ed. ISBN:9781119484745.

[28] Thomas H. (2014) *Delivering with Less – the Impact on Environmental Health Services and Citizens*, Auditor General for Wales, Wales Audit Office, Cardiff.

[29] Turbutt C, Bowering-Sheehan G, Dhesi S. (2014) Supporting environmental health practitioners to evaluate and publish: A review of activity in South-West England. *Journal of Environmental Health Research*, 14(1): 81–87.

[30] Chartered Institute of Environmental Health. (2021) *Environmental Health Workforce Surveyreport: Local Authorities in England*, CIEH, London. https://www.cieh.org/media/5249/cieh-workforce-survey-report-for-england.pdf (Accessed 8 October 2021).

[31] Day C. (2016) Chapter 2: Environmental health – a changing practice. In *Clay's Handbook of Environmental Health*, Battersby S (Ed.), Routledge, Taylor & Francis Group, London, 21st ed. ISBN:9781317382904.

[32] Mullen L, Cowden JM, Cowden D, Wong R. (2002) An evaluation of the risk assessment method used by environmental health officers when inspecting food businesses. *International Journal of Environmental Health Research*, 12(3): 255–260. https://doi.org/10.1080/0960312021000001005.

[33] Jones S, Parry S, O'Brien S, Palmer S. (2008) Are staff management practices and inspection risk ratings associated with foodborne disease outbreaks in the catering industry in England and Wales? *Journal of Food Protection*, 71(3): 550–557. https://doi.org/10.4315/0362-028x-71.3.550.

[34] Fairman R, Yapp C. (2005a) Enforced self-regulation, prescription, and conceptions of compliance within small businesses: The impact of enforcement. *Law and Policy*, 27(4): 491–519. https://doi.org/10.1111/j.1467-9930.2005.00209.x.

[35] Fairman R, Yapp C. (2005b) *Making an Impact on SME Compliance Behaviour: An Evaluation of the Effect of Interventions upon Compliance with Health and Safety Legislation in Small and Medium Sized Enterprises*, HSE Research Report 366, Her Majesty's Stationery Office, London.

[36] Aveyard H, Sharp P. (2017) *A Beginner's Guide to Evidence Based Practice in Health and Social Care*, Open University Press, McGraw-Hill Education, London, 3rd ed. ISBN:978-0335227082.

[37] Thomson H, Thomas, S, Sellstrom E, Petticrew M. (2013) Housing improvements for health and associated socio-economic outcomes. *Cochrane Database of Systematic Reviews*, 2. https://doi.org/10.1002/14651858.CD008657.pub2.

[38] Borley L, Page A (2016) A reflection on the current local authority-led regulation model: Views from small- and medium-sized businesses. *Policy and Practice in Health and Safety*, 14(2): 144–162. https://doi.org/10.1080/14773996.2016.125544.

[39] Farrell J, McConnell K, Brulle R. (2019) Evidence-based strategies to combat scientific

misinformation. *Nature Climate Change*, 9: 191–195. https://doi.org/10.1038/s41558-018-0368-6.

[40] National Institute for Health and Care Excellence (NICE). (2020) *NICE Guideline: (NG149) Indoor Air Quality at Home*, NICE, London. ISBN:978-1-4731-3625-0.

[41] Davis A. (undated) *The Prevention Paradox and Population Level Strategies as Applied to Road Safety: A Public Health Approach.* https://d3n8a8pro7vhmx.cloudfront.net/20splentyforus/pages/419/attachments/original/1539441823/Prof_Adrian_Davis.pdf?1539441823 (Accessed 8 October 2021).

[42] Sumner A, Tribe M. (2008) *International Development Studies: Theories and Methods in Research and Practice*, Sage Publications, London. ISBN:9781412929455.

[43] Toner E. (2010) *The Art of Communication: A Landscape Shared Between Regulator and Ethnic Employer*, Derry City Council, Northern Ireland.

[44] Rudder A. (2006) Food safety and the risk assessment of ethnic minority food retail businesses. *Food Control*, 17(3): 189–196. https://doi.org/10.1016/j.foodcont.2004.10.017.

[45] Harrow M. (2013) *Working Effectively with Minority Ethnic Food Businesses – A Resource Handbook*, Food Standards Agency, London.

[46] Newton I. (1676/1959) Letter to Robert Hooke dated 5 February 1676. In *The Correspondence of Isaac Newton*, Turnbull H (Ed.), Volume 1, Cambridge University Press, Cambridge.

[47] ODPM Office of the Deputy Prime Minister. (2006) *Housing Health and Safety Rating System Operating Guidance (February 2006)*, Office of the Deputy Prime Minister, London.

[48] Horder W. (2004) Reading and not reading in professional practice. *Qualitative Social Work*, 3(3): 297–311. https://doi.org/10.1177/1473325004047174.

[49] Heijen M, Cumming O, Peletz R, Ka-Seen Chan G, Brown J, Baker K, Clasen T. (2014) Shared sanitation versus individual household latrines: A systematic review of health outcomes. *PLoS One*, 9(4): e93300. https://doi.org/10.1371/journal.pone.0093300.

[50] Gibson M, Petticrew M, Bambra C, Sowden A, Wright K, Whitehead M. (2011) Housing and health inequalities: A synthesis of systematic reviews of interventions aimed at different pathways linking housing and health. *Health and Place*, 17(1): 175–184. https://doi.org/10.1016/j.healthplace.2010.09.011.

[51] Thomson H, Petticrew M, Morrison D. (2001) Health effects of housing improvement: Systematic review of intervention studies. *British Medical Journal*, 323: 187–190. https://doi.org/10.1136/bmj.323.7306.187.

[52] Davis, C. (2004) *Making Companies Safe: What Works?* Centre for Corporate Accountability. http://www.corporateaccountability.org.uk/dl/courtreport04/makingcompaniessafe.pdf (Accessed 8 October 2021).

[53] Ormandy D. (Ed.) (2009) *Housing and Health in Europe: The WHO LARES Project,* Routledge, London. ISBN:9781138972001.

[54] Braubach M, Jacobs DE, Ormandy D. (2011) *Environmental Burden of Disease Associated with Inadequate Housing*, WHO Europe, Copenhagen, Denmark. ISBN:9789289002394.

[55] University of Southern California. (2021) *Organizing Your Social Sciences Research Paper: Theoretical Framework.* https://libguides.usc.edu/c.php?g=235034&p=1561763.

[56] Tombs S, Whyte D. (2007) *Safety Crimes*, Willan Publishing, Devon. ISBN:9781843920854.

[57] Lange B. (1999) Compliance construction in the context of environmental regulation. *Social and Legal Studies*, 8(4): 549–567. https://doi.org/10.1177/a010430.

[58] Green J, Thorogood N. (2018) *Qualitative Methods for Health Research*, Sage Publications, London, 4th ed. ISBN:9781473997110.

[59] Robson C, McCartan K. (2015) *Real World Research: A Resource for Users of Social Research Methods in Applied Settings*, Wiley, Chichester, 4th Ed.

[60] Costley C, Elliot G, Gibbs P. (2010) *Doing Work Based Research: Approaches to Enquiry for Insider-Researchers*, Sage Publications Ltd., London. ISBN:9781848606784.

[61] Meyer J. (1993) New paradigm research in practice: The trials and tribulations of action research. *Journal of Advanced Nursing*, 18(7): 1006–1072. https://doi.org/10.1046/j.1365-2648.1993.18071066.x.

[62] Meyer J, Batehup L. (1997) Action research in health care practice: Nature, present concerns and future possibilities. *Nursing Times Research*, 2(3): 175–184. https://doi.org/10.1177/174498719700200304.

[63] Baum F. (1995) Research public health: Behind the qualitative-quantitative methodological debate. *Social Science and Medicine*, 40(4): 459–468. https://doi.org/10.1016/0277-9536(94)e0103-y.

[64] Mays N, Pope C. (2000) Assessing quality in qualitative research. *British Medical Journal*,

320(7226): 50–52. https://doi.org/10.1136/bmj.320.7226.50.

[65] Becker S, Bryman A, Sempik J. (2006) *Defining 'Quality' in Social Policy Research: Views, Perceptions and a Framework for Discussion*, Social Policy Association, Lavenham. http://www.social-policy.org.uk/downloads/defining%20quality%20in%20social%20policy%20research.pdf (Accessed 8 October 2021).

[66] Lincoln Y, Guba E. (1985) *Naturalistic Inquiry*, Sage Publications, Thousand Oaks, CA. ISBN:9780803924314.

[67] Dhesi S. (2013) Reflexivity – researching practice from within. *Journal of Environmental Health Research*, 13(1): 83–87.

[68] Chew-Graham C, May C, Perry M. (2002) Qualitative research and the problem of judgement: Lessons from interviewing fellow professionals. *Family Practice*, 19(3): 285–289. https://doi.org/10.1093/fampra/19.3.285.

69. North R. (1999) *Some Observations on Food Hygiene*, Chadwick House Group Ltd., London.

[70] Stewart J. (2021) Meeting the private sector housing condition and adaptation needs of older people: Responses from London's environmental health and allied services. *Housing, Care and Support*. https://doi.org/10.1108/HCS-03-2021-0009.

[71] Court J, Hovland I, Young J. (2005) Cross cutting issues and implications: Promoting more informed international development policy. In *Bridging Research and Policy in Development: Evidence and the Change Process*, Court J, Hovland I, Young J (Eds.), ITDG Publishing Ltd., Bourton-on-Dunsmore, Warwickshire.

[72] Stewart J. (2013) *Effective Strategies and Interventions – Environmental Health and the Private Housing Sector*, Chartered Institute of Environmental Health, London.

[73] Wright P. (2020) Evidence isn't Enough. In *Kittens Are Evil II: Little Heresies in Public Policy*, Pell C, Wilson R, Lowe T, Myers J (Eds.), Triarchy Press, Dorset. ISBN:978-1-91119377-7.

[74] Rayner G, Lang T. (2012) *Ecological Public Health: Reshaping the Conditions for Good Health*, Earthscan from Routledge, Abingdon. ISBN:9781844078325.

[75] Day C. (2006) Guest columnist: Let's descend to the swamp: EHPs should tackle the messy problems rather than simply manage 'tasks'. *Environmental Health Practitioner*, 25, June.

[76] Schön D. (1991) *The Reflective Practitioner: How Professionals Think in Action*, Ashgate Publishing Limited, London.

[77] Haynes L, Service O, Goldacre B, Torgerson D. (2012) *Developing Public Policy with Randomised Controlled Trials*, Cabinet Office Behavioural Insights Team, London.

ANNEX 4.1: COMMON RESEARCH TERMS (ADAPTED FROM COUCH ET AL. 2012:88–89)

Action research: A qualitative research approach where practitioners and researchers work together to address everyday issues about practice and develop a systematic approach to implement and evaluate change ([36] – page 71).

Case control study: A study where people (or premises, etc.) with a particular disease/condition (cases) are compared to those without the disease/condition (controls) [27]. One environmental health example is Jones et al. [33].

Empirical research: Refers to research based on observation or experience. The opposite is theoretical research which uses ideas to explain phenomena.

Environmental health professional/practitioner: We use this term to refer to all those working to maintain and improve environmental health, not just those with traditional environmental health qualifications (e.g. a degree in environmental health). This inclusivity is driven partly by the authors' own varied backgrounds and in recognition that they have worked with so many other professionals (and others) towards improving public health over the years. This work is also intended to be relevant to EH practitioners around the world. There is also a desire to avoid the insider/outsider politics common to so many professions.

Epistemology: The branch of philosophy concerned with theories of knowledge which include positivism, realism and relativism.

Ethnography: A qualitative research approach where a community is observed in real time to answer questions about how the community behaves [59].

Evidence: Information that indicates whether something is true or valid and can be based on anecdote (e.g. expert opinions, something that's worked before) or, ideally, research.

Evidence-based environmental health: Environmental health policy and practice supported by the best available evidence, taking into account the preferences of citizens and the wider public and the judgement of EH professionals.

Grey literature: These include technical reports and policy statements that have not been subject to formal publication. They are sometimes of the highest quality but might not have been subject to peer review and (as with all literature) should be read with critical eyes.

Grounded theory: A qualitative research approach where data are collected and analysed to generate theory (e.g. explanations of social phenomena) [59].

Hierarchies of evidence: A system concerned with the effectiveness of interventions and used to determine which evidence is the most trustworthy [27].

Methodology: The overall research strategy followed to answer questions/hypotheses which include the theoretical basis for the study and the methods used to collect, analyse and report the data [42].

Methods: The detailed tools and techniques used to collect primary or secondary data.

Mixed methods: Methods incorporating a mixture of quantitative and qualitative tools and techniques to answer research questions. These are not uncommon in environmental health research, for example see Hutter [25] and Fairman and Yapp [34].

Peer review: This is the process used to decide what is published in an academic journal where the editors appoint experts in your field to assess the quality and importance of your research.

Primary data: Data collected by the researcher themselves, in contrast to secondary data collected by someone other than the researcher.

Qualitative research: This tends to use data derived from language (written and oral) [44], not numbers, to explore the meaning and develop in-depth understanding of the research topic as experienced by the participants of the research and the researcher may be involved with the participant who may shape data collection and analysis ([36] – page 68).

Quantitative research: This generally refers to studies that collect and analyse numerical data and often involves high numbers of participants with little or no involvement between the researcher and participant [36].

Randomised controlled trials: A trial where participants are randomly allocated to one intervention or another to determine the effectiveness of the intervention [27]. Environmental health examples do exist and the recent and free publication by Haynes et al. [77] is a good place to start.

Reflexivity: The process of researchers reflecting upon their actions and values during the research process and the effects they might have [59].

Systematic review: A literature review conducted in accordance with a defined approach as exemplified by the reviews of the Cochrane Collaboration.

Theoretical framework: A theoretical framework consists of concepts, together with their definitions, and existing theory/theories that are used for your particular study [55]. It helps to embed research within previously generated knowledge and enables the researcher to make it clear what their contribution to knowledge will be.

ANNEX 4.2: THREE EXAMPLES OF THEORETICAL FRAMEWORKS

Street level bureaucracy – by Michael Lipsky [26]

Political scientist Professor Michael Lipsky developed his theory during the 1970s when the competence of poorly resourced American front-line public services was being called into question – does this sound familiar? By reviewing a vast empirical literature on front-line public officials, including American EH professionals, Lipsky argues that public policy is not best understood as the product of governments or high-ranking policy officials but is instead the product of the crowded offices and daily encounters of front-line workers like EH professionals. Here *the decisions of street-level bureaucrats, the routines they establish, and the devices they invent to cope with uncertainties and work pressures, effectively become the public policies they carry out* ([26] – page xii).

This happens because the uncertainties characteristic of their work gives street-level bureaucrats enormous power over service users and considerable autonomy from their employers. But this power is set against the many dilemmas of being at the sharp end of resource allocation where demand far exceeds supply. Front-line workers therefore find themselves making decisions in circumstances not of their own choosing and devise strategies to protect their working environment. For example, they make decisions back in their private offices or mechanically 'process' clients into categories, whilst reserving the treatment they would ideally like to give all towards those clients more likely to succeed.

One might consider this justification for greater controls on the discretion of EH professionals, but Lipsky is bleak about its effectiveness amidst workplaces with high staff turnover where performance is difficult to measure, and greater supervision can be counterproductive. Clients, particularly the most vulnerable, are also relatively powerless to hold street-level bureaucrats to account, whilst legal systems can be poorly equipped for discretionary decision-making. Professional organisations also do not escape Lipsky's criticism with their 'careerist' tendencies and reluctance to hold fellow professionals to account.

Published research has mentioned the relevance of street-level bureaucracy for describing the work of UK EH professionals [49], but regrettably more than 30 years after

publication Lipsky's work remains largely untapped by EH professionals.

Why EH regulators generally consider prosecution as the last resort? – by Steve Tombs and Dave Whyte [56]

In their book *Safety crimes*, the sociologists Professor Steve Tombs and Dr Dave Whyte explore four competing theories questioning why safety regulators (including EH professionals in the UK) generally consider prosecution as the last resort. They argue that consensus theories of regulation are broadly pluralist (i.e. power is shared between political parties) and based on the belief that the most effective regulatory strategies are those involving persuasion, bargaining and compromise through close relationships between the regulator and regulated that remain dominant in Western societies. Alternatively, in capture theories, such relationships can get too close, and government and regulators become vulnerable to capture by powerful interests like so-called 'big business'. Neoliberal theories of regulation argue that society is overregulated by interventionist states; instead market mechanisms (e.g. competitive advantage, compensation, insurance) could better protect environmental health.

Tombs and Whyte (2007) critique each of these theories before describing their preference for what they call 'critical approaches to regulation' that move beyond struggles between state versus capital only. Their preferred critical analysis argues that regulation is best viewed as a process determined by the product of struggles between states and business and states and the electorate; here power is distributed unequally but spaces for challenging power are not closed down or captured. The role of EH professionals as regulators in managing inevitable conflicts between opposing interests is therefore critical to maintaining social order and a functioning economy.

Environmental health regulation as modern state power – by Tom Crook [23]

The historian Dr Tom Crook applies three theories of modern state power to help us understand why environmental health regulation emerged in the late Victorian/early Edwardian period that could help today's EH professionals better understand why they began doing what they (largely) still do.

The first theory used by Crook associates inspection with the gradual movement away from a laissez-faire (non-interventionist) state in the late Victorian period towards an increasingly bureaucratic and interventionist state as characterised by the emergence of professionally qualified inspectors. They were bound by rules but had considerable discretion and were appointed by new local government structures to carry out their legal environmental health duties. Note that here the term bureaucratic is used not in its derogatory sense but to describe the appointment by the state of professional officials to inspect. The second theory associates inspection with the rise of the bureaucratic surveillance state in which the bureaucratic administration just described is embedded within a broader theory of social power characterised by a belief in the legality of rules and the rights of rule-bound inspectors to issue environmental health commands to discipline and control populations.

Crook accepts that sanitary inspection, as environmental health was then known, can be seen as both a form of bureaucratic intervention and surveillance by the state. But these top-down theories obscure the interpersonal nature of inspection and its operation within a critical and sometimes hostile public sphere with which EH professionals reading this might be all too familiar. Instead he argues that inspection is better understood as a form of liberal surveillance and part of a liberal culture of governance. Here

[p]ower circulates between and inhabits all these agents [state and society, experts and public] as they, by turns, resist and

co-operate with one another . . . in this way freedom is not a goal but a means of liberal governance, a process it works through as a form, however messy, of social ordering . . . governance *was* the struggles inspectors endured and sought to overcome, which informed all aspects of their job, from direct encounters with the public to the ongoing battle for greater professional independence.

([23] – page 393)

Indeed Crook's 'struggles' closely resemble those of Tombs and Whyte's [56] 'critical approaches to regulation' and viewing the work of EH professionals through these theoretical lenses has much utility for describing the complexity of environmental health work today. The nature of environmental health problems is always changing but wherever one works these power struggles are always there and continue to shape the policy and practice of all EH professionals.

ANNEX 4.3: SUGGESTED FURTHER READING

For first-time researchers

Aveyard H. (2019) *Doing a Literature Review in Health and Social Care: A Practical Guide*, Open University Press, London, 4th ed.

Bell J, Waters S. (2018) *Doing Your Research Project: A Guide for First-Time Researchers*, McGraw-Hill Education, London, 7th ed.

Murray R. (2017) *How to Write a Thesis*, Open University Press, Maidenhead, 4th ed.

Routledge Focus on Environmental Health Series. https://www.routledge.com/Routledge-Focus-on-Environmental-Health/book-series/CONENHE.

For more detail on the research process, particularly methods

Bruce N, Pope D, Stanistreet D. (2018) *Quantitative Methods for Health Research: A Practical Interactive Guide to Epidemiology and Statistics*, Wiley-Blackwell, London, 2nd ed.

Costley C, Elliot G, Gibbs P. (2010) *Doing Work Based Research: Approaches to Enquiry for Insider-Researchers*, Sage Publications Ltd., London.

Dytham C. (2017) *Choosing and Using Statistics: A Biologist's Guide*, Wiley-Blackwell, London, 3rd Ed.

Green J, Thorogood N. (2018) *Qualitative Methods for Health*, Research Sage Publications, London, 4th ed.

Greenhalgh T. (2019) *How to Read a Paper: The Basics of Evidence-Based Medicine*, Blackwell Publishing Ltd, Oxford, London.

National Institute for Health and Care Excellence. (2012) *Methods for the Development of NICE Public Health Guidance*, 3rd ed. http://www.nice.org.uk/article/pmg4/chapter/5-reviewing-the-scientific-evidence.

Reed J, Procter S. (Eds.) (1995) *Practitioner Research in Health Care: The Inside Story*, Chapman & Hall, London.

Robertson D, McLaughlin P. (2008) *Looking into Housing: A Practical Guide to Housing*, Research Chartered Institute of Housing, Coventry.

Robson C, McCartan K. (2019) *Real World Research: A Resource for Social Research Methods in Applied Settings*, John Wiley & Sons, London, 4th ed.

For more on evidence-based practice and why it's needed

Aveyard H, Sharp P. (2017) *A Beginner's Guide to Evidence Based Practice in Health and Social Care*, Open University Press, Berkshire, 3rd ed.

Dodd S, Epstein I. (2012) *Practice-Based Research in Social Work: A Guide for Reluctant Researchers*, Routledge, London.

Goldacre B. (2009) *Bad Science*, Harper Perennial, London.

5

Law and practice to achieve outcomes

John Pointing

Common law and civil law systems

The common law originally developed in England (and the duchy of Normandy) in about a century and a half after the Norman conquest of 1066 [1]. Today, it provides the basis for the law operating in the UK as well as in most of the former British colonies, including the USA, Australia, Nigeria and India. Civil law (or civilian law) developed on the continent of Europe from the late middle ages, drawing on Roman law principles. Modern civil law systems owe much to the codification of law established in Napoleonic France by the Civil Code of 1804. Codification means that the domain of law is continually expanding as societies change and develop, so the scope of civil law has expanded far beyond the categories of Roman law. With variations in concept and practice, all the countries of continental Europe have civil legal systems.[1] European Union law draws on this civil law tradition too.

Legal systems have much in common in the regulation of human affairs, and for maintaining order and providing rules for the governance and functioning of institutions.

A fundamental difference arises between common law and civil law systems over the primary sources of law. Although common law systems are dominated by statutes created by the legislative bodies of states, the basic law is judge-made. Any gaps in the law left by statute can be filled by judges when deciding on individual cases coming before them. Law is continually created; never complete and legal cases are open to different interpretations. By contrast, civil law systems do not allow judges to make law. At the heart of civil law is the law of the book, provided by codification which amounts to a complete statement of law. The true meaning of law sets the agenda for legal scholars, whose primary role is interpretative. Here, the 'statutory code is seen as providing the basic gapless law', so the code is forever expanding as new issues arise ([2] at p. 45).

Common law judges possess a law-making function that is absent in the civil law jurisdictions. This ability is circumscribed, so in the UK no judge would be able to declare legislation to be unlawful or unconstitutional, as the US Supreme Court would be able to do.[2] In the employment law case of *Johnson v Unisys Ltd*,[3] Lord Hoffmann (at 37) was

DOI: 10.1201/9781003035640-5

careful to give full recognition to the principle of parliamentary sovereignty.

> Judges, in developing the law, must have regard to the policies expressed by Parliament in legislation. . . . The development of the common law by the judges plays a subsidiary role. Their traditional function is to adapt and modernise the common law. But such developments must be consistent with legislative policy as expressed in statutes.

The role of judges in modernising the common law is limited because this depends on the right case coming before the court in a timely way. The law can become obsolete, requiring reform by statute rather than waiting for a judicial decision by the higher courts. Thus, the common law offence of public nuisance has been extensively criticised over the years for being outdated [3]. Some public nuisances, such as being a 'common scold' or a 'common barrator' (a vexatious litigant), were still offences until their abolition by the Criminal Law Act 1967. Sclerotic regions of the common law can be reformed by putting into effect reforms proposed by the Law Commission, which has reported fairly recently on public nuisance.[4] At the time of writing, the Government is intending to place the common law offence of public nuisance on a statutory footing in the Police, Crime, Sentencing and Courts Bill 2021.

Of great importance in legal practice is whether previous decisions of the higher courts are binding on the way the law is subsequently interpreted. This is crucial not only in relation to judicial decisions made in courts and tribunals but also for lawyers advising clients on the merits of their cases. Generally speaking, only common law systems provide for judicial precedent as a rule for how the law should be interpreted. As Dworkin has suggested the purpose of the doctrine of judicial precedent is to ensure that earlier decisions will be followed in a

sufficiently like later case ([4] at p. 24). There will often be disagreement as to whether a precedent applies to the facts of the new case. Sometimes there are conflicting judicial decisions, as can occur when a statute has been poorly drafted, leaving it to the Court of Appeal or the Supreme Court to decide what a particular section of an Act truly means ([5] at pp. 222–223).

A previous decision may not have binding effect on the instant case heard in a common law jurisdiction – but it can nevertheless be persuasive. The judge in the latter case may accept that the reasoning employed in the earlier decision precisely applies to the case before him. The persuasive effect of cases heard in the higher courts of other common law jurisdictions can also be applied, so an English judge may cite Australian or US cases in his or her decision. The influence of previous decisions of the higher, appeal courts is also felt in civil law jurisdictions, even though no doctrine of judicial precedent applies. In France, for example, no court is legally required to follow a previous decision. But, in practice, French judges follow decisions of the highest appeal court – the *Cour de Cassation* – if they believe a decision of that court represents the law. In doing so, however, they will state that they are following an established legal principle rather than a judicial decision [6].

Duties and powers of local authorities

A local authority is a public body existing as a corporation.[5] It functions by perpetual succession, as successive individuals act on its behalf and then cease to do so once they leave office. Corporate status means that the acts of individual officers become the acts of the local authority, provided that they are acting with actual or ostensible authority. Corporate status provides local authorities with the capacity to act as an individual in some respects. It allows them to own and dispose of property and to enter into contracts and makes them capable of suing and being sued. It also means that local authorities can prosecute in their own name.

As creatures of statute, local authorities can only do things which are expressly or impliedly authorised by Parliament. A local authority is also under a duty not to fetter or to divest itself of its statutory discretions. This means that it is obliged to carry out its statutory duties and will be subject to sanctions by the secretary of state of a government department or become liable in an action for judicial review for a breach of statutory duty. With regard to enforcement duties, as set out in various statutes, a local authority cannot choose not to enforce legislation placing it under a statutory duty, such as to investigate complaints or to take enforcement action. Where a local authority is not placed under a duty to act but has a discretion − as in the case of a power − such discretion should be exercised properly and lawfully.[6]

Local authority decisions

Section 101 of the Local Government Act 1972 gives the local authority wide powers of delegation to discharge any of its functions by a committee, a subcommittee, or an officer of its own or another authority. Delegation may be given effect by Council resolution or standing orders. Additionally, a local authority may make arrangements to discharge its functions by another local authority or to act jointly with it. Any decision taken without proper delegation will be in principle unlawful, whether taken by officers or by an informal group of councillors.[7] A decision cannot be ratified with retrospect effect if the delegation had not been effective at the time it was made.[8] There is no power in section 101 of the Act to delegate to a single member and neither should one be implied. In *R v Secretary of State for the Environment, ex p Hillingdon LBC*,[9] the council delegated, by standing orders, to the chair of the planning committee the power of issuing planning enforcement notices. The court found such delegation had been *ultra vires*.

Standing orders provide a general means whereby local authorities regulate their conduct and proceedings. The terms of standing orders may not be contrary to law and neither do they have the force of law. Central government has the power to require local authorities to include certain provisions or to make or to refrain from making certain modifications to their standing orders.[10] They may be used to define the remit and extent of delegation to each committee, subcommittee, and officer. Standing orders can be changed, but if a local authority does not act in accordance with those in force at the relevant time, such acts will be judicially reviewable.[11]

Members and officers enjoy immunity from personal liability when acting *bona fide* in carrying out statutory duties.[12] The Local Government Act 1972 has the effect that the authority for taking decisions on behalf of a council can be delegated to a properly authorised officer. The power to institute legal proceedings may be delegated to a subcommittee or to an officer with sufficient seniority. Delegation to an individual council member is not permitted since the professional expertise, competence, and accountability of an employee form the rationale for the delegation.[13] Standing orders commonly provide for a power to be delegated to an officer acting in consultation with a specified member, but this should not extend to enabling the member to exercise the power alone or to play the dominant role. It is common practice for the decisions about whether to prosecute or to issue a simple caution against an offender to be delegated to a senior officer acting in consultation with a member.

Limits to local authority powers − As a corporation, a local authority's powers are limited to those expressly given by statute and to those which may be implied by statute.[14] Section 111 of the Local Government Act 1972 enables local authorities to do anything calculated to facilitate, or which is conducive or incidental to the discharge of any of its functions. This statutory limitation applies to all acts carried out by or on behalf of the local authority, including the exercise of its powers and responsibilities. Where a local authority has acted beyond its powers, such acts or decisions are *ultra vires* and subject to judicial review. Implied powers are fairly tightly construed but may include anything 'reasonably incidental' to what the local authority has express or implied authority to do.[15]

A local authority cannot make representations that go beyond its powers, even where this results in a party acting to his or her detriment. In *Southend-on-Sea Corporation v Hodson (Wickford) Ltd*,[16] the borough engineer wrote to a builder stating that certain land did not need planning consent for use as a builder's yard. The builder relied on this representation, purchased the land, and used it as a builder's yard. The authority then took a different view and served a planning enforcement notice. The Divisional Court refused the builder's claim on the basis that an estoppel cannot be raised to hinder the exercise of a statutory discretion conferred upon a public authority. However, where a planning officer's decision, in *Lever (Finance) Ltd v Westminster LBC*,[17] amounted to a representation made within the scope of his ostensible authority, a developer acting on it could render that representation binding on the local authority. The decision of the Divisional Court in *Lever* is difficult to reconcile with the earlier decision in *Southend-on-Sea Corporation*. Later cases have decided that not all representations made within an officer's ostensible authority will bind the council, so the state of the law is not clear.[18] Where an officer has acted beyond the scope of their authority, however, their representations will not be binding, and a contract founded upon those actions will be void.[19]

Legal proceedings

Local authorities can be involved in legal proceedings in various capacities associated with environmental health. These include:

- Appeals against service of administrative notices;

- Prosecution for breaches of administrative notices, regulations or statutes;
- Licensing appeals;
- Planning appeals;
- Applications for injunctions;
- Judicial review;
- Appeals to the High Court and further.

Local authorities may be subject to complaints made by aggrieved persons to the Local Authority Ombudsman Service for maladministration. Where the local authority is the landlord a tenant may complain to the Housing Ombudsman. Ombudsman proceedings are not judicial, but they are quasi-legal, and findings made against local authorities have implications for future policy and decision-making.

Power of local authorities to prosecute

The Local Government Act 1972 empowers local authorities to prosecute and to start (or defend) civil proceedings in their own name. Section 222(1) of the Act provides:

> Where a local authority consider it expedient for the promotion or protection of the interests of the inhabitants of their area –
>
> (a) they may prosecute or defend or appear in any legal proceedings and, in the case of civil proceedings, may institute them in their own name.

Section 222 provides the local authority with a general power to prosecute the full range of regulatory and criminal offences associated with its functions. It can bring public nuisance proceedings in its own name and apply for civil remedies such as injunctions. Section 222 is drafted to give local authorities wide powers, but they must consider whether the proceedings are expedient for the promotion or protection of inhabitants in *their* area (and not elsewhere). Their powers to prosecute are not limited to regulatory functions, so a local authority could prosecute a crime involving criminal intent, such as a conspiracy to defraud.[20]

Appearance in legal proceedings

Section 223 of the Local Government Act 1972 allows any member or officer of a local authority[21] to prosecute or defend in magistrates' court proceedings, provided that they were authorised to do so before the commencement of those proceedings.[22] Proof of authorisation is by production of the relevant council resolution: reliance on the officer's appointment or position is not sufficient. Officers appearing in court should bring with them proof of authorisation, such as a warrant card and a copy of the minutes of the relevant council resolution. Proper authorisation also applies to the service of administrative notices. Where not properly authorised at the time of service, notices may not be cured by subsequent authorisation. A defective notice should not form the basis for any subsequent prosecution.

Section 223 enables employees who are not legally qualified, such as trading standards officers and environmental health practitioners, to appear in the magistrates' courts. Such officers conducting cases are also competent to give evidence in the same proceedings. Nowadays it would be unusual for environmental health practitioners both to conduct a prosecution and give evidence. The usual practice is for legal representation by a solicitor or barrister.

Courts and tribunals

The magistrates' courts hear the vast majority of criminal cases, including those where the local authority is the prosecutor. Some offences are 'summary only' and can only be heard in the magistrates' court – either before a lay bench or before a district judge sitting alone. Some statutes – such as the Food Safety Act 1990, the Health and Safety at Work Act 1974, and the Environmental Permitting

Regulations 2010 – provide for 'either-way' offences. These cases will mostly be heard in the magistrates' courts, but more serious cases can be sent up to the Crown Court for a trial before a jury. The Crown Court has greater sentencing powers than the magistrates' court, including the power to sentence a convicted person to a longer term of imprisonment. Maximum fines available in the Crown Court are usually not limited, neither are they for many environmental health offences heard in the magistrates' courts.[23]

Appeals – Appeals against conviction or sentence can be made to the Crown Court from decisions in the magistrates' courts. If an appeal is on a point of law it should be made to the High Court. Here, the magistrates' court is asked to 'state a case' for the opinion of the High Court on issues of law which it may have applied wrongly in coming to its decision.

Magistrates' courts also have a civil jurisdiction. This will apply when an administrative notice is appealed by the recipient, who commences their appeal by making a complaint to the court, within the time limit specified in the relevant statute. The powers of the court involve either affirming the notice, quashing or cancelling it, or modifying it to make it more appropriate or fairer to the subject. The grounds for appealing a notice will be set down in the relevant statute, or in regulations made to give effect to the right of appeal. Examples of appeals dealt with by the magistrates' courts include abatement notices for statutory nuisances (see Malcolm and Pointing 2011 [7], at chapter 15) and Food Safety Act 1990 improvement notices and emergency prohibition notices (see Malcolm and Pointing [8], at chapter 3).

Appeals against notices can be complicated and technical, and magistrates' courts generally lack experience in dealing with this type of complaint. The Health and Safety at Work etc Act 1974 provides that appeals regarding service of improvement or prohibition notices should be to an employment tribunal. Regulations for hearing such appeals by an expert tribunal have been made, and this probably helps to improve the quality of decision-making compared to proceedings heard in conventional courts [9].

The use of specialist tribunals rather than the courts to hear appeals against administrative notices has increased in recent decades. Appeals to the First-tier Tribunal may be made in respect of the various notices and orders served by local authorities under Part 1 Housing Act 2004 (see Battersby and Pointing [10]). Appeals against licensing decisions in respect of Part 2 and 3 Housing Act 2004 should also be made to the First-tier Tribunal in England[24] or to the Residential Property Tribunal (RPT) in Wales.

A landlord has the right to appeal to the First-tier Tribunal (or RPT) against a civil penalty issued by the local housing authority. The First-tier Tribunal has the power to confirm, vary (increase or reduce) the size of the civil penalty imposed by the local housing authority, or to cancel the civil penalty. If the First-tier Tribunal decides to increase the penalty, it may only do so up to a maximum of £30,000 [11]. A refusal to pay a civil penalty would require the authority to obtain an order in the county court to enforce it.

Appeals against decisions made by the First-tier Tribunal (Property Chamber) can be made to the Upper Tribunal (Lands Chamber).[25] The Upper Tribunal has exclusive jurisdiction to hear appeals from the tribunal below – so this excludes a hearing before the High Court – and can rule on a point of law. Appeals from the Upper Tribunal are to the Court of Appeal.

A move away from using the magistrates' courts for hearing appeals against administrative notices is illustrated by the regulation of contaminated land, under Part IIA of the Environmental Protection Act 1990. When this provision first came into effect, in 2000, appeals against remediation notices served by local authorities were to the magistrates' court. The complex and technical nature of such appeals placed these courts under considerable pressure, so appeals have had to be made to the Secretary of State since 2006.[26] Appeals against decisions of the Secretary of State can be made to the High Court.

More recent regulatory regimes have followed this practice. Thus, appeals against decisions of the regulator with respect to Environmental Permitting should be made to the 'appropriate authority': the Secretary of State, in England, Welsh Ministers in the case of Wales.[27]

Decision to prosecute

The position of local authorities in making the decision to prosecute is more complicated than for conventional crime. This is partly because local authorities are responsible both for investigating and prosecuting offences, so there is not the separation of these functions as there would be in a criminal prosecution involving the police and the CPS. Another reason is because local authorities prosecute mainly regulatory offences in which the operation of businesses and commercial concerns is a major source of offending.

The *Regulators' Code*, published by the Better Regulation Delivery Office [12] in 2014 provides a nuanced approach to enforcement by regulators, including the decision to prosecute businesses for regulatory breaches. Local authorities must have regard to it when developing policies and procedures to guide their regulatory activities. When deciding whether to prosecute, local authorities should consider public interest factors in terms of the *Regulators' Code*. This is particularly important in situations where the regulator provides advice and support in order to secure regulatory compliance, such as when permitting and licensing activities are involved.

There is a discretion to prosecute any type of criminal offence and local authorities are subject to the ordinary principles of administrative law when making such decisions. Councils also need to take into account their enforcement policy. In making the decision whether to prosecute, proper consideration by the person or persons responsible for making the decision on behalf of the local authority is required. Underlying this decision is a requirement for the evidential test in bringing the prosecution to be satisfied before commencement of proceedings. Satisfying the evidential test, followed by proper consideration being given to the public interest factors is the procedure stipulated in the *Code for Crown Prosecutors*, issued by the Crown Prosecution Service (CPS).[28] No matter how serious the offence or the harm resulting from it, if the evidence is not sufficient to provide a realistic prospect of conviction then it would be improper to commence or continue with any prosecution.

Injunctions

An injunction is a civil remedy which can be obtained in the High Court or county court. It is a civil remedy, even though persons found to have breached an injunction are liable to pay an unlimited fine and/or be sentenced for up to two years' imprisonment. An injunction may be the most effective remedy available in some nuisance, housing, or property cases – because it is quick and suitable for use in an emergency. It can be used to order to stop a harmful activity from happening or continuing. It can even be used to reinstate the *status quo*, such as when a building has been unlawfully built without planning permission – in which case the local planning authority can order its demolition, backed by an injunction.

A civil remedy can sometimes be far more effective than using the criminal law. As reported in *The Observer*, the Carlton Tavern – a public house in Westminster – was unlawfully demolished by the developer of the site 'after being denied planning permission to convert it into 10 flats, and two days before English Heritage was due to recommend the pub be granted Grade-II listed status' [13]. Westminster City Council ordered the developer to reinstate the building 'brick by brick', as it had been built in the 1920s – a decision backed by the planning inspector who turned

down the developer's appeal. Enforcement of the reinstatement of the building would be by injunction, though in this case the developer rebuilt the pub to the satisfaction of the Council.

Section 222 of the Local Government Act 1972 allows a local authority to apply for an injunction in its own name, where the purpose of the injunction is to promote or protect the interests of the community. A local authority is not usually required to give an undertaking in damages to the court before an injunction order is made, so any losses sustained by a company as a result of the injunction being granted are not recoverable.[29]

Evidential matters

Environmental health practitioners have extensive powers of investigation, collecting of evidence, and of seizure – whether of documents, stock, or equipment.[30] These are set down in the various statutes governing the use of such powers, which vary depending on the field of regulation and on whether they apply to businesses or to residential premises. Whether such powers are exercised properly and in accordance with the relevant statute is important, for a local authority is a public body and so its decisions are subject to judicial review, or consideration by the Local Authority Ombudsman. Additionally, EHPs should be conscious of the importance of maintaining good practice standards, difficult though this is with restraints on spending imposed by central government.

An essential thread running through all stages of enforcement activity is the quality of the evidence. It is this that should guide decision makers at all stages: from carrying out an investigation, to deciding what action to take, through to the conduct of legal proceedings. Prosecutions should only be taken where there is sufficient admissible, relevant, and reliable evidence that can be presented in court giving a reasonable prospect of conviction. No matter how serious the consequences of a harmful event or state of affairs, cogent evidence concerning who was responsible and in what way for the offence must be obtained.

In criminal cases brought before the magistrates' court or the Crown Court, the prosecution is bound by general rules applying to all types of criminal case, and in particular by the Criminal Proceedings Rules.[31] The prosecution of regulatory offences by local authorities is similarly bound, even though these Rules were drafted with normal criminal cases in mind.

The prosecutor in a criminal matter has to prove their case against the defendant to the criminal standard: 'beyond reasonable doubt'. The evidence needed to bring a prosecution has to be obtained fairly and should also be relevant and probative – focussed on the charges the defendant is required to meet.

The right to a fair trial in both civil and criminal proceedings is enshrined in the common law and in article 6 of the European Convention for the Protection of Human Rights and Fundamental Freedoms. The criminal courts have a discretion to exclude prosecution evidence, including the way the evidence was obtained. Thus section 78 of the Police and Criminal Evidence Act 1984 states:

> In any proceedings the court may refuse to allow evidence on which the prosecution proposes to rely to be given if it appears to the court that, having regard to all the circumstances, including the circumstances in which the evidence was obtained, the admission of the evidence would have such an adverse effect on the fairness of the proceedings that the court ought not to admit it.

A good rule of thumb for investigating officers is to bear in mind that any case they are investigating might end up being appealed or defended in a court or tribunal. Officers might have to justify to a judicial body any decision they have made in the history of a case. Included in this is whether they have acted properly and kept within the scope of their statutory authority. Straying outside these boundaries may mean that cases will be lost and regulation will have failed to protect the public. Where cases are badly managed there is the risk that the reputation of the local authority is damaged, or the professional reputations of enforcement officers may be impugned. To give an example of this from a reported case. In 2004 an abatement notice was served, under section 80 Environmental Protection Act 1990, by the London Borough of Hackney on Rabbi Rottenberg for noise nuisance emitted from premises used as a school and for religious purposes. Noise from these activities had been causing a problem for residents of the adjoining property. Rottenberg's use of the premises – a large, semi-detached house – was compliant with planning permission.

Rottenberg was convicted for a number of breaches of the notice in the magistrates' court but succeeded on all counts in his appeal against conviction at a rehearing before the Crown Court. The council then appealed by case stated to the High Court, where they lost. The High Court judgement is worth reading by all local authority enforcement officers.[32] It shows the dangers for a local authority in prosecuting a case when the evidential base is shaky. The council's case was also misconceived as it was based on the mistaken belief that the court was bound to accept the expert evidence of its investigating officers, when such evidence had not been challenged by an expert giving evidence for the defendant. A number of officers had carried out a series of poorly conducted investigations, based solely on listening to the noise on the complainant's side of the party wall, and without engaging with Rabbi Rottenberg in any way and without taking any noise measurements. The High Court decided that the evidence of the officers fell far short of the standard required for expert evidence, was no more probative than evidence which might have been given by a lay person and was insufficient to justify conviction for a criminal offence.

Expert evidence

A witness appearing in court usually gives evidence as a witness of fact – about what they saw, heard, or did. Only an expert may give evidence of opinion, in which an inference is drawn from a set of circumstances rather than reflecting on something they observed, heard, or did. An EHP will appear in court as a witness to give evidence pertaining to the factual matters of their investigation (what they observed, did, etc.) and also to provide the court with evidence of opinion, based upon their expertise. In the latter situation, technical evidence may be given forming the basis for an opinion; or an opinion might be given as to whether, in a statutory nuisance case for example, the matter was prejudicial to health or otherwise constituted a statutory nuisance.

In environmental health proceedings an issue in dispute will often be resolvable by reference on some form of technical or specialist knowledge, which will be outside the experience of those having to decide the case. Only an expert witness may give an opinion in such cases – either as testimony, or in a report or witness statement. This forms part of the whole body of relevant evidence that enables the court to come to a decision on liability. It is for the court to decide whether the witness is to be treated as an expert, and the qualifications and experience of the witness should be considered before deciding on this. Where an expert witness indicates that he or she does not have sufficient expertise in a particular area, then any opinion evidence given about that area should be disregarded by the court, as they have no more expertise than a layman in regard to it.[33] The court is not bound to accept the opinion of an expert and is placed under a duty to scrutinise all evidence before deciding whether to accept it.[34]

An EHP may be qualified to give an expert opinion in a case because of his or her qualifications and experience. The fact that an officer is employed by the local authority prosecuting the case or acting as a party is not a bar to that person giving expert testimony. However, it is important that they fully understand the role of an expert witness, the necessity for objectivity, and that their overriding duty is owed to the court and not to their employer. In *Field*[35] Lord Woolf opined: 'I would encourage the authority concerned to provide some training for such a person to which they can point to show that he has the necessary awareness of the difficult role of an expert'.

Duty of objectivity – An expert owes a duty of objectivity to the court. This implies that the evidence they give – either as testimony, or in a report or witness statement – would be the same if they were giving that evidence on behalf of the opposing side. In *Whitehouse v Jordan*[36] Lord Wilberforce explained:

> Expert evidence presented to the court should be, and should be seen to be, the independent product of the expert, uninfluenced as to form or content by the exigencies of litigation. To the extent that it is not, the evidence is likely to be not only incorrect, but self-defeating. The expert's responsibility is to the court and not to the client in giving evidence or producing a report. This applies as much to defence as to prosecution experts and applies throughout the duration of the case.

The last sentence in this judgement is worth noting. Should the local authority find that the purported expert evidence being led by the other side lacks objectivity and is tainted, then an application can be made to the court for such evidence or report to be ignored. The court may also decide to report the offending expert to their professional body or regulator for misconduct, or, in an extreme case, to hold them in contempt of court ([14] at paras. 89–92).

Obtaining evidence for use in criminal proceedings

Investigating regulatory offences in order to obtain sufficient evidence to charge the person responsible is bound by a set of rules. These may be found irksome by local authority officers with enforcement responsibilities. It also means that the time taken, from investigating whether an offence has been committed to when the case is finally disposed of by the court, seems excessive. The protections afforded to suspects in criminal cases must be observed conscientiously, otherwise a prosecution might fail, so putting the public at risk by the failure of regulation. The most important provisions are set down in the Police and Criminal Evidence Act 1984 (PACE), which should be read in conjunction with the codes of practice associated with the Act.

Interviewing suspects – A PACE interview conducted with a suspect by the police will often be a significant source of evidence in a normal criminal case. Obtaining a confession that can be used as evidence in court is often the main purpose of such an interview. Evidence of the state of mind or criminal intent of the suspect will often form an important part of the interview. The interview also provides an opportunity for the investigators to put before a suspect other evidence gathered in the course of the investigation in order to elicit a response. However, in an investigation carried out by a local authority, there will usually be other evidence than from an interview which is more important for the offences a suspect may be charged with. Criminal intent is not relevant where the offence is one of 'strict liability', so the state of mind of the suspect should not be explored in any detail during an interview for such offences. Therefore, the purpose of the interview needs to be considered carefully beforehand in an environmental health case. A key reason for conducting the interview might be to explore whether a suspect has a statutory defence, and, if so, to explore what that defence entails. A statutory defence, such as whether 'best practicable means' were employed to avoid commission of an offence or whether there is a 'reasonable excuse' for the offence, might need some time to emerge from the interview. It may also take the local

authority some time to decide whether or not to accept a statutory defence.

A properly conducted PACE interview provides a framework for assessing the evidence provided by the suspect being interviewed and for deciding how the case should be progressed. The answers given, under caution, to carefully worded and fairly put questions, which are recorded in accordance with the requirements of PACE Code 'C', can be used as evidence in a future prosecution [15]. If a company is believed to have committed an offence, then a director, officer, or manager should be interviewed. Such a person must be sufficiently senior to be part of the 'brains' of the company and thus able to make admissions on its behalf.

Code 'C' states that a caution must be given if there are grounds to suspect that the person being questioned has committed an offence ([15] at para. 10.1). The purpose of an interview might be to seek information about the circumstances of the alleged offence, not to obtain evidence to use in a future prosecution. An information-gathering interview should not be carried out under PACE conditions and a caution should not be administered. Issuing a caution at this stage is not only inappropriate but will often result in the person being interviewed ceasing to provide any further information. The purpose of the interview must be made clear to the interviewee, including that they are being interviewed as a witness, not as a person suspected of committing an offence.

Any interview conducted under PACE conditions must give the suspect the choice of refusal to answer any questions put to them: this being one of the purposes of the caution. Interviewing a witness using a statutory power that obliges them to answer questions, such as an information-gathering interview under section 20(j) Health and Safety at Work Act 1974, has the implication that their answers cannot be used against them (or their spouse) in any proceedings.[37]

Powers under the Regulation of Investigatory Powers Act 2000 – The purpose of the

Regulation of Investigatory Powers Act 2000 (RIPA) is to provide a legal framework for ensuring that the organs of the state, including local authorities, do not carry out activities in breach of the protections afforded by the Human Rights Act 1998 (see also Chapter 2). This Act gives effect to the European Convention,[38] in which article 8 states:

(1) Everyone has the right to respect for his private and family life, his home and his correspondence.

(2) There shall be no interference by a public authority with the exercise of this right except such as is in accordance with the law and is necessary in a democratic society in the interests of national security, public safety or the economic well-being of the country, for the prevention of disorder or crime, for the protection of health and morals, or for the protection of the rights and freedoms of others.

Such activities as covert surveillance, in order to be lawful and not breach article 8(2) of the European Convention, should be authorised in accordance with the provisions in RIPA (or other legislation protecting rights under the European Convention). Section 48 of RIPA defines surveillance as including 'monitoring, observing or listening to persons, their movements, their conversations or their other activities or communications'. It includes 'recording anything monitored, observed or listened to in the course of surveillance'.

'Directed surveillance' may be undertaken in some criminal investigations carried out by local authorities. Directed surveillance is defined in section 26 of the Act as covert (where the subject is unaware that it might be taking place) but not intrusive, when undertaken as a specific investigation or operation, and carried out in such a way as to make it likely that private information is obtained about a person. 'Private information' includes any information relating to a person's private or family life. Talking and conversations are included, where the words used can be made

out and have meaning. Arguably, involuntary sounds made by a person with Tourette's syndrome are included. Singing (whether competently or not) and noise from a party generally would not come within the scope of private information. Noisy activities emanating from domestic premises, such as the playing of recorded music, the sound from televisions, noise from the use of domestic equipment do not come within the scope of private information ([15] at para. 3.40).

The scope of authorisation provided by RIPA is limited to criminal investigations. The Code of Practice issued by the Home Office on covert surveillance ([16] at para. 4.44) sets out the effect of amendments made to RIPA by the Protection of Freedoms Act 2012. Firstly, local authority authorisations made under section 28 of RIPA are subject to judicial approval from a justice of the peace and can no longer be given approval by a senior council officer. Justices will be mindful of their duty to consider whether the evidence being sought by the authorisation might be available by some other means than covert surveillance.

Secondly, RIPA has been amended by article 2 of the Regulation of Investigatory Powers (Directed Surveillance and Covert Human Intelligence Sources) (Amendment) Order 2012.[39] The effect of this Order is to restrict the scope of surveillance that can be undertaken by local authorities. The Code of Practice summarises the effect of these changes ([16] at para. 4.44):

- Local authorities in England and Wales can only authorise use of directed surveillance under RIPA to prevent or detect criminal offences that are either punishable, whether on summary conviction or indictment, by a maximum term of at least 6 months' imprisonment **or** are related to the underage sale of alcohol and tobacco or nicotine inhaling products.
- Local authorities **cannot** authorise directed surveillance for the purpose of preventing disorder unless this involves a criminal offence(s) punishable (whether

on summary conviction or indictment) by a maximum term of at least 6 months' imprisonment.

- Local authorities may therefore continue to authorise use of directed surveillance in more serious cases as long as the other tests are met — i.e. that it is necessary and proportionate and where prior approval from a JP has been granted. Examples of cases where the offence being investigated attracts a maximum custodial sentence of six months or more could include more serious criminal damage, dangerous waste dumping and serious or serial benefit fraud.

- A local authority **may not authorise** the use of directed surveillance under RIPA to investigate disorder that does not involve criminal offences or to investigate low-level offences which may include, for example, littering, dog control and fly-posting.

The possibility of a custodial sentence of more than six months being given by the court is, therefore, the boundary line set for authorisation. Provided that the proportionality test is met — that surveillance is justified because of the seriousness of the problem and because gathering sufficient evidence by other means would prove to be too difficult — then food safety, health and safety, and environmental permitting offences can be authorised, subject to approval by a magistrate.

When it comes to noise nuisance, since RIPA was amended, in 2012, it has no longer been possible to obtain authorisation involving covert surveillance for a noise nuisance investigation where a person is suspected of breaching a section 80 EPA 1990 abatement notice. This is because such a breach is not an imprisonable offence.

Investigating officers need to be clear about the circumstances when RIPA is relevant and when it is not applicable, particularly in domestic noise nuisance investigations. As the Code of Practice makes clear, where covert surveillance activities 'are unlikely to result

in the obtaining of any private information about a person, no interference with article 8 rights occurs and an authorisation under the 2000 Act is therefore not applicable' ([16] at para. 2.24). The use of noise-monitoring devices which record sound levels does not result in the listening to or recording of private information, so RIPA is not applicable and such use is allowed.

The position when a recording device records the actual sounds made from talking, speech, or shouting — where the words used can be heard and recorded — is more complicated. An investigation into a noisy party using such a device would be unlikely to result in obtaining private information, in which case RIPA would not be applicable and such use is allowed. If it turned out that the recording did record private information, those parts of the recording could be redacted from the evidence of noise nuisance intended to be used later in court. The redacted private information could then form part of the unused material served on the defendant before the trial.[40]

Where people are holding a private conversation in the home, or in a public space, then they have a reasonable expectation for protection, under article 8 of the European Convention, from those conversations being listened to and a recording being made of them by an agency of the State. A local authority wishing to listen to and make a recording of such private information can only do so if authorised by RIPA. This would be possible for some forms of regulation, but not for a statutory nuisance investigation, because breach of an abatement notice is not an imprisonable offence.

On the recording of verbal content, the Code of Practice states that 'an authorisation is unlikely to be available':

[Where] the recording of verbal content is made at a level which does not exceed that which can be heard from the street outside or adjoining property with the naked ear. In the latter circumstance, the perpetrator would normally be regarded as having forfeited any claim to privacy.

([16] at para. 3.40)

This part of the Code might suggest that where the source of noise – which includes private information – is so loud that it can be heard outside the source premises, then the persons involved have forfeited their rights of privacy. Whereas authorisation would not be available in these circumstances, an assertion of the forfeiture of rights is dubious. This part of the Code is ambiguous, and the forfeiture point is not developed further in the text. It may be that an investigating officer hears verbal content that includes private information whilst outside the source premises, but that does not mean that a recording can be made which is admissible in any future legal proceedings.

The use of regulatory law to achieve positive environmental health outcomes

The purpose of regulatory law is not only for the pre-emptive control of potential harm but also for the retrospective enforcement of a penalty for causing harm [17]. Its scope is thus wider than prosecution or instituting legal proceedings. Making use of formal enforcement powers by EHPs generally occurs after persuasion has failed to bring about resolution of a regulatory problem. Professionalism based on 'soft skills', on good practice considerations is (or should be) central to the regulatory approach of local authorities; with recourse to prosecution for breaches of regulations and statutes taking place only as a 'last resort' [18–19]. One of the risks of this approach is for local authorities to eschew taking any formal enforcement action in nearly all circumstances. The cost and time involved in prosecuting offenders, the lack of resources resulting from the pursuit of austerity policies by governments since 2010, and the deskilling of environmental health roles in

local government all lean towards stasis in achieving positive environmental health outcomes [20].

The deregulation agenda pursued by governments of various political hues have had the effect of curtailing the role of local authorities – a process that started with the enactment of the Deregulation and Contracting Out Act 1994. The changing role of local authorities in the early 1990s and, in particular, the pervasive influence of privatisation, was well captured by Desmond King around this time ([21] at p. 204).

> Instead of envisaging local government as an institution representing a local community and its local tradition, it is to be designed as an institution responsible for overseeing service provision. Local Government is thought of as an enabling institution and not one of direct service delivery. . . . This new role maximises efficiency and profit criteria in local government. It treats citizens as customers of government services. Furthermore, local authorities are viewed as purchasers rather than providers of services.

The Deregulation and Contracting Out Act 1994 gave central government new powers to redirect the scope of interventions by local authorities. In particular, the use of enforcement powers by local authorities came up for criticism – these being considered too inflexible and interfering with the efficient running of businesses. Neil Hamilton, the Deregulation Minister, in addressing the October 1994 Conservative Party conference, described the proper role of local government enforcement officers as 'handmaidens of business – helping them to comply – rather than the local branch of the Gestapo' [22]. The type of regulation being castigated here has been described as 'command and control' – the term employed by Keith Hawkins in his interesting study of the operations of the Health and Safety Executive [19]. Command and control involves both high costs and high visibility. It also requires the regulator to make frequent inspections, relies on formal methods of enforcement and a readiness to resort to prosecution early on for infringements of regulations and statutes. As Ayres and Braithwaite have pointed out, the reality has been more complicated [23]. The practice of local authority regulators, at least since the 1980s, has been to rely heavily on persuasion and encouragement for businesses to comply with regulatory requirements, with formal enforcement powers – particularly prosecution – being used only as a 'last resort' [19]. This 'softer' approach to enforcement has itself received criticism for encouraging a too close and cosy relationship to prevail between regulators and businesses [24].

The purpose of Neil Hamilton's jibe was to promote the superiority of techniques developed in the private sector. Such ideas were (and remain) very attractive to politicians and ministers, and to their policy advisers. The private sector virtues incorporated into the thinking of regulatory agencies have included the setting of explicit standards and measures of performance; adopting private sector styles of management practice; an emphasis on greater discipline and parsimony in resource use [25]. These ideas of corporate governance have been applied to the whole range of public agencies and services, following the Hampton Review of 2005 [26]. The Hampton Review made a number of trenchant and justified criticisms concerning local authority regulation:

- Patchy use of risk assessment;
- Regulators do not give enough emphasis to providing advice in order to secure compliance;
- Too many, often overlapping, forms and data requirements with no scheme to reduce their number;
- Regulators lack effective tools to punish persistent offenders and reward compliant behaviour by business;
- The structure of regulators, particularly at local level, is complex, prevents joining

up, and discourages business-responsive behaviour; and
- There are too many interfaces between businesses and regulators [26].

Implementation of the Hampton Review has been left to a succession of governmental agencies, accountable to the Secretary of State. The Better Regulation Executive (BRE) has provided an administrative layer between government departments and executive agencies to coordinate policy and advice [27]. The Local Better Regulation Office (LBRO), accountable to the Department for Business, Innovation and Skills (BIS), was established following publication of the Hampton Review. After the Regulatory Enforcement and Sanctions Act 2008 was enacted, the LBRO launched its flagship Primary Authority scheme, in April 2009. The Primary Authority scheme was designed to facilitate regulation by a single local authority operating across the different functions and locations of a larger business [28]. Although supported by some local authorities and by the Chartered Institute of Environmental Health [29], the scheme has had little impact outside of the large business sector according to Eccles and Pointing [30].

In 2012, the LBRO was succeeded by the Better Regulation Delivery Office (BRDO). The BRDO was itself replaced, in March 2016, by Regulatory Delivery. This coincided with a change in government department – BIS being replaced by the Department for Business, Energy and Industrial Strategy (BEIS). Regulatory Delivery enjoyed a short life and was succeeded, in January 2018, by the Office for Product Safety and Standards (OPSS). By this time the remit of better regulation had narrowed from the broad scope exemplified in the Hampton Review. Henceforth, the focus would be on trading standards and product safety, with food safety and construction products being looked after elsewhere.[41] The press release launching OPSS proudly proclaimed the hope that: 'The new office will further enhance the UK's world-beating product safety system and give consumers the highest ever levels of protection' [31].

Better regulation

The better regulation agenda calls for the regulator to change its behaviour – by operating strategically and more like a private company – and for the company to cooperate and become a partner with the regulator [32]. This involves deregulation: shifting part of the regulatory burden from local authorities to businesses, with the nature of the burden transitioning from being interventionist and moving towards 'smarter', less intrusive forms of regulation, including self-regulation [30]. The building up of trust between the regulator and company is anticipated to mean that taking on some of the regulatory burden is accepted willingly by the company. Reliance on trust allows self-regulation to flourish, so enabling state enforcement generally to wither away according to Campbell [31].

The role of regulation changes with this shift in burden, with the regulator taking a less direct, more supervisory role, so resulting in cost savings. This change is reflected in the ways in which problems are defined, redefined and 'framed' by policy-makers and decision-takers [34–35]. This process has been promulgated as 'win–win–win' by enthusiastic on-message regulators and better regulation advocates [36]. Issues are often assumed to be uncomplicated, in the words of one influential academic:

> The role of regulation ceases to be primarily about inspectors or auditors checking compliance with rules, and becomes more about encouraging the industry or facility to put in place processes and managerial systems which are then scrutinised by regulators or corporate auditors.
>
> *([37] at pp. 190–191)*

Measuring the success of better regulation is much more difficult than proclaiming its hoped-for benefits. Some commentators such as Braithwaite have taken a broad sweep from an international perspective, arguing that there has been little winding back of regulation in the long term [38]. Evidence of success is patchy, with some larger companies developing effective policies and implementing good regulatory practices, often with the help of regulators. In other cases, a lack of transparency in the way regulatory activity is carried out and in the relations between regulator and the business has meant that nobody takes responsibility when things go wrong [39].

The taking on of a more strategic and supervisory role by local authority regulators has not been easy, particularly during a period of cutting back of financial and human resources. As was predicted by this author several years ago:

> The capacity for local authorities of carrying out this kind of supervisory role has not been tested. It may prove to be too difficult at a time when recession and cuts in public services result in local authorities cutting out technical managerial and supervisory posts. In other words, the individuals being made redundant may be precisely those with the requisite experience and skills suitable for [putting into effect better] regulation.
>
> *([30] at p. 39)*

Cut-backs have also resulted in enforcement officers avoiding using statutory powers and relying, even more than they did previously, on informal routes to compliance. Such over-reliance on informal methods may be the result of a mistaken belief that this is a requirement under section 21 of the Legislative and Regulatory Reform Act 2006.[42] It has been argued that this explains the infrequent use of enforcement powers for dealing with housing hazards under the Housing Act 2004 [10].

A failure of deregulation

The dogmatic, business-friendly ideology pursued by David Cameron's Government fuelled the *Red Tape Challenge* – perhaps the high point of deregulation policy. In December 2014, Matthew Hancock, then the Business Minister, boasted that the 'government's war on red tape has saved business £10 billion over the last 4 years' [40]. Less than three years before the Grenfell Tower disaster of June 2017, Mr Hancock gushed about the future benefits for the construction industry, in transforming: 'housing standards: streamlining hundreds of technical housing standards to just 5 national standards, saving house-builders and councils £96 million per year' [40]. With respect to building standards, particularly fire safety standards as Lord [41] indicates – as highlighted by the Grenfell Tower tragedy – deregulation has proved a disaster. Contributory causes of the tragedy include the type of materials used for cladding the building and the methods used for fixing insulation panels. The testing of materials, conformity with building regulations, the adequacy of building regulations – and their suitability for use in high-rise flats – have also been problematic. A full explanation of how this disaster came about will no doubt be clear once the inquiry chaired by Sir Martin Moore-Bick has completed its work.

The architect Rowan Moore has written about the 'culture of incompetence' that led to the Grenfell Tower disaster, maintaining that this has tainted regulators, private companies, consultants and subcontractors in the construction industry. Commenting on the companies to whom work had been outsourced, he concludes:

> Their primary skills are not necessarily in the quality of their management or products, but in their abilities to drive down costs, navigate procurement processes and manipulate regulations. They become remote from the physical and human consequences of their actions.
>
> [42]

The reaction to Grenfell has included a return to more prescriptive forms of regulation. Preventative measures have been required for existing buildings, such as stripping out flammable insulation and re-cladding of flats above 18 metres in height to replace combustible materials; 'waking watches' being required around the clock in blocks of flats posing a fire risk; and the implementation of more rigorous electrical safety certification in rented dwellings.

The wider picture for deregulation

The broad effects of deregulation have gone far beyond the cutting down of superfluous and obsolete forms of regulation. The fabric and integrity of regulation have been damaged such that the quality of life for many people has deteriorated. Three areas of this corruption of the regulatory system are of particular concern. First, deregulation has compromised the standards of building materials, building quality and design, construction methods, and the supervision of construction projects. Second, building control, which has largely been privatised, has been weakened, leading to worsening housing standards. Third, as Clifford et al. have pointed out, extending the scope of permitted development rights to circumvent local planning regulation has been problematic, particularly for office-to-residential conversions [43].

The first two of these issues were considered in the review of high-rise building, set up by the Government in the wake of the Grenfell Tower disaster, chaired by Dame Judith Hackitt. A number of problems were highlighted in her Final Report. These include the lack of individual accountability and competency among duty-holders, 'silo-thinking' of regulators, a lack of coordination in guidance and documentation, an over-reliance on prescriptive guidance rather than taking a risk-based approach.[43] The recommendations included setting up a coordinating body – the Joint Competent

Authority – so that local authority building standards regulators (to supersede building control) would act in conjunction with fire regulators and the Health & Safety Executive. The recommended new framework would:

> Provide stronger oversight of dutyholders with incentives for the right behaviours, and effective sanctions for poor performance – more rigorous oversight of dutyholders will be created through a single coherent regulatory body that oversees dutyholders' management of buildings in scope across their entire life-cycle.
>
> *[44]*

The Final Report also advocates moving towards a more nuanced relationship between industry and regulators than that exemplified in the *Red Tape Challenge*.

> Moving towards a system *where ownership of technical guidance rests with industry* as the intelligent lead in delivering building safety and providing it with the flexibility to ensure that guidance keeps pace with changing practices *with continuing oversight from an organisation prescribed by government.*
>
> *([44] emphasis in original)*

The third area of concern highlighted here – extending the scope of permitted development rights – has also come up for criticism. Some of the conversions from offices to housing allowed under permitted development have created a toxic nexus of problems for their new residents, such as inadequate internal space, damp problems, noise, poor transport links, remoteness from essential services. Moore has said that many such developments have ended up creating nightmare environments in terms of crime, gang activity, trading in illegal drugs, poor amenities, public health issues and alienation [45]. This is another area where weak regulation has created foreseeable problems. Research sponsored by the Government to compare permitted development projects with comparable ones subject to ordinary planning permission, concluded that: 'permitted development conversions do seem to be more likely to create worse quality residential environments than planning permission conversions in relation to a number of factors vital to the health, wellbeing and quality of life of future occupiers' [46].

To conclude, deregulation has gone too far in many respects – its proponents having allowed themselves to be dazzled by the presupposed virtues of private enterprise with insufficient consideration being given to the drawbacks. There are *inherent* problems to dismantling complex regulatory systems which do not seem to be properly understood by dogmatic, starry-eyed policy-makers in Government. The economist Robert Skidelsky, reflecting on the causes of the 2008 economic crisis, opined: '*Any great failure should force us to rethink . . . the crisis was generated by the system itself and not by some external shock*' [47]. Whether such reflection occurs with deregulation policy is still too soon to tell.

Civil and criminal law and environmental health

It is often asserted by academic lawyers that the role of the criminal law in regulating business activity has been on the wane over the last few decades [17]. However for environmental offences [48] and for food safety and hygiene offences and health and safety offences [49], the publication of sentencing guidelines suggests support for increasing criminal penalties for more serious offences. Publication of such guidelines might suggest that the historic problem of light sentencing, even for fairly serious offences, has been addressed. The Sentencing Council has advocated that the cost to the community of regulatory breaches should be better reflected in sentencing decisions. These should take into account such factors as the extent of harm caused by the breach, the size and resources of the company involved, and the culpability of persons associated with illegal acts. This change in sentencing policy seems to be at odds with the peripheral position of criminal law often attributed by sympathetic commentators on deregulation, such as Dame Julia Black ([50] at p. 70):

> Although criminal law has traditionally been used extensively in the design of regulatory systems in the United Kingdom, in practice criminal law plays a peripheral and indirect role in its implementation. The Hampton Report, and the subsequent Macrory Review, recognise this, and both recommend moving away from criminal law as the primary, and often only, sanction for regulatory non-compliance. The reasons lie in its inappropriateness and ineffectiveness. Regulatory obligations are not seen by businesses, and indeed often courts, as conduct which is *mala in se*, but merely conduct which is *mala prohibita*, due largely to the fact that most regulatory offences are strict liability offences. As a result, the use of criminal law to express moral disapprobation is often misplaced, as it is shared neither by those subject to the regulation, nor by the courts.

This identification of a diminishing role for criminal law does not apply across the field of environmental health regulation. Different types of law come into play for the various areas of environmental health enforcement. These areas have their own legislative histories – some dating back to the latter part of the nineteenth century, when strict liability offences constituted the norm for regulatory enforcement. More recent areas of regulation operate differently, with licensing and permitting frameworks being prominent. These include aspects of administrative and civil law as well as criminal law. Thus regulation has different forms operating across the range of environmental health work. The main fields will next be considered.

Health and safety at work

A number of sources of law are involved with health and safety enforcement. The statutory framework is provided by the Health and Safety at Work Act 1974 (HASAWA), which applies both to local authorities in Great Britain and to the Health and Safety Executive (HSE).[44] Under section 18(4) of HASAWA, local authorities are required to make adequate arrangements for the enforcement in their area of the relevant statutory provisions. Section 18(4) also provides that local authorities should act in accordance with any guidance issued by the HSE.

The statutory provisions include regulations made by the Secretary of State under section 15 of the Act as well as enforcing general duties under sections 2–9. The general duty of care of employers, under section 2(1) of HASAWA, is: 'to ensure, so far as is reasonably practical, the health, safety and welfare at work of all his employees'. This wording of the statutory duty is drawn from common law requirements placed on employers: to provide and maintain a safe place of work with safe means of access and egress; safe appliances and equipment and plant for doing the work; a safe system of work and effective supervision; and competent and safety-conscious employees [9].

Regulations made under section 15 of HASAWA constitute a form of secondary legislation, which does not require parliamentary scrutiny as would primary legislation. The use by the Secretary of State of ministerial powers to make regulations is a convenient way of developing the statutory framework without having to amend the Act. The use by the Secretary of State of section 15 powers allows the legislation to keep up with changing demands and issues, as regulations may be added, modified, or withdrawn in relation to the statutory framework in flexible and responsive ways. Regulations can be very detailed and apply to specific sectors of industry, or apply only to certain parts of Great Britain, or may be made for a limited duration.

The Management of Health and Safety at Work Regulations 1999[45] consists of comparatively general regulations which supplement the provisions of the 1974 Act. For example, regulation 3 requires employers to make suitable and sufficient risk assessments concerning employees at work and other persons affected by the employer's undertaking. The 1999 Regulations have also amended earlier domestic regulations and provide a way of giving effect to EU law. Thus regulation 4, schedule 1 gives effect to the general principles of prevention set out in article 6(2) of the European Council Directive 89/391/EEC.[46]

Legal proceedings brought under section 33(1) of HASAWA include prosecution for failing to discharge a duty of care, contravening any requirement imposed by an improvement or prohibition notice, or for contravening any health and safety regulation. Section 33(1) offences include intentionally obstructing an inspector in the exercise of their powers and duties, making a false statement, or making a false entry in a document. Criminal prosecutions are normally brought for relatively serious or flagrant breaches, or where there is a history of regulatory non-compliance by a company or individual. For 2018–19, the number of prosecutions brought by the HSE and, in Scotland, by the Crown Office and Procurator Fiscal Service (COPFS) totalled 394. The rate of conviction for at least one of the charges brought was 92% [51].

Criminal prosecution is not the primary source of law for resolving health and safety problems. Most formal enforcement is initiated by service of improvement and prohibition notices (i.e. civil law instruments). For 2018–19, the total number of enforcement notices served by the HSE and local authorities in Great Britain was 11,040, of which 66% were prohibition notices. Local authorities served 2,263 of this total [51]. Civil administrative law is, therefore, of great significance in health and safety law. Service of improvement and prohibition notices is a powerful and usually an effective way of resolving health and safety problems. The proportion of notices that are breached and prosecuted in the criminal courts has, over the years, remained low. Appeals against service of improvement and prohibition notices may be brought before an employment tribunal, which has a civil jurisdiction.

Housing conditions

Local housing authorities (LHAs) are required to keep conditions in the housing stock under review and identify any action deemed necessary, including under the Housing Act 2004 and the Regulatory Reform (Housing Assistance) England and Wales Order 2002.[47] Like health and safety at work, regulating housing conditions forms another area where the issuing of administrative notices is the primary response of the enforcement authority ([10] at chapter 5).

Part 1 of the Housing Act 2004 is concerned with housing hazards in dwellings. The Housing Health and Safety Rating System (HHSRS) is designed to provide a means of identifying the greatest threats to health and safety arising from deficiencies in a dwelling. The 2004 Act refers only to Category 1 and 2 hazards, with the HHSRS as the prescribed means for identifying the category into which any hazards fall. Category 1 hazards are those with an HHSRS score of 1,000 or more; Category 2 hazards being those scoring under 1,000. Powers are available to LHA inspectors to serve administrative notices and orders in order to address various hazards. Putting right deficiencies giving rise to the hazard might involve serving an improvement notice (which can be suspended).[48] Where there is an imminent risk of serious harm, an emergency remedial action notice[49] or an emergency prohibition order[50] can be served for a Category 1 hazard. A prohibition order (which can be suspended) may be served to prohibit the use of specified premises or any part of those premises.[51] There is also a power to serve a hazard awareness notice to advise a person about the existence of a hazard, without requiring any steps to be taken.[52] A demolition order is available for a Category 1 hazard.[53] A clearance area declaration can be made where all the dwellings in an area have a Category 1 hazard, or a hazard that is harmful to the health and safety of the inhabitants.[54]

The enforcement framework in Part 1 of the Housing Act 2004 for dealing with housing hazards is based on civil administrative law. The purpose of the legislation is to provide LHAs with statutory powers to protect occupants in respect of twenty-nine HHSRS hazards ([10] at pp. 55–58). These powers are extensive and intended to cover a wide range of circumstances for the protection of occupiers of residential property from hazards – for owner-occupiers as well as for tenants. Appeals against service of notices are civil and may be made to the First-tier Tribunal in England (Residential Property Tribunal in Wales).[55] The scope for criminal prosecutions is limited to situations such as where rogue or criminal landlords seek to evade their responsibilities or place the health and safety of tenants at risk of significant harm.

Officers responsible for resolving housing problems need to be familiar with a number of regulatory approaches besides Part I of the Housing Act 2004 and use of the HHSRS.[56] Other provisions may be more appropriate than the Housing Act 2004 – these are mentioned later.

State of the premises – Premises in such a state as to amount to a nuisance or which are prejudicial to health are a statutory nuisance under section 79(1)(a) Environmental Protection Act 1990 (EPA). This provision requires the premises themselves (which are not restricted to residential property) to have fallen into a state of disrepair, dilapidation, or decay. As a result of such deterioration the property may suffer from damp problems or facilitate the habitation of vermin. Most state of the premises cases, therefore, concern the health limb of statutory nuisance [10]. Enforcement action by local authorities against landlords is available, as is prosecution

by a tenant against their landlord under section 82 EPA 1990. Availability of the nuisance limb means that the premises causing a nuisance to neighbouring property can be dealt with under the EPA 1990, something that is not possible under the Housing Act 2004 which is restricted to the subject property.

Tents, caravans, sheds – Section 268 of the Public Health Act 1936 defines as a statutory nuisance (under section 79(1)(h) EPA 1990) a tent, van, shed, or similar structure used for human habitation:

(a) which is in such a state, or so overcrowded, as to be prejudicial to the health of the inmates; or

(b) the use of which, by reason of the absence of proper sanitary accommodation or otherwise, gives rise, whether on the site or on other land, to a nuisance or to conditions prejudicial to health.

Drainage problems – Various statutory provisions are relevant to drainage problems. The Building Act 1984, section 59 provides a power to local authorities to require an owner to remedy defects in drains, cesspools, soil and rain-water pipes, sinks, and other appliances. Section 60 of that Act is concerned with the use and ventilation of soil pipes. The Public Health Acts of 1936 and 1961 contain various provisions for dealing with drainage problems. Thus, section 17 of the 1961 Act contains a power for the local authority to repair drains, water-closets, and waste or soil pipes.

Pests and vermin – Section 83 of the Public Health Act 1936 provides a power for a local authority to require an owner or occupier to take steps to cleanse and disinfect filthy or verminous premises. The Prevention of Damage by Pests Act 1949, section 4 is also available for local authorities to address problems resulting from rats and mice. Verminous premises could also be considered prejudicial to health – a statutory nuisance under section 79(1)(a) EPA 1990. The EHP needs to make a judgement as to which provision (including the Housing Act 2004) best fits the circumstances of the case in question.

Licensing and HMOs – Part 2 of the Housing Act 2004 provides the framework for the regulation of houses in multiple occupation (HMOs) by local housing authorities. The licensing regime has developed further since the 2004 Act came into force, with the Government using its powers to enact secondary legislation, mainly to extend the scope of regulation. Government has also provided local authorities with non-statutory guidance on licensing powers [52].

The licensing requirements in Part 2 are extensive, and this has not been an area of policy where the deregulation ideology has held sway. The conditions of licences are both extensive and prescriptive – as provided in section 67 and schedule 4 of the Act, setting out mandatory conditions. The Act provides housing authorities with wide discretionary powers, and a licence made under section 67(1):

> may include such conditions as the local housing authority consider appropriate for regulating all or any of the following –
>
> (a) the management, use and occupation of the house concerned, and
> (b) its condition and contents.

Breaches of licence conditions, allowing overcrowding in an HMO, and the failure to hold a licence where this is required expose landlords and their agents to the risk of either a criminal or a civil penalty. Section 72 of the Act sets down summary offences in relation to licensing. Although not imprisonable offences, magistrates' courts can hand down substantial fines on conviction for these offences for which no maximum amount applies.

Under section 70A of the Housing Act 2004, a LHA is under a duty to revoke a licence where a banning order is made under Part 2 Housing and Planning Act 2016. The

introduction in the 2016 Act of banning orders and the keeping of a register of rogue landlords and their agents are important ways of tightening regulation of the private rented sector.

Civil penalties – The power to impose a civil penalty as an alternative to prosecution was introduced by sections 23 and 126 and schedule 9 of the Housing and Planning Act 2016. Statutory guidance provided by the Government is available to local housing authorities concerning the use of civil penalties, including what level of penalty is appropriate for the case under consideration [11]. The LHA may impose a financial penalty of up to £30,000 if satisfied, beyond reasonable doubt, that the person's conduct amounts to a relevant housing offence. A relevant offence is defined in section 249A(2) of the Housing Act 2004 as:

- A failure to comply with an improvement notice under Part 1 of the Act;
- An HMO licensing offence (s.72);
- A licensing offence of houses regulated under Part 3 of the Act;
- A failure to comply with an overcrowding notice (s.139(7)); or
- A breach of management regulation in respect of HMOs (s.234).

Breach of a banning order made under the Housing and Planning Act 2016, section 21 can also be dealt with by imposing a civil penalty.

The same standard of proof (the criminal standard of beyond reasonable doubt) applies where the LHA decides to impose a civil penalty as for a prosecution. Therefore, the quality and sufficiency of the evidence should be the same for both routes. A person served with a notice of intent to impose a civil penalty can make representations to the housing authority; they also have a right of appeal where a final notice has been issued. Perhaps the main advantage, for the regulator, of choosing the civil route is that an appeal will be heard by the First-tier Tribunal, which has experience in dealing with housing matters. Imposing a civil penalty will also offer a quicker resolution than prosecution. The use of civil penalties to levy what can amount to substantial fines without due process – the need to go through the courts system – has raised concerns amongst lawyers and landlords. Interviewing landlords without the protections afforded by Code C of the Police and Criminal Evidence Act 1984 and imposing civil penalties where there is insufficient evidence that the occupation is truly of an HMO have been singled out as abuses perpetrated by some local authorities [53]. Local housing authorities can use the receipts of civil penalties to help fund its housing enforcement responsibilities, and this, as Oatt has said makes it an attractive alternative to prosecution [54].

Statutory nuisances

The statutory nuisance regime – as consolidated in Part III of the Environmental Protection Act 1990 (EPA) – continues to play an important role in the regulation of public health, housing, and nuisance problems.[57] Statutory nuisances are set down in section 79(1) of the Act. Most forms can exist on residential or commercial premises, though some are limited to statutory nuisances arising on industrial, trade, or business premises. Despite its origins in legislation dating back to the mid-nineteenth century and overlapping with more recent statutory provisions, this regime remains a widely used and locally enforced form of environmental regulation [55].

Section 79 EPA 1990 places local authorities in Great Britain under a general duty to inspect their areas from time to time to detect any statutory nuisances. In addition, they are under a duty 'to take such steps as are reasonably practicable to investigate' complaints made by persons living in their area.[58] Where a situation is believed by the local authority to amount to a statutory nuisance – because it is prejudicial to health and/or a nuisance – the

authority is placed under an obligation to serve an abatement notice on the person responsible. In cases where the nuisance arises from a defect of a structural character, service is on the owner of the premises. If the person responsible for the nuisance cannot be found or the nuisance has not yet occurred, the abatement notice should be served on the owner or occupier of the premises. Service of an abatement notice is an administrative action, which can be challenged by the recipient making a civil appeal to the magistrates' court.

The criminal law comes into play for breach of an abatement notice – a criminal offence though not an imprisonable one. Criminal proceedings are summary only, brought by the local authority under section 80(4) of the EPA 1990. An offence is also committed by any person who wilfully obstructs an officer exercising statutory powers to ascertain whether a statutory nuisance exists. Obstructing an officer taking any action or executing any work authorised or required by Part III is an offence.[59] Appeals against conviction and/or sentence in the magistrates' court may be made to the Crown Court.

Other powers available to the local authority under Part III of the EPA 1990 are civil in nature. Where an abatement notice has not been complied with, a local authority may 'abate the nuisance and do whatever may be necessary in execution of the notice' and recover its costs from the person responsible.[60] This power – including doing works in default – is available whether or not there is a prosecution for breach of the notice.

The power to seek a High Court injunction to secure the abatement, prohibition, or restriction of the nuisance is available under section 81(5) of the Act. The local authority can apply for an injunction where it forms the opinion that the procedure for prosecuting breach of an abatement notice would provide an inadequate remedy. It would need to convince the High Court that the perpetrator of the nuisance had been 'deliberately or flagrantly flouting the law or that only an injunction will stop their illegal or potentially illegal activities'.[61]

Is the concept of statutory nuisance still relevant today? – The statutory nuisance regime is based on two separate limbs – common law nuisance and matters prejudicial to health. It originated as a measure of sanitary reform during the latter phases of the Industrial Revolution, at the same period as state regulation to control the most damaging types of industrial pollution came into effect [56]. Both measures were pioneering attempts to mitigate the adverse consequences brought about by industrialisation and the uncontrolled growth of towns and cities (see chapter 3 of [7]).

The nuisance limb is based on two separate, long-established common law torts: private nuisance and public nuisance. The flexibility of private nuisance is derived from the legal concept of unreasonable interference in the use or enjoyment of neighbouring land.[62] Nuisance cases are fact-sensitive and the legal concept resistant to precise definition; as stated in Clerk & Lindsell: 'the essence of nuisance is a condition or activity which unduly interferes with the use or enjoyment of land' [57]. Such flexibility facilitates development in the common law of nuisance, which statute law is able to incorporate as one of the limbs of statutory nuisance, in Part III EPA 1990 [55]. The result may not be elegant, and complexity results from the overlaps between the statutory nuisance regime and more recent legislation. For example, many noise nuisances investigated by local authorities are behaviour driven, sometimes including elements of deliberate or malicious acts intended to punish residential neighbours. Depending on the circumstances, such instances may amount to statutory nuisances and should be regulated under the EPA 1990, or may be dealt with as anti-social behaviour, using powers provided by the Anti-social Behaviour, Crime and Policing Act 2014. Enforcement officers need to be familiar with the legislation as a whole and be fully aware of the extent of their powers under both Acts if they are to carry out their responsibilities competently [58].

The relevance of statutory nuisance to the present day is borne out by the continuing use of some of its earliest forms. Three types of statutory nuisance – arising from the state of the premises, the way animals are kept, and accumulations and deposits – were set down in the 1855 Nuisances Removal Act, and remain on the statute book, using virtually the same wording, in section 79(1) of the EPA 1990. These types of problem form the basis of complaints still being made to local authorities, and the statutory nuisance procedure will often be used in order to deal with them.

Flexibility in the legislative framework is also shown by additions made in more recent times to the list of statutory nuisances set down in section 79(1) of the Act. Noise emitted from premises – originally a statutory nuisance for the whole of Great Britain in the Noise Abatement Act 1960 – was consolidated as section 79(1)(g) in the EPA 1990. Section 2 of the Noise and Statutory Nuisance Act 1993 added noise 'emitted from or caused by a vehicle, machinery or equipment in a street', as section 79(1)(ga) EPA 1990. The viability of employing the noise nuisance provisions of the EPA 1990 is further evidenced by the reluctance among local authorities to utilise the Noise Act 1996 to deal with one-off, night-time noise emitted from domestic or licensed premises [7, 82].[63] More recently, sections 101 and 102 of the Clean Neighbourhoods and Environment Act 2005 added nuisances from insects and from artificial light to the list of statutory nuisances in section 79(1) EPA 1990.

Prosecution by a 'person aggrieved' – Criminal proceedings can be brought in the magistrates' court (in Scotland, the sheriff) by a 'person aggrieved' against the person responsible for a statutory nuisance. In these private prosecutions brought under section 82 of the Act, the court may grant the person aggrieved a nuisance order and may also impose a fine when making the order. A further offence is committed if the nuisance order is subsequently breached (see chapter 18 of [7]). The availability of such an action under section 82 could be seen as demonstrating the continued vitality of statutory nuisance. This action has been criticised, however, with regard to the poor condition of rented housing. The authors of a report prepared for *Shelter* have concluded that prosecution of a landlord by the tenant to improve the state of the premises (section 79(1)(a) EPA 1990) has become obsolete, calling for new legislation to improve tenants' rights. The authors criticise 'the complexity and risks involved in its use, and the fact that it is adjudicated upon by the Magistrates Court which has very limited expertise in housing' [59].

The restricted health limb – The health limb applies to all the statutory nuisances set down in section 79(1) EPA 1990. But in practice the circumstances in which it applies are limited. The state of the premises – section 79(1)(a) EPA – generally engages the health limb, as may cases concerning accumulations and deposits – section 79(1)(e). Statutory nuisances involving insects, atmospheric pollution, or watercourses may also engage the health limb.

But the scope of the health limb has been limited by case law. In *Everett*[64] the Court of Appeal, in a split decision, decided that a steep and dangerous staircase did not come within the scope of the health limb as regards the statutory nuisance pertaining to the state of the premises. The majority decided that public health legislators in the nineteenth century were intent on controlling infectious and contagious diseases, but were not concerned about the risk of physical injury linked to the state of premises. A majority of their lordships were influenced by what they thought Parliament had intended when the wording of this subsection first received parliamentary scrutiny, in 1855.[65] Their lordships were also mindful that a local authority has powers under other legislation to deal with the type of health risk in question. So, in a

similar case today, officers should be looking to the Housing Act 2004 for a resolution of this kind of problem.

The scope of the health limb was subsequently considered in *Birmingham City Council v Oakley*.[66] This case concerned a dwelling rented to the Oakley family by the local authority and due for demolition in an area soon to be redeveloped. The internal arrangement and facilities in the property were below modern standards. A lavatory lacking a wash-hand basin was located next to the kitchen. In order to wash their hands after using the toilet, members of the family would either have to pass through the kitchen into the bathroom, which led off from the other side of the kitchen, or wash their hands in the kitchen sink. A majority of the House of Lords in *Oakley* agreed that there was a fundamental distinction to be drawn between the way the premises are used and the 'state of the premises' in the context of the statute. This was not a case where the existing facilities had fallen into a state of disrepair, so causing a risk to health. It was the use of the premises that posed the health risk, not the actual state of the premises, and this did not come within the scope of the statute. This decision means that problems arising from the internal layout and poor design of premises do not come within the scope of the health limb in section 79(1)(a) EPA 1990. *Oakley* is binding for subsequent decisions of the courts and this decision severely curtails the use of the statutory nuisance regime for improving the quality of housing. Powers available under Part 1 of the Housing Act 2004 should be sufficient to deal with defects in residential property caused by obsolescent layout and facilities.[67]

Noise and the health limb – Use of the health limb for regulating statutory noise nuisances is rare. Local authorities seem loath to use it and there are no reported cases on the use of the health limb for noise nuisances under sections 79(1)(g) or (ga) of the EPA 1990. Is the health limb redundant therefore? During the final reading of the Environmental Protection

Bill, the Government saw the inclusion of an 'injury to health' limb as an important addition to the law of statutory noise nuisance – the earlier noise provision (Control of Pollution Act 1974, section 58) having been solely concerned with nuisance.[68]

The ingredients required to prove prejudice to health may be more problematic compared to nuisance. Noise nuisances will rarely be serious enough to cause actual injury or the risk of injury to health, and as Penn [60] points out proving either of these is likely to be difficult for local authority regulators. An exception to this are cases involving low frequency noise, where nuisance may be difficult to prove where the sound is not readily audible. Sometimes health effects are only discernible over a long period of time. In this situation, proceeding early on in a case when exposure to nuisance can be established, but without any health effects being substantiated, could offer a better approach for the local authority to take. The World Health Organization's definition of health – dating from the 1940s – is extremely broad: 'a state of complete physical, mental and social well-being and not merely the absence of disease or infirmity' [61]. Guidance from Government to promote this positive norm for public health would be needed before local authorities are likely to be tempted to utilise the health limb more often. That said, the Housing Act 2004 defines health as including mental health.

Nuisance and planning – The flexibility of the statutory nuisance regime has meant that it can provide some redress for problems associated with other statutory regimes. Many EHPs will have experience of using statutory nuisance powers to deal with problems resulting from poor decisions made at the planning stage of a development. The grant of planning permission will sometimes result in noise and dust complaints being made during construction. The problems may be exacerbated where the conditions (if any) attached to the planning permission prove to be inadequate or ineffective. Both statutory nuisance

powers and those provided in sections 60 and 61 of the Control of Pollution Act 1974 enable local authorities to control noise emissions from construction sites. But such controls will be more limited than would have been the case had competent decisions been made at the planning stage.

Serious noise problems may become apparent only when the development is completed. Such problems might have been anticipated at the planning stage and might have been avoided with a change in design or specification of the building. But things that might have been or should have been done are sometimes left undone. Where planning regulation has failed to provide sufficient protection and a problem reaches the nuisance threshold, some mitigation might be realised through the use of statutory nuisance powers.

Noise nuisance problems might arise for new occupiers after completion of their housing development, built near to existing commercial or industrial premises. Here, the creation of a nuisance situation by the new development under the 'agent of change' principle might have been avoided had better standards of noise protection been required for the new building. A private nuisance action or the use of statutory nuisance powers may provide some mitigation of the problem. But the standard of protection that can be provided by nuisance actions will usually not be as good as it would have been had competent decisions been made at the planning stage.

Planning law provides only very limited grounds for appealing or judicially reviewing poor planning decisions. Other areas of law can partially fill this void, and rights available in these areas are not extinguished by the grant of planning permission. Common law rights in nuisance are among the most important. The Supreme Court in *Coventry v Lawrence*[69] affirmed the decision of the Court of Appeal in *Allen v Gulf Oil Refinery*[70] that a planning authority has no jurisdiction to authorise a nuisance. An action in private nuisance or regulatory action taken by the local authority under section 80 EPA 1990 to abate, prohibit, or restrict a nuisance is not vitiated by the grant of planning permission.

Anti-social behaviour

The Anti-social Behaviour, Crime and Policing Act 2014 provides various powers enabling local authorities, the police and social landlords to deal with anti-social behaviour (ASB) caused by individuals or by bodies (including businesses). A local authority can exercise discretionary powers under the Act which does not impose compulsory duties on them.

Many councils receive complaints involving behavioural forms of nuisance and this means they should consider possible overlaps between ASB and statutory nuisance. The 2014 Act is also concerned with housing-related nuisances, so there is an important connection with the management of social housing. Practitioners need to be familiar with the legislation in both these areas, not only because of legislative overlaps but also because different remedies are available depending on the statute being utilised. Practitioners should have regard to the latest statutory guidance on the use of ASB powers [62] and the advice published by the Chartered Institute of Environmental Health [58].

Community protection notices – Community protection notices (CPN) are an important enforcement tool that can be used by authorised council officers, police officers, and by social landlords (if designated by the council). These administrative notices are intended to deal with repeated or ongoing conduct – not occasional or 'one-off' incidents. The test for ASB is set out in section 43 of the 2014 Act. This requires that the investigating officer is satisfied, on reasonable grounds, that the conduct of the individual or body (a business, company, or other organisation):

- Is having a detrimental effect on the quality of life of those in the locality; and
- Is unreasonable; and
- The behaviour is of a persistent or continuing nature.

Whilst there is the possibility of conflict between the ASB provisions of the 2014 Act and the statutory nuisance provisions of the EPA 1990, the legal tests applying to the two regimes are different. The threshold for statutory nuisance is a high one. Under the nuisance limb of statutory nuisance, material and substantial interference with personal comfort is the test. Mere annoyance would fall below the threshold for nuisance but would be enough in a case of ASB, provided that all the ingredients of section 43 were made out.

A written warning must be issued to the person or body believed to be responsible for the anti-social behaviour prior to issuing a CPN. Where a written warning has not been heeded, a CPN may be issued to an individual (over the age of 16) or to a body whose conduct meets the criteria specified in section 43 of the 2014 Act.

Whereas issuing a CPN is a civil administrative procedure, breach of a notice is prosecuted in the magistrates' court under its criminal jurisdiction. Failure to comply with a CPN is a summary offence under section 48. The offence is punishable on conviction: (a) in the case of an individual, by a fine not exceeding level 4 on the standard scale, or (b) in the case of a body, an unlimited fine. The offence is one of strict liability subject to the statutory defences that:

- The defendant took all reasonable steps to comply with the CPN; or
- There was some other reasonable excuse for the failure to comply.

Anti-social behaviour and housing – A CPN may be issued in a housing context against those responsible for ASB, such as: tenants, occupiers or other persons, including corporate bodies [10]. Rogue landlords and property agents can be issued with CPNs, provided the section 43 criteria are met. With landlords and agents, if the CPN is breached and the person convicted, a local authority (in England) can apply to the First-tier Tribunal for a banning order under section 15 of the

Housing and Planning Act 2016. The effect of this order is that a person could be banned, for a specified period of at least 12 months, from letting housing or being engaged in letting, agency, or property management work.

Closure notices and closure orders – Authorised council and police officers can close premises, including residential premises, that are causing a nuisance or are associated with disorder. The powers are set down in sections 76 to 93 of the 2014 Act, and explained further in the Home Office guidance [62]. The closure notice procedure can be employed without needing to go to court. A notice is limited in duration to 24 hours, with the possibility of extending it to a maximum of 48 hours. Following the issuing of a closure notice, an application must be made to the magistrates' court for a closure order in order to extend the terms to a maximum of six months. If an application for a court order is not made the closure notice expires.

The test for issuing a closure notice is where the local authority or police are satisfied on reasonable grounds:

- That the use of particular premises has resulted, or (if the notice is not issued) is likely soon to result, in nuisance to members of the public; or
- That there has been, or (if the notice is not issued) is likely soon to be, disorder near those premises associated with the use of those premises; and
- That the notice is necessary to prevent the nuisance or disorder from continuing, recurring, or occurring.

Injunctions – Applications for an injunction under the Anti-social Behaviour, Crime and Policing Act 2014 can be made either to the county court or the High Court. Injunctions to control ASB are intended to be used as a primary remedy under this legislation, rather than where other remedies have been tried but without success [62]. Injunctive proceedings are civil, even though breach of an injunction order invokes criminal penalties:

an unlimited fine and/or up to two years' imprisonment for adults. In respect of individuals aged from 10 to 18, applications should be made to the youth court.

As provided in section 2(1)(a) of the 2014 Act, the test for an ASB injunction in a 'non-housing related' context is: 'conduct that has caused, or is likely to cause, harassment, alarm or distress to any person'. This sets quite a high threshold of harm; it applies to more serious forms of ASB, akin to what would be required in a criminal offence. It is applicable where there is a relatively serious ASB problem in a public place, such as a town or city centre, shopping mall, or local park.

For ASB in a housing context the 'nuisance or annoyance' test will apply, and this sets a lower threshold compared to ASB in a public place. The test for an injunction in a housing context is where the conduct: 'is capable of causing nuisance or annoyance to a person in relation to that person's occupation of residential premises'.[71] This test applies not just to occupiers of residential premises but also to any person affected by the housing-related nuisance or annoyance. In a housing context, therefore, it is permissible to seek an injunction to control behaviour falling below the nuisance threshold. This will be appropriate if there are reasonable grounds to believe that the behaviour is likely to worsen over time unless checked by an injunction.

Food law

As Malcolm and Pointing have said, food law in the UK comprises a complex mix of primary legislation and detailed regulations [8]. Together with associated EU regulations, the General Food Law of the European Union provides the overall framework for food law for most of Europe, including the UK. The General Food Law is set down in Regulation (EC) 178/2002.[72] The free movement of safe and wholesome food, a high level of protection of human life and health, and consistency between Member States regarding food safety requirements form the principal objectives of the General Food Law.

The Food Safety Act 1990 is the primary UK legislation dealing with such matters as enforcement measures, offences, and penalties. Breaches of food safety law are criminal offences based on strict liability, though a due diligence defence is available under section 21. All provisions of this Act have to be consistent with the General Food Law [63]. Section 16 of the Food Safety Act provides a power for the Secretary of State to make regulations pertaining to the production, marketing, labelling, and sale of food, including requirements ensuring food safety and food hygiene. Regulations made under section 16 include the Food Hygiene Regulations.[73] These are comprehensive and deal with the detailed requirements for operating food businesses, as well as setting out the offences and penalties for regulatory breaches.

Fraud offences, requiring proof of criminal intent (*mens rea*), can sometimes be used in serious food crime cases, but as this author has said such prosecutions are rare [64]. Convictions for fraud, where individuals are found to have intentionally or recklessly placed the health of the public at risk, will usually result in prison sentences. The maximum prison sentence for a conviction in the Crown Court under the Fraud Act 2006 is ten years (compared to a maximum of two years in the Crown Court for a Food Safety Act offence).

Most offenders will be prosecuted in the magistrates' courts for food safety offences, which will usually be dealt with by fines. Where offences are 'either-way', trials can be held before a jury in the Crown Court for more serious offences. An upper limit for fines in the magistrates' courts no longer applies for the main food safety offences, including breaches of the Food Safety and Hygiene Regulations.[74] This reflects a recent trend towards increasing the levels of fines in proportion to the harm resulting from the offence, the size of the business, and the extent of the illicit gains from offending [49].

Implications of the UK leaving the EU – The European Union (Withdrawal Agreement) Act 2018 provides that directly applicable EU legislation is retained as UK legislation after exit day – 31 December 2020. The General Food Law is thus retained as UK legislation and continues using the same numbering and year as assigned by the EU. Similarly, associated EU regulations are retained legislation.[75]

The retained EU regulations in force in the UK on 1 January 2021 were amended by the General Food Law (Amendment etc.) (EU Exit) Regulations 2019.[76] Any references made in the legislation to the 'Community' are replaced with the 'United Kingdom'. The 2019 Regulations also repealed parts of the General Food Law referring to EU institutions and inserted new powers for the Secretary of State to make regulatory provisions for England by Statutory Instrument (with similar powers being given to the devolved authorities). Regulation 6(b) of the 2019 Regulations disapplies any effect of the European Food Safety Authority for the UK. Any further amendments to retained EU legislation will have to be made by UK legislation. This means that any subsequent changes to EU food law made by the European Parliament and the Council of the European Union need to be reflected by new UK legislation if consistency between the two jurisdictions is to be maintained.

Selling food not complying with the food safety requirements – Section 8 of the Food Safety Act 1990 remains the most important offence under the Act. It should be read in the light of article 14 of Regulation (EC) 178/2002. Article 14(1) states that 'food shall not be placed on the market if it is unsafe'. Food is deemed to be unsafe if it is injurious to health or unfit for human consumption. Either of these will amount to a breach of the food safety requirements. A person will commit an offence under section 8(2) of the Food Safety Act if he or she sells: 'food [which] fails to comply with food safety requirements. . . [that] is unsafe within the meaning of Article 14'.

Article 14 does not refer to the 'sale' of food but uses a wider concept: food which is 'placed on the market'. This is defined in article 3 of Regulation (EC) 178/2002 as the holding of food or feed for the purpose of sale. It includes offering food for sale or any other form of transfer and applies whether free of charge or not. Sale, distribution, and other forms of transfer are included. All occasions when food is supplied are covered, except private domestic consumption which is exempted under article 1(3).

Adulteration – Under section 7 of the Food Safety Act 1990, an offence is committed by:

> Any person who renders any food injurious to health by means of any of the following operations, namely –
>
> a) adding any article or substance to the food;
> b) using any article or substance as an ingredient in the preparation of the food;
> c) abstracting any constituent from the food, and
> d) subjecting the food to any other process or treatment, with intent that it shall be sold for human consumption.

Not all forms of adulteration come within the scope of section 7. It does not include adding water that is not contaminated to food, as occurs in the processing of chickens to improve palatability (see [65] at chapter 19).

A section 7 offence is committed when the food is adulterated with intent that it be sold (that is placed on the market) for human consumption. The intent required pertains to the sale of food for human consumption; it is not necessary for the prosecutor to prove intention to adulterate the food, which may have occurred accidentally. There is a rebuttable presumption that where food is found in the possession of a person involved in its processing then that person intends it to be sold, or otherwise placed on the market, for human consumption.

Consumer protection offences – Section 14(1) of the Food Safety Act 1990 provides that: 'Any person who sells to the purchaser's prejudice any food which is not of the nature or substance or quality demanded by the purchaser shall be guilty of an offence'. Section 15 of the Act creates an offence of falsely describing or presenting food. Both sections 14 and 15 refer to the 'nature, substance, or quality' of the food. These are separate elements that correspond to separate offences. In drafting an information – the document that sets out the legal and factual basis for the prosecution – these separate ways of committing the offence should not be rolled up by the prosecutor into a single charge.

Article 16 of Regulation (EC) 178/2002 deals with the labelling, presentation, and advertising of food (or feed) and prescribes that these shall not be undertaken so as to mislead consumers. It includes such matters as: the shape, appearance, type of packaging, and packaging materials used; the setting in which the food is displayed; and the information which is made available about the food. This article could be invoked where consumers are likely to be misled by packaging. Examples of this include where the meat was falsely described as Halal. It could also apply where meat was labelled as the product of an animal that had been slaughtered without stunning when in fact the animal had been stunned prior to slaughter [66].

Breaches of the Food Safety and Hygiene Regulations – The Food Safety and Hygiene Regulations form a complex and detailed body of secondary legislation that can be amended in response to the changing requirements of food law. The current regulations in force, in England, are the Food Safety and Hygiene (England) Regulations 2013.[77] Regulation 19 provides that 'any person who contravenes or fails to comply with any of the specified EU provisions commits an offence'. Similar regulations are in force in other parts of the UK, made by the devolved legislative bodies of Scotland, Wales, and Northern Ireland.

An administrative power, under regulation 6, enables an authorised officer to serve a hygiene improvement notice on the operator of a food business. Breach of a notice is an offence. The operator of a food business convicted for an offence under these Regulations may also be subjected by the court to a hygiene prohibition order, under regulation 7. Making such an order depends on whether the health risk condition is fulfilled – a risk of injury to health, whether temporary or permanent. The scope of the health risk applies to the running of the business, the preparation of food, the state of the premises, and the equipment used for food preparation. The effect of making such an order is that, until the problem is rectified, a process, treatment, or piece of equipment, or the premises, cannot be used for a food business. Under regulation 7(4), the court may 'impose a prohibition on the food business operator participating in the management of any food business, or any food business of a class or description specified in the order'. The powers of prohibition available to the court on conviction for an offence are therefore considerable under Food Safety and Hygiene Regulations.

Waste on land – fly-tipping and littering

A report commissioned by the Environmental Services Association has estimated that the economic impact of waste crime in England, in 2015, was at least £604m [67]. It paints a picture of a worsening as well as a complex problem that has proved elusive to effective regulation. The report classifies waste crime into the following categories, in which fly-tipping has had the greatest economic impact:

- Illegal waste sites;
- Illegal burning of waste;
- Fly-tipping;
- Misclassification and fraud;
- Serious breaches of permit conditions, including the abandonment of waste;
- Illegal exports of waste [67].

Waste on land is regulated under Part II of the Environmental Protection Act 1990 (as amended). Responding to the 'polluter pays principle', a duty of care is imposed by the Act on importers, producers, carriers, keepers, treaters, and disposers of waste. The Environmental Agency (in England) is responsible for regulating the more serious problems; local authorities are involved in investigating smaller-scale incidents of fly-tipping, carrying out clear-ups, and prosecuting offenders. Local authorities are also responsible for enforcing the littering provisions, under Part IV of the EPA.

Most waste offences set down in section 33 of the EPA consist of acts carried out without there being an environmental permit in place or as breaches of licensing requirements. This orientation of waste offences towards administrative controls reflects a 'better regulation' policy: to simplify licensing requirements for businesses and to treat them as 'partners' who are willing to cooperate with their regulator [68]. Environmental permitting provides a unified system of licensing, which straddles the control of waste on land and the regulation of industrial processes having the capacity to pollute the environment.

Because of its links to environmental permitting, fly-tipping is considered in more detail in the *Administrative methods of regulation* section. This is not an entirely satisfactory situation because there is not always a sharp divide between fly-tipping and littering – exclusively a local authority concern. Furthermore as Defra reports fly-tipping covers a wide range of activities, ranging from the deposit of a sack of household waste to major incidents of lorry-loads of dumping carried out by organised criminals (who may or may not be operating under an environmental permit) [69]. Eccles and Pointing suggest such crimes do not fit very well with a regulatory framework based on licensing, imbued with the shared values of trust and cooperation presumed by better regulation apologists [30].

Some other forms of regulation are relevant to fly-tipping and littering [70–71]. The oldest of these is the statutory nuisance regime. Section 79(1)(e) EPA 1990 provides that 'any accumulation or deposit which is prejudicial to health or a nuisance' shall be a statutory nuisance. A statutory nuisance may arise on residential or commercial property. If prejudicial to health a statutory nuisance may exist on land open to the public or on the highway. In *Williams* [2001], section 79(1)(e) was used to regulate an accumulation of refuse that included carpets, soil, old furniture, and garden waste which was a potential harbourage for rodents and flies.[78] Where an abatement notice has not been complied with a local authority may 'abate the nuisance and do whatever may be necessary in execution of the notice' and recover its costs of doing any works from the person responsible.[79]

More recent waste regimes usually take preference over statutory nuisance. Section 79(10) EPA 1990 sets a boundary between these. If the activity causing the problem is controlled under Part I of the EPA 1990 or by regulations made under section 2 of the Pollution Prevention and Control Act 1999, summary proceedings may not be instituted under the statutory nuisance regime without the consent of the Secretary of State. This restriction only applies to criminal proceedings: to prosecution for breach of an abatement notice. Section 79(10) is not applicable to service of an abatement notice because this is an administrative process.[80]

Littering – Part IV of the EPA 1990 is concerned with regulating litter deposited in open places. Part IV is focussed specifically on littering, so it will be more appropriate to use it rather than statutory nuisance in most circumstances. It is also more straightforward to use for regulators, as it is based on simple strict liability offences. Under section 87 of the Act (as amended): 'A person is guilty of an offence if he throws down, drops or otherwise deposits any litter in any place to which this section applies and leaves it'. This is a widely drawn offence, which applies to any place 'which is open to the air', unless 'the public does not have access to it, with or without payment'. Litter deposited in water as well as on land is caught by this provision. Section 87 also states that no offence is committed if it is 'done by or with the consent of the owner, occupier or other person having control of the place where it is deposited'. This means that the local authority cannot enforce this provision on private premises, such as supermarket car parks, unless the company owning the premises gives it permission.

Dumping – The Refuse Disposal (Amenity) Act 1978 contains provisions with respect to unauthorised dumping. Section 3 gives the local authority the power to remove abandoned vehicles from land 'open to the air' or 'forming part of the highway'. Section 2(1) is more widely drawn and makes it an offence for any person who:

(a) abandons on any land in the open air, or on any other land forming part of a highway, a motor vehicle or anything which formed part of a motor vehicle and was removed from it in the course of dismantling the vehicle on the land; or

(b) abandons on any such land any thing other than a motor vehicle, being a thing which he has brought to the land for the purpose of abandoning it there.

Offenders may be prosecuted for these littering and dumping offences, or served with fixed-penalty notices.

Other legislation – Littering is an annoying and anti-social activity that could be seen as coming within the scope of the Anti-social Behaviour, Crime and Policing Act 2014. Using enforcement powers provided by the 2014 Act may be considered, particularly where the anti-social behaviour includes additional elements to littering. The power under the 2014 Act to seek an injunction to control the future behaviour of a person might be a reason for choosing this route.

Section 215 of the Town and Country Planning Act 1990 enables local planning authorities to issue a notice to landowners requiring land or buildings to be cleaned up if their condition adversely affects the amenity of their area. Authorities can also enter land and clear it, and recover their costs of doing so from the owner.

Administrative methods of regulation

This section deals with areas of regulation where the emphasis is towards administrative/civil forms of regulation. Environmental health law is not structured with a simple boundary between criminal law and administrative/civil law. Modern legislation, dating from 1974 with the enactment of the Health and Safety at Work Act and the Control of Pollution Act, tends to have a 'hybrid quality', in which aspects of criminal law, administrative law, and civil law are all present. The balance of these is not the same between the different statutes and there may be advantages in moving away from imposing criminal penalties in favour of a more nuanced approach based on civil sanctions. The role of civil sanctions has been described by a leading exponent in the following terms:

> Where regulatory non-compliance occurs, sanctions can ensure that businesses that have saved costs by non-compliance do not gain an unfair advantage over businesses that are fully compliant. Where breaches result in damage or other costs to society, sanctions can assist in ensuring that those in breach provide proper recompense. Sanctions can equally represent a societal condemnation of the regulatory breach, acting as a deterrent to the sanctioned business against future breaches, and sending a wider message to the regulated sector.
>
> *([72] at para. 1.12)*

Since Professor Macrory completed his work on civil sanctions, the recommended levels of fines for unlawful acts and omissions have taken an upwards trajectory, with a wide range to fully reflect the circumstances of infraction. Sentencing guidelines for health and safety, food safety, and environmental offences include recommendations for substantial fines in certain circumstances [49, 73]. Civil penalties, based on a company's annual turnover or on the financial benefit gained by the infraction, may also involve substantial sanctions [72]. But their role is compensatory rather than the more punitive aspect accruing to criminal fines.

Contaminated land

The contaminated land regime, established by Part IIA of the Environmental Protection Act 1990, is based on a civil law approach to regulation. Its guiding principles include the 'polluter pays principle' and the requirement for remediation to be carried out according to a 'suitable for use' approach. Land may have been contaminated by historic pollution – layers of contamination caused or allowed to happen by a number of users over a long period. Allocating who bears responsibility for what pollution over a long period is a complex task, difficult to prove even to a civil standard of proof. With historic pollution, the company responsible for causing the pollution may no longer exist as a corporate entity or in a form which allows the costs of their polluting activity to be paid. Statutory guidance provided by Government for local authorities [74] may assist, but, in practice, liability cannot be established in many cases, in which case the costs of remediation may be borne by the State.

The local authority is the key regulator for implementing the regime for most sites. (The Environment Agency is responsible for 'special sites'.) The central objective of remediation is to bring contaminated land back into a state in which it can serve a useful purpose that avoids the harms caused by the contamination. The powers available to the local authority to achieve this objective go beyond the service of remediation notices on those found responsible for the contamination. Service of a remediation notice may occur towards the end of a long process of investigation and determination of liability by the regulator. But other ways of resolving the problem must be explored first. A local authority will need to give consideration to any voluntary agreement to remediate the land by a developer wishing to redevelop it. Where there is a potential for redevelopment, grants or tax breaks may be available to the developer who is prepared to take on the risks of remediation. The local authority needs to ensure that any scheme of voluntary remediation being proposed is of a sufficient standard and quality. Voluntary agreements to remediate will often mesh with an application for planning permission in order to redevelop a contaminated site, so cooperation between developer and regulator is both desirable and encouraged by this legislation.

Licensing

The exercise of administrative controls – such as registration, prior authorisation, and licensing – over activities that are potentially harmful usually involves the regulator giving some form of permission to a business entity. Compliance with such administrative requirements ensures that the activity taking place occurs within certain boundaries and is lawful. Compliance may also protect the regulated entity from other forms of regulation. An example of this would be the control of noise emitted from a construction site. Construction sites that are subject to section 60 notices served under the Control of Pollution Act 1974 (COPA), or where prior consent is obtained under section 61 of the Act, will be protected from local authority action founded upon statutory nuisance ([7] at pp. 121–122). Should the noise emitted from a construction site reach the nuisance threshold – perhaps because it turned out to be more problematic than envisaged when COPA regulation had been put in place – the local authority will not also be able to serve a statutory nuisance abatement notice to reduce noise levels.

Premises licensing and public nuisance – The Licensing Act 2003 was enacted in the spirit of deregulation, to replace the prescriptive licensing regime first put in place by the Licensing Act 1872. The 2003 Act established a single regime for licensing premises used for the sale or supply of alcohol, including late night refreshment, and the provision of regulated entertainment. Subsequent relaxation of controls has included the suspension of licensing conditions pertaining to musical entertainment in venues having a capacity of less than 500 persons.[81]

Since the 2003 Act came into force, local councils, acting as the licensing authorities, must carry out their functions with a view to promoting the four statutory licensing objectives:

- The prevention of crime and disorder;
- Public safety;
- The prevention of public nuisance;
- The protection of children from harm.

Noise and anti-social behaviour pose significant problems associated with the drinks and entertainments industry, which licensing is intended to keep within acceptable boundaries for local residents and the public generally. With regard to the prevention of public nuisance, as Horrocks and Pointing say, the local authority's environmental health service plays a pivotal role [75]. It can make representations on new applications on the grounds of the noise impact and can object in principle; it can ask for suitable conditions to be attached to any premises licence. The service can, additionally, ask for a review of any licence.

The licensing authority may:

- Grant the premises licence subject to conditions deemed necessary to promote the licensing objectives;
- Exclude any of the licensable activities from the scope of the premises licence;

- Refuse to specify a person as the premises supervisor; or
- Reject the application in its totality.

These powers of the licensing authority are considerable and demonstrate that administrative controls can be an effective way of encouraging the proper running of businesses that avoids causing unreasonable annoyance to local residents and the wider public. Setting out explicit licensing objectives does have its drawbacks, however. Thus, public nuisance, which retains its common law meaning in the Licensing Act 2003 [76] sets a very high evidential threshold for nuisance. This makes it less effective for dealing with noise nuisance problems than the Government may have originally intended; it seems that, at the Bill stage through Parliament, the Government conflated a public nuisance with causing a nuisance to the public [75]. It would, perhaps, be fairer if this licensing objective had been tied to the same test as for anti-social behaviour in the Anti-social Behaviour, Crime and Policing Act 2014, that is, unreasonable annoyance or nuisance to those in the community.

The licensing regime established by the Licensing Act 2003 does not prevent the use by councils (or the police) of powers provided under the Anti-social Behaviour, Crime and Policing Act 2014. These powers can be used against businesses, which tolerate or allow anti-social behaviour amongst their clients or customers, as well as against individuals. Moreover, the statutory nuisance provisions of the Environmental Protection Act 1990, particularly noise emitted from premises, may be resorted to by local authorities when regulating nuisances from licensed premises. Alternatively, the little-used but potentially very effective use of the Noise Act 1996 for dealing with night-time noise from licensed premises is available to local authorities ([7] at pp. 118–121). Enforcement activity grounded in any of these statutory provisions may or may not be linked to licensing hearings in which changes are being sought to licensing conditions.[82]

Environmental permitting

The environmental permitting regime constitutes a detailed and comprehensive regulatory system, based on the licensing of activities and industrial processes having the potential to cause environmental harm if left unregulated. Up to 2007, when the first Environmental Permitting Regulations came into force,[83] separate systems of licensing existed for prescribed processes and waste management. Industrial concerns could be subject to regulation by more than one regulator and this was seen as problematic by policymakers and industry. The Better Regulation Task Force reported, in March 2005, that subjecting industrial concerns to more than one system of permitting resulted in confusion, duplication, and inefficiency. Of particular concern was the duplication of regulators with respect to waste management and Integrated Pollution Prevention and Control (IPPC). The report stated ([68] at p. 37):

> Various licensing requirements are set out in different pieces of legislation and may impose different administrative requirements on industry. The procedures relating to IPPC for an industrial process that might pollute the air, water, or land are different to those required for waste management – yet their objective, to protect the environment, is the same. Many businesses will need to deal with both permitting systems.

The Environmental Permitting (England and Wales) Regulations 2007 and 2010 provide the framework for a unified system of regulation for prescribed processes and waste management. The 'polluter pays principle', drawn from European Union environmental policy, forms a major part of the policy background. Environmental permitting requires the costs of regulation (at least, in part) to be met by licence holders and persons applying for permits are required to pay fees to regulators.

Local authorities have an important role to play as regulators in environmental permitting. The regulated activities and installations that are subject to the permitting regime are set down in the various schedules of the Environmental Permitting (England and Wales) Regulations 2010 (EP).[84] Both Integrated Pollution Control (IPC) and Integrated Pollution Prevention Control (IPPC) processes regulated by local authorities come within the scope of environmental permitting. The Secretary of State has the power, provided under section 2 of the Pollution Prevention and Control Act 1999, to make regulations for the implementation of integrated control measures. This power is given to meet the requirements of European Council Directive 96/61/EC on Integrated Pollution Prevention and Control and other measures intended to prevent and control pollution. Schedules 1 and 2 of the 1999 Act set down the legal framework for permitting, enforcement measures, offences, and for regulatory controls generally.

Detailed guidance is provided by Defra setting out how local authorities should exercise their duties and powers as regulators [77]. Statutory guidance is not fully binding but local authorities must have regard to it, and such guidance should be followed.[85] The guidance also informs businesses subject to environmental permitting what they need to do in order to obtain the required permits and to fulfil the requirements included in their permits.

Environmental permitting is also a dynamic system and regulation 34(1) of EP 2010 requires local authorities to periodically review permits. Permit holders are required to ensure that they employ best available techniques (BAT) in their operation of regulated activities and processes. BAT changes over time, so permit holders need to ensure that new developments are properly addressed. As the guidance states: 'authorities should review permit conditions in the light of new information on environmental effects, available techniques or other relevant issues' ([77] at chapter 26.2).

Once a permit has been issued, the operator must carry out the activity in accordance with permit conditions. Any changes to the activity should be dealt with under the variation notice procedure ([77], in chapter 24). The role of the local authority is to satisfy itself whether this is being done. This is another instance of dynamism in the regulatory system, based on risk assessment. As stated in the guidance:

> The amount of regulatory input should be related to the risks associated with each individual installation, measured in terms of the environmental impact of the installation . . . and the demonstrated capability of the operator.
>
> *([77] at chapter 27.1)*

Permitting thus places significant responsibility for monitoring emissions and other regulatory parameters on operators. Permit conditions should be drafted such that they are enforceable, have clarity both for industry and the public, and are both relevant and workable.[86] Conditions for regulated installations include ensuring that the operator supplies the data needed to check for compliance and informs the authority, without delay, of any incident or accident that is causing or may cause significant pollution [77].

Regulators should carry out inspections based on an assessment of risk; they can also decide to undertake their own compliance monitoring. This proactive aspect of environmental permitting contrasts with reactive regimes, such as statutory nuisance, where regulation is achieved by service of administrative notices enforced in the breach in the criminal courts ([7] at pp. 324–325).

Enforcement – The Regulators' Code, derived from better regulation principles, is of crucial importance in guiding how local authorities should deal with non-compliance with permit conditions [12]. These principles place a requirement on regulators to take a proportionate approach to enforcement which takes into account the circumstances of non-compliance. Giving advice and being persuasive are soft skills needed by enforcement officers where non-compliance is seen as not deliberate, is at the lower end of seriousness, or where other mitigating factors are involved. The next step up the enforcement architecture might involve issuing a warning letter or service of an enforcement notice. Under regulation 36 of EP 2010, an authority has the power to serve an enforcement notice if it believes that an operator has contravened, is contravening, or is likely to contravene any permit condition. Enforcement notices may include steps to remedy the effects of any harm and to bring a regulated facility back into compliance [77]. Before deciding whether to serve an enforcement notice, the local authority should consider whether compliance with permit conditions is more likely to be achieved by issuing a warning letter, with the prospect of court proceedings should this step prove ineffective.

Controlled waste and fly-tipping

Formidable problems face the waste industry and its regulators. There is a serious and worsening problem with waste crime, and a lack of regulation in some sectors; the permitting and enforcement systems are riven with difficulties; much waste crime takes place 'under the radar' and goes unreported. The light-touch regulatory approach fostered by the Government in the Deregulation Act 2015 has probably made a bad situation worse. The *Rethinking Waste Crime* report concluded ([67] at p. 24):

> A regulatory approach that may be laudable in some industries is a poor fit for a sector that in its current form only exists through regulation and enforcement. This has increased the opportunity for illegal activity – ultimately damaging – rather than supporting legitimate businesses.

Waste on land is regulated under Part II of the Environmental Protection Act 1990. These provisions are concerned with *controlled* waste, which is defined in section 75(4) EPA 1990 as 'household, industrial and commercial waste'. Substances or objects need to be discarded in order to come within the definition of controlled waste. There are some exempted waste operations and a separate regime applies to marine licensing. Separate provisions deal with hazardous waste.[87]

Waste management licensing – Persons operating businesses that involve the storage and handling of controlled waste need to be licensed, as required by the environmental permitting regime. Where the activity does not require a permit, because it is exempted as a 'low risk', registration with the Environment Agency is still required.

Applications for permits should be made to the Environment Agency (for England).[88] An applicant needs to be a 'fit and proper' person, that is, with no relevant convictions and able to show sufficient technical competence. A permit may be revoked or suspended if the permit holder is subsequently convicted of a relevant offence. Revocation or suspension can also occur where the management of the concern is found to be incompetent. A failure to comply with a permitting condition within a specified time may also result in suspension or revocation.

A local authority may have responsibility for regulating a waste operation in two situations:

- Where the waste operation is listed in Schedule 3 to the Environmental Permitting (England and Wales) Regulations 2010 (EP) but is not exempt because it is a directly associated activity of a Part B installation;
- Where the non-exempt waste operation is directly associated with a Part B installation and would be separately regulated by the Environment Agency if it were not for a direction made under regulation 33 of 2010 EP to achieve a single regulator for the installation ([77] at chapter 40).

Offences – Section 33(1)(a) EPA 1990 (as amended) provides that a person shall not:

> deposit controlled waste or extractive waste, or knowingly cause or knowingly permit controlled waste or extractive waste to be deposited in or on any land unless an environmental permit authorising the deposit is in force and the deposit is in accordance with the licence.

A distinction is made in this provision between depositing without a permit in place and breaching the terms of a permit. Depositing waste, or knowingly causing or knowingly permitting a deposit to be made – without the relevant permit in place – form three separate fly-tipping offences under section 33(1)(a). The inclusion of 'knowingly' in these offences imports a mental element, which the prosecutor needs to prove to secure a conviction.[89] A due diligence defence and a defence of acting in an emergency, whilst taking reasonable care to avoid commission of the offence and informing the relevant authority about the incident, are available under section 33(7) of the Act.

Section 33(1)(a) offences are triable either in the magistrates' court or in the Crown Court. In addition to the power in both courts to order an unlimited fine, the Crown Court can impose a prison sentence of up to five years. Sentencing powers of the magistrates' courts are limited to up to 12 months' imprisonment, though the court can after conviction refer a case to the Crown Court for sentencing where it believes a longer sentence than 12 months is justified. Authorised officers may, as an alternative to prosecution, issue a fixed penalty notice for a section 33(1)(a) offence. This is an administrative provision intended to be used for relatively minor infractions.[90] For England, in 2019–20, 13,400 fixed penalty notices were issued specifically for small-scale fly-tipping offences – prosecutions for fly-tipping totalling 2,900 during the same period [69].

There is a further fly-tipping offence provided in section 59 of the EPA 1990. A local authority (or the Environment Agency) can serve a notice on an occupier to clear controlled waste that has been illegally deposited by another person on their land. It is an offence not to comply with a notice. The authorities may also enter land and clear it, and recover reasonable costs from the occupier. Section 59 offences could be seen as a rare instance of the victim of crime being made liable instead of the perpetrator. This may explain why the provision seems to be rarely used. Considering the enforcement policy of the Environment Agency, the authors of *Rethinking Waste Crime* concluded ([67] at p. 28):

> Unfortunately this power is used infrequently and, on the face of it, only when there is a serious threat to the environment and human health. This represents a lost opportunity to recover some of the economic damage associated with waste crime.

Sentencing – The Sentencing Council, in its press release issued in February 2014, when its sentencing guidelines were published for environmental crime, proclaimed that: 'custody remains the starting point for the most serious types of individual offenders who deliberately commit a crime that causes significant or major harm' [48]. However, in reality, custodial sentences have rarely been imposed since the sentencing guidelines came into effect. Statistics published by Defra for fly-tipping in England indicate that, between 2015 and 2020, custodial sentences were imposed in 18 to 41 prosecutions per year. Most outcomes were fines, with custodial penalties applying within a range from 0.81% to 1.78% of successful prosecutions brought during this period ([69] at p. 14).

In conclusion, weaknesses in regulators' conduct of enforcement coupled with a pusillanimous approach to prosecution have meant that waste management crimes go largely unpunished.[91] The issue of a warning letter or a fixed-penalty notice are the main actions taken for low-level breaches, and, it would seem, for more serious offences that are not prosecuted but arguably should be.

Notes

1 Civil (or civilian law) law used here has a different meaning from civil law as a system for resolving legal disputes between parties, as distinct from criminal law.
2 The complex interaction between statute law and the common law means that they should not be seen as akin to 'oil and water'. For an illuminating analysis of this relationship, see Burrows ([2] at chapter 2).
3 [2003] 1 AC 518.
4 Modernisation of the law of public nuisance, including placing the crime on a statutory footing, has been proposed by the Law Commission, see *Simplification of Criminal Law: Public Nuisance and Outraging Public Decency* [78].
5 For England, as provided in the Local Government Act 1972, s 2 (as amended); for Wales, the Local Government (Wales) Act 1994, s 2.
6 An example of the exercise of a statutory power, as distinct from a statutory duty, is provided by s 43 of the Anti-social Behaviour, Crime and Policing Act 2014, which provides the power to issue community protection notices.
7 *R v Tower Hamlets LBC, ex p Khalique* (1994) 26 HLR 517.
8 *Barnard v National Dock Labour Board* [1953] 2 QB 18.
9 [1986] 1 WLR 807.
10 Local Government and Housing Act 1989, s 20.
11 *R v Hereford Corporation, ex p Harrower* [1970] 1 WLR 1424.
12 Public Health Act 1875, s 265, extended to legislation (including local Acts) generally by the Local Government (Miscellaneous Provisions) Act 1976, s 39.
13 *R v Secretary of State for the Environment, ex p Hillingdon LBC* [1986] 1 WLR 807.
14 *Baroness Wenlock v River Dee Co* (1885) 10 App Cas 354.
15 *Attorney-General v Great Eastern Railway* (1880) 5 App Cas 473.
16 [1962] 1 QB 416.
17 [1971] 1 QB 222.
18 Such as the decision of the Court of Appeal in *Western Fish Products Ltd v Penwith DC* [1981] 2 All ER 204.
19 *Hazell v Hammersmith & Fulham LBC* [1992] 2 AC 1.

20 *R v Jarrett and Steward* [1997] EWCA Crim 275.
21 Section 223 of the Local Government Act 1972 (as amended) includes in its definition of a 'local authority': the Common Council, a joint authority, an economic prosperity board, a combined authority, a fire and rescue authority created by an order under section 4A of the Fire and Rescue Services Act 2004, the Greater London Authority, and a police and crime commissioner and the Mayor's Office for Policing and Crime.
22 *Bowyer, Philpott & Payne Ltd v Mather* [1919] 1 KB 419.
23 The Legal Aid, Sentencing and Punishment of Offenders Act 2012 (Fines on Summary Conviction) Regulations 2015, SI 2015/664.
24 The rules for appeals are quite complicated and can be found in Schedule 5 of the Housing Act 2004.
25 A flowchart setting out the procedure for appeals from the First-tier Tribunal (Property Chamber) and Leasehold Valuation and Residential Property Tribunals can be found at https://assets.publishing.service.gov.uk/government/uploads/system/uploads/attachment_data/file/718601/t614-eng.pdf
26 Contaminated Land (England) Regulations, 2006, SI 2006/1380, reg 8.
27 Environmental Permitting (England and Wales) Regulations 2010, SI 2010/675, reg 2.
28 *Code for Crown Prosecutors*, Crown Prosecution Service, 8th ed, 2018.
29 *Kirklees MBC v Wickes Building Supplies* [1993] AC 227.
30 Useful guidance for EHPs in exercising their statutory powers is provided in the *Noise Management Guide* [79].
31 Criminal Procedure Rules 2020, SI 2020/759 (L. 19).
32 *R (on the application of Hackney LBC) v Rottenberg* [2007] EWHC 166 (Admin).
33 *Southwark LBC v Simpson* [1999] Env LR 553, 559.
34 *Roper v Tussauds Theme Parks Ltd* [2007] EWHC 624 (Admin).
35 *Field v Leeds City Council* (2000) 32 HLR 618.
36 *Whitehouse v Jordan* [1981] 1 WLR 246.
37 For advice on conducting health and safety interviews, see Penn [9] (at pp.173–194).
38 European Convention for the Protection of Human Rights and Fundamental Freedoms, 1950.
39 SI 2012/1500.
40 The prosecution might decide that the unused material could undermine its case or assist the defence, in which case it should be disclosed under section 3 of the Criminal Procedure and Investigations Act 1996.

41 Food safety had been an important priority in the Hampton Review and for the Better Regulation Task Force, in 2005, see BRTF [68] and [80].

42 The principles set down in section 21(2) of the Legislative and Regulatory Reform Act 2006 for which regulators must have regard are: (a) regulatory activities should be carried out in a way which is transparent, accountable, proportionate and consistent; (b) regulatory activities should be targeted only at cases in which action is needed.

43 Risk-based approaches to regulation are not all the same and the concept of risk is not straightforward. For an analysis of the HSE see Almond and Esbester [81].

44 Except for the provisions to make regulations under sections 15 and 30, the Health and Safety at Work Act 1974 does not apply to Northern Ireland. Section 23, which deals with supplementary provisions pertaining to prohibition and improvement notices applies in Northern Ireland. Part III of the Act, concerning the building regulations, does not apply to Scotland.

45 SI 1999/3242.

46 Council Directive of 12 June 1989 on the introduction of measures to encourage improvements in the safety and health of workers at work (89/391/EEC), OJ No. L183, 29.6.89, p. 1. These principles of EU health and safety law comprise: (a) avoiding risks;
(b) evaluating the risks which cannot be avoided;
(c) combating the risks at source;
(d) adapting the work to the individual, especially as regards the design of workplaces, the choice of work equipment and the choice of working and production methods, with a view, in particular, to alleviating monotonous work and work at a predetermined work-rate and to reducing their effect on health;
(e) adapting to technical progress;
(f) replacing the dangerous by the non-dangerous or the less dangerous;
(g) developing a coherent overall prevention policy which covers technology, organisation of work, working conditions, social relationships and the influence of factors relating to the working environment;
(h) giving collective protective measures priority over individual protective measures; and
(i) giving appropriate instructions to employees.

47 SI 2002/1860.

48 Housing Act 2004, ss 11–19.

49 Housing Act 2004, ss 40–42.

50 Housing Act 2004, ss 43, 44.

51 Housing Act 2004, ss 20–27.

52 Housing Act 2004, ss 28, 29.

53 Housing Act 2004, s 46.

54 Housing Act 2004, s 47.

55 Transfer of Tribunal Functions Order 2013, SI 2013/1036.

56 See Battersby and Pointing [10] for the law and practice regarding Part 1 of the Housing Act 2004.

57 See Malcolm and Pointing [7] for the law and practice on statutory nuisances generally.

58 Similar duties apply in Northern Ireland under s 64 of the Clean Neighbourhoods and Environment Act (Northern Ireland) 2011.

59 Environmental Protection Act 1990, schd 3, para. 3(1).

60 Environmental Protection Act 1990, ss 81(3)(4).

61 *Vale of White Horse DC v Allen and Partners* [1997] Env LR 212, at 214.

62 Per Lord Denning MR in *Miller v Jackson* [1977] QB 966, at 980.

63 The CIEH Noise Survey for England indicates that no notices were served by local authorities under the Noise Act 1996 during 2019/20 [82] (CIEH 2021: 5). Noise notices served in England during this period included: 1655 under section 80 Environmental Protection Act 1990, 993 under section 60 Control of Pollution Act 1974, 339 community protection notices under the Anti-social Behaviour, Crime and Policing Act 2014.

64 *R v Bristol City Council, ex p Everett* [1999] 1 WLR 1170.

65 Nuisances Removal and Diseases Prevention Act 1855.

66 *Birmingham City Council v Oakley* [2001] 1 All ER 385.

67 For a detailed analysis of the health limb, see Battersby, S. and Pointing, J. ([10] at chapter 4).

68 *Hansard*, 6th Series, Vol 178, 31 October 1990, col 1023, per David Trippier, Minister for the Environment and Countryside.

69 *Coventry v Lawrence* [2014] UKSC 13.

70 *Allen v Gulf Oil Refining Ltd* [1979] 3 All ER 1008, subsequently affirmed by the House of Lords at [1981] AC 1001.

71 Anti-social Behaviour, Crime and Policing Act 2014, s 2(1)(b).

72 Regulation (EC) No 178/2002 of the European Parliament and of the Council of 28 January 2002.

73 The current regulations in force for England are the Food Safety and Hygiene (England) Regulations 2013, SI 2013/2996 (as amended).

74 SI 2013/2996.

75 The General Food Law (Amendment etc.) (EU Exit) Regulations 2019, SI 2019/641, reg 4, schedule 2 lists the retained provisions of

EU food law: Regulation (EC) No 178/2002, Regulation (EC) No 852/2004 (food hygiene), Regulation (EC) No 853/2004 (products of animal origin).

76 SI 2019/641.

77 SI 2013/2996.

78 *R (on the application of Knowsley MBC) v Williams* [2001] Env LR 28.

79 Environmental Protection Act 1990, ss 81(3)(4).

80 *R (on the application of Ethos Recycling Ltd) v Barking and Dagenham Magistrates' Court* [2009] EWHC 2885 (Admin).

81 Licensing Act 2003, sched 1, part 2, para. 12ZB(4).

82 According to the CIEH Noise Survey for England, a review of licensing conditions under the 2003 Licensing Act took place in 752 cases involving noise, in 2019/20 ([82] at p. 5).

83 Environmental Permitting (England and Wales) Regulations 2007, SI 2007/3538.

84 SI 2010/675.

85 Environmental Permitting (England and Wales) Regulations 2010, SI 2010/675, reg 64.

86 For useful advice on drafting conditions for environmental permitting, see [77] Annex XI, p. 367ff.

87 Hazardous Waste (England and Wales) Regulations 2005, SI 2005/894. Hazardous waste is estimated by the Environmental Agency to be involved in over 15% of 126 serious fly-tipping incidents reported in 2015 ([67] at p.11).

88 The responsible authority for Wales, the Natural Resources Body for Wales; for Scotland, the Scottish Environment Protection Agency.

89 *Alphacell Ltd v Woodward* [1972] AC 824.

90 A fixed penalty notice can also be issued in respect of a failure to comply with the duty relating to the transfer of household waste.

91 In *Barr v Biffa Waste Services Ltd.* [2011] EWHC 1003 (TCC), reversed by the Court of Appeal but not affecting the point being made here, the High Court was highly critical of the Environment Agency. The case concerned odours emitted from a waste processing plant in which regulatory action taken by the Agency had been largely ineffective over a number of years. Mr Justice Coulson was particularly critical of the weak and 'pusillanimous attitude of the EA' [at paras. 576–580] in failing to take regulatory action against a powerful commercial entity. Biffa were themselves criticised by his lordship for 'intimidation' and for taking an 'unnecessarily aggressive approach' both towards the EA and the residents bringing the action, which spilled over into their conduct at the trial [at paras. 570–575].

References

[1] Glenn HP. (2014) *Legal Traditions of the World: Sustainable Diversity in Law*, Oxford University Press, Oxford, 5th ed.

[2] Burrows A. (2018) *Thinking About Statutes: Interpretation, Interaction, Improvement*, The Hamlyn Lectures 2017, Cambridge University Press, Cambridge.

[3] Pointing J. (2011) Public nuisance: Beyond highway 61 revisited? *Environmental Law Review*, 13(1).

[4] Dworkin R. (1986) *Law's Empire*, Fontana Press, London.

[5] Blake L, Pointing J, Sinnamon T. (2007) Over-regulation and suing the state for negligent legislation. *Statute Law Review*, 28(3): 218–234.

[6] Steiner E. (2002) *French Legal Method*, Oxford University Press, Oxford.

[7] Malcolm R, Pointing J. (2011) *Statutory Nuisance: Law and Practice*, Oxford University Press, Oxford, 2nd ed.

[8] Malcolm R, Pointing J. (2005) *Food Safety Enforcement*, Chadwick House Publishing, London.

[9] Penn C. (2005) *Local Authority Health & Safety Enforcement*, Shaw & Sons, Crayford.

[10] Battersby S, Pointing J. (2019) *Statutory Nuisance and Residential Property: Environmental Health Problems in Housing*, Routledge, Abingdon.

[11] MHCLG. (2018b) *Civil Penalties Under the Housing and Planning Act 2016: Guidance for Local Housing Authorities*, HM Government, Ministry of Housing, Communities & Local Government, London.

[12] BRDO. (2014) *Regulators' Code*, HM Government, Better Regulation Delivery Office, Department for Business Innovation & Skills, London.

[13] Tapper J. (2021) Rising from the rubble: Pub rebuilt brick by brick after illegal bulldozing. *The Observer*, 21 March.

[14] Civil Justice Council. (2014) *Guidance for the Instruction of Experts in Civil Claims*, Civil Justice Council, London. https://www.judiciary.uk/wp-content/uploads/2014/08/experts-guidance-cjc-aug-2014-amended-dec-8.pdf (Accessed March 2021).

[15] PACE. (2018) *Police and Criminal Evidence Act (PACE) – CODE C Revised Code of Practice for the Detention, Treatment and Questioning of Persons by Police Officers*, Home Office, The Stationery Office, Norwich.

[16] Home Office. (2018) *Covert Surveillance and Property Interference – Revised Code of Practice*, HM Government, Home Office, London, August.

[17] Wells C. (2001) *Corporations and Criminal Responsibility*, Oxford University Press, Oxford, 2nd ed.

[18] Ashworth A. (1989) Towards a theory of criminal legislation. *Criminal Law Forum*, 1.

[19] Hawkins K. (2002) *Law as Last Resort*, Oxford University Press, Oxford.

[20] Rose E. (2020) The UK's enforcement gap 2020. *Unchecked.UK*. www.unchecked.UK (Accessed April 2021).

[21] King D. (1993) Government beyond Whitehall: Local government and urban politics. In *Developments in British Politics 4*, Dunleavy P, Gamble P, Holliday I, Peele G (Eds.), Palgrave, London.

[22] Andrews C. (1998) *The Enforcement of Regulatory Offences*, Sweet & Maxwell, London.

[23] Ayres I, Braithwaite J. (1992) *Responsive Regulation – Transcending the Deregulation Debate*, Oxford University Press, Oxford.

[24] Makkai T, Braithwaite J. (1995) In and out of the revolving door: Making sense of regulatory capture. *Journal of Public Policy*, 15: 61–78.

[25] Hood C. (1991) A public management for all seasons. *Public Administration*, 69: 3–19.

[26] Hampton P. (2005) *Reducing Administrative Burdens: Effective Inspection and Enforcement*, HM Government, HM Treasury, London.

[27] BEIS. (2020) *Better Regulation Framework: Interim Advice*, HM Government, Department for Business, Energy and Industrial Strategy, London.

[28] LBRO. (2010) *Primary Authority: Gateway to Better Local Regulation*, HM Government, Local Better Regulation Office, Better Regulation Executive, London.

[29] CIEH. (2008) *Response by Chartered Institute of Environmental Health to the Regulatory Enforcement and Sanctions Act 2008: Consultation on the Primary Authority Scheme*, Chartered Institute of Environmental Health, London.

[30] Eccles T, Pointing J. (2013) Smart regulation, shifting architectures and changes in governance. *International Journal of Law in the Built Environment*, 5: 71–88.

[31] BEIS. (2018) *Government Launches New Office for Product Safety and Standards*, HM Government, Department for Business, Energy and Industrial Strategy, London, press release.

[32] McCrudden C. (2007) Corporate social responsibility and public procurement. In *The New Corporate Accountability*, McBarnet D, Voiculescu A, Campbell T. (Eds.), Cambridge University Press, Cambridge.

[33] Campbell T. (2007) The normative grounding of corporate responsibility. In *The New Corporate Accountability*, McBarnet D, Voiculescu A, Campbell T (Eds.), Cambridge University Press, Cambridge.

[34] Manning P, Hawkins P. (1990) Legal decisions: A frame analytic perspective. In *Studies on Communication, Institution and Social Interaction*, Riggins S (Ed.), Mouton de Gruyter, New York.

[35] Black J. (2007) Tensions in the regulatory state. *Public Law*, 58–73.

[36] BRDO. (2012) *Making It Happen: LBRO and Local Regulatory Reform 2007–2012*, HM Government, Better Regulation Delivery Office, Department for Business Innovation & Skills, London.

[37] Gunningham N. (2009) Environmental law, regulation and governance: Shifting architectures. *Journal of Environmental Law*, 21: 179–212.

[38] Braithwaite J. (2008) *Regulating Capitalism: How It Works, Ideas for Making It Work Better*, Edward Elgar, Cheltenham.

[39] Martinez M, Verbruggen P, Fearne A. (2013) Risk based approaches to food safety regulation: What role for co-regulation? *Journal of Risk Research*, 16: 1–21.

[40] BIS. (2014) *Business Minister Matthew Hancock Announces the Government's War on Red Tape Has Saved Business £10 Billion Over the Last 4 Years*, HM Government, Department for Business, Innovation & Skills, London, press release.

[41] Lord R. (2021) *Fire Safety in Residential Property: A Practical Approach for Environmental Health*, Routledge, Abingdon.

[42] Moore R. (2020a) The culture of incompetence that led to Grenfell still imperils us. *The Observer*, 6 December.

[43] Clifford B, Ferm J, Livingstone N, Canelas P. (2018) *Assessing the Impacts of Extending Permitted Development Rights to Office-to-Residential Change of Use in England*, Royal Institution of Chartered Surveyors, London.

[44] Hackitt J. (2018) *Building a Safer Future – Independent Review of Building Regulations and Fire Safety: Final Report*, Cm 9607, HM Stationery Office, London.

[45] Moore R. (2020b) It's like an open prison. *The Observer, The New Review*, 27 September.

[46] Clifford B, Canelas P, Ferm J, Livingstone N, Lord A, Dunning R. (2020) *Research into the Quality Standard of Homes Delivered Through Change of Use Permitted Development Rights*, HM Government, Ministry of Housing, Communities & Local Government, London.

[47] Skidelsky R. (2009) Where do we go from here? In *Prospect*, Prospect Publishing, London, January.

[48] Sentencing Council. (2014a) *New Sentencing Guideline for Environmental Crimes Brings Higher Sentences for Serious Offenders*, HM Government, London, press release. https://

www.sentencingcouncil.org.uk/news/item/new-sentencing-guideline-for-environmental-crimes-brings-higher-sentences-for-serious-offenders/ (Accessed March 2021).

[49] Sentencing Council. (2015) *Health and Safety Offences, Corporate Manslaughter and Food Safety and Hygiene Offences, Definitive Guideline*, HM Government, London. https://www.sentencingcouncil.org.uk/wp-content/uploads/Health-and-Safety-Corporate-Manslaughter-Food-Safety-and-Hygiene-definitive-guideline-Web.pdf (Accessed January 2021).

[50] Black J. (1997) New institutionalism and naturalism in socio-legal analysis: Institutionalist approaches to regulatory decision making. *Law & Policy*, 19: 51.

[51] HSE. (2019) *Enforcement Statistics in Great Britain, 2019*, Health and Safety Executive, London.

[52] MHCLG. (2018a) *Houses in Multiple Occupation and Residential Property Licensing Reform: Guidance for Local Housing Authorities*, HM Government, Ministry of Housing, Communities & Local Government, London.

[53] Taylor D, Turtle P. (2020) Dramatic increase in use of prosecutions, civil penalty fines and rent repayment orders. *Property 118.com*. https://www.property118.com/dramatic-increase-in-use-of-prosecutions-civil-penalty-fines-and-rent-repayment-orders/ (Accessed March 2021).

[54] Oatt P. (2019) *Selective Licensing: The Basis for a Collaborative Approach to Addressing Health Inequalities*, Routledge, Abingdon.

[55] Malcolm R, Pointing J. (2006) Statutory nuisance: The sanitary paradigm and judicial conservatism. *Journal of Environmental Law*, 18(1): 37–54.

[56] Pontin B. (2007) Integrated pollution control in Victorian Britain: Rethinking progress within the history of environmental law. *Journal of Environmental Law*, 19(2): 173–199.

[57] Clerk & Lindsell. (2020) *Clerk & Lindsell on Torts*, Sweet & Maxwell, London, 23rd ed.

[58] CIEH. (2017) *Professional Practice Note: Revised Guidance on the Use of Community Protection Notices Under Part 4 of the Anti-Social Behaviour, Crime and Policing Act 2014*, Chartered Institute of Environmental Health, London, November. https://www.cieh.org/media/1238/guidance-on-the-use-of-community-protection-notices.pdf (Accessed December 2020).

[59] Carr H, Cowan D, Kirton-Darling E, Burtonshaw-Gunn E. (2017) *Closing the Gaps: Health and Safety at Home*, Kent Law School & University of Bristol Law School. https://england.shelter.org.uk/__data/assets/pdf_file/0010/1457551/2017_11_14_Closing_the_Gaps_-_Health_and_Safety_at_Home.pdf (Accessed December 2020).

[60] Penn C. (2002) *Noise Control: The Law and Its Enforcement*, Shaw & Sons, Crayford, 3rd ed.

[61] WHO. (1946) *Preamble to the Constitution of the World Health Organization as Adopted by the International Conference*, World Health Organization, New York, 19–22 June.

[62] Home Office. (2021) *Anti-Social Behaviour, Crime and Policing Act 2014, Reform of Anti-Social Behaviour Powers: Statutory Guidance for Frontline Professionals*, HM Government, Home Office, London, June.

[63] Pointing J. (2009) Food law and strange case of the missing regulation. *Journal of Business Law*, 6: 592–605.

[64] Pointing J. (2005) Food crime and food safety: Trading in bushmeat – is new legislation needed? *Journal of Criminal Law*, 69: 42–49.

[65] Pointing J, Al-Teinaz Y, Lever J, Critchley M, Spear S. (2020) Food fraud. Chapter 19 in *The Halal Food Handbook*, Al-Teinaz Y, Spear S, Abd El-Rahim I (Eds.), Wiley Blackwell, Chichester.

[66] Pointing J. (2020) The legal framework of general food law and the stunning of animals prior to slaughter. Chapter 17 in *The Halal Food Handbook*, Al-Teinaz Y, Spear S, Abd El-Rahim I (Eds.), Wiley Blackwell, Chichester.

[67] Eunomia. (2017) *Rethinking Waste Crime*, Environmental Services Association, London. http://www.esauk.org/application/files/7515/3589/6448/20170502_Rethinking_Waste_Crime.pdf (Accessed March 2021).

[68] BRTF. (2005b) *Less Is More: Reducing Burdens, Improving Outcomes,* HM Government, Cabinet Office, Better Regulation Task Force, London.

[69] Defra. (2021) *Fly-Tipping Statistics for England, 2019/20*, HM Government, Department for Environment Food and Rural Affairs/ Government Statistical Service. https://assets.publishing.service.gov.uk/government/uploads/system/uploads/attachment_data/file/964062/FlyTipping_201920_Statistical_Release_Acc_checked_FINAL.pdf (Accessed March 2021).

[70] Priestley S. (2017a) *Fly-Tipping – the Illegal Dumping of Waste*, Briefing Paper, No. CBP05672, House of Commons Library, London, 21 June.

[71] Priestley S. (2017b) *Litter*, Briefing Paper, No. CBP06984, House of Commons Library, London, 21 July.

[72] Macrory R. (2006) *Regulatory Justice: Making Sanctions Effective*, HM Government, Cabinet Office, London.

[73] Sentencing Council. (2014b) *Environmental Offences, Definitive Guideline*, HM Government, London. https://www.sentencingcouncil.org.uk/wp-content/uploads/Environmental-offences-definitive-guideline-Web.pdf (Accessed February 2021).

[74] Defra. (2012a) *Environmental Protection Act 1990: Part 2A, Contaminated Land Statutory Guidance*, HM Government, Department for Environment Food and Rural Affairs, London.

[75] Horrocks D, Pointing J. (2012) Licensing and public nuisance. *Acoustics Bulletin*, 37(6): 23–26.

[76] Home Office. (2018a) *Revised Guidance Issued Under Section 182 of the Licensing Act 2003*, HM Government, Home Office, London, April.

[77] Defra. (2012b) *Environmental Permitting: General Guidance Manual on Policy and Procedures for A2 and B Installations*, HM Government, Department for Environment Food and Rural Affairs, London.

[78] Law Commission. (2015) *Simplification of Criminal Law: Public Nuisance and Outraging Public Decency*, Law Commission, LAW COM No. 358, House of Commons, HC 213, London, 24 June.

[79] Defra/CIEH. (2006) *Neighbourhood Noise Policies and Practice for Local Authorities – A Management Guide*. HM Government, Department for Environment Food and Rural Affairs, London.

[80] BRTF. (2005) *Routes to Better Regulation: A Guide to Alternatives to Classic Regulation*, HM Government, Cabinet Office, Better Regulation Task Force, London.

[81] Almond P, Esbester M. (2017) Regulatory inspection and the changing legitimacy of health and safety: Regulatory inspection and legitimacy. *Regulation & Governance*, 12(1).

[82] CIEH. (2021) *CIEH Noise Survey 2019/20: Report on Findings – England*, Chartered Institute of Environmental Health, London.

Cases

Allen v Gulf Oil Refining Ltd [1979] 3 All ER 1008.

Alphacell Ltd v Woodward [1972] AC 824; [1972] 2 WLR 1320; [1972] 2 All ER 475.

Attorney-General v Great Eastern Railway (1880) 5 App Cas 473.

Barnard v National Dock Labour Board [1953] 2 QB 18.

Baroness Wenlock v River Dee Co. (1885) 10 App Cas 354.

Barr v Biffa Waste Services Ltd. [2011] EWHC 1003 (TCC).

Birmingham City Council v Oakley [2001] 1 AC 617; [2001] 1 All ER 385.

Bowyer, Philpott & Payne Ltd v Mather [1919] 1 KB 419.

Coventry v Lawrence [2014] UKSC 13; [2014] AC 822; [2014] 2 WLR 433; [2014] 2 All ER 622.

Field v Leeds City Council [2000] 17 EG 165; (2000) 32 HLR 618.

Hazell v Hammersmith & Fulham LBC [1992] 2 AC 1.

Johnson v Unisys Ltd. [2001] UKHL 13; [2003] 1 AC 518.

Kirklees MBC v Wickes Building Supplies [1993] AC 227.

Lever (Finance) Ltd v Westminster LBC [1971] 1 QB 222.

Miller v Jackson [1977] QB 966.

R (on the application of Ethos Recycling Ltd) v Barking and Dagenham Magistrates' Court [2009] EWHC 2885 (Admin).

R (on the application of Hackney LBC) v Rottenberg [2007] EWHC 166 (Admin).

R (on the application of Knowsley MBC) v Williams [2001] Env LR 28.

R v Bristol City Council, ex p Everett [1999] 1 WLR 1170; [1999] 2 All ER 193; (1999) 31 HLR 1102.

R v Hereford Corporation, ex p Harrower [1970] 1 WLR 1424.

R v Jarrett and Steward [1997] EWCA Crim 275.

R v Secretary of State for the Environment, ex p Hillingdon LBC [1986] 1 WLR 807.

R v Tower Hamlets LBC, ex p Khalique (1994) 26 HLR 517.

Roper v Tussauds Theme Parks Ltd [2007] EWHC 624 (Admin).

Southend-on-Sea Corporation v Hodson (Wickford) Ltd [1962] 1 QB 416.

Southwark LBC v Simpson [1999] Env LR 553.

Vale of White Horse DC v Allen and Partners [1997] Env LR 212.

Western Fish Products Ltd v Penwith DC [1981] 2 All ER 204.

Whitehouse v Jordan [1981] 1 WLR 246.

Legal implications of the UK leaving the EU

Rosalind Malcolm and Amanda Cleary

Introduction

The European Union (EU), an economic and political union of 27 countries, operates as a Single Market allowing free movement of goods, capital, services and people within its internal borders. The current EU Member States are: Austria, Belgium, Bulgaria, Croatia, Cyprus, Czech Republic, Denmark, Estonia, Finland, France, Germany, Greece, Hungary, Ireland, Italy, Latvia, Lithuania, Luxembourg, Malta, Netherlands, Poland, Portugal, Romania, Slovakia, Slovenia, Spain and Sweden.

The UK was a member but withdrew from the EU on 31st January 2020. There then followed a Transition Period as part of the Withdrawal Agreement 2019 which ended within 12 months of that date on 31st December 2020. During this period the UK was no longer a member of the EU but remained a member of the Single Market and the Customs Union[1] and subject to EU rules. The EU-UK Transition Period was supposed to allow time for the UK and EU to reach a deal on their future relationship, but this was imperfectly achieved, and many issues were fudged with loose ends left undone. Today it is possible to state broadly that in principle the UK is no longer bound by EU Treaties, or the EU principles of free movement

or the general principles of EU law, but EU law in the UK has not disappeared, far from it. In fact, post-Brexit legal arrangements are extremely complex, and based on multiple and overlapping legal instruments together with some new legal concepts (such as 'retained law') the impact of which has yet to be established. So, while the purpose of the most important of these legal instruments, the EU (Withdrawal) Act 2018, was to convert EU law as it stood at the end of the Transition Period into domestic UK law, and to preserve laws made in the UK to implement EU obligations, how this is to be applied in practice has yet to be seen. Certainly, the end of the EU-UK Transition Period saw the loss of Single Market access rights and participation in the Customs Union for the UK, but exactly what has replaced it for the purposes of trade, public health and environmental protection is far from clear and it is already apparent that substantial barriers to trade have been created through the failure of the EU and the UK to agree on the operation of essential sanitary and phytosanitary (SPS) measures. To a great extent this is unsurprising as, during the period of nearly 50 years of membership of an organisation with extensive law-making powers, the UK's legal system has become closely aligned with EU law,

DOI: 10.1201/9781003035640-6

and the process of extrication from this legal system created by the European Communities Act 1972 (ECA) will necessarily take many years, giving rise to confusion and litigation. In the meantime, since the practicality of breaking from the EU legal code is not something that could be achieved overnight and was certainly not achieved during the Transition Period, any assumption that the UK is no longer connected with EU law is incorrect and the process of withdrawal does not mean that EU law is no longer applicable. On the contrary, in the areas of environmental health interest, the presumption should be that EU law does continue to apply unless express UK legislation has been adopted to modify or repeal it. The reasons for this will be explained in more detail to follow.

History of EU and of accession of UK to EU

The original forerunner to the European Union, the European Economic Community (EEC), was created in 1957 by the Treaty of Rome and had six members: France, Germany, Belgium, Italy, the Netherlands and Luxembourg. This followed on from the successful establishment of the European Coal and Steel Community (ECSC) in 1951. The EURATOM Treaty was also signed in 1957, providing for a legal framework in the specific sector of atomic energy, whereas the EEC Treaty was more far-ranging in scope and from its inception had objectives that were more than purely economic, despite its name. From the outset it was clear that this was intended as more than just a trade arrangement, and the existence of the highly independent and active European Court of Justice (created in 1951, known since 2009 as the Court of Justice of the European Union (CJEU)) was to ensure that the EEC would develop as a supranational entity, not just another international forum. In a series of key judgments in the 1960s, the European Court of Justice laid down the constitutional principles of the primacy of European law over national law (the doctrine of supremacy), and its direct effect in national law even when not necessarily directly applicable (the doctrine of direct effect).

This was possible because the original treaties all provided a structure for the European institutions to adopt legislation, which today is collectively referred to as EU secondary law, and the CJEU always had the power to review that legislation in such a way that national courts and legislatures were bound by its decisions. It was this 'supranational' quality, along with the treaty-based provisions for enforcement, which makes the EU legal system unique and far more effective than similar intergovernmental systems: it can penetrate directly into the legal systems of its members and citizens of those countries have the right to rely upon most EU law before their national courts and in some (limited) instances to go directly to the CJEU itself.

All the founder members had civil law traditions, with legal systems ultimately derived from Roman law, albeit heavily overlain by Napoleonic, Germanic, canonical and local practices. It was natural, therefore, that EEC law should be modelled upon the same principles, and this has fundamentally influenced the development of EU law. The sources of EU law have expanded and evolved over the years, but the primary source remains the treaties, and all EU action has to be founded on treaty articles. Failure to comply with this basic 'principle of conferral' will render any EU actions void and illegal.

Secondary sources of EU Law are set out today in Article 288 of the Treaty on the Functioning of the European Union (TFEU) as follows:

- Regulations.
  Regulations are legal acts that apply automatically and uniformly to all EU countries as soon as they enter into force, without needing to be transposed into national law. They are binding in their entirety on all EU countries.

- Directives.

  Directives require EU countries to achieve a certain result but leave them free to choose how to do so. EU countries must adopt measures to incorporate them into national law (transpose) in order to achieve the objectives set by the directive. National authorities must communicate these measures to the European Commission.

  Transposition into national law must take place by the deadline set when the directive is adopted (generally within two years). When a country does not transpose a directive, or transposes it incorrectly, the Commission may initiate infringement proceedings. There is also an extensive body of case law as to when citizens can rely on the direct effect of directives.

- Decisions.

  A decision shall be binding in its entirety. A decision which specifies those to whom it is addressed shall be binding only on them.

- Delegated acts.

  Delegated acts are legally binding acts that enable the Commission to supplement or amend non-essential parts of EU legislative acts, for example, in order to define detailed measures.

  The Commission adopts the delegated act and if Parliament and Council have no objections, it enters into force.

- Implementing acts.

  Implementing acts are legally binding acts that enable the Commission – under the supervision of committees consisting of EU countries' representatives – to set conditions that ensure that EU laws are applied uniformly.

- Recommendations.

  Recommendations have no binding force. They allow the EU institutions to make their views known and to suggest a line of action without imposing any legal obligation on those to whom it is addressed.

- Opinions.

  Opinions have no binding force. An 'opinion' is an instrument that allows the EU institutions to make a statement, without imposing any legal obligation on the subject of the opinion. They can however be invoked as an aid in the interpretation of binding law.

- Communications.

  Communications have no binding force but are issued by the European Commission on a wide variety of topics (e.g., EU Commission Communication on the Precautionary Principle 2000 [COM/2000/0001 final]). Communications may include policy evaluations, commentary, or explanations of action-programmes or brief outlines on future policies or arrangements concerning details of current policy. They are frequently accompanied by a Staff Working Paper, which provides more in-depth detail.[2]

A tertiary, but very important, source of EU law are the General Principles of Law (GPLs) which have been evolved by the CJEU over the decades, some of which have now been codified and given treaty status, some of which have not. The GPLs have to be respected both in the making and application of EU law (e.g., fundamental rights, proportionality, legal certainty, equality before the law, subsidiarity). The GPLs have come to include the principles of the EU Single Market, which are the freedom of movement of goods, services, workers and capital between EU Member States without internal barriers (subject to limited exceptions). This notion of the four freedoms represents the very core of the EU Single Market. These arrangements relevant to the Single Market were extended under the European Economic Area Agreement to cover Iceland, Liechtenstein and Norway. The European Economic Area (EEA) was set up in 1994 and extended the EU's internal market to countries in the European Free Trade Association (EFTA). Thus, the EEA comprises the 27 EU

Member States and the three EFTA States of Iceland, Liechtenstein and Norway. Unlike the UK, now that it has left the EU, these three EFTA states may participate in the Single Market albeit they have limited rights in respect of the development of EU legislation, and are subject to the EFTA Court, which is a supranational judicial body. The European Free Trade Association is an intergovernmental organisation which, as its name suggests, is designed to promote free trade and economic integration for its Member States. Founded in 1960 by Austria, Denmark, Norway, Portugal, Sweden, Switzerland and the United Kingdom, and later joined by Finland, Iceland and Liechtenstein, it now only has four members: Iceland, Liechtenstein, Norway and Switzerland. All the other members left at various points to join the EU which leaves the UK, now it has left the EU, as neither a member of EFTA nor the EU.

UK accession and withdrawal

The UK was not an original founder member of either the European Coal and Steel Community (ECSC) or the European Economic Community (EEC) due to widespread political opposition post-war, and instead supported the creation of a larger but less integrated free trade area encompassing all members of the Organisation for European Economic Co-operation (OECD). Hence it initially preferred to support the 'rival' EFTA organisation but for reasons of economic reality applied for EEC membership in 1961. However, it was not until 1969 that its application was successful, and it acceded to membership via the Treaty of Accession signed in January 1972. The European Communities Act 1972 (ECA) was enacted on 17th October 1972, with UK membership of the European Community (EC) coming into effect on 1st January 1973.

UK membership of the EU was frequently uneasy. Mainly for political reasons Britain's difficult relationship with its European partners often led to it being described as 'an awkward partner',[3] and the UK certainly was not a willing participant in what were perceived as 'federal' developments such as the single currency and the Schengen agreement on the free movement of people. From a legal perspective, the twin EU doctrines of supremacy and direct effect always caused controversy and resistance in certain UK political circles and in the early years of membership the UK's differing common law tradition[4] did cause some consternation, probably because the UK and Republic of Ireland were the only members of the European Union with legal systems based on common law. Civil law prevailed (and continues to prevail) in all the other Member States. These two traditions are the basis of quite different ways of approaching EU regulation; and they were at the heart of some of the most critical misunderstandings between the UK and the EU. Arguably they also go some way to explaining the UK's struggle with EU bureaucratic rigidity and with what the Eurosceptics perceived as 'undemocratic regulatory incontinence'.[5] Curiously the difference in the legal systems gave rise to relatively little difficulty in the UK courts, and over the decades the civil/common approaches have tended to dovetail, to the extent that certain principles such as proportionality (EU) and reasonableness (UK) closely resemble each other in practice.

On 23 June 2016 the UK voted by a narrow majority to leave the European Union.[6] Events thereafter were complicated: on 29 March 2017 Article 50 of the Treaty on European Union (TEU) was formally triggered and the two-year countdown to the UK formally leaving the EU began. (Article 50 TEU was introduced by the Lisbon Treaty in 2009 and set out a right for a Member State to leave the EU together with a process for doing that which requires the service of a two-year period of notice.) The UK was therefore expected to leave the EU on 29 March 2019. But following a House of Commons vote on 14 March 2019, the government sought permission from the EU to extend the Article 50 time period and a later Brexit date was agreed, namely 30 June 2019. This was extended to 31 October 2019, and then to 31 January 2020.

On 23 January 2020, the European Union (Withdrawal Agreement) Act 2020 received Royal Assent, thereby completing the legislation to implement the agreement to withdraw that had been negotiated by the UK and the EU in 2018. It sets out the new constitutional arrangements between the EU and the UK following the end of the Transition Period.

The legislative package leading to Brexit

The scope of the terms governing the exit of the UK from the EU were finally agreed on 17th October 2019 and came into legal effect on 1st February 2020. These terms are contained in the Withdrawal Agreement which was made under Article 50 TEU ('the Withdrawal Agreement'). The EU-UK Withdrawal Agreement is an international treaty which, under UK law, has to be ratified into domestic law in order to have legal effect. This resulted in a lengthy legislative process through the UK Parliament and required extensive legislation to achieve that effect. The 'Brexit bundle,' which finally implemented the EU-UK Withdrawal Agreement into UK domestic law, comprises the European Union (Withdrawal) Act 2018, which was subsequently amended by the European (Withdrawal Agreement) Act 2020, and the European Union (Withdrawal) Act 2018 Exit Day Regulations 2019. It was this legislation which withdrew the UK from the European Union. This has in turn been supplemented by the European Union (Future Relationship) Act 2020.

Current law

The European Union (Withdrawal Agreement) Act 2020 is the key constitutional document which ratifies the treaty between the European Union (EU), and the United Kingdom (UK), signed on 24 January 2020 and sets out the terms of the withdrawal of the UK from the EU and Euratom (the Agreement on the Withdrawal of the United Kingdom of Great Britain and Northern Ireland from the European Union and the European Atomic Energy Community 2020). The immediate impact of the 2020 Act was that it provided the domestic legal basis for the UK to ratify the EU-UK Withdrawal Agreement. This enabled it to be ratified without a further vote in Parliament, contrary to the normal rules for the ratification of international treaties as laid down in the Constitutional Reform and Governance Act 2010.

Retained EU legislation (REUL)

From a legal and environmental health perspective it is the impact of the European Union (Withdrawal) Act 2018 and its creation of a new and broad category of UK law which is of the greatest practical importance. It is not always widely understood that the 2018 Act retains all the EU legislation which was part of the UK's legal system on withdrawal day subject to certain provisos. This preserved legislation is now known as 'Retained EU Law' (REUL). A statutory definition of retained law is given in s.6 (7) of the 2018 Act as

> anything which, on or after exit day continues to be, or forms part of, domestic law by virtue of section 2, 3 or 4 or subsection (3) or (6) above (as that body of law is added to or otherwise modified by or under this Act or by other domestic law from time to time).

The 2018 Act protects all such law from immediate demise post-Brexit, both primary (the EU Treaties) and secondary law (the rest). However, by default, any later amendments to EU retained law by the EU will not be applicable in the UK unless expressly adopted in domestic law. This will, of course, be possible, but the potential for confusion is immediately apparent.

This new category of UK law is created under sections 2 to 4 of the European Union (Withdrawal) Act 2018 and came into force on 1st January 2021, following the repeal of the savings to the European Communities Act 1972 (ECA).

The combined effects of Sections 2 to 4 of the 2018 Act are to:

- Retain EU-derived domestic legislation (as saved and modified during the transition period), as it had effect in UK law at the end of the Transition Period. This includes UK legislation that implements EU directives or relates otherwise to the EU or the European Economic Area. This is of considerable importance for all areas of environmental health law which have been heavily reliant on EU legislation. Notable areas are health and safety legislation, some aspects of food law, consumer protection law, air quality and water law.
- Save and convert into UK law most (but not all) directly applicable EU legislation (such as EU regulations), as it had effect in EU law immediately before the end of the transition period, but only so far as it applied to the UK under the UK-EU transitional arrangements.
- Save and convert into UK law most of the EU rights that before the end of the Transition Period were recognised and available in UK law through the savings and modifications to section 2(1) of the ECA 1972, such as directly effective rights in EU Treaties (e.g. a right to a fair hearing).

Retained EU law also includes any post-transition additions and modifications to this body of UK law, and the interpretations of it by the UK courts (s.6 (7) of the 2018 Act).

In more detail, Sections 2 to 6 of EUWA 2018 sets out five categories of retained EU law:

1 EU-derived legislation

Under s.2 EU-derived domestic law is retained. This is any primary or secondary legislation which implemented or related to former EU obligations. This will include much environmental health law where that EU law has already implemented into the UK, in particular, directives implanted via statutory instruments, for example, Directive 2008/50 on ambient air quality and cleaner air.

2 Direct EU legislation

Under s.3 'direct EU legislation' is retained, so far as it is operative immediately before exit day. This means EU legislation which was directly applicable in the UK without implementing legislation, namely EU regulations and decisions, but not directives. One important example is the EU General Food Law (Regulation (EC) No 178/2002). However, many practical issues in the food area have become far more complex as the application of the EU substantive provisions will now depend upon the application of UK regulations in a purely UK context. For example, under Regulation 1151/2021 the EU uses three different protected status schemes for foodstuffs, which provide differing characteristics and levels and types of protection: Protected Designation of Origin (PDO), Protected Geographical Indication (PGI) and Traditional Speciality Guaranteed (TSG). After the Transition Period, the UK initiated a separate scheme governed by the same rules, which applies in England, Scotland and Wales, so that the designations in effect on 31 December 2020 under the EU scheme (from any country), are since 2021 governed by UK law, which is an amended version of Regulation 1151/2021. However, designations applied for under UK law since 1 January 2021 are not recognised in the EU (although an application for registration under EU law can still be made), and the extent to which designations applied for under EU law are recognised in the UK is uncertain and probably limited. Hence a dual system of protection is now in place in practice, despite the substantive provisions being the same.

3   Rights etc. under s2 (1) European Communities Act 1972

Under s.4 'other rights, powers, liabilities, obligations, restrictions, remedies and procedures' which were recognised and available in UK law on exit day will continue to be so. At its widest interpretation this could incorporate almost all EU law into UK law and will doubtless be the subject of much litigation. As it is, the UK has legislated to expressly exclude various aspects of the treaties from the status of 'retained EU law'. Of particular interest is the Freedom of Establishment and Free Movement of Services (EU Exit) Regulations 2019, which will affect the ability of environmental health professionals to live and work in the EU. More excluding legislation can be expected in the future to attempt to introduce greater certainty and clarity in other areas.

As regards rights arising under international agreements, it is thought that directly effective rights arising under treaties brought into domestic law by the European Communities Act 1972, such as international agreements made by the EU with countries outside the EU, will also be retained by virtue of this section.[7]

Unfortunately, there is no official record of which EU treaty rights were incorporated into UK law and equally unfortunately no attempt has been made by the government departments responsible for environmental health matters (or any other) to give any indication as to which pieces of legislation should be considered 'retained' and which not. Instead, an assessment of each EU law right is required to determine if it meets the

European Union (Withdrawal) Act 2018 criteria, and increasingly this will need to be followed by research to determine if it has been amended or revoked by post-Brexit domestic legislation.[8] A conservative estimate is that about 3,000 pieces of legislation are affected.

4   'Retained case law'

Section 6(3) provides that any question as to the validity, meaning or effect of any retained EU law is to be decided so far as this law is unmodified on or after 31st December, and so far, as it is relevant to it (a) in accordance with any retained EU case law, and any general principles of EU law and (b) having regard (among other things) . . . of EU competences. At section 6 (7) 'retained case law' is defined as (a) retained domestic case law and (b) retained EU case law.

However, in England and Wales neither the Supreme Court, nor the Court of Appeal are bound by retained EU case law (see section 6(4) of the 2018 Act and European Union (Withdrawal) Act 2018 (Relevant Court) (Retained EU Case Law) Regulations 2020. The test is the same test as the Supreme Court would apply in deciding whether to depart from the case law of the Supreme Court – 'when it appears right to do so'.[9] Domestic regulations also purport to provide that other courts and tribunals are not bound by retained case law and set different tests for departure (see for example Regulation 23 of the Competition (Amendment etc.) (EU Exit) Regulations 2019[10] regarding the Competition Appeals Tribunal and the High Court). Again, how and when these differing rules will be applied is a matter for conjecture and uncertainty.

5   'Retained general principles of law'

Under s.6 (7) the 'retained general principles of law' are relevant in the interpretation and application of EU retained law, unless expressly excluded. This apparently minor stipulation incorporates a considerable canon of EU law on the General Principles of Law into UK law.

Section 5(2) of the 2018 Act provides that the supremacy of EU law applies to any enactment or rule of EU law *so far as relevant to the interpretation, disapplication or quashing of any enactment or rule of law passed or made before exit day*. This means that the doctrine of supremacy is still relevant and applicable to all retained EU law.

Judicial interpretation of, and divergence from, retained EU law

Given that most environmental health cases are heard before the lower courts (including magistrates' courts), it is important to appreciate that these lower courts are bound to decide any question as to the meaning, validity or effect of EU-derived domestic legislation in accordance with the decisions of the CJEU made prior to the Transition Period completion day. This includes the High Court when dealing with a judicial review or a similar statutory challenge.

This was followed by the Court of Appeal in March 2021 in *Lipton v BA City Flyer*,[11] a case that gives guidance on how this new category of retained EU law should be applied. The Court of Appeal also explains how English law can be altered under the EU (Future Relationship) Act 2020 where it is inconsistent with the EU/UK Trade & Cooperation Agreement (TCA). In *Lipton*, Green LJ established nine basic principles that can be applied in these circumstances.

1. Consider whether the 'old' EU law is retained EU law. This requires looking at the European Union (Withdrawal) Act 2018 to see if the 'old' EU law (e.g., EU Regulations, CJEU case law, general principles of EU law) has been retained.
2. Consider whether the retained EU law has been amended or even revoked.
3. Apply a purposive construction. This requires taking into account recitals and other principles referred to in the body of the retained regulation and in the recitals.
4. To the extent necessary, this process of interpretation would include any provision of international law that has been incorporated by reference in the regulation.
5. Apply CJEU case law made prior to 1 January 2021 to determine the meaning and effect.
6. Apply general EU law principles, as recognised in pre-1 January 2021 CJEU case law and as derived from the Charter of Fundamental Rights and the TFEU.
7. The Court of Appeal (and Supreme Court, but not the High Court) may depart from retained CJEU case law or any retained general principles 'if it considers it right to do so'.
8. The Trade Cooperation Agreement (TCA) and the EU Future Relationship Act (EUFRA) may be relevant to the effect of existing English law if the subject matter of the domestic law overlaps with the subject matter of the TCA and/or EUFRA, and insofar as domestic law does not already cover the subject matter of the TCA.
9. If the English domestic law does not already reflect the substance of the TCA, then the domestic law 'takes effect in the terms of the TCA'. If domestic law already implements the relevant provisions of the TCA, then there is no need for any further transposition.

It is noteworthy that the effect of point 9 is that it rests upon s29 EU Future Relationship Act where the power to modify existing UK domestic law in such circumstances lies. Thus Section 29 EUFRA modifies existing English law where it does not coincide with the TCA and 'modify', includes amend, repeal and revoke. Described by Green LJ as a 'sweeping up mechanism', s 29 means that existing English law can be modified automatically where there is 'inconsistency, daylight, lacuna' between the existing English law and the TCA.

Section 29 is not universally applicable, and a key limitation is that it only applies to 'existing domestic law' which is defined to be such law in existence at the time the TCA came into force, so it does not have any impact on any future domestic law, which it may be assumed would be made in compliance with the TCA. In addition, in *Lipton*, Green LJ states that section 29 only applies in such limited circumstances as where the domestic law is not already consistent with the TCA; and only applies in order to achieve compliance with the international obligations of the UK under the TCA.

Clearly, the judgment in *Lipton* contained in Green LJ's decision is key to the interpretation of the new category of retained EU law. It is worth bearing in mind the range of the TCA which covers many areas such as healthcare, life sciences, public procurement, energy and transport.

The effect of section 29 is to introduce some uncertainty into English law post Brexit. It is necessary to have such a provision to deal with inconsistencies between pre Brexit UK law and the TCA, but the fact that such inconsistencies may emerge over a period of time into the future does render some potential for dispute and argument about the effect of law. Given that the way in which the TCA is read into existing UK law could be open to different interpretations, there is likely to be an area here for litigation into the future.

The decision in *Lipton* may well itself be challenged in the future, especially as the decision concerned a period of time before the Transition Period had expired, thus it is arguable that the judgments should have been confined to the interpretation of the applicable EU law. A High Court decision, *Varano v Air Canada*,[12] saw arguments about the correctness of *Lipton* being made, but judicial precedent meant that the decision in the CA was binding.

Further litigation on the status of EU retained law has arisen in the High Court: (*Pearce v BEIS SoS* [2021] EWHC 326 [Admin]). This was a judicial review of the North Vanguard Offshore Wind Farm Order made under the Planning Act 2008. In *Pearce*, Holgate J. following *Lipton* held that the effect was that the High Court was bound by retained EU case law to apply the stricter EU law test for discretion to decline to grant relief where a judicial review succeeds on an EU point of law ([148]).

This decision confirmed that the Supreme Court and Court of Appeal are not bound by any retained EU case law but can depart from it where 'it appears right to do so' under the test in the Practice Statement (Judicial Precedent) [1966] 1 WLR 1234. This follows from s6(5) which provides that the same test must be adopted as would be applied by the Supreme Court when deciding to depart from its own case law. This position was also taken by the Court of Appeal in *R (on the application of Open Rights Group and another) v Secretary of State for the Home Department and another (Liberty and another intervening)* [2021] EWCA Civ 1573 where Warby LJ at [23] stated that

> a relevant court is not absolutely bound by any retained EU case law: EUWA s 6(4)(ba) and Regulations 1 and 4. It can depart from that law; but the test to be applied in deciding whether to do so is the same test as the Supreme Court would apply in deciding whether to depart from the case law of the Supreme Court: EUWA 6(5A)(c) and Regulation 5. . . . (and this test) . . . is the one laid down by the House of Lords in its Practice Statement [1966] 1 WLR 1234.

The power of the Supreme Court to depart from retained EU case law is set out in s6(4) of the EUWA 2018 whereas the Court of Appeal's power derives from the *European Union (Withdrawal) Act 2018 (Relevant Court) (Retained EU Case Law) Regulations 2020*. In respect of the Court of Appeal, it is bound by the rules of precedent so will only be bound by retained EU case law so far, as there is post-transition case law which modifies or applies that retained EU case law and which is binding on the Court of Appeal.

Since BREXIT day, there has been a plethora of litigation where points are, inevitably, being taken on the role and status of retained EU law. For example, in *TuneIn v Warner Music* [2021] EWCA Civ 44, a case on music copyright infringement relating to an app which allowed access to global radio stations, reference was made to 24 CJEU judgments comprising retained EU case law. The applicants argued that the Court of Appeal should depart from the entire body of CJEU case law on the issue including one CJEU case decided after the end of the Transition Period.

Two of the Lords Justices of Appeal dealt with this question of departing from retained EU case law. Arnold LJ, (at [75]) stated that *this is a power to be exercised with great caution* referring to the earlier decision of Lord Bingham in *Horton v Sadler* [2006] UKHL 27 who said (at [29]) that the power to depart from the CJEU's jurisprudence is one to be exercised only 'rarely and sparingly'.

Arnold LJ gave eight reasons for this which included that Parliament had not changed the domestic legislation and there had been no change in the relevant international legislative framework. In relation to international treaties, he noted (at [79]) that where the matter was regulated by international treaties the courts should strive for 'consistency of interpretation, rather than unilaterally adopting their own interpretations'.

Arnold LJ also recognised the unrivalled experience and developing jurisprudence of the CJEU in dealing with the relevant issues in many different contexts and there was no good reason why the court would create uncertainty by departing from the EU legal order and recreating the regime (see [80] and [83]).

The issue of precedent and the need for certainty was also reflected upon by the Master of the Rolls, Vos LJ, who observed (at [200]) that it was clear that the Supreme Court should not refuse to follow its own earlier decisions simply because the court would have decided it differently.

Part of the judgment referenced the fact that the arrangements for departing from the EU were an aspect of international law and as such the conventions around not departing from international agreements should be observed. Vos LJ, for example (at [198]), commented that the courts of the states that accede to such treaties should, wherever possible, be striving to *achieve harmonious interpretation of them, not individualistic disharmony*. A nation should only depart from the jurisprudence of the CJEU where there was an exceptionally good reason to do so. Vos LJ considered (at [197]) that this case was the perfect example of a situation where it would be inappropriate for the Court of Appeal to exercise its new-found power to depart from retained EU law.

There is a developing abundance of case law and the way in which the courts are likely to interpret retained EU law will become clearer over time. Cases in the environmental law field are rarer at the moment but include the case of *Gladman v SSHCLG & Medway Council*.[13] Gladman upheld the approach that the retained EU law should be followed and applied. In this instance this was the CJEU judgment in *People Over Wind v Coillte* (C-323/17) [2018] Env LR 31[14] – a case which concerned the role of mitigation at the 'screening' stage under the Habitats Directive. The High Court preferred the CJEU case to a domestic decision in *Hart DC v SSCLG*.[15]

There is also the question of the way in which the UK courts will treat decisions of the CJEU after Brexit day. Technically, the courts are not bound by such decisions at all (Section 6(1) of the EUWA 2018). But in *TuneIn*, the Court of Appeal, when considering how to deal with a post-Brexit case, considered that it should be treated as highly persuasive because it was directly relevant to the issues the court had to decide. It would seem that, currently, the general tenor of the UK courts is to adopt an approach of following CJEU jurisprudence unless there is a good reason to depart from it.

To summarise, retained EU case law in the environmental field will continue to apply, provided the domestic legislation remains unchanged. Domestic courts must meet a high threshold for departing from it and to date seem disinclined to do so. This discretion also only applies at the level of the Supreme Court or the Court of Appeal. So, the status quo is likely to be maintained in the environmental law field at least. The extent to which this will be applicable in other fields is entirely a matter of conjecture, but in the interests of free movement of goods it seems likely that the same conservative approach will be taken. Otherwise, there is a risk that there will be the consequence that there may be two versions of the same law in that the retained EU law version may be different from the original EU law on which it is based – a curious result.

Legislative divergence from EU retained law

It was a fundamental objective of Brexit that full parliamentary sovereignty should be restored, and since 1st January 2021 the UK is no longer bound by EU law, claiming an absolute and unlimited legal authority for Parliament which orthodox legal theory says is able to make, amend and repeal any laws it wishes. The UK can therefore now diverge as much as it pleases from EU retained law. However, to date the extent of divergence appears to be limited.

One such example of how divergence is working in practice is in relation to health claims, which were previously covered by EU Regulation 1924/2006 on nutrition and health claims made on foods. Regulation 1924/2006 is retained law but is amended so that, procedurally, authorisations for health claims go to a UK Committee – the UK Nutrition and Health Claims Committee (UKNHCC). Nonetheless, applicants for the authorisation of a health claim should continue to consult EFSA guidance on the scientific requirements for health claims and it is expressly stated that UKHNCC *will take a similar approach to that established by EFSA (European Food Safety Authority).*[16]

Another example relates to the responsibility for assessing food and animal feed safety in the UK which has now been taken over by the Food Standards Agency (FSA) and Food Standards Scotland (FSS). The procedure adopted since leaving the EU is that a risk assessment will be done in consultation with experts from the independent Scientific Advisory Committees[17] and Joint Expert Groups[18] and will be applicable across the UK although there may be nation-based risk assessments. The FSA, in recognising that a greater responsibility now falls on the UK to deal with risk, has taken certain steps such as expanding the role of the Scientific Advisory Committees and establishing new Joint Expert Groups. There is also a new process to advise government ministers on authorising regulated food and feed products for sale in England, Wales and Scotland. This is within the overall context that current food and feed safety rules have not changed because of the status of retained EU law. But when these rules change then the UK will adopt its own process rather than that of the EU.[19] (This, as explained later, is different for Northern Ireland where Annex 2 of the Northern Ireland Protocol means that food or feed marketed in Northern Ireland must follow the whole current body of EU food and feed safety and hygiene regulations (not just EU law which is retained at the day of departure).

In September 2021, the government announced that it would be reviewing the entire corpus of EU law which entered the UK statute book via the European Union (Withdrawal) Act of 2018. The review would remove the special status of retained EU law (REUL), so that it is no longer a distinct category of UK domestic law, but normalised within our law, with a clear legislative status, Lord Frost, on behalf of the government, said. *'Unless we do this, we risk giving undue precedence to laws derived from EU legislation over laws made properly by this parliament'.*

The substantive content of retained EU law would also be scrutinised. Our intention is eventually to amend, to replace, or to repeal all that retained EU law that is not right for the UK. Conceding this would be a '*mammoth task*', he went on to say that the government is looking at '*developing a tailored mechanism for accelerating the repeal or amendment of this retained EU law*'.[20]

So it would appear that the government is determined to follow through with a plan which was originally mooted during the Brexit campaign, but subsequently abandoned as too divisive and ambitious. This decision is against the background of the UK courts having emphasised this year (referred to above) that they will only exercise their new discretion to overturn EU case law in exceptional circumstances because of concerns around legal uncertainty and the state obligation to abide by international agreements. Clearly, any widening of the scope to overturn long-established case law will only encourage more speculative litigation and greater uncertainty as to the meaning and effects of a wide range of legislation, including current environmental health measures.

The details of the reforms that the government has in mind for the status of retained EU law (REUL) in the UK's domestic legal system are as yet unclear. It seems highly probable that the government will target the supremacy and direct effect of some EU legislation in the UK under the Withdrawal Act 2018, which continues to override pre-Brexit domestic legislation if a conflict between the two comes to light (as explained above). Beyond this, changes to the status of retained EU law in UK legislation are more likely to be presentational than substantive, because the reality is that divergence from EU standards in many fields would have serious and damaging economic consequences. Nonetheless, this concern for the presentational has meant that increasingly it appears common practice to put a broad clause into new legislation relating to REUL to give the government the power to amend it without the need for primary legislation.

As of writing in December 2021, in order to make these reforms, the government had flagged its intention to put a general Bill before the UK parliament asking for wide powers to amend retained EU law by secondary legislation. Questions are being asked if the government is looking to give itself a broad superpower – what has been called a Henry VIII clause on steroids[21] –- so it can remove swathes of REUL secondary law where it has failed to already give itself the necessary power.

The parliamentary history of the contentious power in Section 8 of the Withdrawal Act 2018 indicates that such new powers would be highly controversial, especially as even greater discretion would be conferred on government ministers to set new policy in more substantive amendments. This is in contrast to the current situation whereby such substantive reforms generally require an Act of Parliament on each occasion, and the full scrutiny process that involves. There is also the problem that the government's plans to repeal a significant volume of REUL will inevitably bring it into conflict with Scotland which wants to continue to map not just existing but also future EU law in key areas, and Northern Ireland which is obliged to comply with existing and future EU law in the areas covered by the Northern Ireland Protocol (see next section). Given the existing lack of parliamentary mechanisms to scrutinise any secondary legislation, it seems inevitable that parliamentary battles lie ahead as the government's opponents in the Commons and Lords will seek to minimise the discretionary powers of government to re-write consequential legislation in future.

Northern Ireland

One consequence of the withdrawal is that the UK now has a land border with an EU country, the Republic of Ireland.[22] This fact has geopolitical consequences given the history of the UK and Ireland and the necessity of avoiding a hard border between the Republic and Northern Ireland. This was provided for in an annex to the Withdrawal Agreement 2020 known as the Northern Ireland Protocol which came into force on 1 January 2021. This allows Northern Ireland to be part of the UK customs territory so that it will follow UK trading policy, while also being required to apply EU customs law in full. Further, Northern Ireland also remains aligned with the rules of the EU Single Market for goods (Article 5 of the Protocol), even though it is formally outside that market. This to ensure that there are no customs checks or controls between Northern Ireland and the rest of the island. However, because the Single Market provisions require there to be certain customs checks and trade controls at its external borders, in place of an Ireland/Northern Ireland land border the Protocol has created a *de facto* customs border down the Irish Sea for customs purposes, thereby arguably separating Northern Ireland from the island of Great Britain. A consequence of this unique arrangement is that the jurisdiction of the Court of Justice of the European Union (CJEU) will continue to extend to Northern Ireland whereas it no longer has any jurisdiction over the other constituent nations of the UK. Article 13 of the Protocol gives a particular role to the CJEU with regards to procedures in case of non-compliance as well as the possibility and requirement for UK courts to ask for preliminary rulings on the application of EU law and related parts of the Protocol.

The consequences of this unique arrangement continue to unfold and are generally considered to be difficult and unpredictable for both political and legal reasons. This has become even more apparent with the conclusion of the Trade and Cooperation Agreement (TCA) 2020. This agreement does not govern trade in goods between the EU and Northern Ireland, and goods entering NI from Great Britain count as imports (Northern Ireland Protocol). Consequently, such goods must comply with EU product rules and be subject to checks and controls for safety, health and other public policy purposes, as well as SPS controls between the EU and the UK. As yet this has not been fully implemented, but considerable disquiet has been voiced by environmental health professionals about the practical and legal ramifications that full implementation of the Northern Ireland Protocol would entail.[23]

The Protocol gives the Northern Ireland Assembly a decisive voice every four years on whether to continue with these arrangements. The first consent vote will take place in December 2024.

The trade agreements

Trade agreements are of fundamental importance for the economic functioning of the UK, but it is often not appreciated that they affect all aspects of life in the UK either directly or indirectly. So, while their direct impact is to open foreign markets to all UK businesses by modifying trade barriers (such as tariffs, quotas or non-tariff barriers) and creating more transparent conditions for such businesses operating abroad, they may influence the development of broader policies such as the environment, food safety, and health and safety. The importance of trade agreements is difficult to over-estimate, with at least one in five jobs in the UK being entirely dependent on exports (2021 DIT).[24] Membership of the European Union provided a legal framework for UK external trade for 47 years and the removal of that framework poses one of the greatest challenges post-Brexit.

The UK-EU Trade and Cooperation Agreement

On 26th December 2020, the UK concluded a free trade agreement with the EU called the *Trade and Cooperation Agreement between the European Union and the European Atomic Energy Community, of the one part, and the United Kingdom of Great Britain and Northern*

Ireland, of the other part ('the TCA'). This was incorporated into domestic law by the European Union (Future Relationship) Agreement 2020 which received Royal Assent on 31st December 2020. It was provisionally effective from 1st January 2021 and fully enforceable as of 1st May, following ratification by the EU.

The TCA is a new international treaty intended to govern the trading relationship between the EU and the UK. It is intended to create a Free Trade Zone and is therefore subject to the general conditions of the World Trade Organization, of which both the EU and the UK are members. It consists of:

1 A free trade agreement
2 Cooperation on economic, social, environmental and fisheries issues
3 A close partnership for citizens' security
4 An overarching governance framework

The free trade agreement focuses primarily on trade in goods, with limited mutual access for services, and is intended to replace the Single Market regime which was applied to the UK until 31st December 2020. It is an international treaty and seen as the key document which will govern all aspects of trade between the UK and the EU but can be terminated by either party with 12 months' notice.

It sets out preferential arrangements in areas such as trade in goods and in services, digital trade, intellectual property, public procurement, aviation and road transport, energy, fisheries, social security coordination, law enforcement and judicial cooperation in criminal matters, participation in Union programmes and 'thematic cooperation' (TCA). This cooperation extends to economic aspects of the relationship, including energy, environment, health collaboration and social policy. The primary concern is on trade in goods, so the emphasis is on tariffs and quotas (in particular when zero tariffs or quotas can be applied), rules of origin and customs formalities. Of importance from an environmental health perspective are the sanitary and phytosanitary (SPS) requirements: UK agri-exporters are now required to meet all EU SPS import requirements and vice versa. However, the UK was granted 'national listed status' separately from the conclusion of the TCA, which means that UK exports to the EU of live animals and products of animal origin can continue, albeit with greater custom formalities needing to be satisfied. The EU also lifted a number of plant health prohibitions and granted equivalence for certain products (such as some seeds and propagating materials), enabling exports of such products from the UK to the EU to continue.

Also of relevance, particularly in the agri-food sector, are the provisions on technical barriers to trade and product conformity assessments: from 1 January 2021 all products exported from the EU to the UK need to meet the UK's technical product regulations (and vice versa). However, to prevent and reduce unnecessary technical barriers and requirements, the TCA includes a number of provisions related to technical barriers to trade. In particular, the parties have agreed to a definition of international standards which identifies the relevant international standard-setting bodies. This should maximise the extent to which the parties' domestic product standards and technical regulations are compatible. Furthermore, in relation to product conformity assessments, the parties have agreed to maintain self-certification of conformity by the manufacturer where this was applied in both the EU and the UK on the date that the TCA came into force. Finally, sector-specific provisions are included to promote cooperation and reduce barriers to trade in certain sectors, including chemicals, organic products and wine.[25]

These substantive measures are underpinned by provisions designed to ensure a level playing field and respect for fundamental rights. As such it is very much a framework agreement and lacks much of the practical detail necessary to achieve the objectives it envisages. In many keys areas much further negotiation between the parties will be needed to put the necessary agreements and arrangements in place, and it is expected that the TCA will be supplemented in due course with bilateral agreements in other areas. The responsibility for this has been placed on a joint body, known as the Partnership Council. It is this body which will oversee the implementation of the Agreement, and to facilitate this the Partnership Council can meet in different configurations depending on the matter at hand. It will be the forum in which the parties will discuss any issues that might arise, with the power to take binding decisions by mutual consent. It will be assisted in its work by Specialised Committees and Working Groups.[26]

The Northern Ireland Assembly graphic[27] (Figure 6.1) shows the governance structure of the TCA.

The work of the Trade Specialised Committee (TSC) on Sanitary and Phytosanitary Measures, and that of the TSC on Customs Cooperation and Rules of Origin are expected to be of particular relevance to environmental health practitioners, and the TSC on Regulatory Cooperation may also evolve to be important. However, meetings are not expected to be open to the public and proceedings are only being published ex post facto. The decisions of the TSCs will undoubtedly be vital to the practical functioning of the measures to be applied, but the opportunities to influence those decisions appear extremely limited.

The Partnership Council is co-chaired by a member of the European Commission and a UK minister and will meet at least once a year but can meet more often at the request of either party. Decisions have to be taken by mutual consent between the parties, a formula which could well lead to tensions and delays. If a solution to a disagreement cannot be found between the EU and the UK, an independent arbitration tribunal can be established to settle the matter through a binding ruling. This horizontal dispute settlement mechanism covers most areas of the

Agreement, including a level playing field and fisheries. It is accompanied by enforcement and safeguard mechanisms, including the possibility to suspend market access commitments (e.g., by reintroducing tariffs and/or quotas in the affected sector). Both parties will furthermore be able to cross-retaliate if the other does not comply with a ruling of an independent arbitration tribunal. For instance, a breach by one party that concerns a specific economic sector will allow the other party to retaliate with measures in other economic sectors. Finally, any substantial breach of obligations enshrined as 'essential elements' of the Agreement (the fight against climate change, respect for democratic values and fundamental rights, or non-proliferation) can trigger the suspension or termination of all or part of the entire TCA.[28] Crucially, there is no role for the CJEU or the national courts in interpreting the TCA or resolving any disputes relating to it. Disputes are to be resolved exclusively according to the dispute resolution procedures set out in the TCA, in particular in Part 6 TCA. These are loosely modelled on the WTO dispute resolution arrangements. Equally only the EU and the UK are permitted to bring claims under the TCA; no provision is made for natural or legal persons to do so, hence the *locus standi* restrictions are even stricter than those before the CJEU.

Trade agreements with non-EU countries

Previously the UK's trading arrangements with non-EU countries were largely conducted within the scope of the EU's external trade policy and law, whereby its trading agreements were negotiated by the European Commission on behalf of the bloc as a whole. The provisions of these agreements were then incorporated into EU law, which in turn were implemented into UK domestic law. As of 1st January 2021, the EU trade agreements no longer apply in the UK, but on the basis of the principles of EU retained

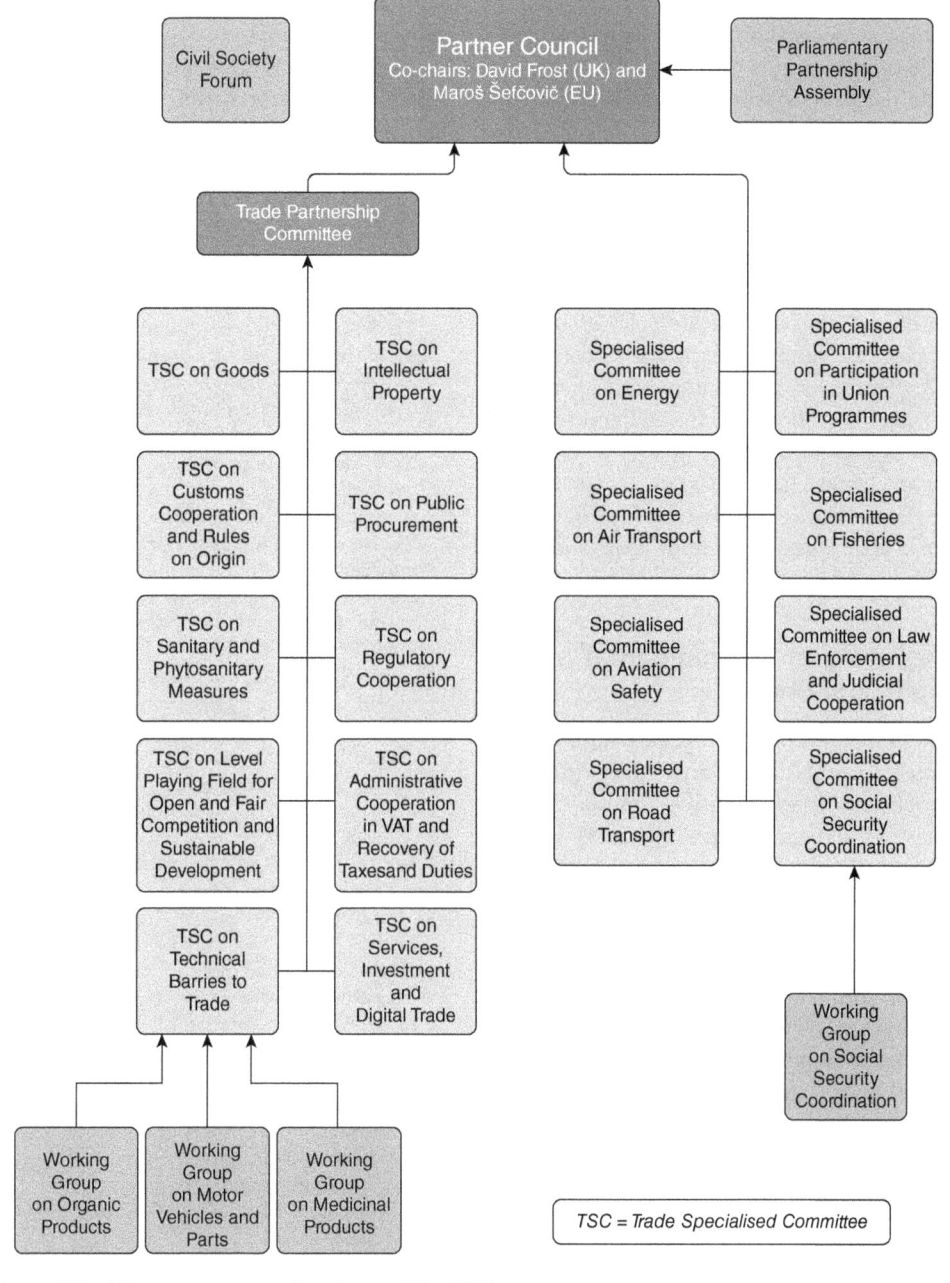

Figure 6.1 The governance structure of the TCA

law, the implementing UK domestic law remains in force.

Although the UK did not actually withdraw from the European Single Market and the European Union Customs Union (and its trade agreements) until 31 December 2020, since 1st January 2020 the UK was free to negotiate its own external trade deals, and it began these negotiations on several free trade agreements to remove or reduce tariff and non-tariff barriers to trade, both to establish new agreements and to replace previous EU trade agreements.

At the time of writing, these negotiations have had limited success. The United Kingdom has concluded only two new major trade agreements, one with Japan; the other with Australia. It has signed trade deals and agreements in principle with 69 countries, mainly by virtue of 35 'trade continuity agreements' but the majority of these are 'rollover' deals – copying the terms of deals the UK already had when it was an EU member, rather than creating new benefits. In addition, it has begun other negotiations, notably to join the Comprehensive and Progressive Agreement for Trans-Pacific Partnership. However, the much-prized goal of a trade deal with the USA appears unlikely in the foreseeable future.

The future for environmental health and Brexit?

The House of Lords European Union Committee published a report *Beyond Brexit: food, environment, energy and health*[29] on the importance of EU membership to the development of food, health and safety, air quality, water quality, product safety and other environmental legislation. As the first attempt post-Brexit to assess the impact of departure from the EU in the environmental health field, its observations are both interesting and instructive. It acknowledges that access to food, energy, healthcare and a safe environment are all being shaped by the EU-UK Trade and Co-operation Agreement (TCA). It finds that even where tariff-free access for many products (including food and agricultural produce) has been achieved by the TCA, a considerable number of substantial barriers have appeared, many of which it fears will be long term.

Overall, the tone of the report is negative. It is clear that, post-Brexit, there is still a considerable gulf between the EU and the UK in the interpretation of the TCA. The Report highlights the importance of the TSCs,

particularly the TSC on Sanitary and Phytosanitary Measures, as a means of resolving the many outstanding issues, and the need for supplementary agreements between the EU and the UK in many of the areas pertinent to environmental health.

Notes

1 The Customs Union means members of the EU have abolished tariffs and quotas between Member States so as to encourage the free movement of goods and services, at the same time adopting a common external tariff on imports from non-member (third-party) countries. The aim is to make the Single Market work effectively.
2 A useful source of information about types of EU law is to be found at: https://ec.europa.eu/info/law_en
3 Stephen George, 'An awkward partner: Britain in the European Community' OUP 3rd ed. 1998.
4 It is important to note that Scottish law is a hybrid or mixed legal system containing civil law as well as common law elements, and has a distinct legal tradition.
5 www.opendemocracy.net
6 How narrow is obvious to anyone who reads William Keegan 'In My Views' in the *Observer*. See https://www.theguardian.com/profile/williamkeegan?page=2 but on a turnout of 72.21% of 46,500,001 registered voters, 51.89% (17,410,642) voted to leave and 48.11% (16,141,241) voted to remain and he constantly has pointed out the decision to leave was made by 37.4% of the electorate.
7 'Converting and preserving law', Factsheet 2, Department for Exiting the EU, July 2017. https://assets.publishing.service.gov.uk/government/uploads/system/uploads/attachment_data/file/714373/2.pdf
8 www.pinsentmasons.com
9 Practice Statement (Judicial Precedent) [1966] 1 WLR 1234.
10 SI 2019 No 93 at https://www.legislation.gov.uk/uksi/2019/93/regulation/23/made
11 [2021] EWCA Civ 454 at [69].
12 [2021] EWHC 1336 (QB).
13 *Gladman v SSHCLG & Medway Council* [2020] Env LR 7 https://www.bailii.org/ew/cases/EWHC/Admin/2019/2001.html
14 *People Over Wind v Coillte* (C-323/17) [2018] Env LR 31 https://www.bailii.org/eu/cases/EUECJ/2018/C32317.html

15 [2008] EWHC 1204 (Admin) available at https://www.bailii.org/ew/cases/EWHC/Admin/2008/1204.html

16 See https://www.gov.uk/government/groups/uk-nutrition-and-health-claims-committee

17 See https://sac.food.gov.uk

18 See https://cot.food.gov.uk/jointexpertgroups

19 See www.food.gov.uk

20 Oral statement by Lord Frost made to the House of Lords September 2021 https://www.gov.uk/government/speeches/lord-frost-statement-to-the-house-of-lords-16-september-2021

21 Barnard C, Retained EU Law and Brexit opportunities at www.euanduk.ac.uk

22 It seems often to have been overlooked that the UK as a member of the EU always had a land border with another Member State.

23 Brexit and Environmental Health, Chartered Institute of Environmental Health Parliamentary Briefing, 2017. https://www.cieh.org/media/1169/brexit-and-environmental-healt.pdf; Smethurst S, Post-Brexit food checks delayed; CIEH backs UK government decision as not all ports ready September 2021; https://www.cieh.org/ehn/food-safety-integrity/2021/september/delay-food-import-checks/

24 Establishing the relationship between exports and the labour market in the UK (March 2021, Department of International Trade) available at https://assets.publishing.service.gov.uk/government/uploads/system/uploads/attachment_data/file/966549/Estimating-the-relationship-between-exports-and-the-labour-market-in-the-UK.pdf

25 Beyond Brexit: trade in goods (March 2021, 24th Report of session 2019-21 HL Paper 249) available at https://publications.parliament.uk/pa/ld5801/ldselect/ldeucom/249/24902.htm

26 See www.eur-lex.europa.eu

27 Available at http://www.niassembly.gov.uk/assembly-business/brexit-and-beyond/governance-of-eu-exit-agreements/

28 See www.ec.europa.eu

29 Beyond Brexit: food, environment, energy and health (March 2021, 22nd Report of Session 2019–21, HL Paper 247) available at https://publications.parliament.uk/pa/ld5801/ldselect/ldeucom/247/247.pdf

7

Business management, environmental health and the EHP/business interface

Roger Pearce

Introduction

Much of the work of EHPs brings them into contact with businesses, and while EHPs have traditionally been employed by governmental agencies (usually local government or municipalities) these days many EHPs work for, or have started businesses, including consultancies and 'outsourced' environmental health services for local government. This chapter aims to provide an insight into business given that EHPs, whether working for regulatory agencies or not, should have some appreciation of the world of 'the regulated' with whom they come into contact.

In the context of this chapter, businesses are generally commercial enterprises or organisations that provide goods or services and generate an income for the business owners. They create jobs (and income) for employees and are essential to modern economies and society, and small businesses are particularly important for local communities. Businesses can create opportunities and innovations. As business becomes increasingly global, commerce works on a macroeconomic scale and plays an integral part in driving better living standards and making resources available. Effective market competition can also provide choice and benefit to the consumer.

Larger businesses usually start as small and medium-size businesses (SMEs) and SMEs are crucial to the UK economy making up 99.9% of the nation's business population. A SME is generally an enterprise with fewer than 250 employees. SMEs are important to all economies, and findings of one study showed that there is a significant relationship between the operation of small and medium-scale enterprises and economic growth in developing nations.[1]

The SME is defined by the EU also as a business with fewer than 250 employees, but also a turnover of less than €50 million, or a balance sheet total of less than €43 million. The EU defines three categories of business within the overall SME definition:

- A medium-sized business has fewer than 250 employees and either a turnover up to €50 million or a balance sheet total of up to €43 million;
- A small business has fewer than 50 employees and either a turnover of up to €10 million or a balance sheet total of up to €10 million;
- A micro-business has fewer than ten employees and either a turnover of up to €2 million or a balance sheet total of up to €2 million.

DOI: 10.1201/9781003035640-7

Given that the UK has left the EU this may not seem relevant; however a number of smaller businesses are relocating wholly or in part to EU countries as the result of difficulties encountered from the UK no longer being a member of the EU (or the Single Market or Customs Union).

Not all enterprises are commercial businesses though; there are charities and not-for-profit organisations, and there are municipalities.

Local authorities are funded through a combination of business rates, central government grants and council tax. In 2018/19, local authorities in England received 31% of their funding from government grants, 52% from council tax and 17% from retained business rates – revenue from business rates that they do not send to the Treasury.[2]

The business rate (payable by non-domestic properties) starts from the rateable value of a business premises as set by the Valuation Office Agency and based on their assessment of the open market rental value. The amount of business rates payable is calculated by multiplying the rateable value by the correct 'multiplier' (an amount set by the government) and then awarding any relevant reliefs and reductions.

Before 2013/14, business rates revenues were collected at a local level but then sent to the Treasury and were then redistributed to local authorities through central government grant funding. This system changed as a result of the Local Government Finance Act 2012 so from 2013/14 authorities kept 50% of the business rates revenues raised locally, while the grant they receive from central government was reduced to compensate. The central share is redistributed to councils in the form of revenue support grant. The aim of this change was in part to give financial incentives to councils to grow their local economies. The government has said that businesses and commercial enterprises are, therefore, key stakeholders for local authorities [1].

The government planned to increase the proportion of gains or losses borne locally to 75% from 2020–21 (from the standard 50%) and has been piloting 100% schemes in local authorities covering half the English population. As a general point the Formula Grant from central government has largely been funded by local business rates income (which is collected for central government). General grant and business rates are added together to make up the Formula Grant, which is then distributed to local authorities using a complex formula. Local government funding is a devolved matter and Scotland, Wales and Northern Ireland have different local government structures and funding from those in England.[3]

The government has frozen the business rates multiplier from 2020–21 to 2021–22, in response to the pressure caused by the Covid-19 pandemic. Normal practice is for the multiplier to rise in line with the consumer price index (CPI). Freezing the multiplier will lead to local authorities collecting less in business rates revenue than they otherwise would.

Unlike central government (and private businesses), local authorities cannot borrow to finance day-to-day spending, and so they must either run balanced budgets or draw down reserves – money built up by under-spending in earlier years – to ensure that their annual spending does not exceed their annual revenue. They also generate income through rents, charges for their services, sales and investments. Whilst not really commercial organisations local authorities are subject to some of the same governance and financial expectations placed on businesses. According to the National Audit Office as the result of the pandemic authorities have reported losses in their sales, fees and charges of £2.1 billion, commercial losses at £523 million and other income streams of £221 million. Authorities have also forecast losses in council tax at £1.3 billion and business rates of £1.6 billion income for 2020–21.[4]

The 2021–22 local government finance settlement was a one-year settlement, based on decisions in the November 2020 Spending Review.[5] The government had planned to implement the Fair Funding Review, and reset the business rate retention scheme, as of April 2021. This was postponed as a result of the Covid-19 pandemic. It is not yet clear whether it will now take place in April 2022 or at a later date.

Councils responded to reductions in their funding by reducing or redesigning services and collecting more local taxes where possible but also by participating in initiatives such as the government's 'city deals' and local enterprise partnerships aimed at driving local economic growth [2].

It is also a time of change for business with increasing regulation and governance played out against a challenging and uncertain global economy and the pandemic. It can be argued that consumers are becoming more socially, environmentally and ethically aware, and through increasing exposure to the work of both environmental health and local authorities in the media, are much more aware of food safety, public health and legal compliance issues.

The Covid-19 pandemic and the lockdowns have affected all businesses to a greater or lesser extent; some have not survived at all. This has affected issues in health and safety and food security. Many businesses have also been affected one way or another by the UK leaving the EU, even if this has only been most obvious by a lack of deliveries of goods or materials.

Most businesses are committed to being responsible and doing 'the right thing', going beyond minimum legal requirements, and protecting their reputation and the relationship with EHPs and other regulatory professionals is widely considered as a partnership rather than 'them and us'. This is reflected in the increase in health and safety, food safety and other environmental health-related employment across business and industry. Senior level safety appointments, including board-level positions, are routinely made and EHPs have applied their specific skills and attributes, or EH-ness, very successfully to such roles.

Business structure and associated legal responsibilities

The legal structure of the business defines the legal responsibilities including registration, taxation, financial reporting and the personal liabilities of the directors.

The main types of business are:

- Sole trader;
- Business partnership;
- Limited company.

Sole trader and business partnerships

Any person working for themselves is classed as a self-employed sole trader. As the exclusive owner of the business, the individual is entitled to keep all profits after tax but retains personal liability for any losses that the business suffers. Sole traders can employ staff.

Typically, sole trader firms are small, require limited capital to set up but are, of course, limited by the trader's available investment capital. This can be overcome by the creation of a business partnership. The partners can be active in the business or 'sleeping' partners who provide investment but otherwise have no involvement in the day-to-day running of the business. A contractual deed of partnership is typically drawn up which sets out the split in ownership of assets and the share of profits; as with the sole trader structure the business owners have unlimited liability for any losses.

As well as additional capital investment, partnerships allow for groups of specialists, for example dentists, solicitors, etc., or business partners with complementary skills to work together. Partnerships in Scotland (known as 'firms') are different and have a legal identity separate from the individual partners.

The legal responsibilities for both sole trader and ordinary or 'unlimited' partnerships include completion of tax returns, payment of income tax on profits and national insurance as well as VAT registration where turnover exceeds the threshold.

Limited companies

A limited company is a business structure used to limit the personal liability of its owners and directors. The basis of limited liability is that all debts incurred by a company are the company's liabilities and are not directly the legal liabilities of the shareholders or of the directors of the company. The company is a separate legal entity from its shareholders and the directors. If the company incurs debts in the course of its business, then only the company is liable for those debts.

In a company limited by shares, the shareholders' obligation is to pay the company for the shares they have taken in it and once the shares are fully paid for, no further money is payable by the shareholders. In the case of a private limited company, the personal liabilities of the owners and directors are limited to paying to the company the amount they have agreed to pay for their shares. This may be a purely nominal amount, for example if the shareholders have each taken one £1 share.

The shareholders of a company limited by guarantee are bound by a guarantee in the company's memorandum of association requiring them to pay the company's debts up to a fixed sum, which is usually £1. The private company limited by guarantee structure tends to be used primarily for non-profit or voluntary organisations, charities, social enterprises and community groups as the organisation still requires a management structure to be in place thus ensuring a democratic approach to achieving the group's objectives.

The directors of limited companies incur no personal liability as all their acts are undertaken as agents for the company. However, there are certain circumstances where liability may be imposed by the court, particularly in respect of wrongful or fraudulent trading. Some potential creditors of a small, limited company may ask the directors to give personal guarantees of the money owed to them, for example a bank loan, overdraft or when taking a lease of premises.

Registered companies

All limited companies must be registered or 'incorporated' with Companies House, which is the United Kingdom's registrar of companies and is an executive agency of the government under the remit of the Department for Business, Energy and Industrial Strategy (BEIS). The United Kingdom has had a system of company registration since 1844 and the Companies Act 2006 covers the requirements placed on registered businesses. England and Wales are treated as a single entity with a unified register while Scotland and Northern Ireland maintain separate registers.

To register, a business must have a company name, registered address, at least one director and one stakeholder, an agreement or 'memorandum of association' of all initial shareholders ('subscribers') to create the company, details of the company's shares and the rights attached to them, known as a 'statement of capital' and written rules about how the company is run or 'articles of association'. Once registered a 'Certificate of Incorporation' is issued confirming that the company legally exists and showing the company number and date of formation.

At the end of June 2021, there were 4,792,548 companies on the total register at the UK Companies House and 4,420,843 on the effective register (which excludes those in the course of removal or liquidation). A company is defined as a specific legal form of business formed under the Companies Act 2006. Incorporation is the process by which a new or existing business registers as a limited company. A company is a legal entity with a separate identity from those who own or run it. The vast majority of companies are limited liability companies where the liability of the members is limited by shares or by guarantee.

In the UK a business cannot operate as a limited company until it has been incorporated at Companies House under the Companies Act 2006. Establishing a business as a company means the directors are required to file certain documents every year such as annual accounts and a confirmation statement. They must also inform Companies House about any changes, such as the appointment or resignation of directors or a change to the company's registered office. Companies are registered at Companies House regardless of whether they go on to trade actively.

During the second quarter of 2021, the total register increased by 76,422 companies (1.6%), whereas the effective register increased by 10,607 companies (0.2%) during the second quarter of 2021. The number of dissolutions during the second quarter of 2021 increased by 100,948 (691.1%) compared with the same quarter of 2020. There were 44,286 more dissolutions in the first quarter of 2021 than in the first quarter of 2012, an increase of over 62%. It is difficult to say to what extent the pandemic has had an effect [3]. As a comparison though in June 2015 there were over 3,260,000 registered businesses and Companies House reported that while there were 52,646 new companies incorporated there were 30,993 dissolved in the same month. This equates to approximately 400,000 companies wound up in the year [4].

Between April and June 2021, there were 190,639 new incorporations and 115,554 dissolutions in the UK. The number of incorporations in the second quarter of 2021 increased by 14,524 (8.2%) compared with the same quarter of 2020 and by 20,663 (12.2%) compared with the same quarter of 2019.

Interestingly the proportion of businesses with employees has fallen since 2000 from around a third, to around a quarter. This decline in the number of employers as a proportion of all businesses is due to the growth in self-employment.[6] As at June 2021 there were just over 5,000 public companies in England and Wales and over 46,000 limited liability partnerships.[7]

There are many different types of limited registered companies, including:

- Public limited companies (PLCs);
- Private companies limited by shares (Limited or Ltd);
- Private companies limited by guarantee;
- Limited liability partnerships (LLPs);
- Limited partnerships (LPs);
- Societas Europaea (SEs);
- Companies incorporated by Royal Charter;
- Community interest companies.

Public limited companies (PLC) – the legal designation of a public limited company or PLC is a company whose shares are traded publicly on a market, such as the London Stock Exchange, and has limited liability. A public limited company's stock can be acquired by any person or organisation and shareholders are only limited to potentially lose the amount paid for the shares.

As can be seen from the figures already mentioned only a small proportion of companies are public companies; these types of businesses tend to be the larger organisations, many of which are household names, and many of which were originally set up as private companies before being floated on one of the stock markets.

The FTSE 100 is the index composed of the 100 largest companies listed on the London Stock Exchange and is traditionally seen as an indication of business performance. The FTSE 100 name originates from when it was owned jointly by the *Financial Times* and the London Stock Exchange, although the FTSE is now a wholly owned subsidiary of the London Stock Exchange. In the UK market there is also a FTSE 250 index, which lists the next 250 largest companies.

A public company must have a minimum share capital of £50,000, of which at least one-quarter plus any share premium must be paid up before the company can obtain its trading certificate from Companies House and start trading. This is the only type of company which may raise capital by offering shares or debentures to the public.

Public companies are subject to more stringent legal requirements than private companies on a wide range of matters, but especially in relation to share capital, directors and accounts. The potential positive and negative implications of listing a company on the stock market can be summarised in Table 7.1.

Private limited companies – the vast majority of trading companies are private companies limited by shares. There are over two million such companies registered at Companies House. A private company limited by shares must have the word 'Limited' or 'Ltd' at the end of its name. The main advantage of trading through a limited company is to limit the financial liabilities of the owners, directors and shareholders.

Private companies are generally smaller organisations with approximately 90% being small or medium enterprises. A private company may not offer shares or debentures to the public.

A private company limited by guarantee is an alternative type of corporation used primarily for non-profit organisations that still need to be a legal entity. This type of company does not usually have a share capital or shareholders, but instead has members who act as guarantors who each give an undertaking to contribute a small nominal amount in the event of the company being dissolved.

Although not the typical purpose of such an organisation it is still possible for a private company limited by guarantee to distribute any profits to its members (subject to the provisions of the company's articles) although distributing profits in this way would prevent a company limited by guarantee from being eligible for charitable status. Converting a limited company to a Community Interest Company (CIC) prevents this as the extraction of profit from this type of company is prohibited.

Limited liability partnerships – in the United Kingdom Limited Liability Partnerships or LLPs are governed by the Limited Liability Partnerships Act 2000 (in England and Wales) and the Limited Liability Partnerships Act (Northern Ireland) 2002 in Northern Ireland. In Scotland the Limited Liability Partnerships (Scotland) Regulations 2001 apply the provisions of the Act. An LLP is a corporate body in that it has a continuing legal existence independent of its members, as compared to a business partnership, which does not have a legal existence.

An LLP is a flexible and relatively unique business entity with LLP members having shared or collective rights and responsibilities to the extent agreed in an 'LLP agreement' but have no individual responsibility for each other's actions. As with a limited company or a corporation, members in an LLP cannot, in

Table 7.1 Implications of stock market listing

Positives	Negatives
• Access to unlimited pool of capital; • 'Cheaper' source of financing; • Diversification of shareholder base; • Perception of lower risk profile; • Increased corporate profile and reputation (customers, suppliers, talent).	• Significant regulatory burden (reporting, corporate governance, CSR); • Significant diversion of CEO and CFO attention to investor relations; • Public visibility of finances, ethics, governance and performance; • Higher expectations of control; • Pressure to hit market forecasts; • Markets' focus on short term can impede long-term priorities and discourage risk-taking; • Risk of hostile takeover.

the absence of fraud or wrongful trading, lose more than they invest.

Since their introduction in 2000 LLPs have been an increasingly popular choice of business entity for both trading and investment businesses. This was, in part, due to some tax advantages in that an LLP is treated as being transparent for tax purposes and so it pays no corporation tax or capital gains tax. Instead, LLP income and/or gains are distributed gross to partners as self-employed persons, rather than as PAYE employees. Partners receiving income and/or gains from an LLP are liable for their own taxation. However, as income tax rates have increased over recent years, corporation tax rates have steadily decreased. There is now such a difference between income tax and corporation tax rates that the tax advantages of LLPs have been reduced and it can be more tax efficient to operate through a limited company than an LLP in many instances.

LLPs have also become common among accountancy, legal and consultancy practices. As of June 2021 there were over 52,000 LLPs registered at Companies House. From 30 June 2016, an LLP could choose to send information usually kept in all or any of certain statutory registers to the registrar of companies to be kept on the public register at Companies House. This choice is an alternative to the obligation to keep those statutory registers at its registered office or a single alternative inspection address.

Other business models – partnerships, co-operatives and franchises – within the principal organisational structures described earlier there are other business models, which are notable.

Employee-owned partnership models operate differently from private-equity backed businesses and stock market-listed companies as instead of profits benefitting shareholders they are shared across the employees in the form of dividends or annual bonuses. The John Lewis Partnership is perhaps one of the best-known employee-owned partnerships although according to the Employee

Ownership Association in 2019 there were more than 400 UK companies with significant employee ownership, a section of the economy that is worth more than £30 bn annually.[8] Other examples include Richer Sounds, Blackwell bookshops, Tiptree conserve maker, Wilkin & Sons and polymers manufacturer Scott Bader.

John Lewis's ownership structure was established by John Spedan Lewis whose father founded the business in 1864. He signed away his ownership rights in 1929 to allow future generations of employees to take forward his *experiment in industrial democracy*. His ideas are set out in the company's constitution that at its heart has the idea of establishing a *better form of business* and *an attempt so to organise and conduct a business that all the advantages whatsoever of owning it shall be shared as fairly as possible by all who are working in it.* The two Trust Settlements made by John Spedan Lewis in 1929 and 1950 established the John Lewis Partnership, to be owned in Trust for the benefit of its members – its employees – who, since 1920, have been known as Partners of which there were 80,000 in 2021 [5].

The partnership as a whole generated total trading sales of over £12 bn although the business made a loss in the year to 30 January 2021 some of which was due to the pandemic and other challenges [5].

As the company itself puts it: *Partners share in the benefits and profits of a business that puts them first.* John Lewis's structure includes a Partnership Board charged with directing the company's commercial activities and supporting the Executive Team and a Partnership Council which shares responsibility for the company's health with the Partnership Board and the Chairperson. Its role is to hold the Chairman to account, influence policy and make key governance decisions such as choosing the Trustees of the Constitution, selecting Board-elected directors, changing the Constitution with the Chairman's agreement and dismissing the Chairman [5].[9]

Co-operatives are another form of business owned and operated by a group of individuals, who may or may not be employees, for their mutual benefit. The International Co-operative Alliance[10] defines a co-operative as an autonomous association of persons united voluntarily to meet their common economic, social and cultural needs and aspirations through jointly owned and democratically controlled enterprise. It is estimated that globally the 300 largest co-operatives or mutuals[11] generate $2,146 billion US.

A simpler explanation of a co-operative is a group of people acting together to meet the common needs and aspirations of its members, sharing ownership and making decisions democratically. Co-operatives are not generally about making big profits for shareholders, but more focused on creating value for customers, and tend to share the same ethical values and principles, for example relating to the environment, animal rights, fair trade or genetically modified food, etc.

There are many co-operative businesses around the world, of which the UK's Co-operative Group is one of the largest, and many other types of co-operative such as housing, building, retailer, utility, worker, credit unions, social, consumer, agricultural and political, amongst others. There are 7237 registered independent co-operatives in the UK covering all industry sectors and owned by almost 15 million people, and in 2021 it is estimated that turnover for the UK's co-ops grew to £39.7 bn. Together those co-operatives contribute more than £37 billion a year to the British economy.

Conventional businesses typically measure performance in profit and return, but social entrepreneurs, charities and the voluntary sector are more focused on making a positive return to society, although profit also may be a consideration for certain social and other enterprises. From an EHP perspective this sector may have fewer resources and require greater input and support in understanding and establishing good practice so far as environmental health needs and legal compliance are concerned.

Although not strictly a business structure, franchises are a business model that EHPs are likely to have significant dealings with. One definition of a franchise is a type of license that a party (franchisee) acquires to allow them to have access to a business's (the franchisor) proprietary knowledge, processes and trademarks in order to allow the party to sell a product or provide a service under the business's name. In exchange for gaining the franchise, the franchisee usually pays the franchisor initial start-up and annual licensing fees.

Franchising is, therefore, simply a method for expanding a business and distributing goods and services through a licensing relationship. There are two different types of franchising relationships; business format franchising is the type most identifiable to consumers and is where the franchisor provides to the franchisee not just its trade name, products and services, but an entire system for operating the business. The franchisee generally receives site selection and development support, operating manuals, training, brand standards, quality control, a marketing strategy and business advisory support from the franchisor.

More than 120 diverse industries use business format franchising as their route to market including fast food, retail and restaurant brands, retail products and services, business services, automotive and property services.

The second type of franchise is traditional or product distribution franchising which focuses simply on the products manufactured or supplied by the franchisor rather than their entire business system. While business format franchises are more obvious in the food and retail sector, product distribution franchising is in fact larger in terms of total sales. Examples of product distribution franchising include bottling, gasoline, automotive and other manufacturing industries.

Franchises are a very popular method for people to start a business, especially for those who wish to operate in a highly competitive industry like the fast-food industry. One of the biggest advantages of purchasing a franchise is that the business has access to an established company's brand name, meaning that franchisees do not need to spend further resources to get their name and product out to customers.

The NatWest and British Franchise Association franchising landscape report in 2018 found that franchise growth had continued to grow with the total contribution of franchising to the UK economy being in excess of £17 billion [6], representing an increase of over £2 billion since 2015. In 2021 there were around 48,000 franchise businesses in the UK. Different brands represented by 39,000 franchised outlets employed over 560,000 people.

Franchisees' own business will be based on one of the business types outlined earlier, including sole traders and limited companies and as with any other business, franchises are subject to the same legal and regulatory requirements. However, franchising is also about protecting the franchisor's brand value and how the franchisee meets its obligations to deliver the products and services to the system's brand standards. This can provide a useful leverage for EHPs where a franchised business is failing to maintain standards or respond to intervention action.

Every registered company, public or private, must file statutory accounts and an annual return with Companies House.

Companies House is also a useful source of information for EHPs who require information about or are investigating a business.[12] Their WebCheck service is still available but registration is required, and it costs £1 to get copies of certain documents. Information that can be obtained includes registered office addresses, previous company names, directors' details, incorporation status (e.g. trading or dissolved), when accounts were filed or due and also a history of a business's filed documents. Filed documents, such as company accounts, annual returns and reports are available for a small charge per document.

Small and medium enterprises (SMEs)

The introduction to the chapter set out the definition of SMEs and these are worthy of separate mention as they are likely to form a significant proportion of the companies with which EHPs are likely to come into regular contact. The business structure of SMEs still follows the models described before.

SMEs are seen as the driver for speeding up economic growth and generating new jobs, and the European Commission developed a range of policy measures and funding to specifically assist this business sector. The Small Business Act for Europe 2008 (SBA) provided an overarching European framework aimed at promoting entrepreneurship through reducing the administrative burden for SMEs, preferential access to finance and supporting access to markets. Following the UK's exit from the EU in March 2021 the UK government opened a scheme where smaller businesses could apply for grants of up to £2,000 to help them adapt to new customs and tax rules when trading with the EU. The £20 million SME Brexit Support Fund is to enable traders to access practical support, including training for new customs, rules of origin and VAT processes. It is unclear what other support will be available to replace European Commission funding to support SMEs. The EC announced further funding to support and provide relief for 100,000 businesses hit by the economic consequences of Coronavirus as of April 2020 with the EC and the European Investment Fund making €8 billion available from the European Fund for Strategic Investments.

Managing a business

There are numerous theories, books and opinions on how to be successful in business. The purpose of this chapter is not to discuss what makes a business commercially successful, but to provide some understanding of the key business activities and functions that need to be in place.

Marketing

Every business needs to have a product or service that meets the needs and requirements of the current market. Most aspects of a business depend on successful marketing, covering advertising, public relations, promotions and sales. Marketing is the process by which a product or service is introduced and promoted to potential customers.

However, marketing is no longer simply about a product. Businesses use comprehensive, integrated marketing strategies to effectively communicate their brand and values to their target audience and build loyalty and reputation. In a world of greater social responsibility and employee movement in the job market, marketing is also a key tool in establishing trust and partnerships in the community as well as maintaining employee loyalty. Brand equity, for both the company and their products, has become an extremely valuable commodity.

Finance

Businesses have three main sources of funding, revenues from their business operations; investment by owners, partners, shareholders or venture capital; and loans. Funds are required for daily operational needs and to meet essential expenses, payments and wages.

Accounting is a very diverse profession, which encompasses financial, management, governmental, taxation, forensic, project, and corporate social responsibility and sustainability accounting. Business accounting would include management accounting, which produces financial information for internal use by the management team, and tax accounting, but for the purposes of this chapter the following information is limited to financial accounting.

In a typical business, depending on its size and complexity, the finance team would provide finance business services, including reporting, analysis and planning, and operational finance support such as administration and processing transactions (invoicing customers, paying suppliers, payroll, etc.). Key roles with a finance structure include the Finance Director or Chief Financial Officer (CFO) and the financial controllers who are responsible for ensuring corporate governance is in place and will set financial policy, controls and delegated authority levels.

Business accounts and accounting

Financial accounting or reporting relates to the production of financial information primarily for external use and to meet reporting requirements. Financial statements represent a formal record of the financial activities of a company, quantifying the financial strength, performance and liquidity of the business. They are based on a set of reporting standards and guidelines known as 'Generally Accepted Accounting Principles' (GAAP) and which accountants in any given jurisdiction must follow.

The four main types of financial statement are:

1   Statement of financial position or balance sheet

The balance sheet is a snapshot of a company's financial position at a single point in time. A standard company balance sheet has three parts: assets, liabilities and ownership equity.

* Assets are something the business owns or controls. Assets are usually listed first and typically in order of liquidity;
* Liabilities are something the business owes to someone else (creditors, bank loans etc.); and
* The difference between the assets and the liabilities is known as net assets. This is the equity of the company – the amount repayable back to the owners.

Assets can be either 'current' which are expected to be sold or otherwise used up within one year (e.g. cash, accounts receivable and inventory), or they can be fixed or 'non-current' assets such as land and buildings, motor vehicles, furniture, office equipment, computers, fixtures and fittings, and plant and machinery which cannot easily be converted into cash, and which will be used over an extended period of time (in excess of a year).

2   Income statement

The income statement is also referred to as profit and loss statement (P&L), statement of financial performance, earnings statement or operating statement. It is a statement of the company's financial performance in terms of net profit and loss and it indicates how revenue is transformed into profit. The income statement comprises income (from sales and revenues, etc.) from which costs including wages, depreciation of assets, charges, etc. are deducted to provide the net profit or loss.

3   Cash flow statement

The cash flow statement is concerned with the flow of cash in and out of the business, capturing both the operating performance and the accompanying changes in financial position. Cash movements are broken down between operating, investing and financing activities. Unlike the balance sheet and P&L, which can be subjective in terms of recognising and valuing assets and liabilities, cash flow is not and the cash flow statement can therefore provide a valuable understanding of a company's performance and financial position.

4   Statement of changes in equity

Also referred to as statements of retained earnings this report details changes in the owners' equity over the period. This is derived from the net profit or loss, share capital issued or repaid, dividend payments and equity gains or losses.

The relationship between the different types of financial report is summarised in Figure 7.1.

Taxation

Individuals and businesses are subject to an array of taxes, the main business-related ones being:

Corporation tax – is a tax on limited companies' taxable income or profits. It is the business's responsibility to assess and pay their

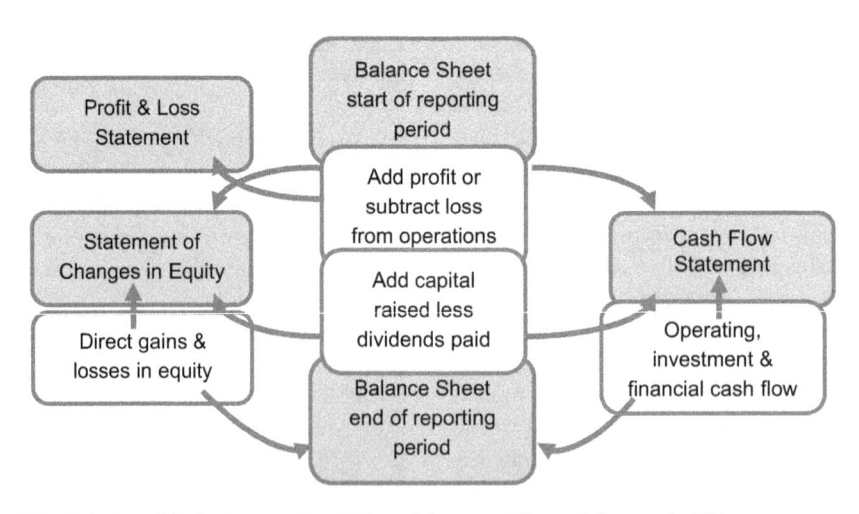

Figure 7.1   Relationship between the different types of financial reports [7]

own tax liability and from April 2021 the main rate for corporation tax was 19%.

Value Added Tax (VAT) – is tax on the final consumption of specified goods and services in the domestic market but is collected at each stage of production, distribution and sale. When a business charges VAT on a sale it is actually collecting that tax on behalf of HM Revenue and Customs (HMRC). If a business is VAT registered it can then reclaim the VAT paid on purchases and the difference between the VAT charged on sales and paid on purchases is paid to HMRC. A VAT return is normally completed every three months.

Companies are required to register for VAT with the HMRC if the business's VAT taxable turnover exceeds the annual threshold; the taxable turnover is the total value of all sales that are not VAT exempt. The threshold for VAT registration changes every year and from January 2021 is £85,000. Businesses must also register for VAT if the predicted sales in the next 30 days are expected to exceed the annual threshold. Companies with a VAT turnover of less than £150,000 can apply for a flat rate scheme under which they pay a fixed annual rate.

For businesses and individual consumers, the standard rate of goods and services increased to 20% on 4 January 2011 (from 17.5%). There is also a reduced rate of 5% and a zero rate as well as some exemptions from VAT. The standard rate applies to most goods and services, with the 5% reduced rate being applied to some children's safety products, mobility aids for the elderly and domestic energy supplies. Zero-rated goods and services include most foods, children's clothes as well as printed materials, some safety equipment, equipment for disabled people and building works for disabled persons and charitable purposes as well as cesspool and septic tank emptying. For supplies relating to hospitality, accommodation or admission to certain attractions the UK government announced at the budget in 2021, that it will be legislating to extend a temporary reduced rate of VAT of

5% until 30 September 2021 and to prepare for a new rate of 12.5% from 1 October 2021 to 31 March 2022.

Most foods remain zero rated with the exception of food supplied in the course of catering, including hot take-away food, ice cream, confectionary, alcohol and other beverages and some snack products. The HMRC produces detailed schedules of VAT rates for goods and services.

There are also some goods and services that are exempt from VAT including insurance, finance and credit, education and training, fund raising, membership subscriptions, and burials and cremations.

Capital Gains Tax (CGT) – CGT is paid by a self-employed sole trader or a business partnership. Other organisations like limited companies pay corporation tax on profits from selling their assets, including land, buildings, plants and equipment. For higher rate income tax payers the rate is 20% and for a basic rate income tax payer the rate will depend on the size of the capital gain. There is, under the entrepreneur's relief scheme (now business asset disposal relief), a reduced rate of 10% up to a lifetime business disposals allowance.

National insurance and *Pay As You Earn (PAYE)* – employers have responsibilities for both National Insurance Contributions (NICs) and deducting PAYE from employees' salaries.

Production and distribution

Every company's success depends on their ability to consistently produce their goods and services at a high enough quality to meet their customer's requirements. Arguably the demand for both quality and consistency is increasing and this is evidenced by the growth of the brands and franchising. An objective of every strong brand is to consistently deliver the business's core values and experience to every interaction with their customers.

A quality management system (QMS) can be defined as the organisational structure, policies, procedures, processes and resources needed to consistently ensure and measure the effectiveness of delivering quality objectives. The purpose of a QMS is to develop and maintain a system and culture, which provides quality assurance and audit across the organisation. This is discussed in more detail in Chapter 8.

The concept of quality first emerged from the Industrial Revolution. The development of mass production changed working practices but also had limitations in terms of variable quality. In the late 19th century manufacturing pioneers, including Henry Ford, started to develop quality departments to oversee the quality of production. Ford also recognised the importance of standardisation of design and component standards to ensure a standard product was produced.

Today the adoption and implementation of quality management standards can extend to all areas of organisational performance and service and is focused on driving efficiency and continuous quality improvement. Customers increasingly expect accreditation to certain quality standards, and in some industries, it has become a pre-requisite for doing business.

ISO 9000 is one of the key standards in that it is the family of standards focused on quality management. The ISO 9000 family of standards cover specific areas, including documentation, training and performance improvements but it is ISO 9001 that is typically referred to by organisations. This is because ISO 9001 is the only standard within the ISO 9000 family that an organisation can become certified against, as it is the standard that actually defines the requirements for a quality management system.

Standards now cover multiple industry sectors, including medical devices, automotive, aerospace, telecoms etc., and a wide range of environmental health related areas such as food safety, occupational health and safety and environmental management. Certification and ongoing compliance with standards is intensively audited and monitored both internally and externally by customers and third party, independent auditors.

The British Standards Institution (BSI) also carries out independent testing of products to ensure they meet the relevant standard or legal requirement. The BSI Kitemark symbol shows that a product conforms to the appropriate British Standard and should be safe and reliable. Many products such as new toys must meet legal requirements before they can be sold within the European Community and must carry CE marking. CE marking attached to a product is a manufacturer's claim that it meets all the requirements of the European legislation.

With regard to distribution the overriding objective is to get products or services to the market in a timely and economic way. Many UK businesses are reliant on road transport and over recent years occupational risks in the road transport sector have, rightly, been subject to increasing focus. Road traffic accidents are a leading cause of workplace death, injury and disability and drivers are regularly exposed not only to the dangers of the road, but to a broad range of other safety issues and hazards including loading and unloading vehicles, working hours, shift work and fatigue, musculoskeletal and vibration-related disorders, lone working and unhealthy lifestyles (for example lack of exercise and poor eating habits). It is important that companies ensure suitable safeguards are in place for their drivers and vehicle fleets.

Research and development

Research and development determines a business's ability to continually innovate and produce new products, services, processes and meet customer and consumer needs. It allows companies to benefit from technological advances, improve production and find new ways of delivering services and develop new products and services. Innovation and product/service development are key elements of most organisations' strategy and objectives.

Depending on the size of the business, research and development may be limited to product improvement due to budget limitations and costs but in larger organisations will include the development of new products or services as well as the improvement of existing products. Research and development is a long-term investment and does not lead to short-term or even guaranteed return on that investment; however effective research and development is essential in ensuring a business stays ahead of their competition. Extensive and current market research is critical to ensuring consumers' and customers' changing requirements are met.

Governance and regulation

Businesses, directors and senior offices are subject to a wide range of regulatory and governance obligations and duties including, but not limited to, corporate governance, financial practices and reporting, and, of course, regulatory compliance with health and safety, food safety and environmental protection.

The executive board is responsible for providing leadership, setting the company's strategic objectives and strategies, ensuring adequate resources are available to achieve those objectives and ultimately setting the standards, ethics and culture of the organisation. Corporate governance and directors' individual responsibilities have perhaps never been under more scrutiny.

In terms of legal obligations, a director is defined by their role within a company rather than job title; 'director' is a widely used title but does not always indicate that the post holder actually sits on the board. Conversely an individual who is not formally appointed to the board can be legally deemed to be a de facto director where their role could be considered as being equivalent to that of a director or if they have acted as a director.

The Companies Act 2006 is an extensive piece of legislation and is the principal source of UK company law. As well as defining the requirements for setting up and registering a business, it sets out the duties and liabilities of directors and other provisions such as the requirement for a company secretary, annual general meetings, annual accounts and reports, and other administrative requirements.

In the case of global organisations, businesses that have legal entities listed on the US stock exchange are required to comply with the Sarbanes-Oxley Act 2002 (SOX) which resulted from governance failings in the US and seeks to achieve the same corporate responsibility and consumer protection as legislation enacted in the UK.

A private limited company is required to have at least one director, who may also be the main shareholder and be responsible for the day-to-day operation of the company. However, as a business becomes larger a board of directors is required to cover the range of directors' responsibilities and provide effective, balanced decision-making. The Financial Reporting Council (FRC) is the UK's independent regulator responsible for promoting high-quality corporate governance and reporting, and the Council maintains the corporate governance code and regulates the activities of the accountancy and actuarial professions. Listed companies are required under the Financial Conduct Authority rules to either comply with the provisions of the code or explain to shareholders why they have not done so. Unlisted companies and those listed on the AIM (the Alternative Investment Market is a sub-market of the London Stock Exchange, allowing smaller companies to float shares with a more flexible regulatory system) are encouraged to adopt the code.

The main principles of the code include:

- Effective leadership with collective responsibility for the long-term success of the company. There should be clear division of responsibilities; the chair is responsible for its leadership and non-executive directors should constructively challenge and help develop strategy;

- Having the appropriate balance of experience, skills, company knowledge and independence to allow the board to meet their responsibilities effectively. Board appointments should be transparent and rigorous and director's performance monitored;
- Accountability for reporting a fair and balanced assessment of the company's position and prospects. This includes establishing formal arrangements for corporate reporting, risk management and internal controls;
- Formal and transparent policy and procedures for determining executive remuneration with remuneration packages designed to promote the long-term success of the organisation; and
- Communication with shareholders to ensure that they understand the company's objectives.

An executive board will meet regularly and although the number of board members depends on the size of the organisation the board structure will typically include a chair, who is often a non-executive director, the Chief Executive Officer (CEO) or managing director, who is an employee and runs the company day to day, at least three executive directors and two or more non-executive directors. The executive directors are also employees and usually manage key business areas such as finance, operations, human resources and marketing. As health, safety and risk management continue to build prominence within organisations there is an increasing number of board directors with specific responsibility for these key areas.

Non-executive directors (NED) are not part of the executive team and typically do not engage in the routine management of the organisation, but they are involved in policy setting. The Institute of Directors describes the role of the NEDs as providing creative contribution to the board by providing objective criticism. Individuals who are appointed as NEDs typically hold other positions either as executive directors with other businesses or multiple non-executive positions.

As such they are appointed to bring additional experience, specialist knowledge, impartiality and independence to a board.

NEDs' responsibilities extend to monitoring the performance of the executive directors and protecting the interests of the shareholders. Even though they are part-time directors of a business there is no legal distinction between NEDs and the executive directors; this means that they still attract the same individual legal duties, responsibilities and potential liabilities. In the UK there is a differentiation between NEDs and independent non-executive directors; the code includes criteria for judging independence based on no material interest (e.g. shares, etc.) in the company, remuneration being strictly limited to director's fees and the NED not being previously employed by the company or serving on the board for more than nine years. Unlike the US where there are stricter controls on independence a survey in 2013 found that only 29% of UK companies declared their NEDs met the independence standards [8]. There is no evidence that much has changed and the existence of NEDs has not prevented some corporate failures such as Carillion.[13]

The UK corporate governance code also requires a board to form three additional groups; audit, remuneration and nomination committees. With clear terms of reference and typically chaired by one of the NEDs other than the chair, each committee will include executive and non-executive directors and other senior employees as required. The audit committee's main role is to review internal financial controls, review and develop external controls including the appointment of auditors, and ensure the integrity of the company's financial statements and announcements. The audit committee typically oversee the activities of any internal audit function and, unless there is a separate board-level risk management or safety committee, the audit committee would also review risk management and safety systems and performance.

The purpose of the remuneration and nomination committees is to ensure that executive directors' and senior managers' pay and benefits are fair and responsible, and that appointments to the board support the strategic aims of the business. The committees usually have delegated authority to set executive pay and make recommendations on appointments. The corporate governance code gives specific guidance on the membership of these additional panels to ensure their independence from the board.

Other governance requirements under the Companies Act include the appointment of a company secretary and need to hold an annual general meeting (AGM), although a private company does not need a company secretary and since the last amendment of the Act may opt out of holding an AGM. The role of the company secretary includes maintaining statutory registers (e.g. registers of shareholders, directors, directors' interests, etc.), giving notice of AGMs and keeping minutes of board and general meetings.

Annual general meetings are the principal annual communication to shareholders and typically include presentation of the annual accounts and directors' reports.

Financial regulatory compliance

The financial services industry has become increasingly regulated following high-profile failures in the markets, with the aim of ensuring this sector is run with integrity and consumers are protected. The majority of business encountered by EHPs will fall outside the control of the Financial Conduct Authority (FCA) but all businesses are subject to some key financial duties. The Companies Act 2006 requires, as previously highlighted, that accounting records are kept and that the directors have a duty to prepare true and fair accounts and annual reports, which need to be published in accordance with the GAPP reporting standards. The legislation also requires that, with the exception of certain small companies, an independent accountant should audit the financial year-end accounts for all businesses.

Other relevant legislation which can apply to either the company or individuals within the business includes the Theft Act 1968, the Fraud Act 2006 and the Bribery Act 2010. Under these Acts not only can the corporate entity be held liable for offences and punished but where culpable the individuals in control of that business can also be held accountable. As with the Sarbanes-Oxley Act, companies listed in the US market are also subject to the Foreign and Corrupt Practices Act 1977 (FCPA), which covers local financial reporting requirements and makes it illegal to bribe foreign officials. Such regulations are becoming globally applicable and carry very significant penalties.

Regulatory compliance

EHPs are responsible for enforcing many, but not all, of the legal responsibilities placed on employers and businesses where the local authority is the relevant enforcing authority.

A company is a legal person or entity capable of being prosecuted for most criminal offences. Liability for serious offences that involve an element of fault (as opposed to strict liability) can attach to corporate entities where the legal breach is attributable to someone who was the 'controlling mind' of the company. An individual who has 'controlling mind or will' over a company is someone who is in sufficient control of the company's affairs that the company can be said to act through them. It would normally only be the directors and senior management of a company, at or close to board level, whose acts can be considered to meet this test and the larger the organisation the more difficult it can be to identify the controlling minds.

As well as the financial and anti-corruption obligations previously described, primary legislative and regulatory requirements where criminal liabilities can also attach to individual directors and senior officers include health and safety, food safety and environmental protection. Table 7.2 outlines the key personal liabilities of directors relevant to

Table 7.2 Key personal liabilities of directors relevant to environmental health and other legislation of commercial significance

Legal statute	Section
Health and Safety at Work Act (HSWA) 1974	**Section 7 General duties of employees at work** Applies to employees who fail to take reasonable care for the health and safety of themselves and other persons who may be affected by their acts or omissions at work. Whilst primarily directed at placing duties on employees other than directors this section may impose liabilities on self-employed individuals and sole traders, although where the employer is primarily responsible action should be taken against the employer alone. **Section 37 Offences by bodies corporate** Applies to all offences created by the Act where that offence has been committed by the body corporate where: (1) that offence has been committed with the consent or connivance of or has been attributable to any neglect of the individual; and (2) the person accused is a director, manager, secretary or other similar officer of the company.
Food Safety Act 1990	**Section 36 Offences by bodies corporate** Where an offence under this Act which has been committed by a body corporate is proved to have been committed with the consent or connivance of, or to be attributable to any neglect on the part of (1) any director, manager, secretary or other similar officer of the body corporate; or (2) any person who was purporting to act in any such capacity, he as well as the body corporate shall be deemed to be guilty of that offence and shall be liable to be proceeded against and punished accordingly.
Environmental Protection Act 1990	**Section 157 Offences by bodies corporate** Applies to all offences created by the Act where that offence has been committed by the body corporate where: (1) that offence has been committed with the consent or connivance of or has been attributable to any neglect of the individual; and (2) the person accused is a director, manager, secretary or other similar officer of the company or other person acting as if he were a director of the body corporate.
Water Resources Act 1991	**Section 217 Criminal liabilities of directors and other third parties** Where a body corporate is guilty of an offence under this Act or under section 4 of the Water Act 2003 and that offence is proved to have been committed with the consent or connivance of, or to be attributable to any neglect on the part of, any director, manager, secretary or other similar officer of the body corporate or any person who was purporting to act in any such capacity, then he, as well as the body corporate, shall be guilty of that offence and shall be liable to be proceeded against and punished accordingly.
Bribery Act 2010	**Sections 1, 2 and 6 General bribery offences and bribery of foreign public officials** Created separate offences of offering, promising or giving a bribe, requesting, agreeing to receive or accepting a bribe, and a distinct offence to bribe a foreign public official. Where a company is guilty of any of these offences its directors can also be held liable with the company if it can be shown the individual consented to or connived in the bribery. The maximum penalty is ten years' imprisonment and unlimited fine with potential disqualification from holding a director's position for up to 15 years.

Table 7.2 Continued

Legal statute	Section
Trade Descriptions Act 1968	**Section 20 Offences by corporations** Where an offence committed by a body corporate is proved to have been committed with the consent and connivance of, or to be attributable to any neglect on the part of, any director, manager, secretary or other similar officer of the body corporate, or any person who was purporting to act in any such capacity, the individual as well as the body corporate shall be guilty of that offence and shall be liable to be proceeded against and punished accordingly.
Data Protection Act 2018	**Section 198 Liability of directors etc.** The Data Protection Act controls how our personal information is used by organisations, businesses or the government. Those responsible for using data must follow strict rules to ensure the information is used fairly and lawfully, is restricted to specifically stated purposes, and is handled and kept safely and securely. Where an offence has been committed by a body corporate, and it is proved to have been committed with the consent or connivance of or to be attributable to neglect on the part of a director, manager, secretary or similar officer of the body corporate, or a person who was purporting to act in such a capacity then director, manager, secretary, officer or person, as well as the body corporate, is guilty of the offence and liable to be proceeded against and punished accordingly.

environmental health and other legislation of commercial significance.

Clearly the impact of enforcement action against a business is, to a degree, proportional to the size and assets of the company concerned; the consequences for a small business or sole trader will be harder felt and be more personal to the business than in the case of a large corporate. However, reputational damage and shareholder reaction are taken extremely seriously and all public limited companies consider compliance a business priority.

With the number of workplace injuries, increasing focus on legal compliance and good business governance, as well as the current move towards higher penalties, businesses and their directors cannot fail to be aware of their duties and responsibilities. Board leadership and involvement in setting expectations and attitudes to safety and other compliance and governance matters is essential. This is clearly reflected in the range of personal liability that now attaches to directors and senior officers of a company as well as the current reviews of sentencing guidelines.

Company directors' can be disqualified if they fail to meet their legal responsibilities, including being convicted of criminal offences, not maintaining proper accounting records, allowing a company to continue to trade when it is insolvent, not paying taxed owed by the company and fraud or theft.

Consent, connivance and neglect

Companies can be held criminally liable for a wide range of offences but as detailed earlier there are a number of relevant statutes under which a director, manager or other senior employee can also be guilty where the offence has been committed with their consent, connivance or is attributable to their neglect.

Consent and connivance have been held to mean that the person had both knowledge of the materials facts and a decision was made on the basis of that knowledge [8]. Connivance has been described by the Courts as wilful blindness in that the person was 'well aware of what is going on, but his agreement is tacit, not actively encouraging what happens but letting it continue and say nothing about it' (*Huckerby v Elliot* [1970] 1 All ER 189).[14]

Neglect is generally taken to mean that the person failed to do something a reasonable person would have done, or if they do something that a reasonable and prudent person would not have done.

It is apparent from case law that there are no fixed rules around establishing consent, connivance or neglect on the part of a director beyond reasonable doubt; the individual's place of activity relative to the workplace and whether the activities leading to the offence were under their immediate direction and control have both been considered by the Courts. However, with greater scrutiny of corporate conduct than ever, Proudlock et al. [9] suggest that the question of what constitutes consent and connivance within a boardroom environment, raises significant questions for businesses and their directors [9].

Corporate manslaughter

Having considered the increasing focus on personal criminal liabilities of directors it would be remiss not to mention the Corporate Manslaughter and Corporate Homicide Act 2007, which has been the subject of significant attention prior to and since its enactment. The Act creates a means of holding an organisation accountable for deaths caused by very serious management failures.

Before the Corporate Manslaughter and Corporate Homicide Act a business could only be held liable for the common law offence of gross negligence manslaughter and for the corporate entity to be convicted it was necessary for a director or senior manager, who could be seen as embodying the 'controlling mind and will' of the company, to be guilty. The 2007 Act put the law on corporate manslaughter (corporate culpable homicide in Scotland) on a new, broader footing, meaning that a company may be convicted where there is a gross breach of the company's duty of care by senior management rather than the

higher test of one or more individuals being guilty of the previous offence of gross negligence manslaughter.

It is important to note that in this case the legislation is only concerned with corporate liability and does not therefore apply to directors and other people who hold senior positions in a company. However, the Act is also not intended to replace the existing gross negligence manslaughter and health and safety offences that continue to apply to individuals. Penalties under the Corporate Manslaughter and Corporate Homicide Act 2007 include unlimited fines, remedial and publicity orders that require a company to openly publish details of the conviction and fine imposed. This is issue is also considered in Chapter 14 in the context of health and safety.

Review of sentencing guidelines

In conjunction with the increased scrutiny on corporate governance the Sentencing Council for England and Wales has issued sentencing guidelines. These include guidelines for environmental offences that were issued on 26 February 2014 and came into force on 1 July 2014; guidelines for sentencing health and safety offences, corporate manslaughter and food safety and hygiene offences were published on 3 November 2015 and came into force on 1 February 2016;[15] these cover offences by organisations and also individuals.

Contrary arguments suggest that the draft guidelines simply aim to increase the level of fines imposed on large and very large companies and there has been concern that guidelines will bring about a significant and disproportionate increase in penalties imposed on all companies prosecuted for health and safety or food safety offences, not just the large and very large organisations. There is no evidence that this has occurred, although this might also reflect low levels of prosecutions.

In December 2020 the Sentencing Code came into effect in England and Wales, consolidating existing sentencing procedure law into a single Sentencing Act. The Code covers sentencing for adults and under 18s and applies to all convictions made on or after 1 December 2020, irrespective of the date on which the offence was committed. From this date, judges and magistrates need to refer to the Code, rather than to previous legislation, although there will be some transitional cases where an offender is convicted before 1 December but is sentenced later. The Code includes general provisions relating to sentencing procedure, the different types of sentences available to the courts, and certain behaviour orders that can be imposed in addition to a sentence. It is a consolidation measure so it has made no substantive changes to the law.

Insurance

A business is required by law to have specified types of insurance cover in place; the requirement extends to employer's liability, public liability and motor vehicle insurance, although there are a number of other classes of insurance policy that most businesses would also buy. It is important to note that a business cannot insure against prosecution for criminal offences including health and safety, and food safety breaches. Financial penalties are intended to deter re-offending and publicise the consequences of non-compliance to other companies and individuals. If the guilty party is not required to pay the fine, because it is covered by insurance, then the intended deterrent would disappear. However, the costs incurred in defending a legal action can be insured.

Insurance does, however, provide protection against civil liabilities and other business risk such as property damage and business interruption but buying insurance is simply a way of transferring all or part of the financial risk connected with a potential event and does not offer protection against reputational damage and other physical impacts of that event on the business.

Employer's liability insurance protects against employee injury and ill health claims and is required by all businesses with the exception of limited companies with a single employee who owns more than 50% of the share capital and sole traders who do not employ anyone or only employ close family members. For most small and medium enterprises (SMEs) employer's liability cover is generally inexpensive and a valid insurance certificate must be displayed where staff can see it.

Public liability insurance is required if members of the public have access to a company's premises, where a business's work activities could injure anyone or if those activities could damage third-party property. While it is not a legal requirement for some businesses it is usually considered essential by most as the potential cost of damages far exceeds typical premiums. Public liability cover also typically provides legal costs for defending claims and is often combined with product liability cover which relates to claims arising from injury or damage caused by products manufactured, repaired or sold by the insured.

Motor insurance for company vehicles is again compulsory, as it is for privately owned cars. The minimum requirement is third-party insurance but most businesses buy fully comprehensive cover as commercial fleet vehicles are more likely to be involved in a claim.

Other commonly held insurance policies include buildings insurance (covering property damage and loss of gross profit or revenue in the event of a business interruption incident), professional indemnity, directors and officers insurance, goods in transit, terrorism, health insurance and fiduciary cover. The basic, mandatory insurance policies are readily available online and directly from insurers but larger businesses usually work with insurance brokers to ensure that adequate and competitive insurance cover is purchased and maintained.

There is often a perception that where cover is in place insurers will cover the entirety of any loss as a matter of course. In reality this is not the case and policies are subject to uninsured deductibles and upper limits of cover and in the event of a claim will investigate or instruct loss adjusters to handle the claim. A claim will only be settled if the terms and conditions of the policy are met and insurers agree with the value of the loss. Claims experience is also a major factor in future insurance costs and organisations should seek to manage their claims.

People and human resources

Finally one of the most important elements of running a successful company are the people required to deliver the products and services offered by the business. Human resource management (HRM) is a complex field subject to extensive employment legislation and is commonly divided into three main areas of people management: staffing, employee compensation and managing the workforce. Essentially, the purpose of HRM is to maximise the productivity of the work force by ensuring the optimum effectiveness of its employees, although most business have long recognised the value of investing in, developing and caring for their employees in order to build a loyal, stable team with the necessary skills and knowledge to ensure business success.

Perhaps most relevant to environmental health and associated legal compliance are employee training and communication (hence Chapter 3 in this publication); both are essential ensuring that people are able to carry out their work effectively and safely. Workforces varying significantly by industry sector, required skill level and geographical location, etc; and by training needs, including language, need to be assessed and effective means of delivery training and assessing learning outcomes implemented.

A company's culture can be a critical determinant of its long-term success and regulatory compliance. Corporate culture is the moral, social and behavioural norms of an organisation, based on the beliefs, attitudes and perceptions of its employees. It is what the employees perceive the organisation really wants and it may differ considerably from what has been stated in writing; culture also dictates whether employees follow procedures when their manager is not around. Communication, both top down and bottom up, are therefore critical to establishing a safe and responsible culture and determining whether a business will be a leader in health and safety or lagging behind.

Factors in business failure

Having reviewed some of the key aspects of running a successful business it is appropriate to look briefly at some of the common reasons for business failure and non-compliance:

- Lack of clear business or management direction, including business plans and agreed strategic objectives;
- Poor cost and financial control, including under investment, failure to achieve required return on investment (ROI), insufficient working capital and budget planning and lack of financial reporting;
- Poor customer relations and service, including inconsistent or poor product quality;
- Loss of sales due to market competition, failure to anticipate changing market trends and uncompetitive pricing;
- Human resource management failures such as employment law issues, industrial action and inadequately skilled workforce; and
- Loss of reputation due to poor performance, incident leading to business interruption, investigation and legal intervention for regulatory or governance non-compliance, and associated media coverage.

Business and the EHP as regulator

There are a number of different enforcing authorities across the UK, including local authorities, the Health and Safety Executive (HSE), the Environment Agency, Trading Standards, the Fire Authorities, the Financial Conduct Authority and, of course, the Police Authorities and national criminal agencies such as the Serious Fraud Office. Of these the local authorities, and particularly EHPs with their health and safety, food safety, environmental protection and licensing responsibilities, have perhaps the widest and most visible contact with businesses. This relationship between business and regulator should therefore be used to work together, educating and influencing businesses to be responsible and compliant as well as building trust in the regulatory bodies.

Local authority EHPs are the main enforcing authority for health and safety within retail, catering and hotel premises, consumer and leisure activities, offices and wholesale distribution and warehousing; the enforcement allocation is set out in the Health and Safety (Enforcing Authority) Regulations 1998. Whilst there are a number of HSE and EHP partnership initiatives, each local authority is an enforcing authority in its own right and must make provision for enforcement.

EHPs are responsible for handling all issues relating to food quality, hygiene and safety but this is carried out under the direction of the Food Standards Agency (FSA) rather than the local authorities alone. The Framework Agreement on Local Authority Enforcement covers the FSA's interaction with enforcement officers, while the Food Law Code of Practice sets out how local authorities should apply food safety law and also how they should work with food businesses.

Business attitudes to compliance and enforcement

There will, of course, be exceptions but the majority of companies, regardless of scale, are committed to the responsible operation of their business although it may seem otherwise, for example the Southern Water case in 2021.[16] There is an established understanding among the business community that good governance allied with strong corporate and social responsibility leads to an enhanced and sustainable financial performance. Most businesses maintain they want to do the right thing by their customers and wider stakeholders.

A company's reputation is one of its most important assets and is, arguably, becoming ever more valuable. Corporate visibility has increased as a result of the proliferation of social media and consumer interest; 'news' is reported almost immediately across the world with incidents and negative messages generating significant attention. Such is the potential impact of these events, which include intervention and enforcement by EHPs and other agencies at local and national levels, that companies are beginning to use social media and negative consumer sentiment to measure customer perception and brand reputation [10]. This 'sentiment analysis' is still developing but it is clear that negative messages are remembered by consumers and put off potential customers, especially where the company concerned fails to respond well or correct the matter. A good reputation, which can take years or decades to establish can be destroyed in hours or days.

Even where a company's response to an incident or crisis is seen as being well managed there can be significant financial and reputational loss. The accident at the Alton Towers theme park in June 2015 saw Merlin Entertainment's Chief Executive personally apologise and accept full responsibility, and other parts of the business were suspended to ensure they were safe. However, the group has since reported substantial reductions in visitor numbers across its sites and has issued a warning that profits may be down by £47 million for the financial year with further impact on results in subsequent years [11]. Unlike Southern Water customers, people can choose not to go to Alton Towers, but there is no choice as to which water and sewerage companies provide those services.

Although the Merlin Entertainments example relates to a large, multi-national business, the impact of a single adverse event was significant in terms of their performance; in the case of smaller companies that EHPs will have more regular dealings with, the effect of more minor incidents and interventions will have relatively equal consequences and, in the case of many small and medium enterprises, may be terminal for the business.

When directors' individual liabilities are also taken into account, the need of good compliance and governance are clear to most directors and companies. This does not mean that businesses always get it right and one of the greatest challenges for smaller businesses is to understand exactly what is required of them in order to achieve compliance and, importantly, ensuring their employees abide by systems put in place to ensure compliance. EHPs also have to be able to consider the extent of culpability and whether there is wilful disregard for the law and public health.

The number of businesses that now have internal, specialist advisors (including commercial EHPs) covering health and safety, food safety and other relevant regulatory areas has increased significantly to become the norm, with a wide range of external consultancy services also available. However, companies should also perceive the EHP as a valued partner in their shared objective of ensuring the business is safe. This is the basic principle of the Primary Authority partnership approach where Primary Authority is based on legal partnerships between businesses and individual local authorities. Businesses can set up their own partnership or belong to a trade association (or other type of group) with an existing partnership.

Similarly, most EHPs in local government and regulatory agencies are open and supportive in their dealings with businesses, finding that this approach is the most effective way of achieving compliance unless the company is uncooperative. The enforcement authorities and EHPs do not always get it right either and poor EHP interventions have been extremely damaging to the level of trust companies sometimes have in the enforcement agencies. It is important for EHPs to recognise that their area of involvement with a company is only one aspect, albeit an important one, of the wider activities and responsibilities a business is dealing with on a day-to-day basis.

The type and size of business may affect both the company's response to the EHP and their approach to dealing with the business and this is reflected in the focus on proportionality in relevant enforcement policies. Whilst larger organisations may represent a higher gross risk this is strongly mitigated by the internal resources, controls and organisational culture that are usually in place and in reality the greater risks are often inherent in the smaller businesses that lack those attributes. This has not perhaps always been reflected by the focus of some EHPs on targeting brands and larger businesses with the resulting publicity, but as corporate transparency and Primary Authority partnerships continue to grow this could change. With more start-up and small businesses being formed, with the added complexities of a multi-cultural and multi-language population of business owners, limited EHP resources should be more directed towards small and medium enterprise.

Working effectively with (responsible) businesses

The HSE's Policy Statement on Enforcement states that HSE take

> enforcement action to prevent harm by requiring duty holders to manage and control risk effectively. This includes

- ensuring action is taken immediately to deal with serious risks;
- promoting and maintaining sustained compliance with the law; and
- ensuring that those who breach the law, including individuals who fail

in their responsibilities, may be held to account (this includes bringing alleged offenders before the courts in England and Wales, or recommending prosecution to the COPFS in Scotland).

[12]

The term 'enforcement' is given a wide meaning and is applied to all the dealings between the authorities and those on whom the law places duties, and is based on the principles of proportionality, appropriate targeting of the highest risk and where the hazards are least well controlled, consistency and transparency.

Local authorities also need to have regard to the revised FSA's Food Law Code of Practice 2021[17] issued under the Food Safety Act 1990 Section 40 and Regulation 26 of the Food Safety Hygiene (England) Regulations 2013. Again, enforcement activity needs to be reasonable, proportionate, risk based and consistent with good practice. Both approaches clearly emphasise the importance of advice, education and informal action with reliance on more formal action only where working with the business informally does not achieve the required effect. There is a need for formal interventions, including prosecution action, but the principles embodied in both also reflect how the majority of companies would expect and want EHPs to work with them.

Identifying and communicating with the right people within a business is essential to achieving the desired outcomes. With the exception of the smallest businesses most companies will delegate responsibility for regulatory compliance to either an internal subject matter expert or a senior manager who has the appropriate level of seniority to ensure the business responds promptly and appropriately to any issues. Additionally, companies will have financial processes in place, including levels of authority delegated from the board. Where material works are required to achieve compliance or meet recommendations relating to best practice it is important that the requirements are effectively communicated to individuals within the company who are at a senior enough level to authorise the expenditure.

The business should also have effective escalation and reporting processes in place to ensure that a visit or intervention by an EHP is notified to the right person internally. However, and particularly with multi-site operations, this should not be relied upon by the EHP who also has responsibility for ensuring that the appropriate management contacts have been identified and contacted. Much more information about individual companies is now available through the internet, with most businesses having some level of web presence. As well as key contacts, these information sources can provide significant insight into an organisation's culture, structure and approach.

The vast majority of responsible companies are keen to work in partnership with the authorities. There really is no commercial advantage in being obstructive; businesses understand that they have legal responsibilities, and many see it as an opportunity to improve their existing practices. However, EHPs should also recognise that some businesses may not come into regular contact with the enforcement authorities and there is therefore the potential for their visits to cause some apprehension, particularly in the event of an investigation rather than routine visits.

More and more companies employ health and safety, food safety and other practitioners, including their own EHPs or consultants, many of whom are highly skilled and qualified. Local authority EHPs should always seek to determine the level of internal competence a business holds and tailor their approach accordingly. The out-dated view that commercial EHPs traded their integrity and environmental health principles when leaving local authority employment was never supportable and working with their commercial peers presents an excellent opportunity for local authority EHPs to not only to encourage best practice but to advance it.

Regardless of the level of relevant in-house competence a company has available, EHPs must, first and foremost, communicate their findings and requirements clearly and effectively as not all the relevant business decision makers will have competence in the specific regulatory areas involved. There are times when an EHP's requirements are not understood by a company or where those requirements are not appropriate or correct; it needs to be recognised that people do not get it right every time and that this applies to local authority EHPs as much as it applies to businesses. It is appropriate therefore that businesses are able to question and challenge EHP interventions and this should be regarded as a key element of the working alliance.

In the case of smaller companies, the role of the EHP as an educator and information source providing best practice advice becomes even more important. This may be more difficult with pressures on resources but should be seen in the context of encouraging businesses and helping the local economy. Working in partnership to share and develop best practice, through business and professional forums, is increasingly common for companies of all sizes and EHPs should be encouraged to become involved in forums wherever possible.

Regardless of verbal reassurances that may be offered by individual front-line staff, most businesses are complex organisations and have established operational procedures in place. Compliance timescales obviously need to be appropriate to the level of risk but need to also recognise that operational and financial sign-off may take time, particularly in the case of capital works.

In summary companies simply want to know what is required of them to achieve compliance or best practice, when it needs to be done by and whether it is a legal requirement or simply a recommendation. Works need to be perceived as proportional to the risk, with informed agreement on the best methods or materials. Identifying and working with the right person in the business,

with the appropriate level authority to make the necessary decisions, is essential to achieving the best outcome.

A business's lasting perception of and response to local authority EHPs will be based on their previous experience of inspections and other interventions. Unfortunately, such experiences are not always positive and, in the same way that the EHP network shares information and views, businesses also communicate with each other and share information. A negative experience with an EHP can therefore impact on the relationship with other businesses in the same area or on the response of other premises within a national network or group of businesses.

EHPs for their part need to develop an understanding of how businesses operate and an appreciation of the different, and often competing, demands placed on businesses. There have already been a number of local authority and commercial sector partnerships offering training placements for student EHPs, but finding opportunities to include some training within businesses as part of their practical training element should be considered good practice and would certainly better equip EHPs for their future dealings with commercial organisations.

Each business is individual and EHPs who adopt a balanced, pragmatic approach are more likely to gain the engagement of businesses. Working in partnership is the most effective way of achieving the compliance goals that both EHPs and businesses seek.

Working as an EHP in the commercial sector

The role of local authority EHPs has changed significantly over the last 25 years with the move away from generalisation. Whilst celebrating the wider transferrable EHP skills, there is essentially little or no difference in the level of technical competence and skills between food safety, health and safety or other discipline specialist EHPs in the public sector and an EHP working in business.

Conversely, many opportunities within business now provide the breadth of role, interest and responsibility that was previously enjoyed by the generalist EHPs of previous years.

Local authorities have increasingly employed staff in a regulatory role whose qualifications may not be the same as qualified EHPs, but they may identify themselves as EHPs to businesses. It is important that local authorities are satisfied that staff are competent to regulate businesses effectively.

The added value of the EHP's core competencies, our EH-ness, has also provided significant opportunities to effectively work with, manage and start very successful businesses. The range of businesses EHPs work within includes almost every possible sector, type and size from consultancy, small and medium enterprise, national and global business covering manufacturing and production, retail and hospitality, healthcare, services industries, facilities management, charities, emergency and armed services to financial and banking.

EHPs in businesses are not only employed to develop and maintain effective policy and procedures, but also hold positions where they can guide, influence and inspire the core values and culture. This level of involvement provides confidence to both internal and external stakeholders that risks are being managed and that the company's credibility and reputation are not being potentially compromised.

Is it gamekeeper turned poacher?

Given the economic and social benefits delivered by business and the increasing size and global reach of businesses, it is disappointing that EHPs working in the commercial sector are sometimes referred to by the less-than-enlightened term 'gamekeeper turned poacher' or similar aspersions relating to their motivation for working in business instead of local authority.

This attitude misses the important point that as a private sector EHP it is possible to

have responsibilities and a positive impact across very significant employee, customer and public populations. Although regulation and its effective enforcement is necessary (and can drive innovation) it does not always go beyond minimum legal compliance. Being a commercial EHP has never just been about a defensive due diligence stance, but more importantly is about setting standards and best practice. Businesses may well be working to higher standards to meet customer expectations and to gain a market (reputational) advantage, so commercial EHPs are still contributing to and in many cases driving the environmental health agenda at a level far above those working within local authority.

Many businesses routinely adopt a culture of continuous improvement and have significantly more frequent and robust internal audits and controls than are placed on them by the enforcement agencies; commercial EHPs and other practitioners have been responsible for developing extremely high standards in safety and hygiene in a number of well-known companies. The work of an EHP in the commercial sector revolves around getting it right the first time and continuous innovation to pre-empt, address and influence business growth and change.

What are the benefits to public health?

Commercial EHPs have not lost sight of their public health origins. Their work directly contributes to a wide range of current and emerging public health issues, often at national and international levels. Some organisations' approach to health and safety extends beyond workplace safety with real emphasis on employee health and wellbeing at work and at home.

EHPs play a leading role in manufacturers' and retailers' efforts to improve the nutritional quality of foods; those involved in reducing salt and fat levels in foods consider their work to have contributed more to improving public health than anything else they have done.

Corporate social responsibility (CSR) and sustainability provide further examples of commercial initiatives in which EHPs have been closely involved. Many companies are successfully striving to reduce water and energy consumption, CO_2 and other emissions as well as investing significantly in recycling and changes in product packaging. Although there are obviously some associated cost savings and brand image benefits, the primary aim of these activities is improved sustainability. The United Nations Industrial Development Organisation (UNIDO) defines CSR as 'a management concept whereby companies integrate social and environmental concerns in their business operations and interactions with their stakeholders' [13]. CSR helps companies achieve a balance of economic, environmental and social imperatives by understanding the impact of their business on local communities as well as the wider world and working to have a positive impact. CSR is firmly embedded in current business culture and there are many excellent examples of local community projects that directly support improvements in public health.

Even if such evidence of contributing to public health fails to persuade the most sceptical public sector EHP, it is difficult to argue that the efforts of the commercial EHP do not, at least, help local authorities to focus their efforts and resources on less enlightened companies or those that cannot yet afford to directly employ their own EHPs or other related expertise.

EHP employment in the commercial sector

Traditionally, the only employment option available to a newly qualified EHP was within local government but those days have disappeared. Today there is a wide range of employment opportunities and professional challenges available to those EHPs who aspire to using their training and experience in broader settings. That said, commercial EHPs will have learned a great deal from their work in the public sector and many continue to

use and further develop those skills once they move into private sector employment. Many have also experienced greater opportunities to improve health and safety, food safety and public health standards, finding greater responsibility and both personal and professional reward than is often possible when working for a local authority.

However, it is important to dispel the popular myth that private sector EHPs enjoy much better remuneration than their local authority peers, and that this in some way justifies the public sector view that commercial colleagues have 'sold out' their integrity; this is not the case. Some private sector salaries can, of course, be attractive but, as with the public sector, very much reflect the associated level of responsibility, and the working hours and employers' expectations on their employees' performance are generally far higher in businesses. Remuneration is also often linked directly or in part to overall business performance.

Salaries in both sectors are also influenced by market forces; there are fewer available local authority jobs for qualifying EHPs and within the commercial job market EHPs are competing against growing numbers of experienced candidates with expertise in food safety, health and safety or other specific environmental health subjects. Despite that, employers do recognise and value the wider range of understanding and skills that most EHPs bring to a role, particularly the ability to assess complex problems, effectively manage communications and mitigate risks. Managing risk is at the very heart of environmental health and public welfare; with our culture of taking a pragmatic approach and working in partnership with others to find the most appropriate solution, the EHP is an ideal risk management professional.

'Risk management' has over recent years become a widely used term and is now used to describe those involved with financial risk management, enterprise or corporate risk management as well as the broader safety-orientated management of risk. With the exception of financial risk, which typically employs people with accountancy or actuarial skills, all other areas of risk management can be considered to fall within our use of the term. As well as risk there has been real growth in other employment areas including sustainability, waste and recycling, carbon management and other 'green collar' jobs, many of which are also suited to EHPs.

Given the very wide range of roles and responsibilities carried out by EHPs in the private sector, and with the number of EHPs having grown enormously over the past 30 years (as evidenced also by the CIEH Consultants' Directory)[18] it is not possible to do justice to their work here. Instead, the following case studies impart a flavour of just some of the different types, extent and primary focus of commercial EHPs jobs.

CASE STUDY 1

Risk Management Director for Moto Hospitality Ltd

I work as Risk Director for Moto Hospitality reporting to the Chief Financial Officer and Audit and Risk Committee. My role covers all the usual areas of regulatory compliance including Health and Safety, Food Safety and Standards, Fire Safety, Environmental Standards and Controls as well as a wider risk remit that includes group insurance, business risk management, business resilience and continuity planning, security, adult gaming, data and cyber security. I was heavily involved in developing Covid-safe operating standards across our business and have led the development of Moto's Environmental, Social and Governance (ESG) Strategy.

Moto operates 64 service areas across the UK, employs over 5000 people and serves over 120 million customers annually. The business operates, under franchise, brands such as M&S Simply Food, Costa Coffee, Burger King, Greggs, Harvester, KFC, Pret, WCP, alongside other retail including a fuel business.

A key skill in commercial environmental health is the ability to innovate to enhance business compliance and reduce cost. We've taken an innovative approach to providing compliance support and auditing across our estate including business risk managers who support the day-to-day compliance management at sites, introducing insurance and customer interaction audits, enhanced site support and introducing an iPad-based audit platform. These steps have significantly reduced time and costs as well as improving compliance visibility. I have overhauled the structure of Moto's insurance programme to reduce cost and improve service delivery, combining the property and casualty insurance programmes into one risk management programme.

Operationally, it is important to work with others in the business. I have supported collegiate working to simplify and digitise incident/accident reporting systems and data analytics to provide an internal online web-based incident reporting portal. This has, alongside a safety culture change programme, led to a 75% reduction in RIDDOR reportable accidents between 2008 and 2014.

Relationships are important and must be founded on a personal approach to leadership that encompasses valuing everyone's opinion, recognising an individual's strengths and weaknesses, honesty and positive communication, whilst retaining the ability to make the correct design and provide direction to the business. The ability to communicate clearly with audiences at all levels within the business is essential.

I am a member of the BEIS Business Reference Panel and the UK Hospitality Food Expert Advisory Group. I also work closely with the Chartered Institute of Environmental Health.

CASE STUDY 2

The RedCat Partnership Ltd Health and Safety Consultants

The RedCat Partnership Ltd is a micro-Environmental Health Consultancy set up by Sarah Daniels and offering Risk Management services (Consultancy and Training in Health and Safety and Food Safety) mainly to businesses in Norfolk. It was founded in 1999, so not ground-breaking, but by keeping the business small the team are able to deliver a completely bespoke service. The company works mainly with SMEs, those businesses that do not have the need for an internal resource and also those whose aspirations are to be the best in their field. However, lately the company have worked with clients that require a very specific skill set.

The team use their fundamental Environmental Health skills of listening and problem solving, finding the best solution to the client's needs and to run our business. The Red-Cat Partnership works to a simple business model; clients are actively listened to and then a solution discussed which fits the culture of the organisation. It is often not a quick fix; clients often require behavioural change or a cultural shift in order to achieve compliance or best practice. Being a consultant or trainer is a trusted position and in a business community it is essential that this credibility and trust remains intact.

Clients respect the fact that RedCat is an SME business too; we fully understand the pressures of running a business in the modern world, and we also own our own five-storey shared office, so facilities management and front-of-house skills have needed to be learned and instigated. Most of our business leads come from recommendation, but RedCat has truly embraced social media, not just as a marketing tool but also to reaffirm its position as a knowledgeable resource in the local and national business community. It is often about helping enquirers with a simple solution.

The synergy of working with like-minded businesses (a business trusted tribe/network) has been a key to RedCat's success and sustainability. The business activities are woven into the local community by our support for networking, events, charities and local social enterprise. This has been further reinforced by RedCat's business home which is specifically designed as an accessible hub for other local businesses; not just those who want to hire out the training/meeting rooms, but also networking and collaboration. The Covid pandemic enabled us to help others, for example by sharing Risk Assessment workshops with local business networks. Essentially, we use our skills and experience to help other businesses to succeed. RedCat has remained true to our core environmental health belief in being friends of the local business community and therefore the human race!

The case studies show the wide ranging responsibilities and achievements of commercial EHPs as well as how key elements of their roles still closely reflect the core environmental health activities undertaken by local government-employed colleagues. EHPs are truly a single profession with a shared common purpose of protecting and enhancing public health.

While local authority EHP numbers appear to have declined there were positive indications that most councils are continuing to provide all environmental health services, and some were adopting different approaches to maintain service delivery. The survey also identified that some authorities are reverting to a more generic EHP role to improve service integration, which also better aligns

CASE STUDY 3

Senior Director, Operational Compliance, Asda Stores Limited

As a Senior Director, Operational Compliance within Asda Stores Limited, the third largest retailer in the UK, I report directly through to the General Counsel and Company Secretary and create an environment to support the senior leadership teams on risk assessment, standards and controls, communication, training, monitoring and successful implementation of a compliance programme to meet the company's Regulatory Compliance Strategy. Asda has 640 stores, 26 depots and two home shopping centres with a turnover of approximately £20 bn. Asda employs 145,000 colleagues and serves over 18 million customers per week.

Strategic support and direction is provided to the business in relation to Compliance, including food safety and hygiene (operational and supply chain), health and safety, fire safety, environmental protection, licensing and permits, pharmacy and optical, labour and employment, trading standards and consumer protection and other compliance-related subjects. In addition, I act as a business partner to Retail and Property and Construction. My team are also involved in managing local and national regulatory and reputational risk through proactive and reactive engagement with local and national regulators. I am a core member of the Incident Management Group whose primary function is Business Continuity in the event of a crisis.

I am a believer in Compliance by design, simplification and automation, and the role of my team is to ensure that we can deliver legal compliance in the most effective way. Having compliance data at our fingertips helps us steer our operators (Stores, Depots and Home Shopping Centres) in the right direction and this then means that they can focus on priority areas whilst continuing to deliver great customer service.

The Compliance team are also a business enabler and have helped deliver tens of millions of pounds of cost savings and efficiencies over the years. Examples of these include simplification of temperature control monitoring within stores, accident reduction directly impacting on claims costs, centralised helpdesk introduction, combining of food safety and cleaning audits and utilisation of a risk-based approach. Although the Compliance team is a support function of the business, the value that the team can provide is recognised by various stakeholders and therefore we are seen as integral part of their operations.

Externally, I am a member of the Department for Business, Energy and Industrial Strategy (BEIS) Business Reference Panel and I am an active contributor on the benefits of the Primary Authority Scheme and the impact on business of regulatory change. Throughout the Covid-19 pandemic I have regularly engaged with governments across all of the UK and provided feedback from a retail perspective on Covid control measures. In addition, I have participated in industry forums such as through the British Retail Consortium (BRC) and Confederation of British Industry (CBI) to share best practice on Covid-19 with other retailers and business sectors.

---

with commercial EHPs' approach and practice [14]. Chapter 2 refers to the 2021 report covering England, where perhaps a demonstration of the value of EHPs was that eight out of ten in local authorities were redeployed in response to the pandemic with the most common activities included liaising with businesses and advising on how to trade safely, developing Covid safe procedures (this also included enforcing restrictions) as well as managing local outbreaks and contact tracing. However, in 2019–2020 it was reported that as few as 455 full-time equivalent posts focused on health and safety at work compared with 1,020 in 2010/11. Some 80% of local authorities were using agency staff to deliver environmental health services [15].

Better regulation and Primary Authority

The Office for Product Safety and Standards (OPSS) is part of the Department for Business Energy and Industrial Strategy and is responsible for the Primary Authority regime. Primary Authority is a means for businesses to receive assured and tailored advice on meeting regulations such as environmental health, trading standards or fire safety through a single point of contact.

The OPSS has the aim of creating a regulatory environment in which businesses have the confidence to invest and grow whilst the community remains properly protected. The OPSS has taken over the functions of what has previously been the Local Better Regulation Office (LBRO) then Better Regulation Delivery Office (BRDO).

Primary Authority was introduced in April 2009 and amended in October 2013, and then the eligibility criteria for Primary Authority was expanded so that from 1 October 2017 all businesses can now benefit from Primary Authority. The partnerships are recognised by other local regulators with the intention that the primary local authority partner provides robust and reliable advice for officers of all other local authorities when they carry out inspections or deal with matters of non-compliance. A central register of the partnerships is maintained by the OPSS who also issue guidance and resolve any disputes that may arise.[19]

Partnerships are available to any type of business, whether starting out or established, as well as other organisations, such as charities and trade associations. Regulators that can become primary authorities include county, district and unitary councils, and fire and rescue authorities. A business can form its own direct partnership. It then receives Primary Authority Advice tailored to its specific needs from its Primary Authority. Alternatively, a business can belong to a trade association (or other type of group) to benefit from a co-ordinated Primary Authority. In this case, the Primary Authority Advice is still from the Primary Authority, but provided via the trade association, and tailored to the general needs of its members. A business can choose the type of partnership best suited to its needs. Most businesses only have one partnership, but it is possible to be in different partnerships for different areas of regulation.

The Primary Authority approach is seen as the route to a simpler, more consistent and effective regulation, giving businesses confidence that they are not going to be subject to different interpretation and regulatory demands by different officers and authorities. The Primary Authority can produce a national inspection plan so as to improve the effectiveness of inspection, avoid repeated checks and enable better sharing of information. Where a problem arises, the Primary Authority can co-ordinate enforcement action to ensure that the business is treated consistently and that intervention is proportionate to the issue.

It is for the business to determine what level of support it needs from its Primary Authority, which can recover the costs involved in entering into such a partnership. Partnerships can cover all environmental health and trading standards legislation, including specific functions such as food safety, occupational health and safety, housing, environmental protection and licensing, as well as fire safety and product safety.

However, offering effective Primary Authority support to a business requires EHPs to develop a good understanding of how those business partners function and operate.

Earned recognition scheme[20]

The principle of earned recognition has been supported by the Food Standards Agency as a sensible approach for improving the efficiency of food regulation and is *at the heart of the Food Standards Agency (FSA) approach to rewarding responsible businesses and encouraging industry to promote the positive role of regulatory standards* [15]. Earned recognition aims to reduce the burden for compliant businesses which allows enforcement activity to concentrate on less compliant businesses. Those who qualify for earned recognition will benefit by receiving less frequent visits by the enforcement authority.

Previously also referred to as earned autonomy, the term earned recognition is more correct as the scheme is based around a business earning recognition of its established systems and management of its risks via approved assurance schemes rather than the organisation working autonomously outside the reach of the legislation. This does reflect the importance of responsible and reputable business but there are a number of possible disadvantages relating to the loss of regular objective inspections, lowering of standards and less confidence in the independence of third-party auditors.

Although effectively established in other countries the implementation of earned recognition in the UK has, so far, been restricted to farming, animal feed but with the economic challenges facing both regulatory authorities and businesses it is likely that ways of further reducing inspection costs will continue to be debated. Smaller and medium-sized businesses will, of course, still welcome the advice and reassurance offered as part of routine EHP inspections.

Conclusion

To be effective in securing necessary behaviour change it is essential that EHPs understand a little about the businesses they are working with, as well as the motivation of those working in commerce. A business-savvy EHP (one who has some better understanding of business) will be better placed to ensure regulatory compliance through advice, guidance and, where necessary and appropriate, formal intervention; in turn the business will engage and respond more favourably where the EHP's approach appreciates what is involved in running a business, is balanced and proportional.

Understanding a little of what is involved in operating a business properly will also help an EHP identify those businesses in which they can have confidence in the quality and competence of management (and identify those which are not well managed and more likely to care less about compliance). The EHP should also have confidence that those that are well managed will implement any necessary changes to meet environmental health needs.

This chapter has attempted to provide a little basic insight into business and management in the commercial world. It has also provided insight into how EHPs working in the commercial world contribute to improved environmental health and hopefully will encourage both public and private sector EHPs to continue to work together as a single profession with a common goal.

Notes

1 See Obi J, Ibidunni AS, Tolulope A, Olokundun MA, Amaihian AB, Borishade TT, Fred P. (2018) Contribution of small and medium enterprises to economic development: Evidence from a transiting economy. *Data in Brief*, 18: 835–839, Also see EU publishes 2021 EU4 Business Report on SME Support in the Eastern Partnership at https://www.euneighbours.eu/en/east/stay-informed/news/eu-publishes-2021-eu4business-report-sme-support-eastern-partnership

2 https://www.instituteforgovernment.org.uk/explainers/local-government-funding-england

3 For more on local government finance see https://researchbriefings.files.parliament.uk/documents/CBP-8431/CBP-8431.pdf

4 https://www.nao.org.uk/wp-content/uploads/2020/08/Local-government-finance-in-the-pandemic.pdf

5 See https://researchbriefings.files.parliament.uk/documents/CBP-9129/CBP-9129.pdf

6 https://researchbriefings.files.parliament.uk/documents/SN06152/SN06152.pdf

7 https://www.gov.uk/government/statistics/incorporated-companies-in-the-uk-april-to-june-2021

8 https://employeeownership.co.uk/wp-content/uploads/EOA-2019-Annual-Review-final.pdf

9 https://www.johnlewispartnership.co.uk/about/how-we-share-power.html

10 See https://www.ica.coop/en

11 A mutual company is a *private company* that is owned by its customers or policyholders. The company's customers are also its owners. As such, they are entitled to receive a share of the profits generated by the mutual company.

12 https://www.gov.uk/get-information-about-a-company

13 See for example "Time to abolish non-executive directors" at https://files.simmons-simmons.com/api/get-asset/EB_Autumn2018_Charles-Mayo_Singles.pdf?id=blte26439a8015af04a

14 A case concerning a prosecution under the Finance Act 1966 in relation to a failure to hold the requisite gaming licence – offence attributable to her neglect.

15 More information can be found at https://www.sentencingcouncil.org.uk/sentencing-and-the-council/about-sentencing-guidelines/about-published-guidelines/

16 See https://waterbriefing.org/home/company-news/item/18550-southern-water-ordered-to-pay-record-£90-million-following-environment-agency-prosecution where the water company pleaded guilty to illegal discharges of sewage which polluted rivers and coastal waters in Kent, Hampshire and Sussex. The HJ Johnson was reported as saying

> each of the 51 offences seen in isolation shows a shocking and wholesale disregard for the environment, for . . . ecosystems along the North Kent and Solent coastlines, for human health, and for the fisheries (including shellfish) and other legitimate businesses that depend on the vitality of the coastal waters. It was necessary to sentence the company for the totality of the offences but that did not reflect the defendant's criminality . . . the offences were aggravated by previous persistent pollution of the environment over very many years.

The court were told Southern Water had deliberately presented a misleading picture of compliance to the Environment Agency, hindering proper regulation of the company. Some commentators suggested that 'fines' were seen as just a 'business cost' or overhead.

17 Note there are separate but parallel versions for England, Wales and Northern Ireland; see https://www.food.gov.uk/about-us/food-and-feed-codes-of-practice#food-law-code-of-practice

18 See https://www.cieh.org/ehn/consultants-directory/

19 Statutory Guidance is available at https://assets.publishing.service.gov.uk/government/uploads/system/uploads/attachment_data/file/707382/primary-authority-statutory-guidance-2017.pdf

20 See https://www.food.gov.uk/business-guidance/earned-recognition-approved-assurance-schemes

References

[1] Department for Communities and Local Government. (2014) *A Guide to the Local Government Finance Settlement in England*, Department for Communities and Local Government, London.

[2] House of Commons Library. (2014) *English Local Government Finance: Issues and Options*, Research Paper 14/43, House of Commons, London.

[3] Companies House. (2021) *Official Statistics Incorporated Companies in the UK April to June 2021*. Crown Copyright. https://www.gov.uk/government/statistics/incorporated-companies-in-the-uk-april-to-june-2021/incorporated-companies-in-the-uk-april-to-june-2021 (Accessed August 2021).

[4] Companies House. (2015) *Statistical Release, Incorporated Companies in the United Kingdom*, Crown Copyright, London, June.

[5] John Lewis Partnership. (2021) *Emerging Stronger John Lewis Partnership PLC Annual Report and Accounts 2021*. https://www.john-lewispartnership.co.uk/content/dam/cws/pdfs/Juniper/ARA-2021/2021-Annual-Report-and-Accounts-Report.pdf (Accessed August 2021).

[6] BVA BDRC. (2019) *2018 Franchise Landscape, for British Franchise Association (BFA) in Partnership with NatWest*. https://rainbow-int-franchise.co.uk/wp-content/uploads/NatWest-Franchise-Report-2018-7.pdf (Accessed August 2021).

[7] Accounting Simplified. (2021) *Financial Statements, Accounting-Simplified.com*. https://accounting-simplified.com/financial/statements/types/ (Accessed August 2021).

[8] Spencer-Phillips C, Lawson R. (2013) *Chair and Non-Executive Director Guidelines*, First

Flight Non-Executive Directors Ltd and ShareSoc UK Individual Shareholders Society Ltd., London.

[9] Proudlock E, Nana C. (2013) Consent and connivance: The criminal liability of directors and senior officers. *The In-House Lawyer.* https://www.inhouselawyer.co.uk/legal-briefing/consent-and-connivance-the-criminal-liability-of-directors-and-senior-officers/ (Accessed 17 August 2021).

[10] Burn-Murdoch J. (2013) Reputation management: Businesses must swallow the bitter pill. *The Guardian.* http://www.theguardian.com/news/datablog/2013/jul/15/reputation-management-business-swallow-bitter-pill (Accessed 17 August 2021).

[11] BBC. (2015) Alton Towers rollercoaster crash causes Merlin profit warning. *BBC.* www.bbc.co.uk/news/business-33672357 (Accessed August 2021).

[12] Health and Safety Executive. (2015) *Enforcement Policy Statement, Health and Safety Executive.* https://www.hse.gov.uk/pubns/hse41.pdf (Accessed August 2021).

[13] United Nations Industrial Development Organization. (2021) *What Is CSR?* https://www.unido.org/our-focus/advancing-economic-competitiveness/competitive-trade-capacities-and-corporate-responsibility/corporate-social-responsibility-market-integration/what-csr (Accessed 17 August 2021).

[14] Chartered Institute of Environment Health. (2014) *CIEH Environmental Health Workforce Survey 2014/15 Phase 1 and 2 Summary Report and Findings,* CIEH, London.

[15] Chartered Institute of Environment Health. (2021) *Environmental Health Workforce Survey Report: Local Authorities in England,* CIEH, London. https://www.cieh.org/media/5249/cieh-workforce-survey-report-for-england.pdf.

[16] Food Standards Agency. (2018) *Guidance on Food Standards Agency Approved Assurance Schemes in the Animal Feed and Food Hygiene at the Level of Primary Production Sectors.* https://www.food.gov.uk/sites/default/files/media/document/FINAL%20FSA%20Guidance%20on%20Approved%20Assurance%20Schemes%20April%202018.pdf (Accessed August 2021).

Further sources of information

UK Government, https://www.gov.uk/business-legal-structures/overview

Companies House, http://www.companieshouse.gov.uk

Employee Ownership Association, https://www.employeeownership.co.uk

International Co-operative Alliance, https://www.ica.coop

British Franchise Association, https://thebfa.org

Bytestart the small business portal, https://www.bytestart.co.uk

The British Standards Institution (BSI Group) https://www.bsigroup.com/en-GB/

HM Revenue & Customs, https://www.gov.uk/government/organisations/hm-revenue-customs

Financial Reporting Council, https://www.frc.org.uk

Institute of Chartered Secretaries, https://www.icsa.org.uk

Institute of Directors, https://www.iod.com

The Sentencing Council, https://www.sentencing-council.org.uk

The Chartered Insurance Institute, https://www.cii.co.uk

8

Quality management systems

Environmental management systems as examples

Stephen Battersby

Introduction

If environmental health is about assessing, preventing or correcting the environmental stressors, including from unsafe food, that adversely impact human health then quality management is something with which environmental health practitioners should be familiar. Much of the work of environmental health officers or practitioners (EHPs) brings them into contact with organisations and enterprises whose activities and products could impact on the wider community, and so an understanding of quality management is necessary. The purpose of quality management is to ensure every time a process is performed, the same information, methods, skills and controls are used and applied in a consistent manner. It is also relevant as EHPs should be able to assist when their employer, be it commercial enterprise or municipality/ council, wishes to improve its performance, particularly environmentally.

The equivalent chapter in the previous edition focused very much on environmental management systems, but here we broaden out the approach a little.

The ISO 9000 series of quality management systems (QMS) is a set of standards that helps organisations ensure they meet customer, client and other stakeholder needs within statutory and regulatory requirements related to a product or service. The standards provide guidance and tools for companies and organisations who want to ensure that their products and services consistently meet customer's requirements, and that quality is consistently improved. The ISO 9000 family addresses various aspects of quality management and contains some of the best-known standards of the International Organisation for Standardisation (ISO). ISO is an independent, non-governmental international (global) organisation with a membership of 164 national standards bodies including the British Standards Institution (BSI) and has its secretariat in Geneva, Switzerland. Its members (or member bodies) have a say in the development of ISO standards and strategy by participating and voting in ISO technical and policy meetings.

There are eight quality management principles, which aim to help organisations to become better managed, more efficient and more customer focused:

- Customer focus;
- Leadership;
- Involvement of people;
- Process approach;

DOI: 10.1201/9781003035640-8

- System approach to management;
- Continual improvement;
- Factual approach to decision-making;
- Mutually beneficial supplier relationships.

The ISO 9001 standard [1] has been adopted by over a million organisations across the world and is used by businesses to continually monitor, manage and improve the quality of their products and services. It is seen as an important tool, providing the framework and guidance an enterprise needs to deliver on a consistent basis and meet their clients' expectations as well as help meet regulatory requirements. ISO has a range of standards for quality management systems that are based on ISO 9001 and adapted to specific sectors and industries.

A QMS can help demonstrate compliance and give the EHO confidence in the management of the operation, be the EHO concerned with health and safety in the workplace, management of waste and emissions and managing the environmental impact of the operation (via an environmental management system EMS), or production of safe food. A QMS can also be used to demonstrate competence in housing management.

ISO 9001[1] promotes the adoption of a process approach when developing, implementing and improving the effectiveness of a quality management system, to enhance customer satisfaction by meeting customer requirements. Understanding and managing interrelated processes as a system within an organisation contributes to the organisation's effectiveness and efficiency in achieving its intended results. This approach enables the organisation to control the interrelationships and interdependencies among the processes of the system, so that the overall performance of the organisation can be enhanced. The process approach involves the systematic definition and management of processes, and their interactions, so as to achieve the intended results in accordance with the quality policy and strategic direction of the organisation. In this structured approach it is possible to

understand the requirements and to meet them consistently. It enables the consideration of processes (and their improvement following evaluation of data and information) in terms of added value and achievement of effective process performance. Key to this approach is the Plan Do Check Act (PDCA) approach (Figure 8.1a). It means that it is easier to identify where and why problems arise and for remedial action to be taken.

The scope of the environmental management system (EMS), considered next, is that area within the broken lines, and the outer area (beyond the dotted lines in Figure 8.1b) is the context of the organisation (such matters as significant environmental aspects, risks and compliance obligations). It should be noted that leadership is a central element.

Environmental management systems (EMS)

An environmental management system (EMS) is a specific quality management system that has much in common with more general management systems. Environmental issues and the impacts of businesses on the environment can also be seen as a fundamental part of environmental health which explains the emphasis here. With increasing recognition of climate heating and the impacts of our activities on the environment, environmental management has become a core issue for many organisations in both public and private sectors. This increased recognition of the environmental imperative and for sustainable development, and also consumer pressure leading to corporate social responsibility has made an EMS a key response for many organisations around the world.[2] Minimising the amount of waste that is produced, reducing energy consumption and making more efficient use of resources can all lead to financial cost savings. In addition, this will help to protect and enhance the environment and improve the image of the company or organisation. Environmental health practitioners should recognise that the quality of the environment

Note: The numbers in brackets refer to the clauses in ISO 9001:2015

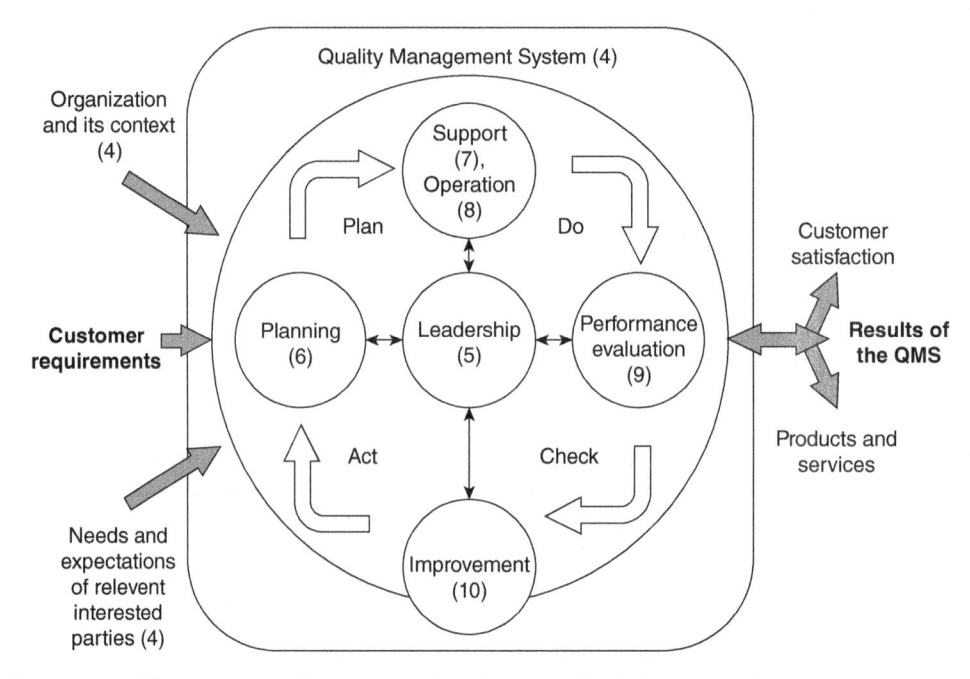

Figure 8.1a The structure of the International Standard ISO 9001 (Quality management systems: Requirements) in the Plan, Do, Check Act (PDCA) cycle.

Source: Permission to reproduce extracts from British Standards is granted by BSI Standards Limited (BSI). No other use of this material is permitted. British Standards can be obtained in PDF or hard copy formats from the BSI online shop: www.bsigroup.com/Shop

locally and globally can influence and impact on the health and well-being of individuals and communities. At the same time EHPs are involved directly in regulating emissions from certain businesses, activities and operations. Although this chapter is largely concerned with environmental management systems, in many ways all quality management systems have much in common, and so this chapter is relevant to all environmental health practitioners regardless of any area of specialism or sector in which they are employed.

The UK Government has also said [2] that an environmental permit requires holders of those permits to have a written management system and *this is a set of procedures describing what [the business] will do to minimise the risk of pollution from the activities covered by [the] permit.* From 7 April 2019, if the company had a waste permit that was granted before 6 April 2008 it has been necessary to manage and operate the waste activity in line with a written management system.

Local authorities themselves are also key players in the local economy. They can be an important influence on the environmental habits of the general public and can make a major contribution to the implementation of the principles of sustainable development at local level by implementing an environmental management system for different activities. For example, Derbyshire County Council's economy, transport and environment department has an environmental management system (EMS) certified to BS EN ISO 14001:2015 [3] (see later).

Businesses and local authorities have environmental impacts through their activities,

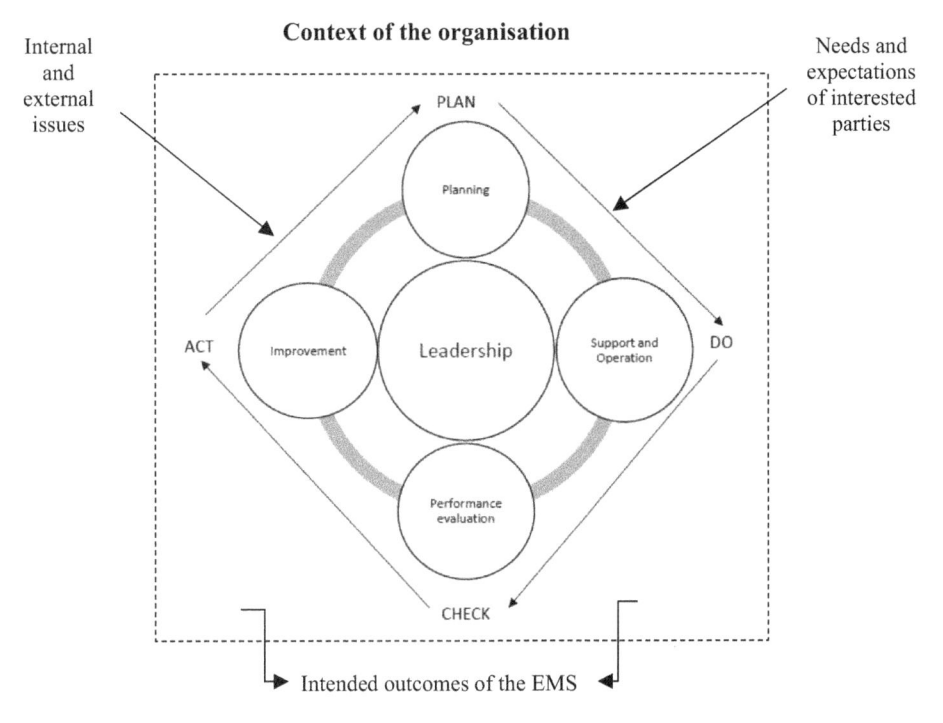

Figure 8.1b Setting out the relationship between PDCA and the framework of ISO 14001:2015 (Environmental management systems – Requirements with guidance for use)

Source: Adapted from Figure 1 in BS EN ISO 14005:2019 (Permission to reproduce extracts from British Standards is granted by BSI Standards Limited (BSI). No other use of this material is permitted. British Standards can be obtained in PDF or hard copy formats from the BSI online shop: www.bsigroup.com/Shop)

products and services. It does not matter whether an enterprise is in the service or manufacturing sector there will be environmental impacts that should be managed. These can range from regulated processes that generate pollution or waste, through to activities that affect resources management and consumption. A key issue of concern to all sectors is the need to mitigate climate change and reduce emissions, but an effective EMS can also help minimise risks of accidental releases to air, land and water. Beyond the benefits of improving regulatory control, an EMS can improve overall environmental performance and business efficiency, helping to reduce costs through more efficient use of materials and other resources.

Whether improved environmental performance is seen as part of corporate social responsibility or is seen as a good thing in itself or a way of improving business efficiency and reducing waste, it is best managed by way of an organised and documented management system.

An EMS works in a similar way to a quality management system. It provides an organisation with a framework through which its environmental performance can be monitored, measured, improved and controlled. When undertaking an inspection or audit of an organisation, an environmental health practitioner will be able to assess how much confidence can be given to the management by ascertaining whether or not the operation has in place a certified and externally validated environmental management system. When determining priorities for inspections and inspection programmes (or managing risk) whether or not a business or site or individual operation has an environmental management system should be a factor that is taken into account. The costs and benefits of the European EMAS scheme were set out in a 2009 study [4], and generally these are applicable to any certified EMS.

The aim of an EMS, as defined in the EU EMAS Regulation [5–6] and the ISO 14001 standard, is to improve overall environmental performance and to ensure that improvements in performance are 'continual'. As ISO 14001 says the aim of the EMS takes a systematic approach to environmental management which can provide top management with information on which to build success over the long term and create options for contributing to sustainable development by:

- Protecting the environment by preventing or mitigating adverse environmental impacts;
- Mitigating the potential adverse effect of environmental conditions on the organisation;
- Assisting the organisation in the fulfilment of compliance obligations;
- Enhancing environmental performance;
- Controlling or influencing the way the organisation's products and services are designed, manufactured, distributed, consumed and disposed by using a life cycle perspective that can prevent environmental impacts from being unintentionally shifted elsewhere within the life cycle;
- Achieving financial and operational benefits that can result from implementing environmentally sound alternatives that strengthen the organisation's market position;
- Communicating environmental information to relevant interested parties.

A commitment to legal compliance is a requirement for ISO 14001 certification and demonstrating compliance is a requirement of EMAS registration. This does not mean that there will never be any incidents that have adverse environmental impacts, but these can be minimised through better management, and an EMS will make it easier to identify where there have been failings in a particular process or operation. Since the revision of the EMAS Regulation Annexes it is easier for an organisation already complying to an environmental management system such as ISO 14001 to step up to EMAS.

Much directly applicable EU legislation, that is, law that applied in the UK without any further legislation by our parliament, such as EU Regulations, was automatically brought into national law by the EU (Withdrawal) Act 2018, as *retained EU law*. In some cases, however, this is not appropriate. Once the UK left the EU the national body no longer has the authority to take part in EMAS and the EMAS Regulation (Regulation (EC) No 1221/2009) is not applicable. However, Article 3(3) of the EMAS Regulation does allow organisations from outside the European Union to register to the scheme through EMAS Global.[3]

An EMS has a similar structure to other management systems, such as those that manage quality or health and safety. It assesses an operation's strengths and weaknesses, helps with the identification and management of significant impacts, provides benchmarks for improvements and helps the business keep track of progress. In addition, it can secure cost savings and internal efficiencies. Some organisations integrate their EMS with other systems such as health and safety management and quality assurance. ISO 14001 [3], and quality management system ISO 9001, for example, now share the same auditing standard ISO 19011 [7], and ISO 14001 is also compatible with health and safety standard ISO 45001 (an internationally recognised standard for Occupational Health and Safety and the world's first global health and safety management system) that also includes guidance for use.[4] Certification to this standard demonstrates quality management in matters of health and safety.

There are specific benefits to an EMS, including better resource management, improved credibility and reliability of environmental information and reduction of any environmental and legal risks and liabilities. Businesses often demonstrate to customers and clients their environmental commitments by having their environmental management system certified, such as through ISO 14001, BS EN ISO 14005:2019 [8], the Green Dragon[5] or EMAS (see later). BS 14005:2019 is a systematic approach to environmental management that allows for a phased approach to implementing an EMS and is discussed later.

Obtaining certification to a recognised standard can increase the credibility of an EMS as well as ensuring commitment from staff and management. This can help to drive improvements, but certification can only be obtained by having the EMS checked on a regular basis (usually annually) by an outside verifier. Businesses with accredited certification to ISO 14001 receive credit under the Integrated Pollution and Prevention and Control legislation resulting in reduced environmental levies for businesses and reduced inspection requirements for the regulator. Whether or not an organisation chooses to have their EMS certified to ISO 14001 or registered under EMAS, the benefits can be summarised as:

- Assuring customers and clients of commitment to demonstrable good environmental performance and management;
- Attracting shareholders and investors with improved access to capital and investment (funding);
- Maintaining good public/community relations;
- Defining clear environmental responsibilities for all staff and shared approach to resolving environmental problems;
- Improving availability of insurance at reasonable cost;
- Identifying opportunities to reduce waste, including raw materials, utility use and waste disposal costs thereby reducing total costs;
- Enhancing image and market share;

- Improving cost control and increased profits;
- Reducing incidents and legal liabilities and risks of legal action and fines for non-compliance with environmental legislation;
- Having procedures in place to minimise environmental impacts, conserving input materials and energy;
- Recording of actual environmental performance against set targets;
- Providing a clear audit trail (that can be used by EHPs on and as part of inspections);
- Facilitating attainment of necessary permits and reduced regulatory oversight;
- Promoting of environmental awareness among suppliers, contractors and others working with the organisation;
- Improving relations with regulatory bodies and government.

Practitioners may wish to advise enterprises on the benefits of EMS as part of their day-to-day work and this chapter should provide the basic information.

When it comes to compliance with environmental law and the management system differences exist between EMAS and ISO 14001. Under EMAS organisations have to demonstrate they can provide for legal compliance with environmental legislation, including permits and permit limits and have procedures in place that enable the organisation to meet these requirements on an ongoing basis. The Environment Agency has in the past said it will use its powers within legislation to suspend companies from their EMAS registration when they break the law, and this would be carried out in accordance with the Agency's enforcement and prosecution policy. Regrettably this now no longer figures in the 2019 policy,[6] presumably linked to the UK exiting the EU, but as we have seen anyway the level of enforcement generally has declined as have staffing levels.[7] The EA suggested it would raise poor legal compliance of companies with the certification organisations for ISO 14001:2015 environmental management systems, but reference to EMS is only made in the context of environmental permits that relate to waste [2]. Legal compliance is a requirement of EMAS, whereas ISO 14001 is more nuanced in that it highlights the need to determine compliance obligations, ensure operations are carried out in accordance with these compliance obligations, evaluate fulfilment of the compliance obligations and correct nonconformities. If results of a compliance evaluation indicate a failure to fulfil a legal requirement, the organisation needs to determine and implement the actions necessary to achieve compliance. A non-compliance is not necessarily elevated to a nonconformity if, for example, it is identified and corrected by the environmental management system processes. Compliance-related nonconformities need to be corrected, even if those nonconformities have not resulted in actual non-compliance with legal requirements. There is of course nothing to prevent local authorities taking the existence of a certificated EMS into account when considering what action to take for non-compliant premises and processes subject to their control.

Terminology and explanations

There are some terms that are often used in the context of environmental management with which those involved in EMS implementation should be familiar. These are set out next, but some are also considered in more detail later in the chapter:

Audit – is to assess whether an EMS conforms to the requirements of the environmental policy and, where the EMS is to be registered to EMAS or certified to the ISO 14001 standard, then it is also audited against the appropriate requirements. The audit should be structured, systematic, documented and undertaken on a regular basis, the frequency of which may depend on the system with which the EMS is being registered. See the following on 'internal audit'. BS EN ISO 19011:2018 [7] sets out the principles of auditing, how to manage an audit programme, how to conduct a management system audit and provides guidance on evaluating the competence of individuals involved in the audit process. This includes the individual(s) managing the audit programme, auditors and audit teams. Unlike previous versions this standard includes a risk-based approach to the principles of auditing and has expanded the generic competence requirements for auditors.

Auditor – in this context is a person with the demonstrable professional competences to conduct an audit of the management system. It can be an individual or group of individuals, belonging to an organisation itself or a person external to that organisation. The role is to assess the environmental management system in place and determine conformity with the organisation's environmental policy and programme, including compliance with the applicable legal requirements (c.f. 'verifier' later).

The skills and competencies set out in BS EN ISO 19011:2018 for auditors are very much those that should be expected of an EHP, for example they should be able to prioritise and focus on matters of significance; communicate effectively, orally and in writing (either personally, or through the use of interpreters); collect information through effective interviewing, listening, observing and reviewing documented information, including records and data; understand the appropriateness and consequences of using sampling techniques for auditing; understand and consider technical experts' opinions and assess those factors that may affect the reliability of the audit findings and conclusions; document audit activities and audit findings, and prepare reports; maintain the confidentiality and security of information [7].

Audit findings – are the results of the evaluation of the collected audit evidence against audit criteria. These findings can indicate either conformity or nonconformity with audit criteria or opportunities for improvement.

BS ISO 14001:2015 also defines 'audit evidence' as consisting of records, statements of fact or other information which are relevant to the audit criteria and are verifiable [3]; and 'audit criteria' are the set of policies, procedures or requirements used as a reference against which audit evidence is compared, as defined in ISO 19011:2018 [7] (the standard provides guidance for any organisation that wants to plan and conduct management systems audits or manage an audit programme).

Environmental aspect – is an element of an organisation's activities, products or services that can interact with or impact on the environment – emissions to air and discharges to water are typical aspects. A significant environmental aspect has or can have a significant environmental impact on the receiving environment, such as pollution of water or contamination of land. These can be direct or indirect.

The EMAS Regulation says that an indirect environmental aspect is one *which can result from the interaction of an organisation with third parties, and which can to a reasonable degree be influenced by an organisation.* Direct environmental aspects are associated with activities, products and services of the organisation itself over which it has direct management control. While a manufacturing or industrial site may be principally concerned with the environmental effects of production activities and emissions and energy use, local authorities and municipalities also have major environmental aspects and related impacts through the way they procure materials or deliver services.

Figure 8.2 is based on the Plan-Do-Check-Act (PDCA) methodology as set out in Figure 8.1a. It is designed to help an

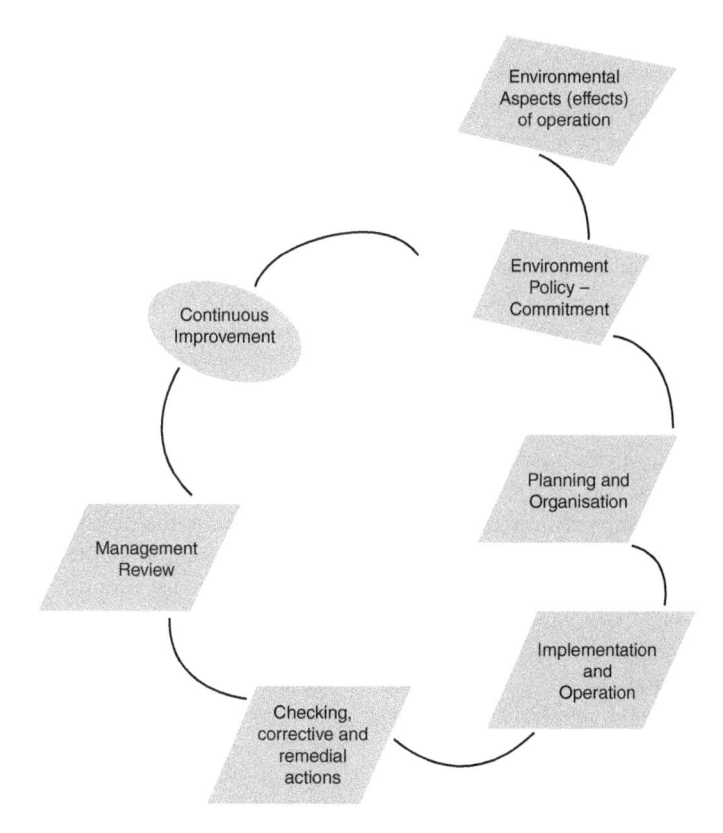

Figure 8.2   Schematic environmental management system

organisation plan, implement and review an EMS. Any EMS should be seen as an organisational framework that should be monitored continually and reviewed from time to time. The purpose is to provide effective direction for an organisation's environmental performance for which there is 'buy-in' at all levels within the organisation and which can take account of changing circumstances.

Environmental impact – specifies any change to the environment, whether adverse or beneficial, wholly or partially resulting from an organisation's environmental aspects.

Environmental issue – can be construed as an environmental impact that reaches wider awareness and concern. For example, cutting down a tree has an environmental impact, but it becomes an issue when that tree has a 'preservation order' and its demise is reported in the media, leading to public awareness and complaints.

Environmental management system – is part of an organisation's management system used to develop and implement its environmental policy and manage its environmental aspects. A management system is a set of interrelated elements used to establish policy and objectives and to achieve those objectives. It includes organisational structure, planning activities, responsibilities, practices, procedures, processes and resources. An effective EMS thus includes:

- A review of an organisation's activities, products, processes and services that might interact with the environment;
- An environmental policy;
- An environmental improvement programme;
- A training and awareness programme;
- Written procedures to control activities with a significant environmental impact;
- Periodic auditing of the system to ensure effective operation;
- A formal review of the EMS by senior management.

Environmental management system audit – is a systematic and documented verification process of objectively obtaining and evaluating evidence to determine whether an organisation's environmental management system conforms to the environmental management system audit criteria set by the organisation, and for communication of the results of this process to management.

Environmental objective – is the overall environmental goal, consistent with the environmental policy, that an organisation or enterprise sets itself to achieve.

Environmental policy – is an agreed documented statement of an organisation's commitment to improve environmental performance, and formally signed by senior managers. An environmental policy is a requirement for both ISO 14001 (paragraph 4.2 in the standard) and EMAS (Annex II). An environmental policy is an organisation's commitment to improve environmental performance by, for example, reducing its carbon footprint, improving recycling, minimising waste and increasing resource efficiency across all operations and in all departments. The environmental policy provides a framework for action and for the setting of environmental objectives and environmental targets. It does not have to be detailed or contain the actual environmental targets.

Environmental performance – is measurable results of an organisation's management of its environmental aspects. In the context of environmental management systems, results can be measured against the organisation's environmental policy, environmental objectives, environmental targets and other environmental performance requirements or criteria. For example, if the policy is to reduce waste year on year per unit of product, performance can be measured against the policy – by comparing waste arising from year to year.

In EMAS an environmental performance indicator is taken to mean 'a specific expression that allows measurement of an organisation's environmental performance'.

Environmental programme – is defined in the EMAS Regulations as a description of the measures, responsibilities and means taken or envisaged to achieve environmental objectives and targets and the deadlines for achieving the environmental objectives and targets.

Environmental performance evaluation – is a process to facilitate management decisions regarding an organisation's environmental performance by selecting indicators, collecting and analysing data, assessing information against environmental performance criteria, reporting and communicating, and periodically reviewing and improving this process. EMAS requires that organisations be able to demonstrate that the management system and the audit procedures address the actual environmental performance of the organisation with respect to the direct and indirect aspects. The environmental performance shall be evaluated as part of the management review process evaluating actual performance against its objectives and targets.

Environmental review – means an initial comprehensive analysis of environmental aspects, environmental impacts and environmental performance related to an organisation's activities, products and services. That is a scoping study to assess the environmental aspects of existing activities and services and provides baseline data for measuring future environmental performance. A review is not mandatory for ISO 14001 but is a fundamental component of EMAS that will be assessed prior to EMS implementation.

Environmental statement – is required by EMAS (not ISO 14001) and outlines the organisation's environmental policy, programme and management system, and summarises its environmental performance with the results achieved and the steps necessary for future improvements, and this has to be validated. The requirements of the statement are set out in Annex IV of the Regulation and the statement has to include such matters as:

- A full description of the organisation and its activities;
- An assessment of all significant direct and indirect environmental issues;

- A description of the environmental objectives;
- A summary of year-by-year data on emissions, waste generation, raw material consumption, energy and water usage and noise emissions and other factors regarding environmental performance including performance against legal provisions with respect to their significant environmental impacts;
- A presentation of the environmental policy, programme and management system;
- The deadline for the next statement;
- The name and accreditation number of the environmental verifier and date of validation.

When the environmental management system has been implemented and the environmental statement prepared, the organisation must have them both validated by an independent accredited verifier. The verification process involves an assessment of the organisation's environmental policy, management system, audit procedure(s) and environmental statement to ensure that they meet EMAS requirements. If the verifier is satisfied that the requirements are met, they validate the information contained in the company's environmental statement.

Environmental target – is a detailed performance requirement, applicable to the organisation or part thereof, that arises from the environmental objectives and that needs to be set and met in order to achieve those objectives.

Internal environmental audit – is defined in the EMAS Regulations as a *systematic, documented, periodic and objective evaluation of the environmental performance* of an organisation, its environmental management system and all processes designed to protect the environment. It is an independent process for gathering evidence and evaluating it objectively to determine the extent to which the environmental management system meets the requirements of the organisation's environmental policy. In many cases, particularly in smaller organisations, independence can be demonstrated by 'first party' auditors – someone from one section or department could audit another section or department. ISO 14001:2015 says that an internal audit can be conducted by the organisation itself (for example one department auditing a different department), or by an external party on its behalf. An audit can be a combined audit (combining two or more disciplines), but independence can be demonstrated by the freedom from responsibility for the activity being audited or freedom from bias and conflict of interest.

Management performance indicator – is an indicator that provides information about the management efforts to influence an organisation's environmental or general management performance.

Significant aspects and impacts – an EMS should include a register (or other recorded information) of aspects likely to have a significant environmental impact. This is one way of establishing priorities for reducing environmental impacts. Determining 'significance' may relate to a number of factors: for example, a regulated aspect such as air emissions may be considered significant because of the potential impacts on the environment or if there are risks of prosecution where regulatory standards are breached. Documentation of 'significant' environmental effects is required for the EMAS Regulation. The ISO 14001 standard requires only the identification of environmental aspects, although these would normally be recorded by some means. 'Significant' aspects could include, for example:

- Emissions to air;
- Discharges to water;
- Solid and other wastes;
- Contamination of land;
- Use of natural resources;
- Energy consumption;
- Noise and vibration, odour, dust and visual impact.

A significant environmental aspect is one that has or can have one or more significant environmental impacts, and a significant environmental aspect will be determined by the organisation applying one or more criteria. The possible or potential effects arising from incidents, accidents and potential emergency or abnormal operating conditions should also be included in the register. This would entail some risk assessment.

Verifier – defined in the EMAS Regulation (where verification of the environmental statement is required) as an accredited body or person which has obtained a licence to carry out verification and validation in accordance with the EMAS Regulation. In the UK, UKAS[8] is the national accreditation body recognised by government to assess, against international agreed standards, organisations that provide certification, testing, inspection and calibration services. UKAS provide EMS accreditation for EMAS environmental verifiers in accordance with the Council Regulation and also accredits certifiers for ISO 14001:2015.

European Accreditation[9] (EA) is a not-for-profit association, registered in the Netherlands. It is formally appointed by the European Commission in Regulation (EC) No 765/2008 to develop and maintain a multilateral agreement of mutual recognition, based on a harmonised accreditation infrastructure. EA currently has 50 members. EA members are National Accreditation Bodies (NAB) that are officially recognised by their national governments to assess and verify – against international standards – organisations that carry out conformity assessment activities such as certification, verification, inspection, testing and calibration. Under Regulation (EC) No 765/2008 – 'Accreditation' means an attestation by a National Accreditation Body that a Conformity Assessment Body meets the requirements set by harmonised standards and, where applicable, any additional requirements including those set out in relevant sectoral schemes, to carry out a specific conformity assessment activity.

Requirements of formalised/certificated environmental management systems in more detail

ISO 14001

ISO 14001:2015 specifies requirements for an environmental management system to enable an organisation to develop and implement a policy and objectives that take into account legal requirements and information about significant environmental aspects [3].

One of the key changes from previously is the requirement for increased importance and prominence to be given to environmental management within the organisation's core business strategy. And the need for environment performance to be integrated into strategic planning and decision-making. This also means there is now greater focus on leadership and commitment by top management with respect to the EMS.

BS EN ISO 14001:2015 is relevant to every organisation, including whether a single site to large multinational companies, high-risk companies to low-risk service organisations, manufacturing, process and the service industries, including local governments, all industry sectors including public and private sectors. There are over 300,000 registrations worldwide. The standard specifies requirements for an environmental management system to enable an organisation to develop and implement a policy and objectives which take into account legal requirements (but does not set compliance as a requirement) and other requirements to which the organisation subscribes, and information about significant environmental aspects. It applies to those environmental aspects that the organisation can control and those that it can influence. It does not itself state specific environmental performance criteria. BS EN ISO 14001:2015 is applicable to any organisation that wishes to establish, implement, maintain and improve its environmental performance, to assure itself of conformity with its stated

environmental policy, and to demonstrate conformity with the standard. Certification to ISO 14001 can be achieved by:

1 Making a self-determination and self-declaration or
2 Seeking confirmation of its conformance by parties having an interest in the organisation, such as customers or
3 Seeking confirmation of its self-declaration by a party external to the organisation or
4 Seeking certification/registration of its environmental management system by an external organisation

This International Standard does not itself establish absolute requirements for environmental performance beyond the commitments in the environmental policy, to comply with applicable legal requirements and with other requirements to which the organisation subscribes, to prevention of pollution and to continual improvement in environmental performance. It is intended that all the requirements in ISO 14001 should be incorporated into the environmental management system; however it is possible to take a step-by-step approach under ISO 14005 [8]. The extent of the application will depend on factors such as the environmental policy of the organisation, the nature of its activities, products and services and the location and the conditions in which it functions. ISO 14001 is a process-driven management system. There is nothing explicitly that requires the products in use or other services to be 'environmentally benign' or good in themselves.

Unlike EMAS, ISO 14001 requires only that an organisation shall establish and maintain a procedure to identify and have access to legal and other requirements to which the organisation subscribes, that are applicable to the environmental aspects of its activities, products or services. Establishing a register is just one way of retaining legal information.

The general requirement of the standard is that an organisation seeking certification should establish, document and implement, maintain and continually improve an environmental management system and environmental performance. The organisation has to define and document the scope of the EMS. The International Standard helps an organisation achieve the intended outcomes of its environmental management system, which provide value for the environment, the organisation itself and interested parties. Consistent with the organisation's environmental policy, the intended outcomes of an environmental management system include:

- Enhancement of environmental performance;
- Fulfilment of compliance obligations;
- Achievement of environmental objectives.

While ISO 14001 sets out the criteria against which an EMS will be certified, BS EN SO 14004:2016 [9] provides further guidance for an organisation on the establishment, implementation, maintenance and improvement of a robust, credible and reliable environmental management system. The guidance is intended for an organisation seeking to manage its environmental responsibilities in a systematic manner so that it contributes to the environmental element of sustainability. This Standard helps an organisation achieve the intended outcomes of its environmental management system and can help an organisation to enhance its environmental performance, enabling the elements of the EMS to be integrated into its core business process.

ISO 14001 refers to an environmental aspect, which the standard defines as an *element of an organisation's activities, products or services that can interact with the environment*. The standard defines environmental impact as any change to the environment, whether adverse or beneficial, resulting from an organisation's activities [3]. Thus a 'significant environmental impact' is generally in line with a 'significant environmental effect' that is in the EU EMAS Regulation. While a register of significant environmental effects is required for EMAS, as we have seen, the ISO 14001 standard requires only the identification environmental aspects. There is now increased emphasis on lifecycle assessment when considering environmental aspects. This also means that there should be proactive initiatives to protect the environment from harm and degradation, such as sustainable resource use and climate change mitigation.

An EMS certified to ISO 14001 can be used as the management system component of EMAS registration, recognising that EMAS also requires legal compliance and a published environmental statement as part of the registration process.

Environmental policy – the starting point for implementation of an EMS is usually the environmental review (see later), although this is not a requirement for ISO 14001. The International Standard requires that top management shall define the organisation's environmental policy, which must be appropriate to the nature, scale and environmental impacts of its activities, products and services. It has to include a commitment to continual improvement and prevention of pollution [3, 9 and 10]. This basic requirement must be formerly acknowledged by the Chief Executive and senior managers, as the continuing commitment and leadership of the top management are crucial to implementing and maintaining an EMS. It sets the level of environmental responsibility and performance required of the organisation against which all subsequent actions should be judged. The criteria for environmental improvement can apply to any number of parameters depending on an organisation's environmental aspects and impacts and stated objectives. It could refer, for example, to a reduction in pollution, greater efficiency in the use of resources, and/or a reduction in waste, and should address mitigation of climate heating. Organisations decide their own performance parameters and set their own rate of continual improvement.

In developing the policy there should be consultation with those working within the organisation. This process should aim to establish a consensus on the principle aims of the policy across the whole organisation. The intention is to be able to get 'buy-in' from all those who contribute to the environmental performance of the organisation. The general aims of the policy should reflect the activities of the organisation, after identification of the environmental aspects of the activities, products and services and should identify environmental performance objectives (e.g., compliance with legal commitments, continual improvement and prevention of pollution), just as is required under EMAS (see [11] and [12]). The policy should be endorsed and actively supported by senior management and accepted by all staff. A written environmental policy makes it easier for management to communicate its aims and objectives to employees and other interested parties, including shareholders, customers and suppliers. It should be consistent with and part of the business strategy. An environmental policy is a voluntary undertaking in the UK, and the structure and content are not regulated under UK law. For companies intending to obtain certification to a formal EMS, including ISO14001 (and EMAS), the environmental policy is a vital document for implementation. The environmental policy should be communicated to all staff and other stakeholders through, for example, annual reports and newsletters; it should also be displayed prominently in reception areas and canteens [13]. Guidelines for general principles, policy, strategy and activities relating to both internal and external environmental communication are provided in BS ISO 14063:2020. It uses proven and well-established approaches for communication, adapted to the specific conditions that exist in environmental communication. It can be used in combination with any of the ISO 14000 family of standards, or on its own.

The policy itself is a public commitment to improve environmental performance and minimise the impacts of pollution and other activities that pose risks to the environment and human health. The policy does not need to set out specific targets but should outline principal aims and objectives, particularly on issues such as legal compliance and environmental impacts.

Initial review – ISO 14001:2015 does not specify an initial environmental review although this is considered a useful exercise to provide baseline information on environmental aspects, impacts and performance. The standard does, however, require a periodic review to determine how the EMS is performing (and revise the process(es) and planned response actions, in particular after the occurrence of emergency situations or tests). An EMS review is likely to cover:

- Identification of environmental aspects both under normal operating conditions and abnormal, in start-up and shut-down situations and accident and emergencies – there is now a stronger focus on managing environmental risks;
- Identification of relevant legal requirements and others applicable to the sector or organisation;
- Examination of any existing environmental practices and procedures including those associated with contractors;
- Evaluation of previous emergency situations and accidents.

Other matters that can be considered include evaluation of performance compared to any other criteria or performance standards (e.g., as set by trade or professional bodies), opportunities for cost reduction, views of other interested parties and any systems that can impede or prejudice performance [10].

Planning – is a critical and ongoing process within the management system. The ISO standard makes clear that it is vital to the fulfilment of the environmental policy, and the establishment and implementation and maintenance of the EMS. It should include the following elements:

- Identification of environmental aspects of the operation(s) and determination of those which are significant (and organisations should select categories of activities, products and services to identify environmental aspects within the defined scope of the EMS, taking account of those which it can control and those it can influence);

- Identification of relevant legal and other requirements including industry and sector norms;
- Setting internal performance criteria;
- Setting objectives and targets that address the aims of the policy and the programmes necessary to achieve them (particularly regarding a commitment to comply with legal and regulatory requirements and the evaluation of significant environmental effects/impacts). The objectives should be specific and measurable wherever practicable. This is common to the ISO standard and EMAS Regulation. ISO 14001 provides guidance on which activities or issues to consider, in terms of prioritising those objectives, as follows:

  - When establishing and reviewing its objectives an organisation shall consider the legal and other requirements, its significant environmental aspects, its technological options and its financial, operational and business requirements and the views of interested parties.
- Publication – there is a requirement the policy should be publicly available. It should also be periodically reviewed [13]. There should be a clear communication strategy relating to the EMS and performance.

Implementation, documentation and operation – senior management has to ensure that the resources (including human resources and skills) are available in order to establish, maintain and improve the EMS. The management programme provides the procedures for ensuring that objectives and targets are achieved. Both the ISO 14001 standard and the EMAS Regulation require that the environmental management programme takes account of the personnel involved.

Roles and responsibilities have also to be defined, documented and communicated to all parts of the organisation. This part of the EMS also covers the requirement to ensure that all personnel have adequate training for the tasks assigned to them within the EMS. Training needs (including awareness training) and competencies need to be assessed and recorded and procedures should be in place to fill any gaps. Both EMAS Regulation and ISO 14001 identify training as an integral part of establishing and implementing an EMS. Employees should be aware of the broader environmental issues that motivated the organisation to develop an EMS. Training can relate to three areas, general environmental awareness, EMS awareness including on the environmental policy, and individual training of staff, particularly those working in an area of the organisation with the potential for environmental impacts (and are subject to legal requirements and where there is potential for legal liabilities).

Procedures for internal communications to and between the various levels and functions of the organisation should be in place. There is also a need for procedures for receiving documentation and responding to relevant communications from external parties. The organisation can itself decide whether to communicate externally about its significant environmental aspects [13]. Whatever the decision, this should be recorded. If the decision is in the affirmative then methods for external communication have to be established.

ISO 14001 requires that documentation for the EMS has to include:

- Environmental policy, objectives and targets;
- Description of the scope of the EMS;
- Description of the main elements of the EMS and their interaction and reference to related documents;
- Documents and records required by the standard; and

- Documents including records determined by the organisation necessary to ensure effective planning, operation and control of processes that relate to the significant environmental aspects.

There is a need for this documentation to be properly controlled, including procedures for approving documents, reviewing and updating documents, identifying any revisions and status of documents, ensuring appropriate and current documents are available at the point of use. Documents of external origin such as industry codes of practice should also be available and updated, as appropriate. Any obsolete documents that are retained must be suitably identified as such. Such records provide tangible evidence that the EMS is working (and in some cases where it needs improving), and that targets are being met. Assessors and auditors need to see the records to determine whether the EMS is performing to the requirements of the Regulation or the ISO standard.

Operational control comprises a series of checks and balances that ensure that activities undertaken by the organisation meet the requirements of the standard and the organisation's environmental policy. Operational controls can relate to documented procedures, work instructions, training and physical controls such as limiting access to key-holders. The choice of the specific control methods depends on factors, such as the skills and experience of people carrying out the operation and the complexity and environmental significance of activities and operations that can give rise to environmental impacts including in emergencies. The operational controls should be those which facilitate management of significant environmental aspects, legal compliance, risk avoidance, objectives and targets and which ensure consistency with the environmental policy.

In the EMS the organisation must address emergency preparedness and response appropriate to its needs. In ensuring preparedness an organisation should take account of a number of factors including:

- The nature of on-site hazards (e.g., flammable liquids, volatile organic compounds, oil storage tanks, compressed gases and any measures to be taken in the event of spillages or inadvertent releases);
- The likely type of emergency or accident;
- Potential impacts and receptors in the case of accidents or emergencies;
- The most appropriate responses to the identified potential accidents or emergencies;
- Appropriate actions to minimise environmental damage;
- Training of personnel for emergency responses, organisation and responsibilities in the case of emergencies or accidents;
- Contact details for relevant organisations such as fire and rescue service, Environment Agency, local authorities;
- Internal and external communication plans for emergency situations;
- Mitigation and response action(s) for different types of accidents or emergencies;
- Testing of emergency response procedure(s) and training plans.

Checking – involves the measurement, monitoring and evaluation of an organisation's environmental performance, compliance with legal requirements and conformance with the EMS. EMS should have defined indicators through which environmental performance can be evaluated as the EMS should be about improving environmental performance.

The aim is to use preventive action to identify and prevent possible problems, rather than correct them after the event. It is a process for identifying what the ISO standard calls nonconformity in the EMS; the corrective or preventive action helps the organisation operate and maintain the EMS as intended. It involves keeping records and managing them effectively, which gives the organisation a reliable source of information on the operation of the EMS. Periodic audits of the EMS help the organisation verify that the system is both designed and operating according to plan.

A range of methods are suggested as a means of assessing conformance, including processes such as

1 Audits
2 Review of documents and/or records
3 Physical inspections of sites and facilities
4 Interviews, with staff
5 Review of projects or working
6 Sampling and analysis
7 Facility tour and/or direct observation

Under ISO 14001 an organisation should establish a frequency and methodology for evaluation of compliance that suits its size, type and complexity (and these should be included in the documentation). Frequency can be affected by factors such as past performance or compliance with specific legal requirements. It can also be beneficial to have an independent review conducted periodically. Whether a phased introduction or otherwise, a management system must maintain and retain all relevant documentation.

Internal audits of an environmental management system can be performed by personnel from within the organisation or by external persons selected by the organisation, working on its behalf. Those conducting the audit should be competent and experienced for the task in hand, including relevant training in the activity or process being audited. An audit of the EMS might involve several staff or external personnel reviewing different parts of the system. The internal audit process of an EMS is one way of periodically identifying what are called 'nonconformities'[10] where a specific activity or function is not performing as it should. As a matter of routine, staff within the organisation should also identify such nonconformities within their areas of work and competencies. Identification of a nonconformity should be followed by further investigation (and records made) so that the cause can be identified, and corrective action taken, including steps to avoid recurrence.

Where a potential problem is identified even though no actual nonconformity exists, the system should allow for preventive action to be taken.

An organisation therefore should establish an audit programme so as to direct the planning and conduct of audits [14]. The programme should also specify the sites or activities that will be covered in the audit, taking account of the environmental aspects and potential environmental impacts. Internal audits do not have to cover the entire EMS at one time but must ensure that all parts of the EMS covering activities, processes and products are audited periodically.

The outcome or results of the internal audit of the EMS are best provided in a report which can then be fed into the management review.

ISO has produced ISO 19011 [7] that provides auditing guidelines suitable for both environmental management systems and quality management systems. It emphasises the need for auditor competencies and the need for auditor independence from the activities or departments being audited. ISO 19011 is applicable to all organisations conducting internal or external audits and ISO 19011 provides guidance on how to run an audit programme and the competence required of the system auditors.

Management review – the management review is the last part of the system as set out in the ISO standard and is an opportunity to consider whether or not the EMS is working effectively and whether it needs amending because some aspects are not working as they were intended or because of new developments. The management review is an opportunity to build on the evidence of the audit findings and to decide on new initiatives to ensure continual improvement in performance, and so should include recommendations for improvement. Senior management will decide on the intervals between reviews of the EMS. The main purpose is to assess the system's current adequacy and effectiveness. It should cover the environmental aspects of activities, products and services that are within the scope of the environmental management system.

A separate standard BS EN ISO 14031:2013 [14] provides comprehensive guidance on environmental performance evaluation (EPE). It sets out a process to help an organisation improve environmental performance. It provides a set of tools to identify, measure, assess and communicate environmental performance using key performance indicators (KPIs), based on reliable and verifiable information. This standard can be seen as companion to BS ISO 14001:2015.

EU Eco Management and Audit Scheme (EMAS)

Although the UK is no longer a member of the EU, there exists the EU Eco Management and Audit Scheme (EMAS) as mentioned previously. This was introduced originally by EC Regulation 1836/93, replaced more recently by EMAS III (EC Regulation 1221/2009).[11] The focus of EMAS may be on 'operational or organisational units' (department, service function or division) or can be site-based such as depots or factories. It is considered to be a premium management instrument for enterprises and other organisations to evaluate, report and improve their environmental performance. EMAS is open to every type of organisation eager to improve its environmental performance. It spans all economic and service sectors and is applicable worldwide. There are elements of the EMAS approach that are worth taking into account when seeking to improve environmental management and these are considered later.

In 2017 Annexes I, II and III of the EMAS Regulation were amended to include the changes associated with the revision of the ISO 14001:2015 standard. The Commission Regulation (EU) 2017/1505 amending these annexes entered into force in September 2017. To provide easier access to the up-to-date EMAS requirements, the Commission has also produced an informal consolidated version of the EMAS Regulation that includes the amended annexes.[12]

Since January 2019, also an amended Annex IV of the EMAS Regulation (EU Commission Regulation EU 2018/2026) has been in place. This amendment includes an update of EMAS's core indicators and the language of the environmental statement. It also allows EMAS organisations new opportunities to report on their environmental performance and to use the organisation's EMAS environmental statement also for other reporting obligations.

EMAS is a voluntary EU-wide environmental management regulation that incorporates an EMS but also which requires organisations to demonstrate legal compliance and produce a public statement. EMAS allows for certified ISO 14001:2015 to meet the EMS component of EMAS and the differences in requirements between EMAS and ISO 14001:2015 are included in Annex II. The scheme is designed to improve companies' environmental performance. The current version, EMAS III strengthened legal compliance and reporting rules but reduced the administrative burden of registration and verification from previous versions. The revisions were intended to allow a wider range of organisations to be eligible to register under EMAS. There is a distinctive EMAS logo to signal EMAS registration to the wider world (Figure 8.3). The system was also

Figure 8.3   The EMAS logo

strengthened by taking into account more strongly indirect effects, such as those related to financial services or administrative and planning decisions. Environmental reporting requirements are also stronger than in the earlier Regulation with 'core' performance indicators mandatory.

In this edition, with the withdrawal of the UK from the EU, less space is devoted to EU EMAS.

EMAS was established by an EU Regulation and so applies directly in law to all Member States. Participation by enterprises and organisations is voluntary and as has already been said, it is possible for organisations outside the EU to participate. Through the EMAS Global mechanism, EMAS is available worldwide. This way, EMAS helps to reduce the environmental impact of (industrial) operations around the world. This type of registration is particularly useful for multinational organisations wishing to implement a unified environmental management system throughout their sites within and outside of Europe.

There are two types of registrations that fall under the EMAS Global category:

- Global Registration: an organisation registers its site(s) in one or more non-European countries;
- Global Corporate Registration: a combination of EU Corporate and Global Registration. It allows an organisation to register its site(s) located in one or more European countries and in one or more non-European countries under one corporate registration procedure in one (leading) country of registration. Organisations can use a single registration number and streamline their auditing and reporting processes.

It is designed to improve the environmental performance of an enterprise or organisation. EMAS recognises organisations that improve their environmental performance on a continuous basis. To achieve EMAS, organisations need to be legally compliant, implement an environmental management system and report on their environmental performance through the publication of an independently verified environmental statement.

EMAS participation is available to any organisation no matter how large or small and regardless of the activity undertaken and can also cover multiple sites. If the organisation comprises one or more sites, each of the sites to which EMAS applies shall comply with all the requirements of EMAS.

Its fundamental aim is to recognise and reward those organisations that go beyond minimum legal compliance and continuously improve their environmental performance. In addition, it is a requirement of the scheme that participating organisations regularly produce a public environmental statement that reports on their environmental performance. It is the intention that the voluntary publication of environmental information, whose accuracy and reliability has been independently checked by an environmental verifier, will give EMAS and the organisations that participate in this scheme enhanced credibility and recognition by the general public.

Member States were required to adopt EMAS and to appoint a 'Competent Body' to oversee the scheme and registration under the scheme. The Institute of Environmental Management & Assessment (IEMA) was the Competent Body in the United Kingdom for EMAS. In Ireland it is Department of Communications, Climate Action and Environment in Dublin. There are also accreditation and licensing bodies for verifiers in each Member State.

What is expected of the environmental review is useful when it comes to an EHP looking at the environmental performance of an organisation even where EMAS itself is not relevant. The scoping of the environmental review should include all the relevant departments and their operations or activities. The EMAS Regulation defines an environmental review as *an initial comprehensive analysis of the environmental aspects, impact and performance related to activities, products and*

Requirements and procedures for implementing EMAS can be summarised as follows:

1   The organisation should start with an environmental review, an initial analysis of all activities the organisation carries out, to identify relevant direct and indirect environmental aspects, and the applicable environmental legislation
2   Then an environmental management system needs to be implemented, in line with the requirements of EN ISO 14001 (Annex II to the EMAS Regulation)
3   The system needs to be checked by carrying out internal audits and a management review
4   The organisation writes an EMAS environmental statement
5   The environmental review and the environmental management system are verified, and an accredited or licensed EMAS verifier validates the statement
6   Once the organisation has been verified, it submits an application for registration to the Competent Body

The European Commission has developed 'Sectoral Reference Documents' for retail, tourism, food and beverage manufacturing, car manufacturing, electronic and electronic equipment manufacturing, public administration (such as local government) agriculture, construction and waste management. Organisations should take these into account when implementing EMAS and specify in their environmental statement how these documents were used.

services. There are five basic requirements of an initial environmental review:

- Identify legal issues – this includes an assessment of environmental legislation and regulations (e.g., consents to discharge effluents and permits for air emissions);
- Identify direct and indirect environmental aspects – examples include emissions and discharges, the use of energy and water, waste management and consumption of materials;
- Determine criteria for assessing the significance of the environmental aspects;
- Examine current policies and procedures – to determine whether current working practices take account of their potential environmental impacts and to assess what those impacts might be;
- Evaluate feedback of abnormal operations, accidents and incidents – an assessment of non-compliance with regulations; accidents which have caused environmental damage and procedures to deal with emergencies.

'Environmental aspect' should be taken to mean an element of an organisation's *activities,*

products or services that has or can have an impact on the environment. Environmental aspects may be input related (consumption of raw materials and energy, for instance) or output related (air emissions, waste generation, etc.).[13]

The environment policy provides the framework for setting and reviewing environmental objectives and targets. It should be fully documented and implemented and maintained and most importantly communicated to all employees. It should be publicly available.

As part of the management system the EMAS Regulation requires top management to define the organisation's environmental policy and ensure that it:

1   Is appropriate to the nature, scale and environmental impacts of its activities, products and services
2   Includes a commitment to continual improvement and prevention of pollution
3   Includes a commitment to comply with relevant environmental legislation and regulations, and with other requirements to which the organisation subscribes

4   Provides the framework for setting and reviewing environmental objectives and targets
5   Is documented, implemented and maintained and communicated to all employees
6   Is available to the public

Implementing environmental improvement programmes should involve workers at all levels of the organisation – starting with top-level management and throughout the structure including those working in administration, maintenance, on production lines or direct service and product delivery. Not only does this help to ensure the success of environmental initiatives, but it also acts as a valuable teambuilding exercise for employees. EMAS specifically requires the involvement of employees in the process of improving the organisation's environmental performance and as a general point is something that organisations should take on-board.

Examples of environment policies

Derbyshire County Council Corporate Environment Policy (v 3 October 2019)

Derbyshire County Council is committed to putting the principles of sustainable development into action in everything the authority does, so that development meets the needs of today without compromising the ability of future generations to meet their own needs. Managing our environment sustainably will be a part of making Derbyshire a place:

- With resilient and thriving communities
- With happy, healthy people and families
- With a strong, diverse and adaptable economy
- Which is great to live in, visit and work.

We recognise the impact we have on the environment and society through the delivery of our operations and are committed to protecting the environment by minimising any adverse environmental impact, while creating opportunities for enhancing positive environmental effects to improve the quality of life for people. Our flagship commitment is to reduce the greenhouse gas emissions from our own estate and operations to net zero carbon by 2032.

We will encourage and enable all our employees to do what they can to translate these commitments into practice. We will also work with our contractors and suppliers to improve our environmental performance. This policy will, therefore, be communicated to all employees and contractors working for or on behalf of the County Council.

We will monitor our environmental performance by setting organisational objectives and targets and report on our progress.

In developing the Environment Policy, the Council is publicly setting out its commitment to continual environmental improvement. The Environment Strategy and Action Plan set out the work the Council will undertake to implement this policy.

In everything we do, Derbyshire County Council is committed to:

Reducing greenhouse gas emissions to net zero carbon by 2032

Identifying, adopting and promoting technologies and practices to reduce the emissions of greenhouse gases, including carbon dioxide, from our estate and operations including Council property, street lighting and fleet and employee travel.

Using water efficiently in the Council's buildings and operations

Using water efficiently in our buildings and operations and ensuring improvements are made to the measurement and monitoring of water consumption across our estate to inform water saving practices.

Reducing waste

Eliminating, reducing, reusing, composting and recycling wastes where possible. Managing our remaining wastes in accordance with our Duty of Care obligations.

Minimising pollution

Minimising, with the goal of eliminating, the release of any pollutant which may cause damage to health or the environment whether from air, land or water.

Protecting the natural and built environment

Protecting, conserving and enhancing the environment, habitats, biodiversity and heritage.

Ensure all staff are able to implement the Corporate Environment Policy

Raising awareness, educating and training employees and those working on our behalf to ensure that all staff have the knowledge, skills and understanding to implement the Environment Policy.

Ensuring that the Council's purchasing power is used positively

Ensuring that the Council's purchasing power is used to reduce negative environmental impacts and to improve the environmental standards and social value of products and services the Council purchases.

We will do this by . . .

Partnership Working

Working closely with employees, other organisations, interested groups and individuals, where appropriate, to further the aims of this Policy.

Objective Setting

Continually improving our environmental performance by setting realistic but challenging objectives and targets and regularly reviewing our progress as set out in the Environment Strategy and Action Plan.

Legal Compliance

Complying with relevant environmental legislation, Council policies and other commitments and striving to deliver best practice.

Environmental Management Systems

Promoting, operating and extending environmental management systems to control, monitor and enhance our environmental performance and communicating this Policy to all employees and contractors.

Policy Review

Reviewing this Environment Policy every three years in view of changes to the Council's activities and priorities in light of new local, national and international developments.

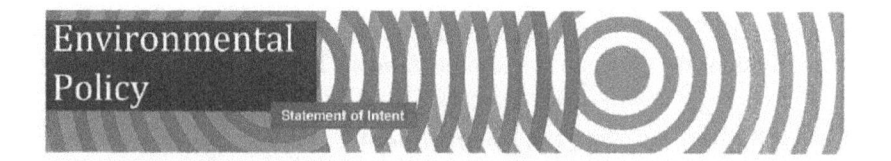

PricewaterhouseCoopers LLP (PwC) seeks excellence in every aspect of our business and is committed to minimising the environmental impacts of our business operations.

Our commitment is to:

- Continuously improve our environmental performance and to integrate recognised environmental management best practice into our business operations.
- Reduce our consumption of resources and improve the efficient use of those resources.
- Measure and take action to reduce the carbon footprint of our business activities to meet our published objectives and targets.
- Manage waste generated from our business operations incorporating reduction, re-use and recycling in accordance with the principles of the waste hierarchy.
- Manage our business operations to prevent pollution.
- Give due consideration to environmental issues (such as biodiversity) and energy performance in the acquisition, design, refurbishment, location and use of buildings.
- Ensure environmental, including climate change, criteria are taken into account in the procurement of goods and services.
- Comply as a minimum with all relevant environmental legislation as well as other environmental requirements to which the firm subscribes.
- Maintain our certification to ISO 14001 through rigorous monitoring and review.

To meet our commitments we will:

- Provide UK Executive Board oversight and review of environmental policies and performance, and allocate resources for their effective direction and implementation.
- Monitor key objectives and targets for managing our environmental performance at least annually.
- Engage with stakeholders, including communicating internally and externally our environmental policy and performance on a regular basis.
- Communicate the importance of environmental issues to our people.
- Work together with our people, service partners, suppliers, landlords and their agents to promote improved environmental performance.
- Promote appropriate consideration of sustainability and environmental issues in the services we provide to our clients.
- Review our environmental policy regularly.

This environmental policy represents our general position on environmental issues, and the policies and practices we will apply in conducting our business. It is accessible via the internal PwC SparkPad and to other interested parties via our website (www.pwc.co.uk) and on request.

Andrew Cope
Partner
PricewaterhouseCoopers LLP
January 2021

pwc

Environmental Policy
January 2021

Data Classification - PUBLIC

It should be noted that an environmental policy statement is in no way an indication of the merits or otherwise of the products or services provided by a business. It relates solely to their environmental performance. So, for example one might disapprove of an arms manufacturer on ethical grounds, but they can still have an environmental policy and management system seeking to minimise the environmental impact of their production.

Indirect environmental aspects are an important consideration for a local authority establishing an EMS. In general, these result from the activities, products and services of an organisation where there may be significant environmental aspects over which it may not have full management control. For example, they may arise from procurement policies and the manner in which services are delivered. As a general point in any organisation these may include, but are not limited to:

1   Product-related issues (procurement, design, development, packaging, transportation, use and waste recovery/disposal)
2   Capital investments, granting loans and insurance services
3   New markets
4   Choice and composition of services (e.g., transport or the catering trade)
5   Administrative and planning decisions
6   Product range compositions
7   The environmental performance and practices of contractors, subcontractors and suppliers

Organisations should demonstrate that the significant environmental aspects associated with their procurement procedures have been identified and that significant impacts associated with these aspects are addressed within the management system.

The organisation should also endeavour to ensure that the suppliers and those acting on the organisation's behalf comply with the organisation's environmental policy within the remit of the activities carried out under

contract. In the case of these indirect environmental aspects, an organisation shall consider how much influence it can have over these aspects, and what measures can be taken to reduce the impact.

In addition to the environment policy the environmental management system element of EMAS requires that there be adequate planning which includes having in place:

1   Procedures for identifying and maintaining up-to-date environmental aspects of the operation
2   Procedures for identifying and keeping up to date with legal requirements on environmental aspects of the operation
3   Procedures for maintaining and documenting environmental objectives and targets
4   A programme for achieving environmental objectives and targets which includes the means and timescale for achieving these and designation of responsibilities

Roles and responsibilities have to be clearly defined and documented so as to facilitate an effective management system. This also entails management ensuring that adequate resources are available to be able to implement and control the EMS. Senior management should appoint specific management representatives with defined roles, including on reporting on performance of the EMS to top management. This reporting enables improvements for the EMS to be identified and prioritised.

In common with any management system, it also important to identify the training needs of staff. All personnel whose work and activities may create a significant impact on the environment should be trained appropriately. They should also understand about the significant environmental impacts from their work activities; their roles and responsibilities in achieving compliance with the policy and procedures within the EMS. In particular this should cover emergency preparedness, and their roles in emergency situations.

Communication is also an important element of the EMS, and organisations should establish and maintain internal communications between the levels and functions in the organisation. There should also be good external communications addressing the significant environmental aspects and record of decisions concerning improvements in performance.

As in all managements systems, records and documentation (preferably electronically) are essential and up-to-date information has to be maintained.

All the operations and activities of the organisation that are linked to significant environmental aspects must be identified. This should include having documented procedures that cover situations that could lead to nonconformity with regulations, or with the environmental policy or where objectives and operating procedures might be compromised. The organisation should also establish procedures relating to significant environmental aspects for goods and services used by the organisation and for communicating relevant procedures to suppliers and contractors.

The organisation shall also establish and maintain procedures to identify the potential for responding to accidents and emergency situations. The procedures should be in place for preventing and mitigating the environmental impacts that may be associated with such accidents and emergencies.

It is necessary to have and maintain documented procedures for monitoring and measuring operations and activities that may have a significant impact on the environment. This also means there should be documented procedures for periodically evaluating compliance with the relevant environmental regulations and permit conditions.

The EMS should include procedures with defined responsibilities for dealing with or investigating nonconformance, and for taking action to mitigate any impacts caused and for completing corrective action. It is necessary to maintain procedures for identifying and maintaining environmental records including on training and the results of audits and subsequent reviews.

The organisation should have a programme of periodic audits which are undertaken to determine whether the EMS conforms to the planned arrangements and has been properly implemented. The audit programme should reflect the environmental importance of the activity concerned and the results of previous audits. Audit procedures should cover the scope, frequency and methodologies as well as responsibilities and requirements for conducting audits and reporting results.

Under EMAS the audit or audit cycle should be completed, as appropriate, at intervals no longer than three years (extending to five years for SMEs). The frequency with which any activity is audited will vary depending upon such matters as the:

- Nature, scale and complexity of the activities;
- Significance of associated environmental impacts;
- Importance and urgency of the problems detected by previous audits; and
- History of environmental problems.

A written audit report of the appropriate form and content should be prepared by the auditors to ensure formal submission of the findings and conclusions of the audit, at the end of each audit and audit cycle. The findings and conclusions of the audit should be communicated to the highest level of management in the organisation.

The environmental statement, which is unique to EMAS, is a publicly available document containing information about an organisation's structure and activities; environmental policy and environmental management system, environmental aspects and impacts; environmental programme, objectives and targets; environmental performance and compliance with applicable legal obligations relating to the environment.

The environmental statement shall be made available and accessible to the public.

Some multi-site organisations registering under EMAS may wish to produce one corporate environmental statement covering the different geographic locations. The intention of EMAS is to ensure some degree of local accountability and so such organisations have to ensure that the significant environmental impacts of each site are clearly identified and reported within the corporate statement.

Other relevant management standards

BS EN 50001:2018

BS EN 50001:2018 (*Energy management systems – Requirements with guidance for use.*) requires the development of an energy policy, identification of an organisation's past, present and future energy consumption as well as the development of an energy monitoring (metering) plan. Analysis of actual versus expected energy consumption will allow businesses to put plans in place to help improve efficiency. The standard aims to help organisations use energy more efficiently and integrate better energy management into business strategy. It does this by outlining how to implement and maintain an energy management system (EnMS) that continually improves the organisation's energy performance and saves money.

BS 8555 and BS 14005

BS 8555:2016 (*Environmental Management Systems – Phased Implementation – Guide*) provided guidance for organisations on the phased development, implementation, maintenance and improvement of an EMS.

It was applicable to any organisation, regardless of the business activity undertaken, size, complexity or level of maturity. Organisations with limited resources and/or organisations new to environmental management might have found the approach useful. The flexibility offered by a phased approach was seen as helpful in driving positive culture change within an organisation and in motivating top management to see the business value of improved environmental performance.

This British Standard has now been withdrawn and replaced by BS EN ISO 14005: 2019 (*Environmental management systems. Guidelines for a flexible approach to phased implementation*). The phased approach was seen as advantageous to many organisations for a number of reasons which are re-iterated in BS EN ISO 14005. Once an organisation has decided to follow the phased approach ISO 14005 suggests it could start with an environmental performance initiative or an environmental improvement project that focuses on one or more of the following options:

- Selected environmental aspects;
- Specific EMS elements, such as the fulfilment of compliance obligations (e.g., specific needs and expectations of interested parties) or communication;
- Increasing the maturity of certain elements, such as ensuring that the organisation's existing environmental policy is taken into account in day-to-day decision-making and operations.

Implementation of an EMS does not need to start from scratch but can build on formal and informal structures already in place.

It provides the means for the management of business risk and demonstrates a high level of environmental commitment without the need for a complete EMS at one time. The aim is for more organisations, particularly small and medium-sized enterprises (SMEs) which lack a formal system to move to a certificated EMS using a phased approach to ultimately meet the requirements of ISO 14001. Each phase incorporates six consecutive stages. The number of phases is flexible. This allows organisations to develop the scope (i.e., the activities, products and services included) and maturity of their EMS, in line with their objectives and available resources.

BS EN ISO 14005: 2019 contains much the same information as was in BS 8555 but is a completely revised and updated version of ISO 14005:2010.

Using a phased approach for implementation, the EMS has a number of additional benefits for organisations. The phased approach offers flexibility that allows an organisation to:

- Develop an EMS at its own pace;
- Decide the scope of implementation and expand it to suit its resources;
- Decide the number of phases it undertakes and the level of maturity it wants its EMS to achieve;
- Start with areas that indicate the greatest potential for environmental improvement and returns on investment;
- Prioritise environmental performance improvement (e.g., improvement with respect to material and energy efficiency, or to a specific waste stream);
- Stimulate a positive culture towards environmental management, expand an existing EMS towards meeting the requirements of ISO 14001.

A phased approach provides familiarity with the basic elements of an EMS, allows an organisation to track legal compliance and experience some of the benefits of managing environmental aspects in a systematic way, and secures management support for implementing it to the full extent.

The six stages of each phase follow the following sequence:

1 Define intended outcomes for the phase
2 Assess EMS status
3 Select areas for EMS improvement
4 Undertake a gap analysis
5 Plan and implement EMS improvements
6 Check and review achievements

The model is set out in Figure 8.4.

As with ISO 14001:2015 itself, this standard is applicable to any organisation regardless of their current environmental performance, the nature of the activities undertaken or the locations. Indeed, the phased approach enables an organisation to develop a system that ultimately can satisfy the requirements of ISO 14001.

Some considerations that are important in moving to a formalised EMS can be summarised as:

- The need to define the environmental management structure, assigning roles and responsibilities with respect to the EMS (building on responsibilities identified in Phase two for legal compliance);
- The need to define the relationships and interactions between different elements of the EMS;
- Structured training and awareness raising programmes, and implement processes for identifying training needs and planning training delivery and review;
- Clear communication practices and processes internally and externally if so decided;
- Effective documentation and records, including control and record-keeping systems. Documentation is an important aspect of the EMS as for any management system, so that the system is not dependent upon any individuals and any problems or infractions can be more easily identified and rectified.

BS ISO 14033

BS ISO 14033: 2019 [15] (*Environmental management: Quantitative environmental information Guidelines and examples*), gives guidelines for organisations on the general principles, policies, strategies and activities necessary to obtain quantitative environmental information for internal and/or external purposes. The purposes, for example, could be to establish inventory routines and support decision-making related to environmental policies and strategies, aimed in particular at comparing quantitative environmental information.

It addresses issues related to defining, collecting, processing, interpreting and presenting

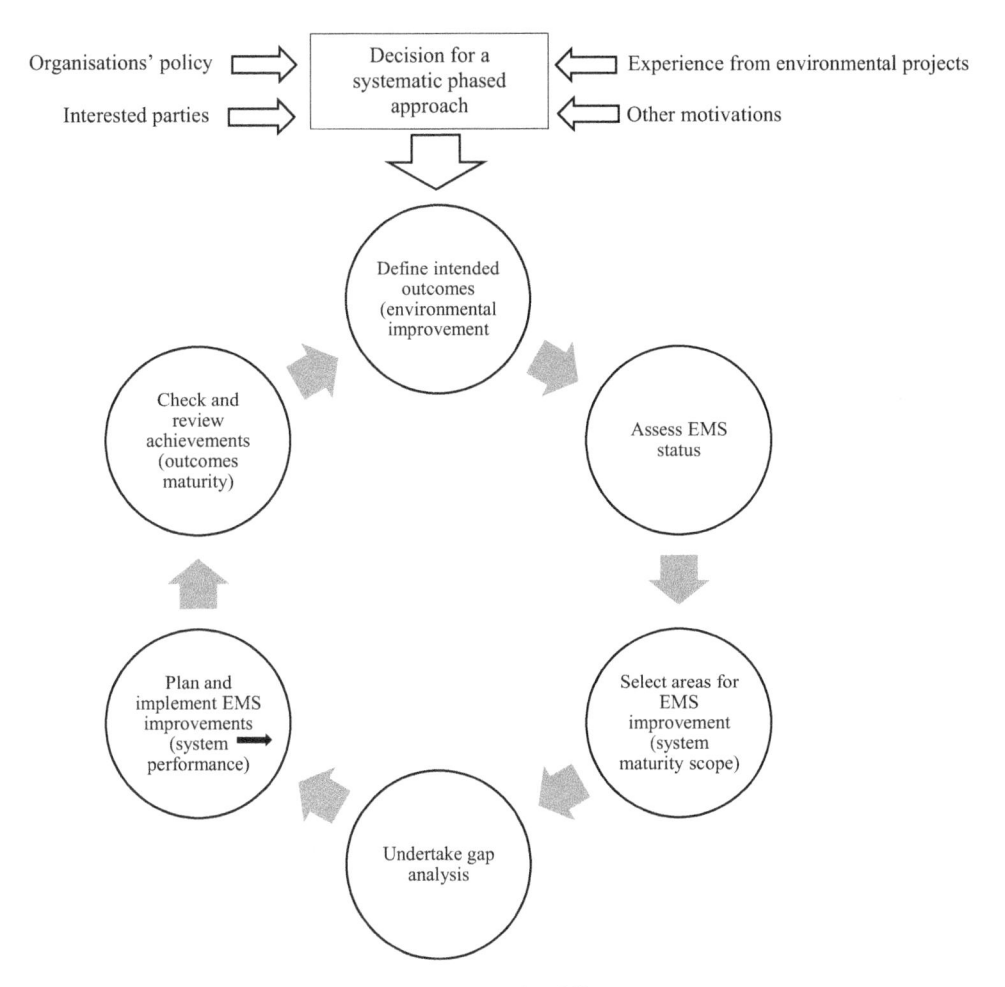

Figure 8.4   Conceptual model for a phase with the different stages

quantitative environmental information. It provides guidelines on how to establish accuracy, verifiability and reliability for the intended use.

Green Dragon scheme

The Green Dragon Environmental Standard[14] is awarded to organisations that can demonstrate effective environmental management and that are taking action to understand, monitor and control their impacts on the environment. It's suitable for organisations of all sizes, in any industry, and is flexible in design, so can be tailored to suitable individual needs. Groundwork Wales is also the only body able to certify an organisation to the Green Dragon Environmental Standard. Groundwork Wales is a UKAS Accredited Inspection Body.

As has been said, for many smaller and medium-sized enterprises it can be difficult and impractical to establish and implement an EMS in one stage. This has been recognised, and to encourage these businesses that may not have the resources, BSI and others have developed a staged approach.

The Green Dragon approach in Wales has taken account of the problems for many smaller and medium-sized companies and organisations.

The attainment of formal environmental management systems such as ISO 14001 can be seen as daunting and may not be or seem relevant to their business needs. Often companies and organisations that are working towards one of the larger schemes such as ISO 14001 do not receive any recognition for the progress made along the way and if ultimately they cannot satisfy all the requirements for certification there is nothing to show for the efforts made.

Within the Green Dragon Standard there are five levels, with each step contributing towards achievement of the Standard, which is structured into five 'Levels', allows organisations to gain a third-party certification that provides recognition of their environmental practices and demonstrates their commitments to sustainability, consideration of environmental impacts, compliance with legislation and environmental protection of the international and European environmental standards ISO 14001 and EMAS.

During the appraisal and audit processes for the Green Dragon Standard, there is an evaluation of costs as well as environmental performance – this means that at each stage the company or organisation will have an outline environmental management system that relates to its bottom line.

The five Levels of Green Dragon are:

Level 1: Commitment to Environmental
    Management
Level 2: Understanding Environmental
    Responsibilities
Level 3: Managing Environmental Impacts
Level 4: Environmental Management
    Programme
Level 5: Continual Environmental
    Improvement.

Level 1 provides the starting point and may be good for any organisation starting out on a journey towards environmental management. If the requirements are met a certificate is issued for one year.

Level 2 is ideal for organisations that have made a commitment to improving environmental performance and are starting to embed an EMS in the workplace. Organisations wishing to register for Green Dragon Level 2 must also achieve the requirements of Level 1. If the standards for this level are met, the organisation will be issued with a certificate and details registered on the Green Dragon website.

Level 3 is aimed at organisations that are actively managing their environmental performance through an EMS. This level is ideal for small to medium-sized enterprises (SMEs), particularly where operations are undertaken on a single site with close management control.

Organisations wishing to register for Green Dragon Level 3 must also achieve the requirements of Green Dragon Levels 1 and 2. There is an audit of the organisation, and this also covers levels 1 and 2. Re-certification visits may require less time. If the organisation has received consultancy support from a Green Dragon consultant, the site audit will not be undertaken by that person.

If the standard's requirements are met the organisation will be issued with a certificate and details registered on the Green Dragon website. Organisations can also display the Green Dragon logo on their corporate literature.

Level 4 is for organisations where environmental performance is embedded throughout the organisation. Level 4 is particularly suited to medium-sized organisations operating different sites; processes or shifts which require documented controls and procedures are assessed through an audit visit undertaken by a Green Dragon Auditor. A Level 4 audit will also cover Levels 1 to 3. Re-certification visits may require less time than the initial audit. Again, if Green Dragon has provided consultancy services that consultant will not be involved in the audit. Once the standard's requirements are met, a certificate will be issued, and details registered on the Green Dragon website. The certificate is valid for one year from the date of issue.

Level 5 is the highest level and demonstrates that an organisation is a leader in its field. Green dragon Level 5 is particularly useful for larger organisations with influence over their supply chain. This level is equivalent to, and in fact requires a greater environmental commitment in respect to public reporting and carbon emissions than ISO14001 and EMAS. Organisations wishing to register for Green Dragon Level 5 must also achieve the requirements of Green Dragon Levels 1 to 4.

Level 5 is assessed through an audit visit undertaken by a Green Dragon Auditor and will also cover Levels 1 to 4. The audits are undertaken by a consultant who has not had any previous involvement with the organisation.

If the standard's requirements are met again a certificate and organisation's details will be registered on the Green Dragon website. The certificate is valid for one year from the date of issue. On achieving this stage of environmental management, the organisation can also display the Green Dragon logo on their corporate literature.

Since its inception, over 1,000 organisations have achieved certification against the Green Dragon Standard. This is significantly more than any other phased EMS schemes operating in the UK. Companies that have achieved the Green Dragon EMS Standard include Argies Coffee (an independent coffee distributor) and Willis Construction (a Cardiff-based construction company) and the Green Dragon website includes case studies.

Companies and organisations that have reached Level 5 may also be successful in obtaining ISO 14001 or EMAS.

Sustainable development management

In addition to the certified environmental management systems there is also BS 8900:2013 (*Guidance for managing sustainable development*), which has been designed as a system framework to help organisations to develop a management approach to sustainable development. It offers practical advice with which to make a meaningful contribution to sustainable development. It provides guidelines on options for sustainability management, managing the balance of the non-financial (i.e., social, environmental and economic) and the financial capital of business, with the aim of continually improving performance and accountability. It applies to all types of organisations with the aim of helping them manage their resources and activities in a more sustainable way.

It is based on the five key and interwoven aspects of sustainable development and the need for enhancing these:

- Human Capital (ensuring that it is contributing positively towards meeting human needs such as subsistence, freedom and security, but also identity, empathy, creativity and leisure);
- Social Capital (supporting the development of the community in which the organisation operates, including economic opportunities);
- Manufactured Capital (using infrastructure, technologies and processes in a way that uses resources most efficiently);
- Natural Capital (limiting and reducing over time the use of substances extracted from planet earth);
- Financial Capital (ensuring that the organisation's financial measures and operations reflect the value and needs of the other four capitals).

Surrounding this is the notion of accountability for actions and use of this capital to society as a whole – being transparent and accountable to stakeholders.

BS 8900 provides a conceptual framework building on the core principles or values for sustainability management. These can be used to assess the sustainability management issues and impacts associated with individual products and services, or to assess the sustainability management issues and impacts associated with the activities of an organisation as a whole.

Organisations should have a framework so that they can take a structured approach to sustainable development by considering the social, environmental and economic impacts of their organisation's activities. It can be used by any organisation including trade unions and other social organisations. It also helps make the connection with existing technical, social and environmental standards, such as the ISO 14001 series of standards. It provides the stakeholders of an organisation with a useful tool to assess and engage in improving organisational performance. It will also help organisations contribute to international-level dialogue in the international standard on social responsibility that is being developed.

BS 8901:2009 (*Sustainable events management system – Specification with guidance for use*) has been developed specifically for the events industry with a purpose of helping the industry to operate in a more sustainable manner. The standard defines the requirements for a sustainability event management system to ensure an enduring and balanced approach to economic activity, environmental responsibility and social progress relating to events.

It requires organisations to identify and understand the effects that their activities have on the environment, on society and on the economy both within the organisation and the wider economy; and put measures in place to minimise the negative effects.

It shares many of the common management principles of other management system standards such as ISO 9001 (*Quality Management*), ISO 14001. The key requirements of BS 8901 include:

- Sustainability policy;
- Issue identification and evaluation;
- Stakeholder identification and engagement;
- Objectives, targets and plans;
- Performance against principles of sustainable development;
- Operational controls;
- Competence and training;
- Supply chain management;
- Communication;
- Monitoring and measurement;
- Corrective and preventive action;

- Management system audits;
- Management review.

This list should take account of all tiers of stakeholders in events, namely. These can be considered as

Tier 1 – Event Owners – This is the person who commissions the event and holds overall responsibility for the event

Tier 2 – Event Organisers – This person or organisation that will have overall responsibility for the management of the event on the ground

Tier 3 – Suppliers – This tier encompasses all people who provide services and products for your events, including couriers, printers, caterers, etc.

This standard should again be applied using the Plan-Do-Check-Act principle.

Quality management around the world

Every country has a system for establishing or adopting standards and accrediting those bodies able to certify compliance. Most of the national bodies are members of ISO and some examples are given next.

Standards Australia is the leading independent, non-governmental, not-for-profit standards organisation in Australia. Through the Accreditation Board for Standards Development Organisations (ABSDO) other Standards Development Organisations can be accredited to develop Australian Standards. *Standards New Zealand* is a business unit within the Ministry of Business, Innovation and Employment. The Standards Approval Board is an independent statutory board set up under a 2015 Act and started work in 2016. It is responsible for approving the membership of standards development committees and standards that are developed by these committees.

Both Australia and New Zealand have adopted the latest edition of ISO 9001. New

Zealand has adopted the new edition of ISO 14001:2015 as AS/NZS ISO 14001:2016[15] and BSI has an Australian site.

As the national standards body, *Enterprise Singapore* administers the Singapore Standardisation Programme through an industry-led Singapore Standards Council. The Council approves the establishment, review and withdrawal of Singapore Standards and Technical References. It also advises Enterprise Singapore on the policies, strategies, initiatives and procedures for standards development and promotion. The Council comprises representatives from the industry, professional bodies, trade and consumer associations, academia and government agencies. It develops Singapore Standards and promotes them so that they are accepted and adopted by stakeholders in Singapore. The Council also actively participates in the development and review of international standards that are important to Singapore.

The Department of Standards Malaysia (*Standards Malaysia*) is the National Standards Body and the National Accreditation Body, providing confidence to stakeholders, through credible standardisation and accreditation services for global competitiveness. The aim is to enhance the quality of Malaysian products and services for the nation and globally. The motive for businesses is to help improve efficiencies, reduce waste and enhance quality for greater marketability of their products and services locally and internationally. A list of accredited certification bodies by Standards Malaysia can be accessed via the website.[16]

In South Africa, the *South African Bureau of Standards* is an autonomous body established as a result of an act of Parliament. The legislation concerning the SABS has been promulgated several times *to cater for changing circumstances and to amend the scope of activities of SABS. The group offers a full spectrum of standards development, information and conformity assessment services.*

Since it was founded in 1918, the American *National Standards Institute* (ANSI) has coordinated the development of voluntary consensus standards in the United States and has represented the needs and views of US stakeholders in standardisation forums around the globe. ANSI is the US member body to ISO.[17]

The *Bureau of Standards Jamaica* (BSJ) is a member of ISO and was created by law under the Standards Act of 1969 to promote higher standards in commodities, processes and practices. The BSJ also administers the Processed Food Act of 1959 and the Weights and Measures Act of 1976. A 14-member Standards Council appointed by the government is responsible for policy direction and the general administration. The National Certification Body of Jamaica (NCBJ) offers certification services to companies in Jamaica and the wider Caribbean and is accredited by the USA ANSI (see earlier) to offer certification to ISO 9001.

Global Reporting Initiative (GRI)

GRI is an independent international organisation, based in Amsterdam, the Netherlands. It operates through regional hubs in Brazil, China, Colombia, India, South Africa and the United States. GRI reports are produced in more than 100 countries. The vision is to establish a *thriving global community that lifts humanity and enhances the resources on which all life depends* with a mission to empower decisions that create social, environmental and economic benefits for everyone.

The GRI Sustainability Reporting Standards (GRI Standards) are the most widely adopted global standards for sustainability reporting. Since GRI's inception in 1997, it has grown from something of a niche practice transformed to one adopted by a growing majority of organisations. It is claimed that 93% of the world's largest 250 corporations report on their sustainability performance.

The argument is that the practice of disclosing sustainability information inspires accountability, helps identify and manage risks, and enables organisations to seize new opportunities. Reporting with the GRI Standards supports companies, public and private, large and small, to protect the environment and improve society, while at the same time thriving economically by improving governance and stakeholder relations, enhancing reputations and building trust.

GRI now work with the largest companies in the world as a force for positive change. As is becoming known some of these companies have revenues larger than the GDPs of some individual countries, with supply chains that stretch the globe. GRI hopes that their work will have a positive impact on social well-being, through better jobs, less environmental damage, access to clean water, less child and forced labour, and gender equality.

Conclusion

Quality Management Systems help to ensure enterprises meet their objectives and are a set of policies, processes and procedures that are necessary for planning and execution in the core business area of an organisation, primarily areas that can impact the organisation's ability to meet customers' requirements. They enable the operation to improve efficiency and identify problems and areas for improvement. Environmental Management Systems are management systems that look specifically at environmental issues arising from the organisations' operations and provide the means of improving environmental performance and may be a means of demonstrating social responsibility. Certified systems allow an organisation to demonstrate this.

Notes

1 There are related standards PD ISO/TS 9002: Guidelines for the application of ISO 90001:2015; ISO 9004: Guidance to achieve sustained success. There are also industry specific quality management standards such as ISO 13485 for the medical devices industry.

2 It does not always follow that a certificated EMS will be used, see for example Waitrose https://

www.johnlewispartnership.co.uk/content/dam/cws/pdfs/Juniper/jlp_cr_report_1819.pdf

3 https://ec.europa.eu/environment/emas/join_emas/emas_global_en.htm

4 For more information see https://www.bsigroup.com/en-GB/Occupational-Health-and-Safety-ISO-45001/

5 The Green Dragon Environmental Standard was originally set up by Groundwork Wales together with ARENA Network to help companies in Wales to develop an EMS in stages.

6 https://www.gov.uk/government/publications/environment-agency-enforcement-and-sanctions-policy/environment-agency-enforcement-and-sanctions-policy#enforcement-options

7 See "The UK's Enforcement Gap – a briefing from Unchecked.uk" at https://www.unchecked.uk/wp-content/uploads/2019/09/The-UKs-Enforcement-Gap-1.pdf

8 https://www.ukas.com/

9 https://european-accreditation.org/

10 That is, under ISO 14001, "a non-fulfilment of a requirement".

11 Full text at https://eur-lex.europa.eu/legal-content/EN/TXT/PDF/?uri=CELEX:32009R1221&from=EN

12 The informal text of the Regulation with the amended Annexes can be found at https://ec.europa.eu/environment/emas/pdf/other/Informal%20consolidated%20version%20of%20EMAS%20Regulation%20-%20post%20amendement%20Annexes%20I,II,%20III%20.pdf

13 See Annex I of Regulation (EC) 1221/2009 for more detail.

14 https://www.groundwork.org.uk/greendragon/

15 https://www.standards.govt.nz/

16 http://www.jsm.gov.my/ms-iso-14001#.Xuy3qJNKgQl

17 https://www.iso.org/members.html

Chapter references

[1] BSI. (2015) *Quality Management Systems Requirements*, British Standards Institution, London, BS EN ISO 9001:2015.

[2] Defra. (2016) *Develop a Management System: Environmental Permits*. https://www.gov.uk/guidance/develop-a-management-system-environmental-permits (Accessed December 2019).

[3] BSI. (2015) *Environmental Management Systems – Requirements with Guidance for Use*, British Standards Institution, London, BS ISO 14001:2015.

[4] Milieu RPA. (2009) *Study on the Costs and Benefits of EMAS to Registered Organisations*

Final Report Study, Contract No. 07.0307/2008/517800/ETU/G.2 for DG Environment, European Commission. https://ec.europa.eu/environment/emas/pdf/other/costs_and_benefits_of_emas.pdf.

[5] Regulation (EC) No. 1221/2009 of the European Parliament and of the Council of 25 November 2009 on the Voluntary Participation by Organisations in a Community Eco-Management and Audit Scheme (EMAS), Repealing Regulation (EC) No 761/2001 and Commission Decisions 2001/681/EC and 2006/193/EC, OJ L342/1 of 22 December 2009.

[6] Regulation (EU) No 2017/1505 of 28 August 2017 Amending Annexes I, II and III to Regulation (EC) No 1221/2009 of the European Parliament and of the Council on the Voluntary Participation by Organisations in a Community Eco-Management and Audit Scheme (EMAS).

[7] BSI. (2018) *Guidelines for Auditing Management Systems*, British Standards Institution, London, BS EN ISO 19011:2018.

[8] BSI. (2019) *Environmental Management Systems: Guidelines for a Flexible Approach to Phased Implementation*, British Standards Institution, London, BS EN ISO 14005:2019.

[9] BSI. (2016) *Environmental Management Systems – General Guidelines on Implementation*, British Standards Institution, London, BS ISO 14004:2016.

[10] BSI. (2010) *Environmental Management – Vocabulary*, British Standards Institution, London, BS EN ISO 14050:2010.

[11] BSI. (2006) *Environmental Management – Life Cycle Assessment – Principles and Framework*, British Standards Institution, London, BS EN ISO 14040:2006.

[12] BSI. (2012) *Environmental Management – Life Cycle Assessment – Illustrative Examples of Application of ISO 14044 to Impact Assessment Situations*, British Standards Institution, London, PD ISO/TR 14047:2012.

[13] BSI. (2020) *Environmental Management – Environmental Communication – Guidelines and Examples*, British Standards Institution, London, BS ISO 14063:2020.

[14] BSI. (2013) *Environmental Performance Evaluation, Guidelines,* British Standards Institution, London, ISO 14031:2013.

[15] BSI. (2019) *Environmental Management: Quantitative Environmental Information. Guidelines and Examples*. BS ISO 14033:2019. https://landingpage.bsigroup.com/LandingPage/Undated?UPI=000000000030349173.

Sources of further information

Journals, websites and other publications:

- *Environmental Management*, Springer, New York, https://www.springer.com/journal/267
- *Journal of Environmental Management*, Academic Press, Elsevier, www.elsevier.com/wps/find/journaldescription.cws_home/622871/description#description
- *Environmentalist Online* https://www.environmentalistonline.com
- *The Environment*, Journal of the Chartered Institute of Water and Environmental Management, Natural Resources Wales https://www.ciwem.org/the-environment/
- Croner-i Environmental Management see https://app.croneri.co.uk/topics/environmental-management

Useful websites:

- European Commission EMAS site http://ec.europa.eu/environment/emas/index_en.htm
- British Standards Institution https://www.bsigroup.com/en-GB/
- Institute of Environmental Management and Assessment www.iema.net/
- United Kingdom Accreditation Service (UKAS) www.ukas.com/
- Environment Agency https://www.gov.uk/government/organisations/environment-agency
- Natural Resources Wales https://natural-resources.wales/?lang=en
- Scottish Environment Protection Agency (SEPA) https://www.sepa.org.uk
- Northern Ireland Environment Agency https://www.daera-ni.gov.uk
- Ireland's Environment Protection Agency (within the Department of Agriculture, Environment and Rural Affairs) https://www.epa.ie/irelandsenvironment/

- Chartered Institute of Environmental Health (CIEH) www.cieh.org
- Business in the Community http://www.bitc.org.uk/
- WRAP (offering a wide range of guidance publications including those from the former Envirowise) www.wrap.org.uk
- NetRegs (NetRegs is a partnership between the Northern Ireland Environment Agency (NIEA) in Northern Ireland and Scottish Environment Protection Agency (SEPA) in Scotland) https://www.netregs.org.uk/about/
- Chartered Institution of Water and Environmental Management (CIWEM) www.ciwem.org
- Health and Safety Executive (HSE) www.hse.gov.uk
- British Assessment Bureau https://www.british-assessment.co.uk/
- Quality Management Systems https://qualitymanagement.co.uk/

9

Constructions and related matters relevant to environmental health

John Bryson and Stephen Battersby

Introduction

This chapter is concerned primarily with some of the key aspects of construction. It looks mainly at buildings and their services (primarily but not exclusively dwellings), but also other constructions and matters relevant to environmental health practitioners when considering matters of hygiene not specifically covered elsewhere in this book.

One of the fundamental functions of any environmental health practitioner is the ability to inspect and identify issues that pose a risk to the health and safety in those locations where people reside, work, enjoy leisure or are part of the food supply chain. In order to be able to fulfil the inspection function adequately and identify problems it is important to have some understanding of the structures with which the EHP comes into contact. Such structures can include different modes of transport and structures associated with sanitation. A book such as this can only provide some basic information, and readers should refer elsewhere for more detailed information (see Further reading). This chapter also gives some pointers on inspection techniques.

As indicated this chapter is relevant to other chapters in the handbook and should be seen as a companion chapter.

Buildings

All buildings should be designed and constructed in a manner that takes account of the purpose, or future use, and intended occupation of the building. The same basic principles apply whether the future occupation is domestic, commercial or industrial. For any new building it is important to consider the location, geological conditions and the prevailing climatic conditions at the design stage. In some cases, the location may be such that the building will have to withstand potentially extreme climatic conditions, for example flooding, severe hot or cold weather or earthquakes.

It should also be designed and finished so that a good standard of cleanliness and hygiene can be maintained. Good design, construction and maintenance are essential to minimise the risk of injuries to the potential occupants and the spread of infections between them. Poor design and layout can affect the psychological and physiological well-being of the occupants including employees. Each element of construction of the building will also have a slightly different purpose.

The hygiene of buildings is an important consideration and as prevention is better than cure, the design and structure should reflect this. Some of these issues are addressed in other specific chapters. At its basic, building hygiene includes provisions to prevent the entry of pests, appropriate ventilation, safe water installations whether for drinking or personal hygiene, ease of maintenance and cleanliness, and effective sanitation and drainage so as to avoid contamination.

Foundations

Foundations, also referred to as "footings", are a fundamental feature of construction as they provide the necessary support for the whole structure. They should take into account the geological conditions in the area and the intended function of the structure. A shallow foundation, which is the most common for housing and smaller commercial buildings is usually embedded a metre or so into the ground. One common type is the spread footing which consists of strips or pads of concrete (or other materials) which extend below the frost line and transfer the weight from walls and columns to the soil or bedrock. The frost line is most commonly the depth to which the groundwater in soil is expected to freeze. The frost depth depends on the climatic conditions of an area, the heat transfer properties of the soil and adjacent materials, and on nearby heat sources.

A deep foundation is used to transfer a load from a structure through an upper weak layer of soil to a stronger deeper layer of soil. There are different types of deep foundations including helical piles, impact-driven (displacement) piles, drilled shafts or replacement piles (using an augur to extract soil and replacing it with reinforced concrete, and piers.

Walls

The functions of external walls in traditional construction carry the load by supporting the floors and roof of the building. This is not the case where the load is carried by framework such as in blocks of flats where there is a steel frame or skeleton frame of vertical steel columns and beams, for example with infill panels or brickwork (or even curtain walling – see later). However external walls also:

- Provide adequate thermal insulation and some forms of construction have better insulation qualities than others;
- Should provide adequate noise insulation;
- Provide an effective barrier against the adverse effects of the weather such as wind and rain (and again some forms of construction are less able to withstand rain than others, e.g. solid walls are more susceptible to penetrating dampness);
- Limit the spread of fire.

For examples of bonding see Figure 9.1 – the bonding used is indicative of the wall construction (whether solid or cavity construction) and also of the age of the building. A solid wall is colder than a cavity whether insulated or not with thermal transmittances (U values considered later) of $2\,\mathrm{W/(m^2K)}$ for a solid wall; uninsulated cavity $1.5\,\mathrm{W/(m^2K)}$; and insulated wall, $0.18\,\mathrm{W/(m^2K)}$ with the lower the figure the better the wall at preventing heat loss from the inside to the outside (see Figure 9.2 for a typical early 20th century form of construction and Figure 9.3 for more modern UK construction).

Other materials used for wall construction include "cob" (or "clom" in Wales) made from clay or earth and straw or other fibre and sometimes lime. In places where the earth used has a high chalk content it is sometimes called chalk cob or wychert. This system of construction has been used in many parts of the world and has various other names including daub, adobe, bauge and torchis in France. Cob and similar type buildings can be found in older buildings in Africa, USA, New Zealand. However, some newer properties that seek to be less damaging to the environment or more sustainable (eco-houses) have been built using cob. Cob is fire-proof but also requires a high moisture content to maintain its strength.

The most common construction found in older housing is the 'Flemish Bond' of brick-work which is used for solid walls and is the usual construction of most houses built before 1919 (1939 in London). The wall is usually one brick thick and is made up by alternating 'headers'(the end of the brick) and 'stretchers' (the long side). This is indicative of a solid wall which by its structure will be colder than cavity construction. Clearly there is no cavity that can be filled to provide additional insulation. English Garden Wall Bond is another solid wall form of construction with three courses of stretchers and one of headers repeated through the height of the wall

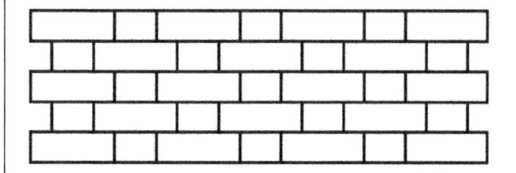

The simplest form of bonding is the 'Stretcher Bond' used for single-thickness walls, including the two individual leaves of a cavity wall. These are usually tied together with metal tie wires. The cavity will now be insulated, and the inner-leaf will normally be blockwork. This bonding is also used for infill panels in steel frame buildings

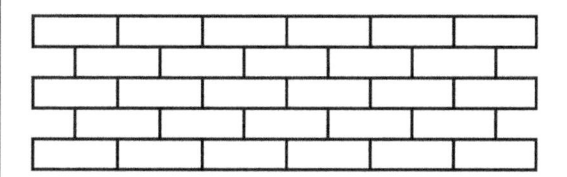

Figure 9.1 Brickwork and bonding

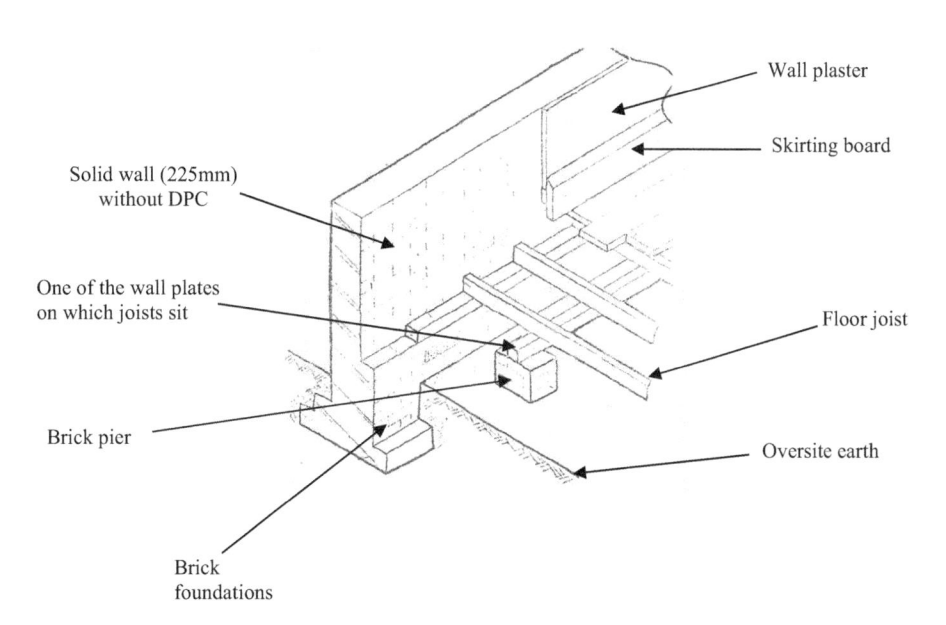

Figure 9.2 Typical early 20th-century construction in England with suspended timber floor
Source: Figure courtesy of Shelter Training

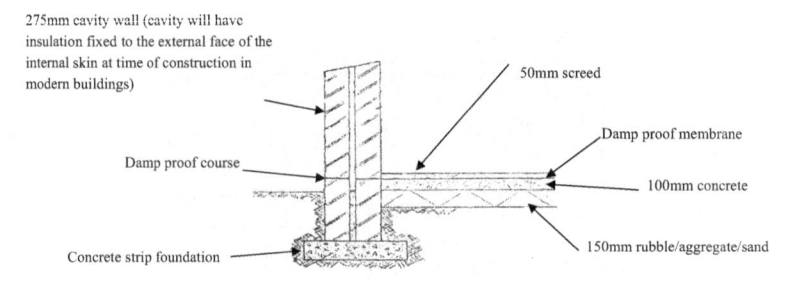

Figure 9.3 Typical more modern construction in UK
Source: Figure courtesy of Shelter Training

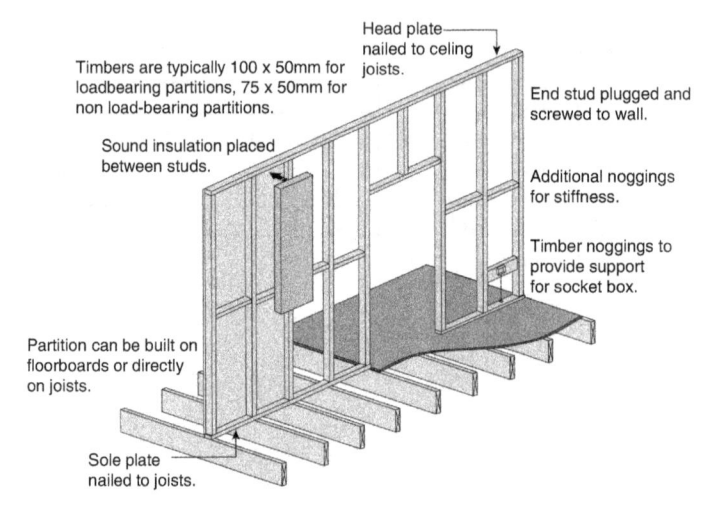

Figure 9.4 Stud partition wall
Source: Marshall et al. (2013) ([1] at page 208), reproduced by permission of Taylor & Francis

Internal Walls should provide:

- Separation into rooms and areas for different functions;
- Some privacy for occupants;
- Some thermal insulation;
- Some sound insulation;
- Some limitation on the spread of fire within the building;
- Wall surfaces in kitchens, food preparation areas, bath/shower and WC compartments should be capable of being easily cleaned and maintained in a hygienic condition.

All internal wall finishes internally should be smooth and even provide a surface that can be readily decorated and cleansed. This is of particular importance in kitchens, bathrooms and WC compartments where hygiene is especially important, and wall surfaces should not provide potential harbourage for any pests such as bedbugs or fleas.

Modern buildings, particularly dwellings, are likely to have mainly non-loadbearing lightweight stud partition walls (see Figure 9.4); older properties may have lathe and plaster partition walls and single block or brick partition have been used over the years.

Roofs

See Figure 9.5 for a pitched roof construction in an older property.

Roofs should provide:

- An effective barrier against the adverse effects of the weather (i.e. wind and rain);
- Adequate thermal insulation;
- Adequate sound insulation;
- Support not only the roofing materials but potential loading stresses from wind and accumulations of snow.

Where there is only one slope to the roof this is a non-pitched roof and can be found commonly in single storey "lean-to" extensions and additions to buildings as well as where it forms the main roof of the building.

Recently constructed or recovered roofs will have felt and also ventilation over the insulation to reduce the risk of condensation in the roof space. More recently another method for controlling condensation in the roof space has been used in new construction, "the breathing roof", in which vapour is allowed to escape via a vapour-permeable underlay which should also include a vapour barrier in the ceiling construction below the roof void and counter-battens to raise the battens used for fixing the roof covering to provide an air space above the breather membrane.

Slates for covering a roof, whether natural or manufactured (usually apparent from their uniformity which is not possible with natural slate) have to be fixed individually. In most instances this will be by nailing to the lath although there are more modern clip systems. Tiles, however, are manufactured, and do not all have to be nailed as they are made with a "nib" which hooks over the lath. It is not necessary (nor desirable) to nail every single tile – nailing as described here will provide a fully functional roof while allowing individual tiles to be removed later if necessary. Every tile along the eaves, up verges and along the ridge should be nailed, but otherwise it is only necessary to nail every fifth tile horizontally and every third tile up the roof.

Other materials have been used to form the roof covering in the UK and around the world, from thatch to corrugated metal and copper and aluminium sheeting.

There has been a move to "green roofs" in new sustainable construction. Green roofs are said to last longer than conventional roofs, reduce energy consumption as there is natural

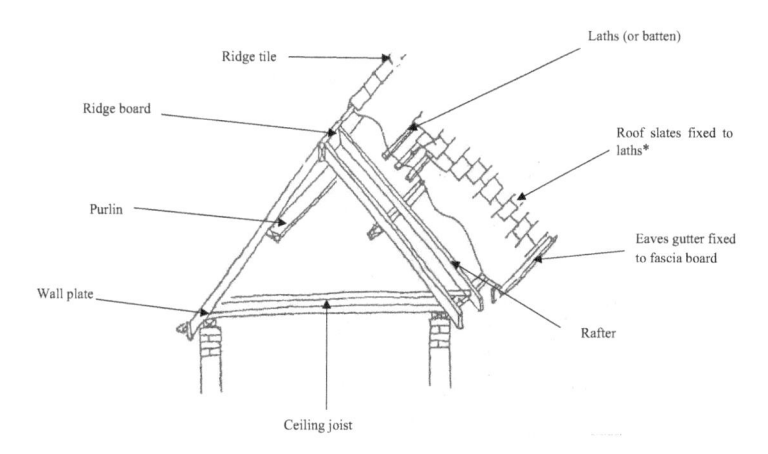

Figure 9.5 Pitched roof construction in an older property with 9 inch (225 mm) solid wall

Source: Figure courtesy of Shelter Training

Note: * In the UK and Europe older roofs that have not been recovered in recent years are unlikely to have any roof or sarking felt and some older forms of roofing felt can perish over time.

insulation. They also have the effect of reducing run-off of rainwater by absorbing storm water, potentially lessening the need for surface water drainage and also reducing the water flow that can lead to flooding. Green roofs also contribute to improved air quality and help reduce the urban heat island effect. The vegetation (e.g. sedum and dry meadow vegetation) is planted on a growing medium, over a waterproof membrane. The roof structure may also include additional layers such as a root barrier and irrigation systems along with drainage for excess water.

Green roofs are increasingly found around the world and in 2019 a report was published by the European Federation of Green Roof and Green Wall Associations (EFB) and Livingroofs.org on behalf of the Greater London Authority[1] supported by the Mayor of London which reported that the total area of green roofs in 2017 in the Greater London Area was 1.5 million m^2 which equates to 0.17 m^2 of green roof per inhabitant. The London Plan is that London currently accounts for around 40% of all green roofs installed in the UK. This has been achieved primarily through the land-use planning process; unlike many other cities around the world, London does not provide financial incentives for their provision.

Flat roofs can be a problem and in the British Isles for a number of reasons they have been considered to have a shorter life span than pitched roofs. Until the 1970s in the UK the "cold" roof was the most common flat roof so that the insulation was placed immediately above the ceiling and between the joists (see Figure 9.6). A vapour barrier was (or should have been) placed below the insulation (between the ceiling plaster and insulation). There should also have been ventilation provided to give an air flow above the insulation to prevent "interstitial" condensation. In practice this ventilation was not always achieved.

More recently "warm" roof systems have been constructed. There can be two versions ("sandwich" and "inverted"). The latter is more frequently found in commercial properties. In both cases the insulation is provided above the deck, so that the temperature of the roof structure is closer to that inside the building than in the "cold" roof. This should prevent movement due to expansion and contraction with the weather and which can cause cracking of the waterproofing layer such as felt on the deck. There is no need for insulation as the risk of interstitial condensation is reduced and indeed allowing ventilation could increase the risk of condensation if the air in the void is cooled. A vapour barrier/layer should be still be laid below the insulation. In the sandwich the insulation is placed above the deck and on top of the vapour barrier, but underneath the waterproof covering, so it is sandwiched between the deck and waterproofing. In the inverted roof the insulation is located above both the deck and the waterproof membrane and is then covered with the final heavy finish which can often be paving slabs that act as ballast. No separate vapour barrier is required in these roofs.

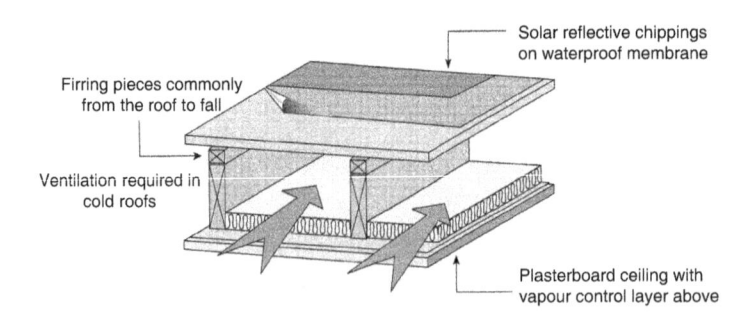

Figure 9.6 "Cold" flat roof construction

Source: Marshall et al. (2013) ([1] at page 192), reproduced by permission of Taylor & Francis

Rainwater goods

Eaves gutters, or eavespouts can be made from a range of materials from lead or zinc-lined timber to plastic, aluminium and cast iron. Asbestos cement was also used for some inter-war houses and those built in the UK shortly after the Second World War. Eaves gutters are intended to collect rain from the roof and carry it to, and connect with, a rainwater pipe. The rainwater, or "fall", pipe should connect to a drainage inlet. In properties this might be via an open gulley grating (gulley pot) or via a back-inlet gulley.

Floors

Most urban housing in the UK without cellars/basements and built in the 19th and early 20th centuries will have ground floors of suspended timber (sometimes with a solid floor in the kitchen or scullery; see Figures 9.2 and 9.3). Since the Second World War houses will usually have a ground floor of solid concrete construction. Some houses such as those built before the 1960s may have a mix of solid and suspended timber floors on the ground floor.

Upper floors in most houses are timber or composition board (or chipboard) sheeting, which is tongue and grooved. Older houses in the UK such as those built in the late 19th century and early 20th century will have close-boarded floors on the ground and upper floor (no tongue and groove) with caulking between the boards which over time may well have been dislodged.

Increasingly, use is being made of suspended concrete floors for ground floor construction. This is chosen as the floors are not so reliant on stable ground conditions as ground-bearing floors. The suspended concrete floors are prefabricated in a factory and consist of a series of T-beams with depth dependent on the span in which they are spaced and suspended on the load-bearing walls so that the infill between them is made up of standard concrete blocks. When the beams and blocks are in position a damp cement and sand grout

is brushed over the surface filling the gaps. The floor is then ready to receive the floor finish. If the air space under the floor is well vented and the ground is effectively drained there is no need for a damp proof membrane (DPM) or over-site concrete unless required for radon (or other gas) or vapour protection.

The floors must provide:

- Separation between different levels;
- Adequate structural support for occupants, their furniture and appliances;
- Resistance to the passage of moisture, heat and sound;
- A safe and even surface to provide easy and safe passage for occupants and visitors;
- In kitchens and food preparation areas, bathroom and WC compartment floors should be impervious, smooth and capable of hygienic maintenance.

Ceilings

Ceilings provide:

- Completion of the vertical separation of the building;
- Thermal insulation;
- Sound insulation;
- Some limitation on the spread of fire within the building – this is particularly important in kitchens and living rooms;
- Smooth and even surface that can be readily decorated and cleansed.

Windows

These provide:

- Weather protection;
- Natural light to the interior of the building;
- Natural ventilation to the interior of the building;
- In domestic buildings, a means for the occupants to view the outside environment;
- A degree of thermal insulation;
- A degree of sound insulation.

Window opening portions should be made to closely fit the window frame. They should be easily operated and prevent draughts which have a cooling effect. They should be fitted with suitable handles and a locking mechanism for security.

Doors

External doors provide:

- Weather protection;
- A means of access to and from the building;
- Privacy and security;

- A degree of thermal insulation;
- A degree of sound insulation.

Doors should be made to closely fit the door-frame and stop. They should be fitted with suitable handles and a secure lock (see Figure 9.7).
Internal doors provide:

- Access between different rooms and parts of the building;
- With the internal walls, completion of the separation between different parts of the building;
- Some privacy;

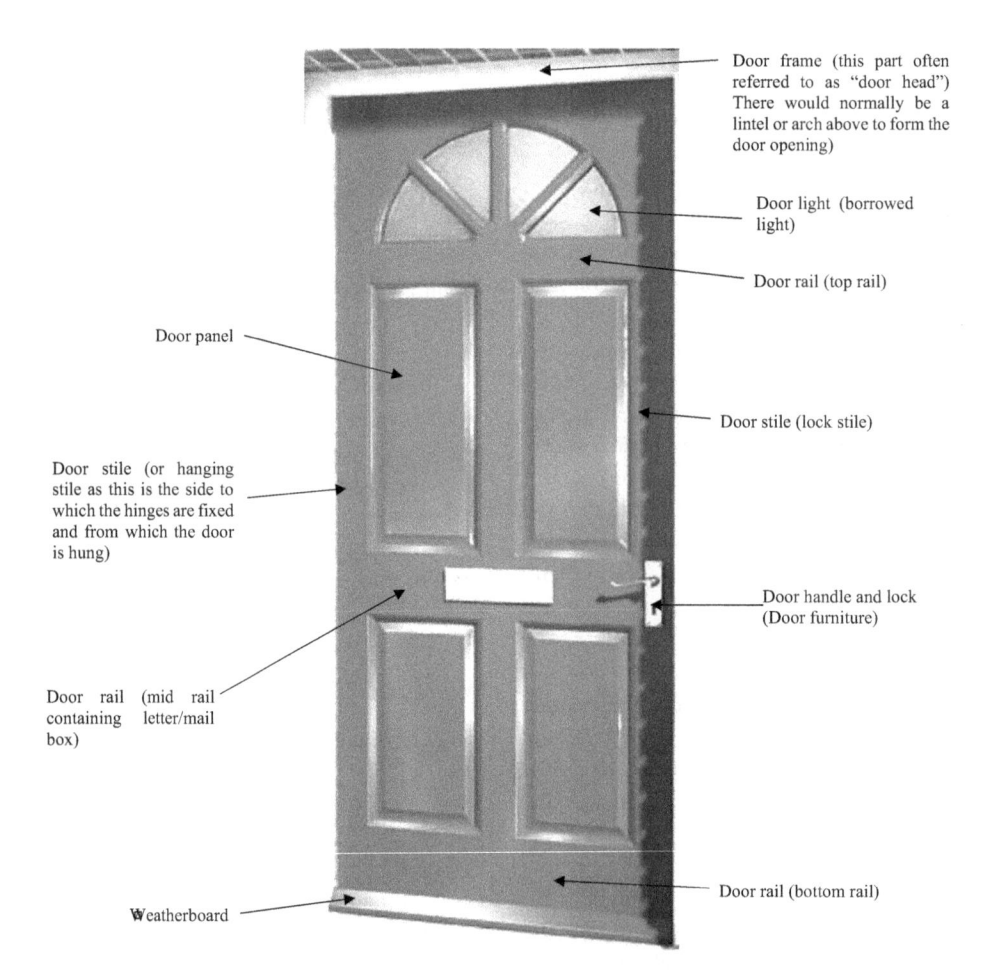

Door frame (this part often referred to as "door head") There would normally be a lintel or arch above to form the door opening)

Door light (borrowed light)

Door rail (top rail)

Door panel

Door stile (lock stile)

Door stile (or hanging stile as this is the side to which the hinges are fixed and from which the door is hung)

Door handle and lock (Door furniture)

Door rail (mid rail containing letter/mail box)

Door rail (bottom rail)

Weatherboard

Below the door opening will be the threshold bar which should be resistant to the weather when the door closed onto it and the frame.

Figure 9.7 Traditional exterior door and frame components identified

- Some limitation on the spread of fire within the building – this is particularly important in kitchens and living rooms.

Fire doors are dealt with in the section Fire safety issues.

The architrave which surrounds openings (most commonly door openings in modern buildings), although partly decorative, also covers the junction between the door casing (frame) and the wall plaster to facilitate maintenance and cleaning. It should be close-fitting and again should not provide potential harbourage for insect pests.

Stairs

Stairs are typically of timber or concrete construction although other materials such as steel can be used in specific circumstances. They should be strong enough to carry the weight of people and the weight of furniture or equipment being moved up or down them. In any situation the staircase should also be able to withstand the effects of fire for a sufficient length of time to allow occupants to escape. As a general matter stairs should:

- Provide a safe and convenient means of accessing different levels within a building;
- Have suitably designed handrails fixed at the correct height;
- Have adequate lighting with artificial lighting being controlled at both the top and bottom of the stairs.

See Figure 9.8 for a simple diagram of stairs. Stairs comprise a series of steps, each step comprise a tread (on which users place their feet) and the riser, which is the vertical face. It is usual for the tread to protrude beyond the riser and this projection is called the nosing. In older stairs, the joint between the riser and the tread under the nosing was sometimes masked by a "scotia mould". Timber members known as stringers usually support the treads and risers on either side. These run the full length of the stairs. The stringer that is

fixed to the wall is known as the wall stringer, the outer stringer is the one remote from the wall. The stringer supports the treads and risers either by housing the ends of the steps in the stringer (this is known as a "closed stringer", or string) or by the stringer being cut on its upper surface to the profile of the steps, and in such a design is known as a "cut" or "open" stringer.

The pitch of a staircase, certainly in domestic premises, should not be greater than 42° and individual going should be not less than 220 mm and the individual rise should be a maximum of 220 mm. All goings and rises should be the same; any variation in a flight increases the risk of a misstep and fall.

It should also be noted that the likelihood of a fall is doubled if there is no wall or guarding to one side of the stair, and guarding and handrails should be at a height of between 900 mm to 1,000 mm. Similarly, the lack of any handrail doubles the likelihood of a fall, even if there is a wall to both sides of the staircase. An accident is also three times more likely on stairs without carpeting, including those stairs designed to be left uncovered such as those built from concrete or with a thermoplastic tile finish and rubber nosing. The minimum headroom on a staircase should be 2,000 mm. The quality of lighting and surface (such as disrepair or frictional quality) will also influence the risk of falling.

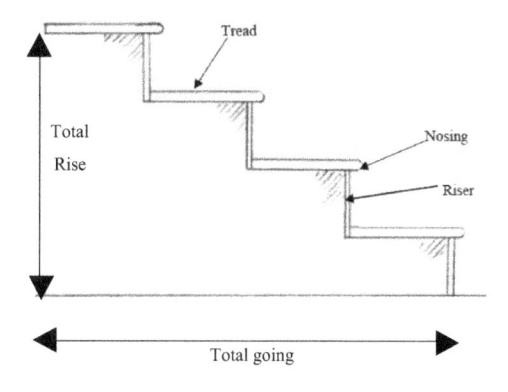

Figure 9.8 Simple diagram of stairs

John Bryson and Stephen Battersby

Non-traditional construction

Timber frame construction is based on factory-made structural elements. The timber framed wall panels carry the loads on the building to the foundations whilst the outer cladding provides decoration and weather protection. Factory production of the timber frame panels ensures that they are accurately manufactured to precise tolerances in a controlled environment away from the vagaries of British weather.

The timber frame panels are rapidly erected on site and, with trussed rafters forming the roof, a weathertight building can be created in a matter of days. This enables work to continue in protected conditions within the building whilst the outer cladding and roof finishes are applied. Cladding is a matter of choice; it can be brick, stone or lightweight claddings, such as timber boarding, tile hanging or render.

Timber is recognised as the only renewable construction material and the softwoods used in timber frame are sourced from environmentally sustainable British and European forests.

The timber frame method of building gives designers flexibility in both layout and external appearance. High levels of thermal insulation are incorporated within the construction, reducing heating costs and conserving energy.

Dry construction not only saves time on site but means that decorations can be carried out soon after completion of the building without risk of cracking and deterioration of finishes.

In Figure 9.9 one section shows brick cladding on the external face, the second diagram shows the external cladding as a proprietary render (coating).

In housing there have, over the years, been a considerable number of other "non-traditional" types of construction with their name often drawn from the company that developed them.

Precast Reinforced Concrete (PRC) types are, essentially, of concrete construction with reinforcing bars. The properties were designed for speedy construction in factories built for wartime production and when there was a shortage of homes as a consequence of bombing and inactivity in house building during the war years. After both World Wars there was a need to build large numbers of homes in the UK very quickly and at both times the majority were built by local councils – social housing. Indeed, very little private housing was built for sale until the mid-1950s. Many of these types were popular and many were bought by the original tenants under "right to buy".

There are a very large number of PRC house types. Distinguishing between them can be quite difficult. Some of the more common ones and features that help identify them are:

- Airey – with "Shiplap" cladding panels, tile hung gable ends;
- Boot – has rough cast rendered externally with front ground floor bay to some houses. Pattern of cracking to external rendering may reveal structure beneath;
- Cornish Unit – has exposed post and panel construction, mansard roofs with tile hung upper portions;
- Orlit – has external walls comprising an outer leaf of paving stabs and an inner leaf of concrete blocks. They may be rendered externally but the panels can give the appearance of blockwork. Roofs may be of flat construction;
- Reema Hollow Panel – in which the panels have a plain or exposed aggregate finish externally. Roofs are either hipped or twin pitched with tiling.

There were also some steel-framed construction dwellings and there are examples of steel frame houses from the early 1920s right through to the late 1960s. Modern steel-framed structures are described in more detail later and are commonly used for high-rise buildings. The most common form for individual dwellings built after the Second World

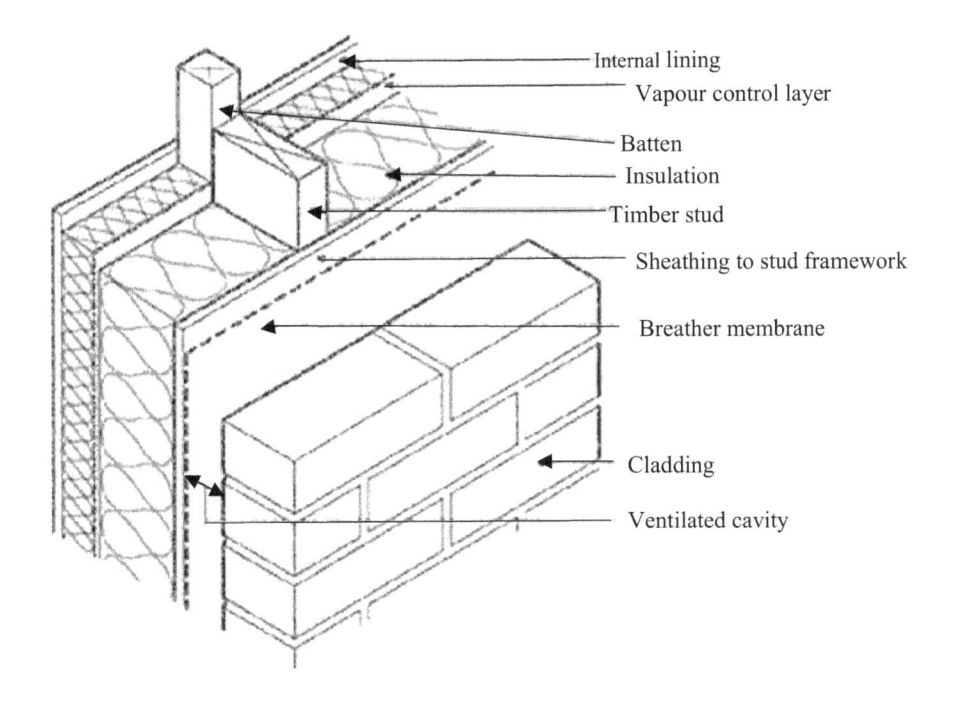

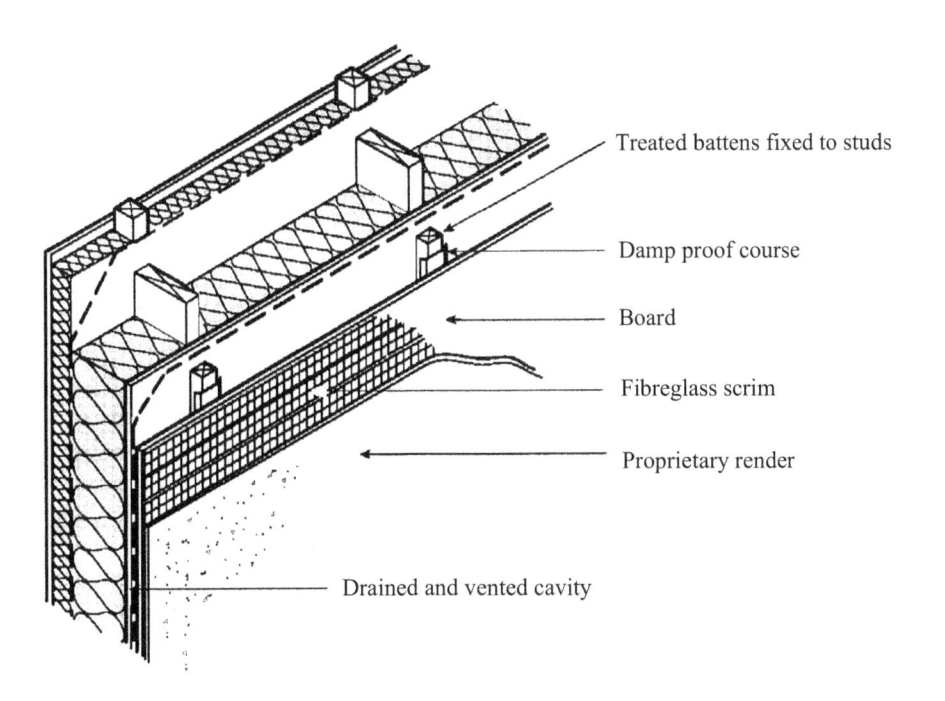

Figure 9.9 Sections through timber framed construction

Source: Figure courtesy of Timber Research and Development Agency

War in the UK was the British Iron and Steel Federation Housing (BISF) house, and these can usually be identified by ribbed metal sheeting to the exterior of the first floor, with metal surrounded windows projecting from the wall face and corrugated asbestos cement roof covering.

Large panel systems (LPS). LPS dwelling blocks typically comprise precast reinforced concrete floor and roof components spanning onto storey-height structural precast concrete wall panels. The precast concrete components made off-site are connected by various forms of joints made on site. Vertical loads are carried to the ground through the structural wall panels, which also provide stability against lateral loads. Walls orientated across the short dimension of the building are usually called cross-walls, or flank walls if they are the exterior walls located at the ends of the building or in re-entrant zones. The structural walls orientated along the long dimension of the building are often referred to as spine-walls.

Most LPS blocks are high-rise and there has been concern in the UK about such system since the collapse of Ronan Point in 1968 (see Figure 9.10 for one example of an LPS). The Building Research Establishment (BRE)[2] has highlighted that hazards within such blocks include internal gas explosions and fire as well as that resulting from impacts. Such blocks have also had problems with dampness either as the result of condensation or penetrating dampness dues to failures at joints between the panels.

The use of concrete with a slow thermal response (but high thermal capacity) and the problem of rain penetration has led to many LPS being clad in a rain-shield to both improve the thermal properties and reduce the risk of condensation and also to prevent rain penetration between the joints. This it seems was what happened at Grenfell Tower in London. The method of installation of the Aluminium Composite Material (ACM) with a polyethylene core used to clad the block contributed to the spread of the fatal fire. ACM cladding has been linked to other

fatal fires such as that at the Lakanal House in south-east London, the Wooshin Golden Suites fire in Busan in 2010, the Lacrosse Tower in Melbourne in 2014 and the Marina Torch and the Address Downtown Hotel fires in Dubai in 2015.

Tower Blocks UK[3] has suggested that the concerns are because the structural design of large panel system blocks is weak: they could collapse in an explosion, high wind or serious fire. There are frequently gaps between the floor and wall panels, and these gaps prevent the flats from containing a fire for one hour and lead to the risk of serious fire spread. Such gaps also allow the easy spread of infestations whether cockroaches, ants or mice.

Steel frames[4] have been used in a variety of structural systems, including in residential apartment buildings and student accommodation. Quite often such buildings will also be of mixed use with commercial or non-residential activities on the ground floor. The structural steel system consists of beams and columns on a regular grid on each floor, and the floor spans between the beams. The floor may be formed from a slab of in-situ concrete poured onto decking or precast concrete units can be installed to form the floor. The façade and internal walls usually comprise light steel infill. The façade is to provide the weather and wind resistance and achieve the required thermal efficiency. The internal walls allow the configuration necessary for the proposed use.

Steel frame systems also include modular construction and a group of them can be arranged to provide a stable building form. Modular construction can also be optimised in steel-frame buildings by designing the modules to incorporate the serviced parts of the unit which will use most services such as bathrooms and kitchens.

Curtain walling – this is a system whereby the outer covering of a building is lightweight, often glass and the outer walls are non-structural, utilised only to keep the weather out and the occupants in. Since the curtain wall is non-structural, it can be made of

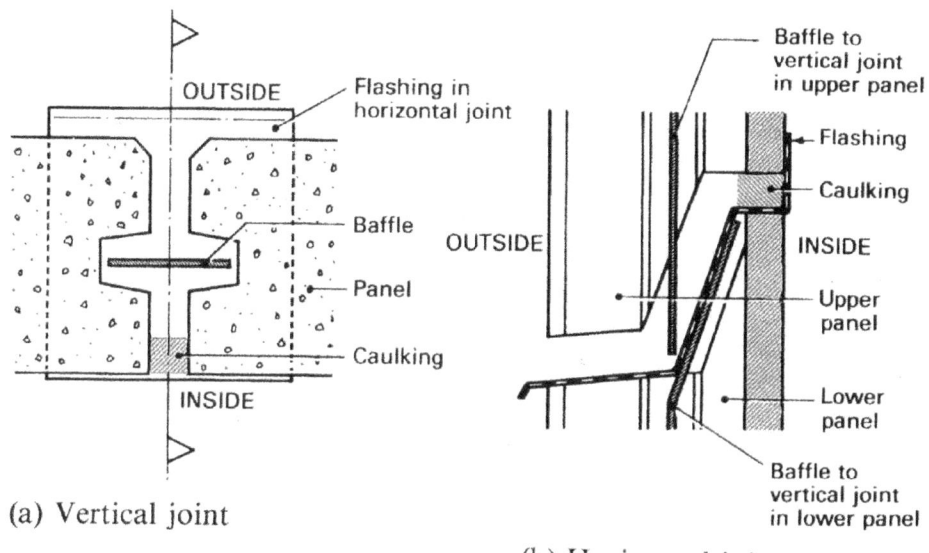

(a) Vertical joint

(b) Horizontal joint

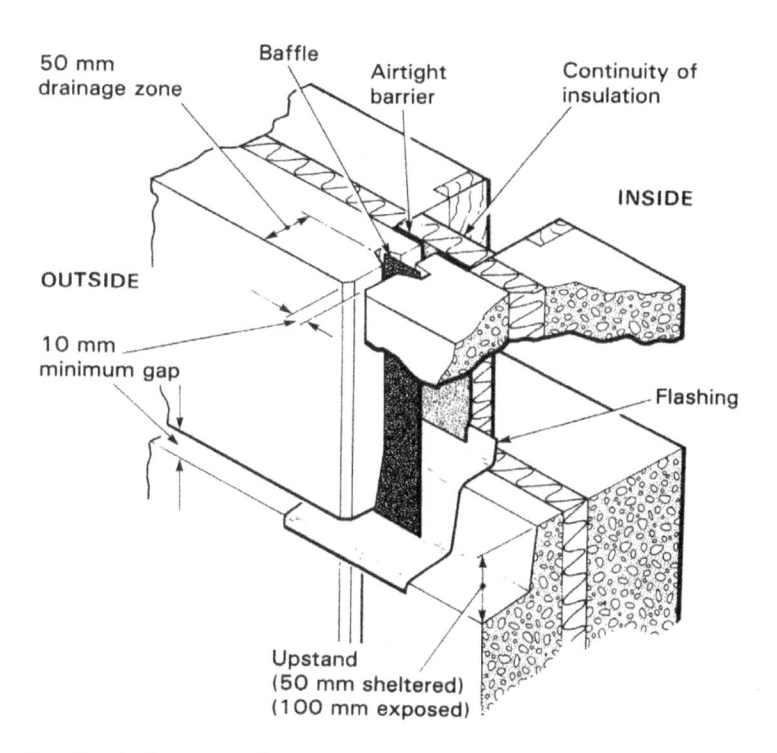

Figures 9.10 Details of a large panel system

Source: © IHS Markit, reproduced with permission from Weatherproof joints in large panel systems: two remedial measures (IP 9/86)

lightweight materials, thereby reducing construction costs. It is usual for the lightweight covering to be held in place via extruded aluminium and gaskets. This system is most often used in office buildings but while allowing the passage of large amounts of natural lighting, there can be a problem of solar heat gain which is more difficult to control when glass is used.

Curtain walling commonly uses the rainscreen principle whereby equilibrium of air pressure between the outside and inside of the "rainscreen" prevents water penetration into the building. For example, the glass is captured between an inner and an outer gasket in the framework. There is a system of ventilation in the framework where the glazing is fitted (the rebate) so that the pressure on the inner and outer sides of the outer gasket is the same. With equal pressure across this gasket, water cannot be drawn through joints or defects in the gasket.

Relative to other building components, aluminium has a high heat transfer coefficient (i.e. aluminium is a very good conductor of heat), so there is a high heat loss through aluminium curtain wall mullions. This can be dealt with in several ways, the most common method being the addition of thermal breaks (barriers between exterior metal and interior metal, usually made of polyvinyl chloride PVC). These breaks provide a point of reduced thermal conductivity in the curtain wall.

Thermal conductivity of the curtain wall system is important because of heat loss through the wall, which affects the heating and cooling costs of the building. On a poorly performing curtain wall, condensation may form on the interior of the mullions.

Mobile homes, park homes and similar structures

Mobile homes are more commonly called "park homes" and although they may have wheels to get them on to a site (or park) these are usually removed and are not really mobile, unlike touring caravans. A park home is a detached bungalow-style home of one or sometimes twin units and usually set in a private estate providing permanent residential accommodation. They are located on plots known as "pitches" and in the UK are cheaper to buy than traditional houses.

Although known as park homes in law, they are "mobile homes" and are covered by the Mobile Homes Act 2013 (in England), as amended by the Housing Act 2004, and there are similar legal provisions in Wales, Northern Ireland and Scotland. There is a British Standard to which park homes should be constructed, BS 3632:2015 *Residential park homes – Specification*. This standard sets standards for thermal insulation and U values (thermal transmittance) for the walls, floor and roof.

There are a number of components to these structures but whether they can be considered as buildings is an arguable point. However, they are usually constructed in a purpose-built factory and full construction can normally be completed in two weeks. The British Standard sets a minimum standard for the accommodation to include a living area, dining area, kitchen area, bedroom, bathroom and/or shower room, and WC.

There will always be a chassis and beams which can have a tow bar (probably detachable) with an axle welded to the main beam; there are also likely to be axle stands. The floor is a timber frame cavity construction (floor cassette) which includes insulation, and each bearer of the floor should be bolted to the chassis. The floor surface will be some form of board such as chipboard and the underside of the floor cassette will be a durable hardboard.

The external walls are constructed of a timber frame which includes insulation between the element of the timber frame (the cavities). The frame can be clad on the exterior with timber or some form of plastic sheets or hardboard with a render similar to that shown previously for more conventional buildings. Internal walls will be plasterboard and should have an integral vapour barrier.

The roof can be flat, mono pitch or double pitch and like flat roofs considered earlier, these roofs can be cold or warm depending on the location of the insulation, and will be of a lightweight construction. The roofs are constructed as one element and then fixed to the walls in the factory. They are covered with a lightweight covering such as roof shingles.

Unlike more traditional accommodation the park home should have an identification number permanently marked on the chassis, on the same side as the main entrance door, in a position that is readily visible. As a minimum, the British Standard requires that identification number identifies the manufacturer, year of manufacture and the individual home.

The British Standard also requires that a fire warning notice of not less than 0.2 m × 0.13 m should be permanently fixed inside the park home in a position that is readily visible, giving simple fire prevention advice and set out the action to be taken in the event of fire. A smoke alarm conforming to BS EN 14604 shall be installed in accordance with the alarm manufacturer's instructions should also be factory-fitted.

Drainage connections should be positioned so as to avoid structural members and be capable of making the connection to the site drainage.

It is common, once the park home has been located on the pitch, for a skirting wall between ground level and the underside of the floor to be put in place and the wheels may even be removed so that the home is no longer "mobile". This may be something that is covered in the site licence conditions and the agreement with the site owner.

Park homes are intended for permanent occupation, but caravans (whether static or towable) and park lodges or cabins are intended only for short-term occupation for holidays and recreational breaks. That said, some lodges which are of timber construction, are luxury properties which meet the BS3632:2015 standard. Most lodges are on sites that are licensed which are to have full residential licences.

The mobile caravans should be built to the BS EN 1647:2018 specification which is a lower standard than that in BS3632; this means that the build quality is different and with less insulation. They are not intended for permanent occupation.

Fire safety issues

Fire doors

All doors should be made to closely fit the door frame and stop and can help prevent the spread of fire even where specific fire resistance is not required. Even where no "fire check" door is justified, a well-fitting door can limit fire spread, but in certain buildings such as multi-occupied houses and buildings it will be necessary to install suitable certified fire check doors (of the appropriate rating) according to the risk of fire. Fire-resistant doors should be fitted with suitable handles and an effective fastening mechanism. Such doors can not only protect the occupant of a letting but also help establish a safe means of escape.

The Building Regulations[5] apply in England and Wales and Approved Document B identifies minimum fire resistance periods for various elements of construction, including fire doors.[6] The guidance recommends doors with a fire resistance period of 20 minutes (FD20) in some instances, and 30 minutes (FD30) in others.

All fire doors must have the appropriate proof of performance for the ratings they carry. All fire doors sold in the UK need to demonstrate proven fire resistance performance through one or more of the following means:

- Independent testing;
- Assessment by the competent authority such as a UKAS accredited test laboratory; or
- Independent third-party certification such as BM Trada Q-Mark product certification scheme.

There is no such thing as a stand-alone, fire rated door frame. The frame can only be considered as "fire rated" when it is machined in accordance with the tested details and installed, with the appropriate door, as a complete doorset. Therefore, if the frame and door leaf (the door itself) are made by different manufacturers, or the frame is to be made on site, its specification must match exactly that of the complete fire tested doorset.

Fire resisting doorsets currently on sale in the UK will have been tested in accordance with either BS 476–22 or EN1634–1 standards. It is not always clear from a visual inspection that existing doorsets comply with the level of resistance required for the particular situation. It may be necessary to require the building owner to provide evidence of the rating of the doorset.

The British Woodworking Federation has a system of certification, BWF-CERTIFIRE. These labels fitted to the doors give the member's name and phone number, and, where applicable, the certification number, a unique serial number and the rating (e.g. FD30 is 30-minutes fire resistance) (see Figure 9.11a). The label provides traceability through the supply chain, ensuring the manufacturer can always be contacted if further information on the fire door is required. The primary manufacturer will put a label on the top edge of the fire door indicating the door's rating and will record where the door has initially been sold.

BM TRADA carry out tests on fire doorsets and hardware. Testing can be undertaken to BS EN 1634–1, BS 476: Pt 22 and ISO 3008, as well as intumescent seal testing to BS 476: Pt 23. The intumescent strip is made of material that swells and chars when exposed to flame and that forms an insulating fire-retardant seal between the door and frame.

A number of manufacturers set a plastic plug into the edge of their doors, indicating the potential fire resistance rating when tested in accordance with British Standard 476: Part 20/22: 1987. This system used by TRADA. It is a circular plug with the design as set out in Figures 9.11b and 9.11c.

It has been suggested that rising butt hinges may be acceptable for use on fire resisting doorsets, providing there is suitable test evidence available. It seems there is no supporting fire test data for the use of rising butt hinges, and it is better if such hinges are not used with fire resisting doorsets other than perhaps in single household dwellings. Another problem with rising butt hinges is that they rely on gravity to close the door over the latch and it is quite common for pressure differentials within a building to prevent the hinges from doing so. This can then lead to the door "bouncing" on its latch and not fully closing within its frame reveal, thus failing to function as an effective fire resisting doorset. Self-closing devices can be defined as a device which is capable of closing the door from any angle and against any latch fitted to the door, and rising butt hinges are unlikely to do this.

Sprinklers and fire suppression

Fire sprinkler and fire suppression systems provide active (as opposed to passive) protection in case of fire. Sprinkler systems have water supplies to each sprinkler head where the sprinkler is held closed by either a heat sensitive glass bulb or a two-part metal link held with a fusible alloy. The glass bulb or metal link apply pressure to a pipe cap that acts as plug preventing the flow of water until the design temperature is reached of the individual sprinkler head. Normally each sprinkler head is activated independently.

Fire suppression systems are different from sprinkler systems and there is a variety of suppression systems. The most common, and the one often referred to in multi-occupied housing is the water mist system, which now are replacing sprinklers as they do less damage because less water is required. There are both high-pressure and low-pressure water mist systems. They can be used in large data rooms, other large areas and local applications. These systems can be used for flammable liquids and electrical rooms, the mist evaporates

Figure 9.11a   Example of BWF CERTIFIRE label

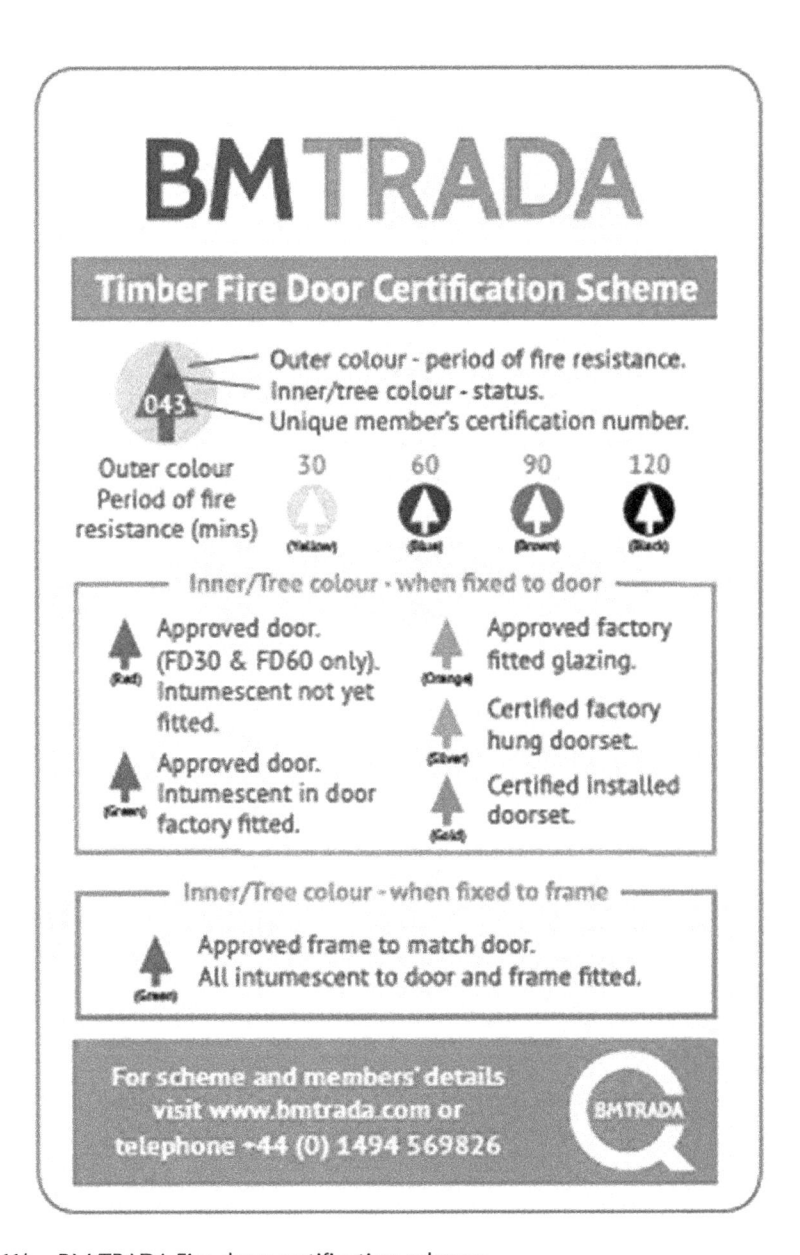

Figure 9.11b   BM TRADA Fire door certification scheme

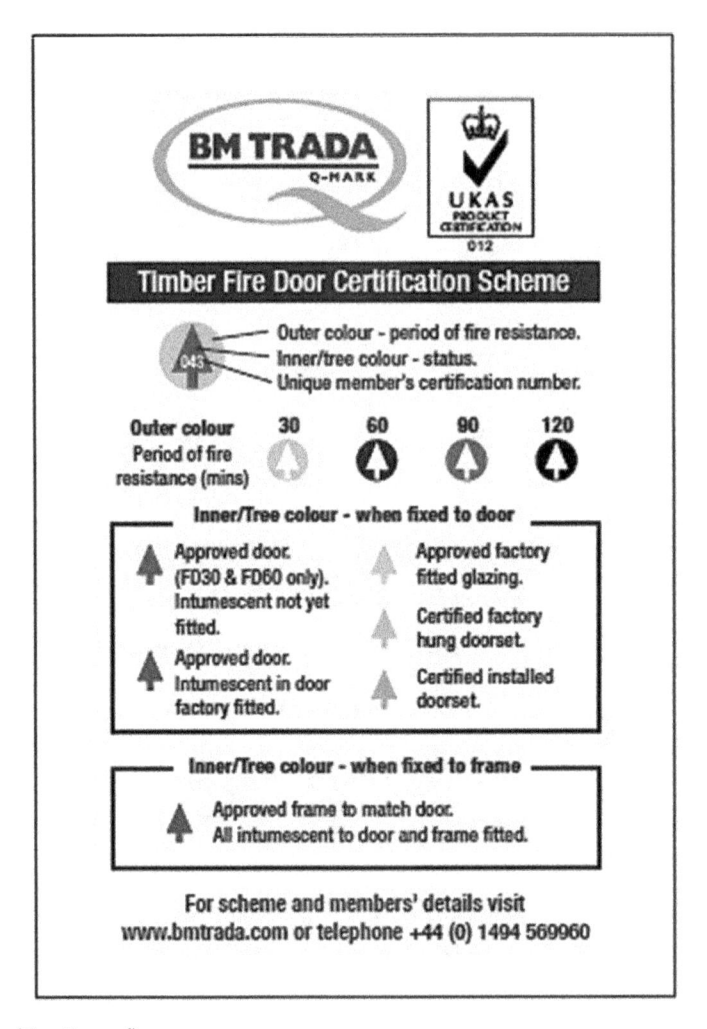

Figure 9.11b   (Continued)

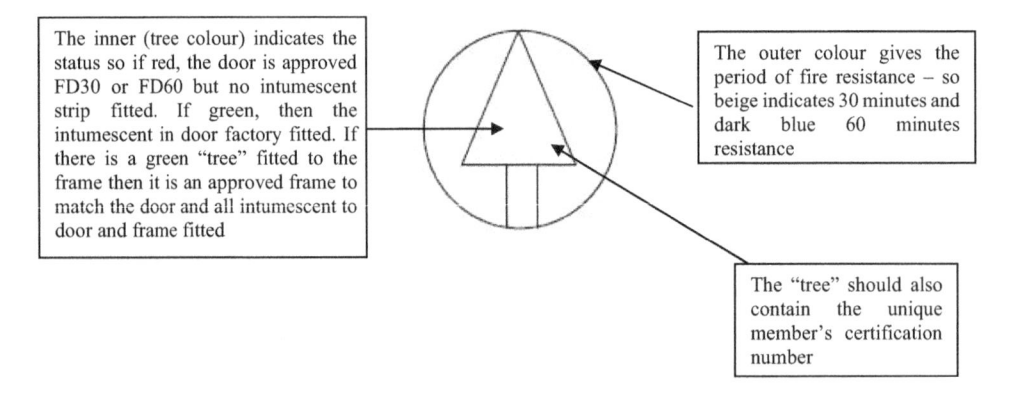

Figure 9.11c   Plastic plug design of indicating fire resistant quality of door

and causes a starving of oxygen effect rather than cooling. There are several relevant British Standards including: BS 8458:2015 *Fixed fire protection systems. Residential and domestic water mist systems. Code of practice for design and installation*; BS 8489–1:2016 *Fixed fire protections systems. Industrial and commercial water mist systems. Code of practice for design and installation*. The BS 8489 series provides recommendations and guidance on the design, installation, commissioning and maintenance of water mist systems for industrial and commercial hazards. The standards use a structured approach to the selection, application and deployment of water mist systems in buildings. They also form a part of the necessary protocols needed for the testing of water mist systems.

There are also gas systems. These use an inert gas such as argon or a blend of nitrogen and argon or carbon dioxide as less detrimental to the environment than HCFCs that have been used. These extinguish a fire by removing the oxygen and/or heat content in the atmosphere. Gas systems are stored as liquid, with nitrogen used to pressurise it. This type of system is said to be best suited to data rooms, switch rooms or communication rooms.

There are also systems that use a foam. One system has been designed for use in commercial kitchens. Nozzles are placed under the cooker canopies/hoods and will propel a water-based agent with a chemical foam-type mix over the fire. The trigger is usually a heat link or manual pull switch. Foam deluge systems are mainly suitable for large applications where you cannot use water or gas. These tend to be external such as transformers, oil tanks and oil storage silos. A simple sprinkler type application disperses a foam concentrate mixed with the water to provide the typical expanding agent (the same as with extinguishers).

Pneumatic heat detection tubes are in essence an extinguisher with a valve and a length of heat detection tube which acts as the detection and propellant feed for the agent. When the temperature reaches a

certain level around the pipe, it blows a small pressurised hole in the pipe which then propels the agent directly onto the fire risk. This suppression system is ideal for boats, vehicles, small machinery, electrical switch cabinets and fume cupboards. This system aims to deal with the fire at the "smouldering stage" but is not appropriate for larger areas or where there could be a larger fire.

Dry risers and wet risers

A dry riser system is a network of pipework and valves that allow the fire service to easily and quickly deliver water to upper floor levels. The maintenance of these pipe networks and associated valves is essential to ensure they do not leak, can maintain the pressures required for firefighting and are accessible for quick and simple connections.

Wet risers are used to supply water within buildings for firefighting purposes and should be in fire-resistant shafts. The provision of a built-in water distribution system means that firefighters do not need to create their own distribution system with which to fight the fire and avoids the breaching of fire compartmentation by running hose lines between them. Although it should be noted that the risers themselves (as for any other services) should be properly fire stopped as they pass through the building; otherwise they could compromise the compartmentation. Wet risers should be inspected and tested regularly to ensure the equipment is functioning correctly and ready for use.

Wet risers are permanently charged with water, but dry risers do not contain water when they are not being used but are charged with water by the firefighters when necessary. Dry risers are used when the water pressure in a building would be insufficient for fire suppression and in unheated buildings where the pipes could freeze. Dry risers have an inlet connector at the emergency service vehicle access point and with "landing valves" at locations on each floor of the building. As with all risers they should be within a fire-resistant shaft and with adequate "fire stopping" when passing through floor/ceiling structures.

Part B of the Building Regulations requires that fire mains are provided in all buildings over 18 m in height.

Fire safety in high-rise residential and other buildings

Traditionally EHOs in the UK have been more concerned about fire safety in more traditional houses that have multiple tenants and single dwellings. In other parts of the world these buildings have often been relatively new developments where it has been assumed modern and adequate standards of fire prevention and protection are in place and fire safety is wholly the responsibility of the Fire Department. In the UK and England and Wales in particular, responsibility is split and there are grey areas, so for example while fire safety issues within a flat might be addressed by the EHO, those in the common parts are dealt with by the relevant fire and rescue service. It may also have been because tower blocks were originally owned by the local authority that employed the EHO. The fires in Grenfell Tower and others since then have highlighted that fire safety is both a real risk in high-rise buildings but also in other buildings even below 18m in height such as at the Cube in Bolton in 2020[7] and Richmond House, Worcester Park Surrey in 2019, either as the result of materials used or deficiencies in construction.[8]

In England, as part of the move to "deregulation" the Regulatory Reform (Fire Safety) Order 2005 (SI 2005 No 1541) placed the responsibility on individuals within an organisation to carry out risk assessments to identify, manage and reduce the risk of fire, rather than on inspection and enforcement by the fire service. Prior to the Order, all public and commercial buildings, and all non-single-household domestic dwellings apart from houses in multiple occupation had to have valid fire certificate issued annually by the fire service. There is a difference between "common parts" in the Housing Act 2004[9] as interpreted by EHOs and definitions used in the 2005 Order which instead describes areas "used in common by the occupants of more than one such dwelling". Under the Housing Act common parts can include the exterior of the building. As the Addendum to the Operating Guidance to the Housing Health and Safety Rating System points out there is a need for consistent and coherent joint working arrangements between the local housing authority and the fire and rescue authority when applying the two sets of legislation, and there should be a local protocol in place on liaison between the two authorities.[10]

The Fire Safety Act 2021 amends the Regulatory Reform (Fire Safety) Order 2005 (the "FSO") for England and Wales enforced by the relevant fire and rescue service and addresses some of the confusion about responsibility. This now makes clear the external walls of a building and the fire doors to individual flats/apartments must now be assessed as part of the requirement for a fire risk assessment. The Act requires the owners and managers of such multi-occupied residential buildings to ensure that the fire risk assessments for such buildings are reviewed and updated to encompass the structure, external walls and flat/apartment entrance doors. The Act is not dependent on the height of the building.

There are four main factors involved in assessing the fire safety (and under the HHSRS assessing the fire hazard). First, the ignition of a fire. A fire is most likely to start within a dwelling, although it could start within common parts where rubbish has been allowed to accumulate or, in other forms of imposed fire loading, such as wood used to box in services passing through common areas. Second, the design, materials, construction and maintenance of the building should limit the spread of fire, containing it within the residential premises (whether the common parts, or the individual flat). This function includes the design, etc. of the entrance door to the dwelling, which should limit the possibility of the fire spreading into common parts; this includes the whole doorset which should fit properly to the adjoining compartment wall. It also includes windows (the frames and openable lights) and ventilation systems, which should reduce the possibility of a fire spreading outwards to ignite possible combustible materials in any cladding system. In addition, if fire takes hold in the cladding it could spread into other flats through a weak point such as windows. Should the cladding system catch fire there is also the risk of structural failure and burning materials becoming airborne.

Third, if a fire starts and begins to spread, residents need to be made aware of it. This means there should be a working smoke/fire alarm system (in the UK complying with BS 5839–1) that alerts the affected residents (alarm systems may be zoned so that residents on the floor above and below are alerted to evacuate the building). It should be recognised that any assessment should not be limited to whether there was compliance with past or previous requirements in the Building Regulations or Codes. Compliance with Building Regulations or Building Codes may not be sufficient, and some stair systems permitted under the Regulations can be considered as hazardous when considering the vulnerable age group.

Finally, and relevant to any assessment, there should be a means of escape and a fire safety plan with which occupiers are familiar. Depending on the particular dwellings and building, this may involve staying within an unaffected dwelling (with the door shut) until rescued although "stay put" policies have been criticised and often ignored once residents are alerted to a fire. Should the procedure and plan require zonal or full evacuation, with or without instruction from the fire service, occupiers need to be able to get out as quickly as possible via a safe and protected route (with provision made for those who cannot evacuate unaided). It should be noted that once a fire is being tackled in a flat this will necessitate the entrance door being open so smoke will enter the escape route.

TowerBlocks UK has a checklist[11] intended for residents but this list includes many matters that an EHO will need to consider. These are not only for individual flats but also common parts, not just structural matters but also matters relating to management of the building and individual apartments.

There should be fire protection between each unit within the building and between the accommodation and the route of escape (common landing and stairway). There should be adequate compartmentation (fire separation) and no obvious disrepair to walls or gaps around service entry points, pipework or ducting, this includes where services such as wet or dry risers, or other services pass through the floor. Nor should the compartmentation be compromised, for example around door frames, window frames, or around service entry points or ducting. One of the most important aspects of fire protection is "fire stopping" which is the sealing or closing of an imperfection of fit between elements, components or constructions of a building, or any joint, so as to restrict penetration of smoke and flame through the imperfection or joint. This would also apply within individual units, and ventilation "penetrations" such as extract fans should be considered insofar as this would permit fire to reach the exterior including cladding. To confirm that all compartmentation is adequate would entail inspecting every unit or apartment.

There should be some system of communicating the occurrence of a fire within the common parts such as a detection and alarm system (including fire alarm break glass buttons) along with clear instructions of what (and what not) to do in the case of fire. Such instructions should be appropriate for the residents of the particular building and will depend on the building fire strategy which is a matter of management of the building. It may be that on any inspection the surveyor should speak with residents to assess the awareness of actions to take in case of fire and any escape strategy should take account of those occupiers who are unable to escape unaided, including those who are unable to hear alarms or who are visually impaired and where there is a need for a personal evacuation plan.

Any surveyor should also consider the current fire risk assessment and its adequacy. There are four types of fire risk assessment for purpose-built blocks of flats/apartments:

Type 1: Common parts only (non-destructive)
Type 2: Common parts only (destructive)
Type 3: Common parts and apartments (non-destructive)
Type 4: Common parts and apartments (destructive)[12]

Type 4 is the most complete and reliable although Type 1 seems to be most common.

The British Standards has published a publicly available specification on fire risk assessments PAS79:12 *Fire risk assessment – Guidance and a recommended methodology* and provides guidance although it is not a British Standard. As the TowerBlocks UK document makes clear in contrast with the approach to compliance with Building Regulations, every fire risk assessment must give thorough attention to fire safety management and, therefore, to matters such as the fire safety strategy for the premises, fire procedures staff training, fire drills, testing and maintenance of fire protection equipment, inspection of means of escape, etc. Good fire safety management also contributes to the prevention of fire by incorporating policies and measures that reduce the likelihood of fire.

Services

Adequate services and utilities should be provided to the building of a type and nature suitable for the intended purpose of the building. The general principles are:

- A constant and wholesome supply of water should be properly conveyed to the building and, in most circumstances, this will be from a mains water supply.
- An electrical supply should be provided via an electrical installation that is safe, adequate and suitable. In England and Wales, the Building Regulations in Approved Document: Part P require that domestic electrical installations are designed and installed safely according to the "fundamental principles" given in British Standard BS7671 Chapter 13. Now these are similar to the fundamentals defined in the International Standard IEC 60364–1 and equivalent national standards in other countries. The legal requirements for electrical work are BS 7671 2018 Requirements for Electrical Installation, IET Wiring Regulations 18th Edition, known commonly as the "electrical (IEE) regs". See the Institution of Engineering and Technology webpage for more information.[13]

- In January 2020, the Government laid in Parliament, the Electrical Safety Standards in the Private Rented Sector (England) Regulations 2020. The new Regulations require inspection and testing, including any necessary remedial work to be carried out on electrical installations within the private rented sector, at intervals not exceeding five years. These Regulations apply to private landlords for all new specified tenancies from 1st July 2020 and all existing tenancies from 1st April 2021. An "existing specified tenancy" means a specified tenancy which was granted before the coming into force of these Regulations; a "new specified tenancy" means a specified tenancy which is granted on or after these Regulations come into force.

- A gas supply, where required, must also be provided in a safe and suitable manner with gas outlets sited in appropriate locations for cooking and heating appliances. Care must be taken in any such locations to provide adequate natural ventilation to provide fresh air to ensure the safe combustion of gas and the removal of waste products of combustion.

- The provision for ventilation should allow adequate, but not excessive, air changes for biological and hygienic reasons and is particularly important in view of the Covid-19 pandemic. Wherever possible, natural ventilation is preferable particularly in domestic situations. Where there are areas of high moisture generation, such as kitchens and bathrooms, additional ventilation will be required.

- Natural lighting is generally preferable for aesthetic, safety and hygienic purposes. Sunlight contains ultra-violet radiation that has sterilising properties which can provide a general benefit but more particularly in kitchen, bathroom and WC areas.
- Artificial lighting is necessary throughout buildings, particularly for winter months, to ensure that occupants can have safe passage around the building and carry out their tasks whether they are domestic, commercial or industrial.
- Certain areas of buildings require specific attention in terms of health, safety and hygiene. Kitchens must be designed so they provide a suitable environment for the safe preparation of food.

In particular:

- There should be sinks of adequate size (the number depends on the circumstances) with impervious surfaces that are suitable for the preparation of food, cleansing of utensils and washing of hands;
- There should be work surfaces suitable for food preparation which are smooth and impervious, located adjacent to sinks and cooking installations;
- The floor should have a non-slip surface that can be readily cleansed;
- Additional artificial lighting may be required including at work surfaces/stations;
- In addition to natural ventilation, mechanical ventilation and/or an extraction system may be required over the cooker(s).

Bathrooms must also be carefully designed to provide all the amenities required with a layout that facilitates their safe use. In particular:

- The bath or shower should have a non-slip surface with a suitably located grab-rail;
- The floor should have a non-slip surface that can be readily dried and cleansed;
- In addition to natural ventilation, a mechanical ventilation and/or extraction system may be required;
- Specific care must be taken with the electrics. The light switch and, if a shower is provided, the shower isolation switch, should both be fitted with a pull cord.

Waste management in high-rise and multi-occupied buildings

There should be a comprehensive waste management system for the transfer of waste from the home or the office to a central deposit point which also meets the principles for high-performance recycling. That is, recycling of materials should be "designed in". Refuse chutes should be of adequate design and preferably allow for separation of waste in recyclables and non-recyclables and be adequately sealed from common parts of the building both to prevent smells and to resist the spread of smoke in case of fire. It is usual for dwellings above the fourth storey to share a single waste container for non-recyclable waste fed by chute, with separate storage for any waste which can be recycled. Alternatively, storage compounds or rooms could be provided for recyclables at each level of the building. In such cases satisfactory management arrangements should be in place to ensure the material is moved to the storage area.

In some cases, non-recyclable waste can be fed by chute into a compacter. Chapter 18 deals in more detail with waste management. Communal storage areas should have provision for washing down and draining the floor into a system suitable for receiving a polluted effluent. Gullies should incorporate a trap which maintains a seal even during prolonged periods of disuse (see later on drainage). Any room for the open storage of waste should be secure to prevent access by vermin. Any compound for the storage of waste should be secure to prevent access by vermin unless the waste is to be stored in secure containers with close-fitting lids. Where storage rooms are provided, separate rooms should be provided for the storage of waste which cannot be recycled, and waste which can be recycled.

Good waste management should also take account of fire risks and security of refuse storage, to reduce the risk of accidental or even deliberate fires in the refuse storage bins. Provision should also be made for the secure storage and collection of bulky items including electrical equipment that cannot be included in the household waste and recyclable streams.

For multi-occupied buildings including houses in multiple occupation, boarding houses etc., there should be adequate secure waste storage for the number of occupants, and the number and type of bins used, including for recyclables, will depend on the frequency of collection. This should be such as to prevent "overfilling" with lids not closing fully.

Air conditioning and heat pumps

These are increasingly being installed in the domestic setting, but air conditioning is found in offices, shops and other commercial buildings. Air conditioning is the process of cooling indoor air to provide thermal comfort. This is particularly used in warm climates, and increasingly used in the UK in offices and domestic premises in periods of hot weather. It has been used commonly where there is need to maintain constant cool temperatures such as in hospitals or other business premises, to counter the elevated temperatures where computing facilities contribute to the elevated indoor temperature. Air conditioning is also found on many modes of public transport. In its broadest sense, the term can refer to any form of cooling, heating, ventilation, humidification (or dehumidification) or disinfection that modifies the indoor air. There are several stages or elements to a simple air-conditioning system.

The indoor unit is where a fan blows the warmer indoor air over a heat exchanging coil through which cold refrigerant flows. The cold refrigerant absorbs the heat from the air and cooled air is blown into the room.

The refrigerant circulates through the units and copper piping takes the heat from the indoor unit to the outdoor unit.

In the outdoor unit as a result of compression, the refrigerant gas is heated and its boiling point increases. In the outdoor unit the heat obtained through compression is released to the outdoor air by means of a fan which blows the outdoor air over a heat exchanging coil. The liquid refrigerant flows back to the indoor unit.

In essence an air conditioner works similarly to a refrigerator. The refrigerant flows through the system and undergoes changes in state. There are four processes in the "refrigeration cycle". The compressor which pumps the refrigerant around the system is the heart of the air conditioner. Before the compressor, the refrigerant is a gas at low pressure. In the compressor, the gas is put under high pressure and heats up and flows towards the condenser. At the condenser, the high temperature, high-pressure gas releases its heat to the outdoor air and becomes sub-cooled high-pressure liquid. The high-pressure liquid goes through the expansion valve, which reduces the pressure, and thus the temperature goes below the temperature of the refrigerated or cooled space. This results in cold, low pressure refrigerant liquid. The low-pressure refrigerant flows to the evaporator where it absorbs heat from the indoor air through evaporation and becomes a low-pressure gas. The gas flows back to the compressor where the cycle starts all over again.

In case of a heat pump the cycle can be reversed. Again, the heat pump consists of the compressor, expansion valve, and two heat exchangers, one to absorb heat from the heat source (e.g. the sun or ground) and one to reject the heat.

The first stage is where the heat transfer medium (refrigerant) is colder than the heat source such as the sun/outside air. As the outside air passes across the first heat exchanger (evaporator) the liquid refrigerant absorbs the heat and evaporates. This vapour then passes to the compressor and is compressed. When compressed the pressure is increased and the temperature of the vapour rises, effectively concentrating the heat. The hot vapour then passes to the second heat exchanger (the condenser) where the heat is rejected and the vapour condenses back into a liquid. That heat can be passed into the hot water of the central heating and hot water system of the building ready for use. The final stage is where the liquid refrigerant passes through the expansion valve, reducing its pressure and temperature before passing to the evaporator for the whole cycle to start again.

A poorly maintained air-conditioning system can occasionally promote the growth and spread of microorganisms, such as *Legionella pneumophila*, which causes Legionnaire's disease, or thermophilic actinomycetes. If the air conditioner is kept clean these health hazards should be avoided (see also Chapters 12, 13 and 15). Conversely, air conditioning, including filtration, humidification, cooling, disinfection, etc., can be used to provide a clean,

safe, atmosphere in hospital operating rooms and other environments where an appropriate atmosphere is important for occupier safety.

Cooling towers have been used as part of air conditioning and are heat removal devices used to transfer process waste heat to the atmosphere. They may either utilise the evaporation of water to remove process heat and cool the working fluid to near the wet-bulb temperature or rely on air to cool the working fluid to near the dry-bulb temperature. Wet-bulb temperature is the lowest temperature that can be reached by the evaporation of water only. It is the temperature felt when your skin is wet and is exposed to moving air. Wet-bulb temperature is an indication of the amount of moisture in the air. Dry-bulb temperature is the temperature of air measured by a thermometer freely exposed to the air but shielded from radiation and moisture. It is the temperature usually considered as air temperature and is the true thermodynamic temperature.

There are two main designs, "crossflow" and "counterflow". In the former the dry air enters through the vertical faces of the tower and passes into the fill material (foam and metal) and the warm water passes down through the fill material by gravity cooled by the air and drops into the sump or collecting basin and then out of the tower for circulation. The air is drawn through across the warm water flow and leaves the unit as moist

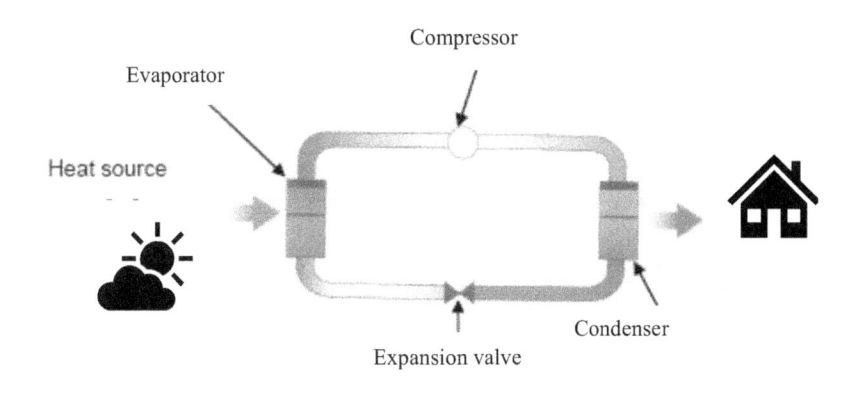

Figure 9.12 Heat pump system

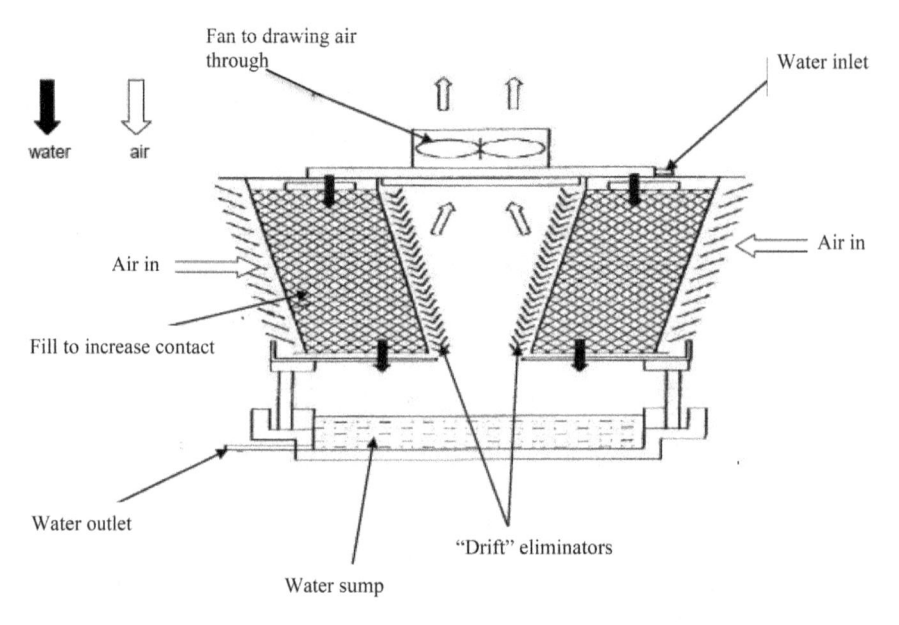

Figure 9.13 Example of an induced draught cooling tower using water that can be found on an office building or similar

warm air (see Figure 9.13). In the counterflow design, the dry air passes up through the fill from the bottom of the unit while the warm water drops through the fill in the opposite direction to be cooled.

One of the issues with cooling towers is "drift" which is when water droplets are carried out of the cooling tower with the exhaust (warmer) air. Drift droplets will have the same contaminants as the water entering the tower plus others from within the tower.

Drainage

A suitable drainage system must be provided and is made up of a series of pipes that remove the waste and other water from a building. In modern plumbing, a drain-waste-vent (or DWV) is a system that removes sewage and wastewater from a building and vents the gases produced by said waste. Waste is produced at fixtures such as toilets, sinks and showers, and exits the fixtures through a trap, a dipped section of pipe that always contains water. All fixtures must contain traps to prevent gases from backing up into the house.

Through traps, all fixtures are connected to waste lines, which in turn take the waste to a soil stack, which extends from the building drain at its lowest point to above and out of the roof. Through the building drain the waste is removed from the building and taken to a sewerage line which leads to a septic tank or, in most circumstances, a public sewer.

Below-ground drainage and systems are often where problems occur either due to age or to poor installation. The pipeline below ground draining a building can be constructed of different materials. The most common for near house drainage these days is plastic of various forms, but older properties will have clay pipes. There are two aspects to below-ground drainage relevant to EHPs, the form of construction and terminology, and the legal status of that pipeline.

A water carriage system of drains and sewers has been fundamental to the improvements in public health (see Chapter 1). However even these days there are many homes, usually in rural areas of the UK (but also in other countries) that cannot be connected to the public sewerage system – the

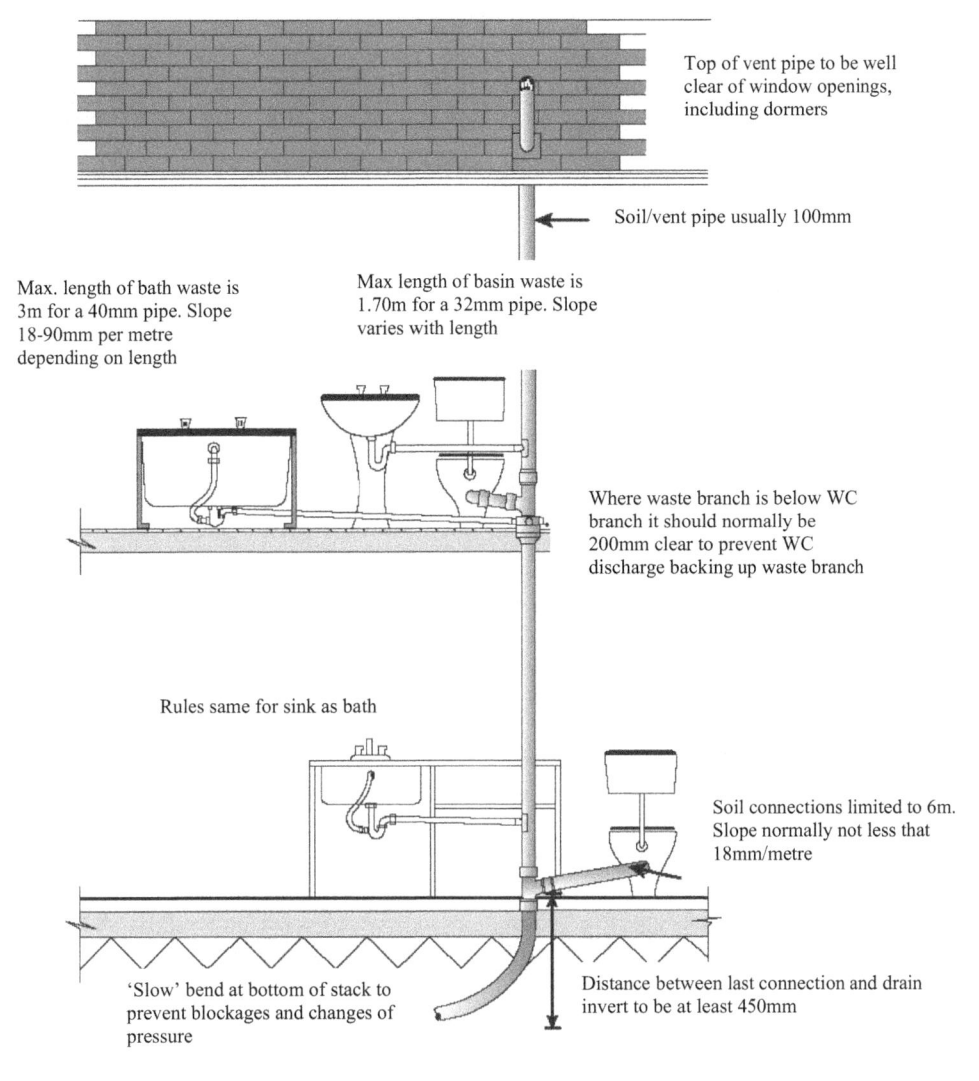

Top of vent pipe to be well
clear of window openings,
including dormers

Soil/vent pipe usually 100mm

Max. length of bath waste is
3m for a 40mm pipe. Slope
18-90mm per metre
depending on length

Max length of basin waste is
1.70m for a 32mm pipe. Slope
varies with length

Where waste branch is below WC
branch it should normally be
200mm clear to prevent WC
discharge backing up waste branch

Rules same for sink as bath

Soil connections limited to 6m.
Slope normally not less that
18mm/metre

'Slow' bend at bottom of stack to
prevent blockages and changes of
pressure

Distance between last connection and drain
invert to be at least 450mm

This diagram shows some of the design parameters for single stack systems

Figure 9.14   Single stack drainage system
Source: Marshall et al. (2013) ([1] at page 401), reproduced by permission of Taylor & Francis

buildings are not connected to "mains drainage". They have to utilise other systems such as septic tanks or cesspools (see Figure 9.15). More on the principles of sewage treatment can be found in Chapter 13.

A "cesspool" is no more than a large holding tank into which all the sewage from a building drains. It holds both liquid and

solid waste and needs to be emptied on a regular basis by a tanker lorry. Householders should maintain records of when and by whom emptying is undertaken. This is necessary in light of waste management requirements and will enable the owner to demonstrate to EHPs (and the appropriate environment agency) that the cesspool is

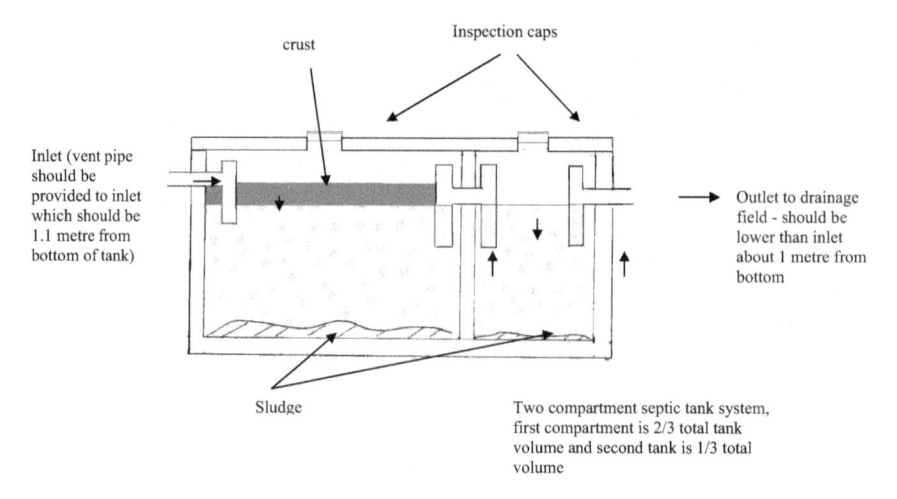

Figure 9.15   Septic tank (not to scale) and how it works

correctly managed whenever investigations are undertaken.

A septic tank is again a holding tank into which the sewage from a dwelling drains but the liquid part of the waste is separated and then leaves the tank via a system of land drains while the solids are held within the tank. As the tank holds only the solid waste it can be smaller than a cesspool, and while it will still need to be emptied, this will be required less frequently. Some septic tanks are fitted with grease traps before the inlet, and may have an additional filtration system, for example via volcanic rocks furnace clinker or other filtration medium before the discharge to the system of land drains or "drainage field" – a system of field drainage.

An alternative to the traditional septic tank is a system known as the "Bio-disc". Central to the operation of such a system is a rotating biological contactor (RBC), which supports the establishment of a biologically active film of aerobic micro-organisms naturally found in sewage. The RBC comprises banks of vacuum-formed polypropylene media supported by a steel shaft. A low energy consumption electric motor and drive assembly slowly rotates the contactor. This facilitates the natural breakdown of the sewage. The media is partially submerged and, as it rotates, the biomass is alternately immersed in the liquor for adsorption and digestion of waste matter and exposed to the atmosphere for oxygenation. The wastewater and sewage flows into a primary settlement zone where solids are settled out as in any traditional septic tank and these are retained and drawn off periodically. Partially clarified liquor containing fine suspended solids then flows upwards into the first stage "biozone" for bacterial activity and breakdown to occur on the RBC. Suspended solids then return to the primary settlement zone and the liquid is transferred to the second stage of bacteriological activity ("biozone") for further treatment. Any solids remaining are then settled out in the final settlement tank. There is a higher quality of effluent to go to the drainage field from this system by comparison with traditional septic tanks, and in some cases the effluent can be discharged to a watercourse, although there should be consultation with the Environment Agency in England, Natural Resources Wales, SEPA in Scotland or the Northern Ireland Environment Agency. The Republic of Ireland advice by the manufacturer is that consultation should be with the Local Authority Public Health Department.

A drainage field serving a wastewater treatment plant or septic tank should be located at least 10 m from any watercourse or permeable drain and be at least 50 m from the point of abstraction of any groundwater supply and not in any groundwater protection zone. It should also be at least 15 m from any building and sufficiently far from any other drainage fields or soakaways so that the overall soakage capacity of the ground is not exceeded. Drainage fields should be designed and constructed to ensure air contact between liquid waste and the subsoil. Drainage fields should be constructed using perforated pipe laid in trenches of a uniform gradient which should not be steeper than 1/200. In order to set the pipes securely, granular fill material should be carefully spread around the infiltration pipes along the entire length of the trench so that it reaches the top of the pipe. Pipes should be laid on a 300 mm layer of clean shingle or broken stone graded between 20 mm and 50 mm with the apertures installed facing downwards. Trenches should be filled to a level 50 mm above the pipe and covered with a layer of geotextile to prevent the entry of silt. The remainder of the trench can be filled with soil; distribution pipes should be laid at a minimum depth of 500 mm below the surface. Drainage trenches should be from 300 mm to 900 mm wide, with areas of undisturbed ground 2 m wide being maintained between in line trenches. An inspection chamber should be installed between the septic tank and the drainage field. Drainage fields should be set out as a continuous loop fed from the inspection chamber.

For more details see BS 6297:2007+A1: 2008 *Code of practice for the design and installation of drainage fields for use in wastewater treatment*. Also, there is BS EN 12566_2016 which is a several part standard for small wastewater treatment systems for up to 50 PT (population total).

Reform of the regulatory system to control small sewage discharges from septic tanks and small sewage treatment plants in England (with equivalents for the other countries in the UK) was brought in so that there are general binding rules for small sewage discharges (SSDs)[14] with effect from January 2015 (made under Environmental Permitting (England and Wales) (Amendment) (England) Regulations 2014).

All septic tanks that discharge into surface water or water courses will have to be either replaced by a sewage treatment plant with full BS EN 12566–3 Certification, or the discharge into the water course will need to be stopped and diverted.

Grease traps are installed to prevent fats oils and greases (FOG) entering the sewerage system. There is good reason for this; there are about 200,000 blockages of public sewers every year and around 75% are caused by FOG. Even where there is not a complete blockage FOG reduces the hydraulic capacity of the sewer and increases the risk of sewer flooding.[15]

Figure 9.16 shows a simple form of grease trap that can prevent fats oils and greases

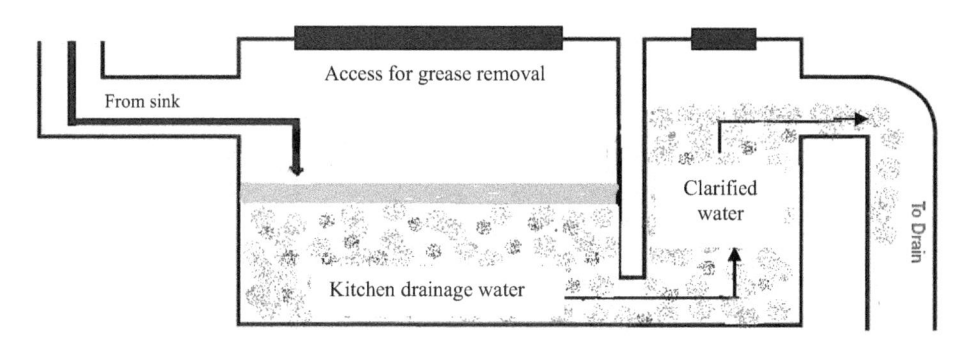

Figure 9.16   An "under the sink" grease trap
Source: Courtesy of Water UK

entering the sewerage system. It allows the fat oil and grease to be separated out from the rest of the wastewater. The grease is retained in the trap and can be collected by a registered waste carrier at regular intervals. Correctly installed, serviced and maintained they can be very effective. A written record of maintenance should be kept. A system such as this is probably preferable to a food macerator, which encourages bad practice, and still allows fat to enter the sewerage system and cause blockages.

There are no also enzyme dosing systems which can be used to break down fats, oils and grease. Some systems are used in conjunction with a grease trap.

As in all aspects of environmental health, drainage has its own jargon. For a start although we are talking about the "sewerage system", sewers and drains are different legal terms. In essence legally a "drain" drains only one premise, whereas a sewer drains two or more premises. New build properties will also have "public laterals", that is the drain once it leaves the boundary will have an access point and from there it is the responsibility of the water and sewerage undertaker (WaSC). It should also be noted that any existing drain

once it has left the boundary of the premises served, becomes the responsibility of the WaSC (another "public lateral"). The transfer arrangements apply only to foul waste and not to surface water drainage.

The following few figures are intended to provide some basic information. More detailed information can be found in the texts listed in the Further Reading section.

Although a rare event, brown rats (*Rattus norvegicus*) are able to climb the vertical ventilation and soil pipes into dwellings with relative ease and the absence of traps within the system increases the risk. A rat can climb by splaying out all four legs and using its back to support itself against the pipe. The Rat Barrier provides a chamber too high for a rat to jump up and too broad for a rat to vertically ascend. The curved shape of the base section excludes the presence of a suitable platform to leap from, whilst the conical shape of the top adaptor section offers no method for the rat to grip or lodge itself. The height of this product may require the use of a deep inspection chamber.

Drainage saddles have been developed to reduce the chance of "bodged connections" to existing pipelines. The saddles have an

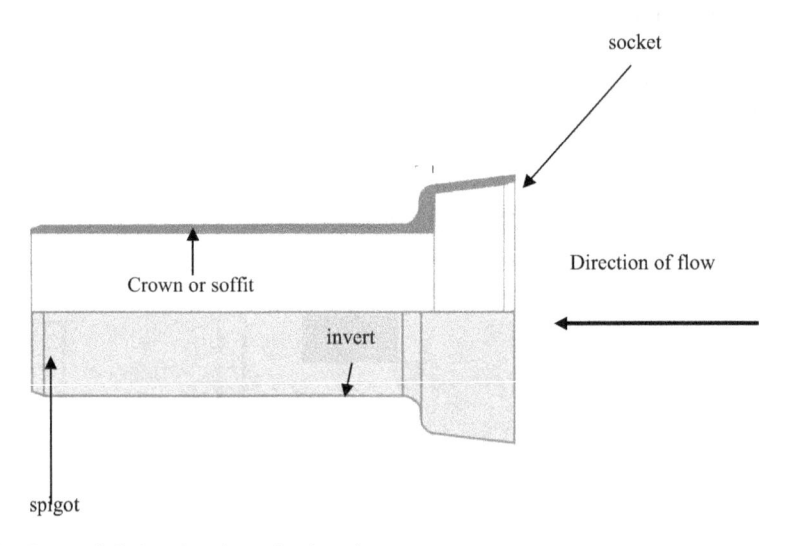

Figure 9.17 Parts of drain pipe (terminology)
Source: Figure courtesy of Wavin Ltd

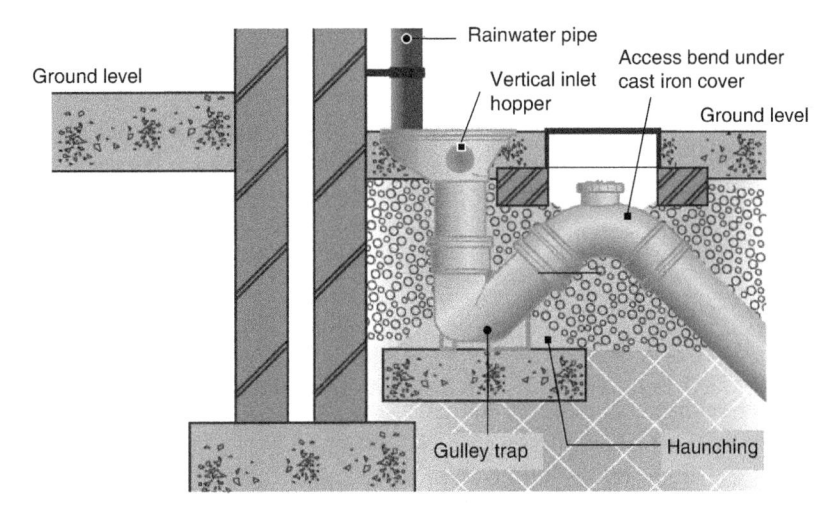

Figure 9.18 Example of a paved area gulley

Source: Figure courtesy of Wavin Ltd

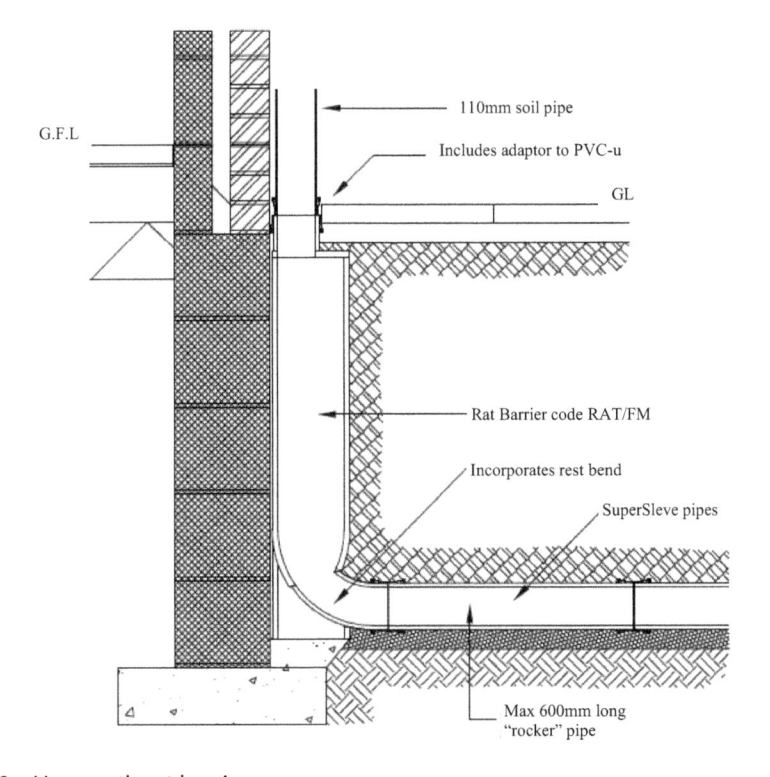

Figure 9.19 Hepworth rat barrier

Source: Figure courtesy of Wavin Ltd

Note: *A rocker pipe is a piece of pipe cut to less than one metre which is placed at the inlet/outlet of a solid structure, such as a manhole or building, to accommodate differential settlement between the structure and the drainage system.

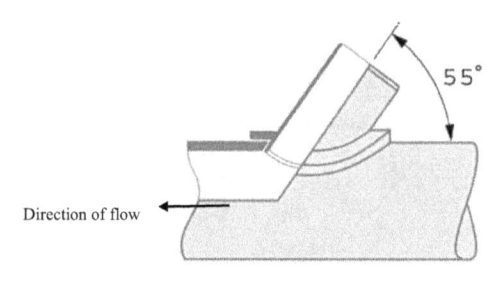

55°

Direction of flow

Figure 9.20 Drainage saddle
Source: Figure courtesy of Wavin Ltd

outer flange to prevent the connection from protruding into the outfall pipe and have been produced in angles from 45 to 90 degrees. The notion is that a template is used to carefully cut a correct-sized hole in the pipeline before fitting the saddle, traditional concrete/sand and cement is used to seal the pipe on the outside and a fast-setting mortar is used to create a smooth and water-tight seal on the inner wall. This will be adequate as long as the hole is cut using a small disc grinder or by stitch drilling around the template (on site this may not be the case). With plastic and more modern systems there are proprietary methods for making new connections. Also available are systems where the saddle is already fitted to a length of pipe of the appropriate diameter that can be inserted into the existing pipeline from which an appropriate length section has been cut.

Building design issues

Room sizes and space

Currently in the UK there are no compulsory national space standards for new dwelling construction, which contrasts with most European Countries and North America where they do have them. Of all EU countries England has the smallest homes by floor area. This is not an issue addressed by the EU but it is unlikely that leaving the EU will lead to increased room size standards. According to Which? (Consumers Association) UK homes built since 2010 offer an average of 67.8 m²

of living space. Home sizes grew steadily over the early part of last century, hitting a peak of 83.3 m² in the 1970s.[16] The company Find Me a Floor has assessed the overall average floor area in England's homes at 71.6 m²; this compares to Italy's average of 108.2 m² and Germany's average of 92.7 m². In France the average was assessed as 79.6 m². Canada has the largest homes in the world measuring overall 150 m². The USA has an average of 130.7 m².[17] A result of the UK housing shortage is that the average size of new homes has continued to shrink, and with that increased risks to health with inadequate space including storage space. The lack of space increases the risk of spread of infection and risks to mental health.

In 2015 the UK Government created a new approach for the setting of technical standards for new housing in England. This sought to rationalise the differing existing standards into a simpler, streamlined system. The new system comprises new additional optional Building Regulations on water and access, and a new national space standard (hereafter referred to as "the new national technical standards").

In October 2015, the nationally described space standard set out detailed guidance on the minimum size of new homes. According to this the minimum floor area for any new home should be 37 m².[18] This system was intended to complement the existing set of Building Regulations and Approved Documents which set out mandatory minimum standards. To implement this new regime, a ministerial statement set out the Government's national planning policy on the setting of technical standards for new dwellings. This statement should be taken into account in applying the National Planning Policy Framework, most recently that of 2019.[19] The rules for local authorities are only optional and must be introduced through the land-use planning system, not the Building Regulation system, so that any policy should be in the local plan.

The standard in the nationally described space standard deals with internal space within new dwellings and was intended for

Table 9.1 Minimum gross internal floor areas and storage (m²)

Number of bedrooms(b)	Number of bed spaces (persons)	1-storey dwellings	2-storey dwellings	3-storey dwellings	Built-in storage
1b	1p	39 (37) *			1.0
	2p	50	58		1.5
2b	3p	61	70		2.0
	4p	70	79		
3b	4p	74	84	90	2.5
	5p	86	93	99	
	6p	95	102	108	
	5p	90	97	103	3.0
4b	6p	99	106	112	
	7p	108	115	121	
	8p	117	124	130	
5b	6p	103	110	116	3.5
	7p	112	119	125	
	8p	121	128	134	
6b	7p	116	123	129	4.0
	8p	125	132		

Notes:

The minimum floor to ceiling height should be 2.3 m for at least 75% of the Gross Internal Area.

Any area with a headroom of less than 1.5 m is not counted within the Gross Internal Area unless used solely for storage.

*Where a 1b1p has a shower room instead of a bathroom, the floor area may be reduced from 39 sq m to 37 sq m, as shown.

application across all tenures. It only sets out requirements for the Gross Internal (floor) Area (GIA) of new dwellings at a defined level of occupancy as well as floor areas and dimensions for key parts of the home, notably bedrooms, storage and floor to ceiling height. The GIA includes built-in storage and includes an allowance of 0.5 m² for fixed services or equipment such as a hot water cylinder, boiler or heat exchanger.

The requirements of this standard for bedrooms, storage and internal areas are relevant only in determining compliance with this standard in new dwellings and have no other statutory meaning or use.

In England, for HMO licensing under the Housing Act 2004 the Government brought in minimum room sizes by way of a mandatory condition of such licences so that licence holder has to:

- Ensure that the floor area of any room in the HMO used as sleeping accommodation by one person aged over 10 years is not less than 6.51 m²;
- Ensure that the floor area of any room in the HMO used as sleeping accommodation by two persons aged over 10 years is not less than 10.22 m²;
- Ensure that the floor area of any room in the HMO used as sleeping accommodation by one person aged under 10 years is not less than 4.64 m²;
- Ensure that any room in the HMO with a floor area of less than 4.64 m² is not used as sleeping accommodation.

Any part of the floor area of a room where the height of the ceiling is less than 1.5 m is not to be taken into account in determining the floor area of that room.

For comparison, in Australia this type of accommodation is known as a "rooming house" and in the State of Victoria of which a rooming house is one type, an owner must not permit a room in the prescribed accommodation to be used as a bedroom if it has a floor area of less than 7.5 m^2. If persons are accommodated in prescribed accommodation for a period of more than 31 days, the maximum number of persons permitted to occupy a bedroom after the 31st day is, in the case of a bedroom with a floor area of less than 12 m^2, one person, and for a bedroom with a floor area of 12 m^2 or more, two persons with an additional person for every 4 m^2 of floor area that exceeds that area.[20]

There is no prescribed method of measuring rooms but whatever the standard against which the property is being assessed, the EHP should be able to measure rooms accurately and determine useable space. One method has been provided by the RICS.[21] This guide includes guidance on measuring commercial as well as domestic property.

Other approaches to space have included the Housing Quality Indicator System (2007), and the *Metric Handbook – Planning and Design Data* (sixth edition published in 2018).[22] In 2007, the Housing Corporation's Scheme Development Standards were replaced by "Design and Quality Standards" (D&QS) but these in turn were withdrawn. The *National Housing Federation Housing Standards Handbook* published in 2016 can be used as one piece of evidence when EHPs are considering the adequacy of space within a dwelling. A history of space standards has been written by the authors of various publication and can be a useful source of information[23] for environmental health professionals.

More relevant to the EHP is to measure room sizes in existing buildings to ascertain their suitability for the purpose intended.

To measure your rectangular room, use a tape measure (a traditional or digital type[24]), pencil and paper to record the length and width of the room including any part covered by fixed cupboards. Once these two measurements have been obtained and recorded, they should be multiplied together to obtain the total area of the room.

If the room being measured has a recess or protrusion, such as a closet or bay window, the shortest length and width of the room should be measured first. Then multiply them to get the main area of the room. After that number has been recorded, the area of the individual recesses should be calculated by measuring their length and width. Lastly, all of the areas should be added together to get the total area of the room. The same principle applies when measuring T-shaped and L-shaped.

Where there is a rounded (rather than square) bay in a room measure the depth at the deepest point from the edge of the opening and the width of the bay at the opening, multiply the width by the depth and divide the product by two. Then multiply this total by π (3.14). The area in a bay window recess should only be included as part of the area of the room if it has a floor (rather than a seat) and the ceiling is of adequate height. In some countries this is 2.13 m high; in England this seems to be 1.5 m, taking the HMO licensing standard as the indicator, but in the nationally described space standard the minimum floor to ceiling height is 2.3 m for at least 75% of the Gross Internal Area.

As a basic example of a simple approach see Figure 9.21 for a rectangular room with a recessed doorway.

Area of the room = (Width A × Length B) + (Width C × Length D)

When calculating the area of U-shaped rooms, use the same method of adding the individual areas together, except when measuring the area of a round recess. To calculate the area of a round recess, measure the longest width and length of the recess, normally through the centre of the space first. Next, divide the length in half and multiply that

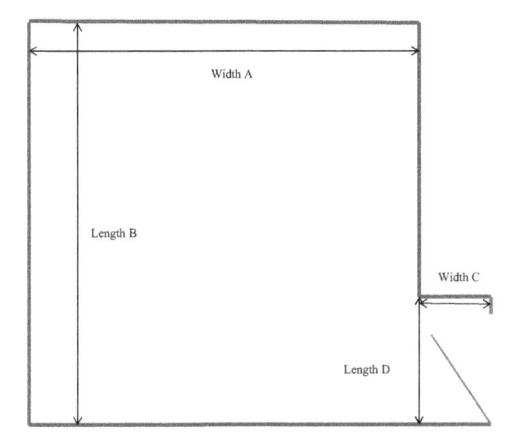

Figure 9.21 Simple approach to measuring room sizes

number by the width then multiply the total by "n" (3.14). This will give the area of the complete circle. However, only half of the circle extends beyond the wall, so the area should be divided in half. This gives the area of the U-shaped protrusion in the room. This number can be added to the area of the other measurements in the room for the total area.

Measuring ceiling height is not quite so clear; indeed for rooms with sloping ceilings and low headroom it can be quite complicated in determining what might be termed useable space. The former 1937 Regulations on measuring rooms were repealed in the 1980s but, in general, they seemed sensible. If the ceiling of an existing room is reasonably level and, as a minimum, is more than 2.1 m in height then that is acceptable although most new build dwellings have a minimum of 2.4m. Measurements for useable space include any area taken up by fixed cupboards in addition to any area such as bat windows. The former Regulations said any measurement should also be made to the back of any skirting but it would seem that this is not really compatible with modern expectations and for useable space then it should be taken from the face of any skirting.

As mentioned earlier, rooms with sloping ceilings and low headroom can be quite complicated when determining what might

be termed useable space. The former Regulations said that you exclude any part of the floor where the ceiling height is less than 5 ft (1.524 m). However, to reduce the likelihood of collision and entrapment it would now seem reasonable to exclude any part where the height is less than 1.7 m.

Although there are no national standards for room sizes the First-tier Tribunal (Property Chamber) (Residential Property)[25] made a determination, which provides some guidance when considering room sizes and their suitability. In consideration of their decision regarding an appeal against a Prohibition Order they came to the view that, in relation to an intervention for the hazard of Crowding and Space,[26] the minimum floor space is 6.5 m². They also added that this does not imply that accommodation including larger rooms cannot also be an actionable hazard if unsatisfactory due to factors such as the shape and ceiling heights.

At a local level the Greater London Authority has introduced minimum space requirements based on daily activities and the space needed for them.[27] Although they do not specify individual room sizes the London standards set a minimum gross internal floor area against the typology of the home (number of bedrooms and storeys) and the designed occupancy level (the number of people the home is designed to accommodate comfortably).

In 2011 the Royal Institute of British Architects produced a critical report, *The Case for Space: the size of England's new homes* [2], which compared the size of new homes with existing ones and also other Western European countries. Their research concluded that newly completed homes were considerably smaller than those existing and those newly built in Europe.

Adaptation and mitigation of climate change

Based on events and future forecasts, we can expect future changes to seasonal rainfall (wetter winters and drier summers), higher temperatures, rising sea levels and coastal erosion. It is also expected that the UK is likely to experience increased extreme events, such as high winds, heavy prolonged rainfall, flooding, drought and heat waves. All of these have direct impacts – but also secondary ones, such as the stability and moisture of soils. It has become clear that it is important to build the potential for adaptation into design and construction methods whether this is new development, refurbishment or regeneration.

Not only is it important to design and adapt buildings to cope with climate change but positive steps should be taken to minimise their contribution to climate change. Careful consideration should be taken in the choice of building materials, energy efficiency measures and sources of energy.

Part L of the Building Regulations deals with the conservation of fuel and power and there are Approved Documents for both new and existing dwellings and for new and existing buildings that are not dwellings. There is an approved target CO_2 emission rate (TER) as the minimum energy performance requirement. It is expressed in terms of mass CO_2 in kg per m^2 of floor area per year as a result of the provision of heating, hot water, ventilation and internal fixed lighting. See Approved Document L1A for more information.

The Code for Sustainable Homes (CSH) is the national standard for the sustainable design and construction of new homes. The CSH is an environmental assessment method for rating and certifying the performance of new homes based on BRE's Ecohomes scheme. It is a Government-owned national standard intended to encourage continuous improvement in sustainable home building.

By way of a written ministerial statement in 2015 the Code was withdrawn (in England) so local authorities should no longer require it as a planning condition for new approvals and the Building Regulations incorporated some of the provisions (although the original technical guidance was still available at the time of writing).[28] Where there were existing

contractual arrangements, for example with Registered Social Landlords under the Affordable Funding Programme 2015–2018, it will be possible to continue to register and certify against the Code.

The new dual level Building Regulations came about because of the Deregulation Act 2015 which also brought in an amendment to the Planning and Energy Act 2008 to prevent local authorities from requiring higher levels of energy efficiency than Building Regulations.

The Building Research Establishment (BRE) has introduced a Home Quality Mark as a standard for new homes, using a 5-star rating to provide impartial information from independent experts on a new home's design and construction quality and running costs. The Mark has been developed based on years of building standards experience. It is part of the BREEAM[29] family of quality and sustainability standards. It will also show the impact of the home on the occupant's health and well-being; as buildings become more airtight, respiratory conditions rise and our population gets older. It will demonstrate the home's environmental footprint and its resilience to flooding and overheating in a changing climate. In addition, the Mark will evaluate the digital connectivity and performance of the home as the speed, reliability and connectivity of new technology becomes ever more critical.

The Home Quality Mark enables housing developers to demonstrate the quality of their new homes and identify them as having the added benefits of being likely to need less maintenance, cheaper to run, better located, and more able to cope with the demands of a changing climate.

There is also the issue of excess heat and there is an increasing chance of heatwaves as climate change will lead to extreme weather events. Some matters that improve energy efficiency such as thermal insulation also prevent solar gain and the interior of the building heating. This is a particular issue in attic flats. Matters such as the orientation of the building and the amount of glazing will influence heat gain. There should also be adequate and controllable ventilation (that does not compromise security). Although not standard in the UK, it is surprising that more use is not made of external shutters (as commonly found on mainland Europe) as these can both reduce heat gain and improve security.

"Cool" roofs (usually light in colour) are roofing systems that can deliver high solar reflectance (albedo) and have the ability to reflect the visible, infrared and ultra-violet wavelengths of sunlight and reduce heat transfer to the building. They also have high thermal emittance (the ability to radiate absorbed or non-reflected solar energy). "Green roofs", sometimes known as "living roofs", are roofs that are partially or wholly covered with vegetation and a growing medium, planted over a waterproofing membrane. It may also include additional layers such as a root barrier and drainage and irrigation systems. They serve several purposes for a building, such as absorbing rainwater, providing insulation, creating a habitat for wildlife, and helping to lower urban air temperatures. Both forms of roof counter the heat island effect. An urban heat island is an urban area that has significantly higher temperatures than surrounding rural or less densely populated areas.

Dilapidations

With the passage of time buildings are likely to deteriorate mainly due to natural wear and tear, the effects of weather and external factors, with extremes of weather as the result of climate change causing increasing rates of deterioration. Regular repair and maintenance will limit the extent of the dilapidation but even so periodical refurbishment of elements and replacement of fixtures and fittings will be required as they reach the end of their natural life. It is important that defects in buildings are identified and remedied as soon as possible after they become apparent. Some defects will be readily observed to have come from an obvious source whilst others will require a more extensive investigation. Some of the more common defects found in

buildings are discussed here. EHOs are likely to be faced most often with dilapidations and defects in housing and these have been dealt with a number of texts; see for example Marshall et al. [3].

Disrepair

Although many problems can arise from design or materials used in construction, or some other inadequacy, such as an undersized heating system, lack of maintenance can result in disrepair. In housing law disrepair has specific meaning, and there has been substantial Case Law on this. Put simply disrepair means that some element of construction does not fulfil its intended function (does not work) and there is damage to that element. So, the mere absence of a damp proof course (dpc) itself would not be disrepair if one had never been present. If it led to perished wall plaster it would be the wall plaster that was in disrepair. However, if there had been a dpc in the wall that had failed then the dpc would be in disrepair as would the wall plaster. That said it should not be taken that repeated patch repairing of the wall plaster would be an adequate response where the absence of the dpc leads to perished plaster (on this see *Elmcroft Developments Ltd v Tankersley-Sawyer* (1984) 15 HLR 63, CA). This, perhaps, highlights the importance of an adequate and detailed knowledge of both construction and the application of that knowledge when inspecting a property.

In the case of *McDougall v. Easington D.C.* (*The Times*, February 2 1989, 21 H.L.R. 310, C.A) under the Landlord and Tenant Act 1985 (which sets out the repairing obligation on landlords) it was said that there were tests as to whether work to deal with poor conditions was work to put the property into repair. It was said that there were three different tests that could be applied separately or concurrently as to the question whether works constituted repairs:

1   Whether alterations were to the whole or substantially the whole of the structure

or only to a subsidiary part – what was the nature of the works

2 Whether the effect of the alterations was to produce a building of a wholly different character than that which had been let

3 What was the cost of the works in relation to the value of the property before the works, and what was the effect of the works on the value and future life of the property – was the substantial alteration in these matters?

From an environmental health perspective the issue is whether the disrepairs give rise to a hazard to the health of the occupiers or others who may be affected.

Dampness

Dampness is probably the most common defect found in buildings and identifying the cause of this is not always straightforward as there are a number of potential sources. Dampness not only affects the occupants but also leads to the deterioration of the fabric and in particular can lead to the rotting of timber and wall plaster becoming perished.

Penetrating damp – This is most commonly a direct result of disrepair allowing rain to enter through the structure principally through the roof, walls, windows and doors. A poorly maintained roof can result in various defects that can contribute to internal dampness including slipped, cracked and displaced tiles/slates or defective flashings to chimneys, windows, soil and vent pipes or walls of adjoining buildings. Defective brickwork and/or mortar to chimneys may allow water to access into the main stack and penetration into the main fabric of the building. Blocked or broken eaves gutters can result in high concentrations of rainwater being discharged on to the masonry. Flat roofs can pose considerable problems particularly where water pools rather than drains away. All too often the materials used for covering flat roofs are easily damaged and extreme temperatures cause expansion and contraction resulting in damage to the integrity of the surface. Flat roofs require more regular maintenance and

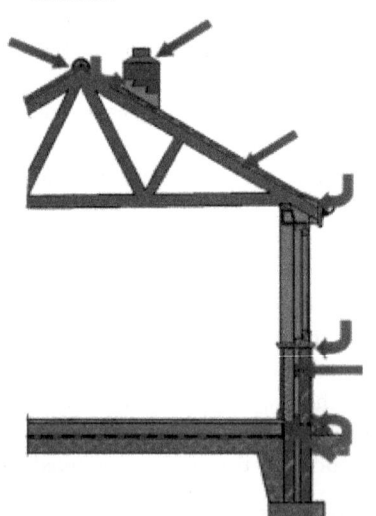

Routes include via cracked and missing flaunching and open pointing, displaced and missing flashings to chimney stack, broken and missing ridge and roof tiles/slates

Routes include from blocked and leaking eaves gutters

Routes around window and door openings where perished or open pointing or missing fillits, lack of damp-proofing where cavity closed or otherwise as the result of 'bridging' of cavity

Bridging or failure of damp proof course

Figure 9.22   Detail showing some of the causes of damp penetration

attention but, by their very nature, flat roofs are generally less visible and therefore any deterioration may go unobserved.

Rising dampness — This is where moisture rises up through the walls by capillary action from the sub-soil and is usually associated with the absence or failure of a damp proof course/membrane. Most materials used for the construction of walls will draw water from the ground and that is why it is necessary to provide an impervious barrier (the damp proof course) to prevent rising dampness affecting the structure and, especially, living space. Broken or damaged damp proof courses, particularly in older buildings where it may have been formed from brittle materials such as slates or dense engineering bricks, will allow moisture to pass through. Dampness can also by-pass a damp proof course by bridging caused by soil or render or debris in the cavity — where present. Rising dampness can also occur where there is a solid floor and the damp proof membrane fails to connect the damp proof course in the walls.

Traumatic dampness — A leaking water supply or drainage pipes can become a source of dampness. Quite often this is due to the failure of the seal to joints between sections of pipe. Another common cause is where old water supply pipes, particularly those made from lead or copper, have failed due to years of corrosive action by some water sources.

Condensation — This is just another form of dampness. It is wholly wrong to say that a problem is "not dampness it is condensation". It is equally wrong to say it is only a matter of "lifestyle". The only difference between most of the other common forms of dampness and this, is that most of the water (H_2O) is generated from within the dwelling rather than coming from the external environment.

The amount of water vapour that can be held by a given volume of air is expressed as a percentage of the total and depends on the temperature of the air — the warmer the air the more water vapour can be held. The percentage of the maximum (saturation) is called relative humidity (RH or sometimes $\varnothing$). It is important to understand this.

If a volume of air of 180 m^3 at 15°C can hold as much water vapour as it can, some 1.6 litres, then RH is 100%. If the temperature is raised to 20°C then the RH will drop to 75% although there are still the 1.6 litres of water vapour present; potentially though, it could carry 2.2 litres. If the air is cooled to 10°C then it would only be capable of carrying 1.1 litres and 0.5 litres must be given up in the form of moisture. The air has passed below its "dew point", that is, the temperature at which moisture starts to form and condenses out. The moisture thus formed is condensate.

In housing the condensation that causes most concern to occupiers is when warm moisture-laden air approaches a cold surface, the wall or ceiling, and cools down so that it passes the dew point. Condensation can also occur in parts of the building not normally visible in the structure, this is "interstitial condensation" and while not visible to the occupiers can lead to deterioration of the fabric. Condensation can also occur in the roof space if there is inadequate ventilation over the insulation; this can lead to an assumption of a leaking roof if it shows as a stain on the ceiling. It certainly will reduce the effectiveness of the insulation in the roof space.

Understanding the physics is important in starting to understand how to investigate a problem. In essence the problems arise from the high RH, and this is the fundamental issue.

Temporary condensation on windows is not a problem, but mould can form on wall plaster as mould spores can germinate in the moisture on the walls. Mould spores are allergens. Surfaces that are damp provide optimum conditions for mould growth and atmospheres where RH is at least 70% together with temperatures between 15°C and 20°C are conducive to the growth of most moulds. More important allergens are house dust mites, which are barely visible to the naked eye. The faecal pellets (detritus) of these mites, which can be found in most homes, are more potent allergens than the mould spores, but it is the occurrence of mould which is more obvious and causes stress and upset too. House dust mites do not breed below an *RH* of 60% but when this is exceeded they start to breed and there are increasing amounts of allergen in the environment.

Exposure to high concentration of these allergens over a prolonged period will cause sensitisation of atopic individuals (those with a predetermined genetic tendency to sensitisation). Once a person is sensitised, relatively low concentrations of the airborne allergen can trigger an allergic response. Symptoms are those typical of allergic responses and can include rhinitis, conjunctivitis, eczema, cough and wheeze (asthma and asthma-like symptoms). It is important to remember that merely removing the mould does not resolve the problem nor reduce the health risks (see Chapter 12).

The World Health Organization has also said that while groups such as atopic and allergic people are particularly susceptible to biological and chemical agents in damp indoor environments, adverse health effects have also been found in non-atopic populations.[30]

The occurrence of condensation is dependent on four inter-related factors:

1 Moisture generated within the dwelling
2 The space heating system provided and the use made of it
3 The ventilation system provided (i.e. the means of getting rid of moisture laden air), and the use made of it
4 The thermal capacity, response and insulation of the dwelling

The designer and/or landlord has influence over factors 2 to 4. The occupier has influence over factors 1 to 3.

Moisture generation in a normal three-bedroom house a family can produce around 10 litres of water per day, and this is without taking into account any washing and drying of clothes. Obviously, the more occupiers there are the more water vapour will be generated because breathing generates moisture

and even when we are asleep we are exhaling water vapour.

Also, if the occupiers are out during the day less water will be generated than a household where some if not all members are present throughout the day. As can be imagined a household with very young children and babies present will not only have someone at home all day, but there is also likely to be more washing of clothes, etc. required. It should not be forgotten either that the presence of other forms of damp will also increase the amount of moisture in the dwelling.

Clothes drying, which produces substantial amounts of water vapour, can be directly affected by adequate and effective ventilation, and tumble driers and clothes drying cupboards should be vented directly to the external air. However, any investigation should take account of the clothes drying facilities and acknowledge that in the winter it may be impossible to dry clothes outside the dwelling, and this is also true of flats in high-rise buildings.

Cooking and washing generate much of the water vapour in the home, so primarily in the kitchen and bathroom areas, warm moisture-laden air will migrate from those areas even around closed internal doors to colder parts of the dwelling. There as the air temperature cools, dew point can be reached and condensation forms. Two factors affect this migration: lack of effective ventilation ensuring that moisture-laden air is safely extracted as close to the point of arising as possible and vapour pressure forces the air of high *RH* to areas where the air is of low *RH*.

Water vapour is a gas, and most building materials are gas permeable unless there is a vapour barrier (e.g. the foil that can be seen on the back of some plaster board).

Relative humidity can be measured quite simply as there are electronic thermos-hygrometers that measure air temperature and *RH*. The dew point can then be read by reference to the psychometric chart or by reference to a simple table based on the chart [3].

Space heating provided in a dwelling is the responsibility of the designer or owner and should take account of the thermal capacity and response of the structure, and the financial resources of the intended occupiers.

It is worth noting that the statutory operating guidance of the Housing Health and Safety Rating System says in connection with thermal efficiency that "*a dwelling should be provided with adequate thermal insulation and a suitable and effective means of space heating so that the dwelling space can be economically maintained at reasonable temperatures*". It is important to check the capacity of any central heating system and that it has the capacity to heat the accommodation effectively. The state of repair and maintenance of the heating system is also a matter for the owner and nothing to do with the occupier.

Dwellings built of materials which have a high thermal capacity will have a slow thermal response and are therefore unsuited to intermittent heating systems, or systems that heat the occupants and air rapidly but which may not be operated long enough to warm the structure. Heating systems that leave parts of the dwelling cold will increase the risk of condensation as warm moist air migrates to the cooler parts.

The cost of running the system is a material consideration and the cost of using the system as well as the way it is used should be identified in any investigation. Obviously how the system is used is a matter for the occupier, but that is only part of the story. If the energy costs are 10% or more of income, then in the past the occupier has been considered to be in fuel poverty (energy precariousness).[31] The factors that influence affordability are the suitability of the heating system, taking account of the thermal capacity and the amount of heat loss from the dwelling. In any assessment it can be necessary to carry out a heat loss assessment, remembering also that draughts have a cooling effect.

It is a general rule that long-period, low-level background heating together with top-up heating is more likely to prevent condensation and mould growth than the same total amount of heat introduced over short periods. However, it can be difficult to persuade occupiers, particularly those on low incomes, of this without demonstration.

Ventilation of any dwelling is necessary to regenerate used air, remove smells and to remove excess moisture from the air within the dwelling. While air changes of 0.5 per hour should be sufficient to prevent the air becoming stale and for hygienic purposes, much higher rates will be required to remove excess water vapour during periods of high moisture generation. In such locations 15 or more air changes an hour could be required to remove the moisture laden air.

Opening windows can, in cold weather or where the dwelling is in an exposed position, result in excessive and uncontrolled ventilation. Excessive ventilation causes draughts, wastes energy and reduces the air temperature, which also means the air is incapable of holding as much water vapour, and leads to cooling of the fabric and internal surfaces. Natural ventilation is affected by wind strength and direction. If a window is opened on the windward side of the dwelling, then moisture-laden air could be forced further into the dwelling rather than being removed from the dwelling. Thus, telling someone simply to open the windows could make the situation worse. That is, excessive and uncontrolled ventilation can increase the likelihood of condensation occurring. Ventilation alone will not prevent condensation occurring and opening the windows could also have the effect of cooling the air, increasing the risk of condensation, so one air change per hour could have little effect on *RH* but a marked effect on temperature [3].

If mechanical extraction is to be used to remove moisture-laden air from a room such as a kitchen or bathroom, then it is better with a humidistat switch as it will switch on automatically when the *RH* reaches a level well short of that which will give rise to condensation. Heat recovery ventilation, also known as HRV, or mechanical ventilation heat recovery (MVHR), is an energy recovery ventilation system using equipment known as a heat recovery ventilator, heat exchanger, air exchanger, or air-to-air heat exchanger which employs a counterflow heat exchanger between the inbound and outbound air flow. Such an approach gets rid of the moisture while warm dry air is returned to the dwelling thus saving energy. With extractor fans the warm air is removed along with the moisture.

Thermal capacity/thermal response of building elements is a major consideration and is important to identify the form of construction. A material that responds slowly to heat input will remain cold even as the air temperature increases; concrete has a slow thermal response (high thermal capacity). The heavier and denser the material, the slower the thermal response and the longer the time lag before the structural temperature reacts to changes in air temperature. These factors affect the type of heating system suitable for the dwelling.

The thermal insulation relates to the ability of the structure to conduct heat and is considered in terms of the thermal transmittance coefficient (U values) measured in $W/(m^2K)$. The U value is the reciprocal of the R value (thermal resistance). Thermal resistance (R) is a measurement of the overall resistance to heat transfer of a material or combination of materials and is expressed in square metres Kelvin per Watt (m^2K/W). The U value of a building element can be calculated from the sum of the R values of the components that make up the element plus its inside and outside surface thermal resistances (Ri and Ro). In the UK while U values are still used in the Building Regulations to set standards for the elements of the building fabric, the overall thermal performance of buildings is now more often assessed using a modelling process – the Standard Assessment Procedure (SAP).[32]

The thermal efficiency of the building and the heating system installed are totally in the control of the designer and not the occupier. However, although U values and SAP can be used to set standards, design criteria or objective as well as allowing comparison of alternative solutions, they are simplifications of reality. Actual performance rarely matches that predicted because of poor workmanship on site (poor supervision), poor detailing and the presence of water in insulation materials.

With adequate thermal insulation warm moist air normally found in an occupied dwelling cannot come into contact with cold surfaces, and so cannot be cooled below the dew point. Poor design and detailing can lead to what are called "cold bridges". They can be a particular problem in well-insulated buildings or modern systems. They occur in localised spots where the nature of construction allows heat to escape through the structure. These can occur where there are gaps in insulation so that these areas have a low internal surface temperature. This can lead to localised damp patches that might be confused with penetrating damp.

Other examples of cold bridges include lintels to door and window openings and the projection of the solid floor through an external wall, such as to form a balcony. Also, it should be noted that in a cube because of the surface to volume ratio, the surface temperature at the corners will be lower than in the middle of the sides. That is, heat is lost from the corners of rooms quicker than through the middle of the walls. This is why mould growth may first be noted at the junction of walls and ceilings. Corners are a form of cold bridge. At cold bridges such as at corners the surface temperatures may be some 5°C colder than adjacent surfaces.

Heat losses are another consideration and there are two main areas of heat loss from a dwelling: first through ventilation where heated air is replaced by cold air, and the second is through the fabric where heat is conducted away from the structure. If the ventilation rate of a dwelling of 180 m^3 volume is reduced from two air changes per hour to one, the heat loss will be cut by around 2,074kW over a heating season. Total heat losses through the fabric depends on the U values of the different materials, that is, the rate at which heat is lost through the material. The smaller the U value the less heat is transmitted, which is a measure of the insulation quality of the material.

Ways of identifying whether the dampness is condensation. It:

- Tends to be seasonal (from October to April);
- Is on the surface of the structure that is damp (wall surface is what is generally noticed, as interstitial condensation will not be seen by the occupier);
- Has associated mould which will normally start near the junction of walls and ceilings;
- Does not always occur where the moisture is generated (moisture-laden air will tend to move away from the relatively warm to relatively cold areas and from relatively wet to relatively dry areas).

Interstitial condensation is condensation that does not occur on the surfaces within the dwelling but within the fabric, such as in a flat roof structure. It usually occurs because of a lack of vapour barrier so that water vapour passes from the dwelling into the structure where the air is cooler and dew point is reached. An illustration of this would be in the case of an insulated roof space, where the water vapour passes through the ceilings (and around any loft access cover) into the roof space because there is no vapour barrier. If there is also no airflow to carry the water vapour away, then the cold air above the insulation cannot carry as much water vapour as that in the dwelling. Dew point is reached where the air approaches the underside of the roof for example and condensation is formed. This can lead to the insulation becoming wet and therefore less effective and there may be staining of the ceilings that appear to indicate a leaking roof, when in fact the dampness is occurring within the roof space. Interstitial condensation can be a problem in timber-framed buildings and flat roofs that have not been constructed

correctly and with the vapour barrier in the correct place.

The critical factor in interstitial condensation is that the internal surface of the structure such as the wall (which is porous) is above dew point but the interior temperature of the wall is below dew point. The water vapour condenses inside the wall rather than on its surface. Interstitial condensation can be a problem with one-brick thick solid walls. It is not normally a problem in a brick and block cavity wall, but it may lead to rotten timber in timber-framed buildings.

Timber elements

Many defects to timber floors are caused by timber and damp problems resulting in rot or woodworm infestation. Through time they cause the deterioration of the timberwork and cause the floor to become springy or, in extreme circumstances, fail completely. Careful examination of the floor structure is required to determine the precise cause of the timber deterioration. Dry rot is often identified, initially, by the fungus which appears as off-white, felt or cotton-wool like sheets on brickwork and timber. In later stages, fungal strands can develop as thick as a finger. The fungus is usually associated with sources of dampness and insufficient ventilation. Where the fungus is exposed to light, it often has a lemon-yellowish tinge. Damage is often confined to timber, but large flat mushroom-like fruiting bodies can easily grow through finishes such as plaster or paint. These fruiting bodies may be the first visible sign of a problem, and they produce numerous spores which are normally brick red in colour. Entirely decayed timber can be crumbled between the fingers, hence the name dry rot – even though it is always associated with the presence of dampness. The fungus leaves deep cracks running across the grain, and there is often evidence of off-white sheets of the fungus on the wood.

If dry rot is not identified at a relatively early stage, then, with the right conditions for growth, it can spread to other timber components of buildings such as skirting boards, door frames, window frames, staircases and even into roof structures.

Wet rot is basically the timber decaying naturally in the presence of high levels of moisture. There is almost always a structural defect causing the problem; it may be that the wall adjacent to the timber is suffering from damp, or water collecting on the timber. Any structural problem must be tackled at the same time as the timber is treated otherwise the problem is likely to reoccur. The problem may just be damaged paint finish on the timber allowing the actual wood to absorb excessive moisture. Damage is normally limited to the timber although the original structural problem may also cause other areas to be affected by damp (such as plaster or just decorations).

Woodworm refers to the larvae of any wood-boring beetle, rather than one particular species. Wood can be infected with eggs or larvae without it being noticeable, and a woodworm infestation may not be discovered for several years. It's a common misconception that woodworm only affects old properties but it can cause damage to any building – even newly constructed ones. Telltale signs of woodworm include:

- Small round holes in your woodwork, similar to the holes in a dart board;
- Fine, powdery dust around these holes (this is known as *frass*);
- Crumbly edges to boards and joists;
- Adult beetles emerging from the holes or present around the house.

Windows and window openings can present various problems which are usually associated with wear and tear, or lack of maintenance. Many 19th- and early 20th-century dwellings and offices were constructed with "sash" windows consisting of an upper and lower sash that slide vertically in separate grooves in the side jambs or in full-width metal weather-stripping. This type of window provides a maximum face opening for ventilation of one-half the total window area. Each sash is provided with springs, balances or compression weather-stripping to hold it in place in any location. To facilitate operation, the weight of the glazed panel is usually balanced by a heavy steel, lead, or cast-iron sash weight or counter-weight concealed within the window frame. The sash weight is connected to the window by a sash cord (more usually) or chain which runs over a pulley at the top of the frame, although spring balances are sometimes used. Traditional problems with wooden sash windows include rot, swelling or distortion of the woodwork, rattling in the wind (due to shrinkage of the wood), and problems brought on by careless application of paint. The sliding mechanism makes sash windows more vulnerable to these problems than traditional casement windows. Sash windows are relatively high maintenance, and the sashcords require checking regularly for evidence of fraying to enable them to be renewed before they break altogether.

The main defects found in timber stairs are worn treads and nosings; creaking steps; and loose, cracked and broken balusters and handrails. Worn treads are often the result of leaving the stair uncovered and exposed to excessive wear and tear. Creaking steps may be caused by a number of things. The most common are loose or missing glue blocks, loose wedges, and a defective joint between a tread and the bottom of a riser. Staircases over 900 mm wide often creak because the centres of the wide steps are not supported well enough. To identify the actual cause (or causes) of the problem, examination of the underside of the staircase is required. More obvious defects are missing, broken or loose handrails and/or balusters which are extremely dangerous and should receive immediate attention.

Sustainable urban drainage (SUDS)

It is now widely recognised that the effects of climate change and the increase in the built

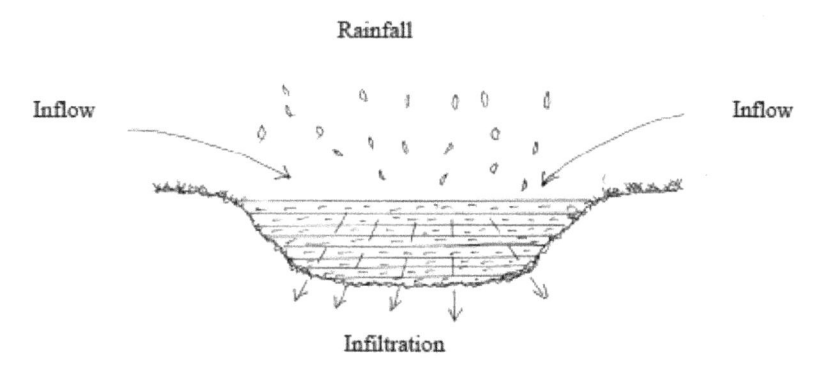

Figure 9.23   A cross-section of an example of a SUDS infiltration basin or swale

environment have necessitated changes to the way in which storm water is dealt with. Sustainable drainage systems, sometimes called sustainable urban drainage systems (SUDS) provide an effective way of mimicking natural drainage before development took place. SUDS can be used to counteract the effect of overloading gravity pipelines and watercourses, which can contribute to flooding downstream; or conversely, dealing with rainwater run-off on site to replenish ground water levels, particularly in times of water shortage. SUDS are increasingly used to mitigate excessive flows from storm water and reduce the potential for pollution from run-offs in urban areas [4].

Susdrain[33] is a grouping that provides a range of resources for those involved in delivering SUDS. Extremes of weather and flooding are being experienced more frequently, as the result of global heating means the need for SUDS is increasing. Susdrain was created by Construction Industry Research and Information Association (CIRIA) to provide an independent and authoritative source of information on SUDS. The website provides up-to-date guidance, information including case studies and videos to assist in the planning, design, approval, construction and maintenance of SUDS.

Well-designed SUDS store and slow surface and floodwater and can treat and immobilise pollutants such as motor oil in run off.

Conservation benefits by protecting or even enhancing landscapes and habitats, as they include grassland, ponds and wetland, which support a variety of plant and wildlife species. They can also assist in providing additional open space and green corridors (helping to deliver on biodiversity action plans). In that sense they not only reduce physical risks from flooding but can have a positive mental health effect.

Infiltration basins are vegetated depressions designed to store runoff on the surface and infiltrate it gradually into the ground. They are dry except in periods of heavy rainfall.

Swales, which in effect are narrow and long infiltration basins, can be installed in housing developments installed alongside roads to replace conventional kerbs. They may be used as conveyance structures to pass runoff to the next stage of treatment and can be designed to promote infiltration where soil and groundwater conditions allow.

Porous pavements are made from materials like concrete blocks, crushed stone or porous asphalt. The water may infiltrate directly into the subsoil or where the ground is not suitable be stored in an underground reservoir such as a crushed stone layer before slowly seeping into the ground. Where infiltration is not possible or appropriate such as where the land is contaminated, an impermeable membrane should be installed and an overflow

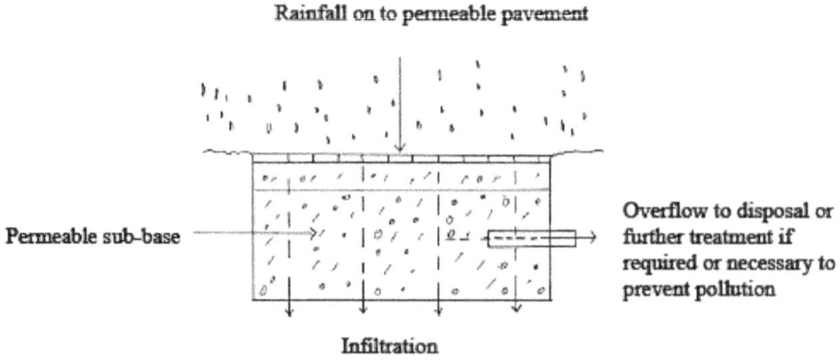

Rainfall on to permeable pavement

Permeable sub-base

Overflow to disposal or further treatment if required or necessary to prevent pollution

Infiltration

Figure 9.24 Permeable pavement as part of SUDS

installed to prevent the pavement itself from becoming flooded. The removal of pollutants occurs either within the surfacing or sub-base material itself, or by the filtering action of the reservoir or subsoil.

The Environment Agency has worked in partnership with the Construction Industry Research and Information Association (CIRIA) to develop a free tool to evaluate the benefits of SUDS available from 2021. The BeST (Benefits of SuDs Tool) provides a means of evaluating the benefit largely based on the overall performance of the drainage system.[34]

In March 2015 Defra issued non-statutory technical standards[35] for sustainable drainage systems, to be read in conjunction with the National Planning Policy Framework and Planning Practice Guidance.

Sanitation provision at events or in emergencies – field or camp sanitation

Among the public health issues associated with the use of camps of all kinds including in emergencies,[36] none is more important than the provision of sanitary accommodation. Large permanent camps require piped water supplies; they also need a water-carriage system of drainage. Where, in addition, a sewage disposal plant is necessary, the types of plant already described are suitable and can

be adapted to meet the needs of most sites. In small camps and those of a temporary nature, some form of conservancy system alone is practicable. In open-air conditions, the provisions described in the following subsections can be quite satisfactory but in every case they must be sited, constructed and maintained under the direction and supervision of the environmental health officer of the district concerned. All latrines must be fitted with seats and covers and otherwise made fly-proof.

Chemical closet

Chemical closets have improved greatly in appearance over the years, although they are now more generally used for camping, caravanning and boating. They all operate with a deodorising, liquefying or sterilising liquid that is basically a solution of formalin. Some manufacturers claim waste matter that has been treated with their chemicals can be safely disposed of down a drain, but unless there are specific disposal points, such as on camp sites, at marinas, etc., the views of the controlling authority should be sought beforehand. The closets are usually constructed of strong plastic or fibreglass and range from an enclosed bucket to semi-permanent recirculating units with filtration units and electric "flushing" pumps. Similar built-in units are used in aircraft, long-distance coaches and in some trains (other trains however, merely dump the sewage effluent on the track).

Earth latrines

The simplest method of disposing of human excrement where there is no access to a sewerage system is a dug latrine, but it is necessary to minimise all danger of spreading disease. Surface pollution must be avoided. Earth latrines, soakaways, etc. may be used with safety only where pollution of subsoil water is not a risk. Faecal organisms reaching the groundwater from a point source, for example, a latrine, do not travel evenly in all directions, but are carried with the groundwater flow. Pollution has been found to travel a distance of 25–30 m "with the stream", but did not reach 3 m in any other direction. Usually, the ground water is flowing in a definite direction and, provided the direction is known, an earth latrine or a soakaway can be located safely. In ordinary soils it may be assumed that the area outside a radius of 6 m, and extending round one-half of a circle "up-stream" or above the latrine, is safe from danger of pollution. In chalk and similar formations, however, the water is in fissures (karstic), which form subterranean streams running long distances in all directions. No earth latrine may be used with safety in such formations.

The deep trench

In temporary camps, where only field sanitation is practicable, a properly constructed and supervised earth latrine is a sanitary and satisfactory provision. The latrine in most common use is of the deep trench type which, owing to its size, is generally arranged for communal use and, because of that, is open to many objections. The trench latrine should be 1 m to 1.5 m wide and as deep as possible, not less than 2.5 m deep and may be lined with timber where there is danger of collapse. It may be of any convenient length, usually between 5 and 10m. It should be provided with a well-constructed riser or seat,

the openings of which should be arranged over the lateral centre of the trench, the front and back being constructed to prevent the ingress of surface water. Self-closing covers should be provided to the openings, and all necessary measures should be taken to render the trench fly-proof when it is closed. The contents of a deep trench latrine should be covered with about 75 mm of fertile top soil (not the earth previously dug out) at frequent intervals. The nitrifying organisms in top soil rapidly neutralise and break down faecal solids, rendering them innocuous. The action is biological and quite efficient. Only where a latrine is sunk into dense watertight earth and contains foul liquid matter should a disinfectant be used. Where it is not possible to install seats the trench is spanned by pairs of wooden boards on which the users squat.

The bore-hole latrine

This simple form of earth latrine has been adopted with marked success in the tropics, but is equally suitable for use elsewhere. By means of a hand-operated land auger of simple design, a hole 400 mm in diameter is bored in the earth to a depth of 4.5–6 m. Being circular and undisturbed, the walls of the hole are, in ordinary soil conditions, self-supporting.

Container closets

The best of these are the earth closet and the chemical closet already described; they are both suitable for camp purposes.

Earth closets

Now little used, these depend upon the power of dry earth to "neutralise" faecal matter. The receptacle is essentially a stout, galvanised iron bucket which, in order to ensure frequent emptying should not be of more than 55 l in capacity. It should be movable, but held in position by a suitable guide, and fitted closely into a suitable enclosure. The earth should be clean, dry, fertile top-soil sifted to exclude particles smaller than 6 mm. Sand or ashes are not suitable for the purpose.

Composting toilets

Unlike conventional toilet systems, there is no flushing water involved and composting toilets can be used in a wide range of locations from farms to golf courses and other recreational areas and can be used as an alternative in buildings that are aiming for environmental sustainability or where there is no access to a water carriage system. Compost toilets depend on aerobic bacteria to break down waste, similar to that of outdoor composting. Rather than flushing, waste is composted with carbon-rich sources like wood shavings, coffee grounds, bark mulch, leaves, or even shredded paper (this is bulking material that helps the waste breakdown ensure that the resulting product is safe to handle and use). The end product, as with any compost, is a soil-like material similar to humus.

There are an increasing number of composting toilets on the market, can be easy to build, and most systems ensure there are no unpleasant odours. Some involve separation of urine and faecal matter or some compartments come with a separate urinal as well. There are even fully accessible units on the market.

A composting toilet must perform three completely separate processes:

1 Compost the waste and toilet paper quickly and without odour
2 Ensure that the finished compost is safe and easy to handle
3 Evaporate the liquid

Where units have a separate waterless urinal much of the liquid avoids the composting chamber and either is allowed to evaporate or be removed manually when the chamber is full. A ventilation system would permit the evaporated liquid to escape into the air outside.

The systems are designed to create the best possible environment to promote quick, efficient break down of organic waste. Once a unit is full, it will need to be emptied but if the system is working properly, this should be a safe, nutrient rich soil, like garden compost, with no real signs of human waste to be seen nor smelled. Normally it should need emptying only once or twice a year.

There are several one- and two-chamber systems and also three-chamber systems. The basic design allows the waste to enter through the toilet into a composting chamber. Multi-chamber designs divert urine into a separate evaporation chamber. The solid waste is then mixed with one of the bulking agents discussed earlier.

In a three-chamber system each of the three main functions take place in its own chamber that allows finished compost to be removed from a finishing drawer, while unfinished compost stays in the bio-drum to be broken down further. This system prevents compost that is further along in the breaking down process from being contaminated with fresh waste. In another system there is liquid separation before mixing with the compost. All liquids drain directly into a front-mounted urine bottle to be emptied when full. All the solid waste is deposited into the main composting chamber where it is mixed with bulking material by turning a mixing handle after each use. In a single chamber design, the waste may need to be transferred to another location to dry.

The interior of the composting toilet in Figure 9.2b shows the bin containing sawdust (left) that should be thrown down the through the pan each time toilet is used.

Sullage water

On no account should crude sullage (bath, shower, kitchen, etc. liquid waste) be discharged into a river, stream, ditch or lake, or over unprepared ground. The simplest means of disposal is by soil absorption through a soakage pit. Even in the most favourable soil conditions its ability to absorb sullage water is lessened as time goes on, and it may be reduced to a point at which complete disposal is not possible. The effects of climate change and periods of excessive rainfall will also impair the effectiveness of soil to absorb the sullage

Plate 9.1a The exterior and Plate 9.1b the interior of a simple composting toilet on an allotment

Plate 9.2a Exterior of fully accessible composting toilet on a nature reserve and Plate 9.2b interior of accessible composting toilet (instructions on use behind the pan) and also urinal which is located to the right of the doorway which goes to a soakaway.

Source: Courtesy of Thames Water

Plate 9.2b (Continued)

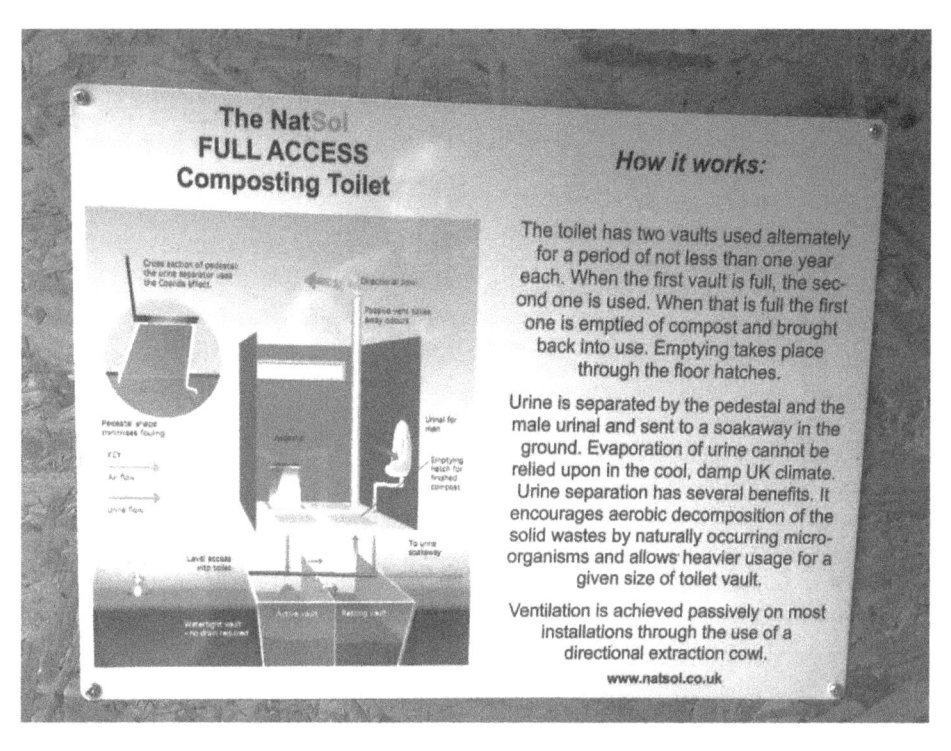

Plate 9.2c Instructions on how the accessible composting toilet works (covers to the two vaults referred to can be seen in Plate 9.2b located on the floor of compartment)

water. In order to preserve soil absorptivity, grease and oils should be removed by passing sullage water through a grease trap before it is discharged into a soakage pit.

For field use, a grease trap is designed to ensure that sullage water passing through it has a long journey at low velocity between inlet and outlet, so that grease may separate from the water and float to the surface, where it is retained. Only two "baffles" are necessary, one to form the inlet and the other the outlet chamber; these should be deep rather than shallow and long and narrow rather than square. A length-to-breadth proportion of 3:1 is recommended. A capacity of 225 litres is suitable for most general purposes.

Soakage pits receive waste liquids as and when they are produced, and act as reservoirs from which they may soak continuously into the surrounding ground. The pit should be filled with coarse rubble to support the sides and cover, while leaving the maximum void. It should be covered with at least 300 mm of earth. Apart from geological conditions, efficiency is dependent upon two factors: water content and the size of the soakage surface. Generally, the water content should be equal to one day's production of sullage. The soakage surface is the area of the perimeter walls plus the base; the shape should provide the maximum perimeter by comparison with the volume, that is, rectangular rather than square. Where the surface stratum is dense, it should, if possible, be pierced and a more permeable stratum be brought into use. A form of vertical drainage, useful for both sullage and surface water, is thereby obtained.

Chemical precipitation can be used in large camps, or where the soil has become non-absorbent, sullage must be purified or, more correctly, clarified; it may then be disposed of in the same way as ordinary surface water.

A simple method of purification that is used in military camps consists of treating sullage with two chemicals: ferrous sulphate and hydrated lime. The quantities required vary with the sullage, but effective precipitation depends upon obtaining a correct alkalinity (pH of 9); the lime is used for this purpose. It is added in equal parts to or slightly more than, the ferrous sulphate. The sullage is collected in a tank. The ferrous sulphate, after being dissolved in water, is then added; the lime is similarly dealt with. Then the whole contents of the tank are thoroughly agitated. A heavy floc results, which precipitates rapidly, forming a closely packed sludge and leaving the supernatant water clear. The process does not completely remove grease; water from a cookhouse should therefore be passed though a grease trap before treatment. The clarified water is comparable to an ordinary purified sewage effluent, except that the oxygen demand is high: about 10 parts per 100,000. This can be corrected by aeration, for example by causing the effluent to pass through an open channel and, if possible, over weirs before discharging into a stream. A sullage purification plant is easily constructed.

Two precipitating tanks are usually necessary, to be used in rotation; they can be formed from precast concrete rings or pipes placed on end in a suitable excavation, the bottoms being concreted in the form of an inverted frustum of a cone, into which the sludge can gravitate. The outlet for clarified effluent is fixed not less than 300 mm above the bottom of the tank and at a level from which the effluent can gravitate to a suitable place of disposal. Two outlets fixed at different levels afford useful flexibility of working. Sludge is dealt with in shallow lagoons. It is usually at too low a level to run by gravity to these; the hydraulic head provided by the supernatant water when the tank is full is utilised to eject and lift it to the required level. It dehydrates rapidly, forming an innocuous residue that may be disposed of on land. The ferrous sulphate and lime contained in the sludge produced from the first two or three charges is used to assist in the precipitation of subsequent charges. For this reason, desludging is carried out only when the sludge level rises to within a few inches of the clear water outlet.

Water and sanitation facilities, etc. on modes of transport

This part is concerned primarily with on-board water systems and sanitation.

Vacuum toilets

Vacuum toilets are toilets employing a vacuum for flushing, rather than gravity alone. There are a number of different types of vacuum toilets, ranging from toilets connected to vacuum sewer systems to toilets with a vacuum assist, which creates pressure to help flush the contents of a toilet with minimal water usage. These are used on aeroplanes, trains and sea-going vessels, and can also be found in other modes of transport and mobile homes. Vacuum toilets use air to drive waste through the toilet and vacuum piping to the treatment tank or intermediate collection tank. By using as little as 1 litre of water, the amount of wastewater is reduced. When the toilet is flushed it opens a valve in the waste line, the vacuum energy stored in the system is released, and the vacuum in the line sucks the contents out of the bowl and into a tank. The bowl is instantly cleared and the waste is moved through the vacuum pump at something of the order of 2 metres per second.

An integral pressure switch activates the vacuum pump. After the flushing mechanism is released, the vacuum pump continues to run until the vacuum level is recharged in the system.

- They use very little water;
- They can use much smaller diameter sewer pipes;
- They can flush in any direction, including upward as a vacuum system does not use gravity to move the water; there is

nothing to stop the sewer pipe from going vertically.

- That the pipe does not have to go downward also means it is possible to avoid cutting the construction to install new toilets.

The toilet is connected into a vacuum sewer system, which may consist of a single sewage tank or a series of sewer lines, or a number of toilets can be linked to a single tank.

All new ships or ships undergoing major conversion should now be equipped with facilities for treating wastes from toilets and urinals, faecal material from hospital facilities and medical care areas, and wastes from food refuge grinders. Holding tanks, properly equipped with pumps and piping, may be installed in place of treatment facilities. Wastes from holding tanks may be discharged to shore connections or to special barges for the reception of these wastes. The design of treatment facilities and holding tanks should be based on 30 gallons (114 litres) per capita per day. The appropriate authority of the country of registration should have approved the design.

On board ships there may be marine sanitation farms, which siphon out the water, treat it until it is clean water, then pump it into the ocean. Aerobic bacteria digest the sludge in storage tanks until it's all offloaded ashore when in port. On some cruise ships "grey water" from galleys, laundries and bathrooms is first mixed in measured proportions with the lavatory waste before bio reactors low down in the ship deal with the waste that is digested by bacteria; the liquid is disinfected by UV radiation. The effluent is monitored for any remaining bacteria such as faecal coliform, and it is discharged into the sea as clean water.

Potable water on ships

Generally, the ship drinking-water supply and transfer chain consists of three major components according to the WHO [5]:

- The source of water coming into the port;
- The transfer and delivery system, which includes hydrants, hoses, water boats and water barges. This water transfer process provides multiple opportunities for the introduction of contaminants into the drinking-water; and
- The ship water system, which includes storage, distribution and on-board production of drinking water from overboard sources, such as seawater.

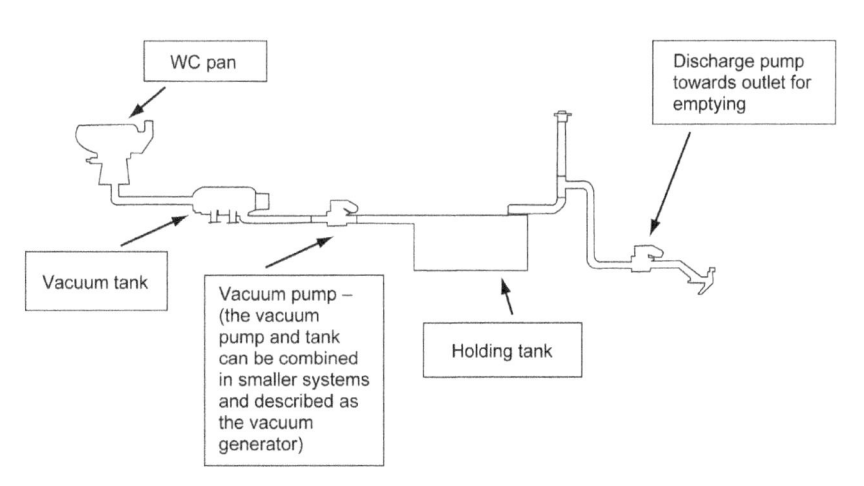

Figure 9.25 Vacuum toilet system and the basic method of operation (there can be multiple toilets attached in some systems, e.g. on board ships)

The WHO Guide to Ship Sanitation [5] says that reference should be made to six international standards in relation to sanitary design and construction of ship water supplies and potable water quality assessment. These are:

- ISO 15748–1: 2002 – Ships and marine technology – Potable water supply on ships and marine structures – Part 1: Planning and design;
- ISO 15748–2: 2002 – Ships and marine technology – Potable water supply on ships and marine structures – Part 2: Method of calculation;
- ISO 19458:2006 – Water quality – Sampling for microbiological analysis;
- ISO 14726:2008 – Ships and marine technology – Identification colours for the content of piping systems;
- ISO/IEC 17025:2017 – General requirements for the competence of testing and calibration laboratories;
- ISO 5620–1/2: 1992 – Shipbuilding and marine structures – Filling connection for drinking-water tanks. (Part 1 is general requirements and Part 2 Components).

Construction of water tanks has to be considered carefully because potable water on ships needs to be stored in tanks constructed and located so as to ensure protection from any contamination from inside or outside the tank. They should be designed so that cross-connections between them and tanks holding non-potable water or pipes containing non-potable water are prevented. Ideally, potable water tanks should be located in rooms that have no sources of heat emission or dirt.

Potable water tanks must be constructed of metal or other suitable material that is safe for contact with potable water and must be robust enough to exclude contamination. Proper maintenance of anticorrosive coatings in water tanks is important. Ideally, potable water tanks would not share a common wall with the hull or other tanks containing non-potable liquids. No drainage line of any kind or any pipe carrying wash water, salt water or other non-potable liquid should pass through potable water tanks. It is also preferable for toilets and bathroom spaces not to extend over any part of a deck that forms the top of a potable water or wash-water tank.

Every potable water storage tank will need to be provided with a vent located and constructed to prevent the entrance of contaminating substances and vectors. Ventilating pipes should not end directly above the water surface, to avoid substances dripping into the water body. A potable water tank vent should not be connected to the vent of any tank holding, or intended for holding, non-potable liquid, as cross-contamination may occur.

It is important that the potable water tank be provided with an overflow or relief valve, located so that the test head of the tank is not exceeded. The overflow must be constructed and protected in the same manner recommended for vents. An overflow may be combined with a vent, but the provisions described for the construction and protection of both vents and overflows must be observed.

The potable water tank should be so designed as to be capable of being completely drained. The end of the tank suction line should be no closer than 50 mm above the tank bottom, to avoid the intake of sediment or biofilms. Potable water tanks need to be equipped with facilities to read the filling level of the tank from outside. This construction should not produce areas of stagnating water that could become a source of contamination.

All potable water tanks need to be clearly labelled with their capacity and words such as "potable water tank".

The potable water tank will need an inspection cover giving access for cleaning, repair and maintenance. To avoid contamination when opening the cover, the opening should not give direct access to the unprotected water surface. An inspection of the empty tank should be performed periodically (e.g. once per year). If tanks are entered by people, clean protective clothing should be worn. Sample cocks should be installed directly on each tank to allow tests to be taken to verify water quality and must point downwards to avoid contamination.

Cold potable water should always be stored at temperatures below 25°C. More detailed information about technical requirements of potable water tanks can be found in ISO 15748–1.

Potable water tanks and any parts of the potable water distribution system shall be cleaned, disinfected and flushed with potable water:

- Before being placed in service; and
- Before returning to operation after repair or replacement; or
- After being subjected to any contamination, including entry into a potable water tank.

Potable water tanks are required to be inspected, cleaned and disinfected during dry docks and wet docks or every two years, whichever is less. On sea-going vessels and marine structures, it has been recommended that two potable water tanks be installed for good reliability of the water supply system.

Ship construction

While most of this chapter has been concerned with constructions on land, some EHPs will have cause to board a ship so we conclude with a little on ship construction, although most of their port health work will involve inspections of facilities in the same way as they would on shore.

Ocean cruising is increasingly popular, and this includes "fly-cruises" where passengers fly to a port to board a cruise ship and fly back to their home from that or another port. The speed of global travel does have implications for the spread of infectious diseases, and we saw in the Covid-19 pandemic and other outbreaks cruises can be susceptible to infectious diseases. The most frequently reported cruise ship outbreaks involve respiratory infections, gastro-intestinal infections (such as norovirus) dealt with elsewhere in this book.[37]

Cruise ships are a particular form of passenger ship. They always require electrical power, including when docked. This is normally provided by diesel generators, although an increasing number of new ships are fuelled by liquefied natural gas (LNG). When docked, ships must run their generators continuously to power on-board facilities, unless they can use onshore power. There have been criticisms of the polluting emissions from the diesel engines of docked cruise ships.

Modern cruise ships typically have aboard a range of facilities and these can include several restaurants and food outlets, casinos, shops, spas, hot tubs and fitness centres, cinemas and theatres, swimming pools and water slides, sports facilities and lounges.

Some ships have bowling alleys, ice skating rinks, rock climbing walls, sky-diving simulators, miniature golf courses, video arcades, ziplines, surfing simulators, basketball courts, tennis courts, chain restaurants, ropes obstacle courses, and even roller coasters.

Principles of inspection

Although this chapter is aimed primarily at providing information on constructions, the corollary of that is this information is for use when carrying out inspections. This last brief section sets out the basic principles of an inspection regardless of the type of premises, that is whether domestic or commercial. Remember this inspection is about collecting evidence even if it is not a matter of legal action and understanding what goes on in the premises. In the first instance it is important to be correctly dressed for the inspection with suitable protective clothing and also to have the appropriate and correct equipment, even if the inspection is non-intrusive and cannot damage the property or its contents.

Any inspection requires use of all the senses; for example it is often possible to smell dampness or mould such as dry rot, or even a pest infestation before you have seen it. Also when speaking with occupants of the premises, listen carefully to their replies

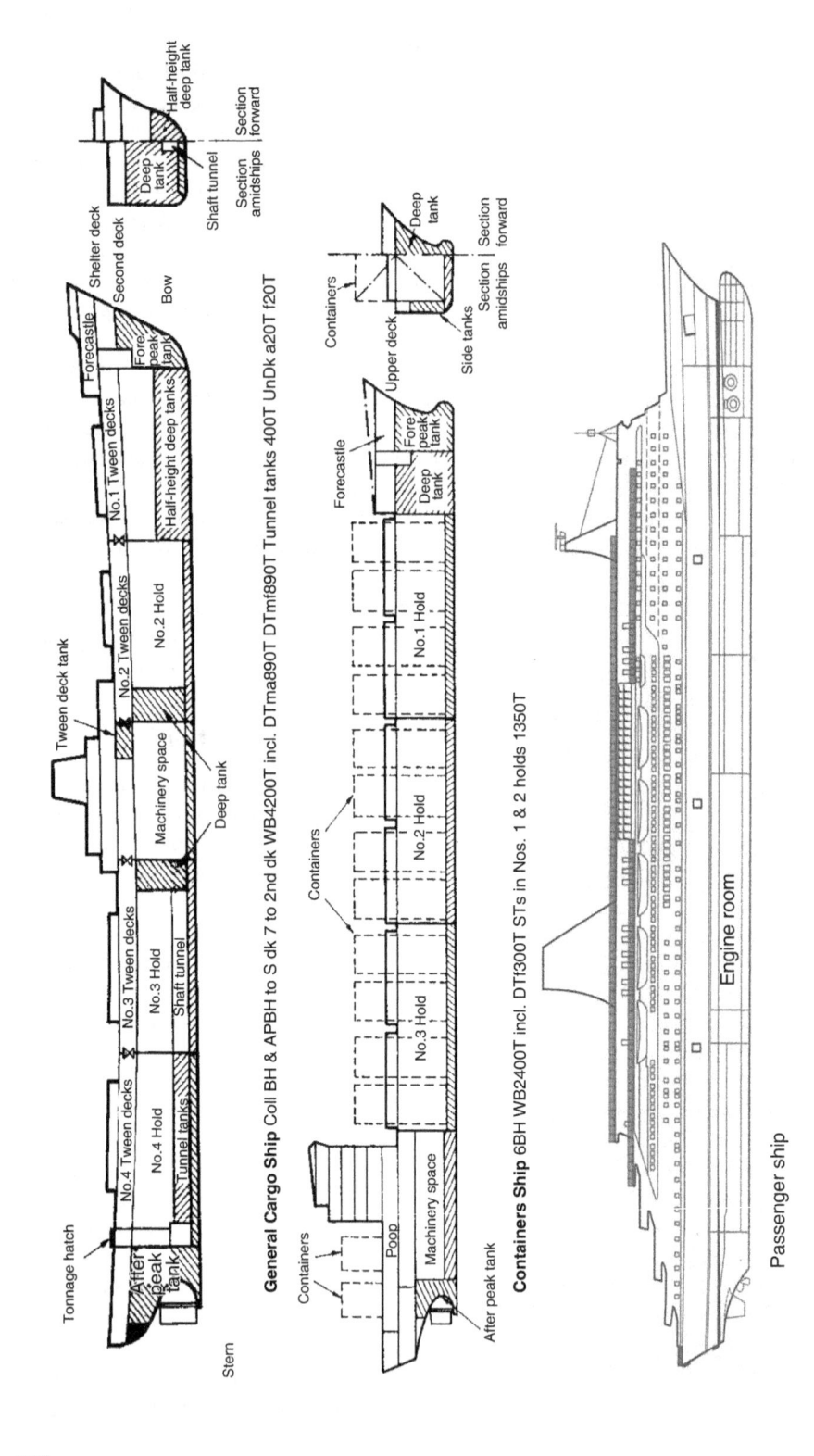

Figure 9.26 Examples of ship construction

Source: Reproduced by kind permission of Lloyd's Register Foundation Library and Archive London

to questions and whether they provide useful pointers to issues, or matters that can be tested by the other evidence collected during the inspection. It is important also to put the inspection into context, and that means looking at the environment in which the premises are located – are there any issues that might impact the premises subject of the inspection.

The principles of inspection:

1 Be accurate and precise as to what you see (or cannot see) – the limitations of any inspection should be noted
2 Be truthful in what you see and record this on site or as soon as possible after (this can be via notebook or voice recording/memo)
3 Work methodically through the premises and in each room or functional space and common parts and then the exterior of the building as well as the immediate surroundings which should include waste storage and management. While a pro-forma inspection sheet can help, this can also lead to unused pages or navigation difficulties (whether paper or on hand-held device) as it is difficult to have a single pro-forma that suits all premises.
4 Take your time and do not rush – (leave time) – and take care as you must also look after your own health and safety
5 Listen to the occupier/operator and what and how they speak to you; this applies whether for example asking about food preparation procedures or use of the heating system but at the same time remember item 6
6 Don't get distracted by occupiers of the premises under inspection, so speak with them at the start of the inspection to get some insight into how the premises are used and to get an insight into any problems, and then at the end of the inspection after you have looked at the physical condition of the premises, check on any issues that have come to light during the inspection

7 But do not record tenants' or occupants' comments as your own, but speaking with them can tell you how the premises and services are used; observe any activities on the premises
8 Check documentation relating to the premises such as gas and electrical safety certificates, and other records such as lift servicing, fire risk assessments and licences
9 Do not damage the property or contents
10 Try not to inconvenience the occupiers or tenant and recognise the limits to a non-intrusive inspection
11 Don't worry if it is not possible to diagnose causes of a problem and there is a need for further investigation, or if more destructive surveys have to be taken by or on behalf of the building owner
12 When leaving the premises close the inspection and tell the occupiers what the next steps will be taken and timescales for any actions

The environmental health practitioner is there to safeguard and promote public health and so it is important that any person with whom the practitioner comes into contact has confidence in their knowledge and expertise.

Notes

1 https://livingroofs.org/wp-content/uploads/2019/05/LONDON-LIVING-ROOFS-WALLS-REPORT_MAY-2019.pdf
2 Matthews S, Reeves B. (2012) *Handbook for the Structural Assessment of Large Panel System Dwelling Block for Accidental Loading*, BRE, Garston, Watford.
3 https://www.towerblocksuk.com/lps
4 https://www.steelconstruction.info/Residential_and_mixed-use_buildings#Attributes_of_steel_construction
5 The Building Regulations 2010 (SI 2010 No. 2214) came into force in September 2010.
6 Approved Document B Volume 1 Dwellings which supports requirements in the Building Regulations can be found at https://assets.publishing.service.gov.uk/government/uploads/system/uploads/attachment_data/file/832631/

Approved_Document_B__fire_safety__volume_1_-_2019_edition.pdf

7 See https://www.manchesterfire.gov.uk/staying-safe/what-we-do/high-rise-taskforce/cube/

8 See https://www.insidehousing.co.uk/news/news/defective-cavity-barriers-contributed-nothing-to-slow-fire-in-worcester-park-blaze-report-reveals-67823

9 Housing Act 2004 s.1(5)

10 https://assets.publishing.service.gov.uk/government/uploads/system/uploads/attachment_data/file/760150/Housing_Health_and_Safety_Rating_System_WEB.pdf

11 https://www.towerblocksuk.com/firesafetychecklist

12 See LGA. (2012) *Fire Safety in Purpose-Built Blocks of Flats*, 41–50. https://www.local.gov.uk/sites/default/files/documents/fire-safety-purpose-built-04b.pdf and https://fire-risk-assessment-network.com/blog/types-fire-risk-assessment-flats/ and http://www.cfoa.org.uk/19532 (which includes link to guidance on choosing a competent fire risk assessor).

13 https://electrical.theiet.org/?utm_source=redirect&utm_medium=legacyredirects&utm_campaign=2019relaunch&origin=homepage-wiring-regs

14 https://assets.publishing.service.gov.uk/government/uploads/system/uploads/attachment_data/file/397173/ssd-general-binding-rules.pdf

15 See Disposal of Fats, Oils, Grease and Food Waste – Best Management Practice for Catering Outlets, Water UK, London. https://dl.dropboxusercontent.com/u/299993612/Publications/Guidance/Wastewater/Disposal%20of%20Fats%2C%20Oils%2C%20Grease%20and%20Food%20Waste.pdf

16 https://www.which.co.uk/news/2018/04/shrinking-homes-the-average-british-house-20-smaller-than-in-1970s/

17 https://www.findmeafloor.co.uk/where-in-the-world-do-you-get-the-biggest-hom

18 https://assets.publishing.service.gov.uk/government/uploads/system/uploads/attachment_data/file/524531/160519_Nationally_Described_Space_Standard_____Final_Web_version.pdf

19 https://www.gov.uk/government/uploads/system/uploads/attachment_data/file/6077/2116950.pdf

20 Public Health and Wellbeing Regulations 2009 (Sr No 178 Of 2009).

21 See https://www.rics.org/globalassets/rics-website/media/upholding-professional-standards/sector-standards/real-estate/rics-property-measurement/rics-property-measurement-2nd-edition-rics.pdf

22 Buxton P. (2018) *Metric Handbook: Planning and Design Data*, CRC Press, Routledge, Abingdon, Oxon, 6th ed.

23 Park J. (2017) *One Hundred Years of Housing Space Standards – What Now?* http://housingspace-standards.co.uk in March 2020; see also Carr H. (2017) Statutory overcrowding standards and England's crisis of housing space. In *Decentring Urban Governance: Narratives, Resistance and Contestation*, Bevir M, McKee K, Matthews P (Eds.), Routledge, London. ISBN 978-1-138-22937-2. E-ISBN 978-1-315-38972-1.

24 It is now possible to download specific "apps" for use on smart phones.

25 Case reference LON/OOBE/HPO/2013/0021.

26 Housing Act 2004 Section 9, Housing Health and Safety Rating System – Operating Guidance.

27 https://www.london.gov.uk/sites/default/files/Interim%20London%20Housing%20Design%20Guide.pdf

28 https://www.gov.uk/government/uploads/system/uploads/attachment_data/file/5976/code_for_sustainable_homes_techguide.pdf

29 See https://www.breeam.com

30 World Health Organization. (2009) *WHO Guidelines for Indoor Air Quality: Dampness and Mould*, WHO, Copenhagen, Denmark.

31 In other countries of the UK this is still the case but in England fuel poverty is now measured using the Low Income Low Energy Efficiency (LILEE) indicator. Under the LILEE indicator, a household is considered to be fuel poor if: they are living in a property with a fuel poverty energy efficiency rating of band D or below and when they spend the required amount to heat their home, they are left with a residual income below the official poverty line.

32 See https://www.gov.uk/guidance/standard-assessment-procedure

33 www.susdrain.org

34 See https://www.ciria.org/ItemDetail?iProductCode=W047AF&Category=FREEPUBS&WebsiteKey=3f18c87a-d62b-4eca-8ef4-9b09309c1c91

35 See https://www.gov.uk/government/uploads/system/uploads/attachment_data/file/415773/sustainable-drainage-technical-standards.pdf

36 See WHO Emergency sanitation – technical options (Technical Note 14). http://www.who.int/water_sanitation_health/hygiene/envsan/sanitationtechoptions.pdf

37 See for example Kak V. (2015) Infections on cruise ships. *Microbiol Spectrum*, 3(4): IOL5-0007-2015. doi:10.1128/microbiolspec.IOL5-0007-2015.

Chapter references

[1] Marshall D, Worthing D, Dann N, Health R. (2013) *The Construction of Houses*, Routledge, Abingdon, Oxon, 5th ed.

[2] RIBA. (2011) *The Case for Space – the Size of England's New Homes*, London. https://www.architecture.com/-/media/gathercontent/space-standards-for-homes/additional-documents/ribacaseforspace2011pdf.pdf.

[3] Marshall D, Worthing D, Heath R, Dann N. (2014) *Understanding Housing Defects*, Routledge, Abingdon, Oxon, 4th ed.

[4] NHBC Foundation. (2010) *A Simple Guide to Sustainable Drainage Systems for Housing*, London. https://geosmartinfo.co.uk/wp-content/uploads/LA-Guidance-SuDS/England/NHBC_An-introduction-to-Sustainable-Drainage-Systems.pdf; https://www.netregs.org.uk/environmental-topics/water/sustainable-drainage-systems-suds/.

[5] WHO. (2011) *Guide to Ship Sanitation*, Geneva, 3rd ed. https://apps.who.int/iris/bitstream/handle/10665/43193/9789241546690_eng.pdf;jsessionid=CD856DDE60BCEEC89A7D734DA481E6EA?sequence=1.

Further reading and sources of further information

Charlett AJ, Maybery-Thomas C. (2013) *Fundamental Building Technology*, Routledge, Abingdon, Oxon.

Chudley R, Greeno R. (2020) *Building Construction Handbook*, Kovac K (Updated and Ed.), Routledge Abingdon, Oxford, 12th ed.

Chartered Institute of Building and Service Engineers (CIBSE). https://www.cibse.org

Chartered Institute of Building (CIOB). https://www.ciob.org

CIEH. (2008) *Guidance on Fire Safety Provisions for Certain Types of Existing Housing*. https://www.cieh.org/media/1244/guidance-on-fire-safety-provisions-for-certain-types-of-existing-housing.pdf and addendum giving some clarification CIEH. (2009) https://www.cieh.org/media/2561/guidance_on_fire_safety_provisions_for_certain_types_of_existing_housing_-_march_2009_update.pdf.

Construction Industry Research and Information Association (CIRIA) http://www.ciria.org/service/Home/AM/ContentManagerNet/HomePages/CIRIA_1502_20080929T115140HomePage.aspx?Section=Home.

Halliday S. (2018) *Sustainable Construction*, Butterworth-Heinemann (Taylor & Francis), Abingdon, Oxford, 2nd ed.

Highfield D, Gorse C. (2009) *Refurbishment and Upgrading of Buildings*, Spon Press, London, 2nd ed.

Mayor of London. (2016) *Quality and Design of Housing Developments*. https://www.london.gov.uk/what-we-do/planning/london-plan/current-london-plan/london-plan-chapter-3/policy-35-quality-and also the Supplementary Planning Guidance – Housing. https://www.london.gov.uk/what-we-do/planning/implementing-london-plan/planning-guidance/housing-supplementary.

Osbourn D, Green R. (2006) *Mitchell's Introduction to Building*, Mitchells Building Series, Pearson Education Ltd., Prentice Hall, Edinburgh, 2nd ed.

Roys M. (2013) *Refurbishing Stairs in Dwellings to Reduce the Risk of Falls and Injuries*, HIS/BRE Press Bracknell, Berks.

Stephenson, J, London District Surveyors Association. (2012) *Spon's Building Regulations Explained: 2012 Revision*, Foulger B (Ed.), Spon Press, London, 8th ed.

Approved Documents under the Building Regulations 2010 (SI 2010 No. 2214). Planning Portal (http://www.planningportal.gov.uk/buildingregulations/approveddocuments/) documents are:

- A Structure (2013) (Incorporating 2010 and 2013 amendments);
- B Fire Safety Vol 1 Dwellings (2019) (Incorporating 2020 amendments), Vol 2 Buildings other than Dwellings (incorporating 2020 amendments);
- C Site preparation and resistance to contaminates and moisture (2013) (Incorporating 2010 and 2013 amendments);
- D Toxic substances (2015) (Incorporating 2010 and 2013 amendments);
- E Resistance to the passage of sound (2015) (amendments in 2010, 2013 and 2015);
- F Ventilation (2015) (Incorporating amendments made in with amendments in 2010 and 2013);
- G Sanitation, hot water safety and water efficiency (2015) (amendments in 2016);
- H Drainage and waste disposal (2015);
- J Combustion appliances and fuel storage systems (2015) (Incorporating 2010 and 2013 amendments);

- K Protection from falling, collision and impact (2013 for use in England) (This includes provisions on glazing previously in AD N);
- L–L1A Conservation of fuel and power (new dwellings) (2014) (with 2016 amendments); L1B Conservation of fuel and power (existing dwellings) (2010) with amendments in 2011 2013, 2016 and 2018); L2A Conservation of fuel and power (new buildings other than dwellings) (2014) (with 2016 amendments); L2B Conservation of fuel and power (existing buildings other than dwellings) (2010) (with 2011, 2013 and 2016 amendments);
- M Access to and use of buildings Vol 1 Dwellings (2015 with 2016 amendments) and Vol 2 Buildings other than dwellings (2015 with 2020 amendments);
- N Glazing (applies in Wales only) (1998 incorporating 2000 and 2010 amendments) https://gov.wales/sites/default/files/publications/2019-05/building-regulations-guidance-part-n-glazing-safety.pdf

- P Electrical safety – dwellings (2013 for use in England);
- Q Security – dwellings (2015);
- R Physical infrastructure for high speed electronic communications networks (2016);
- AD Regulation 7 (Materials and workmanship) (2013 incorporating 2018 amendments).

Information Papers, Digests and Good Repair Guides on Construction matters from the BRE (Building Research Establishment Bookshop see http://www.brebookshop.com/).

BM TRADA https://www.bmtrada.com (BM TRADA is part of the Element Group, a global network of more than 6,700 experts, operating out of almost 200 laboratories, located in North America, Europe, the Middle East, Africa, Asia and Australia)

WHO, 2011, Handbook for Inspection of ships and Issuance of Ship Sanitation Certificates (accessible at http://apps.who.int/iris/bitstream/10665/44594/1/9789241548199_eng.pdf?ua=1)

10

Human physiology, hazards and health risks

*Revati Phalkey, Naima Bradley, Alec Dobney, Virginia Murray,
John O'Hagan, Mutahir Ahmad, Darren Addison, Tracy Gooding,
Timothy W Gant, Emma L Marczylo and Caryn L Cox*

Chapter introduction

If environmental health is concerned with the impact of environmental stressors on human health it is necessary to have some understanding of basic physiology and anatomy as well as the risks to human health and physiological impacts. This chapter examines some of the relevant aspects of physiology, but also looks at the effects of some of the stressors from toxins to radiation in various forms. It is hoped that this will help EHPs explain to people why intervention is necessary.

This chapter is subdivided into six sections with different authors; the first looks at the human body and is a reviewed version of this section in the last edition. The other sections look at chemical hazards and risks, ionising and non-ionising radiation, radon as a particular issue, epigenetics and infectious diseases.

SECTION 1: THE HUMAN BODY IN HEALTH AND DISEASE

David J Baker as reviewed by Revati Phalkey

Introduction

Humankind has evolved from life forms that were originally just single cells. Cell structure and function determines life, disease and death. Almost all cells have fundamental activities, such as the generation of energy, which are necessary for maintaining cell integrity and function. The body as a whole provides each of its 30 trillion cells with a constant suitable environment. This was termed the milieu interior (internal environment) that is the fluid that is inside and outside the cells. This is the environment in which the cells function optimally where there is an appropriate balance of electrolytes, ions, temperature, hormones and all other factors necessary for normal cell function. The essential function of the systems of the human body is to make the necessary adjustments as and when necessary to ensure that this internal environment is preserved and protected from disease or changes in the external environment.

DOI: 10.1201/9781003035640-10

Cell structure

Membranes made up of a double layer of lipid (fat) molecules in which proteins are embedded, surround each cell and small functional units within the cell called organelles. They act as a selective barrier to the passage of molecules, detect chemical messengers arriving at the cell surface and link adjacent cells together to form tissues and organs. However, there exists a small space of about 20 nanometres between the opposing membranes of adjacent cells.

Almost all cells have a single nucleus (exceptions being skeletal muscle cells with multiple nuclei and red blood cells with none) whose primary function is the transmission and expression of genetic information. The most prominent organelle within the nucleus is the nucleolus which is composed of DNA (deoxyribonucleic acid) and RNA (ribonucleic acid) and proteins from which ribosomes (organelles found in the cytoplasm) are assembled. Chromatin is composed of DNA and protein occurs in the nucleus as coiled threads which condense to form chromosomes within the nucleus that store genetic information which is transferred from cell to cell when cells divide. The transference of genetic information from the nucleus to other parts of the cell is done by messenger RNA.

Cytoplasm is the area outside the nucleus within the cell wall and contains the endoplasmic reticulum which is involved in packaging of proteins to be secreted by the cell and as a site for lipid synthesis. In addition, there are ribosomes which are composed of a large number of proteins and several RNA molecules and synthesise proteins from amino acids using genetic information sent by messenger molecules from DNA in the nucleus. The other important organelles in the cytoplasm are the mitochondria, which are spherical or elongated rod-like structures. There are as many as 1000 are present in cells that use large amounts of energy, whilst smaller numbers are present in less active cells. Mitochondria are primarily concerned with the chemical processes by which energy is made available to the cell in the form of adenosine triphosphate (ATP) by a process that uses oxygen and results in formation of carbon dioxide. Therefore, the mitochondria are the powerhouses of the cells. In all cells, the transfer of energy from reactions to cell function is through ATP (adenosine triphosphate). The energy resulting from the hydrolysis of ATP is used for functions/processes in the cells that require energy. Cells use ATP to transfer but not to store energy.

Cell function and metabolism

The mitochondria contain iron-containing proteins called cytochromes whose function depends on oxygen. If oxygen is not available to the cytochrome system due to either decreased oxygen intake into the human body (decreased oxygen concentration in air due to environmental pollutants or by toxic chemicals such as cyanide which block the function of cytochromes), ATP cannot be formed normally. A process called anaerobic metabolism which does not use oxygen takes over which is much less efficient than aerobic metabolism and leads to the build-up of lactic acid, ultimately cell death.

Lysosomes are organelles that break down bacteria and dead cells that have been taken into the cell. This is an intracellular 'digestive system' which plays an important role in specialised cells which make up the defence system of the body.

Metabolism – the thousands of chemical reactions taking place within the human body all the time are collectively referred to as metabolism. In the body there is a balance between synthesis of substances for maintaining cell structure and function (anabolism) and breakdown to excess of or waste products (catabolism). Many of these processes require enzymes to speed up chemical reactions in the cells. The route taken to produce essential chemicals in the body is termed a metabolic pathway.

Cell Division – During the formation of the 30 trillion cells in the human body, from the basic fertilised egg following conception at least 30 trillion cell divisions must occur. During the duplication of genetic material, errors may occur and any alteration in the genetic message carried by DNA is referred to as a mutation. Factors in the environment (e.g. some chemicals, ionising radiation and X-rays) that increase the rate of mutations are called mutagens. Inherited diseases due to gene mutation are referred to as inborn errors of metabolism.

Agents including viruses, radiation and some chemicals which alter or activate various genes involved in cell growth and division are called carcinogens. This results in cells that have lost the ability to respond to the normal control mechanisms that regulate cell growth. Cancer cells have the capacity for unlimited multiplication and spread often via the circulatory system to other parts of the body to form multiple tumour sites (metastases).

Differentiation and Grouping of Cells – Some cells become specialised to perform different functions such as muscle cells for movement, nerve cells for transmission of impulses from the brain and liver cells for metabolism. This transformation of cells to perform specialised functions is referred to as cell differentiation. Approximately 200 distinct types of cells can be identified in the human body, as regards structure and function.

The next feature is migration of cells to different locations to adhere to each other and form multi-cellular structures or tissues (e.g. muscle tissue, nerve tissue, epithelial tissue and connective tissue) which may combine with other types of tissues to form organs. Organs in general have varying proportions of tissues, often arranged differently (i.e. in layers or bundles, etc.). Most organs possess small similar functional sub-units, each performing the functions of the specific organ. Organs performing similar functions are often grouped together as systems. There are ten organ systems in the human body divided on the basis of both structure (anatomy) and function (physiology). Each of these systems can be affected by exposure to environmental toxic agents whilst some play an essential role protecting the human body or minimising harm following toxic environmental exposures.

Systems of the body

The systems of the body are as follows:

1 *Nervous system* – This consists of the brain, spinal cord, peripheral nerves. These are subdivided into voluntary (controlling muscle function) and involuntary (the autonomous nervous system which is further subdivided into the sympathetic and parasympathetic nervous systems). These control the subconscious functions of the body such as heart rate. In addition, the nervous system has ganglia (these are effectively amplifiers in the conduction of nerve impulses using chemicals termed neurotransmitters. The neuromuscular junction controlling voluntary muscle function is an example. There are also special sense organs (eyes, hearing, smelling).

   The principal roles of the nervous system are in determining the state of consciousness, learning, cognition (act or experience of knowing or acquiring knowledge) and regulation of many activities of the body (e.g. central control of respiration, blood pressure, temperature).

2 *Respiratory system* – This comprises the nose, pharynx, larynx, trachea, bronchi and alveoli, where oxygen exchange to the blood takes place. Respiration is essential for intake of oxygen from the air and removal of carbon dioxide from the body, and regulation of the hydrogen ion concentration in the body (the pH).

3 *Circulatory system* – This comprises the heart, blood vessels, blood, lymphatic vessels. The flow of blood through the body is often assessed by skin colour, body warmth, blood pressure and pulse rate.

4 *Gastrointestinal system* – This consists of mouth, pharynx, oesophagus, stomach, small and large intestine, salivary glands, pancreas, liver, gall bladder. Its function is the digestion and absorption of nutrients, water and salt.

5 *Urinary system* – This comprises the kidneys, ureters, bladder, urethra. Its function is the regulation of the concentration in plasma of electrolytes (e.g. sodium, potassium), the regulation of body water, hydrogen ion concentration and excretion of waste products.

6 *Endocrine system* – This comprises 11 glands, secreting hormones (pituitary, hypothalamus, pineal, thyroid, parathyroid, thymus, adrenal, pancreas, testes, ovaries, intestine) which are involved in the regulation and coordination of several important activities of the body.

7 *Immune system* – This comprises the white blood cells, lymph vessels and nodes, spleen, thymus and lymphoid tissue in other organs/tissues. They are involved in defence against foreign organisms, toxins, return of extracellular fluid to the blood and production of white blood cells.

8 *Reproductive system* – In the male this comprises the testes, penis, and associated ducts and glands. In the female this is the ovaries, uterine tubes, uterus, vagina, mammary glands. The system controls the production of sperm and transfer of sperm to the female, production of eggs (ova) and the provision of a suitable environment for the developing embryo (foetus), and the nutrition of infants.

9 *Integumentary system* – This is the skin which provides protection against injury, dehydration, defence against foreign invaders and the regulation of body temperature.

10 *Musculo-skeletal system* – This is comprised of cartilage, bone, ligaments, tendons, joints, skeletal muscles. It is involved in the support, protection and movement of the body.

The nervous system and associated organs

Hearing (Figure 10.1)

1 Ear lobes amplify and direct sound waves to ear canal.

2 The tympanic membrane (ear drum) stretched across the inner end of ear canal vibrates at the same frequency as the sound wave (i.e. it vibrates slowly to low frequency waves, rapidly to high frequency waves).

3 The middle ear cavity (cavity in temporal bone of skull) is connected to the exterior (i.e. atmospheric pressure) by the Eustachian tube (auditory tube) – the cause of earache whilst flying in aeroplanes, etc.

4 Sound energy is transmitted from the tympanic membrane through the middle ear cavity to the inner ear, the cochlea, a spiral passage in the temporal bone, which also contains the semi-circular canals (these contain the sensory organs for equilibrium and movement).

5 Three small bones – the malleus, incus and stapes act as pistons and couple the motions of the tympanic membrane to the membrane-covered oval window which separates the middle and inner ears. As the oval window is smaller in size compared to the tympanic membrane, the pressure caused by the sound waves on the tympanic membrane is increased nearly 20 times.

6 The cochlea contains the Organ of Corti with the ear's sensitive receptor cells (hair cells) that transform pressure waves to receptor potentials (i.e. transducers). Hair cells are easily damaged following exposure to high-intensity noises.

7 Hair cells synapse with afferent fibres of the cochlea nerve. Hair cell depolarisation leads to release of chemical transmitters that activate neurones in the cochleal nerve to generate a nerve impulse. Cochleal nerves reach the brain stem and then on to the thalamus and finally to the auditory cortex.

The hair cells in the semi-circular canal make up the vestibular system which detect changes in motion and position of the head. Information from these hair cells in semi-circular canals are conveyed by the eighth cranial nerve to the cortex. (See Figure 10.1.)

Vision

1 The iris is smooth muscle and the sympathetic nervous system enlarges the pupil while the parasympathetic nervous system constricts the pupil.
2 The ciliary muscle controls the shape of the lens and is controlled by parasympathetic nerves.
3 The cornea focuses the image on the retina and light rays are bent passing from air into the cornea.
4 The lens is elastic and makes adjustments for distance by changing shape (accommodation). Opacity of the lens is called a cataract.
5 The retina is a thin layer of neural tissue with light receptors termed rods and cones. Light rays hit the rods, cones, and transmit signals via bipolar cells – contained in the retina to the topic cortex of the brain.
6 The aqueous humour maintains the intraocular pressure and there may be a role in immune response to defend against pathogens. Its main function is to provide dioptric power to the cornea. Glaucoma is a condition characterised by increased intraocular pressure (pressure within the eye) either through increased production or decreased outflow of aqueous humour.
7 The optic nerve is connected to the thalamus and then the visual cortex within the brain (see later). (See Figure 10.2.)

The role of the nervous system

The nervous system monitors, integrates (processes) and responds to information from inside and outside the body. It controls or regulates many body functions essential to life such as breathing (respiratory centre), circulation (vasomotor centre), hormonal secretions, temperature

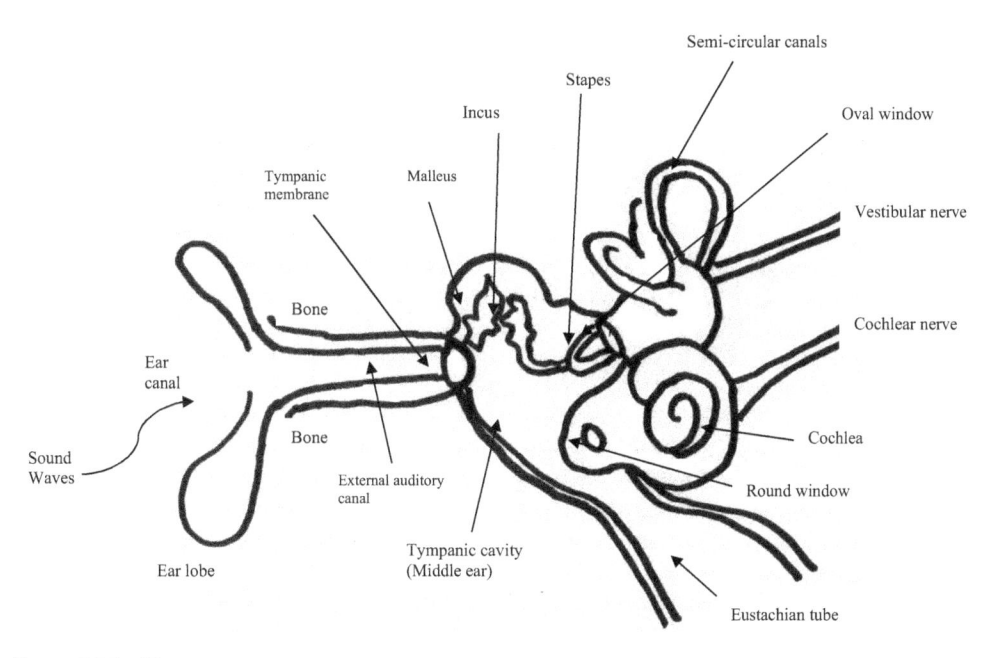

Figure 10.1   The ear

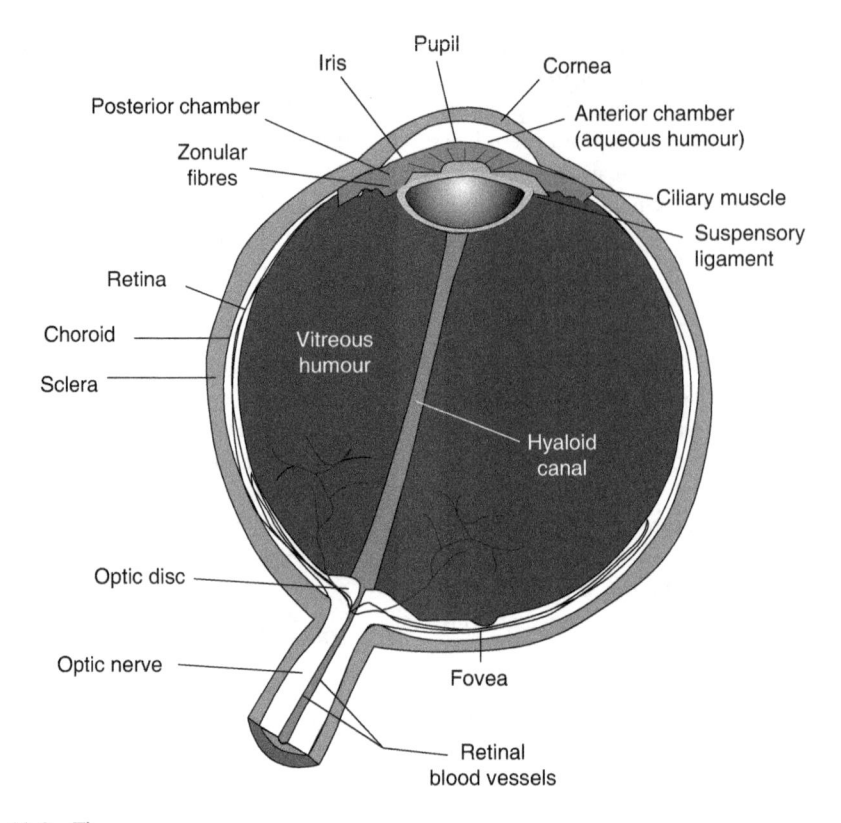

Figure 10.2 The eye

and indirectly the activity of the lungs, blood vessels, heart, kidneys and several other organs. The role of the nervous system is to maintain normal body function by making the necessary adjustments or responses to changes which may eventually cause harm or ill health.

This is done by receiving information from the outside world by means of transmission of electrical impulses passing along nerve cells (neurons). In various parts of the system these signals are passed from cell to cell by special amplifying relay stations called synapses, where the signal is passed by chemical messengers called neurotransmitters. Special synapses are found at the end of nerves which pass a signal directly onto an organ, also through a neurotransmitter.

Various parts of the nervous system are interconnected but may be conveniently divided into the central nervous system

(CNS), comprising the brain and spinal cord, and the peripheral nervous system (PNS), which is composed of nerves extending to and from the CNS to all parts of the body.

Neurons and transmission of impulses

The basic unit of the nervous system is the nerve cell and its processes, which is called the neuron.

The branched outgrowths from the nerve cell, the dendrites and the cell itself constitute the specialised junctions where electrical impulses are received from other neurons. The single long extension from the nerve cell, the axon (sometimes referred to as a nerve fibre) propagates the signals from the cells ending in an axon terminal which transmits chemical signals to cells of other body systems.

The functional classification of neurons is

1 Afferent neurons, which convey information from tissues and organs to the CNS, and
2 Efferent neurons, which transmit signals to cells that perform different functions (e.g. muscle cells and glands). Interneurons (which lie wholly within the CNS) connect the afferent and efferent neurons.

A nerve impulse is a transient event, which for a very short period of time alters the permeability of the cell membrane and allows sodium ions to enter the cell. These positively charged sodium ions change the potential inside the cell from -70mV to +40mV, a process called depolarisation. This sudden change inside the cell is termed an action potential which is propagated along the cell membrane. Each action potential corresponds to a nerve impulse or message which is conveyed to its destination by the sequential changes in permeability of the membrane.

A nerve impulse lasts about one millisecond and each nerve impulse in both motor and sensory nerves (i.e. nerves that carry messages or impulses to the spinal cord and brain from the peripheral tissues as regards pain, temperature, touch) is associated with sodium entering and potassium leaving the cell momentarily.

The rate of propagation of the nerve impulse or nerve conduction varies with the size of the nerve fibre. The large nerve fibres with a diameter of 20 μm have a velocity of conduction of 120 metres per second. These large fibres have a sheath made up mainly of fatty material called myelin (i.e. the nerves are myelinated), and the gaps in this sheath called nodes of Ranvier enable a nerve impulse to 'leap-frog' down the nerve as the exchange of sodium and potassium ions only occur at these interruptions in the myelin sheath. This allows transmission down the axon.

The smaller fibres such as those that convey pain impulses to the brain are about 1 μm in diameter and are not individually myelinated. Therefore, they are able to conduct impulses at about 5 metres per second.

The central nervous system

This is the part of the nervous system consisting of the brain and spinal cord. The brain lies within the skull and the spinal cord within the vertebral column, the nerve tissues and the bony structures being separated by three membranes, the meninges. The dura mater lies adjacent to the bone, the arachnoid in the middle and the pia adjacent to the nervous tissue, which effectively protects this system and also covers the nerves entering or leaving the CNS to varying lengths. The space between the arachnoid and pia is filled with cerebrospinal fluid (CSF), a watery liquid similar, but not identical, in composition to blood plasma, which supports, cushions and nourishes the brain. The meninges also form partitions within the skull and brain. Inflammation of these lining membranes gives rise to the serious condition called meningitis whilst inflammation of the brain tissue itself (nerve cells) is referred to as encephalitis.

The major regions of the brain are the cerebrum, diencephalon, midbrain, pons, medulla oblongata and the cerebellum, which are shown in Figure 10.3.

The cerebrum and diencephalon constitute the forebrain. The cerebrum consists of the cerebral hemispheres (the upper most parts of the brain) making up approximately 83% of total brain mass which is responsible for conscious behaviour containing three different functional areas: the motor areas, sensory areas and association areas. It has an outer layer (2–4 mm thick) of grey matter. Located internally in the cerebral cortex is the white matter, which is responsible for communication between cerebral areas and between the cerebral cortex and lower regions of the CNS, as well as the basal nuclei (or basal ganglia), involved in coordinating muscular movement and several bodily functions.

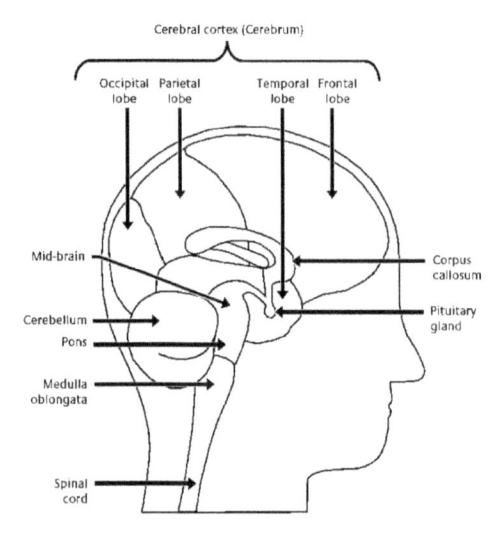

Figure 10.3 The brain

The diencephalon is located centrally within the forebrain (the anterior or front part of the brain). It consists of the thalamus, hypothalamus and epithalamus, which together enclose the third ventricle (a sac containing cerebrospinal fluid found within the brain which is connected to the lateral ventricles in the cerebral hemispheres and to the 4th ventricle in the brain stem). The thalamus acts as a grouping and relay station for sensory inputs (inputs such as pain, touch and temperature from the periphery), ascending to the sensory cortex and associated areas. It also mediates motor activities, cortical arousal or wakefulness and memories. The hypothalamus, by controlling the autonomic (involuntary) nervous system, is responsible for maintaining the body's homeostatic balance by maintaining the concentrations of ions and pH of the internal environment ('the milieu interior'). Moreover, the hypothalamus forms a part of the limbic system, the 'emotional' brain. The epithalamus consists of the pineal gland and its connections.

The midbrain, pons and medulla together form the brain stem which is literally the stalk of the brain through which all nerve fibres that relay signals between the cerebrum and cerebellum and spinal cord pass. Running through the core of the brain stem is the reticular formation, which is a network of nerve cells and is one part of the brain absolutely essential to life.

This part of the brain contains collections of neurons which are referred to as 'nuclei' or 'centres', which control several vital functions such as cardiac (heart) activity and respiration (breathing) and also centres associated with visual and auditory reflexes.

The nuclei in the medulla oblongata regulate respiratory rhythm (the respiratory centre), heart rate (cardiac centre), blood pressure (vasomotor centre) and contain the nuclei of several cranial nerves.

Cranial nerves comprise 12 pairs of nerves which arise directly from the brain and not from the spinal cord and leave the brain through apertures or foramina in the skull.

The cerebellum located behind the pons and medulla accounts for about 11% of total brain mass and has a thin outer cortex of grey matter, internal white matter, and small, deeply situated paired masses (nuclei) of grey matter. The cerebellum processes impulses received from the cerebral motor cortex, various brain stem nuclei and sensory receptors in order to 'fine-tune' skeletal muscle contraction (the muscles that control movement-walking, running, etc.), thus giving smooth, coordinated movements.

The spinal cord is the direct continuation of the brain from the brainstem into the vertebral column and contains the nerve pathways through which messages are sent from the brain (efferent pathways) and to the brain (afferent pathways). Like the brain itself, the spinal cord is composed of central grey matter containing nerve cells and white matter which comprises nerve fibres. Not all messages controlling the body have to go up to the brain and return. The nerves of the grey matter in the cord can act on their own, in response to a sensory stimulus. This action is known as a 'spinal reflex' and the knee jerk is perhaps the best-known example.

The peripheral nervous system

The peripheral nervous system contains motor nerves that control voluntary movement together with sensory nerves that carry information concerning touch, pain, temperature and position to the brain.

The autonomic nervous system is the part of the peripheral nervous system that supplies smooth muscles (in contrast to the striated or skeletal muscles found in our limbs and other parts of the body). The autonomic nervous system also controls the heart, the digestive and urinary systems and the secreting glands such as sweat and salivary glands. In general, the autonomic nervous system is concerned with *involuntary* nerve impulses. The part of the peripheral nervous system that controls *voluntary* actions, such as movement, is known as the somatic nervous system.

The peripheral nervous system is subdivided into

- The sympathetic nervous system;
- The parasympathetic nervous system.

These systems differ in the chemical transmitter involved in the transmission of impulses at synapses. In the sympathetic nervous system, the chemical transmitter is predominantly noradrenaline, whereas in the parasympathetic nervous system the chemical messenger is acetyl choline. The actions of these two sections of the autonomic nervous system are shown in Table 10.1.

The sympathetic nervous system is active in states of emotional excitement and stress and gives rise to what has been called the 'flight or fight' reaction. Increased sympathetic nerve activity causes the heart to beat faster and increases the force of contraction of the ventricles of the heart (the force with which the heart muscle contracts) which result in an increase in the output from the heart (cardiac output). Therefore, the blood pressure increases. In addition, the pupils of the eye dilate, the air passages increase in diameter allowing an individual to breathe in more air, and also contracts the muscles associated with sweat glands and skin causing the hair to 'stand on end' and form goose pimples. The rate at which breathing occurs also increases due to excitement of the respiratory centres in the brain. In addition, sympathetic stimulation slows down the contractions of the digestive tract.

The chemical transmitters found in the sympathetic nervous system are adrenaline (epinephrine) and noradrenaline (norepinephrine). Chemically, noradrenaline and adrenaline are amines of the benzene derivative catechol and are referred to collectively as catecholamines. Noradrenaline is rapidly removed after release, mainly by re-uptake into the nerve so that the target organ is capable of responding to further nerve impulses.

Over-activity of the sympathetic nervous system leads to narrowing of blood vessels (vasoconstriction) and consequently a reduction of the blood supply to the organ or tissue. If it is widespread, the narrowing of the blood vessels will lead to an increase in blood pressure (hypertension), profuse sweating and dilatation of the pupils.

Sympathetic nerve fibres can cause either contraction or relaxation of the smooth muscle of the innervated structure. The receptors, which cause contraction as a response are alpha receptors and those that cause dilatation are beta receptors. These can be blocked selectively by either alpha or beta blockers. Blockade of beta receptors in the heart decreases the force with which the heart muscle contracts and is used to reduce blood pressure. Blocking the effects of alpha receptors on blood vessels, arterioles, where there is maximal resistance to blood flow would cause a fall in blood pressure. Alpha blockers are thus used to treat high blood pressure. Stimulation of beta receptors in the airways is used in the treatment of asthma to dilate or increase the lumen through which air can move in the small airways or bronchioles of the lung. In addition, there are ganglion-blocking drugs that block transmission at the sympathetic ganglia, which are also used to treat very high blood pressure in cases of emergency.

Table 10.1 Actions of the autonomic nervous system

Organ supplied	Sympathetic activity	Parasympathetic activity
Pupil of the eye	Dilates	Constricts
Air passages, bronchi and bronchioles	Dilates	Constricts
Salivary glands	-	Increased salivary secretion and dilatation of blood vessels
Heart	Speeds up, increases force of ventricular contraction	Slows heart rate
Digestive tract	Reduces motility	Increases motility
Sphincters of the digestive tract	Constricts	Relaxes
Rectum	Allows filling	Empties and relaxes internal anal sphincter
Bladder	Allows filling	Empties and relaxes internal sphincter
Blood vessels	Vasoconstriction	Nil (except salivary gland and external genitalia-vasodilation)
Sweat glands	Sweating	Nil

The fibres of the parasympathetic nervous system originate in the cranial nerves and from the lower end of the spinal cord (the sacral region). The 12 cranial nerves have the cells of origin in the brainstem. The third, seventh, ninth and tenth cranial nerves all contain parasympathetic fibres. The tenth cranial nerve or the vagus nerve is the principal parasympathetic nerve and its stimulation causes a slowing of the heart amongst many other effects, such as those on the stomach and stomach secretions, oesophagus and the small airways of the lungs.

The neurotransmitter at ganglions (synapses), which are junctions between the pre-ganglionic fibres and the post-ganglionic fibres, in both the parasympathetic nervous system and the sympathetic nervous system is acetylcholine. Acetylcholine is also the neurotransmitter at the post-ganglionic nerve endings of the parasympathetic nervous system and acts as the neurotransmitter at some post-ganglionic nerve endings of the sympathetic nervous system.

Acetylcholine (ACh) has two distinct actions within the autonomic nervous system and at the neuromuscular junction, which are described as nicotinic or muscarinic.

Like noradrenaline, the effects of acetylcholine are terminated very quickly after its release. However, with acetylcholine it is due to the activity of the enzyme acetylcholine esterase (AChE). Inhibition of this enzyme is the basis of poisoning by organophosphate pesticides and nerve gases.

In most parts of the body, the action of the parasympathetic nervous system is the opposite of that of the sympathetic nervous system (Table 10.1). Thus, it slows the heart rate, lowers blood pressure, constricts the pupils and constricts or narrows the airways. In addition, the parasympathetic nervous system speeds up digestion and plays an important role in defecation and emptying of the bladder and increases secretions from several glands such as the salivary glands and tear glands.

There are many chemicals that act as neurotransmitters. The more common neurotransmitters and those of particular interest in toxicology are:

- Acetylcholine;
- Norepinephrine or noradrenaline;
- Epinephrine or adrenaline;

- Dopamine;
- Serotonin;
- Histamine;
- Gamma amino butyric acid (GABA);
- Glycine;
- Glutamate aspartate.

Dopamine, norepinephrine and epinephrine are a group of neurotransmitters called 'catecholamines'. Norepinephrine used to be known as 'noradrenaline' and epinephrine as 'adrenaline'. Each of these neurotransmitters is produced in a step-by-step fashion by different enzymes.

The cardiovascular system

The cardiovascular system comprises the heart and blood vessels and is involved in the transport of blood around the body. Blood vessels that carry blood away from the heart are called arteries and vessels that bring blood to the heart are called veins. Arteries carry oxygenated blood away from the heart into capillaries supplying tissue cells. The exception is the pulmonary artery which carries venous blood from the right side of the heart to the lungs. Veins collect the deoxygenated blood from the capillary beds in the body tissues and carry it back to the heart, but the exception is the pulmonary vein which brings oxygenated blood from the lungs to the left side of the heart.

The circulatory system is divided into:

- The pulmonary circulation, which takes deoxygenated blood from the right side of the heart to the lungs and returns oxygenated blood to the left side of the heart;
- The systemic circulation, which takes oxygenated blood (oxygen diffuses into the blood whilst it passes through the alveoli – the small units in the lung which contain 'fresh' air taken in from outside) from the left side of the heart to all cells in the body and returns deoxygenated blood (i.e. blood from which the cells have extracted the necessary oxygen and

carrying the carbon dioxide returned by the cells) to the right side of the heart.

The circulatory system also carries food components from the digestive tract to the cells to provide nutrition for growth and energy and waste products from cells in the body to the kidneys to enable the body to get rid of (excrete) these unwanted products in the urine. The circulatory system transports hormones from the glands that produce them (endocrine glands) to other organs of the body and is also involved in ridding the body of surplus heat by increasing loss through the skin.

The heart consists of two pumps, right and left, which circulate blood round the body. Each pump has two chambers, the atrium which collects the blood either from the lungs (the left atrium) or from the tissues (right atrium) and then passes through valves to the major pumping chambers, the ventricles. Thus, the left ventricle pumps the blood with oxygen whilst the right ventricle pumps the blood that has returned from the cells or tissues (after extraction of oxygen and nutrients and containing greater amounts of carbon dioxide than in the blood which reached the cells from the left ventricle) to the lungs to get rid of carbon dioxide and pick up more oxygen.

Each time the heart beats, each ventricle pumps out about 70 ml of blood, and this volume is termed the stroke volume. The heart beats about 70 times a minute, which is termed the heart rate; this is usually measured in patients by counting the pulse rate at the wrist.

From the left side of the heart, oxygenated blood is pumped by the left ventricle to the aorta, the main artery in the body. From the aorta arise the large arteries or branches which carry the oxygenated blood to all parts of the body. For example, the carotid arteries arising from the aorta supply the brain, and the renal arteries arising from the aorta supply the kidneys. From the aorta, blood goes to the arteries and then to the smaller arterioles as it nears organs or cells to which the blood supply is intended. From arterioles the blood flows into the very thin-walled capillaries which travel between cells and give out oxygen and allow carbon dioxide to come into the blood vessel. Now the blood, which has provided oxygen and nutrition to the cells, passes from the end of the capillaries to the venules, veins and then to the larger veins – the superior and inferior vena cava.

The capacity of the venous side of the circulation is much greater than that of the arterial side and often accommodates 75% of the blood volume. This is because the walls of the vessels on the venous side are not as thick and contain less muscle than the vessels on the arterial side.

The left ventricle contracts each time the blood is pushed into the aorta, which produces a wave of flow commonly referred to as a pulse. This pulse can be felt only in arteries which are often not visible. On the contrary veins are often visible and a pulse cannot be felt in veins as the flow in them is not pulsatile.

Blood pressure is the force applied to the wall of the arteries as the heart pumps blood to the cells in the body through the arterioles and capillaries. There are two components to a measurement of blood pressure. The first is the maximum pressure exerted when the heart is pumping the blood – the systolic pressure. The second represents the pressure in the arteries when the heart is at rest – diastolic pressure. In 'normal' people these measurements are recorded as 130/80 mm of Hg. The pulse is due to the fluctuation between systolic and diastolic blood pressure.

The measurement of blood pressure is dependent on the amount of blood pumped by the heart (cardiac output), which depends on the contraction of the heart muscle and the volume of blood in the left ventricle. The latter is determined by the volume of blood circulating and the return of blood to the right side of the heart. It follows that if the heart has been damaged, for example, following a heart attack (injury to heart muscle

caused by the lack of blood flow in the arteries supplying the heart muscle, coronary vessels, most often due to blood clot (thrombus) in the small coronary arteries), or if there is depletion of the blood volume either due to massive blood loss following trauma or due to conditions such as dehydration (lack of fluid) due to severe diarrhoea or vomiting, the blood pressure would fall. In addition, there are some toxic substances and drugs used in the treatment of different disease states that are capable of causing either direct damage to the heart muscle or decreasing the force with which the heart muscle contracts. Another consideration is that for normal cardiac function, the heart has to beat rhythmically – mainly regulated by nerve impulses from the cardiac centre ion the medulla. Some toxins and drugs can alter heart rhythm making the pumping action of the heart relatively ineffective. Such disorders are referred to as arrhythmias.

The resistance to blood flow is mainly in the small arterioles as it is the smaller diameter vessels that offer the greatest resistance to flow through them. The arterioles have smooth muscle in their walls, which is arranged circularly. When the muscle contracts it makes the blood vessels smaller. The sympathetic nervous system provides a nerve supply to these vessels and stimulation of this system causes contraction of the smooth muscle in the walls and thus vasoconstriction and an increase in peripheral resistance and hence of blood pressure. When the arterioles are relaxed or when the sympathetic nerve stimulation is not present, the vessels are said to be vasodilated. The diameter of these vessels is directly under control of the sympathetic nerve outflow regulated by the vasomotor centre in the medulla oblongata of the brain. Blood pressure is determined as follows:

$$\text{Blood pressure} = \text{cardiac output} \times \text{peripheral resistance}$$

There are many nerve cell groupings referred to as receptors that influence the activity of the vasomotor centre, best known being the baroreceptors found in a special area, the carotid sinus, of each carotid artery (arteries supplying blood to the brain). The higher centres in the brain (regions of the brain where conscious thoughts occur) also influence the activity of the vasomotor centre. Emotional stress and excitement cause stimulation of the vasomotor centre and an increase in blood pressure. Carbon dioxide content in the blood also influences vasomotor centre activity. When the carbon dioxide content and tension is low, as in patients who are breathing rapidly, the activity of the vasomotor centre is reduced. A shortage of oxygen, in contrast, would increase the activity of the vasomotor centre.

Blood pressure is not constant and there are several factors affecting it. Some of these occur in healthy persons related to psychological factors such as stress, anger and fear. Blood pressure may also be raised by pathological conditions which include disease of arteries such as thickening or loss of muscle fibres (arteriosclerosis, which usually occurs with ageing), and kidney disease. Certain hormonal disorders are also associated with a high blood pressure. An increase in blood pressure may occur during pregnancy in some individuals.

The haematopoietic system is the body organs and tissues involved in the formation and functioning of blood elements and includes the bone marrow and spleen. An adult has approximately 5 litres of blood in the body. A new-born baby has only 300 ml (80 ml per kg body weight). Blood is composed of cells (45%) and plasma (55%). Blood cells are formed in the bone marrow, which is found in cavities of bones. Blood cells can be divided into red and white blood cells and platelets.

Red blood cells – There are approximately 5 million red blood cells per cubic mm of blood that contain the pigment haemoglobin, which is bright red in colour when combined with oxygen and is purple-blue in colour when no oxygen is present. Every 100 ml of blood has about 15 g of haemoglobin.

Haemoglobin plays an important role in the carriage of not only oxygen but also of carbon dioxide. The life span of the red cell is about 120 days. It has no nucleus. The main requirements for the formation of red blood cells are iron, folic acid and vitamin B12. The hormone erythropoietin which is formed in the kidney also plays a role in the production of red blood cells.

White blood cells – These are grouped as granulocytes (neutrophils, eosinophils and basophils), monocytes and lymphocytes.

Neutrophils are the most common white blood cells, and their main function is to recognise, ingest and destroy foreign particles and micro-organisms (e.g. bacteria). Eosinophils constitute 1–6% of the circulating white cells and are involved in allergic reactions. Basophils which are less than 1% of the circulating white cells bind certain antibodies (IgE) and when the specific antigen enters the circulation, there is degranulation (breakdown) of the cells releasing histamine, leukotrienes and heparin. Thus, they play an important role in hypersensitivity reactions. Monocytes are the largest white cells and are best known for their ability to act, like neutrophils as macrophages. Lymphocytes produce antibodies, which are able to react with foreign substances called antigens and destroy them. Thus, they are an important protective mechanism of the blood and of the body. Up to the age of 7 years, lymphocytes are the most abundant white cells in the body.

Haemoglobin – This is a protein specially adapted for gas transport to and from the lungs. The element in the haemoglobin to which oxygen binds is iron. Anaemia occurs either when the total number of red blood cells are decreased in number or when there is a diminished concentration of haemoglobin in each red cell or when both these occur. Though carbon dioxide is more soluble in water than is oxygen, some carbon dioxide is also carried, bound to haemoglobin as carbamino-haemoglobin.

Platelets and blood clotting – The circulation of blood is a transport system that requires protective mechanisms to prevent loss of blood (haemostasis) and also to prevent inappropriate cessation of blood flow. Haemostasis depends on the interaction between vessel walls, platelets and clotting factors. When a vessel is damaged, the vessel contracts and platelets aggregate at the site of injury to form a plug to arrest blood loss (haemorrhage). Then the clotting factors group to form a fibrin mesh to secure the platelet plug.

Platelets have two main functions in the human body. They are able to clump together and block small holes in the blood vessels by forming platelet plugs. The breakdown of platelets causes the release of a factor called thromboplastin, which converts a component of the blood called prothrombin to thrombin in the presence of calcium ions. Thrombin,

when formed, acts on another component in the blood called fibrinogen, and changes the fibrinogen to fibrin which is actually the blood clot – the protective mesh at the site of injury of the vessel wall.

The coagulation system consists of soluble proteins designated by Roman numerals which are capable of activating one or more components to initiate a cascade of events that result in the formation of the fibrin mesh.

There are a few inherited diseases where the clotting of blood is affected due to lack of specific substances in the person's blood. These diseases include haemophilia and Christmas disease.

Plasma – Plasma is the straw-coloured fluid in which the blood cells are suspended. It consists of a watery solution of plasma

---

Common medical terms used to describe symptoms and signs of heart and blood vessel disorders

Anaemia: There are several different types of anaemia and each one has a different cause. Types of anaemia include vitamin B12-deficiency, folate-deficiency and sickle cell disease, but the most common is iron-deficiency anaemia. It is possible to get anaemia when the bone marrow does not make enough good-quality red blood cells, for example due to a lack of essential vitamins such as vitamin B12 or folate, or as the result of a serious bone marrow disorder such as leukaemia or aplastic anaemia (where the bone marrow fails to produce blood cells correctly) or as the result of a chronic disease such as rheumatoid arthritis. The kidneys produce erythropoietin, which stimulates bone marrow to make red blood cells. Long-term kidney problems may cause anaemia due to a lack of erythropoietin.

Hypotension: is an abnormally low blood pressure which can occur following blood loss, failure of the heart to pump efficiently (heart failure) or as a result of toxic effects on the arterioles causing a decrease in peripheral resistance. Abnormally low blood pressure is seen in the clinical state referred to as shock.

Hypertension: is defined as a sustained normally high blood pressure. Transient high blood pressures are part of a normal circulation. Hypertension causes a strain on the left ventricle which has to pump the blood against a higher pressure. Initially the heart muscle increases in size (left ventricular hypertrophy). This can then lead to failure of the heart to maintain normal function (i.e. heart failure). High blood pressure may also cause blood vessels to burst and this may occur in the blood vessels to the brain, causing a stroke. If this occurs, it may result in paralysis on the side opposite to the site where the blood vessel burst, due to the anatomy of the nervous system and brain circulation.

Ischaemic heart disease: is a disease of the blood vessels resulting in an insufficient supply of the heart muscles with oxygen, which is severe enough to cause temporary strain, or even permanent damage and death of the muscle fibres.

Myocardial infarction: is a term used to describe irreversible injury to heart muscle, which results in loss of function or inability of the muscle to pump blood. Common symptoms include crushing central chest pain that may radiate to the jaw or the left arm. Chest pain may be associated with nausea, sweating and shortness of breath.

Angina: is a chest pain that occurs secondary to the inadequate delivery of oxygen to the heart muscle. When the blood supply to the heart muscle is reduced, due usually to a partial or complete obstruction of the blood flow in the arteries supplying the heart muscle (coronary arteries), the pain that arises is called angina pectoris. It is a tight constricting pain round the chest but may also be felt on the inside of the arms or in the neck.

Dyspnoea: is difficult or laboured breathing; shortness of breath. Dyspnoea is a sign of serious disease of the airways, lungs, or heart.

Oedema: is the presence of abnormally large amounts of fluid in the intercellular tissue spaces of the body, usually applied to demonstrable accumulation of excessive fluid in the subcutaneous tissues. Oedema may be localised, due to venous or lymphatic obstruction or to increased vascular permeability (which may follow stings from insects), or it may be more widespread due to heart failure or renal disease. Collections of oedema fluid are designated according to the site where they are found, for example ascites (peritoneal cavity), hydrothorax (pleural cavity) and hydropericardium (pericardial sac). Oedema due to heart failure is usually first detected as a swelling around the ankles (ankle oedema). Oedema may also occur at the back in front of the end of the spinal cord (sacral area), where it is referred to as sacral oedema. In more serious cases oedema forms in the lungs, a condition known as cardiogenic pulmonary oedema.

proteins and plasma electrolytes and all the substances transported in the blood. It also contains the factors or ingredients necessary for blood clotting. If the factors necessary for blood clotting are removed from plasma, the remaining fluid is called serum.

The respiratory system

Respiration or breathing has three main functions:

1 To deliver oxygen to the cells
2 To eliminate carbon dioxide and
3 To regulate the pH of the blood (Figure 10.4)

Oxygen is a necessity for cells to live and function. This oxygen is obtained from the outside air which we breathe in (inhalation)

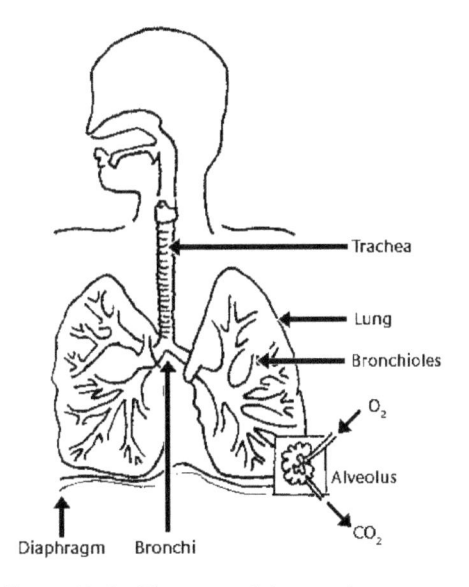

Figure 10.4 Diagram of the respiratory system (not to scale)

through the nose/mouth, windpipe (trachea) and the smaller branching tubes from the trachea – the bronchi, bronchioles to the small air sacs – the alveoli. The alveoli have very thin walls and are surrounded by the smallest of blood vessels, the pulmonary capillaries. These too have very thin walls allowing the diffusion of oxygen and carbon dioxide out of and into the alveoli. Normally the oxygen concentration in the alveoli is higher than that in the blood. After reaching the alveoli in the capillaries during respiration oxygen diffuses easily across to the blood. Similarly, as the concentration of carbon dioxide in the venous blood reaching the alveoli is higher than that in air, carbon dioxide diffuses into the alveoli.

Inspiration and expiration – This is the process by which air enters the lung (inspiration) and the process by which air leaves the lungs (expiration). The rhythmical activity of inspiration and expiration is controlled mainly by the respiratory centre in the brainstem (medulla) which changes the rhythm or pattern, or rate influenced by various impulses reaching the centre from the cortex and other receptors.

Inspiration is an active process, initiated by impulses from the centre to cause contraction of the diaphragm (the muscular/tendinous structure which separates the abdominal cavity, which contains the intestines, liver, kidneys, spleen, etc.) from the thoracic cavity. Sometimes, as with increased exercise, other muscles are involved in inspiration (e.g. the intercostal muscles which consist of internal and external layers between the ribs). Contraction of the diaphragm and other muscles causes expansion of the chest, creating a negative pressure which allows air from outside to enter the lungs. Expiration is essentially a passive process as the lungs regain their original form which results following relaxation of the muscles of respiration, though in some disease states, muscular activity may be required to expel air from the lungs.

If the respiratory centre is damaged (e.g. following a head injury) or its activity is suppressed by drugs such as morphine and other related drugs such as heroin, respiratory activity will deteriorate and may even cease resulting in respiratory failure which stops the supply of oxygen to the cells.

A decrease in oxygen content available to tissue is referred to as hypoxia whilst total lack of oxygen is anoxia.

Cells in the body vary in their sensitivity to lack of oxygen, the brain and heart cells being most sensitive. If brain cells are deprived of oxygen for more than four minutes, there is a great probability of death of brain cells. At rest, 250 ml of oxygen are absorbed per minute during breathing to satisfy the metabolic requirements of the body. The energy requirements depend on the level of activity of the individual. For example, during vigorous exercise, the oxygen requirement may be as high as 5000 ml of oxygen per minute.

Gases diffuse across membranes in amounts that are determined by the difference in the partial pressures of that gas between the two compartments. With the partial pressure of 100 mm of Hg in the alveoli, oxygen diffuses into the blood which is brought by the pulmonary artery with a lower oxygen tension, and this continues until the partial pressure in the blood reaches equilibrium with that the alveolar gas. The blood leaves the lungs with a tension of 100 mm of Hg and arrives at the capillaries with a similar tension of oxygen. As the blood flows through the tissue capillary, it comes in to contact with the tissue fluid with a much lower oxygen tension (around 40 mm Hg). The tension of oxygen in the tissue fluid is low because cells are continually taking up oxygen for metabolism.

As blood flows through the capillary, the oxygen tension falls to 40 mm of Hg (that of the surrounding tissue fluid) and returns to the right side of the heart (to the right atrium and ventricle). Then it comes into contact with alveolar air with an oxygen tension of 100 mm of Hg and equilibrium occurs when the tension of oxygen in the blood reaches 100 mm of Hg and thus becomes referred to as arterial or oxygenated blood.

Tensions of gases in the blood are now measured routinely. This process is referred to as blood gas analysis and indicates the partial pressure of oxygen and carbon dioxide in the blood and also measures the pH of the blood. These measurements provide valuable information about disease processes and the effects of treatment.

The quantity of oxygen carried in the blood depends on the affinity of oxygen to haemoglobin, found in red blood cells. Haemoglobin allows far more oxygen to be carried than would be possible than in solution alone. One gram of haemoglobin has the ability to combine with 1.34 ml of oxygen. Thus, a person who has 15 grams of haemoglobin in every 100 ml of blood would theoretically be able to carry approximately 20 ml of oxygen in every 100 ml of blood. This is termed the oxygen capacity of the blood. In venous blood that is returning from the tissues via the veins to the right side of the heart, only 14 ml of oxygen are present in every 100 ml of blood. As noted, the tension has fallen to 40 mm Hg. Thus, as blood passes through the tissues 5–6 ml of oxygen is taken up by the tissues and during exercise, this amount is much more.

The tension of carbon dioxide in the lungs is 40 mm Hg; whilst in the tissues it is 46 mm Hg. Therefore, as the blood passes through the capillaries in the lung, carbon dioxide diffuses into the alveoli due to the difference of tension of the carbon dioxide in the venous blood and that in the alveoli. Therefore, carbon dioxide moves in a direction opposite to that of oxygen in the alveoli of the lungs. The carbon dioxide content of the blood leaving the lungs is 48 ml carbon dioxide per 100 ml blood. As it passes through the cells and tissues, the carbon dioxide content increases to 52 ml of carbon dioxide per 100 ml of blood.

What emerges from this information is that the changes in oxygen content are greater than the changes in carbon dioxide content. This is because carbon dioxide is not only a waste product. An adequate level of carbon dioxide has to be maintained in the blood in order to maintain an acceptable blood pH to enable cells to function normally.

Carbon dioxide is carried in the blood in three ways: firstly in simple solution as carbonic acid and secondly, as sodium bicarbonate in the plasma and potassium bicarbonate in the red blood cells. Thirdly, it is carried as neutral carbamino protein, mainly with haemoglobin in the red cells.

The transport of the acid-gas carbon dioxide in the blood is closely associated with the maintenance of the blood pH at 7.4. Carbon dioxide dissolves in water, or plasma, and forms carbonic acid. This weak acid is in equilibrium with its salt, the bicarbonate ion. The ratio of bicarbonate ions to molecules of carbonic acid defines the acidity of the blood. Under normal circumstances, this ratio is 20:1. Bicarbonate ions are produced in red blood cells in systemic capillaries as carbon dioxide diffuses into these cells and forms carbonic acid. The formation of carbonic acid is catalysed by the enzyme carbonic anhydrase. Bicarbonate ions diffuse from the red cells to the plasma being replaced by chloride ions moving into the red cells. Hydrogen ions produced during the formation of bicarbonate ions are buffered by haemoglobin in its deoxygenated state. This process is precisely reversed in the capillaries of the lung: bicarbonate ions enter red cells, chloride ions leave red cells and hydrogen ions released from oxygenated haemoglobin combine with bicarbonate ions to produce carbonic acid; carbonic anhydrase catalyses the formation of carbon dioxide and water from carbonic acid and the carbon dioxide diffuses out of the cell into the plasma. Note that carbonic anhydrase catalyses both the formation and breakdown of carbonic acid.

Physiologically, four phases of respiration are recognised:

1 Ventilation: the movement of air to and from the lungs
2 Distribution: air entering the lungs is distributed to all parts including the small

air sacs (alveoli) where gas transfer to and from the blood takes place

3 Diffusion: the oxygen from the air diffuses through the walls of the alveoli to the adjacent blood vessels and carbon dioxide from the blood vessels diffuses back in to the alveoli

4 Blood rich in carbon dioxide and low in oxygen is pumped to the lungs via the pulmonary arteries, by the right ventricle of the heart. Blood low in carbon dioxide but loaded with oxygen is returned to the heart via the pulmonary veins. Matching of ventilation and perfusion (i.e. blood supply to the alveoli – air sacs) within the lung ensures normal gas exchange.

Common signs, symptoms and terms used for disorders of lung or respiratory function follow.

Asphyxia – This is a state in which there is excess of carbon dioxide and lack of oxygen in the body. This occurs when respiratory function or activity is insufficient to meet the demands of the body or when there is obstruction to respiration, for example during strangulation, or when an individual is breathing in a confined space when the expired air has to be inhaled. Asphyxial states stimulate respiration or breathing, as carbon dioxide is a potent stimulus of the respiratory centre, as is a lack of oxygen.

Hypoxia – This is defined as a shortage of oxygen supply to the cells of the body. It is encountered when there is insufficient oxygen in the inspired air or when the tissues are deprived of the normal amount of oxygen due to possibly poor circulation or a blood clot. In this case, tissue hypoxia results and if this occurs in the heart muscle it results in heart attacks or ischaemic heart disease. Hypoxia also depresses the brain as the brain is dependent on sufficient oxygen supplies for the proper functioning of nerve cells. When hypoxia of the brain occurs, a person becomes disorientated, loses all sense of danger, loses consciousness and coma sets in. Hypoxia may be due to decreased amount of oxygen in the

air breathed in (inspired oxygen) as at high altitude, or may be due to lung disease when the oxygen cannot enter the red blood cells in the blood that flows through the lungs.

In contrast to asphyxia, where an individual will struggle to breathe with all the available resources in the body, in hypoxia, the individual will soon lose control and become unconscious.

If the supply of oxygen to the brain cells is interrupted for more than four minutes (as seen when the heart ceases to pump blood effectively, commonly referred to as cardiac arrest), the nerve cells in the brain may be irreversibly damaged and may lead to 'brain death'. If there is a deficiency of haemoglobin to transport the oxygen, as is seen an anaemic patient due to either poor nutrition or prolonged blood loss as with heavy menstruation or bleeding piles, the term used is anaemic hypoxia. The hypoxia associated with carbon monoxide poisoning is an anaemic hypoxia as there is insufficient haemoglobin to transport the oxygen due to its preferential binding with carbon monoxide, which has an affinity about 250 times greater for haemoglobin than oxygen.

If blood flow is very slow, there will be insufficient oxygen for the cells to function. This is referred to as stagnant hypoxia.

Finally, the cells may be unable to utilise the oxygen brought to them by the blood due to the enzymes within the cells being inactive or destroyed. This occurs in cyanide poisoning, where vital cytochrome enzymes situated in the mitochondria (the power units of the cells) are inactivated by the cyanide and normal cellular energy production fails.

Pulmonary oedema – In the lungs, the pulmonary arterial pressure is usually 25 mm Hg compared to about 130 mm Hg in arteries arising from the aorta from the left ventricle. When the pressure in the left atrium or pulmonary veins is elevated (for example when the atrium cannot empty its contents to the left ventricle either because the atrium muscle is not contracting in the normal manner or when there is an obstruction to the flow of blood from the atrium to the ventricle such as narrowing stenosis of the valve between the two chambers (commonly referred to as mitral stenosis), the pressure in the pulmonary capillaries is exceeded to such an extent that fluid would pass from the capillaries into the alveoli. This fluid build-up interferes with diffusion of gases. The presence of fluid in the alveoli or in the lung is referred to as pulmonary oedema. Pulmonary oedema leads to difficulty in breathing (as there is insufficient oxygen and accumulation of carbon dioxide, which are both stimuli for the respiratory centre) and the patient becomes dyspnoeic. This is a state of being conscious of breathing and having difficulty in breathing. Pulmonary oedema may also occur due to the action of poisonous gases such as chlorine or phosgene, where the structure of the alveolar walls and capillaries are damaged leading a leakage of fluid into the alveoli. If the left side of the heart fails, the heart muscle on the left side of the heart does not function or contract adequately and there is a build-up of pressure in the pulmonary veins bringing blood from the lungs. This also causes pulmonary oedema.

Cyanosis − is a bluish discolouration, of the skin and mucous membranes, caused by an excessive concentration of deoxygenated haemoglobin (haemoglobin not bound to oxygen) in the blood. A point of interest to those particularly in countries where anaemia is very common: the haemoglobin levels may be so low, that the amount of haemoglobin without oxygen (deoxygenated haemoglobin) would be insufficient to cause cyanosis.

The gastrointestinal system

Food or water entering the body through the mouth passes down the oesophagus and then to the stomach. From the stomach, partially digested food passes on to the duodenum, jejunum and ileum (small intestine). That which has not been absorbed proceeds to the caecum, ascending, transverse, and descending and to the sigmoid colon (large intestine), the rectum and finally the anal canal (Figure 10.5).

In the mouth, saliva is produced by three paired salivary glands: parotid, submandibular and sublingual. Saliva is secreted by these glands, usually in response to the thought, sight, taste or smell of food. Secretion of saliva is under control of the parasympathetic nervous system. When there is over-activity of the parasympathetic nervous system and there is more secretion of the neurotransmitter of that system (acetylcholine), there will be excessive secretions, for example as in organophosphorus insecticide poisoning. Dryness of the mouth results if drugs such as atropine block the action of acetylcholine on the receptors.

The food formed into a bolus in the mouth passes down the oesophagus due to propulsive contractions of the muscle of the oesophagus, which are controlled by another parasympathetic cranial nerve, the vagus. Then the food enters the stomach which not only acts as a storage organ but also promotes digestion by secretions (pepsin and hydrochloric acid for digestion of proteins) from the cells lining the stomach wall. The secretions of the stomach are also under control of the vagus nerve and also a hormone called gastrin.

At regular intervals of minutes, small quantities of food pass through an opening at the distal end of the stomach called the pyloric sphincter to the duodenum. The contents of a stomach usually empty in about four

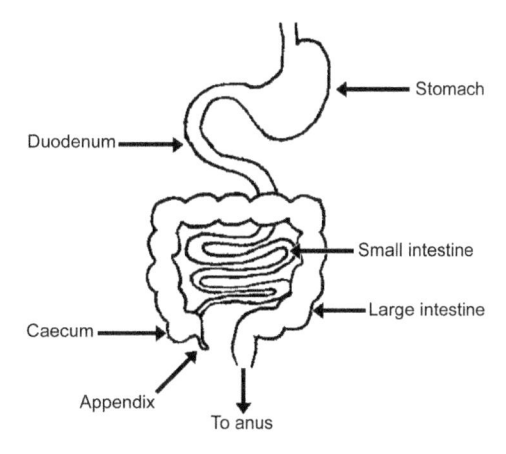

Figure 10.5 The gastrointestinal system (not to scale)

hours. However, with a very fatty meal, the emptying of the stomach becomes much slower due to the release of a hormone called enterogastrone.

Stomach ulcers and duodenal ulcers, commonly referred to as peptic ulcers, occur because pepsin, which aids in digesting proteins, acts on the cells of the stomach wall. Destruction of the stomach wall cells is aided by the presence of hydrochloric acid also secreted by the cells lining the stomach.

The stomach also has a role in vitamin B12 metabolism. Loss of the stomach or a large part of it can lead to a condition known as pernicious anaemia, which is due to lack of vitamin B12.

The pancreas – is a gland which sends its secretions into the blood stream and also into the duodenum. The secretion of insulin and glucagon, which are necessary for the control of blood sugar in the human body, is from the pancreas. The juices from the pancreas which enter the duodenum, known as the pancreatic juices, contain the enzymes trypsinogen and chymotrypsinogen, which are precursors of the protein splitting enzymes trypsin and chymotrypsin. These promote the digestion of food in the small intestine after leaving the stomach.

Pancreatic secretion is also under control of the vagus nerve. Insulin secretion is from collections of specialised cells known as the Islets of Langerhans. Failure to produce sufficient insulin results in diabetes mellitus.

Secretions from the liver also enter the duodenum via the bile duct. The bile is stored and concentrated in the gall bladder, which also contracts due to the action of the vagus nerve and can also contract due to the action of some hormones. The bile constituents may concentrate in the gall bladder and give rise to gall stones. The associated inflammation of the gall bladder is termed as cholecystitis.

The small intestine – is concerned primarily with the absorption of sugars or carbohydrates and produces the related enzymes maltase, sucrase and lactase. Though the nerve supply to the small intestine is both

from the parasympathetic and sympathetic nervous systems, these nerves regulate motility or contractions of the small intestine (peristalsis) and have no role in the production of the digestive enzymes. The absorption of food takes place mainly in the small intestine. Amino acids and fats are also absorbed here.

The large intestine – this mainly absorbs water so that the water content that reaches the end of the large intestine is ultimately reduced by about two-thirds. The large intestine is not essential to life. However, the bacteria present in the large intestine are important in the provision and production of vitamins, particularly those of the vitamin B group.

The liver – this is the chemical factory of the body, both producing essential molecules and modifying and detoxifying ingested toxic substances (Figure 10.6).

The main functions of the liver are:

- Production of essential proteins such as albumin.
- Synthesis of factors that are involved in the clotting of blood.
- Maintaining the level of sugar in the blood. The excess carbohydrate absorbed by the blood from food is converted to glycogen, which is also formed from excess fat and protein. The liver glycogen maintains the normal blood glucose level in the blood when glucose is used up by the cells.
- Formation of urea from the ammonia, which collects after amino acids have been used up (deaminated). The urea is eliminated through the kidney.

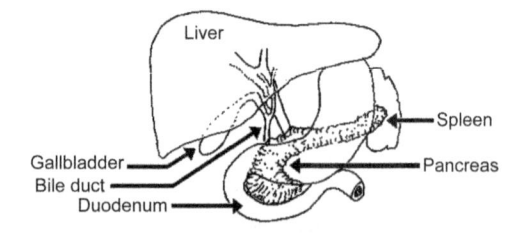

Figure 10.6 The liver and biliary system (not to scale)

- The bile salt produced by the liver along with products from an anatomically closely related organ, the pancreas, plays a vital role in digestion and absorption of fat. It also stores fat soluble vitamins (A and D).
- The destruction of used red blood cells and removal of the breakdown product of haemoglobin (bilirubin) in the bile via the bile duct to the intestine (duodenum). The liver stores vitamin B12 which is necessary for the maturation of blood cells.
- In relation to toxicology, the liver plays a vital role by modifying the toxicity of foreign substances (including alcohol and drugs) that gain entry into the body by any route. Some drugs used in medical treatment are administered as pro-drugs which depend on the liver to produce the active drug through the action of liver enzymes. The liver also has the ability to bind toxic substances to other compounds to make them more water soluble and thus enable the kidneys to excrete them from the body.

Jaundice – one of the important roles of the liver, as noted, is the formation of bilirubin from the red cells that die. Water-soluble bilirubin is responsible for the characteristic colouration of the faeces. The failure to excrete bilirubin gives rise to yellowish discolouration of the whites of the eyes, the skin and nails and mucosal membranes, which is called jaundice. Jaundice is essentially a sign of liver failure. Several toxic compounds such as some pesticides, solvents such as carbon tetrachloride and dry-cleaning fluids, damage the liver cells and prevent them from functioning normally to bind the bilirubin to the glucuronide. There are many drugs used in the treatment of disease which also damage the liver cells and are termed hepatotoxic. For example, very high doses of the common drug paracetamol can cause liver damage, liver failure and jaundice. Another common and important cause of damage to liver cells which often results in liver failure is excessive alcohol (ethanol) consumption.

Therefore, in liver failure, jaundice occurs, the blood urea falls (as urea is no longer formed from ammonia), there is insufficient production of proteins, of which albumin is the most important, which may lead to swelling of ankles or oedema (as proteins are essential to maintain plasma osmotic pressure which keeps fluid within capillaries) and blood clotting will be impaired. The most important effect is that the blood will not have sufficient glucose for the cells to function normally.

In addition, when the liver cells fail to function properly, their ability to make foreign substances less toxic by metabolic enzymes fails and the toxicity of some drugs used in medicine such as morphine is increased.

The liver's role in metabolism of xenobiotics – humans are constantly and unavoidably exposed to chemicals that are foreign to the body, or xenobiotics. These include both manufactured and natural chemicals such as medical drugs, industrial chemicals, pesticides, pollutants, plant alkaloids and plant metabolites and toxins produced by moulds, plants and animals.

The physical property that enables many xenobiotics to be absorbed through the skin, lungs or gastrointestinal tract is their fat solubility or lipophilicity. Lipophilicity is also an obstacle to their elimination as they can be readily reabsorbed. Another important consideration is that lipophilicity facilitates the entry of toxic substances into cells. Therefore, the elimination of xenobiotics often depends on their conversion to water soluble compounds by a process called biotransformation which is catalysed by enzymes in the liver and other tissues. An important result of biotransformation is the conversion of a lipophilic substance to one that is more water soluble (hydrophilic) (see Phase 1 and Phase 2 reactions later).

This transformation is probably one of the most important defence mechanisms of the body. Xenobiotics such as drugs exert beneficial effects and others may cause deleterious effects as in the case of poisons. The effect a xenobiotic produces in the human body is dependent on its physicochemical properties and thus the results of xenobiotic exposure is altered by this process of biotransformation.

Some drugs must undergo biotransformation to be effective because the metabolite of the drug and not the drug itself produces a therapeutic or beneficial effect. Similarly, some xenobiotics undergo biotransformation to produce their harmful or toxic effects. However, in the vast majority of situations, biotransformation terminates the effectiveness of the xenobiotic in the human body, whether beneficial or harmful. In the context of toxicology, this means that many potentially toxic substances are made relatively innocuous by biotransformation by liver enzymes. These are predominantly the cytochrome P450 group of isoenzymes which are responsible for the majority of oxidation reactions which xenobiotics undergo.

The enzymes catalysing biotransformation reactions often determine the intensity and duration of the action of drugs and play a key role in chemical toxicity. The xenobiotic biotransforming enzymes catalyse two types of reactions. These are:

Phase I reactions: addition of a functional group (e.g. -OH. -NH2, -SH or -COOH) to produce a slight increase in water solubility or hydrophilicity.

Phase II reactions: include glucuronidation, sulfation, acetylation, methylation, conjugation with glutathione and conjugation with amino acids (e.g. glycine, taurine, glutamic acid). These reactions cause a large increase in water solubility and thus increase the excretion of the xenobiotic, for example, morphine, heroin and codeine are all converted to morphine-3-glucuronide. In the case of morphine, this is a result of direct conjugation with glucuronide. In the case of heroin (diamorphine) and codeine, conjugation with glucuronic acid is preceded by Phase I biotransformation (hydrolysis) or deacetylation, with heroin and demethylation

involving oxidation by cytochrome P450 isoenzymes with codeine.

The cytochrome P450 enzyme system – the liver is the organ with the highest concentration of enzymes catalysing biotransformation reactions. These enzymes are also located in the skin, lungs, nasal mucosa (mucosa of the nose), eyes and gastrointestinal tract.

In the liver and in most other organs, they are located in the cells, primarily in the endoplasmic reticulum (microsomes) or in the soluble fraction of the cytoplasm, with a smaller concentration in the mitochondria, nuclei and lysosomes.

Amongst the Phase I biotransformation enzymes, the cytochrome P450 system is responsible for most oxidation reactions and is probably the most versatile, detoxifying more xenobiotics than any other enzyme system.

In humans, about 40 different microsomal and mitochondrial P450 enzymes play a key role in catalysing reactions in the following areas:

- The metabolism of drugs, environmental pollutants and other xenobiotics;
- The biosynthesis of steroid hormones;
- The oxidation of unsaturated fatty acids to intracellular messengers; and
- Metabolism of fat-soluble vitamins.

The liver microsomal P450 enzymes involved in xenobiotic biotransformation belong to three main P450 gene families: CYP 1, CYP 2 and CYP 3. The level and activity of each P450 enzyme varies from individual to individual due to genetic and environmental factors.

It is important to remember that the activity of the CYP isoenzymes can be altered by several agents. For example, there are many drugs that increase the activity of the isoenzymes called enzyme inducers. Similarly, some xenobiotics can inhibit the activity of CYP450 isoenzymes. These are called enzyme inhibitors. The induction of CYP450 isoenzymes by drugs such as phenobarbital (a barbiturate used primarily in the treatment of epilepsy) or rifampicin (an antibiotic used in the treatment of tuberculosis) can prevent the effectiveness of the oral contraceptive drug ethinyl oestradiol.

The kidneys – are two bean-shaped organs about the size of a fist lying below the rib cage. Though small in size (about 0.5% of the body weight), the kidneys receive approximately 20% of the blood that is pumped out from the heart via the renal arteries. The main parts of the kidney are an outer lightly coloured cortex and an inner darker medulla (Figure 10.7). The renal (kidney) pelvis is the funnel which collects the urine from nephrons and enables the urine to flow to the ureters. The ureters are the tubes that carry the urine from the kidneys to the urinary bladder.

The blood brought to the kidneys by the arteries is filtered under pressure by a part of the basic functional unit of the kidney, the nephron (Figure 10.8). There are approximately 1 million nephrons in each kidney. Each nephron is a thin long convoluted tube, surrounded by capillaries, which is closed at one end, where the filtering takes place. The filtered fluid is then absorbed from within the nephron, according to the needs of the body.

The cells of the tubules also have the ability to secrete substances into the lumen of the tubule of the nephron and eliminated them from the body in the urine. These are usually unwanted substances and include many toxins and drugs.

The kidney is one of the most important organs involved in the elimination of waste products and unwanted substances from the body. The blood supply, blood vessels and the structure of the kidney enable the entire blood volume of an individual to be filtered 20–25 times a day. The purified blood is returned to the circulation by the renal veins.

Essentially, three basic processes take place in the nephron: filtration, absorption or reabsorption and secretion or excretion.

The principal function of the kidney is to produce urine, which is excreted from the body and ensures the maintenance of the correct chemical environment (milieu interior) for body cells, water balance, electrolyte balance and the pH of the blood. It has other functions such as producing substances necessary for the formation of red blood cells (erythropoietin), converting vitamin D to an active form which promotes the absorption of calcium from the intestine, and also

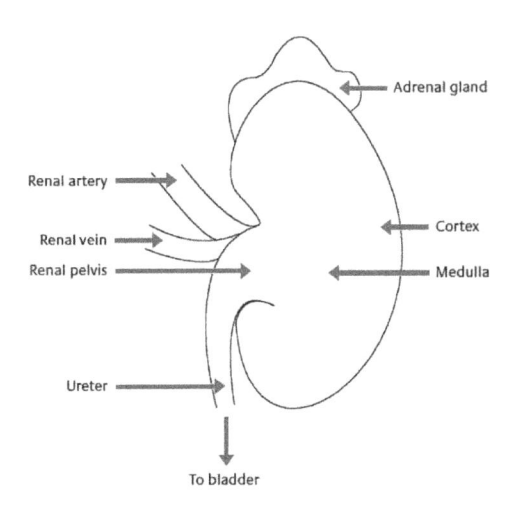

Figure 10.7 Diagram of the kidney (not to scale)

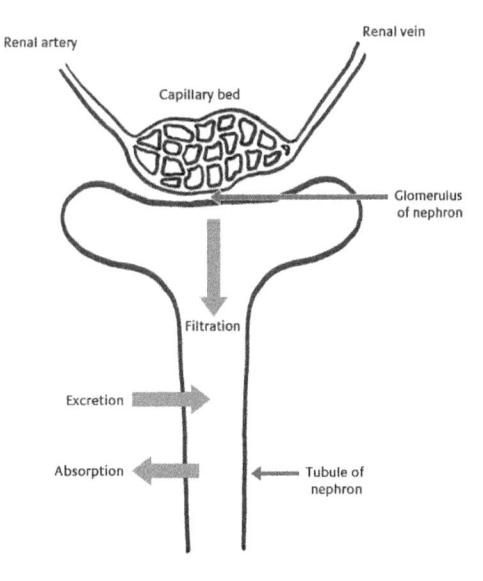

Figure 10.8 The nephron (not to scale)

producing some hormones associated with the regulation of blood pressure.

Urine normally contains surplus water and electrolytes, waste products such as urea, which is formed from the amino acids in the liver, uric acid, which is produced from nucleic acids and purines, and creatinine from muscle. Urine also gets rid of excess amounts of acids or alkalis that have either been ingested or are formed following metabolic processes.

It is important to note that most toxic substances and even beneficial substances such as drugs that are used in treatment are eliminated in the urine and the measurement of the relevant constituents can confirm either the use of drug(s) or the exposure to toxic compounds. Thus, examination and analysis of the urine is an important investigation in toxicology and acts as a biomarker of exposure. The amounts excreted in the urine may also be an indicator of the severity of exposure to a toxic substance.

The endocrine system and the production of hormones

The activities of the organs of the body are primarily controlled by nerve impulses. The other important mode of control of activity is by hormones. These are chemical substances produced by endocrine glands and which act as messengers from one part of the body to another. They enter the blood stream directly following secretions by the glands and are brought to all parts of the body by the cardiovascular system. Most organs are under the influence of both nerve impulses and hormones.

The word 'hormone' is derived from the Greek word *hormao*, which means to excite. A hormone is a naturally occurring substance secreted by specialised cells that affect the metabolism or behaviour of other cells possessing functional receptors for the hormone. Hormones may be hydrophilic, like insulin (from the pancreas), in which case the receptors are on the cell surface, or lipophilic, like steroids (from the adrenal cortex), where the receptor can be intracellular. Hormones are substances that circulate in the blood and bring about an effect on distant organs.

The endocrine glands are:

- Pituitary;
- Thyroid;
- Parathyroid glands;
- Adrenal glands;
- Ovaries in the female;
- Testes in the male;
- Placenta during pregnancy;
- Pancreas, which is both exocrine and endocrine (secretions of exocrine glands reach the blood stream through ducts while those of endocrine glands reach the blood stream directly).

The *pituitary gland* lies in a bony cavity in the skull called the pituitary fossa. The posterior part of the pituitary is suspended by a structure called the pituitary stalk from a part of the brain called the hypothalamus. The gland has two parts: the anterior pituitary or adenohypophysis, and the posterior pituitary or neurohypophysis (Figure 10.9). While the posterior pituitary has neural or nerve connections with the brain, the anterior pituitary has vascular connections with the brain.

The *posterior pituitary* is composed mainly of nervous tissue descending from the hypothalamus and produces two hormones:

- The antidiuretic hormone (vasopressin);
- Oxytocic hormone (oxytocin).

The antidiuretic hormone (ADH) is involved intimately with water balance as it controls or regulates the amount of water that is reabsorbed in the kidney. Large doses of ADH cause high blood pressure or hypertension since the hormone causes vasoconstriction or contraction (narrowing) of blood vessels. The amount of ADH secreted by the posterior pituitary is controlled by the amount of water in the blood. If the body is short of water, more ADH will be secreted, and more water

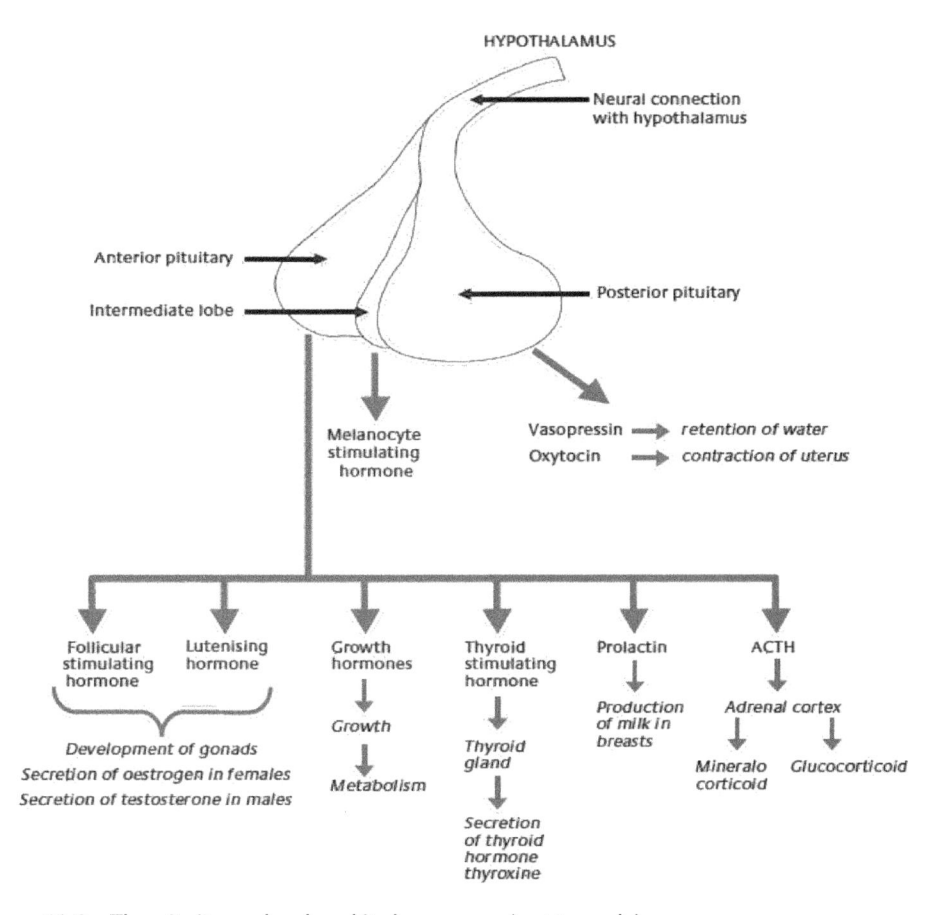

Figure 10.9 The pituitary gland and its hormones (not to scale)

will be reabsorbed by the kidney tubules and less urine will be formed.

The posterior pituitary may fail to produce ADH, producing diabetes insipidus. In this condition excessive amounts of urine are formed and lost from the body and the patients are always very thirsty.

The oxytocic hormone is only important during pregnancy. It causes contraction of the pregnant uterus and facilitates the ejection of milk during lactation.

The *anterior pituitary* releases the following:

- Growth hormone;
- Thyrotrophic hormone;
- Adrenocorticotrophic hormone; and
- Gonadotrophic hormones.

Growth hormone stimulates the growth of bone and muscle tissue during childhood. If there is excessive production of this hormone before puberty (when the long bones fuse with the growing ends or epiphyses and no further increase in growth can occur), gigantism results. Insufficient production of the hormone causes dwarfism. Increased production after puberty leads to an increase in the size of facial bones and also the bones of the hands and feet. This is a condition called acromegaly.

Thyroid stimulating hormone and the thyroid gland – the thyroid stimulating hormone acts on the thyroid gland in the neck and stimulates the release of the thyroid hormones, thyroxine and triiodothyronine. The thyroid hormone stimulates metabolism by acting on the cells to speed up the rate at which food is used up and converted to heat and energy. The thyroid gland is unique in that it stores its hormones as a colloid in small vesicles in the gland. The other glands store their secretions in the cells themselves. The formation of the thyroid hormone requires iodine which has to be in the diet. In regions where populations may encounter a deficiency of iodine in their diets the addition of small quantities of iodine to salt (iodised salt) has helped in the prevention of enlargement of the thyroid or goitre which occurs in iodine deficiency. Deficiency of the thyroid hormone (hypothyroidism) in a child causes cretinism where the development of the nervous system is affected, and the child is mentally retarded. In an adult, deficiency of thyroid hormone causes myxoedema where the body temperature is low, the heart rate is slow, brain activity is sluggish and there is deposition of fluid-like material under the skin. The face and eyelids become puffy.

If there is increased production of thyroid hormone (hyperthyroidism), metabolism is stimulated and more heat is produced, the heart beats faster, the heart excitability is increased which may give rise to disorders of heart rhythm (cardiac arrhythmias). In this situation, which is called thyrotoxicosis, the person is irritable, anxious and nervous. They lose weight though the appetite is good. They often have protrusion of the eyeballs (exophthalmos). In addition, the thyroid gland produces calcitonin, which is important in the regulation of calcium balance.

Parathyroid glands – these are adjacent to the thyroid gland but are not controlled by the secretions of the anterior pituitary gland. There are four parathyroid glands, two on either side of the thyroid. Parathyroid hormone (PTH) from the parathyroid gland controls the calcium levels in the blood. Another factor affecting the blood calcium level is calcitonin, which as noted earlier, is also a secretion of the thyroid gland. Calcitonin acts by trapping calcium in the bones.

Another important factor determining the level of blood calcium is vitamin D.

With increased activity of the parathyroids (hyperparathyroidism), the plasma level of calcium increases to about 20 mg calcium per 100 ml of blood from a normal of 5.5 mg of calcium per 100 ml of blood. This calcium comes from the bone and the kidney gets rid of it from the body. Bones can become thin and fragile and likely to fracture more easily than normal bones with sufficient calcium.

With decreased production of parathyroid hormone (hypoparathyroidism), the blood calcium level falls, which causes increased excitability of the nerves and of the neuro-muscular junctions leading to a condition called tetany (this has to be distinguished from the disease tetanus which follows infection with a bacillus *Clostridium tetani*). In tetany, there is spasm of the hands and feet (carpo-pedal spasms). Increased excitability of the nerve cells in the brain may lead to convulsions.

Adrenocorticotrophic hormone (ACTH) and the adrenal gland – the adrenal glands are located above each kidney (see Figure 10.7). Each adrenal gland consists of a central medulla and an outer cortex.

The adrenal medulla releases the hormones adrenaline and noradrenaline (epinephrine and norepinephrine) in response to nerve stimuli that enter the medulla from the sympathetic nervous system. The adrenal medulla is an integral part of the sympathetic nervous system and is intimately involved in the fight or flight responses to stress, where increased sympathetic activity is lifesaving. It is common to state that a person should have sufficient adrenaline to perform well. The secretions of the hormones from the adrenal medulla are not under the control of the anterior pituitary hormone ACTH.

The adrenal cortex has three layers, and each layer produces a different hormone and at least two of the layers are controlled by ACTH from the anterior pituitary gland.

The outer layer of the adrenal cortex, the zona glomerulosa, produces aldosterone, which is necessary for reabsorption of sodium in the kidney. An excess of aldosterone causes salt and water retention. The secretion of aldosterone is regulated by rennin, a hormone secreted by the kidney.

The inner two layers of the adrenal cortex, the zona fasciculata and the zona reticularis, produce hormones collectively known as corticosteroids. The main corticosteroid secreted is cortisol (hydrocortisone). The corticosteroids have several actions which are as follows:

- Promoting the utilisation of proteins for the production of heat and energy in preference to the use of carbohydrates;
- Anti-allergy;
- Anti-inflammatory;
- Aldosterone-like effects, causing retention of sodium and of water and loss of potassium. The salt and water retention may lead to oedema and/or high blood pressure (hypertension).

Cortisol reduces the utilisation of carbohydrates for energy; thus they increase the blood glucose level (a diabetogenic effect). When body proteins are broken down, wound healing is impaired and the effect on suppression of immune or inflammatory response can lead to 'masking' of infections (which may cause delays in diagnosis) and also an increased susceptibility to infections.

Over-production of the adrenal cortex hormones leads to Cushing's syndrome, which is characterised by 'moon face', due to redistribution of fat and swelling of the face. Redistribution of body fat leads to 'an egg on match sticks' appearance: a expanded abdomen and chest with 'skinny' limbs, particularly lower limbs, diabetes and increased blood pressure. The skin tends to bruise easily, and purple striae appear on the skin and females develop hirsutism. There may also be psychological changes.

Changes similar to Cushing's Syndrome follow treatment with corticosteroids over a period of time in several common disease states. An excessive production of aldosterone leads to Conn's disease, which is associated with muscular weakness, increased loss of potassium and water in the urine.

Decreased activity of the adrenal gland leads to Addison's Disease. This was common following tuberculosis affecting the adrenal gland when both the medulla and cortex were affected. In Addison's disease, sodium and water are lost from the body causing a lowering of blood pressure, muscle weakness, nausea and vomiting. The production of catecholamines (e.g. adrenaline and noradrenaline) is affected and there is increased production of melanin instead. This leads to increased pigmentation of the skin, particularly of exposed parts. Episodes of low blood sugar may occur as adrenaline plays an important role in mobilising glucose, particularly in times of stress.

Immune system

The immune system is a complex defence network made up of a considerable number of different cell types and chemical messengers which work together in a coordinated manner to protect us from a range of threats including bacteria, viruses and also cancer cells within the body (cells which are reproducing in an uncontrolled way) by identifying and destroying them.

The immune system is very efficient in differentiating between harmful and harmless substances. Following an exposure to a harmful substance for the first time, defence mechanisms often destroy this substance. In addition, the system develops a memory, which is often life-long, so that it can recognise and attack this harmful substance if it enters the body subsequently. This memory is in the form of antibodies, each being unique and tailor-made for a specific hazard.

There are five classes of antibodies:

- Immunoglobulin A (IgA) is found in secretions such as saliva, tears and protects against organisms that may invade gastrointestinal and respiratory tracts.

- Immunoglobulin M (IgM) which is formed initially and provides a temporary protection following exposure of the body to a new threat until immunoglobulin G is made.
- Immunoglobulin G (IgG) takes over from IgM to provide long-lasting protection against a specific threat.
- Immunoglobulin E (IgE) (sometimes called the 'allergy' antibody) is responsible for allergic reactions. IgE is usually produced against harmful substances but in some cases with an inherited disorder, IgE is formed in excessive amounts to substances that are usually not harmful to the majority of the population. These 'atopic' individuals thus react abnormally or disproportionately to a substance that should be harmless.
- Immunoglobulin D (IgD) is a unique immunoglobulin with a concentration in serum far below those of IgG, IgA and IgM but much higher than that of IgE. IgD's function has long been a conundrum and is still incompletely understood.

Several chemicals are produced by the body in allergic responses, of which histamine is the best known and causes the well-known symptoms of itching, swelling, redness and increased production of mucus.

An allergen is a substance that triggers an allergic response in atopic individuals (persons who are liable to asthma, eczema, hay fever and urticaria). Amongst the common allergens are house dust mites, pollens, foods such as peanuts, egg protein, moulds, spores, dog dander (scales from hair or fur of dogs) and cat dander (combination of scales from skin and saliva). All allergens have one common feature – they are all proteins. As peanut oil does not contain any protein, individuals who are allergic to peanuts are unlikely to react to peanut oil. The best known non-protein allergen is the antibiotic penicillin.

Skin allergies – the commonest is probably eczema, which is an inflammation of the skin (dermatitis) occurring in atopic individuals.

Skin allergies are attributed to the lack of formation of a protein called filaggrin, which is due to a malfunctioning gene which has been inherited (due to a genetic mutation). Filaggrin forms a protective layer at the surface of the skin that keeps water in and foreign substances out.

Contact dermatitis is dermatitis that occurs following contact of the skin with certain substances. Approximately 20% of cases of contact dermatitis are of an allergic nature in atopic individuals and follow exposure to substances such as nickel, formaldehyde, plant and rubber products. The other 80% follow exposure to irritant chemicals, such as those that are used in hair-dressings, printing, catering and construction.

Urticaria (nettle rash, hives) may occur in both atopic and non-atopic individuals, and is triggered by a wide range of stimuli, not all of which are allergens. Examples of allergic causes are bee and wasp stings, foods such as nuts, milk, beans, fish (shellfish), penicillin.

Non-allergic causes include cold, heat, water, pressure, sunlight and medications such as ibuprofen, aspirin, paracetamol and some other anti-arthritic drugs. Others are food dyes (e.g. tartrazine – E no 102) and preservatives such as ascorbic acid, sulphides and some other anti-oxidants.

Allergies at work are a major concern for workers and employers and there are over 200 substances that can act as allergic sensitisers to cause occupational asthma. Some other substances act as irritants.

Allergies at work are encountered during the manufacture of latex products (gloves, condoms, balloons), use of formaldehyde in glues, carpentry and manufacture of fabrics, newspapers, fertilisers, furnishing foams, smoke and exhaust fumes.

Immune responses – Immunology is the study of the physiological responses by which the body destroys or neutralises foreign matter or xenobiotics, living and non-living, as well as cells of its own that have become altered in certain ways. The ability of the immune response to protect us against bacteria, fungi, viruses and other parasites and other foreign matter is one of the most important defence mechanisms of the human body.

The immune response, the process by which xenobiotics are destroyed or neutralised is therefore essential for a healthy disease-free life. The immune response can also destroy cancer cells that arise in the body and also worn out or damaged cells such as old red blood cells or erythrocytes.

Immune responses can be broadly classified into:

1   Non-specific immune responses, which recognise in a non-selective manner all foreign substances
2   Specific immune responses against substances that are specifically identified and then attacked

Bacteria have the ability to cause damage at their sites of invasion or can release into the body fluids (extra cellular fluids of which blood is the most important) toxins that are carried to other parts of the body to cause damage to cells.

The body also needs protection against viruses, which are essentially nucleic acids surrounded by a protein coat. Unlike bacteria, which have their own metabolic processes and can multiply independent of other cells, viruses lack the enzyme processes and other cell constituents such as ribosomes for their own metabolism and energy production. Therefore viruses can only multiply whilst living inside other cells whose biochemical apparatus they use. The nucleic acids in the viruses cause the production/manufacture of proteins required for the viruses to multiply and also the energy to multiply.

The cells and associated components that carry out immune responses are collectively called the immune system. Though called a system, it has no single anatomical basis but consists of diverse collections of cells found both in the blood and tissues throughout the body.

Cells mediating the immune response are:

1   White blood cells or leucocytes, including neutrophils, basophils, eosinophils, monocytes and lymphocytes. The lymphocytes are grouped into B cells and T cells (cytotoxic T cells, helper T cells, suppressor T cells). White blood cells can leave the circulation or blood (unlike the red blood cells) and enter tissues where they perform their protective function.
2   Plasma cells found in peripheral lymph organs. Plasma cells are transformed (differentiate) from lymphocytes in the tissue and are not found in the blood as the name suggests.
3   Macrophages are present in almost all tissues and organs. They are large cells but their structure may vary from tissue to tissue. They are derived from monocytes (white blood cells) that leave the blood vessels to enter the tissues. As their main function is to engulf foreign material, they are strategically located at sites where entry of foreign substances or organisms is likely to take place.
4   Mast cells are also found in all tissues and organs. Mast cells differentiate from basophils that have left the blood vessels and a characteristic feature is that they usually contain large numbers of secretory vesicles and they secrete mainly locally acting chemical messengers (e.g. histamine). These cells are involved in allergic responses such as hypersensitivity reactions.

Inflammation – is the local reaction of the body to injury or infection which destroys or inactivates foreign organisms and prepares the body to repair the injury caused. The main cells involved in inflammation are phagocytes which engulf the foreign material by a process called phagocytosis. Once inside the phagocyte, the foreign substance is destroyed. The important phagocytes are the neutrophils, monocytes and macrophages.

The usual clinical manifestations of inflammation are redness, swelling, heat and pain, which are produced by a variety of chemical messengers or mediators. The better known of these mediators are the kinins, histamine, complement and eicosanoids. The kinins are produced from the plasma protein kininogen whilst histamine is released from mast cells.

Two important inflammatory mediators, interleukin 1 (IL1) and tumour necrosis factor are proteins which are released by monocytes and macrophages during an inflammatory response.

Complement is a substance which kills microbes without prior phagocytosis. Complement is always present in the blood albeit in an inactive form most of the time. Activation of complement in response to an infection or tissue injury generates active molecules from inactive precursors and the complement system is comprised of at least 20 distinct proteins. Complement also stimulates the secretion of histamine from mast cells and effectively increases the blood flow to the injured area and facilitates the movement of phagocytes such as the neutrophils to the injured area from within the blood vessels.

Lymphocytes also circulate in the blood but tend to gather in large numbers in groups of organs and tissues called lymphoid organs such as bone marrow and the thymus gland a (gland found in the chest which shrinks in size after puberty), lymph nodes, spleen and tonsils. These cells also concentrate in the lining of the intestine, and those of the respiratory, genital and urinary tracts.

The lymphatic system is a network of lymphatic vessels and lymph nodes found along these vessels, through which lymph, a fluid derived from interstitial fluid, flows. It constitutes a route by which interstitial fluid can reach the blood vessels or the cardiovascular system. This movement of interstitial fluid as lymph to the cardiovascular system is very important because the amount of fluid filtered out of all the blood vessel capillaries (except those of the kidney) exceeds that which is reabsorbed by approximately 4 litres each day. These 4 litres are returned to the blood via the lymphatic system. In the process,

the small amount of protein that usually leaks out of the capillaries is also brought back into the circulation by the lymphatic system. The lymph node cells meet the foreign substances and start off the immune response via the lymph flowing through them. Each lymph node is a honeycomb of sinuses (enlargements or sac-like dilations containing lymph) lined by macrophages with large clusters of lymphocytes between the sinuses. The spleen is the largest of the organs containing lymphoid tissue which lies on the left side of the abdominal cavity between the stomach and the diaphragm.

The other structures with large collections of lymphoid tissue are the tonsils, which are small, rounded structures in the throat which often get inflamed in children resulting in the common condition called tonsillitis.

There are multiple populations and subpopulations of lymphocytes termed B lymphocytes, T lymphocytes, cytotoxic, helper and suppressor T cells. There are two broad categories of specific immune responses. Firstly, a lymphocyte is programmed to recognise a specific antigen. The ability of lymphocytes to distinguish one antigen from another is the basis of specific immune responses. Recognition of an antigen implies that antigen is bound to the lymphocyte which has receptors for that antigen.

Once the lymphocyte has attached itself to the antigen, it divides into different types of cells, and this takes place at the site where the antigen has attached itself to the lymphocyte. Some of the divided cells will attack the antigens, whilst others may influence both the activation and function of these 'attack cells'.

The activated cells attack all the antigens that initiated the immune response. The manner in which lymphocytes attack the 'invaders' is by two methods: antibody mediated or humoral and cell mediated.

Antibodies are proteins that are both present in the plasma membranes of B cells and are also secreted by them. These antibodies travel along the blood stream to all parts of the body, combine with the antigens and direct an attack by phagocytes and complement that eliminates the antigen or the cells bearing them. Antibodies belong to a group of proteins called immunoglobulins.

In cell mediated immunity, the T lymphocytes and natural killer cells travel to the location of cells bearing on their surface antigens that initiated the immune response and directly kill them.

Two broad generalisations can be made. Antibody mediated responses carried out by B lymphocytes have a large range of targets and are the major defence against bacteria, viruses and other microbes and against toxic molecules. Cell mediated killing by T lymphocytes and natural killer cells is against a more limited number of targets, specifically the body's own cells that have become cancerous (have mutated) or infected with viruses. The helper cells activate both humoral and cell mediated immune responses. They cells are essential for the production of antibodies except in the case of a small number of antigens. Suppressor T cells inhibit the function of both B cells and cytotoxic T cells.

Homeostatic systems

One of the most important 'defence' mechanisms of the body rests with the organs by which either excessive amounts are eliminated (e.g. kidney, liver, intestine) or where excessive amounts are altered to a non-toxic state or metabolised (e.g. liver, intestine). The lungs also have an important role in elimination of toxic substances, and such substances may also be eliminated through the skin.

Fundamentally, stability of a variable in the internal environment is maintained by the balancing of inputs and outputs. These are a collection of interconnected cells that functions to maintain a physical or chemical property of the internal environment relatively constant. However, it is not possible to maintain all aspects/constituents of the internal environment at a given time and there is a hierarchy of importance. Thus there are some variables that are preferentially maintained with some 'acceptable' changes in others.

Homeostatic systems may work either by regulating the change in the variable by causing changes in the direction opposite to original change i.e. negative-feedback systems. A good example is the control of body temperature. In contrast, an initial disturbance may set off a train of changes that increase the disturbance even further i.e. positive feedback systems, a good example being blood clotting.

The importance of external environmental changes that can cause changes in the internal environment is that negative feedback control systems are usually unable to return the variable that has changed to its original value as long as the adverse external environmental change persists. An example is exposure to cold, which would cause loss of heat from the warm skin which tends to lower the body temperature. This is minimised by narrowing or constriction of the blood vessels to the skin, decreasing loss of heat from the skin (decreased output). If there is continued exposure to cold, homeostatic mechanisms induce shivering to increase the body temperature (increased input). If the cold external environment persists, a return of the body temperature would cause a cessation of the negative feedback mechanisms and thus body temperature would fall. Therefore, for the homeostatic mechanism to function, there has to be a persistent fall in body temperature until the external environment returns to normal.

Aging causes a decrease in function and the capacity of the body's homeostatic mechanisms to respond to environmental changes. This change, which is often worsened by the presence of disease states, for example of the heart or blood vessels, is attributed to the decrease in the number of cells that occur during aging and a slowing of the communication between different cells or systems that are involved in homeostasis.

Components of a homeostatic system – A stimulus is a detectable change in the environment both internal and external such as temperature [external] or sodium concentration [internal]) and the detector of this change is called a receptor. The stimulus acts on a receptor and to transmit a signal to an integrating centre, which may be receiving such signals from varying sources. The integrating centre transmits, depending on the net effect of the inputs from receptors to a system or groups of cells which provide an overall response to maintain homeostasis. Receptors are however rather specific. In a broad sense they are detectors but have the specificity associated with their structure to respond to a particular chemical messenger, the neurotransmitter. Further, different tissues may have the same receptor but the responses could be varied dependent on the main function the tissues regulate. Thus, a single receptor could produce different responses depending on the type of cell that has the receptor. Receptors are subjected to physiological and pathological influences and their affinity to the neurotransmitter may also be altered. Thus, receptors could vary in the number available on a tissue and also in the degree of affinity to the specific neurotransmitter. The associated terms are down-regulation or up-regulation.

The pathway to the integrating centre is the afferent pathway and the pathway that produces the necessary changes is the efferent pathway. The whole of this system is referred to as a reflex – an involuntary built-in system to protect the human body. In simplistic terms it refers to the withdrawal of a finger from a hot object or shutting of the eye when an irritant is in close proximity. In a wider perspective, it refers to systems which control body temperature, electrolyte balance, blood pressure and blood glucose levels. It is necessary that these reflexes are not only mediated by nerves, but hormones which are essentially blood-borne messengers influence responses from the glands which secrete hormones to produce similar effects. Some integrating centres in the brain have the ability to send messages to glands to produce the necessary responses.

Reflexes do not necessarily involve the brain or 'higher' structures and glands. A change in the external or internal environment can produce changes locally at the cellular level which does not involve nerves or hormones. Such 'local' reflexes include the release of chemicals by cells to prevent further injury to the cells and the increase in blood supply to active muscles by the release of chemicals by exercising muscles.

The necessity for all these homeostatic mechanisms is the transmission of 'messages' from all components of the reflex. This transmission between nerves is by neurotransmitters, though they have the ability to influence function/activity of cells that are not nerve cells. Chemical messengers produced locally by cells are called paracrines, which diffuse into neighbouring cells and produce the desired effects. Paracrines unlike neurotransmitters do not generally enter the blood stream to affect distant cells or tissues. However, there are local 'messengers' that are produced by virtually all tissues and cause a wide range of effects (e.g. in blood clotting, regulation of smooth muscle contraction, modulating release of neurotransmitters, release of hormones, and body defences against injury and infection) – eiconasides which include the prostaglandins, thromboxanes and leukotrienes. These are synthesised in response to a stimulus and released immediately to act locally and are quickly metabolised to inactive forms.

Chemical messengers – Responses from cells due to messengers are dependent on proteins on the plasma membrane, the G protein which interacts with another protein either via an ion channel or an enzyme in the plasma membrane which relays the 'messages' from the plasma membrane to the biochemical mechanisms within the cell by second messengers which produce the response from the cell. The best-known second messenger is cyclic 3'5-'adenosine monophosphate (cyclic AMP). Another is 3'5'-guanosine monophosphate (cGMP). Calcium ions too act as a second messenger in several cellular responses.

The neuromuscular junction (NMJ) – The NMJ is one of the most important sites for the action of a chemical messenger in the body. As its name implies it is the gap between the motor nerve and the skeletal muscle fibre, which it controls. Electric signals from the nerve are transmitted at the NMJ using a chemical transmitter, which crosses the junction and then generates another electric signal on the other side which causes the muscle fibre to contract. The chemical transmitter found at the NMJ is called acetyl choline. Acetylcholine is synthesised at the end of the nerve fibre (the nerve terminal) and is discharged when an impulse reaches the end of the nerve fibre. This chemical messenger reaches specialised parts of the muscle fibre called end plates, which contain specific receptors through which sodium ions pass into muscle fibre and produce a change in membrane potential, which is called the end plate potential. This movement of ions or depolarisation spreads to the whole muscle fibre, causing calcium ion release, and muscle contraction follows.

Acetylcholine stays only for a very brief period at the neuromuscular junction as it is quickly inactivated by the enzyme acetylcholinesterase. This enables the next impulse to release acetylcholine again and cause another muscle contraction.

The neuromuscular junction functional activity is vulnerable to many toxic substances. Firstly, toxic substances can interfere with the production and release of the chemical messenger acetylcholine. The end plates may be damaged or altered in disease states such as myasthenia gravis. The enzyme cholinesterase, which restricts acetylcholine in time and space, can be inactivated by several toxic substances, of which the best known are the pesticides belonging to the class of compounds called organophosphates. Some chemical warfare agents (nerve agents) such as sarin, tabun and soman are organophosphates and produce the same effect.

Importantly in medical practice, particularly in the speciality of anaesthesia, drugs are used to prevent transmission at the neuromuscular junction and thus produce muscle relaxation (the drugs used are called muscle relaxants) to facilitate surgery. Historically the Indians of South America used a substance as an arrow poison to paralyse their prey during hunting. This arrow poison was refined to become one of the best-known muscle relaxants: curare or tubocurarine.

Every muscle in the body consists of muscle fibres which are the units that cause muscles to contract and enable us to move, run, talk or do whatever we wish to do. Every muscle fibre needs a nerve supply in order to contract. The origin of these nerves to muscle are from cells in the spinal cord (anterior horn cells) and as there are more muscle fibres than nerve cells, each nerve cell or anterior horn cell innervates more than one muscle fibre. For example, in the leg, as many as 200 muscle fibres may share a single anterior horn cell. Where eye muscles are concerned, only about five muscle fibres would share one anterior horn cell. The nerves leaving the anterior horn cell are called axons or motor nerves and branch to supply a group of muscle fibres on reaching the muscle. The anterior horn cell and the muscle fibres supplied by this neuron are called the motor unit. The motor unit forms the basis for voluntary movement (movements which are intentional). If the motor nerve or axon is cut or damaged, paralysis of muscles occurs.

Nerves from the anterior horn cells carry nerve impulses or messages which enable the muscle fibres to contract. If an anterior horn cell discharges slowly or at a very low frequency, the muscles tend to be relaxed or flaccid. When the rate of discharge from the neurons increases, one may see coordinated contractions. However, these motor neurons are capable of discharging at very fast rates, usually in disease states leading to either tremulous contractions (clonus) or sustained contractions (tetanus). Tetanus is also the name given to an infection (lockjaw) caused by the tetanus bacillus (*Clostridium tetani*), which occurs when wounds become contaminated with soil/faeces which contains the bacteria. The infection releases tetanus toxin which acts on the anterior horn of the spinal cord producing repeated

activation of the motor units. Tetanus was a serious sequel of accidents and war injuries before immunisation against the disease became available.

SECTION 2: PRINCIPLES OF ENVIRONMENTAL PUBLIC HEALTH RISK ASSESSMENT AND APPLICATION TO CHEMICAL INCIDENT RESPONSE

Naima Bradley, Alec Dobney, Virginia Murray

Introduction

The following section in this chapter describes the principal concepts of environmental and public health risk assessment and their application in the management of chemical risks in all environmental compartments, air, water and land. These concepts will be expanded and their applications to specific risk assessment methodologies will be discussed in subsequent sections of this book. Chemicals are essential to modern life. They are used daily in food production and preservation, water sanitation, housing, clothing, cleaning products and healthcare, etc.

The worldwide Chemical Abstract Service Register [1] contains more than 186 million unique organic and inorganic chemical substances, such as alloys, coordination compounds, minerals, mixtures, polymers and salts. Though a large number, it is only a small fraction of the total possible chemical mixtures. With the increasing development, manufacture and use of chemicals comes an inherent risk that through, for example accidents or deliberate releases, chemicals that enter the environment can pose a threat to human health.

The World Health Organization (WHO) estimates that 24% of global deaths (and 28% of deaths among children under 5) are due to modifiable environmental factors [2]. The disease burden varies across the globe from 14% in Western Europe to over 29% in Africa [3]. In Europe, schemes such as the REACH Regulations (Registration, Evaluation, Authorisation and restriction of Chemicals) [4] require companies manufacturing or

marketing industrial chemicals to identify and manage the risks they pose. However, due to the large number of chemicals, there is limited information about their environmental behaviour, their toxicity or their subsequent adverse impacts on health.

Toxicology is the study of the nature and mechanism(s) of toxic effects of substances on living organisms and other biological systems [5]. It is a speciality in science which studies the manner in which toxins cause harmful effects to living organisms, the amounts (doses) that cause such harm, the consequences of harm (e.g. disordered function, disease, death), the manner in which such harm can be prevented and the methods by which harmful effects can be treated.

General concepts of environmental public health risk assessment

Environmental public health risk assessment provides an interdisciplinary approach to the study of environmental hazards and their impacts on human health. This requires a sound knowledge of toxicology, chemistry, environmental epidemiology, environmental science and public health principles [6]. Assessing a risk involves an analysis of the consequences of, and the probability that a hazard will result in harm or cause adverse health effects under specific circumstances. Table 10.2 provides some key definitions of environmental risk assessment. The key stages for risk assessment and risk management are shown in Figure 10.10.

Table 10.2 Key definitions of environmental risk assessment

Hazard	A situation, chemical, radiological, biological or physical agent that may lead to harm (impact) or cause adverse health effects.
Risk	The potential consequences of a hazard combined with their likelihoods/probabilities.
Risk assessment	The formal process of evaluating the consequences of a hazard and their likelihoods/ probabilities.
Risk management	The process of appraising the options for responding to risk and deciding which to implement.
Uncertainty	Limitations in knowledge about the impacts and the factors that influence them. Uncertainties originate from randomness as well as incomplete knowledge.

In human health risk assessment, the problem formulation stage involves a clear statement of what is to be risk assessed and establishes the scope and objective of the assessment. This has several stages including hazard analysis, hazard characterisation and exposure assessment. This is followed by the risk characterisation stage where estimated exposures are compared with health-based guidelines to estimate the severity of the exposure and to inform the risk management options [8].

A simpler model in environmental public health risk assessment takes into consideration the nature of the chemical hazard (source – S), the route by which the chemical reaches the receptor (pathway – P) and the individual characteristics of the people affected (e.g. susceptibility) (receptor – R). This assessment can be qualitative, quantitative or semi-quantitative. Without this source-pathway-receptor (S-P-R) linkage, harm to health will not occur.

As long as the S-P-R linkage exists, a sound, systematic methodology for assessing

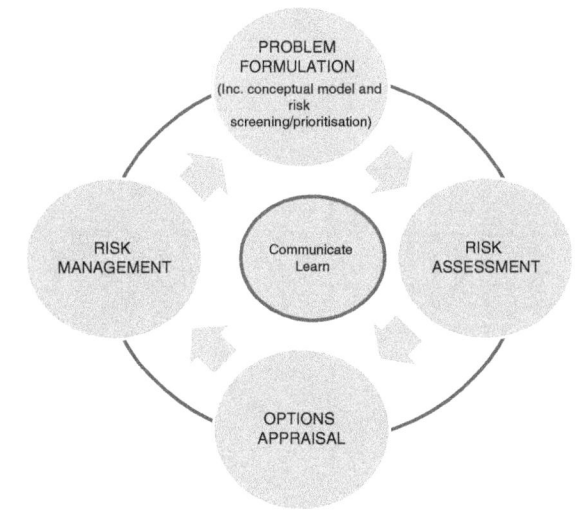

Figure 10.10 Framework for environmental public risk management

Source: Adapted from Green Leaves III, Guidelines for Environmental Risk Assessment and Management. Defra (Department for Environment, Food and Rural Affairs) and Cranfield University, Figure 2, p. 9, © Crown Copyright 2011,[1] licensed under the Open Government Licence v1.0, available from https:// www.gov.uk/government/uploads/system/uploads/attachment_data/file/69450/pb13670-green-leaves-iii-1111071.pdf

Table 10.3 World Health Organization classification of hazards

GENERIC GROUPS[1]	1. NATURAL						2. HUMAN-INDUCED2, 3		3. ENVIRONMENTAL
GROUPS	1.2 HYDRO-METEOROLOGICAL						2.1 TECHNOLOGICAL	2.2 SOCIETAL	3.1 ENVIRONMENTAL DEGRADATION17
SUBGROUPS	1.1 GEOPHYSICAL[4]	1.2.1 HYDROLOGICAL[4]	1.2.2 METEOROLOGICAL4	1.2.3 CLIMATOLOGICAL[4]	1.3 BIOLOGICAL[5]	1.4 EXTRATERRESTRIAL[4]			
Main types – subtypes [sub-subtypes]	Earthquake: – ground-shaking Tsunami Mass movement (geophysical trigger): – landslide – rock fall – subsidence Liquefaction Volcanic activity: – ash fall – lahar – pyroclastic flow – lava flow	Flood: – riverine flood – flash flood – coastal flood – ice jam flood Mass movement (hydro-meteorological trigger): – landslide – avalanche (snow) – mudflow – debris flow Wave action: – rogue wave – seiche	Storm: – extratropical storm – tropical cyclone [cyclonic wind, cyclonic rain, cyclone (storm) surge] – convective storm [tornado, wind, rain, winter storm, blizzard, derecho, lightning, thunderstorm, hail sand/dust storm] Extreme temperature: – heatwave – coldwave – severe winter condition [e.g. snow/ ice, frost/ freeze, dzud][6] Fog	Drought Wild fire: – land fire [e.g. brush, bush, pasture] – forest fire Glacial lake outburst (flood)	Airborne diseases Waterborne diseases Vector-borne diseases Foodborne outbreaks[7] Insect infestation:[4] – grasshopper – locust Animal diseases Plant diseases Aeroallergens Antimicrobial resistant micro-organisms Animal-human contact – venomous animals [snakes, spiders]	Impact – airburst – meteorite Space weather – energetic particles – geomagnetic storms – shockwave	Industrial hazards:[8] – chemical spill – gas leak – radiation [radiological, nuclear] Structural collapse: – building collapse[8, 9] – dam/bridge failures Occupational hazards – mining Transportation:[8, 11] – air, road, rail, water, space Explosions Fire[8] Air pollution:[9] – haze[10] Infrastructure disruption: – power outage[11] – water supply – solid waste, waste water – telecommunication Cybersecurity Hazardous materials in air, soil, water[12, 13] – biological, chemical, radiological Food contamination[7]	Acts of violence Armed conflicts:[14] – international – non-international Civil unrest Stampede Terrorism: – chemical, biological, radiological, nuclear, and explosives[15,16] Financial crises: – hyper-inflation – currency crisis	Erosion Deforestation Salinisation Sea-level rise Desertification Wetland loss/ degradation Glacier retreat/ melting Sand encroachment

Source: World Health Organization (2019) Health Emergency and Disaster Risk Management Framework [7]

The footnote numbers refer to the sources of information as in the original document (Annex 1 of [7]) which should be consulted

and managing that risk is required. In most instances, detailed, quantitative assessment will require the collation of information on the source and the pathways often using tools such as environmental monitoring and dispersion modelling. Gathering very detailed information is possible when dealing with longer-term problems such as chronic exposures to contaminated land or air pollution. Geographic Information Systems (GIS) can be used to provide graphical representations of the S-P-R scenarios with relevant layers added to provide detailed location-specific information such as the extent of residential areas, the presence of schools, hospitals and other locations where vulnerable parts of the population may be located.

Risk assessment during chemical incidents presents unique challenges due to the paucity of information in the earlier stages of the incident and the speed with which risk management interventions have to be implemented to protect public health.

Chemical incidents and public health protection

Since chemical incidents are so varied in nature, they are difficult to define. The World Health Organization defines them as *the uncontrolled release of a toxic substance, potentially resulting in harm to public health and the environment* [9]. Chemical incidents usually trigger a public health response, which include an assessment of exposure and risk and provision of advice to authorities and/or the public.

The potential for chemicals to be involved in accidents or deliberate releases has been realised through a number of familiar historic events such as Seveso, Buncefield, Bhopal and the Tokyo sarin attack and more recently the explosion in Beirut. Accidents causing major devastation are increasing throughout the world but are still rare. Natural events can also trigger chemical-related incidents (e.g. gaseous and ash emissions from volcanic activity). Industrialisation and new technologies have produced new hazards. Many smaller chemically related incidents are dealt with on a daily basis, such as chemical spills, fires or explosions. Chronic exposure to chemicals in everyday use is often a top public concern and continued vigilance is required, as we better understand the interactions between chemicals and biological systems (toxicodynamics).

Consider first *the source* – the identity of the chemical(s) involved in an incident requires identification; however, this information may not be immediately available. Various properties and characteristics of chemicals and a variety of sources of information can be used to identify unknown substances. More than one chemical could be present, or there may be reactions between chemicals resulting in a complex mix of by-products or decomposition processes such that the chemical of concern may not be the original involved. The quantity of the chemical present is extremely important: for some chemicals (e.g. cyanide salts) only a very small amount is required to cause a significant health or environmental effect.

Chemicals may occur as solids, liquids, aerosols, vapours or gases depending on their innate characteristics and their environment. Of those in use, many exert acute or chronic toxic effects, which may range from being immediate and obvious (e.g. cyanide poisoning) to insidious and/or long term such as the development of cancer (e.g. chronic exposure to inorganic arsenic). A competent understanding of chemistry is required when responding to chemical incidents to ensure correct information is gathered and appropriate actions are taken. For example, the physiochemical properties and health effects of phosgene and phosphine are different; however, due to their similar chemical names they can be easily mixed up in transcriptions leading to erroneous assessments. When dealing with chemicals it is important that chemical classification numbers such as Chemical Abstracts Service (CAS) registry numbers or UN numbers are used to ensure the correct chemical information is being recorded.

The physical and chemical properties of the chemical must be considered when evaluating or characterising the potential extent of the contaminated environment during a chemical incident. For example, a chemical with low aqueous solubility such as oil will behave differently within a water environment than a water-soluble chemical which quickly dissolves and disperses. For example, highly toxic water-soluble chemicals could be a significant hazard if released into a lake due to the ease with which they disperse, whereas certain organic materials present in land (dependent on environmental conditions) are able to penetrate buried plastic water pipes and have the potential to contaminate drinking water supplies. Knowledge of the chemical's physicochemical properties is essential to determine how the chemical will behave in the water environment (e.g. whether the substance floats on the surface, dissolves, hydrolyses or settles out within sediments).

Pathways are the second consideration – the behaviour of contaminants released into the environment is controlled by a complex set of processes, which include various forms of transport and cross-media uptake and will determine how the public are exposed to the chemical. Consequently, when one environmental medium is directly contaminated, there is the potential for secondary (indirect) contamination of other media if the contaminant source is not contained and mitigated in an appropriate and timely manner. Environmental media include air (ambient and indoor), land (soil) and water (including groundwater, surface water, coastal waters, rivers, lakes, streams and aquifers). Examples of direct and indirect cross-media contamination are indicated in Figure 10.11. An incident (chemical input) may result in an airborne chemical plume that contaminates outdoor air (direct contamination) and subsequently through ingress contaminates the indoor air environments (indirect contamination). In addition, chemical contamination from the plume may deposit to land or water (or both) and result in subsequent contamination of buildings, farmland and watercourses (indirect contamination) [10].

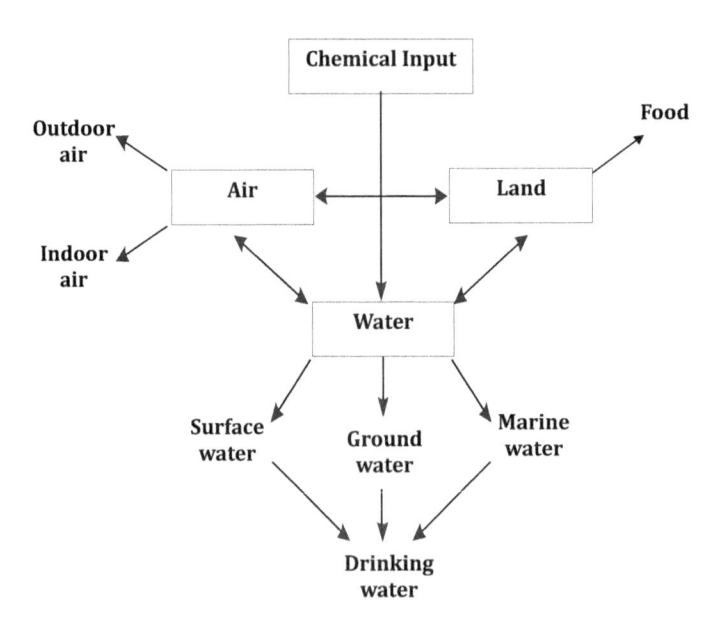

Figure 10.11 Direct and indirect contamination of air, land and water following a chemical incident

Source: Adapted from Wyke-Sanders et al. [10]

Once released, a chemical disperses in the environment, and dependent on the media affected can cause harm through several routes as shown in Table 10.4.

Receptors are the final consideration – in environmental and public health risk assessment, receptors are people potentially or actually affected by environmental pollutants. They may be nearby residents, people in hospitals, schools, care homes, recreational areas, etc. Sensitive or vulnerable receptors are people who may be impacted by lower level of pollution due to their age (the young and elderly), habits (pica behaviour) or health condition (immunosuppressed, people with existing respiratory conditions).

The existence of a chemical contaminant does not always lead to human exposure. Individual exposure through breathing, eating or drinking a chemical substance or by skin contact is required and the adverse health effects will depend on several factors, including the amount to which an individual is exposed (dose), the way they are exposed, the duration of exposure, the form of the chemical and if there are simultaneous exposures to any other chemicals.

Table 10.4 The main route of exposure – the environment/human interface

Entry into the body	Route of exposure
Inhalation	This is a common route of exposure, particularly in occupational settings. The substances are inhaled through the nose/mouth and breathing tubes (bronchi and bronchioles) into the lung and finally into the thinly lined air cells (alveoli), which are surrounded by blood vessels. The substances or particles may cause allergenic responses or depending on their physical/chemical properties, diffuse across the thin lining of the alveoli into the blood stream. **Exposure to a chemical by inhalation can occur from:** re-suspension of materials from contaminated surfaces via air/vapour, or aerosols, or direct inhalation of the chemical or of contaminated material/dust.
Dermal	Dermal (across the skin) – One of the most common routes of exposure is when the potentially harmful substance is absorbed through the layers of the skin. Absorption will be dependent on the properties of the material – in some instances, the vehicle or substances in which the toxin is dissolved can influence the rate of absorption facilitating or decreasing penetration of the skin layers. If the chemical can react with protein then it can be allergenic resulting in the activation of dentritic cells in the skin (the watchkeepers) and subsequent inflammatory reaction. A good example is nickel. **Exposure to a chemical by dermal contact can occur from:** Direct contact with chemical or contaminated environmental media (e.g. water, soil)
Ingestion	Ingestion (orally) – absorption usually occurs in the gastro intestinal tract (i.e. stomach, intestines). On occasions, absorption may occur through the mucous membrane of the mouth. **Exposure to a chemical by ingestion can occur from:** Ingestion of contaminated drinking water or food, also consider water used in food production or inadvertent ingestion of contaminated material.
Mucous membrane	Through mucous membrane – substances may enter the body following absorption through the mucous membrane of the eye. **Exposure to a chemical via the mucous membrane can occur from:** Direct contact with water/aerosol/spray/liquids, solids and atmospherically dispersed materials.

Chemical hazards and toxicology

Chemicals have the ability to cause harm or damage (toxicity) to living organisms. Hazardous substances vary in their origin, chemical structure, and physical properties and most importantly in the manner in which they cause toxicity. Toxicology is the study of the effects of chemical substances on living systems.

Dose response – when a reference to toxicity of a substance is made, it is fundamental to consider the concept introduced by one of the founders of toxicology, Paracelsus, in the 16th century:

> No substance is a poison by itself. It is the dose that makes a substance a poison.

All substances therefore have toxic properties and have the potential to cause harm. Once the dose-response relationship is understood, the exposure characteristics will determine the likelihood of harm arising. This is best summarised by the following equation:

[Dose × exposure duration] (Single or repeated) = toxic effect

This means the dose (e.g. weights or volumes or the concentrations in air, water or land taken at a particular time [with a single exposure]) or the doses in total taken over a specified period of time relate to the toxic effect.

Thus, the quantity of the chemical and the duration during which the exposure has taken place (orally, dermally or by inhalation) is critical in the assessment of the harmful effects a substance would produce in the human body. Following exposure to a potentially harmful substance, the routes of exposure and pathways of absorption, distribution and excretion of toxic chemicals in humans are summarised in Figure 10.12.

With our now much greater understanding of biology, we need to introduce a third factor into this equation, and that is an individual's susceptibility. For various reasons, which can be physical (size, age, sex) and/or genetic, individuals will respond differently to the same cumulative dose:

[Dose × exposure duration] (Single or repeated) × susceptibility = toxic effect

The possible fates of a chemical substance in the body are of concern. The harmful effect of a potentially toxic substance to an organ is determined primarily by the dose of the substance or key metabolite that reaches the *target organ* (such as the lungs, liver, kidneys, heart or brain). The dose of the ultimate toxicant that reaches the target organ is known as the *target dose*. The target dose depends not only on the dose which reaches the body by the exposure route, but also on the processes that take place once the potentially toxic substance gains entry into the body. When a chemical substance enters the human body, it is subjected to a number of processes, which results either in the total elimination of the chemical substance with no ill health effects, ill effects from altered function (of biological systems), death of cells or inhibition of enzyme systems.

These processes include:

- *Absorption* – to be harmful, a chemical substance needs to be transferred from the environment to the circulatory system. This mostly occurs through the skin, alimentary tract or lungs. The substance also needs to be transferred from the blood to the target tissues.
- *Distribution* around the body for excretion by organs such as the liver (via bile), kidneys, lungs, to the sweat glands, to breast milk or for storage in fat, bone or other tissues. This will also include binding to proteins (especially in the blood), fat and other tissues.
- *Metabolism* – whereby a substance is made more water soluble to facilitate excretion. This can result in the metabolite becoming less toxic (detoxification) or more

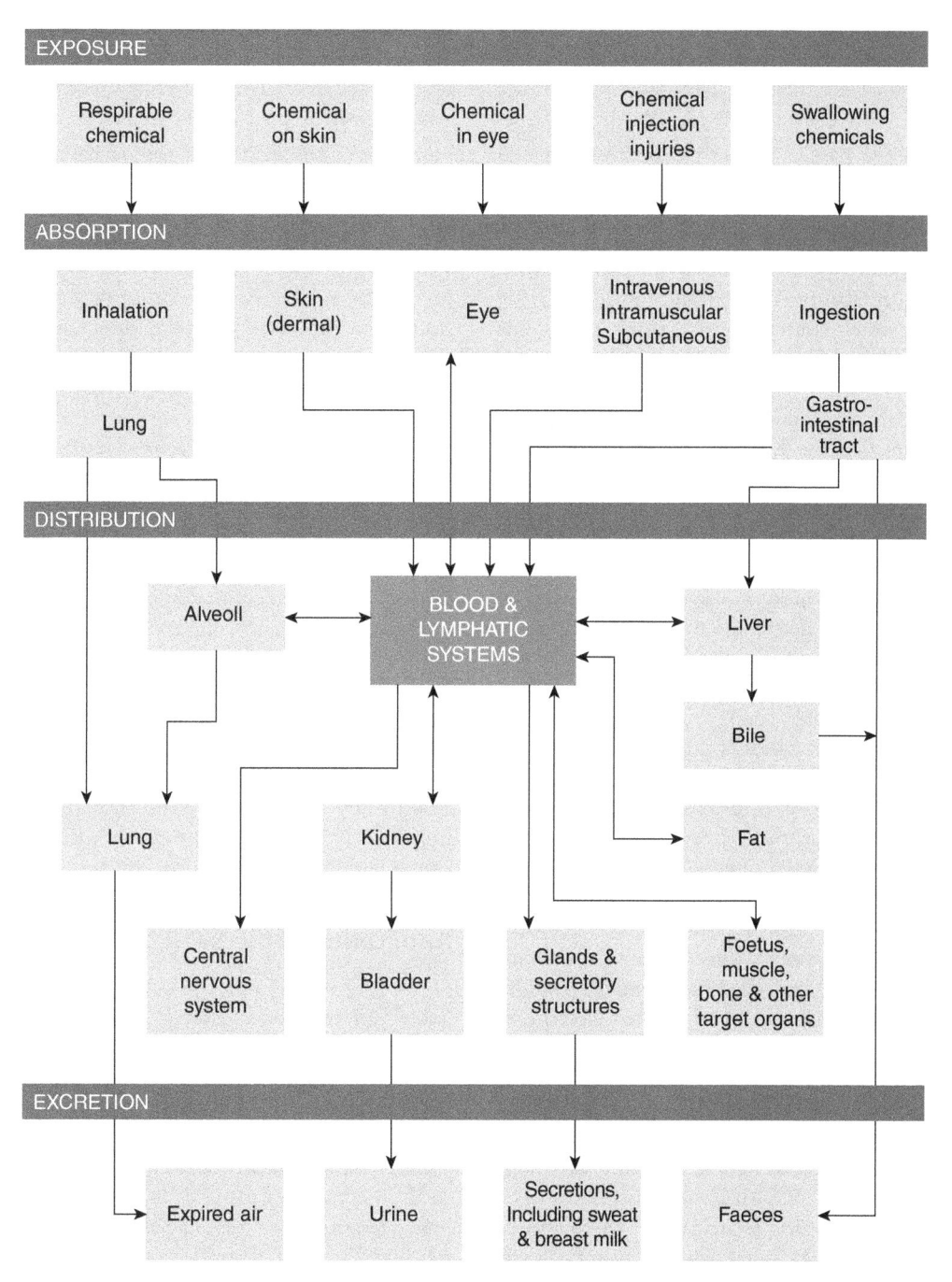

Figure 10.12   Summary of routes of exposure, absorption, distribution and excretion of toxins in the body

Source: Adapted from Baker et al. [5]

toxic (activation). Metabolism mainly takes place in the liver, but can also occur in significant amounts in the kidneys, skin and the intestine. Metabolic systems are affected by an individual's genetic factors and also by foods, alcohol and drugs. The importance of these systems is that they influence the amount or concentration of chemical absorbed.

- *Excretion*, the removal of the substance and/or its metabolites from the body which is usually in the urine, faeces, bile or breath.

The fate of the chemical substance in the human body (i.e. the processes to which a substance is subjected, e.g. metabolism, excretion or binding to tissues or cells or blood components such as proteins) is referred to as toxicokinetics.

The chemical and physical properties of a substance determine the manner in which they are processed by the human body, and the nature of the harmful effects they produce (see Table 10.5).

Two main routes of exposure are examined next.

Inhalation – A volatile substance (i.e. has a strong tendency to evaporate), will usually be preferentially absorbed by inhalation. Substances that are lipophilic (dissolve easily in fat) will enter cells more easily than those which are water soluble, because cell membranes have high lipid (fat) contents. The size of a molecule (usually described in terms of molecular weight) influences both its absorption and excretion (e.g. large molecules can be absorbed less readily than small molecules) and is usually poorly excreted by the kidneys, and may instead be excreted into the intestines via the bile. This latter process can result in enterohepatic recirculation where resorption of excreted molecules occurs. Some volatile or gaseous chemicals (such as carbon monoxide) may be excreted unchanged via the lungs.

Ingestion – when a substance enters the body orally, it is usually absorbed through the gut (gastrointestinal system) and this may occur in the stomach or in the small or large intestines. Once absorbed, the substance is taken by the blood vessels draining the intestine to the liver where enzymes present may metabolise the substance to a more water-soluble species and this may result in a less toxic (detoxified) or more toxic (activated) substance.

Harmful effects may not be immediately visible or detectable. Certain health impacts such as the ability to cause cancers or abnormalities in the development or function of organs may only occur months or years after exposure to the substance. This is known as the *lead time*.

Vulnerable and susceptible populations – it is known that there are susceptible subgroups of the population who are potentially at greater risk of harm from chemical exposures. They can be divided into three main groups based on biological, sociocultural or ethnic characteristics that may affect their vulnerability to adverse effects resulting from environmental exposure to a particular hazardous chemical, shown in Table 10.6. One or more of these risk factors may be present in individuals at any one time and should be considered when making any assessment of toxic harm. Additionally, it is important to remember that susceptibility will vary dependent on chemical.

Application of the S-P-R to chemical incidents

Table 10.7 provides an example of the application of a source-pathway-receptor model to a large fire where all available information is brought together and analysed to establish the severity of the potential exposures and inform the intervention measures. The model is often used for emergency planning as well as emergency response purposes. The availability and reliability of the information relating to the exposure improves as the incident progresses.

This process is part of the multi-agency incident response – the larger the incident, the more complex the response model. Local, regional, national or international responders

Table 10.5 Chemical classification of toxic agents based on toxic effects

Class of substance	Toxic effects
Irritant	Causes inflammation of the skin and mucous membranes (skin, eyes, nose or respiratory system). Skin (dermal) irritants cause contact dermatitis (acute inflammation of the skin), symptoms of which include itching and skin changes ranging from reddening to blistering or ulceration. Examples: dilute solutions of acids, alkalis and some organic solvents.
	Respiratory irritants cause injury to the nose, mouth, throat and lungs. Materials that are very water soluble affect mainly the nose and throat. Less water soluble materials act deeper in the lungs.
Corrosive	A material that can destroy human tissue. Includes both acids and alkalis and may be a solid, liquid or gas.
Asphyxiant	A material that deprives tissues of oxygen and causes suffocation by displacing oxygen or interfering chemically with oxygen absorption, transport or utilisation.
	Simple: A simple asphyxiant displaces oxygen from the atmosphere, which prevents its absorption. Examples: carbon dioxide, methane and nitrogen.
	Chemical: A chemical asphyxiant prevents the uptake of oxygen by the cells. Examples: carbon monoxide combines with haemoglobin and prevents it transporting oxygen to the cells; hydrogen sulphide prevents uptake of oxygen by the cells by inhibiting the action of enzymes – cytochromes – which are necessary for this process.
Sensitisers	A chemical that causes an allergic reaction, such as urticaria or breathing problems. Examples: nickel, colophony.
Systemic effects	A response to chemical exposure that affects the whole body. Systemic illnesses may cause symptoms in one or two areas, but the whole body is affected.
Pulmonary toxin	Irritates or damages the lungs. Examples: asbestos, silica, ozone, chromium.
Neurotoxin	Affects the nervous system. Examples: mercury, lead, carbon disulphide.
Cardiotoxin	Affects the heart. Example: carbon disulphide.
Hepatotoxin	Causes liver damage. Example: carbon tetrachloride, dimethyl sulphate, chloroform.
Haematopoietic toxin	Affects the function of the cellular components of blood. Examples: benzene, ethanol, 2, 4, 6-trinitrotoluene.
Nephrotoxin	Causes kidney damage. Examples: chloroform, mercury, lead.
Carcinogen	A material which can cause cancer. Examples: asbestos, benzene, acrylonitrile, 1–2 nathphylamine, vinyl chloride monomer, aflatoxin B1.
Mutagen	Anything which causes a change in the genetic material of a living cell. Many mutagens are also carcinogens.
Reproductive toxin	Causes impotence or sterility in men and women. Examples: lead, dibromodichloropropane.
Teratogen	A material which interferes with the developing embryo when a pregnant female is exposed to that substance. Examples: lead, thalidomide.

Source: Adapted from Baker et al. (2012) [5] – see sources for further information

regularly plan for, and exercise different incident response scenarios.

The most common way to assess the level of exposure is by comparing measured and modelled levels of the contaminant with health-based standards developed for that purpose. These standards are derived depending on the environmental matrix being considered (e.g. air pollutants, pollutants in soil or in drinking water).

Table 10.6 Factors affecting vulnerability to an environmental hazard

Biological	Sociocultural	Ethnic	Genetic
Age group (e.g. infant, elderly)	Diet	Social (e.g. diet)	Inheritance from parents
Sex	Smoking status		Epigenetic adaptation to environment
Disease state/medication	Alcohol, drugs		
Genetic susceptibility	Socio-economic position 'status'		
Pregnancy (e.g. foetal development) Physiological variation (e.g. height, weight)	Religion		
	Housing		
	Quality of housing		
	Location of occupation		

Source: Adapted from Risk Assessment and Toxicology Steering Committee 1999 – *Exposure assessment in the evaluation of risk to human health* http://www.iehconsulting.co.uk/IEH_Consulting/IEHCPubs/IGHRC/cr5.pdf (Accessed 2 November 2020)

For chemical incidents the most common guidelines are:

- The Acute Exposure Guideline Levels (AEGLs) [11] are based primarily on acute toxicology and derived for risks to humans resulting from once-in-a-lifetime, or rare, exposure to airborne single chemicals. They are designed to protect the general population including vulnerable groups such as children and the elderly. Three levels are derived for a range of exposure times called AEGL1, AEGL2 and AEGL3 representing the level of the chemical in air, at or above which the general population could experience discomfort, serious long-lasting effects and life-threatening health effects respectively.
- Emergency Response Planning Guideline (ERPG) Values [12]. Three levels are defined ERPG1, 2 and 3 as being the maximum airborne concentration below which it is believed that nearly all individuals could be exposed for up to 1 hour without experiencing other than mild transient adverse health effects, experiencing serious health effects/symptoms which could impair an individual's ability to take protective action and developing life-threatening health effects respectively.
- There is no acute exposure guideline for particulate matter contained within AEGL or ERPGs; furthermore, there is no safe threshold that has been identified beneath which there are no health effects for exposure to particulate matter. In the UK there is an Air Quality Strategy objective for PM10 of 50 µg/m³ that should not occur more than 35 times per year (concentration measured as a 24 hour mean) in any one location and 40 µg/m³ (concentration measured as an annual mean). The UK target objective for PM2.5 is 25 µg/m³ (concentration measured as an annual mean) [13] although the UK has set out new policies in their Clean Air Strategy 2019 to reduce concentrations further [14]. The World Health Organization (WHO) have set a series of 24-hour average interim targets for developing countries around the world that experience higher levels of particulate matter. The highest target on their list is 150 µg/m³ measured as a 24-hour average. This is described as *relating roughly to a 5% increase in mortality, an impact that would be of significant concern and one for which immediate mitigation actions would be recommended* [15].

Carboxyhaemoglobin (COHb) is the best marker available at present to confirm exposure to CO and is most often measured in the blood (CO can also be measured in breath). However, this measurement rarely reveals the

Emergency planning and response stages	Source	Pathways considered	Receptors	Risk management
Emergency planning and preparedness stage Before the incident –	Information on the potential chemicals which could be released, products of combustion and their toxicological impacts [10] (e.g. COMAH plans, REACH). Identification of potentially high-risk waste sites by environmental regulators or major accident hazards – chemical storage and manufacturing sites (e.g. Seveso sites) [12]. Deliberate release (CBRN) emergency plans.	Development of air dispersion models for use during air pollution emergencies [11]. Development of protocols for the deployment of air pollution monitoring during emergencies. Modelling of worse-case scenarios is undertaken as a requirement for Seveso for major accident hazards.	Availability of GIS maps for emergency response, including layers with information of the type and number of individuals likely to be in that area (e.g. COMAH plans, public information zone [PIZ] at night and during the day). Location of susceptible populations (e.g. schools, hospitals). Proximity to residential areas when planning location of new industrial, waste depots and large chemical storage and manufacturing sites (e.g. Seveso sites).	Development of protocols, planning and exercising of emergency plans with partner agencies and emergency services. Plan and exercise protocols for implementing multi-agency communicating for population sheltering and evacuation options. Plan and exercise protocols for hospitals and those located in the PIZ. Develop communication plans in case of drinking water contamination. Plan and exercise emergency mass decontamination plans.
Emergency response – initial assessment	**Site information** Location of the site. Obtain information on the type and quantities of material burning or stored at site (e.g. quantity of tyres, type of chemicals stored). Identification of the most likely combustion products and short-and long-term toxicity/health effects (e.g. tyre fires emit a complex mixture of chemicals in gaseous, liquid and solid form). The most important from a public health perspective are irritant gases – sulphur dioxide and acrolein – and particulate matter.	**Environmental data** Availability of simple air dispersion maps giving the direction of the smoke plume and approximate areas affected. Obtain information from fixed monitoring sites (air and water) where available.	**Population information** Identification of population locations Presence of susceptible populations (e.g. people with existing respiratory diseases, children, the elderly and those in nearby hospitals, schools, care homes, open spaces [parks, recreational waters]).	**Options** Consider: Shelter in place (go-in, stay-in, tune-in) Evacuation Health messages and medical treatment Other important considerations impacting on the effectiveness of mitigation measures:

Human physiology, hazards and health risks

(Continued on next page)

Table 10.7 (Continued)

Emergency planning and response stages	Source	Pathways considered	Receptors	Risk management
	Obtain information on the characteristics of the fire (e.g. low temperature, smouldering fires will emit more complex organic compounds), its severity and likely duration.		Consider impact on sources of drinking water.	Communication with the public through local, national, social media. Communication with partner agencies.
Emergency response – as the incident progresses	Update information on the above as the incident progresses. Establish fire fighting strategy (e.g. whether the fire is actively tackled or a control burn technique is adopted), and impact on off-site emissions.	Air – deployment of real-time, mobile air pollution monitoring capabilities, analysis of cata and comparison with health-based standards. Undertake complex air dispersion modelling for short-, medium- and long-term scenarios. Potential for longer term impacts: Water runoff affecting abstraction points for drinking water and soil contamination/ contamination of crops due to deposition.	Consider impact on nearby crops/gardens used for food production. Consider biomonitoring.	Consider sheltering times and potential for ingress indoors. If the incident lasts weeks or months, sheltering may no longer be an option.
Recovery Post incident	Fire extinguished. Site clearance.	Potential chronic exposures due to contamination of soil and water. Management of residual waste remaining at the site.	Health surveillance, post incident epidemiological studies.	Environmental decontamination may be necessary (e.g. asbestos).

CASE STUDY 1

Carbon monoxide – use of barbeques indoors

The Fire and Rescue Service attended a number of serious incidents where families have been hospitalised due to disposable barbeques (BBQs) being taken inside holiday tents to provide warmth during cold weather whilst on holiday.

On one occasion the fire service carried out tests using a combination of different types of disposable BBQs and tents. Results showed that small disposable barbeques with warm coals can produce lethal levels of carbon monoxide (CO) within minutes in a sealed tent: even when the aluminium tray containing the charcoal is cold to touch, CO levels recorded ranged from over 400 parts per million (ppm) to 957 ppm, levels which can result in symptoms including severe headache, weakness, dizziness, nausea, vomiting and even seizure and coma.

Modern tent manufacture is constantly utilising new and improved materials and designs that have improved waterproofing and comfort. More frequently they incorporate integral ground sheets and zips designed to inhibit the flow of drafts through the structure improving warmth. However, this has played an important part in contributing to such incidents.

actual concentration of CO to which an individual has been exposed due to the natural dissociation of CO from the haemoglobin and relatively short half-life. This dissociation is accelerated by the administration of oxygen, often immediately after an incident by ambulance staff. There is a good example of susceptibility here as well because the foetus is more susceptible than the mother as foetal haemoglobin has a higher affinity for CO than the maternal haemoglobin.

Alternatively, environmental sampling may be more indicative of toxicological risks from CO. However, it is vital to remember that there are health and safety hazards in measuring environmental levels and it is very important not to be put at risk if personal protective equipment (PPE) is unavailable. Table 10.8 summarises data on CO concentration in parts per million (ppm) and milligrams per cubic meter (mg/m^3) with symptoms likely to occur at these levels over such exposures and where available, exposure time.

Although much is known about CO poisoning, people are still regularly exposed to CO. To address this, Public Health England and the Chartered Institute of

Table 10.8 Threshold toxicity values and IK guidelines for exposure to carbon monoxide by inhalation

Exposure via inhalation		
ppm	mg m-3	Signs and symptoms
~ 100	~ 115	Slight headache, flushing of skin (indefinite exposure)
200–300	230–345	Headache (5–6 hour exposure)
400–600	460–690	Severe headache, weakness, dizziness, nausea, vomiting (4–5 hour exposure)
1100–1500	1265–1840	Increased pulse and breathing rate, syncope, coma, intermittent seizures (4–5 hour exposure)
5000–10000	5750–11500	Weak pulse, depressed respiration/respiratory failure, death (1–2 minutes exposure)

Environmental Health produced a flowchart to help environmental health practitioners

---

CASE STUDY 2

Organophosphate exposure

Small amounts of harmful chemicals stored in residential garden sheds and garages are often implicated in harmful human exposure. Paramedics were called to two adults exposed to a 1 litre container of a chemical substance that had been stored in their garage for over 20 years. They had recently moved and whilst storing the chemical in the new garage the lid had become damaged. Over a period of two weeks, they noticed fumes, which they thought were from the substance so decided to move the container and, in the process, spilt some of its content on their hands.

The chemical was determined to be a synthetic organophosphate insecticide (demeton-S-methyl), which is highly toxic via dermal contact and inhalation by blocking the action of a key enzyme (acetyl cholinesterase) involved in nerve signal communication. It also causes mucus membrane irritation and chest tightness. The adults were decontaminated at the scene and transferred to the local emergency department for observation.

The emergency services entered the garage in gas protective suits to locate, retrieve and safely remove the chemical for its appropriate disposal.

(EHP) establish whether occupants of residential premises might be at risk from CO [16].[2] As well as helping EHPs identify residential premises where CO could be present, it recommends actions to take if the presence of CO is suspected, detailing key telephone numbers for services providing further advice and practical assistance to protect occupants and others who might be at risk.

Neurotoxins – sources, mode of action and effects – an overview

The nervous system is the communication and information storage system of the body, coordinating and integrating the functions of nearly all organ systems. Consequently, damage to the nervous system by chemicals such as organophosphate has a major impact that goes beyond the better-known functions and activities of this system such as thinking and memory.

Neurotoxicity has been defined as any adverse effect on the structure and function of the central and peripheral nervous system produced by a biological, chemical or physical agent.

Toxic damage to the nervous system occurs mainly via the following mechanisms:

1. Direct damage and death of neurones and glial cells
2. Interference with electrical transmission – usually following disorders at ion channels
3. Interference with chemical neurotransmission (chemical messengers)

The Office of Technology Assessment (OTA) and the US EPA defines neurotoxicity or a neurotoxic effect as an adverse change in the structure and function of the nervous system following exposure to a chemical agent – a definition that hinges on the interpretation of the word 'adverse' where there is considerable disagreement between scientists as to what constitutes an 'adverse effect'.

Due to the diverse functions attributed to the nervous system, assessment and quantification of neurotoxicity is beset with confounders and variations in methods of measurement/assessment including individual biases and skills. In addition, an adverse effect on the CNS may be a secondary effect of the action of a toxin on another organ or system (e.g. liver, kidney).

The broad health, social and economic significance of neurotoxicity is underscored by recent studies that have linked neurotoxicants to anti-social behaviour in adults, and thus an association with crime. An early evaluation of neurotoxicity necessitates a history of drug taking, exposure to industrial or environmental chemicals, inspection of patient's work environment and the awareness of:

1 Non-specific symptoms (which often are associated with other common pathologies) such as headache, vertigo, dizziness, swinging mood, fatigue and memory loss that may be suggestive of neurotoxicity from chemical exposure
2 Changes in breathing and heart rates, motor reflexes (e.g. knee jerk), perspiration and gastrointestinal function are early signs of injury to peripheral nerves

However, the vulnerability of the nervous system to chemicals is a reality, can be explained and is of concern because:

1 Many chemical substances have affinity for fat which makes up about 50% of the dry weight of the brain compared to 6–20% of other organs.
2 Nerve cells have limited capacity for regeneration and thus adverse effects would often be permanent.
3 The average human brain contains 86 billion neuronal cells [17]. These cells have a high metabolic rate and together with a less active antioxidant defence system tend to be more sensitive to toxic insult than many other cells types in the body.
4 The developing nervous system is particularly vulnerable to some neurotoxic agents because:

  a It is actively growing and establishing cellular connections

  b The blood-brain barrier is not completely formed
  c Detoxification systems within the system and cells are not fully developed

5 Substances with the potential to cause sensory and motor dysfunction may interfere with learning and memory processes and cause detrimental behavioural effects.
6 Some psychoactive drugs (e.g. heroin, cocaine, MDMA – ecstasy, phencyclidine) may cause permanent damage to the nervous system. For example, MDMA destroys nerve fibres.
7 Abuse during pregnancy of psychoactive drugs, alcohol, etc., poses a serious threat to the nervous system of the foetus.

Examples of signs and symptoms reported in humans exposed to neurotoxicants:

1 *Sensory changes* – changes in smell, vision, taste, hearing, balance, proprioception (position of body in relation to outside world and state of contraction of muscles), feeling, pain
2 *Motor weakness* – decreased strength of muscles, in coordination or ataxia (unsteady gait) – speech defects, abnormal movements (myoclonus, fasciculations), behavioural seizures, hypermotor or hypomotor activity, Parkinsonism symptoms-tremor, rigidity, poverty of spontaneous movements
3 *Sensorimotor* – paraesthesia, numbness in feet, pain in soles, muscle weakness similar to Guillain-Barré syndrome
4 *Autonomic* – body temperature changes, 'cholinergic crisis', changes in size of pupils (miosis-constricted, mydriasis-dilated)
5 *Cognitive* – learning, memory deficits
6 *Sensorium* – hallucinations, delusions, apathy, stupor, coma

Alcohol consumption is likely to increase the neurotoxic potential of most chemicals.

Conclusion

This section of the chapter has concentrated on the principles of environmental and public health risk assessment and their application in the management of chemical incidents. It covers such topics as routes of exposure to toxic substances, possible fates of a chemical in the body and has identified some of the vulnerable and susceptible populations. The applications of the source-pathway-receptor concept is discussed. The value of using these concepts in chemical incident management are summarised and finally focused by considering case studies, namely an acute carbon monoxide poisoning and organophosphate exposure.

Sources of further information plus a little about some of the references

- Baker D, et al. [5]

  Summarises toxicology found to be of value for frontline professionals and is organised into four sections and an appendix covering:

  Fundamentals of toxicology – provides a general introduction to the subject and explains how toxicological information is derived.

  Applications of toxicology – addresses exposure assessment, susceptible populations and the medical management of chemical incidents. It also provides valuable pointers to sources of toxicological data.

  Environmental toxicology – considers pollutants in air, water and land, and food contaminants and additives. In Occupational Toxicology, it considers exposures to toxic agents in the workplace.

  A review of some toxic agents – addresses in detail a selection of important toxic agents: carbon monoxide, pesticides, heavy metals and trace elements, combustion toxicology as well as the emerging issues of traditional medicines and the deliberate release of toxic agents in warfare.

- Bradley, N. et al. [6]

  Provides practical guidance on the technical aspects of environmental and public health investigations and practical, expert advice on a range of topics from key concepts and framework for investigation to contaminated land and waste management. Case studies are used to aid learning and understanding of the topics discussed.

- Ayres JG, Harrison RM, Nichols GL, Maynard RL. (Eds.) (2010) Environmental Medicine. Hodder Education.

  This book provides information on the investigation, diagnosis and treatment of a wide variety of environmentally acquired disorders.

- Duarte-Davidson R, Griffiths M, Wyke S, and Bradley N. (Eds.) (2014) Recent Developments in Assessing and Managing Serious Health Threats – Environmental International Special issue. Environment International Volume 72, pp. 1–186.

  This special edition contains a number of articles on the recent developments for identifying, assessing and managing serious cross-border health threats.

- WHO (2010), Human Risk Assessment Toolkit – Chemical Hazards.

  The Toolkit contains road maps for conducting a human health risk assessment, identifies information that must be gathered to complete an assessment and lists electronic links to international resources from which the user can obtain information and methods essential for conducting the human health risk assessment. See http://www.who.int/ipcs/methods/harmonization/areas/ra_toolkit/en/

Internet sites:

When accessing toxicology resources on the Internet care must be taken as there are numerous sites, and it is important to access sources which can be considered to be reputable or 'accredited'.

Compendia of Chemical Hazards, Public Health England – This is a series titled the Compendium of Chemical Hazards. The aim is to produce an online information resource for the public and all professionals who may be involved in advising and responding to chemical incidents, mainly public health professionals and emergency services. Each Compendium entry is split into three sections: General Information that provides background information on the compound, including its uses and 'frequently asked questions'. Incident Management focuses on information such as physicochemical properties, health effects and decontamination that may be needed during chemical incidents. Toxicological Overview provides more in-depth toxicology of the compound.

https://www.gov.uk/government/collections/chemical-hazards-compendium (Accessed 30 October 2020)

Chemical Hazards and Poisons Reports – This is a series of reports published by Public Health England, containing information for health and emergency professionals involved in chemical incident response and preparedness in the following categories: chemical incident responses case studies, emergency preparedness and response and environmental and toxicological research. Later editions have contained articles on natural hazards, extreme events and climate change.

https://www.gov.uk/government/collections/chemical-hazards-and-poisons-reports (Accessed 30 October 2020)

Wireless Information System for Emergency Responders (WISER) – developed by the US National Library of Medicine to assist emergency responders. Information is available for Internet download or to mobile devices and includes: substance identification support, physical characteristics, human health information, and containment and suppression guidance.

https://wiser.nlm.nih.gov/ (Accessed 30 October 2020)

The International Programme on Chemical Safety (IPCS) – produces INCHEM as a co-operative programme and is an invaluable tool for those concerned with chemical safety and the proper management of chemicals. IPCS INCHEM directly responds to one of the Intergovernmental Forum on Chemical Safety (IFCS) priority actions to consolidate current, internationally peer-reviewed chemical safety-related publications and database records from international bodies, for public access. IPCS INCHEM offers rapid access to internationally peer-reviewed information on chemicals commonly used throughout the world, which may also occur as contaminants in the environment and food. The site provides quick and easy electronic access to thousands of searchable full-text documents on chemical risks and the sound management of chemicals.

http://www.inchem.org/pages/about.html (Accessed 30 October 2020)

The US National Library of Medicine (NLM) toxicology data service – has resources for members of the public and healthcare practitioners such as, ChemIDplus, a chemical database dictionary of over 400,000 chemicals (names, synonyms and structures). ChemID-plus includes links to NLM and other databases and resources with access to a range of databases on toxicology, hazardous chemicals, environmental health and toxic releases.

https://www.nlm.nih.gov/toxnet/index.html (Accessed 30 October 2020)

PubChem is an open chemistry database at the National Institutes of Health (NIH). 'Open' means that you can put your scientific data in PubChem and that others may use it. Since the launch in 2004, PubChem has become a key chemical information resource for scientists, students and the general public. This is part of the United States National Institutes of Health (NIH). PubChem can be accessed for free through a web user interface.

https://pubchem.ncbi.nlm.nih.gov/ (Accessed 30 October 2020)

Section 2 notes

1 *Green Leaves* was the latest in the UK. A more recent, 2019, WHO publication can be found here: https://www.who.int/hac/techguidance/preparedness/health-emergency-and-disaster-risk-management-framework-eng.pdf?ua=1

2 Note that *CO and fuel combustion products* are one of the 29 hazards included in the Housing

Health and Safety Rating System, and deficiencies that contribute to the hazard can be addressed under Part 1 of the Housing Act 2004. See https://assets.publishing.service.gov.uk/government/uploads/system/uploads/attachment_data/file/15810/142631.pdf0. The HHSRS is under review at the time of writing

Section 2 References

[1] Chemical Abstracts Service (website) (www.cas.org) *A Division of the American Chemical Society*. https://www.cas.org/content/chemical-substances (Accessed 29 October 2020).

[2] Prüss-Üstün A, Wolf Corvalán J, Bos CR, Neira M. (2016) *Preventing Disease Through Healthy Environments: A Global Assessment of the Burden of Disease from Environmental Risks*, addended updated 2016 data tables WHO. https://www.who.int/publications/i/item/9789241565196 (Accessed 29 October 2020).

[3] Kreis IA, Busby A, Leonardi G, Meara J, Murray V. (Eds) (2012) *Essentials of Environmental Epidemiology for Health Protection: A Handbook for field Professionals*, Oxford University Press, Oxford.

[4] European Commission. (2006) *Regulation (EC) No 1907/2006 of the European Parliament and of the Council of 18 December 2006 Concerning the Registration, Evaluation, Authorisation and Restriction of Chemicals (REACH), Establishing a European Chemicals Agency*. http://ec.europa.eu/growth/sectors/chemicals/reach/index_en.htm (Accessed 30 October 2020).

[5] Baker D, Karalliedde L, Murray V, Maynard RL, Parkinson NHT. (Eds.) (2012) *Essentials of Toxicology for Health Protection – a Handbook for Field Professionals*, Oxford University Press, Oxford, 2nd ed.

[6] Bradley N, Harrison H, Hodgson G, Kamanyire R, Kibble A, Murray V. (Eds.) (2014) *Essentials of Environmental Public Health Science – a Handbook for Health Professionals*, Oxford University Press, Oxford.

[7] World Health Organization. (2019) *Health Emergency and Disaster Risk Management Framework*, World Health Organization, Geneva. Licence: CC BY-NC-SA 3.0 IGO. https://www.who.int/publications/i/item/9789241516181

[8] WHO. (2010) *The International Programme on Chemical Safety (IPCS)*, Harmonization Project Document No. 8, Human Health Risk Assessment Toolkit: Chemical Hazards. WHO, Geneva. https://www.who.int/ipcs/publications/methods/harmonization/toolkit.pdf (Accessed 2 November 2020).

[9] WHO. (2009) *Manual for the Public Health Management of Chemical Incidents*, WHO, Geneva.

[10] Wyke-Sanders S, Brooke N, Dobney A, Baker D, Murray V. (2012) *The UK Recovery Handbook for Chemical Incidents (UKRHCI)*. https://www.gov.uk/government/publications/uk-recovery-handbook-for-chemical-incidents-and-associated-publications (Accessed 30 October 2020).

[11] The National Advisory Committee for the Development of Acute Exposure Guideline Levels for Hazardous Substances (AEGL Committee) Acute Exposure Guideline Levels (AEGLs). https://www.epa.gov/aegl (Accessed 5 May 2021).

[12] American Industrial Hygiene Association, ERP Committee, Response Planning Guidelines. https://www.aiha.org/get-involved/AIHAGuidelineFoundation/EmergencyResponsePlanningGuidelines/Pages/default.aspx (Accessed 5 May 2021).

[13] Defra. (2007) The Air Quality Strategy for England, Scotland, Wales and Northern Ireland (Volume 1). *Defra*. http://www.gov.uk/government/uploads/system/uploads/attachment_data/file/69336/pb12654-air-quality-strategy-vol1-070712.pdf (Accessed 6 May 2021).

[14] Defra. (2019) *Clean Air Strategy 2019 Clean Air Strategy 2019 – GOV.UK*. www.gov.uk (Accessed 6 May 2021).

[15] WHO. (2005) *WHO Air Quality Guidelines for Particulate Matter, Ozone, Nitrogen Dioxide and Sulfur Dioxide*. https://apps.who.int/iris/bitstream/handle/10665/69477/WHO_SDE_PHE_OEH_06.02_eng.pdf;sequence=1 (Accessed 6 May 2021).

[16] Public Health England, Guidelines for Environmental Health Practitioners (EHP) on the Risks of CO for Occupants of Residential Premises. https://www.gov.uk/government/publications/carbon-monoxide-co-residential-inspection-aid (Accessed 6 May 2021).

[17] Azevedo FA, Carvalho LR, Grinberg LT, Farfel JM, Ferretti RE, Leite RE, Jacob Filho W, Lent R, Herculano-Houzel S. (2009) Equal numbers of neuronal and nonneuronal cells make the human brain an isometrically scaled-up primate brain. *The Journal of Comparative Neurology*, 513(5): 532–541, 10 April.

SECTION 3: RADIATION (IONISING AND NON-IONISING) AND HEALTH

John O'Hagan, Mutahir Ahmad, Darren Addison

Ionising radiation

The term radiation applies to all emissions in the electromagnetic spectrum. Only ionising radiation has enough energy to cause ionisation of matter and break chemical bonds in living tissue. Most of this damage can be repaired, but some cells may be killed. Breaks of both strands of the DNA molecule in a cell nucleus may not kill a cell but can be a precursor of cancer [1–2].

Natural sources of ionising radiation include radon gas (see Section 4) and cosmic rays. Human-made sources include X-rays and radioactive isotopes produced in nuclear reactors or particle accelerators (e.g. cyclotrons). Most human exposure is due to natural radiation, followed by medical exposures.

Ionising radiation is used in research, medicine and industry. High doses can cause significant acute illness and death. The risks of long-term (stochastic) effects, particularly cancers, are well quantified from studies of A-bomb survivors and occupationally and medically exposed cohorts. Doses that could lead to acute illness are prohibited by regulation. To prevent stochastic effects, all doses should be kept as low as reasonably achievable [2–7].

Isotopes of some elements are unstable and undergo radioactive decay. The half-life is the time taken for half of a given quantity to decay and is particular to the isotope. Half-lives range from fractions of a second to thousands of years. The unit of radioactivity is the becquerel (Bq). 1 Bq equals one atomic disintegration per second. The amount of natural potassium-40 (^{40}K, half-life 1.250×10^9 years) in an average person is about 4,000 Bq. This means that about 15 million ^{40}K atoms disintegrate inside a person per hour.

Unstable isotopes (radionuclides) decay and in so doing, release energy as subatomic particles (alpha or beta), or X- or gamma rays.

Alpha radiation – This is mainly emitted by the isotopes of heavier elements. It consists of helium nuclei, which are made up of two protons and two neutrons and are tightly bound together to create a particle. This is a relatively large subatomic particle with 2 units of positive charge. Alpha particles are densely ionising and are stopped by the dead layer of the skin, so they only constitute a hazard if taken into the body. Because the radiation is particulate and charged, alpha radiation interacts both physically and electrically with the media it passes through and transfers all of its inherent energy into material it interacts with. This energy excites electrons in the absorbing material, causing electrons to be released from their atomic orbits, to produce ions.

Beta radiation – This is mainly emitted by intermediate and lighter elements. It consists of high-speed electrons that originate in the atomic nucleus and carry 1 unit of negative charge. Beta particles can penetrate the body up to a few centimetres. Beta particles produce a similar ionisation to alpha particles but are much lighter and more penetrative. Beta particles will penetrate a sheet of paper and have a range in air of a few metres.

Gamma radiation – This consists of quanta of energy emitted as an electromagnetic wave. It is non-particulate and uncharged. X-radiation is electromagnetic radiation that differs from the other forms mentioned in that it is non-nuclear in origin. It is normally generated electrically, although it can be generated when atomic electrons undergo a change in orbit, such as when beta particles react with other matter. Gamma radiation and X-radiation are extremely penetrative with a very large range in air. Most X- and gamma rays will pass through the human body. However, if absorbed, they can ionise matter. Shielding with high density material such as lead is needed to protect against X- and gamma rays.

These properties of radiation affect the site and extent of cellular damage after exposure and dictate the protection methods required.

Detecting and measuring radiation

Acute cell damage depends on the energy absorbed per unit mass of tissue (absorbed dose) and is measured in gray (Gy). One gray is equal to one joule (J) deposited per kilogram (kg) of tissue [8].

When considering the long-term (stochastic) effects, weighting factors are used to convert the absorbed dose in gray to an effective dose in sieverts (Sv). This allows exposures from any type of ionising radiation to be integrated into one dose which is directly related to the stochastic risk. The UK average annual individual natural background radiation dose is 2.3 millisieverts (mSv). The typical dose from a chest X-ray is 0.02 mSv and from an abdominal CT scan 10 mSv [9].

Radiation detection devices, as well as establishing the presence of ionising radiation may also allow quantification of radiation levels and exposures and isotope characterisation. Some are suitable for environmental surveys of contamination or dose rate in the air, others more appropriate for personal dosimetry. X- and gamma rays travel long distances in air and through matter so radioactivity can be monitored at a distance from the source. Alpha and beta particles penetrate only small distances through the air so different monitoring equipment, used closer to the source, is needed.

The thermoluminescent dosemeter (TLD) has largely replaced the film badge for personal dosimetry. It uses the properties of certain materials to emit light photons in response to impinging ionising radiation. These devices can give an electronic readout of radiation dose.

The Geiger-Muller tube and other ionisation chamber detectors have an air-filled chamber with a voltage across it. Any ionisation of the air in the chamber generates an electrical current, proportional to the radiation intensity, which can be measured and quantified.

A scintillation counter contains material that absorbs the energy from ionising radiation. When the atoms return to their original state the energy is given out as photons, which are detected and amplified to produce a signal, which can be measured and quantified. Environmental radon measurements rely on the ability of alpha particles to cause physical damage to the surface of common materials, the amount of damage being proportional to the amount of incident radiation [10].

Film badges were the mainstay of personal dosimetry for many years. They rely on the property of radiation to darken photographic film. The degree of darkening can be measured and quantified.

Contamination – External exposures, either whole body or partial, do not render people radioactive and pose no radiation risk to people nearby after the exposure has stopped (for example having a medical X-ray does not cause contamination of the patient). However, if the radioactive material is on the skin or clothing, ingested, inhaled, injected or absorbed through wounds, it may continue to emit radiation and be a hazard to that person and to people in the vicinity. After large intakes, sweat and other body fluids may be significantly contaminated. Decontamination of radioactive material on skin or clothing is often straightforward but should not take precedence over lifesaving procedures.

Health effects of exposure to ionising radiation

There are three types of health effects associated with exposure to ionising radiation. Stochastic effects, psychological effects and acute or deterministic effects [11–12].

Stochastic effects – These are effects where the probability of occurrence depends on the radiation dose. They include carcinogenesis and induction of heritable defects. Radiation-induced cancer is clinically and pathologically indistinguishable from other cancers. A linear no-threshold model is assumed at all levels of dose, though there are scant data for very low exposures. Therefore, there is no 'safe' radiation dose, but very small exposures convey very small stochastic risks. The absolute cancer risk per unit of radiation dose (risk coefficient) is estimated to be 5.5% per sievert.

Psychological effects – These are likely to occur following accidental or malicious exposures. They are not necessarily related to radiation dose. They may be significant and enduring, especially if people have been displaced. Readers are referred to the literature on risk communication.

Acute or deterministic effects – These include acute radiation syndrome and radiation burns. They occur after radiation doses above 1 Gy whole body dose or 20 Gy skin dose, when sufficient cells are killed. The early (prodromal) symptoms of acute radiation syndrome are non-specific, resembling influenza or food poisoning so it may be some time before the true cause is appreciated. The latency, severity and duration of prodromal symptoms depend on the radiation dose. After a latent period of apparent recovery, effects of the killing of stem cells appear. The patient suffers organ failure, its severity depending on the radiation dose, usually initiated by reductions in blood count, infection, haemorrhage and anaemia. If the patient survives this phase, recovery is likely. Radiation burns may also be difficult to diagnose but may show surrounding epilation. Skin injuries evolve slowly, usually over weeks to months, may be very painful and are resistant to treatment [12].

Principles of radiation protection for ionising radiation

The system for protection against ionising radiation throughout the world is based on the work of a non-governmental scientific organisation, the International Commission on Radiological Protection (ICRP) [13]. ICRP recommendations are regularly reviewed to take account of new scientific evidence, most recently in 2007. ICRP recommendations are adopted in the UK on the advice of United Kingdom Health Security Agency (Public Health England). ICRP recommend three central requirements for all human actions that add to radiation exposure: justification of the practice so that the net benefits to society outweigh the risks, optimisation of protection by taking all reasonable steps to reduce exposures and application of dose limits (other than for patients undergoing medical exposures) for individual workers or members of the public.

In the UK a substantial body of legislation, built up since the 1940s is supported by a wealth of codes of practice, licenses, authorisations and other advice. Responsibility for enforcing radiation protection in the UK rests mainly with the Health and Safety Executive. Several other government departments and agencies also have important roles. United Kingdom Health Security Agency advises on the system of protection to be used across the UK.

The methods available to protect the public and workers, applicable in planned and accidental situations, include time restrictions on exposure, increasing the distance between the person and the radiation source and shielding the person from the source.

The Ionising Radiations Regulations 2017 (IRR17) [14][3] and the associated Approved Code of Practice [7] have been made under the provisions of the Health and Safety at Work etc. Act 1974 to implement the requirements of the Basic Safety Standards Directive 2013/59/Euratom in Great Britain (Northern Ireland publishes separate regulations). The IRR17 came into force on 1 January 2018, and they replaced the Ionising Radiations Regulations 1999 (IRR99). They lay down the standards for the protection of workers and the general public from harm caused by use of ionising radiation in work activities.

Working areas must be designated as 'controlled' or 'supervised' areas if procedures are needed to restrict significant ionising radiation exposure or persons working therein are likely to receive an effective dose of more than 6 mSv in a year or an equivalent dose greater than 3/10 of any relevant dose limit. Radiation levels have to be monitored in 'controlled' and 'supervised' areas [7, 13–14].

Examples of worker dose limits [14] are:

- For employees aged 18 years and over, 20 mSv in any calendar year;
- For trainees aged under 18 years, 6 mSv in any calendar year;
- For any person under 16 years of age, 1 mSv in any calendar year.

The radiation doses received by workers must be monitored and recorded by a dosimetry service approved by HSE [7]. Classified radiation workers are defined as workers over 18 who are likely to receive an effective dose of more than 6 mSv in a year or an equivalent dose greater than 3/10 of any relevant dose limit. Workers are also classified if they work with any source of ionising radiation, which is capable of giving rise to an effective radiation dose greater than 20 mSv or an equivalent dose in excess of a dose limit within several minutes. Classified workers are subject to medical screening before they can start radiation work and they also need to receive ongoing medical surveillance.

Non-ionising radiation: definition, mechanism of harm and basis of regulation

Non-ionising radiation includes radiant heat, visible light, ultraviolet radiation, radio waves, microwaves and power frequency electromagnetic fields. Apart from ultraviolet radiation, the long-term health effects of non-ionising radiation, if they exist at all, are not so clearly defined [15–22]. Therefore, the regulatory controls prevent the known, mostly acute

adverse health effects of, for example, electric shock, sunburn.

Optical radiation – This is often defined as radiation with the wavelength range from 100 nm to 1 mm. This encompasses ultraviolet radiation (100 nm to 400 nm), visible radiation or light (380 nm to 780 nm) and infrared radiation (780 nm to 1 mm). It should be noted that these boundaries are not precise and there is an overlap between ultraviolet and visible radiation. The main source of optical radiation is generally the sun. Artificial sources include lasers, optical radiation from welding, sunbeds, lighting and infrared illuminators. The Control of Artificial Optical Radiation at Work Regulations 2010[4] apply to worker exposure from artificial sources. The exposure limit values for optical radiation exposure are derived from guidance published by the International Commission on Non-Ionizing Radiation Protection (ICNIRP) [23–24].

Ultraviolet radiation (UVR) – This affects primarily the skin and the eye [22, 25]. The short-term skin effect is sunburn, which may be followed by increased production of melanin (suntan), and which only offers minimal protection against further exposure. Acute ocular exposure to UVR can lead to photokeratitis and photoconjunctivitis (arc eye, snow blindness). Large doses of ultraviolet-A radiation can damage the retina, although this is not common as the lens and cornea tend to absorb the energy. Children are at particular risk as their eyes transmit much more of the UV-A component (which has a longer wavelength and is more penetrating through body tissues) to the retina than in older people. The principal long-term effect to the eye is the production of lens opacities – cataracts.

Short-term effects on the skin are erythema (reddening of the skin), which is similar to sunburn, as the blood vessels are dilated and the blood supply to the affected tissues increases. UV-A induces rapid tanning of the skin as a result of the existing pigmentation (melanin) being stimulated. This does not give a long-term tan. UV-B and UV-C, which are of higher energy than UV-A, tend to produce long-term tanning by increasing the activity of the melanin-generating cells in the skin. UV-C is absorbed in the atmosphere, so is not present from the sun at ground level.

A beneficial effect of small amounts of UV-B, less than needed to tan, is synthesis of vitamin D in the skin [26]. UV-A can trigger the production of nitric oxide, which dilates blood vessels and therefore can reduce high blood pressure [27]. Exposure to UVR can suppress immune responses but the significance for human health is uncertain.

The most serious long-term effect of UVR is induction of skin cancer. The overall incidence is likely to exceed 70,000 cases per year in the UK. The incidence of malignant melanoma, which is much more likely to be fatal, has increased substantially in white populations for several decades and now causes about 2,000 deaths per year in the UK. Chronic UVR exposure also causes photo-ageing, wrinkles and loss of elasticity of the skin [25].

Sunbeds deliberately mimic sun exposure without the adventitious benefits of being outdoors. There is no evidence to suggest that they are any less damaging than solar exposure [25]. The International Agency for Research on Cancer classifies sunbed use as 'carcinogenic' (Group 1) because there is clear scientific evidence for harm. The UK Committee on Medical Aspects of Radiation in the Environment (COMARE) [25] have recommended that regulations should be applied to all commercial uses of sunbeds including the prohibition of the use of commercial sunbeds by the under 18s and the prohibition on the sale or hire of sunbeds to the under 18s. COMARE and the WHO also recommend the prohibition of unsupervised use and/or self-determined operation of sunbeds in commercial outlets [28]. The Public Health etc. (Scotland) Act 2008 and the Sunbeds (Regulation) Act 2010 (covering England and Wales) restrict the use of sunbeds to those who are 18 years or older. The provision of sunbed services is regulated in Wales,

Northern Ireland and Scotland: regulations for England are being considered. Some local authorities require sunbed operators to obtain a licence.

UVR from welding applications is managed in the workplace through the use of personal protective equipment. A number of incidents have resulted where inappropriate tubes have been fitted to UVR-emitting equipment such as insect killers. For most other workplace UVR sources, the most appropriate protection measure is to enclose the source.

UV-C sources can be used for disinfection. Upper-room sources can be used while the room is occupied. Total room exposure units require care to ensure that people are not exposed to levels of UV-C that exceed the exposure limits. Handheld wand-type units are usually not effective.

Visible light – This can be focused by the cornea and the lens of the eye to form an image on the retina [23]. Humans have natural protection measures to protect them from bright light. For most of our evolution, the only potentially hazardous light source has been the sun, with a broad spectrum of wavelengths. The natural aversion responses, including blinking and head turning, are generally adequate for the sun. However, if these are overcome then it is possible to get retinal damage from sunlight. This is a particular concern during solar eclipses when there is a temptation to stare at the sun.

There are a number of potential sources of visible light that do not produce a broad spectrum of wavelengths. The response of the eye to visible light depends on the wavelength, with the peak at about 550 nm (yellow-green) in daylight. Light at wavelengths close to the extremes of the visible region (380–780 nm) will appear much less bright for the same irradiance (power divided by area) at the surface of the eye. It is possible that the exposure limits could be exceeded without the aversion response being triggered.

Most new lighting systems incorporate energy-efficient lamps, such as light-emitting diodes (LEDs). It is important that the quality of the resultant lighting systems is addressed. Except in domestic installations, replacing just the light source without replacing the light fitting may not be appropriate. Professional advice should be sought to ensure that lighting installations do not result in adverse health effects. In particular, some people experience headaches and migraine from LED lamps that flicker at 100 Hz (twice the mains frequency). Flicker can also be introduced by dimming circuits.

There have been concerns about the amount of blue light emitted by LED lamps and display screens. However, the studies claiming to show adverse effects have usually used levels of exposure considerably higher than the exposure limits. The European Commission Scientific Committee on Health, Environmental and Emerging Risks (SCHEER) published an Opinion on Potential Risks to Human Health of Light Emitting Diodes (LEDs) in 2018 [29].

Infrared radiation – This may be emitted from an intended source, such as an infrared illuminator, but it may be a by-product of heating during use. Infrared radiation from 780 nm to 1400 nm can be focused by the eye in a similar manner to light, but it does not trigger a visual sensation. At higher wavelengths, the infrared radiation cannot generally be assessed in isolation. It is important to consider the environmental conditions and the metabolic activity of the person exposed. ICNIRP has issued a statement on far infrared radiation exposure [30].

Lasers – These are treated separately because they usually produce a well-collimated beam of optical radiation which can present a risk of injury at a large distance from the source [24, 31]. Most lasers tend to produce a single wavelength or a small number of discrete wavelengths. Lasers may be used in a wide range of applications from laser printers and DVD players through alignment devices to laser shows and high-street cosmetic practices.

To guide users, lasers are classified into eight laser classes according to the amount of laser beam that is accessible. If a laser is safe under all conditions, it is exempt from the classification scheme. The classification scheme for other lasers is described in Table 10.9.

There are two issues associated with human exposure to laser beams. Firstly, whether the exposure – to the eye or the skin – exceeds the exposure limit value and secondly, whether eye exposure to a visible laser beam can cause distraction, glare or afterimages, even though below the exposure limit value. These latter effects can be of particular concern where the recipient is not expecting the exposure, the ambient light level is low and the person is carrying out a safety critical task such as operating machinery, driving a vehicle or flying an aircraft. Specific legislation applies where the use of a laser (or any bright light source) compromises flight safety[5] or drivers of vehicles, including aircraft.[6]

Laser shows may range from small village discos through street shows to large concerts inside or outdoors. In the UK the entertainment industry has produced guidance for laser shows [32] supported by a document published by British Standards [33]. In addition, the Civil Aviation Authority and other authorities may need to be notified of the use of lasers out of doors [34]. Directing the laser beam across the faces of the audience is rarely justified due to safety concerns. Audience or performer exposures to laser beams requires a detailed assessment, supported by measurements, for which specialist advice should be sought.

Lasers are used for a range of cosmetic and medical practices in high-street locations. Laser surgery to the cornea to correct refractive errors should be carried out as a medical procedure. Other applications, such as hair or tattoo removal, or skin rejuvenation, may be less-well controlled. These practices are regulated in some parts of the UK.

Radiofrequency electromagnetic waves

The widespread rapid adoption of radiofrequency (RF) microwaves in wireless

Table 10.9 Classification of lasers

Laser class	Basis for classification
Class 1 **Safe** Visible/non-visible beams	Lasers which are safe under reasonably foreseeable conditions of operation. Generally, a product that contains a higher-class laser system but access to the beam is controlled by engineering means.
Class 1C **Safe without viewing aids** Visible/non-visible beams	Lasers which are designed explicitly for contact applications to the skin or non-ocular tissue. During operation ocular exposure is prevented by engineering means to either stop or reduce the beam to below the accessible emission limit for Class 1 when removed from contact with the skin or non-ocular tissue. For operation when in contact with the skin or non-ocular tissue, irradiance or radiant exposure levels may exceed the skin maximum permissible exposure as necessary for the intended treatment or procedure.
Class 1M **Safe without viewing aids** 302.5 to 4000 nm	Safe under reasonably foreseeable conditions of operation. Beams are usually collimated but with a large diameter. May be hazardous if user employs telescopic optics within the beam.
Class 2 **Low power** Visible beams only	For continuous wave (not pulsed) lasers, protection of the eyes is normally provided by the natural aversion response, including the blink reflex, which takes approximately 0.25s. (These lasers are not intrinsically safe.) Maximum power = 1 mW for a continuous wave laser
Class 2M **Safe without viewing aids** Visible beams only	Protection of the eyes is normally provided by the natural aversion response, including the blink reflex, which takes approximately 0.25s. Beams are usually collimated but with a large diameter. May be hazardous if user employs telescopic optics within the beam.
Class 3R **Low/medium power** Visible/non-visible beams	Risk of injury is small and not as high as for Class 3B. Up to 5 times the accessible emission limit for Class 1 or Class 2.
Class 3B **Medium/high power** Visible/non-visible beams	Direct intrabeam viewing of these devices is always hazardous. Viewing diffuse reflections is normally safe provided the eye is no closer than 13 cm from the diffusing surface and the exposure duration is less than 10 seconds. Maximum power = 500 mW for a continuous wave laser
Class 4 **High power** Visible/non-visible beams	Direct intrabeam viewing is always hazardous. Specular and diffuse reflections are hazardous. Eye, skin and fire hazard. There is no upper limit for the power emitted by Class 4 lasers. Treat Class 4 lasers with caution

technology, including mobile phones, Wi-Fi, smart meters etc. has led to concerns about adverse health effects. The frequencies are similar, though not identical to those used for TV and radio broadcasting. Exposure to intense radiofrequency fields can cause thermal burns but heating remains the only recognised effect of exposure. The most recent comprehensive review of the scientific evidence relevant to radio wave exposures and health was published by an independent expert advisory group of the Health Protection Agency, now part of the United Kingdom Health Security Agency (UKHSA), in 2012 [21, 35]. The overall conclusion was that, although a substantial amount of research has been conducted in this area, there is no convincing evidence that radio wave exposures below guideline levels cause adverse health effects in either adults or children. However, these are new technologies, and a cautious approach is appropriate. The current UK precautionary advice is that children should only use mobile phones for important calls. More modern phones tend to emit lower power radiofrequency fields though the average duration of use may be increasing over time. No precaution, apart from information and consultation, is advocated for masts, which, like Wi-Fi and smart meters, produce very much lower personal exposures (thousands or tens of thousands of times below the internationally accepted exposure guidelines) than from using a phone [35].

A large amount of new scientific evidence on radiofrequency fields and health has emerged over the past few years [21, 36]. As the research programmes have been coming to fruition, scientific expert committees have reviewed the evidence and have come to consider judgements at international, European and national levels.

For the general public, the United Kingdom supports European Council Recommendation 1999/519/EC on limiting exposure to electromagnetic fields (EMFs), which include radio waves [37]. This recommendation incorporates the 1998 guidelines from the International Commission on Non-Ionizing Radiation Protection (ICNIRP), as advised by UKHSA [38]. ICNIRP restated the RF parts of these guidelines in 2009 on the basis of its own comprehensive review of the scientific evidence published at that time [36–37]. In March 2020 ICNIRP published new RF exposure guidelines that have been developed to take account of the increased scientific evidence. Like the predecessor (1998) guidelines, the restrictions are based on the avoidance of excessive localised and whole-body heating. A wide range of other biological and adverse health effects have been investigated, and ICNIRP concluded that exposure below the thermal threshold is

unlikely to be associated with adverse health effects. The restriction values in the new guidelines are very similar to those in the previous guidelines, especially at frequencies below 6 GHz, where current mobile communications systems operate [39].

For occupational exposure, the Control of Electromagnetic Fields at Work Regulations 2016[7] transpose the requirements of European Commission Directive 2013/35/EU [40], which are also based on ICNIRP guidelines. This Directive lays down the minimum requirements for the protection of employees from risks to their health and safety arising, or likely to arise, from EMF exposure.

The World Health Organization (WHO) states that the main conclusion from its own reviews is that EMF exposures below the limits recommended in the ICNIRP international guidelines do not appear to have any known adverse consequence on health.

The Scientific Committee on Emerging and Newly Identified Health Risks (SCENIHR) advises the European Commission on the health aspects of EMF exposures. SCENIHR takes account of worldwide studies on EMFs and has produced several reports in which it expresses views broadly in line with those of UKHSA, ICNIRP and WHO. The most recent SCENIHR Opinion was published in March 2015 and contains detailed conclusions on different aspects of the scientific evidence. SCENIHR concluded that the results of current scientific research show that there are no evident adverse health effects if exposure remains below the levels set by current standards [41].

Power frequency electric and magnetic fields (EMFs)

There are concerns that power frequency fields might have adverse health effects at levels below those required to interfere with nerves through induced fields and currents, which form the basis for exposure guidelines. Epidemiological studies have shown a broadly consistent association, not necessarily indicating causation, between unusually high magnetic fields in homes and/or residential proximity to power lines and increased risk of childhood leukaemia (possibly 2–25 attributable cases per year in the UK). The evidence linking other illnesses such as Alzheimer's disease, breast cancer and depression with power frequency exposures is weak and less consistent than that for childhood leukaemia. There is no established biological mechanism for this effect which appears to be decreasing in more recent studies. The International Agency for Research on Cancer has classified power frequency magnetic fields as 'possibly carcinogenic'. In March 2004, the UK Health Protection Agency (now part of the UKHSA) recommended:

> The Government should consider the need for further precautionary measures in respect of exposure of people to EMFs . . . the overall evidence for adverse effects of EMFs on health at levels of exposure normally experienced by the general public is weak. The least weak evidence is for the exposure of children to power frequency magnetic fields and childhood leukaemia.

In 2004 the UK Government initiated SAGE, the Stakeholder Advisory Group on Extremely Low Frequency EMFs (ELF EMFs) [36], to consider possible precautionary measures in relation to EMFs. SAGE has published two reports. The first recommended optimum phasing of high voltage overhead powerlines and more information for the public which was accepted by the UK Government. SAGE also recommended manufacturers to investigate the market for 'low field' appliances. The option of implementing 'no build corridors' was not adopted by government because it was a disproportionate measure in the light of the existing evidence for childhood leukaemia alone. The first SAGE report also made some specific recommendations for house wiring and

inspection practice, which were referred to the industry bodies. The second SAGE report made many recommendations relating to distribution networks, many endorsing existing best practice [42].

Static magnetic fields – Exposure to static magnetic fields can occur in various situations, for instance wherever electricity is used in the form of direct current (DC), such as in some power transmission systems and rail and subway systems, and in occupational settings such as in aluminium production and the chlor-alkali industry. ICNIRP published guidelines in 2009 [43] recommending that acute exposure of the general public should not exceed 400 mT (millitesla), for any part of the body, although the previously recommended value of 40 mT is the value used in the European Council Recommendation [37]. However, because of potential indirect adverse effects, ICNIRP recognises that practical policies need to be implemented to prevent inadvertent harmful exposure of people with implanted electronic medical devices and implants containing ferromagnetic materials, and injuries due to flying ferromagnetic objects, and these considerations can lead to much lower restrictions, such as 0.5 mT. People undergoing magnetic resonance imaging (MRI) are exposed to static magnetic fields of at least 1.5 tesla and often more. Head movements in static magnetic fields of this magnitude can cause symptoms such as vertigo, nausea, a metallic taste and phosphenes (seeing light without light entering the eye). There are insufficient data to know if there are long-term health effects from exposures to large static electric and magnetic fields but the benefit of these scans to medical diagnosis is clear.

Section 3 notes

3 SI 2017 No 1075.
4 SI 2010 No 1140.
5 Air Navigation Order 2009.
6 Laser Misuse (Vehicles) Act 2018.
7 SI 2016 No 588.

Section 3 References and further reading on ionising radiation

[1] Berger ME, Christensen DM, Lowry PC, Jones OW, Wiley AL. (2006) Medical management of radiation injuries: Current approaches. *Occupational Medicine (Lond)*, 56(3): 162–172, May.

[2] Christensen DM, Sugarman SL, O'Hara FM, Jnr. (2011) *The Medical Basis for Radiation-Accident Preparedness: Medical Management*, Proceedings of the Fifty International REAC/TS Symposium on the Medical Basis for Radiation-Accident Preparedness and the Biodosimetry Workshop, Miami, FL, September.

[3] Dainiak N, et al. (2011) Literature review and global consensus on management of acute radiation syndrome affecting nonhematopoietic organ systems. *Disaster Medicine and Public Health Preparedness*, 5: 183–201.

[4] Dainiak N, et al. (2011) First global consensus for evidence-based management of the hematopoietic syndrome resulting from exposure to ionizing radiation. *Disaster Medicine and Public Health Preparedness*, 5: 202–212.

[5] Department of Health England. (2013) *Health Emergency Preparedness, Resilience and Response from April 2013 Summary of the Principal Roles of Health Sector Organisations*. https://www.gov.uk/government/uploads/system/uploads/attachment_data/file/156099/EPRR-Summary-of-the-principal-roles-of-health-sector-organisations.pdf.pdf (Accessed 27 July 2021).

[6] Fliedner TM, Meineke V. (Eds.) (2005) Radiation-induced multi-organ involvement and failure: A challenge for pathogenetic, diagnostic and therapeutic approaches and research. *BJR*, Supplement 27.

[7] Health and Safety Executive. (2018) *Working with Ionising Radiation, Ionising Radiations Regulations 2017. Approved Code of Practice and Guidance*. https://www.hse.gov.uk/pubns/priced/l121.pdf (Accessed 3 August 2021).

[8] Thompson NJ, Youngman MJ, Moody J, McColl NP, Cox DR, Astbury J, Webb S, Prosser SL. (2011) *Radiation Monitoring Units: Planning and Operational Guidance*, HPA-CRCE-017, ISBN: 978-0-85951-690-7. https://www.gov.uk/government/publications/radiation-monitoring-units-planning-and-operational-guidance (Accessed 27 July 2021).

[9] HM Government. (2013) *Nuclear Emergency Planning: Consolidated Guidance*. https://www.gov.uk/government/publications/nuclear-emergency-planning-consolidated-guidance (Accessed 27 July 2021).

[10] Advisory Group on Ionising Radiation (AGIR). (2009) *Radon and Public Health*. Documents of the HPA, RCE 11. https://www.gov.uk/government/publications/radon-and-public-health (Accessed 3 August 2021).

[11] ICRP (International Commission on Radiological Protection). (2006) *Biological and Epidemiological Information on Health Risks Attributable to Ionizing Radiation: A Summary of Judgements for the Purposes of Radiological Protection of Humans*, Committee 1 Task Group Report: C1 Foundation Document (Annex A of Main Recommendations). http://www.icrp.org/docs/biology_icrp_foundat_doc_for_web_cons.pdf (Accessed 27 July 2021).

[12] Kutov V, Buglova E, McKenna T. (2011) Severe deterministic effects of external exposure and intake of radioactive material: Basis for emergency response criteria. *Journal of Radiological Protection*, 31: 237–253.

[13] ICRP. (2007) Recommendations of the international commission on radiological protection ICRP publication 103. *Annals of ICRP*, 37(2–4): 1–332.

[14] HM Government. (2017) *Ionising Radiations Regulations 2017*, SI 2017/1075. https://www.legislation.gov.uk/uksi/2017/1075/contents/made (Accessed 3 August 2021).

[15] Mobile Telecommunications and Health Research Programme Management Committee. (2007, 2012) *MTHR Reports 2007 and 2012*. https://webarchive.nationalarchives.gov.uk/20140910130151; http://www.mthr.org.uk/ (Accessed 3 August 2021).

[16] NRPB (National Radiological Protection Board. (1991) *Principles for the Protection of Patients and Volunteers During Clinical Magnetic Resonance Diagnostic Procedures, Limits on Patient and Volunteer Exposure During Clinical Magnetic Resonance Diagnostic Procedures: Recommendations for the Practical Application of the Board's Statement*, Volume 2, No. 1, Documents of the NRPB. http://webarchive.national-archives.gov.uk/20140722091854/http://www.hpa.org.uk/Publications/Radiation/NPRBArchive/DocumentsOfTheNRPB/Absd0201/ (Accessed 3 August 2021).

[17] NRPB (National Radiological Protection Board). (2004) *Advice on Limiting Exposure to Electromagnetic Fields (0–300 GHz)*, Volume 15, No. 2, Documents of the NRPB. http://webarchive.nationalarchives.gov.uk/20140722091854/http://www.hpa.org.uk/Publications/Radiation/NPRBArchive/DocumentsOfTheNRPB/Absd1502/ (Accessed 3 August 2021).

[18] NRPB (National Radiological Protection Board). (2004) *Review of the Scientific Evidence*

for Limiting Exposure to Electromagnetic Fields (0–300 GHz), Volume 15, No. 3, Documents of the NRPB. http://webarchive.nationalarchives.gov.uk/20140722091854/http://www.hpa.org.uk/Publications/Radiation/NPRBArchive/DocumentsOfTheNRPB/Absd1503/ (Accessed 3 August 2021).

[19] NRPB (National Radiological Protection Board). (2005) *Mobile Phones and Health 2004: Report by the Board of NRPB*, Volume 15, No. 5, Documents of the NRPB. http://webarchive.nationalarchives.gov.uk/20140722091854/http://www.hpa.org.uk/Publications/Radiation/NPRBArchive/DocumentsOfTheNRPB/Absd1505/ (Accessed 3 August 2021).

[20] UK HSA (2021) *Common non-ionising radiation safety concerns*. https://www.ukhsa-protectionservices.org.uk/nir/concerns/.

[21] AGNIR. (2012) *Health Effects from Radiofrequency Electromagnetic Fields. Report of the Advisory Group on Non-ionising Radiation*. Documents of the HPA, RCE-20. https://www.gov.uk/government/publications/radiofrequency-electromagnetic-fields-health-effects (Accessed 3 August 2021).

[22] AGNIR (Advisory Group on Non-Ionising Radiation). (2002) *Health Effects from Ultraviolet Radiation*, Volume 13, No. 1, Documents of the NRPB. http://webarchive.nationalarchives.gov.uk/20140722091854/http://www.hpa.org.uk/Publications/Radiation/NPRBArchive/DocumentsOfTheNRPB/Absd1301/ (Accessed 3 August 2021).

[23] International Commission on Non-Ionizing Radiation (ICNIRP). (2013) ICNIRP Statement on limits of exposure to incoherent visible and infrared radiation. *Health Physics*, 105(1): 74–96. http://www.icnirp.org/cms/upload/publications/ICNIRPVisible_Infrared2013.pdf (Accessed 3 August 2021).

[24] International Commission on Non-Ionizing Radiation (ICNIRP). (2013) ICNIRP optical guidelines ICNIRP guidelines on limits of exposure to laser radiation of wavelengths between 180 nm and 1,000 μm. *Health Physics*, 105(3): 271–295. http://www.icnirp.org/cms/upload/publications/ICNIRPLaser180gdl_2013.pdf (Accessed 3 August 2021).

[25] Committee on Medical Aspects of Radiation in the Environment (COMARE). (2009) *13th Report: The Health Effects and Risks Arising from Exposure to Ultraviolet Radiation from Artificial Tanning Devices*. https://www.gov.uk/government/publications/comare-13th-report (Accessed 3 August 2021).

[26] AGNIR. (2017) *Ultraviolet Radiation, Vitamin D and Health*. Report of the independent Advisory Group on Non-ionising Radiation. https://www.gov.uk/government/publications/ultraviolet-radiation-and-vitamin-d-the-effects-on-health (Accessed 3 August 2021).

[27] Holiman G, et al. (2017) Ultraviolet radiation-induced production of nitric oxide: A multi-cell and multi-donor analysis. *Scientific Reports*, 7: 11105. https://doi.org/10.1038/s41598-017-11567-5.

[28] World Health Organization. (2017) *Artificial Tanning Devices: Public Health Interventions to Manage Sunbeds*. https://www.who.int/publications/i/item/9789241512596 (Accessed 27 July 2021).

[29] Scientific Committee on Health, Environmental and Emerging Risks (SCHEER) Opinion on Potential Risks to Human Health of Light Emitting Diodes (LEDs). https://ec.europa.eu/health/sites/default/files/scientific_committees/scheer/docs/scheer_o_011.pdf (Accessed 2 August 2021).

[30] International Commission on Non-Ionizing Radiation (ICNIRP). (2006) ICNIRP statement on far infrared radiation exposure. *Health Physics*, 91(6): 630–645.

[31] British Standards Institution. (2021) *Safety of Laser Products – Part 1: Equipment Classification and Requirements*, BS EN 60825-1:2014+A11:2021. https://shop.bsigroup.com/ProductDetail?pid=000000000030386301 (Accessed 3 August 2021).

[32] Professional Lighting and Sound Association. (2016) *Safety of Laser Displays*. https://www.plasa.org/wp-content/uploads/2017/11/plasa_laser_guidance-1.pdf (Accessed 27 July 2021).

[33] British Standards Institution. (2008) *Safety of Laser Products: Guidance for Laser Displays and Shows*. PD IEC/TR 60825-3. https://shop.bsigroup.com/ProductDetail?pid=000000000030171733 (Accessed 3 August 2021).

[34] Civil Aviation Authority. (2011) *Guide for the Operation of Lasers, Searchlights and Fireworks in United Kingdom Airspace*. CAP 736. https://publicapps.caa.co.uk/docs/33/CAP736.PDF (Accessed 27 July 2021).

[35] Health Protection Agency. (2012) *HPA Response to the AGNIR 2012 Report*. https://www.gov.uk/government/publications/radiofrequency-electromagnetic-fields-health-effects/health-protection-agency-response-to-the-2012-agnir-report-on-the-health-effects--from-radiofrequency-electromagnetic-fields (Accessed 3 August 2021).

[36] Stakeholder Advisory Group on Extremely Low Frequency Electromagnetic Fields. First

Interim Assessment. https://www.emfs.info/wp-content/uploads/2014/07/SAGEfirstinterimassessment.pdf (Accessed 15 July 2021).

[37] 1999/519/EC: Council Recommendation of 12 July 1999 on the Limitation of Exposure of the General Public to Electromagnetic Fields (0 Hz to 300 GHz). https://op.europa.eu/en/publication-detail/-/publication/9509b04f-1df0-4221-bfa2-c7af77975556/language-en (Accessed 3 August 2021).

[38] International Commission on Non-Ionizing Radiation Protection (ICNIRP). (2009) Statement on the guideline for limiting exposure to time-varying electric, magnetic, and electromagnetic fields (up to 300 GHz). *Published in Health Physics*, 97(3): 257–258. http://www.icnirp.org/cms/upload/publications/ICNIRPStatementEMF.pdf (Accessed 3 August 2021).

[39] ICNIRP guidelines for limiting exposure to electromagnetic fields (100kHz to 300 GHz). (2020) *Published in Health Physics*, 118(5): 483–524. https://www.icnirp.org/en/activities/news/news-article/rf-guidelines-2020-published.html (Accessed 15 July 2021).

[40] Directive 2013/35/EU of the European Parliament and of the Council of 26 June 2013 on the minimum health and safety requirements regarding the exposure of workers to the risks arising from physical agents (electromagnetic fields) (20th individual Directive within the meaning of Article 16(1) of Directive 89/391/EEC) and repealing Directive 2004/40/EC. https://eur-lex.europa.eu/LexUriServ/LexUriServ.do?uri=OJ:L:2013:179:0001:0021:EN:PDF (Accessed 15 July 2021).

[41] European Commission Scientific Committee on Emerging and Newly Identified Health Risks (SCENIHR). (2015) *Opinion on Potential Health Effects of Exposure to Electromagnetic Fields (EMF)*. http://ec.europa.eu/health/scientific_committees/emerging/docs/scenihr_o_041.pdf (Accessed 4 August 2021).

[42] Stakeholder Advisory Group on Extremely Low Frequency Electromagnetic Fields SAGE Second Interim Assessment. https://www.emfs.info/wp-content/uploads/2014/07/SAGESecondInterimAssessment2010.pdf (Accessed 15 July 2021).

[43] ICNIRP guidelines on limits to exposure for static magnetic fields. (2009) *Published in Health Physics*, 96(4): 504–514. http://www.icnirp.org/cms/upload/publications/ICNIRPstatgdl.pdf (Accessed 15 July 2021).

SECTION 4: RADON

Tracy Gooding

Introduction

Radon is a radioactive environmental hazard that crosses boundaries from public health, worker health and safety, to indoor air quality. As a lung carcinogen associated with buildings, those who deal with asbestos often find that radon is added to their portfolio. However, unlike asbestos, there is no particular lung cancer that is strongly associated with past exposure and only measurements with specialist equipment will indicate the degree to which radon is present.

Radon is a colourless, odourless radioactive Noble gas that is formed by the radioactive decay of elements (uranium and hence radium) that occur naturally in rocks and soils [1] and may also be found in certain building materials and water. Radon is the single largest source of radiation exposure to the UK population [2] in both homes and workplaces, is present in all indoor and outdoor air, and is the second leading cause of lung cancer after tobacco smoking. Radon is recognised as a Class 1 carcinogen [3] by the International Agency for Research in Cancer (IARC) and included as one of the 12 ways to reduce your cancer risk in the European Code Against Cancer [4].

In the UK, exposure to radon (including the radioactive solid elements into which it decays) is responsible for more than 1,100 lung cancer deaths each year [5]. Smokers and ex-smokers are at greatest individual risk as there is a strong synergy between radon exposure and tobacco smoking. Indoor radon levels, and the consequent risks to the inhabitants of the building, vary greatly across the

UK according to the underlying geology [6]. Similar variability has been observed in other countries [7]. Indoor radon research in the UK and worldwide over more than four decades has enabled specialist measurement equipment to be widely available to professionals and members of the public, and the means by which exposures can be reduced through minor building works.

A brief history

Radon was one of the first radioactive elements to be discovered. In 1899, Ernest Rutherford's experiments on thorium identified a gaseous 'emanation'; Frederick Soddy eliminated all chemical reactions, and suggested that it was a Noble gas, which they named 'thoron'. In 1900, Friedrich Ernst Dorn repeated Rutherford's experiments and identified a similar gaseous emanation from radium (Rutherford had identified the physical processes of 'radium emanation'). Both groups had identified radon, although different radioactive isotopes of the same element, with different half-lives and parent radionuclides. Over subsequent decades, the high lung cancer rates associated with hard rock mining (which had been described in the 16th century [8] as *pestilential air* causing *their lungs to rot away*) were gradually attributed to radon or, more correctly, the radioactive solids into which it decays. However, despite measurements of indoor radon being made in the 1950s, it was not until 1984 that the Royal Commission on Environmental Pollution recognised the health hazard from radon in its tenth report [9]. At the same time, the International Commission on Radiological Protection (ICRP) published a report [10] on the principles for limiting exposure of the public to natural radiation and radon had been identified as a major source of exposure that was amenable to control.

During the 1980s, the UK conducted a national survey of radon in homes [11]. This survey of around 3000 homes obtained representative national statistics and identified hotspots from targeting by geology, for instance mining areas. Numerous workplaces were measured at same time, especially council buildings such as schools, and research was undertaken on how building works could reduce the indoor radon concentration (*remediation* or *mitigation*). This early programme established the foundations for a radon infrastructure that now includes widely available information, measurement services and companies offering mitigation. Corresponding regulatory requirements have been developed that cover risk assessments for workplaces, housing standards, new building works and conveyancing.

The last essential component was the publication of the European case control study of radon in homes and lung cancer in 2005 [12]. This provided unambiguous evidence that indoor radon exposure increased a person's individual risk of developing lung cancer even at modest levels seen in homes, and also depended on personal smoking history. The results were also consistent with those obtained from earlier cohort studies of miners [13], extrapolated to lower radon levels. These studies are explained in the later section on health risks.

The physical properties of radon

Radon is a Noble gas. This enables it to travel by diffusion or pressure-driven flow through porous matter without being chemically retained.

Radon has multiple isotopes, all of which are radioactive and three occur in nature (it is the only naturally occurring radioactive gas). They have the same chemical properties but differ in their nuclear mass. The most important isotope, radon-222 (also written as Rn-222 or ^{222}Rn) is the decay product of radium-226, which is part of the uranium-238 decay chain shown in Figure 10.13. Uranium-238 has a half-life of 4.5 billion years, which makes it a 'primordial radionuclide', and half the uranium-238 that formed with

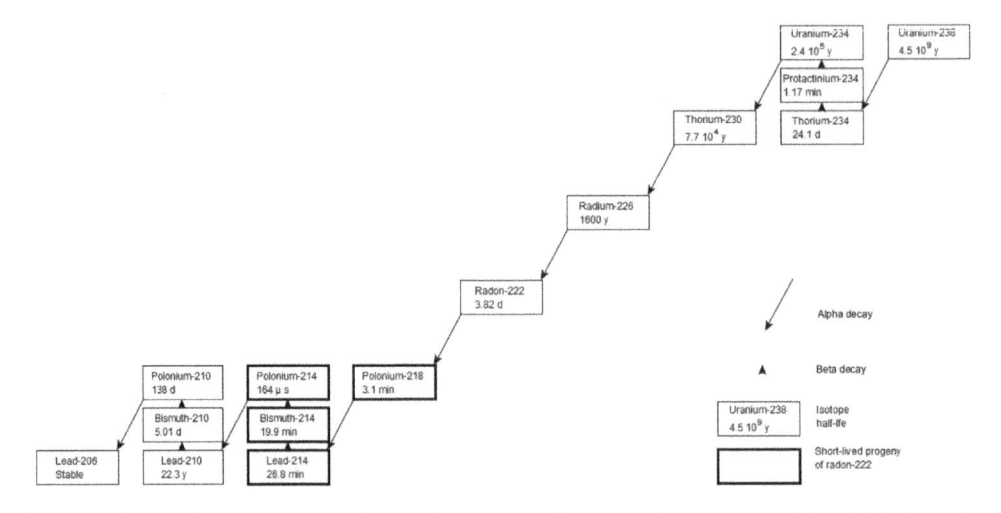

Figure 10.13 Radioactive decay chain of uranium-238 (main branches only), which includes radon-222 and ends with stable lead-206

the Earth is still available. Radon-222 has a half-life of about 3.82 days and decays by the emission on an alpha particle. Several of the isotopes in the radon-222 decay chain are also alpha-emitters and none is a gas. These radioactive isotopes of polonium, bismuth and lead are chemically reactive and, if created by the decay of airborne radon, tend to adhere to other airborne particulates (dust, smoke, water droplets, etc.), which can be inhaled. They may also remain as single atoms or form small clusters; these 'unattached' particles are also chemically reactive and can be inhaled. In general, when referring to 'radon', this means only the isotope radon-222 (and its decay products, as appropriate).

Radon-220 (thoron) is part of the decay chain of the primordial radionuclide, thorium-232 (half-life 14 billion years). Radon-220 has a half-life of 56 seconds and has numerous short-lived, solid, chemically active decay products, some of which are also alpha-emitters.

Radon-219 (actinon) is in the decay chain of the primordial radionuclide, uranium-235 (half-life 700 million years). Radon-219 has a half-life of 4 seconds.

The half-life of radon-222 is long enough to allow some of the gas produced in rocks

and soils close to the earth's surface to be carried in soil gas and migrate. It may either disperse in open air or be trapped under buildings where it can be drawn in through structures with ground contact.

Radon-220, with its much shorter half-life, has a correspondingly smaller capacity for migration. It is generally not a significant source of radiation exposure, apart from in exceptional cases where thorium is present in high concentrations in building materials or open rock faces. Radon-219 has an even smaller capacity for migration.

The concept of radon being transported by carrier gases is important. The partial pressure of radon is exceptionally low – at typical indoor concentrations, radon is only 1 atom in 10^{17}. This means that radon follows the air currents or the rise and fall of other gases, and does not exhibit behaviour due to its relative density, etc.

Sources of radon

The main source of radon is uranium (and hence radium) in near-surface rocks and soils. Uranium is found in a range of geological formations, normally at levels below 1 ppm. However, leaching and deposition can result in significantly higher concentrations. Uranium is found in igneous, metamorphic and sedimentary rocks.

For radon to be a source of radiation exposure, the rock or soil must be sufficiently porous and permeable to allow the migration with soil gas. Radon activity concentrations in the ground are of many thousands of becquerels per cubic metre of soil gas and frequently orders of magnitude higher. Diffusion to the air of open ground allows radon to disperse rapidly; the average outdoor air concentration in the UK air is low: 4 Bq m^{-3} (becquerel per cubic metre of air) [11]. Radon in soil gas exhaling to the ground in contact with the foundations of a building can be drawn into rooms through pressure-driven flow. Processes such as the temperature differential between indoor and outdoor air, heating (the buoyancy of warm air), ventilation and wind action on different faces of the building influence the slightly lower indoor air pressure that drives radon ingress, as illustrated in Figure 10.14. The resulting 'under-pressure' on floors and walls in contact with

the ground may only be of the order of a few pascal (Pa) but is sufficient to produce indoor radon concentrations that range from a few to many thousand Bq m^{-3}, although the average indoor concentration in the UK is 20 Bq m^{-3}. The Action Level for homes [14], the reference level at which householders are advised to reduce the indoor radon concentration, is 200 Bq m^{-3}. The regulatory threshold for work in a radon atmosphere is 300 Bq m^{-3} [15]. These are in line with WHO recommendations [16].

The distribution of indoor radon concentrations in an area corresponds approximately to a log-normal distribution. This is as a result of the previously mentioned processes combining multiplicatively. In a locality with apparently homogeneous geology, neighbouring homes can have markedly different indoor radon levels. Although it is not possible to predict the radon concentration within an individual building, the statistical combination of radon measurements in homes and the underlying geology allow the radon potential to be calculated [17], that is the

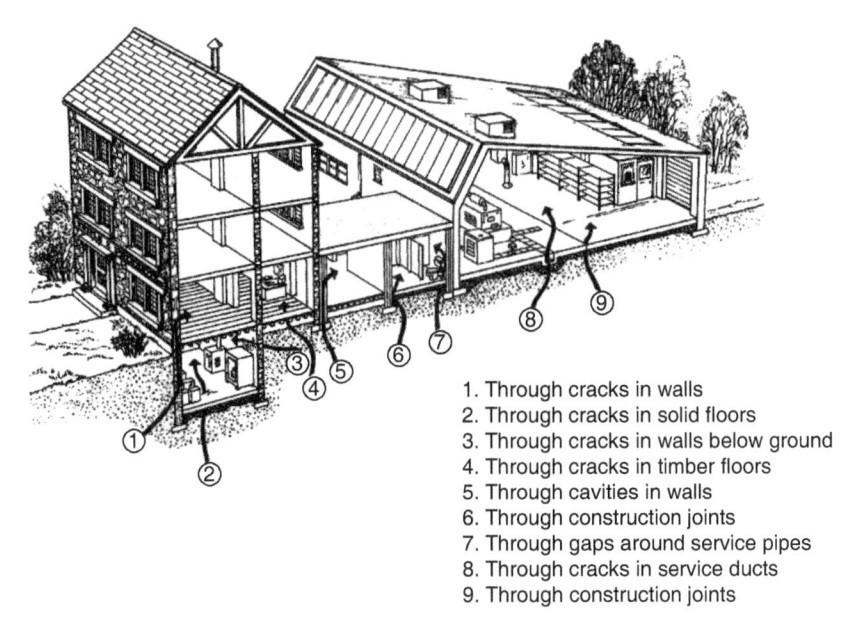

1. Through cracks in walls
2. Through cracks in solid floors
3. Through cracks in walls below ground
4. Through cracks in timber floors
5. Through cavities in walls
6. Through construction joints
7. Through gaps around service pipes
8. Through cracks in service ducts
9. Through construction joints

Figure 10.14 Routes of radon ingress into a building through cracks, cavities, joints and gaps around service entries

percentage of current and future homes in an area that will have radon levels above a specific threshold concentration. UKHSA (formerly Public Health England) and the British Geological Survey have jointly mapped the UK at a resolution of 25 metres. An example map of radon potential is shown in indicative form (the worst-case radon potential at a 1 km resolution) in Figure 10.15. Such maps are revised at irregular intervals when there are major updates to the geological data or significantly more results of radon measurements in homes.

Radon levels vary greatly over time [11]. The indoor radon concentration in a particular building can vary by orders of magnitude over periods from hours to months. This means that any measurements taken over short durations will likely not be representative of the long-term average exposure – the most important quantity when gauging the health impact. For this reason,

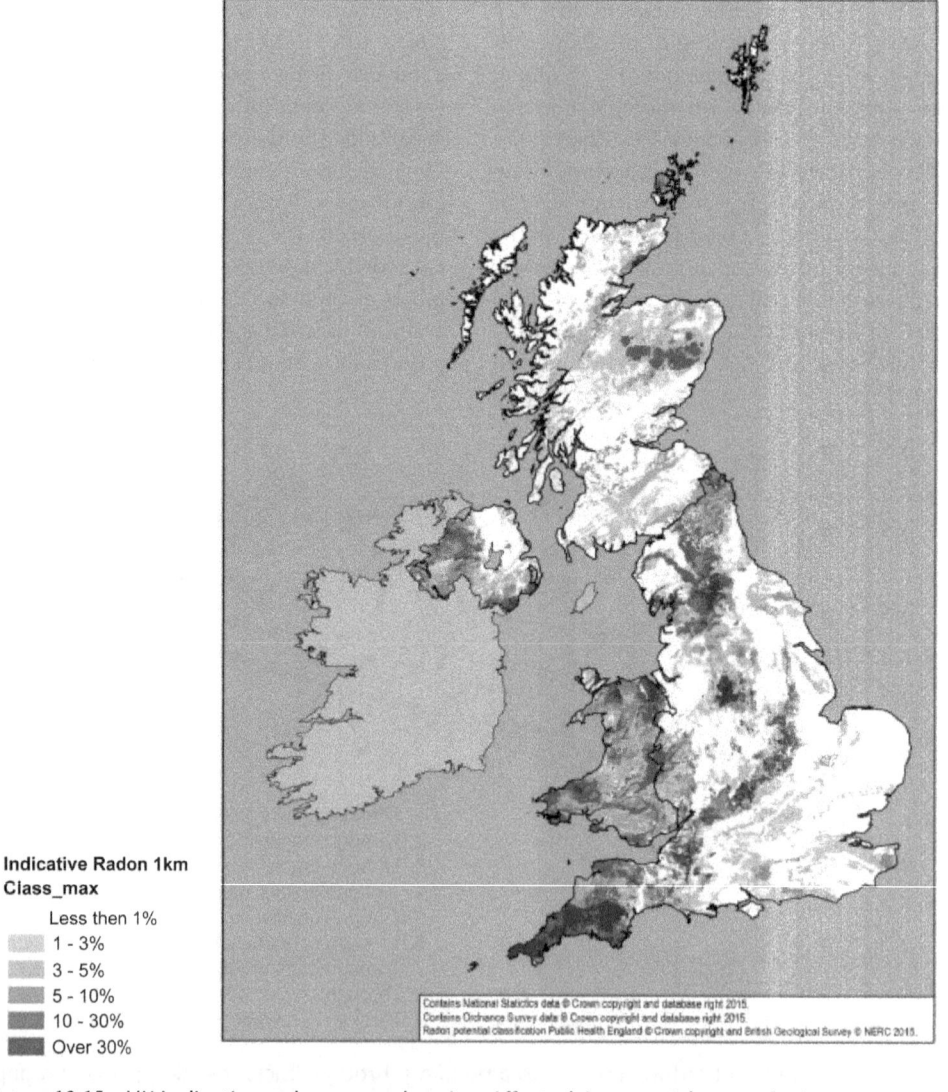

Indicative Radon 1km Class_max

- Less then 1%
- 1 - 3%
- 3 - 5%
- 5 - 10%
- 10 - 30%
- Over 30%

Figure 10.15 UK indicative radon map, showing Affected Areas at 1 km resolution

radon measurements are recommended to cover a continuous a three-month period. Most buildings show a seasonal variation, with radon levels higher during the winter months owing to the greater indoor-outdoor temperature difference and lower ventilation. For the same reasons, diurnal variations are usually observed, with daytime radon levels typically lower than those at night. The seasonal pattern has enabled correction factors to be developed so that the annual average radon concentration can be estimated from a three-month measurement at any time of year [18].

Building materials containing uranium and thorium can exhale radon into buildings. In the UK, this has not contributed sufficient radon to exceed the Action Level for homes. In countries such as Sweden, Hungary and Czechia, high radon levels were attributed to the concrete blocks that included pulverised fuel ash (PFA) and coal slags with high radium levels. International work to develop Europe-wide standards on radon exhalation from building materials is under way.

Building materials containing thorium may exhale radon-220 into buildings. This not only depends on the concentration of thorium (and radium-224) but, owing to the short half-life, the physical matrix of the material must have a sufficiently short path for the radon-220 to escape before it decays. In the UK, radon-220 contributes about 4% of the annual average radiation exposure.

Past human activities involving radium can contaminate buildings and the ground. Radium was used for research and luminising clocks and instruments. The preferred mitigation of such sites is to remove the contaminated material, although this may be impractical and need to be combined with more conventional radon mitigation methods. Although a number of sites have been found to have high radon levels from these sources, the records of many small luminising workshops have been lost.

Radon in water supplies presents a smaller public health hazard than radon in air. The risks and standards are described in a separate section to follow.

Health risks from radon

When radon is inhaled, its 3.82 day half-life means that there is a small probability of the atoms decaying whilst in the lungs, and most are simply exhaled. The decay products, however, not only have much shorter half-lives, but their chemically active nature means that they are more readily caught by the mucous membranes that line the airways; those attached to aerosols tend to deposit in the bronchi, the unattached tend to be deposited more deeply in the lung. Once on the mucous membranes, the isotopes of polonium, bismuth and lead decay radioactively before the body's clearance mechanisms can remove them (to the mouth, nose or GI tract). The range of alpha particles in tissue is particularly important, as it corresponds to the distance from the surface of the mucous membrane to the basal cells that line the airways. Radiation damage to the DNA in the cell nucleus risks mutations that can lead to cancer.

The link between exposure to radon (and its decay products) and lung cancer was firmly established in miners through cohort studies that showed a clear excess of deaths [13]. From the 1940s onwards, epidemiological studies analysed data from uranium miners, which looked at the age at exposure, the dose-response curve, the delay between exposure and effect (latency period) and the interaction of radon exposure and other substances such as tobacco smoke.

The link to lung cancer from the radon levels typical in homes was established in pooled, large-scale case control studies [12]. These showed a linear relationship between the exposure and the excess relative lifetime risk. Overall, this increases an individual's baseline lung cancer risk by 16% for each 100 Bq m^{-3} to which they are exposed over the long term; the baseline risk is strongly dependent on the smoking history and current status, for instance 15% to age 75 for current smokers and 0.4% for those who have never smoked.

The dominance of tobacco smoking on lung cancer risk means that large studies to compare average radon levels and average lung cancer rates over geographical regions, even when accounting for average smoking rates, are fundamentally flawed. Data that show the geographical distributions in England have been published in the Atlas of Variation on risk factors for respiratory disease [19].

There is currently no strong evidence that links radon exposure with any other disease. Calculations of radiation doses to organs other than the lung suggest a small, theoretical risk of cancer, but at rates several orders of magnitude lower. The UK Advisory Group on Ionising Radiation concluded that such effects are *so weak as to be generally undetectable in the published epidemiological studies* [5].

Detecting and measuring radon

Radon cannot be detected by the human senses: it has no smell, taste or colour. Specialist equipment is required that detects airborne alpha radiation with a satisfactory response at the levels typically encountered in homes and at work. In addition, the variability of radon levels favours a long-term measurement if the exposure estimation is to be meaningful, and there is no such thing as an 'instant measurement' for radon and/or its decay products.

Radon detecting instrumentation has separated into two categories: active and passive. Active devices give either an ongoing report of recent radon levels or can be used to provide a 'snapshot' of the radon concentration at some point (but require analysis times ranging from minutes to hours). Passive devices accumulate a signal that is subsequently analysed and interpreted by a laboratory service. Each type of device has optimal circumstances for use, and there is great scope for inappropriate interpretation of results. Most assessments are made with respect to reference levels in terms of annual average radon concentration.

However, the need for a rapid test to determine the effectiveness of a mitigation system or its optimal location favours instruments that can provide reliable results from short-term measurements.

Passive etched-track detectors – This is the preferred type of detector for long-term radon measurements, typically three months. The vast majority of radon measurements in the UK have been conducted with these detectors in more than 600,000 homes and 50,000 workplaces. The number of detectors required for a reliable test is determined by the usage and size of the building (or premises) [20].

The detector comprises a small piece of radon-sensitive material (poly-allyl diglycol carbonate – PADC, or CR-39), which is held in an outer casing that protects it from unintended physical and environmental damage but allows radon gas to diffuse through. The design of the casing excludes the decay products. Alpha particles from the decay of radon gas inside the casing cause the polymer of the PADC to be damaged, which is revealed by chemical etching as visible pits. The density of damage tracks is proportional to the exposure of the detector. Knowledge of the exposure duration enables the average radon level to be calculated. Seasonal correction factors can be applied to estimate the annual average concentration, which can be compared to the Action Level for homes [14] or the Ionising Radiations Regulations 2017 threshold for workplaces [15].

The Validation Scheme [18] provides standards for suppliers of various types of passive detectors for homes and workplace buildings. This covers the detector performance, laboratory procedures and the interpretation of results. A current list of laboratories complying with the Validation Scheme is maintained by UKHSA.

Passive etched-track monitors can also be used for personal dosimetry. However, their use must be carefully controlled to ensure that only doses from radon exposure at work are recorded for regulatory purposes. Personal dosimetry should be conducted by an HSE-Approved Dosimetry Service following advice from a Radiation Protection Adviser (RPA) − see *Managing radon at work*, later.

Diffusion-barrier activated charcoal detectors − This type of detector can be used for short-term measurements, with a limited range of two to seven days. Although they are accurate over this duration, they are fundamentally unsuitable for long-term measurements.

The detector consists of a perforated canister that contains activated charcoal (25–100 g). After unsealing the detector, radon entering the detector will be absorbed into the charcoal. At the end of the exposure, the detector is then resealed and returned to the originating laboratory for assessment. The charcoal retains the radioactive decay products, which enables the absorbed radon to be measured by gamma spectrometry of bismuth-214 and lead-214.

The radon absorbed during the exposure will gradually decay and desorb from the charcoal which, combined with the 3.82 day half-life radon, precludes measurement durations longer than seven days.

Electrets − Electrets are flexible in that they can be configured for short- or long-term measurements (hours to a year). However, the multiple sensitivities available mean that the user must select the most appropriate one(s) in advance, which require an estimate of the likely range of radon exposures (concentration multiplied by duration) that the detector will need to cover.

The detector consists of a Teflon® electret that holds an electrostatic charge, which is placed at the bottom of a conducting plastic chamber − an electret ion chamber. Radon diffuses into the active volume above the electret, which loses charge due to air ionisation produced by the radioactive decay. Measurements of the initial and post-measurement charge can be made by the issuing laboratory, although the reader can be obtained and used by an operator with minimal training; mishandling − such as dropping the electret − can cause significant misreading errors.

Electrets are available in different sensitivities and the chambers in several sizes, including ones that can be turned on and off or are sensitive to radon-220. They are true integrating detectors but have a limited dynamic range; choosing an over-sensitive configuration might 'saturate' and produce an underestimate of the concentration. The devices are sensitive to gamma rays, for which compensation must be made, in additional to altitude (atmospheric pressure) for the most sensitive measurements.

Active electronic instruments − There is an increasing number of instruments that will monitor the ambient radon concentration and produce a time series of radon concentrations. Although they have the advantage over passive monitors in that they can be read without being returned to the supplier, all need to be professionally calibrated before first use and annually thereafter if they are used to demonstrate regulatory compliance [15]. The simpler devices are also routinely misused as their operators might not understand the limitations.

The most sophisticated of these have high sensitivities and a rapid response time (~30 minutes for the first reliable results) but are expensive and are aimed at the professional scientific research community; the data can be downloaded for detailed analysis. These time-resolved data can be useful in determining the radon exposure profiles in workplaces when compared to occupancy patterns or assessing the effects of influences such as ventilation and local weather conditions, e.g. wind direction and temperature.

The electronic instruments available at a lower cost (less than £1000) are orders of magnitude less sensitive than those at the higher end of the market. This means that they take longer to produce the first reliable results (two to five days) and are subject to large statistical variations at radon concentrations around the Action Level or regulatory threshold for workplaces. However, most will display a reading of the radon concentration, for instance the average over the previous day or week, which can be misinterpreted as the current ambient level. Time series can be downloaded for analysis from some instruments and bespoke software may apply an algorithm to the results derived from patterns of the radon level over the measurement period to enhance performance; unfortunately, this may be applied inappropriately to subsequent measurements if the instrument is not reset completely and the memory cleared.

Active 'grab sample' instruments – There is no such thing as an instant radon measurement but grab samples of radon gas and decay products can be taken. These samples are useful when building up a picture of the radon levels in defined locations over a course of visits to a mine or cave, when assessing the possible exposure in a confined space before work is authorised, or to characterise the relationship between the radon gas and its decay products. The time between a radon gas sample and a result is several hours; the corresponding time for a radon decay product result can be a little as 15 minutes. Such measurements are complementary to other, longer-term assessments and alone cannot predict the annual average concentration. The measurements (sampling and/or analysis) are also technically complex and require professional services in consultation with an appropriately qualified and experienced RPA.

Radon decay product concentrations are expressed in terms of $J\ m^{-3}$ or working levels (WL). The relationship between the radon gas concentration and the concurrent radon decay product concentration produces a measure of the equilibrium factor. This unitless quantity is important when estimating doses, and is typically 0.4–0.5 indoors, but can span almost the full range from zero to 1.0 in workplaces such as water treatment works or deep mines [1].

International and national standards

International and national guidance and standards for radon have been developed over

many decades. The International Commission on Radiological Protection (ICRP) includes radon within its system of radiological protection [21], which influences national regulations through international standards. ICRP considers radon exposure to be an 'existing exposure situation' – that is, one at which the exposure is already taking place when control decisions are taken. The recommendations to identify 'radon prone' areas, set reference levels of annual average concentration, and limit individual annual exposures are echoed in international standards and guidance, such as the WHO *Handbook on Indoor Radon* [16]. ICRP also proposes dose conversion factors, which enable the exposure to be converted to a dose (which is analogous to risk) that can be compared with the annual dose limit for workers or members of the public.

The EU Basic Safety Standards Directive (EU-BSS) published in 2013 [22] required each Member State: to establish reference levels of radon in homes and workplaces; carry out measurements of radon in workplaces that are more prone to high radon levels (such as location); take action to reduce radon exposures following the principle of optimisation and apply appropriate measures where the reference levels are still exceeded; and prepare a national action plan for radon for which an indicative list of items was given. The UK already had most of the radon requirements of the EU-BSS in place. However, the existing legislation for radiation at work was updated correspondingly [15], a short set of regulations was created [23] to formalise some of the existing arrangements (including those for public exposure to radon), and the first National Radon Action Plan [24] was published in 2018 after public consultation. The International Atomic Energy Agency (IAEA) published an international BSS in 2014 [25], which contains similar provisions for radon.

Managing radon in the home

Radon in the home contributes around half the average UK resident's radiation exposure [2].

Those people living in homes with high radon levels receive much greater doses overall, of which radon can contribute more than 90%. Maps of radon potential and online services that access the high-resolution detail are available to identify whether individual addresses or locations are more likely to have high radon concentrations,[8] known as radon Affected Areas, that is a 25 metre Ordnance Survey grid square where at least 1% of current or future homes are estimated to exceed the domestic Action Level of 200 Bq m^{-3}. The map in Figure 10.15 shows the widespread nature of Affected Areas. However, being in an area with a low radon potential does not guarantee a low indoor radon concentration in a particular home (and conversely a high radon concentration in an area of high radon potential). UKHSA recommends that all householders in Affected Areas monitor their homes for radon; householders elsewhere may do so if they wish.

To supplement the Action Level, a Target Level of 100 Bq m^{-3} was introduced in 2010 [14]. This was designed as both a target for post-mitigation radon concentrations, and as an action threshold for people at greater individual risk, such as current and ex-smokers.

The Housing Health and Safety Rating System (HHSRS) [26] is a risk-based evaluation tool to help local authorities in England identify and protect against potential risks to health and safety in dwellings. It uses a formula that generates a hazard score based on the possible harm, its likelihood and the spread of outcomes. Radon is described in the HHSRS section on radiation.[9] Category 1 hazards are those where the most serious risk of harm is identified, for example death. Radon concentrations are deemed unacceptable when slightly higher than the Action Level, which means that the landlord has a duty to take action under the Housing Act 2004.

Radon is also addressed in conveyancing. In England and Wales, the standardised CON29 search made of the local authority asks whether the property is in an Affected Area. The standard conveyancing form TA6 used by the Law Society (of England and Wales) also asks the current owner to provide details of radon measurements and any protective measures in the property.

Testing – The routine testing procedure for homes is straightforward. Two passive detectors are sent to the home; one to be placed in a living room and one in a bedroom for a period of three months. Having two detectors means that the most occupied areas in the home are covered and, for two-storey homes, the difference between the upstairs and downstairs is reflected.

The results are combined in an average that is slightly weighted towards the bedroom result, to reflect the typical occupancy pattern. A seasonal correction factor is then applied [18], according to the duration and month in which the measurement started, to estimate an annual average concentration that can be compared to the Action and Target Levels. The results supplied to the householder should include recommendations on further action, as appropriate.

Reducing radon levels (mitigation or remediation) – The techniques for reducing radon levels in the home are informed by knowledge of the mechanisms that encourage radon penetration into the building and by four decades of experience in applying them to UK houses. The preferred method is to prevent radon entering indoor air rather than diluting it by ventilation, although the existing building design will limit options along with the initial radon concentration and reduction factor required.[10] The following methods have been shown to be successful and durable.

The most effective method is a *radon sump*. This is a small void created under a solid floor, which is connected via a pipe to a constantly running fan that draws radon-laden soil gas out from under the building and vents it away from windows and doors, typically above the eaves. This method should achieve a ten-fold reduction in radon levels and is the recommended option when radon levels are over 1000 Bq m^{-3}.

Passive ventilation beneath suspended floors improves the natural air flow and dilutes the soil gas under the building. This can be achieved by clearing, replacing or installing additional air bricks. Its effect is limited and suitable for when modest reductions in radon concentrations are required; however this *passive system* does not require a constant power source.

Active ventilation beneath suspended floors can be used where a passive system is not sufficient. Systems are available to either suck or blow air into the underfloor space – mitigation works installers may need to try both options to see which is the most effective.

Positive ventilation of the indoor space uses a system in the loft space, which slightly increases the indoor air pressure and counters the suction effect on the floor. The system was originally designed to reduce condensation (and mould) in energy-efficient buildings with double glazing.

Sealing cracks and gaps is difficult to achieve effectively, as skirting boards would have to be removed and many floor cracks are not visible. Filling gross voids around pipes and others service entries is useful when combined with other methods.

It is essential to carry out another radon measurement after installation, as before. Mitigation contractors may offer short-term tests, which will show if high radon concentrations persist, but the normal fluctuations in radon levels mean that these cannot give a good indication of the long-term radon levels.

Research has shown that the mitigation systems are durable over many years. Fans have an average lifetime of around seven years and air flow monitors installed with the system should show when problems occur. It is recommended to monitor the radon levels at regular intervals to show that they remain optimal and provide documentary evidence when the home changes ownership.

In order to find a mitigation contractor, householders may consult local listings or contact the self-regulating industry bodies: The Radon Council and the UK Radon Association.

Managing radon at work

The employer has a duty to assess and manage the risks to the health safety of employees. Regulations that specifically cover work with radon have been in place since 1985 and much of the practical implementation has developed in parallel with the knowledge of radon exposure and mitigation in homes.

In Great Britain, there are two key pieces of legislation (both of which have equivalents in Northern Ireland):

1   The Management of Health and Safety at Work Regulations 1999 (with amendments), MHSWR99 [27]
2   The Ionising Radiations Regulations 2017, IRR17 [15]

The MHSWR99 Regulation 3(1)(a) requires the employer to carry out a suitable and sufficient risk assessment of the risks to the health and safety of employees to which they are exposed whilst they are at work, which includes radon as appropriate. The actions resulting from that risk assessment include radon tests, for instance if premises are in an Affected Area, are underground in any location (including occupied basements) or have a source that is likely to emit radon (geological sample, antiques with radium dials, etc.). Caution should be exercised when comparing the working location to the radon Affected Area data: small premises that are comparable with the size of a house can use the online radon potential service; larger workplaces might have a single postal address, but the footprint of the building(s) can be considerably larger than the small parcel of land to which the specific postal address is linked. In the latter case, the employer may use the indicative maps of radon potential to give a 'worst-case' assessment. Alternatively, BGS offers a service to compare the digital footprint of the site with the underlying map of radon potential.

Initial radon testing of workplace buildings should use the same passive detectors over a three-month period as for homes. The performance and reporting of the test results are also covered by the Validation Scheme [18]. As workplaces vary considerably in size and usage, guidelines have been developed and published by UKHSA to help employers determine how many detectors are needed for an adequate test (see www.ukradon. org). Additional precautions can be made for testing in harsh environments such as high humidity.

The IRR17 apply if the radon concentration in any area exceeds 300 Bq m^{-3} annual average radon concentration, as defined in Regulation 3(1)(b). In a workplace building, a three-month measurement is seasonally corrected using the factors published in the Validation Scheme. For environments where the seasonal correction factors are not appropriate, such as mines, caves, tunnels and additional sources such as water, repeated measurements throughout the year (or in opposite seasons) might be needed to obtain a better estimate of the annual average radon concentration.

The IRR17 require the employer to take numerous actions, such as *notification* to HSE through an online form [28]; this is the same for all employers, regardless of their usual health and safety inspector. In addition, the employer needs to consult a suitably qualified and experienced Radiation Protection Adviser (RPA), who can assist with IRR17 compliance.

The preferred method for complying with the IRR17 is mitigation. The building techniques for workplace buildings are the same as those for houses, although multiple systems might be required in large buildings where there are several rooms with elevated radon levels. Forced or controlled ventilation can be optimised in buildings where this is installed already and is an option for underground facilities (working in conjunction

with a ventilation engineer to account for other requirements of the air quality, such as removing pollutants).

Where the radon level cannot be reduced sustainably to below 300 Bq m^{-3}, or it is impractical or prohibitively expensive to do so, the workplace becomes a 'managed exposure situation'. Alternative measures such as limiting the amount of time spent in such areas may be a practical solution for controlling radon exposures to employees, although requiring long-term management and input from the RPA. Regular reviews are also needed to ensure that any special procedures remain in place and that changes in circumstances (such as occupancy, use or building modifications) are taken into account.

New build

It is desirable to install radon preventive measures at the construction stage of new buildings or extensions to existing buildings, especially in areas with high radon potential. In effect, this means that mitigation measures have been built in when it is most cost-effective and convenient to do so. Building regulations [29] and their associated approved documents and guidance are in place in each part of the UK, which identify the criteria for radon prevention as either *basic* or *full* measures, and in which geographical areas they should be installed.

The basic measure consists of a radon-impermeable membrane that covers the footprint of the building, with joints and edges carefully sealed. This is intended to provide a simple barrier to the ingress of soil gas. Experience indicates that these measures offer a degree of protection but are not, in general, able to prevent radon entry completely. This may be due to the building design or the standard of installation [30].

Full measures consist of the basic measure and the provision for a powered mitigation system, but without the electric fan and final pipework, for instance a sump (in which case the sump pipework will be capped outside the building). If radon levels after occupation are found to be high, the additional protection can be completed with minimal disruption and expense.

Full measures should be installed in new homes and workplaces where at least 10% of homes are expected to exceed the radon Action Level. Basic measures should be installed where the radon potential exceeds 1% (Scotland and Northern Ireland) or 3% (England and Wales).

Universal basic measures in all new properties would, over time, reduce the very large numbers of individual exposures that arise from modestly increased radon levels. To date, the cost-benefit of universally installed basic measures has not been sufficient in governmental review of planning guidance to support its adoption. However, the Construction Design and Management Regulations 2015 [31] require that the design of the building should *eliminate, so far as is reasonably practicable, foreseeable risks to the health or safety of any person . . . using a structure designed as a workplace*, which could provide additional impetus for radon preventive measures.

Radon in water

Radon has a relatively low solubility in water. However, significant amounts of radon can be carried in groundwater, especially where the local geology contains elevated concentrations of uranium or radium minerals. Public supplies, which have been treated and stored, and blended with surface water, reduce radon to very low levels (although the airborne radon levels in water treatment works can be high). Private water supplies can retain high levels of radon at the point of entry to the building and at the tap. The consumer may then be exposed through ingestion of radon in the water, and through inhalation of the radon that is de-gassed from water such as when heated or given a large surface area, which happens during showers, etc; the inhalation pathway is dominant in terms of exposure. Anomalously high radon levels in

kitchens and bathrooms, compared to those in living rooms and bedrooms, may reveal unknown high radon levels from the water supply.

Standards for radon in water – Following a 2013 EU directive [32], the Water Supply (Water Quality) Regulations 2016 [33] set monitoring criteria for radon in drinking water and parametric values that should not be exceeded. Regulation 6 requires 'water undertakers' to carry out representative surveys that characterise the likely exposures to radon in water intended for human consumption, considering the groundwater sources, hydrology and geology. Monitoring must be carried out if there is reason to believe that the parametric value might be exceeded. Schedule 2 defines the parametric value for radon at the supply point as 100 Bq l^{-1} (becquerel per *litre*). Remedial action may be taken where the radon concentration exceeds 1,000 Bq l^{-1}. The limit of detection (Schedule 4) is 10 Bq l^{-1}.

Sampling and mitigation – Radon levels in water show short-term, diurnal and seasonal variations. Consequently, grab samples do not reflect the long-term average concentration, exposure or risk. Samples must also be taken with care to avoid air bubbles or an air gap above the water in the container, as this will cause a loss of radon from solution prior to measurement. Samples can be analysed by gamma spectrometry (emissions from the decay products), liquid scintillation (a scintillant is added to the sample and the resulting light signal is recorded) or using an electret (radon is released from the water sample to a closed container containing an electret detector).

Mitigation is achieved either by aeration – allowing the radon to de-gas from the water by creating a large surface area, which is vented to outside air – or by passing the water through activated carbon to adsorb the radon gases from the water. Commercial systems are available; activated carbon has limitations due to concerns over the build-up of radioactivity.

Conclusion

Radon exposures in the home and at work are a proven health risk. More than 1,100 UK residents die annually from radon-related lung cancer. Individual risk depends on long-term exposure to elevated radon levels, combined with the smoking history. The main source is from uranium and radium in rocks and soils, with building materials and water as minority sources. The radon potential of a building can be estimated, and the whole of the UK has been mapped at high resolution, but suitable measurements are required to assess the individual radon levels and exposures. Legislation exists to control exposures in the home and at work, with proven methods available to mitigate existing buildings and limit radon ingress in new construction.

Section 4 notes

8 See www.ukradon.org, www.bgs.ac.uk/georeports)
9 See more on hazards in the home and the HHSRS in Chapter 12 – note also that at the time of writing there is a government project to review the HHSRS.
10 Information on suitable mitigation methods is available on www.ukradon.org, which includes guides written with the Building Research Establishment (BRE) and elsewhere.

Section 4 References

[1] United Nations Scientific Committee on the Effects of Atomic Radiation (UNSCEAR). (2008) *UNSCEAR 2006 Report to the General Assembly, Volumes I and II, with Scientific Annexes*, United Nations, New York.

[2] Oatway WB, et al. (2006) *Ionising Radiation Exposure of the UK Population: 2010 Review*, PHE-CRCE-026, PHE, London.

[3] IARC. (2012) *International Agency for Research in Cancer: IARC Monographs on the Evaluation of Carcinogenic Risks to Humans, Volume 100D. A Review of Human Carcinogenics. Part D: Radiation/IARC Working Group on the Evaluation of Carcinogenic Risks to Humans*, WHO, Lyon, France.

[4] IARC. (2016) *European Code Against Cancer*. https://cancer-code-europe.iarc.fr/index. php/en/.

[5] HPA. (2009) *Radon and Public Health*, Report of the Independent Advisory Group on Ionising Radiation, Documents of the Health Protection Agency, Radiation, Chemical and Environmental Hazards, RCE-11, Chilton, Oxon, June.

[6] The UKHSA Interactive Radon Map. https://www.ukradon.org/information/ukmaps.

[7] EC. (2020) *European Atlas of Natural Radiation*. https://remon.jrc.ec.europa.eu/About/Atlas-of-Natural-Radiation.

[8] Agricola G. (1556). *De Re Metallica*. https://www.gutenberg.org/files/38015/38015-h/38015-h.htm.

[9] R CEP. (1984) *Royal Commission on Environmental Pollution, 10th Report: Tackling Pollution: Experience and Prospects*, Cmnd 9419, HMSO, London.

[10] ICRP. (1984) Principles for limiting exposure of the public to natural sources of radiation. *Annals of the ICRP*, 14(1). ICRP Publication 39.

[11] Wrixon AD, et al. (1988). *Natural Radiation Exposure in UK Dwellings*. Chilton, Oxon, NRPB-R190.

[12] Darby S, et al. (2005) Radon in homes and risk of lung cancer: Collaborative analysis of individual data from 13 European case-control studies. *BMJ*, 330(7485): 223, 29 January. https://doi.org/10.1136/bmj.38308.477650.63.

[13] National Research Council (US) Committee on the Biological Effects of Ionizing Radiations. (1988) *Health Risks of Radon and Other Internally Deposited Alpha-Emitters: Beir IV*, National Academies Press, Washington, DC, Appendix IV, Epidemiological Studies of Persons Exposed to Radon Progeny. https://www.ncbi.nlm.nih.gov/books/NBK218116/.

[14] HPA. (2010) *Limitation of Human Exposure to Radon*, Documents of the Health Protection Agency. RCE-15. https://www.gov.uk/government/publications/radon-limitation-of-human-exposure.

[15] HSE. (2017) *The Ionising Radiations Regulations 2017*. SI 2017/1075. https://www.legislation.gov.uk/uksi/2017/1075/contents/made.

[16] WHO. (2009) *Handbook on Indoor Radon*. https://www.who.int/ionizing_radiation/env/9789241547673/en/.

[17] Miles JCH, Appleton JD. (2005) Mapping variation in radon potential both between and within geological units. *Journal of Radiological Protection*, 25: 257–276.

[18] Daraktchieva Z, et al. (2018) *Validation Scheme for Organisations Making Measurements of Radon in UK Buildings: Revision*. PHE-CRCE-040. https://www.gov.uk/government/publications/radon-in-dwellings-validation-scheme-for-measurements.

[19] The 2nd Atlas of Variation in Risk Factors and Healthcare for Respiratory

Disease. https://fingertips.phe.org.uk/profile/atlas-of-variation.

[20] Workplace Radon Monitoring Density Guide. https://www.ukradon.org/cms/assets/gfx/content/resource_4426cs125c0e943c.pdf and PHE (2018) *UK National Radon Action Plan*, London, Report by the PHE Centre for Radiation, Chemical and Environmental Hazards. https://assets.publishing.service.gov.uk/government/uploads/system/uploads/attachment_data/file/766090/UK_National_Radon_Action_Plan.pdf.

[21] ICRP. (2017) *Occupational Intakes of Radionuclides: Part 3*, ICRP Publication 137, Ann ICRP 46(3/4). https://www.icrp.org/publication.asp?id=ICRP%20Publication%20137.

[22] EC. (2013) *Council Directive 2013/59/Euratom of 5 December 2013 Laying Down Basic Safety Standards for Protection Against the Dangers Arising from Exposure to Ionising Radiation, and Repealing Directives 89/618/Euratom, 90/641/Euratom, 96/29/Euratom, 97/43/Euratom and 2003/122/Euratom*. https://eur-lex.europa.eu/eli/dir/2013/59/oj.

[23] BEIS. (2018) *The Ionising Radiation (Basic Safety Standards) (Miscellaneous Provisions) Regulations 2018. SI 2018/482.* https://www.legislation.gov.uk/uksi/2018/482/contents.

[24] McColl NP, et al. (2018) *UK National Radon Action Plan.* PHE-CRCE-043. https://www.gov.uk/government/publications/uk-national-radon-action-plan.

[25] IAEA. (2014) *Radiation Protection and Safety of Radiation Sources: International Basic Safety Standards*, IAEA Safety Standards Series No. GSR Part 3, IAEA, Vienna. https://www.iaea.org/publications/8930/radiation-protection-and-safety-of-radiation-sources-international-basic-safety-standards.

[26] HHSRS. (2006) *Housing Health and Safety Rating System – Operating Guidance, ODPM.* https://assets.publishing.service.gov.uk/government/uploads/system/uploads/attachment_data/file/15810/142631.pdf.

[27] The Management of Health and Safety at Work Regulations. (1999) SI 1999/3242. https://www.legislation.gov.uk/uksi/1999/3242/contents/made.

[28] The HSE Website for Radiation Notifications. https://www.hse.gov.uk/radiation/ionising/notification.htm.

[29] Approved Document C – Site Preparation and Resistance to Contaminants and Moisture. (2004 Edition Incorporating 2010 and 2013 Amendments). https://www.planningportal.co.uk/info/200135/approved_documents/65/part_c_-_site_preparation_and_resistance_to_contaminants_and_moisture.

[30] Hodgson SA, et al. (2019) *Performance of Basic Radon Protection in New Homes.* PHE-CRCE-046. https://assets.publishing.service.gov.uk/government/uploads/system/uploads/attachment_data/file/792115/PHE_CRCE_new_build_protection_2019_v0.11__1_.pdf.

[31] The Construction Design and Management Regulations. (2015) *SI 2015/51.* https://www.legislation.gov.uk/uksi/2015/51/contents/made.

[32] EC. (2013) *Council Directive 2013/51/Euratom of 22 October 2013 Laying Down Requirements for the Protection of the Health of the General Public with Regard to Radioactive Substances in Water Intended for Human Consumption.* https://eur-lex.europa.eu/legal-content/EN/TXT/PDF/?uri=CELEX:32013L0051&from=EN.

[33] The Water Supply (Water Quality) Regulations 2016 (with amendments). SI 2016/614. https://www.legislation.gov.uk/uksi/2016/614/regulation/1.

SECTION 5: EPIGENETICS AND RELEVANCE TO PUBLIC HEALTH

Timothy W Gant and Emma L Marczylo

Introduction to epigenetics

The foundations for the science of genetics lie with two publications: the *Origin of Species* by Charles Darwin (1859) that laid the basis for the environmental selection of advantageous mutations, and the publication of the structure of DNA by Watson and Crick that laid the biochemical basis for understanding genetic heritage. Less noticed at the time was an earlier work by Jean-Baptiste Lamarck that proposed soft inheritance [1], or the adaptation of species characteristics to their surroundings, adaptation that could then be passed to offspring.

The key difference of Lamarckism compared with Darwinism is that the environment does not select the advantageous characteristic

414

as with Darwin's origin of species, but rather that the organism adapts to a changing environment and then passes that adaptation to its offspring. To understand this further we need to make a leap to Hal Conrad Waddington of the mid-20th century who proposed genetic assimilation, a process by which a phenotype developed as an adaptation to an environmental change could be acquired by future generations [2]. Acceptance of the theories of Lamarck and Waddington had to await a biological explanation, and that is the process we now know as epigenetics. In contrast to the fundamental genetic basis of Darwinian selection, epigenetics is a genetic process by which gene expression is altered in the absence of a mutation of the DNA base sequence. Epigenetic changes can also alter phenotype and can be inherited. Thus, epigenetics provides the mechanism linking altered phenotypic responses to an environmental factor, where the adaptation can be passed to offspring. The important point to note here is that there is no mutation of DNA base sequence which is the fundamental genetic basis of Darwinian

selection. Instead, epigenetics uses one or more of three key biochemical processes in the cell, (a) chemical modification of DNA bases for example by methylation, (b) chemical modification of DNA proteins and (c) qualitative and quantitative change in the expression of non-protein coding RNAs. For a detailed explanation of these three classes of epigenetic modification see Marczylo et al. [3]. The most important DNA modification in epigenetics is methylation of cytosine to make 5-methylcytosine, a base that in the right place in the gene DNA sequence suppresses transcription. These methylation marks are not easily removed, and so if gene reactivation of gene transcription is required this is instead achieved by introducing a hydroxy group on the 5-methyl of the methylcytosine to create 5-hydroxymethylcytosine (Figure 10.16). This modification is not recognised as methylated DNA and so no longer has the effect of suppressing gene transcription and is returned to cytosine in the daughter DNA strand on replication. Later in this chapter we will return to these individually.

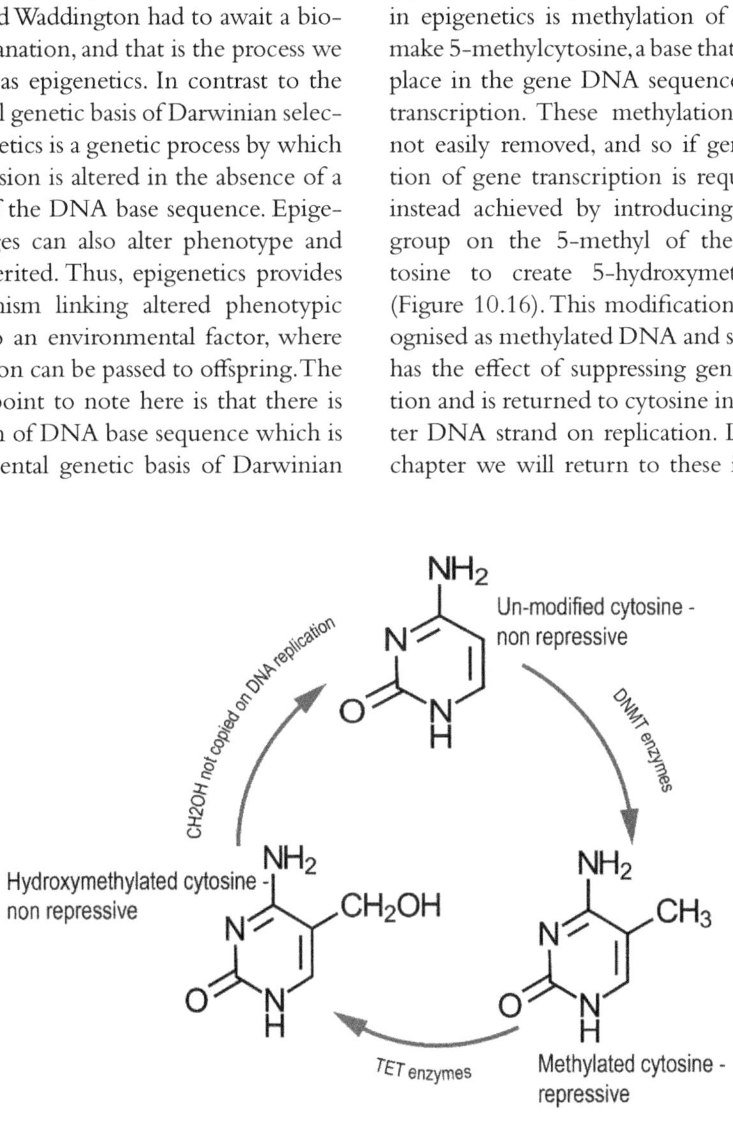

Figure 10.16 DNA methylation and hydroxy methylation at cytosine. The cycle of cytosine methylation by the DNA methyltransferases (DNMT) and conversion to hydroxymethylcytosine by the ten-eleven translocation methylcytosine dioxygenases (TET)

For the moment the important question we need to examine is, does epigenetics matter for biology and the influence of the environment on human health?

The importance of epigenetics for mediating the influence of the environment on human health

For a starting point to look at this subject let us consider a landmark moment in all our lives: conception. At conception two gametes, the oocyte (female) and the spermatozoa (male), fuse to create a zygote. This is the first cell with a full complement of 23 pairs of chromosomes, one of each from mother and father. One chromosome pair, the sex chromosomes, determines biological sex, while the other 22 pairs are autosomes and have no influence on biological sex. From this one cell develops all the trillions of cells that make up the human body, many with specialised functions. All of these cells contain the same complement of genes despite their differing functions as brain cells or liver cells or any other type of differentiated cell. Yet clearly neurones and hepatocytes, for example, perform very different functions. How is this the case when they have the same complement of genes? The answer to this question lies in epigenetics. Epigenetics turns genes on and off and thereby allows neurones to have their characteristic phenotype and the same for all hepatocytes and all other differentiated cells. This process of differentially regulating genes by epigenetic means occurs during differentiation when the cell moves from the totipotent cell (see later for a description of the totipotent cell) to the pluripotent cell and finally the differentiated cell. This is the process hypothesised by Waddington [2] as a cell moving down a valley during its process of differentiation from a stem cell to a differentiated state from which it cannot return (Figure 10.17). Once at the bottom of the valley it cannot return and stays differentiated until it dies and is replaced. The totipotent cell is the founder stem cell of the organism that

is derived from the zygote after epigenetic reprogramming. From the totipotent cells are derived the pluripotent stem cells that form the lineages of differentiated cells that lead to the organs of the organism (Figure 10.17) [4].

To complete the story, we now know that differentiation can be reversed experimentally, and this de-differentiation process can be used to form induced pluripotent stem cells from differentiated cells that can then be re-differentiated into cells of various types. The same processes can be used to make totipotent cells used in the formation of cloned animals, the first of which was famously Dolly the sheep [5]. As this is a specialised laboratory technique further consideration of this de-differentiation process is not necessary here and the interested reader is referred to numerous reviews available, such as Avior et al. [6].

Waddington envisaged a totipotent as the founder cell of the organism being at the top of a valley system and pluripotent cells sitting at the top of separate valleys representing lineages. Differentiation is the process of travel down one of the valleys to a differentiated cell from which there is no return to the pluripotent state. Cells can now be returned to the pluripotent state experimentally, but in nature Waddington's hypothesis holds.

For a second reason why epigenetics matters let's go back to biological sex again. All the gametes from the mother will carry an X chromosome but that from the father will

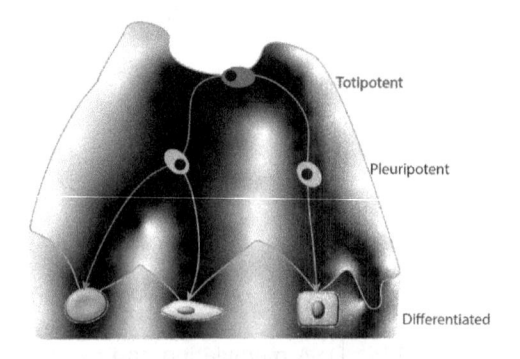

Figure 10.17 The totipotent and pluripotent progenitor cells and Waddington's valleys

carry an X or Y. The sex chromosome from the father that ends up in the zygote will therefore define biological sex, X for a female and Y for a male. Now consider gene dosage. The best way of thinking about this is to consider a disease where gene dosage is abnormal, and for the purposes of this chapter we can consider Down's syndrome as an example. In the common form of Down's syndrome there is an additional copy of chromosome 21, so there are three copies. This means there is differential expression of some genes encoded or controlled by genes on chromosome 21, leading to the characteristic phenotype [7]. Returning to the X chromosome, why is it then that females do not have a gene dosage effect from their additional X chromosome compared to males that have only one X chromosome but are equally dependent on the X chromosome genes? The answer lies in epigenetics. Within each cell of a female one copy of the X chromosome is turned off through the process of epigenetic X chromosome inactivation. This suppression is random so that in any one cell of a female the X chromosome from either the father or the mother is inactivated via DNA methylation [8].

Hormesis, homeostasis and our children

Few words with lots of meaning. To deal with this title effectively we'll break it down into its component parts, starting with homeostasis. Homoeostasis underlies both Darwinian and Lamarckian genetics. Homeostasis is a state of equilibrium with the environment, while hormesis defines the capacity an organism has for adapting to environmental change. Once hormesis is exceeded then an adverse outcome occurs such as toxicity and ultimately death. When an organism is in homeostasis it exists in a balanced state in its environment. When that environment changes, the organism needs to respond. If hormesis is not exceeded the organism can re-achieve a balance with its environment at a different homeostatic level. This is a beneficial

adaptation for the organism and can be transmitted epigenetically to the offspring. If hormesis is exceeded there is an adverse outcome. If this adverse outcome does not result in death, then this adaptation could also be passed to the offspring but may not be beneficial. An example of this outcome is given in Figure 10.21.

Beneficial adaptations where hormesis is not exceeded usually occur at lower levels of the environmental perturbagen and adverse outcomes at higher levels, the classic toxicological dose response curve.

Methylation and the gametes

We have discussed the importance of the methylation marks in the X sex chromosome, but not when this occurs. There are two important periods of life in terms of the establishment of methylation patterns. Before exploring these, it is important to note that while a DNA base mutation in response to an environmental change is permanent, epigenetic changes can be transitory.

The two major periods of demethylation and re-methylation occur following formation of the zygote and during production of the new organism's gametes [9]. Immediately after fusion of the gametes a wave of demethylation takes place that removes most of the cell differentiation marks from the zygote. It is not too difficult to discern why this is the case. The zygote provides the totipotent that gives rise to the differentiated cells of the new organism. We have already discussed that cell differentiation is the result of acquiring epigenetic changes and so for the zygote to give rise to all cells of the body it must start from the beginning. The methylation marks are then re-established as differentiation takes place towards the stage of implantation of the blastocyst. This is a critical state of life and has the potential to be susceptible to environmental exposures [10].

The second major event of demethylation/re-methylation takes place during formation of the gametes and, as such, only occurs in the gametes. This process is primarily to remove and reset maternal and paternal chromosome marks across all chromosomes according to the sex of the offspring. There are some key genes involved in early development that are switched on in males and off in females, and *vice versa*. This is the reason why fertilisation is only successful upon fusion of a male and a female gamete, and not upon fusion of two male or two female gametes. Since half of the chromosomes within a male offspring have been inherited from the mother, they will carry maternal chromosome marks. For successful reproduction of the next generation, these maternal marks must be removed and replaced with paternal marks. The same is true of the paternal marks on the set of chromosomes within a female offspring inherited from the father. Therefore, gametes undergo a second round of demethylation and re-methylation to remove and reset these marks, so they all match the sex of the parent. In males this takes place before birth as the primordial germ cell gives rise to the prospermatogonia, whereas in females this occurs during each menstrual cycle as the primary oocyte develops to the mature oocyte [10–12] (Figure 10.18).

Other mechanisms of epigenetic regulation (Histone modification and miRNAs)

Earlier we have discussed methylation of the cytosine base of DNA as an epigenetic mechanism. This is not the only mechanism of epigenetic regulation and while we do not have space to discuss the other methods it would be remiss if they were not mentioned. As briefly introduced earlier, there are generally considered to be three levels of epigenetic regulation. These involve modification to (a) the DNA itself by methylation of cytosine to form the so-called fifth base – 5-methylcytosine, (b) the histone proteins around which DNA is wound – histone modification, and (c) non-coding RNA species including, but not exclusively, microRNAs (miRNAs). Histone changes are one of the devices the cell has on hand to rapidly and transiently control gene expression. Modification of histone tails is essential to allow unwinding of DNA prior to transcription [13]. As post-translational modifications to histones are more transitory than those to the DNA itself, it has been proposed that they are used by the cell as a trial mechanism to establish if the resulting changes in gene

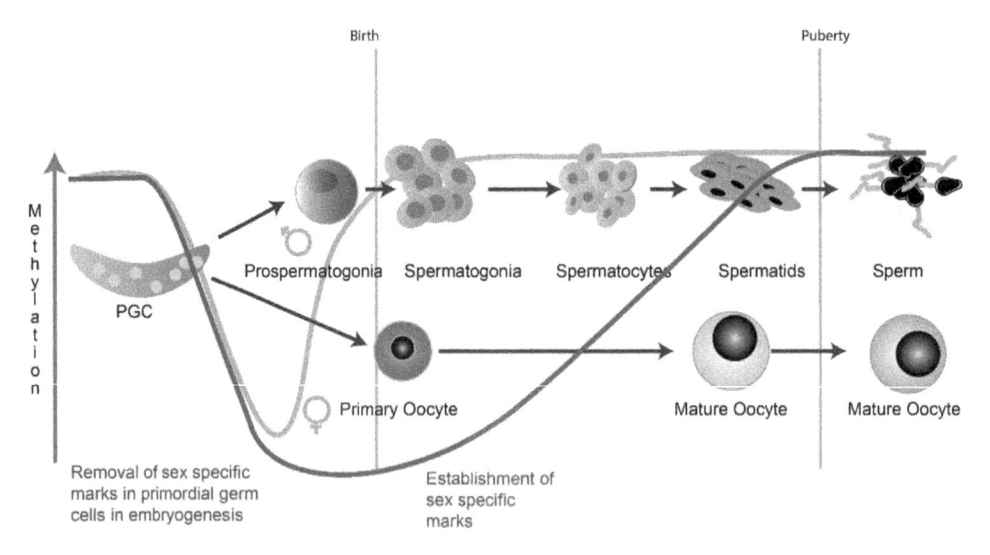

Figure 10.18 Demethylation and re-methylation cycles during formation of the gametes in male and the female offspring

expression are beneficial in response to an environmental factor. If found to be advantageous then the expression change is made more permanent by being converted into a new methylation pattern within the DNA.

Methylation of DNA is carried out by the DNA methyltransferases, all of which are embryonic lethal in mice that have had the genes deleted [14]. There are three genes in the family. DMNT1 is responsible for maintenance methylation where it uses template strand during cell DNA replication to replicate the methylation marks onto the daughter DNA strand. The other two DMNT enzymes, 3A and 3B, are responsible for *de novo* methylation and are particularly important in the context of methylation in the gametes. In human embryonic stem cells the effects of DNMT3 enzyme knockout are less severe than in the mouse embryo, but deletion of the DMNT1 enzyme still causes lethality [15].

miRNA species are one of a series of RNA types known collectively as non-protein coding RNA. These RNAs control gene expression by repressing mRNA translation and thus reducing protein levels [16]. One miRNA can control the expression of many mRNA species operating in a network manner [17]. miRNAs are dysregulated in many pathologies and released into body fluids where they can act as biomarkers of pathophysiology, particularly in cancer but also in response to environmental stress and toxicity [18–20].

Importantly for intergenerational transmission of environmentally induced epigenetic effects, miRNA are transferred in the gametes in easily measurable quantities where they control the early stages of gene expression in the zygote. Transfer takes place in both the ovum and the spermatozoa. In terms of looking at the effects of the environment on the composition of miRNA in the gametes, there have been more studies in the male by virtue of the tissue being more accessible. For example, tobacco smoking has been shown to have effects on the composition of miRNA

in the spermatozoa [21–22]. These are observational studies and so the consequences of the changed composition of miRNA for the zygote the totipotent cells and resulting embryo are not known. Nevertheless, they provide an insight into a potential mechanism by which environmental exposures in one generation may lead to inheritable adaptations in subsequent generations. Qualitative alteration in miRNAs in gametes have also been associated with other environmental exposures [23–24] and diet-induced obesity [25].

What is epigenetically normal?

This is a difficult question to answer given that we have already said epigenetics is a means to respond to the environment, provide hormesis and achieve homeostasis. Inbred animals are often used in safety studies for the very reason that the stable genome provides more homogenous responses to insults and therefore reduces the statistical burden required to test the null hypothesis. To understand whether deviations from baseline for epigenetic changes are important in determining hazard of the test material, it is necessary to understand the baseline and how much, and what type, of deviation from that baseline is deleterious. Realising the importance of this the CEFIC/LRI programme funded a project (C3) in 2014 [26] to look at the baseline methylation patterns across multiple tissues in two rodent strains commonly used for chemical testing. This work showed that there are differences in methylation patterns between gender and strains (Wistar and Sprague Dawley), but within the strains and gender there is relative homogeneity [27]. The study also looked at the 5-hydroxymethylation cytosine modification. As described earlier, the removal of methylation marks begins with conversion of 5-methyl cytosine to 5-hydroxymethylcytosine. Thomson et al. [27] found that the 5-hydroxymethylation cytosine marks are more variable in gene bodies between gender and species and as such might be a better marker for environmentally mediated epigenetic change than the 5-methylcytosine itself.

Human studies

The Dutch hunger winter – The Dutch hunger winter was a period of acute food shortages from September 1944 to May 1945. The result was about 20,000 individuals dying from starvation. Some groups were more vulnerable, including pregnant women and their unborn children. At the peak of rationing, calorific intake in Amsterdam had fallen from about 2000 calories per day to about 400–500 [28]. This decreased calorific availability, particularly during the first trimester, was associated with several chronic disease states in the later life of the resulting children. For example, abnormal glucose handling, leading to higher body mass index (BMI) and an elevated total and low-density lipoprotein (LDL) cholesterol [29–30]. Some of the pathophysiological effects have been associated with differential methylation of key genes, including the Palmitoyl Transferase 1A (*CPT1A*) and Insulin Receptor (*INSR*) genes, which have been tentatively linked to birth weight and lipid metabolism [31]. This was perhaps the first study that has indicated the effect of environmental factors on the epigenome and phenome of the subsequent generation.

Multigenerational, transgenerational and intergenerational inheritance

At this juncture it is important to add a note that the Dutch hunger winter as cited, with effects in the first generation as a result of *in utero* exposure, represents multigenerational and not transgenerational inheritance. What is the difference? In the case of the Dutch hunger winter the unborn child was exposed to the calorific deficiency directly. Therefore, the effect was not transmitted through the gametes of either the mother or the father; it was the result of a direct effect on the embryo (albeit via the placenta, but still a direct

exposure). It is therefore multigenerational in that there was an effect on the mother in terms of lost weight and on her child as a change in later life health. For transgenerational inheritance there must be transmission of the genotype in the gametes with no direct exposure of the affected offspring. In this situation, the exposure leads to an epigenetic change that results in a phenotype that can be passed through subsequent generations. Furthermore, it is important to distinguish the differences between *in utero* and adult exposures. During an *in utero* exposure, the mother (F0 generation), the child (F1 generation) and the gametes that go on to produce the grandchildren (F2 generation) are directly exposed. This means that the effect can only be defined as true transgenerational epigenetic inheritance if the phenotype is found in the great-grandchildren (F3 generation). The great-grandchildren are the first generation to be derived from gametes that have never been exposed. If the exposure occurs in adulthood, then only the F0 generation and the gametes that go onto produce the F1 generation are directly exposed. Here, a phenotype in the grandchildren (F2 generation) would demonstrate transgenerational epigenetic inheritance. For the purposes of the following text, where multi- and transgenerational mechanisms are relevant to health, the term intergenerational will be used to encompass both. (See Figure 10.19.)

Chemical and radiation exposures and intergenerational effects

The potential for chemicals to elicit intergenerational effects is an important one. Risk assessments currently in place generally only consider the exposed organism at the F0 (adult) and F1 (*in utero*) generations (OECD TG416), which is multigenerational but has no consideration of effects beyond this in terms of the germline and subsequent generations (transgenerational). While any generational effect is of potential concern, those

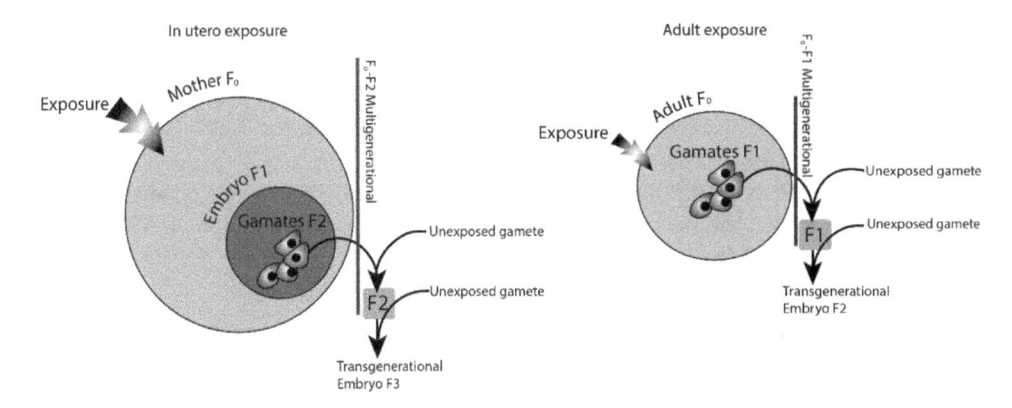

Figure 10.19 Transgenerational verses multigenerational effects. Highlights the mechanism transfer from multigenerational (direct exposure at some level) to transgenerational (no direct exposure) as a result of an exposure *in utero* or adulthood. The collective term used here is intergenerational

exposures that result in an adverse phenotype that can be inherited in the absence of any direct exposure may be of greater concern since the exposure only has to happen once for lasting effects across all subsequent generations. For some chemical types known as endocrine disruptors there may be an intergenerational effect, and an example of the endocrine disruptors vinclozolin and methoxchlor is given later. For this reason Greally and Jacobs [32] have proposed the inclusion of epigenetic endpoints in chemical testing methods that assess hazard.

Apart from Lamarckism several major events seeded an interest in intergenerational effects: (a) thalidomide, which is only multigenerational and will not be considered further here, (b) reports of deaths and defects in the offspring born to fathers exposed to ionising radiation at the Sellafield nuclear plant [33–34], and (c) the effect of diethystilbesterol (DES) on the female offspring of mothers prescribed this drug during pregnancy (also multigenerational) [35].

With respect to Sellafield, a study published by Gardener MJ in 1992 showed an increased incidence of leukaemia in the sons born to men who had been exposed to radiation [36]. This association, which the authors acknowledged had no established causal pathway, has

not been confirmed in larger studies [37–38]. Comparison with other similar construction sites has since suggested an alternative infection related causal pathway [39]. This highlights that epidemiological association does not always translate to causation and further studies are usually required to establish causal pathways. Animal studies can be helpful here, although trans-species relevance of the results is always an issue. For ionising radiation, Dubrova et al. demonstrated instability in a mini-satellite DNA region within the spermatozoa of mice exposed to a high dose (2 gray) of ionising radiation [40] (Figure 10.20). It is important to note that the time between exposure and sampling in this study allowed a full cycle of spermatozoa maturation to take place and so that the sampled spermatozoa were not exposed per se. These results therefore point to an instability in the germ cell, and the transgenerational nature of the instability was indicated in a further study [41] shown in Figure 10.20 and also shown to occur following exposure to chemical DNA mutagens [42].

Similar spermatozoa microsatellite mutations were reported in mice exposed to ambient air near two integrated steel mills and a major highway compared to high-efficiency air particulate (HEPA) filtered ambient air, suggesting that air pollution may also affect germline stability [43]. There are caveats to be added here; (a) a microsatellite instability in the sperm does not indicate an adverse outcome either for the cell or the whole organism, and (b) the dose of ionising radiation used was very high compared to environmental exposures. Thus, although this work provides a potential causal pathway, the data are not likely to be consequential for the Sellafield worker cohort due to the high ionising radiation exposures used [44]. When considered together, these studies indicate the profound difficulty in undertaking this type of work where the generational time in humans is 25 years. Even for mice with a shorter generational time the numbers of animals required becomes prohibitive, and expensive, and the results can be challenging to translate to the human situation.

Reports of the effects of DES on female F1 started to appear in the 1970s and were confirmed in a cohort of 3980 prenatally exposed females in 1984 with an adverse outcome of cervical and vaginal dysplasia [35]. This is a multigenerational effect since the developing embryo is directly exposed. There were early reports that male F1 were also affected, showing increased genitourinary anomalies and increased levels of hormonally related cancers. Further epidemiological studies did not confirm these associations [45–46], though a recent meta-analysis has indicated an increased rate of testicular germ cell tumours in the F1 generation [47]. There have also been reports of adverse transgenerational effects on development of the reproductive organs of sons (F2 generation) born to men prenatally exposed to DES in the absence of any effects in the F1 generation themselves [48–49]. Again, such studies indicate the complexity in assigning causality in intergenerational studies and the importance of establishing a causal biochemical pathway.

The work with DES led to interest in the potential intergenerational effects of endocrine active compounds. The scene was set with a study of the fungicide vinclozolin (antiandrogenic) and the insecticide methoxchlor (estrogenic) from the group of M. Skinner. Publishing in *Science* in 2005, this group showed adverse outcomes in sperm after exposure of the pregnant female. This effect was transgenerational as it occurred to the fourth (F3) generation, and selective mating with unexposed partners demonstrated transfer through the male lineage [50] (Figure 10.21).

Increased apoptosis and functional defects in sperm were transmitted transgenerationally via the male lineage. The latter demonstrated by the presence of adverse effects in the offspring of an F3 male mated with an unexposed female but not in the offspring of an F3 female mated with an unexposed male.

This group have published many subsequent studies of the potential causal pathways leading to these outcomes, and all three epigenetic processes appear to have some involvement, methylation [51], histone modification [52] and non-coding RNA [53]. While this work is mechanistically interesting, its relevance to public health is limited because the doses used were much greater than human exposure from the environment (Figure 10.22). Thus, we are looking at hazard, not a human health risk. Furthermore, differing outcomes in multiple generations has been documented in other studies with endocrine active substances [54]. There is more work

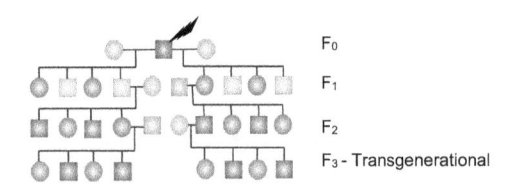

F0

F1

F2

F3 - Transgenerational

Figure 10.20 Study design used by Dubrova et al. to look at the effects of ionising radiation or chemical mutagen exposure on sperm genome stability in subsequent unexposed generations. The founder exposed male is shown at the top, who then sired two lines with two unexposed females six weeks after exposure, the left arm showing paternal transmission and the right maternal transmission

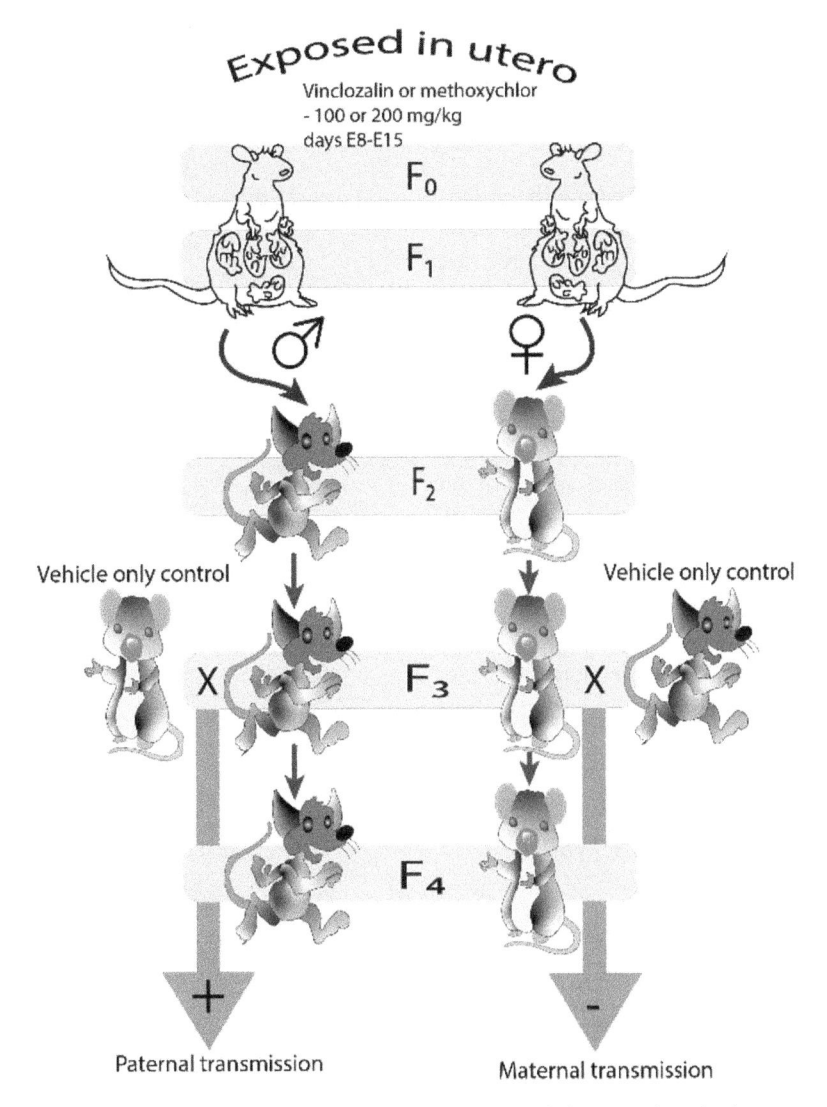

Figure 10.21 Experiment of Anway et al. [50] that showed the transfer of adverse spermatogenic outcomes after exposure to high dose vinclozolin or methoxychlor

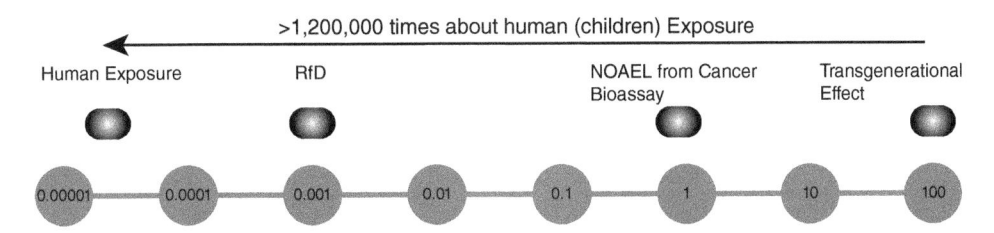

Figure 10.22 Dose at which transgenerational effect of vinclozolin was reported compared to a no observable adverse effect level (NOAEL), reference dose (RfD, an estimate of a daily exposure to the human population) and measured human exposure from biomonitoring data

to be done to fully understand the effects of these substances across generations.

Arsenic

Marczylo et al. [3] have reviewed the potential public health concerns of environmentally induced epigenetic toxicity and the interested reader is referred to this publication. From this work, one example is selected that is of more direct relevance to human health. Arsenic is an environmental contaminant of both natural and human-made origin and is unusual in that an understanding of the potential mechanism preceded an understanding of the associated adverse outcome.

Arsenic is an environmental containment, the levels of which are heavily regulated in most countries. Found in aquifers it has been an issue in regions of Asia, in particular Bangladesh, and parts of India and Pakistan [55–56]. Exposures in these regions from contaminated drinking water wells can be high to very high. Arsenic exposure is associated with lung, bladder, liver, kidney and some forms of skin cancer [57] and for this reason is classified by the International Agency for Research on Cancer (IARC) as a group 1 carcinogen (see also Chapter 16 on air pollution).

Arsenic is metabolised by methylation, first to a monomethylated form before becoming demethylated [58] (Figure 10.23). This involves the methyl transfer molecule S-adenosyl methionine (SAM), which carries the methyl group from methionine for transfer. SAM is also the intermediate required for the methylation of cytosine within DNA, and so we can hypothesise that high levels of arsenic exposure could deplete cells of SAM thereby interfering with normal DNA methylation and thus gene expression [59]. Skin lesions are common in high arsenic areas and some studies

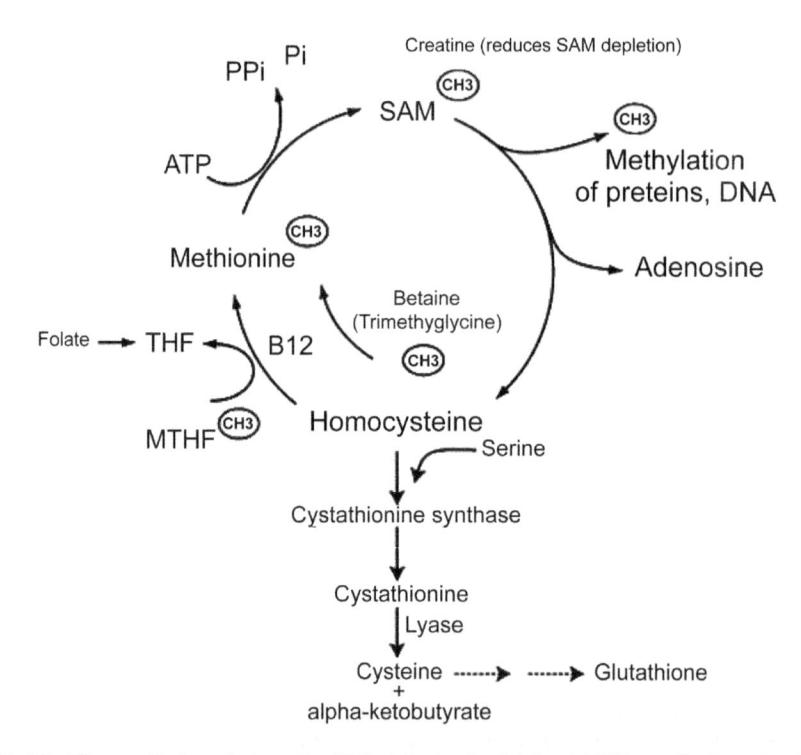

Figure 10.23 The methyl cycle in cells. THF: tetrahydrofolate, MTHF: methyltetrahydrofolate

have linked this to changes in DNA methylation [60]. Another outcome of arsenic exposure is cancer and changes in DNA methylation patterns are highly deregulated in cancers leading to altered gene expression. Given that the metabolism of arsenic via SAM potentially reduces the amount of methyl donor available for DNA methylation, the expectation would be for reduced (hypo) DNA methylation to occur with arsenic exposure. Ren et al. [58] reviewed this area and found that this was indeed the situation, at least in experimental systems. Others have found the same in liver systems [61]. However, although global hypomethylation is indicative of an effect it does not indicate causation. What matters is the altered methylation patterns at the level of the individual genes controlling cellular processes that lead to the adverse outcome, for example oncogenes or cell cycle genes in carcinogenesis.

There is therefore a need not just to measure differences in global methylation, but also methylation changes at the level of genes critical to the development of the adverse outcome. As we have already discussed, this can be challenging because (a) methylation patterns are both cell and organ specific and so it is usually not possible to use a more accessible surrogate biosample such as blood, and (b) changes in methylation patterns occur in response to wide range of environmental factors [62]. It can therefore be difficult to assess the changes occurring in response to one perturbation, particularly in human samples. For further reading, a useful review of metals and the DNA-methylome and carcinogenesis has been published by Brocato and Costa [63].

Diet and epigenetic change

Can the diet of the parents produce phenotypic change(s) that can affect their children? More particularly, can obesity in parents be passed to children [64]? As described, the Dutch hunger winter studies, which detailed the effects of *in utero* malnutrition on their later life BMI, seeded the whole area of intergenerational effects. What about obesity? In male rats fed a high-fat diet there was epigenetic reprogramming of the let-7c miRNA and its mRNA targets. This was associated with a reduced glucose tolerance and imparted ability to gain weight in the female, but not male, offspring of these males [65]. DuPont et al. have provided a review of the area and acknowledge that human studies are scarce, but rodent studies provide evidence of paternally transmitted intragenerational effects resulting from a high-fat diet [25]. For example, increased female body size was transmitted via the paternal lineage to at least the fourth (F3) generation following *in utero* exposure to a maternal high-fat diet, indicating transgenerational inheritance [66].

Aside from diet, there is also some evidence for epigenetic transmission of an obesity phenotype through the gametes following exposure to obesogenic chemicals. This is a contentious area as most of the work is in animal systems where exposures are greater than would be encountered by humans in the normal environment. The interested reader is referred to a published review by King and Skinner [67]. Two points need to be made here:

1 These studies point to areas where further research is required, the outcome of which is likely to have consequences for public health and
2 Some care must be taken with the current evidence

While some of the genes shown to be differentially regulated in these systems have a known role in glucose metabolism, it is not proof they are causally related to the outcome. However, there is enough indication in these studies to justify more work, particularly given the public health consequences of the high level of obesity in the population.

Environment and sperm – fathers' health matters

Small non-coding RNAs are abundant in sperm and are one of two important means by which adaptations in one generation may be passed to the second and further [68]. The first adaptation altered methylation patterns, and we have already discussed this in more detail. The second, spermatozoal miRNAs, we have mentioned only briefly. It is important to re-iterate that after fusion of the gametes to form the zygote, miRNAs control the translation of mRNA, both which are inherited from the spermatozoa and the oocyte. New gene transcription within the zygote does not begin until about the 8-cell stage. For this reason, the composition of these molecules transferred into the early zygote by the gametes is critically important in shaping the early totipotent cells from which all cells of the embryo will be derived [69].

It has already been mentioned several times that phenotypes can be transmitted through the spermatozoa. A lot of current health improvement emphasis has been placed on mothers' health (pre- and post-conception) as being advantageous to a good outcome for the offspring. While not complete, there is sufficient evidence to suggest that a similar emphasis should also be placed on fathers. Small RNA species in the gamete can be affected by diet and other environmental factors. There are several studies in this field, one of which has demonstrated that tobacco smoking can alter the expression of spermatozoal miRNAs associated with biochemical processes that could affect early embryo development [21].

Intergenerational associations from human cohort studies

The Newborn Epigenetics Study Cohort (NEST) [70] found an association between pre-pregnancy obesity and methylation in the TAP binding protein (TAPBP) gene in mothers where the epigenetic change was carried though to both the female and male children [71]. Differential methylation at specific sites in the TAPBP gene were associated with BMI in females and more strongly with systolic blood pressure in females and males. A caveat was that since obesity was present pre-conception, it was also likely carried through pregnancy. Thus, there could have been a similar direct effect on the epigenetic methylation marks of the mother and the developing embryo, and so did not constitute a transgenerational effect. None of the TAPBP CpG sites were replicated in the Avon Longitudinal Study of Parents and Children (ALSPAC) cohort, one of the longest running intergenerational epidemiological studies. An association has though been made in the ALSPAC cohort between paternal smoking before age 11 and elevated body mass index in male (but not female) offspring [72].

Given the observations cited in the text earlier that tobacco smoking can alter miRNAs within sperm [21], this may represent a true transgenerational effect with adverse outcomes for the offspring. Once again, however, reproducibility is an issue. In the Nord-Trøndelag Health (HUNT) study the same association between paternal smoking and altered BMI in sons was not found [73]. A further caveat in respect of this subject. This is an introduction and not an exhaustive review. The two examples are cited to indicate the difficulty of undertaking these studies. The fact that results have not been repeatable does not mean they are wrong in either study. Given that epigenetic marks respond to all environmental influences, there could be other factors involved. They are cited simply to indicate how difficult it is to conduct these studies in the human population. It is only going to be through multiple studies that the weight of evidence will be built to test the hypothesis that environmentally induced phenotypes can be inherited from parent to child through altered epigenetic marks on the DNA or small RNAs in the gametes. Animal studies in faster breeding species will be essential to test and establish causal pathways.

Conclusions

In monozygotic twins, who are genetically identical in terms of DNA base sequence, there is known to be epigenetic drift with time [74–75], and there is clear evidence of epigenetic transfer across multiple generations. This is physiologically necessary as well as a key response for adapting to a changed environment. A review from the Toxicology Department at the UK Health Security Agency (formerly Public Health England) identified examples of epigenetic change in response to environmental factors and considered the implications of these studies for public health and regulatory toxicology [3]. There is more work to do to establish causal pathways, and even more with respect to building this emerging knowledge into a regulatory framework [32]. Nevertheless, evidence suggests that such work is essential to ensure that the health of our children is not affected by the exposures or lifestyle choices of their mothers or fathers. Paternally transmitted outcomes require particular attention due to the intrinsic belief that it is the mother's health that is more important for the child and so fathers are less likely to be targeted in public health improvement drives. There is much work for public health practitioners, epidemiologists, molecular scientists, policy makers and regulators to do in this space to improve health for generations to come [76].

Acknowledgements

TWG and ELM are both partially supported from the National Institutes of Health Research Units; Environmental Exposures and Health held between UKHSA and Imperial College London and the University of Leicester. The input of Dr Miriam Jacobs as the UK National Coordinator for Health to the OECD in this area of work is acknowledged and some of her work is cited. Thoughtful discussion with Prof Naima Bradley (University of Leicester) and colleagues in the Toxicology Department of UKHSA are also acknowledged.

Section 5 References

[1] Lamarck JB. (1809) *Philosophie zoologique ou exposition des considérations relatives à l'histoire naturelle des animaux*, Cambridge University Press, Cambridge, 507.

[2] Waddington CH. (1942) Canalization of development and the inheritance of acquired characters. *Nature*, 150(3811): 563–565.

[3] Marczylo EL, Jacobs MN, Gant TW. (2016) Environmentally induced epigenetic toxicity: Potential public health concerns. *Critical Reviews in Toxicology*, 46(8): 676–700.

[4] Hu K. (2019) On mammalian totipotency: What is the molecular underpinning for the totipotency of zygote? *Stem Cells and Development*, 28(14): 897–906.

[5] Campbell KH, et al. (1996) Sheep cloned by nuclear transfer from a cultured cell line. *Nature*, 380(6569): 64–66.

[6] Avior Y, Sagi I, Benvenisty N. (2016) Pluripotent stem cells in disease modelling and drug discovery. *Nature Reviews Molecular Cell Biology*, 17(3): 170–182.

[7] Pecze L, Szabo C. (2021) Meta-analysis of gene expression patterns in down syndrome highlights significant alterations in mitochondrial and bioenergetic pathways. *Mitochondrion*, 57: 163–172.

[8] Wutz A, Gribnau J. (2007) X inactivation xplained. *Current Opinion in Genetics and Development*, 17(5): 387–393.

[9] Seisenberger S, et al. (2013) Reprogramming DNA methylation in the mammalian life cycle: Building and breaking epigenetic barriers. *Philosophical Transactions of the Royal Society B*, 368(1609): 20110330.

[10] Heard E, Martienssen RA. (2014) Transgenerational epigenetic inheritance: Myths and mechanisms. *Cell*, 157(1): 95–109.

[11] von Meyenn F, Reik W. (2015) Forget the parents: Epigenetic reprogramming in human germ cells. *Cell*, 161(6): 1248–1251.

[12] Gkountela S, et al. (2015) DNA demethylation dynamics in the human prenatal germline. *Cell*, 161(6): 1425–1436.

[13] Bannister AJ, Kouzarides T. (2011) Regulation of chromatin by histone modifications. *Cell Research*, 21(3): 381–395.

[14] Li E, Bestor TH, Jaenisch R. (1992) Targeted mutation of the DNA methyltransferase gene results in embryonic lethality. *Cell*, 69(6): 915–926.

[15] Liao J, et al. (2015) Targeted disruption of DNMT1, DNMT3A and DNMT3B in human embryonic stem cells. *Nature Genetics*, 47(5): 469–478.

[16] Selbach M, et al. (2008) Widespread changes in protein synthesis induced by microRNAs. *Nature*, 455: 58–63.

[17] Ameres SL, Zamore PD. (2013) Diversifying microRNA sequence and function. *Nature Reviews Molecular Cell Biology*, 14.

[18] Starkey Lewis PJ, et al. (2011) Circulating microRNAs as potential markers of human drug-induced liver injury. *Hepatology*, 54(5): 1767–1776.

[19] Taylor E, Gant T. (2008) Emerging fundamental roles for non-coding RNA species in toxicology. *Toxicology*, 249(1): 34–39.

[20] Oda S, Yokoi T. (2021) Recent progress in the use of microRNAs as biomarkers for drug-induced toxicities in contrast to traditional biomarkers: A comparative review. *Drug Metab Pharmacokinet*, 37: 100372.

[21] Marczylo EL, et al. (2012) Smoking induces differential miRNA expression in human spermatozoa: A potential transgenerational epigenetic concern? *Epigenetics*, 7(5): 432–439.

[22] Metzler-Guillemain, C, et al. (2015) Sperm mRNAs and microRNAs as candidate markers for the impact of toxicants on human spermatogenesis: An application to tobacco smoking. *Systems Biology in Reproductive Medicine*, 61(3): 139–149.

[23] Li Y, et al. (2012) A microarray for microRNA profiling in spermatozoa from adult men living in an environmentally polluted site. *Bulletin of Environmental Contamination and Toxicology*, 89.

[24] Meng P, et al. (2018) Maternal exposure to traffic pollutant causes impairment of spermatogenesis and alterations of genome-wide mRNA and microRNA expression in F2 male mice. *Environmental Toxicology and Pharmacology*, 64: 1–10.

[25] Dupont C, et al. (2019) Role of miRNA in the transmission of metabolic diseases associated with paternal diet-induced obesity. *Frontiers in Genetics*, 10: 337.

[26] CEFIC/LRI. (2014) *LRI-C3: Epigenetics – Normality in Toxicologically Relevant Species*. https://cefic-lri.org/request-for-proposals/lri-c3-epigenetics-normality-in-toxicologically-relevant-species/ (Accessed 20 October 2021).

[27] Thomson JP, et al. (2017) Defining baseline epigenetic landscapes in the rat liver. *Epigenomics*, 9(12): 1503–1527.

[28] Painter RC, Roseboom TJ, Bleker OP. (2005) Prenatal exposure to the Dutch famine and disease in later life: An overview. *Reproductive Toxicology*, 20(3): 345–352.

[29] Lumey LH, Stein AD, Susser E. (2011) Prenatal famine and adult health. *Annual Review of Public Health*, 32: 237–262.

[30] Lumey LH, et al. (2021) Overweight and obesity at age 19 after pre-natal famine exposure. *International Journal of Obesity (Lond)*, 45(8): 1668–1676.

[31] Tobi EW, et al. (2014) DNA methylation signatures link prenatal famine exposure to growth and metabolism. *Nature Communications*, 5: 5592.

[32] Greally JM, Jacobs MN. (2013) In vitro and in vivo testing methods of epigenomic endpoints for evaluating endocrine disruptors. *ALTEX*, 30(4): 445–471.

[33] Parker L, et al. (1999) Stillbirths among offspring of male radiation workers at Sellafield nuclear reprocessing plant. *Lancet*, 354(9188): 1407–1414.

[34] Lord BI. (1999) Transgenerational susceptibility to leukaemia induction resulting from preconception, paternal irradiation. *International Journal of Radiation Biology*, 75(7): 801–810.

[35] Robboy SJ, et al. (1984) Increased incidence of cervical and vaginal dysplasia in 3,980 diethylstilbestrol-exposed young women: Experience of the national collaborative diethylstilbestrol adenosis project. *JAMA*, 252(21): 2979–2983.

[36] Gardner MJ. (1992) Leukemia in children and paternal radiation exposure at the Sellafield nuclear site. *Journal of the National Cancer Institute, Monographs*, 12: 133–135.

[37] Wakeford R, et al. (1994) The descriptive statistics and health implications of occupational radiation doses received by men at the Sellafield nuclear installation before the conception of their children. *Journal of Radiological Protection*, 14(3).

[38] Draper GJ, et al. (1997) Cancer in the offspring of radiation workers: A record linkage study. *BMJ*, 315(7117): 1181–1188.

[39] Kinlen LJ, Dickson M, Stiller CA. (1995) Childhood leukaemia and non-Hodgkin's lymphoma near large rural construction sites, with a comparison with Sellafield nuclear site. *BMJ*, 310(6982): 763–768.

[40] Dubrova Y, et al. (1998) Stage specificity, dose response and doubling dose for mouse minisatellite germ-line mutation induced by ionising radiation. *Proceedings of the National Academy of Sciences of the United States of America*, 95: 6251–6255.

[41] Barber R, et al. (2002) Elevated mutation rates in the germ line of first and second generation offspring of irradiated mice. *Proceedings of the National Academy of Sciences of the United States of America*, 99: 6877–6882.

[42] Vilarino-Guell C, Smith A, Dubrova Y. (2003) Germline mutation induction at mouse repeat DNA loci by chemical mutagens. *Mutation Research*, 526: 63–73.

[43] Yauk C, et al. (2008) Germ-line mutations, DNA damage, and global hypermethylation in mice exposed to particulate air pollution in an urban/industrial location. *Proceedings of the National Academy of Sciences of the United States of America*, 105(2): 605–610.

[44] Tucker JD, et al. (1997) Biological dosimetry of radiation workers at the Sellafield nuclear facility. *Radiation Research*, 148(3): 216–226.

[45] Leary FJ, et al. (1984) Males exposed in utero to diethylstilbestrol. *JAMA*, 252(21): 2984–2989.

[46] Strohsnitter WC, et al. (2021) Prenatal diethylstilbestrol exposure and cancer risk in males. *Cancer Epidemiology, Biomarkers & Prevention*, 30.

[47] Hom M, et al. (2019) Systematic review and meta-analysis of testicular germ cell tumors following in utero exposure to diethylstilbestrol. *JNCI Cancer Spectrum*, 3(3): pkz045.

[48] Tournaire M, et al. (2018) Birth defects in children of men exposed in utero to diethylstilbestrol (DES). *Therapie*, 73(5): 399–407.

[49] Gaspari L, et al. (2021) 'Idiopathic' partial androgen insensitivity syndrome in 11 grandsons of women treated by diethylstilbestrol during gestation: A multi-generational impact of endocrine disruptor contamination? *Journal of Endocrinological Investigation*, 44(2): 379–381.

[50] Anway M, et al. (2005) Epigenetic transgenerational actions of endocrine disruptors and male fertility. *Science*, 308(5727): 1466–1469.

[51] Skinner MK, et al. (2019) Transgenerational sperm DNA methylation epimutation developmental origins following ancestral vinclozolin exposure. *Epigenetics*, 14(7): 721–739.

[52] Ben Maamar M, et al. (2018) Epigenetic transgenerational inheritance of altered sperm histone retention sites. *Scientific Reports*, 8(1): 5308.

[53] Schuster A, Skinner MK, Yan W. (2016) Ancestral vinclozolin exposure alters the epigenetic transgenerational inheritance of sperm small noncoding RNAs. *Environmental Epigenetics*, 2(1).

[54] Priyanka, et al. (2020) Gestational and lactational exposure to triclosan causes impaired fertility of F1 male offspring and developmental defects in F2 generation. *Environmental Pollution*, 257: 113617.

[55] Smith AH, Lingas EO, Rahman M. (2000) Contamination of drinking-water by arsenic in Bangladesh: A public health emergency. *Bulletin of the World Health Organization*, 78(9): 1093–1103.

[56] Sohel N, et al. (2009) Arsenic in drinking water and adult mortality: A population-based cohort study in rural Bangladesh. *Epidemiology*, 20(6): 824–830.

[57] IARC. (1994) Some drinking-water disinfectants and contaminants, including arsenic. *IARC Monographs on the Evaluation of Carcinogenic Risks to Humans*, 84: 1–31.

[58] Ren X, et al. (2011) An emerging role for epigenetic dysregulation in arsenic toxicity and carcinogenesis. *Environmental Health Perspectives*, 119(1): 11–19.

[59] Bailey KA, et al. (2016) Mechanisms underlying latent disease risk associated with early-life arsenic exposure: Current research trends and scientific gaps. *Environmental Health Perspectives*, 124(2): 170–175.

[60] Das A, et al. (2021) Depletion of S-adenosylmethionine pool and promoter hypermethylation of arsenite methyltransferase in arsenic-induced skin lesion individuals: A case-control study from West Bengal, India. *Environmental Research*, 198: 111184.

[61] Chen H, et al. (2004) Chronic inorganic arsenic exposure induces hepatic global and individual gene hypomethylation: Implications for arsenic hepatocarcinogenesis. *Carcinogenesis*, 25(9): 1779–1786.

[62] Martinez-Zamudio R, Ha HC. (2011) Environmental epigenetics in metal exposure. *Epigenetics*, 6(7): 820–827.

[63] Brocato J, Costa M. (2013) Basic mechanics of DNA methylation and the unique landscape of the DNA methylome in metal-induced carcinogenesis. *Critical Reviews in Toxicology*, 43(6): 493–514.

[64] Lopomo A, Burgio E, Migliore L. (2016) Epigenetics of obesity. *Progress in Molecular Biology and Translational Science*, 140: 151–184.

[65] de Castro Barbosa T, et al. (2016) High-fat diet reprograms the epigenome of rat spermatozoa and transgenerationally affects metabolism of the offspring. *Molecular Metabolism*, 5(3): 184–197.

[66] Dunn GA, Bale TL. (2011) Maternal high-fat diet effects on third-generation female body size via the paternal lineage. *Endocrinology*, 152(6): 2228–2236.

[67] King SE, Skinner MK. (2020) Epigenetic transgenerational inheritance of obesity susceptibility. *Trends in Endocrinology & Metabolism*, 31(7): 478–494.

[68] Chen Q, Yan W, Duan E. (2016) Epigenetic inheritance of acquired traits through sperm RNAs and sperm RNA modifications. *Nature Reviews Genetics*, 17(12): 733–743.

[69] Yuan S, et al. (2016) Sperm-borne miRNAs and endo-siRNAs are important for fertilization and preimplantation embryonic development. *Development*, 143(4): 635–647.

[70] Hoyo C, Murphy S. (2021) *Newborn Epigenetics Study Cohort (NEST)*. https://tools.niehs.nih.gov/cohorts/index.cfm/main/detail/ids/c178.

[71] Martin CL, et al. (2019) Maternal pre-pregnancy obesity, offspring cord blood DNA methylation, and offspring cardiometabolic health in early childhood: An epigenome-wide association study. *Epigenetics*, 14(4): 325–340.

[72] Northstone K, et al. (2014) Prepubertal start of father's smoking and increased body fat in his sons: Further characterisation of paternal transgenerational responses. *European Journal of Human Genetics*, 22(12): 1382–1386.

[73] Carslake D, et al. (2016) Early-onset paternal smoking and offspring adiposity: Further investigation of a potential intergenerational effect using the HUNT study. *PLoS One*, 11(12): e0166952.

[74] Poulsen P, et al. (2007) The epigenetic basis of twin discordance in age-related diseases. *Pediatric Research*, 61(5 Pt 2): 38r–42r.

[75] Fraga MF, et al. (2005) Epigenetic differences arise during the lifetime of monozygotic twins. *Proceedings of the National Academy of Sciences of the United States of America*, 102(30): 10604–10609.

[76] Lane M, Robker RL, Robertson SA. (2014) Parenting from before conception. *Science*, 345(6198): 756–760.

SECTION 6: COMMUNICABLE DISEASES

Caryn L. Cox

Introduction

As we have seen in Chapter 1, as towns expanded in the United Kingdom and became more remote from their supply hinterland with the Agricultural and Industrial Revolutions, they also became increasingly crowded and insanitary, as the necessary infrastructure including housing was not available or developed sufficiently to support the population growth. For lower-income countries in the 21st century, similar pressures exist, but for different reasons. In addition, there is rapid movement of goods and people around the globe and climate change is having a more significant impact.

Improvements in sanitation, housing, nutrition and other social conditions in the UK in the 20th century led to dramatic reductions in diseases that caused large epidemics such as cholera, typhoid, tuberculosis, diphtheria and dysentery. These improvements from the middle of the 20th century were supported by new developments in antimicrobial therapy and vaccines against common infections, particularly in childhood. These significant medical advances and associated health benefits led to the widespread belief that infectious diseases would be eradicated, particularly in higher-income countries.

However, over the last few decades, a number of infectious diseases have begun to emerge as global public health problems. These emerging infections include zoonoses such as Chikungunya and West Nile Fever viruses which have taken a foothold in human populations where previously they were only identified in animal populations, as well as SARS-associated coronavirus (SARS-CoV) and avian influenza virus (H5N1) that have caused dramatic outbreaks in human and animal populations. The SARS outbreak highlighted the potential for rapid spread of new human pathogens in human populations and reinforced the need for coordinated global response to disease outbreaks.

This was further borne out on 31 December 2019, when the World Health Organization (WHO) was informed of cases of pneumonia with an unknown cause in Wuhan City, China. A novel coronavirus was identified as the cause on 7 January 2020 and was temporarily named 2019-nCoV. A novel coronavirus (nCoV) is a new strain that has not been previously identified in humans.

The new virus was subsequently named the COVID-19 virus.

On 30 January 2020, the WHO Director-General declared the novel coronavirus outbreak a public health emergency of international concern (PHEIC); at that time there were 98 cases and no deaths in 18 countries outside China. On 11 March 2020, the rapid increase in the number of cases outside China led the WHO Director-General to announce that the outbreak could be characterised as a pandemic. By then more than 118,000 cases had been reported in 114 countries, and 4291 deaths had been recorded. By mid-March 2020, the WHO European Region had become the epicentre of the epidemic, reporting over 40% of globally confirmed cases. As of 28 April 2020, 63% of global mortality from the virus was from that Region [1].[11]

The occurrence of outbreaks in poultry and sporadic human infections with avian influenza (H5N1) heightened concerns about the emergence of a pandemic strain of the influenza virus. These concerns were soon realised in the early part of the 21st century with the emergence of the pandemic (H1N1) influenza virus in 2009 – declared a pandemic by WHO in June 2009, which had caused by August 2010, when WHO declared the end of the pandemic, laboratory confirmed infections in 214 countries and overseas territories worldwide, including over 18,398 deaths [2].

Reporting and modelling estimated that between 151,700 and 575,400 died worldwide from the 2009 H1N1 virus infection during the first year the virus circulated – 15 times higher than the number of laboratory-confirmed deaths reported to the WHO and with a disproportionate number in African and Southeast Asian regions [3].

In 2014, the significant Ebola virus disease (EVD) outbreak (formerly known as Ebola haemorrhagic fever) in West Africa led to a global response from the first cases reported in March 2014. The most severely affected countries, Guinea, Liberia and Sierra Leone, had weak health systems, lacked human and infrastructural resources, and had recently emerged from long periods of conflict and instability. In August 2014 WHO declared the West Africa outbreak a Public Health Emergency of International Concern under the International Health Regulations [4]. The outbreak ended in March 2016, two and a half years after the first case was discovered with WHO declaring that the Ebola situation no longer constituted a PHEIC. The 2014–2016 outbreak in West Africa was the largest and most complex Ebola outbreak since the virus was first discovered in 1976 and led to more than 28,600 cases and 11,325 deaths. Subsequent to this, further outbreaks have been declared in Guinea, the Democratic Republic of Congo and in August 2021 Cote d'Ivoire declared an outbreak.

Zika virus disease is caused by a virus transmitted by Aedes mosquitoes. People with Zika virus disease usually have symptoms that can include mild fever, skin rashes, conjunctivitis, muscle and joint pain, malaise or headache and symptoms normally last for two–seven days. The incubation period is not clear but thought to be a few days. There is no specific treatment or vaccine currently available, and the best form of prevention is protection against mosquito bites. The virus is known to circulate in Africa, the Americas, Asia and the Pacific.

During late 2015 it was identified that there was a rise in the spread of Zika virus in Brazil which was accompanied by an unprecedented rise in the number of children being born with unusually small heads – identified as microcephaly. In addition, several countries, including Brazil, reported a steep increase in Guillain-Barré syndrome (GBS) – a neurological disorder that can lead to paralysis and death. In February 2016 following surveillance of the observed increases in neurological disorders and neonatal malformations and the association with Zika infection, WHO declared a PHEIC.

WHO supported countries to control the Zika virus disease by taking actions outlined in the *Zika Strategic Response Framework*; this work included coordination, surveillance, care, vector control, risk communication and community engagement, and research at the global, regional and country level. As a result of key actions undertaken, the PHEIC was ended by WHO on 18 November 2016.

In addition to concerns about emerging infections, other infectious diseases that had hitherto been controlled, such as tuberculosis (TB), have re-emerged as significant public health problems. The re-emergence of previously controlled diseases has been driven in part by a reduction in effective public health control programmes, synergism with new infections – for example Human Immunodeficiency Virus (HIV) and TB co-infection – and changes in a range of socio-economic factors and determinants. These changes include globalisation and the attendant increase in international trade and travel which has resulted in an increase in the spread of diseases across international borders.

Alongside this, changes in food production, processing and transportation practices have driven the emergence of antibiotic resistant strains of common pathogens and provided an opportunity for widespread outbreaks of infectious diseases in animal populations that pose a risk to human health, for example Bovine Spongiform Encephalopathy (BSE) in cattle and Highly Pathogenic Avian Influenza (HPAI) infection in poultry.

In many countries, changes in the patterns of human migration as a result of civil unrest or other factors, for example for economic reasons, have also driven recent changes in the epidemiology of previously controlled endemic diseases.

Growing concerns about the potential impact of changes to the global climate has led to a rise in the development of infectious disease models to predict potential changes in the epidemiology of infectious disease, particularly vector-borne diseases such malaria and dengue fever.

Over time, attention has also been focused on the role of infectious agents in the aetiology of chronic diseases and the growth in

multi-drug resistant infections particularly in healthcare settings.

Diseases such as stomach ulcers, lymphomas, cervical and stomach cancers, and cardiovascular disease (CVD) are now widely accepted to have infectious aetiologies and new preventive and therapeutic approaches have been developed to prevent and control the occurrence of some of these diseases, for example, the use of the Human Papillomavirus (HPV) vaccine to prevent the occurrence of cervical cancer.

The occurrence and spread of drug-resistant infections, particularly multi-drug resistant infections in healthcare settings is a significant public health problem due to the huge burden of preventable morbidity and mortality arising from these infections. The prevention and control of Healthcare Associated Infections (HCAIs) has led to the development of extensive surveillance systems to monitor the occurrence of these infections; policies and guidance to reduce the inappropriate use of antimicrobials and improve infection prevention and control processes in healthcare and community-based settings; and increased research to develop new antimicrobial therapies. Antimicrobial stewardship is one arm to this – where coordinated interventions are designed to improve and measure the appropriate use of antimicrobials by promoting the selection of the optimal antimicrobial drug regimen, dose, duration of therapy, and route of administration.

These observed and predicted changes in the epidemiology of infectious diseases discussed earlier highlight the need for well-resourced, multi-level, public and environmental health systems to support the prompt identification, response and control of epidemics and implementation of long-term measures to prevent and control endemic diseases. Internationally, this has led to important changes in the International Health Regulations (IHR) which requires countries to strengthen their existing capacities for public health surveillance and response as a way of improving national, regional and global public health security.

The combination of increased investment in local and national surveillance systems, laboratory diagnostic facilities, and workforce development including training and education in the use of epidemiological methods and tools, are necessary to ensure the effective prevention and control of infectious diseases.

Basic concepts in infectious disease epidemiology

Infectious diseases are the result of complex interactions between an external infectious agent, a susceptible host and enabling environmental factors. This model of interaction is known as the epidemiological triangle (see Figure 10.24). In order to successfully investigate and control sporadic cases and outbreaks of infectious diseases, an understanding of the infecting organism, and its interaction with hosts and environmental factors is required.

The infectious agent – the infectious agents in the epidemiologic triangle are usually organisms such as viruses, bacteria, fungi or parasites that cause infections in susceptible hosts. An overview of the characteristics of some of these infectious agents is presented in this section.

Viruses – Viruses are very important human pathogens that only replicate inside the living cells of the host organism. Viral particles (virions) have a simple structure that consists of a nucleic acid core (single stranded RNA or double stranded DNA genome) surrounded by a protective protein coat (capsid), with an additional outer glycoprotein envelope in certain viruses.

The viral genome can undergo genetic changes when transferred from one host to another, an adaptive response to any resistance acquired by the hosts over time. This process of adaptation occurs in a number of ways, for example, *antigenic drift*, which entails small changes to the viral genome over time that can confer resistance to a range of antiviral drugs. *Antigenic shift* entails a sudden, major change in the viral genome, resulting in a new strain which hosts have little or no acquired immunity to, for example, changes to the influenza virus leading to epidemics or pandemics. Viruses can be diagnosed using a variety of methods that include the following:

Serology (usually blood) has been a mainstay of laboratory diagnosis of viral infections. Viral serologic testing monitors the immune system's antibody response to viral antigen exposure, including both infection and immunisation. Serological diagnosis involves the use of a variety of techniques which are constantly evolving to improve both the accuracy of the test result and also the speed at which the results can be known.

Direct examination of a specimen can be undertaken to detect viral antigens, for example, the use of immunofluorescence techniques to detect influenza viruses. Other methods of direct examination include electron microscopy to visualise viral particles in specimens such as fluid from vesicles and tissue.

Molecular methods such as polymerase chain reaction (PCR) and nucleic acid-based amplification techniques are increasingly utilised by laboratories for a range of tests. These methods involve the amplification of viral genetic material such as RNA and DNA to enable the qualitative and quantitative detection of the viral genome and during

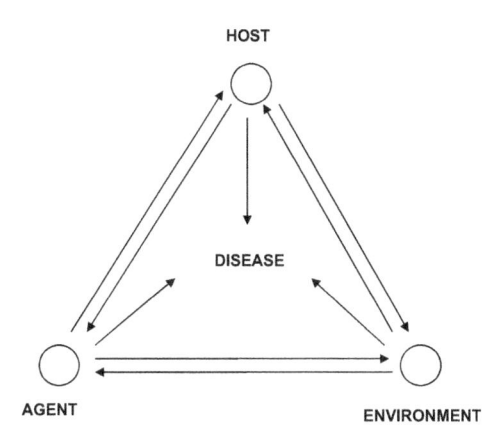

Figure 10.24 The epidemiological triangle

the COVID-19 pandemic has been the test mostly used for the confirmation of a positive case.

Bacteria – These are diverse organisms responsible for many common infections including those of public health significance that have caused significant morbidity and mortality over time – for example the plague caused by *yersinia pestis* and tuberculosis caused by *mycobacterium tuberculosis*. Bacteria are primarily classified according to morphological characteristics such as size and shape, staining, culture and biochemical properties, for example, anaerobic gram-positive *Staphylococcus aureus*. Taxonomic subgroups can be defined on the basis of genotypic characteristics such as strain type and the pattern of plasmids that contain genetic information that code for resistance to antibiotics.

Pathogenic bacteria are polymorphic with apparently identical organisms (even within the same host) exhibiting considerable heterogeneity in their pathogenicity, virulence and growth requirements in host tissue and the environment. Heterogeneity in growth requirements has been exploited to develop diagnostic tools such as culture media that selectively grow bacteria while testing for susceptibility to specific antimicrobial drugs. In addition to supporting the development of new diagnostics, bacterial polymorphism has also been exploited in developing new treatments and vaccines and in understanding the impact of human genetic polymorphism on the development of bacterial disease.

There is no universal mechanism by which bacteria cause disease in humans. A variety of mechanisms have been described such as direct invasion of human tissue resulting in pyogenic infection – for example staphylococcal abscesses; granulomatous infections – for example tuberculosis; and neoplasia – for example HPV causing cervical cancer.

A second mechanism involves the production of bacterial toxins which can be released by the bacteria or be part of the structure of the organism. These toxins can lead to intoxication which affect the host's ability to maintain

physiological processes – for example cholera toxin and poor regulation of salts and fluids; and intoxication that leads to destruction of the host tissue – for example clostridial exotoxin and gas gangrene. Another mechanism of bacterial disease involves the activation of the hosts' immune system resulting in immune-mediated disease processes such as post-infective Guillain-Barré syndrome.

There is also an increasing understanding of the mechanism by which bacterial pathogens cause or contribute to the aetiological pathway of non-infectious disease conditions, for example, increased inflammation of the stomach lining as a result of helicobacter pylori infection and subsequent development of stomach cancer.

The diagnosis of bacterial infection can occur via a variety of methods which includes:

- Direct visualisation of bacteria via microscopy of human samples including body fluids and faeces. Bacteria can also be cultured from human specimen such as blood in specially prepared culture media that provide an ideal environment for these organisms to grow, for example, the use of blood cultures to diagnose infection with *Staphylococcus* and *Streptococcus* species.
- Serology: this involves the use of a variety of techniques to detect and quantify specific antibodies in human samples or provide evidence of raised antibody titres from paired specimen following infection with a specific pathogen.
- Molecular methods: this approach involves the detection of specific molecules or genetic material using techniques such as probe hybridisation, PCR amplification and ligase chain reaction.

Chlamydia, Mycoplasma, Rickettsiae and Coxiellae – These organisms represent a heterogeneous group of bacteria with unique phenotypic characteristic that are dependent on living cells for their survival.

Chlamydiae are intracellular bacteria which are more commonly recognised as a causative of upper and lower respiratory tract infections and sexually transmitted infections.

Mycoplasmas are ubiquitous, small-sized organisms that infect a variety of plant and animal species and have frequently been implicated as cell culture contaminants. Pathogenic mycoplasma infection results in atypical pneumonia.

Rickettsiae are gram-negative intracellular bacteria that are more commonly associated with arthropods such as ticks and fleas. Rickettsiae cause a range of serious zoonotic infections such as spotted fevers, endemic and scrub typhus but are readily treated with antibiotics. It is important to note that they do not cause rickets, which is a result of vitamin D deficiency.

Coxiellae are gram-negative and include *coxiella burneti* which causes human Q fever – which is carried by arthropods (particularly ticks), fish, birds and mammals.

Fungi – These are part of a large group of eukaryotic organisms that live in the environment or may be normal inhabitants of humans and animals.

Yeastlike fungi are round or oval, generally form smooth flat colonies and reproduce by budding. Moulds (moulds) are made up of tubular structures called hyphae, from fuzzy colonies and grow by branching and longitudinal extensions. Virtually all fungi reproduce by forming spores which are widely dispersed into the surrounding environment and are capable of surviving for extended periods of time in often unfavourable conditions.

Pathogenic yeast and moulds cause diseases in humans only when the hosts' defences are compromised, for example, *Pneumocystis* infection in AIDS patients. HIV and the increasing use of immunosuppressive therapy has significantly expanded the range of fungi capable of causing pathogenic infection and occasionally death in humans.

Protozoa – These are morphologically simple, free-living unicellular organisms with a complex lifecycle involving asexual and sexual reproduction. Important examples include *plasmodium* species that cause malaria,

trypanosoma species that cause sleeping sickness, *giardia* and *cryptosporidium* which are common causes of diarrhoeal illness.

Helminths – Helminths are multicellular parasitic worms with a complex life cycle that are recognised as a leading cause of morbidity particularly in developing countries. Helminths that commonly parasitise humans include nematodes (round worms); cestodes (tape worms); trematodes (schistosomes and other flukes); and platyhelminths (flatworms).

Ectoparasites – These are organisms which complete a life cycle and or derive benefit through interaction with the outer surface or skin of the host. These organisms derive nourishment and shelter from their hosts and may cause infestations for years if left untreated. Ectoparasites are more commonly invertebrate arthropods such as lice, ticks, fleas and bugs.

Prions – These are infectious agents composed entirely of protease-resistant proteins called prion protein (PrP). These proteins like viruses do not self-replicate, but rather induce polypeptides in the host organism to take on the abnormal protein structure. Prions are responsible for a group of diseases whose main pathologic manifestations are almost exclusively confined to the central nervous system (CNS), for example, Creutzfeldt-Jakob disease (CJD) and Kuru.

The host – A number of host factors can influence an individual's exposure, susceptibility and response to an infectious agent. These factors may prevent, reduce or increase the likelihood of occurrence of infection or disease following exposure; modify the severity and duration of disease and the likelihood of spread of the infection to other humans. These host factors include non-modifiable factors such as age, sex, ethnicity, immunological and genetic makeup as well as modifiable factors such as socio-economic status (for example income deprivation), lifestyle and behavioural factors such as intravenous drug use and unsafe sexual practices, and diet.

The environment – This refers to environmental factors extrinsic to the infectious agent and host that affect the agent and the opportunity for exposure. These factors include climate, physical surroundings, geography, socio-economic factors and healthcare services. Certain occupational groups are known to be at increased risk of certain infection due to increased opportunity for exposure, for example, sewage workers and leptospirosis. Poor housing quality and overcrowding can increase the transmission and risk of infections such as *Neissera meningitides* and *Mycobacterium tuberculosis*. Climatic differences between tropical and temperate regions favours the occurrence of key vectors required to spread some infections, for example mosquitoes and *Plasmodium falciparium*. Variation in the provision and access to healthcare services can lead to differences in the occurrence of certain infections in subgroups of the population, for example, increased incidence of childhood infections such as measles in groups with poor uptake of vaccines.

The case – In infectious disease epidemiology, a case is a person in a population or group who meets certain criteria that identifies him/her as having the infectious disease of interest. The person who introduces the disease into a population or group, such as a household or school, is known as the *primary case* and the subsequent case(s) that arise as a result of the transfer of the infectious agent from the primary case are known as *secondary cases*. An *index case* is the first case identified by the healthcare system during an outbreak and can be but is not usually the same as the primary case.

Occurrence – Cases are counted to determine the burden of infectious diseases which can vary with time and place. This variation in the pattern of disease occurrence may be *sporadic* with cases occurring or being reported at irregular time intervals. An *endemic* pattern refers to when an infection is always present at low or moderate levels within a given geographic area or defined population while a *hyperendemic* pattern is observed when

infection occurs at high levels and affects all age groups equally.

When the level of infection in the area or population exceeds the expected level for a given time period, it is called an *epidemic (or outbreak)*, or a *pandemic* when the infection spreads across several countries or continents. Outbreaks which are considered disease epidemics with localised increases in disease frequency present in a variety of ways. When the exposure occurs over a relatively brief time period, for example, during a wedding party, a *point source outbreak* occurs; when the exposure is intermittent or continuous, for example, ingestion of contaminated water supplies, a *continuous source outbreak* ensues; and *propagated outbreak* are situations where spread from person to person is apparent.

Two commonly used measures of the occurrence of infections or disease are the *incidence rate*, which is the number of new cases occurring in a defined population (i.e. persons at risk) over a given time period expressed as a proportion of the total population at risk, for example, 10 cases per 10,000 people per year; and *prevalence*, the proportion of a defined population with the disease at a given point in time. Other measures of occurrence used during outbreaks include *attack rate*, the proportion of individuals exposed to an infectious agent who become ill during the outbreak period (a measure of infectivity); and *case-fatality rate* (or inversely the survival rate), the proportion of diagnosed cases of an infectious disease who die (or inversely survive) within a specified time period – this being a measure of virulence.

Reservoir of infection – This is the natural habitat of an infectious agent. Reservoirs can be humans, animals, plants or soil – or a combination of these – in which the infectious agent normally lives and multiplies, and on which it depends for survival. The reservoir does not necessarily have to be the same as the source or vehicle of infection. Humans are the only known reservoir of infections such as measles, polio and hepatitis B, while there are a number of recognised animal and environmental reservoirs for pathogenic organisms such as gastrointestinal bacteria, *Clostridium tetani* and viruses such as rabies amongst others.

Source of infections – A *source* is the person, animal, inanimate object or substance from which an infectious agent is acquired. The source of an infection may sometimes be different from its reservoir. For example, in outbreaks of *Escherichia coli*, the reservoir of the infection is often found to be farm animals, but the source of the infection can be contaminated food or water.

The source of an infection can be described as a *carrier* or representing *colonisation* when found to harbour the infectious agent, with evidence of growth and multiplication in the absence of any discernible clinical disease (i.e. no evidence of illness).

Routes of transmission – This is the pathway or mechanism by which an agent is spread from a reservoir and/or source to a susceptible host. The main routes of spread of infectious agents are direct, indirect (includes vector- and vehicle-borne spread) and airborne transmission.

- *Direct transmission* – This describes the transmission of an infectious agent by direct contact such as touching, biting, kissing, sexual intercourse and by droplet spread. Examples of transmission through direct contact include skin infestations and infections such as scabies and impetigo which are spread by contact with an infected area on another person's body or through contact with contaminated inanimate objects (sometimes known as fomites – for example door handles). Direct transmission via sexual contact is also the route of spread of sexually transmitted infections such as chlamydia and syphilis.
- Direct transmission through the spread of respiratory droplets from an infected person to close contacts can occur through sneezing, coughing and talking. Common examples include measles, influenza and the common cold. The faecal-oral route

of direct transmission describes the transfer of faeces directly to the mouth of a susceptible host as occurs in the spread of a number of gastrointestinal infections and is readily reduced/minimised by effective handwashing.

- *Indirect transmission* – This may involve inanimate material or objects such as fomites that provide a vehicle where the infectious agent may or may not develop in or on before transmission. Examples of such vehicles include toys, soiled clothing, cooking utensils and surgical instruments. Indirect transmission can also occur via vectors such as insects that mechanically transfer the agent to the susceptible host or that support the multiplication and development of the agent before transmitting the infectious form of the agent to humans.
- Indirect transmission of gastrointestinal infections occurs via the faecal-oral route following contact with faecal contaminated food and objects or transfer of faeces by animal vectors such as flies. Another route of indirect transmission is the blood-borne route as observed in the transmission of blood-borne infections such as HIV and hepatitis B where blood or body fluids from an infected person are transferred to a susceptible host through inoculation, injection or transfusion. A third route of indirect transmission is the respiratory route which involves droplets from the nose and mouth contaminating hands and fomites and subsequently resulting in the transfer of the infectious agents to susceptible persons.
- *Airborne transmission* – Airborne transmission entails the spread of suspensions of infectious particles which can remain suspended in air for long periods of time (microbial aerosols). These microbial aerosols are transferred to humans usually via the respiratory route, for example, tuberculosis and legionellosis.

Incubation period – This refers to a time period characterised by subclinical or inapparent pathological changes that begins following exposure to the infectious agent and ends with the onset of symptoms. The incubation period is usually expressed as a range and an average value for a given organism. For example, Legionnaires' disease has a range of two–ten days with a mean of five days whilst rabies has an incubation period of between three and eight weeks but may be as short as nine days or as long as seven years.

Infectious period – This is the time during which the infectious agent may be transferred directly or indirectly from an infected person to a susceptible individual. With some diseases, this period may be lengthy and intermittent such as tuberculosis and rabies or it may involve the development of carriage or colonisation as is observed in staphylococcus infections. Some diseases tend to be more infectious before the onset of clinical illness (i.e. during the incubation period), than during the actual illness – for example measles.

Detecting problems

The identification of trends and patterns in the occurrence of infection, diseases or other health-related events in a defined population and the dissemination of relevant information to those responsible for public and environmental health action describe a variety of activities that constitute surveillance. Surveillance is defined as the systematic collection, collation and analysis of data and the timely dissemination of information to those who need to know so action can be taken.

Public health surveillance plays an important role in identifying and guiding the investigation of clusters/outbreaks of infection; and planning and monitoring the effectiveness of public health control measures and preventive actions. Surveillance information is also used to assess the health status of the population, evaluate programmes and identify important priorities for further surveillance and research.

Data used for surveillance may include information collected primarily for the specific surveillance system (e.g. disease notification data as well as routine data collected for other purposes but then applied to surveillance purposes such as healthcare administration data). The mechanism or process employed in capturing the relevant data may involve a passive process which relies entirely on the initiative of individuals and groups, for example physicians, making these reports in a timely manner (passive surveillance) or a process of actively seeking these reports by the surveillance team (active surveillance). The statutory notification of infectious diseases (NOIDS) in the UK is an example of a passive system and several European and North American countries operate similar systems. The prime purpose of the notification system is speed in detecting possible outbreaks and for this reason, the accuracy (and specificity) of diagnosis is secondary. In the UK, an example of an active surveillance system is that operated by the British Paediatric Surveillance Unit to monitor trends in the occurrence of selected paediatric disorders and infections, for example, congenital rubella.

Some systems described as 'sentinel' surveillance systems do not seek to capture all events but rather focus on improving the completeness and quality of reporting from a dedicated and representative sample of participants, for example in general practice (GP)/primary care. There are also surveillance systems such as 'syndromic surveillance', which are based on information about individual symptoms or aggregations of defined symptoms (i.e. syndromes) rather than laboratory-confirmed diagnoses of disease. These systems may provide some early warning of changes in the occurrence of certain infections, for example influenza which will then need to be corroborated using other systems.

Sources of surveillance information

Death certification and registration – Countries need to know how many people are born and die each year – and the main causes of their deaths – in order to have well-functioning health systems. The only way to count everyone and to track all births and deaths is through civil registration. Civil registration provides the basis for individual legal identity but also allows countries to identify their most pressing health issues.

The WHO receives cause of death statistics from around 100 Member States. However, globally, around two-thirds of 60 million annual deaths and around one-quarter of 140 million births are still not registered.

When deaths go uncounted and the causes of death are not documented, governments cannot design effective public health policies or measure their impact. Information on births and deaths by age, sex and cause is the cornerstone of public health planning [5].

In England and Wales, there has been a legal requirement to register all deaths before a body may be buried or cremated since 1837 and doctors have been required to certify the cause of death for patients under their care since 1841, except for certain categories of death which are referred to the Coroner for investigation. As a result, death certification and registration is virtually 100% complete, but errors in the data can occur at any stage from diagnosis through certification and coding, to processing and analysis. Copies of death entries in the local register are sent to the Office for National Statistics (ONS)[12] by all Registrars in England and Wales. The underlying cause of death is coded in accordance with the codes from the International Classification of Diseases (ICD), and statistics are published weekly, monthly, quarterly and annually in varying detail. Mortality data is not a very useful source when undertaking surveillance of infections that cause mild non-fatal illnesses such as the common cold or gastroenteritis. However, when undertaking surveillance of certain infections such as HIV, mortality data can be usefully combined with morbidity data to provide a comprehensive and representative picture.

Notifications of infectious diseases – In England, health protection regulations outline the requirements for clinical notification of diseases. The legislation adopts an 'all hazards' approach, which specifies that, in addition to the defined list of notifiable infectious diseases, there is also a requirement for clinicians to notify cases of other infections, hazards or contamination which could present a significant risk to human health. Any registered medical practitioner (RMP) suspecting or making these diagnoses is required to notify the proper officer of the local authority, who is often, but not always, a Consultant in Health Protection or a Consultant in Communicable Disease Control (see Table 10.10 for a list of notifiable diseases in England). The list can be amended as notifiable diseases of significance arise and this was most recently seen with a change to legislation with the addition on 5 March 2020 of COVID-19 to the list of notifiable diseases and SARS-COV-2 to the list of notifiable causative agents. This change was made by adding them to the Health Protection (Notification) Regulations 2010.

Statutory notifications are an important way of monitoring trends in infectious diseases, particularly diseases such as whooping cough and mumps, where diagnosis is rarely confirmed by laboratory test. The prime purpose of the notification system is to enable the timely detection of possible outbreaks and epidemics based on the clinical diagnosis of infections with the speed of notification rather than the accuracy of the clinical diagnosis considered more important.

In England, the UK Health Security Agency (UKHSA – formerly known as Public Health England) is responsible for administering the Notification of Infectious Diseases (NOIDS) system (see Table 10.10 for diseases and Table 10.11 for causative agents under this system) and publishes summaries of local and national trends on a weekly basis with regular annual reports. Similar systems operate in the other devolved UK nations.

Other countries worldwide take a similar approach to notification, reporting and surveillance systems. However, countries do differ in the infections that are required to

Table 10.10 Diseases notifiable to proper local authority officers under the Health Protection (Notification) Regulations 2010 (England, UK)

Acute encephalitis	Cholera	Invasive group A streptococcal disease	Plague
Acute infectious hepatitis	COVID-19	Legionnaires' disease	Rabies
Acute meningitis	Diphtheria	Leprosy	Rubella
Acute poliomyelitis	Enteric fever (typhoid or paratyphoid fever)	Malaria	Severe Acute Respiratory Syndrome
Anthrax	Food poisoning	Measles	Scarlet fever
Botulism	Haemolytic uraemic syndrome	Meningococcal septicaemia	Smallpox
Brucellosis	Infectious bloody diarrhoea	Mumps	Tetanus

be reported as these are selected and differ depending on which diseases are of more significance or importance in each country.

Other diseases that may present significant risk to human health are reported under the category 'other significant disease'.

Laboratory reporting systems – Laboratory reporting of infections is a key component of communicable disease surveillance. Laboratory reports represent counts of specified infections which have been confirmed in official and quality-controlled and assured laboratories and captured on a form of database used for routine laboratory reporting. Laboratory reports are then collated, analysed, interpreted and the findings are disseminated via surveillance reports. Clear guidance is provided to laboratories that specify the reporting procedures, list of the priority organisms and illnesses and the level of detail of the required data. Although laboratory data are undoubtedly more specific than clinical reports captured by notification systems, they may still be subject to the same biases observed in notification system reporting, as clinical diagnosis usually precedes laboratory confirmation and any changes in clinical awareness will also affect laboratory reporting. Also, laboratory reports may also

be biased towards the severe end of the illness spectrum and as such may not be truly representative.

Administrative and secondary care consultation data – The collection of data relating to hospital admissions and consultations in secondary care for infectious diseases can be a useful source of administrative data. In England this is known as Hospital Episode Statistics (HES) and is a record level administrative database that contains information relating to hospital admissions including accident and emergency attendances and outpatient appointments in England. The database provides demographic and clinical activity data on a range of conditions including infectious diseases and associated clinical complications for which hospital admission and clinical intervention is required. Other examples of data collected for administrative purposes that may be used for infectious disease surveillance purposes include emergency department attendance and drug prescribing. However, care must be exercised in using these forms of administrative databases (such as HES) as they may introduce bias and be less representative than other data sources, due to their focus on severe cases requiring hospital intervention.

Table 10.11 Causative agents notifiable in England under the Health Protection (Notification) Regulations 2010 (England, UK)

Causative agents notifiable in England under the Health Protection (Notification) Regulations 2010

Bacillus anthracis	*Chlamydophila psittaci*	Ebola virus	Kyasanur Forest disease virus	*Neisseria meningitidis*	SARS-coronavirus
Bacillus cereus (only if associated with food poisoning)	*Clostridium botulinum*	*Entamoeba histolytica*	Lassa virus	Omsk haemorrhagic fever virus	*Shigella spp*
Bordetella pertussis	Clostridium perfringens (*only* if associated with food poisoning)	*Francisella tularensis*	*Legionella spp*	*Plasmodium falciparum, vivax, ovale, malariae, knowlesi*	Streptococcus pneumoniae (invasive)
Borrelia spp	*Clostridium tetani*	*Giardia lamblia*	*Leptospira interrogans*	Polio virus (wild or vaccine types)	*Streptococcus pyogenes* (invasive)
Brucella spp	*Corynebacterium diphtheriae*	Guanarito virus	*Listeria monocytogenes*	Rabies virus (classical rabies and rabies-related lyssaviruses)	Varicella zoster virus
Burkholderia mallei	Corynebacterium ulcerans	*Haemophilus influenzae* (invasive)	Machupo virus	*Rickettsia spp*	Variola virus
Burkholderia pseudomallei	*Coxiella burnetii*	Hanta virus	Marburg virus	Rift Valley fever virus	Verocytotoxigenic *Escherichia coli* (including *E. coli* O157)
Campylobacter spp	Crimean-Congo haemorrhagic fever virus	Hepatitis A, B, C, delta, and E viruses	Measles virus	Rubella virus	*Vibrio cholerae*
Carbapenemase-producing gram-negative bacteria	*Cryptosporidium spp*	Influenza virus	Mumps virus	Sabia virus	West Nile Virus
Chikungunya virus	Dengue virus	Junin virus	*Mycobacterium tuberculosis* complex	*Salmonella spp*	Yellow fever virus
					Yersinia pestis

The control of key communicable diseases[13]

The infectious disease conditions covered in this section have been chosen to reflect existing legal requirements in the UK and/or important epidemiological parameters such as frequency of occurrence, severity, preventability, environmental health focus and public interest and those of interest worldwide identified by the WHO which remain and pose a significant health burden.

Acute encephalitis

This is inflammation of the brain caused by several organisms, the commonest being viruses such as herpes simplex type 1. This is the commonest cause of acute encephalitis in Europe and causes a severe infection with case fatality rates that may be as high as 70%. Encephalitis can also occur as an acute complication of measles and varicella infections.

Incubation periods vary with the causative organism and no specific public health control measures are required in response to cases unless the causative agent is suspected to be rabies.

Acute meningitis

This is inflammation of the lining of the brain (i.e. meninges), caused by a variety of infectious and non-infectious agents. The most common infectious causes are bacteria, viruses, protozoa and helminths. The most common mechanism through which these causative organisms invade the brain is by crossing the blood-brain barrier after establishing a haematogenous focus of infection (i.e. viraemia/bacteriaemia). Incubation periods vary with the causative organism and the clinical illness presents as a syndrome of meningeal symptoms occurring over the course of hours to up to several days. Diagnosis is by isolating the organism from blood or cerebrospinal fluid and also by serological test and scans.

To achieve control and/or prevention all cases of acute meningitis should be promptly notified to public health authorities. Where the causative organism is believed to be a meningococcal infection (caused by the bacterium *Neisseria meningitidis*), the case should be admitted to hospital for immediate assessment, laboratory investigations commenced to confirm their diagnosis and potential antibiotic therapy. Public health action requires the tracing of 'close contacts' who should be offered chemoprophylaxis and vaccination (if the infection is due to a vaccine preventable strain), to prevent early and late secondary cases in the network of close contacts. If an outbreak occurs (for example two or more confirmed or probable cases in a household or educational setting) then wider chemoprophylaxis may be required as a public health action. Many countries have a wide-ranging schedule of vaccination programmes covering serogroups A, C, W135 and Y and most recently introduced in a number of countries for serogroup B.

Acute poliomyelitis

This disease has been eliminated from nearly all countries worldwide, with five out of six WHO regions now certified wild poliovirus free – the African Region, the Americas, Europe, South-East Asia and the Western Pacific. Polio only remains endemic in two countries – Afghanistan and Pakistan. The Global Polio Eradication Initiative, spearheaded by WHO and key partners, has a goal to complete the eradication and containment of all wild, vaccine-related and Sabin polioviruses, such that no child ever again suffers paralytic poliomyelitis.

This is a viral infection of the central nervous system caused by poliovirus types 1, 2 and 3 and mainly affects children under 5 years of age. Most cases are asymptomatic or present with sore throat or diarrhoea and the clinical spectrum can include acute flaccid paralysis, permanent disability or death. Laboratory diagnosis is by isolation of the polio virus in faeces and occasionally nasopharyngeal secretions.

Looking at the source and transmission, humans are the only reservoir. Spread is by the faecal-oral route and is readily transmitted. Long-term carriage does not occur.

The incubation period is usually seven to 14 days but may be up to 35 days and up to 60 days for vaccine-associated cases.

Control and/or prevention requires the implementation of routine primary immunisation of all children. Booster doses are required at ten-year intervals for travel to endemic areas. All suspected cases of acute poliomyelitis should be notified immediately to public health authorities. If confirmed as a sporadic case or outbreak of indigenous polio, then mass vaccination would be required. For vaccine-associated cases, no specific action is required other than a review of vaccine coverage.

Acute infectious hepatitis – hepatitis A

Hepatitis A is an enterically transmitted infection of the liver. The incidence has been decreasing in higher-income countries over the last five decades with the highest incidence in these countries in those aged 15–34 years. The infection leads to a clinical picture that ranges from no symptoms to mild, non-specific nausea and vomiting through to hepatitis (liver inflammation and jaundice) and rarely liver failure. Diagnosis is by serology to identify specific antibodies. There is no specific treatment but passive immunisation with immunoglobulin (HNIG) can confer protection for a limited period.

Humans are the reservoir of infection. Infection is spread primarily by the faecal-oral route from other humans. Secondary transmission in household and educational settings also occurs via this route perhaps aided by transmission via fomites. Outbreaks have been reported in men who have sex with other men (MSM), and intravenous drug users (IVDUs).

The incubation period is 15 to 50 days with a mean of 28 days.

Control and prevention is secured by good personal hygiene especially handwashing, ensuring good toilet hygiene in educational settings such as nurseries and schools, effective sanitary disposal of sewage and provision of safe drinking water and food. There should be vaccination (passive or active) of travellers to countries where the infection is common, relevant household, sexual and other close contacts and individuals with unsafe or high-risk practices such as IVDUs and MSM.

Acute infectious hepatitis – hepatitis B and C

Worldwide, these are important public health issues that cause considerable morbidity and mortality due to the increased risk of chronic liver disease and hepatocellular cancer. The incidence of acute hepatitis B varies considerably across countries worldwide with the UK having one of the lowest rates of acute infection and carriage in the world. Acute hepatitis C infections are largely asymptomatic thus making it difficult to distinguish acute and chronic infections. Chronic carriage of these hepatitis viruses is more common in high-risk groups such as IVDUs, people who regularly change sexual partners, people who receive multiple blood transfusions and renal dialysis patients.

Clinical illness can range from no symptoms to a non-specific prodromal illness, jaundice, fever and malaise. Treatment guidelines exist based on the use of antiviral drugs.

Humans are the only reservoir and so are the source. Hepatitis B and C are transmitted from person to person following contact with blood and body fluids of infected persons including transfusions of blood and blood products, sharing contaminated needles, and unhygienic skin tattooing. Hepatitis B can also be transmitted sexually and from mother to infant during childbirth (vertical spread), but these routes are less effective means of spreading the hepatitis C virus.

The incubation period is six weeks to six months with a mean of 60 to 90 days.

Control is achieved by screening of all blood and blood products including through excluding specific high-risk groups from those who can donate blood. Provide hepatitis B vaccination for infants and/or older children or selective vaccination of high-risk groups and certain occupational groups. There is no vaccine available for hepatitis C, but it is increasingly being treated with a short course (of around 12 weeks) of direct acting antiviral therapy. Provide post-exposure prophylaxis (PEP) using hepatitis B immunoglobulin (HBIG) where indicated. Promote the need to avoid unprotected sexual intercourse and sharing of needles and other injecting paraphernalia.

Anthrax

Anthrax is a potentially serious acute infection caused by *Bacillus anthracis*, an aerobic gram-positive bacterium, which may be fatal. The disease is a zoonosis and has been eliminated in most of Europe but is still endemic in Africa, Asia and countries of the former Soviet Union. Anthrax is an occupational hazard of groups in contact with infected carcases and their products such as skins, hides, hair and wool. There are also concerns that it can be used as a bioterrorism agent. Clinical presentation and disease type is dependent on the route of infection and cases can present with cutaneous, pulmonary and intestinal anthrax. Pulmonary and intestinal diseases are rare. Diagnosis is by a PCR test and culture of swabs from cutaneous lesions, blood and other human samples and treatment is with antibiotics.

All domestic and wild herbivores are potentially at risk of infection. Humans are usually infected from direct contact of cuts/abrasions with infected animal carcases or animal products; inhalation or ingestion of aerosolised spores and in recent years it has emerged that injecting contaminated drugs such as heroin can be a source. Anthrax spores released from infected carcases can remain viable in the soil for decades. There are no known cases of cross infection of inhalational anthrax from human to human.

The incubation period is one to ten days (this varies depending on disease type).

Control and prevention requires that animals (and their products) with confirmed or suspected infections be disposed of safely and effectively. Strict occupational and environmental safety measures should be instituted including the pre-treatment of all animal products.

Anthrax vaccination is recommended for individuals at risk of occupational exposure. In the event of a bio-terrorist release, post-exposure antibiotic therapy is recommended, and post-exposure vaccination can be considered.

Cholera

Cholera is a major public health problem caused by the bacterium *Vibrio cholerae*. It is rare in Europe, but endemic in many developing countries, with seasonal outbreaks of diarrhoeal illness in Sub-Saharan Africa and Southeast Asia. The clinical illness is characterised by abrupt onset of copious amounts of watery diarrhoea, vomiting and abdominal discomfort. Death can occur in severe, untreated cases (up to 50% case fatality rate in severe cases). Diagnosis is by direct microscopy of stool specimen and culture of stool and rectal swabs (although PCR can also be used). Treatment is principally via rapid and appropriate rehydration with oral and/or intravenous fluids. Antibiotic therapy may be used when considered appropriate.

Humans are the only known host and primary reservoir. It is commonly transmitted through the faecal-oral route through ingestion of contaminated water and food by susceptible individuals. Seafood has been the most commonly implicated vehicle in food-borne cholera outbreaks worldwide. Direct person-to-person transmission (without the medium of contaminated food or water) through close contact is possible, but is very rare.

The incubation period is a few hours to five days (usually 24 to 48 hours).

Control and prevention is secured by improved water supply, adequate sanitation and food hygiene measures which are the basis of public and environmental health interventions to prevent and control cholera. Hygiene advice should be provided to cases and contacts. Food handlers should be excluded from work until microbiological clearance is confirmed. Safe and effective oral vaccines have been developed, but are not yet recommended for use in mass vaccination campaigns but can be used where travel takes place to areas where cholera is common and/or visiting remote places without access to medical care.

Diphtheria

Diphtheria is caused by toxin-producing (toxigenic) aerobic bacteria *Corynebacterium diphtheriae* and less commonly by *Corynebacterium ulcerans*. The disease is rare in most countries with well-established immunisation programmes. It may present as classical respiratory diphtheria, nasal and cutaneous diphtheria. Extensive organ damage can occur resulting in neurological and cardiac complications and death if left untreated. The case fatality rate is 5–10%. Diagnosis is by culture from throat and nose swabs, nasopharyngeal/oral secretion and serology from blood. Treatment is with appropriate antibiotics and diphtheria antitoxin.

Humans are the only known reservoir, and transmission occurs via respiratory droplets, contact with nasopharyngeal secretions, or wound exudates in cases of cutaneous disease; rarely from contact with articles contaminated by discharges from infected lesions and contaminated food (raw milk has been known to be a potential vehicle).

The incubation period is two to five days but occasionally is longer.

Control and prevention requires:

- Routine programmes of immunisation for all infants with diphtheria toxoid vaccine;

- Isolation and strict barrier nursing of suspected and confirmed cases; identification and screening of all close contacts; and
- Provision of antibiotic prophylaxis and primary or booster vaccination (depending on vaccination status) to prevent secondary cases;
- Contacts should be kept under surveillance for seven days.

Escherichia coli – Shiga toxin-producing

There are different strains of *E. coli* associated with gastrointestinal illness, but the most serious illness is that caused by Shiga toxin-producing *E. coli* (STEC), also referred to as Verocytotoxin-producing *E. coli* (VTEC). STEC belonging to serogroup O157 (STEC O157) is the commonest serogroup and the most likely to cause bloody diarrhoea and haemolytic uraemic syndrome (HUS) in the UK and other European countries. Infection may cause no symptoms, bloody diarrhoea (haemorrhagic colitis), acute renal failure, HUS, particularly in children; and problems with the blood-coagulation system (thrombocytopenic purpura), particularly in adults. Diagnosis is by isolation of the organism from stool cultures with serology used for retrospective diagnosis, although PCR testing is now regularly used in many countries. Recommended treatment is adequate fluid and electrolyte replacement. Antibiotic therapy is not usually recommended.

The natural reservoir of *E. coli* O157 is the gastrointestinal tract of many mammals, particularly cattle, sheep, goats and horses. Most cases are sporadic, but outbreaks do occur in with links to farms (i.e. open or 'petting' farms), nursery and household settings, particularly in the summer months. Transmission is mainly through ingestion of contaminated food and/or water, direct contact with infected animals, and person to person faecal-oral spread.

The incubation period is from six hours to ten days (typically two to four days).

Control and prevention is achieved by:

- Minimising contamination of animal carcases at abattoirs;

- Effective food preparation practices, particularly ensuring that raw meat is thoroughly cooked;
- Separation of raw foods from foods which are ready to eat to reduce cross-contamination;
- Storage of foods at low temperatures (<5°c);
- Pasteurisation of milk;
- Adequate hygiene practices in all settings where there is a potential for exposure;
- Adherence to health and safety standards by operators of petting farms, which includes providing appropriate facilities and advice to visitors; and
- Exclusion for those identified as being in a high-risk group – see Figure 10.25 – until normal stools have resumed and two consecutive stool samples, taken 24 hours apart are received and test negative.

Giardiasis

This is a common cause of diarrhoea caused by the cyst producing protozoan *Giardia lamblia* (also known as *G. intestinalis* or *G. duodenalis*). Cases occur throughout the year with slight peaks in spring and autumn and the incidence varies across Europe with higher rates in some countries in Eastern Europe. The majority of cases may be asymptomatic carriers or have no evidence of infection. Symptomatic cases tend to present with diarrhoea which may be prolonged, malaise, foul-smelling greasy stools, abdominal cramps and weight loss. Diagnosis is by identification of cysts by microscopy or tests for *Giardia* antigen in stool samples, with PCR also available. Treatment with metronidazole is clinically effective.

When it comes to sources and transmission, humans are the main reservoir for *Giardia lamblia* transmission to other humans although zoonotic transmission is possible. The mode transmission is faecal-oral, either directly or by the contamination of water or food. Direct person-to-person transmission through the faecal-oral route is particularly

common in children and can lead to secondary transmission within families.

The incubation period is usually five to 16 days (but extremes of one to 28 days have been reported).

Control and prevention is achieved by:

- Good personal and food hygiene;
- Exclusion of those in high-risk groups (see Figure 10.25) until 48 hours after the first normal stool;
- Adequate and effective treatment of drinking water supplies including flocculation, sedimentation and filtration processes should prove effective in removing giardia from the water supply.

Food poisoning (see also Chapter 13)

This is a non-specific term used to describe gastrointestinal illness typically characterised by nausea, vomiting and diarrhoea arising from the ingestion of food and/or water contaminated by bacteria, viruses, parasites, toxins or chemicals. Clinical illness is usually mild and resolves in a few days without any specific treatment, but in some instances, severe disease requiring hospitalisation and intensive management may occur. Diagnosis will vary depending on the cause and treatment is not always available or indicated. Some common causes of food poisoning include viruses such as the Norovirus (also known as small round structured viruses [SRSV] or Norwalk virus); bacteria such as *Campylobacter jejuni*, *Listeria monocytogenes*, *Shigella* and *Salmonella* species; and toxins produced by organisms such as *Bacillus*, *Clostridium perfringens* and *Staphylococcus aureus* (see Table 10.12 for more details).

The effective prevention and control of sporadic cases and outbreaks of food poisoning requires the prompt identification and response to incidents and outbreaks and early implementation of appropriate public and environmental health control measures.

Other actions that will be beneficial include promoting the safe handling and processing of food in commercial settings through the adoption of systems such as Hazard Analysis Critical Control Point (HACCP) (see Chapter 13) and the use of legal powers. Statutory powers can also be used in some countries, to exclude people who belong to a high-risk group (see Figure 10.25).

Education on safe handling of food and good and effective personal hygiene practices is needed to reduce faecal-oral transmission, and adequate infection control measures should be instituted in premises such as hospitals and educational settings like nurseries and schools.

Legionellosis

This is a bacterial disease which can be caused by a bacterium *Legionella pneumophila* which can lead to an atypical pneumonia known as Legionnaires' disease and a milder

---

Group A
Any person who is unable to perform adequate personal hygiene due to their lack of capacity or ability to comply, OR lack of access to hygiene facilities.

Group B
All children aged 5 years old or under (up to sixth birthday) who attend school, pre-school, nursery or other similar childcare or minding groups.

Group C
People whose work involves preparing or serving unwrapped ready-to-eat food (including drink).

Group D
Clinical, social care or nursery staff who work with young children, the elderly or any other particularly vulnerable people, and whose activities increase the risk of transferring infection via the faecal-oral route.

Figure 10.25 Groups at risk for ongoing transmission of gastrointestinal infections

form called Pontiac Fever. Legionnaires' disease was first identified after an outbreak of pneumonia among delegates attending an American Legion Convention for service veterans in Philadelphia in 1976, hence the name. Diagnostic tests were developed, and by testing stored specimens it was identified that the disease could be traced back to the 1940s. The infection had escaped detection because the causative organism did not grow on conventional culture media.

Its public health importance lies in its ability to cause outbreaks in community and healthcare settings with a high case fatality rate (10%) in susceptible groups such as the elderly, patients with multiple co-morbidities and the immunosuppressed. Diagnosis is by urine antigen testing and culture of appropriate respiratory specimen on selective media, and serology may also be used. Treatment is with antibiotics.

The source and transmission are environmental. The reservoir is environmental water in which the organism occurs in low concentrations, for example, free standing water, water stored in hot water systems, cooling towers, air conditioning units, hot tub spas, fountain/sprinkler systems and medical equipment-producing aerosols. Humans are infected through inhalation of aerosols or droplet nuclei containing infective doses of the organism. Travel is a major risk factor for Legionnaires'. The infectious dose is unknown, but the risk of disease may be related to the amount of time exposed to the source. Person-to-person spread does not occur. *L. longbeachae* has also been found in gardening composts.

The incubation period is usually two to ten days (usually five days).

The key to control and prevention is to ensure that the appropriate temperature is maintained in domestic and commercial water systems (store hot water at 60°C and deliver above 50°C; store and deliver cold water below 20°C), adequate disinfection, cleaning and changing of water in fountains and whirlpool spas, maintain wet cooling systems in line with recommended standards.

Table 10.12 Selected examples of causative food poisoning organisms*

Organism/Toxin	Clinical features	Incubation period	Transmission	Diagnosis	Treatment	Control measures
Bacillus species (includes *B. cereus* and *B. subtilis*)	An emetic syndrome which presents as nausea, vomiting and abdominal pain OR a diarrhoeal syndrome which presents as diarrhoea and abdominal pain. The symptoms are usually short-lived (approx. 1 day)	1–6 hours OR 6–24 hours	Usually from ingestion of contaminated food items such as rice subjected to inadequate temperature control during storage. Person-to-person spread does not occur.	Cultured from stool or vomit samples. Serotyping possible.	Symptomatic treatment.	Food preparation and storage advice. Consider excluding cases in high-risk groups but microbiological clearance is not required before return to work or school.
Campylobacter jejuni	Sudden onset of nausea, vomiting, foul-smelling watery diarrhoea followed by bloody diarrhoea, abdominal pain and fever.	Within 2–5 days (avg. 3 days)	Usually from ingestion of raw or undercooked meat (particularly poultry), unpasteurised milk and untreated water. Person to person is possible but rare.	Culture of the organism from faecal samples. Serotyping possible. PCR Testing available.	Symptomatic treatment. Consider antibiotics for invasive disease.	Adequate treatment of water supplies, pasteurisation of retailed milk, food preparation and food safety advice, and personal hygiene advice.
Clostridium botulinum – botulism	There may be some initial vomiting and diarrhoea but the main symptoms are neurological with the risk of respiratory failure following muscle paralysis.	Usually 12–72 hours (extremes of 2 hours to 10 days)	Ingestion of preformed toxins in food items contaminated after canning or bottling. Transmission may also occur following ingestion of spores or the introduction of spores into wounds as occurs following the use of contaminated drugs by intravenous drug users (IVDUs). Person-to-person spread has not been reported.	Toxin detection from human samples such as faeces, serum, vomit and gastric fluids. Culture of the organism from faeces, food and gastric fluid. Serotyping. Rapid PCR testing available.	Hospital admission of all cases for investigation and treatment with antibiotics and botulinum antitoxin.	Undertake rigorous public health investigation including a detailed review of food histories. Consider recalling commercially produced food products. Provide appropriate food safety advice to the public and other public health advice to IVDUs if contaminated injectable drugs are suspected as a source of infection.

Table 10.12 (Continued)

Organism/Toxin	Clinical features	Incubation period	Transmission	Diagnosis	Treatment	Control measures
Clostridium perfringes	Watery diarrhoea, abdominal pain, nausea and fever. Cases usually recover within 24 hours.	Usually 8–18 hours.	Usually from ingestion of contaminated food items such as meat and meat products subjected to inadequate temperature control during storage. Person-to-person spread does not occur.	Culture of faecal samples and serotyping.	Symptomatic treatment.	Food preparation and storage advice. Consider excluding cases in high-risk groups but microbiological clearance is not required before return to work or school.
E. coli O157 (see later also)	Abdominal pain/cramps, diarrhoea (sometimes bloody). Infection may lead to a life-threatening disease, such as haemolytic uraemic syndrome (HUS) characterised by acute kidney failure. Low infectious dose.	12 hours–10 days.	Faecal-oral route. Primary sources of EHEC (Enterohaemorrhagic E. coli) outbreaks are raw or undercooked meat and meat products, raw milk and faecal contamination of vegetables. Faecal contamination of water and other foods as well as cross-contamination during food preparation. Acid tolerant and will persist in frozen and chilled conditions.	Culture of faecal samples and serotyping. PCR Testing is also available.	Fluids and ensure re-hydration. Anti-biotics not recommended.	Good personal hygiene and hygienic food preparation. Advice on hygienic food preparation.
Hepatitis A	See later					
Norovirus	See later					

(Continued on next page)

Caryn L. Cox

Table 10.12 (Continued)

Organism/Toxin	Clinical features	Incubation period	Transmission	Diagnosis	Treatment	Control measures
Salmonella (excluding typhoid and paratyphoid)	Variable severity with most cases presenting with loose non-bloody diarrhoea, fever abdominal pain, nausea and malaise.	Usually 12–36 hours (range: 6 to 72 hours)	Ingestion of contaminated food items particularly undercooked poultry and eggs. Person-to-person spread via the faecal-oral route Rarely, direct contact with infected animals including exotic pets and contact with fomites.	Culture from faecal samples, rectal swab or blood. Serotyping.	Symptomatic treatment.	Implement a 'farm to fork' prevention strategy that includes vaccination of poultry flock to reduce infection and carriage in farm animals, use of Hazard Analysis and Critical Control Point (HACCP) systems to ensure food safety in commercial food processing sites, and provision of adequate personal hygiene and food preparation advice to the public. Exclude cases in risk groups until 48 hours after first normal stool. Enteric precautions for cases admitted to hospital.
Shigella	Abdominal pain and diarrhoea (initially watery), nausea, fever and vomiting. Occasionally the diarrhoea may become bloody (dysentery).	Usually 12 to 96 hours.	Person-to-person spread via the faecal-oral route. Ingestion of contaminated food or water.	Culture from faecal samples. Serotyping. PCR Testing is available.	Symptomatic treatment. Moderate to severe cases may need antibiotic treatment.	Safe disposal of human faeces and adequate treatment of drinking water supplies. Reinforce personal hygiene and food preparation advice. Exclude cases from work or school until well (preferably until no diarrhoea for 48 hours).
Staphylococcus aureus	Abrupt onset of nausea, cramps, vomiting, diarrhoea and prostration. Illness 1–2 days.	Usually 2–4 hours with a range of 1–7 hours.	Ingestion of enterotoxins in contaminated food items such as cooked meats and sandwiches.	Culture of food samples, vomit and faecal samples. Detection of enterotoxins in food items. Serotyping.	Symptomatic treatment.	Provision and adherence with food preparation and storage advice. Personal hygiene advice and exclusion of food handlers with purulent skin lesions.

* Note: This is not an exhaustive list.

Leptospirosis

This is a zoonotic disease caused by bacteria of the genus *Leptospira*. In the UK, the most commonly isolated serovars are *L. Hardjo* and *L. icterohaemorrhagiae*, which are associated with cattle and rats respectively. Although all persons are susceptible, the disease is mainly seen in men as a result of greater occupational and recreational exposure to infected animals and the contaminated environment.

The clinical course of the disease is highly variable and ranges from asymptomatic disease to an abrupt onset of headache, myalgia, and fever with meningism, renal and vasculitic manifestations observed in the later phase of disease. A more severe form of leptospirosis characterised by liver and kidney impairment is known as Weil's disease. Diagnosis is via serology, but culture and direct visualisation of the organism is possible with PCR techniques now available. Early treatment with antibiotics may reduce the severity of the illness.

The main reservoirs of infection for transmission to humans are wild animals and domestic animals, with rats being regarded as the most common reservoir worldwide. Transmission to humans occurs following contact with food, water, soil and other material contaminated with the urine of infected animals. Person-to-person transmission is rare.

The incubation period is usually from seven to 13 days (extremes of two to 30 days have been reported).

Control and prevention is achieved by good control of rodents and educational campaigns to avoid exposure to contaminated areas. Suitable and adequate protective clothing to occupational risk groups (e.g. sewage workers) should be provided. Vaccination of people with specific high-risk occupations may be available in some countries.

Leprosy

Leprosy is a chronic infectious disease caused by *Mycobacterium leprae*. The disease occurs in almost all tropical and warm temperate countries with several countries but there are only a few pockets of hyperendemic disease occurrence in Africa, Asia and South America. The disease mainly affects the skin and peripheral nerves causing localised or extensive damage that can result in deformation and disability. Diagnosis is mainly by careful clinical examination. Laboratory diagnosis involves identifying the organism in skin smears or biopsy material. Treatment is with long-term multidrug antimicrobial therapy. Supportive care and rehabilitation are also important components of leprosy management.

Humans are the only known reservoir. The exact mechanism of transmission of leprosy is not known and the most widely held belief is that transmission is via person-to-person contact but recently, the possibility of transmission by the respiratory route is gaining ground.

The incubation period varies from a few months to several years.

Control and prevention is achieved by prompt identification, notification and treatment of cases. Multi-drug therapy treatment is effective and readily undertaken, which has significantly changed the management of people with leprosy and removed the need to isolate people. All household contacts should be examined as well as first- and second-degree relatives because early identification and treatment of cases is the main preventative measure.

Norovirus

Norovirus (NoV, also known as small round structured virus [SRSV] or Norwalk-like virus) is the most common cause of gastroenteritis in Europe and a common cause of viral food poisoning. Outbreaks tend to occur in semi-closed settings such as hospitals, nursing or residential care homes, cruise ships and educational settings such as schools and nurseries. In the UK, hospital outbreaks are more common in the winter (colder) months which has led to the media calling this the 'winter vomiting bug'. All age groups are affected with severe and significant infection more common in the elderly. The clinical illness is relatively mild and characterised by nausea, diarrhoea, vomiting and fever. Cases usually recover fully within two–three days with no long-term complications. Diagnosis is now usually undertaken by antigen testing or PCR testing of a stool specimen.

The source of the disease is humans who are the only known reservoir of this highly infectious organism. Transmission may occur via ingestion of contaminated food and water, direct person-person spread via the faecal-oral route or indirect spread from contaminated environmental surfaces. Transmission is also possible via aerosolised vomit.

The incubation period ranges from six to 72 hours, but usually 24 to 48.

Control and prevention requires:

- Good standards of personal and food hygiene;
- Adequate infection control measures in affected settings including effective environmental cleaning;
- Exclusion of all cases in the high-risk groups in Figure 10.25 until 48 hours after resolution of illness (diarrhoea and vomiting).

Plague

Plague is a serious bacterial infection caused by *Yersinia pestis*, which historically caused large epidemics of what was known as the 'Black Death'. Today, human plague occurs mainly in parts of Africa, South America and Asia but cases are still reported in some parts of the states of the former Soviet Union. Depending on the type, plague is an illness with rapid and severe onset of high fever, bleeding, breathing problems and death if left untreated. Diagnosis is by direct microscopy of clinical specimen and serology with the definitive test being phage lysis of culture isolates. Treatment is with antibiotics.

The source and route of spread is wild rodents and in particular rats are natural reservoirs as the result. Bubonic plague results from the bite of an infected flea (an ectoparasite of rats). This is most frequently *Xenopsylla cheopis* but other species of flea have been implicated. Pneumonic plague is transmitted via the respiratory route from person to person.

The incubation period is one to seven days (but ten to 15 hours for pneumonic plague).

Cases will be infectious until 48 hours after commencing treatment and they have responded to that treatment. Control requires the identification of close contacts and the use of antibiotic chemoprophylaxis. Consideration should be given to vaccinating high-risk workers such as rodent control and laboratory workers. During outbreaks, intensive rodent and flea control programmes should be instituted.

Rabies

A viral infection of the central nervous system caused by a lyssavirus. Rabies in animals currently occurs in all continents, except Antarctica. Individual countries, including most in Western Europe, are reported to be rabies free in terrestrial animals; however, bats are a potential source of infection in the UK and other rabies-free countries. Rabies is almost always fatal once symptoms appear. Diagnosis is only possible after the onset of symptoms and is usually based on the clinical history and examination. However, serology and isolation of the rabies virus or viral antigen can also be used. Once a rabies infection is established, there is no effective treatment.

Rabies is enzootic in animal populations and animal reservoirs include dogs, cats, foxes, bats and skunks. Transmission to humans usually occurs from contact with saliva, through a bite, scratch, or licks of mucous membranes or broken skin. Transmission is affected by the size of the inoculum and is more likely if bites occur through bare skin, rather than through clothing. Airborne spread has been demonstrated in bat caves, but this is unusual. Person-to-person transmission has only been observed in cases of organ transplantation, including corneal and solid organ transplants.

The incubation period can vary from around four days to 19 years (but is usually around three to 12 weeks) – the significant variance in the incubation period is due to the amount of virus introduced, severity of the wound and the proximity to the central nervous system.

For control and prevention environmental control measures are required. These include vaccination of domestic animals and those imported from endemic areas (including the use of effective quarantine measures with sufficient length of time) and in enzootic areas, culling of stray animals is sometimes undertaken.

Pre-exposure vaccination of travellers to endemic areas and those at occupational risk assists control. Early provision of PEP with vaccine, the prompt and thorough cleansing of wounds, and where indicated, rabies immunoglobulin also contributes to effective prevention and control.

Influenza (including pandemic and avian influenza)

The influenza virus (or 'flu' virus) is a ubiquitous and highly infectious cause of acute respiratory infection. The virus is categorised into influenza A, B and C with influenza A and B considered to be the main pathogens responsible for most influenza illnesses. Influenza virus infection has a worldwide distribution with seasonal epidemics. In the UK, these seasonal epidemics (seasonal influenza) occur between October and March (the colder, winter months) and affects all age groups and last on average six–ten weeks, with a peak seen around week four.

The influenza virus is also capable of causing severe pandemics (pandemic influenza) following an antigenic shift which results in the development of a new subtype to which there is little or no pre-existing immunity globally. Influenza A also causes infections in animals such as birds (avian influenza), horses (equine influenza) and pigs (swine influenza) which can occasionally be transmitted to humans leading to large and severe outbreaks and pandemics.

Influenza infection in humans presents as an acute illness characterised by a rapid onset of fever, cough and malaise with additional symptoms such as headache, myalgia, sore throat, runny nose and anorexia. Infrequently, vomiting and diarrhoea may occur. The clinical illness is usually more severe in the elderly, who account for most hospitalisations and deaths, although this pattern may be reversed with a pandemic strain of the virus. Diagnosis is usually based on clinical features, particularly during periods of seasonal epidemics and pandemics with rapid laboratory confirmation of diagnosis usually undertaken by PCR testing of nasopharyngeal aspirates, nasal or throat swabs. There is no definitive treatment, but antiviral drugs can be used if given early enough, to shorten the duration and lessen the effects of the illness.

Seasonal influenza is a human disease arising from influenza A or B subtypes that are already in circulation in human populations. Infection can occur all year round but there is a clear seasonal pattern with varying epidemic cycles in the northern and southern hemispheres.

Pandemic influenza occurs when a novel influenza virus is in circulation in human populations to which there is little or no pre-existing immunity. Pandemics are capable of causing widespread and significant illness worldwide but can also be either mild or severe in the illness and death they cause, and the severity of a pandemic can change over the course of that pandemic. The first of the 21st century was in 2009 and declared a pandemic by WHO in June 2009; by November worldwide more than 206 countries and overseas territories or communities had reported 503,536 laboratory confirmed cases of pandemic influenza, including over 6,250 deaths. The pandemic was caused by a novel virus H1N1, that originated from the animal virus (swine influenza) and caused an illness of mild to moderate severity with most deaths occurring among younger people, including those who were otherwise healthy.

Previous pandemics occurred in the 20th century including:

- Spanish pandemic between 1918 and 1919 with an estimated mortality of between 50 and 100 million people;
- Asian pandemic (H2N2) between 1957 and 1958 with an estimated mortality of between 1 to 4 million people;
- Hong Kong pandemic (H3N2) between 1968 and 1969 with an estimated mortality of between 0.75 and 1 million.

Avian influenza – This is a disease of birds caused by influenza A and the natural reservoirs are wild aquatic birds such as ducks and geese. Domestic and commercial poultry flock are vulnerable to avian influenza infection, with several large outbreaks in the last two decades in several countries worldwide. It is of concern to humans due to the potential to jump the species barrier and infect humans, thus triggering a pandemic. Transmission to humans has rarely occurred with the few reported cases being infected with certain subtypes, particularly highly pathogenic A/H5 and A/H7 subtypes. Highly pathogenic avian influenza (HPAI) A/H5N1 has received a lot of attention in the last decade following the occurrence of large poultry outbreaks that resulted in the culling of millions of poultry and the occurrence of associated human infections with a high case fatality rate (~ 60%) in previously healthy young adults.

The source of influenza B is humans as only humans are affected. The reservoirs for influenza A are animals, particularly wild aquatic birds with the potential for direct spread of new strains to humans or through intermediaries such as pigs. Transmission within humans is via the respiratory secretions of cases mainly via airborne droplet spread but may also occur via direct or indirect contact with cases.

The incubation period is short – usually one to three days (but can be up to five days); the average for type A is 34 hours and 14 hours for type B.

For control and prevention, a WHO-recommended seasonal influenza vaccine is developed each year and used to immunise persons judged to be at increased risk of serious illness from influenza. For example, this should include the elderly, pregnant women and persons with chronic underlying diseases. Immunisation is also offered to health and social care workers to protect patients and staff. In 2009 a vaccine was developed quickly to control the H1N1 pandemic strain that year.

Cases should be encouraged to adhere to basic infection control and personal hygiene measures such as handwashing and safe disposal of contaminated material. During epidemics and pandemics, antiviral drugs may be used as post-exposure prophylaxis and social distancing measures such as exclusion from work and school closures may be considered where appropriate.

Severe Acute Respiratory Syndrome (SARS)

SARS is an acute viral respiratory infection caused by a previously unrecognised coronavirus named the SARS-coronavirus (SARS-CoV). SARS was the first severe new disease to emerge in the 21st century and with a global outbreak in 2002/2003 that predominantly affected mainland China, Canada, Hong Kong, Singapore, and Taiwan and led to 8098 cases and 774 deaths. The clinical presentation of SARS is consistent with that of an atypical pneumonia with rapid respiratory deterioration requiring hospitalisation and isolation. Commonly reported symptoms include fever, cough, headache, myalgia, malaise, chills and rigor. Diagnosis is via serology or PCR detection of RNA. There is currently no definitive drug treatment, but support is given, for example a ventilator to assist with breathing.

Preliminary research indicates that SARS-CoV may have originated in livestock or small mammals. The virus is transmitted by close person-to-person contact and large respiratory droplets. Asymptomatic patients are not considered infectious, and cases are no longer considered infectious ten days after fever resolution.

The incubation period is two to ten days (average five days) but periods of up to 14 days have been noted.

Control and prevention requires prompt implementation of public health measures for control of infectious diseases that include:

- The early identification and isolation of all cases;
- Optimising infection control measures particularly in hospitals;
- Monitoring close contacts of cases; and
- Population level surveillance.

COVID-19

Coronaviruses are a large family of viruses that can cause illness in animals or humans. In humans there are several known coronaviruses that cause respiratory infections, and COVID-19 is caused by the SARS-CoV-2 virus. The symptoms are variable, but can include fever, cough, sore throat, headache, fatigue, runny nose, muscle aches, diarrhoea, breathing problems and loss of smell and taste. Symptoms may begin one to 14 days after exposure to the virus. At least a third of people who are infected do not develop noticeable symptoms.

The estimated incubation period is between two and 14 days with a median of five days. It is important to note that some people become infected and do not develop any symptoms or feel unwell.

SARS-CoV-2 spreads from person to person through close communities. Modes of SARS-CoV-2 transmission are now categorised as inhalation of virus, deposition of virus on exposed mucous membranes, and touching mucous membranes with soiled hands contaminated with virus. When people with COVID-19 breathe out or cough, they expel tiny droplets that contain the virus. These droplets can enter the mouth or nose of someone without the virus, causing an infection to occur. The most common way that this illness spreads is through close contact with someone who has the infection.

So far as control of the spread is concerned (and the existence of new variants) there remain knowledge gaps. However, available evidence seems to demonstrate that existing recommendations to prevent SARS-CoV-2 transmission remain effective. These include physical distancing (six feet or 1.8 m), community use of well-fitting masks (e.g. barrier face coverings, procedure/surgical masks), adequate ventilation, and avoidance of crowded indoor spaces. These methods will reduce transmission both from inhalation of virus and deposition of virus on exposed mucous membranes. Transmission through soiled hands and surfaces can be prevented by practicing good hand hygiene and by environmental cleaning.

Vaccines (including boosters) are available but globally, availability, distribution and take-up have been variable. Vaccines are important for reducing the severity of the illness and allied with other public health measures reduce spread.

Toxoplasmosis

This is a zoonosis with worldwide distribution that is caused by the protozoan parasite *Toxoplasma gondii*. Human exposure is common with the risk of severe complications highest in pregnant women and immunocompromised individuals. Infection is usually asymptomatic with symptomatic cases presenting with features that range from non-specific flu-like illness and painless lymphadenopathy to severe illnesses such as ocular, congenital and acute disseminated toxoplasmosis. Diagnosis is via serology.

The source and route of transmission is via the domestic cat (and other wild felines) which is the definitive host and transmits the infection through faecal shedding of an oocyst, with humans acting as intermediate hosts. Transmission occurs following contact with an oocyst in the environment for example in the soil and ingestion of contaminated food. Vertical transmission to the foetus occurs in women who acquired their primary infection during pregnancy. Direct person-to person spread has not been observed.

The incubation period is usually ten to 25 days.

Control and prevention requires

- The provision of information and education, particularly to pregnant women to avoid raw or undercooked meat and exposure to contaminated environments;
- Treatment of pregnant women to reduce the risk of vertical transmission; and
- Provision of chemoprophylaxis in immunosuppressed patients.

Play areas where cats may defecate, for example sandpits, should be covered to prevent their use by cats. Hands should be washed after handling cat litter with faeces and faeces disposed of regularly and effectively.

Typhoid and paratyphoid (also known as enteric fever)

A severe systemic infection caused by *Salmonella typhi* or *Salmonella paratyphi (A, B or C)* which causes a similar, but less severe illness. Infection is worldwide with the highest incidence reported in south-central and southeast Asia. Most UK cases are imported from these endemic regions and cases are more common in late summer and in spring. Fever is the earliest symptom and is soon followed by a headache, abdominal pain/tenderness, anorexia, myalgia and fatigue. Definitive diagnosis is via culture from a normally sterile site although serology can also be used. PCR testing of faeces is also available in many laboratories but requires follow-up stool cultures. Untreated typhoid has a case fatality rate of

10 to 15%, effective and timely treatment reduces this to below 1%.

So far as the source and transmission of infection is concerned humans are the only reservoir for *S. Typhi* and the main reservoir of *S. Paratyphi*. The infection is spread through the ingestion of food or water contaminated with faeces from a human case or carrier. Transmission can also occur from person to person via the faecal-oral route and rarely, between MSM.

The incubation period is from three to 60 days, with an average range of eight to 14 days. For paratyphoid it is one to ten days.

Control and prevention requires the maintenance of adequate levels of sanitation, particularly around the safe disposal of sewage and provision of safe drinking water. Enteric precautions and good personal hygiene for cases is to be encouraged. Exclude all cases until clinically well for 48 hours with formed stools. Cases that belong to a risk group (see Figure 10.25) should be excluded until three consecutive negative faecal specimens taken at 48-hour intervals and commencing one week after completion of antibiotic treatment. Identify and undertake a risk assessment relating to close contacts of cases; they may, depending on the risk assessment then need screening through faecal specimens but this depends on the risk assessment and whether contacts belong to risk groups.

Viral haemorrhagic fevers (VHF)

These describe a group of distinct and highly infectious RNA viruses that cause severe, life-threatening febrile illnesses. These infections are endemic in Africa, South America, some parts of Asia, the Middle East and Eastern Europe. The clinical course of VHF infections range from an abrupt onset, rapid progression and clinical deterioration seen in Marburg and Ebola infections, to the multiphasic clinical pattern observed in Crimean Congo Haemorrhagic Fever (CCHF) infections, and the gradual onset, less severe haemorrhagic illness caused by the Lassa fever virus. Case fatality is between 2% and 90% depending on the virus. Definitive diagnosis is based on virus isolation, electron microscopy, virus antigen detection and serology. The use of antiviral drugs is only recommended for Lassa fever and CCHF.

So far as the source and spread of the disease is concerned the origin and natural reservoir of Marburg and Ebola viruses still remain uncertain with current research efforts focusing on the role of fruit bats. Transmission to humans can occur through direct contact with infected non-human primates with person-to-person spread identified as the main mode of transmission of infection during outbreaks.

CCHF is a tick-borne zoonotic infection and Lassa fever is transmitted following the ingestion or inhalation of food, and other materials contaminated with the infectious droppings or urine of rodents which are the primary reservoir. Person-to-person transmission occurs through direct contact with the blood and other bodily secretions of infected persons, aerosolised spread, sharing of food, drink, clinical instruments, objects and utensils.

The incubation period is two to 21 days, depending on the causative organism.

Control and prevention requires:

- Education of the affected population about the nature of the infection and necessary outbreak containment measures;
- Strict infection control measures, for example barrier nursing, isolation of suspected cases; and
- Safe disinfection or disposal of contaminated equipment and material;
- Case contacts or individuals with exposure in laboratories should be placed under health surveillance for 21 days after the last exposure and undergo a risk assessment should they become symptomatic.

Viral gastroenteritis

This is caused by a variety of viruses including Norovirus (see previous), Sapovirus (formerly known as human classic calicivirus), Rotavirus, Adenovirus and Astrovirus. These viruses are a major cause of acute gastroenteritis in children and are frequently identified in outbreaks occurring in hospitals and elderly care institutions. Commonly reported symptoms include abdominal pain, nausea, vomiting and diarrhoea. Diagnosis is by electron microscopy, serological tests (such as ELISA), PCR.

So far as the source and transmission are concerned, humans are the only known reservoir for the common strains of Norovirus, rotavirus, sapovirus, astrovirus and adenovirus that commonly infect humans and the viruses are excreted in faeces and vomit. Transmission is from person to person, either directly by the faecal-oral route, or indirectly by contamination of the environment, food or water. Infected food handlers can spread the infection for up to 48–72 hours after they recover and secondary household transmission following food-borne outbreaks is common.

The incubation period is six to 72 hours (up to a few days in some instances).

For control and prevention good standards of personal and food hygiene and effective disinfection of contaminated surfaces and areas is required. Good standards of infection control in hospitals and care homes is also necessary. Food handlers should be excluded from work immediately and for at least 48 hours after symptoms subside. Vaccination is now available in some countries for rotavirus.

Yellow fever

This is an acute viral infection that is endemic in Central Africa and parts of Central and South America. The clinical illness ranges from an inapparent illness to mild, severe and malignant presentations with the distal end of the spectrum occasionally resulting in death. Diagnosis is via isolation of the virus and serology, with PCR available in some laboratories.

Yellow fever is a zoonosis, and the natural reservoirs are wild primates and humans.

Transmission of the virus occurs via the bite of forest canopy mosquitoes (sylvatic yellow fever) or via transmission from a viraemic case following the bite of the *Aedes aegypti* mosquito in urban areas (urban yellow fever).

The incubation period is three to six days.

Control and prevention is achieved by controlling mosquito vectors in urban areas and mass immunisation with the yellow fever vaccine in endemic countries. Travellers to these countries should be vaccinated before travel.

Yersiniosis

An important cause of intestinal infection in several countries is *Yersinia enterolitica*. The infection affects all ages with the highest number of cases reported in children aged less than 5 years of age and peak incidence in autumn and winter. Clinical illness includes diarrhoea which is occasionally bloody, fever, vomiting and abdominal pain with the severe cases developing complications such as arthritis and septicaemia. Antibiotics are not usually required but may be helpful in cases that develop complications.

So far as the source and route of transmission is concerned the reservoir is the gastrointestinal tract of many mammals, particularly pigs who are asymptomatic carriers. Transmission occurs via ingestion of contaminated food and water and direct contact with infected animals, particularly in certain occupational groups such as abattoir workers. Blood-borne transmission from contaminated blood and person-to-person spread can occur.

The incubation period is usually three to seven days but can be one to 12 days.

Control and prevention is achieved by excluding cases and symptomatic contacts of cases in high-risk groups (see Figure 10.25) until 48 hours after normal stools. Standard enteric precautions should be used by cases. Good infection control measures are required in abattoirs, effective food preparation and hygiene, protection and treatment of drinking water supplies and pasteurisation of dairy products.

Section 6 notes

11 The COVID-19 pandemic is continuing at the time of publication and key data and information can be sought from https://www.who.int/emergencies/diseases/novel-coronavirus-2019
12 The statistics can be found at https://www.ons.gov.uk/peoplepopulationandcommunity/birthsdeathsandmarriages/deaths/datasets/weeklyprovisionalfiguresondeathsregisteredinenglandandwales
13 The International Health Regulations 2005 (IHR (2005)) (https://apps.who.int/iris/bitstream/handle/10665/246107/9789241580496-eng.pdf?sequence=1) have the purpose of preventing, protecting against and providing a public health response to the international spread of disease. The IHR include procedures for determination of public health emergencies of international concern. There are specific provisions for certain diseases such as yellow fever and the latest amendment provides that the period of protection from vaccination with an approved vaccine against the disease will be for life rather than ten years as previously required.

11

Understanding and responding to the health and climate emergency

Emma L. Gillingham, Helen L. Macintyre,
Raquel Duarte-Davidson and Revati Phalkey

Introduction

What is the difference between weather and climate?

'Weather' refers to atmospheric conditions for a particular place and time in the short term (e.g. minutes to monthly temperature, precipitation, humidity, wind, cloud cover, visibility or sunshine). 'Weather' is what you might see outside on any particular day and is usually related to a specific location. 'Climate' is the average of these conditions over much longer timescales, usually taken over at least a 30-year period, and usually considers a much wider area than the weather does, such as regions, whole countries or even groups of countries.

What is climate change and what is environmental change?

'Climate change' refers to large-scale and long-term shifts in the climate of the planet. The Intergovernmental Panel on Climate Change (IPCC) is the United Nations body for assessing the science related to climate change, and is an internationally accepted authority on the topic, with thousands of scientists and expert reviewers compiling key findings into reports produced periodically (Box 11.1). The IPCC defines climate change as:

> a change in the state of the climate that can be identified (e.g. by using statistical tests) by changes in the mean and/or variability of its properties and persists for an extended period, typically decades or longer.
>
> [1]

Changes in the climate can be due to natural factors such as volcanic eruptions or changes in the orbit of the Earth, as well as human-driven (i.e. 'anthropogenic') activity such as emissions of greenhouse gases (GHGs) or land use changes. 'Environmental change' includes climate change, and some of its impacts, as well as broader changes that may be unrelated or partly driven by climate change. Examples of environmental changes include climate change, biodiversity loss and subsequent ecosystem disruption, urbanisation, freshwater changes, desertification, and may also refer to changes in the social and cultural environment too.

Greenhouse gases and the greenhouse effect

When solar energy reaches Earth, about one-third is reflected back into space whilst the rest is absorbed by the land and oceans, warming them, and being partly re-emitted back to the atmosphere. Some atmospheric gases are strong absorbers of this re-emitted energy and are known as 'greenhouse gases' (GHGs), as they trap energy in a similar way to the glass of a greenhouse capturing sunshine and becoming warmer, known as the 'greenhouse effect'. The most important GHGs include carbon dioxide (CO_2), methane (CH_4), ozone (O_3) and nitrous oxide (N_2O), and these have both natural and anthropogenic sources. A major source of

GHGs is fossil fuel combustion (coal, oil and natural gas) for energy generation for homes, industry/manufacturing and transport; they are also produced from agriculture, land use and forestry changes, and wetlands. If there were no GHGs, the Earth's surface would be too cold to sustain life, but very high levels of GHGs would lead to a much warmer planet. Many GHGs can stay in the atmosphere for decades, centuries or even millennia, and since the beginning of the industrial revolution, anthropogenic activities (e.g. burning fossil fuels) have increased levels of GHGs in the atmosphere (Figure 11.1a), resulting in more energy becoming trapped and causing the Earth's temperature to rise. The IPCC concluded in their recent Sixth Assessment Report that *observed increases in well-mixed greenhouse gas (GHG) concentrations since around 1750 are unequivocally caused by human activities* [2].

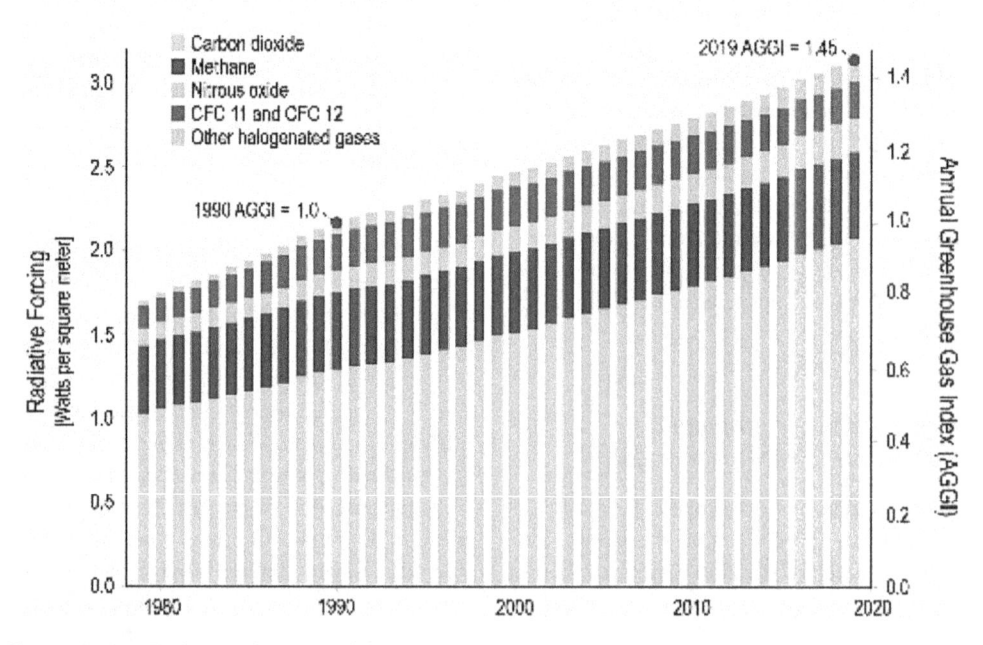

Figure 11.1a Radiative forcing (change in amount of solar radiation trapped by the atmosphere) and annual greenhouse gas index (AGGI), showing that warming influence of long-lived greenhouse gases in the atmosphere increased by 45% between 1990 and 2019[1]

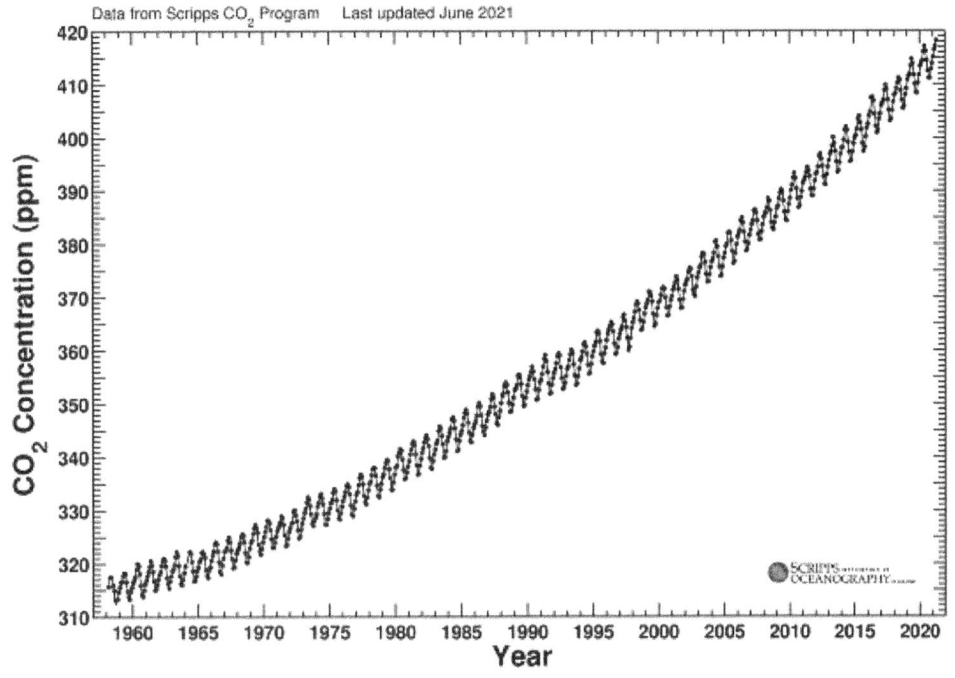

Mauna Loa Observatory, Hawaii
Monthly Average Carbon Dioxide Concentration

Data from Scripps CO_2 Program   Last updated June 2021

Figure 11.1b   Monthly average atmospheric carbon dioxide concentration from 1958–2021 at Mauna Loa Observatory, Hawaii. CO_2 concentration is in parts per million by volume (ppm)[2]

CO_2 levels have been measured continuously since 1958 at the Mauna Loa observatory in Hawaii (Figure 11.1b). When measurements were started, concentrations of 316 parts per million (ppm) were recorded. Pre-industrial levels have been estimated around 280 ppm, and over time, CO_2 levels have risen. As of May 2021, the monthly mean atmospheric concentration was 419 ppm (Figure 11.1b).

Observed changes in our climate

Each of the last four decades has been successively warmer than any preceding decade since 1850 (Figure 11.2). Compared to 1850–1900, global surface temperature was 0.99°C (range: 0.84–1.10°C) higher during 2001–2020, and 1.09°C (range: 0.95–1.20°C)

higher in 2011–2020 [2]. Temperature changes in specific locations can be much greater than the global average; for example, over land (rather than ocean) the increase is already 1.59°C (range 1.34°C to 1.83°C), and there is high confidence that the Arctic will warm at double the global mean rate [2]. In the UK, the top ten warmest years since 1884 have occurred since 2002 [3].

Global mean sea level rose 20 cm from 1901 to 2018 [2], predominantly driven by melting ice (glaciers and ice sheets) and thermal expansion, with smaller contribution from land water storage. The rate of change has accelerated from 1.3 mm yr^{-1} between 1901–1971, to 3.7 mm yr^{-1} between 2008–2018. Unlike some other effects of climate change, sea level rise is hard to reverse and will continue past the end of this century even with strong emission reductions. The

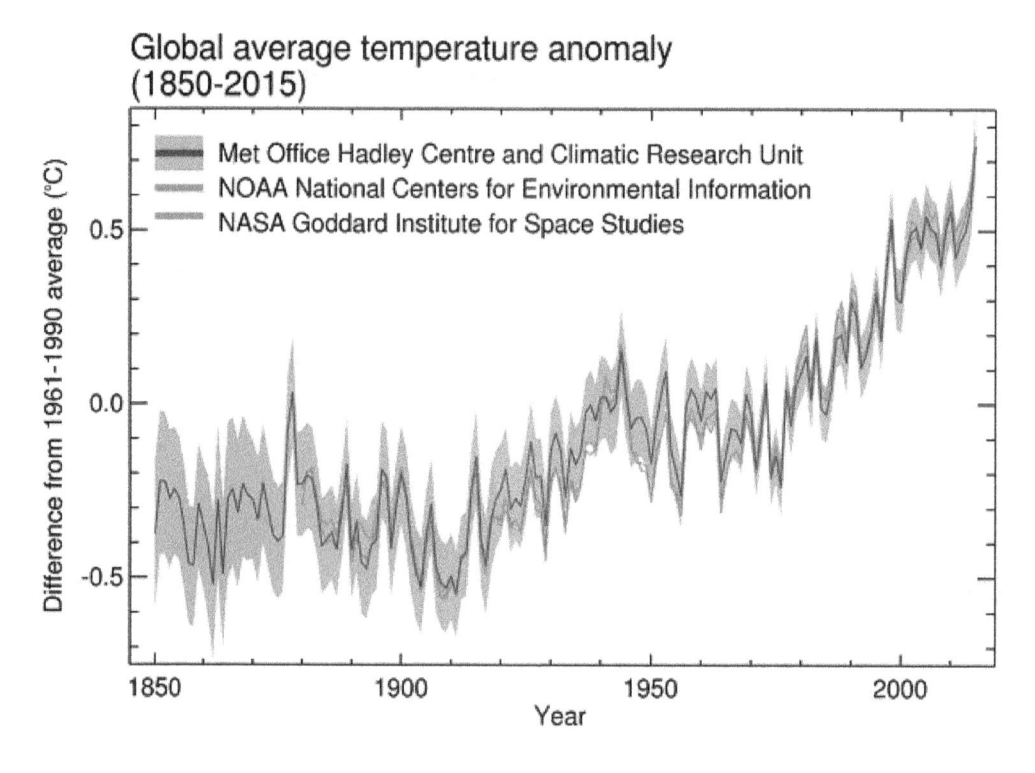

Figure 11.2 Global average temperature anomaly compared to the 1961–1990 mean[3]

upper ocean has warmed since the 1970s, CO_2 emissions are driving acidification of the upper ocean and oxygen levels have dropped in the upper ocean in many regions since the 1950s [2].

Mid-latitude storm tracks have shifted poleward since the 1980s, and global mean precipitation over land has increased, with implications for flooding. In the UK, winters have been 12% wetter (2009–2018 compared to 1961–1990), and rainfall on extremely wet days has increased by 17% [3]. In parallel, some areas globally have become drier and more at risk of drought and subsequent wildfires (e.g. the extreme drought in the western US). Growing seasons have lengthened up to two days per decade since the 1950s in the Northern Hemisphere extra-tropics, and climate zones have shifted poleward in both hemispheres, with implications for changes in ecological systems [2].

Climate scientists have found that it is not possible to explain the observed changes in our climate from natural effects only, and that human activities like emissions of GHGs and land use changes are overwhelmingly responsible for observed climate changes. The IPCC states

It is unequivocal that human influence has warmed the atmosphere, ocean and land. Widespread and rapid changes in the atmosphere, ocean, cryosphere and biosphere have occurred.

[2]

Future climate change

Different emission scenarios are used to project likely future levels of GHGs, and subsequent changes in climate. These are called 'Representative Concentration Pathways'

(RCPs) and describe different climate 'forcing' emission pathways. These are often combined with 'Shared Socioeconomic Pathways' (SSPs), describing how different broad societal choices affect future GHG emissions. The RCPs and SSPs are used in combination to project possible future climate change outcomes under different scenarios. The IPCC has produced an 'Interactive Atlas' tool for flexible analyses of observed and projected climate change information, including regional synthesis for Climatic Impact-Drivers.[4] The frequency and intensity of extreme heat events, heavy precipitation and droughts are projected to increase in future, with the magnitude of change dependent on future global warming level (Figure 11.3).

The UK Met Office has produced a set of bespoke climate projections for the UK

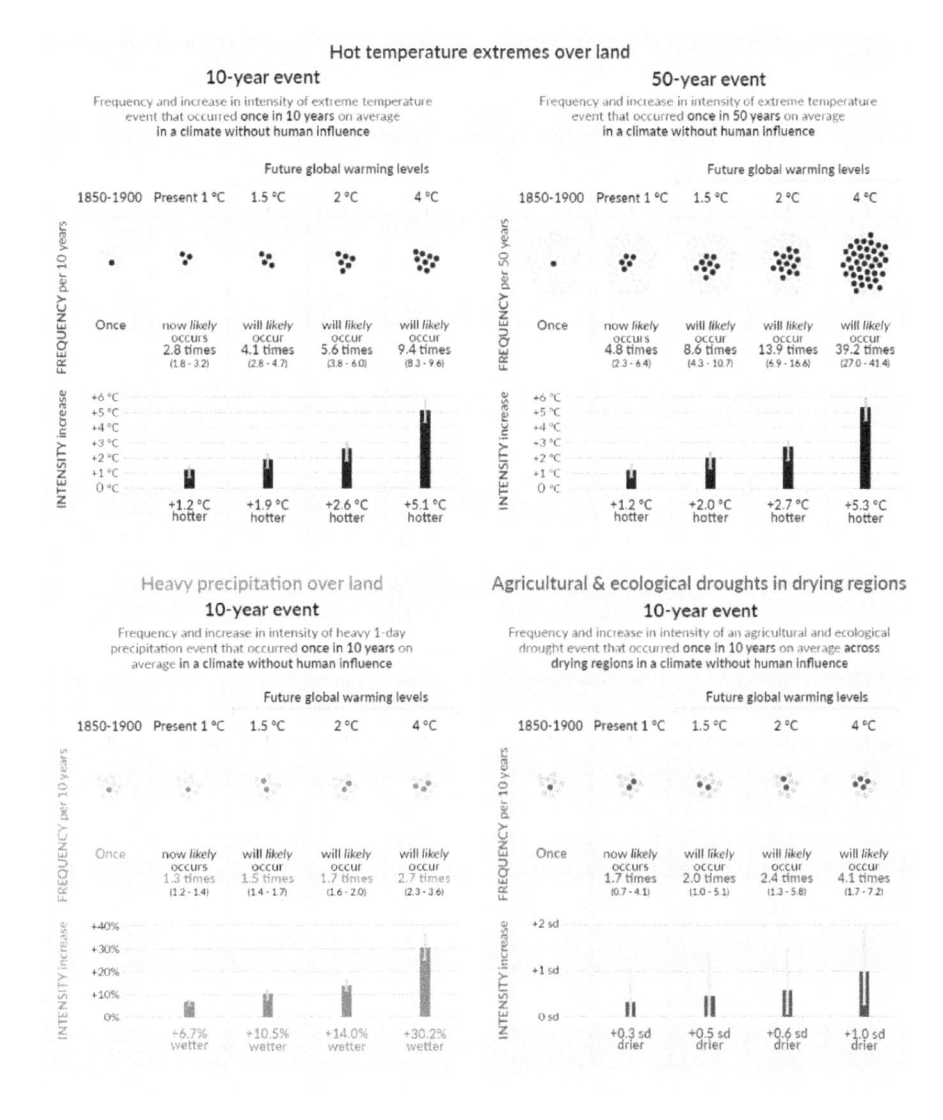

Figure 11.3 Projected changes in the intensity and frequency of hot temperature extremes over land, extreme precipitation over land, and agricultural and ecological droughts in drying regions

Source: Figure SPM.6[5] from IPCC AR6, page SPM-23

(UKCP18 project). These projections show *an increased chance of warmer, wetter winters and hotter, drier summers along with an increase in the frequency and intensity of extremes* and despite the drying summer trend, increases in the intensity of heavy summer rainfall events are projected [3].

Why is climate change a health emergency?

Climate change is the greatest threat to global public health [11]. Without interventions to minimise effects, the geographic distribution and burden of many climate-sensitive health outcomes will be altered, impacting health-care systems delivery. Regions of the world will be differentially affected: the countries who have contributed the least to climate change will be most impacted by the effects, and poverty and inequalities will be exacerbated. Since 2000, there has been an average of 130 per million deaths in disaster-affected areas in low-income countries compared to 18 per million deaths in high-income countries [12]. Climate change has increased economic inequalities, with many low-income countries having a lower per capita GDP than if climate change was not happening [13]. It is estimated that 26 million people annually are pushed into poverty by natural disasters [14] and by 2030, an estimated 132 million people will be pushed into extreme poverty driven by climate change [15].

Climate change is already affecting health globally. The United Nations Office for Disaster Risk Reduction (UNDRR) estimated that during 1998–2017, climate-related and geophysical disasters killed 1.3 million people, with a further 4.4 billion injured, homeless, displaced or requiring emergency assistance [12]. Over 90% of all disasters were caused by floods, droughts, heatwaves, storms or other extreme weather events, and global direct economic losses from climate-related disasters were valued at US$2,245bn, an increase of 68% (US$895bn) reported during 1978–1997 [12]. Between 2030 and 2050, there will

be an estimated 250,000 deaths annually due to climate change, including heat exposure in the elderly, increased incidence of diarrhoeal disease, malaria, dengue and undernutrition [16]. This figure is likely to be an underestimate of the true figure, however, as it only includes a subset of potential causal pathways.

Climate change is already contributing to human displacement. People are displaced due to impacts of food, water, land and increased frequency of extreme weather events, with women, children, older people, people with disabilities and indigenous communities disproportionately affected. Internal displacement, where people move to a different part of the same country, is most likely to occur, although international displacement is becoming increasingly common. As a result of displacement, people can end up living in overcrowded accommodation with inadequate sanitation and result in disease outbreaks. In 2019, weather-related hazards resulted in ~24.9 million displacements in 140 countries and territories, three times the number of displacements caused by conflict and violence [17]. By 2050, it is predicted there will be 143 million climate migrants in Sub-Saharan Africa, South Asia and Latin America alone [18].

The urban population of the world has increased from 751 million (30% of the global population) in 1950 to 4.2 billion (55%) in 2018 and is predicted to increase to 6.7 billion (68%) by 2050 [19]. Population growth rates will be significantly higher in low- and middle-income countries than in high-income countries, with India, China and Nigeria accounting for 37% of the predicted growth between 2018 and 2050 [19]. Urban areas account for 71–76% of CO_2 emissions and consume two-thirds of the world's energy. Increasing urbanisation can affect air quality and water availability and quality. Some effects of climate change, such as heatwaves (page 465), can be more apparent in cities than in rural areas, and large numbers of people will be exposed to the effects of climate change in urban areas.

How does climate change affect health?

Next is a review of exposure pathways of climate change and potential health outcomes. While these are described individually, exposure to multiple hazards and pathways to impact can occur, often simultaneously, leading to overlapping and cascading risks. For example, prolonged dry periods of heat can lead to drought, increasing risk of wildfires, which can affect air and water quality, reduce crop yields and impact food availability.

Heatwaves

Heatwaves are periods of unusually hot weather that persist for at least two consecutive days and are one of the most dangerous natural hazards. Heatwaves can acutely impact many people in a short time, resulting in public health emergencies, excess morbidity and mortality and socioeconomic impacts. Heatwaves affect delivery of health services, causing blackouts and disrupting health facilities, transport and water infrastructure [20]. Those most vulnerable to heatwaves are the poor, displaced or homeless people, children, older people, pregnant women, those with pre-existing conditions (e.g. cardiovascular or respiratory illness), athletes, and outdoor and manual workers. Heat-related illnesses occur when a person is exposed to high temperatures and their body cannot cool itself sufficiently. As the body attempts to cool down, regulatory mechanisms such as heart rate and function are pushed harder to release internal heat, resulting in dehydration and chemical imbalances increasing the risk for heat-related illnesses. Symptoms range from mild swelling, heat cramps, dehydration, heat stroke and hyperthermia. Chronic conditions such as cardiovascular and respiratory diseases may be exacerbated when exposed to extreme temperatures. Hospitalisations and deaths can occur due to heatwaves, and illness or death may be accelerated in those who are already infirm, although there may be a lagged effect (occurring several days later). Between 1998–2017, more than 166,000 people died due to heatwaves [12], with more than 70,000 of these deaths occurring because of the 2003 heatwave in Europe [21]. A study of heat deaths across 732 cities in 43 countries between 1991–2018 found that 37% of heat deaths worldwide can be attributed to anthropogenic climate change, with increased mortality evident on every continent [22]. Heatwaves can affect mental health; an Australian study identified an increase in hospital admissions for mental and behavioural disorders of 7.3% during heatwaves compared with non-heatwave periods [23]. The effects of heatwaves can be greater in cities, due to the urban heat island (UHI) effect. The UHI is driven by urban materials such as concrete, bricks and metal absorbing heat in the day and releasing it at night, increasing temperatures in urban areas compared with surrounding rural areas. The effect is particularly acute in areas where there is a lack of trees and vegetation, and people living in urban areas are likely to be more exposed to higher temperatures than those living in rural areas. With increasing urbanisation, future populations will be more exposed to urban environmental factors. Heatwaves can affect health service delivery, including critical infrastructure, such as energy, transport and water. Blackout events may occur during heatwaves as electricity demand for cooling devices increases, exposing 68–100% of urban populations to elevated risks of heat exhaustion and/or heat stroke [24]. Urban hospitals are particularly vulnerable during periods of high energy demand as they may be asked to shift to emergency power systems to free up resources. As cooling systems are not on emergency power systems, hospitals are at risk of overheating, exposing already at-risk individuals to extreme temperatures.

Hot temperature extremes have become more frequent since the 1950s, whilst cold temperature extremes are less frequent and less severe, and there is high confidence that anthropogenic climate change is the main driver of these changes; it is extremely unlikely that extreme hot temperatures observed during the 2010s would have occurred had it not been for anthropogenic climate change [2]. During 2000–2016, the number of people exposed to heatwaves increased by ~125 million [25]. Economic losses due to extreme temperatures during 1998–2017 was US$61bn [12] and this figure is likely to increase over time as incidence of heatwaves increases. It is very likely with climate change that the intensity and frequency of heatwaves will increase [2]; under a high-emission scenario (RCP8.5), extreme temperatures could result in more than 9,000 additional premature deaths per year by the end of 2100, with annual financial damages estimated at US$140bn compared with US$60bn under RCP4.5 [26].

Floods

The most common flood types are:

a   Flash floods, caused by extremely heavy rainfall over short time periods
b   River floods, where heavy rainfall or rapidly melting snow results in rivers breaking the banks
c   Coastal floods, caused by storm surges often associated with tropical cyclones, hurricanes and tsunamis

Floods were the most frequent climate-related disaster occurring between 1998–2017, comprising 43.4% of all recorded events with economic losses of US$656bn [12]. Between 1998–2017, floods affected more people than any other disaster type, with over two billion people experiencing flooding and 142,088 related deaths [12]. Most flood-related deaths are from drowning (75%), with low- and middle-income countries most vulnerable, as many people live in flood-prone areas and often have more limited capacity to respond to flooding.

Floods can increase transmission of many infectious agents, including water-borne infections such as diarrhoeal diseases. During 1948–1994, more than half of US waterborne disease outbreaks occurred following a period of extreme rainfall [27]. There is evidence of increased incidence of disease following floods; for instance, there were higher cases of *Escherichia coli* in children in flooded areas of Pakistan compared with non-flooded areas [28]. In Dhaka, Bangladesh there were higher cases of cholera (*Vibrio cholerae*), enterotoxigenic *Escherichia coli*, *Shigella* and *Salmonella* species in flooded than unflooded areas [29]. There is mixed evidence in relation to rotaviruses and flooding, with some studies showing increased risk at both low and high rainfall levels [29–31]. In addition to water-borne disease impacts, floods create breeding grounds for vectors which transmit vector-borne diseases and are reviewed later (see also Chapter 17 on vector-borne diseases).

During the winter of 2013/2014, widespread flooding occurred across south-west England, and The National Study of Flooding and Health was established to investigate long-term impacts of flooding on mental health and well-being [32]. Surveys were sent to addresses in affected neighbourhoods and were defined as 'flooded', where floodwater was reported in at least one liveable room of the home; 'disrupted', where homes were not flooded, but participants experienced evacuation, flooding of gardens/garages, loss of household utilities/communications, interruption to healthcare access; or 'unaffected', where there was no flooding or disruption [32]. One year later, the prevalence of probable depression, anxiety and post-traumatic stress disorder (PTSD) was higher for flooded and disrupted responders than the unaffected group (Table 11.1), with adjusted odds of psychological morbidity 6.0–7.0 times higher for flooded and 1.5–2.0 times higher for disrupted respondents than unaffected participants [32]. Participants who were flooded and had to move out of their home for any duration had over 8.0 times higher odds of probable PTSD than those flooded who remained

Table 11.1 Prevalence of mental health outcomes by exposure groups each year during the English National Cohort Study of Flooding and Health

Outcome	Overall cohort	Time since flooding event	Exposure group		
			Unaffected	Disrupted	Flooded
Probable depression	250/1929 (12.6%)	Year 1	16/278 (5.8%)	102/1058 (9.6%)	125/593 (20.1%)
	59/988 (6.0%)	Year 2	2/137 (1.5%)	21/512 (4.1%)	36/339 (10.6%)
	42/733 (5.7%)	Year 3	1/112 (0.9%)	22/380 (5.8%)	19/241 (7.9%)
Probable anxiety	300/1927 (15.6%)	Year 1	18/278 (6.5%)	113/1052 (10.7%)	169/597 (28.3%)
	83/988 (8.4%)	Year 2	4/137 (2.9%)	33/512 (6.4%)	46/339 (13.6%)
	59/731 (8.1%)	Year 3	4/114 (3.5%)	27/378 (7.1%)	28/239 (11.7%)
Probable PTSD	396/1925 (20.6%)	Year 1	22/278 (7.9%)	160/1056 (15.2%)	214/591 (36.2%)
	129/988 (13.1%)	Year 2	0/137 (0.0%)	46/512 (8.9%)	83/339 (24.5%)
	91/771 (11.8%)	Year 3	3/117 (2.6%)	43/397 (10.2%)	45/257 (17.5%)

Source: Taken from [32–34]

in their home [32]. Two years later, the prevalence of probable psychological morbidity remained higher for flooded and disrupted respondents, although rates were lower than in Year One (Table 11.1) [33]. Three years later, mental health outcomes remained higher in the flooded group (Table 11.1), suggesting long-lasting mental health impacts following flood events [34].

In July 2021, severe flooding in Germany and Belgium was caused by extremely heavy rainfall on already saturated ground, breaking historical records by large margins [35], resulting in 184 deaths in Germany and 38 deaths in Belgium, and considerable damage to infrastructure. Analysis suggested that climate change has increased the intensity of summertime maximum one-day rainfall events in Western Europe by 3–19% [35]. If temperatures continue to increase, the intensity of such flooding events in Western Europe would increase by a further 0.8–6% if global mean temperature rise reaches 2°C [35]. Since the 1950s, there has been an increase in the frequency and intensity of heavy precipitation events, and anthropogenic climate change is likely the main driver [2]. There is high confidence that heavy precipitation events will intensify and become more frequent in future, with extreme daily precipitation events projected to intensify by 7% per 1°C of warming [2].

Droughts

Droughts are slow-onset, long-lasting events covering a large area, characterised by a lack of precipitation, resulting in water shortages. One of the main drivers of episodic droughts globally is the El Niño Southern Oscillation (ENSO) [36]. The strongest correlations with intense drought and ENSO occur in Australia, Brazil, India, Indonesia, the Philippines, parts of eastern and southern Africa, central America, USA and the western Pacific basin islands [36]. Droughts can lead to famine, migration, loss of natural resources, and cause hardship in rural communities and increase global food prices [12]. The occurrence of drought is often coupled with heatwaves, and severe droughts can result in reduced air quality and wildfires. Whilst droughts comprised just 4.8% of climate-related disasters between 1998–2017, the second largest number of people (following floods) were affected, totalling 1.5 billion and resulting in 21,563 deaths [12]. The economic losses caused by drought periods during 1998–2017 was US$124bn [12].

Compared with heatwaves and floods, the health effects of droughts have been understudied, as they are harder to identify and depend on drought severity, population vulnerability, existing health and sanitation infrastructure and available resources to mitigate impacts as they occur [37]. Morbidities and mortalities from drought are felt most by low- and middle-income countries: droughts in the Horn of Africa during 2010–2011 caused 250,000 deaths and resulted in 13 million people requiring humanitarian aid [36]. Droughts cause long-term water shortages and extreme heat stress in crops, damaging yields and result in lower food production and availability [38]. Droughts in Argentina resulted in declines of soybean and maize production, with estimated direct losses to be at least US$12bn [36]. Droughts can have a particularly high impact in countries dependent upon agriculture [12]. Over 34% of crop and livestock production loss in low-income and low-to-middle income countries is drought-related, costing US$37bn overall [39]. During 1961–2018, droughts reduced European cereal yields (wheat, barley, maize etc.) by 9% and non-cereal yields (vegetables, sugar cane, citrus, grapes, etc.) by 3.8% [40]. Droughts resulted in US$14bn production loss in Africa during 2008–2018 and caused US$13bn in crop and production loss in Latin America and the Caribbean [39].

Many health risks associated with droughts are linked to food insecurity, malnutrition, reduced water quality, loss of livelihoods, population displacement, and increased risk of food-, water- and vector-borne diseases [36]. During 2009–2010, the Caribbean experienced one of the most severe droughts in 50 years and was followed by an additional drought in 2014–2016. This led to reductions in crop yields, livestock losses, food price increases (resulting in riots in Haiti), increased plant pests and diseases, water shortages and rationing (low reservoir levels), wildfires, gastrointestinal outbreaks due to poor water storage, proliferation of mosquitoes and

significant numbers of landslides when the rains returned [36]. Increased water temperatures can promote the production of algae and toxic cyanobacterial blooms, making it unfit for drinking. In addition, once water sources are replenished following droughts, mosquito densities may increase due to reduced predator abundance [41], leading to outbreaks of vector-borne diseases. Drought can lead to mental health issues in affected populations, resulting in suicide in extreme cases.

Drought and climate insecurity can lead to instability, violence and conflict. During the 1970s and 1980s, prolonged drought in Sudan lead to famine, deforestation and desertification, forcing four million people to migrate south [42]. As land became less fertile and resources scarcer coupled with increasing demand for farmland, tensions broke out between farmers and pastoralists, leading to war and resulting in the deaths of over 300,000 people and displacing more than 2 million people [36].

Since the 1950s, the global frequency of concurrent heatwaves and droughts has increased, and there is high confidence that this is driven by human influence, and for every additional 0.5°C increase in temperature, the frequency of droughts will increase [2]. The frequency of an agricultural and ecological drought event that occurred on average once in ten years during 1850–1900 will be twice as likely in a 1.5°C world, 2.4 times more likely if temperatures rise by 2.0°C and 4.1 times more likely in a 4.0°C world [2]. In Europe, losses from drought are estimated to be €9bn annually, with the highest losses in Spain (€1.5bn), Italy (€1.4bn) and France (€1.2bn) [43]. It has been predicted that annual losses will rise to €9.7bn with 1.5°C temperature increase, to €12.2bn with 2.0°C rise and to €17.3bn with a 3°C rise [43]. Ecological and agricultural droughts in regions in Africa, South America and Europe are projected to increase in frequency and/or severity with medium to high confidence with 2°C warming [2].

Wildfires

Wildfires are unplanned fire burning in natural areas, such as a forests or grasslands which can be caused by natural phenomenon (e.g. lightning) or human activity. Four elements are necessary for a wildfire to start:

1   Large biomass of fuel, usually leaves, bark litter, dead wood and living foliage
2   Fuel dryness, where fire risk increases in areas where multi-year droughts have occurred and there is dried-out vegetation
3   Ignition, which can be anthropogenic or natural
4   Fire weather, where hot, dry, windy conditions enable fires to spread rapidly through an area

Temperature increases, specifically intense heatwaves, plus precipitation decreases and population growth at the wildland-urban interface has increased human exposure to fires.

Wildfires emit large amounts of pollutants, which can affect the health of residents in the fires' immediate vicinity and surrounding areas. One of the biggest wildfires in the UK to date occurred on Saddleworth Moor in north-west England during June 2018. The fire burned for three weeks and covered 7 square miles (18km^2). Concentrations of $PM_{2.5}$ increased by over 300% in parts of Greater Manchester and up to 50% in areas up to 80 km away [45]. A quarter of the population (2.9 million people) were exposed to moderate $PM_{2.5}$ concentrations of 36–53 μg m^{-3} during 23rd–30th June 2018, and calculations suggested 28 deaths were brought forward, with a mean excess mortality of 3.53 deaths per day during the seven-day period [45]. Over the week, there was up to a 165% increase in excess mortality across the region due to exposure to $PM_{2.5}$ from the fire.

Whilst exposure to smoke and pollutants from wildfires can be short-term, mental health impacts on residents affected by fires can be much more long-lasting. The deadliest Australian wildfires to date occurred in Victoria during February 2009 when extreme heat, high winds, low humidity and severe drought created a perfect storm. The worst fires occurred on Saturday 7th February, known as 'Black Saturday', resulting in 173 fatalities and 3,500 destroyed buildings [45]. Three to four years after Black Saturday, affected communities reported higher PTSD, depression and severe psychological distress linked to the fires compared with medium- and low-impacted communities [46]. All communities reported elevated rates of heavy drinking, but rates were highest in the worst affected communities (46). A follow-up study ten years after Black Saturday found that PTSD levels had reduced since the previous survey, but 20% of people from high-impacted communities had a probable psychological disorder [47].

Wildfires can result in large economic losses, not only due to destruction of property but also due to healthcare costs. Economic costs of mortality due to $PM_{2.5}$ exposure of the Saddleworth Moor fire in the UK were estimated at £21.1m, broken down into human costs (£13.9m), lost output (£7.0m), medical costs (£0.07m) and others (£0.15m) [44]. A study of wildfire events in USA during 2008–2012 estimated that there were 5,200–8,500 respiratory hospital admissions, 1,500–2,500 cardiovascular hospital admissions and 1,500–2,500 deaths from short-term exposure to $PM_{2.5}$ concentrations [48]. The economic value of hospital admissions and deaths associated with short-term exposure were estimated to be US$11–$20bn/year, and the cost of long-term $PM_{2.5}$-exposure-related premature deaths was estimated to US$76-$130bn/year [48]. During 2018, more than 8,500 wildfires burned in California and resulted in 104 fatalities, making it the deadliest and most destructive year on record (until the wildfire season of 2020). The amount of damage caused by the fires was estimated at US$148.5bn, ~1.5% of California's annual GDP [49]. Approximately 31.5% of the cost of the wildfires was related to health:

US$32.2bn was associated with 3,652 air pollution deaths, US$210m related to medical expenses and US$130m in work time lost [49].

There is medium confidence that weather conditions which exacerbate wildfires have become more frequent in southern Europe, northern Eurasia, USA and Australia during the last century [50]. Some of the most damaging wildfires have occurred in Australia: during 2019/2020, south-east Australia experienced its worst fire season to date. The wildfires, termed 'Black Summer' began during September 2019 in Queensland, with hundreds of fires occurring in the forests of New South Wales, the Australian Capital Territory, Victoria and South Australia until rainfall eased conditions in February 2020. The severity of the fires is thought to have been driven by weather conditions, as 2019 was hotter and drier than any year on record, with mean annual temperatures 1.9°C above the 1911–1940 historical mean (reaching 2.0°C above in south-east Australia) [51], coupled with a widespread and sustained drought that began in 2017, and continued with low rainfall during the autumn and winters of 2018 and 2019 [51]. Changes in the climate that have already occurred have increased the risk of weather suitable for causing wildfires in south-eastern Australia by 30% [52], and there is medium confidence that increases in weather conditions suitable for wildfires in Australia are driven by human influences [50]. Analysis of wildfires in Arizona and New Mexico, USA during 1984–2015 found hotter and drier conditions were associated with increasing area burned, and areas burned at high severity had the strongest relationship with the climate variables [53]. In future, it is likely that wildfire risk will increase, with fire weather season predicted to lengthen and intensify in many places [50]. There is medium to high confidence that fire weather will increase throughout Australia, parts of South America, North America, Central America and the Mediterranean [50].

Air pollution

Air pollution (see also Chapter 16) globally causes 7 million premature deaths annually, including more than 5 million caused by non-communicable disease (NCDs, e.g. heart disease, stroke, chronic obstructive pulmonary disease, lung cancer): it is the second leading cause of NCD deaths after tobacco smoking. Globally, 90% of people are exposed to air pollution levels that put them at risk. Evidence demonstrating the linkages between ambient air pollution and cardiovascular disease risk is established and strengthening [54].

People living in low- and middle-income countries disproportionately experience the greatest burden of outdoor air pollution, and this disparity is increasing, mainly because of large-scale economic development and urbanisation that has often relied on inefficient fossil fuel combustion for energy. Anthropogenic $PM_{2.5}$ and NO_2 harm health, and in the UK have an effect equivalent to 28,000 to 36,000 deaths annually, associated with a loss of 328,000–416,000 life years [55–56]. Episodes of O_3 have short-term impacts on health, particularly on respiratory hospital admissions [57–58].

Air pollution levels are driven primarily by emissions rather than changes in climate, so the emissions pathways are key drivers for ambient air pollution exposure in future. Meteorological factors such as weather patterns, temperature and precipitation (which affects wet removal of pollution), however, may change as climate changes. Episodes of elevated air pollution can occur during periods of dry, stable weather ('blocking' events) with impacts on health [59–60], but evidence on how such weather patterns might change with climate change is currently mixed [61]. Meteorological factors are more important for O_3, as this is a secondary pollutant formed by photochemical processes and precursors like NO_2 and volatile organic compounds (VOCs), which may be sensitive to changes in weather conditions, and thus O_3

is considered a pollutant which is most likely to be affected by climate change. Methane (CH_4), an important GHG, may also contribute to O_3 formation through its complex degradation reactions in the atmosphere, and is important in pollution from non-local (transboundary) sources; controls on GHGs like CH_4 can impact levels of O_3, and as controls on local emissions of pollutants become effective, non-local sources become more important. Emission of ammonia (which contributes to $PM_{2.5}$) from agricultural practices may also be higher in a warmer world [62]. Looking at different future emission pathways shows the O_3-health burden over the UK may increase by 16–28% compared to during 2003 [57].

Air pollutants and GHGs often have similar sources (e.g. combustion processes, such as vehicle engines, fossil fuelled power plants, or other fuel burning such as domestic wood stoves). Efforts to reduce emission of GHGs from these sources should also reduce co-emitted air pollutants, thus reducing the burden of ambient air pollution on health. Improvements in health and well-being from improved air quality are important mutual benefits to health from efforts to reduce climate emissions, and as such can be strong motivators for countries to take more ambitious policy action on climate change. Health benefits from improved air quality (due to reduced co-emitted GHG emissions) are one of the most cited and quantified health benefits of climate mitigation actions. Meeting the targets of the Paris Agreement is expected to save over 1 million lives a year from air pollution alone by 2050 [63]. A UK-focused study modelled three UK-specific energy-related scenarios and the subsequent impacts on health due to air pollution changes in 2035 and 2050; reduced NO_2 levels resulted in reduced long-term NO_2-related mortality, leading to ~6.5 million life-years gained in 2050 compared to 2011. Decreases in $PM_{2.5}$ concentrations across the UK result in ~17.8 million life-years gained in 2050

compared to 2011 [64]. The Climate Change Committee's (CCC) Sixth Carbon Budget outlines measures to meet the UKs Net Zero commitments, which will benefit air quality and health [65], and in almost all pathways proposed by the CCC, better air quality outcomes will be seen [66].

Building operations contribute about 28% of global CO_2 emissions [67], making them a key priority for emissions reduction through reducing building energy use and decarbonising energy sources. In the UK, improving building energy efficiency through retrofit (e.g. reducing draughts and improving insulation) will be a key part of meeting climate goals. Such retrofit can reduce building ventilation, which will increase exposure to indoor sources of air pollution, and potentially lead to issues with mould and radon exposure if not designed and implemented carefully [68].

Vector-borne diseases

Vector-borne diseases (VBDs) are caused by parasites, viruses and bacteria that are transmitted by arthropod insects (vectors) such as mosquitoes, ticks, lice and flies. Approximately 17% of the estimated disease burden affecting humans is attributed to VBDs, with conditions such as onchocerciasis causing significant debilitation, and over 700,000 deaths annually are caused by VBDs. The two most important VBDs in terms of people affected are:

1 Malaria, which is transmitted by *Anopheles* mosquitoes and cause ~219 million cases globally and more than 400,000 deaths per year, predominantly in children under 5 years of age and

2 Dengue fever, which is transmitted by *Aedes* mosquitoes, with ~96 million symptomatic cases and 40,000 deaths per year

As the vectors that transmit VBDs are ectothermic (i.e. cold blooded), they are particularly sensitive to changes in the climate.

It should be noted, however, that there are many non-climate drivers of VBD transmission, such as land use changes, globalisation, socioeconomic conditions and improved surveillance and diagnostic techniques that make it difficult to ascertain the effect of climate change alone.

Changes in the climate, particularly temperature, can impact many parts of the life-cycle of both vectors and pathogens. Temperature increases result in faster development, increased survival, and increased feeding rate of vectors up to a threshold, with increased mortality occurring if temperatures are too high. The rate of pathogen development within mosquitoes, known as the extrinsic incubation period, accelerates under warmer temperatures, increasing the rate of disease transmission. Warmer conditions during periods of the year that were previously too cold for vector survival increase the risk of disease transmission. Predicted temperature increases of 3°C in Scotland suggest that late winter and early spring would no longer be disease risk-free as higher temperatures would improve tick survival and a higher proportion of questing ticks would become active earlier in the spring and later in the winter [69]. Modelling has shown that the length of transmission of malaria and dengue will increase in future: malaria suitability will increase by 1.6 additional months in the African tropical highlands, the Eastern Mediterranean and Americas, whilst dengue suitability will increase by 4.0 additional months in lowlands of the Western Pacific and the Eastern Mediterranean, resulting in an additional 4.7 billion people at risk of both diseases by 2050 relative to 1970–1999 [70].

Vector distribution may alter with climate change, as regions historically unsuitable for vector survival, including higher altitudes and latitudes may become suitable. Established populations of the deer tick, *Ixodes ricinus* have been reported at higher altitudes in the Alps [71] and at higher latitudes in parts of Scandinavia [72]. In northern Russia, warmer temperatures have increased the spatial spread

of reported tick bites, with bite frequency increasing 40-fold during 2000–2018 [73]. Changes in the geographical distribution of vectors and VBDs can result in outbreaks in new areas, potentially exposing immunologically susceptible populations to novel infections. Malaria cases are increasing in East African tropical highlands due to mosquito survival at higher altitudes in response to warmer temperatures, and immunologically naïve populations are being exposed to the infection and are vulnerable to severe morbidity and mortality [74]. In 1947, Zika virus was first isolated from monkeys in the Zika forest in Uganda. Several mosquito species are able to transmit the infection, including *Aedes aegypti* and *Ae. albopictus*. During 1950–2014, sporadic outbreaks of Zika occurred in parts of Africa, Asia and the Pacific Islands. During 2015/2016, an outbreak occurred in South America, with cases first identified in Brazil followed by subsequent spread throughout the Americas, Africa and other regions. The outbreak was linked to unusually high temperatures driven by a very strong El Niño event during 2015, where the risk of Zika transmission in South America was higher during 2015 than any other time between 1950–2015 [75]. Furthermore, the lack of immunological response against the virus in South America resulted in a significantly larger outbreak than in Africa, where populations have been previously exposed to the virus and were less susceptible to infection [75].

Schistosomiasis is an acute and chronic disease caused by *Schistosoma* trematodes which use freshwater snails as intermediate hosts. Schistosomiasis is prevalent in tropical and subtropical areas of Africa, Asia, the Middle East, South America and the Caribbean. In summer 2013, an outbreak of urogenital schistosomiasis (*Schistosoma haematobium*) occurred in Corsica, previously considered to be outside of the geographical range for schistosomiasis due to cold winter temperatures. More than 100 cases were detected in both residents and tourists and linked to a popular swimming spot in a river in southern Corsica [76]. It is hypothesised that schistosome eggs were unknowingly spread by an infected person/s into the river, and the availability of competent snail hosts combined with warm summer temperature conditions enabled survival [76]. Laboratory studies have suggested that schistosomes from tropical areas are able to survive and establish in temperate areas if there are locally adapted snail species already established [77], which has important implications in terms of future schistosomiasis distribution. Modelling studies of predicted future distribution suggests that infection risk may reduce in current endemic areas due to exceeding the thermal capacity of snail species, whist infections may become endemic in areas that are currently unsuitable for snail survival and infection-free [78].

Precipitation is vital to vector survival (see earlier on flooding); many vectors including mosquitoes have aquatic developmental stages, and humidity is vital to non-aquatic species such as ticks or sandflies. For ticks, relative humidity levels of ~85% are required as ticks absorb moisture from the atmosphere. Following drought conditions in Illinois, USA, densities of larval *Ixodes scapularis* feeding on mice were significantly lower than the eight-year means, likely driven by the drought conditions severely damaging the vegetation and small mammal populations [79]. Following droughts in 1996 and most of 1997, significant flooding from an El Niño event occurred in eastern Africa during October–December 1997, and in early 1998, malaria epidemics were reported in south-western Uganda [80] and north-eastern Kenya [81], thought to have been driven by the floods. Conversely, in Tanzania there were fewer malaria cases following the floods than the preceding year, possibly because heavy rainfall flushed out larvae from breeding sites and there were fewer adults biting people [82], suggesting geographical setting may play a role, which has since been confirmed by additional work. A study of flooded and non-flooded villages in Uganda found that the risk of an individual having a positive malaria test was 30% higher in villages bordering a flood-affected river, with larger relative impacts on upstream compared to downstream villages [83]. Increased use of containers for storing and collecting rainwater during droughts could provide breeding sites close to people's homes. Drought is considered the primary climatic driver of increased West Nile virus (WNV) epidemics in the USA [84]. WNV is transmitted between *Culex* spp. mosquitoes and birds, and water scarcity reduces mosquito predators and drives birds closer to remaining water sources which are close to human settlements.

Non-communicable diseases (including mental health)

Compared with infectious diseases, the impact of climate change on non-communicable diseases (NCDs) has received little attention, despite the incidence of NCDs in communities most at risk of climate change (low- and middle-income countries and lower socio-economic groups) is increasing. As already mentioned, individuals with cardiovascular and respiratory illness are more at risk from heatwaves and air pollution than healthy individuals. Climate change also has the potential to change the risk of UV-related health issues, such as cancers [85].

The effects of climate change on mental health have been understudied compared with physical health effects. As previously discussed, exposure to disasters linked to extreme weather events such as heatwaves, flooding and wildfires can result in mental health illness, including anxiety, depression, PTSD or psychosis (see previous sections). Individuals at greatest risk from mental health impacts of climate change include those with pre-existing mental illness, migrants, women, youth and groups already affected by social inequalities such as living in low-income countries where there is less likely to be access to mental health resources and support.

As a result of severe, long-term extreme weather events, people may be forcibly displaced or required to migrate seasonally or on a permanent basis, both internally and/or internationally. It is likely that underlying mental health risk factors may be higher already due to the events that caused displacement initially, and those with existing mental health issues are likely to be worst affected. Additional factors that are likely to contribute substantially to poor mental health include separation from family, diminished sense of belonging, economic deprivation and inadequate housing [86].

It is not only populations that experience first-hand the effects of climate change that may experience mental health issues. The term 'eco-emotions' has been used to describe anxiety and concerns about climate change. One such eco-emotion is 'eco-anxiety', a term that is gaining attention and has been used to describe worry or distress about impending environmental change [87]. Eco-anxiety may result in anxiety, hopelessness, sleep disturbance and despair, and is particularly associated with younger adults [88]. Another eco-emotion is 'climate grief' or 'ecological grief', which refers to grief experienced or anticipated because of ecological losses due to environmental change and affects those witnessing environmental destruction such as indigenous communities [89].

Nutrition and food security

By 2050, the global population will likely exceed 9 billion, and demand for food will increase by at least 60% over current levels [90]. The aim of SDG 2 (Box 11.1) is to *end hunger, achieve food security and improved nutrition, and promote sustainable agriculture* by 2030. In 2019, the world was off track to achieve the goal of ending world hunger, and this has been further exacerbated by the COVID-19 pandemic. Approximately 660 million people may still face hunger in 2030, an increase of 30 million people than scenarios where the pandemic had not occurred [91]. During

2020, 720–811 million people (~10% of the global population) faced hunger, a rise of ~118 million from 2019 [91]. More than half of the undernourished populations are found in Asia (418 million), whilst more than one-third are from Africa [91] (see also Chapter 13). At least 155 million people in 55 countries/territories faced food crises or worse during 2020, an increase of 20 million people compared with 2019, with 66% of the people in crisis located in just ten countries/territories (Democratic Republic of the Congo, Yemen, Afghanistan, Syrian Arab Republic, Sudan, northern Nigeria, Ethiopia, South Sudan, Zimbabwe) [92]. In addition, 149.2 million children under 5 years of age were affected by stunting during 2020, 45.4 million suffered wasting and 38.9 million were overweight [91]. Climate change impacts on nutrition and food systems are expected to be widespread, geographically and temporally variable and influenced by social and economic factors [93]. One study estimated that by 2050, climate change will result in reductions of global food availability by 3.2% per person, 4.0% in fruit and vegetable consumption and 0.7% in red meat consumption, resulting in 529,000 climate-related deaths, with most deaths occurring in South and East Asia [95]. The implementation of climate change mitigation methods, however, could prevent between 29–71% of predicted deaths [94].

Extreme weather events can negatively impact the agriculture industry (including food production), particularly in the most vulnerable parts of the world. During 2006–2016, ~25% of damage and loss caused by medium- to large-scale climate-related disasters in low- and lower-middle income countries affected the agriculture industry [38]. In 2017, extreme weather was a leading cause of acute food crises and malnutrition, affecting 59 million people in 24 African countries [95]. The frequency of exposure to extreme weather events is increasing in Latin America, Africa and South Asia, and surveys of more than 5,000 households in 15 countries found that 71% of households had experienced extreme weather in the previous five years, with affected households 1.73 times more likely to be food insecure [96]. An area of particular concern is the effect of climate impacts on childhood nutrition, with severe flooding and drought linked to poor nutrition, particularly stunting, in many low-income countries [97]. Factors such as nutritionally diverse agricultural systems, stable crop production and effective governance increased resilience to extreme weather events, with child nutrition in arid low-income, politically unstable countries with poor governance most likely to be negatively impacted [97].

A few countries produce the most important food crops (e.g. maize, wheat, rice and soybean); extreme weather events in these regions have the largest impact on global food production. Regions not directly affected by climatic impacts on food security may suffer from indirect consequences, such as reduced availability and higher food prices. Argentina, Brazil, China and USA account for 87% of the global maize exports, and whilst the current probability that there will be simultaneous production losses greater than 10% is virtually zero, the risks increase to 7% under 2°C warming and 86% under 4°C warming, resulting in trading instability and price increases [98]. Global crop and economic models have suggested there will be a 1–29% cereal price increase in 2050 due to climate change, with 1–183 million additional people at risk of hunger compared to a no climate change scenario [99]. In 2019, healthy diets were considered too expensive for ~3 billion people, and this is likely to increase in most regions due to the COVID-19 pandemic coupled with climate change [91].

It is not only extreme weather events that can impact food products, as increased CO_2 is projected to be beneficial for crop productivity at lower temperature increases, but it is projected to lower nutritional content and quality, with wheat grown at 546–586 ppm CO_2 having 5.9–12.7% less protein, 3.7–6.5%

less zinc and 5.2–7.5% less iron [99]. If atmospheric CO_2 exceeds 500 ppm by 2050, ~148.4 million people will be at risk of protein deficiency (caused by reduced nitrogen concentration), with an additional 53.4 million people in India, 15.9 million people in South Asia and 24.6 million people in sub-Saharan Africa at risk of protein deficiency [100]. Other models have found that elevated CO_2 in 2050 could cause zinc deficiency in 175 million people and protein deficiency in 122 million, with 1.4 billion women and children under 5 years of age at risk of anaemia due to loss of dietary iron [101]. The quantity of vitamins is also of concern, and the presence of vitamins B1, B2, B5 and B9 in rice decline under elevated CO_2 scenarios yet increases in vitamin E have been reported [102].

Response to climate change

As GHGs have long lifetimes and Earth systems take time to respond, even if emissions stopped tomorrow, the climate would continue to change and affect future populations, meaning adaptation to climate change is unavoidable. The level of required adaptation will depend on emission levels over the coming years, so mitigation efforts to reduce emissions and subsequent impacts of climate change is important. Many mitigation and adaptation actions can have mutual co-benefits for health, and health can be a key driver for motivating changes that both address climate change and improve health. The global COVID-19 pandemic has highlighted the connection between society, the economy and health, and has demonstrated the need for resilient health and care systems.

Mitigation

Mitigation refers to reducing climate change, through reducing emissions of GHGs and switching to renewable sources of energy (often referred to as 'decarbonising'), as well as removing GHGs from the atmosphere, for example through carbon capture and storage/utilisation (carbon capture storage / carbon capture utilization and/or storage). Renewable energy generation costs have fallen sharply over the last decade, and new solar and wind projects are undercutting even the cheapest coal-power plants; analysis suggest that replacing 800 GW of existing coal power plants could reduce costs by US$32bn annually and save 3 gigatonnes of CO_2 emissions each year [103].

Historically, health has not been strongly represented in the NDCs under the Paris Agreement (Box 11.1). Around 10% (18 out of 184) of countries' NDCs emphasise the health benefits of mitigation and only two

---

Box 11.1 Key international legislation and organisations for climate change action

The Conference of the Parties (COP) is the supreme decision-making body of the United Nations Framework Convention on Climate Change (UNFCCC) and was created in 1994 with the aim of stabilising GHGs levels. The COP meets annually, and reviews national communications and emission inventories submitted by Parties to ensure that progress is being made to achieve the objectives of the UNFCCC.

The first COP was held in Berlin in 1995 and occurs annually (except in 2020, when COP26 was delayed until 2021 due to the COVID-19 pandemic). At COP3 (1997), the Kyoto Protocol was established which committed industrialised nations to limit their GHG emissions. The most important COP to date was COP21 (Paris, 2015), where a legally binding international treaty on climate change was established, called the 'Paris Agreement'. In total, 196 nations pledged to keep global mean temperature rise 'well below' 2°C and 'pursue efforts to limit it to' 1.5°C above pre-industrial levels. The Agreement also states that developed nations must provide financial assistance to less-developed countries to mitigate climate change, strengthen resilience and enhance adaptation to climate impacts.

The Paris Agreement works on a five-year cycle of increasingly ambitious climate action, with each country producing a plan for climate action, called 'nationally determined contributions' (NDCs). The NDCs include action pledges to reduce GHG emissions to reach the goals of the Paris Agreement as well as building resilience to adapt to the effects of climate change. The first NDCs were submitted in 2020, and to ensure there is collective progress towards achieving the Paris Agreement goals, the first global stocktake (GST) will take place from 2021–2023 and will be repeated every five years thereafter. Following the GST, countries will then submit the next round of NDCs in 2025.

The Intergovernmental Panel on Climate Change (IPCC) was created in 1988 by the United Nations Environment Programme and the World Meteorological Organization (WMO). The IPCC assesses the science of climate change, provides policymakers with regular scientific assessments, its implications and potential future risk, and proposes adaptation and mitigation solutions in its comprehensive Assessment Reports.

In 2015, the General Assembly of the UN adopted the 2030 Agenda for Sustainable Development: a blueprint for peace and prosperity for people and the planet, now and in the future. To achieve sustainable development, there are 17 'Sustainable Development Goals' (SDGs) [4] to achieve a better and more sustainable world for all by 2030. The goals include ending poverty and hunger, ensuring health and well-being and taking urgent action on climate change (see also Chapter 1.)

---

mention quantifying or monitoring health benefits to inform decision making. Only 3% (5 out of 184) of NDCs emphasise health benefits of adaptation, and one in five NDCs refer to National Adaptation Plans (NAPs) or National Adaptation Plans of Action (NAPAs) in relation to public health [104].

Health systems have a leadership role in addressing the impact on health whilst significantly contributing to global GHG emissions and are themselves vulnerable to climate change. Healthcare systems globally are responsible for 4.4% of global net CO_2 emissions [105]. In England, the NHS accounts for 4% of the country's GHG emissions, and it has set an ambitious target to become the first health service to reach net zero emissions by 2040 (Box 11.2).

Box 11.2 Case study of UK action for climate change and health

The UK has made significant, sustained contributions to the global understanding of the health effects of climate change over the last 20 years and was the first major economy to introduce comprehensive legislation with the Climate Change Act 2008. The Climate Change Act requires a national Climate Change Risk Assessment (CCRA), which is an independent assessment of climate risks and opportunities facing the UK. Following the publication of the CCRA, the UK government and the devolved administrations of Scotland, Wales and Northern Ireland then set out their response to the risks and opportunities in their National Adaptation Programmes (NAP). The CCRA and NAP are updated on a five-yearly cycle.

In June 2021, the third CCRA [5] concluded that the most urgent health priorities in the UK for action in the next five years are: risks to health and well-being from high temperatures; risks to people, communities and buildings from flooding and sea level rise; risks and opportunities from summer and winter household energy demand; risks to health from vector-borne diseases; risks to health and social care delivery. Health priorities requiring further investigation are risks to health and well-being from changes in air quality; risks to food safety and food security; and risks to health from poor water quality and household water supply interruptions.

The NAP [6] contains several objectives to address climate change risks to the health of the UK population, as well as research and action priorities identified by the CCRA. The UK Health Security Agency (UKHSA) leads on two of these including a report detailing the evidence base on the direct and indirect health impacts and climate change [7] and publishing the Heatwave Plan [8] and Cold Weather Plan for England [9].

In 2008, the Greener NHS programme within the National Health Service (NHS) was established to promote sustainable environmental, social and financial development in the healthcare system. Since its inception and despite increased clinical activity, there has been significant progress:

- 62% reduction in emissions for the NHS Carbon Footprint, significantly exceeding the 37% requirement by 2020 outlined in the Climate Change Act and 26% reduction in emissions for the NHS Carbon Footprint Plus [10]
- NHS England and NHS Improvement became the world's first health system to commit to become 'carbon net zero' by 2040, with ambition for an 80% reduction by 2028–2032 [10].

Low-carbon health facilities are more cost-effective to run, and renewable energy sources can increase energy availability and access in remote areas, as well as be more stable and reliable than some fossil fuel-based supplies. This can improve access and quality of healthcare, especially in remote settings with limited access to energy [106–107]. An example is using renewable energy in improving cold-chain storage of vaccines and medicines such as anti-venom. In Sub-Saharan Africa, over 70% of health facilities lack reliable access to electricity, impacting storage of medicines and vaccines. The Solar for Health programme focuses on installing solar PV systems in health clinics located in the more vulnerable regions of world. It is estimated that health facilities will see a 100% return on their investment in solar PV cells within 2–3.5 years, which could be reinvested in other health sector priorities [107]. Following Hurricane Irma, a 'Smart Hospitals Toolkit' was used in rebuilding a care home in the British Virgin Islands [106]. Efficient lights and air conditioning units, solar PV, low-flow taps/toilets, and cool roofs helped to reduce energy and water costs, and the carbon footprint of the facility, with solar PV cells alone offsetting 20–30% of the facility's energy use, and combined with other measures reduced running costs significantly [106]. The return on investment through cheaper electricity and less reliance on fuel imports improves energy security, reduces vulnerability to fluctuating fuel prices and dependency on fossil fuel subsidies, and can be reinvested into health systems [108]. Ending fuel subsidies for fossil fuels can be challenging where such subsidies are popular with the public, and their reduction can be met with fierce opposition, but can be acceptable when funds are redirected to visible, tangible benefits such as investing in public health measures [109].

Air pollution is a major environmental risk to health (see earlier and Chapter 16), and often has similar emission sources to GHGs. A major health benefit of reducing GHG emissions, therefore, is improved air quality: ~1 million deaths occur annually as a result of air pollution from coal-fired power, and if improvements in particulate air pollution seen in Europe from 2015 to 2016 are sustained, this would lead to an annual average increase in healthy life-years' worth €5.2bn [110]. The Powering Past Coal Alliance was launched at COP23 to work to advance the transition from coal power generation to clean energy, and encourages all members to commit to phase out coal by 2030 in the OECD and EU, and by no later than 2050 in the rest of the world; this commitment is key to achieving climate targets. The avoided health costs through benefits to air quality of meeting the Paris Agreement goals would outweigh cost of the mitigation measures by 2-to-1, with greatest benefits seen in China and India [63].

In the UK, transport is now the biggest contributor to GHG emissions, contributing 27% of the total in 2019 [111]. Shifting to clean energy and electric vehicles can lower emissions of some harmful pollutants, benefiting health. However, active travel such as walking and cycling can have greater benefits to health through increasing physical activity combined with reducing air pollution. Improving and increasing access to public transport as well as options such as electric bikes (e-bikes) can aid equitable mobility, reducing inequality particularly across vulnerable groups, and improving access to services and employment for those without private vehicles [112]. Mitigation efforts that account for the unequal impact of climate change on health and society will help combat widening health inequalities.

Health impacts linked to nature are of concern in NDCs, with 33 NDCs highlighting VBDs and 27 NDCs highlighting food and nutritional security [104]. Nature Based Solutions (NBS) use natural systems to tackle climate challenges, and include protecting and restoring mangroves (which reduce coastal erosion, sequester CO_2 and provide habitat for fish and wildlife), using natural processes to enhance water availability and quality and for flood management (such as protecting natural wetlands), urban green/blue infrastructure (such as parks, lakes, green roofs/walls), forest protection/restoration, and sustainable agricultural, food and fisheries practices, among many others. NBS often have multiple and varied benefits for climate, health and ecosystems, are realistic and economically viable, and can offer sustainable and scalable solutions. Current land use for agriculture and forestry contributes 23% of global emissions, but land and coastal ecosystems could provide up to one-third of target emissions reduction [113]. Land-based climate change mitigation can support conservation goals, but some measures may have large impacts on biodiversity and ecosystem functions, so should be implemented and monitored carefully [114]. Agriculture is responsible for 80% of global deforestation and 70% of freshwater use, food systems release 29% of global GHGs, and around one-third of food is wasted/lost globally [115]. Mitigation actions that improve biodiversity have benefits for health, and WHO has produced guidance on mainstreaming biodiversity for nutrition and health [115]. The Intergovernmental Science-Policy Platform on Biodiversity and Ecosystem Services (IPBES) has produced critical assessments of the status and trends of the natural world, providing policymakers with objective scientific assessments about the state of knowledge regarding the planet's biodiversity, ecosystems and the contributions they make to people, and actions to protect and sustainably use these vital natural assets [114].

NBS in urban areas include urban green spaces, street trees, sustainable urban drainage systems (SuDS), green roofs and walls, and rain gardens. Green spaces can reduce the UHI, potentially reducing future energy demand for cooling in cities and thus reducing emissions. Cool roofs (paler colour roofs that reflect more sunlight) have also been used to reduce energy consumption. The New York City (NYC) CoolRoofs program, launched in 2009, has coated more than 10 million ft^2 of rooftop with white coatings, providing benefits and savings directly to the building owner by reducing cooling costs by 10–30%, tackling the UHI and reducing GHG emissions, supporting the city's goal of achieving carbon neutrality by 2050, and providing local job seekers with training and work experience [116–117].

Adaptation

Adaptation refers to changing our systems, behaviours and ways of life to protect our homes, health, society, economy and environment from the impacts of climate change. By strengthening resilience to climate-sensitive threats, adaptive capacity can be built to be better prepared to manage the increasing risks of climate change, especially extreme weather (Box 11.3).

Impacts of climate change are already happening on health and health systems, and the cost of inaction will be great. In Pakistan in 2010, over 15 million people were affected by flooding, with 6 million requiring urgent medical care, although healthcare delivery was majorly affected by the destruction of over 200 health facilities [118]. Climate impacts disproportionately affect disadvantaged and vulnerable populations through food and water insecurity, lost livelihood opportunities, adverse health impacts and population displacements. This widens health inequalities, by exacerbating existing inequalities, limiting ability to respond and act, and thus climate change is projected to be a poverty-multiplier [119]. Water security for 80% of global population is under threat, 25% of healthcare facilities lack basic water services and 20% have no sanitation [119]. Currently only 0.5% of climate finance goes to health projects [118].

There can be significant health gains from adaptation actions, which can also support health equity and universal health coverage, as

Box 11.3 Disaster Risk Reduction (see also Chapter 22)

During the 2010s, 83% of all disasters were linked to extreme weather and climate-related events like heatwaves, floods and storms, and these disasters combined led to the deaths of 410,000 people and affected 1.7 billion people [136]. Furthermore, countries most affected by climate-related disasters only receive a small amount of funding available for climate adaptation, leading to a disparity in where the climate risk is greatest and where climate adaptation funding is received [136].

Multi-sector, all-hazard plans can help integrate risk reduction and early-warning/ response systems for health emergencies, ensuring integration into national action plans. Developing heat-action plans (Box 11.2) that provide early alerts, combined with emergency public health measures can reduce heat-related morbidity and mortality. Key global efforts include:

- The Sendai Framework for Disaster Risk Reduction 2015–2030 (see also Chapter 22) aims to achieve substantial reduction of disaster risk and losses in lives, livelihoods and health and in the economic, physical, social, cultural and environmental assets of persons, businesses, communities and countries. It is the first major agreement of the post-2015 development agenda, outlining seven targets and four priorities for action: (a) Understanding disaster risk; (b) Strengthening disaster risk governance to manage disaster risk; (c) Investing in disaster reduction for resilience and; (d) Enhancing disaster preparedness for effective response, and to 'Build Back Better' in recovery, rehabilitation and reconstruction [137].
- The REAP (Risk-informed Early Action Partnership) initiative was launched at the United Nations Climate Action Summit (UNCAS) in 2019, aiming to make one billion people safer from disasters by 2025. REAP includes targets: for 50 countries to have reviewed and integrated disaster risk management and climate adaptation laws, policies and plans; one billion more people to be covered by financing and delivery mechanisms linked to effective early action plans; US$500 million to be invested in early warning systems to target early action in 'last/first mile' communities; for 1 billion more people to be covered with new/improved early warning systems, including for heatwaves [138].
- The UNDRR/ISC Sendai Hazard Definition and Classification Review Technical Report [139] provides a standardised characterisation of hazards that forms a basis for countries to assess and enhance their risk reduction policies and operational risk management practices. The report provides a common set of hazard definitions to characterise the broad range of hazards, and the cascading and complex nature of natural and human-induced hazards, including their potential impact on health, social, economic and other systems. Six recommendations are made: (a) Regular review and update; (b) Facilitate the development of a multi-hazard information system; (c) Engaging with users and sectors for greater alignment and consistency of hazard definitions; (d) Use the hazard list to actively engage policymakers and scientists in evidence-based national risk assessment processes, disaster risk reduction and risk-informed sustainable development, and other actions aimed at managing risks of emergencies and disasters; (e) Conduct further work to operationalise parameters for exposure, vulnerability and capacity, building on the United Nations General Assembly definitions; (f) Address cascading and complex hazards and risks [139].

well as shielding vulnerable populations from the most serious harm and improving preparedness. Early Warning Systems (EWS) can be a key part of adaptation to climate risks. The city of Ahmedabad, India, has a detailed heat action plan which includes improving public awareness, implementing an early warning system and inter-agency coordination, developing warning messaging, capacity building among healthcare professionals and reducing heat exposure and promoting adaptive measures such as the Ahmedabad Cool Roofs Program [120]; since implementation, around 1,000 deaths/year are avoided [121]. Parts of this action plan were based on the Heatwave Plan for England, that the UKHSA (formerly Public Health England) are responsible for (Box 11.2) [8].

The WHO operational framework for climate resilient health systems (2015) identifies ten key components for action to strengthen health systems' resilience and improve their ability to anticipate, prevent, prepare for and respond to climate-related health risks [122]. There is additional guidance on Climate Resilient and Sustainable Health and Care Facilities [123] to enhance capacity to protect and improve the health of their target communities in an unstable and changing climate, and to empower healthcare facilities to be environmentally sustainable. Climate resilient and sustainable health systems are a key component of universal health coverage. The guidance gives an overview of all proposed interventions, as well as a broader framework for achieving climate resilient and environmentally sustainable healthcare facilities [123].

Adapted and resilient agricultural systems have multiple health co-benefits through increased incomes, diversified livelihoods (and hence reduced risks) as well as improved food security amongst others. Climate smart agriculture (CSA) includes crop management, field management, farm risk reduction and soil management practices, agroforestry (farmer-managed natural regenerations), soil and water conservation technologies,

and climate information services. In Kenya, CSA has been used to improve food security, and enhance soil nutrients, with the greatest impact if all categories of practices are included, with adopters 57% more food-secure [124]. In Nigeria, drought-tolerant maize has been used as a 'win-win' adaptation strategy to address food security in rural areas, increasing maize yields by 13%, reducing exposure to drought risk by 81%, and resulting in a reduction in poverty (13%) and chance of food scarcity (84%). Despite these benefits, current adoption rates are low, and future dissemination efforts are crucial, including mainstreaming into national climate adaptation plans [125]. Adoption of CSA is influenced by the gender of household heads and learning through established networks increases probability of technology adoption, as farmers trust more practical experiences demonstrated by peers [124]. Improving resilience of food systems can also help protect against shocks to food prices.

Anticipatory action (AA) pilot initiatives aimed at delivering support to vulnerable communities before disasters strike is growing. Forecast-based early action (FbA), forecast-based financing (FbF) and early warning early action (EWEA) rely on risk information/forecasting to trigger planned, pre-financed actions when a disaster is forecast or before impacts occur. FbA initiatives use forecasts to provide support to at-risk communities before a disaster happens, to reduce the humanitarian impact and consider how aid is spent. Limited evidence on the costs/benefits of these actions suggests that even a false early response is more than offset by the cost of a late response [126]. Overall, evidence suggests that AA at household level is mainly positive, but benefits can be mixed, and the context of actions is key to their success – acting early can be better than doing nothing, but it is unclear whether it is better than other options [127].

'Urban green spaces: a brief for action' (WHO/Europe) contains key findings from research and practical case studies, and lessons learned to inform design of urban green spaces that promote/maximise social and health benefits [128]. Green spaces and other nature-based solutions offer innovative approaches to increase the quality of urban settings, enhance local resilience and promote sustainable lifestyles, improving the health and the well-being of urban residents. Parks, playgrounds or vegetation in public and private places are a central component of these approaches [128]. Spending time in nature promotes better mental and physical health, aids recovery from illness, and biodiversity and nature are often central to cultural well-being [129]. A UK study on associations between contact with nature and health and well-being showed that the likelihood of reporting good health and high well-being was significantly greater if participants reported 120 mins or more weekly contact with nature, with benefits still felt if over several shorter visits or one long visit [130]. Other benefits to health of green spaces in the UK have been reported: in 2017, urban green and blue space in Great Britain removed 27,900 tonnes of air pollution, with avoided health costs estimated at £162.6m, with 70% of the avoided costs due to the positive effects of urban woodland [131]. Additionally, disadvantaged groups gained larger health benefits and have reduced socioeconomic-related inequalities in health when living in greener communities, helping address inequity [131]. Launched in 2021, the EU Biodiversity Strategy for 2030 is a central element of both the EU Green Deal and the EU Recovery Plan, setting ambitious targets for nature protection and restoration which will restore balance with nature and contribute to transformational change that will filter through to all parts of society, ensuring the 'health and prosperity of people and nature' [132].

Adaptation actions can have great benefits to health but need to be carefully designed and implemented to avoid maladaptation, and there may be limits and costs to adaptation. The UNEP Adaptation Gap Report 2018 is health-focused and concludes that unless adaptation efforts are strengthened, health impacts from heat and extreme weather, from infectious diseases, and from risks to food systems and nutritional security will be significant; while health is a priority sector in over half of NDCs featuring adaptation, there are few quantitative targets against which progress can be measured, and current funding for climate change adaptation in health is negligible [133]. Whilst planning has progressed, significant gaps remain, *finance for adaptation must be stepped up urgently, along with faster implementation*, and solutions must be locally appropriate [134]. The first systematic global stocktake of human adaptation to climate change found that despite a growing literature reporting on efforts to adapt, there was negligible evidence of risk reduction, with responses being largely local, incremental, and with little evidence of substantive changes [135].

Chapter references

[1] IPCC. (2018) Annex I: Glossary, Matthews JBR (Ed.). In *Global Warming of 1.5°C. An IPCC Special Report on the Impacts of Global Warming of 1.5°C Above Pre-Industrial Levels and Related Global Greenhouse Gas Emission Pathways, in the Context of Strengthening the Global Response to the Threat of Climate Change, Sustainable Development, and Efforts to Eradicate Poverty*, Masson-Delmotte V, Zhai P, Pörtner HO, Roberts D, Skea J, Shukla PR, Pirani A, Moufouma-Okia W, Péan C, Pidcock R, Connors S, Matthews JBR, Chen Y, Zhou X, Gomis MI, Lonnoy E, Maycock T, Tignor M, Waterfield T (Eds.), In Press. https://www.ipcc.ch/sr15/chapter/glossary/.

[2] IPCC. (2021) Summary for policymakers. In *Climate Change 2021: The Physical Science Basis Contribution of Working Group I to the Sixth Assessment of the Intergovernmental Panel on Climate Change*, Masson-Delmotte V, Zhai P, Pirani A, Connors SL, Péan C, Berger S, et al. (Eds.), Cambridge University Press. https://www.ipcc.ch/report/ar6/wg1/downloads/report/IPCC_AR6_WGI_SPM.pdf.

[3] UK Met Office. (2021) *UK Climate Projections: Headline Findings*. https://www.metoffice.gov.uk/binaries/content/assets/metofficegovuk/pdf/research/ukcp/

ukcp18_headline_findings_v3.pdf: Source: Met Office © Crown Copyright 2021.

[4] United Nations. *Sustainable Development Goals.* https://sdgs.un.org/goals.

[5] Kovats S, Brisley R. (2021) Health, communities and the built environment. In *The Third Climate Change Risk Assessment Technical Report*, Betts RA, Haward AB, Pearson KV (Eds.). https://www.ukclimaterisk.org/wp-content/uploads/2021/06/CCRA3-Chapter-5-FINAL.pdf.

[6] Department for Environment Food & Rural Affairs. (2018) *The National Adaptation Programme and the Third Strategy for Climate Adaptation Reporting.* https://assets.publishing.service.gov.uk/government/uploads/system/uploads/attachment_data/file/727252/national-adaptation-programme-2018.pdf.

[7] Health Protection Agency. (2012) *Health Effects of Climate Change in the UK*, Vardoulakis S, Heaviside C (Ed.). https://assets.publishing.service.gov.uk/government/uploads/system/uploads/attachment_data/file/371103/Health_Effects_of_Climate_Change_in_the_UK_2012_V13_with_cover_accessible.pdf.

[8] Public Health England. (2015) *Heatwave Plan for England.* https://www.gov.uk/government/publications/heatwave-plan-for-england.

[9] Public Health England. (2015) *Cold Weather Plan for England.* https://assets.publishing.service.gov.uk/government/uploads/system/uploads/attachment_data/file/748492/the_cold_weather_plan_for_england_2018.pdf.

[10] NHS England. (2020) *Delivering a "Net Zero" National Health Service.* https://www.england.nhs.uk/greenernhs/wp-content/uploads/sites/51/2020/10/delivering-a-net-zero-national-health-service.pdf.

[11] Atwoli L, Baqui AH, Benfield T, Bosurgi R, Godlee F, Hancocks S, et al. (2021) Call for emergency action to limit global temperature increases, restore biodiversity, and protect health. *Lancet,* 398: 939–941.

[12] UNISDR, CRED. (2017) *Economic Losses, Poverty and Disasters 1998–2017*, United Nations Office for Disaster Risk Reduction (UNISDR) – Centre for Research on the Epidemiology of Disasters (CRED). https://www.undrr.org/publication/economic-losses-poverty-disasters-1998-2017.

[13] Diffenbaugh NS, Burke M. (2019) Global warming has increased global economic inequality. *Proceedings of the National Academy of Sciences of the United States of America*, 116: 9808–9813.

[14] The World Bank. (2017) *Results Brief: Climate Insurance.* https://www.worldbank.org/en/results/2017/12/01/climate-insurance.

[15] Jafino BA, Walsh B, Rozenberg J, Hallegatte S. (2020) *Revised Estimates of the Impact of Climate Change on Extreme Poverty by 2030*, Washington, DC. http://www.worldbank.org/prwp.

[16] World Health Organization. (2014) *Quantitative Risk Assessment of the Effects of Climate Change on Selected Causes of Death, 2030s and 2050s,* Geneva. http://www.who.int/globalchange/publications/quantitative-risk-assessment/en/.

[17] Internal Displacement Monitoring Centre. (2020) *Global Report on Internal Displacement 2020.* https://www.internal-displacement.org/global-report/grid2020/.

[18] Rigaud KK, de Sherbinin A, Jones B, Bergmann J, Clement V, Ober K, et al. (2018) *Groundswell: Preparing for Internal Climate Migration.* https://openknowledge.worldbank.org/bitstream/handle/10986/29461/WBG_ClimateChange_Final.pdf.

[19] United Nations Department of Economic and Social Affairs Population Division. (2019) *World Urbanization Prospects: The 2018 Revision*, United Nations, New York. file:///C:/Users/rocey/Downloads/WUP2018-Report.pdf.

[20] World Health Organization. (2018) *Heat and Health.* https://www.who.int/news-room/fact-sheets/detail/climate-change-heat-and-health.

[21] Robine JM, Cheung SLK, Le Roy S, van Oyen H, Griffiths C, Michel JP, et al. (2008) Death toll exceeded 70,000 in Europe during the summer of 2003. *Comptes Rendus Biologies*, 331: 171–178.

[22] Vicedo-Cabrera AM, Scovronick N, Sera F, Royé D, Schneider R, Tobias A, et al. (2021) The burden of heat-related mortality attributable to recent human-induced climate change. *Nature Climate Change,* 11: 492–500.

[23] Hansen A, Bi P, Nitschke M, Ryan P, Pisaniello D, Tucker G. (2008) The effect of heat waves on mental health in a temperate Australian city. *Environmental Health Perspectives,* 116: 1369–1375.

[24] Stone B, Mallen E, Rajput M, Gronlund CJ, Broadbent AM, Krayenhoff ES, et al. (2021) Compound climate and infrastructure events: How electrical grid failure alters heat wave risk. *Environmental Science & Technology*, 55(10): 6957–6964.

[25] World Health Organization. (undated) *Heatwaves.* https://www.who.int/health-topics/heatwaves#tab=tab_1.

[26] USGCRP. (2018) Chapter 14: Human health. In *Impacts, Risks, and Adaptation in the United States: Fourth National Climate Assessment, Volume II*, U.S. Global Change Research Program,

Washington, DC. https://doi.org/10.7930/NCA4.2018.

[27] Curriero FC, Patz JA, Rose JB, Lele S. (2001) The association between extreme precipitation and waterborne disease outbreaks in the United States, 1948–1994. *American Journal of Public Health,* 91: 1194–1199.

[28] Bokhari H, Shah MA, Asad S, Akhtar S, Akram M, Wren BW. (2013) *Escherichia coli* pathotypes in Pakistan from consecutive floods in 2010 and 2011. *American Journal of Tropical Medicine and Hygiene,* 88: 519–525.

[29] Schwartz BS, Harris JB, Khan AI, Larocque RC, Sack DD, Malek MA, et al. (2006) Diarrheal epidemics in Dhaka, Bangladesh, during three consecutive floods: 1988, 1998 and 2004. *American Journal of Tropical Medicine and Hygiene,* 74: 1067–1073.

[30] Kraay ANM, Man O, Levy MC, Levy K, Ionides E, Eisenberg JNS. (2020) Understanding the impact of rainfall on diarrhea: Testing the concentration–dilution hypothesis using a systematic review and meta-analysis. *Environmental Health Perspectives,* 128: 126001-1–126001-16.

[31] Jones FK, Ko AI, Becha C, Joshua C, Musto J, Thomas S, et al. (2016) Increased rotavirus prevalence in diarrheal outbreak precipitated by localized flooding, Solomon Islands, 2014. *Emerging Infectious Diseases,* 22: 875–879.

[32] Waite TD, Chaintarli K, Beck CR, Bone A, Amlôt R, Kovats S, et al. (2017) The English national cohort study of flooding and health: Cross-sectional analysis of mental health outcomes at year one. *BMC Public Health,* 17: 129. http://doi.org/10.1186/s12889-016-4000-2.

[33] Jermacane D, Waite TD, Beck CR, Bone A, Amlôt R, Reacher M, et al. (2018) The English national cohort study of flooding and health: The change in the prevalence of psychological morbidity at year two. *BMC Public Health,* 18: 330.

[34] Mulchandani R, Armstrong B, Beck CR, Waite TD, Amlôt R, Kovats S, et al. (2020) The English national cohort study of flooding & health: Psychological morbidity at three years of follow up. *BMC Public Health,* 20: 321.

[35] Kreienkamp F, Philip SY, Tradowsky JS, Kew SF, Lorenz P, Arrighi J, et al. (2021) Rapid attribution of heavy rainfall events leading to the severe flooding in Western Europe during July 2021. *World Weather Attribution,* 13: 1–54.

[36] United Nations Office for Disaster Risk Reduction. (2021) *GAR Special Report on Drought 2021,* Geneva. https://www.undrr.org/publication/gar-special-report-drought-2021.

[37] Stanke C, Kerac M, Prudhomme C, Medlock J, Murray V. (2013) Health effects of drought:

A systematic review of the evidence. *PLoS Currents,* 5. ecurrents.dis.7a2cee9e980f91ad7697b570bcc4.

[38] FAO. (2018) *The Impact of Disasters and Crises on Agriculture and Food Security.* http://www.fao.org/3/I8656EN/i8656en.pdf.

[39] FAO. (2021) *The Impact of Disasters and Crises on Agriculture and Food Security,* FAO, Rome.

[40] Brás TA, Seixas J, Carvalhais N, Jagermeyr J. (2021) Severity of drought and heatwave crop losses tripled over the last five decades in Europe. *Environmental Research Letters,* 16: 065012.

[41] Chase JM, Knight TM. (2003) Drought-induced mosquito outbreaks in wetlands. *Ecology Letters,* 6: 1017–1024.

[42] Reuveny R. (2007) Climate change-induced migration and violent conflict. *Political Geography,* 26: 656–673.

[43] European Commission. (2020) *Impacts of Climate Change on Droughts.* https://ec.europa.eu/jrc/sites/jrcsh/files/07_pesetaiv_droughts_sc_august2020_en.pdf.

[44] Graham AM, Pope RJ, Pringle KP, Arnold S, Chipperfield MP, Conibear LA, et al. (2020) Impact on air quality and health due to the Saddleworth Moor fire in northern England. *Environmental Research Letters,* 15: 074018.

[45] Victorian Bushfires Royal Commission. (2009) *The 2009 Victorian Bushfires Royal Commission Final Report Summary.* http://royalcommission.vic.gov.au/Commission-Reports/Final-Report/Volume-1/Print-Friendly-Version.html.

[46] Bryant RA, Waters E, Gibbs L, Gallagher HC, Pattison P, Lusher D, et al. (2014) Psychological outcomes following the Victorian Black Saturday bushfires. *Australian and New Zealand Journal of Psychiatry,* 48: 634–643.

[47] Bryant RA, Gibbs L, Gallagher HC, Pattison P, Lusher D, MacDougall C, et al. (2021) The dynamic course of psychological outcomes following the Victorian Black Saturday bushfires. *Australian and New Zealand Journal of Psychiatry,* 55: 666–677.

[48] Fann N, Alman B, Broome RA, Morgan GG, Johnston FH, Pouliot G, et al. (2018) The health impacts and economic value of wildland fire episodes in the U.S.: 2008–2012. *Science of the Total Environment,* 610–611: 802–809. https://doi.org/10.1016/j.scitotenv.2017.08.024.

[49] Wang D, Guan D, Zhu S, Mac Kinnon M, Geng G, Zhang Q, et al. (2021) Economic footprint of California wildfires in 2018. *Nature Sustainability,* 4: 252–260. http://dx.doi.org/10.1038/s41893-020-00646-7.

[50] Arias PA, Bellouin N, Coppola E, Jones RG, Krinner G, Marotzke J, et al. (2021) Technical summary. In *Climate Change 2021: The Physical Science Basis Contribution of Working Group I to the Sixth Assessment of the Intergovernmental Panel on Climate Change*, Masson-Delmotte V, Zhai P, Pirani A, Connors SL, Péan C, Berger S, et al. (Eds.), Cambridge University Press, Cambridge.

[51] Abram NJ, Henley BJ, Sen Gupta A, Lippmann TJR, Clarke H, Dowdy AJ, et al. (2021) Connections of climate change and variability to large and extreme forest fires in southeast Australia. *Communications Earth & Environment*, 2: 8.

[52] van Oldenborgh GJ, Krikken F, Lewis S, Leach NJ, Lehner F, Saunders KR, et al. (2021) Attribution of the Australian bushfire risk to anthropogenic climate change. *Natural Hazards and Earth System Sciences*, 21: 941–960.

[53] Mueller SE, Thode AE, Margolis EQ, Yocom LL, Young JD, Iniguez JM. (2020) Climate relationships with increasing wildfire in the southwestern US from 1984 to 2015. *Forest Ecology and Management*, 460: 117861. https://doi.org/10.1016/j.foreco.2019.117861.

[54] World Health Organization. (2018) *Fact Sheet: Ambient (Outdoor) Air Pollution.* https://www.who.int/news-room/fact-sheets/detail/ambient-(outdoor)-air-quality-and-health.

[55] COMEAP. (2010) *The Mortality Effects of Long-Term Exposure to Particulate Air Pollution in the United Kingdom.* Committee on the Medical Effects of Air Pollutants, London.

[56] COMEAP. (2018) *Nitrogen Dioxide: Interim View on Long-Term Average Concentrations and Mortality*, Produced by Public Health England for the Committee in the Medical Effects of Air Pollutants. London.

[57] Heal MR, Heaviside C, Doherty RM, Vieno M, Stevenson DS, Vardoulakis S. (2013) Health burdens of surface ozone in the UK for a range of future scenarios. *Environment International*, 61: 36–44. http://www.sciencedirect.com/science/article/pii/S0160412013002043.

[58] COMEAP. (2015) *Quantification of Mortality and Hospital Admissions Associated with Ground-Level Ozone.* Committee on the Medical Effects of Air Pollutants. Produced by Public Health England for the Committee in the Medical Effects of Air Pollutants, London.

[59] Macintyre HL, Heaviside C, Neal LS, Agnew P, Thornes J, Vardoulakis S. (2016) Mortality and emergency hospitalizations associated with atmospheric particulate matter episodes across the UK in spring 2014. *Environment International*, 97: 108–116. http://www.sciencedirect.com/science/article/pii/S0160412016302847.

[60] Fenech S, Doherty RM, Heaviside C, Macintyre HL, O'Connor FM, Vardoulakis S, et al. (2019) Meteorological drivers and mortality associated with O3 and PM2.5 air pollution episodes in the UK in 2006. *Atmospheric Environment*, 213: 699–710. http://www.sciencedirect.com/science/article/pii/S1352231019304212.

[61] Blackport R, Screen JA. (2020) Weakened evidence for mid-latitude impacts of Arctic warming. *Nature Climate Change*, 10(12): 1065–1066. https://doi.org/10.1038/s41558-020-00954-y.

[62] Kirtman B, Power SB, Adedoyin JA, Boer GJ, Bojariu R, Camilloni I, et al. (2013) Near-term climate change: Projections and predictability. In *Climate Change 2013: The Physical Science Basis Contribution of Working Group I to the Fifth Assessment Report of the Intergovernmental Panel on Climate Change*, Stocker TF, Qin D, Plattner GK, Tignor M, Allen SK, Boschung J, et al. (Eds.), Cambridge University Press, Cambridge, New York, 953–1028.

[63] Markandya A, Sampedro J, Smith SJ, Van Dingenen R, Pizarro-Irizar C, Arto I, et al. (2018) Health co-benefits from air pollution and mitigation costs of the Paris agreement: A modelling study. *Lancet Planetary Health*, 2(3): e126–e133. https://doi.org/10.1016/S2542-5196(18)30029-9.

[64] Williams ML, Lott MC, Kitwiroon N, Dajnak D, Walton H, Holland M, et al. (2018) The lancet countdown on health benefits from the UK climate change act: A modelling study for Great Britain. *Lancet Planetary Health*, 2(5): e202–e213. http://www.sciencedirect.com/science/article/pii/S2542519618300676.

[65] Climate Change Committee. (2020) *The Sixth Carbon Budget, the UK's Path to Net Zero.* https://www.theccc.org.uk/publication/sixth-carbon-budget/.

[66] AQEG. (2020) *Impacts of Net Zero Pathways on Future Air Quality in the UK.* https://uk-air.defra.gov.uk/library/reports.php?report_id=1002.

[67] IEA. (2020) *Tracking Buildings 2020*, IEA, Paris. https://www.iea.org/reports/tracking-buildings-2020 https://www.iea.org/reports/tracking-buildings-2020.

[68] Vardoulakis S, Dimitroulopoulou C, Thornes J, Lai KM, Taylor J, Myers I, et al. (2015) Impact of climate change on the domestic indoor environment and associated health risks in the UK. *Environment International*, 85: 299–313.

http://www.sciencedirect.com/science/article/pii/S0160412015300507.

[69] Li S, Gilbert L, Harrison PA, Rounsevell MDA. (2016) Modelling the seasonality of Lyme disease risk and the potential impacts of a warming climate within the heterogeneous landscapes of Scotland. *Journal of the Royal Society Interface*, 13: 20160140. http://rsif.royalsocietypublishing.org/content/8/55/153.full.html#related-urls%5Cnhttp://rsif.royalsocietypublishing.org/content/8/55/153.full.html%23ref-list-1.

[70] Colón-González FJ, Sewe MO, Tompkins AM, Sjödin H, Casallas A, Rocklöv J, et al. (2021) Projecting the risk of mosquito-borne diseases in a warmer and more populated world: A multi-model, multi-scenario intercomparison modelling study. *Lancet Planet Heal*, 5: e404–e414. http://doi.org/10.1016/S2542-5196(21)00132-7.

[71] Garcia-Vozmediano A, Krawczyk AI, Sprong H, Rossi L, Ramassa E, Tomassone L. (2020) Ticks climb the mountains: Ixodid tick infestation and infection by tick-borne pathogens in the Western Alps. *Ticks and Tick-Borne Diseases, Elsevier*, 11: 101489.

[72] Jaenson T, Jaenson D, Eisen L, Petersson E, Lindgren E. (2012) Changes in the geographical distribution and abundance of the tick *Ixodes ricinus* during the past 30 years in Sweden. *Parasites and Vectors*, 5(1): 8. http://www.pubmedcentral.nih.gov/articlerender.fcgi?artid=3311093&tool=pmcentrez&rendertype=abstract.

[73] Vladimirov LN, Machakhtyrov GN, Machakhtyrova VA, Louw AS, Sahu N, Yunus AP, et al. (2021) Quantifying the northward spread of ticks (Ixodida) as climate warms in Northern Russia. *Atmosphere (Basel)*, 12: 233.

[74] Siraj AS, Santos-Vega M, Bouma MJ, Yadeta D, Carrascal DR, Pascual M. (2014) Altitudinal changes in malaria incidence in highlands of Ethiopia and Columbia. *Science,* 343: 1154–1159.

[75] Caminade C, Turner J, Metelmann S, Hesson JC, Blagrove MSC, Solomon T, et al. (2017) Global risk model for vector-borne transmission of Zika virus reveals the role of El Niño 2015. *Proceedings of the National Academy of Sciences of the United States of America*, 114: 119–124.

[76] Berry A, Moné H, Iriart X, Mouahid G, Aboo O, Boissier J, et al. (2014) *Schistosomiasis haematobium*, Corsica, France. *Emerging Infectious Diseases*, 20(9): 1595–1597.

[77] Mulero S, Rey O, Arancibia N, Mas-Coma S, Boissier J. (2019) Persistent establishment of a tropical disease in Europe: The preadaptation of schistosomes to overwinter. *Parasites and Vectors,* 12: 379. https://doi.org/10.1186/s13071-019-3635-0.

[78] Yang GJ, Bergquist R. (2018) Potential impact of climate change on schistosomiasis: A global assessment attempt. *Tropical Medicine and Infectious Disease*, 3: 117.

[79] Jones CJ, Kitron UD. (2000) Populations of *Ixodes scapularis* (Acari: Ixodidae) are modulated by drought at a Lyme disease focus in Illinois. *Journal of Medical Entomology*, 37: 408–415.

[80] Kilian AHD, Langi P, Talisuna A, Kabagambe G. (1998) Rainfall pattern, El Niño and malaria in Uganda. *Tropical Medicine*, 22–23.

[81] Brown V, Issak MA, Rossi M, Barboza P, Paugam A. (1998) Epidemic of malaria in North-Eastern Kenya. *Lancet*, 352: 1356–1357.

[82] Lindsay SW, Bødker R, Malima R, Msangeni HA, Kisinza W. (2000) Effect of 1997–1998 El Niño on highland malaria in Tanzania. *Lancet*, 355: 989–990.

[83] Boyce R, Reyes R, Matte M, Ntaro M, Mulogo E, Metlay JP, et al. (2016) Severe flooding and malaria transmission in the Western Ugandan highlands: Implications for disease control in an era of global climate change. *The Journal of Infectious Diseases,* 214: 1403–1410.

[84] Paull SH, Horton DE, Ashfaq M, Rastogi D, Kramer LD, Diffenbaugh NS, et al. (2017) Drought and immunity determine the intensity of West Nile virus epidemics and climate change impacts. *Proceedings B is the Royal Society*, 284: 20162078.

[85] Hiatt RA, Beyeler N. (2020) Cancer and climate change. *Lancet Oncology*, 21:e519–e527. http://doi.org/10.1016/S1470-2045(20)30448-4.

[86] McMichael C, Barnett J, McMichael AJ. (2012) An ill wind? Climate change, migration, and health. *Environ Health Perspect*, 120: 646–654. http://www.ncbi.nlm.nih.gov/pmc/articles/PMC3346786/pdf/ehp.1104375.pdf.

[87] Clayton S, Cunsolo A, Derr V, Doherty T, Kotcher J, Silka L, et al. (2017) *Mental Health and Our Changing Climate: Impacts, Implications and Guidance*, Washington, DC. https://www.apa.org/news/press/releases/2017/03/mental-health-climate.pdf.

[88] Clayton S, Karazsia BT. (2020) Development and validation of a measure of climate change anxiety. *Journal of Environmental Psychology*, 69: 101434. https://doi.org/10.1016/j.jenvp.2020.101434.

[89] Cunsolo A, Ellis NR. (2018) Ecological grief as a mental health response to climate change-related

loss. *Nature Climate Change*, 8: 275–281. http://doi.org/10.1038/s41558-018-0092-2.

[90] FAO. (2009) *How to Feed the World in 2050.* https://www.fao.org/fileadmin/templates/wsfs/docs/expert_paper/How_to_Feed_the_World_in_2050.pdf.

[91] FAO, IFAD, UNICEF, WFP, WHO. (2021) *The State of Food Security and Nutrition in the World 2021: Transforming Food Systems for Food Security, Improved Nutrition and Affordable Healthy Diets for All.* https://www.fao.org/documents/card/en/c/cb4474en.

[92] FSIN. (2021) *Global Report on Food Crises: Joint Analysis for Better Decisions.* https://www.wfp.org/publications/global-report-food-crises-2021.

[93] Vermeulen SJ, Campbell BM, Ingram JSI. (2012) Climate change and food systems. *Annual Review of Environment and Resources*, 37: 195–222.

[94] Springmann M, Mason-D'Croz D, Robinson S, Garnett T, Godfray HCJ, Gollin D, et al. (2016) Global and regional health effects of future food production under climate change: A modelling study. *Lancet*, 387: 1937–1946. http://doi.org/10.1016/S0140-6736(15)01156-3.

[95] FAO, IFAD, UNICEF, WFP, WHO. (2018) *The State of Food Security and Nutrition in the World 2018: Building Climate Resilience for Food Security and Nutrition.* http://www.fao.org/3/I9553EN/i9553en.pdf.

[96] Niles MT, Salerno JD. (2018) A cross-country analysis of climate shocks and smallholder food insecurity. *PLoS One*, 13: e0192928.

[97] Cooper MW, Brown ME, Hochrainer-Stigler S, Pflug G, McCallum I, Fritz S, et al. (2019) Mapping the effects of drought on child stunting. *Proceedings of the National Academy of Sciences of the United States of America*, 116: 17219–17224.

[98] Tigchelaar M, Battisti DS, Naylor RL, Ray DK. (2018) Future warming increases probability of globally synchronized maize production shocks. *Proceedings of the National Academy of Sciences of the United States of America*, 115: 6644–6649.

[99] Mbow C, Rosenzweig C, Barioni LG, Benton TG, Herrero M, Krishnapillai M, et al. (2019) *Food Security. Climate Change and Land: An IPCC Special Report on Climate Change, Desertification, Land Degradation, Sustainable Land Management, Food Security, and Greenhouse Gas Fluxes in Terrestrial Ecosystems.* https://www.ipcc.ch/site/assets/uploads/sites/4/2021/02/08_Chapter-5_3.pdf.

[100] Medek DE, Schwartz J, Myers SS. (2017) Estimated effects of future atmospheric CO_2 concentrations on protein intake and the risk of protein deficiency by country and region. *Environ Health Perspect*, 125: 087000.

[101] Smith MR, Myers SS. (2018) Impact of anthropogenic CO_2 emissions on global human nutrition. *Nature Climate Change*, 8: 834–839. http://doi.org/10.1038/s41558-018-0253-3.

[102] Zhu C, Kobayashi K, Loladze I, Zhu J, Jiang Q, Xu X, et al. (2018) Carbon dioxide (CO_2) levels this century will alter the protein, micronutrients, and vitamin content of rice grains with potential health consequences for the poorest rice-dependent countries. *Science Advances*, 4: eaaq1012.

[103] IRENA. (2021) *International Renewable Energy Agency: Renewable Power Generation Costs in 2020.* https://www.irena.org/publications/2021/Jun/Renewable-Power-Costs-in-2020.

[104] World Health Organization. (2020) *Health in National Determined Contributions (NDCs): A WHO Review.* https://www.who.int/publications/i/item/who-review-health-in-the-ndcs.

[105] Pencheon D, Wight J. (2020) Making healthcare and health systems net zero. *BMJ*, 368: m970. https://www.bmj.com/content/bmj/368/bmj.m970.full.pdf.

[106] PAHO. (2017) *Smart Hospitals Toolkit.* https://www.paho.org/en/health-emergencies/smart-hospitals/smart-hospitals-toolkit (Accessed October 2021).

[107] United Nations Development Programme. (2018) *UNDP Solar for Health Case Study.* https://www.undp-capacitydevelopment-health.org/en/capacities/focus/solar-for-health/cases/strategy-overview-and-case-studies/ (Accessed October 2021).

[108] LEDS Global Partnership. (2017) *Benefits of Low Emission Development Strategies: The Case of Kenya's Lake Turkana Wind Power Project.* https://ledsgp.org/app/uploads/2017/02/Kenya-Benefits-Case-Study-FINAL.pdf (Accessed October 2021).

[109] Yates R. (2014) Recycling fuel subsidies as health subsidies. *Bulletin of the World Health Organization, World Health Organization*, 92: 547–547A. https://pubmed.ncbi.nlm.nih.gov/25177065.

[110] Watts N, Amann M, Arnell N, Ayeb-Karlsson S, Beagley J, Belesova K, et al. (2021) The 2020 report of the Lancet Countdown on Health and Climate Change: Responding to converging crises. *Lancet Elsevier*, 397(10269): 129–170. https://doi.org/10.1016/S0140-6736(20)32290-X.

[111] BEIS. (2020) *Final UK Greenhouse Gas Emissions National Statistics: 1990 to 2019.* https://

www.gov.uk/government/collections/final-uk-greenhouse-gas-emissions-national-statistics.

[112] Philips I, Anable J, Chatterton T. (2020) E-bike carbon savings – how much and where? *creds.ac.uk*. https://www.creds.ac.uk/wp-content/uploads/CREDS-e-bikes-briefing-May2020.pdf (Accessed October 2021).

[113] United Nations. (2019) *Report of the Secretary-General on the 2019 Climate Action Summit and the Way Forward In 2020.* https://www.un.org/sites/un2.un.org/files/cas_report_11_dec_0.pdf (Accessed October 2021).

[114] IPBES. (2019) *Summary for Policymakers of the Global Assessment Report on Biodiversity and Ecosystem Services of the Intergovernmental Science-Policy Platform on Biodiversity and Ecosystem Services*, Díaz S, Brondízio ES, Ngo HT, Guèze M, Agard J, Arneth A, et al. (Ed.). https://doi.org/10.5281/zenodo.3553579; https://ipbes.net/sites/default/files/inline/files/ipbes_global_assessment_report_summary_for_policymakers.pdf.

[115] World Health Organization. (2020) *Guidance on Mainstreaming Biodiversity for Nutrition and Health.* https://www.who.int/publications/i/item/guidance-mainstreaming-biodiversity-for-nutrition-and-health (Accessed October 2021).

[116] C40 Cities. (2015) *Case Study: The NYC CoolRoofs Program.* https://www.c40.org/case_studies/nyc-coolroofs.

[117] Sustainable Energy for All. (2019) *NYC Cool Roofs Initiative.* https://www.seforall.org/stories-of-success/video-nyc-coolroofs-initiative.

[118] World Health Organization. (2018) *COP24 Special Report: Health and Climate Change.* http://www.who.int/iris/handle/10665/276405.

[119] World Health Organization. (2018) *The 1.5 Health Report: Synthesis on Health & Climate Science in the IPCC SR1.5.* https://www.who.int/publications/i/item/the-1.5-health-report.

[120] Ahmedabad Municipal Corporation. (2016) *Ahmedabad Heat Action Plan 2016.* https://www.nrdc.org/sites/default/files/ahmedabad-heat-action-plan-2016.pdf.

[121] Hess JJ, Lm S, Knowlton K, Saha S, Dutta P, Ganguly P, et al. (2018) Building resilience to climate change: Pilot evaluation of the impact of India's first heat action plan on all-cause mortality. *Journal of Environmental and Public Health*, 7973519. https://doi.org/10.1155/2018/7973519.

[122] World Health Organization. (2015) *Operational Framework for Building Climate Resilient Health Systems.* https://www.who.int/publications/i/item/operational-framework-for-building-climate-resilient-health-systems.

[123] World Health Organization. (2020) *WHO Guidance for Climate Resilient and Environmentally Sustainable Health Care Facilities.* https://www.who.int/publications/i/item/9789240012226.

[124] Wekesa BM, Ayuya OI, Lagat JK. (2018) Effect of climate-smart agricultural practices on household food security in smallholder production systems: Micro-level evidence from Kenya. *Agriculture & Food Security*, 7: 80. https://doi.org/10.1186/s40066-018-0230-0.

[125] Wossen T, Abdoulaye T, Alene A, Feleke S, Menkir A, Manyong V. (2017) Measuring the impacts of adaptation strategies to drought stress: The case of drought tolerant maize varieties. *Journal of Environmental Management*, 203: 106–113. https://www.sciencedirect.com/science/article/pii/S0301479717306497.

[126] Overseas Development Institute. (2018) *Forecasting Hazards, Averting Disasters: Implementing Forecast-based Early Action at Scale.* https://odi.org/en/publications/forecasting-hazards-averting-disasters-implementing-forecast-based-early-action-at-scale/.

[127] Overseas Development Institute. (2020) *The Evidence Base on Anticipatory Action.* https://odi.org/en/publications/the-evidence-base-on-anticipatory-action/.

[128] World Health Organization Regional Office for Europe. (2017) *Urban Green Spaces: A Brief for Action.* https://www.euro.who.int/en/health-topics/environment-and-health/urban-health/publications/2017/urban-green-spaces-a-brief-for-action-2017.

[129] World Health Organization. (2015) *Connecting Global Priorities: Biodiversity and Human Health: A State of Knowledge Review.* https://www.who.int/publications/i/item/connecting-global-priorities-biodiversity-and-human-health.

[130] White MP, Alcock I, Grellier J, Wheeler BW, Hartig T, Warber SL, et al. (2019) Spending at least 120 minutes a week in nature is associated with good health and wellbeing. *Scientific Reports*, 9: 7730. https://doi.org/10.1038/s41598-019-44097-3.

[131] Public Health England. (2020) *Improving Access to Greenspace: A New Review for 2020.* https://assets.publishing.service.gov.uk/government/uploads/system/uploads/attachment_data/file/904439/

Improving_access_to_greenspace_2020_review.pdf.

[132] Directorate-General for Environment (European Commission). (2021) *EU Biodiversity Strategy for 2030: Bringing Nature Back into Our Lives*. https://op.europa.eu/en/publication-detail/-/publication/31e4609f-b91e-11eb-8aca-01aa75ed71a1#.

[133] United Nations Environment Programme. (2018) *Adaptation Gap Report 2018*. https://www.unep.org/resources/adaptation-gap-report-2020.

[134] United Nations Environment Programme. (2021) *Adaptation Gap Report 2020*. https://www.unep.org/resources/adaptation-gap-report-2020.

[135] Berrang-Ford L, Siders AR, Lesnikowski A, Fischer AP, Callaghan M, Haddaway N, et al. (2021) Mapping evidence of human adaptation to climate change. *Research Square*, In press. http://doi.org/10.21203/rs.3.rs-100873/v1 (Accessed October 2021).

[136] International Federation of Red Cross and Red Crescent Societies (IFRC). (2020) *World Disasters Report 2020 Come Heat or High Water*. https://www.ifrc.org/media/8968.

[137] United Nations Office for Disaster Risk Reduction. (2015) *The Sendai Framework for Disaster Risk Reduction 2015–2030*. https://www.preventionweb.net/publication/sendai-framework-disaster-risk-reduction-2015-2030.

[138] REAP. (2019) *Risk-Informed Early Action Partnership*. https://www.early-action-reap.org/.

[139] United Nations Office for Disaster Risk Reduction. (2020) *Hazard Definition and Classification Review*. https://www.undrr.org/publication/hazard-definition-and-classification-review.

Chapter Notes

1 See Globalchange.gov: https://www.globalchange.gov/browse/indicators/annual-greenhouse-gas-index

2 Data from Scripps CO2 Program https://scrippsco2.ucsd.edu/graphics_gallery/mauna_loa_record/mauna_loa_record.html

3 Source: https://www.metoffice.gov.uk/weather/climate/science/global-temperature-records.

4 This is available at https://interactive-atlas.ipcc.ch

5 SPM is an abbreviation of Summary for Policymakers.

12

Housing, health, and the domestic environment

David Ormandy and Véronique Ezratty

Introduction

Ensuring a safe and healthy home for everyone has become one of the most important challenges of our age. The potential effect of housing conditions on health has long been recognised, but the enforced lockdowns and the decrease in mobility as part of the responses to the COVID-19 pandemic in 2020 and 2021 has highlighted the need for healthy housing and that housing is a public health issue (although it is only in the UK that it is seen as a core function of EHPs; elsewhere in the world housing conditions are only dealt with indirectly). The ten-year review of the Marmot Report [1] recorded the more deprived the area, the shorter the life expectancy and this social gradient had become steeper over the last decade, and the 2020/21 pandemic highlighted in-household transmission of viruses which was said to be responsible for a significant number of infections [2].[1]

The reaction of governments to this pandemic included 'lockdowns' (requiring people to remain within their dwelling). These lockdowns emphasised and made everyone more aware of the huge inequities between those living in a spacious house with a garden (even if this may have caused some stress, made many more appreciative of their home) and those living in small, crowded houses and apartments with little or no outside space, where residents could not escape from any threats to health from the conditions. Ageing populations, and climate change causing more frequent and intense extreme events such as heat waves and floods, make the challenges societies are facing to maintain a safe and appropriate housing environment even greater.

Definitions

The words 'dwelling' and 'home' are often used as if they are synonymous, and the word 'housing' (as a noun) used to imply a group of houses. Before discussing the importance of dwelling, home, and housing and the relationship these have with 'health' and 'well-being', we propose definitions for the purposes of this chapter.

The World Health Organization, European Office (WHO) adopted a definition of housing as a broad concept involving four inter-related elements − the dwelling, the home, the immediate environment (or neighbourhood), and the community [2].

The **dwelling** is the physical structure providing shelter with the necessary space,

DOI: 10.1201/9781003035640-12

facilities, and amenities for the occupant(s) or intended occupant(s). A dwelling is also a financial asset, mainly a personal and individual asset, but a national asset as well. As an asset, it will have a psychological significance to the owner (or person purchasing, perhaps through a mortgage), but the degree of this significance will vary depending on whether the house is intended to provide accommodation for personal occupation or is to be let to a tenant. More relevant to this chapter is the primary purpose of a dwelling – that of providing somewhere to establish a home. To this end, it should be designed, constructed, and maintained to provide a safe and healthy environment for the occupant(s) and any visitors. However, it is not possible to provide a completely safe and healthy dwelling – we need some hazards such as electricity, stairs, cooking facilities, windows, and doors – and so any necessary and unavoidable hazards should be as safe as possible. A dwelling should provide protection from the local climate giving an optimum indoor environment for the occupants, it should allow for the normal day-to-day activities throughout the year without problems, and its condition should be such that it does not interfere with the establishment of a home. It should have sufficient space to allow for the inter-relationship between the members of the household and also allow individuals the opportunity for privacy. It should also be affordable, both in terms of the rent or mortgage repayment and in terms of the 'running-costs' – the costs of local taxes, of energy, of water, and of maintenance.

The **home** is the social, cultural, and economic structure created by the individual or household. It is the structure that gives a human refuge from the outside world, enables the development of a sense of identity, and, for a household, a sense of attachment. The home creates an environment where one can be, and develop, oneself.

The **immediate environment** or neighbourhood is the locality where the dwelling is situated. It includes the adjacent dwellings, the walkways and roads, the public services, the shops, the schools, the places for worship, the places for entertainment, and other amenities such as green space and playgrounds. It should be planned and maintained to be safe for use by pedestrians, bicyclists as well as for private and public transport. There should be ready access to the immediate environment for all, with easy connections to other areas.

The **community** is the social, cultural, and economic structure established by those living and working within a neighbourhood.

The WHO definition of health as a state of complete physical, mental, and social well-being and not merely the absence of disease or infirmity [4] is crucial to understanding the impact of housing conditions on the occupiers.[2] The term well-being or quality of life includes feelings of comfort and contentment and so goes beyond what would be considered clinically healthy.

Housing as a determinant of health

Each individual element of housing has the potential to have a direct or indirect impact on physical, social and mental health, and an impact on well-being. Two or more of inter-related elements can, together, have a combined impact.

The evidence base

The last five decades have seen an increased interest in collecting and strengthening evidence on the relationship between housing conditions and the health of occupiers. The evidence base is extensive and growing, and includes reviews on the relationship, with several conferences demonstrating the wealth of international studies, and analyses of data sources. There are references in this chapter to examples of this evidence [5–17] to which readers should refer. The range of intervening and confounding variables (such as life-style and working conditions) means that it can be difficult to demonstrate clear and measurable 'cause/effect'

491

relationships as demonstrated by some studies although Howden-Chapman et al. [18] is an example of a robust study and Thompson et al. [19] provides a systematic review.

There is however, a considerable amount of other evidence (so-called 'grey evidence' or 'grey literature') relating the condition of buildings (including houses) and the indoor environment to the health and safety of users [20]. In addition, many developments are accepted as being beneficial without the need for research or proof, such as child safety locks and restrictors for windows, and cut-off devices for gas and electrical appliances.

The normal day-to-day household activities within a dwelling are well known and so it should be possible to carry these out with unwanted and potentially harmful side effects. However, there remain many examples of housing conditions that can have an impact on the health and/or safety of occupants. These include poor energy efficiency resulting in excess low or high temperatures, dampness and mould growth that can exacerbate respiratory conditions such as asthma and bronchitis, inadequate or inappropriate ventilation reducing indoor air quality and allowing a build-up of pollutants, poor sound attenuation allowing noise to penetrate, poor design making it difficult to maintain a clean and healthy indoor environment, features that increase the likelihood of accidents such as falls, poor design and layout of kitchens increasing the possibility of accidents with hot liquids and equipment, and increased likelihood of a fire starting and spreading.

The World Health Organization LARES project

The WHO organised and co-ordinated the Large Analysis and Review of European housing and health Status (LARES) project in 2002/03 [3] (see also Chapter 17). This project gathered data on the health of 8,519 individual residents and the condition of the 3,373 dwellings they occupied. It covered eight European cities – Angers (France),

Bonn (Germany), Bratislava (Slovakia), Budapest (Hungary), Ferreira do Alentejo (Portugal), Forlì (Italy), Genève (Switzerland), and Vilnius (Lithuania). LARES made a major contribution to the evidence on the links between housing conditions and the health and well-being of the occupiers through the amount and range of data collected and the unique cross-disciplinary approach to the analyses of that data.

Data from each individual city were analysed and a preliminary report presented to the public, the politicians, and the officers. This preliminary report provided the cities with evidence to inform their policies and strategies. The data from the eight cities were then combined and analysed by teams of international experts, each focussing on a particular topic. The results from the analyses gave new insights into the association between the characteristics and conditions of housing and the health of the residents. In particular, the results provided evidence of the relationships between health and indoor air quality, dampness, thermal comfort, mental health, noise, and domestic accidents. So, as well as informing policies at local level, the findings on each topic have been used by both WHO and by specialists and academics in their work on those topics.

The US National Center for Healthy Housing (NCHH) proposes ten principles for a Healthy Home [21]. These are:

- **Dry** – Damp dwellings provide an optimum environment for mites, cockroaches, rodents and moulds, each of which are associated with asthma and allergies.
- **Clean** – Clean homes reduce the possibility of pest infestations and exposure to contaminants.
- **Pest-free** – Studies have shown a causal relationship between exposure to mice and cockroaches and asthma in children. However, inappropriate treatment can exacerbate health problems as pesticides residues can pose a risk of neurological damage and cancer.

- **Safe** – The majority of children's physical injuries occur in the home. Falls are the most frequent cause, followed by injuries from objects, burns, and poisonings.
- **Contaminant-Free** – Exposures include lead, radon, pesticides, volatile organic compounds (e.g., formaldehyde), carbon monoxide, oxides of nitrogen, and second-hand tobacco smoke.
- **Ventilated** – Studies have shown that a supply of fresh air improves respiratory health.
- **Maintained** – Poorly maintained dwellings are at risk from moisture and pests.
- **Thermally controlled** – Tenants and homeowners are at risk for various health problems related to prolonged exposure to excessive heat or cold when their homes do not maintain adequate temperatures.
- **Accessible** – Modifications are often necessary in order for occupants to move safely in their homes. Lack of accessibility in and outside the home can result in reduced physical activity, trips, falls, isolation from family and friends, and poor mental health.
- **Affordable** – Households in which more than 30% of the income in spent on housing are considered to be *cost burdened*; if they spend more than 50% of their income on housing, they are considered *severely cost burdened*.

As well as the relationship between the condition of the dwelling and the indoor environment and the health of the occupants, it appears that the social cohesion of the community, and the sense of trust and collective worth, depends to some extent on the quality of the immediate environment, and that the quality of urban design and maintenance can have an impact on social, mental, and physical health. Poorly planned or badly maintained residential areas that lack public services, greenery, parks, playgrounds, and walking areas, have all been linked with a lack of physical exercise, an increased prevalence of obesity, cognitive problems in children,

and a loss of the ability to socialise. All these impacts were confirmed during the lockdowns instituted as part of the responses to the 2020/21 pandemic. (Lockdowns required people to confine themselves within their own dwelling for long periods.) Declining neighbourhoods affect residents through visual mechanisms (such as litter, pollution, and graffiti) and social mechanisms (including isolation, loitering, and increased fear of crime). Poor urban planning and layout can also lead to an increased dependence on private transportation, which, as well as being expensive, results in increased pollution and noise, and endangers or isolates those likely to be more susceptible such as the very young, the elderly, and those with functional limitations.

Thermal discomfort, overheating, and energy precariousness (fuel poverty)

Thermal discomfort – This is not merely dissatisfaction with the ambient temperature, but is a situation where there is a potential threat to health (i.e., when the temperature falls below 18°C or rises above 24°C for a period of time). This range is aimed at protecting health, in particular the health of those most susceptible to low or high temperatures. It is based on the WHO's guidance on thermal comfort for the home environment [22].

The WHO guidance, followed by most European countries, recommends that for the elderly (65 years and over), the lower limit should be 20°C. Public Health England suggests that a bedroom temperature of 16°C is safe for young babies [23].

The WHO guidance is directed specifically at the housing environment. In this it differs from other guidance, such as the Predicted Mean Vote (PVM) and Predicted Percentage Dissatisfied (PPD) geared to the working (office) environment. PVM/PPD approach is based on guidance proposed by the American Society of Heating, Refrigerating and Air Conditioning Engineers (ASHRAE), the Internal Organization for Standardization

(ISO), and the Chartered Institute of Building Services Engineers (CIBSE) [24–26].

The PMV/PPD approach is intended to determine conditions that will be acceptable to a majority of occupants [27],[3] and is aimed at predicting the response of a large group of people. In contrast to the working environment occupied largely by the relatively healthy working population, it is the very young, the elderly, the sick, and others susceptible to extremes of temperatures for other reasons that occupy the home environment. For such potentially temperature susceptible groups, modelling or predicting 'dissatisfaction' does not ensure the protection of their health and well-being.

Thermal discomfort is influenced by a wide range of environmental and individual objective and subjective factors [22]. Environmental factors within a dwelling influencing air temperature include the temperature of the surrounding surfaces, air movement, relative humidity, and the rate of air exchange (ventilation). The design and construction of the dwelling will determine how well it protects the occupants from environmental factors outside the dwelling (the climate and weather conditions).

Thermal comfort (and discomfort) will also be dependent on the activity and the clothing worn by the individual, their age, health status, gender, their adaptation to the local climate, and the duration of exposure to temperatures outside the comfort range. All these factors will vary for the individual and between members of the household during the day and over time and will be affected by household activities such as cooking. It has also been shown that crowding and under-occupation influence thermal comfort.

As none of these factors remain stable, and as it is not practical or possible to assess some, it is necessary to make some assumptions and to suggest safe limits. It is for this reason that the primary focus for health protection has been to give guidance on ambient indoor air temperature. However, measuring indoor temperature is also problematic and canvassing occupants' perception seems to be the most practicable, even though this is not necessarily reliable for some susceptible groups – such as the elderly and the very young.

Overheating – It is predicted that as the result of global heating there will be an increase of extreme weather events, including heat waves in countries that already experience them and in countries with temperate climates. In addition, there are an increasing number of cases where there is a build-up of heat within dwellings during 'normal' temperate climatic conditions as older existing dwellings may provide inadequate protection and rehabilitated and modern dwellings may 'trap' heat.

Heat-related health impacts can be particularly severe for certain population groups. These susceptible groups include the elderly, the very young, those with chronic physical conditions (such as obesity, diabetes, and cardiovascular disease), those with mental health conditions, and those on certain medications [28–30]. It also seems that women are more likely to be at risk, as are individuals with a lack of mobility or bed-ridden [31–32].

The characteristics of the dwelling (or the building containing the dwelling) will affect the risk of exposure to overheating, including the location, structural insulation, orientation, and air changes (ventilation).

Risks from high temperatures are greater for those living in urban areas. Large urban locations create heat islands where the local ambient temperature may be several degrees warmer than that in a rural location a relatively short distance away, as much as 4°C higher. These urban heat islands (UHIs) are caused by heat from surfaces, from refrigeration and air condition equipment, and that emitted overnight by the thermal mass of buildings and roads that absorbed heat during the day. All compounded by the lack of any protective greenery that might absorb or protect surfaces from heat. This lack of variation between day and night temperatures limits the dwelling, and more importantly the residents, from cooling down.

In addition, a noisy location (such as close to busy roads, railway lines, industrial plants, or airports) influences whether occupants open windows at night to benefit from the cooler air. Similarly, a dwelling on the ground floor of a block will influence whether occupants open windows at night (for security reasons and to avoid insect attacks).

The direct heat gains through glazing can dramatically increase the indoor temperatures, particularly where windows face south through to west. Solar radiation through windows will heat the internal surfaces and that heat will be given off into the dwelling causing internal temperatures to remain high over the following night.

Design and construction practice (both old and new) can mean that one dwelling can be at risk of overheating while an identical dwelling facing the opposite direction is not; this is often the case for older rows of houses as well as relatively modern estates. In the case of apartment blocks, dwellings on one side may be more prone to overheating than those on the opposite side.

For modern, highly insulated and relatively air-tight dwellings, the problem can be magnified. Such dwellings limit the heat loss through the fabric and via infiltration, so retaining even more of the internal and solar heat gains within the dwelling.

Properly designed ventilation is necessary for indoor air quality (including for the removal of moisture laden air), but opening windows when the outside temperature is equal to or above the indoor temperature (i.e., during the day) is at best of no benefit and may be detrimental and increase overheating problems.

As well as the more obvious sources of heat generated by day-to-day activities, such as cooking and bathing, there are several sources of heat within dwellings. The human body gives off between 65 to 80 Watts per person per hour, and the generation of hot water and its distribution are both sources of heat liberated into the dwelling, which may be significant.

Minimising the risks of overheating – The causes and dynamics of overheating can be complex, and it is likely that no single solution will solve all cases. In addition, the cost, planning restrictions, etc. may make some solutions on some buildings inappropriate or prohibitively expensive.

The possible solutions include – the provision of additional thermal insulation to the structure, insulation of pipes, shading to windows, shutters (ideally external), brise-soleil[4] or awning, light-coloured reflective finish to flat roofs, and ventilation (ideally, passive to avoid the need for additional energy for air-conditioning) [22].

Limiting heat gain and reducing indoor temperatures requires the active participation of the occupiers. For example, opening windows is only useful where the outdoor temperature is lower than that indoors (during the night and not the daytime). Air movement helps the body cool principally by evaporation, but fans do not replace the requirement for adequate ventilation, and fans may be ineffective where temperatures are above 35°C.

Energy precariousness (fuel poverty) – To maintain a safe and healthy indoor residential environment, energy is necessary for water heating, space heating and/or cooling, lighting, cooking, food storage, and personal and domestic hygiene. Here, we use the term 'energy precariousness' to denote those situations when a supply of clean, affordable, and sustainable energy is not readily available, guaranteed, or certain.

There may be one or more causes for energy precariousness. It may be related to the energy efficiency of the dwelling, or lack of it (so-called 'cold homes'), to the socioeconomic status of the household (often referred to by the pejorative term 'fuel poverty'), and/or by the cost of energy. Particularly in developing countries, energy precariousness can mean the lack of access to clean energy which may lead to the use of biomass fuel and consequential problems of indoor air pollutants. There are also cases in medium- to high-income developed countries where the cause is attributable to under-occupation, that is where a large family dwelling is occupied by one or two elderly persons (the rest of the family having left home). Whatever the cause or causes, it can threaten the social, physical and mental health and the well-being of the occupiers. It also has social, political, and economic implications for society.[5]

Energy precariousness may mean that some occupiers have difficulty meeting the cost of energy they have used or will use. To cope with this, they may try to economise by using less energy, or, in extreme cases, become disconnected (or may self-disconnect if they have a prepayment meter) from one or other energy supply.[6]

The most obvious direct threat is exposure to low temperatures (below 18°C, i.e., below the thermal comfort range) for a period of time sufficient to have an impact on health. A lack of energy for space heating means that the occupier cannot effectively heat the dwelling, increasing the risk of respiratory and cardiovascular conditions, and in extreme cases, may result in death as Wilkinson and others have found [33–35].

As well as low indoor temperatures, attempts to economise by reducing the energy used for heating can lead to dampness from condensation enabling mould growth to become established and proliferate (on which see later). Cold, dampness, and mould growth may also mean occupiers, particularly of older dwellings, may attempt to reduce heat waste and draughts by sealing ventilators and poorly fitted windows. This can add to the problems increasing the likelihood of poor indoor air quality.

An indirect consequence of disconnection or attempts to economise can be the use of alternative and unsuitable means of heating, such as portable unflued gas or oil heaters. Such appliances may increase the risk from fire, and can result in potentially dangerous levels of carbon monoxide and other indoor air pollutants. A lack of a supply of electricity may also lead to the use of other forms of artificial lighting, such as candles and oil lamps, again increasing the risk of fire and burn injuries. Poor lighting can also increase the risk of depression and of falls and other accidents [36].

The lack of means of heating water may interfere with the maintenance of personal and domestic hygiene (personal washing, clothes washing and general laundering, washing cooking and eating utensils, and general cleaning of the dwelling), so increasing the possibilities of the spread of infections.

Energy precariousness may also mean that there is no, or insufficient, energy for the safe storage and cooking of food. This can result in food becoming contaminated by micro-organisms, and in poor nutrition which can lead to obesity.

Finally, the effect on the social and mental health and well-being of the household should not be under-estimated. Managing on a low income and having to economise on energy (and other necessities) can lead to stress and depression. Being able to afford to heat only one room to a reasonable temperature may be one way for a household to economise, but this interferes with normal household life, making it difficult for individuals to have time away from other household members, and can lead to a reluctance to invite friends and relatives into the dwelling. This social isolation can be exacerbated if the dwelling suffers from dampness and mould because of the inadequate heating.

Minimising the risks from energy precariousness – there seem to be three options to try to avoid energy precariousness and the risk of exposure to low temperatures – subsidising the cost of energy, subsidising households on low income, and improving the energy efficiency of dwellings [37].

The first of these is where a country subsidises the cost of the production of energy, making the cost of energy to the consumers lower and more affordable. The downside of this is that little incentive exists for housing owners to improve the energy efficiency of their dwellings, or for households to economise on energy consumption.

The second is where a country provides a subsidy to individual households (usually based on some form of means testing). While this may encourage some owners to improve the energy efficiency of their dwelling, it is an open-ended and uneconomic solution.

The third option, that of improving the energy efficiency of dwellings, is a long-term solution – reducing the energy needed and consumed, reducing the number of 'cold homes', and reducing the risk of exposure to low temperatures (as well as reducing CO_2 emissions). However, this is an option that, in practical terms, can take several years to achieve. Nonetheless, unlike the other options, the improvement of each dwelling is a one-off expense, and the saving is continuous. Thermal renovation should always be associated with checking and maintaining adequate ventilation as improved insulation can alter the ventilation characteristics of dwellings and have a deleterious impact on indoor air quality.

Dampness[7]

Water in the form of moisture is naturally present in many of the materials used in buildings and dwellings. Provided that the amount of moisture remains within certain limits (which depends on the particular material) it presents no problems. However, it is when the amount exceeds the upper limit for the particular material that problems can occur, and it is this excess water content that is usually referred to as 'dampness'.

Water is held in building materials in several ways. First, it will be chemically combined within the material, such as in concrete and plaster. The construction process will involve using large quantities of water (2,000 litres for an average two-storey house) most of which will dry out as the materials 'cure'. Second, all porous materials will exchange moisture with the air and with adjacent materials; these materials are never completely free of moisture but should be 'air-dry'. Third, water will rise through fine holes in materials by capillary attraction, and it is by this action that moisture can rise from the ground into walls and floors – so-called 'rising dampness' – unless prevented by a water-proof membrane (a damp proof course or membrane). Finally, water can penetrate small cracks in an otherwise water-proof material (such as external protective render) to become trapped in the structure.

Defects in constructions that can lead to dampness are also covered in Chapter 9.

Water is also naturally present in the air in the form of water vapour (a gas). Again, within certain limits, water vapour causes no problems. When it exceeds the upper limits or condenses out, it is another form of 'dampness' (condensation) and can lead to problems.

The amount of water vapour a given volume of air can hold depends on the temperature of that air. The higher the temperature of the air, the greater the amount of water vapour it can hold. The ratio between the amount of water vapour held by the air and the amount it is capable of holding at that temperature is referred to as 'relative humidity' at that temperature. Within a room (and a dwelling) the temperature will vary, both horizontally and vertically, although the amount of water vapour will remain relatively constant. This means that the relative humidity will vary with the temperature gradient, being higher where the air is cooler and lower where it is warmer. Problems of dampness start to occur where the relative humidity is above 70% for long periods.

Put simply, 'dampness' is excess moisture in the building fabric or the air, and 'dampness' occurs before any moisture is visible.

Mould (fungal) spores and house dust mites are always present in dwellings. However, both mites and mould flourish in damp or humid conditions. Mould spores are known allergens and can increase the likelihood of respiratory conditions and allergic reactions. Dust mites and their detritus are potent allergens.

As well as the threats to physical health from mould spores and mites, dampness and visible mould can have social and mental health effects. Occupiers can become ashamed of the conditions and reluctant to invite friends and relatives into the dwelling, so becoming socially isolated.

Domestic hygiene and food safety

The design, construction, and maintenance of a dwelling should be such that it is relatively easy to keep it clean and hygienic. Surfaces should be readily cleansable, particularly floors, worktops and walls of kitchens, and personal washing and sanitary facilities (bathrooms and WCs). Where a dwelling is difficult to keep clean and hygienic there is an increased risk of the spread of infections, including gastro-intestinal diseases, and an increased risk from allergens such as mould spores and dust mites. In addition, where a dwelling is difficult to keep clean it may cause depression and anxiety.

Dwellings should also be designed, constructed, and maintained to limit (as far as possible) entry by pests, and harbourage for pests within the structure. As well as provision for sanitation, there should be provision for the safe storage of household refuse awaiting collection and disposal.

Insect pests can cause allergic reactions (see Chapter 17 for more). Dwellings with obvious cockroach infestations can cause high levels of allergic sensitivity in children. Insects can also cause food spoilage, making the food unpalatable if not inedible. Insects are also known to be mechanical vectors of diseases, picking up disease-causing organisms on their bodies from one source (such as rotting refuse and animal faeces) and transferring them to food intended for human consumption and to kitchen surfaces.

Rats and mice are known to be infected with pathogenic organisms (see Chapter 17 for more information), and birds (such as pigeons), as well as causing a nuisance, carry diseases including salmonella, and can harbour insect pests in their nests (such as the Martin bug Oeciacus hirundinis[8]).

Finally, emotional distress and social isolation are commonly associated with pest infestations and accumulations of refuse.

Crime and fear of crime

There are two aspects relating to crime and fear of crime. First, mugging (street crime), which may involve violence, and affect people's perception of safety when returning to their home at night (after dusk). Second, burglary and attempted burglary (sometimes termed 'housebreaking' or 'breaking and entering'), and aggravated burglary where violence is involved.

The health effects of crime and fear of crime are primarily related to mental health and include stress and anguish resulting from burglary or mugging, and the fear of being mugged or burgled. Where violence is involved, there is the physical injury. Fear of crime is often more widespread than the prevalence of the actual crimes. Reports and news of crimes will generate fear in more local inhabitants than there are victims.

The immediate housing environment should be laid out so that possible areas of concealment for burglars and intruders are minimised. There should be both public and private space that the residents relate to as part of their housing environment (sometimes called 'defensible space'). Pedestrian routes should be well lighted and defined. Where possible, dwellings should be sited so as to provide natural views of neighbouring properties, without invading privacy.

Dwellings should be capable of being secured against unauthorised entry. Adequate and appropriate security will both delay and deter intruders and will make occupiers feel safer. The design of dwellings (including any associated yards, gardens, and outbuilding) should include clearly private and defensible space.

Window locks, deadlocks for external doors, burglar alarms, security lights, and window grills reduce the likelihood of unauthorised entry or attempted entry. Spy holes in exterior doors and door chains can make occupiers feel more secure. While fencing and walls around gardens and yards can hinder intruders, they can also provide concealment. The design and layout of the immediate housing environment and the level of crime will affect the level of security appropriate for the dwelling.

There can be negative health effects, however, where a dwelling is made fortress-like by security devices. Also, there has to be a balance between the security features and the need for safe means of escape in case of fire, and ventilation provided for by openable windows.

In multi-occupied buildings, security provided by an entry-phone or concierge system can help both reduce crime and the fear of crime.

Crowding and space

A dwelling should provide sufficient space to allow for the social interaction between members of the household (and with friends), but also enable individuals to have some privacy away from other household members. Individuals, particularly children, also need space for study and recreation, whether inside the dwelling or in areas immediate adjacent to and observable from the dwelling. However, too much space can lead to a sense of physical and social isolation, particular for those living alone.

Crowding (and even 'overcrowding') should not be confused with density. Crowding cannot be measured objectively, whereas density can and is an objective measure of the number of individuals in a given space. Culture, society, and custom will influence what is considered to be crowding (or overcrowding). What may be considered unacceptable by one group of people may not be seen as a problem by another – *the same objective density may or may not be uncomfortable depending on the situation. High density doesn't always lead to crowding* [38].

Nonetheless, countries attempt to define crowding or overcrowding, based on what appears to be arbitrary and subjective provincial concerns.

The current (2021) English statutory definition of 'overcrowding' is a classic density approach.[9] It was first introduced in 1935 and appears to have been aimed at limiting the spread of disease and on moral grounds to constrain the chances of sexual abuse or promiscuity. When introduced, the Minister of Health stated that *It is relevant to point out that this standard does not represent any ideal standard, but the minimum which is in the view of Parliament tolerable while at the same time capable of early enforcement* [39]. This definition refers to the numbers of persons who should be allowed to sleep in rooms of a given size, and the total number of persons who should be permitted to sleep in a dwelling with a given number of rooms. It also treats a child between the age of 1 and 10 years old as 'half a unit' and disregards altogether a child under 1 year old, and that as children grow older they will pass the arbitrary boundaries of 1 and 10 years old.

The Survey of English Housing (now the English Housing Survey) uses a different definition. For this, the Bedroom Standard, the 'number of bedrooms required' is calculated for each household taking account of the age/sex/relationship composition. This 'number required' is then compared with the number of bedrooms available for the use of that household [40].

New Zealand adopted an 'Equivalised Crowding Index'. This is a formula which applied a concept of an 'adult equivalent'. The formula weights each individual in a couple relationship, and each child under 10 years old as one half. The formula gives an equalised number of people per bedroom and any value in excess of 1.0 represents a measure of crowding. However, the definition and approach have been subject to review and to criticism in recent years [41].

Canada has a 'National Occupancy Standard'.[10] This was developed by the Canadian Mortgage and Housing Corporation to help determine the number of bedrooms a dwelling should have to provide freedom from crowding. It is based on the number, age, sex, and inter-relationship of household members [42].

Personal space and privacy needs are important for the individual members of the same household, and for individuals or households sharing common areas in multi-occupied buildings. These needs vary reflecting both individual and cultural perceptions. Adolescents may need more space and privacy than the elderly. Small children will need as much space as an adult; the need for privacy begins to develop from the age of 8, and they need space to play inside and outside.

A lack of space and crowded conditions have been linked to a range of health outcomes [43]. These include negative mental health outcomes such as stress and poor educational achievement. Crowding also seems to have several effects such as increased heart rate, increased perspiration, reduced tolerance, and reduced ability to concentrate.

Crowding is also linked with increased personal hygiene risks, increased risks of accidents as well as the spread of infectious diseases (including tuberculosis [44]). Because they spend more time at home, those more likely to be affected are the elderly, the very young, carers, and those with impaired mobility.

Within the idea of crowding and lack of space can be included 'storage space'. The ability to move safely around a dwelling may also be influenced by a lack of adequate storage capacity so that passageways become restricted. There is also need for safe storage of household cleaning products and medicines. In buildings that are multi-occupied this can also lead to equipment such as push chairs being left in common parts and hindering means of escape in case of fire.

Noise

Intrusive noise as the result of inadequate sound insulation and defects in the structure can have detrimental effects on health.

It is now well known that noise levels below hearing damaging levels can lead to sleep disturbance, cognitive impairment, physiological stress reactions, and cardiovascular disorders. For more see Chapter 20.

The assessment of housing

There are two approaches to the assessment of housing conditions – the detailed assessment of an individual dwelling (an inspection or survey), and local or national sample surveys of the housing stock. The first is to determine the state and condition of the individual dwelling and whether its condition requires remedial action or improvement. The latter is geared to providing information to inform policies and strategies.

What matters (characteristics and deficiencies) are taken into account during an assessment depends on the purpose of the assessment. However, it will usually include (if not focus on) any legal or administrative controls applicable (e.g., building codes, and other standards or requirements).

Individual surveys

The detail, focus, and approach will again depend on the reason for the assessment. For example, an individual dwelling can be assessed to determining whether it is maintaining (or increasing) its financial value, whether it is providing a safe and healthy environment for the occupying household (or potential occupant(s)), or for both purposes.[11] The detail of the assessment will depend on the status of the assessor (the surveyor). If instructed by the owner of the property, then the surveyor may be able to carry out destructive investigations, such as opening up a wall to trace the source of dampness. However, if the survey is being carried out to determine whether state intervention is justified, then the surveyor will not have the right or power to undertake such investigations and must rely on visual evidence.

The public (state) interest in the dwelling as an asset is really only triggered when the dwelling is a threat to the occupiers, is in need of rehabilitation (upgrading) to extend its useful life, or is coming to the end of its useful life such that demolition and replacement becomes the most likely option.

Sample surveys

Sample surveys provide data on local and/or national conditions that can be used to inform policies and strategies, and to monitor the effectiveness of housing programmes. Where the surveys collect the same (or at least, comparable) information on different geographical areas, the data from local surveys can be combined to give a national picture, and national data can be combined to give a wider picture, for example, that for Europe.

While sample surveys may use the same basis (a standard or assessment methodology), they are not intended to providing information for action in respect of an individual dwelling. However, they can be used for other purposes such as investigating the relationship between housing conditions and the health, social, and economic cost to society attributable to the risks posed, and the cost of removing, or at least reducing, those risks.

Standards and metrics

There is a wide range of possible standards and metrics that will influence the assessment methodologies. Non-legal (non-statutory) standards and metrics can be administrative; examples of these are ones that may be targets for, or financial support to, social or public sector housing, such as the Decent Homes Standard in England [45].[12] Others are those factors included in sample surveys and surveys for international databases, such as the EU Statistics on Income and Living Conditions (EU-SILC), and the European Quality of Life Survey (EQLS).

Legal (statutory) standards are those that apply to housing either when it is being created (constructed or converted from an existing building) or to those already existing. How such standards are formulated depends on several factors including the intended situation (whether applicable to new or existing housing), the status and training and competency of the person(s) applying the standards, and the sanctions that may be associated with them.

Standards and requirements can be quantitative, qualitative, or a combination of both. Quantitative requirements state what should or should not be present (e.g., there should be a bath or shower with a supply of hot and cold, or temperature-controlled water; dampness should not be present) and what can be measured or counted. The advantages of such requirements are that they are clear and understandable by those who apply them, and also by those expected to comply with them. They are particularly suitable for application by relatively untrained staff and so are relatively inexpensive to implement. However, they can be problematic to update, refine, and extend and still subject to individual interpretation. Also, they tend to focus on the condition of the structure and fabric and what facilities should be present, rather than the reason why – any underlying health and safety principles may be unclear or obscure. Such quantitative requirements and standards are particularly suited to new buildings (or conversions, refurbished buildings). As quantitative requirements are based on limits (a pass/fail approach), they can become the norm (even if described as 'minimum'), acting as a disincentive to go beyond those limits.

Qualitative standards and requirements focus on what is to be taken into account and may be more generally phrased (e.g., that there should not be any unacceptable threats to health and/or safety; there should be sufficient personal washing facilities), relying on the expert opinion of the surveyor. This means they rely on those applying them to keep themselves up-to-date with current relevant knowledge. However, non-experts may less easily understand them. Unlike quantitative requirements they are more human focussed, looking at the health, safety, and needs of individuals and households.

Controls on housing conditions

Since the mid-19th century, the relationship between the condition and the health of occupiers has been recognised and considered of sufficient public health importance to warrant state intervention. Such state intervention had to be justified, and that justification was founded on a recognition of a relationship between living conditions and health. It was this recognition that provided the motivation to develop a health-based (originally a sanitary-based) approach for assessing dwellings and it is this health-based approach that will be discussed in this chapter.

More often than not health and safety have been the principles underlying controls on housing conditions. In England, it was the Ministry of Health that proposed the introduction of a Standard of Fitness for Human Habitation in 1919 [46], although the standard was not incorporated in the legislation until 1954 [47].[13] The basis of this Standard was clearly health and safety, although the phrasing focussed on the structure and facilities.

Other historical documents dealing with housing and health include the World Health Organization's The Physiological Basis of Health Standards for Dwellings [48], and the American Public Health Association's Basic Principles of Healthful Housing [49]. This latter document was superseded in 1986 by Housing and Health: Recommended Minimum Housing Standards [50] and then the National Healthy Housing Standard [51] issued in 2014.

The Housing Health and Safety Rating System (HHSRS)[14]

Following several studies into the effectiveness of the existing statutory standard for

housing[15] (the fitness standard), in 1998 the UK Government commissioned the development of a new approach for assessing housing conditions in England and Wales. Put simply, the underlying concept was to focus on the potential threats to health and/or safety posed by the characteristics and condition of a dwelling [52].

The development started with an extensive literature review of medical-, architectural-,

---

A. Physiological requirements
Damp and mould growth etc.
Excessive cold
Excessive heat
Asbestos (and manufactured mineral fibre)
Biocides
Carbon monoxide and fuel combustion products
Lead
Radiation
Uncombusted fuel gas
Volatile organic compounds

B. Psychological requirements
Crowding and space
Entry by intruders
Lighting
Noise

C. Protection against infection
Domestic hygiene, pests and refuse
Food safety
Personal hygiene sanitation and drainage
Water supply

D. Protection against accidents
Falls associated with baths etc.
Falling on level surfaces
Falling on stairs etc.
Falling between levels
Electrical hazards
Fire
Flames and hot surfaces
Collision and entrapment
Explosions
Position and operability of amenities
Structural collapse and falling elements

Figure 12.1 The 29 HHSRS potential housing hazards

engineering-, and building-related sources. This produced a list of 29 potential hazards (see Figure 12.1), each of which, to a greater or lesser extent, was attributable to the condition of a dwelling (i.e., none was attributable solely to occupier behaviour). The review also provided details of the potential health outcomes from each of these hazards. Data on these health outcomes was then matched with data from the English House Condition Survey (EHCS, now the English Housing Survey). Analyses of the matched data provided information on the prevalence of each hazard in the housing stock, the likelihood of a hazardous occurrence and the most probable and other outcomes for each hazard. The analyses also provided national averages for the likelihood of a hazardous occurrence and the most probable and other outcomes for each hazard.

The 29 potential hazards differ widely. Some of them can result in illnesses, some in physical injuries, and others may cause or exacerbate health conditions; some, in extreme cases, could be fatal, while others, such as noise, would not be. For some, such as excess cold or dampness, harm would only result from a period of exposure, while for others, such as falling on stairs, harm could be immediate resulting from a single event.

First, to allow comparison of the possible consequences of the hazards, the differing health outcomes were grouped into four Classes of Harm (see Figure 12.2) based on the degree of incapacity caused whether the outcome was an injury, health condition, or illness. (The Classes of Harm were taken from other work by the BRE (in 2000) that developed seven Classes [53], but only the top four were used for this system as these were serious enough for victims to seek medical attention and so likely to provide the records used to support the HHSRS.)

To allow for comparison of the widely differing hazards and the health outcomes, a numerical score seemed the obvious solution – the higher the score, the greater the

Class I covers the most extreme harm outcomes including – Death from any cause; lung cancer; mesothelioma and other malignant lung tumours; permanent paralysis below the neck; regular severe pneumonia; permanent loss of consciousness; and 80% burn injuries.	**Class II** covers severe harm outcomes, including – Cardio-respiratory disease; asthma; non-malignant respiratory diseases; lead poisoning; anaphylactic shock; cryptosporidiosis; Legionnaires' disease; myocardial infarction; mild stroke; chronic confusion; regular severe fever; loss of a hand or foot; serious fractures; serious burns; loss of consciousness for days.
Class III covers serious harm outcomes, including – Eye disorders; rhinitis; hypertension; sleep disturbance; neuro-psychological impairment; sick building syndrome; regular and persistent dermatitis, including contact dermatitis; allergy; gastro-enteritis; diarrhoea; vomiting; chronic severe stress; mild heart attack; malignant but treatable skin cancer; loss of a finger; fractured skull and severe concussion; serious puncture wounds to head or body; severe burns to hands; serious strain or sprain injuries; regular and severe migraine.	**Class IV** includes moderate harm outcomes which are still significant enough to warrant medical attention. Examples are – Pleural plaques; occasional severe discomfort; benign tumours; occasional mild pneumonia; Broken finger; Slight concussion; moderate cuts to face or body; severe bruising to body; regular serious coughs or colds.

Figure 12.2 The four HHSRS Classes of Harm with some examples

	Class of Harm weightings	Likelihood	Spread of harm (%)
S1 =	10,000 X	1	X 0₁
		L	
S2 =	1,000 X	1	X 0₂
		L	
S3 =	300 X	1	X 0₃
		L	
S4 =	10 X	1	X 0₄
		L	
Hazard score = (S1+S2+S3+S4)			

Where –

S = the row product for each Class of Harm

L = the likelihood of an occurrence over the next 12 months

0 = the outcome expressed as a percentage for each Class of Harm

Figure 12.3 The HHSRS formula

Source: Taken from the Housing Health and Safety Rating System (England) Regulations 2006)[16]

risk. To generate a score a formula was devised (see Figure 12.3) using three sets of figures:

1   A weighting for each Class of Harm to reflect the degree of incapacity for each health outcome – 10,000 for Class I, 1,000 for Class II, 300 for Class III, and 10 for Class IV (based on the weightings suggested by BRE, [53]).
2   The likelihood of a hazardous occurrence over the next 12 months, expressed as a ratio. The 12-month period takes account of the differences between an event and a period of exposure, and also of the effect the seasons will have on certain hazards (such as dampness and excess cold).
3   The spread of possible harms resulting from an occurrence, expressed as a percentage for each of the four Classes of Harm, the highest percentage being given to the most probable outcome. For example, the most probably outcome from a fall out of a window on the ground floor will be bruising, then perhaps a fracture, but such a fall would not be fatal; whereas a fall from a window on the fifth floor would probably be fatal.

Based on the literature review and the analyses, for each of the 29 hazards, profiles were produced [54]. These profiles gave:

- A definition of the individual hazard;
- A summary of the potential for harm, including the typical health outcomes, the prevalence of the hazard, and the national averages;
- The causes and in particular the dwelling features and conditions that could contribute to the likelihood of a hazardous occurrence and severity of the outcome;
- What was currently considered to be the ideal, that is, the safest condition that would mitigate the hazard.

The Operating Guidance [54] is an important document as it set outs in the hazard profiles the exact definitions of each hazard and also the definitions for the terminology used. It also highlights that users of the system need to keep up to date with developments in housing and health research, only by doing so can valid assessments of risks arising from the hazards be made. One means of keeping informed about research on the topic is the Housing Health International Research Bulletin.[17]

The HHSRS assessment process – is based on a full inspection of the dwelling and associated parts such as outbuildings yards and gardens, to identify any defect or deficiency to any element or facility, whether it results from the original design, construction, or manufacture, or from deterioration and a lack of repair or maintenance. Once complete, the surveyor determines whether the deficiencies identified contribute to any of the 29 HHSRS hazards, and if so, to which ones (a single deficiency can contribute to more than one hazard, and several deficiencies, perhaps in different locations, may contribute to a single hazard).

The next step is to assess the severity of the hazards. First, the surveyor judges whether the hazards identified are average for that type and age of dwelling or significantly worse. If average, there is no need for further assessment of that hazard, as the hazard score is known.

For those hazards the surveyor considers to be significantly worse than average, she or he uses her/his professional judgement to assess the severity of those hazards. This is a two-fold assessment – first the likelihood of an occurrence over the next 12 months, and then the most likely and other potential outcomes.

This two-fold approach is considered more logical than merely attempting to judge a hazard on a linear scale. It ensures that the severity of a threat which is very likely to occur but will result in a minor outcome can be compared with one which is highly unlikely to occur but if it did would have a major outcome. It also allows differentiation between similar hazards where the likelihood may be the same, but the outcomes very different.

The weightings given to the Classes of Harms are fixed, while the likelihood and the spread of outcomes are used to reflect judgements made by the surveyor inspecting the dwelling. Surveyors are not expected to be exact in their judgements, but to give a figure that represents a range (e.g., '1 in 10' represents the range from 1 in 7.5 to 1 in 13, and '1 in 180' represents the range of 1 in 130 to 1 in 240). This representative figure is used in the HHSRS formula to generate a score.

The numerical score generated by the HHSRS formula can appear very specific and falsely imply that the score is a precise statement of the risk, rather than a representation of the surveyor's judgement. To overcome this, the scores have been put into Bands (see Figure 12.4), Band J being the safest possible, and Band A the most dangerous.

The result of this inspection and assessment process is a hazard profile for the individual dwelling, giving the Bands for those hazards identified by the surveyor as significantly worse than the average for that type and age of dwelling.

The HHSRS was designed to be useable for the assessment of any kind of dwelling. As it focusses on hazards, it is not dependent on the form of construction, nor on whether the dwelling is self-contained (i.e., all necessary facilities and rooms, etc. for the exclusive use of the occupants) or shares facilities and/or space with others. While the statistical information supporting the HHSRS relates only to houses and dwellings in multi-occupied buildings, the approach can be used to assess mobile homes (such as caravans and boats), self-contained and non-self-contained dwellings (including bedsits). This focus on hazards also means that the HHSRS does not indicate the appropriate solution to remove or alleviate the hazard, which depends on the form of construction and the defects and deficiencies contributing to the hazard. However, it might be expected that the deficiencies identified on inspection giving rise to the hazard would be remedied.

Although commissioned by the UK Government as the means for determining enforcement action, it was developed having in mind the other possible uses for an assessment methodology.

The Housing Act 2004[18] introduced into the law the concept of hazards in dwellings with the HHSRS as the basis for local

Hazard Bands	Hazard score range
A	5,000 or more
B	2,000 to 4,999
C	1,000 to 1,999
D	500 to 999
E	200 to 499
F	100 to 199
G	50 to 99
H	20 to 49
I	10 to 19
J	9 or less

Figure 12.4 The HHSRS hazard Bands

authority action to address housing conditions in its area. It was brought into effect in April 2006 so that the HHSRS replaced the English and Welsh Standard of Fitness for Human Habitation as the prescribed method to be used for assessing housing conditions as the precursor to determining whether they should exercise their duties and powers to deal with unsatisfactory housing conditions.

The legislation defines two levels of Hazards – Category 1 and Category 2 Hazards. Category 1 Hazards are those where the HHSRS Score is 1,000 or more. These are totally unacceptable Hazards, and the legislation places a duty on local authorities to take action to deal with them. For Category 2 Hazards that could be reduced and are still unacceptable, the legislation gives local authorities powers to act.

In 2019, the HHSRS Hazards were incorporated into the Landlord and Tenant Act 1985 by the Homes (Fitness for Human Habitation) Act 2018, which placed an obligation on all landlords to ensure that rented houses and flats are 'fit for human habitation', and are safe, healthy, and free from matters that could cause serious harm. This also provides a means by which tenants can seek a legal remedy.[19]

The US Healthy Homes Rating System

In 2010, the English HHSRS was adopted by the US Department of Housing and Urban Development (HUD). The only adaptions made to the HHSRS were to rename it the Healthy Homes Rating System (HHRS), and to replace the English house/dwelling classifications and technical terms with those used in the US [55].[20] The HHRS was used as a condition of a grant award to an organisation involved in housing work; HUD required that the organisation use the HHRS. While the HHRS has retained the English national averages for each of the 29 Hazards, it was intended that the averages will be replaced when local data had been built up by the organisations as they use the HHRS.

The New Zealand Healthy Housing Index

The Healthy Housing Index (HHI) was developed by the Housing and Health programme at the University of Otaga, Wellington. It is a number formed from housing factors providing a measure of how likely it is that occupiers will suffer ill health or accidents as a result of housing factor(s). The Index was modelled on the concept of the English HHSRS, with modifications made to take account of the unique New Zealand environment and the nature of data available. Like the English HHSRS, the HHI focusses on the structure and condition of the house; there are no indicators of behaviour within the home [56].

The HHI is designed to be a tool to understand the link between housing and health at both a community and individual household unit level. It is intended to be used by the NZ Accident Compensation Corporation, local government, district health boards, primary care providers, and other agencies involved in the housing and/or health sector. The on-going application of the HHI should allow for the identification of high need homes, families, and communities and provide a basis on which to target resources to reduce inequalities in health. The HHI also informed the development of the Rental Warrant of Fitness used in particular areas to 'certify' that a dwelling offered for renting is satisfactory for occupation [57].

The French Domiscore

A tool, the Domiscore, has recently been developed in France aimed at identifying housing conditions and shifting the emphasis to their potential effects on the health and well-being of residents [58]. A pilot testing of this multicriteria grid using a scale ranging from favourable to unfavourable for the residents' health and well-being by 11 expert and non-expert housing assessors was performed on 28 dwellings to assess the ease of handling and to suggest improvements. The feedback of various profiles of potential users of the Domiscore (Conseillers Médicaux en Environnement Intérieur [59], social and medico-social workers, municipal agents, landlords, real estate agents, etc.) should ensure that the tool is practical, understood, reliable, and effective in protecting residents and tackling substandard habitat.

The cost of unsatisfactory housing and the cost benefits of remedial action

As well as being used to assess the condition to determine the severity of threats to health and safety, the HHSRS also can be used to judge the effectiveness of remedial action

by the assessment of the condition after the completion of remedial action. These two assessments, the pre- and post-remedial action assessments, form the basis of the cost benefit analysis developed in this study.

Once it is accepted that unsatisfactory housing conditions can have a negative effect on health, it is then logical to assume that there will be a cost to society. Work by Ambrose (Figure 12.5) highlighted this cost to society [60] and based on such work the Audit Commission stated in 2009 that *Every £1 spent on providing housing support for vulnerable people can save nearly £2 in reduced costs of health services, tenancy failure, crime and residential care* [61].

Work by the UK Building Research Establishment (BRE) developed a methodology

A matrix of costs whose levels can be related to poor living conditions		
	Costs to residents	**External costs**
Systemic – capital	high annual loss of asset value on owner-occupied property	high annual loss of asset value for landlords of rented properly
Systemic – revenue	poor physical health	higher health service costs
	poor mental health	Ditto
	social isolation	higher care services costs
	high home fuel bills	high building heating costs
	high insurance premiums	high insurance payments
	uninsured contents losses	spending on building security
	spending on security devices	high housing maintenance costs
	living with repairs needed	extra costs on school budgets
	under-achievement at school	homework classes at school
	loss of future earnings	loss of talents to society
	personal insecurity	high policing costs
	more accidents	high emergency services costs
	poor 'hygienic' conditions	high environmental health costs
	costs of moving	disruption to service providers
	adopting self-harming habits	special health-care responses
Formalised – capital		Government and EU programmes, SRB, New Deal, etc.
Formalised – revenue		Local authority 'statements of need' Education, police, and NHS funding formulae
		Fire and ambulance services funding formulae
		Housing investment programmes

Figure 12.5 Cost matrix developed by Ambrose [60]

that provided the means to compare the cost of housing interventions with the potential savings to the health services. The methodology used HHSRS data from the EHS, and, relating that data with the health outcomes (see Figure 12.2), calculated the cost to the National Health Service (NHS) of unsatisfactory housing in England using health cost data from various sources including the Department of Health [62]. The BRE also calculated the cost benefits of housing interventions – that is, the estimated annual savings to the health sector attributable to the one-off cost of removing potential housing hazards. The BRE has subsequently updated the work and in 2015 calculated the cost of poor housing to the NHS at £1.4bn per annum [63].[21] This methodology based on the HHSRS was adapted and used in a subsequent study on the cost benefits of upgrading energy inefficient dwellings in France. Results suggest that investment in a suitable programme of energy-efficiency improvement would lead to cost savings for the health sector, and such savings would be greatest for dwellings occupied by low-income households. For those households, the health costs avoided would be of the same order as the renovation costs [64].

Pandemics and HHSRS

The 2020/21 pandemic and the stay-at-home requirement (so-called lockdown) imposed by governments as part of responses meant that some HHSRS hazards took on greater significance, such as the hazards of crowding and space, personal hygiene, sanitation and drainage, noise. If more time is spent within the dwelling, then the greater exposure there is to any of the hazards present, with a consequent increased risk. Although lockdowns and the issue of safety for those carrying out inspections create difficulties, there is also the need for action to ensure that occupiers are not put further at risk as the result of the conditions in which they are confined.

The issue of crowding has been highlighted during this pandemic whether it be migrant workers living in dormitories in Singapore or people living in crowded and multi-occupied accommodation in the UK during the lockdown. In the UK the resulting mortality rates in areas with higher levels of deprivation were partly related to household crowding and lack of space. This is shown with multi-generational households which increases the risk of transmission within the households and between generations. Also crowding and lack of space may lead to an increase in the severity of outcomes as close proximity between people led to a higher viral load. It is also the case that crowded households are also likely to be found in housing in the most deprived areas (often older and/or smaller houses) and there is more crowding among low-income households (sometimes also a consequence of high housing costs).

Marmot [1] has shown how the 2020/21 virus (COVID-19) mortality rates follow a similar inequality gradient to non-COVID mortality. It is logical to argue that environmental health practitioners, given the resources, have an important role in addressing this inequity. As a report for the Northern Housing Consortium [65] reported, many households were living with longstanding repair and quality issues, some of which got worse or more obvious during lockdown. *Lockdown had ultimately worsened such conditions and impaired people's ability to live with those conditions.* Longstanding issues with conditions worsened as carrying out repairs became more problematic because social distancing measures. Furthermore, renters were unwilling to report repairs or press their landlords for fear of eviction even though there was a ban on evictions in England. The report says that 'the COVID-19 lockdown has shown in the starkest of terms that rundown homes are resulting in rundown people'.

Besides, teleworking has been boosted during the pandemic, which means that some workers are exposed to some health risks, physical and mental, for longer periods (for example, to overheating during a heat-wave if their home does not provide enough protection from high temperatures).

Building for the future: towards positive energy buildings with a low carbon footprint

Ensuring that buildings are more energy efficient is part of the European Green Deal which encourages the implementation of an ambitious environmental standard for new buildings. However, it should be noted that in energy-efficient buildings, airtightness can cause problems in terms of indoor air quality, noise, and thermal discomfort. This is so if the ventilation is not adequate, which unfortunately is often the case as it is difficult to ensure and check that guidance on 'adequate' ventilation is followed. (The 2020/21 pandemic highlighted the importance of ventilation, a main issue to guarantee indoor air quality, limiting the spread of infectious diseases, as well as avoiding dampness and mould growth.)

Now and in the future, it is and will be necessary to adopt an integrated multidisciplinary approach, from the design through construction and completion, and to put in place proper maintenance programmes for all dwellings. This should include ensuring that the different parameters of the indoor environment (indoor air, light, noise, temperature) are taken into account. Such an integrated approach is necessary to protect the well-being and health of all occupants, particularly the most vulnerable (elderly, small children, those with chronic diseases, or otherwise compromised). In addition, they will have a significant impact in terms of cost savings to society.

In 2018, the WHO published guidelines on housing and health that included a systematic review and analysis of the health risks associated with unsafe, unhealthy, and otherwise substandard housing [66]. These guidelines emphasise the importance of collaboration between the health sector and other sectors in the promotion of healthy housing.

Multi-occupied buildings/houses

While in many countries the equivalent of the UK environmental health practitioners (EHPs) deal with housing only tangentially, such as drainage or pest management issues or only with certain types of housing (e.g., lodging houses) housing is a key component of EHPs' work in the UK.

Reference has already been made to how the 2020/21 pandemic highlighted and indeed magnified inequalities (and inequity) in health. The US CDC recognised the problems of living in shared housing and provided specific advice.[22,23] So far as shared kitchens are concerned, the advice was to limit the number of people allowed in the kitchen and dining room at one time so that everyone can stay at least 2 metres apart from one another while those who are sick should eat or be fed in their room. It was also advised that dishes, drinking glasses, cups, or eating utensils should not be shared. Further advice was also given but how this could be implemented or even enforced remains problematic.

In the UK, it is only the local authority (and most often the environmental health department) that will have any clear idea of where houses in multiple occupation (HMOs) or private rented homes are located, as it is not unusual for such houses to be operated by those wishing to 'stay under the radar'. That said, most HMOs in England and Wales will be subject to the licensing regime set out in the Housing Act 2004. Any form of multi-occupied housing will have particular problems such as fire safety and management of common parts. It is for this reason that, after many years of campaigning, provisions for the licensing of such housing were introduced. Given the existence of the HHSRS (which is used for Part 1 of the 2004 Act) it is unclear why licensing also addresses some aspects of conditions as well as management.[24] The Act also introduced 'selective licensing' which gave English and Welsh local authorities the power to require all privately rented dwellings to be licensed (see Chapter 5).[25],[26]

Other countries have different descriptions for similar types of properties and shared accommodation and may also have a system of registration or licensing. For example, in Australia such housing falls within the category of 'rooming' or 'boarding' houses.[27]

Management

The health and well-being of occupiers will be affected by the quality of management of a dwelling and/or the building containing the dwelling. This is particularly so for multi-occupied buildings, whether HMOs or blocks containing self-contained flats or apartments. To try to ensure proper standards of management of HMOs, the UK has legal provisions including Management Regulations and for Management Orders.[28]

The term 'management' can be defined as the day-to-day organisation, control, and maintenance of both the building and the individual dwellings. In particular, it is concerned with the moral and legal obligations owed to occupiers and the management of financial, physical, and other risks. Effective management of a multi-occupied building takes account of:

- Residents;
- Individual units of accommodation;
- Internal shared space (e.g., maintenance and cleanliness of kitchens, bathrooms, lounges, stairs, passages, lifts, etc.);
- The structure and exterior of the building;
- Structure such as roof spaces above accommodation;
- Shared external space (e.g., gardens, yards, paths, parking spaces, refuse storage areas, etc.) and the surrounding neighbourhood and community.

Ideally, management procedures should reflect the size and layout of the building,

the number of residents, and the degree of sharing of common parts. For instance, the greater the number of residents, the more intensive will be the management needs.

The management of buildings providing accommodation for particular groups of people should take account of the characteristics of residents. If the building provides accommodation for a particular vulnerable group, such as those with physical or mental disabilities, or those with alcohol or drug dependencies, then management procedures will need to take this into account. Effective management also relies in part on residents being fully aware of their responsibilities.[29]

Retaliatory eviction

While not specific to multi-occupied housing, one concern in the UK has been where an unscrupulous landlord seeks to evict tenants who complain to the statutory enforcement agency (local authority) about conditions.[30] In an effort to avoid this, various provision were adopted to provide some protection to the tenants.

Conclusion

This chapter has focussed on housing and health and how conditions affect health. Housing conditions also have clear relevance when it comes to tackling an epidemic or pandemic; the 2020/21 COVID-19 pandemic highlighted how housing is fundamental to public health.

The chapter has not looked at procedures to remedy unsatisfactory conditions, but has focussed on how threats to health and safety can be assessed. Detailed procedures on ensuring remedial are dealt with elsewhere in other works.[31] It should also be noted that other aspects of environmental health and the work of EHPs are dealt with in other chapters of this book, for example Chapter 9 looks at matters of construction and dilapidations.

Chapter notes

1 See also https://www.bmj.com/content/370/bmj.m3181 where a study is reported as saying contact within households was thought to be responsible for roughly 70% of SARS-CoV-2 transmission when widespread community control measures were in place.

2 It is worth noting that the Housing Act 2004 defines 'health' as including 'mental health' so that in England and Wales, there is a de facto acceptance of the relevance of the WHO definition of health in the context of housing.

3 See also http://ceae.colorado.edu/~brandem/aren3050/docs/ThermalComfort.pdf

4 A horizontal louvred screen to protect windows and walls from the sun.

5 According to NEA the collective impact on society is significant with £1.3bn spent each year on health services in England on treating illness caused by cold homes; and 20% of the UK's carbon emissions come from housing see https://www.nea.org.uk/articles/what-is-fuel-poverty/

6 See also Leng G. (2012) *Poor Homes, Poor Health – to Heat or to Eat – Private Sector Tenant Choices in 2012: An Exploratory Study of the Health Impacts of Welfare Reform on Tenants Living in the Private Rented Sector*, Report for the Pro-Housing Alliance. http://www.sabattersby.co.uk/documents/GLHSreportfinal23-10-12W2007.pdf (Accessed September 2021).

7 For more see Ormandy D, Ezratty V, Battersby SA. (2022) *Dampness*, Routledge Focus on Environmental Health Series, London, in print.

8 Martin bug (*Oeciacus hirundinis*), is closely related to the bedbug (*Cimex lectularius*); although not strictly a bedbug it primarily affects birds but has been reported to affect humans.

9 Housing Act 1985, Part X.

10 See http://www.statcan.gc.ca/eng/concepts/definitions/dwelling06

11 It should be noted that in England a 'private rented sector (PRS) offer' is an offer of a fixed term assured shorthold tenancy of at least one year, offered by a private landlord and arranged by a local authority in order to end its main housing duty under homelessness legislation but that offer must be of suitable accommodation including its physical condition. When determining the suitability of accommodation, housing authorities should, as a minimum, ensure that all accommodation is free of Category 1 Hazards. In the case of an out-of-district placement it is the responsibility of the placing authority to ensure that accommodation is free of Category 1 Hazards under the HHSRS and so EHPs may need to carry out surveys of such properties. See

https://www.gov.uk/guidance/homelessness-code-of-guidance-for-local-authorities/chapter-17-suitability-of-accommodation

12 See https://www.gov.uk/government/publications/2010-to-2015-government-policy-rented-housing-sector/2010-to-2015-government-policy-rented-housing-sector

13 Housing Repairs and Rents Act 1954, s9.

14 At the time of writing a review of the HHSRS is being undertaken by the Government following a scoping study. See Ormandy D, Battersby S. (2020) The HHSRS scoping review – what a dog's breakfast. *Journal of Housing Law*, 23(1): 107–110.

15 See for example http://www.sabattersby.co.uk/hhsrs/1998_Controlling_Minimum_Standards_Main_Findings.pdf

16 SI 2006 No 3208.

17 See http://www.housinghealth.com/

18 Part 1 of the 2004 Act, and Housing Health and Safety Rating System (England) Regulations (2005 SI 2005 3208) (Wales has its own SI), and the Operating Guidance [47].

19 See sections 9A and 10 of the Landlord and Tenant Act 1985 as amended; see https://www.legislation.gov.uk/ukpga/1985/70/contents

20 See http://portal.hud.gov/hudportal/HUD?src=/program_offices/healthy_homes/hhrs

21 In 2017 the cost to the NHS as the result of Category 1 Hazards in housing in Wales was estimated at £95 million per year and taking the cost of remedying these hazards implied a payback period of 6.1 years – see Nicol S, Garrett H, Woodfine L, Watkins G and Woodham A. (2019) *The full cost of poor housing in Wales*, Building Research establishment Ltd, Public Health Wales, Welsh Government. In November 2021 the BRE issued an updated report (*The Cost of Poor Housing in England*) which confirmed the estimated cost of poor housing in England (that containing a Category 1 Hazard under the HHSRS) at £1.4 billion, with more than half of this attributable to defects which give rise to excess cold.

22 https://www.cdc.gov/coronavirus/2019-ncov/community/shared-congregate-house/guidance-shared-congregate-housing.html

23 https://www.cdc.gov/coronavirus/2019-ncov/daily-life-coping/shared-housing/index.html

24 Somewhat strangely the Act at s.67(4) says,

> the authority must proceed on the basis that, in general, they should seek to identify, remove or reduce category 1 or category 2 hazards in the house by the exercise of Part 1 functions and not by means of licence conditions,

but given that the house has to be suitable before grant of a licence the authority is not prevented from imposing licence conditions relating to the installation or maintenance of facilities or equipment within subsection even if the same result could be achieved by the exercise of Part 1 functions. 'The fact that licence conditions are imposed for a particular purpose that could be achieved by the exercise of Part 1 functions does not affect the way in which Part 1 functions can be subsequently exercised by the authority'.

25 See also Oatt P. (2020) *Selective Licensing: The Basis for a Collaborative Approach to Addressing Health Inequalities*, Focus on Environmental Health series, Routledge, Abingdon, Oxon.

26 More detail on HMO licensing will be found in a forthcoming publication in the Focus on Environmental Health series.

27 An example is in New South Wales where there is the Boarding Houses Act 2012 which defines 'boarding premises' as premises (or a complex of premises) that:

'(a) are wholly or partly a boarding house, rooming or common lodgings house, hostel or let in lodgings, and

(b) provide boarders or lodgers with a principal place of residence, and

(c) may have shared facilities (such as a communal living room, bathroom, kitchen or laundry) or services that are provided to boarders or lodgers by or on behalf of the proprietor, or both, and

(d) have rooms (some or all of which may have private kitchen and bathroom facilities) that accommodate one or more boarders or lodgers'. These have to be registered where there are five or more residents.

28 The Management of Houses in Multiple Occupation (England) Regulations SI 2006 No 372 and Interim and Final Management Orders in Part 4 of the Housing Act 2004.

29 Useful information, which although specific to student accommodation, can be found in the accreditation codes issued by Unipol/ANUK. See https://www.unipol.org.uk/the-code

30 See Deregulation Act 2015. In England and Wales (it is different in Scotland and Northern Ireland) from 1 June 2021 all section 21 Housing Act 1988 notices must give at least four months' notice. See https://assets.publishing. service.gov.uk/government/uploads/system/ uploads/attachment_data/file/465275/Retaliatory_Eviction_Guidance_Note.pdf for guidance brought in by the 2015 Act and https:// www.gov.uk/evicting-tenants/section-21-and-section-8-notices where only certain actions under Part 1 of the Housing Act prevent a s.21 Notice being valid. Service of a Hazard Awareness Notice or making of a Prohibition Order

does not prevent the use of s.21 and accelerated possession.

31 See Deveaux T, Bassett WH. (2019) *Bassett's Environmental Health Procedures*, Routledge, Abingdon, Oxon.

Chapter references

[1] Marmot M, et al. (2020) *Health Equity in England: The Marmot Review 10 Years on*, Institute of Health Equity, London. http://www.instituteofhealthequity.org/resources-reports/marmot-review-10-years-on/the-marmot-review-10-years-on-full-report.pdf.

[2] Denford S, Morton KS, et al. (2020) *Preventing Within Household Transmission of COVID-19: Is the Provision of Accommodation Feasible and Acceptable?* Pre-print paper. https://doi.org/10.1101/2020.08.20.20176529; https://www.medrxiv.org/content/10.1101/2020.08.20.20176529v2.

[3] Ormandy D. (Ed.) (2009) *Housing and Health in Europe: The WHO LARES Project*, Routledge, London.

[4] Preamble to the Constitution of the World Health Organization as adopted by the International Health Conference, New York, 19–22 June, 1946; signed on 22 July 1946 by the representatives of 61 States (Official Records of the World Health Organization, no. 2, p. 100) and entered into force on 7 April 1948.

[5] Ranson R. (1991) *Healthy Housing: A Practical Guide*, E & FN Spon, London.

[6] Burridge R, Ormandy D. (1993) *Unhealthy Housing: Research Remedies and Reform*, E & FN Spon, London.

[7] Ineichen B. (1993) *Homes and Health: How Housing and Health Interact*, E & FN Spon, London.

[8] BMA. (2003) *Housing and Health: Building for the Future*, British Medical Association, London.

[9] AJPH. (2003) Special issue – built environment and health. *American Journal of Public Health*, 93(9).

[10] Howden-Chapman P, Carroll P. (2004) *Housing and Health: Research, Policy and Innovation*, Steele Roberts, Wellington, NZ.

[11] RenvH (*Reviews on Environmental Health*) (2004) Special Issue. *Housing, Health and Wellbeing*, 19(3–4).

[12] Papers from Unhealthy Housing conferences at the University of Warwick. (1986, 1987, 1991, 2003, and 2006) http://www2.warwick.ac.uk/fac/cross_fac/healthatwarwick/research/devgroups/healthyhousing/healthhousing_papers/.

[13] WHO. (2002) *Proceedings of the International Symposium on Housing and Health*, Forli, Italy,

21–23 November and WHO. (2004) *Proceedings of the 2nd International Symposium on Housing and Health*, Vilnius, Lithuania, 29 September–1 October, WHO Regional Office for Europe, Copenhagen, Denmark. http://www.ihealthbank.org/Portals/0/Environmental%20Health%20in%20Emergencies/e87878_pt1.pdf.

[14] Marmot M, Davey-Smith G, Stansfield S, Patel C, North F, Head J, White I, Brunner E, Feeney A. (1991) Health Inequalities among British civil servants: The Whitehall II study. *Lancet*, 337: 1387–1393.

[15] Sandel M, et al. (1999) *There's No Place Like Home: How America's Housing Crisis Threatens Our Children*, Housing America, San Francisco.

[16] Attanasio O, Emmerson C. (2001) *Differential Mortality in the UK*, Working Papers (W01/16), The Institute for Fiscal Studies, London.

[17] Ridge T. (2009) *Living with Poverty: A Review of the Literature on Children's And Families Experiences of Poverty*, Research Report No. 594, Department for Work and Pensions, London.

[18] Howden-Chapman P, et al. (2008) Effects of improved home heating on asthma in community dwelling children: Randomised controlled trial. *BMJ*, 337: 852–855.

[19] Thompson H, et al. (2002) *Housing Improvement and Health Gain: A Summary and Systematic Review*, MRC Social and Public Health Sciences Unit, Occasional Paper 5. For Examples of Robust Studies, Glasgow.

[20] CLG. (2008) *Review of Health and Safety Risk Drivers, BD2518*, Department for Communities and Local Government, London.

[21] National Center for Healthy Homes. (2016) *The Principles of a Healthy Home*. https://nchh.org/information-and-evidence/learn-about-healthy-housing/healthy-homes-principles/ (Accessed September 2021).

[22] Ormandy D, Ezratty V. (2012) Health and thermal comfort: From WHO guidance to housing strategies. *Energy Policy*, 49: 116–121. https://doi.org/10.1016/j.enpol.2011.09.003.

[23] PHE. (2014) *Minimum Home Temperature Thresholds for Health in Winter – A Systematic Literature Review*, Public Health England, London.

[24] ASHRAE. (2009) *ASHRAE handbook – fundamentals: Chapter 9*. In *Thermal Comfort*, American Society of Heating, Refrigerating and Air-Conditioning Engineers Inc, Atlanta. http://www.ashrae.org.

[25] ISO. (2005) *Ergonomics of the Thermal Environment – Analytical Determination and Interpretation of Thermal Comfort Using Calculation of the PMV and PPD Indices and Local Thermal Comfort Criteria*, ISO 7730:2005. International Organization for Standardization, Geneva.

[26] CIBSE. (2006) *Guide A: Environmental Design*, Chartered Institute of Building Services Engineers, London.

[27] AREN 3050. (2005) *Environmental Systems for Buildings I*. http://ceae.colorado.edu/~brandem/aren3050/docs/Thermal-Comfort.pdf (Accessed July 2015).

[28] Kovats RS, Hajat S. (2008) Heat stress and public health: A critical review. *Annual Review of Public Health*, 29.

[29] Stafoggia M, et al. (2006) Vulnerability to heat-related mortality: A multicity, population-based, case-crossover analysis. *Epidemiology*, 17(3): 315–323.

[30] IOM. (2011) *Climate Change, the Indoor Environment, and Health*, Institute of Medicine, The National Academies Press, Washington, DC.

[31] Fouillet A, et al. (2006) Excess mortality related to the August 2003 heat wave in France. *International Archives of Occupational and Environmental Health*, 80(1): 16–24.

[32] Vandentorren S, et al. (2006) August 2003 heat wave in France: Risk factors for death of elderly people living at home. *European Journal of Public Health*, 16(6): 583–591.

[33] Wilkinson P, et al. (2004) Vulnerability to winter mortality in elderly people in Britain: Population based study. *BMJ*, 329: 647.

[34] Healthy J. (2003) Excess winter mortality in Europe: A cross country analysis identifying key risk factors. *Journal of Epidemiology and Community Health*, 57: 784–789.

[35] Bhattacharya J, et al. (2003) Heat or eat? Cold-weather shocks and nutrition in poor American families. *American Journal of Public Health*, 93: 1149–1154.

[36] Brown M-J, Jacobs D. (2011) Residential light and risk for depression and falls: Results from the LARES study of eight European cities. *Public Health Reports*, 126(supp 1): 131–140.

[37] WHO. (2007) *Housing Energy, and Thermal Comfort: A Review of 10 Countries Within the WHO European Region*, World Health Organization, Copenhagen.

[38] Jazwininski C. (1998) Crowding: Cited in Gray A. (2001) *Definitions of Crowding and the Effects of Crowding on Health: A Literature Review*, Prepared for the Ministry of Social Policy by Gray Matter Research Ltd, for The Ministry of Social Policy, Wellington, NZ. https://www.msd.govt.nz/documents/about-msd-and-our-work/publications-resources/archive/2001-definitionsofcrowding.pdf.

[39] Ministry of Health. (1935) *Memorandum B: The Prevention and Abatement of Overcrowding*, Ministry of Health, London.

[40] Department for Leveling Up, Housing and Communities (2021) *English Housing Survey: Headline Report, 2020–21.* https://assets.publishing.service.gov.uk/government/uploads/system/uploads/attachment_data/file/1039214/2020-21_EHS_Headline_Report.pdf

[41] Goodyear RK, Fabian A, Hay J. (2011) *Finding the Crowding Index That Works Best for New Zealand*, Statistics New Zealand Working Paper No 11-04. https://www.stats.govt.nz/research/finding-the-crowding-index-that-works-best-for-new-zealand-applying-different-crowding-indexes-to-census-of-population-and-dwellings-data-for-19862006.

[42] Statistics Canada. (released 2017) *Housing Suitability*, website https://www12.statcan.gc.ca/census-recensement/2016/ref/dict/households-menage029-eng.cfm (Accessed March 2022).

[43] ODPM. (2004) *The Impact of Overcrowding on Health & Education: A Review of Evidence and Literature*, Office of the Deputy Prime Minister, London.

[44] Baker M, et al. (2008) Tuberculosis associated with household crowding in a developed country. *Journal of Epidemiology and Community Health*, 62(8): 715–721, August.

[45] Ministry of Housing, Communities and Local Government. (2006) *A Decent Homes: Definition and Guidance for Implementation* https://www.gov.uk/government/publications/a--decent-home-definition-and-guidance

[46] Ministry of Health. (1919) *Manual of Unfit Housing and Unhealthy Areas*, MoH, London.

[47] Housing Repairs and Rents Act 1954. See https://www.legislation.gov.uk/ukpga/Eliz2/2-3/53/contents.

[48] WHO. (1968) *The Physiological Basis of Health Standards for Dwellings*, Public Health Papers No. 33, World Health Organization, Geneva.

[49] APHA. (1938) *Basic Principles of Healthful Housing.* American Public Health Association, Chicago.

[50] APHA/CDC. (1986) *Housing and Health: Recommended Minimum Housing Standards*, American Public Health Association and US Center for Diseases Control, Washington, DC.

[51] APHA/NCHH. (2014) *The National Healthy Housing Standard*, American Public Health Association and National Centre for Health Housing, Washington, DC.

[52] Ormandy D. (2009) The right to healthy housing: Putting health at the centre of English housing policies. In *Conseil d'Etat. Rapport public 2009. Droit au logement, droit du logement*, Paris: La Documentation Française,. 439–454.

[53] A Risk Assessment Procedure for Health and Safety in Buildings. (2000) CRC, London.

[54] ODPM. (2006) *The Housing Health and Safety Rating System: Operating Guidance*, Office of the Deputy Prime Minister, London.

[55] HUD Healthy Home Rating System: Operating Guidance. (2014) Department of Housing and Urban Development, USA. https://www.hud.gov/sites/documents/OPERATINGGUIDANCEHHRS_1-14.PDF (Accessed December 2021).

[56] Keall M, et al. (2010) Assessing housing quality and its impact on health, safety and sustainability. *Journal of Epidemiology and Community Health.* https://doi.org/10.1136/jech.2009.100701.

[57] http://www.healthyhousing.org.nz/wp-content/uploads/2016/09/RHWoF-Summary-19jul2016.pdf and also https://www.healthy-housing.org.nz/our-research/past-research/rental-housing-warrant-fitness

[58] Haut Conseil de la Santé Publique (HCSP). (2020) *Elaboration d'un outil de caractérisation d'un habitat du point de vue de la santé – « Domiscore »*, Rapport de faisabilité, Haut Conseil de la Santé Publique (HCSP), Paris, France. available at https://www.hcsp.fr/Explore.cgi/avisrapports?Langue=en and https://www.hcsp.fr/Explore.cgi/avisrapportsdomaine?clefr=803 (accessed March 2022).

[59] Conseillers Médicaux en Environnement Intérieur (2021) website https://cmei-france.fr (Accessed March 2022) and also see https://www.hcsp.fr/Domiscore.cgi/Debut and https://diagmag.fr/2020/06/26/domiscore-une-premiere-etape-vers-un-diagnostic-habitabilite/ (Accessed March 2022).

[60] Ambrose P, et al. (1996) *The Real Cost of Poor Homes*, Royal Institute of Chartered Surveyors, London.

[61] Audit Commission. (2009) *Building Better Lives: Getting the Best from Strategic Housing*, Audit Commission, London.

[62] Davidson M, Roys M, Nicol S, Ormandy D, Ambrose P (2009) *The Real Cost of Poor Housing*, IHS BRE Press, Bracknell.

[63] Nicol S, Roys M, Garrett H. (2015) *Briefing Paper: The Cost of Poor Housing to the NHS*, Building Research Establishment, London.

[64] Ezratty V, Ormandy D, Laurent MH, Duburcq A, Boutière F, Cabanes PA. (2019) Chapter 2: Health cost benefits of energy upgrades in France. In *Designing for Health & Wellbeing: Home, City, Society*, Jones M, Rice L, Meraz F (Eds.), Vernon Press, Wilmington, DE.

[65] Brown P, Newton D, Armitage R, Monchuk L. (2020) *Lockdown. Rundown. Breakdown.: The COVID-19 Lockdown and the Impact of Poor Quality Housing on Occupants in the North of England*. The Northern Housing Consortium. https://nationwidefoundation.org.uk/wp-content/uploads/2020/10/Lockdown.-Rundown.-Breakdown.-2.pdf.

[66] WHO. (2018) *Housing and Health Guidelines*, World Health Organization, Geneva. Licence: CC BY-NC-SA 3.0 IGO. https://www.who.int/publications/i/item/9789241550376.

Sources of further information

1. The Academic-Practitioner Partnership for Healthier Housing has produced two substantial reports (Good Housing: Better Health and Delivering Healthier Housing) and other submissions to Government. See https://www.healthierhousing.co.uk

2. Care and Repair: http://careandrepair-england.org.uk/health-housing/ https://www.careandrepair.org.uk/en/ and http://www.careandrepairscotland.co.uk

3. Public Health England (as was)– https://www.gov.uk/government/collections/housing-for-health

4. Australian Housing and Urban Research Institute (AHURI). https://www.ahuri.edu.au

5. BRE Trust at https://www.bretrust.org.uk

6. House of Commons Library see https://researchbriefings.parliament.uk

7. National Institute for Health and Care Excellence (NICE) and reports, for example, on Liverpool Healthy Homes (Case Study) in 2017 and Indoor Air Quality at Home (NG 149) 2020 and Excess Winter Deaths NG6 in 2015. See https://www.nice.org.uk/search?q=Housing

8. JRF. See https://www.jrf.org.uk/housing

9. European Network for Housing Research. See https://enhr.net

10. The UK Collaborative Centre for Housing Evidence (CaCHE). See https://housingevidence.ac.uk

13

The influence of society on the UK's food and food regulatory systems

Sian Buckley and Tony Lewis

Introduction

It is often said that people get the society that they deserve! In the same vein, it might also be said that society gets the food that it deserves. So, what does food mean to you within our society? Does food mean pounds and ounces and an expanding waistline, or does it mean a street market with fresh fruit and vegetables for sale? Does it mean watching food-related TV such as the great British Bake Off or does food mean McDonald's golden arches or a packet of Walkers crisps? Or is food in British society of 2022 something else? Is it about the exponential rise in the use of food banks? Is it about the inexorable rise of the food-tech platforms such as Just Eat and Deliveroo, or is food represented in the culmination of years of Brexit negotiations that have delivered little other than shellfish produc-ers in Scotland being unable to export their product to the European Union? Or does it mean empty shelves in our supermarkets as a result of heavy goods vehicle drivers from the European Union having returned home as a result of Brexit?

Whatever food means to you as an indi-vidual, one thing is for certain – your view of food has been formed by the society in which we all live and for the United Kingdom in mid-2021 with a COVID-19 pandemic in full swing, that society is complex, difficult to define, in a state of flux and is largely uncer-tain of its future. Similarly, the type of society that we live in and its relationship with food largely directs the extent and type of controls that it puts in place to manage the 'risks' that people, society and their food present.

If describing a society's relationship with food is difficult to grasp, perhaps it will make life easier if we compare and contrast just four families – the first family lives in Ethiopia, a country ravaged by war and ongoing famine; the second family lives in urban Germany; the third in rural Peru, whilst our final family is typically British.

The family from Ethiopia consists of three children and their parents and to survive for a month they rely on food aid that consists of little more than 1/3 of a sack of maize, 1/4 of a sack of rice and a small selection of veg-etables, supplemented by a small amount of bottled, clean drinking water. This compares significantly with our family from Peru who typically consume over the same period of one month half a sack of sweet potatoes, half a sack of rice, half a sack of yams, several bags of flour, sugar, salt; large supply of all fresh vegetables that include turnips, carrots, green vegetables, plantains, limes and bananas.

DOI: 10.1201/9781003035640-13

Even the comparative food wealth of the Peruvian family contrasts significantly with the amount that our German family will consume over the same period of one month. The typical German family consumes large amounts of fresh meat, a small number of fish, a large amount of fresh vegetables, eggs, butter, cheese, other dairy goods; a large amount of beer and wine to supplement a comparatively small amount of processed foods such as pizza. Finally, comparing again, our British family consumes a relatively small amount of fresh fruits and vegetables and meat but a significant amount of dairy products including milk, butter, cheese and flavoured yoghurt, all of which are supplemented by a large amount of processed foods including sausages, bacon, crisps, pizza, ready meals and confectionary.

Just by looking at the diets of the four typical families we can build a picture of the societies in which each sit. Our Ethiopian family is not atypical; some 795 million people in the world do not have enough food to enable them to lead a healthy active life [1]. That's approximately one in nine people on earth. Whilst our minds are naturally drawn to African societies as being typically undernourished, more than half of all undernourished people (418 million) live in Asia [2].

This position of hunger and malnourishment contrast significantly with the situation of comparative plenty within Western societies; however, modern Western diets are, in many respects, unhealthy. The Western diet is high in fat, salt and sugar and low in the consumption of fruits, grains and vegetables and is strongly linked to excess morbidity and mortality from conditions such as coronary heart disease, stroke, some cancers and type 2 diabetes [3].

This picture of a diet, spawning significant non-communicable disease, is a function of a society within which many Britons live. Furthermore, the reasons why this society has delivered a people that follow such an unhealthy diet are complex but include such things as the influence of our history as well as of 21st-century life.

The history of food not only includes information about population, but it also provides an insight into what people lived on and how they managed to create a food supply, often in difficult circumstances. The history of food also helps explain the changing face of the landscape of an area over the short and long terms. Several major historical events have also been dictated by changing tastes in food; for example, the growth in the consumption of sugar-fuelled slavery.

The history of Britain and its invaders has played a large part in its traditions, its culture and in its food. for example, the Romans brought cherries and stinging nettles to Britain – the latter to be used as a salad vegetable. The Romans also brought us cabbages and peas and improved the cultivation of our corn. They also brought us wine. The Saxons were excellent farmers and cultivated a wide variety of herbs that were used to make food go further and not just as flavour enhancers as we see them today. The Vikings and Danes brought us the techniques for smoking and drying fish and, even today, those techniques continue in the parts of Britain where the Vikings and Danes first landed (i.e. the East Coast of Scotland and the North East Coast of England. The Normans invaded Britain in 1066 and changed our eating habits forever. Normans encouraged the drinking of wine and gave us words for common foods such as mutton (mouton) and beef (boef). In the 12th century the Crusaders were the first Britons to taste oranges and lemons whilst in the Holy Land and they brought them back with them at the end of the Crusades. Saffron was first introduced into Cornwall by the Phoenicians when they came to Britain to trade for tin. In addition, the importation of foods and spices from abroad has extensively influenced British food. For example, in the Middle Ages wealthy people were able to cook with spices and dried nuts from as far away as Asia. In Tudor times, new kinds of foods started to arrive due to the increase in trade and the discovery of new lands. This led to spices being imported from the Far East, sugar from the Caribbean, coffee and cocoa from South America and tea from India. Potatoes from America began to be widely grown at this time. Turkeys were bred in the UK almost exclusively in Norfolk up until the 20th century because they were first landed in that part of the country.

The growth of the empire in the 18th century brought new tastes and flavours – kedgeree for example is a version of the Indian dish Khichri and was first brought back to Britain by members of the East India company and has remained a traditional dish at the British breakfast table since the 18th and 19th centuries.

Whilst the influence of our history on our diet and food has been significant, the influence of 20th- and 21st-century life in Britain is also a matter to be considered. The 20th and 21st centuries can be characterised by a lack of time, pressure of work, national and local planning policy, access to local and world markets, the integration of agrochemical and food manufacturing industries and the large-scale, industrialisation of food production.

So, take a step back and choose a meal that you ate over the last couple of days, perhaps it was the meal that you ate last night, and consider the following: what was the meal, what did it consist of, why were you eating that particular food or product? Do you know where the product came from; do you know where the ingredients came from; do you know how the food was produced; do you know who produced it and do you know about the conditions under which it was produced? More importantly, do you understand the hazards to your health and the health of others that are intrinsic within the consumption of that food?

These are all questions of which, at the very least, environmental health professionals ought to be aware and thereby be able to understand and make sense of the checks and controls that society puts in place in an attempt to manage the hazards and risks presented by the current food system.

Society and its influence on the foodscape

The term 'foodscape' is simply a combination of the terms 'food' and 'landscape' [4]. The places and spaces where you acquire food, prepare food, talk about food or generally gather some sort of meaning from food – this is your foodscape. The concept originated in the field of geography and is widely used in urban studies and public health to refer to food environments. Sociologists have extended the concept to include the institutional arrangements, cultural spaces and discourses that mediate our relationship with food.

The foodscape centres around the food environment. For example, where I live my foodscape includes a supermarket, a community garden, a bakery, two cafes, two pubs, a food bank, two school breakfast and lunch programmes, a farmer's market and several fast-food takeaways. Although these are all within walking distance of my front door, foodscapes are not always local spaces. In many rural and urban areas, shoppers must drive or catch a bus to get to the nearest supermarket, restaurant, café, pub or farmers' markets. In the suburbs of major towns and cities, shoppers can find chain supermarkets that offer large discounts and substantial choice, whereas, in more rural areas, consumers probably have only got access to small village stores with minimal choice and relatively high costs and prices. For middle-class shoppers, foodscapes extend to online spaces that allow consumers to order food from distant warehouses through grocery delivery services or websites like Amazon or Ocado.

In short, the foodscape is never fixed; its boundaries shift depending on how the food environment expands and contracts. There is also a multiplicity of foodscapes in any given space. For example, a town may encompass a gourmet foodscape and a deprivation-led foodscape at the same time. Sociologist Anthony Winson has written critically about the vast amount of processed food and fast-food restaurants clogging our foodscapes and reported by MacKendrick [5]. Reflecting on the obesity epidemic he identifies the profit-seeking motives of agribusiness rather than consumer demand as a primary force that he says led to the dominance of 'pseudo foods' in many foodscapes; consequently, healthy eating becomes the burden of the individual shopper who must learn to navigate their grocery and neighbourhood in search of healthy food.

Institutions such as breakfast or lunch clubs and organisations like the major supermarkets play a fundamental role in determining the food we eat, as well as who eats and who does not. For children from low-income households, free school meals and breakfast clubs are integral to their foodscape. Likewise, food banks and luncheon clubs are vital components of the foodscapes for many low-income individuals, particularly the elderly. The institutional and organisational dynamics that contribute to a foodscape are often reflected in what kind of food is available; for example, social activists consistently document the struggles of poor inner-city residents who they see as being at the mercy of convenience stores and fried chicken shops that serve their community, and which are unlikely to carry affordable fresh produce or healthier food items.

Foodscapes reflect not just profit motives but also racial prejudices. In her research on African American health [6], human ecologist Naa Oyo Kwate explored the ways in which fast-food chains tended to target predominantly low-income, black neighbourhoods in the United States. Kwate found that grocers often need a lot of convincing to open stores in poor communities, especially those with high proportions of black and Asian residents. Referred to as 'retail red-lining', this is a form of spatial discrimination whereby businesses choose not to locate a fresh food supermarket in an area because of racial prejudice – they see black neighbourhoods as having poor retail viability and high incidents of crime. Foodscapes may therefore be classified as racialised environments.

Society influences foodscapes in additional ways too. The contours of foodscapes may also be shaped by what is referred to as cultural politics and trends that relate to the meaning and significance of food. For example, the slogan of 'voting with your fork' is a powerful articulation of food politics. This slogan draws from the ideals of the organic food movement of the 1960s that suggested that individual food purchasers have the potential to transform the industrial food system into one that is more ethical healthy and environmentally sustainable. This argument believes that consumers can irrevocably change their communities by supporting local farmers, organic production and fair-trade practices.

In their book 'Foodies' sociologists Josee Johnston and Shyon Baumann explore the narratives of the gourmet foodscape [7]. Whether high- or lowbrow, they noted that the gourmet foodscape requires the creation of boundaries that mark certain foods as 'authentic' or 'artisan' and this then reinforces a dichotomy of 'good' versus 'bad' food – good being authentic or artisan whilst supermarkets or restaurants deliver bad food. The gourmet foodscape is also constructed through popular culture, including magazines like *BBC Good Food* and food television. UK television broadcasts dozens of shows in any given week from *Diners, Drive-ins and Dives* to *Masterchef.* Food trucks and food festivals are also part of this foodscape; their menus and branding blend the highbrow with the lowbrow and produce messages about the authenticity of their dishes whilst being inaccessible to many and to particularly the socially disadvantaged.

Our local surroundings have a large impact on us, and we live in an environment that can inadvertently encourage unhealthy behaviours, including eating more and exercising less. The factors affecting our choices can include access to active travel and availability of green spaces, as well as the density of fast-food outlets. Many of our streets are saturated with fast-food outlets selling food such as chips, burgers, kebabs, fried chicken and pizza and the sheer density of these outlets may make it easier for us to consume too much, too often. The fact that such outlets have no or limited nutrition information available makes it even more difficult for us to make informed choices. In addition, such premises are likely to have the lowest food hygiene rating scores and are often associated with areas of higher deprivation.

Local authorities acting in conjunction and under advice from health professionals such as EHPs, have powers to help shape such environments including tackling the growth of new fast-food outlets in neighbourhoods that are already saturated with such development. For example, some local authorities have focused on the provision of safe spaces for children and young people to congregate away from fast-food saturated areas; others have also recently introduced more cycling lanes (and been heavily criticised for doing so by the press and Conservative Party politicians [8]) and made improved use of open green spaces in an effort to counter obesogenic environments.

In addition, Public Health England produced a toolkit to help local authorities to work with local businesses to provide healthier alternatives to their biggest sellers, for example, by using less salt, sugar, saturated fat and offering smaller portions [9].

Sustainability and food choice

The most-often quoted definition of sustainability (and in several places in this book – an indication of its importance to the different aspects of environmental health) comes from the United Nations World Commission on Environment and Development:

> Sustainable development is development that meets the needs of the present without compromising the ability of future generations to meet their own needs.

The definition acknowledges that human civilisation consumes resources to sustain our modern way of life [10].

In 2005 the World Summit on Social Development [11] identified three core areas that contribute to the philosophy and social science of sustainable development; these are:

- Economic development;
- Social development;
- Environmental protection.

Economic development is about giving people what they want or need now, without compromising the quality of future life – especially in the developing world. This is the issue that proves the most problematic, as most people disagree (on the basis of political ideology) what is and what is not economically sound, how it will affect business and, by extension, jobs and employability.

In respect of social development, there are many facets to this pillar. Most importantly is awareness of the need to protect the health of people from pollution and other harmful activities of modern life. It is also about maintaining access to basic resources without compromising the quality of life. The final element is education – teaching people about the importance of environmental protection as well as warning of the dangers if we fail to achieve our goals.

Environment protection is the third pillar and, to many, the primary concern of the future of humanity. It defines how we should protect ecosystems, including air quality, along with the integrity and sustainability of our resources and focuses on elements that place stress on the environment.

In 2012, the United Nations Conference on Sustainable Development [12] met to discuss and develop a set of goals to work towards. The Sustainable Development Goals (discussed in Chapter 1) that emerged from the conference include, amongst other things:

- The end of poverty and hunger;
- Better standards of education and healthcare;
- Gender equality;
- Sustainable economic growth while promoting jobs and stronger economies;

with all to be undertaken whilst tackling the effects of climate change, pollution and the other environmental factors that can harm and do harm to people's health, livelihoods and lives.

Sustainability acknowledges modern needs; but, in the developed world we live in and in a consumerist and largely urban existence where we consume natural resources every day at a rate that is unsustainable, climate change is a major challenge in 2021 and we know unsustainable food systems are substantially implicated in that. On both the world stage and within the UK, we face a strange but dangerous paradox – waste versus hunger. Our society generates large amounts of food waste whilst at the same time many of our people face hunger. When you overlay all of this with the challenges generated in the UK by Brexit, then it is easy to see that we in the United Kingdom not only live in a dangerously unsustainable world, our food systems of production, supply, retail and consumption are equally unsustainable and present a range of risks to our society.

Consequently, the major question that we in the UK face is how can we achieve a dietary pattern that provides us with the many nutrients we need for health, in appropriate amounts, but that is also equitable, affordable and sustainable? How do we produce more food with fewer resources such as land, water and fuel to feed the growing population of the United Kingdom? These are just some of the key questions that we face in 2021 for which we need to find answers and quickly.

Sustainable eating is a global challenge that needs to be considered alongside malnutrition which is another global problem. The term 'malnutrition' includes under-nutrition (wasting, stunting, underweight and micronutrient deficiencies) and over nutrition (overweight and obesity). Many countries like the UK are currently experiencing a double burden of malnutrition, where a combination of obesity and hunger issues exist together.

Worldwide, over 150 million children under the age of 5 suffer from stunting, while more than 50 million are affected by wasting. In contrast, 38 million children in this age group are believed to be overweight or obese. Similarly, over 1.9 billion adults globally were defined as overweight or obese in 2016, in contrast to the 462,000,000 adults who were underweight. In the UK, 64% of adults living in England are considered to be overweight or obese [13]. Although obesity is the result of consuming more calories than required in the context of lifestyle and activity levels, it is often associated with a generally poor-quality diet (i.e. too high in sugars, saturated fat and salt, and too low in fibre).

Addressing the global imbalance of nutrition and its causes is a central aim of the Sustainable Development Goals. Sustainable Development Goals also incorporate the need to address climate change and protect our marine and terrestrial ecosystems at the same time. Improvements in nutrition are recognised as playing a pivotal role in accomplishing all of the Sustainable Development Goals, with the period 2016 to 2025 declared a decade of action on nutrition by the United Nations.

Bearing all of this in mind, how do we feed both the UK and the world's growing population?

An estimated 37% of the world's land mass is farmland [14] and there are huge pressures acting on the current global food system which is responsible for producing enough food to supply the global population. Worldwide, demand for food continues to increase, with the growing global population that is expected to rise to close to 10 billion by 2050 [15]. Over half of the world's population now live in urban environments. This growing urbanisation, in combination with rising prosperity in some low- and middle-income countries, together with associated changes in dietary habits, have together led to a growing demand for meat and dairy products being placed on the available land.

There are additional and growing pressures on our water resources. Globally, agricultural irrigation uses 70% of all extracted freshwater. Agriculture, forestry and other land use together account for 24% of worldwide greenhouse gas emissions [16]. Such factors, as well as the likely effects of climate change on what can be grown in the future, must be considered when developing policies that impact on food consumption as well as in the development of food-based recommendations for populations.

Bearing all of this in mind, what is a sustainable diet?

Defining a sustainable diet requires consideration of multiple environmental, social and economic factors, which together form the three pillars of sustainable development.

> Sustainable diets are those diets with low environmental impacts which contribute to food and nutrition security and to healthy life for present and future generations. Sustainable diets are protective and respectful of biodiversity and ecosystems, culturally acceptable, accessible, economically fair and affordable, nutritionally adequate, safe and healthy; whilst optimising natural and human resources. [17]

How do we achieve a sustainable diet and sustainable food systems?

The complexity of the issues surrounding sustainability make it almost impossible for us to know, at the point of picking up a food item in the supermarket, what its impact on the planet to date has been. However, there is enough evidence to support some general principles that we can adopt in the UK in order to eat more sustainably:

- Eat a more plant-based diet;
- Choose fish from sustainable stocks;
- Reduce food waste in the home.

Food-based dietary guidelines around the world encourage the consumption of a more plant-based diet and the UK's guidelines are no different in this respect.

The UK's *Eatwell Guide* [18] is designed to help all those aged over 2 years of age to eat a healthy, balanced diet. The *Eatwell Guide* recommends that diets should be based on starchy foods (e.g. bread, potatoes, rice and pasta), as well as plenty of fruits and vegetables. A variety of foods from these two food groups should make up just over two-thirds of the food that we eat. Good plant-based sources of protein (and vitamins and minerals) are emphasised such as lentils, beans and pulses, alongside sustainably sourced fish, eggs and lean meat as well as some dairy foods. A small amount of unsaturated oils and spreads are included. Foods and drinks high in fat, sugar and/or salt (EJ cakes, chocolate, sweets, biscuits, sugar-sweetened drinks, crisps) are not needed in the diet and, if eaten, should be included less often and only in small amounts.

An analysis performed by the Carbon Trust, and commissioned by Public Health England, reported that consuming a diet in line with the *Eatwell Guide* would have a 32% lower environmental impact (in terms of greenhouse gas emissions, water and land use) than the current average UK diet [19]. A cost analysis of the *Eatwell Guide* concluded that the price of a diet in line with these recommendations (£5–99/adult/day) would be very similar to the typical diet already consumed by the UK population.

Modelling work undertaken for the Carbon Trust has estimated that if everybody in the UK consumed a diet in line with these recommendations, this would lead to 780,000 fewer new cases of type 2 diabetes by 2026, as well as significantly fewer new cases of coronary heart disease, stroke and colon cancer.

In order to adopt a diet that is more in line with this guidance, most of us in the UK need to do the following:

- Eat more plant-based foods such as fruit, vegetables, potatoes, pulses and wholegrains;
- Eat fewer foods that are high in fat, sugar or salt such as cakes, sweets, chocolate,

biscuits, crisps, processed meats and ice cream;

- Consider our meat consumption, choosing plant-based protein sources such as lentils, beans and nuts as alternatives some of the time, as well as fish from sustainably managed resources.

When choosing fish from sustainably managed resources it is also important to select species that have not been overfished and this means moving away from those species on which in the UK there has been too much reliance, namely cod, haddock, tuna, salmon and prawns.

The right to food and the Government's duty to provide food

As a society, we generally believe in justice and compassion. We should have the same chance to get on and succeed in life whether from the Cheshire countryside or urban communities in Stoke on Trent. For most of us, this starts with having enough to eat, proper clothing and a safe place to call home. But what happens when we can't put enough food on the table? Who can, and should, step in to help?

In Britain, one in five children suffer from what UNICEF called food insecurity [20]. This means that their families lack secure access to sufficient safe and nutritious food. More than 8,000,000 British adults struggle to get enough to eat and almost 5,000,000 of us have gone whole days without eating [21].

Food poverty, or household food insecurity, has a myriad of faces:

- It can affect children who lack free school meals outside of school term time; see the campaign recently spearheaded by the Manchester United and England footballer, Marcus Rashford;
- Parents on low incomes going without food so that their children can eat;
- It even affects working people whose low wages leave them struggling to buy healthy food;

- Food poverty can be triggered by a crisis in financial or personal circumstances – but it can also be a long-term, grinding experience of not being able to afford to eat well.

The figures referred to here are quite shocking. Despite our uncertain access to the most basic resources, we need to survive, UK welfare policy is taking support away, rather than giving more out to people in severe need. And whilst elected governments should be free to enact the laws that they please; once these policies start to impact upon people's ability to access adequate and healthy food for themselves and their families this becomes an issue of rights more than policy.

What are our rights?

Our right to food is protected by a number of international standards including:

- Article 25 of the Universal Declaration of Human Rights;
- Article 11 of the International Covenant on Economic Social and Cultural Rights, where States are required to ensure economic, social and cultural rights, such as the right to food, *to the maximum of its available resources.*

The idea is that developing States, that may be genuinely unable to provide food for everyone (until they can build up their economic resources), should not be penalised under international law. However, in a wealthy developed state like the United Kingdom, the Government can hardly argue that it cannot afford to provide all individuals within its jurisdiction with basic nutrition.

What is Government legally required to do?

Government must act to secure the human right that:

Whenever individuals or groups are enabled, for reasons beyond their control, to enjoy the right to food by the means at their disposal, States have the obligation to fulfil (provide) it, for example by providing food assistance or ensuring social safety nets for the most deprived. [22]

What does all this mean for health?

According to the Department of Health malnourished people see their GP twice as often, have three times the number of hospital admissions[1] and people in deprived neighbourhoods live on average seven years less than people in wealthier neighbourhoods. Additionally, in 2015, there were 84,528 hospital bed days taken by patients who were suffering from malnutrition.

People in food poverty often have high levels of stress and depression with poverty being both a potential causal factor and consequence of mental illness health. People with a disability or mental health condition are far more likely to receive emergency food aid than any other group. Furthermore, articles 25 and 28 of the Convention on the Rights of Persons with Disabilities, which the UK has signed and ratified, stipulates that States have a duty to provide people with disabilities with an adequate standard of living and the highest attainable standard of health.

Children have special provisions, such as those in article 24 of the Convention on the Rights of the Child, to protect their right to food, because child malnutrition can result in serious long-term health consequences, and poverty in childhood can lead to an increased susceptibility to cardiac disease and certain types of cancers. Recently, 87% of schools surveyed by the Child Poverty Action Group said poverty affected their pupils' learning to a significant extent [23].

Elderly people are seen as especially vulnerable to malnutrition and there are special provisions in place that ought to protect them. The United Nations' 'Principles for Older Persons', adopted by the General Assembly of the United Nations, says that older persons should have access to adequate food, water, shelter, clothing and healthcare. Yet, an all party parliamentary group on hunger report showed that a shocking 1.3 million older people in the UK are malnourished or at risk of malnourishment [24].

What can be done?

Despite being signatories to all of the conventions referred to already, the UK Government currently does little to ensure that it meets its commitments. This is in sharp contrast to the action being taken by many governments around the world and even to action being taken by the Scottish Government that is currently in the process of working on proposals for a Good Food Nation Bill, which is likely to incorporate the right to food in order to tackle deep-rooted social and economic inequalities in Scotland. Some of the measures under consideration by the Scottish Government include the requirement to have health warnings on what some label as 'junk food', and the provision of holiday hunger programmes for children suffering from food poverty.

Does the Government have a duty to provide food?

Official Government documents reveal that Government feels no legal responsibility for securing our food supplies in an emergency – neither fire, flood, disease epidemic, conflict, supply chain disruption, nor Brexit. However, when faced with COVID-19, the UK Government has taken limited measures to ease the path for the supermarkets to get food to their stores. The UK Government places the main responsibility for securing our food supply in an emergency, squarely with the commercial food industry. However, Government also recognises that private-sector business is governed by commercial considerations; consequently, in a food crisis we are therefore caught between a rock and a hard place.

In response to recent parliamentary questions concerning the Government's legal duty to secure a food supply in a post-Brexit emergency situation, Defra's official statement says,

> Defra is not responsible for the supply of food and drink to the population in an emergency. Local authorities do not have a general duty to provide food but have duties to provide food to particular groups in particular circumstances including schools and care settings. The expertise capability and levers to plan for and respond to food supply disruption lie solely within the industry. The food industry is experienced in dealing with scenarios that can affect food supply.

This response is in sharp contrast to the requirements of the Civil Contingencies Act 2004. The Civil Contingencies Act 2004 is one of the most relevant pieces of legislation to emergency planning in respect to civil emergencies. The Act lists local authorities, the Environment Agency and emergency services as 'Category 1 responders' to emergencies and it places duties on these organisations to undertake risk assessments, manage business continuity, carry out emergency planning, and warn and advise the public during times of emergency. The key question is, how can local authorities and others fulfil their obligations under the Civil Contingencies Act 2004 in respect of a major national emergency when the Government that took decisions that led to that emergency has washed its hands of responsibility and, more importantly, how will the people be fed?

Brexit and food

The European Economic Community (EEC) also initially known as the 'Common Market' was formed in March 1957 and from that time it became a subject of some debate as to whether Britain should become part of the European Economic Community and enjoy the benefits of being part of one of the world's largest trading blocs. Unfortunately, Britain did not achieve membership of the EEC until January 1973 when, finally, the UK became a signatory to the Treaty of Rome.

Whilst successive UK Governments saw the benefit of Britain being part of the EEC, large numbers of British people remained deeply sceptical about membership. Many believed that the country paid too high a financial, social and political price for that membership, and this led to a referendum taking place in 1975 on Britain's continuing membership of the EEC. Whilst 67% of the British people voted to remain in membership, the opposition to the UK being part of the EEC never went away and culminated in a further referendum in 2016 that saw a narrow majority vote to leave. Despite the status of the referendum being merely 'advisory' and not mandatory, the Conservative Government pursued a policy of Brexit that culminated in Britain formally exiting the EU (as it was now called) in January 2021, albeit with transition arrangement in place until January 2022.

Despite many in the UK falsely believing that exiting the European Union was simply a matter of closing one door and opening another, the reality is very different. For almost 50 years all aspects of life within the UK had become increasingly integrated within those of the wider European Community and this was particularly true for the UK's agriculture, food production, distribution, retail and associated regulatory systems. It should, therefore, have been obvious to all that disentangling the UK from the EU in respect of food would prove to be particularly challenging and would be likely to generate significant disruption to the UK's food security, and food regulatory systems as well as to food-related business. However, in order to make sense of the challenges generated by Brexit, it is first of all important to understand the extent to which, over 50 years of EU membership, Britain's food system had become integrated within that of Europe.

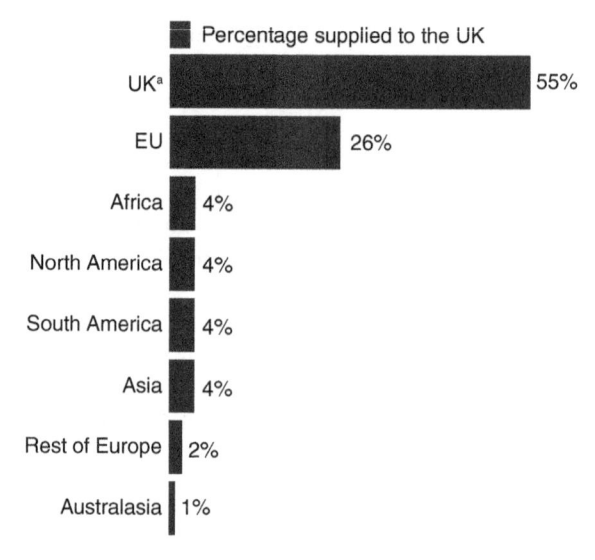

Figure 13.1 Sources of unprocessed food

Where does UK food come from?

The UK does not feed itself and the UK's domestic production of food has been steadily declining for the last 40 years; self sufficiency is now, by value, only around 55% [25]. Furthermore, if we consider the flow of unprocessed as opposed to processed foods, the UK supplied slightly more than half (55%) of its unprocessed food in 2019. The sources by region are given in Figure 13.1.

The UK has had a rising food trade gap for many years. In 2017, the UK imported food, drink and animal feed worth £46.2bn. It exported only £22bn worth, of which whisky accounted for nearly a fifth! Ironically, the UK, famed by many as a land of roast beef, actually imports more meat in total than it exports [26]. In 2019, the UK was only 81% self-sufficient in cattle meat, 61% for pig meat, 90% for poultry, 101% for sheep and lamb, 105% for milk and dairy and 86% for eggs [27]. It's also likely that the actual UK import dependency is greater than financial figures suggest. Taking the available data as a whole, it is clear that the UK currently imports over 50% of its food and animal feed by value.

Consequently, the UK has a serious imported food dependency problem and has what used to be called an ongoing 'balance of payments' crisis [28].

Although there have been vociferous calls in the Conservative Party for the UK to do 'big trade deals' with the USA, the USA has made it clear that this would only happen for financial services if the UK abandoned its attachment to food standards which the US Government, under both Presidents Trump and Biden, view as unnecessary [29]. Indeed, Wilbur Ross, the former US Commerce Secretary, stated that abolishing EU-derived food standards was essential for any UK-USA trade deal [30].

Very few seem to have recognised that the USA would have to fill a huge gap in foodstuffs if it were to replace the food the UK currently imports from the EU. US exports to the UK are currently proportionately small – the reality is that it is tenth out of the top ten food exporters to the UK. For the USA to replace the combined food imports from the other nine of the top ten would require a vast food fleet and a mammoth logistics operation that exceeded that of the 1940–45 Atlantic Convoys. As one senior food industry manager said in a personal communication in 2017: *this is dangerous fantasy* [31]. However, we might reasonably expect that at some time in the future the US will sell hormone-assisted, cheap beef to the UK; but this occurrence will require the UK to abandon EU standards (which do not permit the use of such hormones); this will also blow apart the UK's 'third nation' status in respect of its ability to export to the EU and is likely to be hugely damaging to Britain's agri-food industry.

It is, therefore, perhaps fortunate that the relatively new US President (Joe Biden) is in no hurry to conclude any kind of trade deal with the UK, preferring instead to focus his efforts on maintaining stability in Ireland and specifically protecting the Good Friday Agreement that is currently being threatened by the Northern Irish Protocol to the Brexit agreement.

Threats to UK food security arising from Brexit

Over the last 50 years, British supermarket shelves have been filled with a wide range of reliable, fresh and affordable foods and the public believe this is the norm. However, the vast majority of people in this country are unaware of the performance of the complex, logistically sophisticated, evolving and unstable system on which our food supply and supermarkets depend. We may see the large trucks on British roads, but what we don't see are the satellites and the computers that have been integral to the 'just in time' logistics revolution. We see the cash tills that tally the customer's purchases at the checkouts, but we don't see that the same tills also directly communicate with the supply chain to order replacement stock. Consequently, at any one time, much of the UK's supermarket's 'stock and storage' is, in fact, moving in trucks on the UK's motorways and Europe's autoroutes; very little is held in storage warehouses and that, of itself, presents a serious risk to the food security of the UK.

The UK's Cabinet Office conducts a regular review of national resilience. Food is one

of the 13 sectors considered in those reviews. According to the published summary of a review undertaken pre-Brexit, *The UK Food sector has a highly effective and resilient food supply chain, owing to the size, geographic diversity and competitive nature of the industry* [32]. However, such a conclusion reflects 50 years of the UK integration of our food system with that of the EU. The real position in the summer of 2021 is very different.

A relatively recent study has also suggested that the UK has 'adequate supply' of all key nutrients except fibre [33]. At one level, this was and is almost reassuring, but, putting COVID-19 to one side, legitimate questions exist with regard to sourcing the 'just-in-time' nature of current food logistics and the reliance of food infrastructure on other sectors such as fuel, transport and trade routes that have all been significantly affected by labour shortages generated by Brexit.

In 2018, the British Retail Consortium (BRC) reminded the Prime Minister that about 10,000 containers of food come into the UK from the EU daily, among the 3.6 million containers that pass through our ports annually, supplying about 50,000 tonnes of food to the UK food trade [34]. Overlying this position has been the EU and UK logistics industry and what we have now learned, in the first eight months since Brexit became a reality, is that many EU-based hauliers rejected contracts to move goods to the UK because of difficulties generated by new customs arrangements and those same hauliers are equally concerned by the phyto-sanitary inspection systems that will be introduced at the end of 2021 following the ending of post-Brexit transitional arrangements. In addition, all of this has been compounded by the fact that UK-based hauliers have been decimated by the loss of drivers due to the return of EU nationals to the EU.

In a similar way, what little UK-grown horticulture there is, has also become massively dependent on EU migrant labour. The previous Seasonal Agricultural Workers Scheme (SAWS) was abolished in 2013 by the

Conservative/Liberal Democrat Coalition, on the basis that the Government was confident that Eastern EU workers would fill any gaps [35]. Events have unfolded surprisingly rapidly. Having lost EU-based labour, some UK-based horticultural companies have now started new subsidiary enterprises in Africa to grow foods for the British market, whilst some UK harvests have gone ungathered [36]. Moving production abroad is unlikely to resolve the UK's fundamental food labour problem and is likely to compound the logistics problem. Her Majesty's Government oscillates between denying there are problems, insisting that market forces can and will resolve them, and fantasising about a labour-free future relying on artificial intelligence and computerised phyto-sanitary inspections systems that simply do not exist at present. Such signals do little to provide industrial certainty or to resolve the empty shelves that have become an everyday fact of life in UK supermarkets in mid-2021.

What Government has consistently failed to grasp is that, like all systems operating to finely tuned specifications, the UK food system is fragile and highly vulnerable to disruption. There were previous early warnings that we should have noted. Back in 2000, blockades of depots by a few hundred fuel delivery drivers massively disrupted UK food distribution within 24 hours [37].

More recently, the shortage of industrial quantities of CO_2 in June 2018 not only exposed the food system's dependence on this ingredient (from stunning animals for slaughter to the production of crumpets and beer) but it also showed how intermeshed the north European food sectors are, with CO_2 being sourced across Europe.

European integration has been a key to the creation of this complex food logistics system. A transformation of the UK's food system began in 1967–73 while the UK was negotiating to join the Common Market, which, in turn, became the EU. The transformation accelerated when, in 1975, a majority of UK voters chose, in a Referendum, to stay in the Common Market. It was significantly boosted by the creation of the European Single Market in 1992. That facilitated a wave of change that swept through the EU food system, with many parts of the food chain experiencing rapid concentration through mergers and acquisitions. A small number of dominant firms became pre-eminent in sectors such as dairy products, confectionery, beverages and frozen foods.

UK vulnerabilities, therefore, seem clear and, as we have come to learn during the first eight months of 2021 (at the time of writing and likely to continue), there are several active and serious threats to UK food security, especially in relation to the following issues.

Business continuity

Contracts for food supplies are typically set 12 months ahead. UK food comes via a complex logistics system run on a just-in-time basis (i.e. three- to five-days' supply). There are only tiny food stocks, commercial or public, held in the UK's food distribution chain.

Food safety

Inspections of food took place 'at source' while the UK was in the EU. Now the UK has left the Single Market/Customs Union, inspections have to take place at our ports (including airports for air-freighted foods). Until the end of 2021, port inspection of EU-derived food amounts to paperwork completion that takes an average of about two minutes. Once customs and food safety clearances on goods imported into the UK from the EU become fully operational, the development of any delay over and above the existing two minutes/truck has the potential to generate significant lorry tailbacks at the ports of entry into UK.

Home-produced UK food supplies

Though slowly declining overall since a high point in the early 1980s, the potential for home production is in the process of being eroded further by the signing of new trade deals such as that concluded with Australia in June 2021. Previous research has shown that farm incomes would more than halve if the UK opened its borders to a low-cost regime, would drop by less than half if it adopted a unilateral protectionist regime, and would rise if a free-trade deal was struck with the EU [38].

It would be wrong to consider food security simply as a matter of tonnage on shelves. *"It doesn't matter where food comes from or how"* a Government advisor infamously once said; *all that matters is that it's there.* That advisor was wrong then and is completely wrong now, but this sentiment is not uncommon, and is now in the process of playing out as gaps appear on UK supermarket shelves. The food system is one of the biggest sources of greenhouse gas emissions, biodiversity loss, social injustice and diet-related ill-health [39]. Notwithstanding Brexit, almost all the prognoses about our food supply to 2050 point to a serious squeeze on resources, land, water and people [40]. Irrespective of Brexit, the UK needs to change its food system considerably and Brexit should not undermine that goal.

The UK is a signatory to the UN Sustainable Development Goals [41] (see also Annex to Chapter 1), the Paris Climate Change Accord [42] and the Convention on Biodiversity [43], all of which require food system changes nationally and globally. Those targets are not likely to be achieved without a radical shift in UK and in other high-income countries' food systems and patterns of consumption [44]. What matters is not just that the UK has enough food to eat, but rather what is eaten and how it is produced, processed, distributed and consumed. Sustainability requires more than cutting emissions of CO_2 or cutting waste. It will require a multi-dimensional approach that connects public and environmental health with culture, economics, societal values and governance. The UK needs to reverse the damage

being caused by poor diets, particularly for low-income consumers. The UK diet had the highest proportion of ultra-processed food (high in fat, salt and sugar) consumption in 1998–2011 across 19 EU countries [45]. The average household availability of highly processed foods ranged from 10.2% in Portugal and 13.4% in Italy to 46.2% in Germany and 50.4% in the UK. The marketing and availability of such products distort eating patterns and impose long-term ill-health costs on the NHS. The current rates of obesity in both children and adults, as well as the rising trends, are hugely damaging and are unsustainable.

Social inequalities in the UK diet are lamentable. Low-income consumers have already been affected by higher food prices since the pound dropped in value, following the Referendum and the emergence of COVID. Food Banks have been normalised; they are no longer short-term crisis management. British eating patterns are socially polarised and increasingly so [46].

It is also important to remind ourselves that pre-EU membership, UK food was famous for being brown, over-cooked and plain. In culinary terms, membership of the EU has contributed to a transformation of UK food culture in both positive and negative directions. The UK now has more varieties of artisanal cheese, for example, than France.

In recent years, we have tended to take the view that this country is food secure because its people can afford to buy food sourced from abroad. There is a long history to this policy, which became particularly important in UK politics in debates over the Corn Laws in the early 19th century. With the 1846 Repeal of the Corn Laws, the policy became one of increasing reliance on external (often colonial) sources for imports. By the late 19th century UK agriculture had been run down, only for shocks to occur during World Wars 1 and 2. Learning the lessons from WW2 produced new, bi-partisan support for rebuilding the UK's domestic food supplies, which culminated in the 1947 Agriculture Act. Domestic food production grew to a high point in the early 1980s, supported initially by the UK's deficiency payments subsidy scheme and then by the Common Agricultural Policy subsidy schemes. Some (such as in the Treasury) see food as either of little interest or as a sub-issue of macro-economics that should be left to the market (ignoring how markets are dependent on and framed by politics).

Food can and should be a unifying theme in the UK. Polling also suggests that the British public is overwhelmingly hostile to any lowering of food standards. A large majority of the British public would prefer to keep current food safety standards (82%) over lowering standards for a trade deal with the USA (8%). More people back alignment with EU consumer, environmental and employment standards (49%) than back the weakening of these standards post Brexit (28%) [47].

In the months prior to Brexit, a senior Government advisor leaked the information that plans had been prepared to 'suspend food controls' if there are any delays to imports of perishable foods at our borders. We learned too that other policy commentators had been told the same by senior Defra personnel. Senior industry people have publicly indicated in no uncertain terms that this would be folly and must be avoided, not least because it would threaten exports from the UK to the EU. If the UK were to suspend food safety controls, others might block exports from a country taking such a cavalier approach to public health. It would go completely against all the protestations of commitment to high consumer and health standards. Yet this appears to be what Defra envisaged prior to Brexit and may be forced to consider again if the ending of the transitional period at the end of 2021 goes badly. There had also been several hints that other parts of the Government endorsed that proposal. On 28 February 2018, George Eustice MP, then Minister of State at Defra, informed the House of Lords EU Environment Committee that, in the event of a no-deal Brexit, Defra envisaged operating on the basis of a risk-based 'mutual recognition' regime. This assumed that food

in the EU is acceptably safe to eat and therefore would be safe to import and distribute in the UK, without introducing any safety checks [48].

In March 2018, Transport Secretary of State Chris Grayling stated on BBC TV that the UK will not 'in any circumstances' create a 'hard border' at Dover by imposing lorry checks after Brexit [49]. He said: *We will maintain a free-flowing border at Dover, we will not impose checks at the port, it is utterly unrealistic to do so. We don't check lorries now, we're not going to be checking lorries in the future.* He was incorrect both then and now: there are checks, an average of two minutes per lorry, if the border authorities have concerns. And these are important for health. This thinking that in a growing emergency imported food controls could be put to one side might be presented as 'emergency planning', but, if implemented, it would be catastrophically counterproductive. Firstly, it would contradict the Government's explicit commitment to maintaining high standards. Secondly, it would threaten the UK's food exports. Thirdly, it could consign the UK to pariah status in the eyes of the EU.

If, for example, border checks rose to four minutes in January 2022, there would be 20-mile or so (possibly even 29-mile) lorry tailbacks within one day [50], hence, the fallback of suspending food controls to allow all traffic to be waved through. This should not be seen as 'taking back control'; rather it is abandoning it.

Threats to food safety in the UK from imported products arising from Brexit

In July 2017 Lang, Millstone and Marsden in their Brexit Briefing paper for the Foord Research Collaboration highlighted four examples of potential food safety risks that might be a consequence of the UK agreeing a free trade deal with the USA [51]. They were, beef hormones, bovine somatotropin (BST) used in milk production, genetic modification (GM) of crops and the use of chlorinated disinfectants to reduce bacterial contamination of poultry carcasses. Of those four topics, 'chlorine-washed chicken' was the one most widely discussed in the print and broadcast media, and in Parliament, and has entered the discourse as a symbol of whether standards rise or are subverted. It is also an issue on which important new evidence emerged. In early 2018 it was learnt that while chlorinated water was not permitted in the EU for use to disinfect contaminated poultry, its use is permitted to disinfect leafy vegetables and horticultural products.

In May 2018 a team of scientists based at the University of Southampton published a paper showing that spraying leafy vegetables with chlorinated water did not reduce the presence or virulence of the bacteria, but it changed them into a "*viable-but-nonculturable (VBNC) state*", which means that while the bacteria remained in place and infectious, their presence could not be shown using the standard technique for detecting bacteria [52]. Those tests involve trying to grow, or 'culture', samples of microbes in glass dishes on suitable substrates. The researchers used a different technique enabling them to detect those bacteria on the vegetable leaves, after chlorinated water was used; the treatment blocks and so invalidates conventional culture tests, but as a disinfectant it is entirely ineffective. When the researchers said: "*These data emphasize the risk that VBNC food-borne pathogens could pose to public health should they continue to go undetected*", they understated the importance of their findings. Their new findings imply that current practices in the UK, and even more importantly in the USA, need to change, and to change urgently. Given that US food producers use chlorinated water far more extensively than do UK and European producers, as they apply them not just to vegetables and poultry but also on fish, fruit and non-leafy vegetables, those facts might help explain why the rate of food

poisoning in the USA is approximately ten times that in the UK [53].

There is an evident and urgent need for similar studies to be conducted across the entire range of anti-bacterial washes that are used on food in the UK, the EU, the USA and the rest of the world, to examine their effects, and effectiveness, on the full range of foods which are treated with them. Are any of them ever effective? Or do they only block and so invalidate the conventional tests? Furthermore, how can production standards be improved so that our food supply is not contaminated with pathogenic bacteria, and so does not need to be disinfected? The UK therefore now has even stronger reasons for rejecting chlorine-washed US poultry meat than we had three or four years ago. For the other three issues highlighted in July 2017, we have no reason to think that the risks they pose are any lower than we thought they were. On the other hand, several other problematic features of US food production practices and standards have been recognised. They include the fact that pesticides are sprayed more frequently and more widely in US agriculture than is the case in the UK and EU. Moreover, the maximum permitted residue levels of pesticides in the USA are often substantially higher than those permitted in the EU. Importing US produce would therefore entail a marked increase in the amounts of pesticides residues that UK consumers would ingest.

The standards of animal welfare in US meat, egg and dairy sectors are significantly lower than those in the EU, and as UK standards are in some respects stricter than those in other EU countries, the gap between what is deemed acceptable in the UK and the USA is especially wide [54]. In the EU, it is unlawful to administer a drug called ractopamine to pigs. Pigs fed ractopamine develop more muscle tissue and less fat than similar pigs fed a similar diet but without the drug. A meatier and less fatty carcass is more profitable but less safe. There is extensive evidence indicating possible adverse effects to both pigs and pork eaters. In 2009 the European Food Safety Authority concluded diplomatically there *"were not enough data to show that it is safe for human consumption at any level"*. In the USA, however, the Food and Drug Administration deems the 'benefits' of using ractopamine to outweigh the risks, but some 160 other countries disagree [55]. Another practice deemed acceptable by the US authorities is to incorporate what is politely referred to as 'chicken litter' (i.e. poultry bedding material mixed with chicken faeces), as an ingredient in animal feed products, including those intended for poultry. The UK's experience with BSE was sufficient to persuade us of the risks that can arise when animal wastes are incorporated into animal feeds. Feeding cattle wastes back to cattle massively increased the numbers of animals that were infected with BSE and accelerated the rate at which the disease spread from herd to herd.

The US food supply also contains a wider range of food additives than are permitted in the EU, and at higher levels of usage than are authorised in the EU. For example, potassium bromate and azodicarbonamide are authorised for use in US bread-making as 'dough improvers' but deemed unacceptable in the EU. Where the lists of substances permitted in both the USA and the EU as food additives overlap, often the USA accepts higher levels of usage than has been deemed acceptable or necessary in the EU. Furthermore, US food labelling standards provide consumers with far less information than is the case in the EU. US food industry representatives have told the US Government that in any US-UK trade deal, the UK should be obliged to accept any food product that complies with current US legislation and regulations.

On a few occasions UK ministers have indicated that they would not accept any reduction in UK food safety standards in exchange for a UK-USA trade deal, but different ministers have been saying different things, and it is far from clear if they can be relied upon fully to deliver on their promises. UK food producers also have a direct interest in these matters.

If US foodstuffs enter the UK's market in the same conditions as they enter the US market, the export of foodstuffs from the UK to the EU will at best become very difficult, or at worst completely impossible. Very few UK producers, and no British retailers whatsoever, are keen on the UK accepting US products that fail to comply with prevailing EU standards, or on the introduction of US practices into the UK. This is therefore a relatively rare topic on which the representatives of UK consumers and of the UK food industry are in full agreement. The position of the British Government on these matters, as so many others, is harder to discern.

The national food strategy

In the wake of Brexit, COVID-19, an increasing use of food banks, farming practices that are damaging the environment, population growth and a growing obesity crisis driven in part by the low-quality food that many families have no choice but to eat, the national food strategy [56] is the plan to transform England's food system. The strategy was published in two parts in July 2020 and July 2021, and it follows a landmark review *from farm to fork*, launched in 2019 and undertaken by a panel chaired by businessman Henry Dimbleby, the founder of the Leon restaurants chain. The strategy's professed objective is to stop families going hungry and align the nation's diet with UK climate goals and was the first independent review of England's food system in 75 years.

Dimbleby and his team hope that the strategy will transform the way England produces, sources and consumes food in a bid to cut down on poverty and improve health across the country, as well as maintaining UK food standards after Brexit. A supporting objective is to make healthy food accessible and affordable for everyone, protect the food system from future shocks — such as pandemics or the climate crisis — and ensure the way we get our food does not damage the environment.

What does the national food strategy have to say and what are its implications?

Dimbleby's review lasted a year, with his team speaking to people involved in all parts of the food system, from farmers through to chefs and food manufacturers. The team consulted academics and other 'experts' and gathered evidence from marginalised groups, including low-paid workers in food production and people with health issues caused by diet.

The first part of the strategy was published in July 2020 and focused on the links between poverty, obesity and the high number of people who died with COVID-19 in the UK. The report referred to the "*slow-motion disaster of the British diet*" and, rather shockingly, found that 36% of England's most disadvantaged people are obese.

"*Clearly, the best way to tackle food poverty is to tackle poverty*" the report commented, and it went on to forcefully state that "*there is no dignity in people having to rely on food banks, food stamps or emergency grants from councils*".

In the second part of the strategy, Dimbleby acknowledges that the consideration of economic measures designed to tackle (food) poverty were beyond the review's remit; however, somewhat surprisingly, he risked the wrath of Government by commenting that "*the true cost of eating healthily should be calculated into benefits payments*".

Building on the campaigning work of Marcus Rashford, the report suggests that money raised by new taxes could fund the extension of free school meals to families earning less than £20,000, bringing an additional 1.1 million children into the free school meals scheme. The report also suggests that Government should also increase its spending on schemes to improve community-based food education and proposes helping low-income families access good quality food, by establishing what it refers to as a 'Community Eatwell' programme enabling GPs to prescribe fruit and vegetables to people on low incomes.

The report also advised Ministers to create a tax for salty and sugary foods and to nudge people towards plant-based foods in order to offer protection to the environment and improve the nation's health after the high COVID-19 death toll proved a "*painful reality check*".

It is probably worth reflecting that whilst the national food strategy is a hard-hitting report that has, to some extent, backfired on Michael Gove who commissioned it, it is, at the moment, unlikely to generate significant change. Nothing will be acted upon until Government has taken its time to review the report, the supporting evidence, and, ultimately, decided what makes sense – both economically and politically, with the latter point being really important.

There is no doubt that the report renders a wake-up call to us all; however, whilst COVID-19 continues to dominate national news and Government business we are unlikely to see much change and there is a real danger that the strategy will simply become lost to COVID! Whilst there is a widespread feeling that the system is broken and needs fixing, the UK's food and drink industry will argue that it is doing just fine on its own and that it cannot be expected to take on board the failings of wider Government policy that have become embedded within the system. The food and drink industry will, no doubt, also continue to argue that it has already taken steps to voluntarily make the products it produces better for consumers and that in the wake of COVID it simply does not have the resources to make the structural and cultural changes necessary to help deliver the strategy's objectives.

Food supply systems

The world today is facing a major food challenge of how to feed a growing population. There will be a need to increase food production by about 70% from its current levels (2021) in order to meet the world population growth of 9.6 billion by 2050 [57]. The challenge is not just about population growth, it's also about how we produce food sustainably, whilst managing our ecosystems and all whilst considering the impact of climate change.

The food supply chain is composed of a wide variety of companies that operate in a range of markets worldwide and produce a diverse range of products. Consequently, there is a world-wide need for intertwining regulatory frameworks that control the whole food supply chain at a multiplicity of levels from the primary producer right down to the retail sector and even embracing the consumer.

When considering the chain from farm to fork, there are four components to the food supply chain that need to be considered:

- Primary producers – regarded as the agricultural and fishery sector;
- The food processing industry;
- The distribution sector;
- The retail sector.

If we consider each sector and think about the activities that they include, then primary producers include those engaged in crop production, the raising of livestock and the operation of fisheries; yet, even within these activity sectors, a diverse range of products are included before we even begin to consider how these products may be enhanced, altered or combined by the food processors. Distribution channels are equally diverse with some primary producers selling and distributing their output to the food processing industry (and sometimes back to itself in relation to animal feed) whilst others sell direct to retailers and even the final customer. Some of the food supply chains are complex and long whilst others are short. For example, consider the fishermen of Cornwall – even some of the larger corporate producers (such as W.H. Stevenson) sell their output directly to retail restaurants and even the final consumer. These very short chains can be compared with the chains involved in the production and shipping of North Atlantic prawns, which can be incredibly long and embracing a multiplicity of different producers, processors and brokers.

If we consider the food processing side of the industry; then we can see that once again the sector is heterogeneous with a number of varied activities that include refining products such as sugar, the cutting and drying of fruits and vegetables, slaughtering and meat processing, through to the production of a diverse range that includes canned food, frozen foods and preserving, utilising other means such as curing or smoking, etc.

The food supply chain isn't a linear process and, when considering it as a whole, we also have to think about the inputs and the outputs at each and every stage, along with the wider impact on the environment, population health and even equality. It's also important to recognise the growing competition for land, energy and water; along with the need for concomitant imports such as energy, pesticides and fertilisers and the implications of their use both on human health and on wider ecosystems.

As we have seen, the food supply chain is highly complex and vulnerable and, along with increasing pressure to increase efficiency and decrease waste, the COVID-19 pandemic has brought about unprecedented stresses on the chain by both a shortage of labour and the need to maintain a level of social distancing that has impacted on all sectors from farms, through distribution and logistics and on to retail. Not only has this impacted to some extent on the availability of food in UK supermarkets, it has also led to food price inflation which will, inevitably, have its greatest impact on the poorest in society. COVID-19 alone has demonstrated the importance of the food supply chain having the capacity and flexibility to respond to sudden and unforeseen challenges.

Provenance within the food supply chain

Food business operators and to a lesser extent environmental health professionals need to be mindful of a couple of things in relation to the food supply chain – the vulnerability of the chain and the provenance of food within the chain. Provenance is about knowing where the foodstuff originates from and being able to trace foods and ingredients one step back and one step forwards within the chain. Traceability is also one of the ways in which food business operators can help to ensure that the food they produce or process is safe. It also helps them to assure that when they, in turn, pass food onto another processor, retailer or the final consumer, it is going to be completely safe.

The legal requirement for traceability within the food chain is, on the face of it, quite minimal – one step back and one step forward; however, consider the production of a pizza! A pizza has a wide variety of ingredients associated with it – the dough for the base is made up from flour, oil, water and possibly some additional ingredients. Then there's the toppings to be considered. There's going to be a tomato sauce – what does that contain? Certainly, it will contain more than just tomato and almost certainly garlic, herbs and possibly a little salt and sugar. Then there's likely to be some cheese included along with a range of other toppings. What if you're using meat as a topping – has that meat in turn been processed? If not, where has it come from and can it be traced back to its source? It is crucial that the pizza producer knows where each and every ingredient involved in the production of the pizza originates. There is then the additional responsibility on others within the chain to demonstrate that they know where each element in the ingredient that they have used or processed has come from and where their product goes to. In this way, it should always be possible to trace products and each individual contributory product back to their originating source.

The reality, however, is very different as demonstrated in recent years by the 'Horsegate' scandal of 2013. 'Horsegate' was a food industry scandal in parts of Europe in which foods advertised as containing beef were found to contain undeclared or improperly declared horse meat – as much as 100% of the meat content in some cases. A smaller number of products also contained other undeclared meats, such as pork. The issue came to light on 15 January 2013, when it was reported that horse DNA had been discovered in frozen beefburgers sold in several Irish and British supermarkets. The analysis stated that 23 out of 27 samples of beef burgers also contained pig DNA; pork is a taboo food in the Muslim and Jewish communities.

While the presence of undeclared meat was not a health issue, the scandal revealed a major breakdown in the traceability of the food supply chain, and the risk that harmful ingredients could have been included as well. Sports horses, for example, could have entered the food supply chain, and with them the veterinary drug phenylbutazone which is banned in food animals. The scandal later spread to 13 other European countries.

'Horsegate' clearly demonstrated the complexity of food supply chains and the vulnerability of those chains to criminality. 'Horsegate' showed that all it needs is for one breakdown in the system – especially where we're dealing with complex food items – to deliver complete obscurity, fraud and failure of food safety within that system. 'Horsegate' was sadly not the only example of system failure within the meat industry. Over the last 30 years there have been a litany of similar failures that result from a high value product and system that is difficult to fully scrutinise.

These types of large-scale food system failures are reflections of a system that squeezes the commodity's unit price to such an extent that it doesn't allow for any margin of profit for individuals further back within the supply chain; consequently, temptation arises for individuals to adjust some of their ingredients or to utilise cheaper items to maximise that margin. The knock-on, however, ultimately falls on the consumer, either in terms of an unsafe product or one that is not of the nature, substance or quality expected.

The food system failures referred to are examples of food crime, and such incidents increase when pressure increases within a food system. Bearing that in mind, and as the UK and the wider world emerges from the COVID-19 pandemic, we can reasonably expect to see system stresses, shortages and tight margins, all of which may once again open up the market for food crime. This is likely to include the sale of food which is unfit and potentially harmful; the recycling of animal by-products back into the food chain; packing and selling meat of an unknown origin; knowingly selling goods which are past their use by date and the deliberate misdescription of food to enable a product to be substituted with a cheaper alternative (e.g. farmed salmon sold as wild, or basmati rice adulterated with cheaper varieties). It may also encompass the use of meat from animals that has been stolen or illegally slaughtered; the poaching or killing out of season of wild game animals such as deer and the deliberate mislabelling of foods to avoid the declaration of allergens within the foodstuff. Furthermore, with the great desire for vegan or vegetarian food, we've recently seen plant-based foodstuffs adulterated with undeclared animal-based products, and examples such as this are becoming increasingly common and are likely to present significant challenges to regulators and to public protection and safety in the post-COVID world where we will need to think in a more holistic manner in order to deliver a more sustainable, equitable, safe and healthy food system for the future.

Food hazards

The term 'food safety hazard' refers to any agent with the potential to cause harm or adverse health consequences for the consumer. An awareness of food hazards and whether they can contaminate or are inherent within food, together with how they might be controlled is an important aspect in relation to the work of an environmental health professional. In certain circumstances, this is a relative rather than an absolute concept and examples of this are the presence of an allergen or the presence of *Listeria monocytogenes* in food. Both the allergen and Listeria may make the food unsafe for some people but not for all; so, it's important to avoid thinking about hazards in isolation; but, instead, to think about the probability of those hazards existing within the food AND of causing illness once the food has been consumed. In addition, some foods may also be or become intrinsically unsafe and this, too, is a factor to be considered. A good example of this is oily fish such as mackerel, where autolysis starts as soon as the fish dies, leading to the naturally occurring histidine in the fish breaking down into histamine which is known to generate an allergic response in some individuals. Furthermore, because there is a diversity of impact on different consumers this may then necessitate taking a different approach towards hazard and risk control.

In general terms, food hazards can be classified as biological (including microbiological), chemical, physical or allergens.

Most biological hazards that are of interest to the environmental health professional can be sub-classified as bacteria, viruses, parasites, algae, fungi, yeasts and moulds; and, whilst EHPs tend to focus on the pathogenic nature of such organisms, the most significant risk that these organisms present is to food producers and retailers in terms of shelf-life, food spoilage and customers being sold products that are not of the quality expected.

Whilst some chemical hazards can be naturally occurring (such as solanine in potatoes) EHPs tend to focus on those chemicals that have been intentionally added such as food colourants and additives. These are chemicals that, in the main, come from natural products and are added to foods to enhance the colour, flavour and/or their taste. We also have to consider those chemicals and chemical products that are added deliberately such as antibiotics or those that are added with the specific intention to defraud. We also have to consider those chemicals that may be added accidentally, such as residues from cleaning materials or from the maintenance of machinery.

Physical hazards include any potentially harmful extraneous matter not normally found in the food and these tend to be much more obvious than say microbiological hazards which we can't see or smell. Appropriate examples would include such things as glass or metal packaging material, perhaps even naturally occurring objects such as bones, grit or associated pests.

Allergens include chemical compounds on or in foods (usually proteins) that can trigger an undesirable and sometimes fatal immunological response in people who are sensitive to them. There are 14 allergens that must be controlled from a regulatory perspective [58] and these include:

- Celery;
- Cereals containing gluten – namely wheat, barley, rye, oats, spelt and sub-species of wheat and their hybrid strains and products thereof;
- Crustaceans including prawns, crabs, crayfish and lobster;
- Eggs;
- Fish;
- Lupin;
- Milk – including lactose;
- Molluscs including mussels, oysters, clams and squid;
- Mustard;
- Nuts including almonds, walnuts, hazelnuts, brazil nuts, cashews, pecans, pistachios, macadamias or peanuts;
- Sesame seeds;
- Soybeans;
- Sulphur dioxide and sulphites that are used as preservatives in some foods and drinks and are at a level above 10 milligrams per kilogram;
- Sulphides.

All sectors of the food industry have the same legal and moral obligation; namely to produce or sell food that is safe for all, including those that have allergies.

A food allergy is an adverse immune reaction that occurs upon exposure to specific food and symptoms range from mild skin reactions to severe and anaphylactic shock. Each year in the UK ten patients die from food-induced anaphylaxis due to undeclared allogenic ingredients [59].

Allergens in food pose an important health challenge, especially in view of the increase in prevalence of food allergies in both developed and developing countries [60]. With the development of worldwide manufacturing and more efficient ways to transport products at low cost around the world, what we now eat is increasingly provided by a food system that is global in scope. This has a significant impact on both food quality and safety as different countries are governed by different manufacturing regulations and guidelines. There is a wide disparity between developed and developing countries with regards to the current role and regulation of food labelling and, even among developed countries, significant differences exist. This is of major importance to those with food allergies who need to know with a degree of certainty whether or not a food they consume contains an allergen, perhaps as a result of cross-contamination during production and is therefore likely to trigger an adverse reaction.

Allergen information on product labels is crucial in food allergy management. The inadequacy in current labelling practices is one of the major causes for accidental reactions upon consuming pre-packed food products.

The legal requirement for allergens in food to be labelled is set out in the EU's Food Information for Consumers Regulation 1169/2011 and is supported in the UK by the Food Information Regulations 2014. All packaged foods for immediate consumption as well as pre-packed foods are to be labelled with allergen information.

Regulations in many countries require the labelling of allergenic food ingredients used in food manufacturing. Despite all management measures taken, food manufacturers sometimes cannot guarantee the absence of unintended allergens in the final product and may apply a precautionary allergen labelling.

There are various types of mandatory food labelling relating to food allergens as previously mentioned. Codex Alimentarius established guidelines for all countries that outlined the main foods that should be considered for allergen labelling [61]. Subsequently food labelling legislation has been introduced in many countries. Changes in legislation to introduce what is known as 'Natasha's Law' comes into effect in the UK from October 2021 requiring food businesses to provide a full ingredients lists and allergen labelling on foods that are pre-packaged for direct sale on the premises. The legislation is produced to protect allergy sufferers and give them confidence in the food they buy.

Whilst measures to prevent the development of a food allergy and the development of therapies to cure food allergies are promising areas of research; ultimately, the best option for allergic consumers is still is to avoid ingestion of a specific food.

EHPs and Trading Standards Officers need to be able to support and advise businesses in relation to their responsibilities when dealing with food that does not comply with the law. It is important that such advice is focused on helping business to embed a comprehensive allergen risk management framework. The following are a range of examples of the sort of challenges faced by business along with possible solutions:

1   Risk – ingredients and allergens are neither included nor identified correctly by a supplier (e.g. a pre-prepared sauce doesn't have a full breakdown of ingredients)

    Solution – advise businesses to work with suppliers to make sure they understand the full ingredient and allergen information they are required to provide under law.

2   Risk – ingredients are not correctly imported into the recipe management system by an employee

    Solution – ensure all staff are fully trained on their responsibilities on why ingredient accuracy is so important. Allow appropriate time for ingredients to be updated on the recipe management system.

3   Risk – labels have the incorrect size and layout to meet the requirements of the legislation

    Solution – ensure that the product contains the right size label and the product name using the correct size font. Ensure that the label lists ingredients in order of weight and that allergens are clearly indicated in bold, underlining or in a different colour.

4   Risk – ingredients or formulations are changed from that which have been used in a standard recipe

    Solutions – every delivery should be checked by an employee to ensure that changes should be noted and, where appropriate, updated in your recipe management system. Errors should be queried with the supplier.

5   Risk – recipe changed to take account of seasonal variations of products

    Solution – where there are recipe changes these should be immediately captured in the recipe management system. Allergens and nutrition information have to be re-calculated and updated on the label.

Food contamination

It is important for any food business operator (FBO), environmental health professional and even society as a whole to understand food contamination and how it can be prevented. Such awareness needs to extend to the various hazards – whether they be microbiological, physical, chemical or allergenic – and the likelihood of their occurrence within the food or in/on food premises and/or food handlers, together with the severity of the potential impact that such hazards may pose to individuals which can vary tremendously from person to person. If Food Business Operators, professionals and the public have a reasonable level of awareness of these factors, then the majority will be capable of coming to a reasonable conclusion about the level of risk that is posed in each case and this in turn enables them to instigate measures to protect their business and the public's health.

There are two types of contamination – direct and indirect cross-contamination. Direct sources include raw meat or soiled vegetables whilst indirect sources include surfaces, hands, equipment that has been used with raw food, including food handlers' clothes.

Decomposition

When certain fish, especially scombroid fish such as tuna and mackerel start to decompose, histamine is formed. Histidine is a naturally occurring amino acid and is converted into histamine by an enzyme produced by certain bacteria during the decomposition process – autolysis. Histamine in small doses is necessary for the proper functioning of the human immune system; however, histamine in higher doses may trigger severe reactions when consumed, similar to those seen in allergic reactions (i.e. rash, nausea, vomiting, diarrhoea, headache, dizziness, burning throat, stomach pain and itchy skin). The presence of high levels of histamine indicates that decomposition has occurred, even if the decomposition is not obvious. Toxic amounts of histamine can form before a fish smells or tastes bad and the only way that it can adequately be controlled is through proper temperature control hence the reason why fish, especially oily fish, needs to be put under some form of temperature control under ice as soon as possible because then this controls the process so any abuse of that cold supply chain can lead to more rapid deterioration in the fish but not obviously seen.

Food-borne illness (food poisoning)

When we look at the type of hazards that can generate food poisoning, over and above a range of pathogenic bacteria, viruses and chemical contaminants, we also have to consider a wide variety of foodstuffs that include naturally poisonous plants, some fish; fungi and mushrooms; rhubarb leaves; plum, peach and apricot kernels; red kidney beans etc.; and even some herbal teas which may have been contaminated or where the leaves have been picked and mistaken for safe species.

It is also important to consider chemically induced food poisoning, and these can include additives and pesticides and even substances like acrylamide (a human carcinogen) that can naturally form on some foods during high temperature cooking such as frying, roasting and baking. However, for most people and for many environmental health professionals the term 'food poisoning' is synonymous with microbiological food poisoning.

Microbiological food poisoning

According to the Food Standards Agency in 2020, approximately 2.4 million Britons get sick and circa 270 die each year from food poisoning.

Bacteria, viruses and parasites are the sources of many food poisoning cases, usually due to improper food handling. Some bacteria, in small amounts, are not harmful to most healthy adults because the human body is equipped to fight them off. Problems, however, occur when certain bacteria and other harmful pathogens multiply and spread, which can happen when food is mishandled. Foods that are contaminated may not look, taste or smell any different from foods that are safe to eat. Symptoms of food poisoning vary and develop as quickly as 30 minutes to as long as several days after eating food that's been infected.

As identified by the FSA, five known pathogens (bacteria, viruses and parasites) account for the majority of food-borne illness, hospitalisation and death in the UK [62].

Salmonella – Salmonella is the name of a group of bacteria that causes the infection salmonellosis. It is one of the most common bacterial causes of diarrhoea and the most common cause of food-borne-related hospitalisations and deaths. Salmonella is more severe in pregnant women, older adults, younger children and those with a weakened immune system. Because Salmonella bacteria can live in the intestinal tract of humans and other animals, it can spread easily unless you use proper hygiene and appropriate cooking methods.

Salmonellosis can be contracted by consuming raw and undercooked eggs, undercooked poultry and meat, contaminated raw fruits and vegetables (such as sprouts and melons), as well as raw milk and other dairy products that are made with unpasteurised milk. It also can be transmitted through contact with infected animals or infected food handlers who have not washed their hands after using the bathroom.

Prevention can be achieved by cooking foods, such as eggs, poultry and minced beef, thoroughly, to their recommended internal temperatures. Wash raw fruit and vegetables before peeling, cutting or eating. Avoid unpasteurised dairy products and raw or undercooked meats, poultry and seafood. Wash hands often, especially after handling raw meat or poultry. Clean kitchen surfaces and avoid cross-contamination.

Clostridium perfringens – also known as *C. perfringens*, is very common in our environment. It can multiply very quickly under ideal conditions. Infants, young children and older adults are most at risk.

Illness usually occurs by eating foods contaminated with large numbers of this bacteria that produce enough toxin to cause sickness in the form of abdominal cramping and diarrhoea. *C. perfringens* is sometimes referred to as the 'buffet germ' because it grows fastest in large portions of food, such as casseroles, stews and gravies that have been sitting at room temperature in the danger zone. If food isn't originally cooked, reheated or kept at the appropriate temperature, live bacteria may be consumed and cause illness.

Prevention can be achieved by cooking food thoroughly and keeping it out of the danger zone, above a temperature of 140°F (60°C) or below 40°F (4.4°C). Practice leftover safety by dividing roasts and stews into smaller quantities for faster cooling and refrigerate right away. Leftovers should be reheated to an internal temperature of 165°F (74°C) or higher before serving. However, any foods left out at room temperature for more than two hours should be thrown out and after only one hour if it's 90°F (32.2°C) or warmer.

Campylobacter – is a common cause of diarrhoea. Most cases of campylobacteriosis, the infection caused by Campylobacter bacteria, are associated with eating raw or undercooked poultry and meat or from cross-contamination of other foods by these items. Freezing reduces the number of Campylobacter bacteria on raw meat but will not kill them completely, so proper heating of foods is important. Campylobacteriosis occurs more frequently in the summer and is most common in infants and young children.

Sources include consuming raw and undercooked poultry and other meats, unpasteurised dairy products and untreated water or contaminated produce.

Prevention can be achieved by cooking all foods thoroughly to their appropriate internal temperatures, preventing cross-contamination by using separate cutting boards when handling raw and cooked foods, not drinking unpasteurised milk or untreated water and washing hands frequently. Wash raw fruits and vegetables before peeling, cutting and eating.

E. coli O157:H7 – *Escherichia coli*, better known as *E. coli*, are a large group of bacteria. Although most strains of *E. coli* are harmless, some can make you very sick. One strain, *E. Coli O157:H7* (STEC) is commonly associated with food poisoning outbreaks because its effects can be extremely severe.

Sources include eating raw or undercooked ground beef or drinking unpasteurised beverages or dairy products.

Prevention is achieved by proper handwashing; cooking meat (especially ground meat) and poultry thoroughly to their appropriate internal temperatures; avoiding unpasteurised dairy products, juices or ciders; keeping cooking surfaces clean and preventing cross-contamination. Also, swimmers should not swallow water when playing or swimming in lakes, ponds, streams or pools.

Listeria monocytogenes – eating food contaminated with *Listeria monocytogenes* bacteria causes listeriosis – a serious infection that primarily affects individuals who are at a high risk for food poisoning: older adults, pregnant women, young children and people with weakened immune systems. Listeria can grow at refrigerator temperatures where most other bacteria cannot grow.

Listeria is found in refrigerated, ready-to-eat foods such as hot dogs, deli meats, unpasteurised milk, raw sprouts, dairy products and raw and undercooked meat, poultry and seafood.

Prevention is achieved by cooking all foods to proper internal temperatures and reheating precooked foods to 165°F (74°C); washing raw fruits and vegetables before peeling, cutting or eating; separating uncooked meats and poultry from foods that are already cooked or ready-to-eat; washing hands thoroughly; storing foods safely by making sure the temperature in the refrigerator is at or below 40°F (4.4°C); maintaining a clean refrigerator and kitchen area; and washing reusable grocery totes/bags regularly.

Norovirus – is one of the leading causes of food poisoning and often results in symptoms similar to stomach flu such as stomach cramping, nausea, vomiting and diarrhoea. Norovirus spreads easily by coming in contact with someone who is infected, especially in crowded areas. Foods, drinks and surfaces also can become contaminated with the norovirus. Anyone can get sick with norovirus, but the illness can be especially serious for young children and older adults. You can contract norovirus many times in your life.

Sources include fresh produce, shellfish, ice, fruit and ready-to-eat foods, especially salads, sandwiches and cookies that have been prepared by someone who is infected.

Prevention is achieved by not cooking, preparing or serving foods or beverages while sick. Frequent handwashing with soap and water for at least 20 seconds is crucial. Foods and utensils should be kept clean by washing all fruits and vegetables, cutting boards, knives, kitchen surface areas, table linens, cloth napkins and reusable grocery bags. Foods that are classified as high risk in terms of their propensity to carry pathogenic bacteria include:

- All cooked meat and poultry;
- Cooked meat products including gravy meat pies, patté, cook-chill meals and stock;
- Milk, cream, artificial cream, some custards and dairy produce, as well especially unpasteurised milk products from unpasteurised milk;
- And cooked eggs and egg products, especially those made from raw eggs shellfish; or
- Shellfish and other seafoods, for example cooked prawns, oysters and mussels;
- Cut fruits such as melon and cut salad vegetables such as tomatoes;
- Cooked rice and pasta.

Therefore, for environmental health professionals it's important to understand the entire route of the food from production to consumption and to be able to identify the points at which that food might become contaminated with pathogenic bacteria. Such knowledge encompasses such things as whether:

- The food is intrinsically contaminated with bacteria;
- The food has not been sufficiently temperature controlled;
- The food has not been adequately heated, thawed or stored;
- The food has become cross-contaminated via the food coming into contact with people, surfaces or equipment that are already contaminated with pathogenic bacteria.

Food preservation techniques

Whilst it is important that the environmental health professional has a good understanding of the points in the food system at which food may become contaminated with pathogenic bacteria; it's also important to have an understanding of the various techniques involved in preserving food; because, only by understanding these techniques will the environmental health professional be able to fully appreciate possible points of failure that might lead to contamination, spoilage and potential harm to persons consuming the food.

Food preservation may be described as that part of food science that deals with the process of prevention of decay or spoilage of food thus allowing it to be stored in a safe condition for possible future use/consumption. Preservation ensures that the quality, edibility, and the nutritive value of the food remains intact. Preservation, therefore, involves preventing the growth of bacteria, fungi and other microorganisms as well as retarding the oxidation of fats in order to reduce rancidity. The process may also ensure that discolouration or aging are reduced. Preservation also involves sealing to prevent re-entry of microbes and ensures that food remains in a state where it is not contaminated by pathogenic organisms or chemicals and does not lose optimum qualities of colour, texture, flavour, and nutritive value (Table 13.1).[2]

Table 13.1 Food preservation techniques

Preservation technique	Description
Drying	This is the oldest method of food preservation. This method reduces water activity which prevents bacterial growth. Drying reduces weight so foods can be carried easily. Sun and wind are both used for drying as well as modern applications like bed dryers, fluidised bed dryers, freeze drying, shelf dryers, spray drying and commercial food dehydrators as well as household ovens. Meat and fruits including apples, apricots, dates, figs and grapes may all be preserved by this method.
Chilling and freezing	Keeping fruit, vegetables, prepared meats, dairy products along with other foodstuffs in chillers or freezers.
Smoking	This process cooks, flavours and preserves food by exposing it to the smoke from burning wood. Smoke is antimicrobial and an antioxidant and most often meats, some cheeses and fish are smoked. Various methods of smoking are used including hot smoking, cold smoking, smoke roasting and smoke baking.
Vacuum packing	Creates a vacuum by making bags and bottles airtight. Since there is no oxygen in the created vacuum, aerobic bacteria die.
Salting and pickling	Salting, also known as 'curing', removes moisture from foods like meat. Pickling means preserving food in brine (salt solution) or marinating in solutions containing vinegar (acetic acid) and in Asia, oil may be used to preserve foods. Salt kills and inhibits growth of microorganisms at 20% of concentration. There are various methods of pickling that include chemical pickling and fermentation pickling. In commercial pickles sodium benzoate or EDTA (Ethylenediaminetetraacetic acid) is added to increase shelf life.
Sugar	Used in syrup form to preserve fruits or in crystallised form if the material to be preserved is cooked in the sugar until crystallisation takes place (e.g. candied peel and ginger). Another use is for glazed fruit that gets superficial coating of sugar syrup. Sugar may also used with alcohol to preserve luxury foods like fruit in brandy.

(Continued)

Table 13.1 (Continued)

Preservation technique	Description
Canning and bottling	Sealing cooked food in sterile bottles and cans. The container is boiled and this kills or weakens bacteria. Foods are cooked for various lengths or time depending upon the process being utilised. Once the can or bottle is opened the food is again at risk of bacterial growth and spoilage.
Jellying/ Jamming	Preserving food by cooking in a material that solidifies to form a gel. Fruits may be preserved as a jelly, jam or marmalade. The jellying agent is pectin that is naturally found in fruit. Sugar is also added.
Potting	A traditional British way of preserving meat or shellfish by placing it in a pot and sealing it with a layer of fat.
Modified atmosphere	This preserves food by operating on the atmosphere around it. Salad crops that are difficult to preserve are packaged in sealed bags with an atmosphere modified to reduce the oxygen concentration and increase the carbon dioxide or nitrogen concentration.
Controlled use of organisms	This may be used on cheese, wine and beer. The method uses benign yeasts or moulds to preserve food by introducing them to food where they form an environment within which it is not suitable for harmful pathogens to grow.
Modified atmosphere packaging	This extends the shelf life of fresh food products. The atmospheric air inside a package is substituted with a protective gas mix (typically carbon dioxide or nitrogen) which ensures that the product will stay fresh for as long as possible.
Pasteurisation	The process of heating food to kill pathogenic bacteria and to stop fermentation. There are two possible pasteurisation methods: 1 Low temperature holding method – the food is kept at a low temperature of 62.8 degrees C for 30 minutes 2 High temperature short time method – the food is held for 15 seconds at a high temperature of 71.7 degrees C
Sous vide	Sous vide is a French term meaning 'under vacuum' – in the USA the word 'cryovac' is often used. The process involves freshly prepared foods being vacuum sealed in individual pouches and then pasteurised at time-temperature combinations sufficient to destroy vegetative pathogens, but mild enough to not lead to a deterioration of the organoleptic characteristics of the product. After cooking, the products are chilled, stored, refrigerated and reheated prior to consumption.

Food governance

Food governance in the UK is not set in isolation; it is very much influenced and informed by a wider framework of organisations both nationally and globally. The framework has evolved over time and has not proved to be responsive to new or specific challenges that present themselves such as environmental sustainability and the many forms of malnutrition. Consequently, what we have are global and national governance structures for food that lag behind what we know about contemporary food-related challenges.

Safe food is a primary determinant of human health, and it is a basic human right to have access to safe and healthy food. In seeking to guarantee this right, Government must ensure that available food meets accepted standards of safety. This task is not easy as the world is now more interconnected and food systems are changing faster than ever. Foods are produced, managed, delivered and even consumed in ways that could not have been anticipated two decades ago and these factors call for a fresh approach to improve food safety via the strengthening of national food safety systems whilst improving international and national collaborations as necessary.

While recognising that food safety is a shared responsibility among multiple stakeholders, food safety policies need to ensure that each stakeholder knows and correctly plays their part. However, access to sufficient, safe and wholesome food for all remains an elusive goal. Economic disparities in the country, including marked differences in the strength of the national food safety system and the complex dynamics that operate within food systems, have significantly slowed progress towards achieving this goal.

Food safety is a public health and social economic priority

Food-borne diseases have a significant impact on public health. Unsafe food, containing harmful levels of bacteria, viruses, parasites and chemical or physical substances can cause acute or chronic illness. Unsafe food disproportionately affects vulnerable groups in society [63], particularly infants, young children, the elderly and the immunocompromised. As well as the impact on the most vulnerable, society also needs to give consideration to the disproportionate impact of unsafe food on the economically deprived.

Nutrition, food safety and food security are closely interlinked in delivering health-related outcomes from food systems. Simply put, there is no food security and adequate nutrition without food safety and unsafe food can lead to a vicious cycle of disease and malnutrition that particularly impacts infants, young children, the elderly.

Unsafe food not only negatively impacts health, but it also negatively influences economic growth via agribusiness trade and even tourism. Consider global agricultural value chains – they've become extended and complex with food products often being grown, processed and consumed in a range

of different countries. Whilst such trends have contributed to increasing the quantity and diversity of foods available to consumers throughout the world, the increased volumes of traded foods increase the safety risks to consumers who have the right to expect that both domestically produced and imported foods are safe.

The introduction of international food safety standards for application at domestic levels and in international trade can be demonstrated through the Codex Alimentarius. Failure to ensure compliance with internationally recognised regulations and standards not only leads to economic losses but also to loss of confidence in business. Today, single governments cannot act to significantly reduce standards applicable at the domestic level without the whole sector's access to high-value international markets being put at risk, resulting in expensive export rejections and damage to markets and to national and brand reputation, with a concomitant knock-on to employment and the health of the wider economy.

Global governance of food safety

Global governance of food safety is delivered by a wide variety of partners and interrelated global organisations. The system of global institutions and agreements that touch and concern food is dominated by the United Nations including the United Nations Food and Agricultural Organisation (FAO); the World Food Programme (WFP) and the World Health Organization (WHO). These organisations, along with the domestic governments and agencies that are influenced by their activities, drive and regulate the production, distribution, retail and to some extent the consumption of food – 'from farm to fork'.

At the time of writing this report, both the World Health Organization and the Food and Agricultural Organisation are currently reviewing their food safety strategies [64].

The rationale for the World Health Organization's new strategy is that food systems demonstrate a need to rethink the place of food safety within sustainable development. They also recognise that there needs to be greater relevance of food safety to societal economic development and that safe, sustainable food systems need to be better understood. The emergence of microbial resistance, emerging zoonotic diseases, climate change, agricultural intensification, new technologies, innovation, food fraud, digitalisation of food systems, food waste and circular economies also need to be considered and addressed within the new strategy. The COVID-19 pandemic also demonstrated the increased relevance of food safety within the need to secure emergency food assistance and humane humanitarian food aid. This too deserves consideration.

Codex Alimentarius

The WHO and FAO have jointly produced a food standards programme known as the Codex Alimentarius or food code. It is a collection of standards, guidelines and codes of practice adopted by the Codex Alimentarius Commission. The Commission is a joint intergovernmental body of FAO and WHO with 188 member countries and one member organisation. Codex has worked since 1963 to create harmonised international Food Standards to protect the health of consumers and ensure fair food trade practices.

The international trade in food has existed for thousands of years; however, until comparatively recently, most of the food was mainly produced, sold and consumed locally. However, over the last century, the amount of food traded internationally has grown exponentially and a massive quantity and variety of food, never before thought possible, travels the globe on a daily basis. For the UK, this means that its food choices have never been so extensive and affordable for the public.

The Codex standards contribute to the safety, quality and fairness of this international food trade and, in so doing, build the trust of the consumer. It is important to Codex that this trust is maintained and enhanced, and this has led to the organisation promoting scientific knowledge and prompting global debates on veterinary drug residues in food, pesticides, food additives and contaminants, etc. Although the adoption of Codex standards is voluntary, the application by Member States serves, in many cases, as a basis for national legislation. This is the situation in the UK. It was the situation prior to the UK joining the EU and has been a huge influence on EU regulations that have now been transposed into UK legislation. So, it is reasonable to conclude that UK legislation on food safety has been informed by Codex.

The benefits of Codex informing the regulatory processes in the UK and around the world means that Codex can be extremely powerful in resolving trade disputes. World Trade Organisation members that wish to apply stricter food safety measures than those set by Codex may be required to justify these measures scientifically. So as Britain moves away from being part of the European Union and enters into trade agreements within their own right (as they have done recently with Australia) they can do so knowing in advance that potential partners broadly share their standards.

UK food safety governance

As we have seen, the UK's food governance processes are very much influenced by external global organisations of which we are part – namely, WHO and FAO, Codex as well as the World Trade Organisation (WTO). Furthermore, despite our exit from the European Union we remain strongly influenced by the European Food Standards Authority (EFSA). Over the years those external influences have led to the development of a domestic system for food governance that involves a large number of Government departments and agencies in addition to a number of other food organisations spanning the private, third, financial and philanthropic sectors that work closely and sometimes in formal partnership with others to ensure food safety. A good example of this is the British Retail Consortium and UK Hospitality.

In more recent times, the UK Government has become increasingly reliant on such partnerships to provide additional resources and knowledge, effectively blurring hitherto clear system demarcations. This, in turn, can make it challenging to understand where public sector decision-making stops and where the research and policy development of others begins. A recent good example of this is the Eat Lancet Commission on Food, Planet and Health.

In the UK there are 16 national Government departments, agencies and related bodies that are responsible for food policy making. Together, these cover diverse issues from transportation and the transport infrastructure through to Social Security payments that help tackle food insecurity and diet-related illness. They embrace the Department of Health and Social Care that looks at aspects of diet-related public health including obesity, along with some aspects of food labelling; though the Department of Education takes responsibility for standards for school food and school meals, the provision of school milk and fruit and veg schemes, etc. Two separate Government departments (Department of Health and Social Care and the Department of the Environment, Food and Rural Affairs) provide the oversight of food security, food access, food safety, food opportunities and nutritional opportunities. The same departments develop both agricultural, fisheries and environment policies, climate change policy, the policy relating to food packaging and labelling, and they also provide the UK input to Codex.

The Food Standards Agency (FSA)

The Food Standards Agency is an independent food safety watchdog set up in 2000 to protect the public's health and consumer interests

in relation to food [65]. It was originally proposed in 1997, amid concerns about food poisoning, intensive farming methods and Bovine Spongiform Encephalopathy (BSE).

Food policy experts wanted a new food agency responsible to the Secretary of State for Health, independent of the food industry, but at 'arms' length' from ministers.

Many parts of the UK food industry are, however, highly critical of the FSA in the way that it has been perceived to develop an unduly cosy relationship with the food companies, bio-tech companies and large retailers [66].

A shake-up in 2013 led to the FSA handing over some of its responsibilities to other Government departments so it could focus solely on food safety policy and enforcement. Whilst the Department of Health and the Welsh Government took over nutrition policy in England and Wales, in Northern Ireland the FSA remained in control. The Department of Health and Department for Environment, Food and Rural Affairs (Defra) also took over food labelling in England but the FSA remains in charge of labelling in Wales, Scotland and Northern Ireland.

The FSA is accountable to Parliament through health ministers, and to the devolved administrations in Scotland, Wales and Northern Ireland for its activities within their areas.

The FSA's remit includes:

- Conducting negotiations on behalf of England, Wales and Northern Ireland (by legal experts and policy officials) to ensure that international food law reflects the interests of UK consumers;
- Removing unsafe food from sale, in conjunction with other enforcement organisations;
- Ensuring good food hygiene, through various initiatives delivered in partnership with local authorities (LAs);
- Ensuring meat hygiene in approved slaughterhouses and meat establishments throughout Great Britain; in Northern Ireland, this role is carried out by the Department of Agriculture Environment and Rural Affairs (DAERA).

The FSA has been credited with raising awareness of good eating habits and persuading food manufacturers to reduce their use of salt and sugar and FSA campaigns have included pushing for stricter rules on TV advertising to children of junk foods and the Agency was at the heart of a mass recall of foods containing the dye Sudan 1 in 2005 [67]. The Agency was, however, sorely tested by the horsemeat scandal that broke in 2013 and the report that was subsequently commissioned from Queen's University Belfast Professor and produced by food security expert Chris Elliot. As a result, the FSA created a new National Food Crime Unit to strengthen consumer confidence [68].

Many commentators believe that that culture of 'backing up' the food industry continues and that the FSA sees its role as reassuring consumers that food is safe and downplaying the risks. The Food Standards Agency sees itself as being a science- and evidence-led organisation and that it seeks to gather and utilise good evidence to support the actions that it takes.

Some of the Agency's strategic goals include ensuring "*foods produced or sold in the UK are safe to eat*" and "*that consumers have the information and understanding they need to make informed choices about where and what they eat*".

The FSA's relationship with local authorities in England, Wales and Northern Ireland is one of goal setting and oversight through a framework agreement which provides for local authorities making publicly available local service delivery plans that indicate how a local authority intends each year to deliver its services regarding food safety and food standards enforcement.

Local authorities

Much of the FSA's inspection and enforcement regime is dependent upon delivery partners, principally local authorities. FSA also audits enforcement activity with respect to local authority food and safety standard controls. Local functions are divided between EHPs and Trading Standards Officers (TSOs). Food EHPs oversee food safety and food hygiene, enforcing law across all forms of retail food business organisations (restaurants, takeaways, shops) as well as food processing and food manufacturing outlets. TSOs have responsibilities relating to food labelling and trading standards.

The main duties of the EHPs in the enforcement of food safety law within local authorities include:

- Implementing and maintaining a documented programme of food hygiene and food standards inspections. Premises posing a 'high risk' to the consumer are inspected more frequently than those posing a 'low risk'.
- Implementing a microbiological and chemical food sampling programme. Priority is given to sampling food produced locally.
- Investigating complaints about food including complaints about the hygiene of premises.
- Investigating cases of food poisoning.

EHPs use a range of tools and interventions in order to secure improvements and gain compliance. They make judgements about the most appropriate interventions from education and encouragement through to warnings and the service of enforcement notices. These compel improvements within a certain period and/or prohibit all or some aspects of the business. In exceptional cases legal proceedings can be instituted and fines or imprisonment imposed. When inspecting food premises, EHPs will normally consider: the food safety management system; food safety training; cleaning standards in the premises; the condition of the structure and equipment; source of food and ingredients; storage conditions and temperature control; and the personal hygiene of staff. EHPs have been the driving force in developing hygiene ratings and encouraging food businesses to openly display them. In 2013, Wales became the first country in the UK to introduce a mandatory scheme requiring food businesses to openly display their hygiene rating. This resulted in consumers being provided with more information about where they eat or buy food, helping to drive up businesses' food hygiene standards. A national discretionary scheme operates in the rest of the UK.

COVID-19

In COVID-19, both the world and the UK's food system has just endured its biggest stress test since the Second World War. COVID-19 swept through the world at the end of 2019 and the start of 2020 leading to the rapid recalibration of the entire machinery of food supply and distribution. In general, and with a few exceptions, the recalibration process has been managed reasonably well by the global food industry where a reasonable level of resilience has been demonstrated. In the UK, however, workers in the food production, retail and hospitality sectors have suffered some of the highest death rates from COVID-19 together with the highest numbers of furloughed staff and redundancies.[3] For the UK, the impact of waves of COVID cases have been exacerbated by Brexit that has seen a large number of its non-UK origin workers in its food system return home, generating significant food shortages on the supermarket shelves at various points during 2021.

EHPs and COVID-19[4]

With the impact of COVID and the first lockdown of the country on March 23rd 2020 EHPs had a vital role in protecting the public's health and in enforcing the business closure and social distancing measures announced at the time by the UK Government. On 24th March 2020, the Government required all retailers selling non-essential goods to close with immediate effect along with all other non-essential businesses.

Despite the national lockdown, it was nationally important for the food production,

distribution and retail industry to remain operational; food still need to be provided to people. People continued to need to purchase food and consume food. Many food businesses adapted to the situation and what quickly became evident was the innovative nature of both the food industry and its principal regulators – EHPs. Food businesses moved to produce and deliver takeaway meals and some unlikely players, including community and voluntary groups, became involved at the local level with food distribution. Interestingly, EHPs were still faced with ensuring that key elements of the food system functioned and did so safely – the provision of advice and support became a priority.

Along with the changes that COVID brought to environmental health practice and to everyday activities designed to protect the public's health, local authorities had to find alternative ways to engage with businesses. One of the things that became normal in the pandemic's abnormal times was the introduction and use by local authorities of remote assessment. Hitherto, this approach to delivering interventions was something that had been resisted by the FSA but which has become increasingly used during the pandemic in an attempt to ensure the ongoing protection of food safety and the consumer at a time when it was considered unsafe for EHPs to visit and inspect food businesses.

Role of the EHP in Food

Environmental health practitioners work hard to ensure that the food we eat is safe and is what it says it is. EHPs can be employed in a variety of environments including the public, private and third sectors. They can be involved in developing food policy, providing advice to food businesses and consumers and in enforcing the law. They engage right across the food supply chain from primary producers through to processors, distributors and to the final consumer. Ultimately, they are there to ensure that the consumer's confidence in the food supply system is protected and enhanced. EHPs are also perfectly positioned to spot, investigate and address food crime which is extremely harmful to public safety, economic development and to ensuring fair business competition. Another area in which EHPs are integral, involves working in partnership with Government agencies such that when food safety controls fail and food-borne infections arise, EHPs are there to deliver on-site investigations and the containment of any food-related infection.

The main function of many EHPs engaged with food safety and food integrity lies in relation to inspecting and/or auditing food businesses across the food supply chain. Most of this work is proactive and is based on a UN-recognised system of risk rating. The risk rating of food businesses falls under the governance process that is applied by the FSA and is set out in the latest version of the Food Law Code of Practice (currently March 2021).[5] The intervention risk rating process allows local authorities to prioritise their intervention-based activities based on those businesses that present the greatest risk in terms of food safety.

Local authorities provide information annually (Local Authority Enforcement Monitoring System (LAEMS) data) to the FSA on the activities that each has undertaken to meet their obligations under the law and to deliver food safety within their community. The annual return to the FSA also provides an evaluation of what has been delivered during the year, how it's been delivered, and a summary of the issues may be impacting on the local authority as well as wider issues that may be emerging from a food risk perspective.

For 2021, having been severely impacted on by COVID-19, FSA recognised that the standard annual return was not going to be possible and instead they have utilised the annual LAEMS data to assess the current state of the food system and to actually understand the backlog of interventions that had built up. In turn, this helps FSA inform policy and issue on guidance and advice to local authorities and to focus on recovery planning in the medium term.

EHPs carry out much of their work in food through what is termed 'official controls'. This term has its origin in EU legislation and, despite our exit from the EU, official control-based legislation is still in place and is operational.

The current competency requirements in respect of individuals undertaking this work is set out in the March 2021 Food Law Code of Practice [69]. The majority of EHPs undertake inspections or audits of food businesses across the supply chain and focus strongly on the risks that are not only generated by hygiene practices within the business, but also on the management controls that are in place (or not) within that business. This results in EHPs delivering interventions that are specifically designed to address food safety-related risks within business.

Just as the frequency of inspection and audit of food businesses is based on the level of risk that the business poses, so the interventions undertaken to address breaches of legislative requirements are also undertaken based on the level of risk posed by the food business to the health of the public. This means that interventions range from the provision of advice and guidance, through to informal action such as sending a strongly worded letter or email, through to more formal action such as the service of hygiene improvement notices or even prosecution and possible business closure.

The continued delivery of the official controls system of food control does, however, present some challenges for local authorities in England, Wales and Northern Ireland. These challenges arise from the paucity of financial resources and, more importantly, appropriately competent EHPs who are able to undertake this work in an ever-increasing supply of operational food businesses that require regulating. This situation has been exacerbated by COVID-19 and the enforced hiatus on proactive work, due initially to lockdown and, more latterly, to the need to maintain social distancing.

At the time of writing, it remains unclear whether, beyond the short term, the official control system of regulating food business will continue now that the UK has left the EU. Whilst there is a need to ensure that a measure of alignment exists between the EU and the UK from the point of view of easing the path of food exports from the UK to the EU, in July 2021 the UK Government quietly announced a review of regulation and perceived red tape [70] in the wake of Brexit and such a review could see a very different system of UK regulation emerge into the future.

Food inspection practicalities

Irrespective of the system of official controls that will operate post Brexit and COVID-19, one thing is certain: the physical inspection of food premises will remain a key component to ensure that an appropriate food safety management system is in place and operational and delivering a verifiable Hazard Analysis Critical Control Point (HACCP) system. It, therefore, follows that the primary purpose of the food control inspection undertaken by environmental health professionals is to review food safety systems and control measures and contribute to their improvement. That said, it must be remembered that food safety systems and control are primarily, though not solely, the responsibility of FBOs who profit from their products. However, because the public expects food to be safe, food control legislation demands that the FBO is made accountable for any system failures and inspection remains the primary method of ensuring that happens.

It must, however, be also recognised that the confirmation that the FBO is observing appropriate hygiene and safety practices is just the start in ensuring food safety.

Historically, the objectives of inspection were mainly to determine whether the FBO was complying with legal requirements and to obtain correction of food safety breaches that already exist at the premises. Such an

approach focused on 'walls, floors and ceiling' within the business and was often limited to verifying that regulations were being complied with. There was little focus on system improvement designed to ensure that future contraventions were prevented, and overall levels of food safety risk were reduced. This contrasts sharply with the current method of prioritising inspections using a risk-based approach that focuses on risk factors that are likely to lead to food-borne disease.

Historically, non-compliance and contraventions have usually been dealt with via the service of notices, the imposition of fines and the requirement for corrective action. However, such a regime did not necessarily deliver either the corrective action required or any sense of improvement due to the regulatory structures and enforcement policies (or lack of!) being pursued by the regulator. Furthermore, there was no assurance after this type of intervention that there would not be future recurrences of the contravention addressed. Lasting improvement was not, therefore, an outcome of the system.

It, therefore, follows that by focusing inspections on hazards and risk associated with food-borne disease, environmental health professionals are not only more efficient, but they are also more effective in delivering real improvements over time. By focusing on key hazards and food safety risks the inspection undertaken by the environmental health professional serves to verify that a good hygiene and safety plan is being systematically and effectively implemented. This effectively means that the focus of the inspection has now shifted away from premises (walls, floors and ceilings) and products to overarching hazards, risks and safety.

Risk-based inspection starts with the consideration of hazards associated with the food and a review of the control measures in place to determine if they are adequate. Some of the food-borne disease risk factors associated with the food operations of a typical restaurant are shown in the example.

Practically speaking, the objectives of assessing the hazards, risks and controls during an inspection are to:

- Determine that the FBOs safety management system adequately addresses all identified hazards and risks associated with the operation of the business;
- Identify options for improving the (HACCP) systems in place;
- Deliver ongoing improvements to the food safety management system.

How is an inspection actually undertaken?

Where the premises being inspected is small, and the FBO has neither a formal nor informal food safety management system in place, then the focus of the inspection is to address immediate the immediate food safety risk AND to take the time to assist and encourage the FBO to invest time and effort into the development and implementation of a food safety management system. If, on the other hand, if a food safety management system (HACCP) is in place, then an audit of that system should be undertaken by the inspector.

Such an audit begins prior to any visit to the premises and by consulting premises records to determine the type of premises, the food being produced, the history of compliance. It also goes without saying that in advance of any site visit the environmental health professional should ensure that he/she has access to and adequate supplies of appropriate inspection wear, tools and equipment. It is also important that the inspector's authority extends to cover both the premises to be inspected and the risk rating of that business. The environmental health professional will also need to consider whether there is a need to pre-notify the FBO of the intended visit, recognising that the current iteration of the Food Law Code of Practice recommends unannounced inspections as the norm. However, no pre-notification should take place in response to a complaint from

the public or where urgent matters have to be addressed.

Prior to undertaking the inspection, the environmental health professional should plan for an opening discussion to take place with the FBO, the purpose of which is to:

- Meet the FBO;
- Discuss the objective of the inspection;
- Explain the procedure to be followed during the inspection;
- Review relevant business records including the food safety management system;
- Ask questions.

At this opening meeting, environmental health professionals should answer any questions from the FBO about any aspect of the inspection including the laws and regulations they are authorised to enforce. This should be followed by an observation of the premises and of the food production/retail procedure in order to assess unmanaged hazards. It would be normal for the inspector to ask the premises manager to accompany him/her at this point, so that questions arising throughout the period of the observation can be quickly addressed. It is also perfectly normal for the visiting inspector to talk to the staff and to ask questions about what they have observed and to obtain clarifications or corrections from those who are present and fully engaged in the food process. However, as is the norm with other types of environmental health investigations, the visiting inspector should seek to 'triangulate' the responses to their questions with both management, supervisors and other frontline workers. Such a process ensures that answers are consistent and can be checked for their veracity.

If the facility utilises a HACCP-based food safety management system, it should be referred to by the inspector and its contents should drive the questions posed to both management and employees.

Throughout the inspection, it's important that the inspector records his/her key observations and documents both non-compliances and contraventions. The use of digital photographs, audio and/or video recordings is particularly useful in this respect and not only provides records that are likely to be admissible in court, but can also be quickly forwarded to colleagues or other 'experts' if other opinion needs to be sought on a particular matter.

Upon completion of the inspection, a closing meeting should be held with the FBO to discuss the findings of the inspection. The tone of this meeting is largely determined by the findings of the inspection. Unless the inspection identified serious areas of non-compliance, this meeting is not only an opportunity to gather final information or seek final clarification, but it is also an opportunity to discuss contraventions, areas of concern and areas and means of possible improvement. However, where the inspection discovers major breaches and areas of food safety risk then the meeting's tone will tend to be more formal. Either way, the inspector's professionalism is extremely important.

Where areas of non-compliance and/or contraventions are observed, then a corrective action plan should be discussed and agreed, detailing both the actions necessary and an appropriate timeline for their completion. An appropriate and written inspection report should then be drafted and passed to or sent to the FBO. Such a report must include the corrective action plan where one is necessary. The environmental health professional should then schedule all follow-up visits and other interventions.

Hazard Analysis Critical Control Point (HACCP) essentials

Article 5 of Regulation (EC) No 852/2004 of the European Parliament on the hygiene of foodstuffs requires food business operators to put in place, implement and maintain a permanent procedure based on Hazard Analysis and Critical Control Point (HACCP)

principles. The Regulation allows for flexibility regarding the application of HACCP principles, taking into account:

- The nature of the operations; and
- The size of the business.

HACCP is a management system that is used worldwide that helps to prevent and minimise food safety risk and provide assurance that the food business operator (FBO) has an active food safety programme in operation. It was first developed in the 1950s by the Pillsbury Company, the Natick Research Laboratories and NASA to ensure food safety for the space programme. The system gives a structure which allows the FBO to thoroughly analyse and control any hazards which could take place in the production of food products.

In essence, the system categorises three types of hazards:

1 Biological
2 Chemical
3 Physical hazards

Once identified, the system allows the FBO to develop ways to monitor and control these hazards during the entire manufacturing process from the initial raw material stages of production, procurement and handling of food through to distribution and consumption of the finished product.

Many of the world's largest manufactures use HACCP as a base for their food management safety system and, in the UK, food production businesses are required to have a HACCP-based food safety system by the Food Safety and Hygiene (England) Regulations 2013 and its associated legislation in Scotland, Wales and Northern Ireland. However, a food safety system does not stop with HACCP; to be effective and to have a good overall food safety system, the FBO needs to design, implement and monitor the following additional systems as well:

1 Pest control
2 Traceability
3 Hygiene

Other food safety practices should be considered and merged into the systems, such as the Good Manufacturing Practices (GMP) "Five P's" and the "Four C's" approach developed by the Food Standards Agency (FSA).

The Five P's

People – All employees are expected to strictly adhere to manufacturing processes and regulations. This means employers must provide up-to-date training for all employees to understand their roles and responsibilities.

Products – All products must undergo constant testing, comparison and quality assurance before being distributed to customers.

Procedures – The creation of guidelines for undertaking a critical process is required to achieve a consistent result, which must be followed by all employees.

Premises – Any premises, and the contents within them including equipment, should champion cleanliness at all times to avoid cross-contamination, accidents or even fatalities.

Processes – The processes and controls should ensure the hygiene of the facility and the safety of the food produced and this includes identification and management of critical control points such as time, temperature, humidity, pH, flow rate.

The Four C's

This approach covers key topics such as:

- Cleaning;
- Cross-contamination;
- Cooking;
- Chilling.

HACCP is based on seven key terms, known as 'The Seven Principles of HACCP' which are set out in Regulation (EC) No 852/2004 and in Table 13.2.

Table 13.2 The Principles of HACCP

HACCP principle	Description
Conduct a hazard analysis	The first stage of the hazard analysis process is identifying where risks can be introduced into the food manufacturing process. The risks could be physical (i.e. a screw from a machine or hair from a worker's head), chemical (i.e. a cleaning product contaminating the product) or biological (a virus from someone working as part of food production). *Flexibility*: hazards – generic descriptions of hazards may be sufficient
Identify the critical control points	The second step is to look at which stages in the process controls need to be put into place to prevent or even eliminated the potential hazard that have been identified. These are the critical control points. For each control point the most appropriate preventive measures to reduce or eliminate the risks needs to be identified. As an example, controlling the temperature of the food throughout the food chain can significantly reduce the risks, either by keeping high risk ingredients cold to inhibit bacteria growth or by heating to a level that kills bacteria. *Flexibility*: CCPs – generic guidance may include pre-determined CCPs in the preparation, manufacturing and processing of food

Table 13.2 (Continued)

HACCP principle	Description
Establish critical limits	The next step is to create a critical limit, which is the point where a risk has become present, requiring the establishment of specific criteria for each critical control point that separates acceptability from unacceptability for the prevention, elimination or reduction of identified hazards. This could include things like setting minimum dilution levels on cleaning or setting specific maximum acceptable temperatures for chilled deliveries and minimum holding temperatures for hot food. *Flexibility*: critical limits – it is not always necessary to fix a numerical value, especially where monitoring procedures are based on visual observation
Establish monitoring procedures	The establishment of monitoring procedures to check if a critical control limit has been reached and ensuring that it is measured in a consistent way. Preferentially, a continuous monitoring method should be utilised for the control point, such as a temperature alarm on a freezer or using a post-mix cleaning product dispenser to ensure correct dilution. If this is not possible, as is often the case, a decision needs to be taken on how often the monitoring needs to be performed to show that the risks are under control. *Flexibility*: monitoring – may be a simple procedure, for example, a visual observation to monitor whether the correct procedure is being applied during the food production process
Establish corrective actions	The establishment of actions that are appropriate if the critical limit is exceeded. This includes detailing who needs to be informed; for example if a screw is found in a bag of flour the FBO would need to inform the supplier; if the FBO is the producer of that flour, then the FSA may need to be informed of the chance of contamination on a wide scale and issue a product recall.
Establish record-keeping procedures	Records need to show that the critical limits have been met. Records are the only real proof that the required checks have been completed, and this will have an impact on the score given under both the Food Hygiene Rating Scheme and the food hygiene risk rating scheme. Records need to exist covering every stage of this process and will include written details of each hazard and a breakdown of identified control points, critical limits for that hazard along with details of the monitoring and the established corrective actions. Additionally, staff need to be trained to complete the monitoring and be aware of where the records are held. *Flexibility*: recording – in the case of visual monitoring procedures it can be acceptable to record results only when there is a problem and the corrective action that has been taken; that is, 'exception reporting'; a diary can be a suitable method of record-keeping
Establish verification procedures	The HACCP plan must be verified. This involves the FBO putting the HACCP process in place and then making sure it is effective at finding and preventing hazards. This means constantly reviewing the plan, ensuring any unaccounted issues that emerge are added and the whole system is constantly reviewed and improved whenever possible. *Flexibility*: verification – checking all aspects of the HACCP plan can be spread throughout the year so that all aspects are verified at least once a year to meet the requirement for 'regular' verification

The future?

What does the future hold for the UK food system and the EHPs that regulate it? Because of the rate of change and of the unexpected emergence of disruptors such as COVID-19, it is impossible to look into a crystal ball and envisage a future with any clarity. It is, however, clear that, over and above Brexit, there are a range of additional challenges that will need to be faced by those considering the shape of the UK's future food regulatory system and by those professionals such as EHPs that make the system work, and these include:

- *Unregulated online sales* – in January 2021, the FSA, recognising an issue that had been growing in scale for many years, finally issued a warning about potentially unsafe meat sold via Facebook by an unregistered and unapproved Wiltshire-based vendor [71]. In 2021, such use of social media to sell food and food products has become extensive and this is leading to the sale and consumption of foods that are not compliant with food safety, labelling and traceability requirements and which could be unsafe with the consumer being unaware that such products lie outside of the food regulatory system.
- *Frequency of food inspections in relation to risk* – the food hygiene intervention frequencies for food business that are rated as being Category A in terms of risk is usually every six months compared to 12 months for those in the second highest bracket and 18 months to two years for the lower-risk businesses that are broadly compliant. 'A-rated' premises include businesses with a history of hygiene or compliance problems and those that are supplying vulnerable groups such as care homes or children, large food manufacturers and those handling raw meat or involved in a process with a high danger of contamination. Whilst the frequency of inspection is nominally based on risk, the scheme itself tends to be a rather blunt

instrument and does not currently allow any autonomy for those businesses that can demonstrate that they are capable of handling their risks well. In the same vein, some businesses that are nominally rated as being C or above, because of failures in management and control could find themselves presenting unacceptable levels of risk. Brexit presents an opportunity for such systems to be reappraised.

- *The use of technology and remote inspection* – in the wake of COVID-19 it is recognised that there is a need to review and reset the food regulatory system to allow local authorities to try and catch up with their backlog of inspections/audits and emerging new businesses. Some local authorities during the pandemic also took on the responsibility of introducing arms'-length inspection utilising technological solutions to screen key aspects of businesses' compliance. We have learned a lot during such experiments and remote inspections have demonstrated their ability to play a part in maintaining the food system. The question going forward is whether such innovations could be more widely used to alleviate the pressure on the system brought about by a shortage of EHPs, a rising number of food businesses and a significant post-COVID backlog of inspections.
- Although it is currently recognised that the use of remote assessments cannot and should not fully replace physicals inspections, the FSA has come to the point where it is prepared to at least consider utilising technological solutions to help deliver the food safety system. It is recognised that there are limitations with virtual assessments. For instance, the use of such systems may prove challenging in terms of spotting pest infestations, contamination or temperature control issues but, as a triage system or as a course screening tool, such innovations have a place in the future.
- Furthermore, on average, each inspection undertaken by a person costs the taxpayer

about £150 [72] and there are half a million food businesses across England, Wales and Northern Ireland. The costs of delivering a face-to-face system are, therefore, high and given the growth in the number of food businesses, this is likely to prove unsustainable and hence act as another driver in terms of resetting this aspect of the current food regulatory system.

- *The continued use of sampling as a screening tool to warn of problems* – even prior to COVID, the amount of food sampling undertaken by local authorities as a means of determining compliance or signalling future system problems has decreased significantly because of budget and resource pressures [73]. Paradoxically, there is a rising need for sampling given the growth in novel and imported foods, the increased focused on allergens and potential changes to the standard of imported food now that the UK has left the EU. Resetting the food system in the wake of both Brexit and COVID presents an opportunity to address this disparity.

- According to a National Audit Office (NAO) report, the food regulation system is complex, has come under increasing financial pressure and has elements that are outdated [74]. The report goes on express concerns about the ability of the current regulatory system to even deliver food safety, food-borne disease control (approximately 1,000,000 people in the UK suffer a food-borne illness each year with the total cost being estimated at roughly £1 billion annually (in 2015)) and achieve value for money following the UK's exit from the EU and the emergence of both COVID-19 and a range of additional risks. Emerging risks include climate change, population growth, crop disease, food fraud and potentially importing more food from non-EU countries.

- The NAO went on to make a series of recommendations to the FSA for change in respect of its sampling strategy; the introduction of mandatory hygiene rating display; verifying the impact that the National Food Crime Unit is having on food fraud; the performance of local authorities and the identification of key resource gaps.

- Somewhat inevitably, the old chestnut of whether the UK's food system post-Brexit and COVID should be centralised continues to raise its head and, at the mid-point in 2021, the FSA has once again fuelled this debate by the introduction of a centralised and online food business registration system. This is a considerable departure from the old system where food businesses registered with their local authority and, going forward, with registrations being undertaken centrally, the FSA will inevitably know more about the food system at the local level than the local authorities themselves.

- Interestingly, the FSA has said that they are pressing ahead with developing indicators to assess local authority performance and to ensure that the NFCU is effective. The imposition of further indicators, on which local authorities are obliged to report, is likely to prove inflammatory to local authorities already reeling under the pressure generated by a lack of resources and in the wake of Brexit and COVID.

Regulating online sales of food

Over and above the challenges posed by Brexit and COVID to the ongoing regulation of the UK food system and the work of the EHP, the regulation of online sales of food is probably the most significant. This is, however, not a new problem. As far back as 2017 the European Commission mounted a fact-finding mission to Denmark, France, Germany, Ireland, Portugal, Sweden and the UK by the Directorate-General for Health and Food Safety of the European Commission (DG Health and Food Safety) to investigate online food sales and the regulation of then [75]. Most of these countries did not have specific national legislation for this type of sale and the Commission found that official controls covering food hygiene (temperature requirements, transport and traceability), labelling, health and nutritional claims made in respect of food sold online remain 'limited'.

In addition to the fact-finding mission, in 2017, the Commission organised the first EU coordinated check on food offered via the Internet. Authorities from 25 Member States, Switzerland and Norway checked nearly 1,100 websites and found around 740 non-compliant offers (e.g. 425 offers of unauthorised novel foods and 315 of food supplements with medicinal claims).

The Commission's report found that existing controls are limited and are primarily directed at existing registered Food Business Operators (FBOs). It was also found that non-compliances were mostly related to labelling and health-claim requirements and that the online marketing of dangerous substances as food supplements was occurring in a few cases.

With Internet sales of food increasing at an exponential rate both within the UK and across the EU, there is a growing necessity for controls to be urgently reviewed and enhanced. Evidence gathered by the Commission found that strong enforcement delivered together with cooperation with non-EU countries is the main constraint to effectively control food sold via the Internet.

The Commission found that identifying non-registered food business operators with an online presence has proved challenging due to a number of factors and these include:

- Limited resources and a lack of digital expertise on the part of regulators;
- Operators able to easily and rapidly enter and exit the online marketplace;
- Operators being unaware that EU food safety regulatory requirements (hygiene standards and labelling) also apply to online sales;

- Lack of cooperation by website providers, domain owners and domestic and international Internet regulators;
- The presence of online sellers who actively try to avoid official controls by frequently changing their digital identity.

The Commission's report found that authorities in the Member States visited recognised the need to enhance controls for online sales of food but, in the main, have not taken steps to amend traditional inspection and sampling activities to ensure food supplied online is safe and subject to an appropriate level of controls. In many instances the authorities within the Member States also felt that tackling such sales is not a priority.

The report also found that in two Member States, national legislation gives official staff increased investigation powers which enable them to use assumed identities to control the sale of goods and the supply of services over the Internet. In another Member State, authorities could access private dwellings in the case of ongoing investigations in order to gather a full picture of Internet sellers' on-site activities.

The Official Control Regulation (EU) No 2017/625 provides the legal basis for authorities to shop online without revealing their identity and to use the products received as official samples. However, and despite this level of support for such activities, not all countries have domestic legislation permitting the use of assumed identities and some, the UK included, have legislation to actively prevent covert surveillance.[6] Furthermore, the Commission found that domestic systems covering the registration of food businesses do not ensure correct identification of all food businesses operating online. In most cases, food business registration forms do not explicitly take into account the online dimension and such information is not considered for the business's risk rating and the setting of the frequency of controls by local authorities.

Online controls are in most cases done in conjunction with physical ones. If operators sell exclusively on the Internet and have no physical food-related activities, then food safety controls are mainly non-existent and such other controls that take place are largely on their websites.

The Commission also found that some domestic regulators did target the online sale of foods and supplements due to the importance attributed to their online sale and distribution. Depending on the risk identified, enforcement activities by authorities have included requirements for removing claims, revising product labelling, removing the product from the market, recalling it from consumers and seizing items.

Finally, the Commission found that the online sales of products containing such things as aflatoxin are particularly challenging to regulate. Sampling of such products presents one such challenge. For example, how can sampling take place from suppliers who appear to have no physical premise and where physical premises cannot be identified? Authorities are not able to take samples of food products identified during online investigations because they are 'no longer available' by the time they visit the premises.

The Commission's report concluded that *the enforcement of EU food chain legislation on online sales is cumbersome, especially concerning entities based in non-EU countries with which there is no established cooperation.*

Deliveroo, Uber Eats, Just Eat and myriad of 'food in a box' suppliers

In the UK, the FSA, as far back as 2010, developed and implemented a scheme to provide the public with information on the current state of hygiene within public-facing food businesses. The Food Hygiene Rating Scheme (FHRS), developed by the Food Standards Agency and operated by local authorities in England, Wales and Northern Ireland (Scotland runs their own scheme)

reflects the standards of food hygiene found within premises on the date of an inspection by a local authority. The food hygiene rating is not a guide to food quality.

The scheme helps customers choose where to eat out or to shop for food by giving them information about the businesses' hygiene standards and gives businesses a rating from 5 to 0 which is displayed at their premises and online so customers can make more informed choices about where to buy and eat food.

5 – hygiene standards are very good
4 – hygiene standards are good
3 – hygiene standards are generally satisfactory
2 – some improvement is necessary
1 – major improvement is necessary
0 – urgent improvement is required

FHRS and online food delivery services

The physical food environment is becoming increasingly diverse and fragmented, particularly in terms of the channels through which individuals purchase food. As their disposable incomes increase, they work longer hours and engage in ever-longer commuting times, consumers are increasingly using online services to purchase both pre-prepared, ready-to-eat foods and groceries [76]. The growth in online delivery has been fuelled by the rapidly expanding use of communication technology including smartphones, etc.

Online food delivery services deliver pre-prepared food direct from the producer to the consumer. The online grocery market is also increasingly popular and has been made more so as a result of the COVID pandemic. A hybrid system, sitting between pre-prepared food and grocery delivery, is also growing in popularity with groceries and fresh produce, including fish and meat, being delivered to customers together with a recipe for the customer to follow, leading to the production of a specific meal. Examples of these types of 'meals in a box' can be found from companies such as the 'Mindful Chef', 'Hello Fresh' and 'Amazon Fresh'.

Over the last five years, there has been exponential growth in pre-prepared, ready-to-eat food that is ordered online. This expansion has been underpinned by integrated online food delivery platforms such as 'Uber Eats', 'Deliveroo' and 'Just Eat'. During the COVID pandemic the use of such online services became the norm for many, as, unable to go out, consumers also abandoned cooking at home and ordered 'takeaways' in order to re-create the 'eating out' experience.

These changes in food opportunities, food choices and food access has enabled rapid and convenient access to a myriad of food choices, but also fuelled a concomitant and questionable consumption of unhealthy foods high in fat, salt and sugar and presenting a challenge to the public's health.

Societal inequalities make access to the newer and wider food opportunities more complex for some individuals. Growing unemployment and poverty mean that many are unable to benefit from this growing diversity of food opportunities. At the same time there has been a proliferation of the numbers of people who need to use food banks. This growth has also been fuelled by the COVID pandemic with large numbers of people being furloughed, made redundant or actually finding that access to food became difficult due to food shortages. These inequalities remain and are growing at the time of writing this report.

The online food industry has impacted on the traditional restaurant and takeaway industry and many businesses have had to change how they operate in order to stay in business. Continuing food business viability puts pressure on food businesses to sign up to the various food delivery companies, which, although initially beneficial and profitable often become less so. Criticism has also made of the food-tech platforms such as Just Eat and Deliveroo for not providing food hygiene rating scores, either on their platforms or with any order that the consumer makes from them. Whilst this situation is now changing, it initially meant that the consumer had little awareness of the hygiene and safety standards within the business that provided their order. In the early days of the food-tech platforms, the companies were happily registering food businesses with 0, 1 and 2 star food hygiene ratings, and customers were content to purchase food from them. We subsequently learned that customers of these platforms place a greater value on the immediacy of food access than they do to the food hygiene rating of the food provider.

At the mid-point in 2021, new restaurants on the Just Eat platform must be registered with their local authority and have a food hygiene rating score of 3, a 'pass' in Scotland or be 'waiting inspection', before they are added to their platform. Deliveroo allows new restaurants 'awaiting inspection' to be listed on the platform where it can verify that the outlet is a legitimate business but may not yet be registered with the local authority. That said, the Deliveroo platform has indicated that it will continue to work with existing restaurants that receive a score of 1, meaning major improvement is necessary in certain circumstances.

Research found only one in five British people check the hygiene rating of a takeaway restaurant before ordering and one in ten say that the food hygiene rating does not affect their decision on placing an order [77].

The same research found that Birmingham, at one time, had the most 0 star hygiene rated takeaways followed by Liverpool, Bristol, Manchester and Nottingham. Furthermore, the research revealed that people in Bristol are the least likely to check hygiene ratings with 36% admitting that they don't check because 'the outcome doesn't bother them'! The research also showed that along with immediacy of access, the menu is the biggest influence on people's choice of where to order from, with nearly 50% saying it's an important factor and only 15% that said that online recommendations influence their takeaway decision.

Research has also shown that it's the most deprived areas that are less likely to meet hygiene standards and takeaway sandwich shops and convenience retailers are significantly less likely to meet hygiene standards compared to other types of premises.

The presence of non-white ethnicities is also negatively associated with the probability of food establishment compliance whilst results of modelling show that outlets in areas with a higher percentage of white people have an increased probability of compliance. Outlets in areas with high percentages of individuals without access to a car and zones with a high rate of overcrowded households also have food premises where there is a decreased probability of compliance.

Supermarkets and hypermarkets are up to three times more likely to meet hygiene standards than restaurants cafes and canteens [78]. Takeaways and sandwich shops are 50% less likely to be compliant when compared with restaurants.

The display of the FHRS can be seen as one way of influencing improvement in food hygiene. Customers provide conflicting information regarding such scores. Research indicates that the most common reason to look at scores is customer reassurance. A consumer survey in Oct 2019 with adults in England, Wales and Northern Ireland delivered a majority of participants that felt that companies providing an online ordering service should display their food hygiene ratings where they can easily be seen [79]. Two out of three people associate businesses that don't display a food hygiene sticker with poor hygiene standards. However almost half would buy food from businesses that did not display that food hygiene rating. The same survey showed that the lowest acceptable food hygiene rating was 3 and 4, both at about 39%, and most of these surveys said they would not consider purchasing from a business that had a rating lower than they considered acceptable. More than half actually take quality and type of food into account when deciding where to eat out or purchase takeaway food, closely followed by price; but only one in five considers hygiene standards and food safety.

Further research by Which?, undertaken in April 2020, showed that consumers felt that the mandatory display of food hygiene ratings across the UK should now be in place [80]. They also felt that local authorities should take strong action against sites displaying incorrect ratings that mislead consumers. In Wales and Northern Ireland it's possible to take action against those businesses that display the wrong or inappropriate scores, whereas in England, because the scheme is not mandatory, this cannot be done.

The 2018 LAEMS data show that 95% of businesses which were inspected in England, Wales and Ireland were broadly compliant or better [81].[7] The data also shows that local authorities inspect high-risk businesses more frequently and that over 99% of these businesses (categories A and B) were inspected on time.

The National Audit Office report previously referred to found spending on food hygiene by local authorities fell an estimated 19% between 2012/13 and 2017/18 and some local authorities were failing to meet their legal responsibilities to ensure food businesses comply.

The number of overdue food visits raises questions about whether the UK's food safety regime is ready to move forward and meet the challenges ahead. It has long been argued that local authority enforcement teams are just not being given the tools they need; however, arguments remain about the effectiveness of inspections in actually reducing food-borne illness.

In addition to the challenges to the food system that were highlighted by the NAO report and by both societal and technology development, the following are also questions that need to be considered and which are likely to have an impact on the future functioning of the food system for the UK:

1   What lessons are to be learned for food safety because of the impact of

conditions created by COVID-19 and changes that it imposed to food production, retail and consumption?

2 To what extent will Brexit lead to a reduction in food safety standards and/or a transference of risks to the consumer?

3 In the wake of COVID and Brexit is domestic and international food crime likely to increase and, as a result of Brexit, have we limited our ability to fight it?

4 What impact will Brexit have on shopping habits and will more consumers buy local produce and support British farmers even if such products are at higher price points?

5 Is the FSA still reluctant to embrace 'Permit to Trade' as opposed to food business registration?

6 Should we finally allow highly compliant and well-managed businesses to regulate themselves and, if so, to what extent?

7 In the wake of several high-profile court cases, should we finally recognise that Primary Authority, as currently constructed, is contra to food safety and public protection?

8 In respect of allergen risk reduction, will the changes to the law to embrace the labelling of foods that are pre-packaged for direct sale be proved to have been an unnecessary additional burden on small business that delivers little additional benefit in terms of public protection?

9 Is it time to have a national conversation about the exponential growth in food banks that has taken place in the last few years and should we also consider their regulation as 'food businesses'?

The answers to these questions along with the future shape and direction of our society will inevitably help determine the UK's future food regulatory regime. One thing is, however, certain; given the pace of change and the inevitability of the emergence of significant and unexpected events, the skills of the EHP are more important than ever. It is,

therefore, vital that EHPs play a key role in shaping that future and do not simply react to it!

Notes

1 More information at https://www.malnutrition taskforce.org.uk/resources/malnutrition-england-factsheet and https://blog.ons.gov.uk/2018/02/14/deaths-involving-malnutrition-have-been-on-the-rise-but-nhs-neglect-is-not-to-blame/

2 See for example https://foodsafetyhelpline.com/about-us/ for more

3 See Coronavirus and redundancies in the UK labour market: September to November 2020; Office for National Statistics at https://www.ons.gov.uk/employmentandlabourmarket/peopleinwork/employmentandemployeetypes/articles/labourmarketeconomicanalysisquarterly/december2020 and Coronavirus and changing young people's labour market outcomes in the UK: March 2021; Office for National Statistics and also https://www.ons.gov.uk/employmentandlabourmarket/peopleinwork/employmentandemployeetypes/articles/labourmarketeconomicanalysisquarterly/march2021

4 For more on EHPs and COVID-19 see Day C. (2021) *Covid-19: The Global Environmental Health Experience*, Routledge Focus on Environmental health Series, Routledge, Abingdon, Oxon.

5 See https://www.food.gov.uk/about-us/food-and-feed-codes-of-practice for the Codes covering England, Wales and Northern Ireland. Food Standards Scotland has the 2019 Food Law Code of Practice at https://www.foodstandards.gov.scot/publications-and-research/publications/food-law-code-of-practice-scotland-2019

6 See the Regulation of Investigatory Powers Act 2000, Regulation of Investigatory Powers (Scotland) Act 2000, Data Protection Act 2018, General Data Protection Regulation and Guidance at https://assets.publishing.service.gov.uk/government/uploads/system/uploads/attachment_data/file/786444/Guide_to_the_Regulation_of_Surveillance.pdf

7 The LAEMS Report covering the year to March 2019 reported the percentage of planned food hygiene interventions undertaken by local authorities increased to 86.3% in total compared with 85.1% in 2017/18 and the percentage of food establishments that are 'broadly compliant' with food hygiene law – this means their standards are equivalent to a food hygiene rating of 3, 4 or 5 – increased slightly to 90.7% from

90.2% in the previous year https://www.food.gov.uk/news-alerts/news/fsa-publishes-latest-annual-report-on-local-authority-food-law-enforcement

Chapter references

[1] World Food Programme. (2020) *World Hunger Map.* https://www.wfp.org/publications/hunger-map-2020 (Accessed October 2021).

[2] FAO, IFAD, UNICEF, WFP, WHO. (2021) *The State of Food Security and Nutrition in the World 2021: Transforming Food Systems for Food Security, Improved Nutrition and Affordable Healthy Diets for All*, FAO, Rome. https://doi.org/10.4060/cb4474en; https://cdn.who.int/media/docs/default-source/nutritionlibrary/publications/state-food-security-nutrition-2021-en.pdf?sfvrsn=84e0ae0c_12&download=true.

[3] WHO. (2009) *Global Health Risks – Mortality and Burden of Disease Attributable to Selected Major Risks*, World health Organization, Geneva. https://apps.who.int/iris/bitstream/handle/10665/44203/9789241563871_eng.pdf?sequence=1&isAllowed=y).

[4] Vonthron S, Perrin C, Soulard CT. (2020) Foodscape – a scoping review and a research agenda for food security-related studies. *PLoS One*, 15(5): e0233218, 20 May. https://doi.org/10.1371/journal.pone.0233218. PMID: 32433690, PMCID: PMC7239489. https://pubmed.ncbi.nlm.nih.gov/32433690/.

[5] Mackendrick N. (2014) Foodscape. *Contexts*, 13(3): 16–18. https://doi.org/10.1177/1536504214545754.

[6] Kwate NO (2014) 'Racism still exists': A public health intervention using racism 'counter-marketing' – outdoor advertising in a black neighborhood. *Journal of Urban Health: Bulletin of the New York Academy of Medicine*, 91(5): 851–872, 1 October. https://doi.org/10.1007/s11524-014-9873-8.

[7] Johnston J, Baumann S. (2015) *Foodies – Democracy and Distinction in the Gourmet Foodscape*, Routledge, Abingdon, Oxon.

[8] MacIntyre N. (2020) 'Cycling revolution' at risk as local conservatives lobby to remove funded cycle routes. *The Guardian*, 15 July. https://www.theguardian.com/world/2020/jul/15/english-councils-backpedal-on-cycling-schemes-after-tory-backlash.

[9] Public Health England. (2017) *Health Matters: Obesity and the Food Environment*. https://www.gov.uk/government/publications/health-matters-obesity-and-the-food-environment/health-matters-obesity-and-the-food-environment-2.

[10] UN. (1987) *Our Common Future*, Report of the World Commission on Environment and Development (WCED). Oxford University Press. https://sustainabledevelopment.un.org/content/documents/5987our-common-future.pdf.

[11] Resolution adopted by the General Assembly on 16 September 2005 [without reference to a Main Committee (A/60/L.1)] 60/1.2005; UN World Summit Outcome, October 2005.

[12] UN. (2012) *Proceedings of the United Nations Conference on Sustainable Development, Rio+20.* https://sustainabledevelopment.un.org/rio20.

[13] Global Nutrition Report. (2020) *The 2020 Global Nutrition Report – Action on Equity to End Malnutrition.* https://globalnutritionreport.org/reports/2020-global-nutrition-report/ (Accessed 3 September 2021).

[14] World Bank. (2019) *Agricultural Land (Percentage of Land Area).* https://data.worldbank.org/indicator/AG.LND.AGRI.ZS (Accessed 3 September 2021).

[15] United Nations. (2017) *Department of Economic and Social Affairs.* https://www.un.org/development/desa/en/news/population/world-population-prospects-2017.html (Accessed 3 September 2021).

[16] IPPC. (2019) *Climate Change and Land*, An IPCC Special Report on Climate Change, Desertification, Land Degradation, Sustainable Land Management, Food Security, and Greenhouse Gas Fluxes in Terrestrial Ecosystems. https://www.ipcc.ch/srccl/.

[17] FAO. (2012) *Proceedings of the International Scientific Symposium: Biodiversity and Sustainable Diets United Against Hunger 3–5 November 2010. Sustainable Diets and Biodiversity Directions and Solutions for Policy, Research and Action*, FAO, Rome. http://www.fao.org/docrep/016/i3004e/i3004e.pdf.

[18] Public Health England (2016) *The Eatwell Guide.* https://assets.publishing.service.gov.uk/government/uploads/system/uploads/attachment_data/file/528193/Eatwell_guide_colour.pdf.

[19] Carbon Trust. (2016) *The Eatwell Guide' – A More Sustainable Diet.* https://prod-drupal-files.storage.googleapis.com/documents/resource/public/The%20Eatwell%20Guide%20a%20More%20Sustainable%20Diet%20-%20REPORT.pdf.

[20] Pereira AL, Handa S, Holmqvist G. (2017) *Prevalence and Correlates of Food Insecurity Among Children Across the Globe*, Innocenti Working Paper 2017–09, UNICEF Office of Research, Florence.

[21] The Food Foundation. (2016) *Too Poor to Eat – Food Insecurity in the UK.* https://

enuf.org.uk/sites/default/files/resources/foodinsecuritybriefing-may-2016-final.pdf.

[22] FAO. (2011) *Right to Food – Making It Happen, Progress and Lessons Learned Through Implementation*, United Nations. https://www.fao.org/3/i2250e/i2250e.pdf.

[23] CPAG. (2018) *Child Poverty and Education: A Survey of the Experiences of NEU Members*, Child Poverty Action Group. https://cpag.org.uk/sites/default/files/files/policypost/Child%20poverty%20and%20education%20-%20A%20survey%20of%20the%20experiences%20of%20NEU%20members%20dated.pd.

[24] Forsey A. (2018) *Hidden Hunger and Malnutrition in the Elderly*, Report for the APPG on Hunger, House of Commons. https://feeding-britain.org/wp-content/uploads/2019/01/Hidden_hunger_and_malnutrition_in_the_elderly-1.pdf.

[25] Origins of Food Consumed in the UK. (2019) https://www.gov.uk/government/statistics/food-statistics-pocketbook/food-statistics-in-your-pocket-global-and-uk-supply.

[26] Defra. (2019) *Agriculture in the United Kingdom 2019*, Table 13.2. https://assets.publishing.service.gov.uk/government/uploads/system/uploads/attachment_data/file/950618/AUK-2019-07jan21.pdf.

[27] Defra. (2019) *Agriculture in the United Kingdom 2019*, Table 13.3. https://assets.publishing.service.gov.uk/government/uploads/system/uploads/attachment_data/file/950618/AUK-2019-07jan21.pdf.

[28] Defra. (2019) *Agriculture in the United Kingdom 2019*, Figure 13.1. https://assets.publishing.service.gov.uk/government/uploads/system/uploads/attachment_data/file/950618/AUK-2019-07jan21.pdf.

[29] Newton-Dunn T. (2018) Trump's Brexit Blast. *The Sun*, 13 July. https://www.thesun.co.uk/news/6766531/trump-maybrexit-us-deal-off/.

[30] Partington R. (2017) Trump adviser Ross says UKUS trade deal will mean scrapping EU rules. *The Guardian*, 7 November. https://www.theguardian.com/business/2017/nov/06/trump-ross-says-uk-us-trade-deal-eu-brexit-chlorinated-chicken.

[31] Personal Communication to Prof Tim Lang, 2 July 2018.

[32] Cabinet Office. (2017) *Public Summary of Sector Security and Resilience Plans 2017*, Cabinet Office, London.

[33] Macdiarmid JL, Clark H, Whybrow S, de Ruiter H, McNeill G. (2018) Assessing national nutrition security: The UK reliance on imports to meet population energy and nutrient recommendations. *PLoS One*, 13(2): e0192649.

[34] BRC. (2018) *Letter to Prime Minister May and Mr Barnier*, British Retail Consortium, London, 4 July. https://brc.org.uk/news/2018/there-will-be-food-supply-issues-in-theevent-of-a-cliff-edge-brexit.

[35] McGuiness T, Garton-Grimwood G. (2017) *Migrant Workers in Agriculture*, Briefing Paper Number 7987, House of Commons Library, London, 4 July. https://researchbriefings.files.parliament.uk/documents/CBP-7987/CBP-7987.pdf.

[36] Daneshku S. (2017). Migrant Labour Shortage Leaves Fruit Rotting on UK Farms. *Financial Times*, 3 November. https://www.ft.com/content/13e183ee-c099-11e7-b8a338a6e068f464.

[37] BBC News. (2000) Countdown to crisis: Eight days that shook Britain. *BBC News*, 14 September. http://news.bbc.co.uk/1/hi/uk/924574.stm.

[38] Agriculture & Horticulture Development Board (AHDB). (2017) *Brexit Scenarios: An Impact Assessment*, Market Intelligence, Stoneleigh Park, October. https://ahdb.org.uk/brexit/documents/Horizon_BrexitScenarios_11oct17.pdf; https://ahdb.org.uk/Search?q=Brexit+scenarios.

[39] Mason P, Lang T. (2017) *Sustainable Diets*, Routledge, Abingdon.

[40] Gladek E, Fraser M, Roemers G, et al. (2016) *The Global Food System: An Analysis – Report to WWF*, Metabolic & WWF, Amsterdam.

[41] UN. (2015) *Sustainable Development Goals, Agreed at the UN Summit*, 27–29 September, https://sustainabledevelopment.un.org/post2015/summit.

[42] UNFCC. (2015) *Paris Climate Change Accord*, United Nations Framework Convention on Climate Change, Paris. http://ec.europa.eu/clima/policies/international/negotiations/paris/index_en.htm.

[43] UN. (1992) *Framework Convention on Biological Diversity [and Additions e.g. Cartagena Convention]*, Food and Agriculture Organisation, Rome, 3rd ed. http://www.fao.org/fileadmin/templates/soilbiodiversity/Downloadable_files/cbd-handbook-all-en.pdf.

[44] Garnett T. (2016) Plating up solutions: Can eating patterns be both healthier and more sustainable? *Science*, 353(6305): 1202–1204.

[45] Monteiro C, Moubarac JC, Levy R, et al. (2018) Household availability of ultra-processed foods and obesity in nineteen European countries. *Public Health Nutrition*, 21(1): 18–26.

[46] Public Health England. (2017) *Health Profile for England*, PHE, London, Chapter 5. https://www.gov.uk/government/publications/

health-profile-for-england/chapter-5-inequality-in-health.

[47] Morris M. (2018) *Have Your Cake or Eat It? New Findings on Public Attitudes to Brexit (Part Two)*, IPPR, London. https://www.ippr.org/research/publications/have-your-cake-or-eat-it (Accessed October 2021).

[48] Eustice G. (2018) *Hearing at House of Lords EU Environment & Energy Committee*, House of Lords, London. https://parliamentlive.tv/Event/Index/040d0036-10ec4dab-a693-8abb5f3f2a4d.

[49] Grayling C. (2018) BBC TV Question Time. *BBC*, 15 March. https://www.bbc.co.uk/news/av/uk-politics-43425055/chris-grayling-no-post-brexit-lorry-checks-atdover.

[50] BBC. (2018) Post-Brexit border checks 'may triple queues' to port. *BBC Newsonline*, 12 March. https://www.bbc.co.uk/news/uk-england-kent-43318258.

[51] Lang T, Millstone E, Marsden T. (2017) *A Food Brexit: Time to Get Real: A Brexit Briefing*, SPRU University of Sussex, City University & SPRI Cardiff University, Brighton. http://sro.sussex.ac.uk/id/eprint/69300/1/Food%20Brexit%20Briefing%20Paper%20LangMillstoneMarsden%2016July2017.pdf (Accessed October 2021).

[52] Highmore CJ, Warner JC, Rothwell SD, Wilks SA, Keevil CW. (2018) Viable-but non culturable Listeria monocytogenes and Salmonella enterica serovar Thompson induced by chlorine stress remain infectious. *mBio*, 9(2): e00540-18, March–April. https://doi.org/10.1128/mBio.00540-18.

[53] Sustain. (2018) *Fears New Trade Deals with US Will Increase UK Food Poisoning*, Sustain, London, 21 February. https://www.sustainweb.org/news/feb18_us_foodpoisoning/.

[54] Stevenson P. (2018) *A Better Brexit for Farm Animals: What the Government Must Do to Protect Welfare Standards*, London: Food Research Collaboration and Guildford: Compassion in World Farming, June 27. http://foodresearch.org.uk/publications/a-better-brexit-for-farm-animals-what-thegovernment-must-do-to-protect-welfare-standards/.

[55] Pacelle W. (2014) Banned in 160 nations, why is ractopamine in U.S. pork? *LiveScience*, 26 July. https://www.livescience.com/47032-time-for-us-to-ban-ractopamine.html; https://en.wikipedia.org/wiki/Ractopamine.

[56] Defra. (2020) *National Food Strategy for England; An Independent Review of England's Food Chain from Field to Fork; Department for Environment, Food & Rural Affairs*, Published 29 July 2020 and July 2021. https://www.gov.uk/government/publications/national-food-strategy-for-england.

[57] FAO. (2009) *Global Agriculture Towards 2050*, High Level Expert Forum, Rome, 12–13 October. http://www.fao.org/fileadmin/templates/wsfs/docs/Issues_papers/HLEF2050_Global_Agriculture.pdf.

[58] EU. (2011) *Food Information for Consumers Regulation (EU FIC) Regulation (EU) No 1169/2011 of the European Parliament and of the Council*. https://eur-lex.europa.eu/LexUriServ/LexUriServ.do?uri=OJ:L:2011:304:0018:0063:en:PDF#:~:text=matter%20and%20scope-,1.,functioning%20of%20the%20internal%20market.

[59] Yarham R. (2021) *Food Anaphylaxis in the UK – What We've Learnt by Analysing National Data*. An FSA Blog, 18 February. https://food.blog.gov.uk/2021/02/18/food-anaphylaxis-in-the-uk-what-weve-learnt-by-analysing-national-data/.

[60] Boyce JI. (2012) Food allergies in developing and emerging economies: Need for comprehensive data on prevalence rates. *Clinical and Translational Allergy*, 2: 25. https://doi.org/10.1186/2045-7022-2-25.

[61] Codex Alimentarius. (2020) *Code of Practice on Food Allergen Management for Food Business Operators*, CXC 80–2020, Adopted in 2020. https://www.fao.org/fao-who-codexalimentarius/sh-proxy/zh/?lnk=1&url=https%253A%252F%252Fworkspace.fao.org%252Fsites%252Fcodex%252FStandards%252FCXC%2B80-2020%252FCXC_080e.pdf.

[62] Daniel N, Casadevall N, et al. (2020) *The Burden of Foodborne Disease in the UK 2018*, Food Standards Agency Report Prepared by FSA and LSHTM. https://www.food.gov.uk/sites/default/files/media/document/the-burden-of-foodborne-disease-in-the-uk_0.pdf.

[63] Food Research and Action Centre. (2017) *The Impact of Poverty, Food Insecurity, and Poor Nutrition on Health and Well-Being*. https://frac.org/wp-content/uploads/hunger-health-impact-poverty-food-insecurity-health-well-being.pdf.

[64] WHO. (2021) *Draft Who Global Strategy for Food Safety 2022–2030 – Towards Stronger Food Safety Systems and Global Cooperation*, WHO Department of Nutrition and Food Safety. https://cdn.who.int/media/docs/default-source/food-safety/public-consultation/draft-who-global-strategy-for-food-safety-13may2021.pdf?sfvrsn=ac480bb9_5.

[65] What is the food standards agency? (2013) *BBC News*, 15 February. https://www.bbc.co.uk/news/uk-21476813.

[66] What is the food standards agency? (2013) *BBC News*, 15 February. https://www.bbc.co.uk/news/uk-21476813.

[67] Putting the Consumer First – A Departmental Report; FSA. (2005) https://assets.publishing.service.gov.uk/government/uploads/system/uploads/attachment_data/file/272105/6525.pdf.

[68] FSA. (2021) *The national food crime unit (NFCU) – a law enforcement capability within the FSA*, FSA. https://www.food.gov.uk/about-us/national-food-crime-unit.

[69] FSA. (2021) *Food Law Code of Practice*, Food Standards Agency, March. https://www.food.gov.uk/about-us/food-and-feed-codes-of-practice#food-law-code-of-practice.

[70] Cabinet Office. (2021) *UK to Seize Brexit Opportunities and Unleash Innovation by Overhauling Approach to Red Tape*. https://www.gov.uk/government/news/uk-to-seize-brexit-opportunities-and-unleash-innovation-by-overhauling-approach-to-red-tape-22-july-2021.

[71] FSA. (2021) FSA issues warning about safety of meat sold on Facebook. *Food Safety News*, 22 January. https://www.foodsafetynews.com/2021/01/fsa-issues-warning-about-safety-of-meat-sold-on-facebook/.

[72] Virtual food hygiene inspections could reduce backlog. (2020) *Food Safety News*, 21 July. https://www.foodsafetynews.com/2020/07/virtual-food-hygiene-inspections-could-reduce-backlog/.

[73] Research into food sampling policies and approach. (2020) *FSA*. https://www.food.gov.uk/research/research-projects/research-into-food-sampling-policies-and-approach.

[74] NAO. (2019) *Ensuring Food Safety and Standards*, HC 2217 Session 2017–2019. https://www.nao.org.uk/wp-content/uploads/2019/06/Ensuring-food-safety-and-standards.pdf.

[75] EC DG Health and Safety. (2018) *Review of Official Controls on the Internet Sale of Foods in EU Member States, DG(SANTE) 2018–6537*. http://ec.europa.eu/food/audits-analysis/overview_reports/details.cfm?rep_id=127.

[76] Deloite. (2019) *Future of Food – How Technology and Global Trends Are Transforming the Food Industry*, A Report Prepared for Uber. https://www2.deloitte.com/content/dam/Deloitte/au/Documents/Economics/deloitte-au-economics-future-food-uber-eats-100719.pdf.

[77] Food Safety News. (2020) Most people in UK don't check takeaway hygiene ratings. *Food Safety News, News Desk*, 12 September. https://www.foodsafetynews.com/2020/09/most-people-in-uk-dont-check-takeaway-hygiene-ratings/.

[78] Oldroyd RA, Morris MA, Birkin M. (2020) Food safety vulnerability: Neighbourhood determinants of non-compliant establishments in England and Wales. *Health & Place*, 63. ISSN:1353-8292. https://doi.org/10.1016/j.healthplace.2020.102325.

[79] FSA. (2020) *Food Hygiene Rating Scheme – Consumer Attitudes Tracker (Wave 8, October 2019)*. https://www.food.gov.uk/sites/default/files/media/document/fhrs-consumer-tracker-report-wave-8.pdf.

[80] Whitworth J. (2020) *Which? Renews Mandatory Hygiene Rating Call*, 28 April. https://www.foodsafetynews.com/2020/04/which-renews-mandatory-hygiene-rating-call/ (Accessed September 2021).

[81] FSA. (2018) *Annual Report on Local Authority Food Law Enforcement for England, Northern Ireland and Wales 1 April 2017 to 31 March 2018*. http://www.foodlaw.rdg.ac.uk/pdf/2018-FSA-LAEMS-2017-18.pdf (Accessed October 2021).

Web links for further reading

British Nutrition Foundation: www.nutrition.org

Codex Alimentarius: www.codexalimentarius.net/web/index_en.jsp

Eat Lancet: https://eatforum.org/eat-lancet-commission/

Eatwell: www.eatwell.gov.uk

EFSA: www.efsa.europa.eu/en/

FAO Food Security Statistics: www.fao.org/economic/ess/ess-fs/en/

Food and Agriculture Organisation of the United Nations (FAO): www.fao.org

Food Research Collaboration: https://foodresearch.org.uk/

Food Standards Agency UK: www.food.gov.uk/

FSA Local Authority Audit & Policy Branch (2009) Making every inspection count: internal monitoring advice for LA Food and Feed Law Enforcement team managers, www.food.gov.uk/news/newsarchive/2010/jan/everyinspection

Information relating to the SFBB system and devolved variations is available from the UK Food Standards Agency website: https://www.food.gov.uk/business-industry/caterers/sfbb

Royal Society for Public Health: https://www.rsph.org.uk/

Sustain: The alliance for better food and farming Sustainable Food Guidelines: www.sustainweb.org/sustainablefood/

UK Food Standards Agency: HACCP in Meat Plants: www.food.gov.uk/foodindustry/meat/haccpmeatplants/

US Food Safety and Inspection Service (FSIS): www.fsis.usda.gov/

WHO – food safety: www.who.int/topics/food_safety/en/

World Health Organization (WHO): www.who.int

World Summit on Food Security: www.fao.org/wsfs/world-summit/en/

World Trade Organization (WTO): www.wto.org

The work and leisure environments

Jonathan Hayes and Stuart Wiggans

SECTION 1: OVERVIEW OF HEALTH AND SAFETY IN THE UK

The Environmental Health Practitioner's role in health and safety

Although the role of the Environmental Health Practitioner (EHP) has changed significantly since the introduction of the Health and Safety at Work etc. Act in 1974, the challenges affecting today's EHP in health and safety are still as great as they always have been.

Health and Safety Statistics for 2019–20 and 2020–2021 [1] revealed that 144 and 142 workers were killed whilst at work in each of those years. In 2019–20 some 65,427 injuries to workers were reported under RIDDOR (see later in Section 6) and 38.8 million workdays were lost through workplace ill health and injury. Also 693,000 workers sustained a non-fatal injury according to self-reports from the Labour Force Survey in 2019–20 (LFS). Injuries and cases of ill health resulting largely from current working conditions cost society an estimated £164.2 billion in 2019–20.

The regulatory landscape has changed beyond recognition in the last decade. Many of the changes will be explored later in this chapter; however, first we should look back at the formation of the modern health and safety framework.

The first piece of legislation aimed at worker protection was an Act to protect children and young workers in mills and factories in 1833, which appointed inspectors of factories to enforce the provisions of young worker protection and set the framework for the regulation and enforcement of health and safety that we still see today. The range of hazards and challenges in health and safety have changed drastically with the demise of the industrial and manufacturing base in the United Kingdom resulting in a change of focus from the dominance of mechanical and physical stressors, to a focus on 'softer', although no less important, occupational health stressors such as stress and the impacts of new technology, particularly the use of computers, mobile communication devices, automated and robotic systems and the potential impacts of nanotechnology. There is a growing importance in considering behavioural approaches rather than a pure enforcement approach to health and safety and this presents modern-day practitioners with a different set of challenges from their

DOI: 10.1201/9781003035640-14

predecessors and requires a more open and flexible mindset and the adoption of a risk-based, solutions-focussed approach to safety compliance.

Legislation introduced to give greater protection to workers, such as the 'Six Pack' which will be discussed later, will include Corporate Manslaughter legislation which changes the game in relation to criminal actions for duty holders. However, the Health and Safety at Work etc. Act 1974 remains the cornerstone of the health and safety regulatory framework in the UK.

This introduction to the role of the EHP in health and safety will explore the founding principles of worker protection, new developments in legislation and practice spanning the last decade and the author's views on the future of the role of the EHP in health and safety.

We will also explore developments in the way that health and safety is managed within commercial organisations and enforced within local government. We will review health and safety management strategies, organisational approaches to safety management, links to risk management and explore in some detail current and emerging risks for the environmental health practitioner.

The modern health and safety framework

Worker protection has been at the cornerstone of the regulatory system in the United Kingdom for over 160 years with regulations existing since the Act of 1833 which introduced factory inspectors to protect young workers. This Act also helped shape the role of local authorities in regulatory enforcement. This role is explored in other chapters; however, this role came to prominence with the report of the Committee on Safety & Health at Work, chaired by Lord Robens in 1972 [2].

The Committee was asked to consider whether changes were required to the laws relating to occupational health and safety and

to review the measures for protecting the public against industry-based hazards. The prevailing view of the time was that the current system of regulation and worker protection was not achieving a sufficiently robust degree of worker protection and therefore not fit for purpose. Never before had a such a comprehensive review of the existing system been undertaken, emphasising the lack of trust there was in the regime in place at the time.

The statistics of the time make alarming reading in so far as that in the region of 1000 people a year were being killed at their work, 500,000 suffered injuries and 23 million working days were lost annually to industrial disease and physical injury.

Robens diagnosed apathy as the primary cause for the high level of illness and injury being caused to the country's workers; however, statistics of the time may be unreliable. His conclusion was that the health and safety system required fully overhauling and that a new framework for health and safety in the United Kingdom was required.

He identified four strands to the problem:

- That there existed too much law, most of it poorly drafted and ineffective in its application;
- That worker health and safety could not be effectively provided by having too much legislation enforced by an expanding cadre of inspectors;
- That the primary duty for ensuring the safety and protection of workers should be the responsibility of those who created the risks in the first place alongside those who worked with them;
- That the role of regulatory law and government intervention should focus on influencing attitudes and creating a framework for better organised safety management rather than in detailed prescriptive legislation and enforcement thereof.

In many ways the problems that Robens diagnosed are still the same that affect the delivery of workplace safety and public protection today.

The result of the work of the Robens Committee was the introduction and enactment of the Health and Safety at Work etc. Act 1974, which sets out the general duties that employers have towards employees and members of the public, those that employees have to themselves and each other, and the duties of the self-employed person. The Act also established the HSC and HSE and confirmed the pivotal role local authorities have in health and safety enforcement.

The Health and Safety at Work etc. Act 1974 has stood the test of time, which shows that it is a good example of a well-drafted piece of enabling legislation which permits duty holders freedom to explore ways of achieving compliance appropriate to their individual needs and circumstances. Section 2 of the Act says it shall be the duty of every employer to ensure, so far as is reasonably practicable, the health, safety and welfare at work of all their employees. Section 3 places a duty on every employer to conduct their undertaking in such a way as to ensure, so far as is reasonably practicable, that persons not in their employment who may be affected thereby are not thereby exposed to risks to their health or safety. There was also a duty on every self-employed person to conduct their undertaking in such a way as to ensure, so far as is reasonably practicable, that they and other persons (not being the individual's employees) who may be affected thereby are not thereby exposed to risks to their health or safety.

The Deregulation Act 2015 and particularly section 1 of that Act has been subject to much debate as it provides the means to limit the scope of the duty on self-employed people and exempt some 1.8 million self-employed jobs in occupations that present no potential risk to others from health and safety law. In order to ensure that low-risk businesses that take a responsible attitude to health and safety are not subject to unnecessary health and safety inspections, the Government shifted the focus of our health and safety regulators to concentrate their efforts on higher-risk industries. The Secretary of State will have the power to make Regulations under section 1 of the 2015 Act to bring other self-employed persons within the scope of the general duty.

The exemption for turban-wearing Sikhs from wearing a safety helmet is extended by the 2015 Act from construction sites to all workplaces (except in urgent response to hazardous situations such as fire or riots, or if the individual is a member of Her Majesty's Forces and taking part in a military operation). This therefore includes visitors to a workplace as well as workers.

A particularly important point that is often forgotten, is the 1974 Act provides the cornerstone of the framework that Robens envisaged for the United Kingdom; that of a self-regulating system based on an enabling Act supported by goal setting legislation.

The use of goal setting-based legislation enables the government of the day to set flexible legislation appropriate to the needs of the day. Goal setting legislation sets out the principles of what should be achieved but does not detail what compliance looks like or how this should be achieved. It details what the legal requirements are and provides good practice examples to demonstrate compliance.

Other health and safety legislation is by its nature detailed and very prescriptive and clearly spells out what must be undertaken to achieve compliance. At the foundation of health and safety legislation are some core principles which are explored in more detail next.

In March 2015 the Department for Work and Pensions published the final progress report on implementation of health and safety reforms[1] as part of its deregulation agenda.[2]

Risk assessment

Risk assessment as a principle is explored in greater detail in other sections of this book (there is an introduction to the principles of risk assessment in Chapter 1); however, environmental health practitioners must keep in mind that risk assessment is key to our health and safety regulatory framework and to the vision Robens had of a self-regulating system for safety and health at work, whereby a duty holder took the initiative in identifying and controlling work-related hazards.

The concept of assessing risk is a simple one and often put more simply as 'common sense'. However, knowledge and understanding of the concept of risk assessment is a key skill for the EHP in evaluating workplace safety and is found across the range of environmental health disciplines that the modern environmental health practitioner faces. The key elements of risk assessment in isolation, namely, hazard, risk and probability, should be easily understood by the environmental health practitioner; however, when taken in totality in respect of health and safety at work they are a complex system with many variables.

The key to successful risk assessment in the workplace is a disciplined, structured approach to the process, undertaken by a team of employees who both understand the hazards associated with a particular task and with sufficient health and safety experience to appropriately encapsulate the findings and ensure that these are communicated. Assessments, once documented, (if required) should become part of the organisation's conscience, lived daily by the workforce and reviewed as appropriate dictated by changed in circumstances, task design or other material changes.

Duty holders are required to determine for themselves the level of risk that occupational and workplace hazards present to their workforce. The environmental health practitioner should, when reviewing risk assessments or indeed writing them, be mindful that the level of perceived risk that a hazard presents is highly subjective; and that individual will accept a higher level of risk where they are in either full or partial control of their circumstances than if these are prescribed.

The Management of Health and Safety at Work Regulations 1999[3] provide the regulatory basis for risk assessment in health and safety legislation. These Regulations make explicit the requirement on employers to ensure that a risk assessment of work activities has been carried out; that significant risks are documented; control measures to mitigate the impact of the risk manifesting itself are in place; and that these control measures are communicated to the workforce, via clear instructions, supervision and training.

The role of the HSE and the environmental health practitioner is to ensure that duty holders undertake suitable and sufficient risk assessments and that the control measures and methods of safe working have been communicated to employees. The EHP should work with and support duty holders to ensure these obligations are met to a satisfactory standard.

Risk acceptability – This can be considered as the degree to which risk becomes acceptable and is a judgement call based on fact but also individual circumstances and experiences. Indeed, public perception of risk can skew acceptability.[4] A line of tolerability exists between the extremes of totally safe and totally

---

CASE STUDY 1

A multi-national retail and catering business undertook an exercise to review and update its risk assessment process. The risk management team in the organisation worked with employees from a range of levels in the organisation to develop the assessments and the resultant safe working methods that the process generated. These were tested at a number of sites and modified before being implemented as national generic assessments. The process was underpinned by specific training to enable the site managers to amend the assessments according to their local circumstances. The process was supported by engagement with technical experts and the organisation's Primary Authority partner.

unsafe, where an individual draws the point at which the level of risk becomes acceptable or 'safe enough' is down to the individual. Much has been written on this subject especially in the last decade as the general public have become more aware of public risk [3].

The HSE has historically defined risk tolerability based on a range of risk classifications – intolerable, tolerable and broadly acceptable, based on research within the nuclear power industry [4], although more recently this concept has been extended to other industrial sectors. Risk acceptability decision-making has to be born in mind against a range of factors:

- Facilitating a regulatory approach to enable transparent decision-making and scrutiny;
- Enabling duty holders to manage risks in a way that is compatible with the regulatory framework;
- Providing public reassurance.

The outcome of the work into risk acceptability and tolerance is the development of the Enforcement Management Model by the HSE [5] which acts as a tool to assist both duty holders in decision-making regarding risk acceptability and also for enforcing authorities to rate how effective a duty holder's risk control measures are against a matrix of tolerance and therefore to determine whether an enforcement intervention is required and what this intervention should be. The Enforcement Management Model will be discussed in greater detail later; however, the module provides in essence:

- A framework for consistent risk-based decisions regarding enforcement interventions;
- The ability for non-enforcers or duty holders to understand enforcement decisions;
- Assists in training for duty holders and less experienced inspectors.

"So far as is reasonably practicable" – The test of reasonableness underpins the provisions of the Health and Safety at Work etc. Act 1974, and subsequent legislation. This test is qualified by the determination of 'so far as is reasonably practicable' or (SFAIRP). This concept has been determined by case law since the introduction of the Act and, in essence, implies a balance between the degree of risk that an activity or hazard could present and the output of resources such as time, money and trouble needed to overcome the risk and reduce it to a safe level.

The expectation is that having applied SFAIRP the level of risk a hazard presents should be as low as is reasonably practicable, enabling the activity to be untaken without the assurance of absolute safety.

SFAIRP and "ALARP" ("as low as reasonably practicable") mean the same thing in practice and at their core is the concept of "reasonably practicable"; this involves weighing a risk against the trouble, time and money needed to control it [2]. In essence, making sure a risk has been reduced SFAIRP is about weighing the risk against the sacrifice needed to further reduce it. The decision is weighted in favour of health and safety because the presumption is that the duty holder should implement the risk reduction measure. To avoid having to make this sacrifice, the duty holder must be able to show that it would be grossly disproportionate to the benefits of risk reduction that would be achieved. Thus, the process is not one of balancing the costs and benefits of measures but, rather, of adopting measures except where they are ruled out because they involve grossly disproportionate sacrifices.

SFAIRP is the term most often used in the Health and Safety at Work etc. Act 1974 and in Regulations. ALARP is the term often used by risk specialists. It is the view of the HSE that the two terms are interchangeable except when drafting formal legal documents when you must use the correct legal phrase.

The view of the Court of Appeal was set out in 1949[5] as:

> 'Reasonably practicable' is a narrower term than 'physically possible', a computation must be made by the owner in which the quantum of risk is placed on

one scale and the sacrifice involved in the measures necessary for averting the risk (whether in money, time or trouble) is placed in the other, and that, if it be shown that there is a gross disproportion between them – the risk being insignificant in relation to the sacrifice – the defendants discharge the onus on them.

There is no simple formula for calculating what is ALARP, as it can be complex. More detailed information and guidance is provided by the HSE.[6]

The application of the SFAIRP principle is long established in United Kingdom legislation and through the development of established case law precedent; however, the European Union (EU) had difficulty in accepting it as principle in terms of public safety.

Communication and consultation with workers – Communication and consultation with workers is a key principle which has been enshrined in health and safety legislation since Robens. He believed that effective health and safety management with organisations should be from the board room to the shop floor and vice versa. Advocating positive worker involvement would secure long-term safety improvements if workers felt they had contributed to the system rather than having it imposed.

Existing legislation requires that an employer consults with their employees on health and safety matters. This consultation ensures that safety is put into practice as the employer listens to and enacts worker comments. Safety representatives should be involved in safety decision-making to ensure that worker perspectives are not ignored.

Risk management in the work environment

The concept of risk management is explored elsewhere in this publication; however, the EHP should understand the overlap and links between health and safety and risk management. In many commercial organisations

health and safety management sits within a larger risk management, legal, compliance or trading law team charged with overseeing all regulatory impacts upon a business [6].[7]

The risk management team may be multidisciplinary and include environmental health practitioners and subject area technical experts. The role of the risk management team is to provide the overall strategy for managing business risks working in concert with board members and the other departments found in commercial organisations.

Risk management is a systematic approach to ensuring that risks affecting an organisation are identified and systematically addressed, using risk assessment and risk acceptability principles to determine the organisational attitude to the level of acceptable risk. Risks can be categorised into four risk categories:

- Financial risks (e.g., interest rates, ability to obtain credit);
- Strategic risks (e.g., demand for product, customer marketplace);
- Operational risks (e.g., regulatory, organisational culture, competition);
- Hazard risks (e.g., employees, public, property, environment).

The risk management team would determine the risks, assess their impact using risk assessment and risk acceptability and then determine the priority the organisation gives to the risks on a value basis of likely impact on the organisation. A successful organisation would have in place a means whereby the board members own the risk register and take responsibility for managing risks within their spheres of influence [7].

Managing for health and safety (HSG 65) [7][8] – This sets out health and safety management standards and processes for organisations. First published in 1991, the guide was updated and republished in 2013 and this latest version saw the HSE advocate a shift away from the POPMAR (Policy, Organising, Planning, Measuring performance, Auditing and Review) modules on which versions 1 and 2 were based to a new model – the Plan, Do, Act, Check approach (see Figure 14.1) and also appears in Chapter 8 when discussing quality management systems.

These principles in the context of health and safety are explained:

Plan – is about assessing the current level of H&S performance and setting a clear direction in terms of where the organisation wants to get to – what are your goals/targets. A safety policy and plan needs to be determined and effectively communicated across the organisation so that everyone can share and adopt the safety goals.

Consideration should be given to safety and occupational health hazards which may impact the organisation and success criteria and monitoring objectives may need to be developed.

Do – is about identifying accurately the risks and hazards within an organisation, organising resources to deliver the safety plan objectives, and putting in place the right support tools to enable its delivery such as training, support, equipment, etc.

Check – is all about measuring safety performance – this may be through auditing, checklists, toolbox talks, visual inspections, etc. The key to this is engagement with people who manage or deliver a particular task to ensure that the controls that have been implemented are effective and adequately control the risk. This may also include accident investigation and follow up.

Act – reviewing safety performance, accidents and incidents to learn from them and to make effective changes as necessary.

The updated version of HSG 65 makes simple clear reading and sets out a well-structured approach to effective health and safety management. It identifies some core elements for effective health and safety management. These are:

- Leadership and management;
- A trained/skilled workforce;
- An environment where people are trusted and involved.

Development of health and safety and enforcement of legislation

The main influence on the development of legislation in the United Kingdom since the early 1980s has been the EU. Historically EU directives made under the Framework Directive on Health & Safety required Member States to enact national regulations to bring them into force. This enabled Member States to enact the principles of the new legislation in a way that was sympathetic to their existing legislature and legal system. This disparity across the EU, also reflecting differences in legal systems, meant that in the UK principles of risk assessment and SFASRP have been fundamental to our law.

Recently EU directives have moved towards a greater level of prescription, creating some conflicts in the UK with the principles of 'self-regulation' enshrined with the Health and Safety at Work etc. Act 1974. The 'Six Pack' of regulations (see later) are an example of this as the legislative requirements are prescriptive in their nature.

More recently introduced EU legislation, especially in food safety, has bypassed the need for Member States to enact separate legislation as the EU started using Committee Regulations which have a direct impact.

Many environmental health practitioners working in commercial organisations have responsibility for health and safety management both within and outside of the UK. Similarly, many UK PLCs and mid-sized organisations are now ultimately owned or operated by companies who are not based in the UK, therefore the EHP must appreciate that the organisational mindset and approach may differ to that of a traditional British owned and operated company.

Uniquely in the United Kingdom enforcement is split between the Health & Safety Executive (HSE) and local authorities (LAs).

However, in the majority of EU Member States, health and safety enforcement responsibility rests with a centralised government or labour inspectorate, responsible for labour and employment matters beyond pure health and safety.

Common in continental health and safety frameworks is the role played by the insurance industry, which provide compensation payments for employees who suffer accidents. Their role also extends in some cases to issuing guidance on compliance to which members must follow and applying sanctions.

In the UK, the role played by the insurance industry is more advisory, in so far as although insurance providers may undertake risk audits of clients and make recommendations of good practice, these generally do not impact on an organisation's ability to obtain insurance and have no statutory binding role.

Continental labour inspectorates coordinate their activities through the Senior Labour Inspectors Committee (SLIC), on which the UK is represented by the HSE. The SLIC collate and coordinate information on enforcement approaches and interventions throughout the EU and act as a point of reference for information [8].

While health and safety enforcement responsibilities throughout the EU are often administered at central or regional level, the expectation is fire safety, which is more frequently administrated at the local level.

Role of the enforcing bodies – Within the UK, enforcement of the Health and Safety at Work etc. Act 1974 is split between the HSE and local authorities. The exact nature of the division of premises for enforcement responsibility is set out in the Health and Safety (Enforcing Authority) Regulations 1998[9] on the basis of 'main activity' undertaken at the premises.

The group of premises that each enforcing authority assumes responsibility for is set out in the Schedules attached to the Regulations. Historically the HSE has taken heavy industry and higher risk major installations, whilst local authorities have focussed on lower risk sectors such as retail, catering and leisure premises and offices. However, the distinctions and divisions have become increasingly blurred in recent years with the advent of initiatives such as joint warranting and joint working.

The HSE website provides guidance for enforcing authorities as to the types of work activities that should be prioritised for inspection.

Within the HSE, inspection and enforcement is undertaken by competent inspectors working within distinct divisions or teams such as FOD (Field Operations Directorate) and specialised divisions such as the Chemical and Hazardous Installations Division. The HSE now levies charges against their regulated business for non-compliance and for enforcement actions. It is yet to be seen whether local authorities will adopt this practice.

Local authority inspectors may be environmental health practitioners or other specialist technical staff. Within government, there is a drive to reduce the distinctions between differing types of inspectors and this can reflect in the quality of enforcement undertaken.

All officers, whether local authority or HSE have the same powers as set out within section 20 of the Act; however, the level of authorisation an individual officer has may vary depending upon their experience and qualifications.

Enforcement policy – Health and safety enforcement should be based on the core principles identified and set out within the Enforcement Policy Statement[10] to which the HSE and local authorities have signed up. The Enforcement Policy Statement is issued in accordance with the principles of good regulation set out in the Legislative and Regulatory Reform Act of 2006.

The Enforcement Policy Statement identifies a number of principles on which enforcement decisions and action should be based.

These principles are:

- Proportionality;
  Proportionality is about relating the enforcement action to the level of risk.

Some degree of consideration needs to be given by the EHP has to how far the duty holder has fallen from an acceptable standard of what the law requires. Proportionality is necessary to ensure that enforcement action is related to risk. Clearly, some health and safety requirements are mandatory whereas others rely on the concept of proportionality built into the principle of "So far as is reasonably practicable".

An EHP who can provide a duty holder with a sense of proportionality and who understands these themselves will be effective in achieving compliance.

- Targeting;
  LAs must prioritise their resources and activities according to the level of risk an activity presents or where the least well-controlled hazards are identified. In some cases, this may mean providing some businesses with a 'light touch' approach to enforcement; however, other duty holders might require a much more hands-on and direct approach, utilising the range of options open to the practitioner.

  HSE expects local authorities to have in place a system for prioritising its activities according to risk. This, in essence, means ensuring that inspection activity is targeted at those activities that give rise to the most serious risks or where hazards are least well controlled. Similarly, action should be focussed on the duty holders responsible for the risk. This may mean targeting action at employers, employees, manufacturers, suppliers or others who are in the most appropriate position to control the risk. LAs are required also to take account of the HSE's strategic plan, the HELA strategy and HELA guidance.

- Consistency;
  Ensuring a consistent enforcement approach and recognising that applying a 'one size fits all' approach to enforcement may not always be the right strategy. Dialogue between enforcers and the regulated helps to achieve this consistency.

Consistency should not be interpreted as uniformity. The HSC believes that consistency of enforcement means taking a similar approach in similar circumstances to achieve similar ends. Both the HELA and local authorities have devised a number of feedback loops to ensure dialogue among and between enforcers. In addition, the HSE has conducted further research into the management of consistency within local authorities to enable local authorities to identify the processes that may lead to inconsistency, and the management tools that might assist in the control of the many variables that make the achievement of consistency so complex. The results of this research are being used by the Local Authority Unit (LAU) to advise local authorities.

- Transparency;
  Duty holders need to understand what is expected of them and what they should expect from enforcement officers so that they can protect employee and public safety. This includes ensuring that legislation is clear in drafting and meaning and that guidance is clear and sets out what a duty holder must do to achieve compliance.

  The HSE believes that effective control of risk to health and safety is achieved if duty holders understand what is expected of them and what they should expect from enforcing authorities. Duty holders should understand what they are expected to do and not do by being able to distinguish between statutory requirements and advice or guidance.

  Transparency also includes ensuring that duty holders understand how they can access legislation and guidance and that this access is provided in a clear simple way.

- Accountability;
  The HSE requires that regulators are accountable to the public for their actions. Regulators must be able to provide sound justifications for their decisions based on evidence.

Enforcing authorities must have policies and standards against which they can be judged, and an effective and easily accessible mechanism for dealing with comments and handling complaints.

In 2006 the Legislative & Regulatory Reform Act[11] was introduced which superseded the Enforcement Concordat with the Regulators Compliance Code; however, the principles of good regulation (outlined previously) remained the same.

The Regulators Compliance Code was superseded by the Regulators Code which came into force on the 6th April 2014, issued by the Better Regulation Delivery Office (BRDO). The Code[12] provides a clear, flexible and principle-based approach for how regulators should engage with those that they regulate.

Each local authority and the HSE has a statement of enforcement policy setting out the general approach and principles of enforcement. All inspectors whether HSE or local government should have regard to the HSE Enforcement Policy Statement when making an enforcement decision.

The HSE has also introduced the Enforcement Management Model, which assists practitioners in decision-making with regard to making enforcement decisions. The EMM allows practitioners to work through their decision chain and to identify the correct enforcement action based on the circumstances of the case.

Six regulations on health and safety at work, the 'Six Pack' – These are six regulations on health and safety at work which explain in detail employer duties to ensure they comply with the requirements of the Health and Safety at Work etc. Act 1974 and are relevant to EHPs. The six regulations are often referred to as the 'Six Pack'.

The 'Six Pack' is:

1   The Management of Health and Safety at Work Regulations 1999 (amended 2006) require employers to carry out risk assessments, make arrangements to implement any necessary measures, appoint competent people and to arrange for appropriate information, instruction and training.

2   The Workplace (Health, Safety and Welfare) Regulations 1992 cover a wide range of basic health, safety and welfare issues, to include ventilation, heating, lighting, workstations, seating and the provision of welfare facilities.

3   The Provision and Use of Work Equipment Regulations 1998 require that equipment provided for use at work, including machinery, is safe and maintained in an efficient state, in efficient working order and in good repair.

4   The Manual Handling Operations Regulations 1992 (as amended) cover a wide range of manual handling activities involving the transporting or supporting of a load. This includes lifting, lowering, pushing, pulling, carrying or moving, whether by hand or other bodily force.

5   The Health and Safety (Display Screen Equipment) Regulations 1992 as amended by the Health and Safety (Miscellaneous Amendments) Regulations 2002 cover the requirements to protect the health and safety of employees who work with display screen equipment.

6   The Personal Protective Equipment at Work Regulations 1992 cover the provision and use of personal protective equipment and the requirements for employers to provide appropriate protective clothing and equipment for their employees.

Coordination of regulatory activity – HELA (Health & Safety and Local Authorities Enforcement Liaison Committee) was established in 1975 to provide a means of communication and cooperation between the two enforcing authorities. HELA comprises of both senior figures from local authorities and the HSE (this is also considered in Section 2).

In 2000, The Revitalising Health & Safety Strategy Statement [9] identified that the

HSE should work with local authorities to identify and propose an indicator against which local authority enforcement and promotion could be measured. This has gone further in recent times with local authority action plans becoming embedded within the overall HSE Strategic Delivery Plans.

Coordination of activity is supported by the HSE's Local Authority Unit (LAU)[13] which comprises of a mixture of EHPs who work within the HSE, EHPs seconded to work with the LAU in regional capacities and HSE employees. This blend of backgrounds gives LAU a useful voice in coordinating LA activity with that of the HSE. The LAU is considered in more detail in the next section.

Better Regulation – The Local Better Regulation Office (LBRO) was created as a government owned limited company in May 2007. The Office was brought into existence through the Regulatory Enforcement and Sanctions Act on 1 October 2008. In turn this was replaced by Better Regulation Delivery Office (BRDO). This is also covered in Chapter 7 when discussing the business environment. This was part of the Department for Business, Innovation and Skills (BIS) and was charged with working towards creating a regulatory environment in which businesses have the confidence to invest and grow and citizens and consumers are protected. It operated the Primary Authority scheme and worked to improve the consistency of front-line regulators, providing a platform for businesses to have their say of effective regulation [12].[14] This has now been replaced by the Office of Product Safety & Standards (OPSS) from January 2018 and works with local authorities and others to take forward responsibilities, including Primary Authority and Better Business for All. The OPSS is part of the Department for Business, Energy & Industrial Strategy (BEIS).

Regulatory developments – Since the start of this century a number of key regulatory developments occurred which have influenced the way in which health and safety regulation is delivered by enforcement authorities and managed by businesses.

Philip Hampton was commissioned by the then Chancellor of the Exchequer Gordon Brown to carry out an independent review with regard to improving the efficiency of regulation. Hampton reported to the Government in March 2005 [10]. Hampton reported that much good practice and some excellent practice was identified within national and local regulators. However, he reported that the regulatory system as a whole was uncoordinated, with significant overlap in regulator responsibility. He identified that in an average year, 3 million inspections are undertaken with 2.6 million forms sent out for completion each year.

Hampton recommended that risk assessment should underpin all regulatory interventions and that there should be no intervention or enforcement without a risk-based justification. He also recommended that the 31 national regulators should be reduced to seven thematic inspectorates. In all, Hampton helped shape Government policy with regard to the value of risk assessment and the importance of a risk-based decision-making approach. The Government committed to implementing Hampton's recommendations in full.

The Government published the *Rogers Review* as part of the 2007 Budget [11]. Rogers was asked to deliver on some of the foundations laid in the Hampton Report and was tasked with setting national enforcement priorities for local authority regulators.

The five priorities he identified were

- Air quality;
- Alcohol licensing;
- Hygiene of food businesses;
- Improving health in the workplace;
- Fair trading.

In addition to these, Rogers also recommended the formation of the LBRO.

Macrory sanctions – Richard Macrory[15] built further on the work of Hampton by developing in his report a flexible set of alternative enforcement tools that are consistent with the principles of risk-based regulation.

Historically for health and safety as with other areas of environmental health, enforcement sanctions have been based around criminal prosecution. The Macrory review[16] [12] recommended that the Government agreed to the introduction of a different suite of sanctions based on civil law, including:

- Fixed and variable monetary penalties;
- Extension of the statutory notice scheme to other regulators who do not currently have them;
- Enforcement undertakings;
- Stop notices.

The alternative enforcement sanctions recommended by Macrory were introduced through the Regulatory Enforcement and Sanctions Act 2008.

Regulatory Enforcement and Sanctions Act 2008 – This contained four parts. Part one established the Local Better Regulation Office. Part two made provisions for more consistent and coordinated regulatory enforcement by local authorities through the establishment of the Primary Authority scheme. The scheme, discussed later, enables a business that trades across two or more local authority boundaries to request one local authority to act as their Primary Authority.

Part three introduced the ability for national regulators to apply for and obtain new enforcement sanctions (Macrory sanctions).

Part four created a duty upon regulators not to place additional burdens upon businesses and to remove unnecessary legislative burdens.

Common Sense. Common Safety – in October 2010, Lord Young of Graffham published the 'Common Sense, Common, Safety', report by Lord Young of Graffham to the Prime Minister following a Whitehall-wide review of the operation of health and safety laws and the growth of the "compensation culture"' [13].

The Report contained a number of recommendations addressing issues around insurance and the compensation culture, accident investigation, combining health and safety and food safety inspections, risk education.

A significant number of the recommendations have been acted upon and reflect the changes and updates to this chapter.

Reclaiming Health and Safety for All – In November 2011, Professor Ragnar E Löfstedt published his report *Reclaiming Health and Safety for All*, an independent review of health and safety legislation on behalf of the Secretary of State for Work and Pensions [14]. This report has had a bearing on some of the proposed changes considered earlier.

The report was intended to identify the scope for reducing the burden of health and safety legislation on business, whilst maintaining high standards of consumer and worker protection. Löfstedt concluded in his summary that wholesale changes to the legislator system were not needed; however, he identified some areas where improvements could be made.

His recommendations included exempting from health and safety law the low-risk self-employed, a review of all HSE ACOPs, a change in law to enable HSE to direct the resources of local authorities and a clarification as to the status of Pre-Action Protocols and reviewing the application of strict liability within the civil justice system. The Government, in its response undertook to enact the recommendations expediently.

Corporate Manslaughter legislation – The introduction of the Corporate Manslaughter and Corporate Homicide Act 2007 introduced a new dimension to health and safety enforcement and management. For the first time, corporations and organisations can be found guilty of corporate manslaughter as a result of serious management failures resulting in a gross breach of a duty of care.

The introduction of the Act has meant that employee fatalities are subject to a joint investigation by all the relevant regulatory bodies, including the police. The investigation will follow the guidance set out within the Work-Related Deaths Protocol [15], which is discussed later.

The Act includes provisions for unlimited fines, publicity orders and remedial orders. The Sentencing Council for England and Wales published new definitive guidelines in early 2010 and published a consultation in November 2014[17] on health and safety offences, including corporate manslaughter and food safety and hygiene offences guidelines.

The consultation proposes a significant increase in the level of fine which can be applied in the sentencing of these offences. The aim being to ensure that sentencing reflects the seriousness of the offence and negates any potential economic advantage an organisation may gain through wilful and deliberate non-compliance with the law.

Working together to achieve better regulation

Key to achieving compliance in the modern regulatory framework is joint working and working in partnership with other regulators and with business. Some examples of this type of working are set out next.

Joint warranting – Here local authorities in a specific geographic area have agreed to utilise 'flexi warrants' within their own county boundary and with HSE field operates. This has meant that local authorities have been able to share expertise and resources, not only with each other but with local HSE inspectors, improving officer knowledge and competence.

Primary Authority – The Primary Authority scheme is a statutory partnership between a single or group of local authorities, a business or trade association and the OPSS who administer the scheme.

Primary Authority was created under the Regulatory Enforcement and Sanctions Act 2008, its purpose to provide a business with a consistent source of advice on regulatory and enforcement matters, provided by a single local authority working in close partnership with the business. The scheme aims to reduce the amount of inconsistent advice provided to businesses.[18]

The interface of workplace and public safety

The work of the modern EHP in health and safety links to a number of different regulatory areas and colleagues within a local authority. In particular, close ties exist with food safety, licensing and trading standards colleagues where there are overlaps and common interests in workplaces.

The role of the EHP in Public Safety should not be underplayed as increasingly, the EHP will be asked to comment on plans for public events organised through either another Council department or by a third party. The CIEH has generated some guidance to assist the EHP in making appropriate risk-based decisions regarding such events.[19]

The Licensing Act 2003 includes public safety and the prevention of public nuisance as two key objectives of the Act alongside protecting children from harm and reducing crime and disorder. The EHP will place a key role in the decision-making process regarding the issuing of a premises licence and the circumstances under which a licence may be called in for review.

EHPs are increasingly working outside of local government in businesses, third-party organisations such as Primary Care Trusts or Charities or for themselves as consultants. In these roles EHPs need a different skill set to achieve compliance and influence organisational behaviour as the enforcement tools of regulatory compliance inspections and enforcement powers do not exist. Safety improvements are made through communication with employers and employees and by empowering employers and employees to make the right decisions with regard to workplace health and safety and risk management through consultation and engagement.

In commercial organisations, the EHP will need to work closely with colleagues in other departments such as operations, property, finance and legal. The EHP needs to understand the culture and decision-making processes within an organisation to be able to achieve effective levels of influence.

Key to understanding the social and economic drivers of the organisation is an awareness of the Financial Conduct Authority publication – Risk Control[20] on financial risk management and understanding the financial impact that decisions have on organisational finance. Improvements and investments must be able to show a defined 'payback' period or a return on investment to be commercially attractive.

The HSE has published documents to assist the commercial practitioner influence Boards of Directors, etc.; these include the Leading Health and Safety at Work [16] and also of interest is the publication by the institute of directors – 'Risk Management – helping directors to control the risks that threaten you and your business'.[21]

A key partner of EHPs working outside of local government is the insurance industry. Many commercial EHPs will find themselves either managing insurance within their wider risk remit or heavily involved in civil claims resolution given the close association with health and safety. The standard of proof in civil claims is the lower 'on balance of probabilities' standard which means that the EHP will be involved in frequent investigations to determine cause and liability.

EHPs within local government may become involved in civil claims by virtue of investigating a RIDDOR reportable accident and often their investigation is requested by the claimant as part of their submission of evidence against the company. RIDDOR is considered later in this chapter.

Section 1 Notes

1 See https://assets.publishing.service.gov.uk/government/uploads/system/uploads/attachment_data/file/415692/final-progress-report-h-and-s-reform.pdf

2 See for instance speech from the then Secretary of State for Business, Innovation and Skills https://www.gov.uk/government/speeches/getting-government-off-your-back-our-commitment-to-cutting-red-tape and Van Lerven F, Welsh M (2018) *A Deregulatory Agenda Is Sweeping Across Europe*, New Economics Foundation, https://neweconomics.org/2018/08/deregulatory-agenda-sweeping-across-europe.

3 SI 1999 No 3242.

4 See for example Slovic P (1987) Perception of risk. *Science*, 236(947990): 280–285, 17 April. http://www.ncbi.nlm.nih.gov/pubmed/3563507.

5 *Edwards v. National Coal Board* [1949] 1 All ER 743 (CA) – in which the Court of Appeal held that 'reasonably practicable' has a narrower meaning than 'physically possible'. If a defendant can show a gross disproportion between them, the risk being insignificant in relation to the sacrifice, the duty holder discharges the onus that is upon him or her.

6 See https://www.hse.gov.uk/simple-health-safety/risk/index.htm

7 The Risk Management Standard was originally published by the Institute of Risk Management (IRM), the Association of Insurance and Risk Manager (AIRMIC) and the Public Risk Management Association (Alarm) in 2002. It was subsequently adopted by the Federation of European Risk Management Association (FERMA). Despite the publication of ISO 31000:2018 *Risk Management – Guidelines*, IRM retained its support for the original risk management standard because it is a simple guide that outlines a practical and systematic approach to the management of risk for business managers.

8 See also https://www.hse.gov.uk/pubns/books/hsg65.htm

9 SI 1998 No 494.

10 See https://www.hse.gov.uk/pubns/hse41.pdf (Accessed August 2021).

11 Legislative and Regulatory Reform Act 2006, c51.

12 See https://www.gov.uk/government/publications/regulators-code

13 See https://www.hse.gov.uk/lau/working-together.htm and also https://www.hse.gov.uk/lau/guidance.htm

14 See more at https://www.gov.uk/government/organisations/better-regulation-delivery-office

15 Professor of Environmental Law at University College, London, Professor Macrory is a barrister and a member of Brick Court Chambers, London, and is currently director of the UCL

Centre for Law and the Environment and the UCL Carbon Capture Legal Programme.

16 See https://www.regulation.org.uk/library/2006_macrory_report.pdf – this website includes some useful information too at https://www.regulation.org.uk/deregulation-1948_to_2006.html

17 See http://www.sentencingcouncil.org.uk/wp-content/uploads/Health_and_safety_corporate_manslaughter_food_safety_and_hygiene_offfences_consultation_guideline_web1.pdf

18 For more information see https://www.gov.uk/government/publications/primary-authority-overview

19 See for example guidance on outdoor catering etc. at https://www.cieh.org/media/1254/cieh-national-guidance-for-outdoor-and-mobile-catering.pdf

20 See https://www.handbook.fca.org.uk/handbook/SYSC/7.pdf

21 See https://www.iod.com/services/information-and-advice/publications/directors-guides/risk-management and also https://iodglobal.com/handbooks.html

Section 1 References

[1] HSE. (2020) *Health & Safety Statistics Annual Report for Great Britain 2019/20*. HSE. http://www.hse.gov.uk/statistics/overall/hssh1920.pdf; https://www.hse.gov.uk/statistics/fatals.htm; and https://www.hse.gov.uk/statistics/causinj/index.htm (Accessed September 2021).

[2] The Committee on Safety and Health at Work (Robens Committee). (1972) *Command 5034*, HMSO, London.

[3] HSE. (2001) *Reducing Risk, Protecting People – HSE's Decision Making Process*, HMSO, London. ISBN:0717621510.

[4] Health & Safety Executive. (1988, revised 1992) *The Tolerability of Risk from Nuclear Power Stations*, HSE Books, Sudbury, Suffolk.

[5] HSE. (2013) *Enforcement Management Model (EMM) (Operational Version 3.2)*, Amended to Accommodate Changes in RIDDOR, October. www.hse.gov.uk/enforce/emm.pdf (Accessed August 2021).

[6] Institute of Risk Management. (2002) *Risk Management Standard*, London. https://www.theirm.org/what-we-do/what-is-enterprise-risk-management/irms-risk-management-standard/.

[7] HSE. (2013) *Successful Health & Safety Management (HSG 65)*, HSE Books, Sudbury, Suffolk.

[8] European Commission's Senior Labour Inspectors Committee. (1997) *Labour Inspectorate (Health & Safety) in the European Union*, EU, Luxemburg.

[9] HSC. (2000) *Revitalising Health & Safety Strategy Statement*, Health & Safety Commission and Department of the Environment, Transport and the Regions, HMSO, London, June.

[10] Hampton P. (2005) *Review on Regulatory Inspections and Enforcement, HM Treasury*. http://webarchive.nationalarchives.gov.uk/20130129110402; http://www.hm-treasury.gov.uk/bud_bud05_hampton.htm.

[11] Rogers P. (2007) *Rogers Review – National Enforcement Priorities for Local Authority Regulatory Services*. http://webarchive.nationalarchives.gov.uk/+/bre.berr.gov.uk/regulation/reviewing_regulation/rogers_review/.

[12] Macrory R.B. (2006) *Regulatory Justice: Making Sanctions Effective (Macrory Review)*. http://www.regulation.org.uk/library/2006_macrory_report.pdf.

[13] Lord Young of Graffham. (2010) *Common Sense, Common Safety*. HM Government. https://www.gov.uk/government/uploads/system/uploads/attachment_data/file/60905/402906_CommonSense_acc.pdf (Accessed September 2021).

[14] Löfstedt R.E. (2011) *Reclaiming Health and Safety for All*, Presented to Parliament by the Secretary of State for Work and Pensions by Command of Her Majesty, November. Cm 8219. Crown Copyright. https://www.gov.uk/government/uploads/system/uploads/attachment_data/file/66790/ Löfstedt-report.pdf (Accessed September 2021).

[15] HSE. (2011) *Work Related Deaths Protocol*. HSE Books, Sudbury, Suffolk. http://www.hse.gov.uk/enforce/wrdp/.

[16] HSE. (2013) *Leading Health & Safety at Work: Leadership Actions for Directors, Board Members, Business Owners and Organisations of All Sizes (INDG 417)*, Bootle, HSE. http://www.hse.gov.uk/pubns/indg417.pdf.

SECTION 2: STRATEGIES FOR THE MANAGEMENT OF HEALTH AND SAFETY AND FOR ACHIEVING COMPLIANCE

Review of strategies and HSE plans

On 1 April 2008 Health and Safety Commission (HSC) and Health and Safety Executive (HSE) merged to form a single entity known as the HSE. The HSE is the national regulatory body responsible for promoting the cause of better health and safety at work within Great Britain and continues to work in close partnership with local authorities. One of the first undertakings of the new HSE Board was to reset and reaffirm the direction of health and safety.

The HSE contains policy groups who work on both fundamental and politically sensitive issues; part of their undertaking is ensuring that HSE policy is in line with European policy, and this is still the case even following Brexit. A fundamental aim of future HSE policy has been to simplify and reduce administration without compromising protection. As part of the changes the HSE published a new *Strategic Plan, 'Be Part of the Solution'* [1] to take the original 2000 strategy forward and support with a business plan. Critically the document was aimed beyond the scope of the HSE and incorporates the involvement of other partners to achieve change.

The Strategy document set out a ten-point framework for action:

1   To investigate work-related accidents and ill health and take enforcement action to prevent harm and secure justice when appropriate

2   To encourage strong leadership in championing the importance of, and a common-sense approach to, health and safety in the workplace

3   To motivate focus on the core aim of health and safety and, by doing so, to help risk makers and managers distinguish between real health and safety issues and trivial or ill-informed criticism

4   To encourage an increase in competence, which will enable greater ownership and profiling of risk, thereby promoting sensible and proportionate risk management

5   To reinforce the promotion of worker involvement and consultation in health and safety matters throughout unionised and non-unionised workplaces of all sizes

6   To specifically target key health issues, and to identify and work with those bodies best placed to bring about a reduction in the incidence rate, and number of cases of work-related ill health

7   To set priorities and, within those priorities, to identify which activities, their length and scale, deliver a significant reduction in the rate and number of deaths and accidents

8   To adapt and customise approaches to help the increasing numbers of small medium enterprises in different sectors comply with their health and safety obligations

9   To reduce the likelihood of low-frequency, high-impact catastrophic incidents while ensuring that Great Britain maintains its capabilities in those industries strategically important to the country's economy and social infrastructure

10 To take account of wider issues that impact on health and safety as part of the continuing drive to improve Great Britain's health and safety performance

The supporting business plan established five key areas of work to change behaviour:

•   Improving the working environment;
•   Sharing the responsibility;
•   Transforming the approach;
•   Enabling delivery;
•   Implementing performance measures and targets.

A new strategic approach – In 2010, HSE recognised that a change of approach was needed again if they were to continue pushing down the rate of death, injury and work-related ill health. Over the past 35 years there has been a decline in the rates; however this has remained almost static in the last five years, a situation that is morally, legally and financially unacceptable. An understanding of the reasons is required to help revitalise the process of improvement. The merger into one body provided a prime opportunity to change the direction of health and safety strategy going forward. Under the 2009/10 business plan the HSE supported the Government's Health, Work and Well-Being Agenda as well as undertaking a number of its own initiatives. The emphasis was very much on partnership working, primarily between the HSE and Local Authority, but also involving other working groups.

Areas for action as part of this plan included:

- Targeting key health issues;
- Reducing low-frequency, high-impact incidents;
- Encouraging leadership;
- Enabling people to determine real issues from trivial ones;
- Increasing competence;
- Promoting worker involvement;
- Customising the approach for differing sectors.

The HSE leads the health and safety system through research, identification of new risks, information and advice, and promotion of training. These are managed in conjunction with other government departments and, through the policy group, consultation to ensure that they follow a joined-up approach. HSE intends to develop this further through consultation with other expert groups to ensure a coordinated approach.

Local authorities will continue their partnership with the HSE to ensure that duty holders manage their workplaces with due regard to the health and safety of their workforce and those affected by their work activities. This will also be done through the provision of advice and guidance and the requirements of the law, supported by inspections, investigations and enforcement where required.

Emergence of occupational health as a priority – In 2008, the HSE board agreed that they should undertake work to produce a new long-term strategy for occupational health with a wide-ranging remit to consider both the effect of work on health and the effect of health on work. This culminated in the publication of 'Be Part of the Solution', mentioned earlier.

In 2008 there were 1.2 million people at work who believed they were suffering from a work-related illness; this roughly equated to a loss of 24.6 million working days. The overall trend over the following decade had been downward but by 2019/20 the figure was 1.6 million people suffering from work-related ill health, so further work is required to meet the targets set down in the Revitalising Health and Safety strategy of 2000.

While the health and safety system has always dealt with occupational ill health, recent health initiatives and the promotion of health in the workplace indicate a shifting balance between ill health and accidents. Arguably, the systems for dealing with accident prevention, the management of risks and the control of hazards have reached relative maturity, whereas the approaches to health and lifestyle issues such as stress, to which the working environment may contribute, are still in their infancy.

Setting targets and actions to address work-related ill health can be complicated. Some ill health is clearly work-linked, though may have a long latency, but other causes are not related to work alone, being partly caused, or exacerbated, by external factors. In setting suitable priorities and effective delivery there must be collaboration between the HSE and other bodies to determine who should address specific issues. One of the main areas for discussion is how to manage the interface between the work and external factors that may affect a person's health.

The HSE has highlighted three occupational health priorities as part of its current strategy: occupational lung disease, musculoskeletal disorders and work-related stress. These have been chosen to the impact on society [2].

- Occupational lung disease – Leads to an estimated 12,000 deaths each year.
- Musculoskeletal disorders – The most common reported cause of occupational ill health in Great Britain, accounting for 41% of all work-related ill health cases and 34% of all working days lost due to ill health.
- Work-related stress – The second most commonly reported cause of occupational ill health in Great Britain, It has been estimated that over 11 million days are lost at work a year because of stress at work according to HSE.[22] In 2019/20 stress, depression or anxiety accounted for 51% of all work-related ill health cases and 55% of all working days lost due to work-related ill health.

Developing strategies – The changes to HSE structure have provided an opportunity to develop new strategies for the future. Ongoing there will be increased focus and consultation with stakeholders at all levels and this message is being passed down to business. These stakeholders will include:

- HSE;
- Employers and their representative bodies;
- The self-employed;
- Workers and their representative bodies;
- Local authorities;
- Government, through its departments and agencies, etc;
- The devolved administrations and their agencies, etc;
- Professional bodies;
- Voluntary and third-sector organisations.

The HSE mission remains to prevent death, injury and ill health in Great Britain's workplaces by addressing the constantly changing health and safety issues.

What the HSE currently does and will continue to do:

- It protects people by providing information and advice; by promoting and assuring a goal-setting system of regulation; by undertaking research; and by enforcing the law where necessary;
- It influences organisations to embrace high standards of health and safety and to recognise the social and economic benefits;
- It works with business to prevent catastrophic failures in major hazard industries;
- It seeks to optimise the use of resources to deliver its mission and vision.

What the HSE aims to do:

- It will take into account wider issues impacting on health and safety;
- It will renew momentum to improve health and safety performance;
- It will respond to a wide range of risks from small business, new sectors and new technologies as well as traditional industries and risks;
- It will find new ways to engage workforces across all workplaces, based on previous findings;
- It will guide practical leaders to implement health and safety based on business benefits;
- It will re-affirm the identity of health and safety and challenge its use as a synonym for bureaucracy and an excuse for not doing things.

The reviews by Lord Young and Professor Löfstedt referred to previously found that too often businesses felt they must go beyond what health and safety law required. The issues identified were:

- There were too many inspections of relatively low-risk and well-performing workplaces;

- An overly complex structure for regulation existed;
- Businesses too often received poor health and safety advice from badly qualified consultants;
- A 'compensation culture' led businesses to a fear of being sued for accidents, even where they were not at fault [3].

The following actions were a result of the review and are consistent with the strategy of promoting a common-sense approach to health and safety in the workplace and concentrating health and safety enforcement on higher risk areas:

- A National Local Authority Enforcement Code sets out the risk-based approach to targeting health and safety interventions that local authority regulators should follow [4];
- The Health and Safety Executive (HSE) has reduced the number of proactive inspections it does each year by a third, from around 33,000 to 22,000;
- The Better Regulations Delivery Office has strengthened the Primary Authority scheme, so it is accessible to more small businesses, covers more regulations and has strengthened inspection plans [5].[23]

The Health and Safety of Great Britain/*Be Part of the Solution*, published in January 2014, set out HSE's mission − to prevent death, injury, and ill health to those at work and those affected by work activities [1].

The strategy covered 12 key areas:

- Resetting the direction;
- The pressures to improve;
- Everyone has a role;
- Investigations and seeking justice;
- Building competence;
- Involving the workforce;
- Creating healthier, safer workplaces;
- Customising support for SMEs;
- Avoiding catastrophe;
- Taking a wider perspective;
- Driving change for the better.

The HSE has previously recognised that employee consultation is a big part of achieving positive change in health and safety and this is picked up in the strategy document. Workplace research provides evidence to suggest that involving workers has a positive effect on health and safety performance. Where managers and workers develop a partnership based on trust, respect and cooperation, this creates a positive culture in which concerns can be raised and problems solved.

The HSE has moved away from the using the POPMAR (Policy, Organising, Planning, Measuring performance, Auditing and Review) model of arranging health and safety to a Plan, Do, Check, Act approach already discussed in the overview and highlighted in Chapter 8.

The move towards the Plan, Do, Check Act (see Figure 14.1) achieves a balance between the systems and behavioural aspects of management (compare with Figure 8.1a and PDCA in the context of quality management systems). It can be seen that good health and safety management is an integral part of good management generally, rather than as a stand-alone system [6].

Building on the success of previous strategies − In 2016 The HSE launched Helping Great Britain Work Well, and this built on the success of the previous strategies [7]. The strategy was built on six themes:

- Acting together: Promoting broader ownership of health and safety in Great Britain;
- Tackling ill health: Highlighting and tackling the costs of work-related ill health;
- Managing risk well: Simplifying risk management and helping business to grow;
- Supporting small employers: Giving SMEs simple advice so they know what they have to do;
- Keeping pace with change: Anticipating and tackling new health and safety challenges;
- Sharing our success: Promoting the benefits of Great Britain's world-class health and safety system.

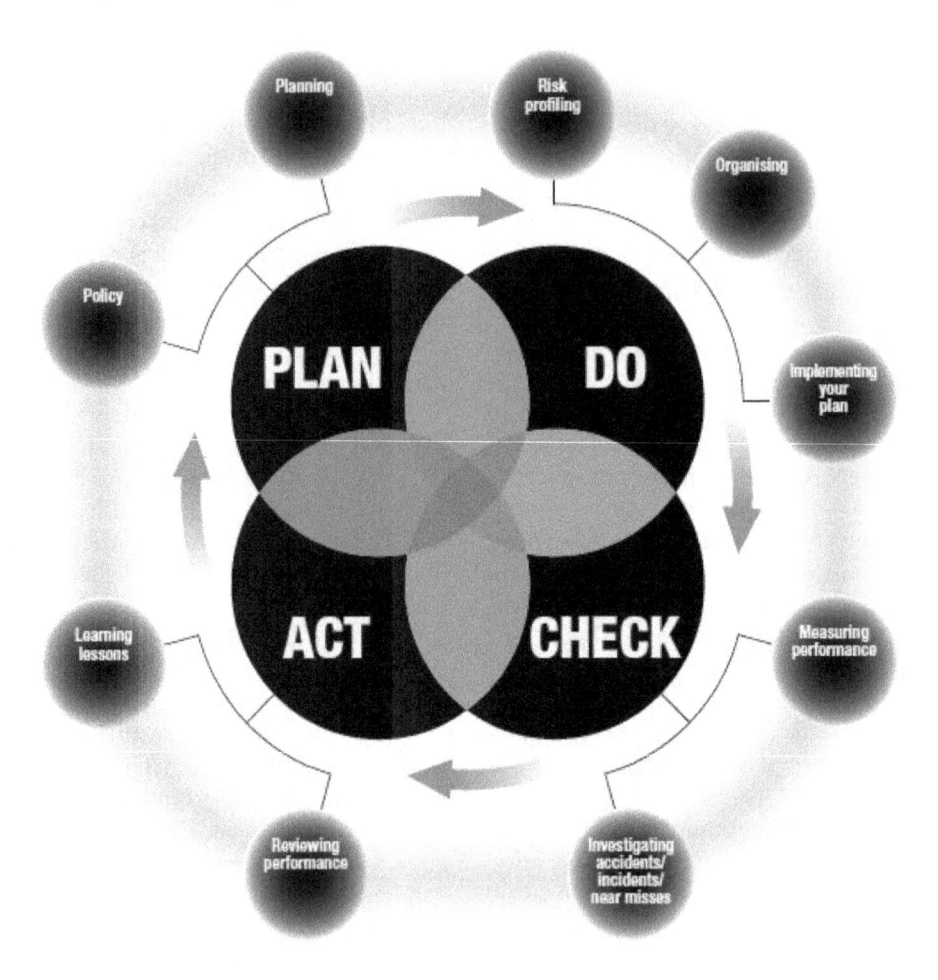

Figure 14.1 Plan, Do, Act, Check

The aim of the strategy is to ensure that all those involved play their part in improving health and safety in the workplace. This is not defined as one single group of individuals, but covers everyone from employers, employees, industry and trade bodies, supply chains, third-sector bodies, insurance and legal bodies, workers' representatives, professional institutions and government, as well as co-regulators and many others.

The strategy is also supported by two different focus areas: Sector Specific Plans and Health Priority Plans [2, 8].

Nineteen sectors, detailed in Table 14.1, are covered based on industry type and risk profile and the plan is broken down for each sector as follows:

- A plan covering its health and safety performance;
- The top three strategic priorities for the next three to five years;
- Actions that the HSE proposes to take.

Local authorities will continue their partnership with the HSE to ensure that duty holders manage their workplaces with due regard to the health and safety of their workforce and those affected by their work activities. This will also be done through the provision of advice

Table 14.1 HSE sectors

• Agriculture	• Explosives	• Manufacturing	• Quarries
• Bioeconomy	• Fairgrounds and theme parks	• Mines	• Sports and leisure
• Chemicals	• Film broadcasting, theatre and events	• Offshore energy	• Utilities
• Commercial consumer services	• Gas and pipelines	• Onshore oil and gas wells	• Waste and recycling
• Construction	• Logistics and transport	• Public services	

and guidance and the requirements of the law, supported by inspections, investigations and enforcement where required.

Supporting the work of the HSE, the International Organisation for Standardisation (ISO) has developed a number of accredited standards, which give guidance or requirements on good management. The ISO 9000 series focusses on quality management through operational process and can be used for health and safety within the workplace. It does this by explaining what standards should be met but without specifying the method that must be taken to meet them, allowing individual businesses to determine an appropriate method for them. One of the requirements of conformity to the ISO 9000 series is the documentation of the management system, including:

- Control of documents;
- Internal audit;
- Corrective action;
- Preventative action.

Conformity lends credibility to a company's efforts to manage health and safety.

Employment and accident trends

The Office for National Statistics outlines the following as of August 2021 [9] in the UK:

- There were 32.2 million people in work;
- The proportion of people aged from 16 to 64 in work (the employment rate), was 75.1%;

- There were 27.8 million employees (1.6 million are temporary);
- There were 4.2 million self-employed;
- There were 24.4 million people working full-time;
- There were 7.8 million people working part-time;
- Self-employed people equalled 4.2 million;
- There were 1.6 million unemployed people;
- The unemployment rate was 4.7% – the unemployment rate is the proportion of the economically active population (those in work plus those seeking and available to work) who were unemployed;
- There were 8.07 million people aged from 16 to 64 who were out of work and not seeking or available to work (known as economically inactive);
- The economic inactivity rate was 21.1%.

Some other key statistics for 2019/20 are as follows [10]:

- 1.6 million working people suffering from a work-related illness;
- 2,369 mesothelioma deaths due to past asbestos exposures (2019);
- 693,000 working people sustain an injury at work according to the Labour Force Survey;
- 65,427 injuries to employees reported under RIDDOR;
- 38.8 million working days lost due to work-related illness and workplace injury;
- £16.2 billion estimated cost of injuries and ill health from current working conditions (2018/19).

The number of fatal injuries to workers fluctuates year on year but the long-term trend continues downwards, with some stabilisation over the past five years [11].

- The number of workers fatally injured in 2020/2021 was equivalent to a rate of fatal injury of 0.43 per 100,000 workers;
- There were 60 members of the public fatally injured in accidents connected to work in 2020/21 (excluding railways-related incidents);
- Of the main industrial sectors, agriculture, forestry and fishing and waste and recycling have the highest rates.

The picture for non-fatal injuries is similar [12].

- For non-fatal injuries reported by employers under RIDDOR, the rate showed a generally downward trend but has been broadly flat in recent years.
- There were 65,427 non-fatal injuries to employees reported in 2019/20. The most common kinds of accident causing these injuries were slips and trips (29%), handling, lifting or carrying (19%), and being struck by moving objects (11%) and these make up 59% of all reported RIDDORs.

Enforcement [13]

- Across Great Britain, 325 cases were prosecuted for health and safety breeches in 2019/20 which is over 50% less than in 2012/13;
- 7,075 notices were served in 2019/20;
- £35.8m fines were issued but the average is down from 2018/19 (£110k versus £150k).

Enforcement strategies – Uniquely, in the UK, both the HSE and the local authorities are responsible for the enforcement of health and safety under the Health and Safety at Work Act 1974 and other relevant statutory provisions. The HSE takes the lead in this partnership and can direct local authority activity to reflect the needs of the health and safety system as a whole. The need for coordination and liaison between the HSE and local authorities to ensure a consistent approach has been a feature of the health and safety system since the 1972 Robens report into the legislative situation regarding health and safety at work recommended a continued role for local authorities. Various mechanisms exist to cement the relationship between the HSE and local authorities and the partnership approach they have adopted to their common responsibilities.

Memorandum of Understanding – The statement of commitment forms part of a wider picture of cooperation between central and local government. Agreed in 2009, the partnership arrangement sets out the commitment of local authorities to improve standards of partnership working with one overriding aim – to prevent the death, injury and ill health of those at work and those affected by work activities – through coordinated enforcement, joint planning and recognition of the national/local balance. It commits them to working together in way that avoids duplication of effort, in which experience and expertise is shared and where appropriate operational policy is developed and agreed. It also expresses specific support for the Health and Safety executive/Local Authorities Enforcement Liaison (HELA).

Enforcing health and safety – The allocation of enforcement responsibilities between the HSE and local authorities is set out in the Health and Safety (Enforcing Authority) Regulations 1998. These regulations maintain the broad allocation of service sector businesses to local authority enforcement, and manufacturing to the HSE.

The HSE Local Authority Unit (LAU) and HELA – HELA provides strategic oversight of the co-regulatory relationship between HSE and local authorities, with a view to maximising effectiveness, improving health and safety outcomes and ensuring delivery of the HSE strategy. HELA meets at least twice a year and is chaired jointly by a senior representative of the Health and Safety Executive

(HSE) and the Chair of the Local Authority Health and Safety Policy Forum.[24]

The (LAU) is an arm of the HSE dedicated to facilitate HELA. It is run by an HSE team, some of which have a local authority background, and ensures that both HSE and local authority perspectives are considered in development and innovation.

In practice this is achieved by:

- Maintaining an effective channel of communication between local authorities and the HSE;
- Aligning planning processes and timescales on major initiatives;
- Promoting consistency, transparency, targeting and proportionality of enforcement;
- Promoting and monitoring the effectiveness of HSE policies and objectives;
- Enhancing the standing of local authority health and safety enforcement among local authorities, the HSE, other government departments, business, trade unions and consumers.

The LAU's yearly work plan flows directly from HELA's strategy and the HSE's key objectives and ongoing aims.

Guidance – Section 18 of the Health and Safety at Work Act 1974 places a duty on local authorities to make adequate arrangements for enforcing relevant statutory provisions. The same section enables the HSE to issue guidance to local authorities, to which they must have regard. There have been very few occasions when the HSE has exercised its right to issue such guidance. Its general approach is to work by consensus, and this is no less vigorously applied in its dealing with local authorities. The HSE believes that the cooperation and goodwill achieved by this approach may be undermined if the HSE relies too heavily on its statutory guidance for the support of local authorities in pursuit of its objectives. The Section 18 Standard sets out the requirements with which HSE and LAs will eventually be obliged to comply in making 'adequate arrangements for

enforcement'. It was given legal effect both as an HSC direction to HSE under section (4)(b) of the Health and Safety at Work etc. Act 1974, and as HSC guidance to LAs under section 18(4)(b) (section 18 guidance) [14].

The Enforcement Policy based on the principles of proportionality, consistency, targeting and transparency was discussed in Section 1.

Good enforcement practice – The LA National Enforcement Code sets out the risk-based approach to targeting health and safety interventions to be followed by LA regulators [15].

The Code provides a principle-based framework that recognises the respective roles of business and the regulator in the management of risk, concentrating on four objectives:

- Clarifying the roles and responsibilities of business, regulator and professional bodies;
- Outlining the risk-based approach to regulation that LAs should adopt;
- Setting out the need for training and competence of LA and Health and Safety regulators;
- Explaining the arrangements for collection/publication of LA data and peer review to give assurance on meeting the requirements of the Code.

The Health and Safety Executives Policy Statement on Enforcement – Is in accordance with the Regulators Compliance Code and the regulatory principles required under the Legislative and Regulatory Reform Act 2006. It sets out the general principles and approach which the health and safety enforcing authorities (mainly HSE and local authorities) are expected to follow. All local authority and HSE staff who take enforcement decisions are required to follow HSEs Enforcement Policy.

Section 18 of the policy states that the HSE expects enforcing authorities to have systems for deciding which inspections, investigations or other regulatory contacts should take priority according to the nature and extent of risks posed by a duty holder's operations. The duty holder's management competence is important, because a relatively low-hazard site poorly managed can entail greater risk to workers or the public than a higher-hazard site where proper and adequate risk control measures are in place. Certain very high-hazard sites will receive regular inspections so that enforcing authorities can give public assurance that such risks are properly controlled [15].

Training and competence of Local Authority staff – Competence of enforcement staff is an essential ingredient in the delivery of appropriate enforcement. The HSE considers that, as part of an enforcing authority's duty to make adequate arrangements for the enforcement of relevant statutory provisions, each local authority must satisfy itself that inspectors authorised to exercise some or all their enforcement powers meet standards of competence set out in the HSE's guidance.

Lead Authority Partnership Scheme (LAPS) – This was an extension of the LOPP scheme in which businesses with multiple sites or national membership were able to form a partnership with a local authority in order to raise the company's health and safety management and encourage consistency from all local authorities dealing with that company. Although closed to new entrants the LAPS support continued until April 2011.

Better Regulation Delivery Office (BRDO)[25] – This became part of Regulatory Delivery in 2016 then replaced by the Office for Product Safety and Standards. The Primary Authority scheme is intended to ensure consistent regulation, improving the professionalism of front-line regulators.

Primary Authority – Through Primary Authority, local authorities and fire and rescue authorities are providing businesses with robust and reliable regulatory advice [16]. This supports growth by enabling them to invest with confidence in products, practices and procedures, knowing that there is endorsement of the resources they devote to compliance. Participating businesses cover a wide variety of sectors and over half of them have less than 50 employees. In 2015 there were 6733 businesses in partnerships and 32 trade

associations in partnerships and 155 local authorities in partnerships [17] and [18] and the number of businesses continues to grow.[26]

The scheme gives companies the right to form a statutory partnership with a single local authority, who will then provide robust and reliable advice for other councils to take into account when inspecting or investigating non-compliance. Before other councils impose sanctions, including formal notices and prosecutions, they must contact the Primary Authority to determine whether their action is consistent with previous advice. This is exempt only where persons are at immediate risk.

For local authority the benefits include increased credibility with both business and other authorities, avoidance of duplication of effort across authorities, and the confidence that business gains will encourage investment, which in turn, reduces the burden on the authorities. They retain the ability to make the decisions but with the national information that enables a proportionate response.

For the business the benefits include clarity on compliance, detailed and tailored advice, and confirmation through one regulatory body that processes are compliant and proportionate responses from other authorities. This enables them to make the right investment to manage health and safety effectively and quickly but also enables them to reduce the amount of repeat information requests and checks.

Managed well the Primary Authority relationship can provide cost and time benefits to all parties and encourages an open attitude to business performance.

Earned recognition – the principle of earned recognition suggests that businesses who are broadly compliant should be exempt from routine inspections. This could be applied where companies file audit reports that demonstrate compliance is being achieved.

Permitting compliant business to self-regulate allows local authorities to concentrate efforts on less compliant business. They would retain a degree of monitoring through complaints and accidents, which allows for review of the earned autonomy status should the situation change.

For the compliant business the ability to self-regulate offers a financial benefit as less time is spent dealing with routine inspection visits and more time can be spent on the business objective and maintaining a safe workplace.

Working groups and strategies for engagement

HSE working groups – As mentioned, the HSE sets up working groups where a health and safety topic arises that requires discussion and the development of a plan to increase awareness. Previous topics have included:

- Slips, trips and falls;
- Asbestos;
- Vehicle loads.

Working groups will contain HSE divisions with interest in the subject, together with external bodies from both enforcement and private backgrounds who can provide expert opinion.

British Retail Consortium and Trade Association Groups – The British Retail Consortium (BRC) is the lead trade association for UK retail industry. It develops and manages policy through Policy Action Groups (PAGs) which are composed of key members from within industry. The PAGs meet regularly to discuss and respond to government policies to ensure that they are not burdensome to the retail sector. The PAGs are broken down into working groups focussing on specific issues within the PAG. As well as the PAG the BRC holds regular meetings with members on a number of topics, including health and safety issues, helping industry to shape the future of regulation.

Instruction and supervision – Are keys to good health and safety, whether from an enforcement perspective for local authority or from a management perspective for business. The Health and Safety at Work etc. Act 1974

(HASWA) requires businesses to provide their employees with information, instruction, training and supervision to ensure the health and safety of their employees. The HSE recognises this importance and offers several pieces of guidance to assist businesses in creating competent employees operating in a safe culture. Similarly, the HSE provides training and support to local authorities on a number of topics to improve skill levels, drive consistency and support the HSE's programmes. Joint training programmes keep the partnership alive and give local authorities access to additional areas of expertise such as accident investigations. Local authorities can pass back information on working with local businesses. Competent supervision by experienced people in both sectors then reinforces the information received during training.

Fit3 – This was a strategic delivery programme developed by HSE and targeted a number of sectors and within each sector a number of priority areas. HSE created a plan to raise awareness on each issue through activities during inspections and investigations, and through publications and briefings directed at the specific sector. For example, within the retail sector the focus topics included slips and trips, moving goods safely, working at height, manual handling and occupational asthma.

Institute of Occupational Safety and Health (IOSH) are another source of health and safety advice to businesses and individuals. Like the HSE they draw up guidance that assists business in achieving compliance through risk identification, competence and training. Their policies are developed through research and consultation with members and focus on addressing key issues within health and safety.

British Safety Council/Royal Society for the Prevention of Accidents – The British Safety Council is one of the leading health and safety organisations with a mission to keep people healthy and safe at work. They provide training courses at all levels on health and safety topics and support the work of the HSE through information, qualifications and audits.

Similarly, the Royal Society for the Prevention of Accidents carries out events, training, and campaigns to raise awareness. It will give accreditation to businesses who can demonstrate good health and safety management and improving accident results.

Section 2 Notes

22 See https://www.hse.gov.uk/stress/
23 The Primary Authority Handbook can be accessed at https://www.gov.uk/government/uploads/system/uploads/attachment_data/file/421841/13-1310-pa-handbook.pdf
24 More information about HELA is available at https://www.hse.gov.uk/lau/hela-structure.htm
25 See https://www.gov.uk/guidance/business-regulation-guidance-and-tools and https://www.gov.uk/guidance/local-regulation-primary-authority
26 There are some 1994 partnerships according to https://primary-authority.beis.gov.uk/par

Section 2 References

[1] HSE. (2014) *The Health and Safety of Great Britain: Be Part of the Solution*, HSE Books, Sudbury, Suffolk. http://www.hse.gov.uk/strategy/document.htm.

[2] HSE. (2016) *HSE Health Priority Plans: Health and Safety Executive*. https://www.hse.gov.uk/aboutus/strategiesandplans/health-and-work-strategy/index.htm.

[3] DWP. (2010) *2010 to 2015 Government Health and Safety Reform*. Policy Paper. https://www.gov.uk/government/policies/improving-the-health-and-safety-system.

[4] HSE. (undated) *National Local Authority Enforcement Code: Health and Safety at Work, England, Scotland & Wales*, Health and Safety Executive. http://www.hse.gov.uk/lau/la-enforcement-code.htm.

[5] BRDO. (2015) *Primary Authority and Growth*. https://www.gov.uk/government/uploads/system/uploads/attachment_data/file/446550/pa-and-growth.pdf.

[6] HSE. (2013) *Successful Health & Safety Management (HSG 65)*, HSE Books, Sudbury, Suffolk.

[7] HSE. (2016) *Helping Great Britain Work Well*, HSE Books, Sudbury, Suffolk. https://www.hse.gov.uk/strategy/assets/docs/hse-helping-great-britain-work-well-strategy-2016.pdf.

[8] HSE. (2016) *HSE Sector Plans: Health and Safety Executive*. https://www.hse.gov.uk/aboutus/strategiesandplans/sector-plans/index.htm.

[9] ONS. (2021) *Labour Market Overview*, Office for National Statistics, London, August.

[10] HSE. (2021) *Health and Safety at Work Summary Statistics for Great Britain 2020: Health and Safety Executive and Office for National Statistics*. Crown Copyright. https://www.hse.gov.uk/statistics/overall/hssh1920.pdf.

[11] HSE. (2021) *Workplace Fatal Injuries in Great Britain, 2021: Health and Safety Executive*. Crown Copyright. https://www.hse.gov.uk/statistics/pdf/fatalinjuries.pdf.

[12] HSE. (2021) *Non-Fatal Injuries at Work in Great Britain: Health and Safety Executive*. https://www.hse.gov.uk/statistics/causinj/index.htm.

[13] HSE. (2021) *Enforcement in Great Britain: Health and Safety Executive*. https://www.hse.gov.uk/statistics/enforcement.htm.

[14] HSE. (undated) *UK National Local Authority (LA) Enforcement Code*. http://www.hse.gov.uk/lau/national-la-code.pdf; supplementary guidance. http://www.hse.gov.uk/lau/supplementary-guidance.pdf.

[15] HSE. (2009) *Enforcement Policy Statement, HSE41(rev1)*, Crown Copyright. http://www.hse.gov.uk/pubns/hse41.pdf.

[16] OPSS. (2017) *Primary Authority Overview, Office of Product Safety and Standards*. https://assets.publishing.service.gov.uk/government/uploads/system/uploads/attachment_data/file/913514/pa-overview-2019A.pdf.

[17] Better Regulation Delivery Office (BRDO). (2015) *Primary Authority Nurturing Partnerships for Growth*, BIS, Birmingham.

[18] Better Regulation Delivery Office (BRDO). (2015) *Primary Authority and Growth*, BIS, Birmingham. https://assets.publishing.service.gov.uk/government/uploads/system/uploads/attachment_data/file/407983/pa-and-growth.pdf.

SECTION 3: THE WORKING ENVIRONMENT

Health safety and welfare

As far as enforcing officers are concerned, control over workplace health and safety can now be thought of as being embodied in the Health and Safety at Work, etc. Act 1974 and in the Workplace (Health, Safety and Welfare) Regulations 1992.[27] The regulations bring together the legislative controls that existed prior to the 1974 Act. Detailed guidance on the regulations is contained in the Approved Code of Practice [1] and on the HSE website.

The regulations contain a general requirement for every employer and others to ensure that any workplace under their control, and where their employees work, complies with any applicable requirement. This duty is extended to those who have control of a workplace in connection with any trade, business or other undertaking (whether for profit or not).

Maintenance of workplace, equipment, devices and systems

Regulation 5 requires that the workplace and certain equipment, devices and systems be regularly maintained in an efficient state, in efficient working order and in good repair. The equipment etc. covered by this regulation is any equipment etc. in which a fault is liable to result in a failure to comply with any of the regulations.

The Approved Code of Practice [1] suggests that any system of maintenance should ensure that:

- Regular maintenance (including, as necessary, inspection, testing, adjustment, lubrication and cleaning) is carried out at suitable intervals;
- Any potentially dangerous defects are remedied, and that access to the defective equipment is prevented in the meantime;
- Regular maintenance and remedial work is carried out properly;

- A suitable record is kept to ensure that the system is properly implemented and to assist in validating maintenance programmes.

The frequency of any maintenance should clearly depend on the nature of the equipment and its age. Regard should be had to advice from, for example, the manufacturers and the HSE.

Examples of equipment covered by this regulation include emergency lighting, fencing, fixed equipment used for window cleaning, anchorage points used for safety harnesses and escalators.

Ventilation – By virtue of Regulation 6, every enclosed workplace must be ventilated by a sufficient quantity of fresh or purified air. This requirement does not extend to work carried on in confined spaces where breathing apparatus may be necessary. In most cases, openable windows or other openings will provide satisfactory ventilation. However, if found to be necessary, mechanical ventilation should be provided. It is important that such systems be regularly maintained, particularly because of the possible risk of Legionnaires' disease.

Temperature in indoor workplaces – Many working environments can be uncomfortable because of excessive heat or cold. Under such conditions, it is possible for workers to suffer from heat or cold stress. Regulation 7 requires that, during working hours, the temperature in all workplaces inside buildings should be reasonable. The Approved Code of Practice [2] suggests a sedentary minimum temperature of 16°C, and 13°C where there is severe physical effort. The only exception to this rule is where processes such as cold storage require lower temperatures. In such circumstances the employer should take all reasonable steps to provide protective clothing and sufficient rest periods for employees.

The regulations also contain a general requirement for employers to provide a sufficient number of thermometers to enable employees to ascertain the temperature.

Gill [2] argues that to obtain a correct assessment of the thermal environment, four parameters need to be measured together:

- The air-dry bulb temperature;
- The air wet bulb temperature;
- The radiant temperature;
- The air velocity.

A whirling hygrometer can be used to measure wet and dry bulb temperature, while a simple globe thermometer can be used to measure the radiant temperature. To measure the air velocity either a kata thermometer or an airflow meter can be used. A detailed description of how to use these measurements to assess the thermal environment is found in Gill [3]. In addition, 'thermal comfort meters' are now available. These not only carry out many of these measurements automatically but are also capable of being linked to recorders and computers to enable rapid analysis of findings.

Lighting – Adequate lighting is an important prerequisite for ensuring the safety and comfort of people at work. While natural light is the most desirous form of lighting, in many work situations this has to be either supplemented or completely replaced by some form of artificial lighting.

The Regulations require that suitable and sufficient lighting is provided. In addition, emergency lighting must be provided where failure of any artificial lighting could pose a risk to health and safety.

The amount and type of lighting provided in any work situation quite clearly depends upon the type of work being carried out there. The HSE provides guidance on different lighting levels on their website. The Chartered Institute of Building Services Engineers has also produced a code for interior lighting [4] to which EHPs should refer when considering if the lighting is suitable and sufficient.

Measurement of lighting levels can be simply carried out using a proprietary light meter. Where possible a light meter with a remote sensing device should be used, as this removes the risk of readings being affected by shadows from the operator.

Cleanliness – Every workplace, and the furniture, furnishings and fittings contained in it, are to be kept sufficiently clean. In addition, the surfaces of the floors, walls and ceilings of all workplaces shall be capable of being kept sufficiently clean, and all waste material must be kept in suitable receptacles.

It is important to bear in mind that only an adequately planned cleaning programme will satisfy this general requirement.

Space – It is a requirement that every room that people work in shall have sufficient floor area, height and unoccupied space for the purposes of health, safety and welfare.

The Approved Code of Practice [1] establishes that personal space should be at least 11 m^3. In calculating personal space, it is assumed that there is a minimum notional ceiling height of 3 m.

Workstations and seating – Workstations must be arranged so as to be suitable for any person who is likely to work there, and for any work likely to be done there. For workstations that are located outside, there must be protection from adverse weather conditions. These workstations must also be arranged so as to enable a swift means of escape in cases of emergency.

Where necessary, a suitable seat must be provided. Suitability is defined as being not only suitable for the operation being carried out, but also for the person it is provided for. Where necessary, a footrest must also be provided [5].

Workstations where visual display units, process control screens, microfiche readers and similar display units are used are covered by the Health and Safety (Display Screen Equipment) Regulations 1992[28] (see later).

Conditions of floors and traffic routes – This covers another of the priority areas of work set for enforcement officers with regard to the prevention of slips, trips and falls. Floors and traffic routes should be of sound construction and must be constructed so as to be suitable for the purpose for which they are used. A traffic route is defined in the regulations as: 'a route for pedestrian traffic, vehicles or both and includes any stairs, staircase, fixed ladder, doorway, gateway, loading bay or ramp'.

The regulations specify that there should be no holes, slope, unevenness or slipperiness that could pose a risk to health or safety. In addition, such routes should be kept free from obstruction, and, where necessary, be sufficiently drained.

Handrails should be provided where necessary. Open sides to staircases should be protected by an upper rail at 900 mm or higher, and a lower rail.

Falls or falling objects – In many workplace situations, for example, construction sites, factories or offices, work has to take place at height. Such work will involve activities such as maintenance operations, installation of plant and equipment or window cleaning. The risk of falling with often fatal results is a common hazard of such operations and because of reported accident statistics, this has also been set as a priority topic for enforcement officers to pay attention to. There are two primary causes of this accident: the first relates to the means of access to the work situation, and the second is the system of work adopted once that working position has been reached.[29]

In many cases employers will have considered the risks to employees of working at height. Exercises such as risk assessments, safety surveys and safety audits will reveal hazards, and meetings of safety committees and discussion with safety representatives will also lead to the establishment of safe systems of work.

When considering the risks of working at height, factors such as the provision of safety harnesses and belts, including the necessary anchorage points for their safe use, and protective clothing such as suitable head covering, gloves and safety footwear for employees need to be borne in mind.

The Working at Height Regulations 2005[30] have superseded most of the provisions relating to falls within the original Regulation 13; however Regulation 13 does still extend to the protection of persons from being struck by a falling object likely to cause personal injury. In addition, every tank or pit shall be securely fenced, and every traffic route, where there is a risk of falling into a dangerous substance, shall also be securely fenced. The regulations define a dangerous substance as:

- Any substance likely to scald or burn;
- Any poisonous substance;
- Any corrosive substance;
- Any fume, gas or vapour likely to overcome a person;
- Any granular or free-flowing solid substance or any viscous substance that, in any case, is of a nature or quantity which is likely to cause danger to any person.

The Approved Code of Practice [1] also suggests the precautions taken to ensure the minimisation of risk to health within racking systems, especially within warehouses. In addition, the HSE has published guidance on warehouse stacking [6].

Windows, transparent or translucent doors, gates and walls – All windows, doors, gates, walls and partitions glazed wholly or partially so that they are transparent or translucent must, where necessary for health and safety, be of safe material and be appropriately marked. The Code of Practice [1] defines safe materials as:

- Materials that are inherently robust, such as polycarbonates or glass blocks;
- Glass that, if it breaks, breaks safely;
- Ordinary annealed glass that meets the thickness criteria laid down in the Approved Code of Practice.

As an alternative to using safe materials, transparent or translucent surfaces may be protected by the use of screens or barriers.

Openable windows, skylights or ventilators should not expose people opening them to a risk to their health and safety, and when opened such windows etc. should not expose people to a risk to their health or safety.

Finally, it is important that windows and skylights be kept clean in order to be effective but in advice that should be given to those employed in any work at height including the cleaning of windows, attention should be drawn to the various sector-specific guidance given by the HSE and the contents of the British Standard 8213–1 (2004) [7] and contractors should be able to demonstrate knowledge of, and compliance with the standard.

Organisation of traffic routes etc. – This is a further priority area of work for enforcement officers due to the seriousness of accidents involving pedestrians and vehicular traffic. The Regulations also address the safe circulation of pedestrians and vehicles in the workplace. There must be sufficient separation between vehicles and pedestrians. All routes must be adequately signed.

All doors and gates are to be suitably constructed to include the fitting of any necessary safety devices. In particular, the regulations require that:

- Any door or gate has a device to prevent it coming off its track during use;
- Any upward opening door or gate has a device to prevent it falling back;
- Any powered door or gate has suitable and effective features to prevent it causing injury by trapping any person;
- Where necessary for reasons of health or safety, any powered door or gate can be operated manually unless it opens automatically if the power fails;
- Any door or gate that is capable of opening by being pushed from either side is of such a construction as to provide, when closed, a clear view of the space close to both sides.

All escalators and moving walkways function safely, be equipped with any necessary safety device and be fitted with sufficient emergency stop controls. There is guidance issued by the HSE [8].

Welfare facilities – Regulations 20–25 deal with the provision of sanitary conveniences, washing facilities, adequate wholesome drinking water, suitable and sufficient facilities for changing clothing and suitable and sufficient facilities for rest and to eat meals. The Approved Code of Practice [1] gives details about the minimum number of sanitary conveniences etc.

Smoking – Smoke-free laws are enforced across Scotland, Wales, NI and England. The laws make all enclosed public places and workplaces smoke free to protect employees and the public from the harmful effects of second-hand smoke. The smoke-free law is enforced by local authorities and failure to comply is a criminal offence with enforceable penalties and fines.

Employers, managers and those in control of premises have a legal responsibility to display no-smoking notices and take reasonable steps to ensure that staff, customers, members and visitors are aware of the new law and do not smoke in buildings. The law also applies to vehicles used for business purpose such as light and heavy goods vehicles and public transport. The law currently does not apply to personal cars.

Display Screen Equipment (DSE)

DSE is covered by a specific set of regulations, the Health and Safety (Display Screen Equipment) Regulations 1992[31] as amended by the Health and Safety (Miscellaneous Amendment Regulations 2002.[32] The regulations implement a European Directive (90/270/EEC) on DSE.

There is some degree of overlap between this set of regulations and the Management of Health and Safety at Work Regulations 1999, particularly with regard to issues such as risk assessment. The guidance on the regulations [9] points out that employers are required to comply with both the specific requirements of the DSE regulations, and the general requirements of the management regulations. However, carrying out a risk assessment on a workstation under the DSE regulations will also satisfy the requirements of the management regulations, but only for that workstation.

Health effects of display screen equipment – DSE has been blamed for a whole range of adverse health effects. One of the most significant is that of musculo-skeletal injury.

This has also been established as a priority enforcement action topic for local authorities. There is also some epidemiological evidence to suggest that DSE can have some impact on other aspects of health status. These effects are linked to the visual system and working posture and include:

- Repetitive strain injury, or work-related upper limb disorders;
- Eye or eyesight defects;
- Photosensitive epilepsy;
- Fatigue and stress.

For a detailed consideration of the health effects of DSE, attention is drawn to the HSE website and the International Labour Office [14].

The requirements of the Display Screen Equipment Regulations – DSE is defined as 'any alphanumeric or graphic display screen, regardless of the display process involved'. An operator or user is someone who 'habitually uses display screen equipment as a significant part of his or her work'. The regulations suggest that generally an individual will be classified as a user or operator if most or all of the following apply:

- The individual depends on the use of DSE to do the job as alternative means are not readily available for achieving the same results;
- The individual has no discretion about using or not using the DSE;
- The individual needs significant training and/or particular skills in the use of DSE to do the job;
- The individual normally uses DSE for continuous spells of an hour or more at a time;
- The individual uses DSE in this way more or less daily;
- Fast transfer of information between the user and screen is an important requirement of the job;
- The performance requirements of the system demand high levels of attention and concentration by the user, for example, where the consequences of error may be critical.

The ACOP gives the following examples of users:

- Word-processing operators;
- Secretaries or typists;
- Data input operators;
- Journalists;
- Air traffic controllers.

Every employer should carry out a suitable and sufficient analysis of workstations used by users or operators, to assess the possible risks to health and safety. The assessment must be reviewed if there is a significant change or if the assessment is no longer valid. The employer, having identified the risks, is then required to reduce them to the lowest extent reasonably practicable.

All workstations should meet the detailed requirements of the schedule to the Regulations. The schedule deals with issues such as the display screen, the keyboard, the work desk or work surface, the work chair, space requirements, lighting, reflections and glare, noise, heat, radiation, humidity and the interface between the computer and the operator/ user. The guidance document gives detailed advice on the interpretation of the schedule.

Employers are required to plan the activities of users so as to ensure that sufficient rest periods are provided. The guidance document suggests that short frequent breaks are more satisfactory than longer breaks, for example, 5–10 minutes for every 50–60 minutes of work.

There is a general provision made in the Regulations for eye and eyesight tests to be made available on request and at regular intervals. If any 'special corrective appliances' (generally speaking these are spectacles/ lenses) are required, then currently these have to be paid for by the employer.

It is a requirement for the provision of health and safety training for DSE users. In addition, employers have to make available for users' information about breaks, eye and eyesight tests and training, both initially and when the workstation is modified.

Section 3 Notes

27 S.I. 1992, No 3004.
28 S.I. 1992, No 2792.
29 See http://www.hse.gov.uk/SLIPS/preventing.
    htm for the slip prevention model.
30 SI 2005 No 735.
31 SI 1992 No 2792.
32 SI 2002 No 2174.

Section 3 References

[1] HSE. (2013) *Workplace Health, Safety and Wel-
    fare – Approved Code of Practice to the Workplace
    (Health Safety and Welfare) Regulations 1992,
    L24.* http://www.hse.gov.uk/pubns/priced/
    l24.pdf (Accessed September 2021).
[2] Gill FS. (2003) Workplace pollution, heat and
    ventilation. In *Safety at Work*, Channing J, Rid-
    ley, J (Eds.), Butterworths, London, 6th ed.
[3] Gill FS. (1995) The thermal environment. In
    Occupational Hygiene, Harrington MJ, Gardiner
    K (Eds), Blackwell, Oxford.
[4] Chartered Institute of Building Services
    Engineers (CIBSE). (2002) *Code for Lighting*,
    CIBSE, London.
[5] Health and Safety Executive. (1994) *A Pain in
    Your Workplace – Ergonomic Problems and Solu-
    tions HS(G)121*, HSE Books, Sudbury, Suffolk.
[6] Health and Safety Executive. (1992) *Health
    and Safety in Retail and Wholesale Warehouses
    HS(G)76*, HSE Books, Sudbury, Suffolk.
[7] British Standards Institution. (1991) *BS 8213
    Part 1 Windows, Doors and Rooflights: Code of
    Practice for Safety in Use and During Cleaning of
    Windows*, BSI, London.
[8] Health and Safety Executive. (1984) *Safety in
    the Use of Escalators PM34*, HSE Books, Sud-
    bury, Suffolk.
[9] HSE. (2003) *Work with Display Screen Equip-
    ment.* Guidance on Regulations, L26 2nd
    Edn, HSE, Crown Copyright. http://www.
    hse.gov.uk/pubns/priced/l26.pdf (Accessed 8
    January 2015).
[10] International Labour Office. (1994) *Visual
    Display Units: Radiation Protection Guidance*,
    ILO Publications, Geneva. http://www.ilo.
    org/wcmsp5/groups/public/-ed_protect/-
    protrav/-safework/documents/publication/
    wcms_107821.pdf.

SECTION 4: WORKPLACE HAZARDS

Slips, trips and falls

Employers have a duty to ensure that slip and
trip risks are controlled and therefore minimise
the risk of people slipping or tripping in order
to fulfil their general duty under the Health
and Safety at Work etc. Act 1974. Employers
must ensure that employees, and anyone else
who may be affected by their work, are kept
safe from harm and their health is not affected.

Slips and trips can result from a combination of a number of different factors or may
be caused by just one. Some of the most
common are discussed next.

Floor surface – Consider the suitability of
flooring for the type of work activity that will
be taking place on it and also:

- Adequacy of fitting and correct maintenance regime – no trip hazard from holes,
  curled up edges, etc.;
- The correct cleaning programme to
  maintain slip resistance properties;

- Any changes of level avoided (e.g., raised
  platforms), or where these cannot be
  avoided, they must be highlighted and
  adequately illuminated;
- Stairs must have a suitable handrail and
  steps should be of equal height and width;
- Materials used for floor surfaces should be
  suitable for the environment that they are
  intended and where necessary have good
  slip resistance.

Contamination/Obstacles – Once a floor
becomes contaminated it is the most likely
cause of a slip, therefore preventing surface
contamination can reduce or even eliminate
the risk of slipping.

Contamination includes things such as
rainwater, oil, grease and wrapping. But in
its most general terms contamination is anything that ends up on the floor and can be
a by-product of work processes or adverse
weather conditions.

Elimination of contamination is the first option. Repairing a leaking machine or changing a system of work could achieve this for example. If the contamination can't be eliminated, then it should be controlled. As an example, having large mats to dry shoes at the entrance to the building or having drip trays for leaks could do this.

If contamination still occurs after these controls, then there must be an effective cleaning schedule in place to remove the contamination as quickly as possible.

Poor housekeeping can contribute to a large number of all trip accidents. Therefore, improving housekeeping could help eliminate many accidents. This can be done by:

- Ensuring that the workplace has suitable walkways throughout;
- Keeping all walkways clear of obstructions such as trailing wires;
- Helping employees keep their work areas tidy by making sure they have adequate storage space.

Encourage people to clean up after themselves by providing bins and other cleaning materials.

Cleaning – Cleaning needs to be carried out in every workplace and affects all employees, so everyone must be responsible for cleaning up their own spillages, etc.

However, there also needs to be a cleaning schedule in place which identifies problem areas and manages the risks appropriately. Cleaners need to be advised on the most effective cleaning methods for the surface that they are cleaning as well as why they need to clean certain areas at certain times.

Other things to consider for effective cleaning are:

- Using the correct cleaning product in the correct concentrations;
- Following the guidelines on the manufacturer's instructions, such as time period for the cleaning product to be left on the surface;

- Maintaining cleaning equipment;
- Spot cleaning where possible;
- Identifying areas where certain items are more likely to cause an issue and increase the frequency of cleaning in these priority areas.

The actual process of cleaning can create a slip or trip hazard, either from damp floors as a result of mopping or from trailing wires attached to a vacuum cleaner. Temporary signs should be used to make people more aware of the hazards and to reduce the chances of either people walking into the cleaning area or being made aware that a cleaning operation is in process if the area cannot be cordoned off.

People – Individuals' behaviour can affect the health and safety of a workplace, including slips and trips. If people have a positive attitude to health and safety, then it can reduce the chances of people slipping or tripping at work. For example, people will be more likely to clean up a spillage if they have a positive attitude towards it and understand the benefits of cleaning the spillage as soon as possible.

The physical attributes of a person can also affect the chances of them slipping or tripping. For example, if someone has poor vision then it could reduce the chance of them seeing a tripping hazard such as a loose wire and increase the likelihood of them tripping over it.

Environment – Environmental issues such as lighting and the weather can increase the risk of slips and trips. For example, too much light on a shiny floor could create a glare which could stop people from seeing trip hazards. Cold weather could also cause snow, frost and ice, which may create a slippery surface.

Footwear – This can play a big role in either contributing to a slip or preventing one. This is even more of an issue if you do not have control over the type of footwear people are wearing, for example if the public can walk through the workplace. In this situation it makes all other controls such as flooring and cleaning even more important.

Having a sensible footwear policy in workplaces where management has control over footwear will help reduce the risk. Some workplaces that are at high risk from slips or trips may require employees to wear a certain type of footwear, such as slip-resistant, as part of their personal protective equipment.

Manual Handling/Manual Handling Assessment Chart (MAC)

The Manual Handling Operations Regulations 1992[33] require that employers manage the risks caused by manual handling. In the guidance on the regulations issued by the HSE [1], it is claimed that more than one-third (38%) of the accidents reported to the enforcing authorities each year are associated with manual handling (i.e., the transporting or supporting of loads by hand or bodily force). Although manual handling accidents are rarely fatal, the majority are major injuries such as multi-skeletal disorders.

Sprains and strains arise from the incorrect application or prolongation of force. Poor posture and excessive repetition of movement can be important factors in the development of such injuries. The main sites of injuries are the back (45%), finger and thumb (16%) and arm (13%). It should be noted that the injuries may occur over a long period.

The Regulations establish a clear hierarchy of measures which, in the first case, requires that employers should, as far as reasonably practicable, avoid the need for employees to undertake any manual handling operations at work that involve a risk of injury. For example, can the activity be brought to the load, instead of moving the load to the activity? This may require the redesigning, automation or mechanisation of tasks but this is not an absolute duty; instead employers must adopt reasonably practicable solutions. To comply with this requirement, a 'general assessment' carried out under the Management of Health and Safety at Work Regulations 1999[34] must be carried out. However, under these regulations there is an additional, specific duty to carry out risk assessments of all manual handling operations where it is not reasonably practicable to avoid them.

Where manual handling operations cannot be avoided, the employer must take appropriate steps to reduce the risk of injury to the lowest level reasonably practicable.

The HSE have produced guidance for employers to assist in the management of manual handling, and have suggested lifting guidelines for males and females, depending on where the lift is carried. The guidelines are not limits, but instead illustrate the point where the risk of injury is likely to increase. The flowchart in Figure 14.2 illustrates how to follow the regulations.

To simplify the assessment process, the Manual Handling Assessment Chart Tool (MAC) has been developed. The tool can be used to assess the risks posed by lifting, carrying and team manual handling activities, based on observation of the task as it occurs in the workplace [2]. The MAC incorporates a numerical and a colour coding score system to highlight high risk manual handling tasks using both physical and environmental factors. See Figure 14.3 for an example of one of the graphs from MAC and Figure 14.4 for the MAC score sheet [2].

Workplace transport

In any workplace where you have moving vehicles there is the potential to cause injury to people or damage to property. Workplace transport safety, as with all hazards in the workplace, must be risk assessed and suitable controls put in place to reduce those risks. Workplace transport covers moving vehicles used in a work setting, including lift trucks, cranes, cars, and delivery vehicles amongst others. It does not extend to transport on a public highway, air, rail or water. Risk assessments should be completed to assess the need for suitable controls, as detailed in the Management of Health and Safety at Work Regulations 1999 and taking account of the Lifting

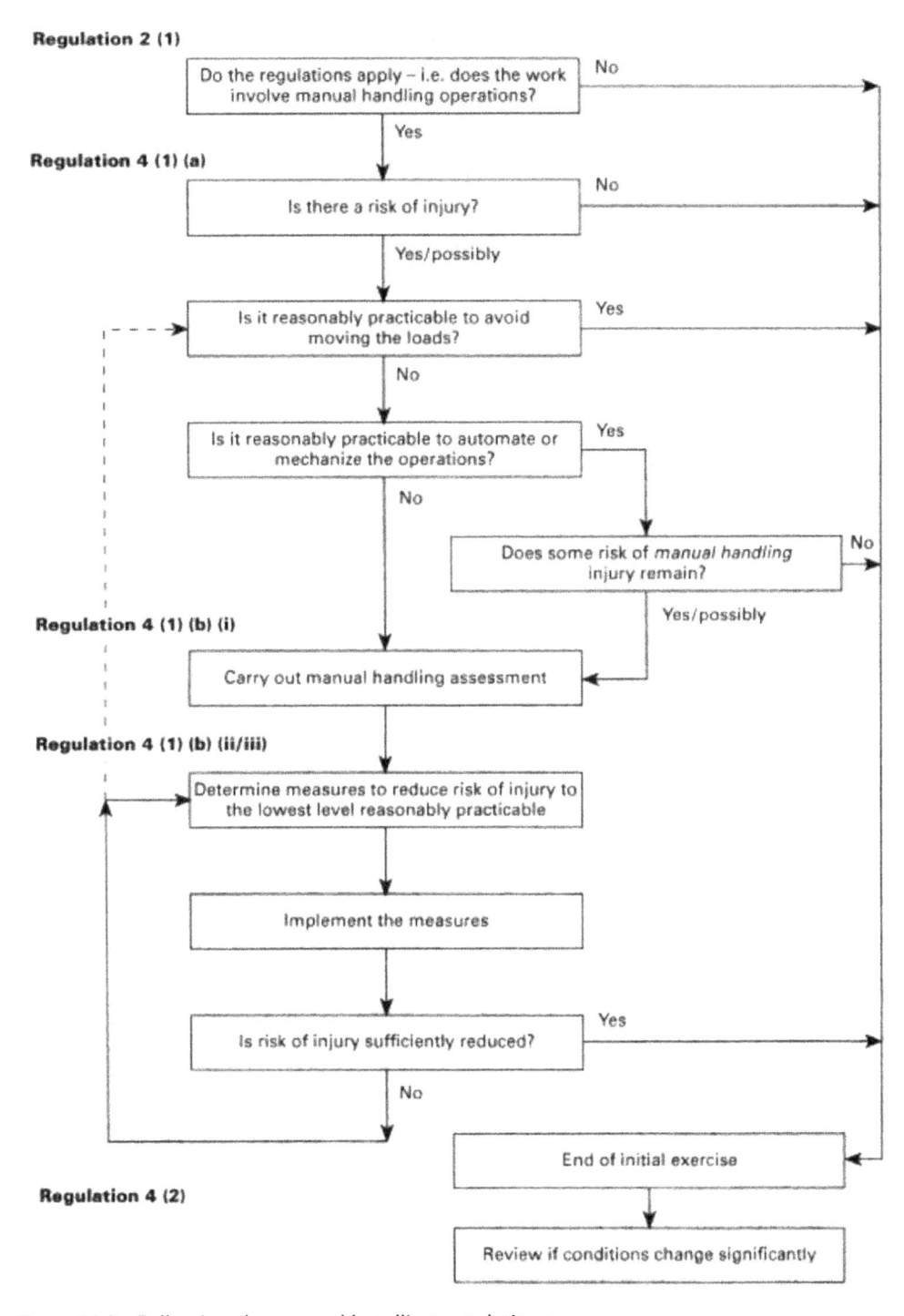

Figure 14.2 Following the manual handling regulations

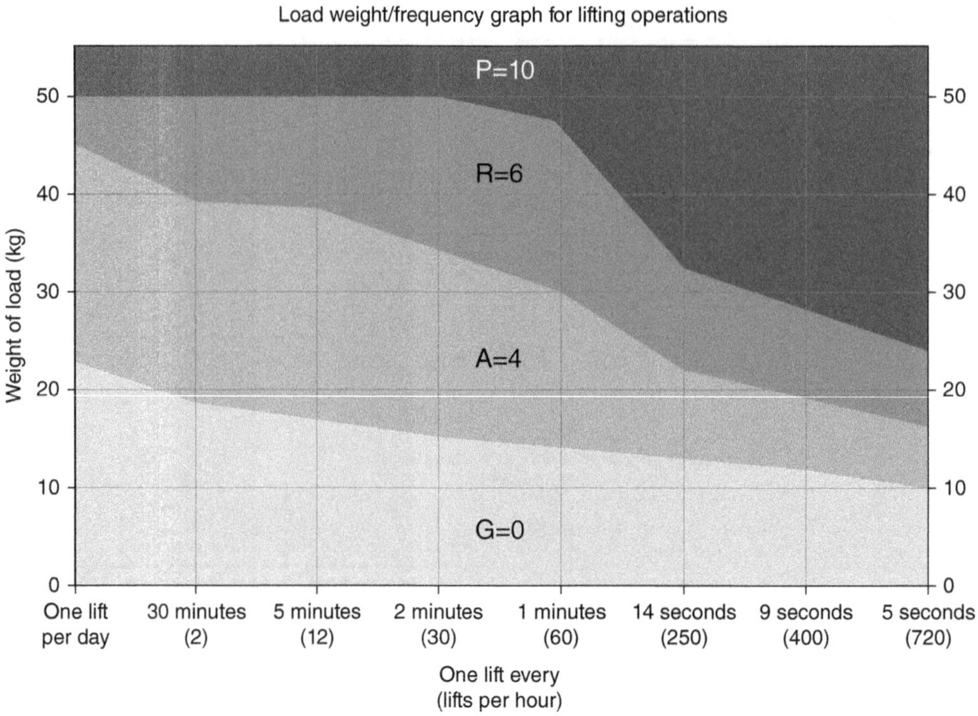

Figure 14.3 Example of graph from MAC

***MAC** score sheet*

Company name:

Task description:

Are there indications that the task is high risk?
(Tick the appropriate boxes)

- ☐ *Task has a history of manual handling incidents (e.g., company accident book, RIDDOR reports).*
- ☐ Task is known to be hard work or high risk.
- ☐ *Employees doing the work show signs that they are finding it hard work (e.g., breathing heavily, red-faced, sweating).*
- ☐ *Other indications, if so what?*

Date:

Signature:

Insert the colour band and numerical score for each of the risk factors in the boxes below, referring to your assessment, using the tool.

Risk factors	*Colour band (G, A, R or P)*			*Numerical score*		
	Lift	Carry	Team	Lift	Carry	Team
Load weight and lift/ carry frequency						
Hand distance from the lower back						
Vertical lift region						
Trunk twisting/sideways bending Asymmetrical trunk/load (carrying)						
Postural constraints						
Grip on the load						
Floor surface						
Other environmental factors						
Carry distance						
Obstacles en route (carrying only)						
Communication and coordination (team handling only)						
Other risk factors (e.g., individual factors, psychosocial factors, etc.; see website – address on page 12)	*TOTAL SCORE:*					

Figure 14.4 MAC score sheet

Operations and Lifting Equipment Regulations 1998.[35]

The HSE has produced detailed guidance in HSG136 on how best to reduce the workplace transport risks to a suitable and sufficient level. Guidance produced by the HSE is split conveniently into the following areas [3]:

- Organising for safety;

  It is extremely important to ensure there are clear controls in place to manage the interaction of moving vehicles and pedestrians. This means it is important to ensure organisations have clear accountabilities and responsibilities for managing workplace transport risks and that these are communicated clearly and enforced by local management. Consideration should be made for contractors or visitors to the workplace. It is also important to ensure where the workplace is shared between organisations those organisations cooperate with each other. Additional controls will be required where the general public has access to areas of the workplace where there is vehicle movement, for example a shopping precinct.

- A safe site;

  A well-designed and maintained workplace will help reduce workplace transport risks. Considerations that will help reduce this risk may include, for example, clearly defined and segregated traffic routes, site speed restrictions, signs, signals and road markings, housekeeping, and adequate lighting amongst other controls. If changes occur throughout the year which could have an impact on the vehicle manoeuvring area, then temporary risk assessments should be undertaken to reflect these changes (e.g., temporary storage is erected in a delivery yard for the Christmas period, and this could impact safe vehicle manoeuvring and the safety of those accessing the temporary storage).

- Safe vehicles;

  Ensuring vehicles are selected that are suitable for the tasks being undertaken, will help reduce workplace transport risks. If necessary, it may be required to ensure the vehicle selected has suitable manoeuvrability, reversing sensors or other aids to reduce risks to an acceptable level. Ensuring vehicles are maintained will also help reduce workplace transport risk.

- Safe drivers.

  As with any piece of work equipment it is important to ensure any employees using workplace transport (cranes, lift trucks for example) are trained to a competent level. Many vehicles such as forklift trucks require the completion of an appropriate training programme [4].

Workplace deliveries

Traffic law is enforced by the police and the Driver and Vehicle Standards Agency (DVSA) [5]. It places duties on transport managers, operators and consignors to ensure that:

- Vehicles are in good mechanical condition;
- Drivers are fit and competent to drive their vehicles;
- Passengers or loads carried on or in those vehicles are carried safely.

There may be specific law and operator licence requirements on HGVs or public service vehicles that take priority over the general guidance on these pages.

The police lead investigations into road traffic incidents on public roads. If issues occur in the delivery area, then EHPs can investigate under health and safety legislation.

Noise

Noise in the workplace is also addressed in Chapter 20, which considers noise as an environmental health issue in detail. The Control of Noise at Work Regulations 2005[36] is concerned with the levels of noise within a workplace and the required measures to protect the hearing of employees. From 6 April 2008, these regulations were extended to cover the music and entertainment industry. The aforementioned regulations require employers to prevent or reduce the risks to health associated with noise in a number of ways:

- Assessing the risks to employees.
  As reinforced in the Management of Health and Safety at Work Regulations 1999, a risk assessment must identify where there is a risk and what must be done to comply with the law. A noise risk assessment is more complex in the sense that it must contain an estimate of employee exposures compared to the exposure action values and must identify any employees who require health surveillance.
- Taking action to reduce the noise exposure that produced risk.
  When a risk assessment has identified the need to control noise, suitable alternative processes should be explored, such as using different equipment or exposing colleagues for shorter periods of time. The principles of the hierarchy of control should be adopted when managing noise also, and so where possible noisy activities should be:
  - Eliminated;
  - Engineered out;
  - Enclosed;
  - Relocated;
  - Limited;
- Your employees should be provided with hearing protection (if the noise cannot be reduced through other methods).
  Hearing protection should be issued where extra protection is needed above what has been achieved using control measures, or as a short-term method whilst these controls are being developed.
- Ensure that legal limits on noise exposure are not exceeded.
  The 2005 regulations list the lower-level exposure action values and these are given in Chapter 20.
  If any area of the workplace is exposed, or is likely to be exposed to noise at or

above the upper exposure action value, the area must:

- Be designated a Hearing Protection Zone;
- Be demarcated and signposted, so that individuals entering that area are aware of the fact that hearing protection must be worn;
- Have restricted access, where practicable;
- Carry out health surveillance where there is a risk to health.

Health surveillance (hearing checks) must be provided to all employees who are likely to be frequently exposed to noise above the upper exposure action values or have been identified as at risk due to hearing loss, for example. The health surveillance should warn employees if they are suffering from the early signs of hearing damage and will give employers the opportunity to check against their current process and proactively improve.

Burns and scalds

Many statutes work to prevent employees suffering burns and scalds at work. With regards to the risk of chemical burns, the Control of Substances Hazardous to Health Regulations 2002 (COSHH)[37] work to ensure that hazardous substances are assessed and that suitable control measures are introduced. In other organisations, it may not be chemicals that pose the risk of burning and scalding. In the catering industry, for example, hot water, hot fat and cooking equipment all pose a risk of burning employees. Where possible, the hierarchy of control should be adopted when looking to reduce risks.

In a kitchen, it would not be possible to remove the need to heat food, but instead the layout should be considered to reduce the level of congestion. Encouraging employees to clean as they go will reduce the likelihood of slips and trips occurring, which could result in contact with hot objects. Training will give the employees the understanding of the safe working procedures, whilst protective clothing should be utilised when all the other control measures do not result in an inherently safe operation.

Electrical safety

Electric shocks in the workplace kill approximately 50 people every year. In addition, poor electrical standards often lead to the occurrence of fires, for example, as a result of overheating electrical conductors. Also, electrical installations and appliances can act as ignition sources and lead to explosions. It must be noted that in flammable atmospheres, static electricity from movement of plant materials or even clothing can be a potential hazard.

It should also be noted that, with prompt action, employees with a basic training in first aid can save the life of a person who has received an electric shock. The HSE publishes a poster containing appropriate guidance for dealing with cases of electric shock (available for HSE http://www.hse.gov.uk/pubns/books/electric-shock-poster.htm). This can be displayed in potentially dangerous areas and brought to the attention of employees. In order to ensure that precautions are taken to prevent injury or fatality, the Electricity at Work Regulations 1989[38] have been made under the Health and Safety at Work Act 1974 and the HSE has published guidance on them.[39] Although intended to assist engineers, technicians and managers to understand the nature of the precautions to be taken, the guidance is also of use to those enforcing the provisions.

When dealing with electrical installations, reference is frequently made to the Institution of Electrical Engineers Regulations for Electrical Installations (IEE Regulations). These are non-statutory regulations evolved by the IEE that relate to the design, selection, erection, inspection and testing of electrical installations. They form a code of practice that has been widely recognised and accepted in the United Kingdom, and compliance with them is likely to achieve compliance with the statutory regulations. Scadden has produced a useful commentary which can assist with understanding of the Regulations [6]. In addition to the information provided by IEE, the HSE has published their own guidance [7].

While carrying out an inspection of premises, the electrical installation and the electrical supply to both fixed and portable machines should be examined. The inspector may be able to recognise simple visual faults that may indicate that the electrical system requires examination by a competent electrician. It is stressed that such an examination must be carried out without risk to other employees or non-employees. Depending upon the circumstances, the employer may be required to provide an electrician's report, or the inspector may engage their own electrician (who must be properly authorised to carry out investigations on the premises).

Consideration must be given to ascertaining whether or not a particular business involves the use of portable electrical equipment. If such equipment is used, then its condition should be ascertained, and enquiries made into how often the equipment is tested and the standards used. Any records kept should be examined. Where portable equipment is used in an external environment, then an isolating transformer or residual current device should be used to protect operators and others in case of an emergency. Some equipment can run on lower voltage supplies than mains voltage. In the case of some parts of electrical installations, for example, display lighting, supplies can be as low as 12 volts.

Some basic points that should be considered are listed in the following.

Flexible leads and cables – The correctly rated 'flex' or cable should be used for the application. Flexes and leads should be positioned and fixed correctly and be in good repair. Common faults include incorrect fixing to plugs or equipment, cuts, abrasions or heat damage. Joints should always be made using a suitable connector or coupler – joints made with insulating tape or 'block connectors' are normally dangerous. Care must be taken to ensure that cables are joined up with the correct polarities in each half of the connector.

Electrical socket outlets – Sockets should be sufficient in number and they must be positioned correctly to avoid tripping hazards from trailing flexes and overloading of the circuit due to the use of multipoint adapters.

Plugs – The flexible lead must be held firmly in the grip or clamp incorporated into the plug, and the individual wires must be connected firmly to the correct plug terminals. A suitably rated fuse must be installed. Some tools and lamps are 'double insulated', and these have only two wires in the flex. Such appliances are recognised by the box-in-box symbol marked on them.

Switches – Every fixed machine must have a switch or isolator beside the machine that can switch off the power in an emergency. Power cables to machines must be armoured, heavily insulated or installed in a conduit. Equipment may also be fitted with a residual current device (RCD) which will trigger and can automatically switch off the device if there is a fault.

Explosive atmospheres – Where the atmosphere of a workstation is either dusty or potentially flammable, then specially protected equipment must be used. Inspectors must consider this risk, for example, when they are taking monitoring equipment into premises. The HSE has produced a range of guidance documents for those working with electrical devices in explosive atmospheres.

Construction Design and Management (CDM)

The Construction (Design and Management) Regulations 2015[40] are concerned with maintaining health and safety in the construction industry. Duties are placed on a number of interested parties and guidance is available from the HSE [8]:

- Clients – anyone having a construction project carried out including individuals and organisations.

- Domestic clients – are people who have construction work carried out on their own home, or the home of a family member that is not done as part of a business, whether for profit or not.
- Designers – those who prepare or modify designs for a building, product or system related to construction work.
- Principal designers – are designers appointed by the client in projects involving more than one contractor. They can be an organisation or an individual with sufficient knowledge, experience, and ability to carry out the role.
- Principal contractors – contractors appointed by the client to coordinate the construction phase of a project where it involves more than one contactor.
- Contractors – a business involved in construction work and can be either an individual or a company.
- Workers – anyone who works for or under the control of contractors on a construction site. All persons involved in the construction are competent in the task for which they have been employed and all parties must be aware of their duties under the Regulations.

The Regulations require:

- The client –

Makes suitable arrangements for managing a project. This includes making sure:

- Other duty holders are appointed;
- Sufficient time and resources are allocated.

Makes sure:

- Relevant information is prepared and provided to other duty holders;
- The principal designer and principal contractor carry out their duties;
- Welfare facilities are provided.
- The domestic client –

Domestic clients are in scope of CDM 2015, but their duties as a client are normally transferred to:

- The contractor, on a single contractor project; or
- The principal contractor, on a project involving more than one contractor.

However, the domestic client can choose to have a written agreement with the principal designer to carry out the client duties.

- The designer –

When preparing or modifying designs, eliminates, reduces or controls foreseeable risks that may arise during:

- Construction; and
- The maintenance and use of a building once it is built.

Provides information to other members of the project team to help them fulfil their duties.

- Principle designer –

Plans, manages, monitors and coordinates health and safety in the pre-construction phase of a project. This includes:

- Identifying, eliminating or controlling foreseeable risks;
- Ensuring designers carry out their duties.

Prepares and provides relevant information to other duty holders. Provides relevant information to the principal contractor to help them plan, manage, monitor and coordinate health and safety in the construction phase.

- Principle contractor –
  Plans, manages, monitors and coordinates health and safety in the construction phase of a project. This includes liaising with the client and principal designer, preparing the construction phase plan, organising cooperation between contractors and coordinating their work.
  Ensures suitable site inductions are provided and reasonable steps are taken to prevent unauthorised access; workers are consulted and engaged in securing their health and safety and welfare facilities are provided.
- The contractor –
  Plans, manages and monitors construction work under their control so that it is carried out without risks to health and safety.
  For projects involving more than one contractor, coordinates their activities with others in the project team – in particular, complies with directions given to them by the principal designer or principal contractor.
  For single-contractor projects, prepares a construction phase plan.
  Workers must:
- Be consulted about matters which affect their health, safety and welfare;
- Take care of their own health and safety and others who may be affected by their actions;
- Report anything they see which is likely to endanger either their own or others' health and safety; cooperate with their employer, fellow workers, contractors and other duty holders.

The HSE have identified five key elements to help deliver construction safety and includes:

1 Managing the risks by applying the general principles of prevention
2 Appointing the right people and organisations at the right time
3 Making sure everyone has the information, instruction, training and supervision they need to carry out their jobs in a way that secures health and safety
4 Cooperating and communicating with duty holders and coordinating their work

5 Consulting workers and engaging with them to promote and develop effective measures to secure health, safety, and welfare

Health and safety on construction sites includes the following:

- Safe working environment with safe access and egress and a secure perimeter;
- Stable structures, both permanent and temporary;
- Safe demolition or dismantling of the site;
- Safe storage, transportation and use of explosives;
- Excavations are protected from collapse, falling items and prevent persons from falling in; any supports are inspected by a competent person;
- Cofferdams and caissons are safe with shelter or escape provided, and inspected by a competent person;
- Inspection reports are carried out and provided to the client/contractor;
- Any electrical supply is safely located and may be isolated or segregated;
- Where water is present, the risk of drowning must be reduced, and suitable safety equipment must be available;
- Suitable traffic routes and protection of pedestrians;
- Safe operation of vehicles, including driving and loading;
- Prevention of fire, explosion, flooding or other asphyxiant risks;
- Emergency procedures in place suitable for the site and any materials on the site, and these are communicated to everyone and tested;
- Emergency routes are present, signed and free from obstruction;
- Suitable fire detection and alarm systems and fire-fighting measures are available, tested and maintained, and people are trained to use the systems;
- Fresh or purified air is circulated on site and any plant providing it has an effective warning system;

- Site is at a reasonable temperature and any equipment to protect against adverse weather is provided;
- Lighting is sufficient.

Fire safety

The main legislation is the Regulatory Reform (Fire Safety) Order 2005 (for England and Wales) as amended[41] and the Fire Scotland Act 2005 (Scotland).[42] It should be noted that the Fire Precautions Act 1971 has been repealed.[43]

There are associated pieces of legislation that link heavily to the previously mentioned but will not be covered in any detail in this chapter. They include the following:

- The Dangerous Substances and Explosive Atmospheres Regulations 2002;[44]
- The Control of Major Accident Hazards Regulations 2015;[45]
- The Manufacture and Storage of Explosives Regulations 2005.[46]

Fire is the rapid oxidation of a substance that generates heat, light and smoke. In order for a fire to be sustained the following three factors are required: oxygen, heat (or an ignition source) and the fuel itself. This can be illustrated in the commonly used Fire Triangle shown in Figure 14.3.

Figure 14.5 **The Fire Triangle**

Table 14.2 Examples of the elements of fire included in the Fire Triangle

Ignition sources	Fuel	Oxygen
Naked flames (matches)	Flammable liquids (e.g., petrol oil)	Air (contains 21% oxygen)
Hot surfaces (e.g., cookers)	Flammable solids (e.g., cardboard, wood, plastics)	Oxidising agents (e.g., fertilisers)
Electrical sparks	Flammable gases (e.g., LPG)	
Friction generating a heat source		

Simple examples of each of these elements are listed in Table 14.2.

In most cases if you remove or eliminate one of these elements a fire will be extinguished or fail to start. To reduce the risk of fire, significant ignitions sources and fuel (combustible items) should be kept separately. Controls on how this is managed will be documented in a fire risk assessment as required in the 2005 Fire Safety Order and is enforced by the Fire and Rescue Services.

Classes of Fire – Fire can be broken down into six classes and are mainly categorised by fuel type. This will help determine the type of fire extinguisher that can be used to extinguish a fire.

- Class A;
  These fires involve primarily solid materials of a cellulosic nature such as wood, paper, cardboard, coal and natural fibres, but do not include metals. Generally, the greater the overall surface area of the solid material the quicker a fire will spread. For example, a solid piece of wood will burn slowly when compared to wood shavings as the surface area is greater.
- Class B;
  These fires involve primarily liquids or liquefiable solids and can be sub-divided into liquids that will mix with water and those that will not.
  B.1 – Liquids that mix with water, for example methanol, acetone and acetic acid
  B.2 – Liquids that do not mix with water for example waxes, petrol and solvents

- Class C;
  These fires involve gases or liquefied gases such as those contained in cylinders such as methane, propane and butane, more commonly referred to as LPG.
- Class D;
  These fires are the most hazardous type of fires and involve certain flammable metals such as aluminium and magnesium.
- Electrical Fires (sometimes referred to as Class E fires);
  These are fires that are either caused by electrical faults or involve equipment that have live electricity passing through them. Electrical fires can be caused by overheating of an electrical circuit or arcing (electricity jumping from one live circuit to another, e.g., lightning is arcing on a large scale). The difficulty in extinguishing this type of fire is that if an extinguishing agent that conducts electricity is used (e.g., water or foam) it introduces the risk of electrocution.
- Class F.
  These fires are a relatively newly classified type of fire and coincide with more widely available wet chemical fire extinguishers. Class F fires involve hot fat and cooking oil.

Fire Spread – Fires can spread in three main ways:

1 Convection is the most dangerous method by which a fire can spread and causes the largest number of injuries and deaths as a result of fire. Smoke

Table 14.3 Fire extinguishers

Class of fire	Description	Extinguisher type	Colour
A	Solid materials (not metals)	Water	Red
		Foam	Cream
		Dry powder	Blue
		CO2	Black
B	Liquids and liquefiable solids	Foam	Cream
		Dry powder	Blue
		CO2	Black
C	Gas or liquefiable liquids	–	
D	Metals	–	
E	Electrical equipment	CO2	Black
F	Hot fat and cooking oil	Wet chemical	Yellow

is generated by combustion, and this will rise and become trapped within buildings. This smoke can become superheated and sometimes reach temperatures in excess of 1000 degrees Celsius and cause spontaneous combustion.

2 By conduction, as some materials will conduct heat extremely well, for example metals. Fire can spread throughout a building by conduction along, for example, metal ducting between rooms.

3 By radiation where radiant heat is transferred through the air in a similar way to an electric bar heater. This can heat combustible items and cause them to smoulder and then burn.

Extinguishing fires – There are three main ways in which fires can be extinguished through removal or reduction of one of the three elements of the Fire Triangle. Therefore, by removing oxygen, the ignition source or fuel (combustible material) a fire will be extinguished.

1 Starvation – this is where the fuel source is removed, for example switching off a gas supply

2 Smothering – limiting the concentration of air or removing the concentration of oxygen removes the ability of the combustible substance from oxidising. For example, wrapping a person on fire in a rug or covering a fire in sand

3 Cooling – by removing the energy, heat or ignition source. This is by far the most common method of extinguishing a fire. Water is the most common method used to remove the energy from a fire

Fire fighting equipment – Fire extinguishers are provided in most workplaces to extinguish small fires and provide a method for people to help escape from a building should they become trapped. Fire extinguishers should not be used to tackle larger fires as this can put individuals at risk. For each type of fire, the appropriate fire extinguishers should be used, each one using one of the three methods of extinguishing a fire listed previously. Table 14.3 highlights which types of fire extinguisher are suitable for which class of fire. Although all red in colour, each type of fire extinguisher can be identified by specific colour coding.

Principles of fire risk assessment – As with other risk assessment and under the Regulatory Reform (Fire Safety Order) 2005, all work premises are required to have a fire risk assessment in place that details the preventative measures and emergency procedure in the event of fire. Similar to other risk assessment processes the main steps of a fire risk assessment are detailed here, and further information can be found in various industries guides issued by the Department for Communities and Local Government (DCLG).[47]

Step 1 – Identifying the Hazards –
    A – Identify sources of ignition
    B – Identify sources of fuel
    C – Identify sources of oxygen
Step 2 – Identifying People at Risk –
    A – Who is at risk (employees, visitors, contractors, neighbours and the disabled)
    B – Why they are at risk

Step 3 – Evaluate, remove, reduce and protect from risk – Having identified the hazards (ignition sources, combustible items and sources of ignition) an organisation must decide what controls to put in place to reduce the risk so far as is reasonably practicable. Controls an organisation may consider are listed here:

A – Remove or reduce sources of ignition, fuel or sources or oxygen (e.g., oxidising agents)
B – Remove or reduce the risks to people by introducing fire protection systems and processes

Fire protection measures – include a range of options that need to be examined:

- Fire warning and alarms systems, including smoke and heat detectors (means of raising the alarm);
- Fire equipment and facilities (e.g., fire extinguishers, sprinkler systems, fire hoses, fire suppressant systems);
- Fire exits and escapes and emergency escape routes – making sure people know what to do and where to go in the event of an emergency, and all escape routes are clear from obstructions and clearly indicated and protected;

- Emergency escape lighting – making sure people are able to clearly see the emergency escape routes in the event of a power cut;
- Signs and notices clearly indicating the nearest emergency escape route or exit are clearly visible, especially where members of the public have access;
- Installation, testing and maintenance – regular maintenance of the fire protective measures should be in line with the relevant British Standards, for example Emergency Lighting, and Portable Fire Extinguishers.

Step 4 – Record, Plan, Inform, Instruct and Train – Significant hazards should be identified together with a clear action plan on how these hazards will be managed.

Emergency evacuation plans should be developed that fit with the protective measures, and hazards identified so all employees, visitors and contractors are able to safety evacuate the premises. This should include provision for people with disabilities where a Personal Emergency Evacuation Plan (PEEP) should be developed.

All employees should be provided with clear information, instruction and training on what to do in an emergency. This may include employees who have specific tasks to carry out in the event of an emergency, for example fire marshals, wardens or incident controllers.

Step 5 – Review – As with any risk assessments it is necessary to review the fire risk assessments on a regular basis. Most importantly, it must be reviewed when significant changes are made, for example an extension to a building.

Personal protective equipment

The Personal Protective Equipment at Work Regulations 2002[48] place duties on employers and the self-employed, providing a framework for the provision of personal protective equipment (PPE) in circumstances where any assessment has shown a need for such protection. The regulations do not apply to the provision of most respiratory protective equipment, ear protectors and some other types of PPE because requirements are already laid down in regulations such as the Control of Substances Hazardous to Health Regulations 2002[49] (as amended) (COSHH), the Control of Noise at Work Regulations 2005[50] already mentioned and the Control of Asbestos at Work Regulations 2002[51] covering the use of PPE in particular circumstances. The main aim of the regulations is to ensure the proper provision of PPE following a risk assessment. Further consideration is given next.

PPE is defined as

> all equipment (including clothing affording protection against the weather) which is intended to be worn or held by a person at work and which protects him against one or more risks to his health or safety, and any addition or accessory designed to meet that objective.

As well as there being specific legal requirements, one must not forget the general duty of care owed to every employee by an employer. All employers must protect their workforce from the risk of reasonably foreseeable injury. Therefore, as part of this general duty, it is reasonable to suggest that not only must PPE be provided free of charge (see section 9 of the Health and Safety at Work, etc. Act 1974), but it must also be readily available at all times. In the guidance to the regulations [9–10], PPE is seen as a last resort. Rather, employers should seek to introduce changes to the process or safe systems of work to reduce the risk to the health and safety of their employees. Any PPE provided by the employer must be 'suitable'. The regulations state that PPE shall not be suitable unless:

- It is appropriate for the risk or risks involved and the conditions at the place where exposure to the risk may occur;

- It takes account of ergonomic requirements and the state of health of the person or persons who may wear it;
- It is capable of fitting the wearer correctly, if necessary, after adjustments within the range for which it is designed;
- So far as is practicable, it is effective to prevent or adequately control the risk or risks involved without increasing the overall risk;
- It complies with any enactment (whether in an act or instrument) which implements in Great Britain any provision on design or manufacture with respect to health or safety in any relevant European Union (EU) directive listed in schedule 1 which is applicable to that item of personal protective equipment.

Before choosing any PPE, any employer or self-employed person must carry out an assessment to ensure that the proposed PPE is suitable. The need to assess the suitability of PPE follows on from, but does not duplicate, the risk assessment requirement under the Management of Health and Safety at Work Regulations 1999. These regulations are concerned with the whole range of hazards present in the workplace and the evaluation of such hazards to determine the necessary controls. One of the controls may be to use PPE; however a separate assessment is required to ascertain the suitability of any PPE. In the simplest cases which can easily be explained, the assessment need not be recorded. In more complex cases, the assessment will need to be documented and kept available. The assessment shall include:

- An assessment of any risk or risks to health or safety which have not been avoided by other means;
- The definition of the characteristics which personal protective equipment must have in order to be effective against the risks referred to in the first bullet, taking into account any risks which the equipment itself may create;

- Comparison of the characteristics of the personal protective equipment available with the characteristics referred to earlier;
- An assessment as to whether the personal protective equipment is compatible with other personal protective equipment which is in use and which an employee would be required to wear simultaneously.

A specimen risk survey table is contained in Appendix 1 of the Guidance to the Regulations [9]. The regulations further require that all PPE be maintained, and that those who use it be adequately trained in its use and the limits of the PPE involved. Furthermore, employers are under an obligation to ensure that employees use any PPE provided in an approved way. Finally, employees are under an obligation to notify their employer of any shortcomings with the PPE they have been provided with.

Risks to health at work from hazards and their management

It may be convenient to divide the hazards to health into three categories. These are physical, chemical and biological hazards. Other relevant information can be found in Chapter 10.

Physical hazards – Include the well-documented hazards produced by, for example, light, heat, cold, noise and vibration. In addition, it may also be convenient to think of hazards such as ultraviolet, infrared, and ionising radiation under this heading.

Chemical hazards – This heading includes the whole range of substances and compounds found in organic and inorganic chemistry. From an occupational health perspective, these hazards should be controlled by the regime imposed by the COSHH Regulations. In addition, substances such as asbestos would also be classified as a significant chemical hazard.

Biological hazards – Those at risk may be involved in the handling of bacteria, viruses, plants, animals or animal products. At one time there was concern about people coming into contact with those infected with HIV (Human Immunodeficiency Virus) but the CIEH produced guidance [11], and it could be argued that SARS-CoV-2 also falls into this category.

Having made this distinction between various hazardous substances generally, it should be made clear that the toxicity of any substance depends upon a number of factors such as:

- The nature of the substance;
- The amount taken in, compared with the weight of the person;
- The physical condition and age of the person when exposure takes place;
- The sex of the person exposed to the substance.

Therefore, before one can attribute the term toxic to any substance and hence determine the need for protection of any worker against exposure to that substance, one must establish some of those factors listed.

Modes of entry of harmful substances – Having established the nature and range of some of the harmful substances that can affect the human body, it is now necessary to look at the ways in which these substances can gain access to the body. It is convenient to think of these 'modes of entry' as falling under four main headings, although it should be borne in mind that some substances may have more than one mode of entry.

- Inhalation;
  A wide range of substances is carried into the human body on the breath. While the nose, airways and lungs behave as fairly efficient filters against many substances, there is a critical size range where penetration can occur along the complete length of the respiratory tract. Coates and Clarke [12] estimate that particles larger than 10μm in diameter are filtered off by the nasal hairs. Particles that are between 5μm and 10μm tend to settle in the bronchi

and bronchioles, caught up in mucus. The muscociliary escalator, formed by ciliary hairs, moves this mucus upwards to the throat where they are either coughed out or swallowed into the stomach. Particles that are smaller than 5μm are able to reach the lung tissue. Coates and Clarke point out that fibres which predispose to disease, have a length to diameter ratio of at least 3:1, and a diameter of 3μm or less. Therefore, the longer the fibre the more damaging it may be.

- Ingestion;

Many substances pass into the body via the digestive system. If the substance is absorbed into the body, then it may be passed to the liver where it is rendered less toxic (detoxification) before being excreted. Some substances, such as bacteria and some chemicals, however, can cause harm without leaving the digestive tract.

- Absorption;

The skin is a very substantial defence against many substances. However, some substances are able to pass directly through the skin and into the underlying tissue leaving the skin intact. In other cases, the substances pass into the epidermis but not through it, resulting in conditions such as dermatitis and some forms of cancer. Some substances, particularly solvents, can reduce the ability of the skin to protect the body against attack.

- Irradiation.

This is the term used to describe the exposure of the body to both ionising and non-ionising radiation. Exposure may result in body surface penetration.

Defence against harmful substances – As well as the protective equipment available to protect the worker, the body also possesses its own defence mechanisms. These are:

- Respiratory filtration;
- Cell defence mechanisms;
- Inflammatory response;
- Immune response;
- Thermoregulation;
- Metabolic transformation.

It is suggested that the reader refer to more specific texts dealing with human physiology for a comprehensive discussion of these factors.

Principles of protection – As has already been made clear, the use of PPE to give protection against a particular hazard should not be seen as a substitute for other methods of dealing with the danger. For example, at a drilling machine with an exposed rotating spindle where there is a risk of entanglement, the aim should be to eliminate the risk by proper guarding of the machine, and not to rely on the operator wearing suitable head covering. The hierarchy of control summarised this way of thinking, and should be applied to all hazards wherever it is reasonably practicable:

- Eliminate the risk;
- Reduce the risk;
- Contain the risk;
- Substitute the process for a safer one;
- Engineer in controls;
- Introduce safe systems of work;
- Use PPE.

It should be borne in mind that personal protection is not an easy option, and it is important that the correct protection is given for a particular hazard. In addition, one must be satisfied that the equipment being used is of sufficient quality in order to afford the worker the required protection. If one is to fulfil legal and moral obligations, it is essential that a programme looking at all aspects of the provision and use of personal protective equipment is in place. Else [13] gives three key elements of information required for a personal protection scheme. These are:

- The nature of the danger;
- Performance data of personal protective equipment;
- The acceptable level for exposure to danger.

Nature of the danger – It is important to know some details about the hazards to be faced. For example, with regard to a physical hazard such as noise, one would need to know the sound level and frequency characteristics of the noise. In addition, information could be gained from recorded accident/incident experiences, safety representatives, safety audits or surveys and medical records.

Performance data of personal protective equipment – The choice of equipment is extremely important. Its quality, durability, suitability and lack of interference with the user's faculties and movements are important considerations affecting choice. In the United Kingdom, the British Standards Institution (BSI) has traditionally conducted assessment of equipment and produced British Standards.

As well as the Personal Protective Equipment Regulations 2002 there is the EU PPE Regulation 2016/425.[52] All PPE supplied for use at work must be independently assessed so as to ensure that it meets 'basic safety requirements'. Satisfactory testing results in a 'certificate of compliance' and entitles the manufacturer to display the 'CE' mark on its product. The other effect of these regulations is to make it illegal for any supplier to sell PPE unless it carries the 'CE' mark.

Acceptable level for exposure to danger – This is an extremely important factor. Quite clearly for some dangers, such as exposure to potentially carcinogenic substances, the only acceptable level can be zero. Both the employer and the enforcing officer must have a sound working knowledge of fixed legal standards, such as 'hygiene limits', and also of those standards of protection that are merely advisory in nature.

These suggested considerations must be viewed in conjunction with the general requirement to engage in a risk assessment contained in the Management of Health and Safety at Work Regulations 1999, and the more specific requirements with regard to PPE contained in the Personal Protective Equipment at Work Regulations 2002.

Selection of personal protective equipment – The general discussion of the Personal Protective Equipment at Work Regulations 2002 earlier has highlighted the need for any PPE chosen to be both 'suitable' and capable of protecting against the risks to health and safety identified. Part 2 of the guidance [9] lays down some general principles about the way equipment should be chosen.

In addition, reference should be made to the information available from suppliers. It should be remembered that all PPE must carry the 'CE' mark [10].

Types of PPE may be divided into the following broad categories:

- Hearing protection;
- Respiratory protection;
- Eye and face protection;
- Protective clothing;
- Skin protection.

Hearing protection – hearing protection can be divided into two main types, earplugs and earmuffs. Earplugs are designed to be inserted into the ear canal. Those that are designed to be disposed of after use are usually made from either mineral down, which is an extremely fine glass down, or from polyurethane foam. Reusable earplugs are made of soft rubber or plastic. They must be thoroughly washed after use. Reusable plugs should be fitted in the first instance by a trained person who should provide advice to the wearer about the correct method of inserting the plugs. BS EN 352–2:2020[53] gives the general requirements and specification for earplugs. In particular, it specifies requirements regarding the sound attenuation of the earplugs, measured in accordance with EN ISO 4869-1:2018.[54]

Earmuffs are designed to cover the external ear. They consist of rigid cups that fit over the ears and are sealed to the head with soft cushion seals. They have several advantages over earplugs. One size will usually fit a wide range of people, they tend to offer greater protection and they are easy to remove and replace. However, they are not without their disadvantages. They tend to make the ears hot and are bulky. BS EN 352–1:2020 deals with the required specification for earmuffs.[55] It is imperative that any type of hearing protection is of sufficient quality to reduce the noise level at the wearer's ear to below any recommended limit. The use of data from octave band analysis should be compared with design data provided by the manufacturer in order to ensure maximum protection. BS EN ISO 4869–1:2018 specifies a subjective method for measuring sound attenuation of hearing protectors at the threshold of hearing.

The Control of Noise at Work Regulations 2005 already referred to lay down a requirement for employers to provide hearing protection in certain circumstances. Where an employer carries out work which is likely to expose any employees to noise at or above a lower exposure action value, personal hearing protectors are to be made available upon request. Where noise cannot be reduced to below an upper exposure action value, hearing protectors must be provided by the employer. The advent of the regulations resulted in the HSE producing a useful practical guide [14], which specifies the criteria for the selection of hearing protection. Attention should be paid to:

- The level and nature of the noise exposure;
- The job and working environment;
- Compatibility with any other protective equipment or special clothing worn;
- The fit to the wearer;
- Any difficulty or discomfort experienced by the wearer.

Respiratory protection – As has been emphasised, personal protection is a form of last resort protection. In the case of respirable dusts and fumes, every effort must be made to try and enclose the process and provide exhaust ventilation. Where this is not possible, suitable protection must be provided. Respiratory protective equipment may be divided into two broad categories: respirators that purify the air by drawing it through a filter, thereby removing the contamination, and breathing apparatus, which supplies clean air from an uncontaminated source.

There are five basic types of respirators.

1 *Dust respirators* are the most commonly available form of respirators. They afford protection against solid particles of matter or aerosol sprays, but not against gases. They generally cover the nose and mouth of the wearer. BS EN 529:2005[56] provides guidance on the best practice for establishing and implementing a suitable respiratory protective device programme.

   As well as dust respirators, there are light, simple *face masks*, which protect the wearer from the effect of nuisance dusts or non-toxic sprays. It should be remembered that while these are a very popular form of respirator, the suitability needs to be assessed against the task being performed.

2 *Cartridge type respirators* give protection against low concentrations of relatively nontoxic gases and vapours. This is achieved by the use of replacement filter cartridges. Care should be taken to change the cartridge regularly to ensure maximum protection.

3 *Canister type respirators* incorporate a full facepiece connected to a replaceable filter canister. This type of respirator offers far more protection than the cartridge type respirator. It is important that the correct canister is fitted to the equipment to ensure maximum protection for the wearer.

4 *Positive pressure-powered respirators* can either cover the nose and mouth or the whole of the face. Air is drawn in through a battery-powered suction unit, through filters and fed to the facepiece at a controlled flow. The excess of air escaping around the edges of the facepiece prevents leakage inwards. This type of protection is mainly used when working with disease producing dusts, for example, asbestos.

Breathing apparatus – The choice of breathing apparatus is extremely complex and should be made by those well versed in the use and limitations of such equipment. BS EN 529:2005 also gives advice on the selection, use and maintenance of such equipment. There are three basic types of breathing apparatus.

Closed circuit systems, which supply either oxygen or air from a cylinder carried by the wearer. BS EN 13794:2002[57] gives a specification for this type of apparatus. The air is supplied via a demand valve. The system gets its name from the fact that the wearer breathes the same air repeatedly. When the wearer exhales, the exhaled air is purified and passes into a breathing bag where it is enriched with fresh oxygen from the cylinder. This results in the apparatus being suitable for use over a longer period of time.

Open circuit systems provide compressed air or oxygen from a cylinder worn by the worker. However, there is no breathing bag, hence the wearing time is greatly reduced.[58]

Airline breathing apparatus has a full facepiece connected to a source of uncontaminated air by a hose. The equipment allows the operative to work in most types of toxic atmospheres, but the trailing tube can limit the movement of the wearer.[59] It should be emphasised that although a breathing apparatus provides the most effective protection against risks, because of its complexity its use requires specialised training and supervision. Detailed guidance has been published by the HSE [15] on the selection and use of respiratory protective equipment.

Eye and face protection – Eye injuries are extremely common. The main hazards are solid particles, dust, chemical splashes, molten metal, glare, radiation and laser beams. Injuries often result in severe pain and discomfort and, in many cases, long-term impairment of vision. It is extremely important that hazards are fully examined before any form of eye protection is chosen.

Industrial eye protectors, such as goggles, visors, spectacles and face screens, are required to satisfy the minimum requirements of BS EN 166:2002[60] and BS 7028:1999 gives advice on the selection, use and maintenance of eye protection.[61] In addition, part 2 of the HSE's guidance [10] gives some advice on the choice of eye protection.

Protective clothing – There is currently a wide range of protective clothing available for use in most industrial situations. This ranges from well-known items such as safety footwear and overalls, to the more specialised gloves and aprons worn in certain work situations. There is a whole range of British Standards and proposed European Norms dealing with the full range of protective clothing. Once again, the HSE's guidance [10] is useful in appreciating the criteria to be employed in selecting protective clothing.

It is important, as pointed out earlier, that those who wear protective clothing should be fully aware of the limitations of its use.

Skin protection – In some circumstances, it may not be possible or desirable to use gloves. In such situations, a proprietary barrier cream may be an alternative. Hartley [16] divided skin protection preparations into three broad groups.

Water miscible – Protects against organic solvents, mineral oils and greases, but not metalworking oils mixed with water.

Water repellent – Protects against aqueous solutions, acids, alkalis, salts, oils and cooling agents that contain water.

Special group – These cannot be assigned to a group by their composition. They are formulated for specific applications. It should be remembered that these creams are of only limited use as they are rapidly removed during the working day.

Petroleum

Dispensing petrol fuel poses a number of risks to health, which must all be identified and controlled through a risk assessment. The

Dangerous Substances and Explosive Atmospheres Regulations 2002[62] require employers to protect employees from fire and explosion and include petrol storage within its remit. Employers must carry out a risk assessment and put in place control measures to remove or reduce that risk. In addition, an emergency plan must be developed, and training provided to employees.

Petrol is a highly flammable liquid which can give off a flammable vapour at low temperatures. As a result of this, a fire or explosion risk is always present where there is source of ignition.

Petrol will float on the surface of water and can therefore travel a long way from the source, carrying those risks away from the initial leak or spillage. Petrol vapours can be harmful if inhaled and liquid petrol must not be swallowed. Employees should be trained in the safe use of site equipment and should also have a sound understanding of the risks associated with petrol.

Consideration must be given to personal protection and hygiene, including the need to wear gloves, goggles and footwear, and the provision of a suitable soap and hot water. Risks should be brought to the attention of customers through correct signage and fire precautions.

Petrol forecourts can also be dangerous places due to the volume of traffic and the interaction of vehicles and pedestrians. Tankers should have clearly demarcated parking, whilst consideration at the design stage should ensure that pedestrians are kept away from vehicle traffic, where possible. Spilt fuel poses a slip hazard, so sand (or a suitable alternative) should be made available for customers to apply themselves. Employees should monitor spillages so that they can intervene when necessary.

Hazardous equipment

Many serious accidents at work involve some type of machinery and most machinery must therefore be regarded as being intrinsically

dangerous. Where equipment is identified as potentially hazardous there is a duty on the employer through Health and Safety at Work to manage that hazard through controls on design and use, and through implementing a safe system of work. However, certain pieces of equipment are subject to more stringent controls, and these are defined by their use. The Health and Safety Executive (HSE) has published a helpful guide that summarises many of the key points that must be considered by employers, employees and enforcement officers [17] and [18].

Injuries may result from any of the following:

1 Trapping:

    a Trapping between moving parts and fixed parts, for example, guillotine blades, garment pressers and sliding tables.

    b Trapping between moving parts, for example, rollers, cogs, drive belts, food beaters and mixers. This category includes trapping by entanglement of hair or clothing, resulting in part of the body being brought into contact with the dangerous machine part.

    Common sources of injuries of this type are 'nips' and 'pinches'. Of particular concern are 'in-running nips'. Examples of this form of trapping and other hazards posed by a wide variety of dangerous machinery can be found in BS EN ISO 12100:2010.[63]

2 Contact with moving parts, for example, cutting blades, abrasive wheels, gear wheels – even slowly rotating parts must be considered to be potentially dangerous. This is particularly true in the case of gears or beaters such as those found in food processing machinery.

3 Burns, for example, from hot exhausts and deep fat fryers.

4 Electrocution, for example, from exposed electrical conductors.

5 Contact with moving workpieces, for example, lathe.

6 Striking by machine parts or work-pieces thrown from the machine, for example, fractured abrasive wheels and non-secure cutter blades. Other non-mechanical hazards and environmental considerations are dealt with in the following sections.

Principles of Machine Guarding – These were established as a result of the introduction of the Factories Act 1961. The established standards were supported by a considerable body of case law and the Provision and Use of Work Equipment Regulations 1998 (PUWER)[64] (see later) have amplified the requirements for the guarding of equipment and applied them across all industrial, commercial and service sectors.

PUWER implemented European Directive 89/655/EEC, which addresses minimum health and safety requirements for the use of work equipment by workers at work. The regulations place duties on employers and others, including certain people who control work equipment, people who use, supervise or manage its use or how it is used, to the limit of their control.

The regulations address the following aspects of work equipment:

- Suitability of work equipment;
- Maintenance;
- Inspection;
- Specific risks;
- Information and instructions;
- Training;
- Conformity with European Union (EU) requirements;
- Dangerous parts of machinery;
- Protection against specified hazards;
- High or very low temperatures;
- Controls for starting or making a significant change in operating conditions;
- Stop controls;
- Emergency stop controls;
- Controls;

- Control systems;
- Isolation from sources of energy;
- Stability;
- Lighting;
- Maintenance operations;
- Markings;
- Warnings.

When selecting a control system, employers must ensure that they consider the failures and faults that might be expected to occur during its normal use. Where the safety of work equipment depends on its installation, employers must also ensure that it is inspected by a competent person after each installation, prior to being put into service for the first time, or after assembly at a new site/location. The results of such inspection must be recorded and kept until the next inspection. If any work equipment is used outside the business site, records of the last inspection must accompany it [19].

The regulations also cover mobile work equipment such as tractors and lift trucks. The provisions relate to:

- Employees carried on mobile work equipment;
- Rolling over of mobile work equipment;
- Overturning of forklift trucks;
- Self-propelled work equipment;
- Remote-controlled self-propelled work equipment;
- Drive shafts.

The Lifting Operations and Lifting Equipment Regulations 1998 (LOLER)[65]

These regulations known as LOLER support PUWER and lay down the health and safety requirements for lifting equipment. The regulations apply to employers and the self-employed but not to suppliers of lifting equipment. Lifting equipment is defined as: 'work equipment for lifting or lowering loads, including any attachments used for anchoring, fixing or supporting it'.

Employers must ensure that:

- Lifting equipment is of suitable and sufficient strength and stability for each load, having consideration to the stress placed on mountings and fixing points;
- All parts of a load, anything attached to it and anything used in lifting is of adequate strength;
- Lifting equipment for lifting people will prevent passengers from being crushed, trapped or struck, or from falling from the carrier;
- Lifting equipment for lifting people, so far as is reasonably practicable, will prevent anyone who is using it to undertake activities from the carrier from being crushed, trapped or struck, or from falling from the carrier;
- Lifting equipment for lifting people has suitable devices to prevent the carrier falling; if these risks cannot be prevented, the carrier must have an enhanced safety coefficient suspension rope or chain, which should be inspected daily by a competent person; if anyone becomes trapped in a carrier, he should not be exposed to danger and should be able to be freed;
- Lifting equipment must be positioned/installed so as to minimise the risk of the equipment/load striking a person, or of the load drifting, falling freely or being released unintentionally; there must also be suitable devices for preventing people from falling down shafts/hoist ways;
- Machinery and accessories must be clearly marked to indicate their safe working loads (swls); separate swls must be given for different configurations; accessories must be marked so that the characteristics needed for their safe use may be identified; lifting equipment for lifting people must be clearly and correctly marked, and equipment that is not meant for lifting people but that could be used as such should be clearly marked to the effect that it should not be used for such a purpose;
- All lifting operations involving lifting equipment are properly planned, supervised and undertaken in a safe manner;

- Before any lifting equipment is used for the first time it must be thoroughly examined; where safety depends on its installation, lifting equipment must be thoroughly examined after installation and prior to first use, after assembly and prior to use at a new site/location;
- Lifting equipment and accessories used for lifting people are thoroughly examined at least every six months, and other lifting equipment at least every 12 months; it must also be examined when an incident has occurred that may adversely affect the safety of the equipment;
- Records of the last thorough examination must accompany lifting equipment used outside the business.

The regulations outline the requirements regarding the making and keeping of reports and exemptions for the armed forces.

The application of LOLER must be risk-based and take into account not only the equipment performing the lift, but also the nature of the goods being lifted and the height to which they are raised [20–21]. It should be noted that equipment with more than one function, for example a forklift truck, may be subject to both sets of regulations. The truck as a whole would fall under PUWER, but the lifting mechanism would be regulated using LOLER.

Machinery protection

There are three basic principles of machinery protection that should be taken into consideration when applying occupational safety legislation.

Intrinsic safety – Intrinsically safe machinery is that which is safe without any further additions or alterations. This means, for example, that when dangerous parts are in motion, or in use in some other way, they are not accessible to the operator or any other person.

This principle cannot always be applied. However, it is essential that potentially dangerous machinery should be made intrinsically safe wherever possible by virtue of design. This is an important aspect of BS EN ISO 12100:2010 referred to earlier (see Note 45).

Safety by position – Care should be taken concerning this aspect of machine safety. A machine should not be assumed to be safe because of its position; for example, just because it is only approached infrequently as it is 'out of the way'. In such cases, a judgement must be made of the circumstances involved, bearing in mind the requirements of the 1974 Act.

Fencing or guarding of machinery – Ideally, guards should be designed as an integral part of the machine and not added on as an afterthought. Essentially, fixed guards must be robust, able to withstand severe treatment, adjustable and safe to use. They should also be capable of protecting operators or people in the vicinity against injury. This may mean that as well as forming a physical barrier, the guard may have additional functions, such as assisting the removal of toxic fumes or reducing noise to a safe level.

Many different types of guard are available but, in general, there is an order of preference, which is given in the following:

- Fixed guard;

  This type of guard is usually preferable, as it is securely attached to the machine and permanently guards the dangerous part or parts. The guard should be sufficiently large to stop anyone reaching over it, and it should be sufficiently distant from the dangerous parts so that it does not form a trapping zone between itself and any moving parts.

  This type of guard is not suitable in many circumstances, for example, where workpieces have to be inserted and removed regularly, or where regular maintenance is necessary.

  It is often designed with a necessary opening in it, for example, for a workpiece to contact a cutting blade. In such circumstances, the opening must be as limited as possible. In some circumstances,

the operator is removed from the dangerous part by the incorporation of a sleeve or tunnel guard. This topic is dealt with in greater detail in BS EN 292–1.

When examining fixed guards, several points should be considered. The guard must be of sound and substantial construction, in good repair and trapping or entanglement should not be possible. Any openings should be acceptable only if the trapping area is not accessible through them.

- Interlocked guard;

  There are two types of interlocked guard.

  The electrical interlock consists of an electrical switch mechanism, which can be installed in a variety of ways. In one form, when the guard is moved, to give access to the dangerous parts, then the power to the machine is cut off, thus rendering the machine relatively safe. This is not acceptable if a dangerous part such as a cutting blade can still be contacted. By corollary, if such a guard is open, then the machine cannot be started up.

  It is important that if such systems fail, then they 'fail to safety', that is, the electrical interlock will ensure that the machine remains incapable of being operated until the interlock is replaced or repaired.

  The mechanical interlock is a moveable guard where, when it is open, the machine cannot be operated, and when the machine is in motion, it is not possible to open the guard. An interlock guard should satisfy the tests for a fixed guard. In addition, the extent to which the guard can be opened can be examined. Also, as the guard is closed, the operation of any limit switches should be noted, that is, the extent to which the guard is in its correct position, before the machine will operate. Because it is a moveable guard, signs of wear should be looked for.

- Automatic guard;

  This is one where the dangerous parts are enclosed while the machine is prepared for operation. The guard may open to an extent where the work can be removed when a work cycle is completed. When the work is removed, the guard closes again. Normally, the guard is closely fitted around the workpiece.

Examination of this type of guard should extend to the tests for fixed guards. In addition, when the moving part of the guard is closed, it should not be possible to reach the dangerous parts. Finally, the extent of wear of any of the guard components should be assessed.

- Trip guard.

These may be moveable mechanical guards, which are operated when a person approaches a dangerous machine beyond a safe distance. A variation on this is where the guard incorporates either photo-electric cells, where a light beam can be disrupted, or an induction field detection coil, where the presence of a person will alter an electrical field. In both cases, when the sensor device is triggered, the machine can be shut off, to render it safe. Trip guards must satisfy the tests outlined here for the other guards.

It should be remembered that machine safety does not stop with the presence of a guard. Any guards must be made from suitable materials and must be maintained regularly. If maintenance can only be carried out by removing guards, consideration must be given to providing some other form of protection to ensure a safe system of work.

All machine controls must be located in such a position that they cannot be operated unintentionally. Also, emergency cut-off switches should be clearly identified, readily accessible and in working order. People operating machines should understand any risks involved and be capable of rendering the machine safe in any emergency.

A common source of potentially dangerous machines is commercial kitchens. The hazards are associated with a variety of machines, including food slicers,

food processors, planetary mixers and waste-compactors, pie- and tart-making machines, and should incorporate some of the devices referred to here. There is section of the HSE website for catering and hospitality.[66]

Lift trucks

The legislative controls relating to these are contained in the LOLER (see earlier).

Forklift trucks represent a special type of mobile machine that has become commonplace in factories, warehouses and large shops. They are especially useful for moving and storing goods, but they do feature predominantly in accident statistics. Some injuries involving lift trucks are fatal. The key aspects involved in the safe use of these machines are dealt with in [22].

Frequently, forklift truck accidents are associated with a lack of suitable operator training. Therefore, the need for employers to provide suitable training in this field should be recognised in order to comply with their legal duties under the Health and Safety at Work, etc. Act 1974.

It is important to appreciate the characteristics of forklift trucks, so that the hazards that arise during their operation can be understood. It is also important to ensure that the racks used to store and unload many of the materials where lift trucks are employed can carry imposed loads, are in a good state of repair and are fixed securely in position.

Forklift trucks are designed to lift relatively heavy loads, most commonly at the front of the vehicle, although there are some side-loading lift trucks. If the load is too heavy, the truck can be tipped over. Also, if the vehicle is unevenly loaded, driven on sloping or uneven ground, or cornered at excessive speed, the stability of the vehicle can be affected, resulting in either shedding of the load or turning over of the truck.

The truck is more likely to turn over if the load is raised to a high level, as the centre of gravity of the vehicle is raised.

The construction and use of forklift trucks is controlled by a variety of British Standard specifications. Operators and inspectors should be familiar with the information that is required to be displayed on forklift trucks, which is as follows:

- The manufacturer's name;
- The type of truck;
- The serial number;
- The unladen weight;
- The lifting capacity;
- The load centre distance;
- The maximum lift height.

These details must be recorded when investigating any accident or dangerous occurrence involving forklift trucks.

Lift trucks are mobile machines, powered either by electric batteries or internal combustion engines. There are risks associated with each of these types. Electric batteries must be recharged, and this process involves the liberation of hydrogen gas. Battery charging should therefore only be carried out in a well-ventilated area where smoking is prohibited, and sources of ignition are adequately protected or removed.

In buildings where forklift trucks are driven by internal combustion engines, effective ventilation is necessary to remove exhaust gases, which are toxic. In addition, areas used for refilling vehicles with petrol or diesel fuel should be located in the open air. Liquefied petroleum gas (LPG) cylinders should also be changed in a well-ventilated area. There must be no ignition sources in the vicinity of refuelling areas. These battery changing areas should also have access to eye wash/shower facilities, where necessary, so that they can be used in the event of a battery spill on an individual.

Because of the risk of causing an explosion, forklift trucks should not be used in premises where flammable vapours, gases or dusts are likely to be present, unless they have been specially protected for such use.

Operator training – Because of the extent and nature of accidents involving forklift trucks, the HSE has published an Approved Code of Practice [22]. This document essentially provides practical guidance for employers with regard to their duties under section 2 of the Health and Safety at Work, Etc. Act 1974: 'to provide such information, instruction, training, and supervision as is necessary to ensure, so far as is reasonably practicable, the health and safety at work of his employees'.

One important point of this Approved Code of Practice is that operator training must only be carried out by instructors who have themselves undergone appropriate training in instructional techniques and skills assessment.

The training provided should be largely practical in nature and should be provided 'off-the-job', so that trainees and instructors are not diverted by other considerations.

Testing of trainees should be carried out by continuous assessment as well as by a test or tests of the skills and knowledge required for safe lift truck operation. The employer should keep records of each employee who has completed the basic training and testing procedure. In the case of an accident, the availability of this type of record could be helpful to the outcome of the investigation.

Violence and aggression

Work-related violence is defined by the HSE as 'any incident in which a person is abused, threatened or assaulted in circumstances relating to their work' [23].[67]

Violence and aggression towards employees is a common issue that workplaces can face, especially where there are interactions between employees and members of the public. Not only is it unacceptable but it has physical and psychological health effects on those involved.

The risk of being exposed to violence depends on the employee's occupation and the interaction with the public and other employees. Employees are at a higher risk of violence especially if they handle money, work with violent people, provide care, give advice, deal with complaints, work in education, work unsocial hours or have the power to act against the public, such as the police.

It is important for both employers and employees to reduce violence and aggression at work as the consequences can lead to concerns in employees' physical health and psychological well-being in the form of stress and anxiety.

Employers can suffer from poor morale amongst colleagues, poor image and reputation for the organisation and colleague recruitment and retention problems. Extra costs can also be seen by the organisation with higher absenteeism, insurance premiums and compensation payments.

There are five main pieces of health and safety law which are relevant to violence at work:

- The Health and Safety at Work Act 1974;
- The Management of Health and Safety at Work Regulations 1999;
- The Reporting of Injuries, Diseases and Dangerous Occurrences regulations 2013;
- Safety Representatives and Safety Committees Regulations 1977 (a) and the Health and Safety (Consultation with Employees) Regulations 1996 (b).

Under the provisions of all these regulations employers are responsible for identifying and managing the risk of violence at work. They should provide clear policies in relation to violence detailing their own responsibilities as well as those of their workforce and set standards for workplace behaviour. Employers should encourage employers to raise any concerns about the risk of violence, discuss precautions and report any attacks. The HSE has produced a guidance note [23], which provides guidance for employees on violence at work.

Stress

Stress can occur in a variety of forms. Some would argue that it is a desirable part of our everyday existence. However, prolonged

exposure to stress can be extremely harmful, leading to coronary heart disease, hypertension and gastrointestinal disorders. The Health and Safety Executive (HSE) defines stress as: '*The adverse reaction people have to excessive pressures or other types of demands placed on them*'.

At a social level, excess stress can result in the disruption of relationships. Within the workplace, people suffering from excessive stress may be more likely to lose concentration, possibly at vital moments, thereby increasing the risk of death by injury to themselves or to a fellow employee. Prolonged exposure to stressors can lead workers to the point of nervous breakdown or serious adverse impact on the health of individuals.

Given that stress can be extremely harmful to many people, it is important for both employers and enforcement officers to recognise those factors, including the physical factors discussed here, that may result in stress.

In response to the reported adverse effects of stress in the workplace, the HSE have developed a number of useful guides and policy documents for employers, managers and employees, which are available via the HSE's website.[68]

There is a growing body of case law on this topic, combined with examples where employees' claims have been settled out of court. Perhaps the most significant case is that reported in the Court of Appeal in the case of *Sutherland and others v. Hatton and others* (2002).[69] The court set out a number of practical propositions for future claims concerning workplace stress.

1   Employers are entitled to take what they are told by employees at face value unless they have good reason to think otherwise. They do not have a duty to make searching enquiries about employees' mental health.
2   An employer will not be in breach of duty in allowing a willing employee to continue in a stressful job if the only alternative is to dismiss or demote them. The employee must decide whether to risk a breakdown in their health by staying in the job.

3   Indications of impending harm to health at work must be clear enough to show an employer that action should be taken, for a duty on an employer to take action to arise.

4   The employer is in breach of duty only if they fail to take steps, which are reasonable, bearing in mind the size of the risk, the gravity of harm, the cost of preventing it and any justification for taking the risk.

5   No type of work may be regarded as intrinsically dangerous to mental health.

6   Employers, who offer a confidential counselling advice service, with access to treatment, are unlikely to be found in breach of duty.

7   Employees must show that illness has been caused by a breach of duty, not merely occupational stress.

8   Compensation will be reduced to take account of pre-existing conditions or the chance that the claimant would have fallen ill in any event.

Work at height

In many workplace situations, for example, construction sites, factories or offices, work has to take place at height. This work will involve activities such as maintenance operations, installation of plant and equipment or window cleaning. The risk of falling is a common hazard of such operations and because of reported accident statistics, this has also been set as a priority topic for enforcement officers to pay particular attention to as it can sometimes result in fatal accidents. Two outstanding routes to this type of accident have been typified, first relates to the means of access to the work situation, and second to the system of work adopted once the working position is reached [24].

The Work at Height Regulations 2005[70] control the risk of working at height, and is not confined to construction and factory work. In fact, a place is determined to be 'at height' if a person could be injured falling

from it. The regulations detail a hierarchy for managing and selecting equipment for working at height [25].

Employers must:

- Avoid work at height where possible;
- Use work equipment or control measures to prevent falls;
- Use work equipment or other control measures to minimise the distance and consequences of a fall (fall arrest equipment).

Exercises such as risk assessments, safety surveys and safety audits will reveal hazards, and meetings of safety committees and discussion with safety representatives will also lead to the establishment of safe systems of work. When considering the risks of working at height, factors such as the provision of safety harnesses and belts, including the necessary anchorage points for their safe use, and protective clothing such as suitable head covering, gloves and safety footwear for employees need to be borne in mind. If work at height cannot be avoided and fall arrest equipment is required, a fall protection plan should also be developed to ensure that the process is managed and remains effective. A fall protection plan should include:

- A policy on working at height;
- Risk assessments;
- A consideration of control measures;
- Training;
- Inspection and maintenance procedures for fall arrest equipment.

The plan should also be reviewed on a continual basis to allow for a system which constantly improves and becomes safer.

Vibration

There are two classifications of vibration, both of which are governed by the Control of Vibration at Work Regulations 2005.[71] These are hand-arm vibration and whole-body vibration.

Hand-arm vibration – Is caused using hand-held power tools, hand-guided power equipment and powered machines which process hand-held materials. Hand-arm vibration is caused as the vibration is carried from the process and into the worker's hands and arms, and as such is commonly associated with pneumatic hammers, lawnmowers and grinders. Ill health is often acute, with symptoms including tingling in the fingers and a temporary loss of strength. These symptoms can be an indicator of problems later in life if exposure to vibration is frequent and intense. Chronic symptoms include pain, sleep disturbance and reduced grip strength. As these symptoms develop, individuals lose the ability to do fine work and the ability to perform normal social activities can be lost [26–27].[72]

HSE guidance outlines the responsibilities of employers under the regulations. Following on from the requirement under the Management of Health and Safety at Work regulations, employers shall assess the vibration risk to employees,[73] and from that assessment determine whether or not colleagues shall be subject to vibration above the daily exposure action value of 2.5m/s^2 $A(8)$ based on an average over an 8-hour day. Where this is the case, a programme of controls shall be introduced. Employers must also identify if employees are likely to be subject to vibration above the daily exposure limit value of 5 m/s^2 $A(8)$. Should this be the case, action must be taken immediately to reduce the value.

Risks from hand-arm vibration can be effectively controlled in several cost-effective ways, including:

- Adapting alternative work methods;
- Selecting suitable equipment;
- Replacing old equipment;
- Considering workstation design;
- Maintaining equipment to a high standard;
- Adjusting work schedules;
- Providing protective clothing.

Whole body vibration – Involves the shaking or jolting of the body through a surface, most often a seat or the floor. Whole body vibration is common amongst construction workers and those working in agriculture who use vehicles on rough terrain or operate earth-moving machines. Back pain is the most likely symptom from whole body vibration, and it is often attributable to:

- Poor design;
- Incorrect adjustment of seat by driver;
- Sitting in one position for long periods;
- Poor driver posture;
- Repeated performance of manual handling tasks by the driver.

As with hand-arm vibration, employers must assess the risk to employers, and should also consider vulnerable workers, such as older people, young people, pregnant women and those that have already suffered back and neck pain. The assessment shall determine whether or not colleagues are subject to vibration above the daily exposure action value of 0.5 m/s^2 $A(8)$. Where this is the case, a programme of controls shall be introduced. Employers must also identify if employees are likely to be subject to vibration above the daily exposure limit value of 1.15 m/s^2 $A(8)$. Should this be the case, action must be taken immediately to reduce the value. It is highly likely that many off-road machines will exceed the exposure limit value at some point, but this will depend on the speed of the vehicle, the ground conditions and the driver's level of skill, which all need to be managed [28].

The risks can be controlled in a number of ways, including:

- Training and instructing operators and drivers;
- Choosing suitable machinery;
- Maintaining the machinery and roadways.

Section 4 Notes

33 SI 1992 No 2793.
34 SI 1999 No 3242.
35 SI 1998, No 2307.
36 SI 2005 No 1643.

37 SI 2002 No 2677.
38 SI 1989 No 635.
39 See http://www.hse.gov.uk/pubns/books/hsr25.htm
40 SI 2015 No 51.
41 SI 2005 No 1541 as amended by the Fires Safety Act 2021.
42 See https://www.legislation.gov.uk/asp/2005/5/contents and also SI 2006 No 456 and also https://www.gov.scot/policies/fire-and-rescue/non-domestic-fire-safety/
43 See https://www.hse.gov.uk/toolbox/fire.htm
44 SI 2002 No 2776.
45 SI 2015 No 483.
46 SI 2005 No 1082.
47 See https://www.gov.uk/workplace-fire-safety-your-responsibilities/fire-risk-assessments
48 SI 2002 No 1144.
49 SI 2002 No 2677 and see https://www.hse.gov.uk/coshh/index.htm
50 SI 2005 No 1643.
51 SI 2002 No 2675.
52 The PPE Directive 89/686/EEC was repealed with effect from April 2018 and the PPE Regulation (EU) 2016/425 came into force (see https://www.legislation.gov.uk/eur/2016/425/contents); in addition, the WHO has prepared technical specifications for PPE for COVID-19. See https://www.who.int/publications/i/item/WHO-2019-nCoV-PPE_specifications-2020.1
53 BS EN 352–2:2020 Hearing protectors. General requirements – Earplugs
54 BS EN ISO 4869–1:2018 Acoustics. Hearing protectors – Subjective method for the measurement of sound attenuation.
55 BS EN 352–1:2020 Hearing Protectors. General requirements – Earmuffs.
56 BS EN 529:2005 Respiratory protective devices. Recommendations for selection, use, care and maintenance. Guidance Document (currently under review).
57 BS EN 13794:2002 Respiratory protective devices. Self-contained closed-circuit breathing apparatus for escape. Requirement, testing, marking (currently under review).
58 BS EN 1146:2005 Respiratory protective devices. Self-contained open circuit compressed air breathing apparatus incorporating a hood for escape. Requirements, testing, marking (currently under review).
59 BS EN 14594:2018 Respiratory protective devices. Continuous flow compressed air line breathing devices. Requirement, testing and marking.
60 BS EN 166:2002 Personal eye protection. Specifications.
61 BS 7029:1999 Eye protection for industrial and other uses. Guidance on selection, use and maintenance.
62 SI 2002 No 2776.
63 BS EN ISO 12100:2010 Safety of machinery. General principles for design. Risk assessment and risk reduction (under review).
64 SI 1998 No 2306.
65 SI 1998 No 2307 and see https://www.hse.gov.uk/work-equipment-machinery/loler.htm
66 See https://www.hse.gov.uk/catering/guidance.htm
67 See also Work-related violence case studies – Managing risk in smaller businesses at https://www.hse.gov.uk/pUbns/priced/hsg229.pdf
68 See https://www.hse.gov.uk/stress/what-to-do.htm and https://www.hse.gov.uk/stress/risk-assessment.htm
69 *Hatton v. Sutherland* (2002) EWCA Civ 76 (2002) PIQR P241.
70 SI 2005 No 735.
71 SI 2005 No 1093.
72 See also https://www.hse.gov.uk/VIBRATION/hav/index.htm for more information
73 See https://www.hse.gov.uk/vibration/hav/source-vibration-magnitude-app3.pdf

Section 4 References

[1] Health and Safety Executive. (2004) *Manual Handling – Guidance on Regulations*, L23, HSE Books, Sudbury, Suffolk, 3rd ed. http://www.hse.gov.uk/pubns/priced/l23.pdf.
[2] HSE. (2008) *Manual Handling Assessment Charts*, INDG383, HSE Books, Sudbury, Suffolk. http://www.hse.gov.uk/pubns/indg383.pdf.
[3] Health and Safety Executive. (2005) *Workplace Transport Safety: An Employers' Guide*, HSG136. HSE Books, Sudbury, Suffolk.
[4] Health and Safety Executive. (2002) *Safety in Working with Lift Trucks*, HSE Books, Sudbury, Suffolk.
[5] HSE. (2021) *Driving and Riding Safely for Work: Health and Safety Executive*. https://www.hse.gov.uk/roadsafety/employer/the-law.htm.
[6] Scadden B. (2002) *Sixteenth Edition IEEE Wiring Regulations Explained and Illustrated*, Newnes, London, 6th ed.
[7] Health and Safety Executive. (2015) *Electricity at Work Regulations 1989*, Guidance on the Regulations HSR25, HSE Books, Sudbury, Suffolk, 3rd ed.
[8] Health and Safety Executive. (2015) *Managing Health and Safety in Construction – Guidance on Regulations L153*, HSE Books, Sudbury, Suffolk.
[9] Health and Safety Executive. (2013) *Personal Protective Equipment at Work – a Brief Guide, INDG 174 (rev2)*, HSE Books, Sudbury, Suffolk.

[10] Health and Safety Executive. (2015) *Personal Protective Equipment at Work, Guidance on Regulations L25*, HSE Books, Sudbury, Suffolk, 3rd ed.

[11] Chartered Institute of Environmental Health. (1991) *Acquired Immune Deficiency Syndrome: Guidance Notes for Environmental Health Officers*, CIEH, London.

[12] Coates T, Clarke ARL. (2003) Occupational diseases. In *Risk Management – Safety at Work*, Ridley J, Channing J (Eds.), Butterworths, London, 6th ed.

[13] Else D. (1981) *Occupational Health Practice*, Butterworths, London, 2nd ed.

[14] Health and Safety Executive. (2005) *Noise at Work, INDG362*, HSE Books, Sudbury, Suffolk.

[15] Health and Safety Executive. (2013) *Respiratory Protective Equipment at Work – Practical Guide, HSG53*, HSE Books, Sudbury, Suffolk, 4th ed.

[16] Hartley C. (1999) Occupational hygiene. In *Risk Management – Safety at Work*, Ridley J (Ed.), Butterworths, London, 4th ed.

[17] Health and Safety Executive. (1999) *Essentials of Health and Safety at Work*, HSE Books, Sudbury, Suffolk, 3rd ed.

[18] HSE. (2014) *The Health and Safety Toolbox – How to Control Risks at Work, HSG 268*, Crown Copyright, HSE Books, Sudbury, Suffolk. http://www.hsc.gov.uk/pubns/priced/hsg268.pdf.

[19] HSE. (2013) *Providing and Using Work Equipment Safely: A Brief Guide, INDG291 (rev1)*, HSE Books, Sudbury, Suffolk. http://www.hse.gov.uk/pubns/indg291.pdf.

[20] HSE. (2013) *Lifting Equipment at Work – a Brief Guide, INDG 290 (rev1)*, HSE Books, Sudbury,

Suffolk. http://www.hse.gov.uk/pubns/indg290.pdf.

[21] HSE. (2014) *Safe Use of Lifting Equipment – Lifting Operations and Lifting Equipment Regulations 1998: Approved Code of Practice*, HSE Books, Sudbury, Suffolk. http://www.hse.gov.uk/pubns/priced/l113.pdf.

[22] HSE. (2013) *Rider-Operated Lift Truck – Operator Training and Safe Use, Approved Code of Practice and Guidance, L117*, HSE Books, Sudbury, Suffolk, 3rd ed. http://www.hse.gov.uk/pubns/priced/l117.pdf.

[23] HSE. (1996) *Violence at Work – a Guide for Employers, INDG69 (rev10/96)*, HSE Books, Sudbury, Suffolk. http://www.hse.gov.uk/pubns/indg69.pdf.

[24] Multi Author. (2012) *Tolley's Health and Safety at Work Handbook*, Tolley (LexusNexis), London, 25th ed.

[25] Health and Safety Executive. (2014) *Working at Height: A Brief Guide INDG401 (rev2)*, HSE Books, Sudbury, Suffolk. http://www.hse.gov.uk/pubns/indg401.pdf.

[26] HSE. (2012) *Hand-Arm Vibration at Work: A Brief Guide, INDG175(rev 3)*, HSE Books, Sudbury, Suffolk. http://www.hse.gov.uk/pubns/indg175.pdf.

[27] HSE. (2005) *Hand-Arm Vibration – The Control of Vibration at Work Regulations 2005 Guidance on Regulations, L140*, HSE Books, Sudbury, Suffolk. http://www.hse.gov.uk/pubns/priced/l140.pdf.

[28] HSE. (2005) *Control Back-Pain Risks from Whole-Body Vibration, INDG242*, HSE Books, Sudbury, Suffolk. http://www.hse.gov.uk/pubns/indg242.pdf.

SECTION 5: OCCUPATIONAL HAZARDS

Introduction

In this section reference will be made to hazards that are discussed in detail in other chapters, in particular Chapter 10, and so readers are referred to that chapter for more detailed information.

Classification of hazardous substances

Many of the substances used in the workplace are obviously dangerous, but there are

many more that are not obviously hazardous. These hazards may be biological, physical or chemical.

Classification, Labelling and Packaging Regulations (CLP) – In 2009, the European Regulation (EC) No 1272/2008 on classification, labelling and packaging of substances and mixtures came into force. The CLP Regulations, in conjunction with the United Nations Globally Harmonised System (GHS), should ensure that classifying substances and mixtures is easier and cheaper by allowing global flexibility and unification. The Regulations have been amended and retained in Great British law as GB CLP following the UK's exit from the EU [1]. Substances and mixtures placed on the market in Northern Ireland however are subject to the EU CLP Regulation (placed on the market includes import into the territory).

'This Regulation should ensure a high level of protection of human health and the environment as well as the free movement of chemical substances, mixtures and certain specific articles, while enhancing competitiveness and innovation' (Regulation (EC) 1272/2008).[74]

Globally Harmonised System (GHS) – Developed by the UN, GHS is a voluntary agreement; for it to become legally binding 'it has to be adopted through a suitable national or regional legal mechanism' which is what enables CLP throughout the EU [1].

Like CHIP before it, CLP concerns general mixtures and substances. Specific chemicals are subject to their own control measures and legislations (e.g., biocides; carcinogens and pesticides) [2].

CLP has replaced the Dangerous Preparations Directive and the Dangerous Substances Directive, which were implemented through the now-revoked CHIP Regulations. CLP ensures that consumers and workers across the EU are aware of the hazards that a mixture (preparation), or a substance could have by way of clear classification and labelling. Chemical suppliers must 'establish the potential risks to human health and the environment of such substances and mixtures, classifying them in line with the identified hazards' [3].

The hazardous chemicals covered by CLP must have clear labels in line with the GHS to allow the consumer/worker to know what type of chemical it is and what effects it can have on them and the environment [3].

The CLP does not apply to the following:

- Substances/mixtures subject to customs supervision;

- Radioactive substances/mixtures;
- Substances/mixtures for scientific research and development which are not placed on the market;
- Non-isolated intermediaries;
- Waste.

CLP also does not apply to the following categories of chemicals that are in a finished state:

- Cosmetics;
- Food;
- Medicines/medical devices and veterinary medicines;
- Feed stuffs (e.g., used in animal nutrition; flavourings and additives for food) [4].

CLP puts an obligation onto 'manufacturers, importers and downstream users to classify substances and mixtures placed on the market' and 'suppliers to label and package substances and mixtures placed on the market'.

Transportation and CLP – CLP does not cover transport of dangerous goods by sea, road, air, rivers, canals or rail, unless Article 33 applies.

Article 33 of the CLP states

> where a package consists of an outer and an inner packaging, together with any intermediate packaging, and the outer packaging meets the labelling provisions in accordance with the rules on the transport of dangerous goods, the inner and any intermediate packaging shall be labelled in accordance with this Regulation.

Outer packaging can also be labelled as per CLP.

Any outer packaging that does not need to be labelled as per the transport of dangerous goods should be labelled as per CLP (including inner and intermediate packaging). Outer packaging does not need any labels if the labels inside the packaging (either intermediate or inner) can be seen through it.

CLP and Safety Data Sheets – Underlying various chemical regulations is the need for a Safety Data Sheet (SDSs). Suppliers of chemical substances/mixtures are required to provide SDSs, and these are now aligned with the GHS under Article 31 and Annex II of REACH (Registration, Evaluation, Authorisation and Restriction of Chemicals 2007).[75]

Under Article 31 and Annex II of REACH, and the CLP and GHS requirements, SDSs must contain:

1 Identification of the substance/mixture and of the company/undertaking
2 Hazard(s) identification
3 Composition/information on ingredients
4 First aid measures
5 Fire-fighting measures
6 Accidental release measures
7 Handling and storage
8 Explosive controls and personal protection
9 Physical and chemical properties
10 Stability and reactivity
11 Toxicological information
12 Ecological information
13 Disposal considerations
14 Transport requirements
15 Regulatory information
16 Other information as appropriate [6]

CLP has introduced new pictograms, definitions and hazard statements, and this information is guided by the information contained within a chemicals' Safety Data Sheet.

The CLP glossary is as follows:

- Hazard pictogram – graphical indication of the hazard in a red diamond;
- Hazard class – health, physical or environmental;
- Hazard categories – this is a division of the criteria within hazard classes which specifies the severity of the hazard;
- Signal words – two are used:
  - Danger – for more severe hazard categories;

- Warning – for less severe hazard categories.
- Hazard statement – each statement relates to a specific hazard class and category. It describes 'the nature of the hazards of a hazardous substance or mixture, including, where appropriate, the degree of hazard'.
- Precautionary statement – provides the recommended steps to take to prevent or minimise the effects of being exposed to the hazardous substance/mixture during its use or when disposed of [2].

Hazard pictograms – CLP determines the hazard pictograms that must be used, according to Annex 1 – Part 2 covers physical hazards, Part 3 health hazards and Part 4 covers environmental hazards. For substances or mixtures that have been classified as having more than one hazard associated with them, several pictograms may be required. Certain pictograms take precedence over others, and only the more severe hazard pictogram has to be used.

Hazard statements are split by hazard class.
Physical – H200, including:

- H200 unstable explosive;
- H221 flammable gas;
- H242 heating may cause a fire;
- H290 may be corrosive to metals.

Health – H300, including:

- H300 fatal if swallowed;
- H301 toxic if swallowed;
- H302 harmful if swallowed;
- H331 toxic if inhaled;
- H350i may cause cancer by inhalation;
- H373 may cause damage to organs.

Environmental – H400, including:

- H400 very toxic to aquatic life;
- H410 very toxic to aquatic life with long-lasting effects.

GHS01: Exploding bomb – explosive, self-reactive, organic peroxide

GHS02: Flame – flammable gases, liquids and solids, flammable aerosols organic peroxides, self-reactive, self-heating, contact with water emits a flammable gas, pyrophoric

GHS03: Flame over circle – oxidising gases, liquids and solids

GHS04: Gas cylinder – gasses under pressure

GHS05: Corrosion – corrosive (severe eye damage and burns to the skin)

GHS06: Skull and cross bones – acute toxicity, very toxic (fatal), toxic

GHS07: Exclamation mark – harmful skin irritation and serious eye irritation

GHS08: Health hazard – respiratory sensitiser, carcinogen, mutagen, reproductive toxicity, aspiration hazard, systemic target organ toxicity

GHS09: Environment – harmful to the environment [7]

Figure 14.6   Hazard pictograms

CLP labels must include:

1 Contact details of supplier
2 Product identifiers (usually from SDSs), name (can be the trade name) and identification number
3 Hazard pictograms
4 Signal words
5 Hazard statements
6 Precautionary statements
7 Supplemental labelling information [5]

Adaptations to technical progress – CLP is regularly updated to incorporate technical and chemical developments (adaptations to technical progress – ATPs). There are two main ATPs that are used to update CLP:

1 Proposed new harmonised substance classification' which will occur approximately every year
2 Changes to GHS 'classification criteria and technical annexes' which will occur approximately every other year and hopes to 'align GHS more closely to the transport regulations'.[76]

The Control of Substances Hazardous to Health Regulations 2002[77] (As Amended) – These regulations referred to as COSHH, were made under the Health and Safety at Work, etc. Act 1974 and apply to substances that have already been classified under the provisions of the CLP Regulations as outlined earlier. They also apply to:

1 Substances that have an approved workplace exposure limit
2 Substances that are a biological agent
3 A substantial concentration of dust of any kind
4 Any other substance that is not listed in any of the aforementioned categories, but that creates a risk to health due to its chemical or toxicological properties, or the way in which it is used

It must be noted that not all substances that can be a hazard to health are controlled by these regulations. Some examples of these 'exempted' substances include:

- Coal – so far as the Coal Mines (Control of Inhalable Dust) Regulations 2007[78] apply;
- Lead – so far as the Control of Lead at Work Regulations 2002[79] apply;
- Asbestos – as far as the Control of Asbestos Regulations 2012[80] apply;
- Substances that are only hazardous to health, solely by virtue of any radioactive, explosive or flammable properties, or solely because it is at a high or low temperature or a high pressure.

There are a number of substances that are 'prohibited' under these regulations. The substances and the extent to which they are prohibited are contained in Schedule 2 of COSHH.

The critical point of these regulations is that employers shall not carry on any work that is liable to expose any employees to any substance hazardous to health unless they have *made a suitable and sufficient assessment of the risks and of the steps that need to be taken to meet the requirements of the regulations.* Such assessments must be reviewed where a significant change in work occurs, or the original assessment becomes invalid for any other reason.

The assessment itself must include:

- An assessment of the risks to health;
- The steps that need to be taken to achieve adequate control of exposure to hazardous substances;
- Identification of any other action necessary to comply with the regulations.

The assessment should consider what types of substances employees are liable to be exposed to, and this must include consideration of the consequences of the possible failure of measures provided to control exposure, the form in which the substances may be present and their effect upon the body. Of great importance is a consideration of the extent to which other workers or other people (including non-employees) are likely to come into contact with the hazardous substances being assessed.

The employer must make an estimate of exposure levels and compare these to any available, valid standards. If the assessment indicates that control is, or is likely to be, inadequate, then the employer must determine the steps that must be taken to obtain adequate control. It should be remembered that the COSHH regulations require that personal protective equipment should be used as a method of exposure control only after all other methods have been employed as far as reasonably practicable.

An assessment will be considered suitable and sufficient if the detail and expertise with which it is carried out are commensurate with the nature and degree of the risk involved with the work. In some circumstances, it will only be necessary to read manufacturers or suppliers' safety information sheets to ensure that current working practices are satisfactory. In other cases, considerable atmospheric monitoring may be necessary before true exposure levels can be ascertained.

In addition to making an assessment of risks, employers are also required to ensure that the exposure of employees to substances hazardous to health is either prevented or, where this is not reasonably practicable, adequately controlled (by means other than personal protective equipment).

Where an employer provides any control measure to meet the requirements of these regulations, he is also required to ensure that those measures are maintained in an efficient state, in efficient working order and in good repair. Thorough examination and testing of engineering controls is required.

Where monitoring is carried out or is specifically required to be carried out, then a record of that monitoring must be kept, and the record itself or a summary must be kept available. Where the record is representative of the personal exposures of identifiable employees, this must be kept for at least 40 years. In any other case, the record must be kept for at least five years.

Where it is appropriate for the protection of the health of employees who are, or are liable to be, exposed to a substance hazardous to health, the employer must ensure that those employees are subject to suitable health surveillance, including biological monitoring where necessary.

An additional requirement of these regulations is that where any employee is exposed, or may be exposed, to substances hazardous to health, then the employer must provide the employee with such instruction and training as is suitable and sufficient for that employee to know:

- The risks to health created by such exposure;
- The precautions that should be taken.

This would include the disclosure of the results of any monitoring of exposure in the workplace and information on any collective health surveillance (so presented that it cannot be related to any individual).

Any person carrying on a work activity must understand the nature of any hazards associated with the materials being used. It is essential that a hazard data sheet be obtained from the supplier when acquiring potentially hazardous materials. This data must be read and understood and kept in a readily accessible place. The contents of such data sheets will form the basis of instructions and training given to employees, as well as of the selection and provision of any necessary protective equipment.

Safe arrangements must be made for the reception, storage and use of such materials and, as far as reasonably practicable, emergency procedures should be designed in liaison with agencies such as the HSE, local authority and fire authority, as necessary.

Finally, suitable arrangements should be made for any medical examinations (including health surveillance) required or deemed necessary, and first-aid facilities should be made available to the standard required by the enforcing authority of the Health and Safety at Work, Etc. Act 1974.

The COSHH Regulations 2002 provide a comprehensive set of measures that will act to improve the health and safety at work of employees and others affected by work activities. The HSE has published an Approved Code of Practice [6].

Workplace Exposure Limits (WELs) – Workers can be exposed to a hazardous substances at work (e.g., dust, fumes, chemicals and fibres) which can be harmful to their health. Exposure to these hazardous substances can cause ill health if not properly controlled. Harm can be caused by:

- Swallowing;
- Inhalation;
- Absorption through the skin;
- Direct contact with skin.

Some hazardous substances can cause illness long after exposure occurred. Guidance on the controls for absorption through the skin can be found in the COSHH Approved Code of Practice (ACOP) (currently the sixth edition) and HSE website.[81]

WELs are occupational exposure limits which have been established to protect the health of workers from the effects of the inhalation of hazardous substances. WELs are 'concentrations of hazardous substances in the air, averaged over a specific period of time, referred to as a Time-Weighted Average (TWA)' and have two time reference periods:

- Short-term – 15 minutes;
- Long-term – 8 hours.

Short-term exposure limits (STELs) are created to protect workers from effects that may occur after exposure for only a few minutes (e.g., eye irritation).

Substances with a given WEL are subject to requirements under the COSHH Regulations: adequate controls in place if a WEL is not exceeded, good control practice principles in place and exposure to carcinogens, asthmagens and mutagens has been reduced to as low as is reasonably practicable.

Substances that have not been assigned a WEL may not necessarily be safe and should be controlled 'to a level to which nearly all the working population could be exposed, day after day at work, without any adverse effects on health'.

Where there is no short-term limit, 'a figure three times the long-term exposure limit should be used'.

In WELs, airborne particle concentrations are expressed in mg.m^{-3} however, for volatile substances limits are expressed in ppm, therefore a conversion to mg.m^{-3} is required (see Equation 1):

Equation 1

$$\frac{WEL\ in\ ppm \times MWt}{24.05526} = WEL\ in\ \text{mg.m}^{-3}$$

Where MWt is molecular weight of the substance and 24.05526 takes account of normal atmospheric conditions in Great Britain.

The acquired figures for WELs in mg.m^{-3} should then be rounded to:

- Less than 0.1 = 1 significant figure;
- To less than 100 = 2 significant figures;
- 100 or over = 3 significant figures.

It should always be born in mind that other factors can put additional stresses on the human body, including UV exposure, high pressure, temperature and humidity which can all increase the toxicity of a substance on the body.

It should be noted that this guidance does not extend to detailed considerations of asbestos and lead exposure levels, where specific legislation exists, nor does it cover situations where work is below ground or exposure to microorganisms.

Reference periods – Substances that are considered to be hazardous to health may cause adverse effects that range from irritation of the skin or mucous membranes through to death. The effect may be produced either over a very short exposure period or over a much longer period. It is, therefore, important to develop exposure limits that reflect these differences, and this has been done with regard to WELs. These are listed either in the COSHH regulations or Guidance Note EH40 [7] as either 'long-term exposure limits' (eight-hour time-weighted average reference period) or 'short-term exposure limits' (15-minute reference period).

The eight-hour reference period refers to a well-established procedure whereby the sum of individual occupational exposures over a 24-hour period are treated as being equivalent to a single, uniform exposure for an eight-hour period. This is known as the 'eight-hour time-weighted average' (TWA) for the exposure, and may be represented mathematically by Equation 2:

Equation 2

$$\frac{C_1 T_1 + C_2 T_2 + \cdots\cdots\cdots C_n T_n}{8}$$

Where C_1 is the occupational exposure and T_1 is the associated exposure time in hours in any 24-hour period. See the following examples.

Short-term reference period – here exposure should normally be measured over the prescribed period for the hazardous substance concerned: this is normally 15 minutes. Measurements for periods greater than 15 minutes should not be used to calculate the short-term exposure [15].

Long-term exposure limits are thus designed to protect against the chronic effects of exposure, while short-term exposure limits are designed to avoid the acute effects of toxicants. Eight hours was selected to reflect the typical exposure during one working shift averaged over a 24-hour period, while the 15-minute period represents any 15-minute period during the working day. Where only short-term samples are taken, information such as episodic peak concentrations may be missed, whilst if a long-term sample is taken, significant high peak concentrations, which could be harmful, might not be detected as the result is averaged over that longer period. Monitoring systems and equipment must therefore be carefully selected and positioned in order to ensure that the data gathered can be usefully interpreted in the light of the available standards and guidance [7].

Monitoring strategies for toxic substances – Carrying out monitoring can be an expensive and time-consuming exercise; therefore

---

Example 1:

A person works for 7 hours and 20 minutes on a process where they are exposed to a substance hazardous to their health. The average exposure during that period is measured as 0.12 mg.m-3.

$$\text{8-hour TWA} = 7 \text{ h } 20 \text{ min } (7.33\text{h}) \text{ at } 0.12 \text{ mg.m-3}$$
$$\text{Then } 40 \text{ minutes } (0.67\text{h}) \text{ at } 0 \text{ mg.m-3}$$

Therefore (including rounding by 1 significant figure):

$$\frac{(0.12 \times 7.33) + (0 \times 0.67)}{8}$$
$$= 0.11 \text{ mg.m}^{-3}$$

Example 2:

As working with hazardous substances may be periodic, this example takes into account regular breaks a worker may have.

Working period	Exposure (mg.m⁻³)	Duration of sampling (h)
0800-1030	0.32	2.5
1045-1245	0.07	2
1330-1530	0.2	2
1545-1715	0.1	1.5

Exposure is assumed to be 0 in rest periods – equalling 1.25 hours. The 8-hour TWA:

$$\frac{(0.32\times2.5)+(0.07\times2)+(0.20\times2)+(0.10\times1.5)+(0\times1.25)}{8}$$

Therefore:

$$\frac{0.80+0.14+0.40+0.15+0}{8}$$

$$= 0.19 \text{ mg.m-3}$$

---

some form of monitoring strategy must be applied in order to gain sufficient data for the purposes of the exercise in as cost-efficient a way as possible. The data must, however, be accurate and reliable so that the requirements imposed by any health and safety legislation can be fulfilled. Suitable advice is contained in the HSE's Guidance Note HSC173 [8]. This document discusses the factors that influence airborne concentrations of hazardous substances and urges that a structured approach be taken. This procedure involves a number of distinct phrases for the monitoring work:

1 Initial appraisal
2 Basic survey
3 Detailed survey
4 Routine monitoring

With regard to these types of monitoring, a decision has to be made about the level of sophistication that the survey requires, both in terms of the quantity and quality of the data collected. Finally, advice is given concerning the interpretation of results. It is essential that anyone carrying out or recommending an assessment under COSHH is familiar with this document.

Various other documents produced by the HSE are of significance when considering this topic: General Method for the Gravimetric Determination of Respirable and Total Inhalable Dust [9], and General Methods for Sampling Gases and Vapours [10].

Asbestos

The Control of Asbestos Regulations 2012[82] updated the previous Regulations to ensure full implementation of EU Directive 2009/148/EC. The main change from the 2006 Regulations is that some non-licensed work with asbestos now has additional requirements.

The main requirement under 2012 Regulations is the duty to manage for those who manage nondomestic premises:

- Manage the risk from asbestos by establishing the extent of asbestos on the premises;
- Keeping accurate records of location and condition;
- Assessing the risk from the material;
- Prepare a plan showing how asbestos risks are to be managed;
- Implement the plan;
- Review and monitor the plan and the arrangements to put it into place;
- Provide information to those who may work on or disturb it.

This requirement is a key part of the HSE's work to raise awareness of the risks and also reduce risk from asbestos exposure.

When considering the problems associated with exposure to asbestos, it is important to understand the concept of 'the fibre'.

Work was undertaken in the late 1990s to establish a unified methodology for evaluating airborne fibres in the work environment, led by the World Health Organization [11]. The principle of the method is to draw a known volume of air through a membrane filter by a sampling pump. The filter is then cleared and mounted on a slide. The fibres that are visible on a measured area are counted using PCOM – Phase-Contrast Optical Microscopy and then the concentration of the fibres in the volume of air is calculated.

In this document the definition of a 'fibre' is an 'object with a length $>5\mu m$, a width of $<3\mu m$ and a length: width ratio (aspect ratio) $>3: 1$, using a phase-contrast optical microscope [11]. These are the particles that are counted, but it should be noted that, as it is not possible to identify the chemical nature of the fibres collected, all 'fibres' are counted as being asbestiform and used in determining

whether control limits and action levels have been exceeded.

The key features of the 2012 Regulations are as follows:

- Assessments;

An employer must assess adequately any work with asbestos to determine the nature and degree of any exposure, and the steps that must be taken in order to prevent or reduce that risk. In the case of work that consists of the removal of asbestos, a suitable written 'plan of work' must be prepared. This must include details of:

1   The nature and probable duration of the work
2   The location of the place of work
3   The methods to be applied
4   The characteristics of the equipment to be used for the protection and decontamination of those carrying out the work and the protection of other persons on or near the worksite

Employers are required to identify the type of asbestos involved in the work, or to assume that it contains brown asbestos (amosite) or blue asbestos (crocidolite), which are the most hazardous types.

Employers must generally provide the enforcing authority with the particulars specified in Schedule 1 of the regulations at least 14 days before commencing the work, unless the enforcing authority agrees to a shorter period.

- Prevention of exposure;

Employers have a duty to prevent or reduce the spread of asbestos from a workplace to the lowest level reasonably practicable. They must also prevent or reduce employees' exposure by means other than the use of respiratory protective equipment. Other related duties include the need to monitor the air for asbestos fibres and to provide information and training related to the risks involved and the precautions that must be taken.

Licensable work with asbestos is:

- Work on asbestos coating;
- Work whereby the workers exposure to asbestos is neither of low intensity or sporadic;
- Whereby the risk assessment cannot demonstrate the 0.1 f/cm^3 over a 4-hour average will not be exceeded;
- Work on asbestos insulating board/insulation and the risk assessment shows that the work is not of a short duration [12].

Notifiable non-licensed work (NNLW) – All non-licensed work on asbestos must be carried out with the appropriate controls in place, and some work requires additional requirements – this is classed as NNLW. It requires employers to:

- Maintain health records;
- Identify the area/s where work is being undertaken;
- Notify the appropriate authority;
- Make sure medical examinations are undertaken;
- If the work is not licensable, employers must establish if it is NNLW by considering the following:
- Type of work;
- Type of asbestos (e.g., if it is friable it is likely to be NNLW);
- Condition of the material – if it is in poor condition, it is likely to be NNLW [13].

Non licensed work – To be exempt from a licence, the work undertaken must be:

- Carried out so workers are not exposed to asbestos above the legal limit of 0.1 f/cm^3 averaged over a four-hour period;
- Sporadic and of low intensity and therefore not exceeding 0.6 f/cm^3 over a ten-minute period.

Non-licensed work must also meet at least ONE of the following criteria:

- Work that involves ACMs (Asbestos Containing Materials) in good condition which will be encapsulated or sealed and therefore will not be easily damaged in the future;
- Work that involves the removal of ACMs that are found to be in reasonable condition which won't be broken up and the fibres of asbestos are contained within another material such as paint, coatings and cement;
- The work being carried out is to check for fibre concentrations in the air, or collection and analysis of samples;
- The work being carried out is not continuous, is of short duration and is only on materials that are not friable [14].

Control limits – The control limit is the concentration of asbestos fibres measured and averaged over four hours, as referenced in the WHO method earlier; this is currently 0.1 asbestos fibres per cubic cm of air (0.1 f/cm^3). This is not what can be termed a 'safe level' and any work involving asbestos should be undertaken as far below the control limit as possible.

There is also a short-term exposure limit (STEL) identified in the ACOP [15]. Some work can be carried out on lower-risk ACMs if the risk assessment can show the control limit will not be exceeded and that the exposure of workers to any asbestos will be sporadic and of low intensity. Therefore, the control limit for this type of non-licensed work is 0.6 f/cm^3 over a ten-minute period. If the risk assessment cannot demonstrate any of these measures, the work must be carried out under licence by a licensed contractor.

A further useful publication is HSE publication HSG248 [16].[83] Of particular importance is the inclusion of a methodology for 'clearance monitoring', which must be carried out before a site is handed back to the control of its occupiers after asbestos removal operations. Using the specified method, the lowest fibre level that can be reliably detected above background levels in a 480 L sample is about 0.01 fibres/ml, and this level is therefore taken as the 'clearance indicator level'.

Sites should not be considered to be satisfactory until such a test has been successfully completed.

The Control of Asbestos in the Air Regulations 1990 apply currently only in Scotland as they were revoked by the Environmental Permitting (England and Wales) (Amendment) Regulations 2013,[84] in so far as they apply to England and Wales, and which prescribe a limit value for the discharge of asbestos from outlets into the air during the use of asbestos. These regulations also provide for the regular measurement of asbestos emissions from specified types of premises. Provision is also made for the control of environmental pollution by asbestos emitted into the air as a result of the working of products from the demolition of buildings, structures or installations containing asbestos. Finally, the Asbestos (Prohibitions) Regulations 1992 (as amended)[85] implement EU directive 91/659/EEC 'on restrictions relating to the marketing and use of dangerous substances and preparations (asbestos)'.

Legionella

Legionnaires' disease is a pneumonia-like condition caused by the bacterium *Legionella pneumophila*. The symptoms of this disease are similar to those of influenza: headache, nausea, vomiting, aching muscles and cough [17]. The bacterium thrives at 20–45°C but can survive cold temperatures and will source nutrients from rust, scale and other bacteria. Legionellosis is also considered in Chapter 10 (Section 6) and Chapter 12.

Legionnaires' disease was first identified after an outbreak of pneumonia among delegates attending an American Legion Convention for service veterans in Philadelphia in 1976, hence the name. Diagnostic tests were developed, and by testing stored specimens it was discovered that the disease could be traced back to the 1940s. The infection had escaped detection because the causative organism did not grow on conventional culture media.

Inhaling droplets containing viable *L. pneumophila* organisms that are fine enough to penetrate deeply into the lungs cause infection. There is no evidence that the disease can be transmitted from person to person, nor is the dose required to infect a person known. Males are more likely to be affected than females, and most cases have occurred in the 40–70-year-old age group.

During the past few years, there has been a growing awareness of this condition, and some highly publicised cases have heightened public awareness. In England and Wales, about 200–350 cases of Legionnaires' disease have been reported each year, of which between 10% and 20% have been fatal. Public Health England reported 308 confirmed cases between 1 January 2019 and 1 August of that year and 295 cases from January to October 2020.[86]

This infection has caused much concern because of its association with hot water systems and the cooling towers of air conditioning and industrial cooling systems, where there is a generation of fine water droplets. The organism may therefore be present in a large number of workplaces and any risk of infection is clearly within an employer's duties to employees and non-employees under sections 2 and 3 of the Health and Safety at Work Act 1974. The Control of Substances Hazardous to Health Regulations 2002 include provisions for managing biological hazards and require employers to carry out a risk assessment for any 'at risk' water systems. Control measures may include the monitoring of water systems and the provision of information and training.

It is therefore important to ensure that any high-risk areas in workplaces, such as cooling towers, are inspected on a regular basis to ensure that they present no risk to health and safety as far as is reasonably practicable.

The procedures for the identification and assessment of risk of legionellosis from work activities and water sources as well as the measures to be taken to reduce the risk from exposure to the organism are explained in the Approved Code of Practice [17].[87]

Procedures for the sampling of water services in all types of buildings, including domestic, commercial and industrial premises, are provided in BS 7592.[88] This includes sampling a variety of water services such as hot water services, cold water services, cooling systems and associated equipment. It should be noted that the British Standard is not applicable for sampling potable-water services or vending machines. There is a British Standard (ISO standard) for methods for the isolation of *Legionella* organisms and estimation of their numbers in water and related materials.[89]

Where a potential risk exists, it is essential that cooling towers and associated water systems are subject to routine sampling as part of the general water-treatment programme. It is suggested that records of the analytical laboratory employed, the source of the sample and the results obtained should be kept readily available for inspection.

Legionella bacteria can colonise water systems without being associated with infections. The risk of infection can be minimised by good engineering practice in the design, construction, operation, and maintenance of water installations. The best method of controlling *Legionella* is to operate the system in such a way that the growth of *Legionella* is prevented. This can be achieved by preventing stagnant water from collecting in pipes, controlling the release of water spray, treating the water and avoiding temperatures and conditions favourable to growth. Water treatment methods that are effective against *Legionella* include the use of biocides, ultraviolet irradiation, copper or silver ionisation and ozone. Where the water system is designed specifically for hot or cold water growth can be controlled by maintaining a hot water storage temperature of greater than 60°C and movement temperature greater than 50°C, or cold water temperature lower than 20°C.

In the case of systems that are liable to produce a spray or aerosol in operation, or where a spray could be generated incidentally during the cleaning or maintenance, additional precautions should be taken by the person who has responsibility for the plant under the Health and Safety at Work Act 1974.

Cooling towers are of particular concern, and it is important that all parts of the system are thoroughly cleaned and disinfected, normally twice per year (in spring and autumn). In order to achieve adequate disinfection, sodium hypochlorite solution may be used to give a concentration of at least five parts per million (ppm) of free available chlorine. Water should be sampled periodically near the circulation return point. Chlorine levels can be determined using swimming pool testing kits. Regular chlorination of 1–2 ppm is normally advocated, but this is not appropriate in all cases as it may react adversely with other water treatments. Some biocides are known to be effective against *Legionella* bacteria, although resistance has been known to develop. In such cases, shock dosing using chlorination of 25–50 ppm of free chlorine is effective. A level of at least 10 ppm should be maintained in the system for 12–24 hours. The use of biocides can be alternated with the use of chlorination to achieve a satisfactory level of disinfection.

In view of the risk of legionellosis, the Notification of Cooling Towers and Evaporative Condensers Regulations 1992[90] requires any person who has, to any extent, control of nondomestic premises to notify the local authority that such plant is situated on those premises, giving information set out in the schedule to the Regulations. This assists environmental health practitioners in ensuring that appropriate, effective treatment and control systems are in place, as well as being vital in the case of any outbreak.

It is important that a safe system of work is devised in all premises where cleaning, maintenance and testing of the type of plant just discussed is carried out. This system of work, as well as any instruction and training necessary, must be made known to the personnel involved. For example, the HSE recommends that suitable respiratory equipment, such as a high-efficiency positive-pressure respiratory with either a full-face piece or a hood and blouse, be used. Also, such operations must be carried out in such a way as to avoid risks to other people working in the vicinity or members of the public.

In addition to its occurrence in cooling systems and water services, it should also be noted that spa baths have been linked with various infections, including Legionnaires' disease. Spa baths are small pools where water is vigorously circulated by means of water and/or air jets. The water is usually quite warm (typically above 30°C) and is not changed between bathers as it is in the case of whirlpool baths. Careful water treatment is therefore needed for spa baths.

If an outbreak of Legionnaires' disease is suspected, past cases are a valuable source of reference. One useful publication is the *Broadcasting House Legionnaires' Disease*, which charts the investigation carried out by Westminster City Council in 1988 [18]. It is also useful to refer to documents relating to the outbreak in Barrow in Furness [19].

Anthrax

Anthrax is an infection caused by the bacterium *Bacillus anthracis* and can be fatal. It can be transmitted to humans by incidental infection after coming into contact with diseased animals, their secretions, hides, hair or other products. The routes of transmission are through skin lesions (cuts and abrasions) or puncture, inhalation or ingestion. Human cases of anthrax in the UK are very rare (since 1981 only 17 cases have been reported with no fatalities), although there are some jobs that put people more at risk of contracting the infection. These include farm workers, veterinary surgeons, local authority workers, zookeepers and abattoir workers/butchers who could all come into contact with infected animals. Construction workers can also be at risk if they are developing a site where infected carcases were once buried.[91] More information about who is at risk from anthrax infection can be found in the HSE Guidance for Anthrax [20].[92]

In order to comply with the Control of Substances Hazardous to Health Regulations 2002 (COSHH) (see previously), employers must assess the risk of anthrax infection for their employees. This can be done through a risk assessment, which will also outline any controls needed if a risk is identified. These controls can include providing information and training, monitoring exposure and carrying out health surveillance where the assessment shows that these are required.

General measures can also be taken by most people to reduce the chances of them becoming infected by anthrax. These include:

- Use good basic hygiene practices including regular handwashing and avoid hand to mouth/eye etc. contact;
- Cover all cuts, abrasions and other breaks in the skin with waterproof dressings and/or gloves;
- Dispose of contaminated waste safely;
- Take rest breaks, including meals and drinks, in separate accommodation away from the workplace;
- Provide suitable, regular, health surveillance arrangements where applicable [20].

Leptospiridium

Leptospirosis causes two types of bacterial infection: the more serious, Weil's disease (commonly caught from water contaminated with rat urine) and Hardjo (from infected cattle urine). Leptospirosis is not usually spread from person to person. Bacterial infection occurs through cuts and abrasions or through the mucous membrane of the eyes, nose and mouth. The disease presents as mild flu-like symptoms but can become more serious and lead to fatality [21].

At-risk occupations are those dealing directly with animals and sources of fresh water which could be contaminated by animals. These include:

- Weil's disease: farm workers, pest control workers, sewage and wastewater workers, construction workers;

- Hardjo: cattle and dairy farmers, veterinary surgeons, abattoir workers.

General control measures can be put in place including controlling the rat population (see Chapter 17), covering cuts and abrasions, wearing suitable gloves when handling rats and good occupational health practices such as washing with soap and warm water.

Antibiotics are successful in treating most cases of leptospirosis, but more serious cases can require hospitalisation and intravenous antibiotics.

Occupational asthma

When exposed to certain substances, such as flour or wood dust, some people can have an allergic reaction. When this occurs in the workplace then it comes under occupational asthma. The substances that cause the allergic reactions are called 'respiratory sensitisers' and can change people's airways which is known as the 'hypersensitive state'. Becoming hypersensitive to a substance doesn't necessarily mean that the person will develop asthma. However, once they are hypersensitive a very small dose of the substance could trigger an attack.[93]

Certain workplaces and trades are more at risk from occupational asthma; these include bakers, woodworkers, agricultural workers, vehicle spray painters and healthcare workers.[94] There are many ways that the risk of occupational asthma can be reduced, which will vary depending on the trade, workplace and substance causing the risk.

Early detection of the risk of occupational asthma is important in reducing the length of exposure and possible health implications to employees. This can be achieved through health surveillance and HSE provides advice for employers.[95] Regular monitoring of employees' health will help identify the early signs of a work-related illness and allow the employer to put controls in place to remove or reduce the exposure for their employees.

Ionising and non-ionising radiation

Ionising radiation and the health effects are considered in Chapter 10 to which readers should refer. This part of this section looks briefly in the context of health and safety at work and leisure.

EHPs may encounter ionising radiation when dealing with waste disposal matters or with the issue of radon gas and its radioactive 'daughter elements' (the elements produced in the decay of radon) in dwellings and, increasingly, in the workplace.

The potential for ionising radiation to produce damage in living tissues is considerable and is discussed in Chapter 10.

The standards that must be applied may be derived, for example, from standards such as those contained in the Ionising Radiation Regulations 2017[96] made under the Health and Safety at Work etc. Act 1974. Also, the associated Approved Code of Practice [22] is made under the provisions of the Health and Safety at Work, etc. Act 1974 to implement the basic requirements of various European Directives. They lay down basic standards for the protection of workers and the general public against the dangers arising from the use of ionising radiation in work activities.

One of the requirements of the Regulations is that, with certain exceptions, employers and self-employed persons are required to notify the Health and Safety Executive (HSE) that they are intending to work with ionising radiation.

As with the Control of Substances Hazardous to Health Regulations 2002 (COSHH), employers are required to take all necessary steps to restrict the exposure of employees and non-employees to ionising radiation, so far as is reasonably practicable.

Radon and issues relating to exposure to radon are discussed in more detail in Chapter 10.

The nuclear industry is a diverse major hazards industry, with a legacy of nuclear technology development, which currently contributes just under 20% of the UK's electricity needs and provides the country with its strategic defence requirements.

The industry is currently regulated through the Office for Nuclear Regulation (ONR)[97] as a statutory Public Corporation under the Energy Act 2013. ONR regulates nuclear safety and security (including transport) at UK sites and operates independently of the Government.

Non-ionising radiation (NIR) is again considered in detail in Chapter 10. Non-ionising radiation (NIR) is the term used to describe the part of the electromagnetic spectrum covering two main regions, namely optical radiation (ultraviolet (UV), visible and infrared) and electromagnetic fields (EMFs) (power frequencies, microwaves and radio frequencies).

The Advisory Group on Non-Ionising Radiation (AGNIR) was an independent advisory group that reported to Public Health England and had a programme of work that encompassed exposure to electromagnetic fields, ultraviolet radiation and ultrasound and infrasound.[98]

HSE has produced guidance on the Control of Electromagnetic Fields at Work Regulations 2016;[99] it provides guidance on how the provisions of the Regulations should be met.[100] An EMF is produced whenever a piece of electrical or electronic equipment (i.e., TV, food mixer, computer, mobile phone, etc.) is used. EMFs are static electric, static magnetic and time-varying electric, magnetic and electromagnetic (radio wave) fields with frequencies up to 300 GHz. EMFs are present in virtually all workplaces and if they are of high enough intensity, employers may need to take action to make sure their workers are protected from any adverse effects.

The sun, as well as a number of artificial sources such as insect killing devices, mercury discharge lamps, photocopiers and sun beds emit ultraviolet radiation. The effects of exposure depend on a number of factors, such as the energy of the radiation concerned, the dose received, and the part of the body affected. Of particular concern to EHPs is the risk from UV tanning equipment [23]. HSE says the use of any ultraviolet (UV) tanning equipment (e.g., sunlamps, sunbeds and tanning booths) may expose staff and will expose customers to UV radiation. UV radiation can cause injuries and ill health either in the short term (e.g., sunburnt skin or conjunctivitis) or in the long term (e.g., premature skin ageing, skin cancer and cataracts). Exposure to UV radiation tanning equipment before the age of 35 years significantly increases the risk of several types of skin cancer. Younger people seem to be more at risk from the cancer-causing effects of indoor tanning [24].[101] The Sunbeds (Regulation) Act, effective from April 2011, makes it illegal for people under 18 to use UV tanning equipment.

The HSE guidance note [24] gives the scale and nature of the risks and offers advice on how these risks can be minimised. The guidance also applies to premises in Scotland, although operators there are bound by the Public Health etc. (Scotland) Act 2008 (Sunbed) Regulations 2009,[102] which imposes other legal requirements in the use of tanning equipment. Further examination of the risk of skin cancer (cutaneous melanoma) arising from exposure to fluorescent lights and ultraviolet lamps can be found in the work of Swerdlow et al. [25].

Welding is an occupation that provides many risks.[103] In the context of optical radiation, welders, for example, should wear suitable goggles or a mask, and they must also ensure that any passers-by are not exposed to the ultraviolet radiation generated (in addition to other hazards). Specific eye protection including filters for welding operatives (to prevent arc eye) should conform to relevant British and European standards[104] and the following.

Ultraviolet lamps must be of an appropriate type, for example, in insect killing devices. Equipment such as photocopiers must be properly enclosed, and staff using the equipment must be properly instructed in the use of the machines and apprised of any risks involved.

It should, however, be noted that there are some hazards associated with visible light. One of the principal hazards is dealt with in the section on lasers. Unsuitable forms of lighting can introduce an element of hazard, for example, when fluorescent lights are used to illuminate moving equipment. It is possible that since the fluorescent light will 'flicker' at a high rate that cannot normally be detected by the human eye, it may produce a strobo-scopic effect on moving parts of machinery, giving the illusion that a rapidly spinning fan, for example, is static. Clearly, this could prove to be hazardous, and it is therefore important to ensure that the type of lighting used is suit-able for the purpose.

Infrared radiation can produce effects simi-lar to sunburn, that is, a general irritation of exposed skin. It must be noted, however, that as with some other forms of non-ionising radiation, the eye is particularly vulnerable to damage: parts such as the lens have no protec-tive, cooling mechanism and therefore as the heat builds up, damage such as coagulation of the proteins in the lens can occur, giving rise to cataract formation.

Quite a wide variety of infrared sources are encountered in the workplace, such as lights (including some lasers) and heaters. The nor-mal solution to the problem of exposure to infrared radiation is to wear eye protection. This means that employers must be able to select and provide suitable protective equip-ment and supply adequate information to staff who may be exposed to this risk. The British Standard BS EN 175[105] provides fur-ther guidance on this subject.

The HPA (AGNIR) produced a report on the health effects from radiofrequency elec-tromagnetic fields in 2012. It concluded that although a substantial amount of research has been conducted in this area, there is no convincing evidence that RF field exposure below guideline levels causes health effects in adults or children [26].[106]

Lasers are commonplace in work situations such as product launches, night clubs, theatres, outdoor and indoor displays, supermarket checkout systems, beauty therapy studios and used in devices such as compact disc players and DVD players. These are considered in more detail in Chapter 10 including the clas-sification of lasers.

General guidance concerning the assess-ment of safety is available in several places.[107] The British Standard BS EN 60825–1:2014[108] deals with the classification of lasers (see Table 10.8). HSE produced guidance on the safety of laser lighting displays in 2021.[109]

In the case of people who carry out main-tenance work on all types of laser products, there may be a risk. They must receive proper training and may need to be equipped with appropriate eye protection while carrying out servicing operations. They must also ensure that when safety interlocks are overridden, other people are not affected when systems are being tested.

Finally, the risks posed to officers enforc-ing health and safety provisions must not be ignored. It is important that where such peo-ple are likely to be put at risk, they too are equipped with suitable forms of protection. They must also be supplied with, or should be able to obtain the use of, a suitable form of measuring equipment.

The Dangerous Substances & Explosive Atmospheres Regulations 2002[110] (DSEAR) – These require employers to control the risks to safety from fire, explosions and substances corrosive to metals. Dangerous substances can put people's safety at risk from fire, explosion and corrosion of metal. The regulations put duties on employers and the self-employed to protect people from these risks to their safety in the workplace, and to members of the public who may be put at risk by work activity.

Dangerous substances are any substances used or present at work that could, if not properly controlled, cause harm to people as a result of a fire or explosion or corro-sion of metal. They can be found in nearly all workplaces and include such things as sol-vents, paints, varnishes, flammable gases such as liquid petroleum gas (LPG), dusts from machining and sanding operations, dusts from foodstuffs, pressurised gases, and substances corrosive to metal.

Duty holders must:

- Find out what dangerous substances are in their workplace and what the risks are;
- Put control measures in place to either remove those risks or, where this is not possible, control them;
- Put controls in place to reduce the effects of any incidents involving dangerous substances;
- Prepare plans and procedures to deal with accidents, incidents and emergencies involving dangerous substances;
- Make sure employees are properly informed about and trained to control or deal with the risks from the dangerous substances;
- Identify and classify areas of the workplace where explosive atmospheres may occur and avoid ignition sources (from unprotected equipment, for example) in those areas.

The EHP is most likely to encounter the regulations and a DSEAR risk assessment in relation to the operation of petrol filling stations. Guidance on DSEAR is available from the HSE website.[111]

Section 5 Notes

74 See http://eur-lex.europa.eu/LexUriServ/LexUriServ.do?uri=OJ:L:2008:353:0001:1355:en:PDF

75 Regulation (EC) No 1907/2006 of the European Parliament and of the Council concerning the Registration, Evaluation, Authorisation and Restriction of Chemicals (REACH). 18 December 2006 at http://eur-lex.europa.eu/legal-content/EN/TXT/PDF/?uri=CELEX:02006R1907-20140822

76 See http://www.hse.gov.uk/coshh/detail/coshh-clp-reach.htm

77 SI 2002 No 2677 as amended by SI 2004 No 3386.

78 SI 2007 No 1894.

79 SI 2002 No 2676.

80 SI 2012 No 179.

81 https://www.hse.gov.uk/pubns/books/l5.htm

82 SI 2012 No 632.

83 https://books.hse.gov.uk/Health-and-Safety-Guidance-HSG/?DI=654258&CLICKID5923CLICKID – This Analysts' Guide H publication is primarily for analysts involved in asbestos work and the authoritative source of asbestos analytical procedures within Great Britain. The guidance has been updated in 2021 to take account of findings from HSE interventions, and developments in analytical procedures and methodology. It provides clarification on technical and personal safety issues, especially in relation to sampling and four-stage clearances. New information on sampling soils for asbestos is included.

84 SI 2013 No 390.

85 SI 1992 No 3067, SI 1999 No 2373 and SI 2003 No 1889.

86 https://assets.publishing.service.gov.uk/government/uploads/system/uploads/attachment_data/file/833550/Legionella_monthly_LD_Report_August-2019.pdf and https://assets.publishing.service.gov.uk/government/uploads/system/uploads/attachment_data/file/948028/Legionella_Mthly_Rprt-Aggregate_rprt_Jan-Oct20.pdf

87 See also https://www.hse.gov.uk/legionnaires/

88 BS 7592:2008 *Methods for Sampling for Legionella bacteria in water systems. Code of Practice*. British Standards Institution.

89 BS EN ISO 11731:2017 *Water quality – Enumeration of Legionella* British Standards Institution.

90 SI 1992 No 2225.

91 See https://assets.publishing.service.gov.uk/government/uploads/system/uploads/attachment_data/file/364744/Guidance_of_Assessing_Risk_of_Anthrax_on_building_Land_151014.pdf

92 See https://www.gov.uk/guidance/anthrax-how-to-spot-and-report-the-disease and also https://www.hse.gov.uk/agriculture/zoonoses-data-sheets/anthrax.pdf

93 See *About Asthma* available on HSE website on www.hse.gov.uk/asthma/about.htm

94 See *Your Trade* available on HSE website on www.hse.gov.uk/asthma/trade.htm

95 See www.hse.gov.uk/asthma/employers.htm

96 SI 2017 No 1075 which revokes and supersedes the Ionising Radiations Regulations 1999 SI 1999 No 3232.

97 See https://www.onr.org.uk

98 https://www.gov.uk/government/groups/advisory-group-on-non-ionising-radiation-agnir

99 SI 2016 No 588.

100 https://www.hse.gov.uk/pubns/priced/hsg281.pdf

101 https://www.hse.gov.uk/pubns/indg209.pdf see also https://www.hse.gov.uk/pubns/misc869.pdf

102 SSI 2009 No 388.

103 See https://www.hse.gov.uk/welding/health-risks-welding.htm on the health risks from welding.

104 See for example BS7028:1999 Eye protection for industrial and other uses. Guidance on selection, use and maintenance.

105 BS EN 175:1997 *Personal Protection Equipment for Eye and Face Protection during Welding and Allied Processes*, British Standards Institution (under review).

106 See also https://www.gov.uk/government/publications/electric-and-magnetic-fields-health-effects-of-exposure/electric-and-magnetic-fields-assessment-of-health-risks

107 See https://www.gov.uk/government/publications/laser-radiation-safety-advice/laser-radiation-safety-advice

108 BS EN 60825-1:2014 +A11:2021 *Safety of laser products - Part 1: Equipment classification and requirements*, BSI. London.

109 https://www.hse.gov.uk/pubns/indg224.htm

110 SI 2002 No 2776.

111 https://www.hse.gov.uk/fireandexplosion/dsear.htm

Section 5 References

[1] Health and Safety Executive. (undated) *The CLP Regulation, HSE.* https://www.hse.gov.uk/chemical-classification/legal/clp-regulation.htm (Accessed September 2021).

[2] Health and Safety Executive. (undated) *Transition Arrangements, HSE.* http://www.hse.gov.uk/chemical-classification/legal/transition-from-chip-to-clp.htm (Accessed 21 March 2015).

[3] European Chemicals Agency. (undated) *Understanding CLP*, ECHA, Helsinki. http://www.echa.europa.eu/regulations/clp/understanding-clp (Accessed 21 March 2015).

[4] European Chemicals Agency. (2011) *Guidance on Labelling and Packaging in accordance with Regulation (EC) No 1272/2008*, ECHA, Helsinki.

[5] UK REACH Competent Authority. (2012) *REACH and Safety Data Sheets, Information Leaflet Number 13 – Safety Data Sheets November 2012.* http://www.hse.gov.uk/reach/resources/reachsds.pdf (Accessed 21 March 2015).

[6] Health and Safety Executive. (2013) *Control of Substances Hazardous to Health, the Control of Substances Hazardous to Health Regulations 2002 (as amended) Approved Code of Practice and Guidance, HSE.* http://www.hse.gov.uk/pubns/priced/l5.pdf (Accessed 23 March 2015).

[7] Health and Safety Executive. (2011) *EH40/2005 Workplace Exposure Limits, Containing the List of Workplace Exposure Limits for Use with the Control of Substances Hazardous to Health Regulations (as amended).* http://www.hse.gov.uk/pubns/priced/eh40.pdf (Accessed 28 March).

[8] Health and Safety Executive. (2006) *Monitoring Strategies for Toxic Substances*, HSG173, HSE Books, Sudbury, Suffolk. http://books.hse.gov.uk/hse/public/saleproduct.jsf?catalogueCode=9780717661886.

[9] Health and Safety Executive. (2000) *General Methods for Sampling and Gravimetric Analysis of Respirable and Inhalable Dust*, MDHS14/4, HSE Books, Sudbury, Suffolk.

[10] Health and Safety Executive. (1993) *General Methods for Sampling Airborne Gases and Vapours*, MDHS70, HSE Books, Sudbury, Suffolk.

[11] World Health Organization. (1997) *Determination of Airborne Fibre Number Concentrations, A Recommended Method, by Phase-Contrast Optical Microscopy (Membrane Filter Method)*, Geneva. http://www.who.int/occupational_health/publications/en/oehairbornefibre.pdf (Accessed 28 March 2015).

[12] Health and Safety Executive. (undated) *Licensable Work with Asbestos.* http://www.hse.gov.uk/asbestos/licensing/licensed-contractor.htm (Accessed 30 March 2015).

[13] Health and Safety Executive. (n.d.) *Notifiable Non-Licensed Work.* http://www.hse.gov.uk/asbestos/licensing/notifiable-non-licensed-work.htm (Accessed 30 March 2015).

[14] Health and Safety Executive (n.d.) *Non-Licensed Work with Asbestos.* http://www.hse.gov.uk/asbestos/licensing/non-licensed-work.htm (Accessed 30 March 2015).

[15] Health and Safety Executive. (2013) *Managing and Working with Asbestos Control of Asbestos Regulations 2012, Approved Code of Practice and Guidance, L143.* http://www.hse.gov.uk/pubns/priced/l143.pdf (Accessed 30 March 2015).

[16] Health and Safety Executive. (2005) *Asbestos: The Analysts' Guide for Sampling, Analysis and Clearance Procedures*, HSG 248 HSE Books, Sudbury, Suffolk.

[17] Health and Safety Executive. (2013) *Legionnaires' Disease, the Control of Legionella Bacteria in Water Systems Approved Code of Practice*

and Guidance on Regulations, L8. HSE, 4th ed. https://www.hse.gov.uk/pubns/priced/l8.pdf (Accessed September 2021).

[18] Westminster City Council. (1988) *Broadcasting House Legionnaire's Disease,* WCC, London.

[19] HSE. (2007) *Report of the Public Meetings into the Legionella Outbreak in Barrow-in-Furness, August 2002.* The Stationery Office, PO Box 29, Norwich NR3 1GN. http://www.hse.gov.uk/legionnaires/assets/docs/barrow-report.pdf.

[20] HSE. (undated) *Construction Micro-Organisms: Anthrax from Contaminated Land and Buildings.* https://www.hse.gov.uk/construction/healthrisks/hazardous-substances/harmful-micro-organisms/anthrax.htm (Accessed September 2021).

[21] HSE. (undated) *Leptospirosis (Weil's Disease and Hardjo).* https://www.hse.gov.uk/agriculture/zoonoses-data-sheets/leptospirosis.pdf (Accessed September 2021).

[22] HSE. (2018) *Work with Ionising Radiation: Approved Code of Practice and Guidance L121,* HSE Books, Sudbury, Suffolk, 2nd ed. https://www.hse.gov.uk/pubns/priced/l121.pdf.

[23] The International Agency for Research on Cancer and Working Group on artificial ultra-violet (UV) light and skin cancer. (2006) The association of use of sunbeds with cutaneous malignant melanoma and other skin cancers: A systematic review. *International Journal of Cancer,* 120: 1116–1122.

[24] HSE. (2011) *Reducing the Health Risks from the Use of Ultraviolet Tanning Equipment, INDG209 (rev2),* HSE Books, Sudbury, Suffolk. http://www.hse.gov.uk/pubns/indg209.pdf.

[25] Swerdlow AJ, English JFC, Mackie RM, et al. (1988) Fluorescent lights, ultraviolet lamps and the risk of cutaneous melanoma. *British Medical Journal,* 297: 647–650, 10 September.

[26] HPA. (2012) *Health Effects from Radiofrequency Electromagnetic Fields – Report of the Independent Advisory Group on Non-ionising Radiation,* RCE-20 HPA, Chilton, Didcot, Oxon. https://www.gov.uk/government/uploads/system/uploads/attachment_data/file/333080/RCE-20_Health_Effects_RF_Electromagnetic_fields.pdf (Accessed September 2021).

SECTION 6: THE INVESTIGATION OF INCIDENTS, ACCIDENTS AND DANGEROUS OCCURRENCES IN THE WORKING ENVIRONMENT

In this section we explore the role that investigating accidents and incidents plays in helping to manage health and safety.

Some principles underpinning the role that accident investigations can play were documented as early as 1956 when the then Ministry of Labour identified six principles of accident investigation [1]. These principles included recognition that accident investigation is an essential part of a company's overall health and safety management system, that an organisation should have a defined organisational safety policy and that all employees and management should be involved in accident investigations.

The principles became enshrined is legislation by the introduction of the Health and Safety at Work etc. Act 1974 and subsequent legislation. Before we can explore the purpose of accident and incident investigations, we should first understand what an accident is.

Bamber [2] put forward the following definition in 2003, that an accident is

an unexpected, unplanned event in a sequence of events, that occurs through a combination of causes; it results in physical harm (injury or disease) to an individual, damage to property, a near-miss, a loss, or any combination of these effects.

This definition is wide raging and goes beyond a simple incident involving a slip or trip or trip to an employee where a physical injury occurs. The breadth of the definition should act as a reminder to organisations and the EHP of the importance of undertaking thorough accident and near-miss investigations and of the importance of learning from such incidents to improve worker and public safety.

It is widely accepted that accidents do not just happen; there is usually always an underlying cause or sequence of causes that have resulted in the potential for harm being realised and an accident occurs. Indeed, in the HSE argues [3] that 'all accidents, ill-health and incidents are preventable'.

Reporting of accidents and near-miss incidents

An organisation should, as part of its health and safety management system, have a defined procedure for reporting accidents and near miss incidents. The size of the business or organisation will reflect on the complexity of the reporting system as processes will be more defined in large organisations than in small and medium-size enterprises and will involve an investigation process.

Accident reports in many large organisations may utilise internet-based reporting systems and therefore full safety documentation may not be held in paper form in individual trading units. Most large organisations have internal Health and Safety, Trading Law, Compliance or Risk Management teams depending upon the culture and internal design of the organisation. EHPs should not only engage with the sites where the incidents occurred but for larger organisations, they should also liaise with the head office teams and where necessary should consider engaging the Primary Authority.

Organisations should capture internal accident and incident information for several reasons. Firstly, reporting may be required under the Reporting of Injuries, Diseases and Dangerous Occurrences Regulations 2013 (RIDDOR);[112] secondly, as part of internal health and safety management procedures to prevent recurrence or as part of an insurance investigation in support of either an employee or public liability claim.

The RIDDOR came into force from 1 October 2013 and set out the process for reporting work-related accidents and ill health. The regulations define the type of incidents that should be reported, who they should be reported to, and the various categories of people for whom reporting is required. Guidance is available from HSE.[113]

RIDDOR introduced significant changes to the previous requirements. The main changes made were to simplify the reporting requirements in the following areas:

- The classification of 'major injuries' to workers that has been replaced with a shorter list of 'specified injuries';
- The previous list of 47 types of industrial disease has been replaced with eight categories of reportable work-related illness;
- Fewer types of dangerous occurrence require reporting.

RIDDOR does not introduce any significant changes to the reporting requirements for:

- Fatal accidents;
- Accidents to non-workers (members of the public);
- Accidents which result in the incapacitation of a worker for more than seven days.

Recording requirements remain broadly unchanged, including the requirement to record accidents resulting in the incapacitation of a worker for more than three days.

RIDDOR places a duty on all employers and 'responsible persons' who have control over employees and a work premises to report relevant accidents and incidents to the relevant enforcing authority. In some circumstances (e.g., a fatal work-related accident), the report should be made immediately by the quickest possible means. However, normally an employer has ten days to make a report. Reports can be made via the internet, telephone or fax. The Regulations specify injuries that need to be reported for workers and non-workers. Such injuries for workers include fractures, amputations, blinding or reduction in sight, crushing, burns, scalping and loss of consciousness.

Certain dangerous occurrences must also be reported (specified in Regulation 7 and Schedule 2). Examples of these include explosion of pressure vessels, certain fires and the full or partial collapse of a building. Notification procedures are also defined in Bassett's Environmental Health Procedures [4].

Once an enforcing authority has received a RIDDOR notification, it will be assessed and if necessary, a telephone call will be made to confirm details of the incident with the notifier. The authority will then determine whether to carry out an investigation into the cause of the incident with regard to any relevant Accident Investigation Protocol and the National Code. Investigators may have regard to local knowledge if there is additional intelligence to support an investigation into an incident, for example where several incidents have occurred which individually would not require investigation but collectively cause concern. When assessing the incident, the previous inspection history and confidence in management will be considered. Self-regulation is reserved for organisations with good safety management systems.

Investigations should focus not only on identifying the underlying causes or factors that led up to the accident occurring but should seek to establish ways in which similar accidents could be prevented. Close liaison with the senior management of the organisation and their internal compliance monitoring teams will assist in this process as organisations have an explicit duty under the Health and Safety at Work etc. Act 1974 to provide both safe systems of work and a safe place of work.

Section 20 of the Health of Safety at Work etc. Act 1974 provides an EHP with powers to ask questions, the answers to which may be used to form a statement. Practitioners should ensure that the procedures identified within the Police and Criminal Evidence Act 1984 (PACE) are followed. Witness statements may be obtained voluntarily under section 9 of the Criminal Justice Act 1967 and is often used where there is no reason to believe that the individual has committed a crime. However, where there is reason to believe a criminal offence has been committed, then the investigator must caution the individual and obtain the statement under PACE 1984. The individual has the right to remain silent which is detailed within the caution. The caution contains two parts; the first part states their right as an individual and the second and often neglected part is questioning that the individual understands the caution.

Besides taking statements, practitioners can examine and remove documents, take photographs, paperwork, IT or other work equipment from the scene of an accident or the organisation as required. The EHP should establish the cause of the accident as clearly as possible and work with the organisation's management to avoid a recurrence. Any reports produced by the practitioner may form part of a criminal case against the organisation and may also be used in any subsequent claim for damages.

The HSE guidance (HSG 245) [5] was aimed at providing employers, unions and safety representatives clear guidance on how an accident investigation should be undertaken and this provides value to a commercial organisation. Guidance rather than legislation was chosen as the correct route after a wide-ranging consultation exercise. EHPs working in commercial organisations will use internal accident investigations as part of a suite of tools when seeking to influence the decision-making by top management and to encourage safety investment.

Fatal accident investigations

The Work-Related Death Protocol has been developed between the Association of Chief Police Officers, The British Transport Police, the HSE, the Local Government Association and the Crown Prosecution Service. The Protocol for Liaison [6] and the Practical Guide [7] provide a clear framework for practitioners to follow when undertaking a fatal accident investigation. Practitioners should bear in mind that although the police have primacy in such investigations, given the introduction of Corporate Manslaughter legislation in 2008, the purpose of the guidance is centred on the principal of 'joint investigation'. Practitioners should remain up to date in terms of knowledge and actively seek to read major accident investigation reports such as the report into the Buncefield fire [8].

Chapter 6 Notes

112   SI 2013 No 1471.
113   https://www.hse.gov.uk/riddor/

Section 6 References

[1]   Ministry of Labour and National Service. (1956) *Industrial Accident Prevention*, Report of the Industrial Safety Sub-Committee of the National Joint Advisory Council HMSO, London.

[2]   Bamber L. (2003) *Principles of the Management of Risk, in Safety at Work*, Ridley J, Cumming J (Eds.), Butterworths, London, 6th ed.

[3]   HSE. (2013) *Successful Health & Safety Management* (HSG 65), HSE Books, Sudbury, Suffolk. https://www.hse.gov.uk/pubns/priced/hsg65.pdf.

[4]   Deveaux T. (2020) *Bassett's Environmental Health Procedures*, Routledge, Abingdon, Oxon, 9th ed.

[5]   HSE. (2004) *Investigating Accidents and Incidents – a Workbook for Employers, Unions, Safety Representatives and Safety Professionals*, HSG245. HSE Books Sudbury, Suffolk. https://www.hse.gov.uk/pubns/hsg245.pdf.

[6]   HSE. (2016) *Work Related Deaths, a Protocol for Liaison (England and Wales) Fourth version*. http://www.hse.gov.uk/pubns/wrdp1.pdf HSE (Accessed September 2021).

[7]   HSE. (2016) *Work Related Deaths, Practical Guide*, HSE. http://www.hse.gov.uk/pubns/wrdp2.pdf (Accessed September 2021).

[8]   HSE, EA, SEPA. (2011) *Buncefield: Why Did It Happen?* COMAH, Crown Copyright. http://www.hse.gov.uk/comah/buncefield/buncefield-report.pdf.

SECTION 7: EMERGING THREATS TO HEALTH AND SAFETY

Avian/swine flu

Influenza as a disease is considered in Chapter 10. It is caused by a viral infection and primarily affects the respiratory system with associated aches and fever; it can be transmitted from person to person through small droplets released in coughs and sneezes and tends to spread quickly, causing seasonal epidemics.

Normally the viruses are highly species-specific and rarely cross into other species. When they do cross over, they are often mild but occasionally a pathogenic strain will emerge. Over the last couple of decades there have been a number of influenza viruses that have transmitted into the human population and the growth of international travel has brought these viruses to the global population. Two such animal-origin influenza viruses over recent year are the H5N1 and H1N1 variants originating in birds and pigs respectively. However the 2020–21 COVID-19 pandemic caused by SARS-CoV-2 has had the greatest global impact. Although the ultimate source is undetermined at present it is thought to have originated in animal species.[114]

However H5N1 crossed over to humans several times in recent years, each time originating in the Far East, and has caused the largest number of deaths. WHO figures [1] show that, although the total number of confirmed cases have fluctuated since 2003, the average fatality rate from H5N1 avian influenza remains above 50%, primarily across the Middle and Far East.

The 2009 pandemic (H1N1) emerged as a new influenza virus that had never circulated among humans before. The fatality pattern also differs somewhat from seasonal flu, affecting primarily the young and those without existing medical conditions. Because this was a new strain there was no innate immunity and no specific vaccine available, making it a high risk to public health [2].

As the COVID-19 pandemic has shown, employers need to be aware of the risks from such pandemics and there is a risk that they will become more frequent so they will need to take this into account in managing health and safety in the workplace. This is not least because most emerging diseases appear to be zoonotic. It is estimated that more than six out of every ten known infectious diseases in people can be spread from animals, and three out of every four new or emerging infectious diseases in people come from animals.[115]

Nanotechnology

Nanotechnologies involve the creation and/or manipulation of materials at the nanometre (nm) scale. Nanotechnology is the science and application of materials with a size below 100 nanometre. One nanometre is 10^{-9} m or one millionth of a millimetre. By comparison, a human hair is approximately 70,000 nm in diameter, a red blood cell is approximately 5,000 nm wide and simple organic molecules have sizes ranging from 0.5 to 5 nm.

These materials can be either formed unintentionally such as those made through pollution, or they may be engineered for products such as cosmetics or sports gear. Nanoparticles such as zinc oxide have been used within sunscreens and cosmetics for some time. The Friends of the Earth's report *Nanomaterials, sunscreens and cosmetics: Small ingredients, big risks* [4], detailed 116 products that were commercially available at that time.[116]

Due to their very small size, the properties of nanomaterials are very different from the bulk material that they are generated from. This is due to the increase in surface area which makes them more reactive. New characteristics are created at the nanoscale such as reactivity, magnetism and electronic properties. An example of this is Nanosilver which has very different properties from silver. Microbes do not become immune to Nanosilver and so it acts as a bactericide and viricide and therefore can be added to wound dressings and food containers.

There are a number of concerns in the scientific community around this emerging technology, due to the minute size of nanoparticles. These concerns are in relation to:

- Health – due to their size, nanoparticles can be taken up by the human body more easily and can cross biological membranes, thereby entering organs in the body. There is also a concern of cardiovascular disease as the particles can get into the nervous system which can affect the rhythm of the heart.
- Environment – the end of life of the product may be an issue as the particles can work their way up the food chain and once in the environment, they will be very difficult to remove.

On 18 October 2011 the European Commission adopted the recommendation on the definition of nanomaterial as:

> A natural, incidental or manufactured material containing particles, in an unbound state or as an aggregate or as an agglomerate and where, for 50 % or more of the particles in the number size distribution, one or more external dimensions is in the size range 1 nm – 100 nm.
>
> In specific cases and where warranted by concerns for the environment, health, safety or competitiveness the number size distribution threshold of 50 % may be replaced by a threshold between 1 and 50 %.
>
> By derogation from the above, fullerenes, graphene flakes and single wall carbon nanotubes with one or more external dimensions below 1 nm should be considered as nanomaterials.

The definition will be used primarily to identify materials for which special provisions might apply (e.g., for risk assessment or ingredient labelling). Those special provisions are not part of the definition but of specific legislation in which the definition will be used. A material that falls within this definition is not automatically hazardous and a material that falls outside this definition is not necessarily of low hazard.

As HSE has said Nanotechnology is an emerging field.[117] However, along with any new innovation there come uncertainties as to whether the unique properties of engineered nanomaterials pose an occupational health risk. Gaps in our knowledge about the factors that are essential for predicting health risks such as routes of exposure, translocation of nanomaterial once inside the body and the interaction of the nanomaterial with the body's biological systems are not yet fully understood.

Assessment of health risks arising from exposure to nanomaterials or other substances requires understanding of the intrinsic toxicity of the substance, the levels of exposure (by inhalation, by ingestion or through the skin) that may occur and any relationship between exposure and health effects. More data is needed on the health risks associated with exposure to engineered nanomaterials.

The HSE has issued guidance on the use of nanotechnology in the workplace, using nanomaterial at work, HSG 272. The guidance provides information on how to manage accidental exposure to manufactured nanomaterial at work, including the protection of employees. The use of nanomaterial falls within the wider remit of COSHH, which is discussed previously in this chapter [5].

Section 7 Notes

114 See https://www.who.int/health-topics/corona virus/origins-of-the-virus

115 See https://www.cdc.gov/onehealth/basics/ zoonotic-diseases.html

116 See also https://euon.echa.europa.eu/what-kind-of-products-contain-nanomaterials

117 https://www.hse.gov.uk/nanotechnology/ what.htm#regulatory-definition

Section 7 References

[1] WHO. (2021) *Cumulative Number of Confirmed Human Cases for Avian Influenza A(H5N1) Reported to WHO, 2003–2021*, WHO, Geneva.

[2] WHO. (2010) *Pandemic (H1N1) 2009*, WHO, Geneva.

[3] Friends of the Earth. (2006) *Nanomaterials, Sunscreens and Cosmetics: Small Ingredients, Big Risks*. http://www.appletonlaw.com/files/2009/ PDFs/Bio_Kimbrell.pdf.

[4] HSE. (2013) *Using Nanomaterial at Work*, HGS 272 2013, Health and Safety Executive. https:// www.hse.gov.uk/pubns/books/hsg272.pdf.

15

Water and environmental health

Kathy Pond, Tom Bond

Introduction

This chapter is an update of the chapter in the 21st edition written by Kathy Pond and Steve Pedley.

In July 2010 the UN General Assembly adopted resolution 64/292, which explicitly recognises the human right to water and sanitation:

> [The UN] recognizes the right to safe and clean drinking water and sanitation as a human right that is essential for the full enjoyment of life and all human rights.

The formal recognition of the human right to water and sanitation puts the onus on States and international organisations to support efforts to accelerate the provision of drinking water and sanitation for all, in particular to those people in developing (or low-income) countries that lack access to these basic services. But recognising a right to drinking water and sanitation alone is insufficient without the inclusion of definitions of the indicators that can be used to measure progress towards the human right. In this context, the UN Committee on Economic, Social and Cultural Rights offered in its General Comment number 15 of 2002, a series of criteria for drinking water:

> The human right to water entitles everyone to *sufficient, safe, acceptable, physically accessible and affordable* water for personal and domestic uses.

The meaning of each of the entitlements has been debated extensively in order to develop some measure that each can be assessed by. The following definitions have been taken from the UN's Water for Life web pages[1] and are broadly accepted by the international water and sanitation community:

Sufficient. The water supply for each person must be sufficient and continuous for personal and domestic uses. These uses ordinarily include drinking, personal sanitation, washing of clothes, food preparation, personal and household hygiene. According to the World Health Organization (WHO), between 50 and 100 litres of water per person per day are needed to ensure that most basic needs are met and few health concerns arise.

Safe. The water required for each personal or domestic use must be safe, therefore free from micro-organisms, chemical substances and radiological hazards that constitute

DOI: 10.1201/9781003035640-15

a threat to a person's health. Measures of drinking-water safety are usually defined by national and/or local standards for drinking-water quality. The WHO Guidelines for drinking-water quality provide a basis for the development of national standards that, if properly implemented, will ensure the safety of drinking-water.

Acceptable. Water should be of an acceptable colour, odour and taste for each personal or domestic use. All water facilities and services must be culturally appropriate and sensitive to gender, lifecycle and privacy requirements.

Physically accessible. Everyone has the right to a water and sanitation service that is physically accessible within, or in the immediate vicinity of the household, educational institution, workplace or health institution. According to WHO, the water source has to be within 1,000 metres of the home and collection time should not exceed 30 minutes.

Affordable. Water, and water facilities and services, must be affordable for all. The United Nations Development Programme (UNDP) suggests that water costs should not exceed 3 per cent of household income.

In 2016, United Nations Member States committed themselves to ensuring access to safe drinking water and to sanitation in Goal 6 of the 2030 Agenda for Sustainable Development (SDG 6) (see Annex to Chapter 1 for more on the SDGs). By doing this, they reiterated their commitment to the human right to water and sanitation.[2]

The provision of safe drinking water and sanitation has been shown to markedly improve human health and, as a consequence, to increase the prosperity of communities [1–4] and [5]. It is this connection between water and human health that is particularly pertinent to environmental health professionals and will form the main focus of this chapter. In particular water quality will be considered. Water quality is the combination of natural and anthropogenic (arising from human activity) contaminants in the water that define its characteristics. Not all these contaminants are harmful – indeed, some are beneficial – or harmful if present in high concentrations. Understanding where these contaminants enter the water, the factors that control their concentration, how they impact upon human health, how they are measured and how their concentrations are controlled in the water before and after human contact is an essential pre-requisite for being able to assess and manage the risks associated with contact with water.

'Introduction' Notes

1 http://www.un.org/waterforlifedecade/human_right_to_water.shtml (Accessed August 2021).
2 For more see https://www.un.org/sustainabledevelopment/water-and-sanitation/ (Accessed August 2021).

SECTION 1: BASIC CONCEPTS

This section provides a brief overview of the basic concepts in water and health. Some of these concepts will be developed further in the later sections of this chapter.

Water cycle

Looking at it objectively, water will never be in short supply, but there may be issues around its distribution and availability on the planet. Estimates of total reserves are imprecise but the data that is currently available indicates that there are about 1.4 billion cubic kilometres of water on earth. Approximately 3% of this total (35 million cubic kilometres) is available as freshwater, but of this amount, despite global heating approximately two-thirds is locked away in ice, permafrost, and in deep inaccessible aquifers. Consequently, at any particular time, just over 11 million cubic

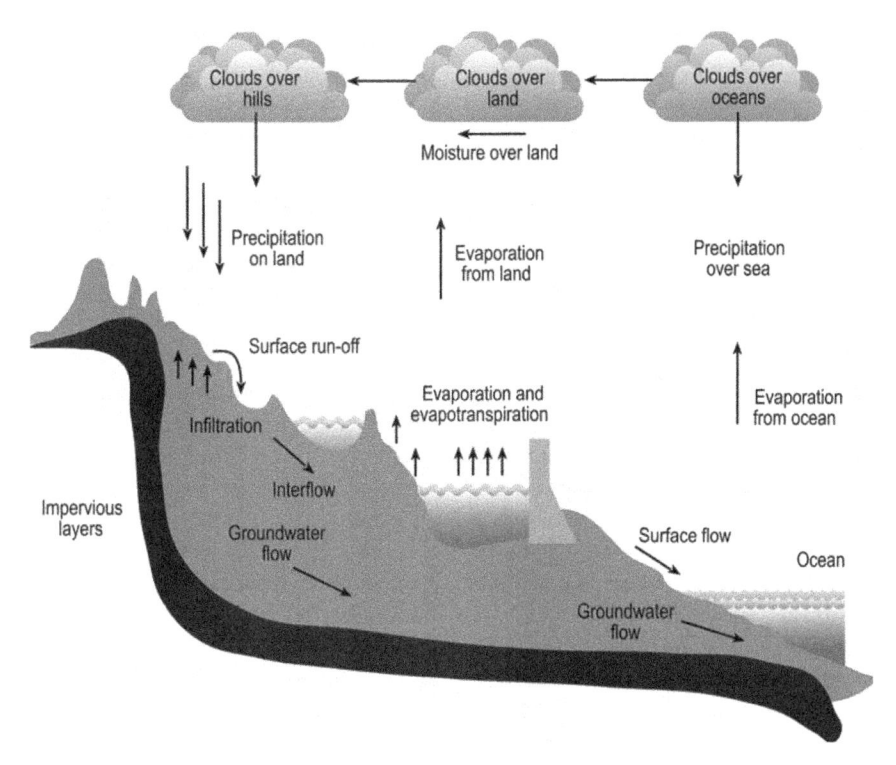

Figure 15.1 The water cycle (CROWN COPYRIGHT)

Source: Scottish Executive, June 2006 (reproduced with permission by Crown Copyright)

kilometres of freshwater are available for us to use.

The water that exists in the oceans, glaciers, rivers, lakes and groundwater is not fixed but moves between them in what is called the water cycle[3] (Figure 15.1).

The various stages of the water cycle are relatively straightforward. Evaporation from the oceans and surface waters, and evapotranspiration by plants, raises water vapour into the atmosphere. The water vapour condenses in the cooler air to form clouds, which, when the conditions are right, release their water as rainfall, snow or hail. Rain falling directly back into the ocean rapidly completes the water cycle, but water falling onto the land can follow any one of a number of routes before it eventually flows back into the ocean.

Some of the rain falling onto land will run across the surface and into surface water systems such as streams, rivers and lakes. The amount of run-off water as a proportion of the total volume of rainfall is not fixed but will vary depending on the conditions in the local environment and on the level of saturation in the soil. Some of the surface water will evaporate back into the atmosphere, some may infiltrate into the ground, but a significant proportion will flow back into the oceans. Water entering surface flows completes the water cycle reasonably quickly; sometimes within days but extending to several years in situations where the river flow is interrupted by lakes and reservoirs. Precipitation falling as snow may also have a long cycle duration if it becomes incorporated into glacier systems, but otherwise will follow the same pathway as water that runs into surface water systems.

Rainwater that does not run directly off the surface of the land may seep into the soils and infiltrate the subsurface to form groundwater. The boundary between the groundwater and the unsaturated surface layers of soil is called the water table. Once the water has entered the subsurface it will continue to drain downwards under gravity until it reaches an impervious layer, such as clay or granite, and will then flow down the gradient created by the impervious layer. Where the groundwater exists in a rock formation in sufficient quantities to be exploited it is called an aquifer.

Different aquifers have different properties and have different yields of water. One property that is particularly important for understanding the potential for groundwater contamination, is whether or not it is protected from the direct infiltration of water from the surface. An aquifer that is not protected remains vulnerable to contamination. These are termed unconfined aquifers. Groundwater which flows between two impervious layers – one above the groundwater and one below – is called a confined aquifer. Confined aquifers are better protected from contamination sources on the surface than unconfined aquifers.

The residence time of water in aquifers can be as short as a few days or as long as several thousands of years. Furthermore, the rate at which water flows through an aquifer can vary considerably depending on the type and structure of the rock formation. For example, some rock formations, such as sandstone, have low permeability and the flow of water may be as little as a few millimetres a week. By contrast some highly fissured rock formations, like limestone, permit the rapid flow of water over several kilometres per day.

Aquifers discharge water at the land surface in a number of ways. Springs are the most obvious example, but groundwater can also feed rivers and other surface water systems by a process called basement flow. This is where the river bed cuts through the water table leading to a direct connection between the surface water and the groundwater. For many river systems a large proportion of the dry-weather flow can come from basement

flow. In coastal areas, the aquifer may continue from the land to under the sea and the groundwater completes the water cycle by emerging at the sea bed.

At the beginning of the water cycle – the point where water vapour rises from the water surface – the water is in a relatively pure form. However, as it passes through the intermediate stages of the cycle the water collects contaminants that can influence its properties, the way it interacts with other environmental systems, and, ultimately, determines its suitability for human use. Even at the point where the water vapour condenses to form clouds it can dissolve airborne contaminants. In extreme cases, such as those that arose in the 1960s and 1970s[4] with acid rain, the contamination of rainwater can have far reaching consequences that includes the destruction of forests and the death of fish in lakes and rivers. Once the water comes into contact with terrestrial systems it begins to dissolve other chemicals and to pick up and transport organic and inorganic particles. Thus, water quality is a dynamic character that changes at every point of the water cycle.

River basins

The river basin is an important concept in the management of water resources. The surface features of an area – its topography – controls the way that precipitation is collected and moves through that area, either as surface water flow or groundwater flow. In simple terms, the topography can act as a funnel, channelling water from a wide area into hydrological features, such as streams and rivers. Each naturally distinct geographical and hydrological unit is called a river basin.

Historically, water resource management had not been organised in the most sensible way, with the areas to be managed being dictated by political imperatives. Thus, single river basins may have been managed by different authorities, with different, perhaps conflicting requirements for the water. These conflicts often are more pronounced at international boundaries, but they can also occur between regions in a single country. Again this is in part why water management is addressed in the Sustainable Development Goals that have been discussed in the Annex to Chapter 1. The limitations of a management unit created from political rather than geographical boundaries was further recognised in Europe by the European Union, and was addressed in the "Directive 2000/60/EC of the European Parliament and of the Council establishing a framework for the Community action in the field of water policy", often referred to as the EU Water Framework Directive or WFD. The WFD was adopted on the 23rd October 2000. A description of the WFD and its implications for water resource management can be found on the European Commission website.[5]

A key initiative in the WFD was to create a unified approach to the management of water resources on the river basin.[6] The WFD then requires that a management plan is created for each river basin – to be updated every six years – which sets out water quality objectives for the river basin and maps out improvements to the water quality to achieve these objectives. In England and Wales, river basin management plans can be accessed directly on the UK Government website[7] that includes links to the management plans for river basins in Scotland and Wales.

Water use

The risks to human health from contact with water is not just a consequence of the quality of the water but is also in the way water is used and how that use dictates the way we come into contact with the water.

Published statistics of the global water use divide the different uses into three broad categories: domestic, agricultural and industrial use. These are classes of convenience because they can be used to estimate the volume of water withdrawn from freshwater resources for different purposes, and any sub-class of use can usually be included in one of these

headings. For the purpose of this chapter, however, there is another use of water that can have both positive and negative impacts on health but falls outside these categories because it does not involve the withdrawal of water. The recreational use of marine and freshwater is an important activity that can offer health benefits but is linked to ill health if the quality of the water is impaired.

At a global scale agriculture dominates in terms of the volume of freshwater it uses (approximately 70% of withdrawals); industry is the second heaviest user of water (approximately 20% of withdrawals); and domestic use takes the lowest volume. However, these averages mask wide variation in the volumes of water used for different purposes at the regional and national levels. Applying a simple division between high-income countries and low- and middle-income countries reveals a staggering difference in the proportion of water used by the different sectors. In low- and middle-income countries the dominant use is agriculture, whereas industry accounts for the greatest volume of water in high-income countries. Interestingly, domestic water use is similar between the two categories of countries (approximately 8% in low- and middle-income countries and 11% in high-income countries).

Domestic use – the uses of water in domestic settings are diverse and include some activities that take place outside the dwelling. Definitions of the domestic use of water are not easy to come by, but the one published by the U.S. Geological Survey (USGS) in the United States is reasonably comprehensive:

> Water used for indoor household purposes such as drinking, food preparation, bathing, washing clothes and dishes, flushing toilets, and outdoor purposes such as watering lawns and gardens. Domestic water use includes potable and non-potable water provided to households by a public water supplier (domestic deliveries) and self-supplied water.[8]

Domestic water consumption can be defined as:

> The quantity, or quantity per capita, of water consumed in a municipality or district for domestic uses or purposes during a given period, generally one day. It is usually taken to include all uses included within the term public supplies (Municipal Use of Water) and quantity wasted, lost, or otherwise unaccounted for.[9]

According to the organisation Waterwise the average person in the UK uses around 150 litres of water per day (127 litres per day for those with a water meter and 160 litres per day for those without a water meter),[10] although the proportion of this volume that is used for consumption, either drinking or for food preparation, is quite small. Analysis of domestic water consumption in the UK has shown that only 4% is used for drinking. A further 33% is used for personal hygiene; 21% is used for washing clothes and dishes; 30% is used to flush the toilet; the remaining 12% is used for outdoor and other purposes, such as watering the garden and washing the car.

Between the years 2000 and 2011, water abstractions in England and Wales were falling year on year but following an increase of 12.6% in 2012 and a 5% reduction in 2013 the volume of water abstracted between 2013 and 2017 rose between 3 and 7%, mainly as a result of an increase in abstraction for electricity generation [6]. It is widely accepted that current levels of water use in the UK and elsewhere cannot be sustained and that a strategy to reduce domestic water use is required. Recycling wastewater to supply some categories of domestic use, such as flushing the toilet, is an option that has been considered on many occasions and may be introduced in the future; however, there are risks associated with using recycled wastewater in the house that will have implications for environmental and human health.

Agricultural use – according to the UN Food and Agriculture Organization (FAO), agriculture accounts for 70% of global water withdrawals.[11] A large portion of the water consumed by crops, an estimated 78%, comes directly from rainfall that infiltrates the soil to generate soil moisture. The other 22% is from surface and groundwater sources [7]. It is estimated that the evapotranspiration required to produce the daily food consumption for one person is 3000 L [7]. The most important driver in water use during the coming decades will be the increase and changes in global food demand due to population growth and changes in diet. Although the world's population growth rate is slowing, the global population is expected to reach 9.7 billion by 2050, and 11.2 billion by 2100 [8]. In addition, poor management of water resources also threaten the resource base on which agriculture depends and the protection of ecosystems continues to be more important and urgent [9].

Several factors are putting pressure on the availability of water for use in agriculture:

- There is increasing competition for water supplies from the domestic, industrial and leisure sectors;
- There is a desire, and regulatory pressure to maintain the ecological diversity of rivers and lakes by avoiding over-abstraction of water; and
- Global climate change models predict reduced summer rainfall and more frequent dry years.

There is clearly a need to reduce the amount of water used globally by the agricultural industry. To maintain water availability or reduce the impacts of drought and flooding farmers may need to adapt their current farming practice. The policies and investment strategies chosen to increase food production will affect water use, the environment, and the extent and depth of rural and urban poverty. Feeding an increasing population will require water management and development strategies that promote improvements in food security while maintaining the productivity of land and water resources and enhancing natural ecosystems.

Industrial use – water is used extensively in all sectors of the economy. The European Environment Agency has estimated that 40% of total water abstraction in Europe is by industry.[12]

Discharges to "controlled waters" in England and Wales apply to industry and are matters for the environmental permitting regime whether as part of the integrated pollution prevention and control regime or as part of the control of water discharge consent regime.[13] It is an offence to cause or knowingly permit a water discharge activity or groundwater activity except in compliance with permit conditions. In Northern Ireland and Scotland there are pollution prevention and control and waste management licensing regimes (see also Chapters 16, 18, 19 and 20).

Recreational use – the use of inland and marine waters for recreational purposes is increasing in many countries. Uses range from body contact sports such as swimming, surfing and diving to non-contact sports such as fishing, walking and bird watching. These activities can have major health benefits to users, but they also present health risks in the form of drowning hazards and exposure to polluted water. Some activities such as boating are also potentially polluting for the environment.

Competition for suitable waters and the popularity of water sports often generates conflicts between activities. Water is important for the tourism industry, which can lead to growing competition for the use of coastal waters and beach areas. This can be most acute in areas where the water has been used traditionally by the local community for fishing, and other commercial purposes. The introduction of tourism to an area creates the need for clear regulations and codes of conduct as well as monitoring waters for health hazards. Water resources are often scarce in prime tourist areas where the very nature of the industry generally makes excessive use of

water through the provision of swimming pools, hotels and related activities.

Water and health

Water-related disease – the WHO Guidelines for Drinking-Water Quality [10] emphasises the risk from the presence of pathogens in drinking water and recommends that eliminating pathogens should take precedence over the removal of toxic chemicals. The reason for making this recommendation relates to the different mechanisms by which pathogens and toxic chemicals lead to disease. Whereas a single exposure to a pathogen in drinking water can lead to infection and the possible occurrence of disease, toxic chemicals tend to have a cumulative effect[14] so that multiple exposures over a longer period of time are required before the disease symptoms emerge. In other words, a person may be able to consume water containing, for example, arsenic for several days or weeks without coming to any harm, but consumption of a single glass of water containing an enteric pathogen can result in an episode of disease. In making this recommendation the WHO is not advocating ignoring toxic chemicals in drinking water, but that the priority should be to provide water that is microbiologically safe before dealing with the chemical contaminants.

There is a similar emphasis on microbiological quality for the management of recreational waters. International guidelines as well as regional and national standards focus on monitoring for the presence and levels of particular microorganisms to determine the risk to health of those who use the water for recreation.

Water-related diseases are of many kinds, and the aetiological agents of these diseases can initiate infection by a variety of routes, not just by ingestion of contaminated water. The diversity of infectious agents and routes of transmission have long created a problem for public health engineers trying to control the spread of these diseases. In order to inform intervention strategies, the concept of classifying diseases according to their ecological and environmental characteristics, rather than their disease characteristics, was developed and is now widely used in the water sector [11]. In its simplest form, the environmental classification of water-related diseases defines four transmission categories: water-borne diseases (including those due to microorganisms in water people drink, for example, cholera and norovirus) that are controlled by improving water quality; water-washed diseases (for example, trachoma and scabies) that are controlled by hygiene interventions, and, therefore, the quantity of water; water-based diseases (for example schistosomiasis and guinea worm) which have part of their life cycle in water and can be controlled by limiting contact with water; and, finally in this group, water-related insect vector diseases (such as malaria and dengue fever) in which the insect vectors require water to complete their life cycle. Outside this group of water-related diseases are additional risks, for example those associated with recreational waters, such as drowning and injuries, and domestic and industrial waters which may pose risks from legionellosis carried by aerosols containing certain microorganisms. The Institute for Health Metrics and Evaluation (IHME) provides regular updates on the contribution to the global burden of disease of a wide range diseases and injuries, including intestinal infections, malaria and other water-related diseases.[15]

Indicator bacteria have been used for many years to assess the microbiological quality of water. These are not typically disease-causing organisms but are found universally in faeces, and ideally nowhere else, and so their presence in water indicates faecal contamination and, therefore, the risk of gastrointestinal disease-causing organisms also being present. The larger the number of faecal indicator bacteria, the higher the risk. The most widely used indicator bacteria are of the total coliform, thermotolerant coliform (formerly faecal coliform), intestinal enterococci groups and *E. coli*. Bacteriological tests are used to assess the sanitary quality of water and the potential public health risk from waterborne diseases. The Fourth Edition of the WHO Guidelines for Drinking-Water Quality [10] provides a summary of the characteristics of each indicator and an assessment of their value for assessing the risk of the presence of enteric pathogens and overall health risk.

One of the challenges for the use of the traditional indicator bacteria in the future is that the spectrum of enteric diseases is expanding and many of the virus and protozoal pathogens do not show any direct correlation with the presence of indicators. However, current advances in analytical techniques are raising the possibility of directly detecting individual pathogens in water as well as identifying new and emerging pathogens. These new methods are forcing a reassessment of the indicator concept and may ultimately provide a better assessment of the risks from water-related disease.

Despite the emphasis on the microbiological quality of water, the presence of chemicals in water can cause serious health problems. The widespread presence of arsenic in water is just one example, with its devastating consequences having been felt most severely in Bangladesh. Fluoride is another natural contaminant of water that is toxic when consumed at high, although not unnatural concentrations. Fluorosis in its mildest form can be seen as dark patches, or mottling, of the teeth, but prolonged exposure to high levels of fluoride can lead to skeletal deformities and death. The WHO Guidelines for Drinking-Water Quality contains recommended maximum concentrations for a large number of chemicals in water that have the potential cause for disease, and in a further publication provide guidance on how to assess priorities for managing the chemical safety of water [12] and [10].

Drinking water – an adequate supply of good quality drinking water has tangible benefits to the health and well-being of individuals as well as to the wealth and productivity of a society [10]. It is a condition that much of the developed world takes for granted; but it is not universal. At the time of writing about 26% of the global population do not have access to safely managed drinking water [13]. Development agencies worldwide acknowledge that the lack of water, together with inadequate provision for sanitation and poor environmental hygiene, helps to maintain many millions of people in a condition of extreme poverty with little prospect of social improvement.

The sequence of events and innovations leading to our modern standards of drinking water and sanitation accelerated during the middle of the 19th century, after the discovery that cholera can be transmitted through water, and that treating water by filtration, and later by disinfection, can significantly reduce the spread of the disease. By the middle of the 19th century, cholera had become established in Britain, with regular outbreaks of the disease occurring in many towns and cities. The prevailing opinion at the time held that cholera, like many other infectious diseases, was spread by foul air (miasmas). This was an understandable mistake given that the incidence of disease would have shown a correlation with the poor, foul-smelling districts of the city, and that it would have risen during the summer months when the smell would have been at its worst. John Snow held an alternative opinion about the transmission of cholera. He proposed that cholera was transmitted by the ingestion of water. His opportunity to test his theory came during an explosive outbreak of cholera in the area around Broad Street (now Broadwick), Soho, London (as considered in Chapter 1). By using careful observation of the distribution of cases, and noting common features between the daily routine of infected households, Snow was able to identify the source of the outbreak as being the water pump on the corner of Broad Street. The remarkable proof of his theory came when he removed the handle from the pump and the outbreak subsided. Thus, the role of water in the transmission of the disease was established.

The next significant advance from the perspective of the control of waterborne disease was made during the investigation of an outbreak of cholera during 1892 in Hamburg, Germany. The outbreak was severe and followed a predictable course, but the striking observation was that the prevalence of disease in Hamburg was significantly higher than in the adjoining town of Altona. Both towns abstracted drinking water from the River Elbe – indeed, the water supply to Altona was taken from downstream of the waste discharges from Hamburg – but differed in their use of treatment to improve the quality of the water before distribution. While the water supplying Hamburg was untreated, the water to Altona was treated by slow sand filtration. This observation provided convincing evidence that water treatment not only improved the aesthetic quality of water, but also had a significant role in the control of waterborne disease.

By the end of the 19th century, advances in the new sciences of microbiology and epidemiology, and in the disciplines of public health engineering provided the evidence to drive forward improvements in drinking water treatment, water quality management and water quality monitoring. At this time, disinfection by chlorine was introduced as a routine component of drinking water treatment, and the multiple barrier principle[16] emerged as a fundamental concept for the control of drinking water quality. The use of indicator bacteria (*Escherichia coli* (*E. coli*) and the coliform group of bacteria) for the monitoring of drinking water quality was also introduced and has continued, relatively unchanged, ever since. This era also saw the introduction of regulations to control the quality of drinking water. The U.S. Public Health Service developed the first drinking water regulations in the United States in 1914. The U.S. Environmental Protection Agency (EPA) assumed responsibility for this task when it was established in 1970. The Safe Drinking Water Act (SDWA) became law in 1974, and was significantly revised in 1986 and 1996. In the European Union, water legislation was accepted by the European Council as early as 1973. Water quality objectives were set for specific types of water (e.g. the Surface Water, Fish Water, Shellfish Water, Bathing Water and Drinking Water Directives) and emission limit values for specific water uses (e.g. Dangerous Substances Directive and the old Groundwater Directive). The quality of drinking water is prescribed for EU Member States in the Council Directive 98/83/EC on the quality of water intended for human consumption.[17]

It is worth noting that for dwellings in England and Wales that Water Supply is also one of the hazards in the Housing Health and Safety Rating System[18] considered elsewhere, although this excludes the quality of water supplied from public mains. It does cover the quality and adequacy of the water supply for drinking and domestic purposes and includes threats to health from contamination.

International efforts to create health-based standards for the quality of drinking water began in 1958 with the publication by the World Health Organization (WHO) of the First Edition of the International Standards for Drinking-Water Quality. Two further editions of the International Standards were published in 1963 and 1971. However, the use of the term Standards implied that the parameter values it contained had to be achieved, irrespective of the resources and capability of a particular country. In 1984, the WHO changed its perspective and published the first edition of the Guidelines for Drinking-Water Quality. The intention of the Guidelines was not to appear to impose maximum concentrations for particular parameters in drinking water, but to suggest parameter values derived from the most current scientific evidence that countries might wish to adopt within national standards. This particular philosophy has continued through to the publication of the current Edition of the Guidelines [10].

New evidence of the links between different constituents of drinking water and human health continue to be published in the scientific literature. The WHO regularly reviews the evidence and where necessary will make changes to the Guidelines. Revisions to the Guidelines are published on the WHO website and we would encourage the reader to visit the site regularly to look for updates and to review the scientific evidence that has been used to support the guideline parameter values.[19]

Case study 1

Cryptosporidium outbreak associated with a groundwater supply (see Willcocks et al., [16])

In 1997 a waterborne outbreak of cryptosporidiosis occurred, affecting 345 people in the North Thames region. The cases were linked by geographical area, which was served by a single water treatment works, which received all its water from eight deep boreholes. Cryptosporidial oocysts were found in one borehole source, the combined raw water entering the treatment works, and in the distribution system at concentrations up to 0.3 oocysts per litre. Despite an extensive investigation of the catchment area around the borehole no obvious source of contamination was found. It was concluded that the oocysts either entered the aquifer directly from the river through interstices in the chalk or from surface water through the adit system (horizontal tunnels below the ground water table) or a defect in the well or borehole lining. This unusual outbreak raises questions about the type of water treatments for boreholes and has implications for our understanding of the quality of water from groundwater sources.

Despite our extensive knowledge of the transmission of disease by drinking water and the toxicology of many of the natural and human-derived contaminants in water, outbreaks of waterborne disease still occur frequently, even in developed countries [14], and many millions of people continue to suffer the often severe consequences of chemical contaminants in water [15].

Recreational water – the first study to review the incidence of disease associated with the use of recreational water was carried out in the early 1920s in the USA. This was followed by two major epidemiological studies. The first of these studies by Stevenson [17] measured the incidence of disease amongst bathers using two inland freshwater bathing sites in the USA. The results showed that there was an appreciably higher level of illness amongst bathers compared to non-bathers. The second study by Moore [18] was conducted at 43 beaches in the UK, but the results of this study conflicted with the findings of Stevenson. Moore concluded that there was a "negligible risk to health" from bathing in sewage-polluted waters even when the beaches were "aesthetically very unsatisfactory" [19]. Indeed, Moore went on to claim that a serious health risk would only exist if the water were so fouled as to be revolting to the senses [19].

There are now a large number of studies that have been published describing the health outcomes arising from exposure to recreational water while participating in a range of different activities, not just swimming. In contrast to the claims made by Moore [18], the overwhelming conclusion from these studies is that exposure to recreational water does increase the risk of illness, and that the risk increases with the level of exposure and the level of contamination as measured by faecal indicator bacteria (for a full review of the subject see Pond [19]). Often the illness is relatively mild – short duration gastrointestinal illness, acute febrile respiratory infection, ear infection – but very occasionally more severe illnesses can be contracted by exposure to contaminated water. An example of the latter is Weil's Disease caused by strains of bacteria from the genus *Leptospira* [20]. Infection by the bacterium is rare and the primary disease symptoms may be mild; however, severe forms of the disease can be fatal if not treated. To put the risk into perspective, in the UK there is an average of 1.9 deaths per year associated with leptospirosis [19], and the British Canoe Union has estimated the risk of a canoeist dying of leptospirosis to be 1:330000 [20].

Legislation and water quality standards to manage coastal and freshwater bathing waters emerged as the link between the contamination of bathing water and the risk of disease was established. Many regions and individual countries have introduced standards for bathing waters that have been informed by the outcomes from major epidemiological surveys, and have selected the indicators for monitoring the quality of bathing water from the correlations found between the different microbial indicators and health outcomes. Throughout the European Union countries are bound by the requirements of the EC Bathing Water Directive.[20]

There is currently an optimistic picture with regards to the quality of marine and freshwater bathing areas in Europe. The latest report by the European Environment Agency, which assesses the bathing water quality in all 28 EU Member States, as well as Albania and Switzerland, showed that 98.3% of the monitored bathing sites in the European Union met minimum standards for water quality in 2020 under the EC Bathing Water Directive.

Climate change poses a threat to the quality. Storm surges, for example, greatly increase the amount of pollutants entering recreational coastal waters, and may resuspend and disperse pathogens that may persist in the environment bound to sediments. Precipitation is predicated to increase in northern Europe at least, and it has been shown that the risk of gastroenteritis and respiratory infections from recreational water use are much higher during the rainy season rather than the dry season [21]. Conversely, extended periods of hot weather can increase the mean temperature of water bodies, which can result in the accelerated growth of algal blooms and pathogenic microorganisms such as *Vibrio* spp bacteria (including *Vibrio vulnificus* and *Vibrio cholerae* non-O1 and non-O139), which are indigenous to the Baltic and the North Sea.

Coastal bathing waters tend to be of better quality than freshwater sites. In 2020, 85% of coastal bathing waters were deemed excellent against the Bathing Water Directive. On the other hand only 78% of inland bathing waters were of excellent quality in 2020.

Artificial recreational waters, such as swimming pools, leisure pools, spas and Jacuzzis have also been implicated in the transmission of disease, despite the water in these systems being treated by filtration and disinfection. A common cause of disease associated with treated swimming pool water is Cryptosporidium – a coccidial protozoan parasite [20] whose infectious stage is an oocyst that is excreted by the infected host – and a number of outbreaks have been reported in the UK and the USA. Cryptosporidium is a significant source of gastroenteritis globally [22]. Oocysts are highly resistant to environmental stress, but the size of the oocysts means that it can be removed by filtration (the same process prevents Cryptosporidium from entering piped water supplies). However, if the filtration step fails, or contaminated filtrate enters the pool water, the oocysts can resist the normal concentration of chlorine used to disinfect the water. Under these circumstances there is a high risk of swimmers becoming infected. A comprehensive discussion of the infections that can be acquired by contact with treated swimming pool water can be found in the manual published by the Pool Water Treatment Advisory Group [23].

Despite the clear need for controls in the pool environment, the regulation of swimming pools is poorly covered. Whilst coastal bathing waters are regulated by a dedicated directive (Bathing Water Directive), swimming pools across the European Community, and further afield, are not. In the United States, for example, there is no federal regulatory authority responsible for disinfected swimming pools, water parks, and so on; all pool codes are developed, reviewed and approved by state and/or local public health officials. As a result, there are no uniform, national standards governing the design, construction, operation and maintenance of swimming pools and other treated aquatic facilities. Thus, the requirements for preventing and responding to recreational water illnesses can vary significantly amongst local and state agencies [14]. Similarly, in the UK, there are no specific regulations governing the design, construction and management of swimming pool facilities; however, designers and operators are liable for their actions under the provisions of the Health and Safety at Work etc. Act 1974.[21] There is an increasing requirement for designers and operators to consider health implications associated with pool use and implement procedures for mitigating the potential risks inherent in the design and use of swimming pools [24].

The European Parliament has passed advisory laws governing pool safety products and procedures relating to both public and private swimming pools. Currently, these have not yet become enforceable directives throughout the whole of Europe. Of relevance to swimming pool operation is the EU Biocidal Products Regulation, which includes disinfectants.[22] This provides a list of approved chemicals complying with the directive.

Section 1 Notes

3 See also https://www.metoffice.gov.uk/weather/learn-about/weather/how-weather-works/water-cycle

4 Although the acidification of rain water by industrial gas emissions was first proposed in the mid-19th century, the problem did not gain prominence until the 1960s and 1970s following news reports of widespread deforestation and the devastating loss of fish stocks in lakes and rivers.

5 European Union. (2000) Directive 2000/60/EC of the European Parliament and of the Council establishing a framework for the community action in the field of water policy. *Official Journal Law*, 327. http://ec.europa.eu/environment/

water/water-framework/info/intro_en.htm (Accessed June 2021).

6   See http://www.eea.europa.eu/themes/water/water-management/river-basin-management-plans-and-programme-of-measures (Accessed June 2021).

7   See https://www.gov.uk/government/consultations/river-basin-planning-working-together (Accessed June 2021).

8   See https://www.usgs.gov/mission-areas/water-resources/science/water-use-terminology?qt-science_center_objects=0#qt-science_center_objects (Accessed August 2021).

9   See Dieter CA, Maupin MA, Caldwell RR, Harris MA, Ivahnenko TI, Lovelace JK, Barber NL, Linsey KS. (2018) Estimated use of water in the United States in 2015: *U.S. Geological Survey Circular*, 1441: 65. https://doi.org/10.3133/cir1441. [Supersedes USGS Open-File Report 2017–1131.]

10 See https://waterwise.org.uk/wp-content/uploads/2019/10/WWT-Report-.pdf and https://assets.publishing.service.gov.uk/government/uploads/system/uploads/attachment_data/file/785567/Water_Abstraction_Statistics_England_2000_2017_Final.pdf (Accessed June 2021).

11 FAO. (2015) *AQUASTAT* website. Food and Agriculture Organization of the United Nations (FAO). http://www.fao.org/aquastat/en/ and reports for example at http://www.fao.org/documents/card/en/c/8dd680fd-70d3-4725-8d9f-30f9a02455a0/ (Accessed June 2021).

12 https://www.eea.europa.eu/themes/water/ (Accessed June 2021).

13 The Environmental Permitting Regulations (England and Wales) 2010 (SI 2010 No 675) came into force on 6 April 2010, replacing the 2007 Regulations. There are three types of surface water and groundwater activities that are eligible for registration as exemptions including small discharges of domestic sewage effluent from septic tanks and package treatment plants. Industrial discharges will require a permit whether a 'standard' permit or 'bespoke'. The water pollution provisions in the Water Resources Act 1991 have been repealed or amended by these Regulations.

14 Lead is an obvious example which has been perhaps not given priority in recent years but lead in drinking water can be an issue. In the UK this is also most commonly the result of lead supply/service pipes where old pipes have not been replaced – for more on lead see PHE Lead Exposure in Children Surveillance System Annual Report 2019 (https://assets.publishing.service.gov.uk/government/uploads/system/uploads/attachment_data/file/967402/

hrp0521_LEICSS.pdf) and EC Joint Research Centre Institute for Health and Consumer Protection (2009) Guidance on Sampling and Monitoring for Lead in Drinking Water (publications.jrc.ec.europa.eu/repository/bitstream/JRC51562/jrc51562.pdf) and also Hayes C, et al. (2010) Best Practice Guide on the Control of Lead in Drinking Water. In: Hayes C, ed. Best Practice Guide on the Control of Lead in Drinking Water. London (United Kingdom): International Water Association Publishing.

15 See http://www.healthdata.org/

16 The multiple barrier principle is an important idea in water treatment. It requires that several processes are used in combination to treat water (see Section 2) such that in the event of a failure of one of the processes, the remaining treatment steps reduce the chance of pathogens being introduced into the water supply.

17 Council Directive 98/83/EC, 1998. Council Directive 98/83/EC of 3 November 1998 relating to the quality of water intended for human consumption. Official Journal of the European Communities No. L 330, 05.12.1998, pp. 32–54 http://eur-lex.europa.eu/legal-content/EN/TXT/?uri=CELEX:31998L0083 (Accessed June 2021).

18 See https://assets.publishing.service.gov.uk/government/uploads/system/uploads/attachment_data/file/15810/142631.pdf

19 https://www.who.int/health-topics/water-sanitation-and-hygiene-wash (Accessed June 2021).

20 Directive 2006/7/EC of the European Parliament and of the Council of 15 February 2006 concerning the management of bathing water quality and repealing Directive 76/160/EEC.

21 http://www.hse.gov.uk/legislation/hswa.htm

22 The Biocidal Products Regulation BPR, Regulation (EU) 528/2012 concerned the placing on the market and use of biocidal products, which are used to protect humans, animals, materials or articles against harmful organisms like pests or bacteria, by the action of the active substances contained in the biocidal product. It has been applicable from 1 September 2013, with a transitional period for certain provisions. It repeals the Biocidal Products Directive (Directive 98/8/EC). After the withdrawal from the EU the existing EU Biocidal Products Regulation (EU BPR) has been copied into GB law and amended to enable it to operate effectively in GB. This means that most aspects of EU BPR will continue in the same way under the new stand-alone regime – the GB Biocidal Products Regulation (GB BPR) came into force at 11 pm on 31 December 2020.

SECTION 2: WATER AND RISKS

Risk assessment

Improving water quality, in conjunction with improvements in excreta disposal and personal hygiene can be expected to deliver substantial health gains in the population. The detection and enumeration of pathogenic microorganisms from contaminated water is both difficult and costly, so for many years the non-pathogenic indicators described in Section 1 have been used. Their enumeration is relatively easy and inexpensive compared to the isolation and enumeration of individual pathogens. Microbial contaminants, however, are not limited to bacteria, and illness may result from exposure to pathogenic viruses or protozoa, both of which have different environmental behaviour and survival characteristics to bacteria. This, coupled with the fact that testing water leaving the treatment plant or within distribution systems (end-product testing) can only identify a potential health risk after the water has been consumed, highlighted the need to adopt additional proactive approaches to assuring water quality and safety. The approach advocated by the WHO is termed Water Safety Planning and implementation of this approach has the potential to greatly improve challenging and resource limited drinking water supplies, both small and large [24]. In water quality management, risk assessment and risk management are essential components of a system for ensuring public health.

Risk assessment informs the development of policies for controlling and managing health hazards. The process involves four very distinct steps: the first step is hazard identification, which involves the collection, organisation, and evaluation of all information pertaining to a pathogen or a nutrient. Second is hazard characterisation, which determines the relationship between a pathogen and any adverse effects. Third is exposure assessment, which involves determining the quantity of pathogens that might be ingested. The fourth, and last step, is risk characterisation, which involves evaluating the risk and related information [25].[23]

Risk assessments can be qualitative or quantitative, with the latter being more expensive, labour intensive and time consuming. Microbial risk assessments are inherently uncertain as bacteria can multiply as conditions change in the water environment. However, researchers are making progress in developing predictive models and other tools that will meet the technical requirements for quantifying estimates of risk.

The objective of risk management is to consider the events that contribute to risk and to focus on barriers to mitigate the risk. An important strategy in risk management to provide safe water for the consumer is the multiple barrier approach, which is generally applied to the water treatment process. Barriers generally act to reduce rather than eliminate risk. Therefore, since events that contribute to risk are linked, the use of multiple barriers provides multiple levels of protection that act together to reduce the total risk by more than the reduction that would be achieved by one barrier. In addition, where the effectiveness of one barrier is reduced, the presence of others helps to maintain a level of protection, or reduced risk throughout the failure. A chain of events involving barrier failures and/or other unusual events, are the key risk sources and therefore targets for risk management [26].

The WHO Guidelines for Drinking-Water Quality [10] and Guidelines for Safe Recreational Water Quality [27] are the starting points worldwide for setting water quality standards, including microbiological standards. The guidelines are, in large part, health risk assessments and are based on scientific consensus, best available evidence and expert opinion. The guidelines advocate that a risk-benefit approach is taken to control public health hazards associated with water. This approach can be applied to both large and small drinking water supplies as well as recreational waters.

Managing risk in real systems requires a systems approach. During the revision of the third edition of the WHO Guidelines for Drinking-Water Quality [28] the value of the Water Safety Plan (WSP) approach (based on the Hazard Analysis and Critical Control Points process; HACCP) was repeatedly highlighted. HACCP has a focus on controlling hazards as close to their source as possible. It has been used to assure food and beverage safety since its codification in 1993 by the United National Food and Agricultural Organization (FAO) and the WHO Codex Alimentarius Commission.

Water Safety Plans are considered by the WHO and throughout the water industry as the most effective means of maintaining a safe supply of drinking water. Their use should ensure that water is safe for human consumption and that it meets regulatory water standards relating to human health. Comprehensive risk assessment and risk management is at the core of these plans, which aim to direct management of drinking water-related health risks away from end-of-pipe monitoring and response.

The principles and concepts of other risk management systems are extensively drawn upon in WSP design, including the multi-barrier approach and HACCP. In order to produce a plan, a thorough assessment of the water supply process from water source to the consumer's tap must be carried out by the water provider. Hazards and risks should be identified, and appropriate steps towards minimising these risks are then investigated.

Practical guidance to facilitate the development is provided by the World Health Organization *Water Safety Plan Manual* [30] and is illustrated in the following pages with examples from domestic water and recreational waters. Water Safety Plans can and should be adapted to address the impacts of climate change and equity issues.

Risk assessment approach to domestic water

The following section provides the broad steps that should be followed to establish a WSP.

Step 1: Set up a team and decide the methodology by which a WSP will be developed.

A team of experts who understand the whole process of water abstraction,

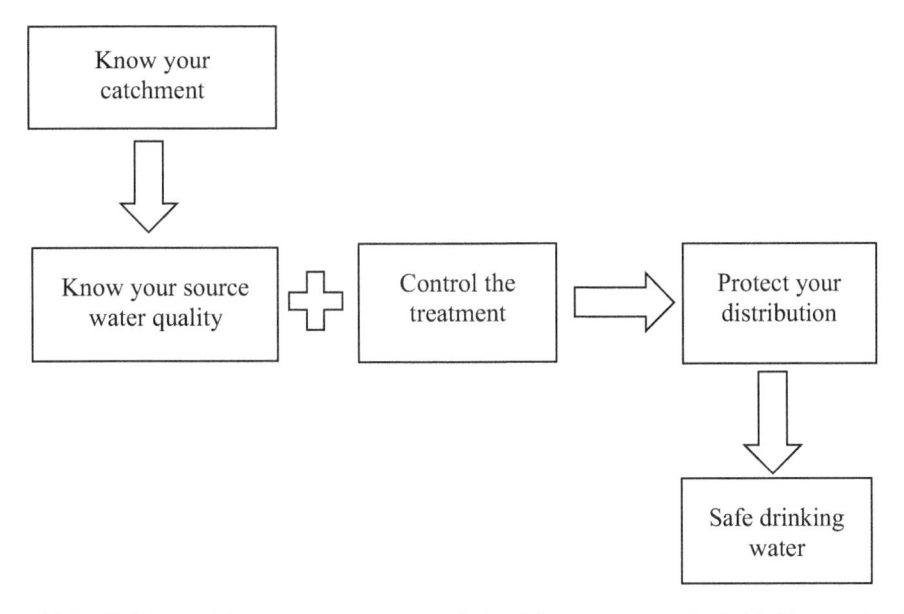

Figure 15.2 Catchment to consumer approach to risk management of drinking water (see Medema et al. [29])

treatment and distribution should develop the WSP. This may require the involvement of expertise from outside the utility; indeed all those with responsibility towards ensuring the safety of the water in question should be involved. This may include agricultural workers, landowners, local government, consumers and others. The first task of the WSP team is to describe the water supply. This should involve field investigations in order to identify the vulnerabilities of the system, relevant types of hazards and control measures. Typical information that is required includes maps showing the distribution systems, types of land use and management in the catchment, details of the industry and risks in the catchment, relevant water quality standards, details of the source of the water, details of any potential contaminants. The description should cover the whole system from the source to the end point of supply. A validated flow diagram helps to identify hazards, risks and current controls.

Specifically, when considering surface waters the type of water such as river (direct abstraction), river (abstraction in storage reservoir), impounding reservoir or lake should be taken into account. In addition, an inventory should be made of point discharges, such as sewage effluents, industrial effluents, water from mining; water quality and how it varies seasonally and with weather patterns; flow and reliability of source and retention time if stored; recreational and other human activity; any existing source protection systems should be developed.

For the assessment of groundwater, the following should be considered: whether it is a confined or unconfined aquifer, hydrology and recharge area; flow rate, direction of flow and dilution characteristics; whether fast or slow response to activities and events on surface; depth of casing and abstraction and any wellhead protection; inventory of activities in the recharge area that could affect water quality.

Step 2: Identify all the hazards and hazardous events that can affect the safety of a water supply from the catchment, through treatment and distribution to the consumers' point of use.

Hazards are defined as physical, biological, chemical or radiological agents that can cause harm to public health. However, hazards in the wider sense should also be considered, such as availability of trained staff, floods, power supplies and laboratory facilities. Identification of a hazard relating to a public supply does not mean the water company is automatically responsible for the cause. Many hazards are naturally occurring – soils, geology and topography of the land can all contribute to hazards, as can hydrology, meteorology and weather patterns; nature of the land and it use, in particular degree of urbanisation, industrial activities, animal rearing and arable farming, degree of natural land and its wildlife, quarrying and mining (that is, uses and activities that could give rise to contamination of raw water sources).

The WSP approach requires water utilities to work with other stakeholders to make them aware of their responsibilities and their impact on the utility's ability to provide safe drinking water. Although the approach will be the same whatever the size of the water system, for owners of small supplies, where funding, expertise and human resources are limited, the most significant risks should be identified. For example, if sheep or cattle have access to a spring or stream that is used as a water supply, it would be obvious during a site visit that faecal contamination is a hazard and this would become a priority.

Step 3: Assess the risk presented by each hazard and hazardous event.

This may be quantitative or semi-quantitative, consisting of an estimation of likelihood/frequency and severity/consequence or a simplified qualitative approach based on the expert judgement of the WSP team to identify the severity of the hazards and to prioritise them.

Step 4: Consider if controls or barriers are in place for each hazard and hazardous event.

Examples of typical control measures are:

- Implementing a catchment management plan that includes measures to protect surface and groundwater such as controls of all point sources of effluent or overflows discharging to the catchment; site-specific protection requirements, such as containment for industrial chemicals and agricultural slurries; designation, limitation and control of specific activities or uses with all or part of the catchment; regular inspection of the catchment;
- Planning regulations including protection of water resources against future activities;
- Protection at the point of abstraction such as security to prevent unauthorised access by humans or animals; appropriate arrangements for surface water abstraction from rivers, reservoirs and lake and for wellhead and casings for groundwater abstraction.

Step 5: Validate the effectiveness of each control or barrier, reassess and prioritise the risk.

Each control measure should be subject to intensive validation monitoring where possible, and to routine monitoring to check that it is working. For example, a sewage works or industrial effluent should be subject to a discharge consent (for example set by an environmental regulator) that specifies quality criteria that must be met. The sewage works or industrial operator should be required to monitor effluent quality on a regular basis and should be required to notify the water supplier and the environmental regulator should the quality criteria not be met. For many control measures the monitoring will be by regular inspection of the catchment and potential polluting sites and regular inspection of the point of abstraction and associated works.

Step 6: Develop and implement an improvement plan where necessary.

The improvements should be monitored to ensure they are working and the WSP updated accordingly. An example of an improvement plan may be where there is a requirement to review the need for reducing the risks from viruses or protozoans from sewage systems. The improvement may be to develop additional sewage disinfection and downstream water disinfection. The actions should always be accountable to a particular person, in this case a water quality officer.

Step 7: Demonstrate that the system is consistently safe.

Monitoring of control points is essential for supporting risk management to show that the control measures are effective and that if a problem is detected, actions can be taken quickly to prevent water quality targets being compromised. Water quality monitoring is covered in Section 3 and the reader is referred to this section for further details. It is sufficient to say here that monitoring data provides important feedback on how the water supply system is working and should be frequently assessed. Accurate records of all results and actions should be kept.

Step 8: Regularly review the hazards, risks and controls.

It is important to regularly review the hazards, risks and controls to ensure that the WSP is working properly. Verification of the WSP involves compliance monitoring, internal and external auditing of operational activities and consumer satisfaction. Audits should be undertaken regularly and may be done by regulatory authorities or qualified independent auditors.

Step 9: Keep accurate records for transparency and justification of outcomes.

Documentation of all aspects of the WSP is essential. This should be done during normal conditions (Standard Operating Procedures) and when the system is operating under corrective actions. Management staff should ensure these are kept up to date to keep operators and management staff involved.

Adequate and effective programmes for training, maintenance, research and development and consumer information support the development of staff skills and knowledge and commitment of the water utility to the WSP approach.

It is often not feasible to monitor supplies each day to ensure the water is of potable quality. A risk assessment allows problems that may result in contamination of the water to be identified and remedied on a continuous basis.

The following paragraphs include an example of a risk assessment for a private water supply in Scotland, which has been adapted to provide an illustration for this publication [31].[24]

The example relates to a surface-derived water supply that supplies a visitor centre (where seven people work and which receives around 24,000 visitors a year) and three dwellings housing ten people. The water supply is a stream that flows into an artificial pond from where the water is abstracted into two tanks before it flows onto a small chlorination system and then into the supply.

The risk characterisation is high as there is evidence of sheep grazing around the stream and dead animals have had to be removed from the catchment in previous years. Other animals have also been seen in the catchment. This hazard is considered permanent. The catchment also includes large areas of commercial forestry. This also constitutes a permanent hazard to the water supply. Other hazards are identified such as the drain and scour lines not having vermin protection at the outlets. The stream shows varying levels and flow throughout the year depending on the prevailing weather conditions and after heavy rains the appearance of the water changes. Recommended interventions focus

on trying to reduce the access of animals to the stream, controlling forestry activity in the catchment, protecting drain and scour pipes from entry by vermin, identifying pipe materials and increasing awareness of the effect of rainfall on water quality. Table 15.1 provides an example of a general site survey where the hazards to water quality can be identified and recorded, and Table 15.2 shows an example of a supply survey.

Overall risk assessment – the overall risk assessment for the source is taken as the highest individual risk category identified from the two surveys. The hazard assessment is quantified and will determine the priorities for remedial work to be carried out.

Generally, a local authority or owner of the supply will become aware of a quality failure after receiving water sample results. Samples must be taken correctly and it should be ensured that laboratories are properly accredited to carry out analysis, as this will have a significant bearing on the success of any formal action. Samples should be taken in accordance with standard requirements, for example as specified in the Private Water Supplies Regulations 2009.[25] Further details on water quality sampling and standardisation are given in Section 3.

When obtaining water samples, as much information as possible should be taken about the site including the location of the supply, surrounding land use and who else shares the supply. This information may already exist as part of the risk assessment for the supply.

Records must be accurate, and that information must be available to the appropriate authorities to enable effective action to be taken. If the supply is considered a risk to human health, measures should be put in place to safeguard public health as quickly as possible. In order for this process to be effective, it is essential that relevant information is available to the authorities. This includes identification of the relevant person(s), details of any legal agreements between owners and users, the number of premises on the supply and the tenure of the properties. Advice and

Table 15.1 Example general site survey. Are any of the following likely to be present and likely to influence water quality at the source?

	Risk characterisation			Hazard assessment		
	Yes	No	Don't know	Likelihood	Severity	Score
History of livestock production						
Evidence of wildlife						
Surface run-off from agricultural activity diverted to flow in the source/supply						
Soil cultivation with wastewater irrigation or sludge/slurry/manure application						
Disposal of organic wastes to land						
Farm wastes and/or silage stored on the ground (not in tanks or containers)						
Remediation of land using sludge or slurry						
Forestry activity						
Awareness of the presence of drinking water supply/source by agricultural workers						
Waste disposal sites						
Disposal sites for animal remains						
Unsewered human sanitation including septic tanks, pit latrines, soakaways						
Sewerage pipes, mains or domestic						
Sewage effluent lagoons						
Sewage effluent discharge to adjacent watercourse						
Evidence of use of pesticides near source						
Evidence of industrial activity likely to present a contamination threat						

Source: Adapted from Scottish Executive [31]

discussions should take place with the relevant person(s) who should ensure, as a matter of best practice, all users of the supply are aware of the advice and are kept informed on progress, in addition to other actions they may be required to take under the Private Water Supplies Regulations. Authorities should consider whether further samples are required, including alternative sampling points, before detailed advice is given.

Authorities responsible for ensuring the risk assessment is carried out should prepare standard letters and methods of distributing results. The relevant person(s) should be provided with general advice and, where appropriate and available, site-specific advice. Results and advice may be given face to face, by telephone, by fax or by email but should always be confirmed by letter, which should detail the results, their interpretation and steps to remedy the supply. A target improvement date should be stated in the letter.

Officers should consider the need to monitor the quality of the supply whilst improvements are undertaken. Consideration should be given to expanding the parameters sampled for. Samples may also be taken following each stage of improvements so that the effectiveness of improvements can be monitored. Relevant person(s) and users of the supply

Table 15.2 Example supply survey. Are any of the following known to occur at the head works site or in relation to the supply?

	Risk characterisation			Hazard assessment		
	Yes	No	Don't know	Likelihood	Severity	Score
Is the supply network constructed from material liable to fracture?						
Are intermediate tanks (e.g. collection chambers) adequately protected?						
Do junctions present in the supply network, particularly supplying animal watering systems, have back siphon protection?						
Has maintenance (including chlorination) been undertaken in the previous 12 months?						
If present, does the header tank within the property have a vermin-proof cover?						
Has the header tank been cleaned in last 12 months?						
Has any point of entry/point of use treatment equipment not been serviced in accordance with the manufacturer's instructions in the past 12 months?						
If present, are UV lamps operating?						
Is there a noticeable change in the level and flow of water throughout the year?						
Is there a noticeable change in the appearance of water (e.g. colour, turbidity) after heavy rain or snowmelt?						

Source: Adapted from Scottish Executive (2009) [31]

should be kept up to date with sample results and progress on work.

Once improvements have been carried out and the water re-sampled, an assessment should be made as to whether further improvements and samples are required.

By working closely with the relevant person(s), a good degree of co-operation may be achieved. If formal action is subsequently taken, the local authority will be able to demonstrate it took all reasonable steps to try to improve the quality of sufficiency of the supply.

Risk assessment as applied to recreational waters

In terms of recreational water quality, assessing the risk associated with human exposure

can be carried out directly with epidemiological studies or indirectly using quantitative microbial risk assessment (QMRA). In the past 20–30 years epidemiological studies have been used to show a relationship between faecal pollution and adverse health outcomes. Pruss [32] reviewed the literature showing a relationship of certain symptoms or symptom groups to the count of faecal indicator bacteria in recreational water. Gastrointestinal symptoms were the most frequent health outcome for which significant dose-related associations were reported. Key studies undertaken by Kay et al. [33] were used by WHO to develop guideline values for coastal recreational waters used by healthy adults [27].

QMRA can be used to indirectly estimate the risk to human health by predicting infection or illness rates given densities of particular pathogens, assumed rates of ingestion and appropriate dose-response models for the exposed population. However, pathogen numbers, as opposed to faecal index organisms, vary according to the prevalence of specific pathogens in the contributing population, and may show seasonal trends. The advantages of QMRA studies are that the potential benefits and limitations of risk management options may be investigated via numerical simulation and that risk below epidemiologically detectable levels may be estimated under certain circumstances.

The WHO Guidelines for Safe Recreational Waters interprets the HACCP approach to risk management as shown in Table 13.3.

Table 15.3 Implementation of the HACCP approach for recreational water management (WHO [27])

Initial steps	Implementation
Assemble HACCP team	Composition of the team should be representative of all stakeholders and cover all fields of expertise as possible.
Collate historical information	Summarise previous data from sanitary surveys, compliance testing, utility maps of sewerage, water and stormwater pipes and overflows. Determine major animal faecal sources for each recreational water catchment. Reference development applications and appropriate and appropriate legal requirements. If no historical data are available, collect basic data to fill gaps.
Produce and verify flow charts	Produce and verify flow charts for faecal pollution from sources to recreational exposure areas for each recreational water catchment. If required, conduct a new sanitary survey. The flowcharts should show what happens to water between catchments and exposure in sufficient detail for potential entry points of different sources of faecal pollution to be identified and any detected contamination to be traced.
Core steps	
Hazard analysis	Identify human versus animal faecal pollution and potential entry into recreational waters. Determine significance of possible exposure risks. Identify preventive measures (control points) for all significant risks.
Critical control points	Identify points or locations at which management actions can be applied to reduce the presence of, or exposure to, hazards to acceptable levels.
Critical limits	Determine measurable control parameters and their critical limits.
Monitoring	Establish a monitoring procedure to give early warning of exceedances beyond critical limits. This should include water sampling and analysis, visual inspection of potential sources of pollution in catchment.
Management actions	Prepare and test actions to reduce or prevent exposure in case exceedances of critical limits.
Validation/verification	Obtain evidence that the envisaged management action will ensure the desired water quality will be obtained or human recreational exposures will be avoided. Obtain evidence from auditing management actions that the desired water quality or change in human exposure is obtained and the good operational practices are being complied with at all times.
Record keeping	Ensure that monitoring records are kept in a format that allows external audit and compilation of annual statistics.

HACCP principally addresses the need for information for immediate management action but can be applied to long-term actions for the management of recreational waters.

Different stakeholders play different roles in the management of swimming pools and other recreational water environments for safety. There may therefore be a number of groups involved in assessing the risks associated with these environments. The areas of responsibility can be aligned with the areas that require risk assessments. These may be divided into four major categories, although there may be overlap between responsibilities: design and construction; operation and management; public education and information; and regulatory requirements (including compliance). The WHO Guidelines for Safe Recreational Water environments [34] provides guidance on potential health hazards associated with these facilities and forms a basis for the development of approaches to controlling the hazards. Numerical values are presented as indicators of safety or good management where possible. The guidelines use a risk-benefit approach, which in the case of recreational waters concerns not only health risks but also benefits associated with the well-being derived from the use of these waters. In developing risk assessment for the protection of public health, competent government authorities should take into account social, economic and environmental factors, including the general education of children and adults. The approach can lead to the adoption of standards that are measurable and can be implemented and enforced.

There has been some debate about the use of faecal indicator bacteria as the main indices of risk to bathers as there are a number of other possible contamination sources. A more holistic approach to risk assessment and management of coastal and inland recreational waters is therefore advocated. The WHO promotes this issue with the use of sanitary surveys [27]. This is designed to encourage managers to know the beach catchment and

the concept has been followed up in New Zealand, Australia and the European Union (EU).

In the EU, the Bathing Water Directive (2006/7/EC) includes a risk assessment in the form of a bathing water profile. The main objective of the 2006/7/EC Bathing Water Directive is to reduce gastroenteritis and other waterborne health risks. Therefore, the Directive requires, in addition to monitoring, the drafting of bathing water profiles. This is to gain an understanding of the faecal sources and routes of pollution and focusses on the indicators for faecal pollution: either *Escherichia coli* (*E. coli*) and intestinal enterococci (parameters of the EU Directive) or thermotolerant bacteria of the coliform group and faecal streptococci.

Information must be available about the route by which bathing water quality is being influenced and its extent. In other words, the manager of the bathing water location will have to give an estimate of which sources of emission negatively influence the bathing water quality and via which dispersion routes.[26] Currently, no such system exists for swimming pools.

Risk assessment for bottled water

Three major types of bottled water can be identified according to Ferrier [35].

Natural mineral water – is, in the European Union, an extremely specific product responding to strict criteria. It is wholesome underground still or aerated water, protected against pollution hazards and characterised by a constant level of minerals and trace elements. This water cannot be treated, nor modified by the addition of exogenous elements, such as flavours or additives. The USA requires for natural mineral water to have a minimum level of 250 ppm total dissolved solids.

Spring water – in Europe is also underground water protected against pollution hazards. It cannot be treated but it doesn't need to have a constant mineral composition. Water from different springs can be sold under the same brand name. In United States, spring water is derived from an underground formation from which water flows naturally to the surface of the earth.

Purified water – is surface or underground water that has been treated in order to be suitable for human consumption. It differs from tap water only by it being distributed in bottles rather than through pipes.

In the UK an official Industry Guide has been developed in accordance with Article 8 of Regulation (EC) No 852/2004 on the hygiene of foodstuffs and follows the Guidelines produced by the Food Standards Agency in 2005. It is a revision of the Food Standards Agency (FSA) Guidelines to Good Hygiene Practice for Bottled Water and was published in 2013.[27] The revised Guide provides advice on how bottled water companies may comply with the EC food law, national implementing legislation and other relevant requirements related to the hygiene of bottled waters.

Food business operators are required (under Regulation (EC) 852/2004) to put in place, implement and maintain a permanent procedure or procedures based on HACCP principles. The Guide provides advice on preparing a HACCP plan looking at a specific food process operation step by step from source of product and supplied materials to the finished product arriving with the consumer. Hazards may occur in three broad categories: microbiological contamination, chemical contamination and physical contamination or damage. Specific hazards should be identified with respect to the health of the workers in the industry, the premises on which the water is to prepared, treated and bottled (which may be temporary premises and hence have different hazards to permanent premises), transport (for example where water is not bottled at source and the water is transported from the source to the bottling/canning facility by road tanker), equipment, waste products, water supply, personal hygiene and packaging.

The food industry utilises large volumes of public water supplies or privately treated water derived from a borehole. Water is used directly in food as an ingredient; for washing food containers; washing raw foods; transport of materials; factory cleaning; and in cooling systems, boilers and steam raising.

All water in direct contact with food must be of potable quality and should be free of pathogenic microorganisms [36].[28] Of concern is the contamination of bottled water and water used in the food industry with Cryptosporidium oocysts. This is a particular concern where the water receives no treatment prior to bottling. The Natural Mineral Water, Spring Water and Bottled Drinking Water (England) Regulations 2007[29] require that the water does not contain:

1 Any microorganism (other than a parameter) or parasite
2 Any property, element or substance (other than a parameter), at a concentration or value which would constitute a potential danger to human health

Special case of *Legionella*

Legionella bacteria are common in natural water courses such as rivers and ponds. Since *Legionella* are widespread in the environment, they may contaminate and grow in other water systems such as cooling towers and hot and cold water services. They survive low temperatures and thrive at temperatures between 20–45°C if the conditions are right (e.g. if a supply of nutrients is present such as rust, sludge, scale, algae and other bacteria). High temperatures kill the bacteria.

In the UK the Health and Safety Executive has published the Approved Code of Practice and Guidance for the control of *Legionella* bacteria in water systems. Approved Code of Practice and guidance L8 contains advice on duties under the law to control *Legionella*, and on technical aspects of the assessment and control of *Legionella* risks.[30] The Code of Practice provides guidance on carrying out a risk assessment to identify and assess the risk from *Legionella* bacteria. The risk assessment should include identification and evaluation of potential sources of risk and the means by which exposure to *Legionella* bacteria is to be prevented; or if prevention is not practicable, the means by which the risk from exposure is to be controlled. The Code of Practice provides a list of some of the factors, which should be considered when carrying out the assessment as well as providing guidance on managing the risk.

WHO advocate the development of a Water Safety Plan (WSP) as the preferred approach to managing specific health risks of exposure to *Legionella* from water systems [28]. Assessment of the water system supports the steps of the plan (as described earlier), allowing effective strategies for controlling hazards to be developed and implemented [37].

Section 2 Notes

23 See https://www.who.int/publications/i/item/ 9789241515719?seque

24 Note the Private Water Supplies Technical Manual has been superseded by that from the Drinking water Quality Regulator for Scotland (DWQRS) although much has the same content and the original "was created to support local authorities when the previous PWS regulations were introduced in 2006. Much of the information contained in it is still relevant". See https://dwqr.scot/private-supply/ technical-information/pws-technical-manual/

25 SI 2009 No 3101.

26 http://ec.europa.eu/environment/water/water-bathing/pdf/profiles_dec_2009.pdf

27 See FSA (2013) Food Industry Guide to Good Hygiene Practice: Bottled Water. The Stationary Office, London. 56pp ISBN 9780117082137 available at https://www.tsoshop.co.uk/bookstore. asp?FO=40152&Action=Book&ProductID= 9780117082137

28 Private Water Supplies Regulations 2016, which came into effect on 27 June 2016, and in Wales through the Private Water Supplies (Wales) (Amendment) Regulations 2010, which came into effect on 27 June 2016, and the Private Water Supplies (Wales) Regulations 2017, which came into force on 20 November 2017. The Private Water Supplies (England) (Amendment) Regulations 2018 came into force on 11 July 2018.

29 SI 2007 No 2785.

30 See Health and Safety Executive (2013). Legionnaires' disease. The control of *Legionella* bacteria in water systems. Approved Code of Practice and Guidance. L8. 4th Edition HSE, Norwich, UK. Available at https://www.hse. gov.uk/pubns/priced/l8.pdf

SECTION 3: WATER QUALITY MONITORING AND ASSESSMENT

Introduction

Throughout this chapter we have referred to the use of various risk assessment frameworks for the management and control of water quality. Despite one intended consequence of using risk assessment to reduce the emphasis on analytical results as a means of informing management decisions, water quality monitoring will remain a vital tool for verifying the quality of water for its intended use. The correct practice of water quality monitoring, therefore, is an important skill.

Water quality monitoring and water quality assessment are different activities despite sometimes being treated as synonymous. The definitions of each, taken from Chapman [38] are:

Water quality monitoring. The collection of information at set locations and at regular intervals in order to provide data, which may be used to define current conditions, establish trends, etc.

Water quality assessment. The overall process of evaluation of the physical, chemical and biological nature of water in relation to natural quality, human effects and intended uses, particularly uses which may affect human health and the health of the aquatic system itself.

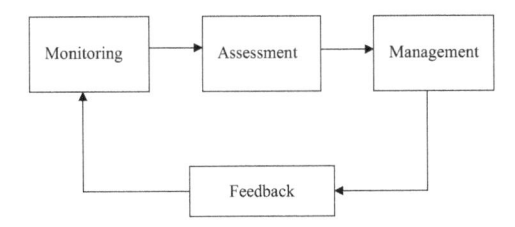

Figure 15.3 Water quality monitoring sequence
Source: Adapted from [38]

Thus, monitoring is the systematic process by which data is collected – the acts of sampling and analysis – and assessment is the interpretation and reporting of the monitoring results in the context of the environment from which the samples were taken. Inevitably, the assessment report will inform the management of the water system, which, in turn, will need to be monitored to verify that management decisions have been effective. The sequence is summarised by the following, simple flow diagram (Figure 15.3, adapted from the text of Chapman [38]).

The key message from the flow diagram is not to view each element as an entity that can exist independently, but to see them as being interdependent components of a management system leading to continuous improvement of water quality.

Environmental health practitioners (EHPs), for example, will be involved with both water quality monitoring and water quality assessment: gathering data, and interpreting the data in the context of the source. Because the latter is influenced by the interplay between several factors that can be unique to the source, it is impossible within the scope of this chapter to be able to provide useable guidance for water quality assessment. For those readers who wish to learn more about the subject, the book *Water Quality Assessment* [38] provides an excellent description of the art and the science that is involved. For the remainder of this section we shall concentrate on the purpose and practice of water quality monitoring.

Water quality monitoring

The saying 'rubbish in, rubbish out' that is used in computing is equally relevant to water quality monitoring. Inadequate consideration of the purpose of the water quality monitoring programme and an inappropriate water sampling plan will generate data that is worthless or dangerously misleading, no matter how well the water samples are analysed by the laboratory. The preference, particularly in the water sector, for using laboratories that have accreditation for their analytical work is understandable, yet the focus on the analytical results has grown to the point where it overshadows the requirement for taking appropriate samples properly. The consequence is that the latter can get overlooked. There is no value in knowing the concentration of a particular contaminant to within plus or minus 0.5% of its true value in the water sample if the sample was taken incorrectly, from the wrong point at the wrong time, in the wrong type of bottle, and then improperly transported to the laboratory. Yet this mistake is widespread, and inevitably leads to costly mistakes being made.

The first step in any water quality monitoring programme is to establish the purpose of the programme. At its simplest this is a decision between the programme being done for regulatory purposes (regulatory/compliance monitoring), to investigate a problem that has been recognised in a water system (investigative monitoring), or to determine the characteristics of a water system at a particular point in time or over a period of time (operational monitoring). The terms in parentheses are ours and you may find that other authors use different terms; nevertheless, the principles remain the same. Once the purpose of the water quality monitoring programme has been established, the subsequent steps in implementing the programme and assessing and reporting the results are more easily defined.

EHPs are often required to take water samples to ascertain if the quality of the water in a system complies with the standards that apply to its use. For instance, in England does the water quality comply with the Private Water Supply Regulations, 2016[31] or the Water Supply Regulations 2010?[32,33] Is the water being used for food preparation potable in accordance with the relevant regulations? The outcome of this type of monitoring programme is a statement that the water does or does not comply with the standard. If the water does not comply with the standard, this may trigger an investigation to find the source and extent of the contamination, for which a different type of monitoring programme will have to be carried out. The investigation may in turn find shortcomings in the regulation, which then have to be addressed by introducing changes to the regulation, which may then feed back into revised recommendations for water quality monitoring. This type of event is uncommon, and for the most part regulatory monitoring is confined to a simple determination of compliance.

In comparison to investigative or operational monitoring, regulatory monitoring is easier to plan and implement because the requirements for the monitoring programme are often stipulated in the regulations. These requirements may include the types of parameters that are to be measured, where the samples should be taken from, how often the samples should be taken, the analytical methods that should be used, and what the admissible concentrations are for each parameter. For example, the Private Water Supplies Regulations 2016, which place a responsibility on local authorities to identify and check private water supplies, define many of the actions that need to be taken in order to determine the quality of these supplies.

Where water is supplied by a water undertaker or licensed water supplier and is then further distributed by a person other than a water undertaker or licensed water supplier, the monitoring must be carried out on the basis of the risk assessment. The regulations specify the monitoring requirements. Two terms used in water supply regulations, such as those applicable to England, Wales and Scotland [26],[34] to describe two different types of monitoring activity are check monitoring and audit monitoring. The difference

between the two is important and EHPs will be required to carry out both activities from time to time.

The descriptions of audit and check monitoring are slightly different between the various regulations to reflect their scope and application; nevertheless, the principles are the same. The following definitions are taken from the Private Water Supply Regulations 2016:

> "Audit monitoring" means sampling for each parameter listed in Parts 1 and 2 of Schedule 1 (other than parameters already being sampled under check monitoring) –
>
> (a) in order to provide information necessary to determine whether or not the private water supply satisfies each concentration, value or state prescribed in those Parts of that Schedule, and
> (b) if disinfection is used, in order to check that disinfection by-products are kept as low as possible without compromising the effectiveness of disinfection.
>
> (3) The local authority may, for such time as it may decide, exclude a parameter from audit monitoring of a private water supply –
>
> (a) if it considers that the parameter in question is unlikely to be present in the supply or system at a concentration or value that poses a risk of the private water supply failing to meet the concentration, value or state specified in Part 1 or 2 of Schedule 1 in respect of that parameter,
> (b) taking into account the findings of any risk assessment, and
> (c) taking into account any guidance issued by the Secretary of State.
>
> (4) A local authority may monitor anything else identified in the risk assessment.

Check monitoring means sampling for each parameter listed in Table 1 in the circumstances listed in that table in order –

(a) to determine whether or not water complies with the concentrations or values in Schedule 1;

(b) to provide information on the organoleptic and microbiological quality of the water; and

(c) to establish the effectiveness of the treatment of the water, including disinfection.

Thus, audit monitoring is carried out with the purpose of ascertaining if a water supply meets all the requirements of the regulation, whereas check monitoring takes a number of key parameters and uses them to determine if the water continues to fulfil the requirements of the regulations. This difference in purpose is reflected in the frequency of sampling recommended for the two approaches in Schedule 2 of the Private Water Supply Regulations 2016 (Tables 15.4 and 15.5).

In addition to the information that is contained in the water quality regulations, there

Table 15.4 Sampling frequency for check monitoring

Volume m³/day	Sampling frequency per year
≤10	1
>10 ≤100	2
>100 ≤1,000	4
>1,000 ≤2,000	10
>2,000 ≤3,000	13
>3,000 ≤4,000	16
>4,000 ≤,5000	19
>5,000 ≤6,000	22
>6,000 ≤7,000	25
>7,000 ≤8,000	28
>8,000 ≤9,000	31
>9,000 ≤10,000	34
>10,000	4 + 3 for each 1,000 m³/day of the total volume (rounding up to the nearest multiple of 1,000 m3/day)

Table 15.5 Sampling frequency for audit monitoring

Volume m³/day	Sampling frequency per year
≤10	1
>10 ≤3,300	2
>3,300 ≤6,600	3
>6,600 ≤10,000	4
>10,000 ≤100,000	3 + 1 for each 10,000 m³/day of the total volume (rounding up to the nearest multiple of 10,000 m³/day)
>100,000	10 + 1 for each 25,000 m³/day of the total volume (rounding up to the nearest multiple of 25,000 m³/day)

is an extensive library of published information that provides guidance on best practice for undertaking regulatory monitoring. In the UK, these documents can be found on the websites of the Drinking Water Inspectorate for England and Wales, the Drinking Water Quality Regulator for Scotland, the Environment Agency for England and Wales, and the Scottish Environment Protection Agency.[35] Outside the UK, regional and national governments publish their own sets of standards and guidelines. Valuable information is also published on the websites of the U.S. Environmental Protection Agency,[36] the European Environment Agency,[37] the environmental protection agencies for the different states of Australia[38] and Environment Canada.[39]

Parameter values will change from time to time as new information becomes available about the health or environmental consequences of their presence in water. In addition, the regulatory requirements for monitoring may also change to reflect an increase or decrease in the importance of the parameter. Frequently, revised guidelines for parameter concentrations emerge from the regular assessments of these parameters carried out by expert groups at the WHO. The results of these consultation exercises and the likely consequences of their findings for water quality monitoring can be found on the WHO website[40] and on the other websites listed previously.

Investigative and operational water quality monitoring require a rigorous approach to planning and implementation because there is no set framework to inform the design of the monitoring programme. For these two types of monitoring, it is particularly important that the whole process from design, through implementation, to reporting is documented. Not only does the act of documenting the process draw attention to the different elements of the programme, their interrelationships and the resources that are required to complete the programme, it also creates a detailed record of the programme that can be used to evaluate the significance of the data in the context of new information about the water system under investigation. This is seldom done well, but in cases where it has been done the quality and value of the data produced is extremely high. There is insufficient space in this chapter to deal adequately with the different approaches to investigative and operational monitoring. For a full discussion of the principles and practice of water quality monitoring the reader is directed to Bartram and Balance [39], Ward et al. [40] and Bartram and Rees [41].

Water sampling

The importance of using the correct procedures for taking water samples cannot be overstated. One of the major sources of error in gathering water quality monitoring data can be the sampling process. Poor sampling practices create problems for those interpreting results and can lead to costly and incorrect decisions. Too often samples are collected or transported incorrectly which renders the results of the analysis as useless.

The International Organization for Standardization publishes standards for water sampling. ISO 5667–5:2006 establishes principles to be applied to the techniques of sampling water intended for human consumption. The guidance given in ISO 5667–5:2006 is confined to those circumstances where water is drawn from treatment works and municipal or similar distribution systems (including individual systems) where prior treatment and/or quality assessment has resulted in the water being classified as suitable for drinking or for other potable uses. Specifically, the standard is applicable to water that is in continuous supply relative to any stage of use up to and including the point of consumption in a distribution system. This includes distribution within large buildings in which additional water quality management might be applicable. ISO 5667–5:2006 is also applicable to sampling situations that can arise relative to the investigation of system defects or emergency situations where the safety of sampling operatives is not compromised.

Methods for the Examination of Waters and Associated Materials, published by the Standing Committee of Analysts, and available as a UK Government publication,[41] provides details of the practices and procedures that should be adopted for taking samples for microbiological analysis. More specific guidance for sampling private water supplies is available online (e.g. on the DWQR website).[42]

The British Standards Institution (BSI) BS ISO 5667–24:2016 provides an audit protocol to monitor conformity with declared, or assumed, practices in all areas of water quality sampling. Specifically, the standard provides guidance on the systematic assessment of sampling practices and procedures in the field, and assessing conformity with those given in the organisation's sampling manual. It is applicable to the audit of sampling activities from the development of a sampling manual through to the delivery of samples to the laboratory.

BS ISO 5667–24:2016 applies to the auditing of sampling practices relevant to the management of water stored in containers, such as temporary supply tanks and bottled supplies.

Other BS standards for water quality include:

BS 1427:2009 Guide to on-site test methods for the analysis of waters

BS ISO 5667–11:2009 Water quality. Sampling. Guidance on sampling of groundwaters
BS 7592:2008 Sampling for *Legionella* bacteria in water systems. Code of practice.

The reader is referred to the BSI webpage[43] for further information and the publications.

It should be stressed that it is not necessary to undertake a large programme of bacteriological sampling of water when the results of previous sampling show that there are no apparent problems. What is more important is to establish a profile of water quality so that quality control staff are instantly aware of any microbiological deviations from the norm and that the sampling regime is adequate to comply with the requirements of the relevant directive or legislation.

Laboratory selection

The use of an accredited laboratory provides a level of assurance that the processes, in terms of methods, traceability, document control, performance of analysts, etc. are under control. The same standard for testing and calibration bodies is applied internationally and laboratories will be subject to the same rigorous audits by their national accreditation bodies. For example, the United Kingdom Accreditation Service (UKAS)[44] is the sole national accreditation body recognised by government to assess, against internationally agreed standards, organisations that provide certification, testing, inspection and calibration services. The general requirements for accreditation are provided in the International Standard general requirements for the competence of testing and calibration laboratories (ISO/IEC 17025: 2005). In the UK drinking water samples taken from consumer taps are generally analysed by a laboratory accredited by UKAS.

Data analysis and interpretation

The analysis of water quality data and the interpretation of the results are the final, and frequently most difficult, stages of water quality monitoring and assessment. Water quality monitoring for the purpose of measuring compliance with water quality standards presents fewer problems of interpretation than monitoring for the purpose of investigating a problem or establishing the characteristics of the water body. For the former, the final assessment may simply record that the parameters tested do or do not comply with the relevant standards for the water system. Over a period of time, the number of times the water fails to comply with the standard may lead to the conclusion that the water is unfit for purpose: the monitoring and assessment of bathing waters according to the EC Bathing Water Directive provides a good example of this (see previously). The interpretation of data from operational and investigative water quality monitoring programmes is more challenging because so many factors have to be taken into account during the analysis. For these types of monitoring programmes, the planning and design of the study are crucial to the interpretation of the data. This is discussed in detail in both Bartram and Balance [39] and Ward et al. [40].

It is not uncommon to isolate *Legionella* bacteria from water systems in buildings and it is important that appropriate actions are taken to control, and if possible eliminate, the organism from the system. The UK HSE publication on the control of *Legionella* bacteria in water systems, provides guidance on the interpretation of the results of *Legionella* tests and actions to be taken to control the risk to public health from the organism.

The legislation applicable to the safety of water in the food industry and for recreational waters has been discussed elsewhere in this publication. In England and Wales the results from the analysis of potable waters are interpreted according to the Water Supply (Water Quality) Regulations 2016 which are informed by EC Directive 98/83/EC. The regulations require that coliforms *E. coli* and enterococci are absent in 100 ml water. Very low levels of coliforms, in the absence of other problems, may be due to inadequate cleaning of the tap or outlet before sampling. In this case, it may be useful to thoroughly sanitise the tap and re-sample. Food premises must have a supply of water sufficient in quantity and that any water shall be clean and wholesome.

In terms of recreational waters the EC Directive on the quality of bathing waters (2006/7/EEC) constitutes the interpretation of recreational water quality results by EHPs. Annex 1 to the EC Bathing Water Directive defines the criteria for labelling water quality as poor, sufficient, good or excellent based on the 90th or 95th percentile evaluation of the number per 100 ml of intestinal enterococci and *E. coli*. The microorganisms listed in the EC Directive 2006/7/EEC are indicators of faecal pollution. Their presence, in excess of the values in the directive, identifies waters which may have received volumes of sewage that have not been given adequate treatment or dilution. Equally, however, large concentrations of sea-birds or agricultural run-off can lead to the presence of these indicators in bathing waters.

Section 3 Notes

31 https://www.legislation.gov.uk/uksi/2016/618/contents/made

32 https://www.legislation.gov.uk/wsi/2010/994/contents/made

33 In Wales the Private Water Supplies (Wales) (Amendment) Regulations 2010, which came into effect on 27 June 2016, and the Private Water Supplies (Wales) Regulations 2017, which came into force on 20 November 2017.

34 https://www.legislation.gov.uk/sdsi/2014/9780111024782/contents

35 http://www.dwi.gov.uk/; https://dwqr.scot; https://www.gov.uk/government/organisations/environment-agency; and http://www.sepa.org.uk/

36 http://www.epa.gov/

37 http://www.eea.europa.eu/

38 See for example http://www.epa.vic.gov.au/; http://www.epa.nsw.gov.au/ and http://www.ntepa.nt.gov.au/

39 http://www.ec.gc.ca

40 http://www.who.int

41 https://www.gov.uk/government/publications/standing-committee-of-analysts-sca-blue-books

42 https://dwqr.scot/private-supply/technical-information/pws-technical-manual/full-technical-manual/

43 http://www.bsigroup.com/en/

44 See http://wwwkas.com/ Other countries have similar systems (e.g. in Australia and New Zealand there is the Joint Accreditation System JAS_ANZ see http://www.jas-anz.org) and in the USA there is the American Association for Laboratory Accreditation – https://www.a2la.org/ – all members of International Accreditation Forum; see https://iaf.nu/en/home/ – the world association of conformity assessment accreditation bodies.

SECTION 4: EMERGENCIES AND INTERVENTIONS

Public health protection is the main aim of a drinking water utility, delivered through supplying safe drinking water. However, it is acknowledged that from time to time risk factors that might compromise this goal can go undetected in organisations, acting as precursors for water quality incidents. Managing risk is more complicated than ensuring compliance or managing chemicals in a catchment: it requires vigilance at all levels within organisations balanced around business goals [42]. The goals and training of water utilities and public health authorities under normal circumstances may be quite distinct and the two probably rarely have the need to collaborate except where there is a public health risk. Emergency preparedness and response to a major drinking water incident with its potential health consequences require effective inter-agency collaboration. Although not always the case, several drinking water safety incidents in the past have been aggravated by the inability of responsible organisations to secure such relationships [43].

This section reviews two high profile drinking water incidents and tries to identify some of the lessons that can be learnt from them. The two incidents are now quite old

Case study 2

Camelford, UK

In 1988 20,000 people in the Camelford area of Cornwall, UK were exposed to high concentrations of aluminium in their drinking water after 20 tonnes of aluminium sulphate was accidentally put into the water at Lowermoor Water Treatment works. The contaminated water flowed into the distribution system and it was several days before the fault was detected and corrective measures implemented. As a result other metal salts, including copper, zinc and lead from the water pipes, were also leached into the water supply. Immediate symptoms experienced by the consumers included nausea and vomiting, skin rashes and mouth ulcers. Some reported that hair, nails and skin were stained brown or blue [44]. The water company responsible for the mistake was slow to react and reportedly gave false reassurances [45]. Following public concern and outrage a commission of enquiry was set up by the Department of Health. The reports coming out of that enquiry concluded that the effect of acute short-lived intoxication was not a cause for concern, even though the report showed that for three days, the EC maximum admissible concentration for aluminium had been exceeded several hundred-fold. This only increased the concern of residents especially as people were still experiencing symptoms. The media reports raised the profile of the situation, resulting in further anxiety to residents who feared a cover-up. Litigation cases were started and local pressure groups set up. Two years later 400 people had illnesses that they attributed to the incident [46].

Lessons

The main issues of concern with this incident relate to lack of co-ordination and delayed public health response which fuelled sustained suspicion and mistrust. Inaccurate information provided by the water utility and large media interest amplified residents' concerns.

First and foremost, communication is essential in such cases. It has been reported that the anxiety and disputes about health effects from this incident would have been less if the public had been better informed from the beginning. Those who have to prepare to manage similar incidents should prepare for high-quality epidemiological assessment and public relations [46].

The plans should identify actions for a variety of unpredictable events such as contamination by chemicals, radiation or microorganisms, and it should not be forgotten that water is only one way of pollutants reaching the human population.

The public often confuse acute incidents and persistent contamination, leading to increased concern. Normally, acute pollution of drinking water is dealt with by the Environment Agency and the water company responsible, and the risk to the public is minimal. Small incidents are relatively common, whereas those on the scale of Camelford are rare. Planning for such incidents is essential as they require the provision of alternative water supplies to large numbers of households, public information and a thorough assessment of the risk to health. This will require the co-ordination and collaboration of a number of agencies.

In the first stage of the incident informing the relevant agencies is vital. The preliminary assessment of the risks to human health is difficult, especially if it is not clear which contaminants are involved. Even if the initial contamination is known, secondary chemical reactions may occur, such as the acidic dissolution of the water pipes in Camelford.

Until the results of analysis are available, the concentrations of the contaminants in the consumers' taps will not be known. The flow of water in the pipes is not constant and often water is mixed from several sources, adding to the uncertainty about health implications. Therefore it may be advisable to issue boil water notices or even provide bottled water to consumers. This will also help to alleviate concerns of the public and provide a positive message that authorities are working on the incident.

Although advice can be obtained from National Poisons Information Service commissioned by Public Health England, a population study on the effects on health backed up by an epidemiological investigation with toxicological support is desirable. Expertise of district health authorities may be useful in such situations as well as independent advisors. Such studies are not instant and clearly expensive. To carry out an epidemiological study after every incident would probably only heighten public concern. However, of high importance in such incidents is the attention given to public relations. A speedy public health response is vital to obviate speculation. In the Camelford incident the media were blamed for provoking the anxiety of the public. The public lost trust in the authorities as there was a lack of information between the incident occurring and people experiencing health effects. It may be advisable therefore to employ the services of a professional public relations company specialising in crisis management.

Case study 3

Walkerton

In May 2000, thousands of residents of the town of Walkerton, Ontario, Canada became ill from drinking municipal water contaminated by the bacteria *E. coli* O157:H7 and *Campylobacter jejuni*. Seven people died, and many suffered debilitating injuries [47]. The water supply had become contaminated with microbial pathogens after manure was spread on a farm near a public supply well. The local government quickly declared the incident was an isolated incident, blaming severe rainstorms and karst hydrology, together with administrative mistakes made by the local water utility management and staff. The incident led to a public enquiry by the Ontario government and criminal charges were brought against staff of the Walkerton public utility. The official enquiry [48] reported a litany of bad practice prior to and during the incident including the failure of operational staff to maintain effective chlorination of the water supply and carry out routine chlorine residual checks, the falsification of water quality records by staff and the failure of treatment managers to respond satisfactorily to positive microbial tests, any one of which may have significantly reduced the impact of the incident or prevented it altogether.

Lessons

The issues of concern with this case relate to ineffective chlorine levels in the water supply. Operators failed to perform simple checks and falsified records. The system was vulnerable to contamination, but previous enforcement was ineffective and managers having learnt of microbial failures in 'treated' waters did not share them with other agencies despite questioning from public health personnel. The results were only disclosed after the water quality problem became obvious leading to the boiled water notice being issued late [43]. The official report into the Walkerton incident [48] recommended that water suppliers adopt a Total Quality Management system based upon:

1. The adoption of best practices and continuous improvement
2. 'Real-time' process control (e.g. continuous monitoring of turbidity, chlorine residual and disinfectant contact time) wherever feasible
3. The effective operation of robust multiple barriers to protect public health
4. Preventative rather than strictly reactive strategies to identify and manage risks to public health and
5. Effective leadership

but they still offer valuable lessons to those responsible for the management and provision of safe drinking water supplies.

Of particular importance is clarity about responsibility in such incidents. Each responsible authority should have a clearly defined role, which has been identified before such incidents occur. An emergency plan can then slot into place with very little delay, helping to reduce the effects of the contamination and installing public confidence in the authorities.

Jalba et al. [43] reviewed 14 incidents involving drinking water safety, including Walkerton and Camelford, and identified critical institutional relationship components that were deficient: proactivity, communication, training, sharing expertise, trust and regulation. The authors note that where there is not quick authoritative advice and the provision of alternative drinking water supplies (where appropriate), water quality incidents can become high profile events with media and customers driving the agenda during and after the event.

In terms of preventative risk management approaches the Water Safety Plan approach discussed in Section 2 provides an opportunity and a challenge to water utilities regarding developing inter-agency relationships. In one-third of these incidents water system or management deficiencies were apparent

before the incident occurred but were not corrected. Environmental regulators and public health authorities did not use their enforcement powers to correct the situation. Regular inter-agency co-operation before incidents occur helps to increase mutual trust and allows expertise to be shared during incidents [49].[45] Other preventative measures that had been overlooked were:

- Failures to control sewage contamination risk for drinking water supply;
- No specific protocols directed at suspected waterborne outbreak scenarios resulting in lack of guidance for public health authority investigations;
- No sentinel surveillance system to ensure timely detection and public health response to indications of an outbreak;
- No structures in place to deal with chemical contamination of drinking water supplies.

In Camelford there was no statutory duty on either side to involve the public health (or environmental health) authority in the drinking water pollution incident response, although the public health authority was the only competent authority to advise the population on health risks.

Lack of regular communication between water utilities and other relevant agencies affects communication of critical information prior to and during incidents. Inter-agency communication should aim to ensure that all parties share a common view of circumstances and co-ordinate their actions. Feedback is essential as is external communication with the media which must be accurate and consistent between all parties.

Training is another area that can help with emergency incidents. Lack of technical knowledge for water operators, insufficient understanding of surveillance responsibilities by regulatory authorities to provide an independent audit of drinking water safety, or a lack of expertise to achieve are significant barriers to preventing and controlling emergency incidents [43]. Specifically, the following have been identified: lack of public health training related to drinking water, waterborne pathogens and failure to recognise public health risks; inadequate understanding of the purpose and significance of regulations and standards, and the reporting of requirements to the regulator; insufficient appreciation of public health responsibilities by water utilities and municipalities providing drinking water, meeting a statutory standard of care; under-funding, understaffing, inadequate risk assessment and public health risk management; inadequate training in internal and inter-agency emergency preparedness and response and incident management; public health authorities failing to address public concerns on drinking water safety due to lack

of expertise to assess health risks and/or risk communication training; failure to learn from previous incidents either local or published.

Unclear or flawed drinking water regulatory regimes leads to confusion over responsibilities, which can have serious consequences, for example unclear criteria for issuing boil water notices during contamination events may lead to residents drinking contaminated water. See also the 2015 problem in Lancashire, England with *Cryptosporidium*.[46]

Therefore, as well as the importance of risk management frameworks, risk analysis tools and organisational structures, the strength of inter-agency co-operation has a major influence on the quality of response during emergency incidents. The Water Safety Plan approach discussed earlier provides the opportunity for water utilities to implement many of these insights, and thus prevent major water quality incidents.

Section 4 Notes

45 See also Tumpey AJ, Daigle D, Nowak G. (2019) Communicating during and outbreak or public health investigation. In *The CDC Field Epidemiology Manual*, Rasmussen SA, Goodman RA (Eds.), New York: Oxford University Press (see https://www.cdc.gov/eis/field-epi-manual/chapters/Communicating-Investigation.html)

46 See http://www.itv.com/news/granada/2015-08-21/lancashire-boil-water-notice-300-000-homes-still-boiling-tap-water-after-cryptosporidium-outbreak/

SECTION 5: EMERGING ISSUES – WHAT EVENTS ARE TAKING PLACE THAT COULD CHANGE THE RISKS TO ENVIRONMENTAL HEALTH THROUGH THE WATER CYCLE?

It is generally accepted that global average surface air temperature and precipitation over land increased during the 20th century [50].[47] Annual variations in climate produce major seasonal changes in environmental conditions in many parts of the globe [51], leading to an intensified water cycle. Some aspects of

this intensification, such as the more frequent occurrence of extreme events, are a threat to at-risk populations.

Some of the most significant impacts of climate change will be experienced through more droughts, floods and less predictable rainfall and water flows [30]. A report by

Howard and Bartram for the WHO [52] stresses the effects that climate change will have on natural water stores, such as mountain glaciers and groundwater. Research into glacier recession makes it increasingly clear that there will be problems for high mountain communities. In the Andes, for example, the cities and towns are reliant upon glacial melt for their water supply. However, the impact of glacial recession is likely to be lower on large rivers, particularly in their downstream stretches where rainfall is more important in driving the hydrological system [53]. There is significant variation in the response of glaciers to climate change; for instance there is some evidence that glaciers in the eastern part of the Himalayan mountain chain may be accreting rather than receding [54]. Climate change impacts on groundwater resources are inevitable but not well understood, mainly as a result of relatively little being known about available groundwater resources in many regions [52].

The impact of climate change on the water cycle and therefore access to drinking water supply will vary both by region (in terms of the specifics of the change in climate expected) and by facilities (in relation to the vulnerability of the facility to the expected change in climate). Problems such as these were highlighted as serious and likely consequences of climate change by the Intergovernmental Panel on Climate Change in 2013 [55] where it said in its Fifth Assessment Report that:

> Warming of the climate system is unequivocal, and since the 1950s, many of the observed changes are unprecedented over decades to millennia. The atmosphere and ocean have warmed, the amounts of snow and ice have diminished, sea level has risen, and the concentrations of greenhouse gases have increased.

Climate change is already being experienced, with observations of increases in global average air and ocean temperatures, widespread melting of snow and ice, and rising global average sea level. The IPPC in 2014 [56][48] reported that the period from 1983 to 2012 was very likely the warmest 30-year period of the last 800 years in the Northern Hemisphere. Measurements taken since the 1980s show that the average atmospheric water vapour content has increased over land and ocean because of warming temperatures. The oceans are heating up, which is causing seawater to expand, and accelerating glacier and snow melt are both contributing to measurable rises in sea level. Global average sea level rise is estimated to have been 1.8 mm per year between 1961 and 2003, but 3.1 mm per year between 1993 and 2003. It is not clear whether the faster rate represents variability or a longer-term trend. These observations were reported by the IPCC [56] with medium confidence based on, but also limited by, the available data. Table 15.6 summarises recent trends and the likelihood of them continuing.

Over the past century, long-term trends in quantities of precipitation have been observed over many large regions, including significantly increased precipitation in the eastern parts of North and South America, in northern Europe, and in northern and central Asia. Additionally, heavy precipitation events have become more frequent over most land areas. In the future, precipitation changes are predicted to follow these observed trends with increases very likely (>90%) in high latitudes and decreases likely (>66%) in most subtropical land regions. Drying has been observed in the Sahel, the Mediterranean, southern Africa and parts of southern Asia, with more intense and longer droughts observed over wider areas since the 1970s, particularly in the tropics and subtropics. Many of these semi-arid and arid areas are projected to suffer a decrease in water resources as a result of climate change [58].

Geographical patterns of warming in the 21st century are projected to be similar to those observed over the past decades,

Table 15.6 Recent trends and projections for extreme weather events for which there is an observed late 20th-century trend

Phenomenon and direction of trend	Likelihood that trend occurred in late 20th century	Likelihood of future trends based on projections for 21st century
Warm spells or heat waves Frequency increases over most land areas	Likely (>66%)	Very likely (>90%)
Heavy precipitation events Frequency (or proportion of total rainfall from heavy falls) increases over most areas	Likely (>66%)	Very likely (>90%)
Area affected by droughts increases	Likely (>66%) in many regions since 1970	Likely (>66%)
Intense tropical cyclone activity increases	Likely (>66%) in many regions since 1970	Likely (>66%)
Increased incidence of extreme high sea level (excludes tsunamis)	Likely (>66%)	Likely (>66%)

Source: Charles, Pond, Pedley et al. [57]

with warming expected to be greatest over land and most high northern latitudes. The IPCC [56] report that impacts from recent climate-related extremes, such as heat waves, droughts, floods, cyclones and wildfires, have highlighted the significant vulnerability of some ecosystems and many human systems to current climate variability. Furthermore, if hot extremes, heatwaves and heavy precipitation events continue to become more frequent it is also likely that tropical cyclones will increase in intensity, with increases in wind speed and more heavy precipitation. This would increase run-off and floods and reducing the ability of water to infiltrate the soil and may affect the regional distribution of surface and groundwater supplies.

Climate change is also expected to affect water quality [58]. Higher water temperatures and increasing run-off from more intense rainfall are predicted to contribute to deterioration in water quality, including increasing algal blooms and higher turbidity. Rising sea levels and temperatures, and decreased groundwater recharge, will increase salinity problems. Other impacts of climate change include coastal inundation, saline intrusion,

vectors of disease, emergency responses and indirect effects of climate change. Increases in the intensity and frequency of storm events linked to climate change could lead to more severe episodes of chemical contamination of water bodies and surrounding watersheds. Changes in salinity may affect aquatic organisms as an independent stressor as well as by altering the bioavailability and in some instances increasing the toxicity of chemicals. The impacts of these changes on water resources are likely to be further compounded by increasing water demand from population growth, increasing affluence and changes in other water demands [58].

The potential consequences of projected climate change for the provision of potable water, particularly in developing or low- and middle-income countries, are particularly significant. It is therefore imperative that the water resources sector and the water supply sectors work together to implement adaptive responses to ensure continuity of water services in a changing climate [59]. As reported by the IPCC [56], adaptation experience is accumulating across regions and governments at various levels are starting to develop adaptation plans and policies and to integrate climate-change considerations into wider development plans.

As well as direct health impacts associated with contamination of drinking water sources, climate events can severely affect the water delivery infrastructure. Flash or high-velocity floods can damage water systems because their physical force can knock out key components such as water treatment works and pumping stations [60]. Water treatment works can be inundated by flood water, potentially causing major disruption. In addition, extreme stormwater events may result in the degradation of materials used to construct water supply pipelines through impacts caused by increased ground movement and changes in groundwater levels [61].

Climate change is also predicted to have an impact on energy supplies [62][49] through its effect on energy demand and energy production. Energy supplies may be affected by extreme events damaging infrastructure, or by lack of water for power generation or cooling. These impacts on energy supplies will also affect water treatment facilities and distribution systems, as well as sewage pumping and treatment facilities.

Different issues arise in situations where water is scarce. Urban water supplies in Africa, Latin America and the Caribbean, and Asia, can operate intermittently during periods of drought and about 66% of the population (4.0 billion people) lives under severe water scarcity at least one month of the year [63].[50] This adversely affects water quality in the supply system. When networks are empty and unpressurised for prolonged periods of time, contaminants enter the pipes through leaks in the supply pipes. The situation is particularly serious in cities with unhygienic excreta disposal where sewage flows in open ditches close to water distribution pipes.

A critical weakness of piped water systems is their lack of flexibility when relatively sudden changes occur to the water source that feeds the system, in particular when the source dries up during prolonged drought.

Prioritising access to groundwater via multiple improved sources and a range of technologies, such as hand-pumped and motorised boreholes, supported by responsive and proactive operation and maintenance, increases water supply resilience especially in rural areas [64]. Slow and predictable changes in water resources can be accommodated in the strategic development plans of large utility water supply facilities, and engineering solutions implemented to mitigate the problem. However, short-term changes in water availability resulting from drought or flood highlight the vulnerability of utility piped water systems and the scale of the impact when the systems do fail.

Increase in the reuse of wastewater

Severe water shortages and dry periods are the main driving forces behind water reuse for some countries, whereas increasing environmental constraints and the fact that water quality discharge regulations have become stricter are also significant drivers. Water reuse practices have also become more technically feasible due to the development of better purification processes. The implementation of reuse for industrial purposes depends, to a large extent, on economic incentives [65]. As water prices rise, there will come a point when existing or developing technologies will make water recycling and reuse a viable commercial operation [66]. In many countries, agricultural irrigation remains the main reuse application for industrial wastewater [67]. Reuse in the food industry has been limited for many years due to strict regulations. Current guidelines and regulations regarding use and reuse of water in the food industry now acknowledge the use of other water qualities than that of potable water, for example, in Council Directive 98/83/EC on the quality of water intended for human consumption − it is not [68].[51] This provides flexibility but at the same time requires a high degree of multidisciplinary knowledge and substantial documentation from industry and regulatory authorities [65]. The concern about reducing hygienic standards has limited research and development in this area. However, due to potential future climate changes these challenges need to be addressed. Reusing process water in the food industry throws up a number of health issues that cannot be ignored. Microbiological quality of the water to be reused must be ensured. Establishing safety tools such as WSPs does this, when correctly implemented (see earlier).

Water conservation

Freshwater is a limited resource, and the treatment and distribution of drinking water is expensive. Where freshwater resources are particularly limited, the conservation of water to reduce the amount that is used is a frequent and valuable strategy. However, water conservation can present its own problems through a reduction in the quality of the water. Limiting the use of household water means that water can stagnate in pipes. Static water in distribution pipes allows suspended particulates to settle and form sediments. It also creates an environment for biofilms to develop [69]. Similarly, the long-term storage of finished water in reservoirs (lasting weeks or several months) can result in waterborne heterotrophic bacteria growing in the sediments and attaching to the inner walls of the reservoir producing a biofilm [70]. Biofilms are a complex mixture of microbes, organic and inorganic material accumulated amongst a microbially produced organic polymer matrix attached to the inner surface of the distribution system [71]. Contaminants, including some pathogens and the total coliform group of faecal indicator bacteria, may attach to or become enmeshed in biofilms on pipe walls in distribution systems. Many pathogens have been found to survive, and grow, in these pipe biofilms where they are protected from the action of disinfectants. Over time, coliform bacteria may detach or slough from the biofilm, causing persistent total coliform detections. Pathogens may also be present in the detached biofilm material raising the risk of an outbreak of waterborne disease. Long retention time can also result in reduction in disinfectant residual and cause the release of ammonia through the decay of chloramines [69].

During periods of severe water restrictions there is a potential for people to reduce the amount of water they use for hygiene or use alternative sources that are not subject to sanitary control. Poor personal hygiene, resulting in transfer of microorganisms from person to person via contaminated food, is often a significant factor in the spread of disease. This route of infection becomes more likely where water supplies are inadequate or interrupted and frequent washing is impractical.

Rainwater harvesting for domestic use

The UN Food and Agriculture Organisation (FAO) have emphasised the importance of fully integrating rainwater into water resource management strategies to cope with water scarcity [72]. They note that water management strategies rarely integrate rainwater, focussing only on surface and groundwater.

Although the opportunities for expanding rainwater collection are clear, there is conflicting evidence in the literature about the safety of stored rainwater for domestic use. In a study of rainwater collection systems in New Zealand, Simmons et al. [73] found widespread microbial contamination of the water and showed that consumption of the contaminated water was associated with symptoms of gastrointestinal infection. Furthermore, a review of health risks associated with the consumption of untreated rainwater identified several cases of infection with bacterial and protozoal pathogens and helminths [74]. In contrast, Dillaha and Zolan [75] have shown that roof-harvested and stored rainwater in Micronesia was suitable for drinking and cooking.

Evans, Coombes and Dunstan [76] have highlighted the importance of local meteorological and environmental conditions in determining the quality of roof-harvested rainwater, which provides a partial explanation for the lack of a consensus about water safety. Several textbooks have been published describing techniques for the collection and management of rainwater (for example Pacey and Cullis [77]) but it is evident that the processes will be site specific, and management and treatment options need to be designed accordingly. Whilst robust assessment techniques are required to deliver this goal, some simple practical adaptations for protecting the quality of the water can be applied. These include:

- Management of the collection area;
- Water collection procedures that discard the first flush of water from the catchment surface; and
- Design, cleaning and maintenance of the storage reservoir.

Despite the reported problems of rainwater collection and storage, in areas of increasing rainfall amounts and pattern variability, strategies to enable communities to directly harvest, store and manage rainwater could significantly improve drinking water supply at the household level, and provide other benefits to the households. This requires the introduction of facilities, and the development of local capacity in skills and knowledge. Facilities for harvesting rainwater include surface or underground tanks, strategically created micro-watersheds (such as impermeable roofs or surfaces), and in-built rainwater purification systems and treatment [78].

The impacts of climate change on water quality need to be managed in order to consistently ensure the safety of a drinking water supply. At a community or utility level this can be done through the use of Water Safety Plans (see previous sections). Different strategies are being applied to combat droughts. These include initiatives to encourage people to use groundwater and domestic water appropriately, and the construction of water reservoirs like the Alqueva Dam in Portugal.

Microplastics in drinking water

Microplastics are now recognised as a ubiquitous environmental pollutant, having been detected in the ocean, freshwater, wastewater, air, food and drinking water, the latter including both bottled and tap water [79–80]. They are typically defined as plastic particles or fragments <5 mm in length. Surface run-off and wastewater effluent are believed to be key entry routes for microplastics in freshwater, though other inputs are also likely. A wealth of scientific literature has accumulated on the occurrence of microplastics in the environment in recent years. However, a lack of standardised sampling and analytical techniques and, in some cases appropriate quality assurance procedures, means making direct comparisons is difficult. In drinking water, polyethylene terephthalate and polypropylene fragments and fibres have been the main types reported [79–80]. Observed microplastic concentrations in drinking water have ranged from 0 to 10,000 particles/L, though these are affected by the experimental methods used, notably the mesh size in sampling. The human health risk posed by the ingestion of microplastics in drinking water is currently uncertain. Three main hazards have been identified: (a) the plastic particles themselves, (b) other chemicals (e.g. persistent organic pollutants, which can sorb to microplastics) and (c) microorganisms/biofilms, potentially including pathogens, which can colonise microplastics. Based on the (limited) evidence available, there is believed to be low risk to public health from (b) and (c), while no reliable information suggests that the toxicity of smaller microplastics (i.e. nanoplastics) is a concern [80].

Section 5 Notes

47 As IPCC says in *Climate Change 2021: The Physical Science Basis. Contribution of Working Group I to the Sixth Assessment Report of the Intergovernmental Panel*

on Climate Change "Climate change is intensifying the water cycle. This brings more intense rainfall and associated flooding, as well as more intense drought in many regions". See https://www.ipcc.ch/2021/08/09/ar6-wg1-20210809-pr/

48 See also IPPC. (2014) *Climate Change 2014 Impacts, Adaptation and Vulnerability – Summary Report for Policymakers.* http://www.ipcc.ch/pdf/assessment-report/ar5/wg2/ar5_wgII_spm_en.pdf

49 See also Wilbanks TJ, Romero Lankao P, Bao M, Berkhout F, Cairncross S, Ceron J-P, Kapshe M, Muir-Wood R, Zapata-Marti R. (2007) Industry, settlement and society. In *Climate Change 2007: Impacts, Adaptation and Vulnerability.*

Contribution of Working Group II to the Fourth Assessment Report of the Intergovernmental Panel on Climate Change, Parry ML, Canziani OF, Palutikof JP, van der Linden PJ, Hanson CE (Eds.), Cambridge University Press, Cambridge.

50 See also Mekonnen MM, Hoekstra AY. (2016) Four billion people facing severe water scarcity. *Scientific Advances*, 2: e1500323.

51 See http://eur-lex.europa.eu/LexUriServ/LexUriServ.do?uri=OJ:L:1998:330:0032:0054:EN:PDF the position is unclear following UK withdrawal from the EU but see https://publications.parliament.uk/pa/cm201719/cmselect/cmeuleg/301-xviii/301-xviii.pdf

SECTION 6: SPECIAL ISSUES RELATED TO WATER AND ENVIRONMENTAL HEALTH IN DEVELOPING (LOW-INCOME) COUNTRIES

An estimated 25% of all preventable illness is caused by environmental factors [81]. Whereas in developed or high-income countries good quality water is often taken for granted, in developing countries (the low- and middle-income countries) poor water quality is a significant health risk, with the poor generally having the least access to clean water sources [82]. Those communities living in extreme conditions of poverty are most affected by the lack of safe water and sanitation. The reasons for this are many and complicated but will include: lack of priority given to the sector, lack of financial resources, lack of sustainability of water supply and sanitation services, poor hygiene behaviours and inadequate sanitation in public places including hospitals, health centres and schools. Providing access to sufficient quantities of safe water, the provision of facilities for a sanitary disposal of excreta and introducing sound hygiene behaviours are of capital importance to reduce the burden of disease caused by these risk factors [83].

Water supplies

Access to safe drinking water is an essential element of sustainable development, and it is central to the goal of poverty reduction within the Sustainable Development Goals for 2030[52] (and first discussed in the Annex to Chapter 1).

Increased access to safe water in adequate quantities can enhance health, enrolment in educational programmes, economic productivity and dignity. It thus played a key role in efforts to reduce poverty – one of the SDG targets for 2030. It is widely recognised both in developing or low-income countries as well as developed nations (the high-income countries) that small-scale systems to provide community supplies, particularly for rural, remote and indigenous communities, are the most vulnerable. Such systems are most liable to contamination and failure, and consequently pose a continuous public health risk. Communities in developing as well as developed (high-income) nations, relying on small systems to supply drinking water, are often not able to overcome the challenges posed by these systems for a number of reasons:

- Isolation and remoteness lead to increased costs associated with accessing supplies;
- The quality of drinking water in small systems tends to be poorer, yet the water is sampled less frequently and often not treated;
- The financial resources available for funding capital and operating expenses are limited;

- The per capita costs for water sampling and testing are high;
- Recruiting and training competent or certified operators is a challenge, especially when funding is scarce;
- Little capacity exists to undertake risk assessments or sampling;
- Owners of very small systems and private wells often lack knowledge about or interest in the relationship between poor water quality and ill-health;
- Operators often lack a support network, standard operating procedures and technical support;
- Training for operators and managers of small systems is inadequate and management expertise is lacking;
- The infrastructure of small systems is often characterised by poor construction and inadequate maintenance;
- Communities often lack the skills or financial means to protect water sources or have little influence on factors that may affect water sources;
- The community perception of risk is often inaccurate;
- Risks and risk factors are often hard to quantify and compare in small systems;
- Surveillance of waterborne diseases associated with small systems is especially difficult because of underreporting of waterborne illness and unsystematic collection of data;
- Communities are often faced with a number of other priorities, such as housing, hygiene and socioeconomic problems, which compete for priority with concerns relating to water;
- Communication to the public is deficient, including about the management of water within the home;
- There is insufficient political engagement;
- Regulatory agencies do not have the resources to adequately regulate small systems that provide community water supplies;
- The perception that there is no ownership of a water supply system and no awareness

of the true cost of water may result in poor decision-making;
- Poor infrastructure in rural areas in general inhibits delivery of safe water.

To overcome some of the challenges facing small systems that provide community water supplies, the following actions have been identified which environmental health practitioners working in developing (low-income) countries may be involved in:

- Better management of community water supplies;
- Management of priorities;
- Information generation and dissemination;
- Bringing communities together to share experiences;
- Development of communication strategies to inform the public and decision-makers about risks;
- Development of tools to ensure that decision-makers at community, regional and national levels are aware of their responsibilities;
- Advocacy and political will at all levels;
- Identification of appropriate regulations for community water supplies;
- Commitment and responsibility of governments to investment;
- Adequate institutional support to ensure outreach mechanisms;
- Capacity-building for water operators and managers, including incentives to stay within the community;
- Promotion and strengthening of community-level capacity to manage water supplies, including the establishment of regional networks to facilitate information sharing and mentoring;
- Investment by small communities in their own water supply systems. [84]

Waterborne and other water-related diseases consist mainly of infectious diarrhoea, typhoid, cholera, salmonellosis, shigellosis, amoebiasis and other protozoan and viral intestinal infections. Some of the pathogens causing these diseases are transmitted by water, although other forms of transmission do occur such as person-to-person contact, animal-to-human contact, transmission through food and aerosols, and by contact with fomites [20]. In addition to the dangers posed by pathogenic microorganisms, chemicals such as nitrates, fluoride or arsenic in water can have toxic effects. People who consume water contaminated with these chemicals may not immediately display symptoms of disease, but the long-term effects on their health can be extremely severe, as shown by the example of arsenic poisoning in Bangladesh [85]. In addition, Santaniello-Newton and Hunter [86] propose a category of diseases that are spread by the daily migration of people to collect water, such as meningococcal disease ('water-carrying disease'). Various non-infectious disorders of the musculoskeletal system resulting from the prolonged carrying of heavy weights, especially during childhood, should also be considered.

A number of studies from low-income countries have indicated that improved access to water – and the resulting increases in the quantity of water or time used for hygiene – are the determining factors of health benefits, rather than improvements in water quality [87]. Indeed, inadequate drinking water and sanitation are estimated to cause 502,000 and 280,000 diarrhoea deaths, respectively. By improving water and sanitation provision *and* improving hand hygiene it is estimated that 361,000 deaths of children under 5 years old could be prevented, representing 5.5% of deaths in that age group [88].

Providing water security can play a wider role in poverty reduction and improving livelihoods by reducing uncertainty and releasing resources that can be used to decrease vulnerability. It has been noted that improved domestic water supplies and improved local institutions can enhance food security, strengthen local organisations and build cooperation between people [89]. A water source may be very close to a village but may be of poor quality or only seasonally accessible. In order to reach a source of good quality it may

be necessary to travel a considerable distance, thus resulting in less time for other activities. In fact, it has been demonstrated that the biggest benefit, in terms of both water and sanitation, is time-saving through better access [90].

In addition to the health benefits and the saving of time and energy, providing safe water can also have an influence on school enrolment and attendance. In many cultures, this particularly affects young school-age girls. For many poor families, the economic value of a girl's work at home exceeds the perceived returns from schooling. On a wider scale, however, the education of girls is widely attested to lead to a fall in fertility rates and in the next generation's mortality and morbidity rates [91]. Clearly, improvements in water supply increase well-being.

Environmental hygiene

Because diarrhoeal diseases are generally of faecal origin, interventions that prevent faecal material entering the environment are likely to be of greatest significance for public health. Therefore, environmental health interventions such as new sanitation facilities, sustainable access to sufficient safe drinking water and personal hygiene are three major factors that contribute to enhancing public health in developing (or low- and middle-income) countries.

There has recently been a refocus on non-engineered interventions, in particular to the importance of behavioural interventions such as community-led total sanitation (an approach which aims to initiate a change in sanitation by a whole community rather than individual behaviours [92]); and household water treatment and safe storage interventions such as boiling water, the use of solar disinfection and ceramic filters [93].

Hygiene interventions act by reducing contamination of hands, food, water and fomites, and have been shown to be as effective as other interventions to reduce diarrhoea in developing (low-income) countries [94]. Curtis and Cairncross [87] showed that handwashing with soap and water after contact with faecal material can reduce diarrhoeal diseases by 42% or more. This work showed that although the most enabling condition for personal hygiene is the availability of water, for behavioural change to occur and be sustained there is a need to continue the hygiene promotion until the behaviour has become entrenched.

There are many different types of sanitation in use in developing (low- or middle-income) countries. The World Health Organization has a wealth of information available online, including a training package intended for managers and planners who are concerned with implementing effective operation and maintenance of rural water supply and sanitation services in developing (low-income) countries.[53] The needs of the users and the resources available should be considered when selecting the most appropriate type. These range from simple pit latrines to costly sewerage systems. Disposing of excreta safely, isolating excreta from flies and other insects, and preventing faecal contamination of water supplies greatly reduce the spread of diseases (see also Chapter 9).

This implies a 'One Health' approach,[54] a concept recognising that the health of humans is connected both to the health of animals and the environment and strongly promotes the collaboration of the human health, veterinary health and environmental health communities to achieve successful public health interventions. This is of crucial importance in developing countries but is applicable globally.

Some of the issues and problems surrounding this have been explored in Ruel-Bergeron et al. [95] and readers with an interest in this aspect of environmental health are referred to this.

Section 6 Notes

52 See http://www.un.org/apps/news/story.asp?NewsID=51968#.VgZr6RNVhHw

53 See https://www.who.int/health-topics/water-sanitation-and-hygiene-wash

54 See http://www.onehealthinitiative.com/

Chapter references

[1] Bartram J, Cairncross S. (2010) Hygiene, sanitation, and water: Forgotten foundations of health. *PLoS Medicine*, 7(11): e1000367.

[2] Evans B, et al. (2013) *Public Health and Social Benefits of at-House Water Supplies Final Report*, University of Leeds, Leeds, DFID, vi, 53. https://www.gov.uk/research-for-development-outputs/public-health-and-social-benefits-of-at-house-water-supplies-final-report (Accessed August 2021).

[3] Hunter PR, Zmirou-Navier D, Hartemann P. (2009) Estimating the impact on health of poor reliability of drinking water interventions in developing countries. *The Science of the Total Environment*, 407(8): 2621–2624.

[4] Hutton G. (2013) Global costs and benefits of reaching universal coverage of sanitation and drinking-water supply. *Journal of Water and Health*, 11(1): 1–12.

[5] Prüss-Ustün A, et al. (2014) Burden of disease from inadequate water, sanitation and hygiene in low- and middle-income settings: A retrospective analysis of data from 145 countries. *Tropical Medicine & International Health*, 19(8): 894–905.

[6] Defra. (2015) Water abstraction from non-tidal surface water and groundwater in England and Wales, 2000 to 2013. *Defra*, 3, 16 April.

[7] de Fraiture C, Wichelns D. (2010) Satisfying future water demands for agriculture. *Agricultural Water Management*, 97(4): 502–511.

[8] UN. (2015) *World Population Prospects: The 2015 Revision. Key Findings and Advance Tables*, New York. http://esa.un.org/unpd/wpp/Publications/Files/Key_Findings_WPP_2015.pdf (Accessed August 2021).

[9] Falkenmark M, Finlayson M, Gordon L. (2007) Agriculture, water and ecosystems: Avoiding the costs of going too far. In *Water for Food, Water for Life: A Comprehensive Assessment of Water Management in Agriculture*, Molden D (Ed.), Earthscan and International Water Management Institute, London and Colombo.

[10] WHO. (2017) *Guidelines for Drinking-Water Quality*, WHO, Geneva, 4th ed. incorporating, 1st Addendum. https://www.who.int/publications/i/item/9789241549950 (Accessed August 2021).

[11] Mara DD, Feacham RGA. (1999) Water and excreta-related diseases: Unitary environmental classification. *Journal of Environmental Engineering*, 334–339, April.

[12] WHO. (2007) *Chemical Safety of Drinking-Water: Assessing Priorities for Risk Management*, WHO, Geneva. https://apps.who.int/iris/bitstream/handle/10665/43285/9789241546768_eng.pdf;sequence=1 (Accessed August 2021).

[13] Anon. (2021) *UN-Water, 2021: Summary Progress Update 2021 – SDG 6 – Water and Sanitation for All*, Version, Geneva, Switzerland, July. https://www.sdg6data.org/ (Accessed July 2021).

[14] CDC. (2021) *Waterborne Disease in the United States*. https://www.cdc.gov/healthywater/surveillance/burden/.

[15] Prüss-Ustün A, et al. (2011) Knowns and unknowns on burden of disease due to chemicals: A systematic review. *Environmental Health*, 10(1): 9.

[16] Willcocks L, Crampin A, Milne L, et al. (1998) A large outbreak of cryptosporidiosis associated with a public water supply from a deep chalk borehole. *Communicable Disease and Public Health*, 1(4): 239–243.

[17] Stevenson AH. (1953) Studies of bathing water quality and health. *American Journal of Public Health*, 43: 529–538.

[18] Moore B. (1959) Sewage contamination of coastal bathing waters in England and Wales; a bacteriological and epidemiological study. *Journal of Hygiene*, 57: 435–472.

[19] Pond KR. (2005) *Water Recreation and Disease: Plausibility of Associated Infections: Acute Effects, Sequelae and Mortality*, IWA Publishing, London.

[20] Hunter PR. (1997) *Waterborne Diseases, Epidemiology and Ecology*, Wiley & Sons Ltd., Chichester.

[21] Dwight RH, Baker DB, Semenza JC, Olson BH. (2004) Health effects associated with recreational coastal water use: Urban versus rural California. *American Journal of Public Health*, 94(4): 565–567.

[22] Ryan U, Lawler S, Reid S. (2017) Limiting swimming pool outbreaks of cryptosporidiosis – the roles of regulations, staff, patrons and research. *Journal of Water and Health*, 15(1): 1–16.

[23] PWTAG. (2017) *Swimming Pool Water: Treatment and Quality Standards for Pools and Spas*, Pool Water Treatment Advisory Group. https://www.pwtag.org/swimming-pool-water-book/.

[24] Herschan, J, Rickert, B, Mkandarwire, T, Okurut, K, King, R, Hughes, S, Lapworth, DJ Pond, K. (2020) Success factors for water safety plan implementation in small drinking water supplies in low and middle income countries. *Resources*, 9(11): 126.

[25] Ross T, Sumner J. (2002) A simple, spreadsheet-based, food safety risk assessment tool. *International Journal of Food Microbiology*, 77(1–2): 39–53, 25 July.

[26] WHO. (2001) Management strategies. In *Water Quality: Guidelines, Standards and Health: Assessment of Risk and Risk Management for Water-Related Infectious Disease*, Fewtrell L, Bartram J (Eds.), IWA Publishing, London. https://apps.who.int/iris/handle/10665/42442.

[27] WHO. (2003) *Guidelines for Safe Recreational Water Environments*, Volume 1, Coastal and Freshwaters, WHO, Geneva.

[28] WHO. (2004) *Guidelines for Drinking-Water Quality*, Volume 1, Recommendations, WHO, Geneva, 3rd ed.

[29] Medema GJ, Payment P, Dufour A, Robertson W, Waite M, Hunter P, Kirby R, Anderson Y. (2003) Safe drinking water: An ongoing challenge. In *Assessing Microbial Safety of Drinking Water – Improving Approaches and Methods*, Dufour A, Snozzi M, Koster W, Bartram J, Ronchi E, Fewtrell L (Eds.), IWA, London.

[30] Bartram J, Corrales L, Davison A, Deere D, Drury D, Gordon B, Howard G, Rinehold A, Stevens M. (2009) *Water Safety Plan Manual: Step by Step Risk Management for Drinking-Water Suppliers*, World Health Organization, Geneva.

[31] Scottish Executive. (2006) *Private Water Supplies: Technical Manual*, The Scottish Executive, Edinburgh.

[32] Pruss A. (1998) A review of epidemiological studies from exposure to recreational water. *International Journal of Epidemiology*, 27: 1–9.

[33] Kay D, Fleisher JM, Salmon RL, Wyer MD, Godfree AF, Zelenauch-Jacquotte Z, Shore R. (1994) Predicting likelihood of gastroenteritis from sea bathing results; results from randomized exposure. *Lancet*, 355(8927): 905–909.

[34] WHO. (2006) *Guidelines for Safe Recreational Water Environments: Volume 2, Swimming Pools and Similar Environments*, WHO, Geneva. https://www.who.int/publications/i/item/9241546808.

[35] Ferrier C. (2001) *Bottled Water: Understanding a Social Phenomenon. An Independent Report Prepared for the WWF*. http://assets.panda.org/downloads/bottled_water.pdf.

[36] Drinking Water Inspectorate. (2010) *Legislative Background to the Private Water Supplies Regulations*, Section 9 (E&W) of the Private Water Supplies: Technical Manual and Regulation 4 and Schedule 1 of SI 2016 No. 573, The Private Water Supplies (England) Regulations 2016, London.

[37] Bartram J, Chartier Y, Lee JV, Pond K, Surman-Lee S. (2007). *Legionella and the Prevention of Legionellosis*, WHO, Geneva. https://www.who.int/publications/i/item/9241562978.

[38] Chapman D. (1996) *Water Quality Assessments: A Guide to the Use of Biota, Sediments and Water in Environmental Monitoring*, E & FN Spon, London, 2nd ed., 626.

[39] Bartram JK, Balance R. (1996) *Water Quality Monitoring: A Practical Guide to the Design and Implementation of Freshwater Quality Studies and Monitoring Programmes*, E&FN Spon, London.

[40] Ward RC, et al. (1990) *Design of Water Quality Monitoring Systems*, Wiley and Sons, London.

[41] Bartram J, Rees G. (2000) *Monitoring Bathing Waters: A Practical Guide to the Design and Implementation of Assessments and Monitoring Programmes*, E&FN Spon, London.

[42] Pollard SJT, Stephenson T. (2008) *Risk Management for Water and Wastewater Utilities*, IWA, London. https://doi.org/10.2166/9781780401980.

[43] Jalba DI, Cromar NJ, Pollard SJ, et al. (2010) Safe drinking-water: Critical components of effective inter-agency relationships. *Environmental International*, 36: 51–59.

[44] David AS, Wessely SC. (1995) The legend of Camelford: Medical consequences of a water pollution accident. *Journal of Psychosomatic Research*, 39(1): 9.

[45] Lowermoor Incident Health Advisory Group. (1989) *Water Pollution at Lowermoor, North Cornwall*, Cornwall and Isles of Scilly District Health Authority, Truro.

[46] Mayon-White RT. (1993) How should another Camelford be managed? *BMJ*, 307: 398–399.

[47] Prudham S. (2004) Poisoning the well: Neoliberalism and the contamination of municipal water in Walkerton, Ontario. *Geoforum*, 35(3): 343–359.

[48] O'Connor DR. (2002) *The Events of May 2000 and Related Events, Report of the Walkerton Inquiry (Part 1)*, Ontario Ministry of the Attorney General, Toronto.

[49] Williams T, Hrudey S. (2007). Public health protection demands effective communication. *International Water Association – Bonn Principles Series*, 3: 1–4.

[50] Folland CK, Karl TR, Christy JR, et al. (2001) Observed climate variability and change. In *Climate Change 2001: The Scientific Basis*, Houghton JT, Ding Y, Griggs DJ, et al. (Eds.), Cambridge University Press, Cambridge, 99–181.

[51] Few R, Laike I, Hunter PH, Gia Tran P, Trong Thien V. (2009) Seasonal hazards and health risks in lower-income countries: field Testing a multi-disciplinary approach. *Environmental Health*, 8(Suppl 1): S16, 5.

[52] WHO. (2010). *Vision 2030: The Resistance of Water Supply and Sanitation in the Face of Climate Change. A Technical Report*, World Health Organization, Geneva. https://apps.who.int/iris/bitstream/handle/10665/70462/WHO_HSE_WSH_10.01_eng.pdf?sequence=1&isAllowed=y.

[53] Rees G, Collins DN. (2004). *An Assessment of the Impacts of Deglaciation on the Water Resources in the Himalayas*. SAGARMATHA Final Technical Report, Volume 2, CEH, Wallingford.

[54] Fowler HJ, Archer DR. (2005) Conflicting signals of climatic change in the upper Indus basin. *Journal of Climate*, 19: 4276–4293.

[55] IPCC. (2013) *Climate Change 2013: The Physical Science Basis. Contribution of Working Group I to the Fifth Assessment Report of the Intergovernmental Panel on Climate Change*, Stocker TF, Qin D, Plattner GK, Tignor M, Allen SK, Boschung J, Nauels A, Xia Y, Bex V, Midgley PM (Eds.), Cambridge University Press, Cambridge, New York.

[56] IPCC. (2014). *Climate Change 2014: Impacts, Adaptation and Vulnerability. A Summary for Policy Makers*, WMO and UNEP. https://ipcc-wg2.gov/AR5/images/uploads/WG2AR5_SPM_FINAL.pdf.

[57] Charles K, Pond K, Pedley S. (2010) *Vulnerability, Adaptability and Resilience of Water Supply and Sanitation Facilities to Climate Change and the Related Changes in Rainfall Patterns*, Report Prepared for Department for International Development, London.

[58] Bates BC, Kundzewicz ZW, Wu S and Palutikof JP. (Ed.) (2008) *Climate Change and Water Technical Paper VI of the Intergovernmental Panel on Climate Change*, IPCC Secretariat, Geneva, 210.

[59] Cronin A, Pond K. (2008) Just how big is the schism between the health sector and the water and sanitation sector in developing countries? *Environmental Health Insights*, 2: 39–43.

[60] McCluskey J. (2001) Water supply, health and vulnerability in floods. *Waterlines*, 19: 14–17.

[61] CSIRO. (2007). *Infrastructure and Climate Change Risk Assessment for Victoria*, Report to the Victoria Government. http://www.cmar.csiro.au/e-print/open/2007/holperpn_c.pdf.

[62] IPCC. (2007) *Industry, Settlement and Society. Climate Change 2007: Impacts, Adaptation and Vulnerability. Contribution of Working Group II to the Fourth Assessment Report of the Intergovernmental Panel on C. Climate Change*, Parry M, Canzian O, Palutikof J, van der Linden P, Hanson C (Eds.), Cambridge University Press, Cambridge. https://www.cambridge.org/pm/academic/subjects/earth-and-environmental-science/climatology-and-climate-change/climate-change-2007-impacts-adaptation-and-vulnerability-working-group-ii-contribution-fourth-assessment-report-ipcc.

[63] WHO/UNICEF. (2004/2021) *World Health Organization and United Nations Children's Fund Joint Monitoring Programme for Water Supply and Sanitation (JMP)*, Progress on Household Drinking Water, Sanitation and Hygiene 2000–2020, Five Years into the SDGs, Geneva.

[64] MacAllister DJ, MacDonald AM, Kebede S, et al. (2020) Comparative performance of rural water supplies during drought. *Nature Communications*, 11: 1099. https://doi.org/10.1038/s41467-020-14839-3.

[65] Casani S, Rouhany M, Knochel S. (2005) A discussion paper on challenges and limitations to water reuse and hygiene in the food industry. *Water Research*, 39: 1134–1146.

[66] Hancock FE. (1999) Catalytic strategies for industrial water re-use. *Catalysis Today*, 53(1): 3–9.

[67] Angelakis AN, Marecos Do Monte MHF, Bontoux L, Asano T. (1999) The status of wastewater reuse practice in the Mediterranean basin: Need for guidelines. *Health & Environmental Research Online (HERO)*, 10, 2201–2217.

[68] Codex Alimentarius. (2019) *Codex Alimentarius Commission: Codex Committee on Food Hygiene. Proposed Draft Guidelines for the Hygienic Reuse of Processing Water in Food Plants*, Fifty-First Session, Cleveland, OH, 4–8 November. http://www.fao.org/fao-who-codexalimentarius/sh-proxy/en/?lnk=1&url=https%253A%252F%252Fworkspace.fao.org%252Fsites%252Fcodex%252FMeetings%252FCX-712-51%252FWD%252Ffh51_09e.pdf (Accessed August 2021).

[69] Brandt M, Clement J, Powell J, Casey R, Holt D, Harris N, Ta C. (2004) *Managing Distribution Retention Time to Improve Water Quality – Phase I*, AwwaRF, Denver, CO.

[70] Geldreich EE. (1996) *Microbial Quality of Water Supply in Distribution Systems*, Lewis Publishers, Boca Raton, FL.

[71] USEPA. (2002) *Health Risks from Microbial Growth and Biofilms in Drinking Water Distribution Systems*. https://www.epa.gov/sites/default/files/2015-09/documents/2007_05_18_disinfection_tcr_whitepaper_tcr_biofilms.pdf (Accessed July 2021).

[72] FAO. (2021) *Water Scarcity*. http://www.fao.org/land-water/water/water-scarcity/en/.

[73] Simmons G, Hope V, Lewis G, Whitmore J, Wanzhen G. (2001) Contamination of potable roof-collected rainwater in Auckland, New Zealand. *Water Research*, 35(6): 1518–1524.

[74] Lye DJ. (2002) Health risks associated with consumption of untreated water from household roof catchment systems. *Journal of the American Water Resources Association*, 38(5): 1301–1306.

[75] Dillaha TA, Zolan WJ. (1985) Rainwater catchment water quality in Micronesia. *Water Research*, 19(6): 741–746.

[76] Evans CA, Coombes PJ, Dunstan RH. (2006) Wind, rain and bacteria: The effect of weather on the microbial composition of roof-harvested rainwater. *Water Research*, 40(1): 37.

[77] Pacey A, Cullis A. (1986) *Rainwater Harvesting: The Collection of Rainfall and Runoff in Rural Areas*, Intermediate Technology Publications, London.

[78] Malley Z, Taeb M, Matsumoto T, Takeya H. (2007) Environmental change and vulnerability in Usangu plain, Southwestern Tanzania: Implications for sustainable development. *International Journal of Sustainable Development & World*, 14: 145–159.

[79] Koelmans AA, Mohamed Nor NH, et al (2019) Microplastics in freshwaters and drinking water: Critical review and assessment of data quality. *Water Research*, 15(155): 410–422. https://doi.org/10.1016/j.watres.2019.02.054. Epub 28 February 2019. PMID: 30861380, PMCID: PMC6449537. https://pubmed.ncbi.nlm.nih.gov/30861380/ (Accessed August 2021).

[80] WHO. (2019) *Microplastics in Drinking-Water*, World Health Organization, Geneva. https://apps.who.int/iris/bitstream/handle/10665/326499/9789241516198-eng.pdf?sequence=5&isAllowed=y.

[81] Gopalan HNB. (2003) Environmental health in developing countries: An overview of the problems and capacities – guest editorial. *Environmental Health Perspectives*, 3, July.

[82] WHO. (2017) *Safely Managed Drinking Water – Thematic Report on Drinking Water 2017*, World Health Organization, Geneva, Switzerland. Licence: CC BY-NC-SA 3.0 IGO.

[83] WHO. (2010) http://www.who.int/water_sanitation_health/hygiene/en/.

[84] Cameron J, Hunter P, Jagals P, Pond K. (2011) *Valuing Water, Valuing Livelihoods: Guidance on Social Cost-Benefit Analysis of Drinking-Water Interventions, with Special Reference to Small Community Water Supplies*, IWA Publishing, London.

[85] Smith AH, Lingas EO, Rahman M. (2000) Contamination of drinking-water by arsenic in Bangladesh: A public health emergency. *Bulletin of the World Health Organization*, 78: 1093–1103.

[86] Santaniello-Newton A, Hunter P. (2000) Management of an outbreak of meningococcal meningitis in a Sudanese refugee camp in

Northern Uganda. *Epidemiology and Infection*, 124: 75–81.

[87] Curtis V, Cairncross S. (2003) Effect of washing hands with soap on diarrhoea risk in the community: A systematic review. *Lancet Infectious Diseases*, 3: 275–281.

[88] WHO. (2015) *Preventing Diarrhoea Through Better Water, Sanitation and Hygiene: Exposures and Impacts in Low- and Middle Income Countries*. http://www.who.int/water_sanitation_health/gbd_poor_water/en/.

[89] Soussan J. (2003) *Poverty, Water Security and Household Use of Water*, International Symposium on Water, Poverty and Productive Uses of Water at the Household Level, Muldersdrift, South Africa, 21–23 January.

[90] Hutton G, et al. (2007) Global cost–benefit analysis of water supply and sanitation interventions. *Journal of Water and Health*, 5: 481–502.

[91] World Bank. (2012) *Gender Equality and Development*, World Development Report 2012, World Bank, Washington, DC. http://siteresources.worldbank.org/INTWDR2012/Resources/7778105-1299699968583/

7786210-1315936222006/Complete-Report.pdf.

[92] WSP. (Ed.) (2007): *Community-Led Total Sanitation in Rural Areas: An Approach That Works*, Water and Sanitation Program, Washington, DC. https://www.wsp.org and https://pt.ircwash.org/sites/default/files/Sanan-2007-Community.pdf.

[93] Clasen T. (2009) *Scaling Up Household Water Treatment Among Low-Income Populations*, World Health Organization, Geneva, 72. https://www.who.int/publications/i/item/WHO-HSE-WSH-09.02.

[94] Fewtrell L, Kaufmann RB, Kay D, Enanoria W, Haller L, Colford Jr, JM. (2005) Water, sanitation, and hygiene interventions to reduce diarrhoea in less developed countries: A systematic review and meta-analysis. *Lancet Infectious Diseases*, 5: 42–52.

[95] Ruel-Bergeron S, Patel J, Kazi R, Burch C. (2021) *Assessing Public Health Needs in a Lower Middle Income Country*, Routledge Focus on Environmental Health Series, Routledge, Abingdon, Oxon.

16

Air quality

Angela Hands

SECTION 1: AIR POLLUTION AND AIR QUALITY AS A PUBLIC HEALTH MATTER

Introduction

Air is essential for life as can be seen in Chapter 10. We breathe from the moment we are born and need the oxygen from the Earth's atmosphere. Humans breathe approximately 432 litres of oxygen per day, and the body needs approximately 352.8 litres of oxygen per day when the body is at rest. Yet globally, the atmosphere is being polluted such that each time we breathe could cause harm. This chapter focusses primarily on local air quality and emissions, with global heating dealt with earlier in the book although there is clearly an overlap as local emissions can contribute to global heating and climate change.

The atmosphere that surrounds the Earth is made up of nitrogen (78%), oxygen (21%) and other gases such as argon and carbon dioxide and helium in much smaller percentages. The atmosphere is stratified into layers of different thicknesses and varying chemical composition. The layer closest to the surface of the Earth is called the troposphere. It is approximately 8 to 15 km thick and is thinnest in the polar regions, and thickest near the equator.

The atmosphere not only provides the gases that are required for respiration, but it also provides protection from harmful radiation and regulates the temperature of the Earth's surface. The troposphere is the layer where the weather, as we experience it, arises and this supplies a means of distributing freshwater across the globe.

Reports over the years, including from the UK Parliament's Environmental Audit Committee (EAC) and the World Health Organization (WHO), have illustrated why air quality is a continuing public health issue. In December 2014 the EAC published its third report on air quality since 2010 [1]. The report said:

> Our main recommendations for the Government in 2010 and 2011 have not been implemented. Meanwhile air pollution continues to be an invisible killer, costing the lives of 29,000 people per year [maybe an underestimate, see later]. . . . The UK Government has been found guilty of failing to meet EU air quality targets in our cities, some of which will not meet the required limits until 2030.

DOI: 10.1201/9781003035640-16

However, meeting EU standards should be the minimum requirement.

Regardless of EU rulings it is unacceptable that UK citizens could have their health seriously impaired over decades before this public health problem is brought under control.

The WHO in 2014 [2] estimated that around 7 million people died in 2012 as a result of air pollution exposure, representing around one in eight of total global deaths. The WHO said this finding confirmed that air pollution, both household and ambient air pollution, is now the world's largest single environmental health risk. As the WHO then said in 2021 [3] globally the toll in deaths and lost years of healthy life has barely changed since the 1990s. While ambient air quality may have improved in high-income countries it has generally deteriorated in low- and middle-income countries.

In 2016 the Royal College of Physicians reported that outdoor air pollution causes an estimated 40,000 deaths in the UK every year [4][1] with more attributable to indoor air pollution. Exposure to outdoor air pollution is also estimated to lead to a loss of 15 minutes of life expectancy each day. In addition to the health implications, air pollution has a significant impact on business and our health services. In the UK, these costs add up to more than £20 billion every year. Exposure to outdoor air pollution is associated with lifelong health implications, including effects on foetal development (particularly in relation to lung and kidney development), and increases in heart attacks and strokes for those in later life. Air pollution is also linked to asthma, diabetes, dementia, obesity, and cancer.

As the CIEH has said "*the death of 9-year-old Ella Adoo-Kissi-Debrah in 2013 having been exposed to toxic levels of $PM_{2.5}$ and a second air pollutant, nitrogen dioxide (NO_2), in excess of limits set by the WHO, has highlighted the need for urgent action*"[2] and highlights why air quality is a public health issue.

Although much of the focus of this chapter is on ambient air pollution, the health impacts from indoor exposure to combustion products from heating, cooking and the smoking of tobacco are also important when it comes to health. As building technologies and materials have changed, including efforts to make new housing more energy efficient has also had impacts on indoor air quality. Such changes can also increase pollutants emitted from building materials, pollutants such as volatile organic compounds (VOCs), which arise from sources including paints, varnishes, solvents and preservatives. The effects of biological particles in the home such as mould and house dust mites are discussed elsewhere, but these also are matters of indoor air quality. Clearly reducing both indoor and outdoor air pollution could save millions of lives.

Air pollution as an environmental health issue – a historical perspective

Chapter 1 considered some of the history relevant to environmental health and air pollution. Most concerns, then and until quite recently, have focussed on visible pollutants. This culminated in the 1950s with the London Smogs, and the Great Smog of 1952, which lasted for four days from 5th to 8th December. This episode was considered responsible for 4,000 extra deaths [5], although a subsequent more recent re-evaluation has suggested that a further 8,000 deaths can also be attributed to the 1952 smog episode [6] as well as many thousands of non-fatal cases with respiratory diseases and longer-term effects.[3] Aside from the illness and misery caused, the main effect of the 1952 episode was to finally provide the necessary momentum to kick-start action to reduce this century-old air pollution problem. This action led to the Clean Air Act 1956, which introduced a number of measures to reduce air pollution, including the introduction of smoke control areas in urban areas, where only smokeless fuels could be burnt. The work of EHPs was crucial to the implementation of these subsequent clean air provisions. This Act and the subsequent 1968 Clean Air Act also included measures to relocate power stations away from cities, to increase the height of some chimneys and to control dark smoke.

These Acts and the work of EHPs and the uptake and widespread use of natural gas to replace coal as a fuel led to air quality improvements in UK urban areas through the reduction in smoke and sulphur dioxide levels. These benefits have continued to the present day.

Smogs similar to those experienced by London, were prevalent in major cities across the world such as Los Angeles, California, USA leading to similar health issues. However, the conditions of smoke and fog were very different in California which led to the term photochemical smog being introduced, describing the promotion by sunlight of a series of chemical reactions in the atmosphere. These reactions involve nitrogen oxides (NO_x), volatile organic compounds (VOCs) and oxygen, which create airborne particles, also known as particulate matter and ground-level ozone.

The recommendations from the Committee[4] set up to investigate the Los Angeles smogs included reducing hydrocarbon emissions from refineries and fuelling operations and establishing vehicle exhaust emission standards for the first time [7]. The subsequent legislation and measures introduced since the 1950s have resulted in reductions in ozone and therefore improvements to air quality. As with the London-like smogs (i.e., based on smoke and sulphur dioxide), the existence of photochemical smog (i.e., based on ozone and particles) is widespread in other parts of the world, including the UK.

The influence of human-made air pollution however also extends above the troposphere into the stratosphere (i.e., that part of the atmosphere 15 to 50 km above ground level). Naturally occurring ozone (O_3) is found within the stratosphere, where it forms the ozone layer. This ozone is produced following the photolysis of molecular oxygen and here it absorbs incoming ultraviolet (UV) radiation from the sun. Thus, it provides an invisible protective screen against this radiation, which is harmful to both humans and other organisms.

Ozone is highly reactive chemically, and in the stratosphere it can be depleted by reaction with other chemicals that may be present. Some of these reactions are with naturally occurring chemicals; however, others are with anthropogenic organohalogen compounds, especially chlorofluorocarbons (CFCs) and bromofluorocarbons. These compounds are not found in nature and were specifically produced because of their highly stable capable nature. CFCs have been widely used as refrigerants, propellants in aerosols applications and solvents. In the atmosphere they can rise to the stratosphere, where chlorine and bromine radicals are liberated by the action of UV light. Each radical is then free to initiate and catalyse a chain reaction capable of breaking down ozone molecules. The consequent breakdown of ozone in the stratosphere results in reduced protective screening, so that increased UV radiation can reach the Earth's surface.

Bans on the CFC-containing aerosol sprays thought to damage the ozone layer were introduced in the late 1970s by the United States, Canada and Norway. However, it was only after the discovery of the Antarctic ozone hole in 1985 by a team from the British Antarctic Survey that international action followed. This led to the Montreal Protocol, an international treaty which sharply limited CFC production from 1987 and phased it out completely by 1996. The Montreal Protocol has been effective in reducing emissions of long-lived CFC gases, but the problem remains with high concentrations still in the atmosphere that to lead to significant ozone destruction in polar regions. The results from the end of 2014 confirmed that the ozone hole over the Antarctica was similar in size to the previous ten years and only slightly smaller than the peak around the turn of the century, according to NOAA satellite observations [8].

The term acid rain has been associated with the effect of increased acidification in lakes and rivers, adverse impacts on forests and soils, as well as damage to buildings since the 1970s. Robert Angus Smith in Manchester, England in the 1850s first described this relationship between acid rain and atmospheric pollution. More correctly the term used should be acid deposition, rather than acid rain, since the acid deposition can be either by wet or dry means and therefore acid deposition is a better term.

Acid gases can be produced naturally; it is those gases, including both oxides of sulphur and oxides of nitrogen, produced by anthropogenic combustion processes that are of greatest importance. Acid deposition refers to the reaction of these gases with water to produce weak solutions of either sulphuric acid or nitric acid. All rain is naturally slightly acidic due to the presence of carbon dioxide, however acid rain normally refers to a pH of less than 5, although the pH can be even more acidic and less than pH 4.

The Clean Air Acts (discussed further in Section 5) in the UK required coal and oil-fired power stations being sited outside of the main urban areas and having much taller chimneys. A direct impact of this was that the gases from these larger power stations were released from much taller chimneystacks, often over 200 m high, leading to acid deposition a considerable distance downwind of the emissions. Examples of acid deposition were observed following a lower than expected pH of local rain in Scandinavia and Canada, as a result of emissions from the UK and the United States respectively.

The UK has also been affected by acid rain,[5] and the Acid Waters Monitoring Network (AWMN) was set up in the late 1980s to assess the condition of waters in the UK. Kernan et al. published a report from the AWMN in 2010 [9] which indicated UK waters were beginning to recover. From 2013 the AWMN became the Upland Waters Monitoring Network,[6] designed to track changes in surface water quality and freshwater biodiversity across all upland regions of the UK.

The reductions of emissions are agreed at international level, such as the EU National Emission Ceilings Directive and the Gothenburg Protocol, which was established in 1999 and revised in 2012.[7] The aim of these agreements was to help create the conditions in which damaged lakes and rivers can recover. Emissions of the air pollutants that cause acid rain are expected to continue to reduce. The 2012 revision of the Gothenburg Protocol agreed to include even more stringent emission reduction commitments for 2020. This agreement included reduction targets for particulate matter (PM) and technical annexes with improved emission limit values.

Current air pollution and air quality issues

This section discusses those air quality issues which are of current concern based on existing knowledge and understanding. These issues are loosely based around geographical scale, as follows: global (i.e., climate change/global heating dealt with in Chapter 11, but over 100 years ago Arrhenius was the first scientist to show that increases in concentrations of carbon dioxide in the Earth's atmosphere would lead to increased temperatures [10]); transboundary (i.e., tropospheric ozone); and local air pollution (nitrogen dioxide, particulate matter and other toxic air pollutants). The World Health Organization published new global air quality guidelines, which are shown in Table 16.1.[8]

Tropospheric ozone – A key pollutant of concern is tropospheric (or ground level) ozone (O_3). The increase in ozone concentrations will lead to a consequent increase in attributable deaths and hospital admissions, as exposure to ozone may also increase the risk of sensitisation to airborne allergens in predisposed individuals. The increased morbidity and mortality could be significant, with up to about 2,000 or more extra deaths

Table 16.1 WHO air quality guidelines

Pollutant	Averaging time	Interim target				AQG level
		1	2	3	4	
PM$_{2.5}$ µg/m³	Annual	35	25	15	10	5
	24 hour[a]	75	50	37.5	25	15
PM$_{10}$ µg/m³	Annual	70	50	30	20	15
	24 hour[a]	150	100	75	50	45
O$_3$ µg/m³	Peak season[b]	100	70	-	-	60
	8 hour[a]	160	120	-	-	100
NO$_2$ µg/m³	Annual	40	30	20	-	10
	24 hour[a]	120	50	-	-	25
SO$_2$ µg/m³	24 hour[a]	125	50	-	-	40
CO mg/m³	24 hour[a]	7	-	-	-	4

Source: WHO Global Air Quality Guidelines [3]
[a] 99th percentile (i.e., three to four exceedance days per year)
[b] Average of daily maximum eight-hour mean O$_3$ concentration in the six consecutive months with the highest six-month running-average O$_3$ concentration

per annum in the UK by the 2030s. This is on the assumption of no threshold for the relationship between eight-hour average ozone and all-cause mortality for all ages and all years [11].[9]

Tropospheric ozone has adverse public health and environmental impacts, including food production. It is one of the principal pollutants in photochemical smog and a powerful greenhouse gas. Climate change could affect local to regional air quality through changes in chemical reaction rates, boundary layer heights that affect vertical mixing of pollutants, and changes in synoptic airflow patterns that govern the movement of pollutant transport [12].[10],[11]

Ozone is found naturally in the troposphere as a result of both incursions down from the stratosphere and in equilibrium with NO$_x$. Additional tropospheric ozone arising as a secondary pollutant formation is formed as a result of photochemical reactions of precursor chemicals (such as nitrogen oxides (NO$_x$), carbon monoxide (CO) and volatile organic compounds (VOCs)) in the atmosphere. Motor vehicle exhaust, industrial

emissions and chemical solvents are the major anthropogenic sources of these chemicals. A key aspect is that these precursors often originate in urban and industrial areas before winds transport the pollution hundreds of kilometres, causing the formation of ozone to occur in rural areas.

The Royal Society report [13] provided the modelled results for future ozone concentrations, taking into consideration global predicted emissions for different scenarios. The results are reported at the global, regional and urban scales. The impacts of climate change are also important, and this shows an increase ozone production in already polluted environments, while decreasing it in unpolluted environments. In Europe, reductions in regional ozone concentrations by 2050 are projected as a result of NO$_x$ and CO emission reductions.

At the urban scale the modelling indicates that ozone will increase in many cities by 2050 due to reductions in local NO$_x$ concentrations, thereby increasing exposure of urban populations. Ground level ozone affects human health, and it is projected that hot summers (and ozone episodes), such as that experienced in 2003, will become more frequent. With an increase in the frequency and magnitude of such events, mortality and morbidity are also expected to increase. Based on these findings the Royal Society advocated that ozone be treated as a global pollutant with appropriate policy frameworks, as without this even current regional level controls are unlikely to achieve their policy objectives.

Particulate matter – Particulate matter or particles are different from other atmospheric pollutants, in that they are not one specific chemical. Rather, particulate matter is a generic term that refers to any material (solid or liquid) found in a gas (or liquid). In this chapter we are specifically concerned with atmospheric pollution and therefore particulate matter refers to dust, soot, aerosols, biogenic material, minerals, etc. found in air.

To characterise particulate matter (PM) we can seek to identify its specific chemical composition (e.g., elemental carbon or nitrate, etc.) or determine its size. Today, most commonly PM is classified on the basis of size. This tends to be in relation to how deeply material can be breathed into the human respiratory system. PM_{10} and $PM_{2.5}$[12] are the most frequently referred to PM pollutants, and health-based standards have been introduced by governments around the world based on these classifications.

Based on the ability of different size PM to penetrate the human respiratory system, $PM_{2.5}$ is sometimes referred to as fine particles. Further divisions as follows:

PM_{10} – thoracic fraction   (i.e., less than 10 µm)
$PM_{2.5}$ – respirable fraction (i.e., less than 2.5 µm)
$PM_{0.1}$ – ultrafine fraction (i.e., less than 0.1 µm
(or 100 nanometres))

The ultrafine fraction can penetrate deepest into the lung and can penetrate lung tissue and enter the circulatory system. The term coarse fraction is also used. This relates to that fraction of PM that is greater than 2.5 µm but less than 10 µm.

Other descriptions of PM relate to its origin and formation. Specifically, primary PM arises directly from its emissions source, be it a combustion source (e.g., domestic wood burning stoves (discussed in Section 5)), diesel engine or other activity (e.g., abrasion (such as grinding by machinery)). Secondary PM refers to the formation of PM from precursor pollutants in the atmosphere. These precursor pollutants can include gaseous pollutants from combustion sources emitting NO_x and SO_2 leading to the formation of secondary nitrate and sulphate particles [14].

Not all sources however are anthropogenic. Natural sources include volcanoes, forest fires and sea salt crystals (arising from sea spray). The behaviour of an atmospheric particle is determined by its size; in general, the smaller and lighter a particle is, the longer it will stay in the atmosphere. Larger particles tend to settle very quickly by gravity within hours (or less). Hence nuisance dust is found close to the source – for example within hundreds of metres, soon after emission from the source. The smallest particles (less than 1 µm) however can stay in the atmosphere for weeks and are mostly removed by precipitation.

As a result, smaller particles can be carried long distances such as many hundreds of kilometres, and be stable with a lifetime of a month or more in the atmosphere. The smaller, secondary, longer residence PM is considered to be a transboundary pollutant, rather than a local pollutant. Examples of the long-distance travel include emissions from volcanoes that can circumnavigate the world, affecting climate and human activities far from their origin: dust blown to the UK from the Sahara and forest fires such as that those in Siberia affecting Europe and others affecting southeast Asia and Australia.

PM affects health by being toxic or by providing a surface for transporting toxic compounds to where they can do harm. PM_{10} and $PM_{2.5}$ are associated positively with cardiovascular, respiratory and cerebrovascular mortality, and there is evidence that both $PM_{2.5}$ and PM_{10} are associated with increased mortality from all causes, including lung cancer [3]. PM can have short-term health impacts over a single day when concentrations are elevated, and long-term impacts from lower-level exposure over the life-course. Effects are amplified in vulnerable groups including young children, the elderly, and those suffering from breathing problems like asthma [14].

The WHO [3] has said that compared to the previous global WHO evaluation, the evidence base has increased substantially. However, studies conducted in low- and middle-income countries (LMICs) are still limited. Associations remained below the current WHO guideline value of 10 μg m^{-3} for $PM_{2.5}$.

In the UK, the Committee on the Medical Effects of Air Pollutants (COMEAP) estimated in 1998 that on average up to 24,000 people in the UK die prematurely every year as a result of short-term exposure to air pollution and thousands more are hospitalised; but later this was thought to be an underestimate and could be nearer to 50,000 [15].[13] Epidemiological and toxicological studies have shown that $PM_{2.5}$ does not only induce cardiopulmonary disorders and/or impairments, but also contributes to a variety of other adverse health effects, such as driving the initiation and progression of diabetes mellitus and eliciting adverse birth outcomes. Of note, recent findings have demonstrated that $PM_{2.5}$ may still pose a hazard to public health even at very low levels of exposure, which may be below national emission standards [16].

The UK government's 2007 Air Quality Strategy also estimated that PM reduces life expectancy by around seven to eight months, averaged over the whole population of the UK. The 2019 Clean Air Strategy [14] refers to COMEAP suggesting the long-term impacts of UK PM concentrations in terms of mortality can be quantified as equivalent to 340,000 life years lost.

Previously the UK Department of Health had commissioned work from the Institute of Occupational Medicine (IOM) [17] to compare the benefits of eliminating anthropogenic $PM_{2.5}$ with the elimination of motor vehicle traffic accidents and the elimination of exposure to passive smoking. The results are showed that a reduction in air pollution would be equivalent to a reduction in both road traffic accidents and passive smoking combined.

PM effects on health are better understood now but other specific effects of PM can include soiling of buildings. Furthermore, PM and other emissions from volcanoes can have deleterious and potentially disastrous effects on human activity. Another effect is that the particles scatter and absorb solar and infrared radiation in the atmosphere, known as solar or global dimming. The pollutants can become nuclei for cloud droplets, thereby creating clouds and reflecting the incoming sunlight back into space so that less reaches the Earth's surface. The result is an increase in the amount of light reflected back into space, which results in a decrease in the amount of solar radiation that reaches the surface. Although at night, clouds cause the re-radiation of heat to the Earth and hence slow heat loss. These clouds also have smaller droplets leading to decreases in the chance of precipitation. If precipitation is suppressed, this results in excess water remaining in the atmosphere. This disruption overall can lead to the extreme effects in weather discussed earlier as a consequence of climate change.

Nitrogen dioxide/oxides of nitrogen (NO_x) – The WHO Guidelines [3] focus on nitrogen dioxide (NO_2) but other oxides of nitrogen should not be ignored. As the 2019 Clean Air Strategy points out [14] the majority of NO_x emitted as a result of combustion is in the form of nitric oxide (NO). When NO reacts with other gases present in the air, it can form nitrogen dioxide (NO_2), which is harmful to health. It is also important in the formation of ozone. NO coverts to NO_2 very quickly and vice versa.

Anthropogenic sources arise from combustion-related activities, such as motor vehicle exhausts, space heating, gas cooking, power generation and industrial activities.

Therefore, the emissions tend to be more concentrated in urban environments, leading to locally higher concentrations. Weather influences NO_2 concentrations, and summer episodes are more likely to occur when the weather is hot and sunny and wind speeds low.

Short-term exposure to concentrations of NO_2 can cause inflammation of the airways and increase susceptibility to respiratory infections and to allergens. Furthermore, the WHO reported an association of short-term exposure with asthma exacerbation. It exacerbates the symptoms of those who are already suffering from lung or heart conditions, shortening their lives.

Between 2006 and 2019 inclusive, the annual mean NO_2 concentration at urban background sites in the UK reduced by an average of 0.9 µg/m^3 each year.[14] Reductions in concentrations were observed at most monitoring sites across the UK. Emissions of NO_2 in the UK and Europe have continued to decrease as newer road vehicles subject to stricter emission standards enter the fleet and power generation moves away from the use of coal, particularly in the UK.

Between 2019 and 2020, the annual mean NO_2 concentration at urban background sites reduced further by 4.5 µg m^{-3} (23%). It is likely that a reduction in traffic because of COVID-19 restrictions was a contributing factor to this relatively large decrease. The average annual mean concentration of NO_2 at the roadside has decreased over the time series to 23.0 µg m^{-3} in 2020, a low since in 1997. The annual mean NO_2 concentration in 2020 is greater at roadside sites compared to urban background sites because of substantial NO_2 emissions from road transport sources, as most concentrations at the roadside come from local transport sources. One example, found in Oxford Street, London, a street that experiences heavy motor vehicle traffic in 2013 a peak hourly concentration of 489 µg m^{-3}.[15]

In the UK an important episode that lasted four days arose at the government's background monitoring site in London during December 1991 when mean concentrations of NO_2 of around 285µg m^{-3} and a peak hourly concentration of just over 800 µg m^{-3} were recorded. These concentrations represented the highest concentrations monitored in the UK since measurements began in the 1970s. Subsequent analysis of the air pollution episode showed that it was associated with an increase in mortality and morbidity, which was unlikely to be explained by the prevailing weather [18], with an additional 160 deaths being attributed to the pollution episode.

This episode however provided an impetus for action on invisible air pollutants. As a result, the UK government set up the Quality in Urban Air Group and in its first report in early 1993, it outlined an initial assessment of the extent of air quality problems and monitoring, including that of NO_2. A Health Impact Assessment for London estimated that air pollution contributed to the deaths of as many as 9,400 Londoners in 2010; around 3,500 from particle pollution and up to 4,900 from nitrogen dioxide [19]. Importantly this was the first time that nitrogen dioxide had been included. The report indicated that most of the deaths linked to NO_2 were because of NO_2 emissions from diesel vehicles and other sources within the capital. The deaths related to PM were from both local and wider sources.

NO_2 is a key precursor of a range of secondary pollutants whose effects on human health were discussed earlier. This is through the formation of strong oxidants that can lead to the eventual conversion to nitrate and sulphate particles that are measured as PM_{10} or $PM_{2.5}$.

Other air pollutants of concern

Other air pollutants with a health impact include Non-methane Volatile Organic Compounds (NMVOC), sulphur dioxide, carbon monoxide, benzene, lead in air, 1,3-butadiene, polycyclic aromatic hydrocarbons (PAHs), various metals.

NMVOCs – NMVOCs are a large group of organic compounds which differ widely in their chemical composition but display similar behaviour in the atmosphere. NMVOCs are emitted to the air as combustion products, as vapour arising from petrol, solvents, air fresheners, cleaning products, perfumes and numerous other sources related to product use. They directly contribute to adverse impacts to health and the environment. NMVOCs can react with other air pollutants outdoors in the presence of sunlight (ultraviolet radiation) to produce ground-level ozone which poses risks to health by triggering inflammation and asthma, but also causes oxidative damage in vegetation such as crops.

NMVOCs also pose a threat to health indoors. Reactions between different NMVOCs and chemicals from combustion processes such as smoking, heating, cooking or candle burning can produce chemicals like formaldehyde, a human carcinogen that also causes irritation to the eyes and upper airways at low concentrations. Formaldehyde as an NMVOC can also be emitted directly from furniture, finishes and building materials such as laminate flooring, kitchen cabinets and wood panels.

The UK government says that emissions of NMVOCs have fallen by 66% since 1970, to 812,000 tonnes in 2019. There was a decrease in emissions of 0.97% between 2018 and 2019. Although emissions from many sources (including domestic solvent use) have been stable for several years, there has been a recent increase in operator-reported emissions related to venting and flaring at oil and gas installations. In the 1990s road transport contributed 30% of total emissions but by 2019 contributed 3.7% (30,000 tonnes).[16]

Sulphur dioxide (SO_2) – This is a colourless gas that is readily soluble in water. The gas was involved in smogs, and acidic deposition discussed earlier. Historically, SO_2 was derived from the combustion of fossil fuels, particularly coal, causing serious environmental health problems in large urban areas of the UK, Europe and other industrialised areas.

SO_2 can irritate the skin and mucous membranes of the eyes, nose, throat and lungs. High concentrations of SO_2 can cause inflammation and irritation of the respiratory system, especially during heavy physical activity. The resulting symptoms can include pain when taking a deep breath, coughing, throat irritation and breathing difficulties. High concentrations of SO_2 can affect lung function, worsen asthma attacks, and worsen existing heart disease in sensitive groups. This gas can also react with other chemicals in the air and change to a small particle that can get into the lungs and cause similar health effects. Children may be more sensitive to the effects of sulphur dioxide due to their smaller size. It is unknown whether sulphur dioxide causes harm to the unborn child.[17] The WHO [3] reported a positive association between short exposure to ambient SO_2 and all-cause and respiratory mortality. Mortality increases on days with higher SO_2 levels.

As a result of the reductions in emissions, ambient concentrations have also decreased, both as annual average and peak levels. Annual mean concentrations in most major UK cities are now said to be well below 35 ppb (100 μgm^{-3}) with typical mean values in the range of 5–20 ppb (15–50 μgm^{-3}). Hourly peak values can be 400–750 ppb (1000–2000 μgm^{-3}) on infrequent occasions. Natural background levels are about 2 ppb (5 μgm^{-3}).[18]

Carbon monoxide – This is a colourless, odourless and tasteless gas, which is slightly lighter than air. It is highly toxic to humans and animals and can be a cause of death in poorly or unventilated areas. It is primarily an issue in indoor environments. It is produced by the incomplete combustion of carbon-based fuels.

CO combines with haemoglobin to produce carboxyhaemoglobin, which reduces the ability of the body to deliver oxygen to tissues. The most common symptoms of CO poisoning resemble other types of poisonings and infections, including symptoms such as headache, nausea, vomiting, dizziness, fatigue and a feeling of weakness. Infants may be irritable and feed poorly. Neurological signs include confusion, disorientation, visual disturbance, syncope and seizures. Low-level carbon monoxide exposure in the at-risk population of individuals with heart disease is an important,[19] but difficult to predict, factor as it may provide an increased risk of sudden death from arrhythmia.

Natural sources of carbon monoxide (CO) include volcanoes, forest fires, plus it is also formed naturally by photochemical reactions in the atmosphere, which are subsequently oxidised. CO emissions remain dominated by road transport activities, although as a result of improved technologies (including abatement measures) CO has reduced by more than 79% since 1970 and also as a result of controls on road transport, agricultural field burning and the domestic sector [33]. The highest ambient concentrations arise in urban areas closest to busy roads during wintertime when dispersion may be reduced.

With an atmospheric lifetime of several months, CO is also important precursor to the formation of the global background concentration of tropospheric O_3. CO further has an indirect radiative forcing effect by elevating concentrations of methane, a greenhouse gas, through chemical reactions with other atmospheric constituents (e.g., the hydroxyl radical (OH)) that would otherwise react and destroy them.

Ammonia (NH_3) – Agriculture is the primary source of ammonia from the storage and spreading of slurries. The main concern with ammonia is the contribution to particulate matter, and Public Health England attributed the 2014 smog in London in part to agricultural emissions. Ammonia produces fine particles when mixed with nitrogen oxides and sulphur dioxide and stays in the atmosphere for just a few hours as a gas. When converted to particulate matter, this extends to several days as the particulates can then travel long distances.

Benzene – Benzene (C6H6) in air arises from the evaporation of or combustion of petroleum products. It is an aromatic liquid that is colourless and sweet smelling that was formerly used as an additive to petrol. It is however still an important industrial chemical and solvent.

Outdoor air contains low levels of benzene from tobacco smoke, petrol stations, motor vehicle exhaust and industrial emissions. Indoor air generally contains levels of benzene higher than those in outdoor air. The benzene in indoor air comes from products that contain benzene such as glues, paints, furniture wax and detergents. The air around hazardous waste sites or petrol filling stations can contain higher levels of benzene than in other areas.

Tobacco smoke contains benzene, and this is estimated to be responsible for approximately one-tenth to one-half of smoking-induced total leukaemia mortality and up to three-fifths of smoking-related acute myeloid leukaemia [20].

As a volatile organic compound (VOC), benzene can contribute to the formation of tropospheric ozone.

Lead – This is a soft metal that is found in air in the form of very small particles. Although it can come from volcanoes, soil erosion and forest fires, industrial sources include smelters, mining operations, waste incinerators and battery recycling. It is a major concern in the USA, particularly from paint.

As the result of the move to unleaded petrol, there was a decline in emissions from the road transport sector, but research from Imperial College London has found that lead from leaded petrol persists in London's air despite its ban in 1999. The study found that up to 40% of lead in airborne particles today comes from the legacy of leaded petrol [21].

Post-1999 the highest contributing sources have been from iron and steel production, although abatement measures reduced these. In addition, emissions have also declined as a result of the decreasing use of coal and due to improved controls on Municipal Solid Waste (MSW) Incinerators from 1997 onwards. The Air Quality Directive[20] provided a Limit Value for lead concentration in air of 0.5 µg m^{-3}, expressed as an annual mean.

Lead can be absorbed if lead-containing dust or fumes are swallowed or breathed in. Although small amounts do not cause any specific symptoms, up to 10% of the total that is absorbed remains within the body and is not excreted; consequently it can gradually build up in the body. Large amounts of lead in the body can cause pain in joints and muscles. Other symptoms of lead exposure include anaemia, nausea, gastric problems, sleep problems, concentration problems, headaches and high blood pressure. In children, the symptoms of lead exposure can be poor development of motor abilities and memory, reduced attention span and colic/gastric problems.

Small children and unborn babies are at most risk because their bodies are smaller, and they are still growing and developing. Any amount of Pb can be a health risk for pregnant woman because the unborn baby is exposed to Pb in the mother's blood. A large amount of Pb in the mother's body can cause premature birth, low birth weight or even miscarriage or stillbirth.

High levels of acute exposure to Pb may also cause brain and kidney damage. Chronic exposure can lead to effects on the blood, kidneys, central nervous system and vitamin D metabolism.

1,3-butadiene – Like benzene 1,3-butadiene is a hydrocarbon used in industrial processes, particularly in the production of synthetic rubber. It is a colourless gas with a petrol-like odour, but in industry is used as a refrigerated liquid. Emissions of 1,3-butadiene arise from its manufacture and use in the chemical industry, and also from the combustion of petroleum products.

1,3-butadiene is not present in petrol and is formed as a by-product of combustion only. It is ubiquitous in the urban environment because of widespread combustion sources.

Therefore, UK emissions of 1,3-butadiene were dominated by the road transport sector. The introduction of catalytic converters in the early 1990s had a significant impact, causing a decline in total emissions of 79% between 1990 and 2008. Emissions from other significant combustion sources, such as other transportation and machinery have reduced, however not as significantly as road transport.

Excessive exposure to 1,3-butadiene may affect blood, the brain, eye, heart, kidney, lung, nose and throat. However, although this is unlikely in ambient air. It is a suspected human carcinogen and teratogen, and animal data suggest the carcinogenic effects of 1,3-butadiene may have a higher sensitivity to women than men when exposed. As a suspected carcinogen, the assumption is made that there is no threshold to such effects and that any exposure results in some increase in risk, albeit this may be very small.

As a volatile organic compound (VOC), 1,3-butadiene can contribute to the formation of tropospheric ozone.

PAHs (polycyclic aromatic hydrocarbons) – Polycyclic aromatic hydrocarbons (PAHs) are a large group of organic compounds that comprise two or more aromatic rings, including both naturally occurring and human-made chemicals. PAHs are widely distributed in the atmosphere. They have low vapour pressures and are found at ambient temperatures in air as gases or often absorbed onto particles of soot. They are relatively insoluble in water but dissolve readily in fats and oils and can accumulate through food chains. They are generated by incomplete burning of fuels, particularly solid fuels.

Human health effects from environmental exposure to low levels of PAHs are unknown. Large amounts of naphthalene in air can irritate eyes and breathing passages. The International Agency for Research on Cancer (IARC) has classified Benzo(a)pyrene (BaP) as carcinogenic to humans. BaP is a common component of combustion products and although emissions of PAHs have decreased 98% since 1990, the use of wood as a domestic fuel has increased in recent years and is estimated to produce 78% of the total national BaP emissions.

PAHs are classed as persistent organic pollutants (POPs) under the UNECE protocol, which aims to introduce strict standards to minimise their release. The following chemicals are specified by the protocol as indicators of the group: BaP, benzo(b)fluoranthrene, benzo(k)fluoranthrene, indo(1,2,3-cd)pyrene.

Heavy metals and arsenic – Those with air quality limit values are arsenic, cadmium, mercury and nickel. In the UK the Heavy Metals Network is currently managed and operated for the Environment Agency, on behalf of Defra and the Devolved Administrations by the National Physical Laboratory. The network monitors the concentrations in air, and the deposition rates of a range of metallic elements at urban, industrial and rural sites. In 2014, the Urban & Industrial Heavy Metals Network and Rural Heavy Metals Network were combined into a single Heavy Metals Network comprising 24 monitoring sites.[21] All stations (except Lough Navar) measure Arsenic (As), Cadmium (Cd), Chromium (Cr), Cobalt (Co), Copper (Cu), Iron (Fe), Manganese (Mn), Nickel (Ni), Lead (Pb), Selenium (Se), Vanadium (V) and Zinc (Zn) in the PM_{10} fraction of air. Additionally, heavy metals in deposition are measured at five rural sites with mercury in deposition additionally measured at four of these stations. The Heavy Metals Network now forms the basis of the UK's compliance monitoring for the Limit Value for lead concentration in air of 0.5 μg/m3, expressed as an annual mean and the target values for arsenic, cadmium, nickel (and polycyclic aromatic hydrocarbons) in the PM_{10} particulate fraction of ambient air.[22]

The heavy metals cadmium, lead and mercury are common air pollutants, emitted mainly as a result of industrial activity. Even low atmospheric levels contribute to build-up in soils, where they persist in the environment and accumulate in the food chain both on land and in water.

Arsenic (As) occurs naturally in the environment and exposure comes by breathing and consumption of contaminated food or water. Arsenic is used by industry for making electronic components, special alloys and in the manufacture of certain glass and ceramic products. Cigarette smoke also contains arsenic and smoking cigarettes can double a person's daily exposure.

Inhalation of air with high levels of arsenic can cause lung damage, shortness of breath, chest pain and cough, which may lead to death in severe cases. However, arsenic levels in the environment are usually not high enough to cause lung damage. Chronic inhalation exposure is associated with irritation of the mucous membranes; it is also strongly associated with lung cancer. Inorganic arsenic compounds are classified as carcinogens.

Emissions in the UK are said to have declined by three-quarters since 1970.[23] Historically the largest source of emission was coal combustion with other sources being very small by comparison.

Cadmium (Cd) combines with other elements, and cadmium oxide is most commonly found in the air. Cadmium is used in a number of industries, such as welding and soldering, photography, production of iron, steel and nickel-cadmium batteries.

Breathing high levels of cadmium can damage the lungs, which may lead to death in severe cases. Chronic effects via inhalation can cause a build-up of cadmium in the kidneys that can lead to kidney disease. Inhalation of cadmium also increases the risk of lung cancer, and it has been classified as a carcinogen.

Again, UK emissions of have declined by substantially since 1990.[24] Historically, the main sources of emission have been industrial combustion including energy production, non-ferrous metal production and iron and steel manufacture. Road transport emissions have recently become more significant due to successful reductions from other sources. Emissions from non-ferrous metal activities include lead-zinc smelting and lead battery recycling plants. The decline in emissions is mainly a result of the general fall in coal combustion and fuel oil combustion in power generation. Large reductions in waste emissions were due to improved controls on Municipal Solid Waste (MSW) incinerators.

Elemental mercury (Hg) evaporates to form mercury vapour, which is the predominant form of mercury in the atmosphere. Again, mercury emissions in the UK have declined post-1990.[25] The main sources are energy and iron and steel production, waste incineration, manufacture of chlorine (using mercury cells) and other industrial combustion. Emissions have declined as a result of improved controls in industry and the decline of coal use. The improved controls on MSW incinerators have also led to a large reduction in emissions.

After breathing elemental mercury vapour, about 80% enters the blood from the lungs. Breathing in elemental mercury vapour for a short time affects the nervous system and lungs leading to tremors, walking difficulties, chest pains and breathlessness, respectively. After longer periods, the lining of the mouth and lungs may be damaged. Kidney damage may also occur as well as stomach irritation, nausea, vomiting and diarrhoea.

Nickel (Ni) is a metal that is widely distributed in the earth's crust, air and water and is used to produce stainless steel and other alloys. Nickel compounds are also used in the production of nickel-cadmium batteries. Human activities including combustion of heavy fuel oil and coal, municipal incineration, steel and other nickel alloy production can all release nickel into the atmosphere.

Inhalation of nickel can cause irritation to the nose and sinuses and also lead to the loss of the sense of smell. Long-term exposure may lead to asthma or other respiratory diseases. Cancer of the lungs, nose and sinuses as well as the larynx and stomach has been attributed to occupational exposure to nickel. It is classified as possibly carcinogenic to humans.

Nickel emissions have also declined post-1990.[26] The decline in use of coal and oil as fuels is largely responsible for the reduction in total emissions. The use of Orimulsion,

a bitumen-based fuel, instead of fuel oil at some power plants led to an increase in emissions in the 1990s. Information on many but not all emissions and pollutants mentioned here can be found in the reports of the National Atmospheric Emissions Inventory.[27]

At the international level, the Gothenburg Protocol and amendments to it set emissions ceilings levels for various pollutants. Its aim is to control long-range transboundary pollution. It is implemented at the EU level through several directives, including the National Emission Ceilings Directives of 2001 and 2016. The 2001 Directive was implemented in the UK through the National Emission Ceilings Regulations 2002. The 2016 Directive set emission ceilings which apply from 2020 and has been implemented by the National Emission Ceilings Regulations 2018.[28] The main pollutants addressed, and which have reduction commitments, are SO_2, NO_x, non-methane VOC,[29] NH_3 and $PM_{2.5}$.[30]

In the UK responsibility for meeting ambient air quality limit values is devolved to the national administrations in Scotland, Wales and Northern Ireland.

Units of air pollutants

Documents on air pollution will often contain different units when considering how much pollutant there is in a given amount of air.

It is important to understand the quantitative terms used for comparative purposes. When we describe the constituent proportions of something, we normally describe these parts in fractional or percentage terms. These descriptions of quantities can also be used for the trace amounts of gases in a given volume of air; however, this soon becomes very cumbersome. Instead, it is better to describe it, in volume terms, as one part (of X pollutant) per million parts of air (i.e., in terms of ppmv (the v is normally dropped)). If this is still too cumbersome (as it is for most pollutants) then we can describe the volumes in terms of ppb or even ppt, where:

ppm = parts per million (10^{-6}) (for context one ppm is equivalent to 1 milligram of something per litre of water)
ppb = parts per billion (10^{-9}) (the equivalent to about three seconds out of a century)
ppt = part per trillion (10^{-12})

Volumes of the same gases however can vary at different temperatures and pressures; the major advantage when using these units is that they are independent of such variations.

Gases also have mass, and this characteristic can be used to determine a unit of pollutant in a given volume of air. This is described as a mass concentration, and it relates the mass of the gas present to a standard volume of air. For pollutants, this is either expressed as milligrams (i.e., mg m^{-3}; or more typically micrograms per cubic metre, i.e., µg m^{-3}). Gravimetric units of pollutants are used for the purposes of environmental standards (e.g., European Limit Values), whereas volumetric units can be used by analysing equipment when determining pollutant concentrations. Consequently, both have advocates for their use.

Gravimetric concentrations vary at different temperatures and pressures so for precise scientific accuracy these conditions are specified. For the EU Limit Values (and UK air quality standards) of gaseous pollutants there are standardised conditions for the volume; these are a temperature of 293K and an atmospheric pressure of 101.3 kPa.[31]

EHPs work to environmental standards and will therefore use gravimetric units. However, there are occasions when it is necessary to convert between the units. The conversion factors used for the standardised conditions referred to earlier are given in Table 16.2. The basic principle is derived from the gas laws that govern chemistry and Avogadro's Law specifically, which states that the volume occupied by an ideal gas is proportional to the amount of moles (or molecules) present. This gives rise to the molar volume of a gas, which at standardised conditions is 24 litres. From the following conversion factors can be derived.

The temperature and pressure under which concentrations of gases are measured are unlikely to be standardised conditions. As a result, a correction is required to convert gravimetric units at one temperature and pressure to gravimetric units at another temperature and pressure (e.g., to the standardised conditions mentioned for the EU Limit Values).

The mass of a sample of gas will not change if the temperature and/or pressure is changed; the volume of the sample however will change if the temperature and/or pressure is changed. This is it necessary to determine the volume of the gas for changes of temperature and pressure. This determination is made

Table 16.2 Conversion factors (volumetric units to mass concentrations)

Gas (pollutant)	Conversion factor	Notes
Carbon monoxide	1.16	ppm to mg m^{-3}
Nitrogen dioxide (NB NO_x is also expressed as NO_2)	1.91	ppb to µg m^{-3}
Nitric oxide	1.25	ppb to µg m^{-3}
Benzene	3.25	ppb to µg m^{-3}
Sulphur dioxide	2.66	ppb to µg m^{-3}
1,3-butadiene	2.25	ppb to µg m^{-3}
Ozone	2	ppb to µg m^{-3}

by using gas laws; specifically rearranging the Ideal Gas law as in Equations 16.1 and 16.2.

Equation 16.1

Where P_1, V_1 and T_1 are the initial pressure, volume and temperature and P_2, V_2 and T_2 are the final pressure, volume and temperature. These can be rearranged as follows to determine the final volume (V_2).

$$\frac{P_1 V_1}{T_1} = \frac{P_2 V_2}{T_2}$$

P_1 = first pressure
P_2 = second pressure
V_1 = first volume
V_2 = second volume
T_1 = first temperature
T_2 = second temperature

Equation 16.2

$$V_2 = \frac{T_2}{P_2} \times \frac{P_1 V_1}{T_1}$$

Section 1 Notes

1 The 2019 Clean Air Strategy [14] refers to a Committee on the Medical Effects of Air Pollution (COMEAP) estimated the long-term exposure to human-made air pollution in the UK leading to a, annual impact of shortening life spans equivalent to 28,000 to 36,000 deaths.

2 In this case, H.M. Assistant Coroner for Inner South London, Philip Barlow, concluded in December 2020 that Ella died at 9 years of age from acute respiratory failure, asthma and "*air pollution exposure*". This is the first time that a Coroner has found that air pollution was a contributory cause of illness and death. See https:// www.blackstonechambers.com/news/inquest-death-ella-adoo-kissi-debrah/ and also https:// www.cieh.org/news/press-releases/2021/ government-must-include-who-air-quality-limits-in-environment-bill/

3 See Phillips DIW, Osmond C, Southall H, et al. (2018) Evaluating the long-term consequences of air pollution in early life: Geographical correlations between coal consumption in 1951/1952 and current mortality in England and Wales. *BMJ Open*, 8: e018231. https://doi. org/10.1136/bmjopen-2017-018231.

4 Dr Arnold Beckman (scientist and scientific instrument manufacturer) helped initiate studies on the sources of photochemical smog, and helped develop control regulations and warning procedures for Los Angeles County. In 1953 Beckman chaired a special technical committee on air pollution appointed by the governor of California. Its scientific findings and recommendation for smog reduction became a standard reference for later control programmes.

5 It was reported that the then Prime Minister Margaret Thatcher only became concerned when she realised that the UK was being affected and that forest dieback as the result of acid deposition was not limited to other countries such as Norway and Sweden.

6 The network is currently co-ordinated by ENSIS/ Environmental Change Research Centre, UCL (ECRC) London (see http://www.ecn.ac.uk).

7 https://unece.org/environment-policyair/ protocol-abate-acidification-eutrophication-and-ground-level-ozone

8 WHO Global Air Quality Guidelines (2021) *Particulate matter (PM2.5 and PM10), ozone, nitrogen dioxide, sulfur dioxide and carbon monoxide*. Geneva: World Health Organization. Licence: CC BY-NC-SA 3.0 IGO. (see https://www.who.int/ news-room/questions-and-answers/item/who-global-air-quality-guidelines and https://www. who.int/publications/i/item/9789240034228).

9 On this see COMEAP. (2015) *Quantification of Mortality and Hospital Admissions Associated with Ground-level Ozone*, Report for PHE. https:// assets.publishing.service.gov.uk/government/ uploads/system/uploads/attachment_data/file/ 492949/COMEAP_Ozone_Report_2015__ rev1_.pdf

10 See also Zoë L, Fleming ZL, et al. (2018) Tropospheric ozone assessment report: Present-day ozone distribution and trends relevant to human health. *Elementa: Science of the Anthropocene*, 6: 12, 1 January. https://doi.org/10.1525/elementa.273

11 See also https://www.epa.gov/air-research/ air-quality-and-climate-change-research

12 This notation refers to its size being less than 10 μm (micrometres) in diameter for PM_{10} and 2.5 μm for $PM_{2.5}$. (A micrometre is one millionth of a metre; it is also the same as a micron, which is still sometimes used, although it is no longer a SI unit.) To put this into comparative terms a human hair is about 100 μm and a red blood cell is around 8 μm; thus a particle of PM_{10} is microscopic in size and not visible to the naked human eye.

13 On differential health effects of particulate matter see also https://assets.publishing.service.gov. uk/government/uploads/system/uploads/ attachment_data/file/411762/COMEAP_

The_evidence_for_differential_health_effects_of_particulate_matter_according_to_source_or_components.pdf

14 See https://www.gov.uk/government/statistics/air-quality-statistics/ntrogen-dioxide

15 See http://www.londonair.org.uk/

16 See https://www.gov.uk/government/statistics/emissions-of-air-pollutants/emissions-of-air-pollutants-in-the-uk-non-methane-volatile-organic-compounds-nmvocs

17 For more see https://assets.publishing.service.gov.uk/government/uploads/system/uploads/attachment_data/file/318234/hpa_Sulphur_dioxide_General_Information_v1.pdf

18 See http://air-quality.org.uk/04.php

19 See for example Samoli E, Touloumi G, Schwartz J, et al. (2007) Short-term effects of carbon monoxide on mortality: An analysis within the APHEA project. *Environmental Health Perspectives*, 115: 1578–1583.

20 2008/50/EC.

21 See https://uk-air.defra.gov.uk/networks/network-info?view=metals

22 As originally set out in the Air Quality Directive (2008/50/EC) and the Fourth Air Quality Daughter Directive (2004/107/EC).

23 See A Review of Arsenic in Ambient Air in the UK. https://uk-air.defra.gov.uk/assets/documents/reports/empire/arsenic00/arsenic.htm

24 See https://naei.beis.gov.uk/overview/pollutants?pollutant_id=12

25 UK Emissions Inventory Team, AEA. (2011) *UK Emissions of Air Pollutants 1970 to 2009*. https://uk-air.defra.gov.uk/assets/documents/reports/cat07/1401131501_NAEI_Annual_Report_2009.pdf

26 Emissions have declined by 63% since 1990. Further information can be seen at https://naei.beis.gov.uk/overview/pollutants?pollutant_id=16

27 The most recent report is the Air Pollutant Inventories for England, Scotland, Wales and Northern Ireland 2005–2019 and can be found at https://uk-air.defra.gov.uk/assets/documents/reports/cat09/2109270949_DA_Air_Pollutant_Inventories_2005-2019_Issue1.1.pdf

28 SI 2018 No 129.

29 Non-methane VOC (NMVOC) are a set of organic compounds that are typically photochemically reactive in the atmosphere.

30 For more see https://commonslibrary.parliament.uk/research-briefings/cbp-8195/

31 European Directive 2008/50/EC.

SECTION 2: LOCAL AIR QUALITY MANAGEMENT

Introduction

Air quality policy in the UK, Europe and elsewhere is driven mainly by health concerns as discussed in Section 1, and to a lesser extent by concerns with the environment. The current issues relating to air quality were first highlighted in the late 1980s. In this chapter, the main focus is ambient outdoor air quality, rather than indoor or occupational air quality.

It should be noted however that indoor air quality is also very important to public health, particularly in the developing world. Concern for air quality is reflected in the Sustainable Development Goals (SDGs discussed in Chapter 1), for example SDG Indicator 11.6.2 (annual mean levels of $PM_{2.5}$) and SDG Indicator 3.9.1 (mortality rate attributable to household and ambient air pollution). For example, in India, poor indoor air quality is thought to be responsible for up to a million deaths per year with the prime contributor being the burning of solid fuel for cooking.[32] The Institute for Health Metrics and Evaluation (IHME) estimated that 1.6 million people died prematurely as a result of indoor air pollution in 2017. This was 3% of global deaths. In low-income countries, it accounted for 6% of deaths [22].

Concerns are also being expressed about indoor air quality as the result of improvements on energy efficiency and reduced ventilation.[33]

Part IV of the Environment Act 1995 was introduced as a result of the health concerns from outdoor air quality (although it did not actually commence until December 1997). It imposed a duty on the Secretary of State to prepare and publish an air quality strategy for the UK setting out standards and objectives for air quality, as well as the measures to be taken by local authorities and others for the purpose of achieving those objectives. The main purpose of Local Air Quality Management was to

Table 16.3 Summary of the health effects of Local Air Quality Management pollutants

Pollutant	Health effects
Benzene	Benzene affects the bone marrow and can cause a decrease in red blood cells, leading to anaemia. It can cause excessive bleeding and affect the immune system. Benzene is carcinogenic to humans. No safe level of exposure can be recommended according to WHO (see https://www.who.int/ipcs/features/benzene.pdf). (Note: benzene has been detected at high levels in indoor air. Some of this exposure might be from building materials (paints, adhesives, etc.), but most is from cigarette smoke in both homes and public spaces.)
1,3-butadiene	Some studies have shown chronic exposure to be linked with cardiovascular disease and cancer but a true causal link is difficult to show because of confounding factors such as exposure to benzene and smoking. The International Agency for Research on Cancer (IARC) has classified 1,3-butadiene as a human carcinogen (see also https://webwiser.nlm.nih.gov/substance?substanceId=137&3-Butadiene&catId=83).
Carbon monoxide	It reduces the oxygen carrying capacity of red blood cells. Symptoms of long-term exposure include headache, flu-like illness, fatigue, nausea, difficulty concentrating and diarrhoea. Exposure above 9 ppm for longer than eight hours leads to COHb levels of above 2.5%; health effects might occur at such levels. (https://oem.bmj.com/content/59/10/708)
Lead	Can lead to elevated blood-lead levels and harm the kidneys, liver, nervous system and other organs. It may cause neurological impairments such as seizures, mental retardation and behavioural disorders. Even at low doses, lead is associated with damage to the nervous systems of foetuses and young children (who are more vulnerable).
PM_{10}	Causes lung disease and an increase in PM_{10} concentration increases daily hospital admissions and premature deaths from respiratory and cardiovascular diseases. Persons with pre-existing cardiovascular and respiratory diseases were found to be most susceptible.
Sulphur dioxide (SO_2)	Sulphur dioxide irritates the eyes and nose. Inhalation of sulphur dioxide causes narrowing of the airways (bronchoconstriction), which people suffering from asthma and chronic respiratory diseases are more sensitive to than other people.
$PM_{2.5}$	$PM_{2.5}$ is a stronger risk factor than the coarser part of PM_{10}. Exposure to $PM_{2.5}$ reduces the life expectancy of the population. It aggravates asthma and also increases mortality from cardiovascular and respiratory diseases and from lung cancer.
Ozone	At ground level, ozone is a highly reactive gas which can irritate the eyes and bring upper and lower respiratory symptoms to healthy people. It may also provoke asthmatic attacks in people having asthma. Ozone can also increase a person's susceptibility to respiratory infection and aggravate pre-existing respiratory illnesses.
Polycyclic aromatic hydrocarbons (PAHs)	PAHs are carcinogenic and may also have an effect on reproduction. (Note also high concentrations of particulate PAH compounds have been reported in indoor environments during the burning of fossil fuels and biofuel for cooking, generally in unvented stoves.)
Nitrogen dioxide (NO_2)	It irritates the mucosa of the eyes, nose, throat and the lower respiratory tract. Causes lung irritation (and reduces lung function) as well as lowering resistance to pneumonia and other respiratory infections. In the presence of sunlight, its precursor (nitrogen oxides (NO_x)) can react to produce a photochemical smog. May cause increased response to allergens in those with asthma and aggravates existing chronic respiratory diseases.

Note: Health effects here relate primarily to chronic exposure, not acute or occupational exposure to the pollutant.

complement policies related to the control of industrial pollution control and vehicle emission standards in seeking to improve air quality standards to acceptable levels. So local authorities have a central role in achieving improvements in air quality.

Through the Local Air Quality Management (LAQM) system local authorities are required to assess air quality in their area and designate Air Quality Management Areas (AQMA) if improvements are necessary. Where an AQMA is designated, local authorities are required to produce an air quality Action Plan describing the pollution reduction measures it will put in place [23].

Local authorities are required to submit an Annual Status Report (ASR) each year. The overall aim of this document should be to report on progress in achieving reductions in concentrations of emissions relating to relevant pollutants below air quality objective levels. It is also where local authorities identify new or changing sources of emissions.

Where there are two-tier local authorities (districts and counties), guidance sets out the responsibilities for each tier [23].

Clean air strategy

The UK Government's Clean Air Strategy was published in 2019 [14] for England. The Clean Air Wales Programme in 2015; the Scottish Government Cleaner Air for Scotland, The Road to a Healthier Future (CAFS);[34] and also the programme in Northern Ireland are discussed.

The Clean Air Strategy air quality objectives are intended as a statement of policy intentions set by government. They are designed to bring air quality as close as possible to the levels at which no significant health effects would be expected in the population as a whole.

In addition, for many pollutants there is a threshold effect (i.e., there can be a level of exposure that can be tolerated for which there is no identified ill effect). The threshold may be set to allow for the most sensitive and vulnerable individuals and normally include an uncertainty factor to allow for an extra margin of safety. In the UK, the terms used are standards and objectives.

Air quality standards are concentrations considered acceptable in terms of effects on health and the environment and are tracked over a specific time period. If the concentration of a pollutant is higher than the standard over this period of time, this is known as an exceedance. Usually, exceedances are reported by the number of days a standard is not met.

Air quality objectives are set as the date by which an air quality standard must not be exceeded by a specific number (e.g., for the 15-minute sulphur dioxide standard the objective level is based on the specified concentration being exceeded more than 35 times in a year).

The following pollutants have air quality standards in the UK:

- Nitrogen dioxide (NO_2)/NO_x (vegetation);*
- Particulate matter (PM_{10});*
- Fine particulate matter ($PM_{2.5}$);
- Ozone (O_3);

Table 16.4 UK key pollutants and target reductions

	2005 baseline (kt)	Reduction required by 2020	Reduction required by 2030	2020 ceiling (kt)	2030 ceiling (kt)
NO_x	1,714	55%	73%	771	463
SO_2	773	59%	88%	317	93
NMVOCs	1,042	32%	39%	709	636
$PM_{2.5}$	127	30%	46%	89	69
NH_3	288	8%	16%	265	242

Source: Defra [14]

- Sulphur dioxide (SO_2);★
- Benzene;★
- Lead (Pb);★
- Carbon monoxide (CO★);
- Benzo[a]pyrene (BaP), a PAH;
- Nickel (Ni);
- Cadmium (Cd);
- Arsenic (As);
- 1,3-butadiene.★

Those marked ★ have a particular relevance to AQMA, while ozone, fine particles ($PM_{2.5}$)

and PAHs are considered to be transboundary and are therefore a matter for national government.

The National Air Pollution Control Programme (NAPCP) is a UK-wide document [24]. It sets out measures and technical analysis which demonstrate how the legally binding 2020 and 2030 emission reduction commitments (ERCs) for five damaging pollutants (nitrogen oxides, ammonia, NMVOCs, PM and sulphur dioxide) can be met across the UK.

Table 16.5a UK objectives and EU Limit Values (for the purposes of Local Air Quality Management)[35]

National air quality objectives and European Directive limit and target values for the protection of human health							
Pollutant	Applies	Objective	Concentration measured as	Date to be achieved by (and maintained thereafter)	European obligations	Date to be achieved by (and maintained thereafter)
Particles (PM_{10})	UK	50 µg/m³ not to be exceeded more than 35 times a year	24 hour mean	31 December 2004	50 µg/m³ not to be exceeded more than 35 times a year	1 January 2005
	UK	40 µg/m³	annual mean	31 December 2004	40 µg/m³	1 January 2005
	Indicative 2010 objectives for PM_{10} (from the 2000 strategy and Addendum) have been replaced by an exposure reduction approach for $PM_{2.5}$ (except in Scotland – see next)						
	Scotland	50 µg/m³ not to be exceeded more than 7 times a year	24 hour mean	31 December 2010	50 µg/m³ not to be exceeded more than 35 times a year	1 January 2005
	Scotland	18 µg/m³	annual mean	31 December 2010	40 µg/m³	1 January 2005
Particles ($PM_{2.5}$) Exposure Reduction	UK (except Scotland)	25 µg/m³		2020	Target value – 25 µg/m³	2010
	Scotland	10 µg/m³	annual mean	31 December 2020	Limit value – 25 µg/m³	1 January 2015
	UK urban areas	Target of 15% reduction in concentrations at urban background		Between 2010 and 2020	Target of 20% reduction in concentrations at urban background	Between 2010 and 2020

(Note: Tables 16.5a,b, & c are in fact one table that has been split for this publication for convenience)
Source: Defra

Table 16.5b UK objectives and EU Limit Values (for the purposes of Local Air Quality Management)

National air quality objectives and European Directive limit and target values for the protection of human health						
Pollutant	Applies	Objective	Concentration measured as	Date to be achieved by (and maintained thereafter)	European obligations	Date to be achieved by (and maintained thereafter)
Nitrogen dioxide	UK	200 µg/m³ not to be exceeded more than 18 times a year	1 hour mean	31 December 2005	200 µg/m³ not to be exceeded more than 18 times a year	1 January 2010
	UK	40 µg/m³	annual mean	31 December 2005	40 µg/m³	1 January 2010
Ozone	UK	100 µg/m³ not to be exceeded more than 10 times a year	8 hour mean	31 December 2005	Target of 120 µg/m³ not to be exceeded by more than 25 times a year averaged over 3 years	31 December 2010
Sulphur dioxide	UK	266 µq/m³ not to be exceeded more than 35 times a year	15 minute mean	31 December 2005	-	-
	UK	350 µg/m³ not to be exceeded more than 24 times a year	1 hour mean	31 December 2004	350 µg/m³ not to be exceeded more than 24 times a year	1 January 2005
	UK	125 µg/m³ not to be exceeded more than 3 times a year	24 hour mean	31 December 2004	125 µg/m³ not to be exceeded more than 3 times a year	1 January 2005
Polycyclic aromatic hydrocarbons	UK	0.25 ng/m³ B[a]P	as annual average	31 December 2012	1.0 ng/m³	31 December 2012

Local Air Quality Management in practice

EHPs have traditionally intervened to control emissions from certain industries and fixed sources, even if until relatively recently the tools have been inadequate to secure reduction of emissions other than smoke. As is clear, less visible pollutants are also an issue, and local air quality is affected by emissions from transport. As a result, the Environment Act 1995 places a duty on local authorities to review the quality of air in their areas, and to assess whether the objectives set out in the AQS are likely to be met.

In Scotland and Wales, powers under Part IV of the Act were devolved to the Welsh Assembly Government and the Scottish Government respectively and in Northern Ireland; separate provisions comparable to Part IV were also made.

Table 16.5c UK objectives and EU Limit Values (for the purposes of Local Air Quality Management)

National air quality objectives and European Directive limit and target values for the protection of human health						
Pollutant	Applies	Objective	Concentration measured as	Date to be achieved by (and maintained thereafter)	European obligations	Date to be achieved by (and maintained thereafter)
Benzene	UK	16.25 µg/m³	running annual mean	31 December 2003	-	-
	England and Wales	5 µg/m³	annual average	31 December 2010	5 µg/m³	1 January 2010
	Scotland, Northern Ireland	3.25 µg/m³	running annual mean	31 December 2010	-	-
1,3-butadiene	UK	2.25 µg/m³	running annual mean	31 December 2003	-	-
Carbon monoxide	UK	10 mg/m³	maximum daily running 8 hour mean/ in Scotland as running 8 hour mean	31 December 2003	10 mg/m³	1 January 2005
Lead	UK	0.5 µg/m³	annual mean	31 December 2004	0.5 µg/m³	1 January 2005
		0.25 µg/m³	annual mean	31 December 2008	-	-

Local authorities also have responsibility for other local government air quality roles (under Clean Air and Environmental Permitting legislation discussed in the following sections of this chapter) as well as for land-use planning functions such as development management and the preparation of Local Plans and other planning policy. County authorities have responsibility for local transport planning and Part IV makes express provision for the contribution of county councils to LAQM in areas where there are two tiers of authority.

In London, there are other special provisions under the Greater London Authority Act 1999. These require the Mayor to produce a London air quality strategy setting out proposals and policies for implementation of the national air quality strategy and for the achievement of air quality standards and objectives in London.

The Mayor is also responsible for many strategic matters in London, including spatial and economic development, plus transport. The London boroughs however have the responsibility for review and assessment and action planning, but are required to work with the Mayor, and vice versa.

The Secretary of State and Devolved Administrations have issued extensive guidance, setting out what is expected of local authorities in fulfilling their duties under Part IV.[36] These include both policy and technical guidance.

Local Air Quality Management – policy guidance 16

The latest Local Air Quality Management guidance for England was published in 2016 [23] and provides details on how local authorities must assess air quality in their area. If improvements are needed, this guidance also provides information on how to declare an Air Quality Management Area.

Local authorities are required to submit an Annual Status Report (ASR) each year. It should be noted that the objectives for pollutants: benzene, 1,3-butadiene, lead and carbon monoxide have been met for several years and are well below limit values. Government accepts that, in the absence of any particular concerns in a local area, national monitoring is currently providing a sufficient basis for the review of these four pollutants under LAQM. On this basis, local authorities are not expected to report annually on these pollutants in their ASRs.

The single ASR replaces the previous three-yearly cycle of reporting, removing the need to compile Updating and Screening Assessments (USAs), Detailed Assessments, Further Assessments Progress Reports and Action Plan Progress reports. Key elements of the ASR are to:

- Report progress on the implementation of measures in the local air quality Action Plan and other measures and their impact in reducing concentrations below air quality objectives;
- Provide a summary of monitoring/modelling data (whether locally retrieved and/or from the national network) to assess the air quality situation in the area and likelihood of air quality breaches, and to provide the necessary evidence base for the impact of air quality measures;
- Report on significant new developments that might affect local air quality; and
- Present information in an executive summary for the public and lay reader so that the local public can more easily engage with local air quality issues and measures taken to improve it.

Table 16.6 Air Quality Objectives contained in the Air Quality (England) Regulations 2000 as amended (reflecting the objectives in Tables 16a,b and c)

Pollutant	Objective	Average period
Nitrogen dioxide – NO_2	200 µg m^{-3} not to be exceeded more than 18 times/year	1 hour mean
	40 µg m^{-3}	Annual mean
Particles PM_{10}	50 µg m^{-3} not to be exceeded more than 35 times/year	24 hour mean
	40 µg m^{-3}	Annual mean
Sulphur dioxide SO_2	266 µg m^{-3} not to be exceeded more than 35 times/year	15 minute mean
	350 µg m^{-3} not to be exceeded more than 24 times/year	1 hour mean
	125 µg m^{-3} not to be exceeded more than 3 times/year	24 hour mean
Benzene	16.25 µg m^{-3}	Running annual mean
	5 µg m^{-3}	Annual mean
1,3-butadiene	2.25 µg m^{-3}	Running annual mean
Carbon monoxide	10 µg m^{-3}	Maximum daily running 8 hour mean
	10 µg m^{-3}	Running 8 hour mean
Lead	0.5 µg m^{-3}	Annual mean
	0.25 µg m^{-3}	Annual mean

If a local authority finds any places where the objectives are not likely to be achieved, it must declare an Air Quality Management Area there. This area could be just one or two streets, or it could be much bigger [23]. Then the local authority will put together a plan to improve the air quality – a Local Air Quality Action Plan ideally within 12 months of the date of designation. AQMAs are only designated in areas where the objectives are not achieved (e.g., from the results of monitoring in new areas). The Action Plan will set out what measures the local authority proposes to take in pursuit of the achievement of air quality standards and objectives.

It is recommended that all local authorities, particularly those that have not had to designate AQMAs or do not expect to designate an AQMA in the future, but who have areas at risk of exceedance, should consider drawing up an Air Quality Strategy.

When determining the boundary of an AQMA the authority should make an appropriate judgement based on the extent of predicted areas of exceedance, the locations of relevant receptors, the nature and location

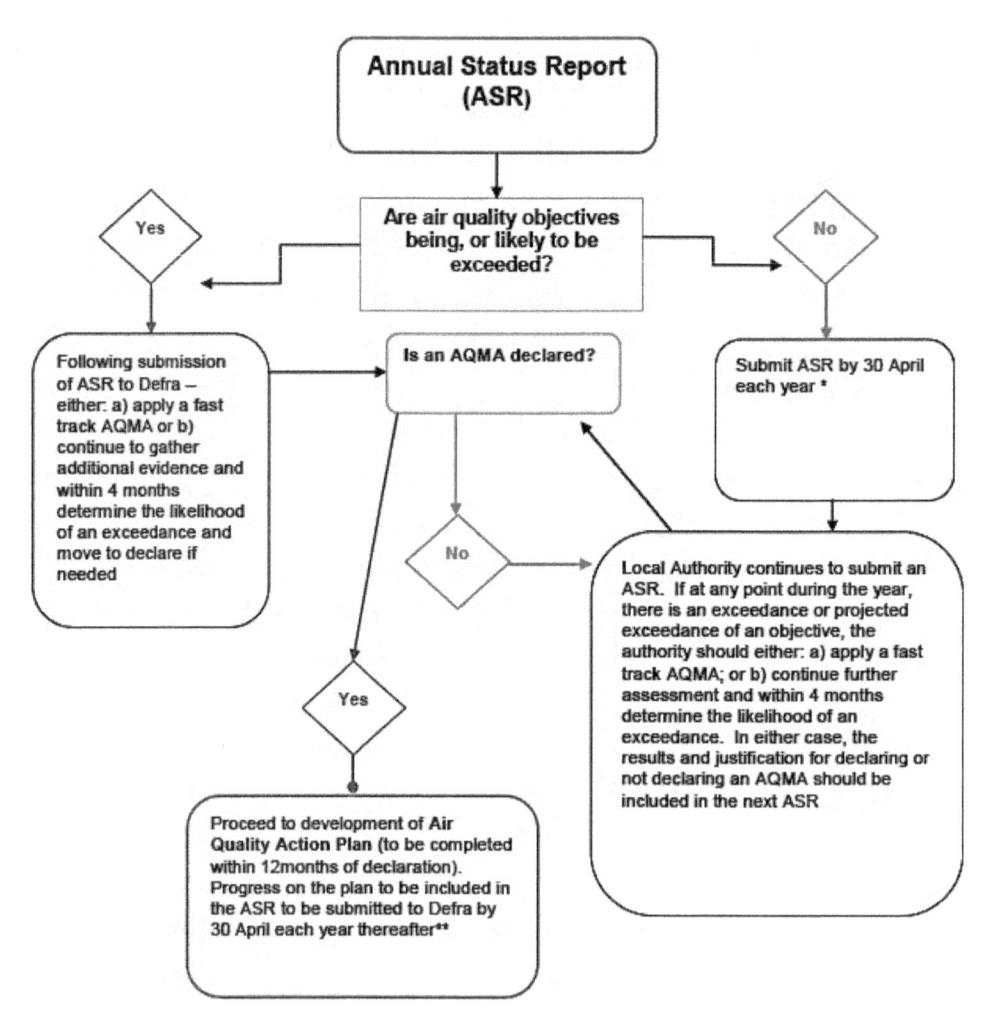

Figure 16.1 The review and assessment (reporting) process

Source: [23] PG16 (https://laqm.defra.gov.uk/documents/LAQM-PG16-April-16-v1.pdf)

of relevant sources, and other local factors. These areas can cover single streets or road networks, a junction, roundabout or even a single dwelling. In many urban and built-up areas, especially where transboundary pollution is an issue, the authority may decide to designate the entire borough as an AQMA. This kind of declaration provides greater flexibility for the local authority to respond to pollution issues as and when they arise and does not prevent officers from then focussing on key areas within an AQMA for taking action.

Alternatively, applying smaller, individual AQMAs for specific hot spot locations may provide a clearer focus for Action Plan measures and allow for strategic targeting based on the best use of resources. Local authorities should not expend significant resources narrowing down the parameters of an AQMA if this is detrimental to the identification of measures and action to improve air quality.

Policy guidance 16 states that Action Plans should include the following:

- Quantification of the contributions from different sources whether fixed and more specifically mobile sources (e.g., HGVs, buses, taxis private vehicles) responsible for the exceedance of the relevant objective. Knowing the source of the problem will allow the Action Plan measures to be targeted.
- Quantification of the impacts of the proposed measures – including, where feasible, data on emissions and concentrations (either locally obtained and/or via national monitoring/modelling statistics). The local authority must show how it intends to monitor and evaluate the effectiveness of the plan.
- Clear timescales should be set, including milestones and expected outcomes which the authority and other delivery partners propose to implement the measures within the Action Plan.
- How the local authority, including transport, planning and public and

environmental health departments, and other external partners such as Highways England, local businesses and Regional Mayors will accept the problem and how they will work together to implement the Action Plan.

The Public Health Outcomes Framework (PHOF) contains an air quality indicator for fine particulate matter ($PM_{2.5}$) which is referred to in PG16. As mentioned in Chapter 1, the PHOF is intended to focus public health action on increasing healthy life expectancy and reducing differences in life expectancy between communities. This is intended to support reductions in health inequality. Public and environmental health professionals and air quality specialists, along with transport and planning officers, should be aware of the public health indictor for air pollution in their local authority area. The public health indicator for PM 2.5 may also be used to raise awareness of the effect of this pollutant on public health.

Local authority assessments of air quality should focus on those locations where members of the public are likely to be present and exposed for a period of time appropriate to the averaging period of the objective only. So, for example, in a busy shopping street one individual does not have to be present more than 18 hours during the year for the one-hour nitrogen dioxide to be relevant. For more information see Local Air Quality Management Technical Guidance (TG16).[37]

Having assessed that pollution levels exceed the objective and that there is relevant public exposure the local authority is required to designate an AQMA. In coming to this conclusion, they should also demonstrate that the uncertainties in all the monitoring data and modelling assumptions have been fully considered. The verification process, particularly if high-quality continuous monitors are used, will aid confidence in decision-making. A final requirement for the assessment is to estimate of the size of

Table 16.7 The number of AQMAs by pollutant in the UK

Pollutant	England	Wales	Scotland	N. Ireland	London
Benzene					
Nitrogen dioxide NO_2	510	43	29	17	33
Particulate matter PM_{10}	37	1	21	3	28
Sulphur dioxide SO_2	6		1		

population exposed to pollutant concentrations above the objective.

Note Table 16.7 does not provide a total count for the number of AQMAs declared. It is a count of the number of AQMAs declared for each pollutant. For example where an AQMA has been declared for both NO_2 and PM_{10} it will have two entries in the table.

Other actions to improve air quality

There are other powers that can be used by departments across local authorities (i.e., not necessarily just EHPs although controlling industrial emissions is crucial to improving local air quality) to try to seek improvements in air quality. These include:

1   Local Transport Authorities have to produce Local Transport Plans.[38] These should not only relate to traffic management such as improving traffic flows and reducing traffic levels, but promotion of alternative forms of transport and movement (safe walking and bicycle use) can also be used to re-allocate space to introduce bus or cycle lanes; create high occupancy lanes; restrict access to a road or area to some or all vehicles at different times of the day (e.g., to create a pedestrianised area); promote cleaner vehicle technology.

2   The Road Traffic (Vehicle Emissions) (Fixed Penalty) (England) Regulations 2002 which enable local authorities with AQMAs to apply to the Department for Transport for the power to conduct roadside vehicle emissions testing (this power has been used by a few local authorities, more to show that they are taking some action. In practice its effectiveness for air quality management is limited and resource intensive).

3   The Highways (Traffic Calming) Regulations 1999 and the Highways (Road Humps) Regulations 1999 allow local authorities to introduce a wide range of physical measures to control traffic at low speeds, although careful design consideration is needed to reduce changes in acceleration and deceleration.

4   Parking controls are useful measures, and the Road Traffic Regulation Act 1997 permits local authorities to determine where motorists can park and how much it will cost them. Authorities can also use the planning process to regulate the amount of private non-residential parking associated with a new development. The use of parking control enforcement has been shown to improve flow in some areas.

5   Other initiatives include park and ride schemes, home zones, clear zones and travel plans (either local, school or workplace based). As with many initiatives the local authority should always be seen as taking the lead and use of travel plans can often be a good example of best practice.

6   The planning and air quality functions of local authorities should be carried out in close liaison. The National Planning Policy Framework[39] sets out national planning policies and

principles for England. The most recent update suggests planning officers take account of air quality and air quality management areas when considering new developments. The NPPF also considers sustainable transport modes such as cycling and walking, which may also benefit improvements in air quality.

Low Emission Zones and Clean Air Zones

Low Emissions Zones, referred to as LEZs (currently in London and Glasgow), and Clean Air Zones (CAZ) (currently in Bath and Birmingham) are areas that councils have defined to limit certain types of vehicles driving through them in a bid to cut harmful pollutants from the air. Though best known for being deployed in London there is also an Ultra Low Emissions Zone (ULEZ). An LEZ is a geographically defined area where the most polluting of vehicles are restricted, deterred or discouraged. The aim is improved air quality by reducing the number of more polluting vehicles in the area, based on particular emission standards (such as Euro vehicle emission standards). In many cases the intention is to bring the use of lower polluting vehicles forward in time. The most significant scheme in the UK is the London LEZ scheme first introduced in 2008, although other smaller examples exist in Europe.

The London Ultra Low Emissions Zone (ULEZ) was introduced in an effort to improve the quality of the air in central London by cutting the number of older, higher-polluting cars taking journeys through the capital. Certain low-emitting cars – predominantly electric vehicles – are exempt from paying any fees; though if the car is not ULEZ compliant, a charge will be levied for entering the zone.

Glasgow city centre was the first area in Scotland to have a LEZ and came into effect on 31 December 2018. Two phases form the plans, with the first having covered local buses. The second will see all vehicles needing to meet certain emissions standards to be able to enter the zone, and will be introduced in 2023.

Clean Air Zones are geographically defined areas allowing action and resources to be targeted to deliver the greatest health benefits. Local authorities can adopt Clean Air Zones as a way of focussing actions to improve air quality. There are different classes of Clean Air Zones. Each successive class includes more vehicle types to bring about a larger reduction in emissions. Vehicle owners will be required to pay a charge if they enter a zone, and their vehicle does not meet the required emission standard.

Bath launched its Clean Air Zone in March 2021. Private cars and motorbikes were not affected initially, but business-use vehicles like taxis and HGVs will have to pay a fee if they do not meet certain emissions standards.

Birmingham's Clean Air Zone covers all roads within the Middleway Ring Road but not the Middleway itself. Unlike Bath, this will cover private vehicles as well and non-compliant cars will be subject to a daily charge for entering the zone and will operate 24 hours a day, 365 days a year. It had been planned to be introduced in 2020, though the Covid-19 crisis set that back.

The Clean Air Zones are areas where only the cleanest vehicles are encouraged to operate to improve air quality. There is guidance issued for England [25] which says, *poor air quality is the largest environmental risk to public health in the UK and investing in cleaner air and doing even more to tackle air pollution are priorities for the UK government.* The devolved administrations have issued their own guidance.

Local authorities can also implement a Clean Air Zone that operates on a voluntary basis (i.e., without charging) to raise public awareness and act as a focus for targeting additional action to improve air quality. Guidance was issued in 2020.[40]

Plate 16.1   Electric and hybrid car re-charging points

Plate 16.2   On-street secure bike store to encourage more cycling

Section 2 Notes

32 See also Ritchie H, Roser M. (2013) Indoor air pollution. *OurWorldInData.org*. https://ourworldindata.org/indoor-air-pollution [Online Resource].

33 For example see Howieson S. (2014) Are our homes making us ill? *Perspectives in Public Health*, 134(6): 318–319.

34 See https://www.gov.scot/publications/cleaner-air-scotland-road-healthier-future/

35 See https://uk-air.defra.gov.uk/air-pollution/uk-eu-limits

36 See for example Local Air Quality Management Policy Guidance (PG09) of 2009. https://assets.publishing.service.gov.uk/government/uploads/system/uploads/attachment_data/file/69348/pb13566-laqm-policy-guidance-part4-090302.

pdf; and Local Air Quality Management – Technical Guidance LAQM.TG(09). https://assets.publishing.service.gov.uk/government/uploads/system/uploads/attachment_data/file/69334/pb13081-tech-guidance-laqm-tg-09–090218.pdf

37 https://laqm.defra.gov.uk/documents/LAQM-TG16-April-21-v1.pdf

38 See https://greenertransportsolutions.com/directory/local-transport-plans-guidance-england-wales/

39 http://planningguidance.planningportal.gov.uk/blog/policy/

40 See at https://assets.publishing.service.gov.uk/government/uploads/system/uploads/attachment_data/file/863730/clean-air-zone-framework-feb2020.pdf from the Department for Transport.

SECTION 3: MEASUREMENT OF AMBIENT AIR POLLUTANT

The need for air quality measurements

Measurement is frequently concerned with providing data for the purposes of monitoring pollutant concentrations against the predetermined standards and objectives. The results can then inform policy decisions at local, regional and national government levels. The Environment Agency produced a Technical Guidance note on monitoring ambient air (M8).[41]

Monitoring provides raw measurements of air pollutant concentrations, which can then be analysed and interpreted. Analysis of continuous measurements allows an assessment air pollution from day to day, and which areas are worst affected (or improving). Measuring different pollutants in the air allows an assessment to be made of how pollutants interact with each other and how they relate to traffic levels or industrial activity. Analysis of the relationship between meteorology and air quality permits prediction of how weather conditions will give rise to episodes with high levels of pollution and this in turn means public health warnings can be given.[42]

Another important use is to validate computer models and their outputs. These models can test different scenarios, including areas where there are no measurements. Where there is good agreement comparing predicted results at a location with measured information for the same location provides greater confidence in the prediction method. The guidance for review and assessment requires this checking and also the verification (calibrating) of the model output against measured data.

Increased awareness in air quality issues has also led to public demand for better and more accessible information.

There are many considerations that need to be taken into account before undertaking any measurement. High amongst these is the requirement to provide measurements that are accurate and precise. Accuracy tells us how close the measurements of the variable are to the true value of the variable and precision tells us how precisely the measurements are made.

In 1992, the then Department of Environment established an Enhanced Urban Network (EUN), and in 1995, all statutory and other urban monitoring was consolidated into one comprehensive programme. During the following five years, over 50 local authority sites were integrated into the resulting network, including 14 of the London Air Quality Monitoring Network sites.[43] In 1998, the previously separate UK urban and rural automatic networks were combined to form the current Automatic Urban and Rural Network (AURN). The AURN is the most important and comprehensive automatic national monitoring network in the country, which is made up of 127 sites, across the UK. Local authorities provide some AURN stations with government support, whereas most other local authorities focus only on those pollutants of most local concern.

Methods for monitoring air quality

There are two types of method for the measurement of air pollution generally in use; one can be termed automatic and the other non-automatic.

Non-automatic monitoring methods – These are generally cheaper and easier to use and therefore are not as robust and reliable as automatic methods. The most commonly used non-automatic method involves the use of passive sampling using Palmes type diffusion tubes. These are widely used in the UK for indicative measurement of ambient concentrations of nitrogen dioxide (NO_2) in the context of Local Air Quality Management because the NO_2 objectives include an annual mean concentration and NO_2 pollution is widespread. However, no UK or international standard method currently exists for diffusion tube preparation or analysis. They do provide a simple and inexpensive method of screening air quality across an area, to give a general indication of average pollution concentrations over long periods. A typical exposure for one tube is two to four weeks. As such their accuracy and precision can be expected to be lower. This is followed by manual handling in the field and subsequent analysis in a laboratory.[44] Extensive guidance has been produced for local authorities, including the LAQM Technical Guidance (see Section 2 Notes 5 and 6).

The tubes need to be sited on a bracket at least 50 mm from the nearest surface. A cap is removed from one end and at the closed end there is an absorbent, which for NO_2 is triethanolamine (TEA) in a preparation (of water of acetic acid). The tubes following exposure of two to four weeks are sent to an analysing laboratory. Hence, they can only ever be used to produce longer-term averages and any study should be at least six months long.

There is further guidance to help overcome some of the problems arising from individual laboratories preparing the diffusion tubes, thus creating additional uncertainties [26].

Similar types of diffusion tubes can also be used for SO_2 and O_3. Other absorbents are used for these gases. But, as diffusion tubes are best for long-term averages, their use is very limited and not recommended for LAQM purposes.

Other non-automatic methods include the use of samplers that actively pump a known volume of air through a filter, absorbent or reagent for period of time. The collected sample (filter or reagent) is then sent for subsequent laboratory analysis. The most common uses of these for LAQM are samplers with size selective filters that are used for gravimetric purposes, typically PM_{10}. Other samplers can be used for lead and PAHs. Pumped samplers are used for benzene. The traditional eight-port sulphur dioxide/smoke bubblers used after the 1950s smogs for several decades

were another example of a non-automatic method.

Other sampling devices can also be used for the assessment of dust nuisance based on deposition of dust (i.e., particulate matter which is not size selected).[45] All of these sampling devices are designed to measure dust deposition, usually quoted as $mg/m^2/day$. There are no UK or European standards for assessing dust; however a mass deposition rate of 200 $mg/m^2/day$ has been widely used in environmental assessments as a "custom and practice" limit, in the absence of any other criteria.

Automatic or continuous analysers – These provide reliable and high-resolution measurements for gaseous pollutants such as ozone, oxides of nitrogen, sulphur dioxide, carbon monoxide and particles. (Hydrocarbons can also be automatically measured, but the costs are even higher.)

A sample of pollutant is drawn into the analysers by a pump and the analysis is undertaken in real time. The resultant measurement data are then stored within the analyser (or on a separate logger); the data may then be downloaded remotely by modem. This enables for rapid dissemination to the public in many different ways while it is still relevant. Data from automatic analysers can be used to form the daily, weekly and monthly reports, as well as annual reports for the purposes of comparison with air quality objectives and EU Limit Values.

Analysers used for measuring nitric oxide (NO), the main component of oxides of nitrogen, use chemiluminescence to determine concentrations. For these analysers ozone is generated within the analyser and reacted with NO in ambient air directly within a sample cell to form electronically excited NO_2, which in turn creates a photon emission that is measured using a photocell. Another sample measures the sum of NO and NO_2, where the ambient air is passed a catalyst to convert the NO_2 to NO. The difference of the two values is then reported as the NO_2 concentration.

For sulphur dioxide (SO_2), a sample of ambient air is irradiated with UV light to stimulate fluorescence at the specific wavelength for SO_2. This is then measured in a photocell. In an ozone analyser, the sample of air passes along a tube that has UV light shone through it. The attenuation of UV at the far end of the tube depends on the ozone concentration and this is compared to that of air, from which ozone has been removed (scrubbed) to determine the concentration. The determination of carbon monoxide (CO) relies on the amount of infrared absorption which is attenuated at a sample passes through the sample tube. This method can also used for many other gases based on the different absorption bands.

The determination of concentrations of particles in real time is more difficult in part because concentrations relate directly to mass and not to volumetric terms. A non-automatic method to determine concentration uses the collection of a physical sample on a filter, which has been extracted from a volume of air, followed by a subsequent gravimetric determination using a balance. This is the basis of the determination required by the EU reference method for PM_{10}. This cannot be undertaken in real time, as the determination of mass requires the sampling of greater quantities of air to obtain a physical sample capable of being weighed.

There is a need to understand precisely the individual source of particles for the purposes of controlling emissions and the mechanism as to why particles are harmful to health and why there is a need to measure particles continuously.[46] There is also the need to relate particle concentrations in comparable and meaningful ways, whilst recognising that particles are non-uniform and heterogeneous, compared to gaseous species, which are homogeneous. In 2012 the Air Quality Expert Group concluded that measurements of $PM_{2.5}$ in the UK depended heavily on the FDMS instrument, which has a relatively short track record in monitoring networks. AQEG recommended that issues including long-term reliability, dryer performance and the handling of semi-volatile components were further investigated.

The measurement of $PM_{2.5}$ remained a challenge, with current measurements falling below the requirements of the Air Quality Directive [27]. In November 2020, AQEG launched a call for evidence on the modelling of future $PM_{2.5}$ concentrations in order to inform their advice to Defra on the development of AQ targets required by the Environment Bill in Parliament at the time of writing.

For gaseous analysers, the Type Approval testing process is managed in the UK by the Environment Agency under its MCERTS scheme with certification provided by SIRA, the appointed certification body. The MCERTS Performance Standards mirror the requirements of the CEN Standard Methods and decisions are made by a certification committee.

Defra has worked with the Environment Agency to develop a new and additional level of certification through MCERTS. This sets out the requirements for instruments to demonstrate equivalence in a representative particulate matter pollution climate for the UK and therefore receive approval from Defra for use in the UK. This certification is called "MCERTS for UK Particulate Matter" and its requirements are set out in an Annex to the Continuous Ambient Measurement Systems (CAMS) performance standard; Defra approved equipment is available on the website.[47]

The techniques used for monitoring within the UK's national compliance monitoring network, the Automatic Urban and Rural Network (AURN), are summarised in Table 16.8. Apart from the automatic PM_{10} analysers, the reference methods of measurement are defined in the relevant EU Directives.

An important and useful document is the 2015 AURN local site operator's manual[48] that deals with selection and maintenance including calibration of equipment. The calibration of the instruments is an important first step. For analysers of gaseous pollutants this will be on an automatic daily basis. In addition, specific calibration should be undertaken at least once every two weeks. Data validation includes checking the calibrated data for spurious and unusual results to identify potential

Table 16.8 AURN measurement techniques

O_3	UV absorption
NO/NO_2	Chemiluminescence
SO_2	UV fluorescence
CO	IR Absorption
PM_{10} and $PM_{2.5}$	• Tapered Element Oscillating Microbalance • Beta attenuation monitor • Gravimetric monitor • Filter Dynamics Measurement System (FDMS) • Optical light scattering • Fine Dust Analysis System (FIDAS)

Table 16.9 Air quality monitoring site types and descriptions

Site type	Description
Urban centre	An urban location representative of typical population exposure in town or city centres, such as shopping areas.
Urban background	An urban location away from specific sources, broadly representative of background conditions (e.g., urban residential areas).
Suburban	A location type situated in a residential area on the outskirts of a town or city.
Roadside	A site sampling typically within 1 to 5 metres of the kerb of a busy road (although distance can be up to 15 m from the kerb in some cases).
Kerbside	A site sampling within 1 metre of the kerb of a busy road.
Industrial	An area where industrial sources make an important contribution to the total pollution burden.
Rural	An open countryside location, in an area of low population density distanced as far as possible from roads, populated and industrial areas.
Other	Any special source-orientated or location category covering monitoring undertaken in relation to specific emission sources such as power stations, car-parks, airports or tunnels.

faults or episodes of high pollution. These can be flagged for further investigation.

The final checking process is known as ratification. For this all measurements are retrospectively recalculated using information on longer instrument performance and calibration histories.

Measurement sites

Determining where to undertake measurements or locate monitoring sites is not straightforward. It is always important to identify the reason for monitoring, which may include the following: to confirm compliance with air quality objectives (or EU Limit Values); to identify areas of high pollution or hotspots; to establish background concentrations (i.e., areas away from local or specific sources of emission); to check in rural areas (the ecosystem guidelines); to check for the grounding of plumes of pollution from large industrial chimneys; to establish the exposure of a population to assess health impact; to set background sites for the exposure reduction targets; to create a network in an area lacking monitoring; or to validate modelled predictions.

Plate 16.3 Automatic monitoring site in a suburban location

Once the purpose for monitoring is con-firmed, the specific site can be selected. This may involve some practical challenges. To overcome these, pre-planning is required and issues to consider include: the presence of utilities such as electricity at a poten-tial site; whether a hard standing is required then checking for subterranean services such as drains is necessary. Furthermore, if a new building is proposed, sight lines for junctions will need to be maintained; confirm access for deliveries, especially gas cylinders and ensuring permissions are obtained. It is also important to ensure any local sources of air pollution are avoided, for example flues from domestic boilers.

Sites can be classified according to the descriptions in Table 16.9 as in the LAQM Technical Guidance of 2021.[49]

Section 3 Notes

41 See https://assets.publishing.service.gov.uk/ government/uploads/system/uploads/attachment_ data/file/301188/TGN_M8_Monitoring_Ambient_ Air.pdf

42 See https://uk-air.defra.gov.uk/forecasting/ for example.

43 See https://londonair.org.uk/london/asp/pub-licstats.asp?region=0&site=&la_id=&network= All&postcode=&MapType=Google&VenueC ode=The London Air Quality Network (LAQN) is run by the of Imperial College London (https:// www.imperial.ac.uk/school-public-health/ environmental-research-group/).

44 See Diffusion Tubes for Ambient NO_2 Moni-toring: Practical Guidance for Laboratories and Users, report to Defra from AEA Group, February 2008. https://laqm.defra.gov.uk/ documents/0802141004_NO2_WG_Practical-Guidance_Issue1a.pdf

45 It is suggested that although many have been developed (British Standard Deposit Gauge; ISO Deposit Gauge; CERL Deposit Gauge) to undertake long-term assessment and identify long-term trends of deposited matter they are not of great value as collectors and also will col-lect rain and snow which undoubtedly impedes results. BS 1747–1:1969 *Methods for the meas-urement of air pollution. Deposit Gauges* has been withdrawn but BS 1747–5 1972 *Methods for the measurement of air pollution. Directional dust gauges* is still current.

46 See https://www.gov.uk/government/statistics/ air-quality-statistics/concentrations-of-particulate-matter-pm10-and-pm25

47 See https://uk-air.defra.gov.uk/networks/ monitoring-methods?view=mcerts-scheme

48 https://uk-air.defra.gov.uk/assets/documents/reports/empire/lsoman/LSO_manual_Oct_2015_Issue_1_Part_A.pdf

49 https://laqm.defra.gov.uk/documents/LAQM-TG16-April-21-v1.pdf

SECTION 4: INDUSTRIAL EMISSIONS

Apart from vehicle emissions, local air quality is affected by industrial emissions from fixed sources. Part One of the Environmental Protection Act 1990 (EPA 1990) established the main system for minimising air pollution from industrial sources in the UK, known as Local Air Pollution Control. This legislation meant that for the first time, local authorities (and EHPs) could regulate emissions to air (other than visible pollutants) in advance, rather than rely on statutory nuisance provisions[50] after the problem had already arisen. For the issue to count as a statutory nuisance, it must do one of the following: unreasonably and substantially interfere with the use or enjoyment of a home or other premises, injure health or be likely to injure health. The Environmental Protection Act 1990 introduced two lists of industrial activities (Part A and Part B) which require a permit to operate. Those processes under Part A cover emissions to air, land and water and are the responsibility of the Environment Agency, whereas Part B only covers emissions to air and are controlled by the local authority. Through this system, operators are authorised through a series of conditions to reduce emissions that may be harmful to human health or the environment. The use of Best Available Techniques Not Entailing Excessive Costs (BATNEEC) is encouraged.

Local authorities regulate Part A(2) installations and Part A(2) mobile plants (including any waste operations, water discharge activities or groundwater activities carried on as part of the installation or mobile plant). This is known as Local Authority Integrated Pollution Prevention and Control (LA-IPPC). Local authorities also regulate Part B installations and Part B mobile plant (LAPPC).

Although regulated by local authorities, the Environment Agency can set the minimum standard for releases to water for a permit applicable to a Part A(2) installation. For any facilities where both a local authority and the Environment Agency are regulators in relation to any one site, they should work together in the permitting process. Powers exist for a direction to be issued by government to change the regulator. This direction can be for a specific regulated facility or for a particular class of regulated facility. This power is used to help make regulation simpler and it has been used in specific cases usually for the Environment Agency to take the regulatory role.

The Pollution Prevention and Control Act 1999 (PPCA 1999) then repealed this section of the 1990 Act to apply a new scheme, Integrated Pollution Prevention and Control. The Pollution Prevention and Control (England and Wales) Regulations 2000 introduced new Part A1 and A2 categories. Part A1 processes remained under the control and Part A2 became the responsibility of local authorities, along with those under Part B.

The regime is now called environmental permitting which requires operators to apply for a permit to operate. The Environmental Permitting (England and Wales) Regulations (SI 2016 No 1154) at Schedule 1 sets out which processes are in Part A(1) and (2) and Part B. The full list of regulated activities is found in Schedule 1 to the Regulations, and any directly associated activities are carried on. An installation is made up of any stationary technical unit where one or more activities listed are carried on. The regulations also apply to mobile plant that is used to carry on either one of the Schedule 1 activities or a waste operation. The Regulations require that to be mobile, a plant (other than a Part A mobile plant) must be designed to move or be moved (by road, rail or water). A mobile plant is also defined as Part A mobile plant, Part B mobile plant or waste mobile plant. In Scotland, the Scottish Environment Protection Agency (SEPA) regulates all IPPC installations. In Northern Ireland, the Environment and Heritage Service regulate around IPPC installations.

In England and Wales Part B processes include smaller foundries, timber activities, crematoria, many solvent-using processes, car refinishing establishments and service stations (in all about 80 different types of installation). These are known as Part C installations in Northern Ireland.

Waste operations are regulated normally by the Environment Agency and SEPA in Scotland. Some waste operations are carried on at Part B installations (see Chapter 18 for more). Also regulated are solvent emission activities as listed in Annex VII of the IED.

A process may be designated as Part A1, Part A2 or Part B, depending upon the activities taking place at the site. To decide which permit is required, Schedule 1 of the Regulations needs to be consulted. For example, under section 1.1 of Schedule 1 combustion activities the type of permit depends on the size of appliances used to carry out the activities. An A1 permit would be required if the process covers the "*Burning any fuel in an appliance with a rated thermal input of 50 or more megawatts*", including "*where two or more appliances with an aggregate rated thermal input of 50 or more megawatts are operated on the same site by the same operator*". Alternatively, a Part B permit would be needed if the process is burning any fuel in a boiler, furnace, gas turbine or compression ignition engine, with a net rated thermal input of 20 or more megawatts, but a rated thermal input of less than 50 megawatts plus burning any waste oil in an appliance with a rated thermal input of less than 3 megawatts.

Similarly, under section 2.1 of Schedule 1 (Ferrous metals) Part A(1)(b),

> Producing, melting or refining iron or steel or any ferrous alloy, including continuous casting, except where the only furnaces used are; electric arc furnaces with a designed holding capacity of less than 7 tonnes, or cupola, crucible, reverberatory, rotary, induction, vacuum, electro-slag or resistance furnaces.

Under Part A2 (local authority IPPC) the Schedule says, "*unless falling within Part A(1)(b) of this Section, producing pig iron or steel, including continuous casting, in a plant with a production capacity of more than 2.5 tonnes per hour*". So, which particular permitting regime (and regulator) depends upon the capacity of the plant being used by the operator.

The LAPPC permitting regime of local authority-controlled facilities is now risk based so that fees and charges are based on environmental risk with the idea that this provides an incentive for business to reduce their risks.

Best Available Techniques (BAT)

The Environmental Protection Act 1990 introduced the notion of Best Available Techniques Not Entailing Excessive Costs (BATNEEC) as mentioned earlier. The approach and basis of control under current legislation is focussed on Best Available Techniques (BAT), which is taken to mean the available techniques which are the best for preventing or minimising emissions and impacts on the environment and should be (according to the EU aim at a "high general level of protection of the environment". "Techniques" include both the technology used and the way the installation is designed, built, maintained, operated and decommissioned.

The UK, despite leaving the EU, is intending to maintain environmental standards and to apply the existing model of integrated pollution control. The EU Withdrawal Act

2018 maintains established environmental principles and ensures that existing EU environmental law will continue to have effect in UK law, including the Industrial Emissions Directive[51] (IED) and BAT Commission Implementing Decisions made under it. If this is to be the case, then the UK will also take into account other aspects of how the EU is implementing the IED.

Article 2(12) of the IPPC Directive (2008/1/EC) defined Best Available Techniques as follows:

> 'Best available techniques' means the most effective and advanced stage in the development of activities and their methods of operation which indicate the practical suitability of particular techniques for providing in principle the basis for emission limit values designed to prevent and, where that is not practicable, generally to reduce emissions and the impact on the environment as a whole.
>
> (a) 'techniques' shall include both the technology used and the way in which the installation is designed, built, maintained, operated and decommissioned;
>
> (b) 'available techniques' means those developed on a scale which allows implementation in the relevant industrial sector, under economically and technically viable conditions, taking into consideration the costs and advantages, whether or not the techniques are used or produced inside the Member State in question, as long as they are reasonably accessible to the operator;
>
> (c) best shall mean most effective in achieving a high general level of protection of the environment as a whole.

Best Available Technique Reference Documents (BREF) are drawn up for defined activities (e.g., food, drink and milk industries) and describe in detail applied techniques, present emissions, consumption levels, Best Available Techniques and BAT conclusions. The European IPPC Bureau (EIPPCB)

organises and co-ordinates the exchange of information between Member States/nongovernmental organisations (NGOs) and the EU Commission that leads to the drawing up and review of BREFs.

Commission Implementing Decisions (CIDs) are made once a BREF has been reviewed and adopted. A CID lists the BAT conclusions and is published for each BREF. The adoption of CID and BAT conclusions is mandatory in the permitting/licensing process within the EU. The CID, for example the food, drink and milk industries, was published on 11th November 2019 by the European Commission. The European Environment Protection Agency's role is to ensure that the relevant industry apply all relevant BAT conclusions as soon as practicable but not more than four years after the CID is published. A Commission Implementing Decision (CID) for BAT conclusions will be published for each BREF reviewed under the IED.

The UK government has made secondary legislation to ensure the existing BAT conclusions continue to have effect in the UK, to provide powers to adopt future BAT conclusions in the UK and ensure the devolved administrations maintain powers to determine BAT through their regulatory regimes. The UK government will put in place a process for determining future UK BAT conclusions for industrial emissions. This will be developed with the devolved administrations and competent authorities across the UK.

So far as the regulators in the UK are concerned, they are all expected to operate in a similar manner as there is no distinction between the mechanisms to regulate installations and facilities that fall within the scope of the Environmental Permitting Regime.

The system is intended to control emissions and/or discharges from facilities that have the potential to either harm the environment or human health or both unless they are controlled. The Environmental Permitting Regime requires operators to obtain permits for some facilities, to register others as exempt and provides for ongoing supervision by regulators. The aims are:

- To protect the environment so that statutory and Government policy environmental targets and outcomes are achieved;
- To deliver the regulation effectively and efficiently, providing increased clarity and minimising the administrative burden on both the regulator and the operators;
- To encourage regulators to promote best practice in the operation of facilities; and
- To continue to fully implement European legislation as incorporated into legislation.

Permits contain conditions that should tell the operator what BAT must be used and may set emission limit values (ELV) or other environmental outcomes, based on BAT. The permit can say that BAT must be followed or "appropriate measures" taken to achieve an outcome or ELV. There are guides available on BAT. An operator can decide which BAT if the permit does not say which BAT to use, and an operator may also need to take additional measures to meet the conditions in the permit. When applying for an Environmental Permit the operator must state whether they are going to follow each BAT that applies to their activity or propose an alternative.

For any BAT that an operator is proposing to follow, they must explain how they are going to either:

- Follow the BAT conclusions and meet the BAT-associated emissions level;
- Follow the BREF note and the technical guidance for activities that don't have BAT conclusions.

For any BAT the operator is not going to follow, they must propose an alternative technique.

Environmental permitting is discussed in other chapters in the context of waste management and land, and the legal framework within the EP regulations includes concepts that set out:

A simplified permitting system has been in place for several years for four sectors – small waste oil burners, petrol stations, dry cleaners and vehicle refinishing activities which use Process Guidance Note PG6/34(11) (Statutory guidance for re-spraying of road vehicles) as guidance.

Not all facilities need an environmental permit due to their low risk, and these are identified in the Schedules to the Environmental Permitting Regulations (England and Wales) 2016. Exempt facilities must be registered as exempt, and details of this process are also included in the above Regulations.

There may be more than one regulated facility on the same site. Conversely a single Environmental Permit can be granted for more than one regulated facility where the regulator and operator are the same and all the facilities are on the same site (unless it refers to a mobile plant, which does not have to be operating on the same site).

Local authority environmental permitting

Applications – We have looked at applications in the context of BAT but there are other matters to consider when an operator[52] applies for a permit. Pre-application discussions between operators and EHPs representing local authorities are encouraged as these can help in improving understanding and therefore the quality of the formal application. The local authorities however should not provide advice that might prejudice its determination of an application. Any advice given may clarify whether a permit is likely to be needed, how to prepare their applications, what key issues to focus on and what guidance is available.

Operators will need to support their application and may use other sources of information such as extracts from Environmental Impact Assessments, externally certified environmental management systems (see Chapter 8) and reports prepared for planning purposes. The operator can make the application at different times but usually should be able to submit an application at an early stage containing all information the needed. An operator however can apply after the construction of the installation at his or her own risk, but this is not good practice as the local authority may require additional works, which can then create expensive delays. If an installation also needs planning permission, it is recommended that the operator should make both applications in parallel. This will allow for early formal consideration.

The requirements for applications are set out in Schedule 5 to the Regulations. The application can also be withdrawn at any time before it is determined but the local authority is not obliged to return the application fee.

A level of detail proportionate to the environmental risk is needed. Examples of sample forms for local authority use are provided in the Part C of the Environmental Permitting General Guidance Manual on Policy and Procedures for A2 and B Installations (GGM) [28]. Any form provided by a local authority for an A2 installation must require the applicant to provide the information specified in Article 6(1) of the IPPC Directive.

Any application submitted should give all the information needed to make a determination. If an operator fails to provide enough information the application may not be "duly made", which means that it cannot be determined. An application may not be duly made if. for example, it has not been submitted on the correct form, the necessary fee has not been paid or it has not adequately addressed a key point in the application form. When coming to this decision it is important to apply tests of reasonableness and common sense to assess whether applications are duly made. The applicant should always be told why the application was not duly made.

A duly made application must be acknowledged and the applicant advised the date it expects to determine the application. If the application is not determined on time, the applicant can consider the application to have been refused and so appeal. Changes can also

be made to a duly made application where appropriate, although this could lead to a new application being required.

If an LA-IPPC (Part A2) installation also needs planning consent, the operator would be wise to make both applications in parallel whenever possible. Although the two systems are separate, they are also complementary and the local authority departments should liaise closely on the two applications but also ensure that one system is not used to achieve goals that should be part of the other system.

Determining applications – From the date an application is determined as duly made the period for determining an application is four months for the grant of an Environmental Permit (the determination period for dry cleaners and small waste oil burners is three months). The determination periods can be extended where further information is required to determine the application or decisions are required as to whether information is sensitive due to commercial or industrial confidentiality and/or national security.

The local authority can serve a notice asking for more information it needs to determine the duly made application; this information is essential to allow the application to be determined. It can relate to assessing whether the proposal meets any directive or other requirements or determining the appropriate permit conditions to impose. A reasonable period should be given for the applicant to provide the information.

The operator must apply to the regulator to vary the permit conditions when proposing a change that would mean that a permit condition could no longer be complied with. This is known as an application for variation.

A permit can be transferred and where an operator wants to transfer a permit to someone else, they must make a joint application. Partial transfers are also possible so that the regulated facility becomes two regulated facilities.

An operator of a Part B installation may simply notify the local authority to surrender his or her permit, but a Part A(2) installation must be made via a formal application.

For all applications made under the Regulations a decision is required whether to grant or refuse the proposal in an application and, where applicable, what permit conditions to impose. The local authority must ensure that its determination delivers provides the required level of protection to the environment and the relevant European Directive and other requirements.

A local authority must refuse a permit application if it is not complete or duly made. An application may also be refused if the local authority has reason to believe that the operator is not competent to run the regulated facility in accordance with the permit, the environmental impact would be unacceptable, the information provided by the operator does not provide a reasonable basis to determine the permit conditions or the requirements of relevant European Directives cannot be met.

Permit conditions – More typically, assuming that the operator has taken full account of advice and guidance available and also has the necessary other permissions, the local authority can grant an Environmental Permit and must include any conditions it sees fit, but all conditions stated in a permit should be both necessary and enforceable. "Necessary" means that the local authority should be able to justify the condition at a subsequent appeal. To be enforceable, the permit conditions may comprise some or all of the following: conditions stipulating objectives or outcomes; standards to mitigate a particular hazard or risk; or conditions addressing particular legislative requirements. The conditions should also avoid any duplication with the requirements of other legislation.

To assist, detailed guidance for classes of installations has been issued to local authorities by government. For LA-IPPC (Part A2) installations these are ten sector guidance (SG) notes.[53] For LAPPC (Part B) installations, these are a series of process guidance (PG) notes, covering seven industrial sectors.[54] The PG notes include relevant information on the process, details of emission and control to be applied, draft application form and a model permit to assist EHPs dealing with the installation. This guidance is reviewed on a six-year cycle. In the coming years more standard rules permits are likely to be issued.

BAT requires that the cost of applying Best Available Techniques is not excessive compared to the environmental protection that they provide. Consequently, the more environmental damage BAT can prevent, the more the local authority can justify requiring the operator to spend on it before the costs are considered excessive. Furthermore, if the emissions would cause serious harm even after applying BAT, the local authority may impose even stricter permit conditions or refuse the permit application altogether.

Local authorities are required to have regard to the guidance notes when determining BAT. In general terms, what is BAT for one installation is likely to be BAT for a comparable installation and each is decided on a case-by-case basis. However, the nature of BAT is that as standards improve overtime, BAT will also improve.

Most importantly local authorities should include in a permit a general condition requiring the operator to use BAT to prevent or reduce emissions that are not covered by more specific permit conditions. This is intended to cover the most detailed level of plant design and operation where the operator will usually be in the best position to understand what pollution control means for an installation in practice. This replaces the implied BAT duty on the operator to use BAT in relation to matters not covered by specific permit conditions.

Standard rule permits – The Environment Agency initially developed a simplified system of permitting "standard rules permits". These were developed for waste operations and used rather than site-specific permit conditions. It was deemed that such facilities had similar characteristics regarding environmental hazards. It is only the Secretary of State, the Welsh Ministers and the Environment Agency that can decide on standard rules.

It is the operator's decision as to whether they wish to operate under standard rules.

Other permitting requirements – Where the local authority grants an application for the variation of an Environmental Permit and there are additional variations needed as a consequence of the application, it should include those necessary variations to the Environmental Permit. A local authority does not have to accept an operator's proposals to vary a permit. If it does, it must impose conditions to secure compliance with the Regulations. The local authority may decide that only some parts of the variation sought should be reflected in revised permit conditions. The local authority may also consider that it needs to impose conditions that go beyond the operator's proposals.

If the application is for a transfer the local authority must refuse it if the local authority considers that the proposed transferee will not be the operator or will not operate the facility in accordance with the Environmental Permit. The primary consideration in transferring a permit is the proposed new operator's competence to operate the regulated facility.

An Environmental Permit remains in force until it is surrendered, revoked or consolidated and the operator therefore remains subject to its conditions until that time. When a regulated facility ceases to operate, an operator should (although not required) seek the surrender of the permit to end regulation and the requirement to pay the associated annual subsistence charges.

The general requirements for permit surrender of a Part A(2) installation are different to Part B installations. As this relates primarily to land see Chapters 18 and 19. For a Part A(2) installation the local authority must accept the surrender of the Environmental Permit if it is satisfied that the necessary measures have been taken to avoid any pollution risk resulting from the operation of the regulated facility (e.g., removing tanks containing pollutants), and to return the site of the regulated facility to a satisfactory state, having regard to

the state of the site before the facility was put into operation.

The local authority must ensure that all necessary measures have been taken to return the site of the regulated facility to a satisfactory state, and in some instances, this may be stricter than could be required under the contaminated land regime.

If the operator satisfies the local authority that it has removed any pollution risks and restored the site to a satisfactory state, then the local authority should accept the surrender and give the operator notice of its determination. The permit then ceases to have effect on the date specified in the notice of determination.

Following an application for the grant or transfer of an Environmental Permit there is a specific duty on the local authority not to grant or transfer the permit if it considers that the operator will not operate the facility in accordance with the permit and permit conditions. Examples may include an inadequate management system, inadequate technical competence, a poor record of compliance with previous regulatory requirements, or financial incompetence.

Compliance – The Regulations place a duty on a local authority to undertake appropriate periodic inspections of installations. There is also a duty to carry out periodic inspections of waste operations exempted by the local authority.

Checking compliance with the Environmental Permit conditions is the principal way in which the operator's performance is assessed. In addition, the operator's systems audit for the management and supervision of the facility should be scrutinised by the local authority.

All operators have significant responsibility for monitoring and record keeping under the Environmental Permit conditions. These need assessment to demonstrate whether they are meeting the conditions of the permit.

The local authority can serve an Enforcement Notice[55] if it believes an operator has contravened, is contravening, or is likely to contravene any permit conditions. Enforcement notices must specify the steps required to remedy the problem and the timescale in which they must be taken. Enforcement notices may include steps to remedy the effects of any harm and to bring a regulated facility back into compliance.

If the operation of a regulated facility involves a risk of serious pollution, the local authority may serve a Suspension Notice.[56] This applies whether or not the operator has breached a permit condition. The local authority may also serve a suspension notice for non-payment of a charge.

The suspension notice must describe the nature of the risk of pollution and the actions necessary to remove that risk and relate to the entire installation or to specified activities only. A compliance deadline must also be specified. When the operator has taken the remedial action required by the notice, the notice must be withdrawn.

Prosecution is usually the last enforcement option; however if an operator has committed a criminal offence under the Regulations, then prosecution should be considered. Conviction in a magistrates' court carries a fine of up to £50,000 and up to 12 months' imprisonment for the most serious offences under the Regulations. Conviction in the Crown court for those offences may lead to an unlimited fine and imprisonment for up to five years.

A local authority can revoke a permit, in whole or in part, by serving a Revocation notice. The local authority may use revocation whenever appropriate. Revocation may be appropriate where exhaustive use of other enforcement tools has failed to protect the environment properly. If an appeal is lodged, the revocation does not take effect until the appeal is determined or withdrawn.

The local authority may vary permit conditions at any time, even if the operator has not requested this. This may follow an upgrade of the installation or permit review, where additional conditions are needed to deal with new matters or where compliance assessment has identified a need to vary the conditions. Permit reviews are usually required following updates to Guidance Notes and to reflect appropriate standards. They are intended to guard against permits becoming obsolete as techniques develop and standards improve.

Operators must also pay subsistence charges to support the local authority's ongoing costs for compliance assessment. If an operator fails to pay a subsistence charge, the local authority may revoke the permit as already discussed.

An operator may appeal against certain decisions made by the local authority. These include when the local authority has refused an application for the grant of a permit, variation of a permit, transfer or surrender of the permit. Appeals can also be entered if there is disagreement with the conditions imposed by the regulator in their permit or an application is deemed to have been refused or the local authority has deemed an application to be withdrawn.

Appeals are made to the Secretary of State or the Welsh Ministers in England and Wales, who usually use the Planning Inspectorate and who have produced separate guidance on appeals for environmental permitting.

Public registers of specified information on environmental permitting have to be maintained and readily and easily available for the purposes of public participation under the Regulations. Each local authority must hold a public register and also include information on facilities regulated by the Environment Agency in their areas. The Environment Agency must also provide this information directly to the relevant local authority. However, no information should be included in a register if, in the opinion of the Secretary of State or the Welsh Ministers, the inclusion of that information in the register would be contrary to the interests of national security.

Information may also be withheld from the public registers where it is judged to be commercially or industrially confidential. When this arises, a statement must be placed on the register indicating the existence of that information.

Section 4 Notes

50 See https://uk-air.defra.gov.uk/networks/monitoring-methods?view=mcerts-scheme

51 Directive 2010/75/EU – See https://ec.europa.eu/environment/industry/stationary/ied/legislation.htm

52 The "operator" of a regulated facility is defined as the person who has (or will have) control over the operation of the installation (or the future operation if it has not started). As such the operator must demonstrably have the authority and ability to ensure the Environmental Permit is complied with. (A triviality exemption exists for Part B installations, where the substances released are of a trivial quantity with insignificant capacity to do harm and the pollution potential is so low as to be inconsequential.)

53 These can be accessed at https://www.gov.uk/government/collections/integrated-pollution-prevention-and-control-sector-guidance-notes and cover: composite wood based board manufacture; glass manufacture; ferrous foundries; non-ferrous foundries; galvanising; surface treatment solvents; ceramics manufacture; rendering; burning recovered fuel oil; animal carcass incineration.

54 Available at https://www.gov.uk/government/collections/local-air-pollution-prevention-and-control-lappc-process-guidance-notes#animal-and-vegetable-processing-sectors – the seven sectors are: animal and vegetable processing; combustion and incineration; minerals; metals; organic chemicals; petroleum and powder coating; and solvents. Within each sector there are up to 23 Process Guidance Notes and each one includes what is considered BAT.

55 Under Regulation 36.

56 Under regulation 37(2) of the Environmental Permitting Regulations.

57 https://commonslibrary.parliament.uk/research-briefings/cbp-8824/ The Bill went through many iterations, and was eventually published as the Environment Act 2021, which can be found at https://www.legislation.gov.uk/ukpga/2021/30/contents/enacted

SECTION 5: CLEAN AIR LEGISLATION

The Clean Air Acts of 1956 and 1968 were introduced as a result of the smogs of the 1950s and 1960s referred to at the start of this chapter and were primarily about visible air pollutants. The Clean Air Act 1993 (CAA1993) consolidated (and repealed) the original acts, and maintained the concept of smoke control through the prohibition of smoke from domestic and industrial chimneys and other sources.

Since the introduction of the original Acts, many large towns and cities in the UK have declared Smoke Control Areas (SCAs).[57] As a result of the 1956 and 1968 Clean Air Acts, smoke and associated levels of sulphur dioxide (SO_2) greatly reduced from the 1950s, along with the widespread uptake and use of natural gas by the domestic and industrial sectors.

Part 1 of the 1993 Act prohibits the emission of dark smoke from chimneys (subject to certain exemptions) and similarly prohibits the emission of dark smoke from industrial or trade premises (unless inadvertent, and all practical steps were taken). Dark smoke is defined as smoke, which, if compared in the appropriate manner with a Ringelmann Chart[58] would appear to be as dark as or darker than shade 2 on the chart.

Other sections of the Clean Air Act 1993 allow the Secretary of State to prescribe emission limits on grit and dust from furnaces other than domestic furnaces and prohibit the use of a furnace, other than a domestic furnace, in a building or outdoors which burns pulverised fuel, solid fuel at 45.4 kg/h or more or liquid and gas fuels at 366.4 kW or more, unless it has grit and dust arrestment plant fitted that have been agreed by the local authority or unless the local authority has been satisfied that the emissions will not be prejudicial to health or a nuisance.[59]

Smoke Control Areas once declared, remain in force. In these areas it is still an offence to emit smoke from a chimney of a building, from a furnace or from any fixed boiler if located in a SCA. It is also an offence to acquire and sell an unauthorised fuel for use within a Smoke Control Area unless it is used in an exempt fireplace, meaning that coal, oil and wood cannot be burnt in these areas unless on an exempt fireplace. More recently the problem has been the use of wood-burning stoves (often inadequately dried wood is used) in SCAs. The main pollutant emitted by burning solid fuels like wood is ultra-fine particulate matter, $PM_{2.5}$, which was discussed earlier. Domestic solid fuel, which includes burning wood and coal, is responsible for 38% of $PM_{2.5}$ pollution in the UK, the largest source. Research has shown the risks of using such stoves [14, 29–30][60] and advice has been issued by Defra.[61]

Exempt fireplaces are appliances that have passed tests to confirm that they are capable of burning an unauthorised or inherently smoky solid fuel without emitting smoke, thus they are exempted from the controls that generally apply in a SCA.

Additional guidance in Part B of the General Guidance Manual [27] for Environmental Permitting has been prepared to clarify the interface between the CAA 1993 and the Environmental Permitting Regime (EPR). The guidance indicates that any exemptions under the CAA 1993 for small-scale incinerators or combustion plant burning waste material do not apply if the activity comes under EPR.

The CAA 1993 is considered the main regulatory control that local authorities have over the new biomass burners that are now frequently installed as renewable energy technologies. As already indicated in an SCA either authorised fuels or exempt appliances must be used. Biomass fuels have sulphur contents below 2% on a dry basis and so easily pass the requirement for authorised fuels, although they normally struggle to pass the smoke test[62] to become authorised fuels. However, a biomass fuel is acceptable in a SCA if it is burnt in an exempt appliance. Exempted appliances have undergone type-approval emission tests to determine if they can operate within prescribed particulate emission limits which are related to appliance capacity/output.

There are provisions in the Act relating to the height of chimneys and prohibiting the use of certain furnaces, unless the local authority has approved the chimney height (a hangover from the high chimney approach to pollution control that sought only to remove the pollutant from the immediate vicinity of the furnace). Pellet appliances burning biomass at larger sizes than wood chips would be caught by both the arrestment plant requirements and chimney heights provisions. The Act also potentially applies controls to particulate matter emissions for wood combustion activities up to 20MWth.[63] Beyond that size the Environmental Permitting Regulations apply unless the fuel is deemed to be a waste (or derived from a waste).

Defra commissioned a report to assess potential change scenarios, including the removal of SCAs. This review was published in 2012 [31]. The findings identified that the removal of constraints in Smoke Control Areas had potential for large increases in emissions from domestic solid fuel combustion including significant impacts on national emissions of Benzo(a)pyrene, PM_{10} and $PM_{2.5}$.

The report also could identify credible quantifiable scenarios relating to the dark smoke and cable burning provisions. It stated that *in the absence of CAA controls/supervision, incidents of dark smoke could be expected to increase and this would lead to an increase in loss of amenity, potentially increase emissions of products of incomplete combustion, and/or potentially nuisance situations.*

Following the publication of this report, Defra launched an inquiry and then published their findings [32]. The outcome was included in the Deregulation Act 2015 which included the provisions only to amend the Part 3 CAA 1993 in England.[64]

Conclusion (and the Environment Bill)

This chapter has sought to provide information that an EHP needs to help tackle a vital issue for our world. Air quality and pollution of the air is very much a public health matter and hopefully EHPs will work with local Directors of Public Health and other agencies to tackle the problems locally. The chapter has sought to set in particular the health risks from poor air quality and particular pollutants. Other chapters in the book provide further relevant information.

At the time of writing in the UK the Environment Bill is in Parliament. Within the UK, the environment is a largely devolved area. The Government has said that the Bill is *part of the wider government response to the clear and scientific case, and growing public demand, for a step-change in environmental protection and recovery*. The Bill makes provision to amend existing environmental legislation and introduce new measures on a range of environmental policy areas within the UK. So readers of this and other chapters will need to check on developments.

The Bill covers two broad themes; first it provides a new domestic framework for environmental governance and second, it makes provision on specific environmental policy areas including waste, air quality, water, nature and biodiversity, and conservation covenants. There is some concern that there might be an attempt to water-down existing standards and that on the face of the Bill there are no legally binding targets relating to the Climate Emergency/Global Heating.

Part 1 specifies a series of environmental principles[65] and requires the publication of a policy statement on these principles setting out how they are to be applied by Ministers during policymaking. It establishes an Office for Environmental Protection (OEP), which will have scrutiny, advice and enforcement functions. It also makes provision for the setting of long-term environmental targets in four "priority areas" of air quality, water, biodiversity and resource efficiency and waste reduction, along with the production of statutory Environmental Improvement Plans.

Part 2 of the Bill relates to environmental governance in Northern Ireland, allowing the OEP to exercise its functions in Northern Ireland, subject to the approval of a restored Northern Ireland Assembly.

Part 3 of the Bill makes provisions for the managing of waste and producer responsibility. It also provides a framework for a deposit return scheme.

Part 4 of the Bill deals with air quality and amends the requirements and management of Local Air Quality Management Frameworks. It also seeks to provide local authorities with greater powers in Smoke Control Areas and includes provision to require the recall of motor vehicles on environmental grounds.

Part 5 of the Bill includes provisions relating to water resources management.

Part 6 of the Bill provides for the creation of a new biodiversity net gain requirement, in England, of 10% for developers though the planning system. Gains will be mandatory and maintained for at least 30 years. It also expands the duty on relevant authorities from "conserving" to "conserving and enhancing" biodiversity; and requires the creation of Local Nature Recovery Strategies to cover the whole of England.

Part 7 of the Bill legislates for the introduction of voluntary legally binding conservation covenants between landowners and "responsible bodies" which conserve the natural or heritage features of the land.

Part 8 of the Bill provides powers to the Secretary of State to amend the EU REACH Regulation (registration, evaluation, authorisation and restriction of chemicals) in order to transfer of the EU REACH Regulation into UK law. (By publication of this book the Bill is now an Act see https://www.legislation.gov.uk/ukpga/2021/30/contents/enacted)

Section 5 Notes

57 The locations declared by local authorities throughout the UK can be seen on the Defra website https://www.gov.uk/smoke-control-area-rules

58 BS 2742C:1957, Ringelmann chart (https://shop.bsigroup.com/products/ringelmann-chart/

preview). The Ringelmann Chart is a paper chart that has five shades of grey with 0 being clear and 5 being black.

59 S.6 of the 1993 Act.

60 See also https://uk-air.defra.gov.uk/assets/documents/reports/cat05/1801301017_KCL_WoodBurningReport_2017_FINAL.pdf

61 See https://uk-air.defra.gov.uk/assets/documents/reports/cat09/1901291307_Ready_to_Burn_Web.pdf

62 BS 3841–1 Determination of smoke emission from manufactured solid fuels for domestic use – Part 1: General method for determination of smoke emission rate.

63 For more information on Clean Air Act procedures see Deveaux T. (2020) *Bassett's Environmental Health Procedures*, Routledge, Abingdon, Oxon, 9th ed.

64 Relate primarily to requirements on the Secretary of State to publish details of "authorized fuels" and "exempting fireplaces".

65 https://commonslibrary.parliament.uk/research-briefings/cbp-8824/ The Bill went through many iterations, and was eventually published as the Environment Act 2021, which can be found at https://www.legislation.gov.uk/ukpga/2021/30/contents/enacted

Chapter references

[1] House of Commons. (2014) *Action on Air Quality*, Sixth Report of Session 2014–15, (Report HC 212) Environmental Audit Committee, House of Commons, The Stationery Office Limited. http://www.parliament.uk/documents/commons-committees/environmental-audit/HC-212-for-web.pdf (Accessed September 2015).

[2] World Health Organization. (2014) *Burden of Disease from the Joint Effects of Household and Ambient Air Pollution for 2012*, WHO, Geneva, Switzerland. http://www.who.int/phe/health_topics/outdoorair/databases/FINAL_HAP_AAP_BoD_24March2014.pdf.

[3] WHO. (2021) *Global Air Quality Guidelines Particulate Matter (PM$_{2.5}$ and PM$_{10}$), Ozone, Nitrogen Dioxide Sulphur Dioxide and Carbon Monoxide*, World Health Organization, Geneva, Licence: CCBY-NC-SA 3.0IGO.

[4] Royal College of Physicians. (2016) *Every Breath We Take: The Lifelong Impact of Air Pollution*, Report of a Working Party, London. https://www.rcplondon.ac.uk/projects/outputs/every-breath-we-take-lifelong-impact-air-pollution (Accessed September 2021).

[5] Ministry of Health. (1954) *Report on Mortality and Morbidity During the London Fog of December 1952*, Ministry of Health No. 95, HMSO, London.

[6] Bell ML, Davis DL, Fletcher T. (2004) A retrospective assessment of mortality from the London smog episode of 1952: The role of influenza and pollution. *Environmental Health Perspectives*, 112(1): 6–8.

[7] California Air Resources Board. (undated) *History*. https://ww2.arb.ca.gov/about/history and also SCAQMD (the air Pollution Control Agency for Orange County and urban portions of Los Angeles, CA) (undated) *The Southlands War on Smog: Fifty Years of Progress Towards Clean Air (through May 1997)*. http://www.aqmd.gov/home/research/publications/50-years-of-progress (Accessed September 2021).

[8] US National Weather Service Climate Prediction Center. (undated) *Stratosphere: Southern Hemisphere Ozone Hole*. http://www.cpc.ncep.noaa.gov/products/stratosphere/sbuv2to/ozone_hole.shtml (Accessed August 2021).

[9] Kernan M, Battarbee RW, Curtis CJ, Monteith DT, Shilland EM. (2010) *UK Acid Waters Monitoring Network 20-Year Interpretative Report to Defra*, Environmental Change Research Centre, UCL, London, ISSN: 1366–7300. https://uk-air.defra.gov.uk/assets/documents/reports/cat13/1206251208_20yearInterpRpt.pdf (Accessed September 2021).

[10] Christianson GE. (1999) *Greenhouse: The 200-Year Story of Global Warming*, Universities Press, New York. ISBN:0094 80030.

[11] Vardoulakis S, Heaviside C. (Eds.) (2012) *Health Effects of Climate Change in the UK*, Public Health England. ISBN:978-0-85951-723-2, HPA (now part of UK Health Security Agency), Didcot, Oxon. https://www.gov.uk/government/uploads/system/uploads/attachment_data/file/371103/Health_Effects_of_Climate_Change_in_the_UK_2012_V13_with_cover_accessible.pdf (Accessed September 2021).

[12] Ebi K, McGregor G. (2009) Climate change, tropospheric ozone and particulate matter, and health impacts. *Cien Saude Colet*, 14(6): 2281–2293, November–December. https://doi.org/10.1590/s1413-81232009000600037. PMID: 20069198.

[13] The Royal Society. (2008) *Ground-level Ozone in the 21st Century: Future Trends, Impacts and Policy Implications*, The Royal Society, London, ISBN: 978-0-85403-713-1.

[14] Defra. (2019) *Clean Air Strategy 2019*, London. https://assets.publishing.service.gov.uk/government/uploads/system/uploads/attachment_data/file/770715/clean-air-strategy-2019.pdf (Accessed September 2021).

[15] Committee on the Medical Effects of Air Pollutants. (2009) *Long-Term Exposure to Air Pollution: Effect on Mortality*, Health Protection Agency, Didcot, Oxon, Crown Copyright. ISBN:978-0-85951-640-2.

[16] Feng S, et al. (2016) The health effects of ambient $PM_{2.5}$ and potential mechanisms. *Ecotoxicology and Environmental Safety*, 128: 67–74, June. https://doi.org/10.1016/j.ecoenv.2016.01.030. Epub 19 February 2016. PMID: 26896893.

[17] Miller BG, Hurley JF. (2006) *Comparing Estimated Risks for Air Pollution with Risks for Other Health Effects*, Research Report TM/06/01 March, Institute of Occupational Medicine, Edinburgh. http://www.iom-world.org/pubs/IOM_TM0601.pdf (Accessed September 2015).

[18] Anderson HR, Limb ES, Bland JM, et al. (1995) Health effects of an air pollution episode in London, December 1991. *Thorax*, 50: 1188–1193.

[19] Walton H, Dajnak D, et al. (2015) *Understanding the Health Impacts of Air Pollution in London. For: Transport for London and the Greater London Authority*, King's College London, 14 July. https://www.london.gov.uk/sites/default/files/HIAinLondon_KingsReport_14072015_final_0.pdf (Accessed September 2021).

[20] Korte JE, Hertz-Picciotto I, et al. (2000) The contribution of benzene to smoking-induced leukemia. *Environmental Health Perspectives*, 108(4): 333–339, April. https://doi.org/10.1289/ehp.00108333.PMID:10753092; PMCID: PMC1638019.

[21] Resongles E, Dietze V, et al. (2021) Strong evidence for the continued contribution of lead deposited during the 20th century to the atmospheric environment in London of today. *Proceedings of the National Academy of Sciences*, 118(26): e2102791118, June. https://doi.org/10.1073/pnas.2102791118.

[22] GBD. (2017) Risk factor collaborators: Global, regional, and national comparative risk assessment of 84 behavioural, environmental and occupational, and metabolic risks or clusters of risks for 195 countries and territories, 1990–2017: A systematic analysis for the global burden of disease study 2017. *The Lancet*, 392: 1923–1994, 8 November. https://doi.org/10.1016/S0140-6736(18)32225. http://www.healthdata.org/research-article/global-regional-and-national-comparative-risk-assessment-84-behavioral-0 (Accessed September 2021).

[23] Defra. (2016) *Local Air Quality Management – Policy Guidance (PG16).* https://laqm.defra.gov.uk/documents/LAQM-PG16-April-16-v1.

pdf There is also a support website at https://laqm.defra.gov.uk.

[24] Defra. (2019) *Air Quality: National Air Pollution Control Programme*, London. https://assets.publishing.service.gov.uk/government/uploads/system/uploads/attachment_data/file/791025/air-quality-napcp-march2019.pdf.

[25] Defra & DoT. (2020) *Clean Air Zone Framework: Principles for Setting Up Clean Air Zones in England.* https://assets.publishing.service.gov.uk/government/uploads/system/uploads/attachment_data/file/863730/clean-air-zone-framework-feb2020.pdf.

[26] Loader A. (2008) *Diffusion Tubes for Ambient NO2 Monitoring: Practical Guidance for Laboratories and Users*, Report to Defra. ED48673043, Issue 1 February. http://laqm.defra.gov.uk/documents/0802141004_NO2_WG_PracticalGuidance_Issue1a.pdf.

[27] AQEG. (2012) *Fine Particulate Matter (PM₂.₅) in the United Kingdom*, Report from the Air Quality Expert Group to the Department for Environment, Food and Rural Affairs; Scottish Government; Welsh Government; and Department of the Environment in Northern Ireland. https://uk-air.defra.gov.uk/assets/documents/reports/cat11/1212141150_AQEG_Fine_Particulate_Matter_in_the_UK.pdf.

[28] Defra. (2012) *Environmental Permitting General Guidance Manual on Policy and Procedures for A2 and B Installations.* https://assets.publishing.service.gov.uk/government/uploads/system/uploads/attachment_data/file/211863/env-permitting-general-guidance-a.pdf.

[29] Chakraborty R, Heydon J, et al. (2020) Indoor air pollution from residential stoves: Examining the flooding of particulate matter into homes during real world use. *Atmosphere*, 11(12): 1326. https://doi.org/10.3390/atmos11121326.

[30] ERG Kings College, London. (2017) *Airborne Particles from Wood Burning in UK Cities*, ERG Kings College London and NPL a report for Defra. https://uk-air.defra.gov.uk/assets/documents/reports/cat05/1801301017_KCL_WoodBurningReport_2017_FINAL.pdf (Accessed September 2021).

[31] AEA. (2012) *Assessment of the Effectiveness of Measures Under the Clean Air Act 1993.* Report for Defra AEA/R/ED46626/3289, AEA, Didcot, Oxon. https://www.gov.uk/government/uploads/system/uploads/attachment_data/file/183198/20072012-AEA-Report-CAA.pdf (Accessed September 2021).

[32] Defra. (2014) *Review of the Clean Air Act 1993 – Call for Evidence Summary of Responses,* Defra, London, Crown Copyright. https://www.

gov.uk/government/uploads/system/uploads/ attachment_data/file/326129/clean-air-act-sum-resp.pdf (Accessed September 2021).

[33] UK National Atmospheric Emissions Inventory (NAEI) Website, (page last modified 28 January 2013) http://naei.defra.gov.uk/overview/ap-overview (Accessed August 2015)

Further reading

- Defra. (2009) *Local Air Quality Management Practice Guidance 1 – Economic Principles for the Assessment of Local Measures to Improve Air Quality*, Defra, 3C Ergon House, Horseferry Road, London, February.
- Defra. (2009) *Local Air Quality Management Practice Guidance 2 – Practice Guidance to Local Authorities on Low Emissions Zones,* Defra, 3C Ergon House, Horseferry Road, London, February.
- Defra. (2009) *Local Air Quality Management Practice Guidance 3 – Practice Guidance to Local Authorities on Measures to Encourage the Uptake of Low Emission Vehicles*, Defra, 3C Ergon House, Horseferry Road, London, February.
- Defra. (2009) *Local Air Quality Management Practice Guidance 4 – Practice Guidance to Local Authorities on Measures to Encourage the Uptake of Retro-Fitted Abatement Equipment on Vehicles*, Defra, 3C Ergon House, Horseferry Road, London, February.
- Defra. (2009) *Local Air Quality Management Technical Guidance (TG09)*, Defra, 3C Ergon House, Horseferry Road, London, February.
- Faulkner M, Russell P. (2010) *Review of Local Air Quality Management*, IHPC Policy Projects for CLG, DfT, DECC and Defra, March. http://webarchive.nationalarchives.gov.uk/20130402151656/http:/archive.defra.gov.uk/environment/quality/air/air-quality/local/documents/laqm-report.pdf

17

Pest management and vector control

Paul Charlson and Stephen Battersby

Recent developments in pest-borne diseases such as West Nile fever in the United States of America and the spread of Lyme disease in both Europe and North America have signalled strongly the crucial need to carefully assess the potential threat of urban pests to public and environmental health.

Dr Robert Bertollini
Director, Special Programme on Health
and the Environment
WHO Regional Office Europe
In the Foreword to *Public Health
Significance of Urban Pests* [1]

Introduction

This chapter is very much an update and review of the chapter in the last edition [2]. As before, the focus is on why pest management is an environmental and public health issue, rather than merely a service that some local authorities have chosen not to provide or not provide fully. The pests we focus on are those of a concern for EHPs and public health because of where we find them – in food premises and other working environments and in and around the home – so primarily urban pests.

The focus of this chapter is on prevention and why pests are a public health issue. While there is some discussion of pest and vector control techniques, this chapter does not consider methods of treatment in any detail. Significant improvements have been made in this regard, particularly for rodent control, where the Campaign for Responsible Rodenticide Use (CRRU) has been instrumental in creating a UK-wide Code of Best Practice in both treatment methodology and the safe use of rodenticide. More detailed guidance on control methodologies can be found using the resources identified at the end of this chapter.

The primary concern relates to pests of public health significance (rather than agricultural or horticultural pests) and pest management strategies including Integrated Pest Management (IPM), which although originally developed in the agricultural context is relevant to urban pests. A key message to emerge from many studies is the under-reporting of rodent zoonoses and that, in many cases, insufficient attention is paid to the diagnosis of these important diseases. It has been suggested that in developed countries, where there is overproduction of food crops and adequate storage, the justification for rodent control is on grounds of hygiene, public health and animal hygiene, rather than for economic reasons, whereas in the tropics and subtropics the opposite is more likely to be true [3].

DOI: 10.1201/9781003035640-17

It should be noted that even where the pest is not implicated in the spread of disease, if it increases the risk of a chronic health condition (for example asthma or allergies) or generally causes stress and adversely affects well-being and mental health then it is still of public health importance. Pests and pest management is also an environmental health issue that cuts across different aspects of the profession and so those other chapters of the handbook should also be consulted. There is a brief discussion of relevant existing legal provisions.

SECTION 1: PESTS AS VECTORS OF PUBLIC HEALTH SIGNIFICANCE

Introduction

The challenges faced by EHPs in attempting to ameliorate the impact of pest species on the health of the public are significant and this chapter will attempt to provide evidence-based advice and guidance on approaches to pest management. Getting the right balance between addressing generic and site-specific issues where pests are found is key. An understanding of pest biology and behaviour is an essential pre-requisite in developing setting specific control programmes and so should not be underestimated.

Despite considerable advances in public health and the substantial investment in the research and development of pesticide products, many of the major public health pests which plagued the UK a century ago continue to cause significant threats to the health and well-being of the population. Many emerging diseases are also zoonotic. Changes in both the distribution of pests and the patterns of vector-borne diseases as a result of many inter-related factors (e.g. global heating, increases in international travel, changes in land use, changes in pesticide efficacy and application, etc.) make it difficult to predict the future challenges which face those attempting to lessen or control the impact of pests. However, if real progress is to be made in dealing with the public and environmental problems caused by pests, there must be a cultural shift away from dealing with the consequences of pests to recognising and addressing the conditions and environments that facilitate the introduction and persistence of infestations.

The full economic consequences of urban pests are difficult to capture comprehensively for higher-income countries. Whilst it may be relatively straightforward to estimate the economic impact of pest damage in agricultural settings by measuring the quantitative and/or qualitative losses when crops are damaged or contaminated by pests, the urban environment provides a more complex setting. The costs in human health and well-being, in addition to food damage/contamination, structural damage to buildings, infrastructure damage and reductions in productivity, the costs of chemical and non-chemical controls applied by both private and public organisations and amateur users in attempting to resolve the infestation can be difficult to establish. The reliability of cost estimates is confounded by the fact that many householders and businesses do not report infestations to anyone. Despite these difficulties, a model was constructed to attempt to estimate the costs of rodent infestations in the UK. Battersby [4] in the early part of this century proposed a figure of between £61.9 million and £209 million (£103.3 million to £348.8 million at 2021 prices) and concluded that, based on the size of the rodent control industry, the higher figure was more likely.

Climate change as well as the ease of global movement of goods and people also increases the threat to public health from pests as vectors of disease. The COVID-19 pandemic has shown how easy it is for diseases to cross the globe very quickly despite the best efforts of many nations, which is something reflected previously by the spread of West Nile virus.

Defining a pest

A number of definitions of a 'pest' exist, such as 'any organism which, in a given circumstance, adversely affects human wellbeing or economy'. Bezant's definition as *an organism in a situation where it is actually or potentially capable of transmitting disease, doing damage, arousing fear, giving offence, producing fouling, causing contamination or just being a nuisance* [5] is probably most apt in an environmental health context.

Many animals have achieved pest status because of their impact on health, aesthetics, economics or damage to material or structures – termites for example. Their presence will not only be unacceptable but will probably lead to some form of chemical or nonchemical control effort. Understanding the characteristics of the setting in which a pest is found, the life cycles of the pests present, an appreciation of the regulatory and legislative frameworks and an understanding of the principles of IPM are fundamental in developing effective and sustainable pest management. Successful outcomes will only be achieved if these factors are incorporated into a strategy which is specific to that situation. Applying generic control strategies may result in a brief reduction in pest numbers, but are unlikely to deliver long-term, sustainable control.

Three groups of animals are likely to account for the vast majority of pest infestations in urban environments: insects, rodents and birds. In some settings it will take significant numbers of the pest species to be present before there is any threat to public health or nuisance. In others, the presence of *any* pest species could provide a significant risk to vulnerable groups. It is therefore important that control measures are justified and proportionate to the risks posed. Table 17.1 provides

a brief summary of the main risks posed by the more common urban pests, with further details provided in the text.

The public's response and expectations also have a bearing on action taken. It would seem that while constraints on local government expenditure have led to a reduction in any pest management service, public expectation has not changed. This can lead to inadequate and ultimately ineffective approaches – in many cases an individualised response will not be successful. As far back as 2012 it was reported that the number of councils still offering free or subsidised pest control had dropped substantially with many having no comprehensive in-house pest control service at all [6]. It will not have improved since, as in 2015 it was reported that pest control was the most commonly discontinued service [7]. Accordingly, the CIEH has always supported the importance of effective pest control and has suggested that more than half of the public see local authorities as the persons responsible for pest control, and a similar percentage would ask their council first for advice on a pest problem.

Urban pests and disease

Most of the diseases transmitted by pests are caused by bacteria, viruses, fungi or protozoa and may be vectored either indirectly or directly. Indirect or passive transmissions occur when pathogens present on the surface of the pest are mechanically transferred, contaminating food and/or surfaces that the pest has come in contact with. For example, cockroaches can move easily from sewers to kitchens and will contaminate the surfaces and food with which they come into contact. In addition to this indirect transfer, the cockroach will regurgitate and defecate as part of its feeding habits and thereby introduce further contamination. Direct transmission occurs when pests either bite or sting. Fleas and mosquitoes have vectored globally important diseases such as plague and malaria. Further pathogens may be introduced because

Table 17.1 Summary of the main risks posed by the presence of pests in food and domestic premises

Food/ domestic	Allergy	Disease/ transmission	Contamination	Structural damage	Blood feeder	Risk rating (internally)	
						Food	Domestic
Mice	Yes	Direct and indirect	√	√	X	High	High
Rats	Yes	Direct and indirect	√	√	X	High	High
Exotic ants (principally Pharaoh ants)	Not known	Indirect	√	Rogers ants will burrow downwards	X	Medium	Low in domestic, but high in hospitals
Cockroaches	Yes	Indirect	√	X	X	High	High
Bedbugs	Yes	Indirect (through irritation from bites)	X	X	√	N/A	Medium/ High
Fleas	Not known	Direct and indirect	X	X	√	No	High
Flies	Not Known	Direct and indirect	√	X	Some	High	Medium
Dermestids	Not reported	Indirect	√	X	X	High/lack of hygiene	Low Medium
Stored product insect pests (beetles, moths, etc.)	Not reported	X	√	X	X	Medium/ High	Low/ Medium
Birds	Not reported	Direct and indirect	√	√	X	Medium/ High	Low/ Medium
Mice	Yes	Direct and indirect	√	√	X	High	High
Rats	Yes	Direct and indirect	√	√	X	High	High
Exotic ants (principally Pharaoh ants)	Not known	Indirect	√	Rogers ants will burrow downwards	X	Medium	Low in domestic, but high in hospitals
Cockroaches	Yes	Indirect	√	X	X	High	High
Bedbugs	Yes	Indirect (through irritation from bites)	X	X	√	N/A	Medium/ High
Fleas	Not known	Direct and indirect	X	X	√	No	High

Table 17.1 (Continued)

Food/ domestic	Allergy	Disease/ transmission	Contamination	Structural damage	Blood feeder	Risk rating (internally)	
						Food	Domestic
Flies	Not Known	Direct and indirect	√	X	Some	High	Medium
Dermestids	Not reported	Indirect	√	X	X	High/lack of hygiene	Low/ Medium
Stored product insect pests (beetles, moths, etc.)	Not reported	X	√	X	X	Medium/ High	Low/ Medium
Birds	Not reported	Direct and indirect	√	√	X	Medium/ High	Low/ Medium

of the intense scratching and irritation which can occur following a bite, resulting in secondary infections.

Asthma and allergens

Allergens have been discussed briefly in Chapter 10 and asthma is a serious global health problem although Stevenson et al. [8] reported that prevalence varies between countries and cities and between ethnic groups within cities. In the USA, cockroach asthma is an important public health problem that affects patients who are the least likely to be compliant with treatment with asthma medications or environmental control [9]. Although there is little UK-based research on the prevalence of pest-induced asthma, it is unlikely to be greatly different.

The Health Survey for England [10] at that time found that 39% of 2- to 15-year olds had received diagnoses of asthma or eczema or hay fever and 2% had been diagnosed with all three conditions. Asthma UK has reported in 2021 that 5.4 million people in the UK were receiving treatment for asthma, with 1.1 million children (1 in 11) and 4.3 million adults (1 in 12) being treated. In 2017 (the most recent data available) 1,484 people in the UK died from an asthma attack in the UK and the

NHS spends around £1 billion a year treating and caring for people with asthma.[1]

International comparisons have ranked Great Britain second highest for the prevalence of eczema symptoms, third highest for asthma symptoms and 13th for symptoms of allergic rhino conjunctivitis. Whilst death from asthma in childhood is a rare occurrence,[2] the 'mountain of morbidity' casting a 'shadow of sickness' represented the impact of asthma on health services and mortality statistics in 2004 [11].

Clearly, exposure to an allergen must occur for allergic sensitisation to develop and Platts-Mills [12] confirmed that decreasing exposure to indoor allergens (such as cockroach or mouse allergens) is a rational and well-accepted part of the treatment of chronic allergic disease. The relationship between the presence of pests and pest allergens and the response of those living in infested dwellings is complex. For example, a study of inner-city homes in the US confirmed that 41% of homes had substantial amounts of the cockroach allergen Bla g 1 present, but only 18% had a current cockroach infestation. This finding confirms the persistence of allergens despite successful cockroach eradication measures. Resident reports of the presence of mouse or cockroach infestations have consistently been associated with high levels of their respective allergens [13–15] and indeed cockroach levels may be used as an indicator or predictor of allergen in dust [14]. Despite the uncertainties around the quantitative thresholds for sensitisation, studies in other countries have demonstrated a clear relationship between increasing domestic exposure to allergens from cockroaches, mice and dust mites and an increased risk of allergic sensitisation and severe asthma [16]. These authors estimated that exposure to pest allergens could affect 4–17% of children living in urban environments.

Rodent pests

Commensal (or perhaps more accurately synanthropic[3]) rodents have demonstrated an impressive ability to adapt to new environments and this, in tandem with their considerable reproductive potential, has made them very successful pests of human environments. Rats and mice have been associated with a wide number of zoonoses [17]. Although in general, bacteria and viruses have caused the most concern about rat-associated zoonoses in urban environments, the role of brown rats as reservoirs for helminth parasites and the associated risk for humans has been less well noted [18]. It has been suggested that globally rat-borne diseases have claimed more lives than all the wars ever fought [19]. There are two main rodent pest species in the UK – the house mouse, *Mus domesticus*, and the brown rat, *Rattus norvegicus*. Although the black rat is still recorded in Britain, its range has diminished to a small number of port areas and tends to be restricted to indoor environments.

People often report strong reactions to the presence of rats and mice in both domestic and commercial premises (and also confuse the two species). Where control programmes have failed to eradicate rodents from domestic dwellings, residents have been known to move because they find their presence intolerable.

The WHO LARES study found that those living in mouse-infested properties were twice as likely to report suffering from depression and eight times more likely to report migraine/frequent headaches than those living in properties that were mouse-free [20]. Indeed, cockroaches, rats and mice were significantly related to asthma and allergies even after allowing for co-founders [21].

Rats and mice can cause significant damage to food in both domestic and commercial premises. In addition to the health risks associated with their presence, they can have significant impact on the built environment through their gnawing and burrowing activities. Both rats and mice can cause direct damage to the structure and fabric of buildings and secondary flood or fire damage by gnawing through water or gas pipes or electricity cables. Rats are also implicated in incidents of subsidence through their burrowing activity [4].

The house mouse (*Mus domesticus*) is a common pest that can be found in both domestic and commercial properties across the UK. They are poor competitors [22–23] and in the urban environment spend most, if not all, of their life cycle indoors where other competitor species such as wood mice will not be present. Interestingly, research indicates that the environment in which they live has an impact on the size of the mice, as some research has shown a correlation of skeletal size with island area where they are located [24].

Once they have gained access to a building, they can spread very rapidly, both vertically and horizontally and are likely to extend their range throughout the building. They can contaminate surfaces and foods both directly and indirectly, including defecation and urination, and are able to gain access through very small holes and exploit un/disturbed environments such as attics and subfloor spaces. This is an important consideration for multi-occupied buildings where a holistic approach to the entire building is required, rather than treating individual residences or units in isolation. Like rats, house mice owe much of their worldwide distribution and prolific numbers to their close association with humans. For EHPs in local authorities, it seems to have become thought that house mice should be classified as nuisance pests rather than public health pests. This division has led to differences in the approaches to treatment in the past and the services provided [25]. It seems that in the domestic setting they were viewed as a nuisance because of what were thought to be modest economic losses associated with food spoilage and structural damage through gnawing. In the apparent absence of research, it may well have been thought that house mice posed no risk to health by comparison with rats. This is wrong as studies have shown that they should be considered as public health pests and researchers have confirmed their ability to vector diseases, either directly or indirectly [26].

Urban house mice are known to vector Lymphocytic Choriomeningitis [27][4] and leptospirosis. Their faeces were found to carry a number of bacterial (e.g. Salmonella, Listeria and pseudomonas) and protozoan (Cryptosporidium and Toxoplasma) infectious agents [3, 28]. However, there is some debate about Salmonella in mice, as Salmonella was not isolated from 222 faecal samples of mice in one study [29] and only rarely in another study [30]. (Although in the latter study it was concluded that mice in buildings can carry *Clostridium difficile* ribotypes that are associated with clinical disease in humans.) Whether the mice are the source or whether they picked up these bacteria from the human environment has not been investigated. Either way, mouse droppings in the indoor environment are a hazard for transmission of *C. difficile* to humans. There might be reasons for this variation in findings that relate to the environment where mice have been sampled. For example, Salmonella itself might be fatal to the mice as another study found *S. enteritidis* in the liver and the intestine of most of the mice on a poultry farm, indicating a systemic infection [31] and would effectively reduce the chance of sampling an infected mouse. Wild mice excreted

S. enteritidis intermittently, and an infected mouse died after intermittently excreting small numbers of *S. enteritidis* in its droppings for 19 weeks. *S. enteritidis* was also found in foetal tissue in a naturally infected mouse suggesting the possibility that the organism might be transmitted vertically. Whilst not conclusive, the research would indicate that it is safer to assume that mice do excrete *Salmonella* spp.

A study by Murphy et al. has clarified their role as an intermediate host in the life cycle of toxoplasmosis in urban areas [32] just as for rats [3]. The public health impact of *T. gondii* lies in its ability to cause spontaneous abortions and foetal abnormalities, and to induce serious illness in immuno-compromised subjects. Cats are the only known definitive host for *T. gondii*, and acquire the infection in one of two main ways: via consumption of infected prey (such as house mice or rats) or via ingestion of oocysts within their food or water [33]. Marshall et al. [34] found that 75% of unborn mouse foetuses tested positive for *T. gondii* infection, confirming vertical transmission from dam to foetus. These researchers concluded that mice were likely to be an important reservoir for the persistence of the infection in that urban area.

A number of studies have also reported the prevalence and exposure risk for mouse allergens. Cohn et al. [15] collected dust samples from a number of locations within domestic dwellings and found detectable concentrations of mouse allergens in at least one sampling location in 82% of US homes. Kitchen floors exhibited the highest prevalence with 22% having concentrations greater than 1.6 µg/g which was a level previously found to be associated with significantly increased mouse allergen sensitisation rates.

Although the brown rat (*Rattus norvegicus*) may invade domestic and commercial premises in urban areas, they are more likely to be found in sewers and drains. Those that do appear on the surface and invade urban premises are likely to mechanically vector a number of pathogens present in the sewers.

Battersby et al. [35] screened urban rats for a number of diseases and compared the disease profile of those rats with an earlier study that had screened rural rats [36] (see Table 17.2).

One theory is that differences in prevalence of zoonoses may reflect different population densities (crowding), with rural populations (on farms) likely to have more crowding of

Table 17.2 Zoonoses associated with commensal rodents

Human disease	Vector, pathogen or both
Ectoparasites	
Bubonic plague	Asiatic rat flea (*Xenopsylla cheopsis*) – *Y. pestis* (the pathogen)
Louse Borne Relapsing Fever (LBRF)	Body louse – *B. recurrentis*
Tick-borne relapsing fever	Ticks (*Ornithodoros hermsi*) – *Borrelia* spp.
Lyme disease	Ticks (*Ixodes* spp.) – *B. burgdorferi*
Rickettsial pox [a]	Rodent mite (*Liponyssoides sanguineus*) – *Rickettsia akari*
Murine typhus [a]	Asiatic rat flea – *R. typhi* Body louse – *R. typhi*
Endoparasites	
Capillariasis	*Capillaria* spp.
Toxocariasis	*Toxocara* spp.
Rat tapeworm infection	*Hymenolepis nana*
Angiostrongylosis [*****]	*Strongyloidea* spp.
Diarrhoeal disease	*Trichuris* spp.
Diarrhoeal disease	*Hymenolepis* spp.
Diarrhoeal disease	*Taenia* spp.
Schistosomiasis [***]	*Schistosoma* spp.
Trichinellosis [*]	*Trichinella* spp.
Cryptosporidiosis [a]	*C. parvum*
Toxoplasmosis [a]	*T. gondii*
Babesiosis	*Babesia* spp.
Sarcosporidiosis	*Sarcocystis* spp.
Coccidiosis	*Coccidia* (*Eimeria* spp.)
Amoebic dysentery	*Entamoeba* spp.
Bacteria	
Leptospirosis [a] (Weil's disease)	*Leptospira* spp.
Listeriosis	*Listeria* spp.
Yersiniosis	*Y. enterocolitica*
Pasteurellosis	*Pasteurella* spp.
Rat-bite fever [a] (Haverhill fever)	*Streptobacillus moniliformis* and *Spirillum minus*
Melioidosis	*Pseudomonas* spp.
Q fever	*C. burnetii*
Salmonellosis [a**]	*Salmonella* spp.
Diarrhoeal disease	*Vibrio* spp.
Tularaemia [*]	*F. tularensis*
E. coli 0157/VTEC [*****]	*E. coli* 0157

Table 17.2 (Continued)

Human disease	Vector, pathogen or both
Virus	
Haemorrhagic fever with pulmonary syndrome (HFPS)	Family *Bunyaviridae*, *Sin Nombre virus*
Hantaan fever (haemorrhagic fever with renal syndrome, HFRS)	Hantaviruses (Seoul, Hantaan, Prospect Hill, Puumala)
Lassa fever	*Arenaviridae* virus
Lymphocytic Choriomeningitis [b] [****]	Lymphocytic Choriomeningitis virus

Source: Taken from [12]

Source: Webster and Macdonald (1995), Battersby (2002) and Battersby, Parsons and Webster (2002), except *[13], **Seguin et al. (1986), Hilton, Willis and Hickie (2002), ***Gratz (1984), ****Lehmann-Grube (1975) and ***** Meerburg B G, Grant R, Singleton G R, and Kijlstra A (2009)

[a] Indicates zoonoses of house mice and *Rattus* spp.; [b] indicates zoonosis of house mice.

Source: Battersby et al. [35] and Webster and Macdonald [36], except 'Nowak [19], ''Seguin et al. [40] and Hilton et al. [41], '''Gratz [42] ''''Lehmann-Grube [43] and '''''Meerburg et al. [44]

rats than in urban areas where there is greater predation including via pest control. Greater crowding facilitates disease transmission [37]. A study found that rat population density varied remarkably over short geographical distances in an inner city, and this could explain observed spatial distributions of rat-associated zoonoses. Season also appeared to influence rat population composition even within the urban environment, which could cause temporal variation in pathogen prevalence [38].

In a recent paper, it was again concluded that there was a tendency for brown rat activity in rural areas to harbour more rat-borne microbes than urban areas, but the opposite could be observed for the black rat. The study clearly indicated that an improved surveillance on wild rats is needed in Europe, and further indicated the pathogens and geographical areas where the major focus is required. A number of important human pathogens that are carried by increasing populations of rats were identified and may constitute significant and/or increasing threats to public health. It is suggested that wild rats in Europe are monitored regularly at least for the six most widespread pathogens identified in this study (i.e. virulent/resistant *E. coli*, pathogenic *Leptospira* spp., the helminths *Hymenolepis*

diminuta, *H. nana* and *Capillaria hepatica*, and the protozoon *Toxoplasma gondii*) [39].

Although these organisms have been identified as infecting rodents, the rodents may also act as a reservoir of infection. The rodents harbour these disease-causing organisms and serve as potential sources of disease outbreaks, which may be via a vector (the flea as in bubonic plague, tick, sand–fly, etc.). Alternatively, they may be a carrier in that they show no or limited symptoms of a disease but harbour the disease-causing agent and can pass it directly to humans via a bite or contamination of food or the environment.

It should be noted that although there may be a high prevalence of ectoparasites such as fleas, mites and lice on rats, they appear currently to represent little risk in the UK in the absence of primary pathogens such as the bacterium *Yersinia pestis* which causes bubonic plague. Ticks can also be a vector for *Borrelia* and *Babesia* in the UK. When considering the impact of climate change however, it has been shown that in Central Asia, where human plague is still reported regularly, the bacterium is common in natural populations of great gerbils. Field data from 1949–1995 and previously undescribed statistical techniques, has been used to show that *Y. pestis* prevalence in gerbils increases with warmer springs and wetter summers. A 1°C increase in spring is predicted to lead to a >50% increase in prevalence. Climatic conditions favouring plague apparently existed in this region at the onset of the Black Death as well as when the most recent plague pandemic arose in the same region, and they are expected to continue or become more favourable because of climate change. Threats of outbreaks may thus be increasing where humans live in close contact with rodents and fleas (or other wildlife) harbouring endemic plague [45].

Battersby et al. [35] confirmed that urban rats had the potential to vector diseases, although the range and prevalence of pathogens was generally less than that found in rural rats and some explanation for this has been given earlier. However, global heating, changes in human activity and behaviour and reduced or less effective pest control and management activities by local authorities and others may lead to changes in the pattern and prevalence of rodent-borne diseases with reduced 'predation' leading to increased rodent densities in some areas.[5] It is also possible that, during the lockdowns associated with the COVID-19 pandemic, rat behaviour changed and populations disrupted given that rat behaviour reflects human patterns of behaviour and activity.[6]

Insects

Insects form a diverse group of animals that have adapted to life in a wide range of habitats. Only a very small proportion of insects (less than 0.5%) are classed as pests and most of those are found in agricultural settings. However, those that are urban/public health pests do cause significant problems. There are a number of useful entomology texts and websites that provide more detailed information on the general biology and development of insects [46]. In addition, the British Pest Management Manual [47] as well as the CIEH National Pest Advisory Panel (NPAP) guides [48] for example, provide useful summaries of the life cycles of the more common urban insect pests.

Cockroaches – Cockroaches are omnivorous insects with mouthparts adapted for chewing, enabling them to feed on a wide range of organic materials. They exhibit paurometabolous metamorphosis (gradual or 'incomplete' metamorphosis) with the adults and nymphs feeding in similar environments and indeed studies on the diet of first stage nymphs confirmed that facultative coprophagy (feeding on faecal matter) assists their development to second stage nymphs by reducing the need to forage [49]. Less than 1% of cockroach species are classified as pests, with the vast majority living in the tropics among dead or decaying leaves or plant debris [50]. However, a small number have developed an anthropophilic habit and are found where people prepare or store food. Whilst they prefer starchy/sugary foods, they will feed on a wide range of substances, including dead insects, blood and faeces. As part of their digestive process, they regurgitate partially digested food and deposit faeces whilst feeding.

A number of authors have provided detailed reviews of the pathogens associated with cockroaches [51–52]. These include *Campylobacter*, *E. coli*, *Listeria*, *Salmonella*, *Shigella*, *Staphylococcus*, *Streptococcus* and *Yersinia* spp.

Cockroaches have been found in sensitive locations such as hospitals, nurseries and care homes and in commercial kitchens where it is highly probable that they are involved in disease transmission. However, it has been difficult to prove categorically in these situations that cockroaches were the sole vector and that all other means of disease transmission could be discounted [53].

Cockroaches evoke strong responses and there is often significant psychological distress and social stigma attached to their presence. In addition, they produce secretions which give a persistent and characteristic odour to areas visited by them [54].

Cockroach allergens are known to be responsible for allergic and atopic reactions. Faecal material and skin casts have been shown to play a key role in the development of human allergies and asthma. Female cockroaches produce more allergen Bla g 1 (25,000–50,000 units of allergen) in their faeces compared to males (2000–3000 units) [55–56]. Cockroaches are also very mobile within premises spreading allergens over large areas. Demark and Bennett [57] and Liccardi et al. [58] outlined the following steps to minimise exposure to cockroach allergens:

- Remove waste food and water leakage which facilitate the growth of cockroach populations;
- The integrity of plasterworks and floors should be controlled to avoid cockroach access indoors;
- Use a range of available chemical agents for cockroach extermination in different formulations;
- Use an intensive vacuum cleaner for removing cockroach materials after extermination with chemical agent;
- Encourage compliance by providing educational material for the public. [58]

Blatella germanica (the German cockroach) and *Blatta orientalis* (the oriental cockroach) are the two most common cockroach species in the UK (for pictures and help in identifying the different species see the CIEH NPAP guides)[7] although several other species may also be found including *Periplaneta americana* (the American cockroach) and *Periplaneta australasiae* (the Australian cockroach). Importation of fruits such as bananas from America and Australia are usually the source of these infestations. *Supella longipalpa* (the brown-banded cockroach) is more common in the South of England, with sightings in more northern regions likely to be associated with people moving from South to North.

Accurate data on the distribution of cockroaches within the UK remains elusive. However, most authors agree that *Blatta orientalis* is more common than *Blattella germanica* [59]. Although both species are reported indoors, there have been a number of reports

confirming that *Blatta orientalis* is capable of surviving outdoors. Brief details about the two common UK cockroach species are provided next.[8]

Blatella germanica is a cosmopolitan pest that has adapted well to living in close association with humans. It is light brown in colour and around 12–15 mm in length and the key identifying feature is the two parallel black stripes on its prothorax (which is most pronounced in the adult). Its preferred temperature range is between 24–33°C and is therefore suited to centrally heated premises. This species can survive adverse temperatures for a short period and may have contributed to its widespread distribution.

In domestic dwellings it is most likely to be found in kitchens, adjoining storage rooms, bathrooms and airing cupboards. In flats they may be found in all rooms, but often congregate around refuse chutes and boiler heating ducts. In commercial food premises, infestations are focused around kitchen equipment (for example congregating around the compressor at the rear of refrigerators) and in the structure (for example, behind stainless steel splash backs or cracked tiles).

Adult German cockroaches live for around three months and during that time females will produce four–eight ootheca (egg cases) containing 37–44 nymphs (and approximately 90% of these will hatch). Adult males mate repeatedly, but females usually mate only once. The ootheca is carried by the female whilst it develops over approximately two weeks, until just before the nymphs hatch. Nymphs moult five–seven times over 30–60 days (depending on the resources available and ambient temperatures). Nymphs and adults are very active and adept climbers. They have been recorded moving up to 75 metres over a five-day period [57]. Their prolific breeding potential can result in significant infestations developing over a short period of time.

Blatta orientalis, the oriental cockroach is larger than the German cockroach at around 20–27 mm in length and is reddish brown to black in colour, with nymphs tending to be darker than adults. It is found in northern, temperate regions and its preferred temperature range is 20–29°C. It is unable to climb smooth surfaces and its distribution in buildings tends to be more limited than that of the German cockroach. It is found in cooler areas such as basements and cellars, drainage pipes, sewers and in subfloor spaces. Adults are less mobile than nymphs and tend to stay around their harbourages.

Adult oriental cockroaches live for around six months, but under favourable conditions may survive for up to two years. Females produce around eight ootheca with 16–18 nymphs in each. Oothecas are dropped shortly after being formed and are usually sited near a food source. Development within the egg case usually takes around 40–50 days, but in colder conditions may remain dormant until more favourable conditions arrive.

Bedbugs – Bedbugs are reddish brown, wingless, dorso-ventrally flattened insects measuring between 5–10 mm in length, depending on how recently they have had a blood meal. Both males and females take blood meals and feeding usually occurs at night. Bedbugs have a very flat body shape and can hide in virtually any crack or crevice, preferring dark, isolated and protected areas [59]. This, combined with their ability to withstand starvation for considerable periods of time, results in them being difficult to control. In the early 1900s bedbugs were common in many homes and tended to be associated with rundown and poorly maintained properties [60].

After several decades of declining infestations, the bedbug (*Cimex lectularius*) has seen an impressive global resurgence in numbers. A survey of Local Authority pest management services in the UK in 2009 found that almost 40% were reporting an increase in bedbug infestations in their area. They cited high rise multi-occupier flats with transient populations as the most common location for bedbugs and the most challenging to treat because of difficulties in negotiating access to each of the flats within the block. The recent resurgence in developed countries appears to have started almost synchronously in the late 1990s in Europe, the United States and Australia [61], with infestations reported regularly in new locations such as upmarket hotels. These infestations can, in part, be traced back to recent overseas travel, but other factors such as insecticide resistance may also be important. Although bedbugs are not considered to be important in the transmission of disease, their biting behaviour can result in secondary infections from the intense itching resulting from their bites. Their presence however can cause significant distress, anxiety and disgust.[9]

Flies – The Diptera are among the most diverse insect orders, with approximately 150,000 described species globally. These insects are diverse not only in species richness but also in their structural variety, ecological habits and economic importance. Around 3000 species are found in Britain, although only a small number are classified as public health pests. They are distinguished from other insects that are sometimes called flies by the presence of only one pair of front wings. The dipteran life cycle includes a series of distinct stages or instars. A typical life cycle consists of a brief egg stage, three or four instars, a pupal stage of varying length, and an adult stage that lasts from less than two hours to several weeks or even months. Several families of Diptera are of major economic importance and involved in the transmission of more disease pathogens to humans and other animals than any other group of arthropods. Biting flies cause annoyance that impacts tourism, recreation, land development, and industrial and agricultural production, whereas their effects on livestock can cause reduced milk, egg and meat production [62].

Flies develop via complete metamorphosis and control measures must focus on correctly identifying the species and addressing *all* stages of the life cycle. Adult flies feed by ingesting liquid, and their mouthparts and feeding habits have evolved to allow them to feed on a diverse array of substances, from mosquitoes feeding on blood to phorid flies feeding on decaying organic matter [63].

From a public health perspective, flies can be split into two main categories:

1   Those that feed on human food (causing harm through contamination of food, indirect transmission of pathogens and statutory nuisance)
2   Those that feed on human blood (causing harm in the UK mainly through biting nuisance, but with the potential to directly transmit pathogens)

Flies that feed on human food are a major concern to the EHP. When an adult fly lands on solid food, it must convert it to a liquid to facilitate ingestion. Flies produce large quantities of saliva together with regurgitated gut contents. This mixture, rich in digestive enzymes, may also include bacteria, viruses and protozoa from their gut. It is vomited onto the surface and then sucked back up.

This may be repeated several times and the fly may also then defecate to reduce its overall body mass before flying off. This feeding pattern underlies the principal mode by which flies contaminate food with disease pathogens and spoilage organisms [60]. Table 17.3 provides a brief summary of the more common flies and the situations where they are likely to be found.

Localised congregations of adult flies often indicate nearby breeding sites. Fly marks on florescent lighting, window and door frames and walls all provide indicators of the presence of flies. They lay their eggs in moist environments and development of the larval and pupal stages requires moisture. Blocked drainage channels and gully traps offer ideal breeding conditions. Accumulations of organic matter such as vegetable/animal waste both internally (around machinery and equipment) and externally (in dustbins, waste skips and pallets) could also provide potential breeding sites.

Table 17.3 Common urban flies and their habitats

Species	Size	Habitat
Common housefly (*Mus domesticus*)	6–8 mm long 3–15 mm wingspan	Rubbish tips, refuse storage areas, pig and poultry farms
Lesser housefly (*Fannia canicularis*)	5–6 mm long 10–12 mm wingspan	Indoors in both domestic and commercial food premises; poultry houses
Fruit flies (*Drospholia replete; Drospholia funebris*)	2–3 mm long 3–4 mm wingspan	Anywhere that contains decaying matter in a moist environment, such as drains, mops, sponges and recycling bins, also breweries, fruit drink manufactures (and anywhere there is ripening or rotting fruit), vinegar factories, public house bar areas, empty soft fruit drink bottles, etc.
Bluebottles or blowflies (*Calliphora* spp) Green bottles (*Lucilia* spp) Flesh flies (*Sarcophaga* spp.)	9–13 mm long 18–20 mm wingspan	Meat processing plants, canteens, food factories, retail food outlets, waste storage areas
Moth flies/sewer flies/owl midges/filter flies (*Psychoda*)	2 mm long characteristic hairs on wings	Sewage works, filter beds, manure heaps, etc.
Phorid/scuttle flies (*Phorida*)	3–4 mm long 9–10 mm wingspan	Faecal/organic matter, blocked or leaking drains/sewers or defective sanitary ware.

In the past, the common housefly was an important pest in domestic dwellings and was associated with the transmission of diseases such as typhoid and dysentery. However, the increasing use of insecticides and improvements in hygiene reduced its importance in domestic settings and it is now more likely to be found in livestock units such as pig and poultry houses. Conversely, the lesser housefly is now more common in houses than the common housefly. It prefers cooler areas within a dwelling and is less likely to settle on foodstuffs and therefore less likely than the common housefly to spread disease.

Fruit flies (family Drosophilidae) are associated with fermenting materials, such as decaying fruit and vegetable and alcoholic beverages. They can contaminate surfaces and food and may carry a number of pathogenic bacteria (such as *Salmonella*, *Shigella* and *E. coli*) and may therefore play a role in disease transmission because they may move from faecal material to other food [63]. Although both red- and dark-eyed drosophila may be found in domestic and food premises, the dark-eyed form (*Drosophila repleta*) is associated with un-cleaned drains and equipment and will feed on faecal matter, posing a greater public health risk than the red-eyed form (*Drosophila funebris*).

Blowflies (family Calliphora) are associated with rotting meat/animal remains and may be found in meat processing plants, canteens, retail food outlets and waste storage areas. They lay eggs in clusters of up to 600 that hatch after 18–48 hours, depending on temperature. Maggots feed for 8–11 days during which time they undergo three larval moults. When fully grown they migrate away from the food source and pupate. Adult flies emerge in around two weeks and live for just over a month. They have been shown to transmit a range of pathogenic bacteria (including *Salmonella*, *Campylobacter*, *Shigella*, *E. coli* and *E. coli* 0157-H7) through contaminating the surfaces they land on.

Phorids or scuttle flies (family Phoridae) are found on wet, damp or decaying organic matter, usually associated with blocked or leaking drains or sewers where faecal matter may have seeped into the soil substrata. Their common name derives from their habit of running or scuttling rather than flying away when disturbed. These flies will readily contaminate surfaces they alight on because they frequent unsanitary locations and have the potential to transmit disease-causing bacteria. Control measures should focus on actively locating and repairing any defective drain or sanitary ware to eliminate the infestation.

Owl midges, drain or filter flies (*Psychoda* spp) are small flies approximately 2 mm long, grey in appearance with hairy wings held tent-like over the body when at rest. Eggs are laid in wet, decaying matter in batches of approximately 200 and take between one–six days to hatch. The larvae grow to approximately 9 mm in length and feed on decaying organic matter, particularly of bacterial or fungal origin in drains or gullies. Development takes between 10–50 days depending on temperature with as many as eight generations in a year. They can be an indicator of poor hygiene as they are strongly associated with drainage issues and the build-up of sludge in drains and sink overflows in domestic and commercial premises. They may be found infesting filter beds at sewage works. They can be seen as a nuisance when numerous.

Flies that feed on human blood are however another public health problem and a number of UK flies feed on human blood. The most important of these are the mosquitoes and the biting midges. The UK has around 30 species of mosquitoes (or gnats) which belong to the family Culicidae (divided into two subfamilies, the *Anophelinae* and *Culicinae*) and over two-thirds of these are known to readily bite humans. In the UK the main problem associated with mosquitoes is biting nuisance. Ague, a form of malaria, was once present in Britain and caused significant morbidity and in some instances, mortality, in marshland and coastal areas, but declined by the late 1800s as a result of draining the marshes, improvements in housing, climatic changes and the availability of anti-malarial drugs [64]. All six species of British *Anopheles* are capable of transmitting malaria and with predicted climate change there are concerns that exotic species or new variants of native species could establish in the UK and vector serious diseases such as West Nile Virus and Chikungunya.[10] Environmental initiatives such as the Great Fens Project which aims to restore over 3000 acres of fenland could re-establish suitable breeding habitats for a number of mosquito species that in the past vectored ague.

The Asian Tiger mosquito (*Aedes albopictus*)[11] has shown an impressive extension in its range and many of the invasions in Europe and the USA have been linked to the trade in used tyres. This species lays drought resistant eggs that remain viable and hatch following local rainfall. *Aedes albopictus* vectored an outbreak of Chikungunya in Italy in 2007. Whilst this infection was known to be endemic in parts of Africa, South East Asia and India, the outbreak in Italy was the first time this tropical disease had been transmitted in a non-tropical area by native mosquitoes. Chikungunya virus was first identified in Tanzania in 1952 and for the following ~50 years was isolated and caused occasional outbreaks in Africa and Asia. Since 2004, Chikungunya has spread rapidly and been identified in over 60 countries throughout Asia, Africa, Europe and the Americas.[12] In 2013 The European Centre for Disease Prevention and Control (ECDC) reported 72 cases, with France, the UK and Germany observing the most cases. The need for effective surveillance and monitoring, particularly at UK ports is paramount (see Chapter 21).

Fleas – Fleas are ectoparasitic on mammals and birds, with their mouthparts modified for piercing host skin and sucking blood. Adult fleas usually live for 6–12 months, and the female will lay around 300–1000 eggs, mostly in small batches of about 3–18 a day, with four–eight eggs laid after each blood meal. Larvae feed on almost any organic debris including the host's dead skin cells, faeces

and partly digested blood evacuated from the digestive tract of adult fleas.

The role of the tropical mouse flea (*Xenopsylla cheopis*) in the transmission of the Black Death has been well documented, with bubonic plague killing around a third of the European population within a few decades [65] in the mid-14th century. Annually, the World Health Organization reports around 1000–2000 cases of plague worldwide each year (with over 95% of cases recorded in Africa). However, this flea is rarely reported in the UK, following the demise of the black rat.

The most common fleas encountered in domestic dwellings are cat fleas (*Ctenocephalides felis*) and bird fleas (*Ceratophyllus gallinae*). Whilst there is little evidence that cat fleas vector disease, they are capable of transmitting *Bartonella henselae* [66]. This tends to be sub-clinical in both pets and people, but immuno-compromised patients may be at greater risk. The main problem associated with fleas in domestic settings relates to the significant discomfort caused by their biting which may result in secondary infections due to the intense itching that occurs after being bitten. In domestic settings, identifying the species involved is important in helping to track down the possible source (e.g. pet bedding for cat fleas; birds' nests in roof spaces for bird fleas, etc.) and planning appropriate control measures which target *all* stages of the life cycle.

Carpet beetles[13] – The adult varied carpet beetle *Anthrenus verbasci* is not dissimilar in size and shape to a ladybird and is attracted to daylight and is often found on window sills. Adults are capable of flight and are mainly nectar feeders. It is the larval stage (known as 'woolly bears') of the varied carpet beetle that causes problems. It is brown with bunches of golden hairs at the tip of the abdomen. They eat any high protein materials and in domestic settings may be found infesting woollen carpets, airing cupboards and wardrobes. In may also be found in commercial premises such as flour mills, warehouses and stored product packaging. Birds' nests are often the source of the infestation and will need to be removed to prevent re-infestation.

Casual invaders and other public health pests

There are a number of pests such as wood-lice, earwigs, ground beetles, slugs and snails which may invade domestic premises at certain times of the year and whilst not of any public health significance, can be disturbing to residents from their nuisance value. The following table (Table 17.4) summarises these pests. Clearing away vegetation from around buildings and proofing obvious entry points should control most of these invasions.

Ticks

Ticks are arachnids (belonging to a group which includes spiders, scorpions and mites) and they feed on vertebrate blood. *Ixodes ricinus* (the deer or sheep tick)[14] is an indigenous hard tick species having a wide geographical distribution and is the most common hard tick in the UK and when unfed is around 6 mm in size. However, following a blood meal, it can increase its body weight by 20–150 fold.

Ixodes ricinus is involved in the transmission of a large variety of pathogens of medical and veterinary importance, including:

- *Borrelia burgdorferi s.l.* causing Lyme borreliosis (Lyme disease);
- Tick-borne encephalitis virus, *Anaplasma phagocytophilum*, causing human *granulocytic ehrlichiosis*;
- *Francisella tularensis* causing Tularaemia;
- *Rickettsia helvetica* and *Rickettsia monacensis*;
- *Babesia divergens* and *Babesia microti* responsible for Babesiosis;
- Louping ill virus and Tribec virus.

Lyme disease is the most common vector-borne disease in the United States. It is transmitted to humans through the bite of infected ticks. Typical symptoms include fever, headache, fatigue and a characteristic skin rash called erythema migrans. If left untreated, infection can spread to joints, the heart and the nervous system.[15]

Lyme disease is monitored in England and Wales through routine surveillance. Data is published in the quarterly Health Protection

Table 17.4 Casual invaders

Species	Description	Habitat
Earwigs	Light to dark brown, 12–20 mm in length with leathery short wing cases and pronounced pincers	Principally a carnivorous outdoor pest of horticulture. Will invade dwellings found in cracks and crevices around door frames, skirting boards, etc. Females are strongly maternal.
Woodlice	Slate grey in colour, ovoid segmented body 15 mm in length with seven pairs of legs	Prefers a damp environment, feeds mainly on plant material, rotting vegetation, decaying wood. Harbourages, cracks and crevices, subfloor spaces, etc.
Ground beetles	Shiny black to dark brown, 6–20 mm in length, depending on species.	Essentially carnivorous, outside pests which may enter buildings looking for shelter.
Slugs and snails	4–10 cm depending on species, grey/fawn in colour. Underside, or sole, whitish with a darker zone along the centre. Mucus colourless or white.	Prefer dark, damp habitats feeding on a variety of living plants as well as on decaying plant matter. Most active at night and during cloudy or rainy days.

Report and annually in the UK Zoonoses Report.

Cases of Lyme disease are not statutorily notifiable by medical practitioners in England, Wales and Northern Ireland. Since October 2010 under the Health Protection (Notification) Regulations 2010, every microbiology laboratory (including those in the private sector) in England is required to notify all laboratory diagnoses of Lyme disease to the UK Health Security Agency (UKHSA which was formed from what was previously Public Health England). It has been estimated that there are between 1000 and 2000 additional cases of Lyme disease each year in England and Wales that are not laboratory diagnosed.

Laboratory-confirmed reports of Lyme disease have risen steadily since reporting began in 1986. Mean annual incidence rates for laboratory-confirmed cases have risen from 0.38 per 100,000 population for the period 1997–2000, to 1.64 in 2010, and to 2.70 cases per 100,000 population in 2017 [67]. There has been over a five-fold increase between 2001 and 2017.

The increase in UK acquired infection may be related to a number of factors, including better diagnostics and enhanced surveillance, successive mild winters, increased recreational travel (such as hiking, trekking and mountain biking) into forests, heaths, moorland and sub-urban parkland where ticks may be present.

Although it is not a notifiable disease in England, Wales and Northern Ireland it is in Scotland. Lyme disease may be asymptomatic, but outcomes can be debilitating.

Clinical presentations include facial palsy, meningitis and radiculopathy (spinal nerve root inflammation) occurring within weeks or months of infection. Up to a third of cases in the UK may present with localised or general symptoms without evidence of a rash. Lyme arthritis is a rare complication of infections acquired in the UK but is more common in patients who have been infected in North America or central Europe

(*Borrelia burgdorferi sensu stricto*, the genospecies strongly associated with Lyme arthritis, is rare in the UK). Acrodermatitis chronica atrophicans (ACA), a skin condition caused by long-standing infection, which occurs in Scandinavia and central Europe, is seen occasionally in the UK.

Current advice to reduce the risk of being bitten includes wearing appropriate clothing, using a DEET-based insecticide repellent and checking thoroughly for ticks on the skin at the end of a day's activities in at-risk locations.

Birds

Wild birds form an important part of both the urban and rural landscape and are protected by the Wildlife and Countryside Act 1981 as amended, which transposed the EC Wild Bird Directive (79/409/EEC) within Great Britain. The basic principle of the Act is that all wild birds, their nests and eggs are protected by law and some rare species are afforded special protection. However, Section 16 (*power to grant licences*) permits the issue of general licences to control protected species of birds for specific purposes (e.g. preventing damage, preserving public health and safety, or air safety).[16] A general licence permits activities that would otherwise be criminal offences under Part 1 of the 1981 Act. In accordance with Section 16 of the 1981 Act, the Secretary of State has consulted with Natural England as to the circumstances in which, in their opinion, general licences should be granted. Only an authorised person (for example a landowner or occupier) can act under these general licences and must act in accordance with the standard licence conditions, but they may kill or take, in certain situations and by certain methods, so-called 'pest species' and destroy or take the nest or eggs of such a bird.

The Class CL03 licence permits owners and managers of food premises (and any persons they authorise to act on their behalf) to catch species of wild birds listed on this licence,[17] that have become trapped in the building.

Any birds captured under this licence must be released unharmed outside the premises. This licence may only be used for the purpose of preserving public health or public safety and only at food premises either that are owned or occupied by organisations registered with Natural England or that have been individually registered with Natural England. An individual food premises can be registered to use this licence or the representative of a group of premises under the same ownership can register. Anyone acting under the licence must be competent to undertake the actions planned to remove the bird or birds.[18]

Birds can act as potential vectors of zoonotic diseases. Feral pigeons have tested positive for a number of pathogens, including *Cryptococcus neoformans*, *Chlamydophila psittaci*, *Salmonella enterica*, *Aspergillus* and *Candida parapsilosis* and directly vector psittacosis and ornithosis (for a more detailed discussion see Haag-Wackerngael and Moch [68]. See also Pellizzari and Loughlin D [69] for a discussion on the humane control of urban pigeon populations which also includes a discussion of some of the health risks).

Birds, such as gulls, can cause nuisance and disturbance through their feeding behaviour or through their calling. The CIEH NPAP has published on bird species which includes the control of urban gulls.[19]

Birds can also cause indirect damage to structures as their droppings may increase the rate of lichen growth and their nests may block drains and gutters, resulting in water damage. Birds' nests and bird guano can act as a focus for insect and mite infestations which then migrate into buildings or dwellings causing biting nuisance. This is often most noticeable after birds have fledged when, for example, mites may migrate down cavity walls and enter bedrooms through air vents to bite occupants. Residents often mistakenly assume that bites are being caused by bedbugs or fleas and this underlines the importance of spending time in identifying the pest(s) present before embarking on pest management measures.

Stored food pests

Many of the common stored product insect pests belong to two orders – the Coleoptera (beetles) and the Lepidoptera (moths). The larval stage is often the most damaging, causing quantitative and/or qualitative losses via their larval by-products – frass and/or webbing. These may cause blockages in food equipment and machinery and can contaminate finished products. Table 17.5 presents some of the more common stored product beetles and moths and the likely locations where they should be found. However, it is not a comprehensive list and where possible, samples must be obtained to confirm identification so that appropriate control measures are applied.

Psocids

Booklice (*Psocids* spp.) are a common pest both in the domestic and commercial settings. They are soft bodied, wingless insects 1–2 mm in length, are fast runners and are pale straw to dark brown in appearance with biting mouthparts. They require an environment with a high humidity of 70% with a preferred temperature of between 20–25°C, although some species may prefer cooler conditions. They feed on moulds, yeasts or mildews growing on cardboard packaging, wooden pallets, etc. and will find harbourage in cracks and crevices that may be found in domestic kitchens, for example between the shelving and chipboard carcass of food cupboards, underneath paper labels of tinned produce and on wooden pallets of produce stored in damp warehouses. They also feed on raw materials such as oilseeds, processed cereals, cereal grains and powdered milk. They produce one generation a year and can overwinter as nymphs. The main damage associated with psocids is contamination and reducing humidity and improving general housekeeping should reduce infestations.

Silverfish (*Lepisma saccharina*) are very primitive insects and are a nuisance pest. They

Table 17.5 Stored food pests

Species	Description	Habitat
Biscuit beetle *Stegobium pancieum*	2–4 mm in length, elongated reddish brown with head tucked under thorax	Bakeries, pantry stored products (e.g. biscuits, pet food, cereals, spices, nuts, etc.)
Larder beetle *Dermestes lardarius*	7–9 mm in length, oval in shape with a white saddle on elytra	Kitchens – poor housekeeping, birds nests
Rust-red flour beetle *Tribolium casteneum*	2.3–4.4 mm in length, elongated mahogany brown with distinct short three-clubbed antennae	Flour mills, warehouses, pantry stored flour, etc.
Confused flour beetle *Tribolium confusum*	3–4 mm in length, elongated red to dark brown body with short clubbed antennae	Flour, cereal products, grain, nuts, dried fruit, etc.
Grain/rice/maize weevils *Sitophilus granarius/oryzae/ zeamais*	3–4 mm in length, dark brown with curved proboscis and elbowed antennae	Maize, barley, rice, millet, biscuits, peas, etc.
Saw-toothed grain beetle *Oryzaephilus surinamensis*	2.5–3.5 mm in length, dark brown with six-toothed projections on thorax	Dried fruit, dried soup, nuts, processed foods, grain, etc.
Mealworm beetle *Tenebrio molitor*	15 mm in length, shiny dark brown to black elongated body, unclubbed antennae	Birds' nests, dead pigeons, cereals, stored food products, etc.
Australian spider beetles *Ptinus tectus*	2.5–4 mm bulbous body with distinct waist, long antennae and legs.	Abandoned birds nests, seeds, grain, spices, nuts, etc.
Warehouse moth *Ephestia elutella*	8–10 mm in length, with wingspan of 14–17 mm, grey brown, speckled or banded wings	Food storage warehouses, cereals, grain, nuts, dried fruit
Mill moth *Ephestia kuehniella*	10–14 mm in length, wingspan 20–22 mm, blue grey forewings with dark navy bars and spotted wingtips	Serious pest of flour milling and baking industry, inside plant dead spaces. Characteristic webbing. Feeds on cereals, bran, nuts, cereal-based products.
Indian meal moth *Plodia interpunctella*	8–10 mm in length, 14–20 mm wingspan, wings yellow, bronze to reddy brown outer halves of wings	Warehouses, dried fruits and nut processing industry. Dried fruit, vegetables, nuts, chocolate.

have a tapered body 15–20 mm in length, are silver in appearance with long antennae and three distinctive bristly tails. Eggs are laid in cracks and crevices such as skirting boards, door frames, etc. and undergo ten nymphal stages before reaching adulthood over a 12-month period. They are nocturnal and found in warm damp environments such as bathrooms. They feed on dead insects, dried powdered foods, book binding gums, textile and leather goods. Reducing the humidity will normally control these insects.

Ants

Black garden ants (*Lasius niger*) are essentially an outdoor pest that nests under walls and paving slabs but will invade both domestic and commercial premises in search of food. Worker ants are dark brown to black in appearance, with a single segmented waist. Worker ants are approximately 3.5 to 5 mm in length and the queens are around 15 mm. Whilst not significant as a public health pest, they can cause contamination of food stuffs and work surfaces when foraging for food. In mid-summer, usually on a hot humid day, queen ants will swarm and mate with males on the wing. On returning to the ground, the mated queen will bite off her wings and dig a cell in the soil in which to overwinter. Eggs will be laid in the following spring and the legless larvae will be fed on secretions from the queen's salivary glands until fully grown. After pupating, the first brood of worker ants hatch. The entire cycle takes approximately two months to complete and a nest may be active for several years.

Lasius neglectus appeared 20 years ago in Turkey and has since spread across the continent. It was first discovered in the UK at a National Trust property in Gloucestershire, where an infestation of around 35,000 was exposed in the Hidcote Manor gardens. Five years later – and 100 miles east – the ants were discovered in a London home and another site was identified in Cambridgeshire. The ants have been discovered in Yorkshire and two sites on the south coast. This ant is in fact smaller than the common native UK variety. It builds 'super' colonies with many queens and interconnected nests which can stretch for miles. The alien species can out-compete native ants commonly found in UK gardens. Native ants form new colonies when a queen vacates the nest and starts her own, competing for food and space with the old colony. By living in extended colonies in large numbers, the super ant is able to out-compete for aphids and space.[20]

Fire ants (*Myrmica rubra*), are a red and brown coloured ant known for stinging and biting people and pets for accidentally disturbing a nest. They are an extremely aggressive ant species with a length of 4 to 6 mm. Fire ants live under stones, piles of pine leaves

and underneath logs. This species of ant is very active all throughout the year. When the weather is humid and hot they begin their mating cycle. During that time of the year, you can see flying red ants, red male flying ants and red queen flying ants. Once mating is complete the new fire ant queens will then find a new home to start a colony.

Perhaps the most important exotic species of ant in the UK is the Pharaoh ant (*Monomorium pharaonis*), which has a worldwide distribution associated with human habitats [70]. These ants originated from tropical regions and in the UK are confined to heated premises, with relatively high humidity. They are most commonly found in hospitals, restaurants and other large-scale food premises, hotels and domestic high-rise flats. Their nests are often in inaccessible locations, which makes their detection and eradication difficult. Edwards and Baker [71] reported that Pharaoh ant infestations were present in 10% of the UK hospitals they surveyed, with London and its suburbs being most heavily infested. Apart from causing nuisance, these ants can cause contamination in food premises and public health problems in hospitals. They are known to spread a number of pathogens, including *Streptococcus* and *Staphylococcus* spp. [72] and were observed feeding on wounds and found inside sterile dressings having bitten through the packaging.

Section 1 Notes

1 https://www.asthma.org.uk/about/media/facts-and-statistics/
2 In England 13 children under 14 years died from asthma in 2016 according to figures from the NHS see https://www.england.nhs.uk/childhood-asthma/
3 A synanthropic species is one that lives near, and benefits from, an association with human beings and the habitats that humans create, while commensal and describes the situation better than commensal which although also means two species living in close association, in commensalism no harm is caused to either which is not strictly true for rodents under discussion here.
4 CDC says for example that Lymphocytic Choriomeningitis (LCM) is a disease directly

transmitted by house mice such as by breathing in dust that is contaminated with urine or droppings or direct contact with mice or their urine and droppings (https://www.cdc.gov/rodents/diseases/direct.html).

5 A recent study in Trento, Italy revealed rodent-borne hanta- and arenaviruses are an emerging public health threat. Two pathogenic hantaviruses (Puumala and Dobrava-Belgrade virus) and one arenavirus (Lymphocytic Choriomeningitis Virus) are known to circulate in rodent reservoirs and the study concluded that the general exposure of residents in the Alps to these viruses has probably increased during the last decade given the increasing human seroprevalence of hantaviruses and LCMV in the Province of Trento. The study recorded an increased circulation of DOBV-Af among rodent hosts, so the probability of transmission of these viruses from rodents to humans has increased. See Tagliapietra V, et al. (2018) Emerging rodent-borne vitral zoonoses in Trento, Italy. *EcoHealth*, 15: 695–704. https://doi.org/10.1007/s10393-018-1335-4.

6 For example, although complaints remained static in Sydney, there was an increase and then decline in captures; see Bedoya-Perez MA, et al. (2021) The effect of COVID-19 pandemic restrictions on an urban rodent population. *Scientific Reports*, 11(1): 12957. https://doi.org/10.1038/s41598-021-92301-0. PMID: 34155237; PMCID: PMC8217515.

7 See in particular Pest Control Procedures Manual – cockroaches. https://www.urbanpestsbook.com/downloads/procedures-pest-specific/

8 See https://www.rentokil.co.uk/cockroaches/species/ for help in identification of the different species.

9 The common bedbug (*Cimex lectularius*) has long been a pest – feeding on blood, causing itchy bites and generally irritating their human hosts. The Environmental Protection Agency (EPA), the Centers for Disease Control and Prevention (CDC), and the United States Department of Agriculture (USDA) all consider bedbugs a public health pest. However, unlike most public health pests, bedbugs are not known to transmit or spread disease. https://www.epa.gov/bedbugs/introduction-bed-bugs.

10 See for example Medlock JM, Leach SA. (2015) Effect of climate change on vector-borne disease risk in the UK. *The Lancet*, 15(6): 721–730. http://www.thelancet.com/journals/laninf/article/PIIS1473-3099(15)70091-5/fulltext (Accessed 7 July 2015).

11 See https://www.cdc.gov/mosquitoes/about/life-cycles/aedes.html for more information.

12 See https://www.who.int/news-room/fact-sheets/detail/chikungunya

13 See https://www.nhm.ac.uk/take-part/identify-nature/common-insect-pest-species-in-homes/carpet-beetles-identification-guide.html

14 See https://www.ecdc.europa.eu/en/disease-vectors/facts/tick-factsheets/ixodes-ricinus for more information

15 See https://www.nice.org.uk/guidance/ng95/chapter/recommendations#clinical-assessment see also NICE Guideline [NG95] of 2018 at https://www.nice.org.uk/guidance/ng95

16 See https://www.gov.uk/government/publications/wild-birds-licence-to-kill-or-take-for-conservation-purposes-gl40/gl40-general-licence-to-kill-or-take-certain-species-of-wild-birds-to-conserve-endangered-wild-birds-or-flora-and-fauna

17 The birds listed are blackbird (Turdus merula), blue tit (Cyanistes caeruleus – formerly Parus caeruleus), dunnock (Prunella modularis), great tit (Parus major), house sparrow (Passer domesticus), pied wagtail (Motacilla alba), robin (Erithacus rubecula), song thrush (Turdus philomelos), starling (Sturnus vulgaris).

18 https://www.gov.uk/government/publications/birds-licence-to-catch-them-on-food-premises/licence-to-take-birds-trapped-in-food-premises-to-preserve-public-health-or-safety-cl03

19 See https://www.urbanpestsbook.com/downloads/procedures-pest-specific/

20 See https://nerc.ukri.org/planetearth/stories/1818/

SECTION 2: PEST MANAGEMENT STRATEGIES AND INTEGRATED PEST MANAGEMENT

The continued shift towards increasing urbanisation combined with the persistence of many pest species, despite considerable efforts to control them, underscores the need for effective and sustainable pest management strategies. The global proportion of urban populations rose from 13% (220 million) in 1900, to 29% (732 million) in 1950. According to the World Bank, by 2020 some 55% of the world's population – 4.2 billion inhabitants – lived in cities. This trend is expected to continue. By 2050, with the urban population more than doubling its current size, globally nearly seven out of ten people will live in cities.[21]

Effective pest management strategies must focus on prevention, be outcome driven and environmentally sustainable. The growing body of evidence-based research work highlights the need for a multi-disciplinary approach to effectively address the risks posed by ever-increasing exposure to pests in urban settings. Emerging from this are strategic, integrated approaches to managing pests in urban settings.

The principles of IPM and their application in environmental health activities

Integrated Pest Management (IPM) was a system developed in the agricultural arena, to provide a more holistic approach to crop protection. Burn, Coaker and Jepson [73] defined it as 'a control strategy in which a variety of biological, chemical and cultural control measures are combined to give stable long-term pest control'. Whilst the organisation of control measures for infestations affecting human health and well-being is somewhat different from those used in crop protection, many of the principles are relevant in urban and rural settings and the approach is highly relevant to EHPs who should take a wider view of any pest problem. As Sarisky et al. [74] stated, the focus of IPM is to eliminate the source of pest problems – that is, the conditions conducive to the establishment, survival and reproduction of pests. These principles should control pest infestations and pest access to people, their food, buildings and other 'at risk' urban settings. The differences between IPM and non-integrated pest control have been summarised by Sarisky et al. [74] (see Table 17.6).

The concepts of IPM are a key element within WHO's Global Strategic Framework for Integrated Vector Management [75]. Its focus is on the enormous benefits for health and socio-economic development that flow from effective control of vector-borne diseases. However, this proactive and holistic approach requires significant initial investment before the long-term benefits are realised. The US Centers for Disease Control and Prevention has produced a manual on IPM and conducting urban rodent surveys.[22]

General principles

EHPs should adopt a risk-based approach that considers the characteristics of each setting.

Table 17.6 Differences between Integrated Pest Management and non-integrated control measures

Pest management programme components	Non-integrated pest control	IPM
Programme strategy	Reactive	Preventive
Customer/public education	Minimal	Extensive
Potential liability	High	Low
Emphasis	Routine pesticide application	Pesticides used when exclusion, sanitation and other means are inadequate
Inspection and monitoring	Minimal	Extensive
Use of non-chemical controls	Minimal	Extensive
Positive identification of pests	Sometimes	Required
Use of pest thresholds	Minimal	Extensive
Outcome evaluation	Sometimes	Required

The most effective approach in all settings is to prevent pests from gaining access and establishing an infestation. The principles of good hygiene, rigorous waste management, sound construction, education and/or training on pest identification and spotting the early signs of infestation and clear guidelines on what to do once an infestation is confirmed are all vital. The visit should attempt to establish:

- The identity of the pest(s) present in that location;
- The size and extent of the infestation(s);
- The food source(s) and resting/nesting sites of the pest(s);
- The characteristics of the environment that may benefit the pest species concerned and/or facilitate/hinder control measures.

In addition to focusing on high-risk areas such as kitchens, the indoor and outdoor environments, adjacent/adjoining properties, accumulations of rubbish and waste disposal arrangements must be considered. When this information has been gathered, the next step is to define the elements and stages of a control programme and the outcomes required – including the pest tolerance/threshold levels. Control methodologies should follow a risk hierarchy approach as part of IPM, which initially focus on those methods that present the least risk to non-target species and/or the environment before more invasive and chemical methods. Therefore at the heart of any successful long-term control programme is careful consideration of the environmental factors that support and/or encourage the presence of pest species and evaluative mechanisms to review the short and long-term effectiveness of the treatments undertaken. In sensitive settings such as hospitals or nursing homes, where vulnerable groups are present, a zero tolerance/threshold level must apply. The next steps therefore are to confirm the desired outcomes (e.g. complete eradication, reduction in numbers, etc.) and then, in consultation with the property owner/occupier

and where appropriate other stakeholders (for example, in domestic dwellings, other residents (through resident associations), a social housing manager, planning, operational services, health and social care sections, etc.) develop a plan to address the key risks identified in that location. The specifics should include:

- A detailed plan of the non-chemical/chemical methods suitable for use at that site and agreed timescales/access, etc. (this may need to encompass the legislative framework);
- Application of the control measures followed by evaluation of the impact;
- A review of the success of the treatments and where control has not been achieved development of a revised control plan;
- A summary report to the owner/occupier and/or other stakeholders on the actions that they must take to ensure long-term maintenance of pest-free premises.

The methods of application, formulation and the range of products available to deal with pests have seen striking changes in recent years, with changes often driven by operator and environmental safety concerns. This has resulted in a number of products being added to, or removed from, the market. Broad spectrum organophosphates such as Fenitrothion have been phased out and replaced by more targeted pesticides such as gels, with precise application achieved through the use of bait guns or placing gels in sealed plastic insecticide bait stations, minimising human exposure. Emulsifiable concentrates have been superseded by formulations such as water dispersible granules and suspension concentrates, again, to reduce operator exposure. There is also concern for the effects on non-target species as well as animal welfare and so there have been changes in the way some products, and particularly rodenticides, can be used. Such changes can have an effect on the efficacy of the control programmes.

Surveying for pests

One of the reasons why pests are often able to increase in numbers is their ability to hide from view, including:

- Nocturnal activity patterns (for example rats and mice, cockroaches);
- Exploiting undisturbed niches (for example stored product pests);
- Human indifference to reporting and managing infestations.

The risks pests pose vary depending on the pest present (public health or nuisance pest), the setting in which they are found (cockroaches found in a hospital kitchen or a single ground beetle in a domestic dwelling), the numbers present (a heavy infestation of rodents in a high-rise block of flats or a single fly in a domestic kitchen) and the length of time they may have been present (chronic or acute infestations).

Settings – Whilst there are countless urban settings that could harbour pests, two main settings have been chosen in this chapter – domestic dwellings and food premises. Many of the principles of IPM used to deal with infestations in these two settings will, in practice, overlap, but there are important differences in the approaches that may be taken.

In both domestic and commercial food premises, pests may be divided into two main groups – those that live in the structure of the building or migrate into the building to feed and those that are resident within a food source. There may be some overlap in this definition, for example, mice could be classified as both visitors (living in various harbourages within the structure, such as roof spaces, under the floorboards, airing cupboards, behind plinths, etc.) and residents (living within raw products that have been introduced into the building in packaging, crates or in the food itself). The course of action taken in attempting to deal with infestations will be influenced by whether the pests are found to be residents and/or visitors.

With the former, the only course of action is to ensure that the infested food is destroyed, as contamination is likely to be the primary concern. In commercial food premises this food may be seized or surrendered and used as evidence to prove the presence of pests.

Where pests are 'visitors' and either entering the building from outside or living within the structure, the source of the infestation should be established and proofing measures instigated to restrict and/or exclude them.

Pests in domestic dwellings – As outlined at the start of this chapter, there is a growing body of evidence demonstrating the impact of pest species on human health. Well-being is a multi-faceted concept, and so the direct impact of pests is not always easy to quantify (although research on allergens has demonstrated the impact of particular pests on health). The results from a large, exploratory pan-European housing survey, commissioned by the World Health Organization are providing crucial evidence on the health-related outcomes of living in infested dwellings. The Large Analysis and Review of European housing and health Status (LARES) study was carried out between 2002–2003 in eight European Cities, obtaining information from 8,519 residents in 3,374 dwellings. Residents completed a detailed questionnaire, followed by a personal interview and then an inspection of their home. The questionnaire contained ten questions related to pests and insects. Six in ten dwellings had been infested in the previous year. The primary pests reported were ants, flies, cockroaches, mice, mites, bedbugs, rats, fleas and mosquitoes. Table 17.7 summarises some of the health outcomes associated with the presence of pests in domestic dwellings.

These results showed that those living in mouse-infested houses were twice as likely as those in pest-free dwellings to report depression and eight times more likely to report migraine/frequent headaches than those living in pest-free dwellings. Whilst pests may not directly cause disease, the LARES study confirmed that those living in infested premises were more likely to report conditions such as

Table 17.7 Health outcomes from living in infested premises

Pest/health problems	Odds ratio (95% confidence limits)
Mice present in the house	
Depression	2.21 (1.3–3.75)
Migraine/frequent headache	1.97 (1.17–3.34)
Mice present in flat	
Migraine/frequent headache	8.06 (4.05–16.04)
Cockroaches present in the house	
Migraine/frequent headache	3.26 (1.78–5.96)
Flies present in the house	
Asthma (diagnosed by physician)	1.73 (1.03–2.90)

depression/migraine and frequent headache/asthma than those living in pest-free dwellings. The growing evidence base of the impacts of pests on health underlines the importance of maintaining pest-free dwellings.

A number of issues require careful consideration when planning pest operations in a domestic setting and the risk assessment will, because of the nature of the setting, be more complex than would be considered in commercial food premises. The presence of vulnerable adults and/or children or the presence of pets (either living in the property or visiting) may influence the formulations used (for example, choosing to use block baits instead of loose grain to reduce accidental spillage), the presentation of the bait (for example, placing bait in a secure tamper-proof bait station, rather than an open tray to minimise accidental access, or using gels in preference to dusts where residents report health-related conditions such as asthma) and the control measures considered (for example chemical or non-chemical methods).

In addition to the information outlined previously it will be useful to confirm the:

- Tenure of the property (to ensure that owners discharge their responsibilities to control infestations within their properties and/or where treatments have been undertaken, to recharge if appropriate;

- Person who made the complaint (e.g. the owner occupier, tenant, landlord/neighbour);
- Pest(s) involved (public health or nuisance pest). Local authorities may provide an in-house pest control service which deals with common urban pests, others may contract out some or all parts of their service, so the pest present may not be covered in the work they undertake;
- Location and type of property. For example, mice have been shown to migrate along terraced properties using the roof space or subfloor space along the block. This will require a block treatment to ensure long-term control.

Where possible, dealing with pests in domestic settings should be undertaken through providing advice and guidance to the owner-occupier about the best methods to control infestations and highlighting the importance of proofing, maintaining good hygiene and ensuring that waste is properly stored and disposed of. Guidance on the services offered by their Local Authority, the importance of early reporting/resolution of infestations, other organisations that may be able to provide treatments and what they can do to treat infestations themselves will be crucial in ensuring that chronic infestations do not establish in domestic settings.

If access to a domestic property where infestations are suspected has been denied or attempts to address the problem have proved unsuccessful or lacking, EHPs may need to resort to the statutory instruments available to address the infestation. A later part of this chapter provides a summary of the powers to deal with infestations in domestic premises.

The HHSRS guidance[23] provides a useful list of preventive measures for domestic settings, which underscore the importance of integrated pest management principles in maintaining pest-free dwellings:

- Design/construction/subsequent maintenance of building should help it to be kept clean preventing build-up of dirt and dust;
- Personal washing/sanitation/food preparation/cooking/storage areas should be capable of being maintained in a hygienic condition;
- Reduction of the means of access by pests into buildings to a minimum;
- All internal surfaces easily cleaned/pest-resistant material to be used where possible;
- Dwelling exterior free of cracks and unprotected holes, otherwise grilles/other methods to be used for protection;
- Service ducting/roof/floor spaces to be effectively sealed but with suitable access if treatment is needed;
- Drain openings, WC basins to be sealed with an effective water tight seal;
- Drainage inlets for waste and surface water to be sealed;
- Any points in walls penetrated by waste, drain or other pipes or cables to be effectively sealed;
- Holes through roof coverings, eaves and verges to be blocked to deny ingress to rats/mice/squirrels/birds. Necessary holes to be covered by grilles;
- Adequate and closed storage for refuse awaiting collection or disposal outside dwelling;
- Suitable storage for refuse within the dwelling;
- Storage to be accessible to occupants but not be a danger to children; and
- Refuse facilities should not cause hygiene problems.

Pests in food premises

Pests are a common aspect in the statutory or voluntary closures of many food premises and are cited as a contributory factor in the majority of food complaints reported to Environmental Health Departments each year. Pests also feature prominently in many of the prosecutions taken under the Food Hygiene legislation. Much of the unfit food seized by or surrendered to EHPs is infested with or damaged by pests. Effective pest management is an integral part of good business practice and the costs associated with that are small in comparison to the distress caused to customers, the potential spread of disease, the damage done to the business reputation and the damage to buildings and fittings.

In devising a pest management programme for food premises, the emphasis must be on preventing pests from gaining access to the site and in reducing the conditions which could encourage or facilitate their presence. Encouraging owners to adopt good levels of hygiene, stock management and continued vigilance is essential. Their engagement in promoting an environment that results in sustained control and is part of the daily processes that all staff understand and act on is key. Reinforcement through regular training and providing information at work stations and sensitive/critical sites should promote the embedding of these principles into good working practices. The CIEH NPAP has published relevant guidance which is available in several languages.[24]

Although the visit/inspection is likely to focus on the inside of the building and pest ingress, checking the exterior environment (specifically arrangements for waste disposal and drainage) and assessing the influence of adjoining properties on pests (for example, a pet shop within a block which could be the focus for an infestation of cockroaches migrating into an adjoining food premises) are important. In preparing to undertake an inspection the following steps will be useful:

1. On arrival, ask for the site plan and pest control report book (if available). Reports that cover the last 12 months may give a clear indication of seasonal variations in the presence of pest species. If this is a return visit, bring the previous inspection report with you.

2. If staff are present, talk to them about the procedures in place to report the presence of pests.

3. Check for the signs of pest activity/infestation (e.g. fly spots, rodent droppings, etc.) or conditions (such as poor hygiene) that could encourage infestations to establish. There are a number of high-risk locations/activities in food premises which should be checked:

   a. Incoming raw ingredients and packaged food (pests may be introduced within the raw ingredients – how are these checked, is there clear separation between raw and final products to prevent cross infestation?)

   b. Storage and stock rotations (is there a schedule to check store rooms for signs of infestation?)

   c. Cleaning schedules (are areas of high risk identified and cleaned more frequently than less risky areas?)

   d. Temperature control (can temperatures be lowered to inhibit insect reproduction?)

   e. Refuse disposal, both internally and externally (how frequently is food waste removed from the site?)

   f. Personal hygiene and training of staff (are staff aware of their responsibilities in maintaining pest-free premises?)

   g. Equipment and services, particularly those that generate conditions that may support pest activity and/or provide harbourage, such as refrigeration, air conditioning or pipes/ducting. (Are these items and locations routinely checked, maintained or cleaned?)

Problems can arise where managers believe they have discharged their legal obligations by engaging a pest management company to control pests at their premises. Using a competent and qualified professional pest controller is important, but the business operator must also act on the advice of that person. In examining the pest control contract or agreement, the EHP should consider:

1. The quality of the company – such as membership of a reputable trade association. BASIS Prompt makes it easy to check whether a professional pest controller is suitably qualified and trained to carry out your pest control requirements in a safe, efficient and sustainable way[25]

2. The range of pests covered by the contract

3. The frequency of site visits

4. The areas of the building (both internal and external) that are covered by the contract

5. The mechanisms in place to report pest sightings

6. The plan of where bait/monitoring stations have been placed

7. The record of bait takes (if applicable)

8. Response times when pests are reported (ideally, this should be immediately, but often it will be rolled into the timing of the next routine visit)

Owners must be able to demonstrate that they are taking reasonable steps to prevent infestations. The record of visit provided by the pest control professional should include:

- Pest sightings log (recording who reported it and when);
- Pest control recommendations (confirmation of the pest, the treatment/active ingredient used, the recommendations

suggested to the site manager by the contractor to mitigate the problem);
- Site paperwork (site plan, COSHH assessments, zero point checklist, etc.).

This information will give an indication of the current approaches to control. During the visit, it may be advisable to check that the contract is active – there have been instances where owners have engaged a pest control contractor and following receipt of the associated paperwork, have then cancelled the contract. Evidence will therefore be available, but contractors are no longer servicing the premises. It is also important to check the records provided by the pest control professional for any pattern(s) in pest activity being reported or remedial works identified that may suggest that the business owner is not responding to the advice provided. Many infestations have resulted from a breakdown in communication between the business and the contractor, or from the business simply assuming that the problem was being controlled by the contractor.

In some food premises the use of pesticides as a control measure may be prohibited and other measures will need to be considered. In these settings it is crucial that comprehensive monitoring is in place to pick up early indications of the presence of pests in tandem with physical control methods. Proofing is likely to be the principal method used to control infestations in these settings and during the visit, it is important to confirm the efficacy of this approach.

Pest and vector control techniques: a case study

Understanding how infestations become established and spread in domestic settings is key in developing appropriate control measures [76]. This section focuses on applying the processes outlined earlier in an urban context. An urban setting that was well known to harbour chronic infestations

of mice was chosen as the study area. The boundaries were drawn to ensure that the study area reflected a 'typical' urban landscape, with a mix of housing types (terraced, semi- and detached housing, flats) and construction ages (from late 1800s to post 1964). The study site contained 253 domestic dwellings. From the outset this work had strong political support and a partnership was developed that included key stakeholders, including the City Council, the residents, healthcare workers and Salford University.

1   Identity of the pest in that location

The research group worked closely with the City Council's pest management division and information regarding rodent infestations was gathered initially from three sources:

a   Resident questionnaires
b   Structural survey reports
c   Complaints received either via local politicians or the City Council's pest management services

It was unclear from this information whether there were rats, mice or both infesting the properties and tracking plates were placed externally and internally to confirm activity.

2   Size and extent of the infestation

Of the 253 properties surveyed, 30% were found to have positive evidence of mouse activity. Only one property had an internal rat infestation (related to a minor drainage defect) and the tracking plates placed externally confirmed another small external infestation outside an empty property where there was a small orchard nearby. Extensive data collection enabled the research team to understand the factors that were influencing the distribution of the infestations within the properties (see Tables 17.8 and 17.9). Results confirmed that tenure, date of construction and dwelling type all had a significant influence on the distribution of infestations.

Table 17.8 General characteristics of domestic properties and their association with the presence of mouse infestations

Variable	N	% infested
Tenure (χ^2 = 9.04, 2 df, P = 0.011)		
Privately rented	23	71
Privately owned	84	48
LA rented	10	20
Date of construction (χ^2 = 42.14, 3 df, P = <0.001)		
Pre 1919	60	77
1919–1939	6	67
1940–1964	3	67
Post 1964	48	15
Dwelling type (χ^2 = 31.7, 1 df, P = <0.001)		
Terraced/flats	74	70
Detached/semi-detached	43	16

Table 17.9 Kitchen hygiene measures and their association with the presence of mouse infestations

Variable	N	% infested
Food storage (χ^2 =15.85, 2 df, P = <0.001)		
Good		23%
Satisfactory	69	58%
Poor	14	79%
Refuse storage (χ^2 = 20.52, 2 df, P = <0.001		
Good	34	23%
Satisfactory	69	55%
Poor	14	93%
Under cupboards voids (behind plinths or kick plates/boards) χ^2 = 9.77, 1 df, P = 0.002		
Yes	73	62%
No	44	32%
Overall hygiene (χ^2 = 14.35, 2 df, P = 0.001)		
Good	34	23%
Satisfactory	64	59%
Poor	19	68%

Results confirmed that mice were most likely to be infesting privately rented, older (pre 1919) terraced properties.

3 The food source(s) and resting/nesting sites of the pest(s)

Detailed internal surveys found that kitchen hygiene was an important determinant of infestation status (see Table 17.9).

Poor food storage, poor refuse storage, the presence of under cupboard voids and poor hygiene overall were each significant factors in determining the distribution of mouse infestations in these dwellings.

4 The characteristics of the environment that may facilitate/hinder control

The survey work confirmed that older, terraced properties were most likely to be infested. However, poor hygiene was also an issue and that without improvements to the standards of hygiene and food storage, in tandem with effective proofing, it was unlikely that long-term control would be achieved. Providing tailored educational materials was an important part of this work and the researchers developed a self-help booklet and attended a number of schools and health fairs in the area.

There was strong support from most of the residents (as a result of regular newsletters (in both English and Urdu) and community meetings), but there was concern about negotiating access to *all* properties within a terraced row to undertake treatments where mouse activity in *any* of the adjoining properties had been confirmed.

5 A detailed plan of the non-chemical/ chemical methods suitable for use at the site and agreed timescales/access (including legislative powers)

Following the survey work, intensive trapping was undertaken in 27 properties where daily access (to comply with home office regulations) was negotiated. DNA analyses of the 200 mice trapped in these properties provided crucial evidence about their population structure and strongly influenced the planning and delivery of the control strategies adopted.

The results confirmed that groups of houses which were attached to others (e.g. a terrace) operated as an 'island' population. Mice caught in *any* of the houses within a terraced block were closely related, which meant they were travelling easily between these individual properties but were not breeding with

mice from other terraced blocks in the area. This aligned to previous studies that found house mice tended to spend most if not all of their time indoors and were unlikely to venture outside. It was crucial to access each property within a block where *any* evidence of a mouse infestation if eradication was to be achieved. Liaison with the key stakeholders during the early stages ensured that the research team had the support of local politicians and resident groups. The vast majority of residents facilitated regular access to trap mice and/or undertake treatments. In a small number of cases, environmental health officers spoke to residents about the powers available to them to gain entry to undertake treatments and these residents then agreed to the research team accessing their property.

6 Application of control measures followed by an evaluation of the impact

Treatments were undertaken over a four-week period. All residents were notified in advance that their property was to be treated, with an explanation of the approach adopted. Strict health and safety rules were adhered to throughout the treatment phase so that all baits were placed safely and did not pose a risk to vulnerable individuals and/or non-target animals such as pets. Around 15 baiting points were used in each property and their position recorded. Treatments were synchronised where possible so that all properties within a block were treated at the same time. Following the initial treatment, further visits were arranged with the occupants on a weekly basis to check and monitor results. Each bait point was checked for takes and replenished where necessary. If no further bait was taken over two weeks, all the remaining bait was removed.

7 A review of the success of the treatments and where control has not been achieved, develop a revised control plan

Following removal of all bait, tracking plates were placed in all properties to monitor mouse activity. Only five properties had positive signs of continued mouse activity and a further treatment was undertaken, which eradicated the infestation.

8   A summary report to the owner/occupier and/or other stakeholders on the actions needed to ensure long-term maintenance of pest-free premises.

Residents were sent information about the success of the eradication programme, but told that they needed to remain vigilant. Each household was given five free snap traps, with instructions about how to use them. They were sent a booklet, providing information on proofing and the importance of good hygiene. A follow-up questionnaire to residents confirmed that these dwellings had remained free of mouse infestations.

Summary

Although this was a labour-intensive project, the results confirmed that block treatments were a key component in the planning of this control programme. The dialogues between the research team, the residents, the Pest Management Services of the City Council and other key stakeholders played a prominent part in ensuring that this was viewed as a shared problem and that ultimately, the residents would have to take responsibility in remaining vigilant and treating re-infestations at an early stage. This approach, in conjunction with focused advice on good hygiene and basic maintenance and repair to prevent pest infestation, resulted in complete eradication of the chronic mouse infestations in this area.

Pest management at ports

Port health is dealt with in detail in Chapter 21 and the risks of importing disease as a result of international travel and trade have long been recognised with measures in place

to protect communities from diseases through restricting access of individuals, livestock and goods.

The International Health Regulations (IHR) were revised by the World Health Organization and came into force in 2007 and are discussed elsewhere. The revisions addressed the need to detect and respond to international disease outbreaks and emergencies caused by non-infectious disease agents. The IHR provides a legal framework for the rapid gathering of information, for determining when an event constitutes a public health emergency of international concern, and for countries seeking international assistance. The new reporting procedures aim to expedite the flow of timely and accurate information to WHO about potential public health emergencies of international concern. With more than half of the world's population at risk of infection by vector-borne disease it is essential that surveillance programmes at airports and seaports are in place to prevent further spread and ensure that appropriate pest management procedures are in place.

The case for aircraft disinsection was well made by Gratz et al. [77]. They postulated that the public health consequences of imported mosquito vectors from countries where vector-borne diseases are endemic include:

- Infected mosquitoes transmitting disease in the country of arrival (e.g. airport malaria);
- Infected vectors establishing and resulting in autochthonous transmission by a local vector;
- Introduced mosquitoes establishing in new areas (e.g. *Aedes albopictus*);
- Costly control measures to eradicate/prevent further expansion of range.

Almost 940 million international journeys were undertaken in 2010. Global travel on this scale exposes many people to a range of health risks – different disease agents and changes in temperature, altitude and humidity are just some of these risks – all of which can lead to ill-health. Many of these risks, however, can be minimised by precautions taken before, during and after travel. There will have been less global travel under relevant restrictions in response to the COVID-19 pandemic of 2020–2021, but disease spread was demonstrated by the emergence of the Delta variant (first detected in India in December 2020 and by April 2021 it was the most common variant in India) that was subsequently found in 96 countries at that time.[26] The WHO has produced a manual on international travel and health[27] for health professionals and others and updates are freely available on the internet.

A number of methods for aircraft disinsection, are set out in the following box.

All insecticide formulations and active ingredients used in the discussed disinsection procedures must be approved for use under the UK Control of Pesticide Regulations (1986) and the EU Biocidal Products Regulations 528/2012 (EU BPR) which is concerned with the placing on the market and use of biocidal products, which are used to protect humans, animals, materials or articles against harmful organisms, like pests or bacteria, by the action of the active substances contained in the biocidal product.[28] The EU Biocidal Products Regulation (EU BPR) has been copied into GB law and amended to enable it to operate effectively in GB. This means that most aspects of EU BPR will continue in the same way under the new stand-alone regime – the GB Biocidal Products Regulation (GB BPR) came into force at 11 pm on 31 December 2020.[29]

The International Health Regulations also require States' parties to establish programmes to control vectors that may transport an infectious agent that constitutes a public health risk to a minimum distance of 400 metres from those areas of points of entry facilities that are used for operations involving travellers (conveyances, containers, cargo and postal parcels). The minimum distance must be extended if vectors with a greater dispersal range are present. Surveillance and monitoring will become a more prominent aspect of port health work, with effective permanent public health measures and response capacities at designated airports, ports and ground crossings in all countries.

Blocks away: This is a procedure where aerosols containing 2% d-phenothrin, a non-residual pyrethroid insecticide are discharged into the cabin by cabin crew once passengers have boarded and the doors have been closed prior to take off. The aerosols, either single or multi-shot canisters have unique reference numbers and are retained by the aircrew following treatment for inspection by the port health authorities at the port of arrival as proof of disinsection.

Pre-flight and top of descent spraying: This is a similar procedure to 'blocks away' except that the aircraft is sprayed with a residual pyrethroid insecticide when on the ground, before passengers board, allowing overhead lockers, toilets, etc. to be sprayed, followed by an in-flight treatment (as detailed earlier) as the aircraft starts its decent to the destination airport.

Residual disinsection: This disinsection procedure involves spraying insecticide on internal surfaces (apart from food preparation areas) of the aircraft cabin, flight deck, aircraft holds, etc. with a residual pyrethroid insecticide containing permethrin. This is carried out at regular intervals (usually every eight weeks) when the aircraft is out of service for maintenance.

Section 2 Notes

21 See https://www.worldbank.org/en/topic/urbandevelopment/overview

22 See https://www.cdc.gov/nceh/ehs/docs/ipm_manual.pdf

23 See Office of the Deputy Prime Minister. (2006) *Housing Health and Safety Rating System: Operating Guidance, Guidance About Inspections and Assessment of Hazards*. https://assets.publishing.service.gov.uk/government/uploads/system/uploads/attachment_data/file/15810/142631.pdf – but note in 2021 a review of the system was being undertaken – similar guidance has been issued by US HUD for the Healthy Homes Rating System. https://www.hud.gov/sites/documents/OPERATINGGUIDANCEHHRS_1-14.PDF

24 See https://www.urbanpestsbook.com/downloads/procedures-sector-specific/

25 See https://www.basis-prompt.co.uk/prompt-verified

26 See https://www.ecdc.europa.eu/en/covid-19/variants-concern and https://www.who.int/emergencies/diseases/novel-coronavirus-2019/media-resources/science-in-5/episode-45-delta-variant (Accessed September 2021).

27 See https://www.who.int/publications/i/item/9789241580472 and continuous updates are freely available on the Internet at www.who.int/ith

28 See also the WHO. (2013) *International Programmes on Chemical Safety (IPCS) Environmental Health Criteria 243 Aircraft Disinsection Insecticides*. https://www.who.int/ipcs/publications/ehc/ehc243.pdf Note all aircraft entering Australia must be treated with disinsection to prevent biosecurity risks as approved by the Director of Human Biosecurity. The Department of Agriculture, Water and the Environment (the department) administers disinsection requirements on behalf of the Australian Department of Health. The Department and the Ministry for Primary Industries (MPI) New Zealand work in partnership to develop and regulate joint aircraft disinsection requirements. Requirements are based on WHO recommendations. A new 'pre-departure disinsection (PDD)' method commenced on 2 August 2021. Procedures for PDD can be found in the Aircraft Disinsection Procedures. The WHO has recommended that the 'Pre-flight and Top-of-descent' aircraft disinsection method be ceased. Pre-flight and Top-of-decent aircraft disinsection will no longer be accepted from 2 August 2021 for aircraft entering Australia and New Zealand.

29 See https://www.hse.gov.uk/biocides/brexit.htm for more information.

SECTION 3: THE USE OF ANTICOAGULANT RODENTICIDES IN THE UK – RISK MITIGATION MEASURES

The EU reviewed second generation anticoagulant rodenticides (SGARs) containing the active ingredients brodifacoum, bromadiolone, difenacoum, flocoumafen and difethialone.

UK scientists found residues of these compounds in the bodies of a number of predatory and scavenging species of birds and mammals. EU and UK regulatory risk assessments concluded that the use of SGARs outdoors presented a higher level of risk to non-target animals such as mammals, predatory birds and the environment than would normally be considered acceptable.

The use of these products under normal circumstances would therefore not be authorised. The UK government, however, recognised that despite these risks, and in the absence of suitable alternatives which are as equally effective and pose less risk to humans, non-targets and the environment, it would be necessary for the continued use of SGARs outdoors provided properly managed rodent control strategies with acceptable stewardship were developed to control rodent infestations that may for example pose a risk to public health.

The Health and Safety Executive (HSE), the Competent Authority for the UK, proposed that SGARs should continue to be authorised for use in the UK by both professional and non-professional users with, however, restrictions in their use outdoors.

The Campaign for Responsible Rodenticide Use (CRRU) was nominated by HSE to develop a Stewardship Proposal for the responsible use of these products across all sector groups using these rodenticides to minimise the impact of these compounds on wildlife and the wider environment and to adopt high-level principles.[30] Five years have passed since the start of the UK Rodenticide Stewardship Regime in 2016. The UK landscape for the use of rodenticides has changed dramatically in that time. Some change has been driven by the European Commission, with introduction of new product label phrases and a new framework of use scenarios. Many new label requirements concern the management of resistance and enhanced risk mitigation. The outcome of regulatory changes implemented in the UK by HSE has been to define use patterns, target species, application methods and risk mitigation procedures more precisely, and in many cases to restrict what products can be used, by whom and where.

The main purpose of the regime is to promote and support best practice and ensure the diligent application by all users of professional rodenticides. Before it can be applied, however, practitioners need to understand what best practice is. The wide experience and knowledge of CRRU stakeholder organisations is harnessed to provide CRRU best practice guidance.[31]

The main sector user group of anticoagulant rodenticides in the UK are:

- Professional pest control services and Local Authority;
- Agriculture;
- Gamekeepers;
- Amateur use.

Proposals were sought from each of these sector groups and a product Stewardship Regime was developed by CRRU addressing the concerns identified by HSE regarding the outdoor use of SGARs.

A timetable for implementation and authorisation of SGARs in the UK was introduced by HSE. All users of SGARs (across all user sectors) needed to demonstrate they had achieved the necessary level of training and competence (currently RSPH/BPCA Level 2 Award or equivalent) to purchase products from suppliers and to carry out treatments in open spaces. This came into effect on 1st June 2016.

The CRRU UK Code of Best Practice: Guidance for Rodent Control & the Safe Use of Rodenticides (CoBP) is endorsed by HSE and the government's Oversight Group. This CoBP gives advice on issues such as rodent control strategies and treatments, risk hierarchy (selecting non-chemical control measures etc. before considering the use of SGARs), avoiding rodent infestations through environmental management, carrying out environmental assessments prior to treatments and rodenticide resistance.

Monitoring of residual SGAR levels in sentinel wildlife species (e.g. barn owls) will continue with the aim of demonstrating significant reduction in anticoagulant residues over a period of time from the date the Stewardship Regime was implemented.

If residues are subsequently found not to be reducing at the prescribed rate set by Oversight Group, then the HSE would implement further restrictions on SGAR use, possibly in the form of authorising the use of the products only for use indoors or in and around buildings.

Accordingly, when the Regime was established, HSE said that it would undergo 'major review' after five years and CRRU has published a report.[32] As CRRU reports, the goal to reduce the exposure of barn owls to anticoagulants has not yet been met and it is hoped that more time is given more time to deliver it. The extension of use outdoors of some more potent anticoagulants however has not resulted in a significant increase in the exposure of wildlife, as once feared. As the report says, less reassuring is the apparent spread of anticoagulant resistance among target rodent species, including new strains that carry two different resistance mutations.[33]

Section 3 Notes

30 See https://www.hse.gov.uk/biocides/rodenticides.htm

31 https://www.thinkwildlife.org/download/crru-uk-code-of-best-practice-2021/?wpdmdl=18095

32 See https://www.thinkwildlife.org/download/five-years-of-rodenticide-stewardship-2016-2020-report/?wpdmdl=18111

33 See https://research.reading.ac.uk/resistant-rats/anticoagulant-resistance-project/ The rodenticide resistance map can be viewed at https://guide.rrac.info/resistance-maps/norway-rat/europe/united-kingdom.html

SECTION 4: LEGAL PROVISIONS AVAILABLE TO ENVIRONMENTAL HEALTH PRACTITIONERS

As the previous edition of this book made clear, the evidence-based research into the impact of pests on human well-being has not necessarily been matched by legislative tools which facilitate and support timely and comprehensive control measures. Whilst education and advice should be the starting point in attempting to resolve pest issues, where human health is at risk, a more urgent course of action may be necessary.

The UK law relating to the control of pest species (as is the case in many aspects of environmental health) is complex and often fragmented and has been developed in response to specific issues. The pest (rodents/birds/insects), the setting (domestic/commercial) in which it is found and the tenure of domestic premises (e.g. public/private/private rented) may influence the legislation (s) available to support the control of pests. The law relating to the control of pests, particularly in domestic settings is, in part, outdated, but despite continued pressure, it has been difficult to instigate a legislative review. Revisions to food safety law have undoubtedly strengthened the legal powers available to local authorities, but anomalies persist. For example in a room storing equipment within a food premises (where food is neither prepared nor stored), authorities must revert to the Prevention of Damage by Pests Act 1949 to ensure that such rooms are included in the control measures required to eradicate infestations. Yet that legislation was brought in to make provision for preventing loss of food by infestations, but was drafted at a time of rationing, food shortages and very different agricultural practices at that time.

Local Authority powers only relate to substantial infestations on land [78]. There is no statutory requirement placed on Local Authorities to provide in-house pest management services, but there are statutory functions and duties placed upon them that relate to inspecting their district and in controlling pests and/or infestations, for example under the Prevention of Damage by Pests Act 1949; Environmental Protection Act 1990; Public Health Acts 1936/1961 and Housing Act 2004.

Pest control in domestic premises

EHPs and/or Pest Control Technicians play a vital role in educating residents about IPM and the role residents must play in maintaining pest-free dwellings. In domestic settings, there may be a number of complex issues which have contributed to the development of chronic infestations and a resident may need the support of other Local Authority services to resolve the factors that may be facilitating the persistence of pests – these may be waste management, social care or housing management. Where possible, attempts to resolve the situation in a sympathetic and sensitive way, in partnership with other agencies and services, without recourse to the legal powers available, should be made. However the health of the individual and/or those living nearby is paramount and a more robust course of action may need to be adopted when other avenues have failed to resolve the situation satisfactorily. Where infestations are confirmed in terraces or flats it may be necessary to undertake control measures in all properties (block treatment) to ensure eradication.

In domestic settings, there are four main acts used to control pests:

- Prevention of Damage by Pests Act 1949;
- Public Health Acts 1936 and 1961;
- Environmental Protection Act 1990;
- Housing Act 2004.

These Acts, used on their own or in combination, provide a range of powers to deal with pest infestations. Local authorities appear to differ in how they use them to resolve domestic infestations, with local practice/preference determining what is used in practice.

In addition, where the infestation arises from deficiencies in a rented dwelling, it is possible for the tenant to take their own legal action under the Landlord and Tenant Act 1985 as amended by the Homes (Fitness for Habitation) Act 2018.[34]

The *Prevention of Damage by Pests Act 1949 (PDPA)* provides local authorities with powers to deal with rodent infestations. Where rats and mice are present, the PDPA facilitates the control of rodents on land (including occupied buildings). As already mentioned this Act was introduced in very different times and its application in urban settings today can be difficult. Section 4 (*Power of Local Authority to require action*) gives powers to local authorities to serve notice on the owner and/or occupier requiring that reasonable steps be taken (as specified within the notice) for the destruction of rats or mice on the land or otherwise for keeping the land free from rats and mice.

- Access to inspect;

  a   Access permitted

Where access to the premises is available, a visual inspection should be undertaken, looking for signs of rodent activity such as the presence of fresh droppings, gnawing or smear marks. If there are no obvious visual signs but an infestation is suspected, then non-toxic baits can be laid and arrangements made to re-inspect the premises for signs of 'takes' a few days later.

b   Access denied

Where access has been denied, a 'notice of intended entry' must be posted and 24 hours given to the occupier as detailed under PDPA 1949 – Section 22 (*Powers of entry*). Unfortunately, the Act does not provide powers of entry and many Local Authorities have introduced local enabling legislation to facilitate their entry or use powers of entry under other legislation, such as the Public Health Act 1936. Notice of entry can be served by hand or affixed to the front door of the premises. The notice specifies the date and time a revisit will be made to gain entry.

If access is not forthcoming, application to a JP/local magistrates' court for a warrant to enter the premises, if need be, by force, is required. If the warrant is granted, the premises can be entered, and a full inspection carried out to determine if any action is required under PDPA 1949. In executing the warrant, it may be necessary to secure the services of a locksmith or alarm engineer and the police. Whether the premises are occupied or not, they must be entered, and inspected/test baited to substantiate the complaint and positive evidence of rodent activity either at the premises or within the immediate area confirmed before formal enforcement action can be taken.

- Evidence of an infestation confirmed;

If a rodent infestation is confirmed, then the Local Authority can serve notice on the owner or the occupier or, where the owner is not the occupier, on both. The notice should specify:

- The nature of the treatment programme;
- The times when the treatment programme should be carried out;

- The carrying out of structural or other works required.

The period for compliance is usually 21 days, to allow for any appeal against the service of the notice. Notices under Section 4 may also be used to secure the removal of refuse, etc. which may be providing harbourage for rodents such as rats, providing there is evidence of active infestation at or close to the site. Details of a statutory notice served are usually entered onto the Local Land Charges Register.

- Non-compliance with the notice;

Once the time period specified in the notice has expired and compliance has not been secured the notice can be actioned in default by the Local Authority and the charges incurred ultimately registered as a charge on the land.

- Other relevant information;

In the domestic setting, establishing ownership of the property prior to serving a statutory notice or if access has been denied or the premises are unoccupied is important. This may necessitate a search via Land Registry or by service of a notice requiring this information under the provisions of the Local Government (Miscellaneous Provisions) Act 1976.

- Outcomes;

Failing to take the steps required within a notice is an offence and those found guilty will be liable to a fine on summary conviction.

It should be noted that Part II of the PDPA (Infestation of Food) does not give local authorities powers because it requires every person whose business consists of or includes the manufacture, storage, transport or sale of food to without delay give notice in writing to the Secretary of State or Welsh Minister of any infestation in any premises or vehicle or any equipment belonging to any premises or vehicle, used or likely to be used in the course of that business for the manufacture, storage, transport or sale of food in any food manufactured, stored, transported or sold in the course of that business, or in any other goods for the time being in his or her possession which are in contact or likely to come into contact with food so manufactured, stored, transported or sold. Similar provisions apply to the master of a sea-going ship used or likely to be used for the transport of food.

The PDPA is a useful tool where block treatment is needed to eradicate infestations from, for example, terraced properties or high-rise flats.[35] Enforcement notices can be served on all properties within the block to control rodent activity and if necessary, specify any structural disrepair such as the need for proofing, to minimise the potential for future re-infestation.

Public Health Acts 1936 and 1961 – These are used to resolve problems associated with insect and bird infestations. Section 90 of the 1936 Act defines *vermin* as – 'in its application to insects and parasites including their eggs, larvae and pupa, and verminous is to be construed accordingly'. Section 74 of the 1961 Act provides Local Authorities with powers to reduce the numbers of pigeons and other birds in built-up areas.

- Access to inspect;

Authorised officers have a right to enter any premises (at a reasonable time of day) to confirm whether or not there has been contraventions of either Act, but for domestic premises they must give 24 hours' notice of their intention. Refusal to allow entry is deemed an offence and a warrant would be required to enter the premises, if need be by force.

- Evidence of an infestation confirmed;

Section 83 (*Cleansing of filthy or verminous premises*) of the Public Health Act 1936 (as amended) facilitates the treatment of verminous premises, infested with, for example cockroaches or bedbugs or other insect pests. Enforcement notices are served on the owner or occupier of the premises requiring them (usually within a 21-day period), to remedy the infestation. Notices may specify the steps to be taken and may include the removal of wallpaper or other coverings, etc. and specify the type of treatment required to destroy the vermin. Provision is also made within the legislation for the use of gas fumigation; however, in practice this is rarely, if ever, used, as the development of more specific residual insecticide technology provides more effective treatments.

- Non-compliance with a notice;

There is no right of appeal against the service of the notice, and where there is non-compliance with a notice, the Local Authority may carry out the works specified in default and recover the costs incurred. However, the defendant is allowed to question the reasonableness of the notice and why the notice was served on them as owner and not the occupier.

- Other relevant information;

The purchase of second-hand appliances and furniture can introduce insect infestations into dwellings. German cockroaches, for example, infesting a second hand fridge (harbourage usually found close to the compressor/heat exchanger) can easily be introduced into households; similarly bedbugs can be introduced through the purchase of second-hand or repossessed beds/mattresses or bedroom furnishings. Section 37 (*Prohibition of sale of verminous articles*) of the Public Health Act 1961 prohibits the preparation/sale of verminous household articles, if, to the knowledge of a dealer the articles were known to be verminous or that if, by taking reasonable precautions s/he could have known them to

be verminous. Local authorities can require disinfestation or destruction of such articles and can recover the costs incurred in so doing. A 'household article' is defined as any article of furniture, bedding or clothing or any similar article.

As with the PDPA, these Acts can be effective in controlling insect infestations in multiple dwellings within a block of premises. In this instance enforcement notices would be served on all owners within the block.

Environmental Protection Act 1990 – There may be situations where the Acts outlined earlier do not address the problems caused by pests and the Environmental Protection Act 1990 (EPA) can be used to control pests or for those situations which may arise where it is felt taking enforcement action under this legislation may be deemed inappropriate. Section 79 defines statutory nuisance as:

- Any premises in such a state as to be prejudicial to health or a nuisance;
- Any accumulation or deposit which is prejudicial to health or a nuisance;
- Any animal kept in such a place or manner as to be prejudicial to health or a nuisance;
- Any other matter declared by any enactment to be a statutory nuisance.

Section 101 of the Clean Neighbourhoods and Environment Act 2005 has placed insects in the statutory nuisance provisions of Section 79 of the Environmental Protection Act 1990. This may have arisen as the result of longstanding problems of fly and smell problems from the Mogden sewage works.[36]

The EPA places a duty on local authorities to inspect its area from time to time to detect statutory nuisances.

- Access to inspect;

Where a complaint of a statutory nuisance is made by a person living within its area, local authorities must take such steps as are reasonably practicable to investigate the complaint.

Authorised officers have powers of entry under Section 69 of the Act.

- Evidence of infestation/statutory nuisance;

Where nuisance has been confirmed, an abatement notice must be served:

(1) On the person responsible for the nuisance, except in a case falling within paragraph (2) or (3) the following
(2) Where the nuisance arises from any defect of a structural character, on the owner of the premises
(3) Where the person responsible for the nuisance cannot be found or the nuisance has not yet occurred, on the owner or occupier of the premises.

- Non-compliance with a notice;

A person served with an abatement notice may appeal against the notice to a magistrates' court within the period of 21 days from the service of the notice. If a person served with an abatement notice, without reasonable excuse, contravenes or fails to comply with any requirement or prohibition imposed by the notice, they are guilty of an offence and liable on summary conviction to a fine not exceeding level 5 on the standard scale (unlimited) together with a further fine of an amount equal to one-tenth of that level for each day on which the offence continues after the conviction. In the case of an offence on industrial, trade or business premises, the fine will not exceed £20,000.

If the magistrates' court is satisfied that the alleged nuisance exists, or is likely to recur on the same premises, the court will make an order:

(1) Requiring the defendant to abate the nuisance, within a time specified in the order, and to execute any works necessary for that purpose
and/or

(2) Prohibiting a recurrence of the nuisance, and requiring the defendant, within a time specified in the order, to execute any works necessary to prevent the recurrence
and/or
(3) May prohibit the use of the premises for human habitation until the premises are, to the satisfaction of the court, rendered fit for that purpose.

The court can impose a maximum fine at Level 5 which is unlimited. Where an abatement notice has not been complied with the Local Authority may carry out the works in default and recover the cost reasonably incurred. In cases where the Local Authority is of the opinion that action taken in the magistrates' court would provide an inadequate remedy, they may institute proceedings in the High Court.

- Other relevant information;

The Clean Neighbourhoods & Environment Act 2005, under Section 101, added insects to the statutory nuisance provisions of Section 79, of the Environmental Protection Act 1990:

> any insects emanating from relevant industrial, trade or business premises and being prejudicial to health or a nuisance.

There are a number of caveats associated with the inclusion of insects as statutory nuisance (see Schedule 5 to the Wildlife and Countryside Act 1981).

'Relevant industrial, trade or business premises' are defined as premises that are industrial, trade or business premises but exclude:

(1) land used as arable, grazing, meadow or pasture land
(2) land used as osier land, reed beds or woodland

(3) land used for market gardens, nursery grounds or orchards

(4) land forming part of an agricultural unit, not being land falling within any of paragraphs (1) to (5)

(5) land included in a site of special scientific interest (as defined in Section 52(1) of the Wildlife and Countryside Act 1981)

For more on Statutory Nuisance see Battersby and Pointing [79].

Housing Act 2004 Part 1 – This Act introduced the concept of Category 1 and Category 2 hazards and a statutory (prescribed) methodology for identifying these was introduced – the Housing Health and Safety Rating System (HHSRS) (for more see Chapter 12). The definition of health in the Act includes mental health, thus the likelihood of severe stress as a result of a hazard can be taken into account. Hazards rated as Category 1 (are those scoring more than 1000 using the HHSRS) place a duty on a Local Authority to take action. Category 2 hazards (scoring 999 or less using the HHSRS) allow discretion as to the actions to be taken. The rating is based on a risk assessment that takes account of the likelihood of an occurrence that could cause harm arising from deficiencies in the dwelling and the possible harm outcomes. The dwelling also has to be capable of providing adequate protection from all potential hazards prevailing in the local external environment. Thus the assessment should take account of those deficiencies which could permit entry of pests as a matter of building hygiene where pests are present in the vicinity or building (e.g. garden or sub-floor space), even if not in the dwelling itself.

The course of action should be that which is most appropriate in all the circumstances, and the hazard rating is just the initial consideration.

The hazards arising from Domestic Hygiene Pests and Refuse can result from:

- Poor design, layout and construction of the dwelling which prevents it readily from being clean and hygienic;

- Access into, and harbourage within the dwellings for pests;
- Inadequate and unhygienic provision for the storing and disposal of household waste.

The conditions which may give rise to pest infestations in domestic premises, and therefore relevant to the likelihood of an occurrence and severity of outcomes referred to in this guidance, include:

- Unprotected or damaged ventilators;
- Design deficiencies (lack of proofing) providing harbourage and access points;
- Inadequately stored/accumulated refuse allowing access to insect/rodent/pests/birds/squirrels/foxes/cats/dogs;
- Service ducts and holes around pipes (e.g. central heating) harbouring insects and providing access between dwellings in blocks;
- Access to open drains by rodents;
- Access for rodents by means of ill-fitting doors and windows;
- Uneven and/or cracked internal walls and/or ceilings allowing access for pests;
- Missing/damaged brickwork including airbricks to external walls and other disrepair to external walls and roof.

The powers available under Part 1 of the Housing Act 2004 are considered next:

- Access to inspect;

Local authorities have powers of entry given under the Housing Act 2004 (Section 239) to carry out an inspection. Generally, at least 24 hours' notice would be given to the occupier and/or owner (if known) of the intention to exercise these powers (and placed by fixing it conspicuously to the premises or building). However, some Residential Property Tribunals have quashed notices where that time has not been given to the owner and/or occupier. Notice is not required where emergency remedial action is needed (see later) – the right of entry may be exercised at any time (Section 40) but a notice must be served on the occupiers before entry, setting out, amongst other things, the action they propose to take. Where entry is refused or the property is vacant an application must be made to a JP for a warrant to enter (Section 240), if need be by force.

During the detailed internal and external inspection (including the garden or rear yard), deficiencies should be identified, and significant hazards arising from those deficiencies (i.e. worse than the average) noted, using the HHSRS. The statutory operating guidance (SOG) gives the factors that contribute to likelihood of an occurrence over the next 12 months and the possible spread of harm should there be an occurrence.

- Evidence of infestation confirmed;

Following the inspection, each case must be decided on its merits and the hazards posed considered carefully and whether the infestation is the result of deficiencies in the dwelling which are the owner's responsibility. What is obvious is that if there is an infestation within the accommodation then under the HHSRS likelihood of an occurrence will be far higher (and hazard rating greater) than if pests are not within the dwelling (such as in the garden) even though there are deficiencies in the structure that could permit entry. It should be noted that as the HHSRS focuses on the structure and ignores occupier behaviour, the HHSRS may not be appropriate if evidence is that the occupier introduced the infestation.

Further detailed information may be found in the NPAP publication, *Pest Control procedures in the housing sector* [60]. A number of actions are available under this Act. Guidance on enforcement is contained within the Statutory Enforcement Guidance [80] which as at 2021, like the HHSRS itself, is under review. Whichever action is taken with respect to a hazard, a statement of reasons must accompany the notice or order. This should set out why that course of action has been taken as opposed to another. The courses of action available (so long as a Management Order made under Part IV of the Act is not in force) are:

Improvement Notices – These may be served when dealing with Category 1 (under s.11) or Category 2 (under s.12) hazards and must, at the very least, remove any Category 1 hazard(s). In addition to detailing any work that may be required as the Act requires that the notice includes 'any remedial action' for the hazard, this can be interpreted more widely and could be more than just 'work' to the premises. It could relate to matters of management which could deal with the hazard, for example of waste storage and collection, or even a requirement to have a pest servicing contract. The notice may be suspended by the authority.

Hazard Awareness Notice – This is an option that provides for advisory action where the Local Authority wants to draw attention to the need for improvements. The notice must give details of the hazard concerned and what is needed to deal with it. It may not be considered appropriate for a Category 1 hazard.

Prohibition Order – This may be judged to be an appropriate response to a Category 1 or 2 hazard where conditions present a serious threat to health or safety. It may prohibit the use or occupation of part or all of the premises for some or all purposes. Any such order made must contain similar information to that required by an Improvement Notice (as earlier) and becomes operative 28 days after it is made.

Emergency remedial action – Where there is an imminent risk of serious harm this may be appropriate. The Local Authority may make an emergency prohibition order and take emergency (immediate) remedial action to remove the imminent risk although it may be that a serious hazard could still remain justifying the service of an Improvement Notice.

- Non-compliance with the notice or order;

Compliance with an Improvement Notice – This is the beginning and completing of the remedial action. The notice must then be revoked. It is an offence to fail to comply with an Improvement Notice once it has become operative (if no appeal, that is, after 21 days, or once an appeal has been finally determined and as set out in the Act). It is possible for the Local Authority to vary the notice either of its own volition or with the agreement of the person on whom it was served. The local housing authority may prosecute in the magistrates' court and/or undertake the remedial action itself in default of compliance and recover the costs involved in doing so. It is also possible to undertake the remedial action by agreement before the notice has expired. The offence of failure to comply with the Improvement Notice is a fine not exceeding level 5 on the standard scale (unlimited).

There is no offence for failing to comply with a Hazard Awareness Notice, but where that action does not resolve the hazard, then the local housing authority can take another course of action such as an Improvement Notice.

Where there a failure to comply with the terms of a prohibition order the offence can be prosecuted in the magistrates' court and a fine not exceeding level 5 on the standard scale (unlimited) can be imposed and a daily penalty as set by the court for each day after conviction.

Food hygiene legislation and pests

Inspecting food premises to ensure the safe production and consumption of food is a vital component of the work local authorities undertake. In tackling infestations in food premises, there are four main Acts used in controlling pests:

- Food Safety Act 1990 and Food Hygiene Regulations 2006;
- Prevention of Damage by Pests Act 1949 (see previous section);
- Public Health Acts (1936 and 1961) (see previous section);
- Environmental Protection Act 1990 (see previous section).

The Food Safety Act 1990 (as amended) is a wide-ranging piece of legislation aimed at food safety and consumer protection in relation to food production and consumption in the UK. Its main aims are to ensure that all food meets consumers' expectations in terms of nature, substance and quality and is not misleadingly presented. It provides legal powers which specify offences in relation to public health and consumers' interests and enables the UK to fulfil its responsibilities in the European Union.

UK law applies the EC Regulation No 852/2004 made by the European Parliament to the United Kingdom. These regulations provide enforcement powers by serving Hygiene Improvement Notices, Hygiene Prohibition Notices and Orders, and Hygiene Emergency Prohibition Notices. In addition, the regulations require that the design and layout of the premises permits good food hygiene practices, including protection against contamination and in particular pest control. Rooms with external opening doors and windows used for the preparation of food should, where necessary, be fitted with insect-proofed screens. Requirements relating to the design, siting, construction, cleanliness and good repair, etc. to avoid contamination by pests, extend to vending machines. The Food Standards Agency has produced guidance for the meat industry but the advice is relevant to other parts of the food industry.[37]

- Access to inspect;

Authorised officers have a right of entry at all reasonable hours to inspect the premises to establish if there is or has been any contravention of the Food Safety legislation. If access is refused the authorised officer may apply to a

JP/magistrates' court for a warrant to enter the premises if need be by force.

- Evidence of infestation;

When carrying out inspections of food premises for pest infestation, having first examined the pest record book for current/previous pest reports, attention should be focused on any visible signs of pest activity, such as fresh rodent droppings, signs of gnawing or grease marks, etc. Although insect monitors can be useful, the absence of pests on these traps does not necessarily mean that there are no insects present. A torch with a strong light and a mirror are useful aids for inspecting behind fixed equipment. Using a small portable flexible fibre optic camera may also prove useful in this regard.

Mouse boxes containing a only few grains of canary seed can be placed at strategic points and checked several days later for signs that the grains have been de-husked, indicating current activity. Similarly, the efficacy of the lure in crawling insect monitors may have declined since the previous visit and it may be advisable to lay new monitors. Sticky boards can be useful for detecting cockroaches.[38] These should be dated and laid at strategic points and arrangements made to carry out a further examination after several days in situ. It is essential when inspecting the premises that any equipment present (e.g. preparation tables) is inspected for signs of cockroach activity, which may be confined to harbourages within the voids of the equipment.

Food found to be infested with pests either directly in the food itself (e.g. grain weevils), or contaminated by stored product pests (e.g. with webbing and/or frass from stored product moths) or rodent droppings, hair, etc. will fail to comply with food safety requirements and may be seized, or surrendered by the food proprietor.

- Evidence of infestation confirmed;

If evidence of an infestation or other reasonable grounds for believing that there are contraventions of Food Safety legislation is found, the officer may serve an *Improvement Notice* on the proprietor of a food business. The Improvement Notice may prohibit or regulate the use of a process or treatment in the preparation of food or more commonly in the case of pests, requiring improvements in hygiene and practices connected with the food. The notice may also include any pest proofing requirements (e.g. the provision of flying insect screens to food rooms, etc.) and requires compliance within a specified period not being less than 14 days.

If, on inspection of the premises the officer believes that an infestation exists and there is a risk of injury to health (for example from an infestation of rodents or cockroaches in food preparation areas), a *Hygiene Emergency Prohibition Notice* may be served imposing an immediate prohibition or closure of the premises as may be specified in the notice. A person served with an Improvement Notice must comply with the requirement of the notice within the time period specified or face prosecution for non-compliance.

Following service of an Emergency Prohibition Notice an application must be made within three days to a magistrates' court to issue an *Emergency Prohibition Order*. If granted, the Emergency Prohibition Order will remain in force until compliance has been achieved. If an order is not granted, compensation must be paid for any loss suffered by the food business operator.

- Non-compliance;

A food proprietor failing to comply with the requirements of notices served under the hygiene legislation will on summary conviction be liable to a fine not exceeding the statutory maximum (£5,000 England and Wales), two years' imprisonment or both. Appeals against a hygiene Improvement Notice or a Hygiene Emergency Prohibition Notice are to the Crown Court.

In all proceedings for offences in relation to Improvement Notices, Prohibition Orders, Emergency Prohibition Notices, etc. it is a defence to prove that the defendant took all reasonable precautions and exercised all due diligence to avoid the commission of the offence by himself or by a person under his control.

- Outcomes.

A weakness of the current food safety legislation is that non-food rooms which may be ancillary to, for example, a food preparation process, fall outside of the scope of the legislation. In practice, this means that rooms with infestations that may be adjacent to food rooms need to be actioned using other legislation. For rodents this would be the Prevention of Damage by Pests Act 1949 and for insects, the verminous legislation of the Public Health Acts 1936 and 1961.

CIEH National Pest Advisory Panel – The National Pest Advisory Panel (NPAP) was formed in May 2001 to advise the CIEH on pest management issues and policy. Its aims and objectives are to:

1 Raise profile of pest management in the UK, leading to better understanding of the need for good pest management
2 Establish channels of communication throughout industry, government, local authorities, academics, leading to greater awareness of problems and need for priorities
3 Improve the standards of pest management throughout the UK by promoting good practice, leading to reduced pest levels and pesticide use
4 Provide expert advice to Government Departments and agencies via CIEH
5 Identify and promote research needs into pest management issues

The panel has published a useful range of documents which can be downloaded from https://www.urbanpestsbook.com/downloads/. These which include:

- Protecting present and future generations – implementing lessons from the WHO/Euro book: public health significance of urban pests (March 2010);
- Pest control procedures in the housing sector (January 2010);
- Good composting practice: guidance on composting without attracting rodents (September 2009);
- The role of pest management in protecting public health (May 2009);
- Public health issues posed by mosquitoes – an independent report (May 2009);
- Pest control procedures manual – rodents (May 2009);
- Pest minimisation – best practice for the hospitality industry (April 2009);
- Pest control procedures in the food industry (January 2009);
- Urban pests and their public health significance CIEH summary (based on a WHO textbook) – 2008 – textbook summary;
- The impact of climate change on pest populations and public health (November 2008);
- Pest minimisation – best practice for the construction industry (September 2008);
- Pest control procedures manual: urban gulls (April 2015);
- Code of practice in the use of vertebrate traps (September 2014);
- Pest control procedures manual: social insects – ants, wasps, & bees (August 2014);
- Urban foxes: guideline on their management (October 2013);
- Pest control procedure manual – cockroaches (April 2013);
- Public perception: statistics that prove the essential value of public health pest control (April 2013);
- National sewer baiting protocol best practice guidance document (April 2013);
- A guide to carrying out an environmental assessment prior to the use of rodenticides (October 2012);

- Impact of climate change on pest populations (May 2012);
- The perfect storm: this need not be the end of a golden age of public health protection (February 2012);
- Pest minimisation: pest management for outdoor & mobile catering (February 2012);
- Pest control procedures in the social care sector (February 2011);
- Pest control procedures manual: bedbugs (April 2011);
- Charging for pest control services – policy briefing (February 2011).

Section 4 Notes

34 This Act amended the standard of fitness in the 1985 Act to include all the HHSRS hazards, so that if a hazard of 'Domestic Hygiene, Pests and Refuse' (risk of harm) arising from deficiencies in the dwelling is so substantial as to make the dwelling not reasonably suitable for occupation, the landlord will be in breach of the obligation to keep the dwelling fit.

35 It should be noted that in *Sharpe v Manchester MDC* (1977) 5 HLR 71 it was held an insect infestation emanating from within the same building could cause a nuisance and that if a landlord failed to take all reasonable steps to remove the nuisance it would be found negligent. On the fact, the council should have used an alternative pesticide and treated the service ducts and was therefore liable to the claimant. From when the claimant took up residence in the property, his flat suffered from a cockroach infestation. The defendant council undertook to remove the infestation with the use of DDT pesticide and was aware it was not the most efficient method of destroying the cockroaches. Furthermore, the council only treated the inside of the claimant's flat and whilst this treatment did successfully kill the cockroaches, it did not address the source of the issue which was the service ducts to the property and which could have been easily treated without disturbing other residents.

36 See *Dobson & others v Thames Water Utilities Ltd* (2011) All ER (D) 125 (Dec) [2011] EWHC 3253 (TCC).

37 See https://www.food.gov.uk/about-us/key-regulations#codes-of-practice and https://www.food.gov.uk/sites/default/files/media/document/pest-control.pdf

38 Note: The Glue Traps (Offences) Bill was introduced Parliament by Jane Stevenson MP in June 2021 and will enable a full ban on the use of glue traps to catch rodents, and has the backing of the government; see https://www.gov.uk/government/news/government-backs-bill-banning-the-use-of-glue-traps-for-pest-control

Chapter references

[1] Bonnefoy X, Kampen H, Sweeney K. (2008) *Public Health Significance of Urban Pests*, WHO Regional Office for Europe, Copenhagen, Denmark.

[2] Murphy G, Oldbury D. (2016) Pest management and vector control. In *Clay's Handbook of Environmental Health*, Battersby SA (Ed.), Routledge, Abingdon, Oxon, 754–789, 21st ed.

[3] Battersby SA. (2015) Rodents as carriers of disease. In *Rodent Pests and Their Control*, Buckle AP, Smith RH (Eds.), CABI, Wallingford, Oxon, 2nd ed., 81–100.

[4] Battersby SA. (2004) Public health policy – can there be an economic imperative? An examination of one such issue. *Journal of Environmental Health Research*, 3: 19–28.

[5] Bezant ET. (1984) *Principles of Pest Control*, Ministry of Agriculture, Fisheries and Food, Rome.

[6] CIEH. (2012) *The Perfect Storm – This Need Not Be the End of a Golden Age of Public Health Protection*, Chartered Institute of Environmental Health, London. (https://www.urbanpestsbook.com/download/perfect-storm-need-not-end-golden-age-public-health-protection/) (Accessed September 2021).

[7] CIEH. (2015) *Environmental Health Workforce Survey 2014/15 Phase 1 and 2 Summary Report of Findings*, Chartered Institute of Environmental Health, London. https://www.cieh.org/media/1262/environmental-health-workforce-survey-2014_15.pdf (Accessed September 2021).

[8] Stevenson LA, et al. (2001) Sociodemographic correlates of indoor allergen sensitivity among United States children. *The Journal of Allergy and Clinical Immunology*, 108: 747–752.

[9] Arruda LK, Vailes LD, et al. (2001) Cockroach allergens and asthma. *The Journal of Allergy and Clinical Immunology*, 107: 419–428.

[10] Prescott-Clarke P, Primatesta P. (Eds.) (1998) *Health Survey for England: The Health of Young People '95–97*, Stationery Office (Department of Health Services HS No. 7, London.

[11] Office of National Statistics. (2004) *The Health of Children and Young People*, Office for National Statistics, London.

[12] Platts-Mills TA. (2004) Allergen avoidance. *The Journal of Allergy and Clinical Immunology*, 113: 388, 391. https://www.jacionline.org/article/S0091-6749(03)02837-9/fulltext.

[13] McConnell R, et al. (2003) Cockroach counts and house dust allergen concentrations after professional cockroach control and cleaning. *Annals of Allergy, Asthma & Immunology*, 91: 546–552.

[14] Mollet JA, et al. (1997) Evaluation of German cockroach (Orthoptera: Blattellidae) allergen and seasonal variation in low-income housing. *Journal of Medical Entomology*, 34: 307–311.

[15] Cohn RD, et al. (2004) National prevalence and exposure risk for mouse allergens in US households. *The Journal of Allergy and Clinical Immunology*, 113: 1167–1171.

[16] Peranowski MS, et al. (2006) Endo toxin in inner city homes: Associations with wheeze and eczema in early childhood. *The Journal of Allergy and Clinical Immunology*, 117: 1082–1089.

[17] Battersby SA, et al. (2008) Commensal rodents. In *Public Health Significance of Urban Pests WHO Regional Office for Europe*, Bonnefoy X, Kampen H, Sweeney K (Eds.), Copenhagen, Denmark.

[18] Galán-Puchades MT, et al. (2018) First survey on zoonotic helminthosis in urban brown rats (Rattus norvegicus) in Spain and associated public health considerations. *Veterinary Parasitology*, 15(259): 49–52. https://doi.org/10.1016/j.vetpar.2018.06.023. Epub 30 June 2018. PMID: 30056983.

[19] Nowak RM. (1999) *Walker's Mammals of the World*, Volume II, Johns Hopkins University Press, Baltimore, MD, 6th ed.

[20] WHO. (2007) *Large Analysis and Review of European Housing and Health Status (LARES) – Preliminary Overview*, WHO Regional Office for Europe, Copenhagen, Denmark.

[21] Annesi-Maesano I, Moreau D. (2009) Potential Sources of Indoor Air Pollution and Asthma and Allergic Diseases. In *Housing and Health in Europe – the WHO Lares Project*, Ormandy D (Ed.), Routledge, Abingdon, Oxon, 111–124.

[22] Berry RJ, Tricker BJK. (1969) Competition and extinction – mice of Foula, with notes on those of Fair Isle and St Kilda. *Journal of Zoology*, 158: 247.

[23] Tattersall FH, Smith RH, Nowell F. (1997) Experimental colonisation of contrasting habitats by mouse mice. *Zeitschrift fur Saugetierkundle*, 62: 350–358.

[24] Lister AM, Charlotte Hall C. (2014) Variation in body and tooth size with island area in small mammals: A study of Scottish and Faroese House Mice (Mus musculus). *Annales Zoologici Fennici*, 51(1–2): 95–110. https://doi.org/10.5735/086.051.0211.

[25] Murphy RG. (2002) Rats and mice: Is there a public health threat? In *Proceedings of the International Housing and Health Symposium*, Bonnefoy X, Rusticali F (Eds.), WHO Regional Office for Europe, Forli, Italy, Copenhagen, 122–123, 21–23 November.

[26] Battersby SA, et al. (2008) Commensal rodents. In *Public Health Significance of Urban Pests WHO Regional Office for Europe*, Bonnefoy X, Kampen H, Sweeney K (Eds.), Copenhagen, Denmark.

[27] Buchmeier MJ, et al. (1980) The virology and immunobiology of lymphocytic choriomeningitis virus infection. *Advances in Immunology*, 30: 275–331.

[28] Healing TD. (1991) *Salmonella* in rodents: A risk to man? *Communicable Disease Report*, 1(10): 114–116.

[29] Pocock MJ, et al. (2001) Patterns of infection by Salmonella and Yersinia spp. in commensal house mouse (Mus musculus domesticus) populations. *Journal of Applied Microbiology*, 90(5): 755–760. https://doi.org/10.1046/j.1365-2672.2001.01303.x. PMID:11348436.

[30] Burt SA, et al. (2018) Wild mice in and around the city of Utrecht, the Netherlands, are carriers of Clostridium difficile but not ESBL-producing Enterobacteriaceae, Salmonella spp. or MRSA. *Letters in Applied Microbiology*, 67(5): 513–519, November. https://doi.org/10.1111/lam.13066. Epub 19 September 2018.

[31] Davies RH, Wray C. (1995) Mice as carriers of Salmonella enteritidis on persistently infected poultry units. *The Veterinary Record*, 137(14): 337–341. https://doi.org/10.1136/vr.137.14.337. PMID:856068.

[32] Murphy RG, Williams RH, Hughes JM, Hide G, Ford NJ, Oldbury DJ. (2008) The urban house mouse (*Mus domesticus*) as a reservoir of infection for the human parasite *Toxoplasma gondii*: An unrecognised public health issue? *International Journal of Environmental Health Research*, 18(3): 177–185.

[33] Dubey JP, Beattie CP. (1988) *Toxoplasmosis of Animals and Man*, CRC Press, Boca Raton, FL.

[34] Marshall PA, Hughes JM, Williams RH, Smith JE, Murphy RG, Hide G. (2004) Detection of high levels of congenital transmission of *Toxoplasma gondii* in natural urban populations of *Mus domesticus*. *Parasiotology*, 128: 39–42.

[35] Battersby SA, Parsons R, Webster JP. (2002) Urban rat infestations and the risk to public health. *Journal of Environmental Health Research*, 1: 57–65.

[36] Webster, JP, MacDonald DW. (1995a) Parasites of wild brown rats (*Rattus norvegicus*) on UK farms. *Parasitology*, 111: 247–253.

[37] Anderson RM. (1993) Epidemiology. In *Modern Parasitology*, Cox FEG (Ed.), Blackwell Science, Oxford, 75–11.

[38] Himsworth CG, et al. (2014) The characteristics of wild rat (Rattus spp.) populations from an inner-city neighborhood with a focus on factors critical to the understanding of rat-associated zoonoses. *PLoS One*, 9(3): e91654. https://doi.org/10.1371/journal.pone.0091654.

[39] Strand TM, Lundkvist Å. (2019) Rat-borne diseases at the horizon: A systematic review on infectious agents carried by rats in Europe 1995–2016. *Infection Ecology & Epidemiology*, 9: 1. https://doi.org/10.1080/20008686.2018.1553461.

[40] Seguin B, et al. (1986) Bilan épidémiologique d'un échantillon de 91 rats (*Rattus norvegicus*) capturés dans les égouts de Lyon. *Zentralblatt für Bakteriologie, Mikrobiologie und Hygiene, Serie A*, 261: 539–546.

[41] Hilton AC, Willis RJ, Hickie SJ. (2002) Isolation of *Salmonella* from urban wild brown rats (*Rattus norvegicus*) in the West Midlands, UK. *International Journal of Environmental Health Research*, 12: 163–168.

[42] Gratz NG. (1984) The global public health importance of rodents. In *Proceedings of a Conference on the Organisation and Practice of Vertebrate Pest Control*, Dubock AC, (Ed.), World Health Organization, Geneva, 413–435.

[43] Lehmann-Grube F. (1975) Lymphocytic choriomeningitis virus. In *Virology Monographs*, Gard S, Hallauer C, Meyer KF (Eds.), Volume 10, Springer-Verlag, New York, 1–173.

[44] Meerburg BG, Grant R, Singleton GR, Kijlstra A. (2009) Rodent-borne diseases and their risks for public health. *Critical Reviews in Microbiology*, 35(3): 221–270.

[45] Stenseth NC, Samia NI, et al. (2006) Plague dynamics are driven by climate variation. *PNAS*, 103(35): 13110–13115, published ahead of print 21 August. https://doi.org/10.1073/pnas.0602447103.

[46] Meaney P. (1998) *Insect Pests of Food Premises*, National Britannia Ltd., Caerphilly.

[47] Meyer A, Allan E, Madge O. (2010) *British Pest Management Manual*, BPCA, Derby.

[48] NPAP. (2015) *Pest Control Procedures in the Food Industry*, CIEH, London. https://www.urbanpestsbook.com/downloads/procedures-sector-specific.

[49] Kopanic Jr RJ, et al. (2001) An adaptive benefit of facultative coprophagy in the German cockroach *Blattella germanica*. *Ecological Entomology*, 26: 154–162.

[50] Rehn JAG. (1945) Man's uninvited fellow-traveller – the cockroach. *Scientific Monthly*, 61(4): 265–267.

[51] Brenner RJ. (1995) Economics and medical importance of German cockroaches. In *Understanding and Controlling the German Cockroach*, Rust MK, Owebs JM, Reirson DA (Eds.), Oxford University Press, New York.

[52] Baumholtz MA, et al. (1997) The medical importance of cockroaches. *International Journal of Dermatology*, 36: 90–96.

[53] Rust, MK. (2008) Cockroaches. In *Public Health Significance of Urban Pests WHO Regional Office for Europe*, Bonnefoy X, Kampen H, Sweeney K (Eds.), Copenhagen, Denmark.

[54] Robinson WH. (1996) *Urban Entomology: Insect and Mite Pests in the human Environment*, Springer, New York.

[55] Gore JC, Schal C. (2005) Expression, production and excretion of Bla g 1, a major human allergen, in relation to intake in the German cockroach, *Blattella germanica*. *Medical and Veterinary Entomology*, 19: 127–134.

[56] Denmark JJ, Bennett GW. (1995) Adult German cockroach (Dictyoptera: Blattellidae) movement patterns and resource consumption in a laboratory arena. *Journal of Medical Entomology*, 32: 241–248.

[57] Liccardi G, Cazzola M, D'Amato, M, et al. (2000) Pets and cockroaches: Two increasing causes of respiratory allergy in indoor environments. Characteristics of airways sensitization and prevention strategies. *Respiratory Medicine*, 94: 1109–1118.

[58] Alexander JB, Newton J, Crowe GA. (1991) Distribution of oriental and German cockroaches, *Blatta orientalis* and *Blattella germanica* (Dictyoptera), in the United Kingdom. *Medical and Veterinary Entomology*, 5: 395–402.

[59] Doggett SL. (2013) *A Code of Practice for the Control of Bed Bug Infestations in Australia*, 4th ed. https://www.pmanz.nz/uploads/5/3/1/0/53106237/bed_bug_cop_edition4march2103.pdf.

[60] National Pest Advisory Panel. (2010) *Pest Control Procedures in the Housing Sector*. https://www.urbanpestsbook.com/downloads/procedures-sector-specific/.

[61] Boase C. (2008) *Bedbugs: An Evidence-Based Analysis of the Current Situation*, International Conference on Urban Pests, Budapest.

[62] Merritt RW, et al. (2009) Chapter 76 – Diptera: (Flies, Mosquitoes, Midges, Gnats). In *Encyclopedia of Insects*, Resh VH, Cardé RT (Eds.), Academic Press, London, 284–297, 2nd ed.

[63] Hogsette JR, Amendt J. (2008) Flies. In *Public Health Significance of Urban Pests WHO Regional Office for Europe*, Bonnefoy X, Kampen H, Sweeney K (Eds.), Copenhagen, Denmark.

[64] Medlock JM, Snow KR. (2008) British mosquitoes. *British Wildlife*, 19(5): 338–346.

[65] Gage KL, Kosoy MY. (2005) Natural history of plague: Perspectives from more than a century of research. *Annual Review of Entomology*, 50: 505–528.

[66] Shaw SE, et al. (2004) Pathogen carriage by the cat flea *Ctenocephalides felis* (Bouché) in the United Kingdom. *Veterinary Microbiology*, 102: 183–188.

[67] Public Health England. (2018) *Lyme Disease Epidemiology and Surveillance*. https://www.gov.uk/government/publications/lyme-borreliosis-epidemiology/lyme-borreliosis-epidemiology-and-surveillance (Accessed September 2021).

[68] Haag-Wackernagel D, Moch H. (2004) Health hazards posed by feral pigeons. *The Journal of Infection*, 48: 307–313.

[69] Pellizzari M, Loughlin D. (2017) Controlling urban pigeon populations humanely. In *Proceedings of the Ninth International Conference on Urban Pests*, Matthew P, Davies, CP, William HR (Eds.), Uckfield, East Sussex. https://www.icup.org.uk/media/42unjh30/icup1202.pdf (Accessed September 2021).

[70] Oi DH. (2008) Pharaoh ants and fire ants. In *Public Health Significance of Urban Pests WHO Regional Office for Europe*, Bonnefoy X, Kampen H, Sweeney K (Eds.), Copenhagen, Denmark.

[71] Edwards JP, Baker LF. (1981) Distribution and importance of the Pharaoh's ant *Monomorium pharaonis* (L.) in national health service hospitals in England. *The Journal of Hospital Infection*, 2: 249–254.

[72] Beatson SH. (1972) Pharaoh's ants as pathogen vectors in hospitals. *Lancet*, 1: 425–427.

[73] Burn AJ, Coaker TH, Jepson PC. (1987) *Integrated Pest Management*, Academic Press, London.

[74] Sarisky JP, Hirschhorn RB, Baumann GJ. (2008) Integrated pest management. In *Public Health Significance of Urban Pests WHO Regional Office for Europe*, Bonnefoy X, Kampen H, Sweeney K (Eds.), WHO, Copenhagen, Denmark.

[75] WHO. (2004) *Global Strategic Framework for Integrated Vector Management*, World Health Organization, Geneva. https://apps.who.int/iris/bitstream/handle/10665/68624/WHO_CDS_CPE_PVC_2004_10.pdf;jsessionid=48

8A11559D3E084C484943A6A29FF6F4?sequence=1 (Accessed September 2021).

[76] Murphy RG, Williams RH, Hide G. (2005) Population biology of the urban mouse (*Mus domesticus*) in the UK. *Proceedings of the 5th International Conference on Urban Pests*, 351–355.

[77] Gratz NG, Steffen R, Cocksedge W. (2000) Why aircraft disinsection? *Bulletin of the World Health Organization*, 78(8), August. http://www.ncbi.nlm.nih.gov/pmc/articles/PMC2560818/pdf/10994283.pdf (Accessed September 2021).

[78] Meyer AN, Murphy G. (2001) *Prevention of Damage by Pests Act'49 Fact or Fiction – Rodent Control: A Modern Perspective*, Conference Abstracts, Pest Ventures Seminar Kegworth, Notts, March.

[79] Battersby S, Pointing J. (2019) *Statutory Nuisance and Residential Property*, Routledge Focus on Environmental health series, Routledge, Abingdon, Oxon.

[80] Office of the Deputy Prime Minister. (2006) *Housing Health and Safety Rating System: Enforcement Guidance*, made under s.9 Housing Act 2004, London. https://assets.publishing.service.gov.uk/government/uploads/system/uploads/attachment_data/file/7853/safetyratingsystem.pdf.

Further reading and other sources of information

Buckle AP, Smith H. (Eds.) (2015) *Rodent Pests and their Control*, CABI, Wallingford, Oxon, 2nd ed.

Busvine JR. (1980) Insects and hygiene. In *The Biology and Control of Insect Pests of Medical and Domestic Importance,* Chapman and Hall, London and New York.

Nowak RM. (1999) *Walker's Mammals of the World*, Volume II, Johns Hopkins University Press, Baltimore, MD, 6th ed.

CRRU. (2021) *UK Code of Best Practice: Best Practice & Guidance for Rodent Control & the Safe Use of Rodenticides*. http://www.thinkwildlife.org/downloads_resources/.

Useful websites

Campaign for Responsible Rodenticide Use www.thinkwildlife.org

CIEH http://www.cieh.org/

Defra http://www.defra.gov.uk/wildlife-pets/wildlife/management/pest-control/index.htm

Commonwealth Scientific and Industrial Research Organisation (CSIRO) (Australia) https://www.csiro.au/en/

Food and Environment Research Agency h t t p : / / www.fera.defra.gov.uk/

Natural England http://www.naturalengland.org.uk/

Association of Port Health Authorities

BASIS Prompt http://www.basis-reg.co.uk/pest-control/promptabout.aspx

British Pest Control Association http://www.bpca.org.uk/

HSE http://www.hse.gov.uk/index.htm and Pesticides Safety Directorate http://www.pesticides.gov.uk/

National Pest Technicians Association http://www.npta.org.uk/

Royal Society for Public Health http://www.rsph.org.uk/

18

Waste and resource management

Jeff Cooper

Introduction

The major influence on the way the UK Government legislates for the control of solid waste has been the European Union (EU) and, despite no longer being part of the EU, this influence will last for the coming decade and beyond because the UK had signed up to many policies and legislative requirements before it had left the EU. In addition, the UK has obligations under international treaties. In the process of leaving the EU there were a significant number of changes made to legislative provisions with reference to EU requirements, but these latest changes are not referred to where they have no substantive effect on the earlier regulations.

In the period from 2000 the United Kingdom's need to conform to the requirements of the Landfill Directive had been instrumental in determining waste management policies and practices during the period to 2020. With the Landfill Directive's requirement to restrict the amount of biodegradable municipal waste to just 35% of the total sent to landfill in 1995 by 2020 (by tonnage, not proportion), the Directive has had a profound influence on the Government's plans and the waste strategies developed for each of the devolved administrations: England, Scotland,

Wales and Northern Ireland. Every aspect of the Directive had influenced waste management by local authorities and businesses.

For the future it is the influence of the EU's policies on the circular economy, because this legislation was passed on 4 July 2018 when the UK was still in the EU, which will have an effect on all of the UK waste management activities. The principles of the Circular Economy Package therefore affect all dimensions of resource management.

Overriding every aspect of policy making and legislative practices in almost every country there is the requirement to tackle the threat of climate change. This is fundamental to the establishment of the European Green Deal and provides the focus for most countries' post-pandemic economic policies. For waste and resource management waste prevention is the key priority and this is referenced later in the chapter in the sections dealing with waste prevention and exemplified in detail for food waste.

Background to waste collection, disposal and administration

Legislative provision for the collection and disposal of waste in the UK before and during the 19th century was extremely patchy. Arrangements for the regular clearance of wastes from streets, privies, middens and ash pits was recognised as fundamental to improving the health of towns, but until late in the century these needs were met by the passage of private Acts of Parliament empowering individual towns to undertake the necessary measures.

Usually the privilege of collection and disposal of waste was sold to the highest bidder because the UK was then a largely agricultural country with the majority of the waste being organic – from horse, human and food waste – and the material was therefore a valuable source of fertiliser, especially as it was frequently left in situ for weeks or months to mature. However, its quality could not always be guaranteed, as the following example from Windsor demonstrates:

> The ditches of which I have spoken are sometimes emptied by carts. And on the last occasion their contents were purchased for the sum of £15 by the occupier of the land in the parish of Clewer, where meadows suffered from the extraordinary strength of the manure which was used without previous preparation. [1]

An attempt made to provide a minimum level of cleansing provision through the Public Health Act 1848 failed because it was largely permissive, not compulsive. Although the Act was adopted by about 200 local authorities, this was inadequate given the need to establish minimum standards for public cleansing, especially after the repeal of the Corn Laws when waste became less valuable as competition from imported grain progressively impoverished the agricultural sector.

The multiplicity of bodies responsible for different aspects of public health and the continuing problem of, inter alia, poor waste collection and disposal practices led to the appointment of a Royal Sanitary Commission in 1869 to resolve the crisis. The resulting Local Government Act 1871 introduced sanitary districts throughout the country in 1872, and the appointment of a medical officer and an inspector of nuisances became obligatory for each district.

The Public Health Act 1875 thereafter consolidated, for areas outside London, the wide spread of laws relating to public health. This prescribed the duties of urban and rural sanitary districts which included, inter alia, responsibility for scavenging. Unlike the period before 1875, when appointed, the municipal scavengers were paid employees, and although clearance of waste was not always sufficiently frequent to avoid putrescible materials becoming a nuisance, the solid waste from houses was taken off by hand- or horse-drawn cart to be disposed of, after any valuable components had been salvaged, by tipping or burning.

This system of waste collection is still often being sought to improve public health in developing countries where there are many risks associated with the poor management of waste. These dangers include:

- Contamination of water sources;
- Blockages in drainage channels; and
- The open burning of waste, which generates air pollution emissions that are much more toxic than they were a century ago.

Although the waste in developing countries is predominantly organic and therefore composting would be possible, often it is the lack of systems for separation that is the main barrier to more effective waste management.

Improving waste management in developing economies

Globally between two and three billion people do not have their waste collected or properly managed. Collection of waste is the essential first step in low- and middle-income countries to establish basic public health protection for waste management. Thereafter the next steps are dependent on the infrastructure that can be provided to further manage the waste: from engineered landfill through to recycling by composting and production of items by manufacturing them from reclaimed materials. Useful guidance can be obtained from the UK arm of Waste Aid International, WasteAid UK.[1] In 2017 WasteAid UK was sponsored by the Chartered Institution of Wastes Management to publish *Making Waste Work: A Toolkit* to show how communities in low- and middle-income countries can manage much of their waste independently and cost-effectively, preventing pollution and creating jobs. This can be accessed through the toolkit on the WasteAid website.[2]

The collection of waste

Prior to the Control of Pollution Act 1974 (COPA), domestic solid waste was officially known as 'house refuse', for which there was no legal definition. However, it was accepted that it was the sort of refuse that arose from the ordinary domestic occupation of a dwelling. The main Act dealing with domestic refuse before the COPA was the Public Health Act 1936. Under Section 72, a Local Authority (LA) could, and if required by the minister had to, undertake the removal of household refuse with respect to the whole or any part of its area. Surprisingly, perhaps, it was only with the introduction of the Collection and Disposal of Waste Regulations 1988, which enforced the provisions of Sections 12–14 of the COPA, that for the first time every collection authority was under a duty to collect household waste in its area.

The variation in the standards of refuse collection services provided by LAs was examined in great detail by the Working Party on Refuse Storage and Collection. Its report, published in 1967, revealed great diversity in the frequency of collection, types of material that were collected or not accepted for collection, and in the types of waste receptacle and equipment used for collection. Concern was also expressed about the lack of provision for the disposal of bulky waste, garden

waste and cars, which were often disposed of in quiet country lanes, ditches or any other spot convenient for the increasingly affluent and mobile population.

The need for facilities for people to dispose of these types of waste was recognised by a Private Member's Bill introduced by Duncan Sandys MP, which became the Civic Amenities Act 1967 and was subsequently revised as the Refuse Disposal (Amenity) Act 1978. These Acts have permitted local authorities to establish civic amenity (CA) sites, now referred to as household waste recycling centres (HWRC) because these sites now incorporate these facilities. Indeed, the duty to manage the site in return for salvage rights can be of financial advantage to the LA. The term civic amenity site is internationally recognised.

The disposal of waste

Reports of the irresponsible disposal of toxic waste became steadily more prominent in the press during the 1960s, and the fear of contamination of water sources prompted government action. In 1964, the Minister of Housing and Local Government, together with the Secretary of State for Scotland, appointed the Technical Committee on the Disposal of Toxic Wastes. Its terms of reference were:

> To consider present methods of disposal of solid and semi-solid toxic wastes from the chemical and allied industries, to examine suggestions for improvement, and to advise what, if any, changes are desirable in current practice, in the facilities available for disposal and in control arrangements, in order to ensure that such wastes are disposed of safely and without risk of polluting water supplies and rivers.

The recommendations of the Key Report, Report of the Technical Committee on the Disposal of Toxic Waste, in 1970, the companion Report of the Working Party on Refuse Disposal produced in 1971, and the earlier Report on the Working Party on Refuse Storage and Collection in 1967 were being incorporated into a comprehensive piece of environmental legislation when several incidents of fly-tipping of drums of toxic chemicals in the Midlands prompted a rapid legislative response.

The Deposit of Poisonous Waste Act 1972 was enacted in 20 days, and was the first piece of legislation anywhere in the world to protect the environment from the dumping of hazardous waste. While it should have been repealed and replaced by the COPA in 1974, the 1972 Act lasted until 1980 when the special waste consignment note system was brought in under the Control of Pollution (Special Wastes) Regulations 1980, now replaced by the Hazardous Waste (England and Wales) Regulations 2005 and equivalent regulations in Scotland and Northern Ireland.

These regulations were designed primarily to comply with internationally agreed obligations under the EU Directive on dangerous and toxic wastes 78/319/EEC. The consignment note system was similar to the notification requirements of the 1972 Act to provide cradle-to-grave notification to ensure that the Environment Agency (EA) is able to monitor movements of hazardous waste and the techniques of disposal. The present arrangements are outlined at pages 846–52.

The administration of waste disposal

Until 1965, responsibility for the disposal of house waste rested with the same authority that collected it, namely the metropolitan borough, county borough, or the urban or rural district council. However, this was recognised as increasingly unsatisfactory in metropolitan areas. Generally, waste disposal had to be undertaken through incineration, where standards of emission control were almost non-existent, or through the use of transfer stations where spillage *en route* accounted for a proportion of the disposal.

In 1963, the London Government Act transferred responsibility for waste disposal to the Greater London Council (GLC), which effectively became the first Waste Disposal Authority (WDA). The new and enlarged London boroughs retained responsibility for collection of household waste, but were directed by the GLC on its ultimate destination for delivery to a landfill site, incinerator or transfer station. Even at its abolition in 1986 the establishment of an environmentally sound system of waste disposal by the GLC had not totally been achieved.

This system for waste management was extended to the rest of England under the Local Government Act 1972, but in Wales, Scotland and Northern Ireland responsibility for waste disposal remained with the much-enlarged district councils rather than being transferred to the upper tier of local government administration (and that position has been retained through subsequent reorganisations of local government). Under the COPA these new local government bodies and English county councils became WDAs after they came into existence on 1 April 1974.

In 1976, under the COPA, for the first time a system of day-to-day regulation of waste disposal and waste transfer facilities was added to the planning conditions and Public Health Act 1936 powers that had been the only means of controlling waste disposal since 1947. Under the Town and Country Planning Act 1947, any change of land use, including use of land for disposal of waste, required planning permission, although use of land for waste disposal prior to 1 January 1948 meant that land could continue to be used for disposal of waste. The local government bodies that administered these waste regulation duties were the new WDAs created under the COPA.

The abolition of the metropolitan counties and the GLC in 1986 prompted a number of changes in the arrangements for waste disposal and waste regulation, with both subsequently being superseded by the provisions of the Environmental Protection Act 1990 (EPA) and the Environment Act 1995, respectively. The post-1986 arrangements led to a series of WDAs, most formed voluntarily by metropolitan district councils covering a former metropolitan county area, or groups of London boroughs, which dealt with both the disposal of waste and the administration of waste regulation. However, under Section 32 of the EPA, the direct administration of waste disposal operations of all LAs in England and Wales were progressively privatised, either through the formation of arms'-length Local Authority waste disposal companies (LAWDCs) or through sale of assets, formation of joint ventures or handing over of management waste disposal facilities to private sector waste management companies [2]. Of the original LAWDCs, all have now been subject either to management buyout or sold to waste management companies.

The formation of the Environment Agency for England and Wales, and the Scottish Environment Protection Agency (SEPA), under the Environment Act 1995 on 1 April 1996 brought together the National Rivers Authority (NRA), Her Majesty's Inspectorate of Pollution (HMIP) and the waste regulation departments of the LAs. From April 2014 there has been a separate regulatory body for Wales, Natural Resources Wales, and also the Northern Ireland Environment Agency (NIEA) has been set up in Northern Ireland.

European perspectives

The EU has had and will continue to have a predominant influence on environmental legislation in the United Kingdom. However, it is commonly acknowledged that the first European statement on waste management, the Framework Directive on Waste 75/442/EEC, was closely modelled on the COPA. The Directive on dangerous and toxic wastes (78/319/EEC) was one of the first daughter directives to be formulated. The Waste Framework Directive 2008/98/EC, which incorporates the Hazardous Waste Directive, and the Waste Oil Directive has amended both these directives. In addition, there are several other directives affecting waste management in the United Kingdom, covering, for example, disposal of polychlorinated biphenyls (PCBs) 96/59/EEC, and a European Waste Shipments Regulation 1013/2006/EC.

In the early 1990s, a number of waste streams were identified by the EU for special attention, some because of their environmental effects and others, such as packaging, mainly because of their political significance. The Packaging and Packaging Waste Directive 94/62/EC, approved on 23 December 1994, set targets for the recovery and recycling of packaging waste and was amended in 2004 (2004/12/EC) with higher targets for recycling. Since 2020 recovery by energy or composting is no longer applicable, the targets can only be met by recycling.

The packaging directive

The directive stemmed from the introduction of legislation covering packaging waste introduced in Germany in 1991, which set the agenda for the whole of Europe. In the United Kingdom, the packaging directive has been implemented through the producer responsibility provisions of the Environment Act 1995, Sections 93–95. The UK has opted for an industry-based shared producer responsibility to raise the levels of recycling of packaging wastes to an estimated 70% and more of the 11.5 million tonnes of packaging used annually in the UK. While most countries take the view that shared producer responsibility for packaging waste will involve a partnership between the consumer, LAs (municipalities) and industry, the United Kingdom has a much more specific and narrower definition. Shared producer responsibility for packaging waste in the United Kingdom refers only to the industries in the packaging chain that produce or use packaging. When LAs and consumers are drawn in it will be to help the packaging producers to fulfil their obligations, but there is no regulatory compulsion.

Shared responsibility has been instituted through imposing specific responsibilities

Table 18.1 EU Directive recycling targets for 2025 and 2030

Material	2025 target	2030 target
Paper	75%	85%
Glass	70%	75%
Aluminium	50%	60%
Steel	70%	80%
Plastic	55%	55%
Wood	25%	30%
Overall recycling	65%	70%

Table 18.2 UK business recycling targets 2022 and 2021 (2007 amended regulations)

Material	2022 target	2021 target
Paper	83%	79%
Glass	82%	81%
(of which re-melt)	72%	72%
Aluminium	69%	66%
Steel	87%	86%
Plastic	61%	59%
Wood	35%	35%
Overall recycling	77%	76%

on all businesses that pass on more than 50 tonnes of packaging or packaging materials and have a turnover of more than £2 million per annum. The share in the achievement of the UK's recycling targets accounted for by each business is dependent on their place in the packaging chain. In addition, in order to determine the recycling targets of businesses, progressively increasing levels of recycling have been set by the Producer Responsibility Obligations (Packaging Waste) Regulations 2007 as amended. The original legislation dated to 1997. Smaller business, those with a turnover of between £2–5 million annually, can take advantage of a standard formula which was 30 tonnes per £1 million turnover in 2020.

Businesses can either arrange for the recycling of packaging waste themselves, but in most cases through agents acting on their behalf by joining a compliance (collective or exempt) scheme, thereby placing responsibility on the scheme to arrange for the recycling to be undertaken on their behalf. Data on their packaging flows still has to be provided by companies.

In future, industry will have to do much more to meet the minimum requirements to improve governance and cost efficiency of EPR schemes set by the EU's Circular Economy Package (CEP). This will mean that industry will need to pay more of the costs associated with the management of packaging waste, including littering. Therefore, once the CEP's obligations for industry are enacted through the Environment Act 2021[3] LAs will receive greater financial assistance from industry, albeit at the time of writing the mechanism was still unclear.

The End-of-Life Vehicles Directive

The End-of-Life Vehicles (ELV) Directive was one of the producer responsibility directives, which were developed as one of the priority waste streams initiatives of the European Commission in the early 1990s, which eventually passed into European legislation through the ELV Directive (2000/53/EC). In the United Kingdom producer responsibility for the treatment and processing of ELVs came into effect in 2007.

The Waste Electrical and Electronic Equipment Regulations 2006

The Waste Electrical and Electronic Equipment (WEEE) Directive (2002/96/EC) is intended to pass the costs of treatment and processing for recovery and recycling of WEEE onto manufacturers and importers (producers) of EEE. Segregation of WEEE from households is not compulsory but MSs are required to encourage consumers to undertake separation for reuse, recovery and recycling. The extent of LA involvement depends on the extent to which LAs offer facilities for consumers to separate WEEE items and treat it separately. Retailers have a responsibility for taking back WEEE items on a like-for-like replacement basis unless they join the Distributor Take-Back Scheme, with enforcement of this aspect of the Regulations by the Department for Business, Energy and Industrial Strategy (BEIS).

There is also the Restriction of Hazardous Substances in Certain Electrical and Electronic Equipment (RoHS) Directive (2002/95/EC), which precludes the use of heavy metals, and certain other materials in new EEE. Most WEEE comes back through LA-managed HWRCs and there is a Code of Practice for Designated Collection Facilities produced under the auspices of BEIS. The latest version was produced in February 2019 and is periodically updated.

The Batteries Directive

The EU Batteries Directive, Directive 2006/66/EC on Batteries and Accumulators and Waste Batteries and Accumulators sets targets not only for the recycling of batteries but also for the proportion of portable batteries, small domestic batteries, for example, to be collected, currently 45% each year. However, the European Commission plans to raise this target to 65% in 2025 and 70% in 2030 in a new Batteries Regulation, which will not affect the UK. There are restrictions on the disposal of automotive and industrial batteries and they all have to be recycled.

There is, however, a problem with the use of lithium-ion (li-ion) batteries that are becoming increasingly common. They are a natural substitute for nickel cadmium batteries which are being phased out of use as required by the Directive. Li-ion batteries have one major disadvantage; they can catch fire if the anode and cathode connect.

The most common reason for fires in waste treatment facilities caused by li-ion batteries is the inappropriate disposal of li-ion batteries through domestic refuse collection systems, both into the residual waste stream but also in the recyclable waste stream. When the anode and cathode come into contact through any normal handling of waste, such as compaction in a refuse collection vehicle, movement by handling equipment in waste facilities or processing of WEEE, for example, a fire may occur. Without stronger guidance to both waste management facilities and also for the public the problem is set to get worse in future.

There was one incident of a refuse collection vehicle fleet in East Northamptonshire that was burnt out due to li-ion batteries, and battery storage and sorting facilities in France, Germany and the UK were destroyed by fire (Figure 18.1).

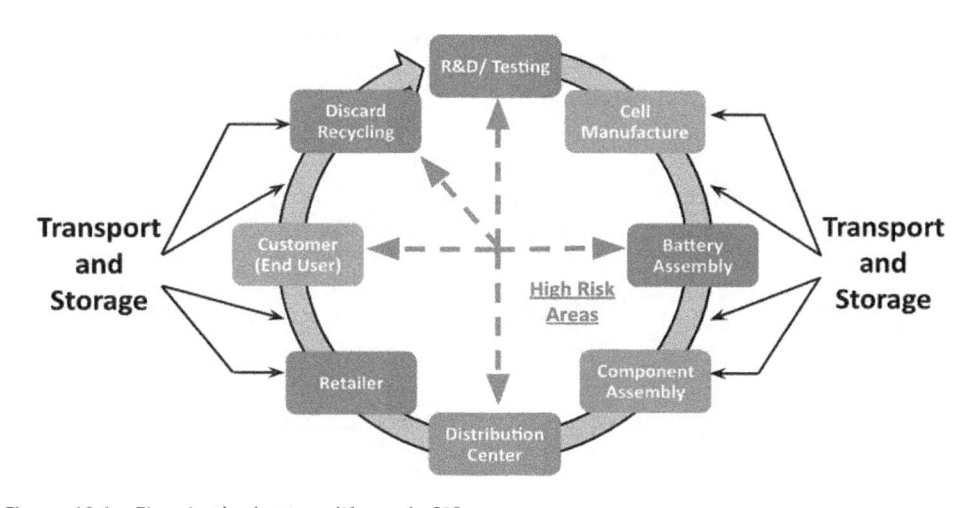

Figure 18.1   Fires in the battery life-cycle [3]

One concern is that consumers have limited guidance about the proper disposal of li-ion batteries because the information provided for recovery of batteries for recycling covers all types of battery chemistry. In the UK citizens have only very indirect incentives to segregate and separately dispose of their portable batteries appropriately and they can still discard them quite legally into their residual waste container, as is the case in many countries, albeit national and LAs provide clear guidance against this practice. Ideally citizens ought to take them to a shop or community facilities, such as libraries, and place the batteries in a container to be collected by a battery compliance scheme directly or aggregated through the retailers' regional distribution centres. Also, batteries, where possible, should be removed from WEEE and segregated for collection via the same route. Alternatively, WEEE items should be taken to a shop offering take-back services or to a HWRC/CA site. With small WEEE items, such as electric toothbrushes and shavers, these items are generally sealed for safety reasons so that battery removal is extremely difficult and mainly these items are discarded into consumers' residual waste containers.

Guidance regarding the issue of the handling of spent li-ion batteries in waste management facilities is often very general and not specific to li-ion batteries. However, there are examples of guidance, including:

In the UK, the WISH (waste industry safety and health) Forum has produced guidance for waste management facilities.[4]

In October 2020 SWANA (Solid Waste Association in North America) produced its *Guide for developing lithium battery management practices at materials recovery facilities*.[5]

With regard to the wider issue of guidance to the public, beyond the current guidance on return to shops, the only instance of more specific guidance with regard to li-ion batteries comes from Norway where consumers are requested to tape over the anodes and cathodes of li-ion batteries. In the UK the Environmental Services Association launched

its campaign to encourage people to recycle their batteries more responsibly in October 2020, *Take Charge*, takecharge.org.uk (#TakeCharge, #ZombieBatteries) but this initiative was not wholly directed specifically to li-ion battery issues.

New EPR systems

Although there are only four EU Directives covering products affected by EPR and the UK government has implemented only those, a number of EU states have instituted EPR programmes for a range of other products, including textiles, clothing, mattresses, tyres, vehicle lubricating oils, cooking oils and agricultural plastic films. There have been calls for the UK to institute EPR systems for clothing and mattresses, but a scheme put forward in 2008–09 for agricultural non-packaging plastic films failed to be endorsed by the farming sector.

The Landfill Directive

While there were lengthy discussions on the production of a landfill directive in the early 1990s, no directive resulted, mainly due to disagreement between the European Commission and the European Parliament. The introduction of a further proposed landfill directive in 1997 led to the Landfill Directive 2000/76/EC.

Overall, the various articles in the directive seek to move waste 'up the hierarchy' away from landfill by:

- Reducing the landfill of biodegradable waste;
- The pre-treatment of all waste prior to landfill; and
- Banning the co-disposal of hazardous industrial wastes and household wastes.

The main difficulty was that for a considerable period the United Kingdom had sought to support the 'flushing bioreactor' model of sustainable landfill, whereby through the promotion of enhanced degradation of biologically active waste materials a landfill could be stabilised within a generation, approximately 30 years. However, there has been no instance of the use of this concept in practice and thus it is difficult to assess whether a landfill could be chemically and physically stabilised within 30 years. One of the main provisions in the Landfill Directive is the progressive reduction in the amount of biodegradable municipal waste (BMW) allowed into landfill sites. This is designed to reduce the potential for generating landfill (mainly methane) gases. The reductions in the UK were to:

- 75% of 1995 amounts by 2010;
- 50% of 1995 amounts by 2013;
- 35% of 1995 amounts by 2020.

The UK and some other Member States were allowed to take four years longer than other Member States to reach these targets because they sent to landfill more than 80% of their municipal solid waste (MSW) in 1995. The Landfill (England and Wales) Regulations 2002[6] were now superseded by the Environmental Permitting Regulations, and the Landfill (Scotland) Regulations[7] introduced most of the requirements of the Directive, including:

- Classifying landfill sites for hazardous, non-hazardous and inert wastes;
- Banning explosive, corrosive, oxidising or highly flammable wastes from July 2002;
- Banning infectious clinical waste from July 2002;
- Banning whole tyres from landfill from July 2003 and shredded tyres from July 2006.

The EU's CEP has now introduced a further target of 10% BMW to landfill in 2035.

The EU's Circular Economy Programme

Although the Circular Economy legislation was only passed in July 2018 the European Commission had introduced an Action Plan to implement its draft Circular Economy Package in 2015. The 2018 Circular Economy Package included a Europe-wide EU Strategy for Plastics in the Circular Economy to transform the way plastics products are designed, produced, used and recycled. By 2030 all plastics packaging should be recyclable. The Commission has also set up a monitoring framework on progress towards a circular economy covering ten key indicators which cover each phase: production, consumption, waste management and secondary raw materials and economic aspects such as investments, jobs and innovation. In addition, there are a number of provisions which will impact the services that LAs provide, including: separate collection obligations for hazardous household waste by the end of 2022, bio-waste by 2023 and textiles by 2025. As of late 2021 the introduction of mandatory bio-waste collections in 2023 appeared doubtful.

Waste prevention plans

The Waste Framework Directive (2008/98/EC) set a legal obligation for Member States to adopt waste prevention programmes. The Waste Prevention Programme (WPP) for England was published in 2013 and it was due to be updated in 2020 but was delayed by the pandemic until 2022. There were a number of projects initiated, and support provided for a wide range of technological and behavioural innovations, but a review conducted by Defra revealed little in the way of long-term results. However, from the annual reviews of MSs WPPs undertaken by the European Environment Agency also showed patchy results from all those WPPs.

Economic and other waste policy instruments

While throughout Europe the landfill or waste disposal tax is the most popular economic instrument, there is a wide range of other waste policy instruments. These include: deposit refunds, product taxes, recycling targets, changing responsibilities, mandatory collection, subsidies, price support, recycling credits and bans.

These policy instruments are often used in combination and thus the UK approach to the EU Packaging and Packaging Waste Directive involves a combination of changing responsibilities and recycling targets. Industry is being forced to take on an enhanced responsibility for packaging waste, partly away from LAs in the case of household packaging wastes, to reach the recycling target levels laid down in the Producer Responsibility Obligations (Packaging Waste) Regulations 2007. A combination of the landfill tax and the aggregates levy has resulted in around 30% of aggregates used in the UK coming from reclaimed materials.

The landfill tax – the United Kingdom's landfill tax was introduced in October 1996 and is applied to all wastes going to landfill. It was applied at two rates, originally at a lower rate of £2 per tonne for inactive (inert) waste, such as brick or concrete waste, and a higher rate of £7 per tonne applying to all other wastes. The landfill tax was raised to £10 per tonne in March 1998, rising by £1 per tonne until 2004/05 after which it was raised at £3 and subsequently £8 per tonne each year until 2013/14 when it reached £80 per tonne and subsequently rises are in line with inflation, which also applies to the inactive rate, reaching £94.15 per tonne in 2020/21 and £3.00 for the lower rate.

The effect of the landfill tax is a reduction in the deposit of waste, especially that from construction and demolition (C&D). More of this waste is being sorted and processed on building sites or at transfer stations and recycling facilities and only an estimated 5% of the C&D waste goes to landfill. A considerable amount of the combustible fraction, together with suitable commercial and industrial waste is exported to be incinerated in European energy recovery facilities, this being a cheaper option than landfill. In 2018 this amounted to over 3 million tonnes but the amounts exported have subsequently been declining as more energy recovery facilities

have been built in the UK and latterly as a result of Brexit.

The aggregates levy – was introduced on 1 April 2002. It is payable on sand, gravel and crushed rock at a rate of £2.00 per tonne. The current rate of exploitation of these aggregates is around 240 m tonnes per annum (tpa), a figure much below the high levels of 360 m tpa reached in the 1980s. As with the landfill tax the revenue raised through the aggregates levy is used for a reduction in employers' National Insurance Contributions.

Climate Change Levy (CCL) – the CCL affects industry by increasing the cost of fossil fuel energy but again the costs are offset against Employers' National Insurance Contributions. The aim of the CCL is to help ensure industry improves its energy efficiency to meet the United Kingdom's commitments under the UN Framework Convention on Climate Change, commonly known as the Paris Climate Agreement. From the perspective of waste management, the significant factor is that substantial savings of CCL can be effected through changes in sources of fuel and/or raw materials in order to improve energy efficiency. Thus, for example, there is an added incentive for glass manufacturers to use cullet from waste sources and for cement manufacturers to use a wide variety of alternative fuels from waste sources, including: tyres, solvents, refuse derived fuel and selected waste packaging products and production residues.

The duty of care for waste

One of the fundamental changes introduced by the EPA was the concept of a duty of care under which

> it shall be the duty of any person who imports, produces, carries, keeps, treats or disposes of controlled waste, or, as a broker, has control of such waste, to take all such measures applicable to him in that capacity as are reasonable in the circumstances.

Waste producers and others responsible for waste have 'to prevent the escape of the waste' on to unlicensed land or in contravention of the conditions of a waste management licence. It also requires the holder of the waste to transfer it only to an authorised person such as:

1   The WCA
2   The holder of a waste management licence
3   A registered carrier under Section 2 of the Control of Pollution (Amendment) Act 1989

Although initially the duty of care did not apply to households for their household waste in 2005 the Waste (Scotland) Regulations 2005[8] introduced a duty of care for householders in Scotland, followed by the Waste (Household Waste Duty of Care) (England and Wales) Regulations 2005[9] and the Waste (Household Waste Duty of Care) (Wales) Regulations 2006,[10] now incorporated into the Waste (England and Wales) Regulations 2011.[11]

Guidance on what actions are 'reasonable in the circumstances' is contained in a Code of Practice produced by Defra (Department of the Environment, Food and Rural Affairs). Under the Environmental Protection (Duty of Care) (England) (Amendment) Regulations 2003[12] WCAs have the power to serve a notice on businesses requiring them to furnish the WCA with their duty of care records. This legal provision is designed to help officers of the WCA to check whether businesses are transferring their waste in accordance with the law. Also it will help WCAs investigate fly-tipping incidents by, for example, being able to track back evidence through information deposited, such as a letter headed paper.

The carriage of waste

The Control of Pollution (Amendment) Act 1989 (COP(A)A) enabled the Secretary of State to make regulations requiring any persons carrying controlled waste in the course of their business, or otherwise with a view to profit, to be registered. The Controlled Waste (Registration of Carriers and Seizure of Vehicles) Regulations 1991[13] came into force on 14 October 1991. Those exempt from registration were:

• WCAs/WDAs;
• People who carry only their own waste (except where building or demolition waste is concerned; all people involved in the carriage of waste from construction, building repair/improvement or demolition must be registered);
• Train operating companies, ferry operators, etc.;
• A charity or recognised voluntary body.

One of the aims of the registration of carriers is to curb fly-tipping activities. Therefore, to ensure compliance with the duty of care it is an offence to transfer controlled waste to an unauthorised person; an authorised person being a registered carrier for these purposes. Registration for people and companies is with the Environment Agency with registration at their principal place of business. A fee is payable with the application for registration, which lasts three years.

The EA must decide whether an applicant should be registered to carry controlled waste. It may refuse to register a carrier if the applicant (or an associate) fails to comply with the application requirements or has unspent environmental convictions for prescribed offences. The prescribed offences relate primarily to environmental matters but also include the offence of non-possession of a vehicle operator's licence. The registration may be revoked if the carrier commits a prescribed offence. Carriers whose applications are refused or whose registration is revoked may appeal to the Secretary of State for the Environment. Carriers who are registered receive a Certificate of Registration that is standardised throughout Great Britain and bears a unique registration number. They may also purchase certificated copies.

Carriers stopped by the police or by EA officers may be required to produce documentary proof of registration within seven days. The COP(A)A also empowers the EA to seize vehicles suspected of being used for the illegal deposit of waste where it proves impossible to trace those in charge of the vehicle(s) at the time of the offence. If not claimed, the vehicle and its load can be sold.

From 2011 the arrangements for the registration of waste carriers were changed in order to ensure that the UK conformed to the requirements of the WasteFD and the outcome of European Court of Justice (ECJ) cases under which it was determined that even if people were carrying their own waste a waste carrier's permit was required. Therefore, the new regime requires any business carrying their own waste to be registered with the Environment Agency/NRW/SEPA/NIEA. However, these newly obligated waste carriers, together with farmers, charities and voluntary groups have a less onerous and less costly registration procedure, a single lifetime registration compared to those long obligated operators that have been trading as waste carriers, where registration every three years continues.

The Transfrontier Shipment of Waste Regulations 2007 provides a system of prior notification and authorisation for the movement of waste consignments between states in the EU and in and out of the EU.

The definition of waste

Article 1 of WasteFD provides that '"waste" means any substance or object which the holder disposes of or is required to dispose of pursuant to the provisions of national law in force'. This provision was given effect in Great Britain by the definition of waste in Section 75 of the EPA. These provisions do not define authoritatively what is and is not waste. What they do is to include within the ordinary meaning of the word 'waste' certain substances about which there might otherwise

have been doubt. Other MSs adopted their own national definition of waste. One of the few aspects of the COPA not changed by the EPA was the definition of controlled waste. The definition used under Section 30 of the COPA is re-stated in Section 75 of the EPA, whereby:

Waste includes:

(a) Any substance which constitutes a scrap material or an effluent or other unwanted surplus substance arising from the application of any process; and

(b) Any substance or article which requires to be disposed of as being broken, worn out, contaminated or otherwise spoiled; but does not include a substance which is an explosive within the meaning of the explosives act 1875. In addition, under Section 75(3), 'anything which is discarded or otherwise dealt with as if it were waste shall be presumed to be waste unless the contrary is proved'.

Controlled waste comprises household, industrial and commercial waste 'or any such waste'.

Household waste means waste from:

- Domestic property, that part of a building used wholly for the purposes of living accommodation;
- A caravan;
- A residential home.

Increasingly waste legislation, such as the Hazardous Waste Regulations 2005, is using the term domestic waste, which only applies to waste from private homes. Therefore waste from premises other than homes, such as hazardous waste, is subject to the controls of the Hazardous Waste Regulations while waste from domestic premises is exempt from hazardous waste controls. However, where the LA collects separately or provides collection facilities for such hazardous wastes the authority becomes the waste generator and needs to provide the appropriate disposal and documentation for these hazardous wastes.

Industrial waste comes from:

- Any factory (within the meaning of the Factories Act 1961);
- Premises connected with transport services;
- Premises used for gas, water, electricity or sewage services; and
- Premises used for postal or telecommunications services.

Commercial waste means waste from premises used for trade and business or for sport, recreation or entertainment, but excludes inter alia household and industrial waste.

The Controlled Waste (England and Wales) Regulations 2012[14] in Schedule1 lists various types of waste as household, industrial or commercial waste. The Regulations do not give an exhaustive list for each category but concentrate on wastes that would otherwise be difficult to classify. The Regulations also list those household wastes for which the Waste Collection Authority may make a charge for collection and disposal.

Defra, through the Controlled Waste (England and Wales) Regulations 2012, allowed LAs to charge premises, such as schools, hospitals and prisons, for the disposal of their waste. Previously they could only charge for collection. Only charity shops, village halls and other premises used for public meetings are now exempt from charging.

Wastes from agriculture, mining and quarrying – certain types of agricultural and mining and quarrying wastes became controlled waste through the Waste Management (England and Wales) Regulations 2006[15] to meet the requirements of the WasteFD. The original UK definition of controlled waste did not correspond with the WasteFD, and was therefore extended to include non-natural wastes from farming, such as plastic sheeting for silage and crop cover and wastes from buildings at mines and quarries.

The Waste (Scotland) Regulations 2005[16] brought wastes from mines and quarries within the definition of industrial waste. Waste from premises used for the breeding, boarding, stabling or exhibiting of animals is classified as commercial waste. Waste from arable farms is not specifically mentioned, but would seemingly qualify as industrial waste. Under the Waste Management (England and Wales) Regulations 2006,[17] agricultural waste and waste arising from mining and quarrying is classified as industrial waste. The extension of regulation to these wastes means that the uncontrolled burning of waste on farms and the use of 'farm tips' are illegal. Farmers require a landfill permit to dump waste on their land.

Municipal waste – municipal waste was first defined in legislation through the Landfill (England and Wales) Regulations 2002 (SI 2002 No. 1559) as 'waste from households as well as other waste, which because of its nature or composition is similar to waste from households'. The definition was significant because the European and domestic legislative duty to divert waste from landfill only applies to municipal waste. The original Defra view of municipal waste was confined to waste collected by the WCA or through WCA contractors. The view in some other Member States is that all similar wastes fit within that definition. This would include at least all commercial waste. In 2009 Defra changed its view to bring the UK in line with the definition of MSW applied in most other Member States.

Following communication with the European Commission, Defra and the DAs reclassified a significant amount of commercial and institutional and a smaller amount of industrial waste which is collected by the private sector as 'municipal waste'. This waste will be subject to the landfill diversion targets under the Landfill Directive. Until then, only waste collected by LAs had been included in the targets. Defra issued guidance in 2012 on the new definition of Local Authority collected municipal waste (LACMW), which included business waste where collected by the LA and was similar in nature to household waste. Local Authority collected waste (LACW) is all waste collected by the LA and is broader in scope than LACMW and could include C&DW when collected. LACW is the definition used for statistical purposes, where municipal waste was previously used and is the information to be submitted to the EA's Waste Data Flow website.

Extension of waste

One of the principal reasons for revising the Waste Framework Directive (WasteFD) started by the European Commission in 2004 was the concern that decisions by the European Court of Justice (ECJ) and in the English domestic courts were leading to the extension of definition of waste. This meant that waste which went through a treatment process, such as composting, might still be regarded as waste at the end of that process. However, from 2008 onwards both the English court system and the ECJ began to consider that the overall objectives of the WasteFD should be the main consideration in their judicial decision process, including the aim of resource efficiency. There were a number of court cases which have been decided on that basis. Therefore in order to provide guidance which would assist decision makers regarding the status of their material and processed waste, Defra issued guidance on the legal definition of waste and its application in August 2012. The content of this document was agreed between Defra, the EA, Welsh government and Northern Ireland's Department of the Environment and NIEA. The document also deals with end-of-waste issues.

End-of-waste criteria

Definition of waste and scope was the most significant issue in reviewing the WasteFD, particularly the end-of-waste criteria. This was to have been clarified through a waste stream approach on a European-wide basis by the European Commission. The priority waste streams covered by this approach were compost, metals and aggregates from waste but there was no consensus among Member States.

Nevertheless the revised WasteFD provides a new definition so that a substance or object resulting from a production process will be regarded as a by-product, and not a waste, only if the following conditions are met:

- Use of the substance or object is certain;
- The substance or object can be used directly without any further processing other than normal industrial practice;
- The substance or object is produced as an integral part of a production process;
- Further use is lawful provided the substance or object fulfils all relevant product, environmental and health protection requirements for the specific use and will not lead to overall adverse environmental or human health aspects.

In the UK, slag from blast furnaces has been recognised as a by-product by the Environment Agency (and subsequently EU-wide), as has gypsum from flue gas desulphurisation abatement at waste incineration plants. Under the WasteFD standards-based waste to product initiatives can be adopted and other changes can be instituted. The Environment Agency and WRAP (Waste and Resources Action Programme) embarked on a series of studies in order to establish the end of waste criteria for a number of waste streams. There are agreements for compost and anaerobic digestate, BSI PAS (publicly available standards) 100 and 110 respectively and BSI PAS 111:2012 covering processed wood waste for a variety of uses, including panel-board, biomass energy generation and animal bedding. Subsequently, there was a LIFE+ project, Equal: Ensuring quality of waste-derived products to achieve resource efficiency to support businesses to reuse and recycle waste materials into new products while protecting human health and the environment.

Hazardous waste

Section 17 of the COPA, which provided enabling powers to the Secretary of State to make regulations on hazardous waste, formerly referred to as special waste, was retained by Section 62 of the EPA. Hazardous waste is controlled under the Hazardous Waste (England and Wales) Regulations 2005[18] and the List of Wastes (England) Regulations 2005[19] which introduced a new system of control for producers of hazardous waste. Producers of hazardous waste have to be registered with the EA but only if they will generate more than 500 kg a year.

Because the European Waste Catalogue defines a wide range of wastes as hazardous, including end-of-life vehicles, cathode ray tubes (old televisions and computer monitors) and fluorescent tubes, a large number of businesses and all LAs are likely to be classed as hazardous waste producers. The EA has produced guidance on the system for hazardous waste classification and control; see later.

Illegal deposit of waste

The key section of the EPA is Section 33. It makes fly-tipping illegal and requires that any other waste disposal activity is undertaken only under and in compliance with the conditions of an environmental permit. Section 33(1) states that a person shall not:

(a) Deposit controlled waste, or knowingly cause or knowingly permit controlled waste to be deposited in or on any land unless a waste management licence authorizing the deposit is in force and the deposit is in accordance with the licence;

(b) Treat, keep or dispose of controlled waste or knowingly cause or knowingly permit controlled waste to be treated, kept or disposed of:

    i.   in or on any land; or
    ii. by means of any mobile plant except under and in accordance with a waste management licence;

(c) Treat, keep or dispose of controlled waste in a manner likely to cause

pollution of the environment or harm to human health.

A contravention of the main offence under Section 33(1), or a contravention of a condition of an environmental permit, is an offence. Section 33 introduces the concepts of 'pollution to the environment' and 'harm to human health'. The definitions are given in Section 29 and reiterate the definitions used in Part I of the Act with specific reference to waste on land. Penalties for offences under Section 33 are unlimited fine and/or 122 months' imprisonment on summary conviction and on indictment an unlimited fine and/or imprisonment for five years. For offences involving hazardous waste the latter prison term is raised to a maximum of five years. These sections have been retained following the introduction of the Environmental Permitting Regulations in April 2008 and April 2010 but much of the waste regulation perspectives of the EPA have been replaced by those regulations.

Regulation of waste facilities

The environmental permitting programme came into force in April 2008, was expanded and revised in 2010 and applies in England and Wales only. The relevant SI is the Environmental Permitting (England and Wales) Regulations 2010 (EPR) (SI 2010 No. 675). The EPR replaced both the Pollution Prevention and Control (PPC) and waste management licensing regimes. Activities listed in the Regulations, which include most waste operations, require an environmental permit in order to operate.

Waste operations, which are unlikely to present a significant risk to the environment have been granted exemptions from the need for a permit. However, they must be registered with the Environment Agency.

The EPR have not changed the standards of environmental protection expected by the regulators, and in general do not require any additional action from regulated industries.

The regime was a 'Better Regulation' initiative, intended to simplify and streamline procedures, cut costs for both industry and regulators and move towards an increasingly risk-based system of regulation. The EPR do not apply in Scotland (see separate section later).[20]

In 2009 the EPR was extended to mineral wastes, implementing the EU Directive on waste from extractive industries. The 2010 EPR brought water discharge consents, groundwater authorisations and radioactive substances regulation within the scope of EPR. Guidance on the permitting of these sectors, together with guidance on permitting requirements for the recovery of waste batteries, is also available. The EA is the regulator for these activities.

The aim of the Environmental Permitting regime is to:

- Protect the environment;
- Fully implement EU legislation;
- Regulate industry in an effective and efficient way;
- Encourage regulators to promote best practice in the operation of regulated facilities.

Under S.12 of the EPR, it is an offence to operate a regulated activity without an environmental permit. The permit will include conditions specifying the measures the operator is to take in order to protect the environment. The regulator will ensure that the permit conditions achieve the objectives of any applicable EU Directives.

The following facilities require an environmental permit:

1   Installations carrying out any of the activities listed in Schedule 1 to the Regulations and any activities that are technically linked (Schedule 1 activities are those formerly regulated under PPC and LAPC).
2   Waste operations (i.e. waste disposal or recovery as defined in the WasteFD).

Some waste operations are exempt or excluded from the need to hold a permit (see Exemptions later).

3   Mobile plant, including mobile plant used to treat or dispose of waste.
4   Mining waste operations.
5   Facilities using radioactive substances.
6   Activities discharging to water or groundwater.

The EPR contain schedules which refer to the following waste-related Directives:

- PPC Part A: Schedule 7;
- Part B Installations under PPC: Schedule 8;
- Waste Framework Directive: Schedule 9;
- Landfill Directive: Schedule 10;
- End-of-Life Vehicles: Schedule 11;
- Waste Electrical and Electronic Equipment: Schedule 12;
- Waste Incineration: Schedule 13;
- Waste Batteries: Schedule 19;
- Mining Waste: Schedule 20.

Other environmental Directives are also covered. Whereas previous UK legislation contained detailed provisions to implement these Directives, the EPR refer back to the text of the Directives themselves. This makes the new regulations less detailed and prescriptive. A particular facility may have to comply with more than one set of Directive requirements (e.g. a large incinerator will be covered by the IPPC Directive, the WasteFD and the Industrial Emissions Directive). Defra has produced guidance for the EPR available on the gov. uk website. Under EPR a single permit can cover more than one regulated facility. A single permit can be granted where:

- The regulator is the same for each facility; and
- The operator is the same for each facility.

Generally, all the facilities must be on the same site. However, a single permit can be granted to an operator for more than one item of mobile plant. A single permit can also be granted to an operator for several 'standard facilities' (see standard rules later) on different sites provided that none of the facilities are covered by the IPPC Directive. A single permit cannot cover Part B installations and other facility types.

The regulator can replace several permits with a single permit where there are several facilities with the same operator. This would normally only be done with the agreement of the operator.

As is shown in other chapters (e.g. Chapter 16) the Environment Agency regulates:

- Part A(1) installations and mobile plant;
- Waste operations;
- Discharges to water and groundwater.

The Local Authority regulates:

- Part A(2) and Part B installations and mobile plant;
- Waste operations carried out as part of the Part A (2) or Part B installations or mobile plant.

Schedule 1 to the EPR lists the types of installations and mobile plant in Part A (1), Part A (2) and Part B. The environmental permit is held by the operator of the facility. PPC permits, water discharge consents and waste management licences automatically became environmental permits. New applications are only required for new facilities or variations to the permit. For many waste operations, planning permission must be in force before the permit can be issued. Schedule 5 to the EPR lays down the procedure for applications.

If the operation of the facility changes, the operator may have to request a variation to the permit. This will always be the case where the operator is no longer able to comply with the permit conditions. If there is a substantial change (see later), public consultation is required. A substantial change is a 'change in operation which, in the regulator's opinion,

may have significant negative effects on human beings or the environment'.

A permit can be transferred completely, or partially, so that another operator takes over part of the operation. For all types of transfer, the two operators must make a joint application. Regulators may need to vary the conditions to reflect a shared operation.

The regulator will decide whether to grant or refuse the permit, and will set the permit conditions. These will include conditions to ensure that the relevant Directives are complied with. The regulator must refuse a permit where:

- The applicant will not be the operator;
- The applicant cannot comply with the conditions. The regulator may decide to refuse a permit where:
  - They believe the operator is not competent to comply with the permit;
  - The environmental impact would be unacceptable;
  - The operator does not provide sufficient information;
  - The requirements of EU Directives cannot be met.

There are two methods for surrender of an environmental permit.

1 *Notification* – an operator can simply notify the regulator if the permit is for a Part B installation or mobile plant. This is because the permit conditions do not extend to the condition of the land.
2 *Application* – for Part A installations, the operator must apply to the regulator.

The regulator must accept the surrender of the permit if it is satisfied that the necessary measures have been taken to:

- Avoid any pollution risk resulting from the operation, for example by removing any tanks containing pollutants;
- Return the site to a satisfactory state.

The regulator needs to ascertain the initial state of the site; any additional contamination found after closure will normally be attributed to the operator. The operator will be expected to restore the site to the condition it was in when the environmental permit, PPC permit or waste management licence was issued. However, the operator is not responsible for historical pollution caused before the issue of the permit/licence. The guidance states that in restoring the site, the operator should:

- Remove any residual waste deposits (except in the case of a landfill);
- Remove any contamination as far as practical;
- Where removal is not practical, treat or immobilise any contamination;
- Remedy any harm the contamination may have caused; and
- Mitigate the effects of any harm.

Standard permits – certain types of activity can be regulated under a standard permit. The purpose of standard permits is to simplify regulation for activities that are relatively uniform and do not present a serious risk to the environment. A standard permit has just one condition, requiring the operator to comply with a set of standard rules. It can cover more than one standard facility operated by the same operator. The fee for a standard permit is lower than that for a site-specific permit, but the operator has no right of appeal against the rules. The operator can decide whether to operate under standard rules. If so, they must apply for a new environmental permit.

Standard rules are available for a range of low- to medium-risk waste operations, including:

- Transfer stations;
- Mechanical biological treatment (mbt) plant;
- Civic amenity sites;
- Metal recycling facilities;
- Composting plant;
- Materials recycling facilities (mrfs);
- Clinical waste treatment and transfer stations;
- Animal carcase incinerators;
- Mobile treatment plant for contaminated soil;
- Pet cemeteries;
- Sites storing scrap metal for recovery, up to 1 million tonnes a year;
- Recovery of scrap metal and the dismantling of de-polluted ELVs.

The 2010 Regulations amended the system of waste exemptions with the effect that some activities which were formerly exempt from permitting now require a standard permit.

Operator competence – in deciding whether to issue a permit, the regulator will consider whether the operator is competent to comply with the conditions. They will take into account the following aspects of competence:

- Environmental management system;
- Technical competence;
- Record of compliance with legislation;
- Financial competence.

The regulators can reassess competence at any time during the regulated facility's life and revoke the permit if they are not satisfied. Requirements to maintain competence can be included in the permit conditions.

Larger, more complex facilities are expected to have an environmental management system certified to ISO 14001, and to have registered for EMAS (see Chapter 8). Simpler facilities are encouraged to obtain certification to BS 8555 and consider 'EMAS easy'. The environmental management system should contain mechanisms for assessing and maintaining technical competence. There are two competence schemes for the waste industry: one has been developed by the Waste Management Industry Training and Advisory Board (WAMITAB) and the Chartered Institution of Wastes Management and the other is the Competence Management System (CMS) developed jointly by the Environmental Services Association and EU Skills (Energy & Utility Skills), which is operated by EU Skills.

The new WAMITAB scheme is more flexible than the one it replaced. It introduces a requirement for 'continuing competence: over a two-year period, candidates must update and demonstrate their knowledge and pass a test'. The CMS Scheme enables operators to demonstrate technically competent management of permitted activities on the basis of both corporate competence and employees' individual competence. It complements established management systems (environmental, quality and health and safety) and so enables a systemic approach to technical competence.

The regulator will not grant a permit to someone who has been convicted of an environmental offence if it believes that it would be undesirable for him or her to hold a permit. This particularly applies to those who have shown a deliberate disregard for the environment by committing repeated offences or have deliberately made false or misleading statements. Regulators will consider an operator's financial solvency if running costs are high relative to the profitability of the activity, or if they doubt the financial viability of the activity. The Agency will expect landfill operators to meet stricter criteria for financial provision.

The EA takes a risk-based approach to regulation that is formalised through its OPRA (Operational Risk Appraisal) methodology. It will target activities that:

- Pose the greatest risk to human health or the environment;
- Operate to poorer standards;
- Fail to comply with permit conditions; and
- Have a greater adverse environmental impact.

The regulator will carry out periodic inspections of regulated facilities and exempt waste operations. They will also review permits periodically, to ensure the conditions remain adequate.

If the operator is found to be contravening permit conditions, the regulator can issue an enforcement notice. Where there is a risk of more serious pollution, a suspension notice can be issued. Both types of notice specify the steps to be taken to remedy the problem and the timescale within which they must be taken. Where other types of enforcement have failed, the regulator can issue a revocation notice to revoke the permit. The revocation notice will list steps that the operator must take to return the facility to a satisfactory state and avoid pollution risk.

Where an activity causes serious pollution, the regulator may carry out remediation and charge it to the operator. The guidance emphasises that restoration of the site at closure cannot justify letting the operator contaminate the site during the lifetime of the facility.

It is an offence to operate a regulated activity without a permit, or to contravene permit conditions. The Agency will prosecute serious offenders. Further guidance on EPR can be found in Defra and the Welsh Government's Environmental Permitting: Core Guidance for the Environmental Permitting (England and Wales) Regulations (revised 2016).[21]

The general principles of environmental permitting described earlier apply to waste operations. Additional requirements for waste management can be found in Schedules 9 and 25 to the EPR and are described in the official EP Guidance.

Landfill sites are subject to the requirements of Schedule 10 to the EPR, considered later in this chapter under the Landfill Directive. Householders are not expected to hold a permit for the disposal of household waste.

Waste Framework Directive

In drawing up permit conditions for waste operations, the regulator must ensure that the requirements of the Waste Framework Directive 2008/98/EC are met. The Directive requires MSs to ensure that waste is disposed of or recovered without endangering human health or the environment. MSs should ensure that there is an adequate network of waste management facilities, and they should encourage the movement of waste up the hierarchy (see earlier). The EPR deals only with those aspects of the Directive that relate to permitting. A permit is required to carry out a waste recovery or disposal operation (see section on definitions of waste). Disposal and recovery operations are listed in the Directive; however, operations that are not listed may still qualify as disposal or recovery. Sometimes it is not clear whether an operation is disposal or recovery: the guidance gives some pointers to help classify such operations:

- Whether the holder pays to get rid of the waste, or is paid for a recoverable material;
- The marginal benefit derived from the operation; a very small benefit might indicate disposal;
- The use of excessive quantities of waste (e.g. in site restoration, which may indicate disposal); and
- Whether the waste is suitable for the intended purpose.

Under the WasteFD, incineration of waste with energy recovery is classed as recovery, provided criteria for heat recovery efficiency are met.

The Directive requires that environmental permits for disposal should cover the:

- Types and quantities of waste;
- Technical requirements;
- Security precautions;
- Disposal site;
- Treatment method.

If the facility requires planning permission, the regulator may not grant a permit for a relevant waste operation unless planning permission is in force (EPR Sch.9). A 'relevant waste operation' is an operation which was subject to waste management licensing prior

to the introduction of the PPC regime (i.e. it does not include incinerators).

Landfills and other large waste facilities are subject to the IPPC Directive 96/61/EC and classified as Part A(1) or A(2) activities under EPR (see later). Furthermore, the IPPC Directive contains provisions related to waste management, which apply to a wide range of industrial installations. Article 3 of the IPPC Directive sets down the following general principles:

(a) All the appropriate preventive measures are taken against pollution, in particular through application of Best Available Techniques (BAT) (see below);
(b) No significant pollution is caused;
(c) Waste production is avoided if possible. Where waste is produced, it is recovered, or where that is technically and economically impossible, it is disposed of, keeping the environmental impact to a minimum;
(d) The necessary measures are taken to prevent accidents and limit their consequences;
(e) When the activities cease, the site is returned to a satisfactory state and any necessary measures taken to avoid pollution risk.

Article 9 additionally provides that permits should:

(a) Aim to minimise trans-boundary and long distance pollution;
(b) Protect soil and groundwater;
(c) Ensure waste is managed properly;
(d) Protect the environment when the installation is not operating normally (e.g. during temporary stoppages);
(e) Ensure the operator carries out site monitoring and remediation, and any other appropriate steps before and after operation;
(f) Specify monitoring requirements;
(g) Require the operator to inform the regulator without delay of any polluting incident.

The regulators will also take into account EU and national Environmental Quality Standards when setting emission limits. These may lead to stricter limits than would be otherwise required.

BAT is 'the most effective and advanced stage in the development of activities and their methods of operation' which will enable compliance with emission limit values, reduce emissions and minimise overall environmental impact. A technique may be rejected as BAT if its costs far outweigh the environmental benefits.

In assessing what is BAT for a particular installation, the regulator will give particular attention to the effects of releases. Annex III to the Directive lists the main polluting substances but this is not an exhaustive list, and in particular carbon dioxide emissions should also be taken into account.

Annex IV of the IPPC Directive requires other environmental issues to be taken into account when setting permit conditions. These are:

1   Consumption and nature of raw materials
2   Energy efficiency
3   Waste minimisation: the assessment of BAT options should cover the amount of waste produced and the possibility of preventing waste, recovering it or disposing of it safely
4   Accidents
5   Site restoration: the regulator will consider whether activities risk polluting the site and should plan ahead for decommissioning and closure

The European Commission has issued guidance on BAT in the form of BAT Reference Documents (BREFs) for 30 different sectors and these are reflected in the EA's domestic guidance documents. However, the regulators will also need to use their expert judgement in weighing up the relative significance of different environmental effects. Once the options have been ranked, that which minimises environmental impact will be BAT unless economic considerations render it unavailable. Normally, the regulator will decide on BAT for a whole sector rather than considering the profitability of an individual business. In 2021 the UK government started a consultation to develop a BAT UK.

Exemptions from waste permitting

The Waste Framework Directive allows Member States to grant exemptions from the requirement to have a permit. Exemptions may be granted to:

*   Establishments which dispose of their own waste at the point of production;
*   Establishments which carry out waste recovery.

An exemption may only be granted to an activity that does not present a risk to the environment or endanger human health. Exemptions which were already registered under the waste management licensing regime have automatically become registered exemptions under the EPR.

The exemptions for England and Wales are listed in Schedule 3 to the EPR and were substantially amended by the EPR 2010. Most of the entries specify limits to the quantity of waste that can be managed, and/or time limits on storage. For larger quantities, an environmental permit is required. Types of waste are identified by the six digit European Waste Catalogue (EWC) code as well as by a written description. These exemptions must be registered with the enforcing authority, which is in nearly all cases the EA. The registration process is very simple and with the exception of activities involving WEEE, free of charge. The exemptions relating to scrap metal and WEEE are very detailed and prescriptive.

Non-WasteFD exemptions

Part 3 of Schedule 25 lists categories of waste storage which do not require a permit and do not have to be registered as exemptions because they are not covered by the WasteFD:

- Temporary storage of waste at the site of production, for up to 12 months;
- Temporary storage of waste on a site controlled by the producer (e.g. a builder's yard where demolition waste from remote sites is stored), for up to three months. This is subject to the following limits:

  - No more than 50 cubic metres of solid waste;
  - No more than 1000 litres of liquid waste;
  - No asbestos or flammable liquids;

- Temporary storage at a collection point of:

  - No more than 30 cubic metres of WEEE;
  - No more than 50 cubic metres of other non-hazardous waste;
  - No more than 5 cubic metres of hazardous waste;
  - No asbestos or flammable liquids.

These exemptions cannot be used on sites where the treatment of waste is taking place, for example, the crushing of fluorescent lamps. However, they can be used if the treatment does not alter the composition of the waste and is merely ancillary to storage (e.g. compacting of boxes or shredding of confidential papers).

Schedule 2 to the EPR describes the procedures for registering an exemption with the EA. Where waste is stored at a site regulated under Part B of EPR (e.g. scrap metal furnaces or boilers burning waste as fuel), it is no longer necessary to register an exemption as the waste storage activities are covered by the Part B permit.

The operator must provide the following information to the regulator:

- Name and address of the establishment or undertaking;
- Nature of the operation;
- Location of the operation.

This information will be entered on the public register. The Agency will carry out periodic inspections of registered exempt activities.

The 2010 EPR implemented substantial changes to the waste exemptions, which are listed in a revised Schedule 3. The aim of the changes was to provide a more risk-based and proportionate approach to exemptions. They also present the exemptions in a more structured and logical order. The Schedule divides the exemptions into four chapters:

1 Use of waste;
2 Treatment of waste;
3 Disposal of waste;
4 Storage of waste pending its recovery.

Various activities, which were formerly regulated under a notifiable exemption such as the use of waste in construction and land spreading, require a standard permit under EPR. At the same time, new exemptions have been created for some low-risk activities that formerly required a permit. Activities that used to rely on an EA 'low risk position' and were not regulated at all now need to register an exemption. This affects a large number of operators. The majority of exempt activities continue to be registered as simple exemptions (about 83% of those formerly registered) but will need to renew their registration every three years. Under the previous scheme of exemptions, there was a category of notifiable exemptions for which additional records had to be kept. This separate category no longer exists, but the requirement to keep records still exists for the following activities:

- Spreading waste to agricultural or non-agricultural land for benefit;
- Recovery of scrap metal;
- Repair or recovery of WEEE.

The records will describe the quantity, nature, origin, destination and treatment method of the waste. They must be retained for two years and made available to the Agency on request.

The Regulations contain detailed provisions regarding the recovery of Waste Electrical Electronic Equipment (WEEE) and end-of-life vehicles (ELVs) in order to implement the requirements of the WEEE and ELV Directives.

WEEE activities must be registered separately from other exemptions. There is a general requirement that the best available treatment, recovery and recycling techniques are used when treating the waste. The operation must meet the technical requirements specified in Annex III of the WEEE Directive. WEEE operations are inspected annually to ensure that they are complying with the technical requirements and safety precautions laid down in the Directive.

The environmental permitting regime does not apply in Scotland. The regulator for waste operations in Scotland is the Scottish Environment Protection Agency (SEPA). The system that applies in Scotland is the same as that which applied in England and Wales prior to April 2008. Waste activities covered by the IPPC Directive, and all landfills, require a PPC permit under the Pollution Prevention and Control Act 1999 and the Pollution Prevention and Control (Scotland) Regulations 2012.[22] Certain activities regulated under PPC are classified as Specified Waste Management Activities (SWMAs):

- All landfills;
- Facilities disposing of hazardous waste with a capacity of more than 10 tonnes per day;
- Treatment plants disposing of non-hazardous waste with a capacity of more than 50 tonnes per day;
- Plants with a capacity of more than 10 tonnes a day recovering hazardous waste through use as fuel, solvent regeneration, oil refining or reuse, recovering components used for pollution abatement;
- Disposal of waste by incineration.

SWMAs must be under the control of a fit and proper person.

A fit and proper person is defined as someone who:

- Does not have past convictions for (deliberate) environmental offences;
- Can demonstrate technical competence through possession of a certificate of technical competence; and
- Can make adequate financial provision for the operation of the site, both during its lifetime and post-closure, so that the site will not present any threat to human health or the environment.

Other waste activities are regulated under the Waste Management Licensing (Scotland) Regulations 2011. Two principal amendments were introduced:

- The abolition of the requirement for Certificates of Technical Competence as a means of demonstrating that an applicant for a waste management licence is suitably qualified;
- The removal of the exemption from waste carrier registration for businesses who only carry waste they produce themselves, as is also the case for the rest of the UK.

Construction waste

Site waste Management Plans (SWMP) under the Site Waste Management Plans Regulations 2008[23] came into force in April 2008. They were abolished in 2013 but ironically at the same time the Wales Assembly Government implemented a parallel set of regulations. However, the guidance in the SWMP procedures is still of value both in preventing the generation of waste on construction sites and also ensuring its effective management when waste is produced.

These regulations now therefore only apply in Wales with Building Standards Departments as the regulatory body. For guidance, netregs.gov.uk, ciria.org.uk and wrap.org.uk are probably the best.

Waste management planning

One of the main concepts that emerged in the 1990s and that continues to influence the planning of waste into the new millennium is sustainability. Following the Earth Summit in Rio de Janeiro in 1992, the UK government produced Sustainable Development: the UK Strategy in 1994. Chapter 23 provided a sustainable framework for waste, including:

- Minimising the amount of waste produced;
- Making the best use of the waste that is produced; and
- Minimising pollution from waste.

It also defined a hierarchy of waste management options:

1   Reduce
2   Reuse
3   Recover

    a   Materials recycling
    b   Composting
    c   Energy recovery

4   Disposal

where this represents the best practicable environmental option (BPEO). This waste management hierarchy has been reiterated through the WasteFD. These waste priorities have been reinforced in the statutory national waste plans established under the EPA, Section 92. The Resources and Waste Strategy 2018 was published in December 2018 as a waste strategy for England. Wales, Scotland and Northern Ireland produced their own revised and updated waste strategies in 2010 with subsequent additional revisions and consultations to move towards zero waste and promote a more circular economy.

The key objectives in the Resources and Waste Strategy 2018 are set within the framework of the government's 25 Year Environment Plan. The Plan aimed: to eliminate avoidable plastic waste over the 25 years of the Plan, to double resource productivity and eliminate avoidable waste of all kinds by 2050. The main tenets of the Strategy are:

Sustainable production
Helping consumers to take more considered actions
Improving resource recovery and waste management
Tackling waste crime
Cutting down on food waste
Making Britain a global environmental leader

Also, soon after the Strategy was published there were four consultation papers published covering some of the most problematic issues to be addressed in order to implement the Strategy, covering: packaging and packaging waste, a deposit return scheme for beverage containers, consistency in Local Authority collection arrangements for recyclable wastes and proposals for taxation of single-use plastics.

As far as the LA sector is concerned these key issues will have profound implications for their operational contracts and finances. However, since 2010 LAs have had no statutory recycling targets.

New infrastructure

New waste management facilities will be needed if the waste diversion and recycling set by the Strategy are to be met. There is a particular need for more energy from waste (EfW) facilities, including anaerobic digestion. Energy from waste is expected to account for up to 35% of municipal waste treatment by 2025 compared with 16% in 2015. In addition, there is encouragement to develop anaerobic digestion (AD) facilities to tackle the most biodegradable wastes from all waste generators. A new standard and protocol for the digestate from anaerobic digestion, PAS110 is helping to secure outlets for this material for land spreading, albeit there are concerns about amounts of plastic particles that can be contained in the composted digestate.

LAs received support including financial help through the Defra Waste Infrastructure Delivery Programme up until 2011 to ensure that the new infrastructure was ready on time. Planning applications are determined by LAs with the exception of larger energy recovery facilities; those greater than 50MW output in England and Wales are regarded as nationally significant infrastructure projects (NSIP), which are examined and determined by the Planning Inspectorate.

Local authorities – two-tier authorities are already working together in waste management. They are also being encouraged to help small businesses to reduce and recycle their waste. There are several very successful and often long-standing examples of successful joint working, including Devon, Hampshire and Dorset. In addition, there is a progressive move in England to establish more and ever larger unitary LAs, such as for Cornwall and for Bournemouth, Christchurch and Poole carved out of Dorset. Also there is an increasing trend towards districts and boroughs working together to provide joint services or arrange for the purchasing of equipment, such as wheeled bins in response to ever tighter budgetary constraints after the financial crisis of 2008.

Public awareness and Plastic Bag Controls – each of the devolved Governments are encouraging schools, community groups and retailers to play their part in waste reduction. For example, in England Defra set up a Household Reward and Recognition Fund in 2010 with a limited amount of funding for local authorities to bid to introduce innovative ways of encouraging households to prevent waste generation and enhance source segregation of recyclable wastes. This initiative has also included the Third (voluntary) sector in many of the projects funded.

In addition, even before the introduction of compulsory charging for plastic shopping bags by supermarkets there had been several local successful campaigns to 'ban the bag'

from shops. In each of the devolved administrations there is now a compulsory charge of 5 pence for each thin supermarket bag passed onto customers. The early results from the early adopters of this policy such as Scotland and Wales showed a dramatic reduction in the use of plastic bags. Most supermarket chains have removed the thin plastic bags in favour of selling reusable bags priced at £0.10p and for Tesco at £0.20p and Waitrose £0.50p. However, these bags contain 75 times more plastic than the thin-walled bags and are often regarded as equally disposable by consumers. In early 2021 it was reported that the ten largest supermarket chains provided 1.58B 'bags for life' in 2020, equivalent to 57 per household.[24]

Local authority (LA) waste strategies

The main focus for LAs in waste in those areas where there are two tiers of local government has been on establishing partnership working for waste management. Consideration should also be given to government statements on land use planning policy, particularly Planning Policy Statement (PPS 10), Planning for Sustainable Waste Management.

In 2003, the Household Waste Recycling Act imposed a duty on Waste Authorities to prepare sustainable waste strategies and to report annually on progress in improving recycling rates. The WCAs were required to collect at least two types of recyclable waste together or separated from the rest of the household waste by 31 December 2010, unless the cost of doing so would be unreasonably high or comparable alternative arrangements are available.

Reporting on waste tonnage data is through the Waste Data Flow (WDF) system, which is run under the auspices of the Environment Agency. The waste data includes information about recyclable wastes collected and residual waste streams. Financial information on costs of running waste-related services is collected and collated through the Audit Commission although the Audit Commission was officially abolished in March 2015.

Waste collection and waste receptacles

Sections 45–47 of the EPA largely replicate the duties and powers of WCAs under Sections 12–14 of the COPA, the regulations for which were finally introduced as part of the Collection and Disposal of Waste Regulations 1988 (17A). The ability of WCAs to establish kerbside source segregation collection schemes, as opposed to the pre-1992 door-to-door schemes, has been permitted under Section 46(1), whereby

> the authority may . . . require the occupier to place the waste for collection in receptacles of a kind and number specified. However, agreement has to be given by the highway or roads authority and provided that arrangements have been made as to the liability for any damage arising out of their being so placed.

Under Section 45, the WCA has a duty to arrange for collection of household waste except for premises that are isolated or inaccessible so that collection costs would be 'unreasonably high', and that adequate arrangements for its disposal can reasonably be expected to be made. In most UK WCAs there is a door-to-door household waste and generally also a recyclables collection. However, in Mediterranean countries households have to deliver these wastes to large collection bins. This system was adopted and progressively rolled out by Brighton and Hove City Council from 2011 in response to budgetary cuts, using 1600 litre and 1100 litre metal waste containers.

There is also a duty for the WCA to collect or arrange collection of commercial waste when requested. As regards industrial waste, the WCA may arrange for its collection if requested by the occupier, but a WCA can only exercise this power with the consent

of the WDA (Section 45(2)). For commercial and industrial waste, the WCA has a duty to recover the costs for both its collection and disposal, 'unless in the case of charge in respect of commercial waste the authority considers it inappropriate to do so'.

Section 45 also deals with the duties of WCAs with regard to the emptying of cesspools and privies. While Section 46 deals with the provision of receptacles for household waste and the collection authority can require the occupier to place such waste for collection in 'receptacles of a kind and number specified', Section 47 covers those for commercial or industrial waste. The authority has considerable discretion about the number and type of containers to be used and their positioning for the collection of waste and recyclables; under Section 46(7) an occupier can appeal to a magistrates' court that the authority's request is unreasonable or that 'the receptacles in which household waste is placed for collection from the premises are adequate'. However, this appeal procedure must be initiated within 21 days of a notice being served on the occupier or the end of the notice period. A similar provision exists for receptacles from industrial and commercial premises, but the grounds for objection are different in that while the requirement of unreasonable provision is repeated the other ground is that 'the waste is not likely to cause a nuisance or be detrimental to the amenities of the locality' (Section 47(7)(b)).

Managing waste arisings

Analysis of waste

The analysis of household waste is concerned with the physical characteristics and composition of that waste, for example, its density, output per household and the amount of paper. Data obtained from such analyses help to predict the changing nature of waste, and is a valuable aid in determining methods of collection, transportation and disposal.

Figure 18.2 shows the national average composition of household waste, although it may vary substantially between different Local Authority areas. Reliable waste analysis can be used for organising collection rounds, calculating storage requirements of blocks of flats and long-term forecasting to enable strategies to be planned and research organised. In seeking improved efficiency and higher standards of waste recycling, treatment and disposal, it is essential that reliable data are established locally.

There are now several analyses of the estimated composition of waste which have been undertaken, including those that consider only household waste (see Figures 18.2 and 18.3). In addition, in Wales WRAP undertook an analysis of MSW in Wales during 2009. These analyses are based on the assessment of waste generated by households or the wider community and are based on statistical sampling so an error factor should be attached.

Waste collection systems for high-rise and high density housing

British Standard 5906:2005 *Waste management in buildings: code of practice*, provides guidance for methods of storage, collection, segregation for recycling, recovery and on-site treatment of waste from residential and non-residential buildings. BS 5906 applies to new buildings, refurbishments and mixed developments including retail and offices and includes healthcare waste from hospitals and other healthcare establishments.

In the UK there has been a lack of longer-term thinking about how we ought to be handling residual and recyclable waste collections in future. There are existing and common engineering solutions that have been adopted in other countries with considerable benefit for public health, environmental gain and public acceptance. Meanwhile, with a few exceptions, Britain has retained its long-standing convention of door-to-door

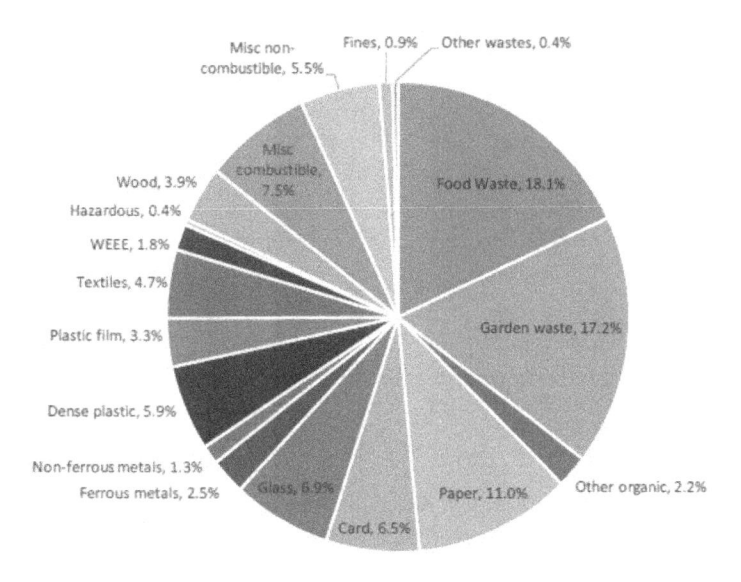

Figure 18.2 Composition of household waste and recycling in the UK 2017 (percentage of total arisings)

Source: WRAP National Household Waste Composition 2017 [4]

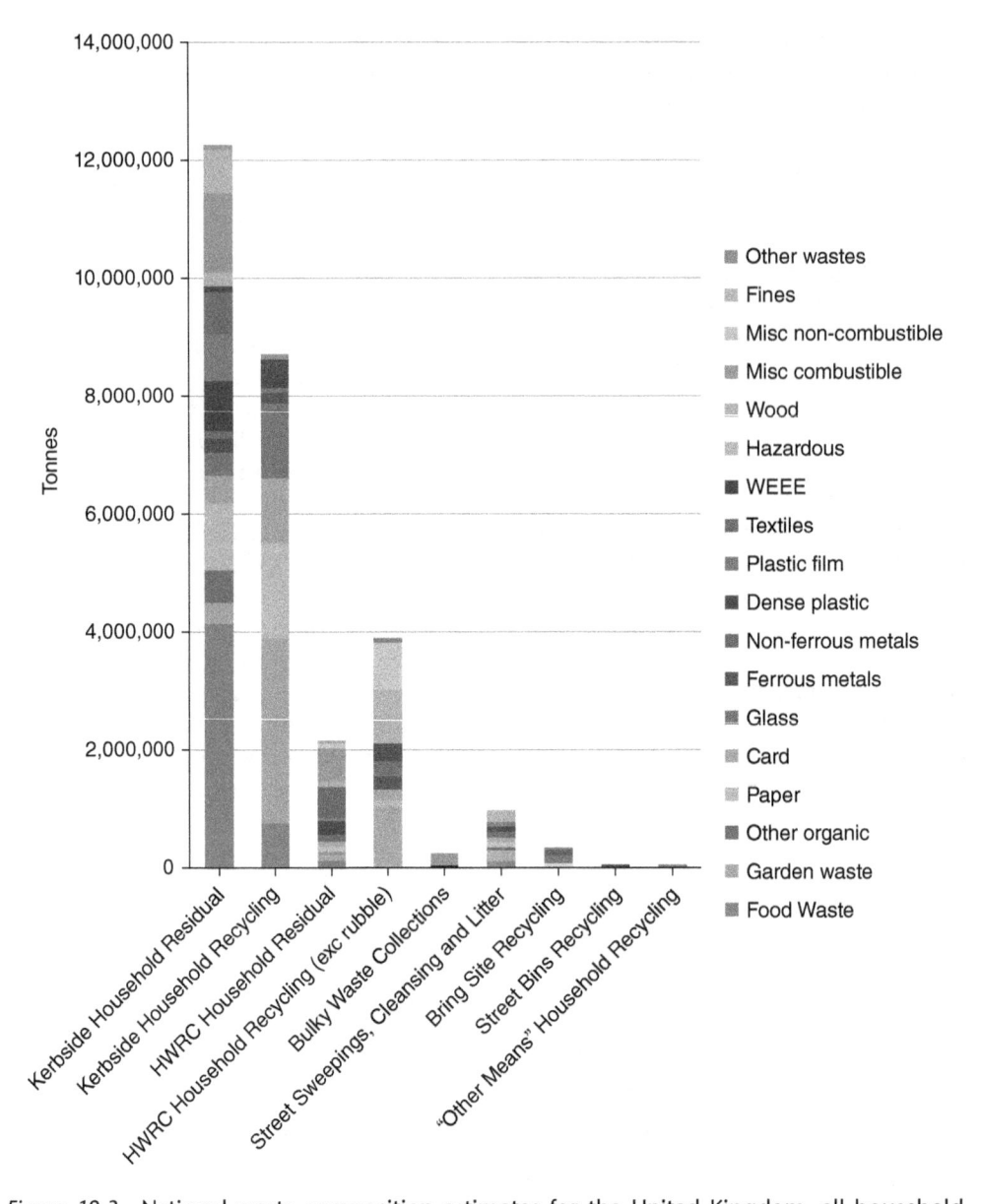

Figure 18.3 National waste composition estimates for the United Kingdom, all household waste and recycling streams, 2017 (tonnes)
Source: [4]

collections of residual waste and added on the segregated collection of recyclable wastes. Three alternative systems for the initial collection of waste from households are detailed:

- Refuse chutes, also redesigned for recovery of recyclables;

- Vacuum collection systems for waste and recyclables;
- Underground storage systems for residual waste and recyclables.

In addition to providing these engineering and technological improvements it also has to be recognised that there are many ways in which these techniques can be further enhanced and improvements made to existing waste management systems through behavioural and public educational programmes. These informational education initiatives are vitally important when introducing new systems and also periodic reinforcement programmes are required in order to maintain public participation at a high level.

In the UK the use of refuse chutes for the collection of refuse from walk-up blocks and for high-rise flats has had a fraught history. In the past, up to the 1950s, all household waste was put down the chute. This was fine when this waste was predominantly ashes and cinders from households' coal fires, where much refuse was also burnt. The waste went straight down the chute into a waste container, predominantly paladin bins had been used, which could then be wheeled to refuse collection vehicles. However, with the change to other methods of heating rather than open coal fires, mainly as a result of the Clean Air Acts, the type of waste put down these chutes changed, with more packaging and food waste being generated. As a result, the number of chute blockages increased, necessitating caretakers clearing the chute and there was more chance of flies and vermin being attracted to the waste. Most chute systems' hopper deposit points in high rise housing from the inter-war and post-WWII periods have increasingly been blocked off, thus forcing residents to walk their waste and recyclables to the ground floor or basement bin stores.

Currently, however, the use of refuse chutes in Britain is being dictated through the increasing pressure on land, which is forcing greater building of high-rise properties in cities and larger towns, a strategic necessity which is common to many countries world-wide. These blocks of apartments are often built on a restricted land footprint and in consequence the amount of space allocated to refuse collection is limited. Therefore, there is a move towards reintroduction of refuse chutes for high-rise blocks, albeit with a range of new controls to ensure that they can be managed effectively. The chute systems also have brushes and sanitising sprays which are located at the top of the unit and can be periodically activated to clean down the chute.

In addition, manufacturers of this equipment have developed more sophisticated options (bi- and tri-separation systems) so that the separation of dry recyclable wastes and residual waste can be assured, provided that residents are adequately informed and educated about the way that they should prepare their waste and use the chute system.

Food waste collections ideally would need a separate chute in order to ensure the integrity of the quality of dry waste recyclables and therefore eliminating the problems that could be caused by split food waste bags. However, in the UK as of 2021 there are no facilities using a separate chute [5]. The blocks currently separating food waste use the tri-separation system with residents provided with suitable bags and clear guidance for their use. However, the tri-separation facilities in operation at the present time have a range of third item options as an alternative to food waste, including glass and plastics.

The glass option is usually used in sites where the chute drop height is more limited, due to potential damage to equipment if used in taller buildings. The issue of incorporation of glass bottles and jars in this type of system is carefully nuanced in each development. In some developments residents are requested to take glass containers down to bottle banks; in others glass is required to be left out by residents' doors for collection by caretaking staff or a combination of both approaches.

The majority of modern chute systems use a steel lining rather than concrete and incorporate control mechanisms that only allow one bag of refuse or recyclables to be placed in a hopper on any floor at any one time. Also, from the system control position in the basement all of the hopper controls can be locked off to permit bin changes and periodic maintenance. Often, given constraints on the bin capacity, the bins are automatically monitored to alert facility operatives to full bins and effect change-over procedures. On average the amount of residual waste and recyclable waste discarded is roughly equal by volume but obviously the residual waste weighs much more than the recyclable waste, typically at least twice or three times the amount.

One of the tallest examples of a tower block with a bi-separation system is The Tower at St George Wharf at Vauxhall, London, on the south side of the River Thames. The Tower is 181 m tall with 52 floors of mainly residential accommodation and was completed in 2014. The system installed used a 600 mm diameter steel-lined chute serving 48 floors with a bi-separator, refuse compactors and an automatic chute cleaning system. There are additional features which have been added to this system, such as digital display panels, a scrolling text for user information and a numeric/scrolling bar countdown to indicate hopper door availability. There are also very tall towers with these chute systems, including Owen Street, Manchester, and South Quay, Canary Wharf, London.

In the UK as of 2021 there were only two vacuum systems for refuse that were operational, at Wembley Park, from 2012, and more recently, starting in 2020, at Barking Riverside. In each case the systems were planned into the development from the start. However, in many other countries around the world the use of vacuum systems for both residual waste and recyclables is commonplace within both new developments and for retrofitting into historic city centres, such as Copenhagen, for example.

The Barking Riverside development covers 440 acres and will have 10,800 homes with a population of 26,000 people when it is completed in 2026–30. It has been built on the site of the old Barking power station.

The vacuum or pneumatic system requires residents to place their bagged refuse or recyclables into an inlet hopper. This inlet is normally accessed with a fob key system, with each household provided with a fob. Where recyclables are added into the scheme there are two separate access inlet hoppers with two different containers at the pumping station for the refuse and recyclable wastes. The bag of waste or recyclables is deposited and drops down to the bottom of the 500 mm pipe which is 2.5–3 m deep and has a flap at the bottom which can be opened for emptying into the main horizontal pipes.

Periodically, generally twice each day, the vacuum pump is activated, and the waste moved to the pumping station from where it falls into a bulk refuse container. Alternative bulk containers are available on site. When full the container is then moved off site for further treatment. The eventual 460 inlet points can be monitored and there are laser sensors which will automatically trigger the pneumatic system, if required beyond the normal clearance timings, due to high loading of any of the inlet pipes.

The planning for food waste collections has also been allowed for with the provision for space and access for a third inlet at the Barking Riverside development. This is because it is planned that the UK government will require mandatory food waste separation in 2023/24 following the lead of the devolved administrations that have already instituted mandatory food collections for households as well as for commercial and industrial premises.

Although less sophisticated than the two other systems, the use of underground refuse storage (URS) systems nevertheless offer most of the advantages of the two other systems but with lower initial costs. Again, while there are only a few isolated examples of the use of these systems in the UK their use overseas is quite commonplace. Initially in Southern Europe they offered an alternative to street sited bulk bins, normally 1100 or 600 litre, for daily collection of residual wastes. Underground storage offered a reduction in smells and a much cleaner environment around the refuse collection points.

Additional containers can be added for each type of recyclable material to be separately collected or a single container for co-mingled collections, albeit glass bottles and containers are generally collected separately and occasionally there are two glass containers for clear and coloured glass. The capacity of the URS containers can be varied, with typical sizes being 3, 4 and 5 cubic metres, and these containers are manufactured by several refuse container manufacturers but all from steel. For refuse alone 80 containers of 5 m^3, from which refuse is collected twice weekly, is sufficient for a population of 10,000 people.

In the UK there are a few examples of their use, with the inner London boroughs of Lambeth and Tower Hamlets and the Borough of Newham being the earliest exemplars but Cambridge, Bedford, Peterborough and Edinburgh also have URS bins. Public access internationally is normally through fob systems, which allows some degree of monitoring of the use of the containers and records can be kept of the number of visits to the containers, which can be used for billing purposes in some cases, and potentially some further checking on the quantity and even quality of discarded materials. This monitoring is generally at a neighbourhood level basis. However, the London schemes allow free access without the use of any form of control.

As with the other systems there are problems, most specifically in finding potential sites for the containers but in addition there needs to be consideration of the space above the containers. Finding space for sets of containers in a crowded urban environment can often be difficult, especially when many services, such as drainage, electrical and telecommunications and other utility services, are positioned in underground channels. The space above the containers also has to offer sufficient room to raise the container from its underground chamber and manoeuvre it into a truck using a double hook on-board crane lift mechanism.

Therefore, the engineering of the container chamber slots into the ground can be tricky, depending on the location and size of the site, sub-surface geology and the ambient water table levels. Choices also need to be made regarding the size of the containers to be used, again with consideration of the overall head space plus the size of the vehicle to be used for clearing the site.

These sites can also be automatically monitored to reduce the number of times that each site is emptied and improve the overall cost and environmental benefit from using such a system.

Commercial and industrial waste storage

Where there is a large volume of waste accumulation, refuse skips or bulk containers of 1100/1300 litres are available for the raw material emanating from the waste industry. There are numerous examples where waste analysis, correctly carried out, has enabled treatment and disposal plant design to be adequately assessed. Similarly, there are many cases where lack of analysis data has not only led to under- or gross over-design, but also to the wrong treatment system being selected. Table 18.4 shows the distribution of waste disposed in the United Kingdom. The EA

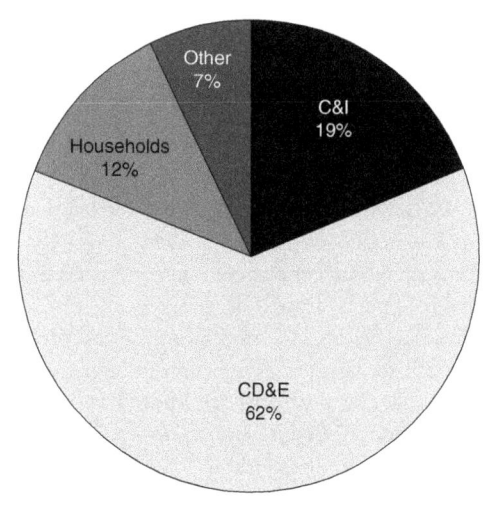

Figure 18.4 Waste generation split by source (UK, 2018)

Source: [6]

undertook a national survey of household waste.

The total UK estimated waste generated in 2018 was 222.2 million tonnes. Percentages may not sum exactly 100% due to rounding; C&I figures presented here differ from those in the C&I section in that they include sewage sludge. However, as these figures are from the WStatR return, which requires sludges to be converted to dry weight for reporting, the C&I figures do not differ greatly from those presented in the C&I section of the report. CD&E figures include excavation waste and dredging. Household figures are based on the WfH measure.

Refuse storage

It is important that during periods of storage, refuse on any type of premises should be kept dry. An accumulation of moist or wet refuse is not only offensive and difficult to remove but forms an excellent breeding medium for flies. The vast majority of domestic residences now have one of two alternative types of refuse storage facility following the phasing out of the traditional galvanised dustbins. Plastic wheeled dustbins have replaced them or black plastic sacks in a small minority of cases. The main receptacle is a 240 litre wheeled bin or increasingly smaller (180 or 140 litre) wheeled bin for residual waste plus a wheeled bin, multiple wheeled bins, or in addition or alternatively boxes or sacks for an increasing range of recyclable wastes.

With the wide adoption of alternative weekly collections (AWC) there has been concern about the possibility of adverse health aspects from the storage of residual solid waste. However, there are several research studies that have been undertaken to show that the problems are limited and can be easily controlled. There are now a few LAs that have introduced three-weekly or even monthly collections of residual waste.

Concern about the dangers of needle stick injury and sharp, protruding refuse from black plastic sacks has added to the impetus to install wheeled bins. Where collection frequency is limited or there is a problem of space, refuse compactors may be installed. These can reduce the volume of waste down to 20% of its original bulk, which also helps to limit odours and litter and reduce fire risk. Waste that is compacted is not likely to attract vermin and insects and gives lower transport costs. BS 5832 deals with the most popular type of compactor waste containers, which have a capacity of 10 m^3. Provision for the installation of the compactor should be made at the design stage so that space, electrical fittings and access to, plus removal of, compacted refuse is 'planned in'. Compacted refuse is heavy and may need wheeled transport, and so the siting of the compactor must take account of steps and ramp access.

External storage – sufficient storage capacity should be provided for the agreed span of time between collections, and consideration should be given to providing an enclosed compound for refuse containers, which encourages hygiene around the building, controls litter, discourages vandalism and helps to avoid the problem of fly-tipping. Enclosed refuse storage compounds must have natural and artificial lighting, appropriate ventilation, smooth, cleanable floor surfaces, hose points and drainage facilities, be easily accessible for deposit and removal of refuse, and have a lock-up entry. The containers must usually be accessible from the ground floor, with vehicular access not more than 25 m from the material without collectors having to pass through a building. Rear access is an advantage, especially where traffic is heavy and parking is a problem.

All access roads for refuse vehicles must have suitable foundations and surfaces to withstand the maximum weights of the vehicles likely to be used. Manhole covers, gully gratings and the like should be of a heavy-duty construction; overhead service cables and pipes should be not less than 7 m from ground level; and there should be sufficient room to allow manoeuvrability of heavy goods vehicles in the 17–24 tonne range. If skips or large compactors are used, consideration must be given to high overhead clearance for the emptying of such equipment.

The screening of waste storage areas by planting creates a good impression. Landscaping should be practical with low maintenance requirements, and should be impact resistant.

Recycling

Recycling of waste is undertaken in order to fulfil a number of objectives but will often not reduce the additional costs of collection of recyclables compared to a single collection of all refuse. At a national level it will provide a positive cost benefit through providing a cheaper alternative to virgin raw materials in the manufacturing process and fulfil other circular economy objectives.

Two methods of waste recovery from households are currently used in the UK: recycling at source, and recycling at an intermediate or at the disposal point. Most LAs base sets of containers at strategic locations around their area where members of the public can bring recyclable commodities, such as glass, cans, plastics and paper, to central collection points. Predominantly LAs have introduced a second wheeled bin or a box or bags so that householders can separate potentially recyclable materials at source. This waste is then taken to an MRF (materials recovery facility) where it can be manually but mainly mechanically sorted. Increasingly artificial intelligence and robots are being deployed in MRFs in order to reduce the need for workers to sort waste and to improve the accuracy of sorting.

In some authorities, in addition, facilities are available to extract certain materials at the waste treatment and/or disposal point, for example, electromagnetic separation of steel cans. If the recycling of household waste material is to become the preferred means for disposal of our waste, it is imperative that suitable markets for reclaimed waste are found. WRAP has the responsibility to collate information about current tonnage and price data for all waste management options and the market prospects for all the main markets for reclaimed materials.

Waste paper

There are many grades of waste paper, the highest being produced by paper mills themselves, almost all of which is recycled. Publishers and printers also create waste, which is worth recycling. In addition, there are the paper products found in households, mainly newspapers and magazines and grey packaging board used to pack cereals. The newsprint is used in the United Kingdom's three newsprint mills and there is still a huge surplus for export. OCC (old corrugated cases/cardboard) and mixed papers are used in board mills. Over half the paper and board produced in the United Kingdom is made from waste paper, and over three-quarters of it is used to make packaging such as cardboard boxes and cartons for cereals, soap powders and other products. Waste paper is not used for cartons that come into direct contact with food because of possible hygiene problems. In the 2020s with increasing emphasis being placed on quality, LAs are being pressed to have segregated paper collections.

Plastics

Recycling of plastics from household waste in the United Kingdom varies more widely in comparison to other materials. The most commonly available material is polyethylene film, which is increasingly recycled, especially from supermarkets. It is used as a shrink-wrapping for pallets of goods and for trays of packaged goods. Plastic bottles made from PET used for carbonated soft drinks and an increasing range of food and non-food goods and HDPE used mainly for milk containers are also recycled and since 2009 these are often processed in the UK to be flaked for the reintroduction back into the production of containers for food and drink products. This 'closed loop' recycling system is now the main use for PET and HDPE containers as the UK has sufficient capacity to recycle all PET and HDPE containers reclaimed from waste. PET can also be reused in the clothing industry as a fibre source. Polypropylene, used in the making of crates and battery cases, can be recycled into drainage pipes. Mixed plastic waste can be reprocessed to make products such as road signs, motorway acoustic insulation walls, fencing, pallets and outdoor furniture. With the introduction of facilities able to segregate pots, tubs and trays and LA encouragement to households to separate these items there are now regional plastic MRFs to enable single polymers to be separated and then reprocessed to produce higher value-added products. (For further information refer to the British Plastics Federation and RECOUP websites.)

The Plastics PACT was agreed in 2019 to progress towards less plastic being used, especially in the food supply chain. Bringing together brands, retailers, manufacturers, producers, recyclers, NGOs, governments and LAs they have developed a roadmap with a set of four objectives that by 2025 should achieve:

- 100% of plastics packaging to be reusable, recyclable or compostable;
- Eliminate single use packaging;
- 70% of plastics packaging effectively recycled or composted;
- 30% average recycled content across all plastics packaging.

However, one major effect of the pandemic was a rise in the use of more plastic packaging as consumers ordered more food for home delivery from both shops and take-away establishments; more products were plastic wrapped and those working had canteen food packaged individually in order to ensure hygiene and social distancing, for example.

Directive (EU) 2019/904 on the reduction of the impact of certain plastic products on the environment, commonly referred to as the Single Use Plastics Directive, banned products for which suitable alternatives are readily available, such as: cotton bud sticks, cutlery, beverage stirrers, straws, plates, sticks for balloons, all EPS food containers, beverage containers and cups and oxo-degradable plastics from mid-2021.

A simple waste reduction measure that can substantially reduce the amount of PET bottles for mineral water is the use of refillable bottles for tap water. Although the water bottle refill campaign suffered a severe setback in 2020–21 when many publicly available refill points were cordoned off, there should be a boost to this trend when a deposit return system (DRS) for beverage containers will possibly introduced in 2025 because the consultation paper of March 2021 did not seem certain as to future plans. However, Scotland appears determined to set up its DRS sooner.

Ferrous metals

All ferrous metals are magnetic and it is therefore not critical for this material to be separated at source if the waste is ultimately to go through MRF for sorting or after incineration or other treatment plant with overband electromagnetic separation after initial processing. However, larger materials, for example, white goods (cookers and washing machines), can normally be segregated at the disposal point for collection. At present, about 50% of food and drinks cans (by number, not weight) are made of ferrous metal and are most easily extracted from mixed waste by magnetic extraction.

Aluminium

Much aluminium waste is obtained from old cars and white goods, with only small quantities from source segregated household waste. Aluminium currently makes up less than 1% of household waste; it is used for about half (in number) of all drinks cans, milk bottle tops and foil food containers. In 1992, a £28 million aluminium can smelting facility was opened in Warrington, which is capable of reclaiming 50,000 tpa of all-aluminium cans. Even in 2021 the Novelis plant still relies on imports of scrap aluminium cans from other countries for ensuring full production.

Glass

Glass bottles had traditionally been re-filled rather than recycled, for example, door-to-door milk delivery, but inevitably there are casualties even in this system. It is difficult to separate glass from other household waste mechanically, although in the UK, following US practice, there is a consistent trend towards the collection of commingled dry recyclable wastes, including glass, rather than requiring consumers to deliver that glass to bottle banks, as is the norm throughout mainland Europe. Ideally every effort should be made to keep different coloured glass separate because two-thirds of the glass produced in the UK is clear containers, and mixed cullet can only be used in green glass furnaces. From 2014, however, colour separation plants have been built in the UK in order to sort mixed coloured glass cullet for further processing.

Since 2001 increasing amounts of green and mixed coloured glass have been used as an aggregate substitute. However, this recycling option relies on additional money being added by packaging compliance schemes through the PRN (packaging waste recovery note) system to supplement the lower value of glass cullet when used in this application.

Food waste

The first UK food waste prevention initiative was started in 2007 with the then government-sponsored Waste and Resource Action Programme's (WRAP) launch of their *Love Food, Hate Waste* campaign. At that time there was 10 million tpa of food waste being generated each year in the UK, predominantly from households (70%). Following that initial promotional effort that utilised a wide range of media there was a decline in the food waste generated by households, a 13% reduction from 2007 to 2013. In one respect the timing of the campaign was opportune because it coincided with the global financial crisis which made everyone more aware of the cost of food they were purchasing and potentially discarding uneaten. The consumer response would therefore also have been reinforced by constraints on consumer budgets to deliver a reduction in many people's unnecessary food purchases.

In addition, at the same time a number of municipalities in the UK introduced segregated food waste collections for their residents. This reinforced information to those residents that they were purchasing food only to discard it into their food waste caddies (kitchen containers) un-eaten. However, more recent unpublished research is ambivalent about the effect of providing segregated food waste collections on the behaviour of residents. While some people might have felt they were wasting food unnecessarily others would regard their discards as environmentally sound because their food waste would be used for productive purposes, such as production of electricity through AD.

Research undertaken by WRAP revealed that food waste generation had decreased in 2015–18 having plateaued in the previous three years and this downward trend has continued to 2020. This result of the WRAP research in 2019 reveals the following estimates:

Households	6.6 million tonnes (of which 4.8m had been edible)
Primary production	1.6 million tonnes
Food product manufacturing	1.5 million tonnes
Hospitality and food sector	1.1 million tonnes
Retail sector	0.3 million tonnes

The total amount of this wastage in monetary terms was around £20B. Faced with continuing high levels of consumer food wastage

WRAP was forced to change the emphasis of its campaigning so that it addressed a wide range of issues that influenced food wastage. These campaigns have included, for example, 'Chill your Fridge' in October 2018, which reached 1.3 million households, because research had revealed that a high proportion of the nation's fridges were operating at too high a temperature, rather than the ideal 0–4°C. This campaign was supported by 21 businesses and 26 LAs. The campaign was reintroduced in the autumn of 2019.

In 2019 the national food waste policy was changed to become more nuanced to address specific issues and groups with a range of low-budget initiatives to address a variety of food waste issues and nudge citizen behaviour towards food waste prevention, including:

- *Eat Me/Freeze Me* to encourage people to freeze any food that remains after initial consumption of an item and that might subsequently have been discarded, especially bread;
- *COMPLEAT* – directed mainly to bread products where many of the crusts and for some people the edges of loaves are discarded and to encourage people to eat all the fruit, such as apple skins, and vegetables, such as carrots and potatoes and broccoli stems without peeling them;
- A number of social media initiatives were undertaken to address the 18–34 year age group, the main demographic cohort with the highest food waste discards;
- *Wasting Food: It's out of date* in October 2020 with the strapline *we won't fix climate change if we don't stop wasting food*;
- *Look, Smell, Taste, Don't Waste* launched in January 2021 by 30 brands, mainly in the dairy sector, to encourage consumer awareness of date labelling guidance;
- Food Waste Action Week 1–7 March 2021.

The initial WRAP *Love Food Hate Waste* campaign focused on consumers but also engaged supermarkets and their suppliers to spread the waste prevention and food reutilisation message. The campaign was assisted by an existing voluntary industry programme co-ordinated through WRAP that was initially directed to the minimisation of packaging waste for both consumer packaging products and the secondary and transit packaging used by retailers and their suppliers. The success of the Courtauld Commitment, as the packaging initiative was titled, has latterly been redirected towards tackling food waste. In conformity with the food waste hierarchy the food retail sector emphasises food surplus redistribution to charitable causes and failing that use for animal feed before those foods that cannot be reutilised are sent for recycling by AD processing.

In addition, the retail sector was encouraged to help their customers to avoid food waste by, for example: cutting out 2 for 1 offers for food products, not giving additional food products when specific items were purchased and through the wider use of re-sealable packaging to extend products' shelf life. The supermarket chains were also invited to use the promotional materials WRAP had devised for the '*Love Food Hate Waste*' campaign. Several supermarkets also stimulated their consumers' food waste prevention through their own in-house leaflets and newsletters.

The Courtauld Commitment food packaging reduction programme was therefore an easy option for extension to food waste prevention, given that the majority of consumer packaging is used for food products. Since 2005 the Courtauld Commitment had gone through several iterations but in 2015 *Courtauld 2025* (C2025) was launched with the aim to reduce UK food and drink waste per head of population by 20% by 2025 compared to 2015 in conformity with the interim UN SDG 12.3 target. C2025 also has ambitious targets for reducing the greenhouse gas emission impacts of food production, processing and final consumption and in addition limiting the water utilised for food production throughout all of its lifecycle stages. In 2021 WRAP announced that there would be a Courtauld 2030 in order to try to fulfil the 2030 SDG 12.3 food waste reduction targets.

In 2012 it was recognised that there was one major food sector outside the conventional food supply chain that also needed to be addressed: the whole range of hospitality companies that provide prepared food for customers, from cafés and restaurants, pubs and clubs through to hotels and specialist hospitality venues, such as conference and event facilities and their supply chains. The food waste generated by these venues amounted to 1 million tpa in 2015. There was also the packaging associated with the production and distribution of this food.

The Hospitality and Food Service Agreement (HaFSA) was therefore set up under the auspices of WRAP in 2012 as a three-year programme in order to stimulate the sector to reduce their overall waste and especially their food waste generation. The participating trade organisations, companies and premises were encouraged to follow the resource management hierarchy in order to achieve an overall reduction in their waste generation to ensure both greater economic and environmental benefits for their businesses.

Food waste costs to the UK restaurant sector were around £800 million tpa from almost 200,000 tpa of food waste. The total cost of this food waste amounted to £4,000 per tonne in 2011, which included: food purchases, labour, utility costs and waste management. Food purchases and labour are more than 90% of the total cost.

After the HaFSA had come to an end in December 2015 the sector subsequently agreed to continue its efforts on waste prevention, reuse and recycling under the auspices of the C2025 which brings together several different strands of food waste prevention in order to rationalise the UK's efforts to more efficiently produce, process and redistribute food in the UK. Specifically, the HaFSA sector agreed to participate in the industry-wide Food Waste Recycling Action Plan. Specifically key associations and companies planned to build on the work of HaFSA

to increase recycling rates by promoting the case for change and encouraging businesses to reduce food waste, redistribute surplus food and to separate their food waste and to reuse and recycle it.

In addition to the HaFSA sector programme, in June 2017 a C2025 your Business is Food Manufacturing Focus Group was set up to address the earlier stages of the food supply chain, from companies taking raw ingredients and converting these to a wide range of food products for onward sale to processors, wholesalers and retailers and the HaFSA sector. In 2018 WRAP joined forces with the IDG to launch the Food Reduction Roadmap with its theme of *Target, Measure, Act* in order to address food waste occurring in the processing and distribution supply chain.

There has also been some work undertaken under the auspices of WRAP with the farming sector, generally linked to supply of materials to their primary customers. This has included assessment and selection of specific types of potatoes, for example, so that the most productive marketable varieties could be identified and grown in future.

While up to 2020 the food waste initiatives in England had been on a purely voluntary basis throughout the food supply chain, the Government's *Resources and Waste Strategy* published in December 2018 and the subsequent early 2019 consultation papers on different aspects of wastes management suggested that there may be government intervention in future. However, it is likely that for the period up to 2022 there would be no legislation and the government would instead rely on exhortation and voluntary action by businesses to reduce food wastage. This laissez-faire approach for England is in stark contrast with the food waste legislation and regulations introduced by all three of the UK's devolved administrations (DAs).

Northern Ireland, Scotland and Wales have adopted their own policies to tackle food waste. In addition to utilising the WRAP-initiated *Love Food Hate Waste* model and its promotional initiatives, the DAs also have their own regulations and campaigns to promote food waste prevention by citizens and businesses.

In Northern Ireland (NI) the Food Waste Regulations (Northern Ireland) 2015 adopted the Scottish approach to regulate the food waste disposal of any business that generates more than 5 kg of food waste each week. This legislation requires these businesses to segregate and present this food waste for separate collection, treatment and recycling. The larger businesses generating more than 50 kg per week had to fulfil this requirement in 2016 and smaller ones from 2017.

From April 2017 there was a mandatory requirement for NI Councils to provide households with containers for food waste for separate collection. These containers are collected on a weekly basis. In itself this change in the NI segregated waste collection system increased the proportion of household waste recycled by 5% (2018 compared to 2017) and also reinforced the recycling message to residents for dry recyclable wastes to be segregated.

Scotland was the first of the UK's DAs to introduce food waste legislation. The Waste (Scotland) Regulations 2012 requires all premises which generate more than 5 kg of food waste each week to have this collected separately and treated, obviously with the larger generators having to conform sooner than smaller ones. However, in recognition of its mountainous topography and the difficult road communication network there are provisions which allow companies not to have a collection if they are in very remote locations. As for household food waste collections, these are being introduced at a slower rate than in NI and Wales because there are so many isolated rural settlements where distances to reprocessing facilities would be both prohibitively expensive and environmentally counter-productive.

In Wales, every household is required to be provided with a separate food waste collection and the Welsh Assembly Government

(WAG) allocated considerable financial resources to enable Welsh councils to purchase equipment, including specialist vehicles, collection bins and kitchen caddies, for their collections. In addition, WAG had provided finance to companies to treat the collected food waste through AD in three plants that serve the whole of Wales.

One of the other kitchen wastes generated in both domestic and commercial premises that should be segregated and treated separately are fats, oils and grease. These present a huge problem for the waste-water utility companies where releases of these items into the sewage system causes blockages ('fat-bergs') and helps to feed rats in the sewer system. Ideally domestic consumers should convert the waste into fat balls to feed birdlife while commercial premises can place these waste materials in waste containers used initially for the delivery of the oil, to be passed back through the suppliers of the new fats and oils and they will pass on the waste oils for conversion into bio-diesel.

Animal by-products

The English Animal By-Products Regulations and the EU Animal By-Products Regulation restrict the ways in which animal wastes can be treated and spread on land. The EU Animal By-Products Regulation (ABPR) 1069/2009 came into force on 4 March 2011 and like its predecessors affects a wide range of catering establishments because raw meat and fish and their associated packaging is banned from landfill. Owners and operators of these establishments have to find alternative disposal outlets for these wastes, such as rendering plants or incinerators. The 2005 Regulations were replaced in 2011 by the Animal By-Products (Enforcement) (England) Regulations 2011 which, following representations to the European Commission, rescinded the planned landfill ban on cooked meat and fish products. Defra's *Guidance for the animal by-product industry* was published in September 2014 and updated in March 2020.

However, as with the 2005 Regulations, raw meat and fish are banned from landfill together with any immediate packaging materials. Therefore, any commercial premises handling these materials need to have alternative disposal arrangements in hand to ensure that this type of waste does not end up in landfill. These restrictions do not affect domestic consumers.

LAs, usually through trading standards, are responsible for the enforcement of ABPR legislation in England, except licensed slaughterhouses, cutting plants and cold stores.

Meat and fish waste that may NOT be disposed of to landfill:

- Raw meat and raw fish and seafood in packs or loose that needs to be cooked before eating;
- Raw sausages;
- Raw bacon and gammon;
- Raw meat, poultry, fish or seafood items in any coating (such as breadcrumbs or batter) or sauces (such as barbeque, garlic, etc.), for example chicken kiev, spare ribs or chicken wings in barbeque sauce, southern fried chicken, scampi, calamari, fish fingers or fish portions;
- All raw meat and fish that come from food production, even if they were not meant to be cooked before human consumption;
- Partially cooked meat;
- Sashimi (raw fish) or sushi containing sashimi;
- Smoked dried or cured meat (e.g. bresaola, beef jerky, salami, Parma ham, chorizo, etc.)
- Raw burgers;
- Raw eggs.

Meat and fish waste that may be disposed of to landfill:

- Meat and fish, and meat and fish products, that are pre-cooked;
- Cooked meat and cooked fish items such as cooked cocktail sausages and frankfurters (ready to eat or ready for re-heating);

- Dips with cooked meat or fish;
- Ham including Parma and Serrano, salami, pate, etc.;
- Cooked prawns, dressed crabs and lobsters, seafood sticks, cooked mussels (ready to eat);
- All ready meals (including meals for re-heating where the meat or fish is already cooked) such as roast dinners;
- Pies and pasties, including ocean pie, pizza and shepherd's pie where the meat is pre-cooked;
- Shells from cooked seafood and eggs;
- Highly processed meat and fish products in cans, jars, pouches or shelf stable aluminium packaging;
- Dried animal products, such as powdered soups;
- Flavourings or other highly processed animal products;
- Stock cubes made using meat or fish extracts;
- Fats or dripping that came from cooking meat or poultry.

Refuse collection

The collection of refuse from domestic, and some commercial and industrial premises is undertaken by LAs themselves, using their direct labour organisations, or through contracts with commercial waste collection and disposal companies. While there had been a significant increase in the involvement of the latter organisations since the introduction of compulsory competitive tendering to the UK waste management industry in the late 1980s there are now very few changes made at the end of contractual periods, usually 5–7 years.

The traditional transport system for refuse collection is now compaction vehicles, ranging from 16-tonne single rear axle through to 24-tonne and 32-tonne twin rear axle vehicles. Where the topographical nature of the round is such that large vehicles cannot negotiate the area, then smaller derivatives are used, but in order to ensure maximum economic payloads and work content, these should operate within a short distance of the disposal point, or should have a transfer facility to ensure that there are no long, unproductive hauls to the transfer site.

The optimum arrangement is to have two or, at the most, three trips to the disposal point, allowing rest periods and lunch breaks to coincide wherever possible, thus ensuring faster working when the work is resumed. The vehicles used tend to be compaction vehicles with power-operated pressure plates producing compaction ratios in the region of 4:1. Typical payloads of a 17-tonne gross vehicle weight (GVW) vehicle would be in the order of 8 tonnes per load. Although there are few variations between the major manufacturers of refuse collection vehicles (RCVs), there are specialist variants for unusual situations, such as a smaller, non-HGV type vehicle that may be used in rural locations where there is considerable inter-premises travel; narrow-bodied vehicles are used in inner cities where access and manoeuvrability is limited, and some authorities are experimenting with one-person crews using automatic lifting devices and/or side loading lifting devices on the vehicles, which are commonly used in suburban locations in Australia and North America. Where the wheeled bin system is in operation, a lifting rig is fitted to the rear of the vehicle. Other lifting devices can be fitted to the rear of large-capacity compression vehicles to enable them to lift refuse compactors, refuse skips and a variety of bulk containers. Split-bodied RCVs allow for the collection of both residual and recyclable wastes at the same time. For certain industrial applications, front-end loading vehicles may be used where access is particularly difficult but these vehicles are being phased out of use.

Health and safety perspectives

There are serious and long-standing health and safety implications involved in the operation of refuse compaction vehicles and collection vehicles for recyclable wastes, plus the operation of CA sites and the operation of waste treatment and disposal facilities. Between 1977 and 1984, 28 fatal and 64 major injuries and 320 accidents were notified to the Health and Safety Executive (HSE). Of these total accidents, 77% occurred during collection rounds. Subsequently the position had grown worse. The HSE in 2009 reported that 'recycling has the highest 3-year average rate of fatal injury rates at 15.1 deaths per 100,000 employees compared to the average across all industries of 1.3 per 100,000'.

HSE issued a safety alert in March 2006 when nine fatalities occurred in just eight weeks. A HSE report in 2020 showed that in 2019/20 there were five fatalities, which compares favourably to the average number of nine for the period 2015/16–2019/20. Nevertheless, there is continuing emphasis on trying to reduce fatalities and accidents through an industry-led initiative, the WISH (waste industry safety and health) Forum which produces guidance and information for the sector (wishforum.org.uk).

A Fire Prevention Plan (FPP) is an important risk mitigation measure required by the UK's environmental regulators to ensure safety in waste management and recycling sites. An FPP is an essential aspect of an environmental permit to meet three objectives:

Minimise the likelihood of a fire happening

Aim for a fire to be extinguished within four hours

Minimise the spread of fire within the site and to neighbouring sites

Litter

In the United Kingdom the cost of managing litter is estimated to amount to £1 billion, according to Keep Britain Tidy. Demand for action against litter is consistent and hardening as exemplified by its prominence in the waste management objectives detailed in the government's Resources and Waste Strategy in 2018 and preceded earlier that year by its Litter Strategy for England.

There remains the paradox that despite the activities of Keep Britain Tidy and the ever-widening legislative provision against littering in the early 1990s onwards, most people perceive that the problem is becoming worse. Therefore, in 2005 the Government introduced the Clean Neighbourhoods and Environment Act complementing provisions in the Anti-Social Behaviour Act 2003, amended in 2015, partly in order to tighten up on the problem of local social disruption, one of the outcomes being environmental deterioration due to irresponsible discarding of waste. The Act introduced a range of measures that local authorities could utilise, including: action against abandoned and nuisance vehicles, graffiti and fly-posting, dogs, noise and a range of fixed-penalty notices.

Although Part IV of the Environmental Protection Act 1990 (EPA) did not repeal the whole of the Litter Act 1983, it did take over its main provision (Section 1), creating an offence of leaving litter (Section 87 of the EPA). The 1983 Act itself consolidated the previous Litter Acts of 1958 and 1971, and repealed the Dangerous Litter Act 1971.

The essence of Part IV is provided in Section 87. If any person throws down, drops or otherwise deposits in, into or from any place to which this section applies, and leaves anything whatsoever in such circumstances as to cause, or contribute to, or tend to lead to, the defacement by litter of any place to which this section applies, he shall (subject to subsection 2) be guilty of an offence. Subsection 2 says that no offence is committed under this section where the depositing and leaving of the thing was:

(a) Authorized by law, or
(b) Done with the consent of the owner, occupier or other person having control of the place in or into which that thing was deposited.

The maximum fine for littering was raised to £2500 from £400 under the 1983 Act, somewhat academic given that fines in 2015 averaged less than £100, although the level of costs awarded for offenders was often greater. Section 88, however, extended to all litter authorities the fixed penalty notice system originally introduced by Westminster City Council through the City of Westminster Act 1988 introducing on-the-spot fines of up to £80, raised to £150 in 2019. These powers cannot be exercised by county councils (Section 89(a)), except for areas designated by the Secretary of State, but do include National Park Committees and Boards and the Broads Authority, which are not principal litter authorities under Section 86(2).

Members of the public can take court action to compel authorities to remove litter from public land in their control (Section 91, see later), although the five-days' notice requirement (Section 91(5)) prior to the institution of proceedings does provide ample time for remedial action to be taken. There is also encouragement to institute proceedings in that the authority would have to reimburse the expenses of the complainant in bringing the complaint and proceedings before the court where magistrates are satisfied that the complaint is justified.

Local authorities have the power to establish litter control areas under Section 90, which would apply to private landowners subject to the Secretary of State setting the necessary regulations (Sections 90(1)). Also street litter control notices can be issued by the LA (Section 93) on occupiers of premises to curb litter or refuse on the street or adjacent land.

The most significant change in litter control under the EPA is that LAs have to reach a quality standard for cleanliness, or absence of litter, from an area (see the Code of Practice later). This is in contrast to the traditional specification of service input for street cleansing and litter removal services (Section 89(7)).

The final section of Part IV, Section 99, provides powers to district and borough councils with regard to abandoned shopping and luggage trolleys. The extent of their powers to curb the measure of abandoned trolleys is given in Schedule 4, which includes the power to seize and remove trolleys and to exact payment for their return or to arrange for their disposal after a period of six weeks.

The EPA placed a new duty on the Crown, local authorities, designated statutory undertakers and the owners of some other land to keep land to which the public has access clear of litter and refuse, as far as is practicable. (In the case of designated statutory undertakers, the duty may apply additionally to land to which the public does not have access.) The duty also applies to the land of designated educational institutions.

The Code of Practice – the Act requires the Secretary of State, under Section 89(7), to issue a Code of Practice to which Section 119 of the Local Government Act 2003 allows LAs to retain the amounts of fine received and use them to pay their litter control functions. Those under the duty are required to have regard. The original code was issued in 1991, revised in 1999 and in 2006 and modified in 2019. The objective of this Code of Practice on Litter and Refuse is 'to provide guidance on the discharge of the duties under Section 89 by establishing reasonable and generally acceptable standards of cleanliness which those under the duty should be capable of meeting'.

It will immediately be apparent that this code, in its approach to litter clearance, is innovative in at least two ways. First, it attempts, by defining standards of cleanliness that are achievable in different types of location and under differing circumstances, to ensure uniformity of standards across Great Britain.

Second, the code is concerned with output standards rather than input standards – that is to say, it is concerned with how clean land is, rather than how often it is swept. Indeed, this code does not suggest cleaning frequencies at all – it simply defines certain standards which are achievable in different situations. This may mean that an area which all but escapes littering will seldom need to be swept whereas a litter black-spot may need frequent attention.

It will be seen that the code offers considerable scope for local authorities and others to target their resources to areas most in need of them, rather than simply sweeping a street because of the dictates of an arbitrary rota. Expressed in its simplest terms: 'if it isn't dirty, don't clean it'.

For Scotland there is the Scottish Government's Code of Practice on Litter and Refuse (Scotland) 2018 issued under Section 89 of the Environmental Protection Act 1990.

Statutory duties – Section 89(1) of the Act places on the Crown and (in England and Wales) county, district and London borough councils, the Common Council of the City of London, and the Council of the Isles of Scilly and (in Scotland) district or island councils, and joint boards (collectively known as 'principal litter authorities') a duty to ensure that all land in their direct control which is open to the air and to which the public has access is kept clear of litter and refuse, so far as is practicable. In addition, where the duty extends to roads, they must also be kept clean – again, so far as is practicable.

Section 86(9) transfers the responsibility for cleaning all roads except motorways (which remain with the Highways Agency) from the highways authorities to the district and borough councils. This duty may be transferred to the highway authority by the Secretary of State. See, for example, the Highway Litter Clearance and Cleaning (Transfer of Responsibility) Order 1998. A similar duty is placed on designated statutory undertakers. One difference between the duty as it applies to statutory undertakers and the duty applying to principal litter authorities is that the duty on the statutory undertakers might cover some land in the direct control of a statutory undertaker to which the public has no right of access, such as railway embankments.

The duty also applies to land in the open air, and which is in the direct control of the governing body or local education authority of designated educational institutions.

Similar duties may be imposed by principal litter authorities (other than county councils or joint boards) on owners of other land by designating their land as a 'litter control area'.

The Act allows the Secretary of State to specify descriptions of animal faeces to be included within the definition of refuse (see the Litter (Animal Droppings) Order 1991) xx. The Secretary may also, by regulation, prescribe particular kinds of things which, if on a road are to be treated as litter or refuse. (See the Controlled Waste Regulations 2012.)

Practicability – the caveat in the summary of the duty concerning practicability is very important. It is inevitable that on some occasions, circumstances may render it impracticable (if not totally impossible) for the body under the duty to discharge it. It will be for the courts to decide, in all cases brought before them, whether or not it was impracticable for a person under the duty to discharge it, but certain circumstances are foreseeable in which the discharge of the duty may be considered by the courts to be impracticable.

Enforcement – in the great majority of cases those under the duty will wish to achieve the highest possible standards of cleanliness. However, the Act makes provision for the occasion when a body under the duty may not discharge it adequately. Under Section 91, a citizen aggrieved by the presence of litter or refuse on land to which the duty applies may, after giving five days' written notice, apply to the magistrates' court (or, in Scotland, the sheriff) for a 'litter abatement order', that requires the person under the duty to clear away the litter or refuse from the area which is the subject of the complaint. Failure to comply with a litter abatement order may result in a fine (with additional fines accruing for each day the area remains littered).[25]

Any person contemplating enforcement action should not just consider the presence of litter but is advised to consider whether the body in question is complying with the standards in the code before notifying them, since, under Section 91(11), the code is admissible in evidence in any court proceedings brought under that section. Similarly, LAs can act against any other body under the duty, which appears to them to be failing to clear land of litter and refuse.

Street cleansing

There are many different types of street cleansing systems depending on the type of location, be it urban, rural, heavily industrialised, inland, seaside, seasonal population changes, etc. No particular system is wrong or right, it is very much a case of 'horses for courses' and tailoring the best option for each particular situation.

It is true to say that there has been a shift away from the manual emphasis of street cleansing towards mechanical systems in the last three decades. The simplest conventional form of street cleansing is the road sweeper with a street orderly barrow.

This is unquestionably the most thorough method of cleansing, but by virtue of the considerable element of walking, and the limited capacity of the barrows involved, the distance travelled will not be great in a working day. The merits of this system are local knowledge, community identity and thoroughness. In inner town areas where access is restricted but where large accumulations of sweepings and waste are anticipated, pedestrian-controlled vehicles are often used. Their capacity may be as much as that of eight bins each of 1.13 m³. As an alternative, the vehicle may have a side- and end-loading tipping body, and the facility exists for a crew to work in a gang of three or four people, thus covering a larger area.

The conventional mechanical sweeping vehicle is used extensively on trunk, principal and main arterial roads. The machines are available in single sweep or dual sweep options to cater for single carriageway and dual carriageway. The single version is left-hand drive to enable the driver to position themselves with good sight of the channel that they are sweeping. These machines sweep at between 2 and 5 mph (3 and 8 kph). Smaller, scaled-down versions can be used in pedestrian precincts and areas of restricted manoeuvrability. Increasingly, these units are now being electric powered.

There are a number of miscellaneous vehicles that can be used for street cleansing purposes; for example, refuse collection vehicles or light vans, once they have fulfilled their useful life on a regular round, can often be used to service manual street cleansing crews, picking up bags or emptying street cleansing bins from strategic locations.

In some of the larger cities in the UK, street orderly boxes are set into the pavement, and at first glance sweepers appear to be sweeping litter and detritus down a street gully. These boxes are often emptied by a nightshift street cleansing crew.

Street washing is common in some large cities in the United Kingdom. It can be undertaken by attachments fitted to mechanical sweepers or gully emptiers, or in connection with a bowser and reel or stand-pipes. This practice is not, however, as universal as it was 20 years ago, although open-air markets often still use it.

Seasonal variations – winter brings snow, frost and ice as additional hazards, spring and summer the growth of weeds on footways and channels, and autumn leaf deposits. Pavement gritters, knapsack and road sweepers, mounted weed control kits and pedestrian-operated or vehicle-mounted leaf blowing units have been developed to improve operator performance in the field of street cleansing. Other non-elemental factors that must be taken into account are tourism and student populations, often requiring seasonal staff to cope with fluctuating service demands and workload.

Street cleansing – prevention or cure? – the whole question of street cleansing and its 'bedfellow', litter abatement, hinges on the question, does it provide prevention or cure? Unquestionably, the cure is easier

to manage and simpler to perform but is more expensive to achieve. However, in the mid-1990s a new approach to the problem of an untidy country emerged. The Tidy Britain Group's 'People and Places Programme' is aimed at increasing the awareness of all sections of the community of littering and its effect on society. The programme is designed to change attitudes to the problem and promote a long-term systematic approach to litter abatement and environmental improvement. Nationally, around 190 councils have adopted the programme.

Waste disposal and treatment

The principal methods are:

- Incineration or other energy recovery techniques;
- Controlled landfill;
- Composting and anaerobic digestion (AD).

Development of other techniques, principally mechanical-biological treatment (MBT) is being promoted because of the opposition to the more conventional methods of recovery and disposal but ultimately most of these newer techniques rely on the pre-treatment of wastes in order to utilise these three conventional methods. Therefore, mechanical-biological treatment (MBT), for example, provides processing options for separating of certain recyclable wastes, processing wastes for fuel use and treating organic wastes by aerobic or anaerobic methods, mainly as a pre-treatment technique prior to landfill. The potential market for SRF (solid recovered fuel), which has a recognised specification and also RDF (refuse derived fuel) should improve as the use of conventional fossil fuels decreases and more incentives are provided for renewable energy sources. Indeed, there are a few UK energy recovery plants which use RDF as their feedstock rather than mixed residual waste. However, the potential market opportunities for CLO (compost-like output) from this type of treatment are limited due to potential contamination problems.

Considerations that influence the adoption of a particular method of disposal are:

- The physical characteristics of the district;
- The situation of disposal sites;
- Daily yield and average composition of refuse;
- Ultimate disposal of residual wastes;
- Capital outlay and running costs.

Energy recovery through incineration

The recovery of energy from that fraction of the MSW, which has not been set out for reclamation by recycling through incineration or other techniques such as gasification, is not popular. Incineration lost favour with LAs in the 1960s mainly on the grounds of its higher cost compared to long distance transport and landfill in large regional landfill sites. Subsequently due to public opposition, mainly based on unfounded fears of the health impact of air emissions, it became politically difficult to approve the building of incineration plants. However, because of a lack of landfill space and increasing distances for haulage, landfill tax and the restrictions on BMW deposit under the Landfill Directive, an increasing number of areas have built and are adopting incineration as their means of disposal. Plants using the optimal systems for emission control should cause no nuisance, as confirmed by an assessment of the health impacts of waste management published by the UK Government and in many other countries.

The limits on emission of pollutants are becoming more stringent under EU directives and were initially implemented through Part I of the Environmental Protection Act 1990. The former EU controls on the incineration of waste in 89/369/EEC and 89/429/EEC were replaced and extended by the Waste Incineration Directive 2000/76/EC (WID). The Waste Incineration (England and Wales) Regulations 2002[26] came into effect on 28 December 2002. The government decided to bring all incinerators and all the other operations covered by the WID, such as co-incineration of waste in cement kilns and waste oil combustion under the Pollution Prevention and Control (PPC) regime. Facilities, such as fossil fuelled power stations, previously had less tight emission standards than incinerators because they were regulated under the replacement to the large combustion plant directive. However, the Industrial Emissions (integrated pollution prevention and control) Directive 10/75/EU (IED) forced all plants by 2016 to improve their emissions to the higher standards set for all larger waste burning facilities. The IED sets stringent emission limits for pollutants such as dioxins and acid gases. It also covers matters such as discharges to water and the recycling of ash. Compliance is achieved through conditions in the PPC permit. These limits increase the cost of disposal by this method.

Incineration of household waste is only viable when it is used to generate energy. Therefore, even after 1996 in the UK, only incinerators that reclaimed energy from waste continued to operate.

It is possible to utilise the incinerator bottom ash (IBA) from the incineration process, especially when the ferrous metal content has been removed by magnetic extraction and non-ferrous metals by eddy current separation. With both the landfill tax and aggregates levy and improved control of incinerators which allows for the maximum burn-out of combustible wastes, there is ever-increasing processing IBA for aggregate use, mainly as a fill material for the sub-base of roads, car parks or similar civil engineering projects.

A variety of combustion systems are used for waste incineration, some being more suited to liquid chemical wastes or sewage sludge and others being most suited to municipal waste. More recent developments have mainly been located at sites which previously housed old incineration plants and are generally under 200,000 tpa capacity. However, the Belvedere plant in SE London, for which it took 17 years to obtain planning permission, and which was originally designed to process 1.2 m tpa of waste, incinerates 585,000 tpa, very similar in capacity to the North London Waste Authority's Edmonton plant, opened in 1970 and still functioning after several refurbishments but due to be replaced with a new facility in 2028. Also in the Belvedere industrial zone a planning application was submitted by Cory Environmental for its Riverside Industrial Park in 2016 and was approved by the Secretary of State under the NSIP procedures to process 655,000 tpa of waste by incineration and 40,000 tpa of green and kitchen waste by AD.

Mass burn incinerators are used to burn municipal solid waste but can also be used to burn suitable commercial and industrial wastes. The capacity of existing plants in the United Kingdom varies according to the number of furnace units incorporated into the design (up to five). The size of each unit can vary from around 10 tonnes per hour up to 30 tonnes per hour.

Waste is fed by crane from storage bunkers to the feed hopper, from where it flows down, usually aided by a mechanical stoker, into the combustion chamber. Air is introduced through a moving grate in the chamber, which agitates the waste promoting its thorough exposure to the air. Waste is fed on to the grate and during the initial drying stage, at 50–100°C, volatile compounds are released. The wastes burn above the grate, where secondary air is introduced to facilitate complete gas phase combustion. The remaining waste moves down the grate and continues to burn slowly. After about an hour the residues are discharged from the end of the grate, where they are usually quenched. In most cases the ferrous fraction, and increasingly the non-ferrous fraction, is magnetically extracted and in addition the incinerator bottom ash is being processed to produce a secondary aggregate. Although there is a considerable variety of design for mass burn incinerators, the most efficient appear to be those with reciprocating or roller grates and the combustion chamber in the form of a vertical shaft.

Controlled landfill

The term 'controlled landfill' or 'sanitary landfill' describes a system of disposing of refuse by depositing in a methodical, as distinct from a crude or indiscriminate, manner. The greater part of all residual controlled waste is disposed of in this manner and, if properly supervised, the system provides a satisfactory method of disposal. Mineral excavations, low-lying or open-cast sites are selected for the purpose, and, by the utilisation of refuse in the manner described later, are reclaimed for useful purposes. In many districts, land reclamation alone justified this method of refuse disposal. The main points about the deposition of the crude refuse are as follows:

1   Where the base of the site cannot support the weight of vehicles, a preformed base is required.

2   Waste should be deposited at the top or base of a shallow sloping working face, and not over a vertical face. The face of the tip should slope at an incline that is no greater than 1:3.

3   The deposit of waste should be in thin layers. The use of a compactor enables a high density to be activated. Each layer should not exceed 0.3 m in thickness.

4   At the end of the working day, all exposed surfaces should be covered with an inert material to a depth of at least 0.15 m.

5   The management and workforce should be fully aware of site safety regulations and of the need to observe them.

6   An effective system of litter control is essential.

7   Measures need to be taken to control birds, particularly gulls and crows, where

sites receive quantities of putrescible matter. Control measures include:

- Bird scarers;
- Distress calls;
- Falcons;
- Nets.

8 Effective measures to ensure good pest control are necessary. Good compaction of the material will reduce the likelihood of infestation by both insects and rodents, as will the daily covering of waste. However, regular systems of both rodent and insect control need to be established on a programmed basis.

9 Waste should not be burnt on the landfill site.

The main elements of the site restoration plan following the cessation of landfill are as follows:

1 A plan for restoration should form an integral part of the landfill operation.

2 The intended final levels and contours should be indicated within the plan, as should the systems for leachate control.

3 There must be a clear and efficient system for the management of landfill gas.

4 Landfill site capping should be constructed of material with a permeability of 1×10^{-6} cm/s or less, and should extend over the whole site to increase water run-off. A basic aim of the cap is to minimise leachate production by minimising water entry to the tipped area. The cap should be covered with an appropriate thickness of soil to protect the cap from damage by, for example, agricultural machinery [6].

The modern landfill containment site is a site where the base, sides and top (once the landfill has finished) are sealed with a suitable mineral or synthetic impermeable liner. In the United Kingdom, clay is extensively used to seal sites. In the United States and some Continental European countries, synthetic plastic liners are used. Depending on the country and geology of the site, a single liner, a composite liner or a double liner is incorporated. Further research is being undertaken to evaluate the advantages and disadvantages of various liners in different conditions.

Within a landfill site, aerobic conditions prevail initially, but anaerobic conditions are rapidly established. The biodegradation of various components of refuse in landfills is extremely complex, but three main stages can be distinguished. In the first stage, degradable waste is attacked by aerobic organisms, which are present in the oxygen in the air trapped in the waste, to form more simple organic compounds, carbon dioxide and water. Heat is generated and the aerobic organisms multiply.

The second stage commences when all the oxygen is consumed or displaced by carbon dioxide and the aerobic organisms die back. The degradation process is then taken over by organisms that can thrive in either the presence or absence of oxygen. These organisms can break down the large organic molecules present in food, paper and similar waste into more simple compounds such as hydrogen, ammonia, water, carbon dioxide and organic acids. During this stage, carbon dioxide concentrations can reach a maximum of 90%, but usually reach about 50%.

In the third and final stage, which is anaerobic, methane-forming organisms multiply and break down organic acids to form methane gas and other products. Under the Landfill Directive all landfills must conform to the detailed technical requirements of the directive. These cover matters such as lining systems, leachate and gas controls, siting of landfills and waste acceptance criteria.

Although disposal by engineered landfill is at the bottom of the waste hierarchy, for developing countries it is an essential first step towards the sound management of waste and a necessary basic component of a waste management strategic plan in order to protect public health.

Composting

As a waste treatment technique in the United Kingdom, composting was almost eliminated in the 1970s. Up to that time several schemes were producing compost from a feedstock of ordinary household waste, usually using Dano drums or similar slowly rotating trommels where the waste was retained for between two days and one week before being placed in windrows for further processing. The production of a growing medium made from mixed refuse became increasingly difficult to market and the cost of this method of waste processing was considerably higher than that of landfill.

In contrast, in most Continental European countries composting had continued to be used for treatment of up to 10% of household waste, although in most cases the resulting material has been used in land reclamation or as landfill cover. While the average bin contains around 20–35% of putrescible and largely compostable waste, there is an even greater amount of readily compostable material available from residents' gardens and the councils' parks departments. Although traditionally composted, until recently increasing amounts of these materials were going to landfill for disposal, partly due to the introduction of wheeled bins and relaxation on the restriction which prohibited the inclusion of garden waste into traditional dustbins and plastic sacks.

A national collection of compostable waste would therefore yield around 1.5–2 million tpa of compost from garden waste, most of which is currently taken to civic amenity sites initially and thence to specialised composting plants.

Composting systems in the United Kingdom takes two main forms:

- Centralised composting of green wastes;
- Promotion of home composting systems.

The main priority of LAs promoting home composting is the need for a comprehensive coverage of composting units in those areas chosen for this type of waste reduction strategy to ensure that there will be an effective diversion of the putrescible waste fraction.

The development of markets to accommodate increasing amounts of compost is an essential aspect of the development of centralised composting systems. The greatest barrier to the expansion of green waste composting schemes was the requirement to develop markets to take the growing media, mulches and clippings produced throughout the year, as the market is highly concentrated into the spring months. However, the introduction of Publicly Available Standard 100 (PAS 100) and other quality control measures has meant that the markets for waste-based composts are expanding.

In addition, there had been increasing concern with regard to bio-aerosols from composting plants but for normal well-managed operations the EA has accepted that the cordon sanitaire should be 250 m. Therefore, most composting facilities are now following the lead of mainland European countries building enclosed composting facilities, which obviously increases the cost of this method of waste treatment. Increasingly UK facilities are being enclosed, both to avoid the bio-aerosol issues and also to allow the inclusion of kitchen waste, where required.

LAs, faced with budgetary constraints, are increasingly introducing charges for the separate collection of garden waste from households. This has had a negative effect on the English recycling statistics from 2013 onwards. The previous rising level of collections of waste for recycling stabilised and there were concerns that England would be unable to reach the EU's 50% target for the recycling of household waste in 2020.

In contrast, the establishment of home composting facilities attempts to reduce waste at source. These initiatives included the free or heavily subsidised provision of conventional garden composters, rotating units and Green Cone digesters. In a minority of instances wormeries have also been offered, but these have greater operational limitations than other types of unit. There can also be pest management issues from home composting (see Chapter 17).

One of the main difficulties in assessing the success of home composting schemes is that there is often little data on what proportion of the population is already practising composting at home and what proportion of the materials generated in the home are composted. How many more people will be composting as a result of the provision of the new composting units is difficult to determine.

Now, following extensive research by WRAP there are methods to calculate the waste prevention impact of introducing home composting schemes and a wide range of other waste prevention initiatives.[27]

Anaerobic digestion (AD)

AD has constantly changed its focus within waste management treatment preferences: as a means of generating renewable energy, producing a digestate which can be utilised as a soil amendment and/or also to provide a liquid fertiliser. In the UK the emphasis is currently on its role as a renewable energy source. However, until 2007 the Government had totally neglected this technology.

There are several different techniques for processing biogenic wastes, including paper and cardboard through the use of the AD process. The main distinction is between the mesophilic (35–55°C) and thermophilic processes, which operates at a higher temperature. There is also a distinction between high solids AD and normal AD processing but this is largely academic as high solids is only 25–30% waste in water.

The methane produced by the degradation of the waste is mainly used within the plant, including the pasteurisation of feedstock, which might contain wastes affected by the ABPR. The surplus is generally used for the generation of electricity but it can also

be utilised as a fuel for transport use or with upgrading put into the natural gas grid. PAS 110:2014 *Specification for whole digestate, separated liquor and separated fibre derived from the anaerobic digestion of source segregated biodegradable materials* has enabled AD plants to generate a quality digestate that can be applied to agricultural and other land as a soil amendment and growing medium. Although the separate collection of biodegradable kitchen wastes is expensive, in order to reach the 50% household waste target set by the EU for 2020 higher targets beyond that date means that the collection of kitchen waste to be processed by AD is regarded as essential.

Mechanical-biological treatment (MBT)

MBT now has limited applicability but its development was based on the premise that energy facilities would increasingly be built in order to utilise the combustible fraction as an energy source or feedstock for fuel. At present the largest UK facility to accept RDF as a feedstock is the 420,000 tpa combined heat and power (CHP) waste to energy (WtE) plant built as part of the Greater Manchester Waste Disposal Authority's Public Finance Initiative contract. The WtE CHP plant for Ineos Runcorn TPS Ltd, a special purpose vehicle which was set up for the procurement, operation and maintenance of the plant.

Hazardous waste

Waste that is hazardous for human health or the environment has long been of particular concern as the effects of illegal depositing of such waste are greater than household waste, and inappropriate mixing can lead to environmental problems. In addition, the further the distance of travel to final disposal the greater the chance of waste escaping into the environment. The important matter is the continuity of control from arising to final disposal. EHPs dealing with industrial sites may need to be familiar with the provisions, particularly those controlled by LAs under the IPPC provisions. The Hazardous Waste (England and Wales) Regulations 2005[28] introduced new duties for producers of hazardous waste and modified the consignment note system which applied under the previous Special Waste Regulations (now revoked). The Regulations apply mainly to England with similar provisions for Wales, and there are provisions relating to cross-border consignments with Scotland. The EA enforces the Regulations, and now has additional powers to use civil sanctions in attention to criminal prosecution (see end of section).

Hazardous waste should not be mixed with a different category of hazardous waste and this is prohibited under Regulation 19. However, mixing by disposal or recovery operations is not illegal if it is allowed under the conditions of their permit. The Agency's 2006 guidance on treatment of hazardous waste for landfill advises when mixing is illegal. Where hazardous wastes have been illegally mixed, the holder has a duty to separate them, provided separation is:

- Technically and economically feasible; and
- Necessary to comply with the Waste Framework Directive (e.g. to protect human health and the environment).

The EA has issued guidance on the mixing of hazardous waste. Mixing is banned if:

- An undesired reaction or phase separation would occur;
- Mixing compromises the treatment of one of the waste streams;
- It dilutes or hides the identity of a hazardous waste;
- It prevents recycling and recovery.

The Agency allows mixing of the following wastes:

- Hazardous and non-hazardous oil/water mixes;
- Hazardous waste from sewer interceptors and grease traps;
- Similar wastes from different sources (e.g. oily wastes).

Segregation of mixed wastes should be carried out where feasible – for example, taking a TV set from the top of a skip. It must not pose a threat to health, safety or the environment.

Notification of premises

Producers of hazardous waste have a duty to notify their premises to the EA. This duty also applies in Wales. The requirements are that:

- All premises that produce more than 500 kg of hazardous waste per annum must register with the EA/NRW.
- All places where hazardous waste is produced are considered to be premises; however, the EA has published a table of locations for which notification will not be enforced. These mainly apply to infrastructure such as roads, waterways and railways.

The following information must be given:

- Name and address of waste producer;
- Address of premises;
- SIC[29] classification of premises;
- Any other information the Agency may reasonably require.

The Agency must be notified every 12 months, and a fee is payable. After receiving the notification, the EA will issue a premises code.

It is an offence to remove waste from premises that are neither notified nor exempt. Exempt premises are still required to ensure that waste is removed only by an authorised person, such as a registered carrier and use the consignment system, explained later.

All premises that produce more than 500 kg of hazardous waste per annum must register with the EA. The Agency has issued *A guide to the Hazardous Waste Regulations: Site premises registration*, available on the Agency's part of the gov.uk website.[30] The easiest way to notify is online but every alternative is available.

The Agency has a duty to inspect hazardous waste producers and carriers (Regulations 56–57). The waste producer is responsible for giving each consignment of hazardous waste a consignment code, using a system set up by the Agency. Under the 2005 Regulations, there is no longer a requirement to pre-notify the Agency of consignments of hazardous waste. Three or four copies of the consignment note are required, depending on whether the producer and consignor are the same person. The consignment note is reproduced in Schedule 4 to the Regulations.

1 The waste producer fills in Parts A and B of the consignment note, and gives all the copies to the carrier

2 The carrier completes Part C, and gives all three copies back to the waste producer (or consignor)

3 The waste producer or consignor completes Part D, then

- One copy is kept by the producer;
- One copy is kept by the consignor (if different);
- Two copies go to the carrier.

4 The two carrier's copies travel with the waste, and the carrier gives them to the consignee (i.e. waste management contractor)

5 The consignee completes Part E, then

- One copy is kept by the consignee;
- One copy goes back to the carrier.

This applies where more than one carrier transports the consignment. The consignor prepares copies of the schedule for:

- The waste producer (if different from the consignor);
- The consignor;
- Every carrier;
- The consignee.

The consignor gives all the copies to the first carrier. The copies travel with the consignment. The first carrier passes all the copies to the second carrier. The second carrier completes the certificate on each copy, returns one copy to the first carrier and keeps the rest. This same procedure is followed for all subsequent carriers. When the waste reaches its destination, the last carrier keeps one copy of the schedule. The other copies are given to the consignee.

Multiple collections

Regulation 38 applies where one carrier collects several consignments of hazardous waste, which all go to the same consignee. A multiple consignment collection note is required. The carrier prepares:

- Two copies for the carrier;
- One copy for each waste producer;
- One copy for each consignor (if different from the waste producer).

The carrier completes Parts A and B on each copy.

Before the waste is collected:

- The waste producer completes the Annex on each copy;
- The consignor and carrier sign declarations to the Annex on each copy;
- The carrier returns a completed copy to the producer (and consignor if separate).

After the last consignment has been collected, the carrier completes the relevant parts of Section C on both remaining copies of the note.

On delivery, the carrier gives both copies to the consignee. The consignee completes the relevant parts of Section C and completes the certificate in Part D, on both copies. The consignee keeps one copy and passes the other back to the carrier. At a landfill site, the contractor must keep records to show where hazardous waste has been deposited. The deposits are identified by:

- EWC code and description;
- Description of composition;
- Consignment note;
- A site plan showing the location of the waste.

If the waste is deposited at the producer's premises, instead of a consignment note, the producer refers to the quarterly return made to the Agency (see later).

At transfer stations and sites where waste is recovered or disposed of other than by landfill, similar records are required. The records should include information on the quantity, nature and origin of the waste, its hazardous properties, the recovery method and the location of the waste. At transfer stations, the records must be updated a maximum of 24 hours after the consignment is received, stored or removed. All these records must be kept for a minimum of three years, or until the permit is surrendered or revoked.

Records

The producer or holder and (if different) the consignor shall keep a record of the following information regarding the hazardous waste consignment:

- Quantity;
- Nature;
- Origin;
- Destination;
- Frequency of collection;
- Mode of transport, including identification of carrier;
- Treatment method;

and keep the records for at least three years after the transfer of the waste.

The various records listed previously must be kept in registers, which should also include copies of (where applicable):

- Consignment notes received;

- Consignee's returns received;
- Carrier's schedules received.

Most of these records are now processed and recorded digitally.

If requested, contractors and others must supply a previous holder of the waste with information to confirm that the disposal or recovery operation has been carried out. The request for information must be in writing and must allow seven days for reply. Every consignee must make a consignee quarterly return to the EA regarding all consignments received during the quarter.

The consignee must send a return to the producer to prove that the waste has been disposed of or recovered. The format for these returns is shown in Schedule 8 to the Regulations. Alternatively, copies of consignment notes can be returned to the producer. A return should be sent to the producer within one month of the end of the quarter in which the waste was accepted.

The EA can request documentary evidence to show that, for example, disposal has been carried out. The Agency is entitled to request any of the records kept under these Regulations.

Part 9 of the Regulations covers the actions to be taken when hazardous waste presents a threat to the population or the environment. The waste holder is allowed to breach the Regulations if this is necessary to avert or mitigate an emergency, for instance when there is no time to draw up a consignment note, or when waste must be removed from a site that has not been notified. The Agency must be notified of the emergency. It is the waste holder's duty to take all reasonable and lawful steps to avert or mitigate grave danger.

Offences and penalties

Offences relating to notification and documentation are punishable by a fine or by a fixed penalty of £300. The Hazardous Waste Regulations were one of the first of a set of new civil sanctions introduced through the

Environmental Civil Sanctions (England) Order 2010.[31]

These are one of an increasing number of civil sanctions for minor environmental offences to be introduced following the introduction of the Regulatory, Enforcement and Sanctions Act 2008, which in turn followed Professor Richard Macrory's report on regulatory sanctions for environmental offences (considered also in Chapter 14).

The proposals introduce the following civil sanctions for regulators to use:

- A range of enforcement notices, such as a compliance notice, restoration notice and stop notice;
- Fixed monetary penalties for lesser examples of non-compliance;
- Enforcement undertakings – by which an operator could offer voluntary commitments to comply, and to fully make amends for the effects of their non-compliance. The regulator may decide to accept an undertaking instead of imposing other sanctions;
- Variable monetary penalties – enabling regulators to impose a proportionate financial penalty for moderately serious non-compliance, including removal of financial benefit from non-compliance.

The following more serious offences are punishable by the statutory maximum fine, or by a maximum of two years' imprisonment, or both:

- Mixing hazardous waste (unless allowed by a permit);
- Failure to avert an emergency, or to notify the Agency in the event of an emergency;
- Giving false information in response to a request.

Classifications

The Environment Agency has issued updated guidance on hazardous waste consignment notes. The definition of hazardous waste

comes from Directive 91/689/EEC (the Hazardous Waste Directive, or HWD). The Hazardous Waste Regulations (England and Wales) 2005[32] and List of Wastes (England) Regulations 2005[33] brought the EU definition fully into effect in England. Although domestic wastes such as batteries, DIY and cleaning products may have hazardous properties, they are not regulated as hazardous waste. Radioactive waste, which does not display any other hazards, is also excluded.

In summary, a waste is hazardous if its entry in the European Waste Catalogue (EWC) is marked with an asterisk. For many wastes (for example, hydrogen peroxide) there is only one entry and it is marked with an asterisk. These wastes are hazardous regardless of the quantity or concentration. However, for many other wastes there are two 'mirror entries' in the EWC, for example:

04 02 16★ dyestuffs and pigments containing dangerous substances

04 02 17 dyestuffs and pigments other than those mentioned in 04 02 16

The assessment procedure is summarised in the flowcharts in Figures 18.5 and 18.6. The Agency produced a comprehensive guidance document, WM2 Hazardous Waste: Interpretation of the Definition and Classification of

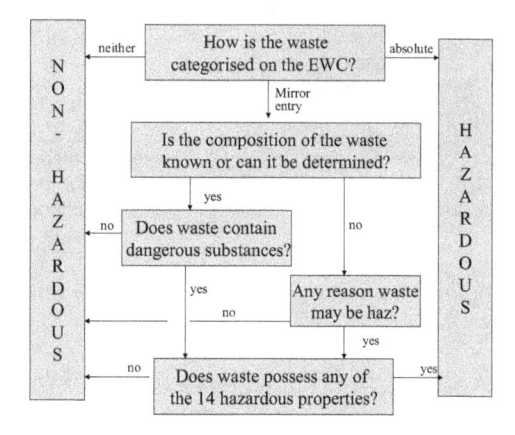

Figure 18.5 Determining whether the waste is 'hazardous'

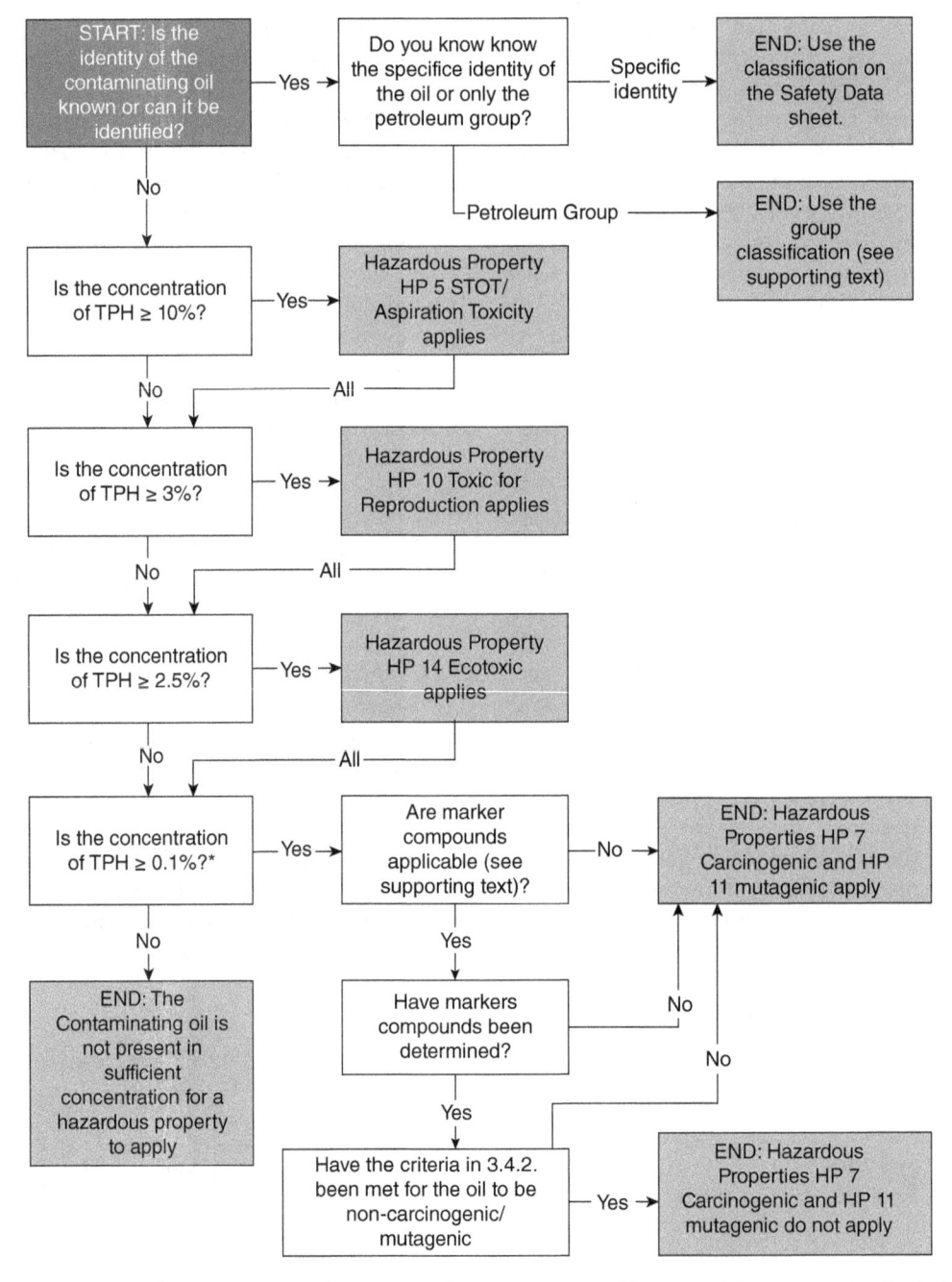

Figure 18.6 The assessment of wastes, other than waste oils, containing or contaminated with oil

Hazardous Waste (Technical Guidance WM2, 2008 edition), which in 2015 was superseded by a joint UK environmental agencies publication *Technical Guidance WM3: Waste* *Classification – guidance on the classification and assessment of waste (1st edition 2015).*[34] (See Figure 18.5.) The definition of hazardous wastes includes commonplace items such as

TVs, computer monitors, fluorescent light tubes and end-of-life vehicles.

Every business will produce at least a small quantity of hazardous waste and must be aware of their legal duties.

Hazardous wastes may only be placed in landfill at a specialist hazardous waste landfill site. Even then, it will only be accepted for landfill if it meets the Waste Acceptance Criteria established by the Landfill Directive (see earlier). Where WEEE and other hazardous items are sent for recycling and recovery, the waste producer must ensure that the storage or treatment facility is permitted to receive hazardous waste.

Dangerous Goods Regulations – those who transport hazardous waste by road and rail should be aware of the ADR (road) or RID (rail). ADR is the European Agreement concerning the International Carriage of Dangerous Goods by Road implemented in the UK by the Carriage of Dangerous Goods and Use of Transportable Pressure Equipment Regulations 2009.[35] Compliance is normally the responsibility of the carrier, but waste producers will need to classify, package and label dangerous goods correctly prior to transport. Waste oils, chemicals, clinical waste and other dangerous substances should be transported in UN-approved packaging and labelled with the relevant hazard warning diamond and UN number. For details see the dangerous goods section of the HSE website www.hse.gov.uk.

Healthcare waste

The World Health Organization (WHO), European Commission and United States Environmental Protection Agency (EPA) in the course of the last few years have all emphasised the need for healthcare waste to be safely managed from the point of production via various methods of movement to its final disposal. The concern expressed by these bodies reflects the growing awareness of the threat to public health that can arise from the improper handling and disposal of such wastes. This was emphasised during the 2014 outbreak of Ebola in West Africa when guidance from the WHO had to be improved in order to protect the health workers who cared for the sick and those responsible for burying the dead.

In the UK the management of healthcare waste is regulated under several different regimes: waste management, health and safety and carriage of dangerous goods. In February 2007 the Department of Health (DoH) (now Department of Health and Social Care and Social Care) published new guidance: *Health Technical Memorandum 07–01: Safe management of healthcare waste* and updated in 2013.[36] The term 'healthcare waste' is used to encompass a broad range of wastes including offensive hygiene wastes, which do not present a risk of infection.

Like any other controlled waste, healthcare waste is subject to the general duty of care, and is regulated under EPR. WEEE from healthcare premises is subject to the WEEE Regulations (see earlier). However, there is an exemption for infected products and active medical implants.

Healthcare wastes can be hazardous on account of their infectious nature or their chemical composition. The DoH guidance creates a unified methodology for the classification and assessment of healthcare waste. The procedure is described in detail in the guidance. In brief, the assessment methodology will place healthcare waste into one of the following categories:

Healthcare waste is presumed to be infectious until assessment has taken place. Infectious waste is hazardous waste. It should be classified for transport under ADR.

Under the Hazardous Waste Regulations 2005, hazardous property H9 (Infectious) is described as 'Substances containing viable microorganisms or their toxins which are known or reliably believed to cause disease in man or other living organisms'. Note that under the EWC entry 18 01 03★, an

Table 18.3 Categories of healthcare waste

Healthcare waste		Municipal waste
Clinical	Offensive	Offensive
Infectious		
Medicinal		
Chemical		

Table 18.4 Segregation and colour coding of healthcare waste

Colour	Description
Yellow	Requires incineration
Orange	May be treated or incinerated
Purple	Cytotoxic and cytostatic waste: to be incinerated
Yellow with black stripe	Offensive/hygiene waste. Can be landfilled
Black	Municipal waste. Remove recyclables and landfill
White	Amalgam waste: for recovery

infectious substance will only be a hazardous waste if special requirements (for example, separate storage and collection) are necessary to prevent infection. Infectious waste must be segregated from other waste streams.

Implanted medical devices, which have been contaminated by infectious bodily fluids are classified as infectious waste. Infectious waste from non-healthcare activities (e.g. drug litter, body piercing and tattooists), is assigned the code 20 01 99. This is not hazardous waste but will need to be classified for transport under ADR.

'*Offensive waste*' is non-clinical waste that is non-infectious and does not contain pharmaceutical or chemical substances but may be unpleasant to anyone who comes into contact with it. Examples include outer dressings and protective clothing (masks, gowns and gloves that are not contaminated with body fluids); hygiene waste and sanitary protection (nappies and incontinence pads); and sterilised (autoclaved) laboratory waste.

'*Medicinal waste* includes expired, unused, spilt and contaminated pharmaceutical products, drugs, vaccines and sera that are no longer required and need to be disposed of appropriately.' It also includes discarded items contaminated from use in the handling of pharmaceuticals, such as bottles or boxes, gloves, masks, tubing, syringe bodies and drug vials. Medicinal waste is subdivided into three groups.

Waste cytotoxic and cytostatic medicines must be separated from other medicinal wastes, otherwise the entire waste stream will be classed as hazardous. It may also be necessary to segregate different kinds of 'cyto' waste where different treatment and disposal routes are envisaged.

The relevant EWC codes are 18 01 08★ (human healthcare), 18 02 07★ (animal healthcare) and 20 01 03★ (separately collected municipal waste). Pharmaceutically active, but not cytotoxic and cytostatic, although not hazardous waste may possess hazardous properties, which should be identified for the purpose of the duty of care. The materials safety data sheet provided by the supplier should identify these hazards. The relevant EWC codes are 18 01 09, 18 02 08 and 20 01 32.

Healthcare wastes not pharmaceutically active and possessing no hazardous properties include saline and glucose products, for example. EWC codes are 18 01 09, 18 02 08 and 20 01 32. This waste can either be disposed of with other non-hazardous medicinal waste or, in small quantities, discharged to sewer and the containers placed in the offensive waste stream, 18 01 04.

For non-infectious medical devices the most suitable code may not be one of those given in Chapter 18 of the EWC. The description on the consignment note must adequately describe the waste and any hazardous characteristics, even if the waste is not hazardous waste. For example, 16 02 13 Discarded equipment containing hazardous components other than those mentioned in 16 02 09 to 16 02 12. Hazard: Corrosive (H8).

Table 18.5 Recommended treatment and disposal methods for healthcare waste

Packaging	Waste types	Treatment/disposal
Yellow with radioactive sticker	Healthcare waste contaminated with radioactive material	Incineration
Yellow with purple band	Infectious waste contaminated with 'cyto' products	Incineration
Sharps receptacles with purple lid	Sharps contaminated with 'cyto' products	Incineration
Yellow bag or box	Infectious waste, anatomical waste, chemical waste	Incineration
Yellow sharps receptacle	Partially discharged sharps not contaminated with 'cyto' products	Incineration
Yellow box	Medicines	Incineration. Hazardous waste incineration if not in original packaging
Orange bag or box	Infectious waste, potentially infectious waste, e.g. soiled dressings, and autoclaved laboratory waste	Treatment facility
Sharps receptacle with orange lid	Sharps not contaminated with medicinal products, or fully discharged sharps contaminated with medical products other than 'cyto'	Incineration or treatment
Yellow bag with black stripe	Offensive/hygiene waste	Landfill
Black or clear bag	Domestic (municipal) waste	Landfill
White container	Dental amalgam waste	Recovery

Assessment of chemical waste properties – healthcare waste often contains small quantities of chemicals, for example:

- Samples used for diagnosis in laboratory areas;
- Diagnostic vials or reagent kits;
- Therapeutic chemicals used for treatment or disinfection;
- Disinfectants, cleaning agents and hand-washing materials;
- Body fluids containing metabolites from medicines.

The normal procedure for assessment of hazardous waste should be carried out. Hazardous waste will be assigned the code 18 01 06★ (human) or 18 02 05★ (animal). If the waste is not hazardous, it will be assigned the code 18 01 07 or 18 02 06.

Dental amalgam is a hazardous waste, which contains mercury. All dental practices are required to install amalgam separators.

Offensive/hygiene waste does not require specialist treatment or disposal and does not need to be classified for transport under ADR. If a healthcare waste does not possess infectious, medicinal or chemical properties, it is offensive/hygiene waste.

Notes

1 WasteAid is an independent UK charity (non-profit), set up by waste management professionals to share practical and low-cost waste management know-how with communities in low-income countries.

2 https://wasteaid.org/toolkit/

3 Environment Act 2021 see legislation.gov.uk/ukpga/2021/30/contents/enacted

4 See https://www.wishforum.org.uk/wp-content/uploads/2020/05/INFO-08.pdf

5 See https://community.swana.org/HigherLogic/System/DownloadDocumentFile.ashx?DocumentFileKey=ef833fcf-7d02-6f76-c496-910881cae25e

6 SI 2002 No .1559.

7 SSI 2003 No. 208.

8 SSI 2005 No. 22.

9 SI 2005 No. 2900.

10 WSI 2006 No. 123; 6a SI 2011 No. 1988.

11 SI 2011 No. 988.

12 SI 2003 No. 63.

13 SI 1991 No. 1624.

14 SI 2012 No. 811.

15 SI 2006 No. 937.

16 SI 2012 No. 811.

17 SI 2006 No. 937.

18 SI 2005 No. 894.

19 SI 2005 No. 1806.

20 The full text of the EPR can be downloaded at www.opsi.gov.uk

21 SI2016No.1154:https://www.gov.uk/government/uploads/system/uploads/attachment_data/file/211852/pb13897-ep-core-guidance-130220.pdf

22 SI 2012 No. 360 came into force in January 2013 to implement the requirements of the Industrial Emissions Directive as well as consolidating the PPC Regulations SI 2000 No. 323.

23 SI 2008 No. 314.

24 As a comparison in 2002 Bangladesh became the first country in the world to ban thinner plastic bags altogether, after they were found to have choked local drainage systems during floods and many other developing countries have followed suit. 17 A SI 1988 No. 819.

25 xx SI 1991 No. 961.

26 SI 2002 No. 2980.

27 See http://www.wrap.org.uk/content/home-composting-guidance-and-information

28 SI 2005 No. 894.

29 The current Standard Industrial Classification (SIC) used in classifying business establishments and other statistical units by the type of economic activity in which they are engaged.

30 See https://www.gov.uk/dispose-hazardous-waste

31 SI 2010 No. 1157.

32 SI 2005 No. 894.

33 SI 2005 No. 895.

34 See https://www.gov.uk/government/publications/waste-classification-technical-guidance

35 SI 2009 No. 1348.

36 See https://www.gov.uk/government/uploads/system/uploads/attachment_data/file/167976/HTM_07-01_Final.pdf

Chapter references

[1] Poor Law Commissioners. (1842) *Report on the enquiry into the Sanitary Condition of the Labouring Population of Great Britain*, HMSO, London, 14.

[2] DoE. (1986) *Landfilling Wastes, Waste Management Paper No. 26*, TSO, London.

[3] Timpane MR. (2018) https://www.epa.gov/sites/production/files/2018-03/documents/timpane_epa_li_slides312_ll_1.pdf.

[4] WRAP. (2020) *National Household Waste Composition 2017 in Quantifying the Composition of Municipal Waste*, WRAP, London.

[5] Cooper JC. (2020) *Food Waste Policy, chapter 14 of Sustainable Food Waste Management: Resource Recovery and Treatment*, Wong J et al. (Eds.), Elsevier, New York.

[6] Defra. (2021) *UK Statistics on Waste, Government Statistical Service*. https://assets.publishing.service.gov.uk/government/uploads/system/uploads/attachment_data/file/1002246/UK_stats_on_waste_statistical_notice_July2021_accessible_FINAL.pdf (Accessed October 2021).

Further reading and sources of information

The Chartered Institute of Waste Management (professional body for waste and resource management): www. ciwm.co.uk (publisher of bi-monthly magazine *Circular*).

Environment Agency: www.gov.uk/environment-agency

The International Solid Waste Association has a Knowledge Hub, a large data bank of reports and guidance available for free through www.iswa.org

NetRegs: www.netregs.gov.uk/

Waste & Resources Action Programme (WRAP) www.wrap.org.uk

19

Contaminated land and land use

Roger Braithwaite

Introduction

This chapter is divided into five sections. The first gives the historical perspective, the second a look at the legislation, the third considers the role of land-use planning, the fourth considers sources of advice and assistance for environmental health practitioners (EHPs), and finally the fifth section briefly considers remediation techniques. The notes for each section and quotation references are all at the end of the chapter.

We live in a world dominated by 'sound bites'. These are defined as a short clip of speech used to promote the full-length piece. This made me wonder what sound bites I might use here to promote this work and encourage you to read more.

I have found many and chosen a handful from a wide range of individuals from all walks of life in the hope they may form the seed that blossoms into a rewarding career, that of caring for this essential part of our environment.

My favourite word in the English language is, without doubt, foundations. Whether they be physical, practical, educational, emotional or spiritual, everything and everyone needs sound foundations to develop and succeed. The land and the soils form the foundation

of life on Earth. As President Franklin D Roosevelt once said:

> A nation that destroys its soils destroys itself. [1]

Environmental health is necessarily an integrated discipline, on the one hand deeply complex, dealing with all the physical elements of the environment that sustain us. On the other, quite simply in that we merely need clean air, water, land and food, which nature has provided in abundance. Land is fixed in supply and location, but as with the more mobile elements of our environment, it is a 100% sustainable resource. It can go on sustaining life *ad infinitum* but it is necessary to look after it.

It is true, and in light of the 26th Conference of Parties on Climate Change (COP26), which at the time of writing is a couple of months away, what Chief Seattle is reported to have said in 1854:

> The earth does not belong to man, man belongs to the earth. All things are connected like the blood that unites us all. Man did not weave the web of life; he is merely a strand of it. Whatever he does to the web, he does to himself. [2]

DOI: 10.1201/9781003035640-19

This chapter has been prepared with the aim of providing sufficient information to enable the reader to form a basic understanding and first reference on the subject.

'Brownfield' colloquially refers to land which is not 'greenfield', and therefore believed to be contaminated in some way. Successive definitions of 'brown', 'green' and 'contaminated' have become a bone of contention over the years. Which dog wins the bone depends from which angle you are viewing the problem.

Inevitably there is a plethora of guidance available on the subject and it is the wont of many commentators to simply direct students to this. However, this only leads to confusion as there are so many volumes, a lot of it conflicting, invariably repetitive, and so complex as to put anyone completely off the subject. There is also a presumption that the most recent is the best, but experience shows that this is certainly not always the case. Some of the earlier guidance was succinct and to the point with many wise comments, probably the best of these being from a piece of Government guidance now over 40 years old:

> A choice will often have to be made from options ranging from low cost/high risk, to high cost/low risk. The consequences of failure, which may not occur for 15 to 20 years or more, and the steps which must be taken to remedy failure, should be consciously considered. The building and construction industries are only too aware that failures do occur – for a variety of reasons including lack of knowledge, failure to allow sufficiently for bad workmanship, a lack of quality control and 'Acts of God'. [3]

This is such an important statement; therefore this chapter must commence with a brief history of contaminated land in the UK. The author has worked as an Expert Witness for over 20 years, yet never purported to be an expert in any of the minutiae of the subject, only to possess a general knowledge. This has been found to be particularly valuable by those sitting in judgement in highly complex cases, in helping them form an informed view. If someone is needed to evaluate the individual parameters of an algorithm used in a unique exposure scenario, then that should never be the EHP; there will always be someone out there who can help with that.

In Section 4 on 'suitably qualified' or 'competent' persons, reference is made to the 'Land Contamination General Practitioner' or LCGP. It should be the aim of every local authority officer working on contaminated land to achieve such status.

To commence that process, a basic knowledge of the duties of the contaminated land professional over the last 50 years or so is essential, especially when dealing with the thorny issue of identifying liabilities for past wrongs.

Therefore Section 1 is a simple history of contaminated land, which began in earnest in the 1970s. Section 2 considers the rarely used legislation which was created specifically to deal with, the 'legacy of contaminated land', known simply as Part IIA. In Section 3 I seek to explain how the 'legacy of contaminated land' is in reality largely dealt with via Town & Country Planning. Throughout the guidance for both statutory processes, the UK Government are at pains to point out that all work undertaken on this broad and highly complex subject must be undertaken by 'suitably qualified' or 'competent' persons. No one individual is suitably qualified and competent in all aspects of contaminated land and land use, therefore Section 4 covers the difficult, but essential subject area of how to find the best people to help you make the right decisions. Section 5 provides a brief overview of remediation techniques.

There is a long list of laws and guidance available from Governments and NGOs in each of the devolved areas of the UK. These are being amended and updated constantly; therefore the reader should identify these appropriate to their own particular circumstances. This chapter is a starting point.

SECTION 1: CONTAMINATED LAND IN THE UK – A BRIEF HISTORY

Industrial change and demographic shift during the 20th century resulted in the need for large scale re-organisation of our towns and cities. Industries moved out or disappeared altogether leaving large, 'brownfield', gaps in our urban landscape. At the same time, changes in heating methods, and the advent of the consumer society had a significant effect on the type and volume of refuse it has been necessary to landfill. Inevitably, these changes left their mark in the form of 'alien substances', in and on the land, which in some cases, have been found to be potentially harmful.

Ever since Victorian times there has been legislative provisions to deal with contaminated land, as a statutory nuisance. The Handbook of Hygiene and Sanitary Science dated 1898, states within its 800 pages:

> Frequent limited outbreaks of disease have been traced to the pan manure mixed with ashes which is conveyed in canal boats from large towns and deposited in heaps. . . . Such heaps, if allowed to remain in close proximity to houses, are not only extremely offensive, but may give rise to diarrhoea, ill defined typhoid symptoms, ulcerated sore-throat of a diphtheritic character, and the risks are intensified if the children are allowed to play on the heaps. [4]

The Public Health Acts

The Public Health Act of 1875 defined two of the statutory nuisances at section 91 as:

> (1) Any premises (includes land) in such a state as to be a nuisance or injurious to health;
> (4) Any accumulation or deposit which is a nuisance or injurious to health.

This was, in effect, the first definition of contaminated land, which continued virtually unchanged (as a nuisance) for a further 120 years.

In 1990 the Environmental Protection Act definitions in section 79 were as follows:

> a) Any premises (includes land) to be in such a state as to be prejudicial to health or a nuisance;
> c) Fumes or gases emitted from premises (only relates to private dwellings) so as to be prejudicial to health or a nuisance;
> e) Any accumulation or deposit which is prejudicial to health or a nuisance.

The definition of *prejudicial to health* was, and still is, *injurious or likely to cause injury to health*. The addition of the word '*likely*', a seemingly minor amendment, has become a key word in environmental law allowing a judgement to be made on what may and may not be *likely* in any particular circumstances.

Hazardous waste

Despite the existence of nuisance legislation to deal with any dangerous 'accumulation or deposit', a serious incident in Nuneaton, Warwickshire in 1972 resulted in a degree of panic in Government circles. The following from Hansard:

The Secretary of State for the Environment (Mr. Peter Walker):

> Preliminary investigations suggest that between 3pm on 23rd February and 9am on 24th February, 36 drums containing sodium cyanide ash were dumped on a site forming part of a disused brick clay workings near Bermuda village, Nuneaton. The drums were found by a local resident and the police were informed. The drums were guarded while police investigations were commenced and arrangements made by the local authority with a firm of waste disposal contractors for the dumped material to be removed.

In the light of this incident that could very easily have become a terrible tragedy, the issue of responsible treatment of hazardous waste was given a high priority by Parliament, resulting only days later in the Deposit of Poisonous Waste Act 1972. This commenced a new era in waste regulation, which it should be noted has its origins in the disposal of waste, as opposed to the 'legacy of contaminated land'. For completeness I explain in brief how hazardous waste regulation then developed.

The 1972 Act was repealed in 1981 and replaced by the Control of Pollution (Special Waste) Regulations 1980 made under section 17 of the Control of Pollution Act 1974. This sought to implement European Directive 78/319/EEC on toxic and dangerous waste. This established a list of 27 substances which in certain circumstances might cause waste to be toxic and dangerous.[1]

Part I of the Control of Pollution Act 1974 was the principal legislation governing the collection and disposal of waste and introduced for the first time the concept of waste management licensing (commencing in 1976). The main objective of this part was to ensure that licensed activities (such as landfill sites) did not cause pollution of water, danger to public health or detriment to local amenities. Within a few years however, the provisions were considered inadequate. There were institutional conflicts of interest with waste disposal authorities taking part in the operations that they regulated; the standards applied in licensing varied across the country and were not strictly enforced; and licence holders could surrender their licence in their discretion and thereby shed their responsibilities for the condition of their land.

Detailed guidance for waste disposal authorities on the implementation of the Control of Pollution Act 1974 Part I was contained in several departmental Circulars, and subsequently, Waste Management Papers. All of this guidance arose from the then Department of the Environment (DOE) which was well staffed and, at the same time, working on guidance to deal specifically with contaminated land.

Early advice on contaminated land

Within the then DOE were several relevant sections including:

- Central Directorate on Environmental Pollution (CDEP);
- Interdepartmental Committee for the Redevelopment of Contaminated Land (ICRCL), and later;
- Hazardous Waste Inspectorate (HWI).

The ICRCL was set up in 1976 as a multi-disciplinary team developed to consider risks associated with the redevelopment of contaminated sites and offer advice to local authorities and others. In June 1977 DOE Circular 49/77 announced the setting up of the committee, and this was sent to all local authorities. It is important to note the word, 'redevelopment', indicating the desire to make use of despoiled land in an attempt to preserve useful agricultural land and other greenfield sites. These were, in effect, planning documents.

HWI was set up in December 1982 with the primary objective of encouraging consistent standards of waste management across the country. Terms of reference were as follows:

> To examine the management of hazardous waste at all its stages from the point of arising to that of final disposal by visiting facilities being used to handle, store, treat, process and dispose of such waste either separately, or in conjunction with other controlled waste; to advise Waste Disposal Authorities on the execution of their duties under Part I of the Control of Pollution Act 1974; to make recommendations with the object of ensuring that standards of operation, site licensing and enforcement are both adequate to protect health and the environment

and also equitable and consistent across the country; and periodically to publish a report.

Information relating to the setting up of the HWI was sent to all local authorities in the form of DOE Circular 20/83. From 1978 onwards there were many documents advising on contaminated land sent out to waste disposal authorities and local councils for consultation and guidance.[2]

Advice was also given to local authorities on a site-specific basis at no cost. This series of guidance introduced the concept of 'trigger', 'intervention' and 'action' levels, in an attempt to standardise and simplify the approach to evaluating risks associated with the redevelopment of contaminated land. These can be found (in some form) in ICRCL 18/79, ICRCL 59/83 and ICRCL 79/90. In addition, 'normal' ranges in soils can be found in ICRCL 23/79, ICRCL 42/80 and ICRCL 79/90.

The 'trigger level' guideline values were formally withdrawn in December 2002 as they were considered to be incompatible with the risk-based approach set out in the new Part IIA.

Of particular importance in relation to contamination at all coal carbonisation sites was the document *Problems Arising from the Redevelopment of Gas Works and Similar Sites* published by the then Department of the Environment (DoE) in 1981.[3] This was followed by the second and final edition in 1988 [5]. This was colloquially known as the *Yellow Book* and key to the assessment and decontamination of the many thousands of old gas and coking works that fuelled the 20th century. The fact that this book listed around 70 references demonstrates how much information and advice was available on contaminated land by this time.

In 1979 a paper *Site Investigation and Material Problems* by RT Kelly was widely circulated [6]. This paper was well known and became of national significance due to the inclusion of a table – 'Guidelines for Contaminated Soils – Suggested Range of Values', to become known throughout the industry as the 'Kelly Tables'. These were subsequently reproduced in Health & Safety Guidance

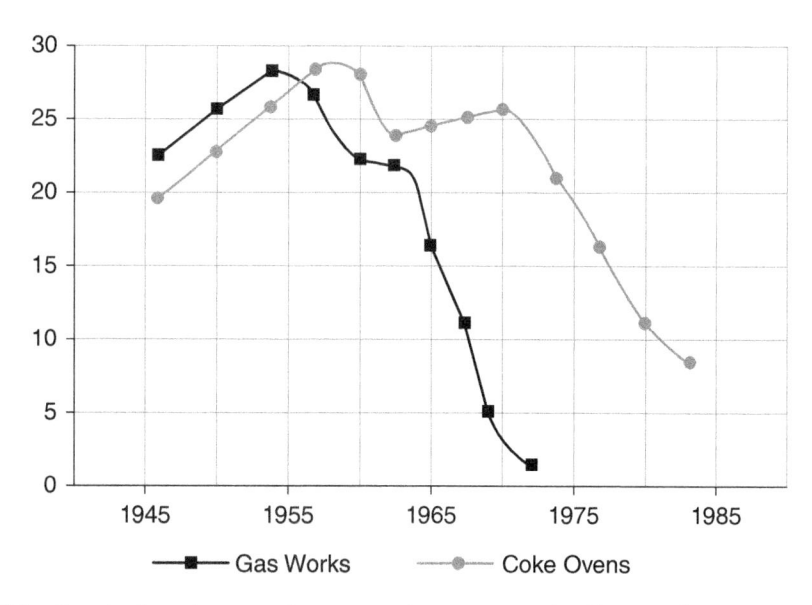

Figure 19.1 Graph of post-war coal consumption at gas and coking works
Source: [5]

Plate 19.1   Old gasworks site for the manufacture of town gas. (Following the introduction of North-Sea gas many thousands of old gasworks sites were decommissioned and subsequently redeveloped.).

[7]. The tables were derived on the basis of empirical evidence arising from the analyses of hundreds of soil samples from a range of sites across the Greater London Council area. The criteria were developed mainly to assist in the characterisation of contaminated soils for disposal purposes and whilst they were not specifically intended to be used in risk assessment criteria for assessing land that was to be redeveloped, they provided a useful indication of the severity of contamination. They also included a broader suite of metals than were included in the ICRCL guidance.

In 1988 the British Standards Institute produced DD 175: 1988 *Code of practice for the identification of potentially contaminated land and its investigation*. This was a draft in development from April 1988 but was subsequently accepted as a full British Standard without amendment in April 1992. This quickly became the accepted standard to which

industry worked, but has now been withdrawn and replaced [8].

Around this time much more guidance was published, examples of some of the most useful include:

- *Industry Profiles*, published by the DOE, there were 47 Industry Profiles describing processes and wastes in over 50 industries – these have proved invaluable;
- *Contaminated land*, Published by the World Health Organization – 1997;
- *Code of Good Agricultural Practice for the Protection of Soil*, Published by MAFF – 1998;
- *Code of Good Agricultural Practice for the Protection of Water*, Published by MAFF – 1998;
- *Guidance for the safe development of housing on land affected by contamination*, R&D Publication 66, published by the Environment Agency – 2000 Later updated in 2008 (Collaboration between the NHBC,

Chartered Institute of Environmental Health and the Environment Agency).

To put the reasons for concern about the state of soil or land into context the then Ministry of Agriculture Fisheries and Food (MAFF) said:

> Soil is a basic, limited resource that will continue to be essential for many human activities. It includes both topsoil and subsoil to depth of at least one metre. The biological, physical and chemical characteristics of soil need to be protected for it to perform its important functions, including the production of food, raw materials and energy. Soils provide a filtering and buffering action to protect water and the food chain from potential pollutants; they help maintain gene pools and wildlife populations. All soils should be managed sustainably in the long term. [9]

Similarly, the current President of the National Farmers Union argues that:

> Quite simply, soil is our greatest asset. Good soil health is central to food security, carbon sequestration, biodiversity, and key to building our resilience to events such as floods and droughts. [10]

The development of law to deal with the 'legacy of contaminated land'

Political pressure in the mid-1980s demanded purpose-made legislation in an attempt to deal with the many known contaminated and potentially dangerous sites. The Government, in its response to the 11th report of the Royal Commission on Environmental Pollution in 1985, announced that the DOE was preparing a circular on the planning aspects of contaminated land. The draft of the circular (The Development of Contaminated Land, DOE October 1986) stated that:

> Even before a (planning) application is made, informal discussions between an applicant and the local planning authority are very helpful. The possibility that the land might be contaminated may thus be brought to the attention of the applicant at this stage, and the implications explained.

This suggested that it would be advantageous for the planning authorities to have available a list of potentially contaminated sites.

In 1988 the Town & Country Planning (General Development) Order required the local planning authority to consult with the waste disposal authority (WDA) if development was proposed within 250 m of land which had been used to deposit refuse within the last 30 years. This resulted in a concerted effort on the part of WDAs and Districts to identify all old closed landfills and produce the beginnings of a contaminated land register.

In January 1990 the House of Commons Environment Committee published its first report on contaminated land.[4] This document expressed concern that the Government's *suitable for use approach* 'may be underestimating a genuine environmental problem and misdirecting effort and resources'. The committee produced 29 recommendations, including the proposals that:

> The Department of the Environment concern itself with all land which has been so contaminated as to be a potential hazard to health or the environment regardless of the use to which it is to be put, and;
>
> The Government bring forward legislation to lay on local authorities a duty to seek out and compile registers of contaminated land.

This could reasonably be considered as the official starting point of an attempt to try to legislate specifically for the legacy of contaminated land.

Immediately following the House of Commons report the Government introduced legislation which sought to deal with just one of the committee's 29 recommendations. This was in the form of a particularly ill thought-out provision of the Environmental Protection Act 1990, titled 'Public registers of land which may be contaminated' (section 143). If enacted, this would have required local authorities to maintain registers of land which was, or may have been, contaminated as a result of previous specified uses. A list of the types of uses that could require entry in the register was produced in July 1990 with the publication of a pilot study carried out in Cheshire for the Department of the Environment titled *A Methodology for Identifying Contaminated Sites* [11].

The list was subsequently reproduced in a DOE consultation paper of May 1991 called *Public registers of land which may be contaminated*. This received a mixed response, with local authorities very concerned at the resource implications of the new proposals, but more significantly, the landowners and financial institutions even more concerned at the potential reduction in land values and the resultant economic implications on a national scale. Despite this, the then head of the contaminated land branch of the DOE, in December 1991, emphasised that the introduction of section 143 registers was key to the Government's strategy on contaminated land.

The following year, the Dutch had estimated that there may be as many as 110,000 contaminated sites in Holland with an estimated clean up cost of $27 billion. The Dutch at that time were working on a multi-functional strategy, proposing to clean up all land to 'as new' condition. Figures like this set alarm bells ringing all around Europe and the world. Suddenly the UK's *suitable for use approach* was looking more attractive.[5]

In March 1992 the concern about registers began to register with the Government, and a press release was published by the Secretary of State delaying the introduction of section 143 stating:

> The Government are concerned about suggestions that land values would be

unfairly blighted because of the perception of the registers.

Subsequently in July 1992, draft regulations were released with significantly reduced categories of contaminative uses, 'to those where there is a very high probability that all land subject to those uses is contaminated unless it has been appropriately treated'. It was estimated that land covered by the registers would be only 10 to 15% of the area previously envisaged. Unfortunately, however, this still did not satisfy financial institutions, so on the 24th of March 1993 the new Secretary of State (Michael Howard) announced that the proposals for contaminated land registers were to be withdrawn and a belt and braces review of land pollution responsibilities was to be undertaken.

This resulted in the DOE consultation paper, *Paying for Our Past* in March 1994, which elicited no less than 349 responses. The outcome of this was the policy document, *Framework for Contaminated Land*, published in 1994 [12]. This useful review emphasised a number of key points:

1 The Government was committed to the 'polluter pays principle' and 'suitable for use approach'
2 Concern related to past pollution only (there were effective regimes in place to control future sources of land pollution)
3 Action should only be taken where the contamination posed actual or potential risks to health or the environment and there are affordable ways of doing so
4 The long-standing statutory nuisance powers had provided an essentially sound basis for dealing with contaminated land

It was also made clear that the Government wanted to:

- Encourage a market in contaminated land;
- Encourage its development; and

- That multi-functionality was neither sensible nor feasible.

These conclusions were significant and would ensure:

1 The UK would not identify many contaminated sites (unlike the Dutch);
2 Past polluters and developers would (or should) bare the cost;
3 There would be minimal central costs;
4 Any new legislation would be much like the old (statutory nuisance powers);
5 Most contaminated land would be dealt with by development rather than enforcement.

The new legislation received Royal Assent in July 1995, and although long and cumbersome, the simple definition of contaminated followed, almost to the word, the statutory nuisance provisions.

Part II of the Environment Act 1995 (section 57) inserted a new 'Part IIA' at the end of section 78 of the Environmental Protection Act 1990, titled 'Contaminated Land'. The new definition read as follows:

land which appears to the local authority in whose area it is situated to be in such a condition, by reason of substances in, on or under the land that –

(a) *significant harm is being caused or there is a significant possibility of such harm being caused; or*
(b) *pollution of controlled waters is being, or is likely, to be caused.*

This in effect meant that *land which is injurious, or likely to be injurious to health or a nuisance* was replaced by *land which is causing significant harm, or there is a significant possibility of causing significant harm.*

On the face of it these seem similar, and on the face of it they were. However, it was then realised that instead of the courts deciding which land might fall within the definitions, the

Government was attempting to do it. The end of section 78A(2) included the following statement:

and, in determining whether any land appears to be such land, a local authority shall, subject to subsection (5) below, act in accordance with guidance issued by the Secretary of State in accordance with section 78YA below with respect to the manner in which that determination is to be made.

Much of the plethora of guidance has resulted from this perhaps noble, but potentially ill-fated attempt to consider what may be considered *harm*, *significant harm* and *significant possibility*, in each and every case. The complexity of the task should not be underestimated and those who understood the subject knew from the outset it would always be impossible. It took several years for Statutory Guidance and secondary legislation (Regulations) to appear, which, it could be argued, resulted in more questions than answers.

At this point it is worth considering some non-statutory definitions of the word 'contaminated' as these would have been rigorously debated in the event of the matter being left to the courts to decide. The term 'contamination' it could be said has been commonly confused or used interchangeably with the word 'pollution'. Although several interpretations of these terms exist, the definition given by Holdgate in 1979 was widely accepted. This states that 'pollution' is:

The introduction by man into the environment of substances or energy liable to cause hazards to human health, harm to living resources and ecological systems, damage to structures or amenity, or interference with legitimate uses of the environment. [13]

Other definitions use the term 'contamination' where the anthropogenic inputs do not appear to cause obvious harmful effects.

'Pollution' is applied only where toxicity has occurred, but this raises the question, at what point does toxicity occur?

The Royal Commission on Environmental Pollution in 1984 drew the following distinction between the two terms:

> 'Pollution' can be defined as the introduction by man into his environment of substances or energy liable to cause hazards to human health, harm to living resources and ecological systems damage to structures or amenity, or interference with legitimate uses of the environment. Substances introduced into the environment become pollutants only when their distribution, concentration or physical behaviour are such as to have undesirable or deleterious consequences.
>
> For comparison 'contamination' can be defined as the introduction or presence in the environment of alien substances or energy, on which we do not wish or are unable to pass judgement on whether they cause, or are liable to cause, damage or harm. 'Contamination' is therefore a necessary, but not sufficient condition for 'pollution'. [14]

Becket and Simms only considered contaminated land in the context of its safe development:

> Contaminated land is that land which, because of its former uses, now contains substances that give rise to principal hazards likely to affect the proposed form of development and requires an assessment to decide whether the chosen development may proceed safely, or whether it requires some form of remedial action, which may include changing the layout and form of the development. [15]

Becket was the former secretary to the Inter-departmental Committee on Contaminated Land (ICRCL) and formative on Government thinking at that time.

Land – is it Green or Brown? A significant proportion of all land is contaminated with something, rarely will it be pristine and unaffected by anthropogenic inputs. It has generally been believed that a perfectly natural *greenfield* site should, if possible, be left to agriculture and/or to sustain biodiversity. This results in two other types of site which hinge around the previous definitions, and both of which can be considered *brownfield*. The first contaminated with things from previous uses, but not causing any of the *harms* specified in Part IIA. The second likewise, but causing *harm* as defined. Let us call these Brownfield Sites 1 and 2:

- Brownfield Sites 1 – These are contaminated land but with a small 'c'. They subsequently become to be known as *land affected by contamination*. They can be large and badly contaminated industrial sites, but a judgement has been made that they are not causing *harm* to any person, any thing or any water (so defined).
- Brownfield Sites 2 – These are contaminated land with a capital 'C'. It is not necessary for them to be large or badly contaminated, but if they are affecting the health of an individual, contaminating groundwater or causing any other 'harm' to a defined receptor (again as defined), then they fall within the legal definition of *Contaminated Land*.

For future reference, and in an attempt to minimise confusion, the legal terms *land affected by contamination* for Sites 1, and *Contaminated Land* for Sites 2 will be used.

Since it became clear to the Government that land is a finite resource there has been an ever-increasing pressure to develop both of these types of sites. The former (*land affected by contamination*) form the vast majority of contaminated sites in the UK. An example would be a derelict industrial site where all the contamination lies undisturbed below the surface, no crops being grown, no access to people or animals, the contamination not affecting ground or surface water. It is not, therefore *Contaminated Land*, but as soon as the site is developed three things potentially occur:

- Receptors are introduced;
- New pathways are created;
- Contaminants are mobilised.

This is the way the most *land affected by contamination* is dealt with, not just in the UK, but throughout the world. This confirms that the *suitable for use approach* is the only financially sustainable option. The land was suitable for the use of a vacant/derelict site; if you choose to change the use, then you have to make it suitable for that new use. Put simply this means that the Development Control Regime, through the Town & Country Planning legislation in the UK, must not create *Contaminated Land*.

Development control

As was shown in Chapter 1 and previously in this chapter, public health law was a product of the 19th century, but it did not deal with the factory built in the garden suburb, or workers houses next to an array of offensive trades.

In contrast, planning law was a product of the 20th century and commenced with the Housing, Town Planning etc. Act 1909. Part II of that Act dealt with Town Planning and was the first enactment ever to introduce the concept of 'development control'.

During the post-war period there was prospect of large building programmes throughout the UK, so to help direct this effectively the Town & Country Planning Act 1947 was introduced on the 1st July 1948. This, *inter alia*, prohibited development without the consent of the local planning authority.

The Mineral Workings Act 1951 was enacted to establish a fund for the purpose of financing the restoration of worked-out ironstone land. The fund was under the control of the Secretary of State and operators were required to pay an annual contribution calculated according to the tonnage of ironstone worked in the year. The Secretary of State also made a contribution. Payments were then made out of the fund to restore fertility to the land.

Prior to the 1970s then, the local planning authority would have sought to ensure polluting industries such as gas works, steel works, etc were properly located, and regulated to a degree, by the use of planning conditions. The same would apply to minerals extraction and landfill sites which had to apply for planning permission before they could operate. It is fair to say, however, controls may have been minimal at that time. The Town & Country Planning (Minerals) Act 1981 amended the Town & Country Planning Act 1971 with respect to the winning and working of minerals and the reinstatement of land after mineral working had ceased.

In 1986 a special regime was introduced to control the storage (not disposal) of hazardous substances on land. The introduction of hazardous substances to a site alone did not constitute 'development', therefore traditionally avoided control. A new part inserted into the Town & Country Planning Act 1971 by the Housing and Planning Act 1986, rectified this. This eventually was taken out of the principal Act in the form of the Hazardous Substances Act 1990. As time went on the results of case law and public enquiries contributed more and more to the planning process and a series of Circulars arose from various Government departments.[6]

In determining an application for development of a contaminated site there were two primary 'material' planning considerations to be taken into account:

- Contamination or the potential for it and its effect;
- The stability of the ground in so far as it affects the land use.

Government advice was until 1994 contained in circular 21/87, The Development of Contaminated Land, and Circular 17/89, Landfill Sites and Development Control. The advice in these was then consolidated into planning policy guidance note PPG23 – Planning and Pollution Control. Stability was dealt with PPG 14, Development of Unstable Land.

PPG 23 – Planning and Pollution Control

This guidance was completed at the same time as the *Framework for Contaminated Land* [12] and therefore, not unexpectedly, used much of the same language. Part 4 and Annex 10 of this guidance is of particular relevance here.

Great emphasis was again placed on the need to deal with any risks to health or the environment, taking into account its actual or intended use, though this was not to preclude the local authority from requiring very thorough remedial works where the circumstances justified. It also pointed out that very few sites were so badly contaminated that they could not be re-used at all.

The aims of the *suitable for use approach* were three-fold:

- Deal with actual or perceived threats to the health, safety or the environment;
- Keep or bring back such land into beneficial use; and so
- Minimise avoidable pressures on *greenfield* sites.

As contamination, or the potential for it, was (and still is) a 'material' planning consideration, local authorities should consider:

1   Whether there is or may be a contamination hazard on the site
2   What information they would need to decide that question
3   Whether the proposed use would give rise to unacceptable risks to health or the environment
4   If so, what steps should be taken to reduce those risks

It was emphasised that responsibility clearly lay with the developer to provide information on whether a site may be contaminated.

Paragraph 14 of Annex 10 also stated that where the presence of contamination was known or suspected, a separate notice should be issued to the applicant stating that the responsibility for the safe development and secure occupancy of the site rests with the developer. It should also warn the applicant that the local authority has determined the application on the basis of the information available to it.

Investigations into the proposed use and likely impact of the development on surrounding land and the environment, would need to be carried out before the application was decided. Alternatively, it may have been appropriate to make a detailed investigation a condition of planning permission, or a reserved matter following granting of outline permission.

The advice on when comprehensive site information should be required prior to granting permission, or after, was detailed in paragraphs 6 to 10 of Annex 10. Generally speaking, where it was known, or strongly suspected that a site was contaminated, a comprehensive report would be required with the application. This should include:

1   A phased site investigation
2   Hazard identification
3   Site-specific risk assessment
4   Development options and proposals
5   A reclamation programme
6   A working plan

This work has always been costly; therefore developers must instruct their consultants to conduct a phased investigation to enable them to withdraw at an early stage if the development was not feasible. Investigation should always be undertaken to a recognised standard, such as that detailed in BSI Code of Practice DD175, *The Identification of Potentially Contaminated Land and its Investigation* [8].

Development options and the reclamation programme should be justified on sound technical grounds. If these requirements were met, not withstanding any other material planning considerations, the EHP would have no objection to the development, and planning permission may be granted subject to planning conditions, or a planning obligation (section 106 agreement, Town & Country Planning Act 1990).

Item 6 on the previous list, the 'working plan', should be submitted prior to the development taking place. This was to identify how the reclamation would be managed, quality procedures and safe systems of work in accordance with Health & Safety Executive Guidance Note HS(G)66 *Protection of Workers & the General Public During the Development of Contaminated Land* [7].

Site evaluation and risk assessment

Very little was mentioned in PPG 23 about risk assessment, though paragraph 12 of Annex 10 did emphasise that the assessment of the significance of the contamination and the associated risks requires careful professional judgement.

The Government's strategy hinged on the *suitable for use approach* and the need for the careful evaluation of risks to known targets and cost-effective solutions. As a result there has been considerable investment in the development of both compliance cost and risk assessment procedures.

In November 1996 the Government consulted on draft regulations, preliminary compliance cost assessment and preliminary risk assessment. All of which was derived from the September consultations. The draft Regulations dealt with:

- What are Special Sites;
- How to write remediation notices;
- How to appeal against remediation notices;
- Compensation to other affected persons (grantors); and
- What goes in registers.

The Oxford English Dictionary talks of risk as 'exposure to mischance or peril'; of assessment as 'the action of assessing (quantifying) the amount'. So as has been discussed in the Annex to Chapter 1, risk assessment is a process of quantifying how likely we are to be exposed to mischance or peril and effects of that exposure. We are all very familiar with the concept of risk from childhood and learn to modify our behaviour according to our experience of risk exposure although perception may not match an objective assessment.

The word risk is probably derived from the Greek word, 'rhiza', the hazard of sailing too near to the cliffs. The idea of risk management has been well understood by ancient Greeks and Romans, who identified many common hazards and potentially effective ways of minimising their capacity to cause harm. The Roman writer Vitruvius observed well before the birth of Christ that workers exposed to the fumes of molten lead suffered disorders of the blood and concluded that water should not be carried in lead pipes. It was not until the mid-17th century, however, that Blaise Pascal laid the foundations of the theory of probability and scientists began to take responsibility from the soothsayers. This was fortunate in that the scientists' technical approach was thorough, but rather remote from the intuitive concept we all come to develop through our lives. Pascal turned the subject into a complex science of probability of the likelihood that something unpleasant may happen.

More specifically in calculating toxic risks, Rodericks defined risk as:

> The likelihood, or probability, that the toxic properties of a chemical will be produced in populations of individuals under their actual conditions of exposure. [16]

He goes on then to describe the types of information required to evaluate the risk of toxicity occurring:

1   The types of toxicity the chemical can produce, its targets and the forms of injury they incur

2 The conditions of exposure (dose and duration) under which the chemical's toxicity can be produced

3 The conditions (dose, timing and duration) under which the population of people whose risk is being evaluated is or could be exposed to the chemical

It is not sufficient to understand any one or two of these; no useful statement about risk can be made unless all three are understood, and these factors form the basis of the whole of Rodericks's writing.

The BMA (1990) defined risk evaluation (qualitative assessments), as:

> Relying on social and political judgements to determine the importance of hazards and the risk of harm, from the point of view of the individual and the community. This aspect of risk assessment includes the perception of risk, and the trading of perceived risks and benefits.

And risk estimation (quantitative assessments) as:

> Relying on scientific activity and judgement. Statistically significant numbers of previous incidents can be used to predict both the magnitude and the likelihood (the risk) of harmful events in the future. [17]

This is interesting because the system that set the wheels of the *Contaminated Land* merry-go-round in motion in the UK, was strictly *risk evaluation*, with particular emphasis on the political element. When looking at *risk estimation*, however, it becomes apparent that statistically significant numbers of previous incidents are in short supply. It is a fact that people are rarely killed, or harmed in any significant way, as a result of *Contaminated Land*. That is not to say that there have not been cases of ill-health as the result of land contamination.[7]

Cairney emphasises that:

> Reporting and interpretation is arguably the most important part of any site investigation, since it represents the culmination of all the work undertaken and will probably receive the most attention. . . . Execution of the site investigation on a logical and rational basis will enable the maximum value to be drawn from the information obtained. This will lead to economy in the expenditure of resources and confidence in the end result. [18]

These are the words of a practical person, the logical and rational leading to economy and confidence. EHPs are trained to be practical and pragmatic when applying science. There are always many cases requiring judgement relating to a range of risks, and these have to be evaluated quickly, rather than quantified with exacting precision. This way deadlines are met and costs remain manageable.

In a nutshell, risk is about the link between hazards and harm (Figure 19.2).

In 1996 this became formally known as the *pollutant linkage*. These three things rarely exist together, but in the few instances where they do, in 90% of cases it will be obvious that there is going to be no *significant risk of harm*, as defined (or pollution of controlled waters).

Section conclusion

In this section an attempt has been made to demonstrate the thought processes behind the development of legislation to deal with the 'legacy of *Contaminated Land*'. While the industrial revolution was a creative process and progress romped along at a pace, it was also careless, leaving a mess behind in its wake.

HAZARD >>>>>>>>>>>> **RECEPTOR**
Pathway

Figure 19.2 Pollutant Linkage

When this was realised, two forms of *prevention* were necessary to stop this continuing:

1 Legislation to control what people did with wastes in the future
2 Legislation to separate people in their homes from new polluting industry

Then in the UK legislation was necessary to deal with the 'legacy' and:

- The *suitable for use* approach would form the mainstay of the Government's new policy on *Contaminated Land*.
- *Land affected by contamination* would be left alone, if identified as doing no *harm* or, if the *harm* not considered *significant* (then considered *suitable for* existing *use*).
- Development of this category of *brownfield* land would be actively encouraged in an attempt to protect *greenfield* sites.
- *Land affected by contamination* would not be left if identified as doing, or *likely* to do, *significant harm* to one of a series of specified receptors. The land then becoming formally defined as, *Contaminated Land*.
- Action required would then be merely to break the *pollutant linkage* (i.e., it may not be necessary to do anything with the contamination or polluting substances).

As the Prince of Wales said in Venice in 2009:

> The industrial base which underpinned success in the Twentieth Century cannot transfer itself, un-adapted, to the Twenty-First Century . . . dealing with our industrial legacy in a way that enhances both our natural environment and makes vibrant, healthy new communities, is a global problem. Regeneration of brownfield sites in particular is an area of great interest in both our countries. In my own country over sixty per cent of all new development happens on brownfield or formerly used sites, many

of them former industrial sites such as the ones you are discussing today. [19]

Section 1 Notes

1 At the same time the issue of contaminated land in the USA was raised to a national issue by the Love Canal case which led to the Superfund and the Comprehensive Environmental Response, Compensation and Liability Act of 1980. See https://www.epa.gov/superfund/superfund-cercla-overview and https://www.nationalgeographic.org/article/superfund/

2 Examples are given here: ICRCL 3/78 Redevelopment of Contaminated Land – general notes, October 1979; ICRCL 16/78(6) Redevelopment of Contaminated Land: Acceptable levels of contaminants in soils (consultation paper), October 1979; ICRCL 17/78, Redevelopment of Contaminated Land: Notes on the development of landfill sites, 3rd edition December 1978, 4th edition June 1979, 5th edition December 1983, 8th (and final) edition December 1990. ICRCL 18/79, Redevelopment of Contaminated Land: Gas works sites, 3rd edition January 1979, 5th (and final) edition April 1986. ICRCL 19/79 – Progress Report of the Interdepartmental Committee on the Redevelopment of Contaminated Land, 1979, October 1979. ICRCL 23/79, Notes on the re-development of sewage works and farms, 1st edition October 1979, 2nd (and final) edition November 1983. ICRCL 38/80 Redevelopment of Contaminated Land: Tentative guidelines for acceptable levels of selected elements in soils, March 1980. ICRCL 42/80, Notes on the re-development of scrap yards and similar sites, 1st edition October 1980, 2nd (and final) edition October 1983. ICRCL 59/83, Guidance on the assessment and redevelopment of contaminated land, 2nd (and final) edition July 1987. ICRCL 61/84, Notes on the fire hazards of contaminated land, 2nd (and final) edition July 1986. ICRCL 64/85, Asbestos on contaminated sites, 2nd (and final) edition October 1990. ICRCL 70/90, Notes on restoration and aftercare of metalliferous mining sites for pasture and grazing, 1st (and final) edition February 1990.

3 The authors were DC Wilson and C Stevens of AERE Harwell Laboratory, Environmental and Medical Sciences Division.

4 House of Commons Environment Select Committee First Report on Contaminated Land. HoC paper 170. HMSO 1990.

5 See Swartjes FA, et al. (2012) State of the art of contaminated site management in The Netherlands: Policy framework and risk assessment tools. *Science of the Total Environment*, 427–428: 1–10; also Bauw E. (1996) Liability for contaminated land: Lessons From the Dutch experience. *Netherlands International Law Review*, 43(2): 127–141. https://doi.org/10.1017/S0165070X00004903.

6 Following is a selection published by the DOE over approximately 20 years to demonstrate how the sector developed. Science of the Total Environment, Circular 26/71 Report of the working party on refuse disposal Circular 35/71 Alkali &c. Works Regulation Act 1906: Alkali &c. Works Order 1971 Circular 60/71 Town & Country Planning (Minerals) Regulations 1971 Circular 1/72 Development Involving Use or Storage in Bulk of Hazardous Material. Circular 70/72 Deposit of Poisonous Waste Act 1972. Circular 118/72 Third Report of the Commission for Environmental Pollution. Circular 55/73 First Report of the Standing Committee on Research into Refuse Collection, Storage and Disposal. Circular 97/74 Development Control & Hazardous Uses. Circular 1/76 Control of Pollution Act 1974 – Part I (Waste on Land) Commencement Order number 4. Circular 49/77 Redevelopment of contaminated land (announced the setting up of the Interdepartmental Committee on the Redevelopment of Contaminated Land). Circular 79/77 Control of Pollution Act 1974 – Part I (Waste on Land) Licensing of Waste Disposal (Amendment) Regulations 1977. Circular 1/78 Report of the Committee on Planning Control over Mineral Working. Circular 28/78 Control of Pollution Act 1974, Part I: Waste on Land. Circular 14/79 Town & Country Planning (Minerals) Regulations 1971: Time Limited Planning Permissions.

Circular 4/81 Control of Pollution Act 1974 – Control of Pollution (Special Waste) Regulations 1980. Circular 1/82 Town & Country Planning (Minerals) Act 1981. Circular 4/82 Directive on the protection of groundwater against pollution caused by certain dangerous substances (80/68/EEC). Circular 26/82 Hazardous Substances. Circular 20/83 Hazardous Waste Inspectorate. Circular 9/84 Planning Controls over Hazardous Development. Circular 17/84 Water and the environment – The implementation of Part II of the Control of Pollution Act 1974. Circular 13/85 Discharges to the water environment: Public registers. Circular 18/85 Water and the environment. Circular 25/85 Reclamation and Re-use of Derelict Land. Circular 28/85 Mineral Workings – Legal Aspects Relating to Restoration of Sites with a High Water Table. Circular 11/86 Town & Country Planning (Minerals) Act 1981. Circular 21/87 Development of Contaminated Land. Circular 15/88 Environmental Assessment. Circular 7/89 Water and the environment. Circular 17/89 Landfill Sites: Development Control. Circular 20/90 EC Directive on the protection of groundwater against pollution caused by certain dangerous substances (80/68/EEC): Classification of listed substances. These Circulars and Waste Management Papers were later followed by Planning Policy Guidance Notes (PPGs) and Minerals Policy Guidance Notes (MPGs).

7 See https://www.epa.gov/report-environment/human-exposure-and-health on experiences in the USA and also UWE. (2013) Science for environment report, *In-Depth Report Soil contamination: Impacts on Human Health*, for European Commission DG Environment. https://ec.europa.eu/environment/integration/research/newsalert/pdf/IR5_en.pdf and also see http://ec.europa.eu/science-environment-policy

SECTION 2: UNDERSTANDING PART IIA ENVIRONMENTAL PROTECTION ACT 1990

Rachel Carson, author of *Silent Spring*, a book that that inspired many environmentalists, wrote in 1953:

> The real wealth of the Nation lies in the resources of the soil, water, forests, minerals, and wildlife. To utilise them for present needs while ensuring their preservation for future generations requires a delicately balanced and continuing program, based on the most extensive research. Their administration is not properly, and cannot be, a matter of politics. [20]

I mention in the 'history' section previously that legislation was developed with the intention of creating a new regime for the identification and remediation of *Contaminated Land*. Section 57 of the Environment Act 1995 inserted a new Part IIA at the end of Part II of the Environmental Protection Act 1990. This was a substantial piece of new legislation (sections 78A to 78YB comprising 33 pages). It is still current and still difficult to understand. The wording is unforgiving, and regularly refers to Statutory Guidance to be issued by the Secretary of State (sections 78A(2) and (5), 78B(2), 78F(6) and (7)).

Statutory Guidance was not published until the end of March 2000 (within DETR Circular 02/2000) so Part IIA did not come into force until the 1st of April 2000. At the same time secondary legislation was issued in the form of the *Contaminated Land* (England) Regulations 2000.[8]

The Regulations provided for procedural matters of the regime such as description of Special Sites, public registers, remediation and appeals. Six years later they were replaced by the *Contaminated Land* (England) Regulations 2006,[9] which added radioactivity to the list of 'harms' that could result in land being declared *Contaminated Land*, and a Special Site (Scotland has its own Regulations and Guidance). They also changed the appellant authority to the Secretary of State from the Magistrates' Court. In August 2006 the Statutory Guidance was updated to address the changes resulting from the new Regulations (within DEFRA Circular 01/2006). The *Contaminated Land* (England) (Amendment) Regulations 2012[10] considered (mainly) the impact of *Contaminated Land* on Controlled Waters.

Finally in April 2012 the Statutory Guidance was revised again with the aim of simplifying and to a degree, toning down. This quote from the Ministerial Forward:

> It has been refined in order to give greater clarity to regulators as to how to decide when land is not actually contaminated land. It is shorter, simpler and more focused towards achieving optimum results in terms of dealing with sites most in need of remediation. . . . To enable

local authorities to take a more targeted approach which remains precautionary rather than a blanket approach which is over-cautious.

The 2012 Guidance did not include radioactivity; this was subject to a separate document published the same year, and later updated in 2018. The two guidance documents together comprise 142 pages, while the 2006 version was 110 pages.

One of the omissions from the supposed 'shorter and simpler' 2012 Guidance was a glossary of terms. A small selection of definitions is included at the end of this section. Readers should always refer also to the current Guidance and the legislation itself when necessary.

It was the Government's intention in 2012 to achieve savings to business through more land being cleaned up more quickly.

A few definitions were changed and, interestingly, contaminant linkage replaced with pollutant linkage. It is for this reason these terms were discussed in Section 1. The Guidance does make clear, however, that a lot of the earlier and extant guidance still uses the term pollutant linkage.

Included in this section are matters which should have been dealt with at the outset, such as strategy formulation, as inspection strategies must be reviewed every few years. As time goes by this onerous task will inevitably be left to new officers.

The Government's aim was to provide an improved system for the identification and remediation of *Contaminated Land* within the scope of the legislation. The regime was designed to reflect the approaches already in place relating to statutory nuisance (Environmental Protection Act 1990 Part III) and pollution of controlled waters (Water Resources Act 1990 section 104). The original objectives were:

- To improve focus and transparency of controls ensuring authorities take a strategic approach to land contamination (this relates to the *Contaminated Land* strategy that all local authorities must produce, publish, review and maintain);

- To enable all problems resulting from contamination to be handled as part of the same process; previously separate regulatory action was needed to protect human health and the water environment (this emphasises that only one enforcement agency, local authorities, are empowered to declare land as *Contaminated Land*);
- To increase the consistency of approach taken by different authorities (given that local authorities and the Environment Agency, have all had differing views, levels of expertise, and attitudes to enforcement, relating to soil and water contamination); and
- To provide a more tailored regulatory mechanism, including liability rules, better able to reflect the complexity and range of circumstances found on individual sites (as sites may have had a series of owners/tenants therefore liability for past contamination is rarely clear).

The regime has five main stages:

1   Strategy formulation
2   Inspection
3   Investigation
4   Declaration
5   Enforcement

Strategy formulation

Section 78B (1) requires local authorities to inspect their areas from time to time to identify *Contaminated Land*. Environmental health staff will be familiar with this terminology as it has been used in relation to statutory nuisances for over a hundred years. Tours of the district to seek out and remedy nuisances have been a luxury not usually afforded to local authorities whatever the legislation might imply. The significant difference in this enactment follows in paragraph (2) in that the authority must act in accordance with guidance issued by the Secretary of State. This demanded a new approach in an

attempt to satisfy the need for 'transparency and focus', including a requirement to produce, formally adopt and publish, a written strategy within 15 months of implementation (by June 2001).

The strategy must:

- Be rational, ordered and efficient;
- Be proportionate to the seriousness of any actual or potential risk;
- Seek to ensure the most pressing and serious problems are identified first;
- Ensure that resources are concentrated on investigating areas where *Contaminated Land* is most likely to be found; and
- Ensure that the local authority efficiently identifies requirements for the detailed inspection of particular areas of land.

The document must also 'merit individual inspection'. There are no statutory provisions for formal approval of the strategy, but it must consider liaison with other bodies including the Environment Agency, who must be consulted and receive a copy following publication. The authority must also keep the document under periodic review.

Whilst the strategic approach was seemingly a reasonable requirement, when local authorities began to write their strategies, they found some policies rather difficult to formulate.

Strategies must relate to local circumstances – This means no 'off the shelf' strategies, and authors must be familiar with the geology, hydrogeology and hydrology of the area. This knowledge should extend to the industrial past, potential sources of contamination including natural contamination and background levels, location of all potential specified receptors, the quality of remediation already carried out on previously developed sites, the likelihood of significant pollutant linkages being identified, and their location.

Each authority must explain its aims and objectives, how the characteristics of the area have influenced the strategy and propose time scales for inspection and investigation.

Authorities must carry out extensive consultation – While producing the strategy the author must consult with the Environment Agency, County Council (where appropriate), neighbouring authorities, statutory regeneration organisations, English Nature, English Heritage, the relevant Government department (Department for Environment Food and Rural Affairs) and any other relevant bodies who may have information regarding the possibility of significant pollutant linkages.

Detailed arrangements and procedures must be included for (*inter alia*):

- Land which the Council owns (or did own);
- Obtaining information on pollutant linkages;
- Evaluating information;
- Evaluating risk;
- Liaison;
- Dealing with representations and complaints;
- Planning and reviewing inspection programmes;
- Site investigations;
- Reviewing and updating assumptions and managing new information;
- Data handling.

At the time of writing the original strategies it quickly became evident that writers of strategies must not only be very knowledgeable about the general subject of land (affected by contamination), the legislation and the procedures, but must also have detailed knowledge of the area in question. Some of the more difficult decisions could be illustrated by some practical examples.

Property is built on known contaminated sites but levels of remediation are unknown – Every local authority has properties built on sites known to be of a type normally associated with contamination. Common examples are sewage works, scrap yards, wood yards, landfill sites, town gas works, etc. In many cases records relating to how the site was prepared prior to development were either absent or incomplete. In such circumstances the authority should consider how it would

approach these sensitive cases. After the usual desktop procedures have been exhausted, investigations should commence on site.

Property built on known contaminated sites and levels of remediation known to be unsatisfactory – Many thousands of contaminated sites were developed with the knowledge and blessing of local authorities. Controls may have been imposed via the town and country planning development control system. With the publication of new guidance, these controls were often found to be inadequate.

Liaison with the Agency – Formal arrangements had to be made with the Environment Agency to enquire, in every potential case of *Contaminated Land*, whether the contaminants would be likely to form a risk to water resources. This placed an enormous burden on local Agency offices and councils were advised to work together with the Agency to ensure officers were not overwhelmed with requests.

On the question of who investigates potential Special Sites, only a local authority can declare land statutorily contaminated, but authorities were, and still are, expected to authorise Agency officers to undertake investigations on their behalf on defence and other sites which would be Special Sites if found to be significantly contaminated. This can result in inevitable difficulties when seeking to ensure consistency and a quality assured approach.

Hazard identification and risk evaluation – What if local authorities get it wrong? There will always be considerable pressure on local authority officers not to make mistakes, as a result there will be a desire to undertake investigations to the letter in accordance with published guidance. If these requirements are stated in the strategy, and subsequently not followed due to the cost implications, investigating officers will be laying themselves open to criticism in any future litigation.

Initially statistical risk evaluation was difficult but is not much better now. Site-specific risk assessment may have been a noble aim but can be very expensive and requires someone, at the end of the process, to make the difficult final decision.

Data handling – Imagine a housing estate built on a potentially contaminated site, no records of remediation. Following the strategic inspection of the district the site is given priority for a detailed inspection into the potential for significant harm arising from the occupation of the site. The site is identified on a database and GIS system which is available to the public with the rest of the district wide survey. A potential purchaser of a house pulls out because of this information that was provided to them by their solicitor. The householder is likely to complain that his or her property has been blighted because of data held by the local authority. So how data is handled, used and stored must be carefully considered in the strategy.

How to deal with complaints – In certain circumstances it may be in the interests of owners and/or occupiers to have the land which their property occupies declared *Contaminated Land*, and an individual or company found liable. This will apply particularly where sites become devalued as a result of actual or perceived risks. Conversely it will be in the interests of many to have their land taken off the 'at risk register' to encourage the market. These issues will always be difficult to resolve and require extensive historical knowledge.

Appointment of 'suitable persons' – Throughout *Contaminated Land* guidance there are references to suitable, suitably qualified or competent persons. Councils have specific powers to authorise 'suitable persons' to undertake inspections and investigations of land (Environment Act 1995 section 108). The question is who decides who is suitable and who is not? The tasks are onerous and the responsibilities great. Specifications need to be tight and how contractors are chosen and quality assurance dealt with must also be considered in the strategy.

This is a very important but much neglected area, so a separate section in the chapter (Section 4) deals with getting the right help.

Inspection of the district

Following publication of the formal *Contaminated Land* Strategy each local authority must survey its district to identify potentially contaminated sites which merit further individual inspection. An ordered and efficient method of prioritisation must be employed. This is known in the Guidance as the 'strategic inspection'.

The manner of the initial survey should have been clearly dictated by the strategy. Surveying the whole of a district is a long and time-consuming task, with the results then digitised and held on geographical information systems. Upon completion, the first on-site investigations would commence to decide whether pollutant linkages exist.

The local authority must consult fully with all parties who may have information on the condition of the land and the potential for impact on known relevant receptors. One of the principal aims of the Act has been to protect water resources. In the UK a significant proportion of potable water is drawn from ground sources. Each time an aquifer becomes contaminated there is potential for it become unusable. Local authorities and the Environment Agency must work closely together to identify land that has the potential to impact on relevant waters.[11]

Information held by past or present owners, potential developers and other local authorities (where contamination may cross boundaries) must also be sought in an attempt to minimise the impact of the investigation process on the surrounding community. At some stage, however, it will be clear that the land is contaminated to some degree, there are clearly relevant receptors, and they have the means of coming together to form a contaminant linkage. An on-site investigation of some sort must then take place to establish whether one or more linkages really do exist, and if so, whether they may result in significant pollution of a controlled water or there is a significant possibility that they may cause harm.

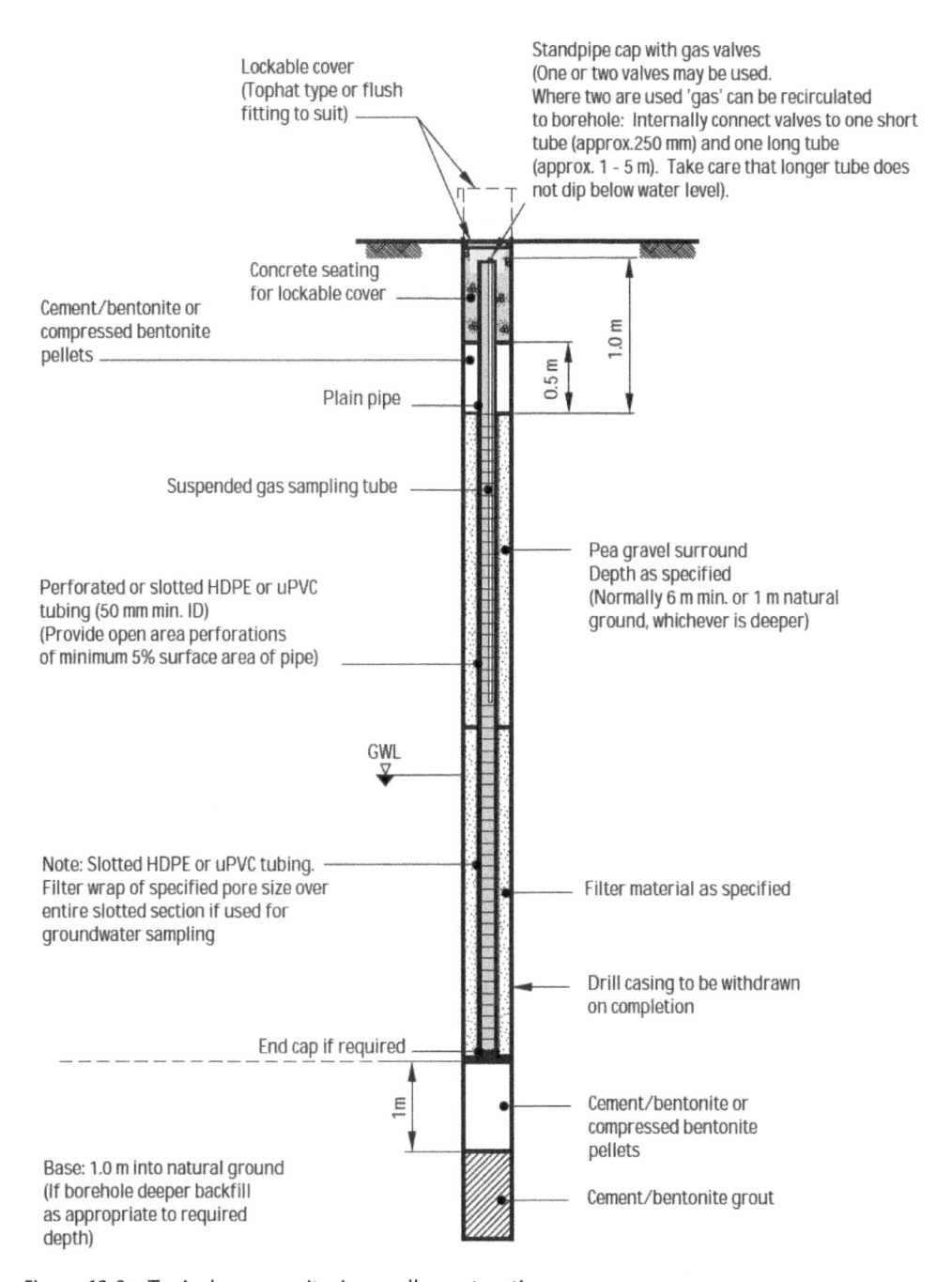

Figure 19.3 Typical gas monitoring well construction

Source: From Figure C1 in BS 10175:2001 [8], which has now been replaced by BS 10175:2011 + A2:2017 *Investigation of potentially contaminated sites – Code of Practice*. Permission to reproduce extracts from British Standards is granted by BSI Standards Limited (BSI). No other use of this material is permitted. British Standards can be obtained in PDF or hard copy formats from the BSI online shop: www.bsigroup.com/Shop

The site investigation process

Site investigations (known as detailed inspections) should be carried out incrementally, the most likely linkages investigated first. After desktop studies have been exhausted, these may include walk-over surveys, limited on-site testing, geophysical analysis and ultimately intrusive investigations such as boreholes or trial pits.

If at any stage a single contaminant linkage is established and the land declared to be *Contaminated Land*, the local authority is advised that it may stop there and commence the enforcement process. There may be many more contaminant linkages but further investigation may be required as part of a remediation notice.

Who should undertake investigations? – Investigations must be carried out by 'suitable persons', the definition of which is never clear. Local authority and local Environment Agency officers should always work closely together to ensure no embarrassing misunderstandings occur.

Local authority strategies should have considered this aspect of the process and whether they would regard one or more of their own officers as 'suitable', and where not, what level of training and experience may be appropriate to enable them to undertake investigations themselves. Selection criteria for outside contractors should also have been included in the strategy.

With regard to sites which, if contaminated, would be declared Special Sites, the local authority must always identify these and authorise a suitable Environment Agency officer to investigate on its behalf. For example, where an authority is aware that part of its district overlies lower Permian basal breccias, conglomerates and sandstones, and that these may be affected by substances on the land which possess carcinogenic, mutagenic or teratogenic properties, they should notify the relevant Environment Agency immediately. These definitions are to be found in Regulation 3 and Schedule 1 of the *Contaminated Land* (England) Regulations 2006 and give an indication as to the complexity of the detail. Although most local authority officers would perhaps struggle with these issues it is a local authority duty.

Where the local authority wish to authorise a third party (consultant/contractor) to undertake investigations, as mentioned earlier, it has specific powers under section 108 of the Environment Act 1995 to do so. This can involve entering premises (includes land), taking samples or carrying out related activities for the purpose of enabling the authority to determine whether any land is *Contaminated Land*. It should be noted that the Agency does not have this power, so, where the local authority and the Agency both agree that there is a potential for a contaminant linkage to exist, and if it did, the resulting *Contaminated Land* would be declared a Special Site, the Agency must apply to the local authority to appoint a third party. Clearly then, there must be broad agreement on what constitutes a 'suitable person' in these circumstances.

How should investigations be undertaken? – Local authorities must always carry out investigations in accordance with appropriate technical procedures and take all reasonable precautions to avoid creating contaminant linkages during that process. Specifications for investigative work must therefore be prescriptive and unambiguous and require strict adherence to a quality assured procedure which will bear close scrutiny in the event of dispute (such as an appeal against a declaration that land is *Contaminated Land* within the terms of the legislation). Local authorities may devise their own site-specific procedures or require compliance with existing approved guidance or British Standards. The problem with this, however, is that the guidance itself is, by necessity, open to considerable interpretation.

As mentioned in the introduction, there is a plethora of guidance available, and the aim here has been to provide a broad process outline without going into the minutiae.

Similarly, it is important to work only with the guidance with which the EHP or officer is comfortable. As a contaminated land Expert Witness for over 20 years the author has never ventured into any of the areas where a high level of knowledge in soil phase transformations, stochastic modelling or groundwater flow through semi-permeable aquifers, is required. A general knowledge of such a broad and complex subject requires considerable expertise in itself. The key to success is to be aware of the boundaries of one's knowledge. Bearing this in mind a list of a few essential guidance documents is included later.

The key aspect of any investigation is to know why it is being undertaken. Clear objectives must therefore be formulated relating directly to the potential contaminant linkages identified. The most straight forward investigations will arise where there is only one potential receptor. Complexity will increase as additional receptors and exposure routes (pathways) are identified. For example, land contaminated by sewage sludge with the potential to affect property (in the form of crops or livestock), ground water, surface water, ecosystems, as well as human targets.

Local authorities should identify all the potential contaminant linkages and consider conceptual models to justify the investigation process. The specification must require the investigator to obtain data incrementally and terminate the process immediately if it becomes clear that linkages do not exist. It would not be appropriate to extend the process to consider the impact of the site on any new receptors that may be introduced in the future as a result of certain forms of development. That should be undertaken at the town (land use) planning stage.

Conceptual modelling – Is used across many fields, for example in the sciences, economics and IT. It invariably involves cartoons, flowcharts and diagrams and as such can sometimes become over-elaborate. Practitioners should understand that the overall aim must always be to simplify what may be quite a complex concept. To this end the contaminant

(pollutant) linkage diagram is a useful way of explaining as simply as possible what may be happening at any particular time.

A simple example would be a closed petrol station suspected to have lost fuel from underground tanks. The preliminary desktop survey may have identified the area underlain by mudstone which should be relatively impermeable (though coarse sandstone lenses possible). Contamination of a minor sandstone aquifer at depths of 180 to 200 feet by light non-aqueous phase liquids (LNAPLs) would be unlikely as they would not penetrate to depth. The site is closed and all down to hard cover. The most significant receptors being property, in the form of underground services, which are known to run through the site, and human, via plastic water mains.

The objectives of any investigation may be to:

1 Identify subterranean structures and pathways (natural and artificial)
2 Identify all potential receptors
3 Create an accurate conceptual model of the site
4 Identify contaminants in the soil, their spatial distribution and concentration

Some local authorities may choose to present contractors with a simple list of objectives like this and ask them to produce a quality assured investigation strategy for them. If that were to be the case, officers must ensure they have the time and the ability to evaluate proposals carefully.

Ideally, they should produce detailed specifications themselves considering the potential linkages in some detail, the contaminants of most concern, sampling methods, frequency and techniques, and laboratory analysis. It is important that there should be a high level of confidence in the data produced before the next process commences.

Risk assessment – Once the quality assured data has been obtained and hazards identified, it must be assessed in respect of each linkage as set out in Figure 19.2.

```
              Transformations
                 physical
                chemical
                biological
HAZARD >>>>>>>>>>>> RECEPTOR
                 Pathway
                   air
                  water
                   soil
                  dust
                  biota
```

Figure 19.4   Pollutant Linkage.

This simple model in Figure 19.2 can readily be elaborated to demonstrate its underlying complexity as set out in Figure 19.4.

The local authority has the sole responsibility for determining whether any land appears to be *Contaminated Land*. Having identified hazards in the form of contaminants, having identified specified receptors, and any ways the two may come together, a basic risk assessment has already been undertaken. This process does, however, need to be extended somewhat to carry out a formal risk assessment that will satisfy the needs of the Act.

Some of the difficulties inherent in the process can best be demonstrated by examples. Some may seem somewhat idiosyncratic but there is a purpose which the reader will realise at the end of these examples.

Is broken glass Contamination? It is a substance, it is on land, and it may undoubtedly cause harm to known specified receptors (human beings). It is the duty of the risk assessor to decide whether the harm, *the human health effect*, if it occurred, would be significant, and whether the potential for such an effect would also be significant.

Significant harm to a human being includes death or serious injury; a child running and falling on glass could very easily result in serious lacerations of the hands and arms, which could result in rapid loss of blood, and potentially, death. The potential for significant harm is therefore undoubtedly confirmed in this acute risk.

Significant possibility of significant harm being caused in respect of human health effects includes:

> The amount of pollutant which the human receptor might be exposed which would represent an unacceptable medical risk assessed on the basis of relevant information on the toxicological properties of that pollutant.

With glass we are not considering toxic effects, but there are 'other human health effects', which must be taken into consideration. These relate particularly, but not exclusively, to fire and explosion, so could it be appropriate to consider other physical effects such as lacerations? This is not clear, but all assessments must be made on authoritative and scientifically based information consistent with providing an appropriate level of protection from the risk.

If, ultimately, the assessor considers the chances of the significant harm occurring are significant, the land must be declared *Contaminated Land* under the legislation. If not, it may be declared, *land in a contaminated state*, which is defined as, *land where harm is being caused or there is possibility of harm being caused*. The important aspect to note here is the absence of the word *significant* (in relation to both harm and possibility). Such land is then exempt from statutory nuisance action and must be left as it is until (and if) it becomes *Contaminated Land* as defined in the legislation, although other remedial action may be available to enforcing authorities.

Are Yew berries Contamination? Poisoning of cattle by Yew is not uncommon. Yew berries are extremely toxic to animals including humans; they also look like jelly sweets. Yew berries are substances as defined as being able to very readily cause significant harm (including death) to humans or property (includes livestock). If a large Yew tree in the village churchyard deposited hundreds of toxic berries on the land next door placing known receptors at significant risk, could the land be declared *Contaminated*?

Is asbestos Contamination? A more conventional example includes perhaps an old scrap yard that has taken all types of metals and scrap over decades. It is in a residential area and has been left abandoned for many years. The public have open access to it and children and youths use it as a BMX and motorcycle scrambling track. Residents have complained of the dust blowing off the site in dry weather, especially when the scramblers are racing.

Initial investigations reveal that the yard took a lot of steel that was fire protected with friable asbestos. The men used to scrape this off by hand and, 'dispose of it' on site. A few exploratory soil samples on the site reveal significant counts of blue, white and brown asbestos fibres at the surface. How would this be assessed?

Again, seemingly very simple, one contaminant (asbestos fibres), one receptor (humans), clear pathways (soil to air to lungs), an undoubted contaminant linkage. There is ample authoritative information on the human health effects of airborne asbestos, but can the risks of significant exposure in the open be accurately quantified? It must be 'Contaminated Land' surely – or are there circumstances when it may not be? Would it be possible to suggest that significant harm could occur, but the possibility of it occurring would not be significant?

Unfortunately, risk assessment will rarely be simple and there will be temptation to employ costly modelling processes in an attempt to deliver accurate scientific judgements. Where, however, there are several dozen potential contaminants on a site, accuracy may have to be sacrificed to sound judgement due to a shortage of reliable human toxicity and exposure data.

The same applies to water and ecological risks where expertise on hydrology, hydrogeology, leaching potential, soil chemistry, environmental fate and impact will always be at a premium for complex mixtures of chemicals on a very wide range of targets.

Local authorities will consult their partners such as Defra, the Forestry Commission, the National Parks Authorities, English Nature and the Environment Agency and so on, but few will have specialist risk assessors to help.

Risk assessment for land contamination is still a very confusing assemblage of concepts and methods borrowed from various sociological, psychological and technical disciplines. No one process provides all the answers, and there are no answers to many of the questions. As a result, when it comes to judging a risk, many people would rather trust the opinion of a friend than take the word of a scientist.

Scientific risk assessment is slow, cumbersome and expensive. It attempts to achieve a degree of precision that is invariably unachievable and often unnecessary. The results are at best suspect, at worst, dangerously wrong. At the end of the process the pragmatic local authority officer will have to utilise the wisdom of Solomon to make the final decision on whether the *land affected by contamination* is *Contaminated Land*. The outcome of which could on the one hand blight a whole housing estate in perpetuity, on the other, leave a generation of children at significant risk.

After the first decade of the 'new' regime the Government became aware of these problems and produced the 2012 Statutory Guidance. It is worth referring the reader to the Ministerial Foreword quoted in the early part of this section. It is very much focused on not being 'over cautious'. In my experience this has been a great help, leaving most sites to be considered merely *land affected by contamination*, and then to remain (hopefully) stable and undisturbed until they are proposed for development.

The Guidance requires local authorities to use a 'precautionary rather than blanket approach'. I would consider it more *pragmatic* than precautionary as indicated by the extensive use of words which require considerable judgement on the part of the final decision maker; these include 'reasonable' and 'robust'.

The following are a selection of examples from the Guidance:

- Remediation requirements should be reasonable;
- Where uncertain the local authority should use its judgement to strike a reasonable balance;
- The LA may decide the land is not *Contaminated Land* on the basis of information from the land owner . . . provided they are satisfied with the robustness of the information;
- Where the local authority considers there is a reasonable possibility of *Contaminated Land* . . . it should inspect the land;
- Local authorities should have regard to good practice . . . and make robust decisions in line with the Statutory Guidance;
- Inspections should be carried out quickly with as little disruption as reasonably possible but be sufficiently robust;
- Decisions should be based on what is reasonably likely, not what is hypothetically possible;
- At the same time LAs should ensure that the time and resource is sufficient to provide a robust basis for regulatory decisions;
- External expertise may be required to conduct a robust risk assessment;
- Where experts are not sure, it is for the LA to use its judgement to form a reasonable view based on a robust assessment;
- Decisions should be supported by robust scientifically based evidence;
- Generic assessment criteria (GACs) etc. should have been produced in an objective, scientifically robust and expert manner.

There are very many more statements like these which ultimately place a heavy burden of responsibility on local authority officers who will have varying degrees of knowledge and experience. Several times in the Guidance we are told that these particularly onerous duties are 'the sole responsibility of the local authority'. Yes, they may seek external help from suitably qualified persons, but then

they have to have the ability to decide who may be suitably qualified to make a 'robust' scientific assessment, and who may be suitably qualified to make a 'reasonable' judgement on risk in line with the Guidance without being over cautious.

Returning to the three earlier examples:

The broken glass – The Authority may consider that similar physical risks are commonplace and it is not reasonable to consider such as *significant* in every case. This balances the hypothetically possible versus the reasonably likely. If action were deemed necessary, it may be considered more appropriate to use waste or littering legislation in the particular circumstances.

The Yew berries – The scale of the problem renders this matter impractical to ever deal with. There are Yew trees in every town and village, but more importantly there are many more plants and trees that have potentially poisonous elements. It is a fact of nature that such risks exist and cannot be overcome by removing the risk, only by managing it.

The asbestos – This is more of a realistic contamination problem than the other two. It is likely that an old scrap yard would have many different contaminants, 'in on or under the land', but under normal circumstances they would not be bothering any person or any thing. In this case it is the receptor/pathway generator that needs to be removed from the linkage. It is likely this would be dealt with informally.

Declaration

Once a decision has been made that the land is Contaminated by definition, the authority must prepare a written record of the determination including:

- Description of the contaminant linkage(s) identifying all three components (contaminant, pathway, receptor);

- Summary of the evidence (factual) upon which the evidence is based;
- Summary of the relevant assessment of this evidence; and
- Summary of the way the authority considers that the requirements of the Statutory Guidance have been satisfied.

So, for example, if the contaminants are hydrocarbons from petrol (see Figure 19.5), the factual
evidence would be obtained from the site investigation explaining the technical procedures, the identification of the pollutant, how it is spatially distributed, concentrations on the site and its proximity to the property in question, usually PVC services.

The assessment will be the scientific basis for the officer's decision and should include a conceptual model for why this pollutant has been considered significant as opposed to others present on the site or likely to be present (which will most likely relate to its relationship to the receptor), and why other linkages have been discounted, or require further investigation, such as those to water or humans.

Finally, all the relevant aspects of the Statutory Guidance must be shown to be satisfied including matters relating to:

- Significant harm;
- Significant possibility of significant harm;
- Pollution of controlled waters (PCW);
- The likelihood of the PCW;
- Consultation procedures;

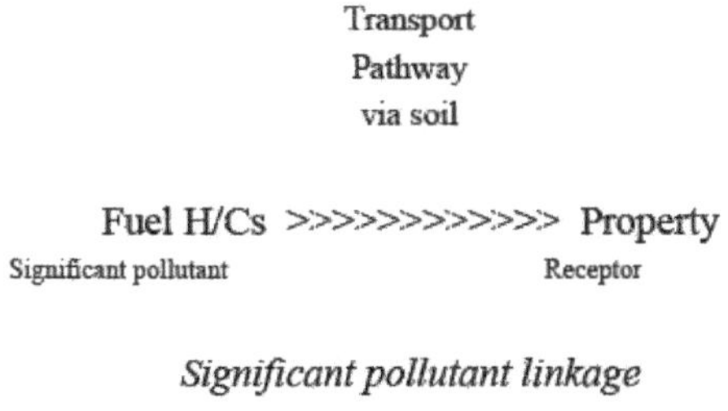

Transport
Pathway
via soil

Fuel H/Cs >>>>>>>>>>>>>> Property

Significant pollutant   Receptor

Significant pollutant linkage

Figure 19.5 The pollution linkage

- Agreements and consistency with other regulatory bodies (or if otherwise, why);
- Determination of the physical extent of the land which may be affected;
- Why the land is or is not considered to be a Special Site.

Notification – The local authority must at this stage initiate the formal notification procedure which has the effect of commencing the consultation process on what remediation might be appropriate.

In respect of each part of the land in question, the authority must establish who is the owner, the occupier(s) and the appropriate person(s), and notify them, together with the Environment Agency, that the land has been identified as being *Contaminated Land*. This is a formal notification which must be in writing. In many cases the authority may not have been able to establish with certainty who falls into these categories, particularly the appropriate person. It must, however, act without delay on the best information available to it at the time, whilst continuing investigations.

Included in the notification may be:

- The written record of determination;
- Site investigation reports (or their availability);
- Statements relating to who appears to be appropriate persons and why;
- Details of all other parties notified;
- Information on tests for exclusion and apportionment.

Special Sites – Having identified the land as *Contaminated Land* under the legislation the authority must decide whether it is a Special Site. Where they are of the view that it is, they need to ask the Environment Agency if they agree. Where the Agency do agree, or they fail to notify the local authority that they disagree, within 21 days, the land is formally designated a Special Site and the Agency becomes the enforcing authority. Where there is a dispute the local authority must refer the decision to the Secretary of State together with written statements from both parties explaining the reasons for their respective positions. The local authority still has the duty to notify the relevant persons where Special Sites are determined.

Liability – When all significant contaminant linkages have been identified, the procedure relating to the apportionment of liability must commence. This has five distinct stages as follows:

1   Identifying potential appropriate persons and liability groups
2   Characterising remediation actions
3   Attributing responsibility to liability groups
4   Excluding members of liability groups
5   Apportioning liability between members of a liability group

Space does not permit detailed consideration of these complex issues; some key points are therefore identified next. The process commences with the establishment of liability groups. All appropriate persons for any one linkage are a liability group. These may be Class 'A' or Class 'B' persons.

Class 'A' appropriate persons – Are, generally speaking, the polluters, but section 78F(2) includes persons who 'knowingly permit'. So should a developer choose to leave contamination on a site which subsequently results in the land being declared *Contaminated Land*, s/he may become an appropriate person and thereby liable for remediation. These are known as Class 'A' persons.

The test of 'causing or knowingly permitting' has been used as a basis for establishing liability in environmental legislation for more than 100 years, primarily in relation to water pollution. A comment by Earl Ferrers from the House of Lords in July 1995 whilst debating the Environment Bill is relevant (Hansard 11th July 1995):

> The test of "knowingly permitting" would require both knowledge that the substances in question were in, on or under the land and the possession of the power to prevent such a substance being there.

As such innocent owners with knowledge, it is suggested, should never become Class 'A' persons.

The matter of appropriate persons must be considered for each linkage; therefore where a site has had a series of contaminative uses over the years, each significant contaminant linkage must be identified separately, and liability considered for each.

Class 'B' appropriate persons – Where no Class 'A' persons can be found, liability reverts to the owner or the occupier. These are known as Class 'B' persons. The local authority must, however, make all reasonable enquiries to identify Class A persons first.

Apportionment of costs – Generally speaking the members of a liability group will have the total costs falling on the group as a whole apportioned between them. It may also be necessary to apportion costs between liability groups.

There are three basic principles which apply to exclusion and apportionment tests:

1   The financial circumstances of those concerned should have no relevance.
2   The local authority must consult persons affected to obtain information (on a reasonable basis having regard to the cost). If someone is seeking to establish an exclusion or influence an apportionment to its benefit then the burden of providing the authority with supporting information lies with it.
3   Where there are agreements between appropriate persons the local authority should try to give effect to these agreements. This is subject to the caveat that an agreement must be disregarded to the extent that it would increase costs borne by someone who would benefit from limitation on cost recovery (adding the cost to the public purse).

Limitation of costs to be borne by appropriate persons – There are six tests specified to identify Class 'A' groups who should be excluded from liability. These should be applied in sequence and separately for each linkage. The exclusion of Class 'B' persons is much less complex with the purpose of the single test merely excluding those who do not have an interest in the capital value of the land. Therefore, tenants etc. who occupy under licence with no marketable value, or who pay rack rent with no other beneficial interests, are excluded.

Where a local authority has apportioned the costs of each remediation action and before serving remediation notices, they must consider whether any of those liable may not be able to afford it. If, after taking into consideration the Statutory Guidance, it decides that one or more of the parties could not, it cannot serve a remediation notice on any of the parties. It must, instead, consider carrying out the work itself and produce and publish a remediation statement.

The enforcement process

Before remediation notices can be served, as mentioned earlier an extensive consultation process must be completed and ample encouragement given to arrive at an informal solution. The local authority must make reasonable endeavours to consult the appropriate person(s), owners, occupiers, etc. about their views on the state of the land. This is unusual and could result in some real nightmares and delays for local authority officers. Where a housing estate is affected for example, it would not be unreasonable to expect house owners, landowners, lenders, insurers, builders, geotechnical engineers, residents' groups, etc. all to have differing views according to their position in the jigsaw.

It should be noted, the Government requires that remediation notices are served only as a last resort, and then only after this lengthy consultation process.

Notices can be authorised after two tests are satisfied:

1   That the remediation actions will not be carried out otherwise
2   That the local authority has no power to carry out the work itself

If these are met the authority must serve a remediation notice on each appropriate person. It cannot be served less than three months after formal notification that the land is contaminated unless urgent action is deemed necessary (where there is imminent risk of serious harm).

Specifying remediation – Local authority officers have the task of specifying remediation measures. These have to be reasonable, appropriate and cost effective. The aim of any remediations should be to ensure that the land is no longer considered *Contaminated Land*, taking the shortest and lowest cost route. This means in most cases attention will be focused on the pathway, rather than the contaminant or receptor. The standard for remediation should relate to best practicable techniques, an interesting hybrid between best practicable means and best available techniques.

Reasonableness will, however, always be the key. This is determined in relation to the cost of carrying out the remediation against the cost of failing to. Like hardship, which must be considered at the outset of the enforcement process, the matter of reasonableness and complex cost benefit analysis will always be a major stumbling block to progress in remediating *Contaminated Land* by enforcement action.

Remediation by the local authority – Before notice(s) can be served the local authority must determine first whether it has the power to carry out any of the remediation actions. There are five specified circumstances where this may be the case:

•   Where urgent action is required;
•   Where no appropriate person can be found;

- Where one or more appropriate persons are excluded (on grounds of hardship);
- Where the local authority has made an agreement with the appropriate person(s) that it should carry out the remediation;
- In default of a remediation notice.

Urgent remediation action should be authorised where there is imminent danger of serious harm or serious pollution of controlled waters.

Section conclusions

This section provides a brief overview of the legal process relating to the remediation of *Contaminated Land* and demonstrates clearly why over the 20 or so years since its inception, the law has rarely been used for the purpose. There was a slow realisation that any land with contamination (in, on or under) should be left undisturbed until proposed for development. Only if it the contamination was likely to cause a real (and significant) problem should the local authority intervene. This has meant that the vast majority of contaminated sites would be dealt with via the Town & Country Planning process known as *development control*.

Glossary of terms

As with many aspects or strands of environmental health there is considerable range of terminology. The 2006 Statutory Guidance to the legislation (Part IIA of the Environmental Protection Act 1990 as amended) had a ten-page glossary defining terms. The 2012 Guidance (current) which replaced it has been said to be 'shorter and simpler', and to that end has no glossary. To assist, next are a few definitions largely taken from the body of the 2012 document but paraphrased for simplicity.

Readers should consult current guidance and the Environmental Protection Act 1990 Part IIA for full definitions, when necessary.

Appropriate person – a person determined to bear responsibility for anything which is to be done by way of remediation of *Contaminated Land*.

Class A person – a person who is an appropriate person because he or she has caused or knowingly permitted a pollutant to be in, on or under the land.

Class B person – a person who is an appropriate person because he or she is the owner or occupier of land and no Class A person can be found.

See also liability group and orphan linkage.

Conceptual model – part of the process of identifying risks centred around the understanding of contaminant linkages.

Contaminated Land – is land which meets the Part IIA definition. This is necessarily complex and requires consideration of the terms 'harm', 'significance', 'significant possibility' and pollution of controlled waters (PCW). These are described at length in the Statutory Guidance including the four 'Categories of Harm'.

Contaminant/Pollutant/Substance – have the same meaning for the purposes of the Guidance. Something with the potential to cause significant harm to a relevant receptor, or to cause significant PCW.[12] See also 'normal' later. Typical contaminants are heavy metals, asbestos, tar and other hydrocarbons including pesticides/herbicides, volatile metals such as mercury and organic chemicals including solvents and the results of waste disposal to landfill.

Contaminant (or pollutant) linkage – the relationship between a contaminant, a pathway and a receptor. This becomes a significant linkage when the risk is such that the land meets the definition in Part IIA. (See Figures 19.2 and 19.4.)

Controlled waters – relevant territorial waters/coastal waters/inland freshwaters/ground waters within the saturated zone.

Current use – includes any likely future use not requiring planning permission, also informal use such as children playing, even if unauthorised. Importantly does not include any agricultural use of land used for crops or stock 'not habitually grown or reared'.

Detailed inspection – see inspection types.

Duty to inspect – Part IIA places a duty on local authorities to inspect their districts to identify *Contaminated Land* and Special Sites in accordance with the Statutory Guidance. This duty is the sole responsibility of the authority.

Enforcing authority – the local authority in who's area the *Contaminated Land* is situated, except when the local authority declares the *Contaminated Land* a *Special Site*. The Environment Agency then become the enforcing authority.

Escaped substances – an exclusion test for Class A persons.

Inspection types:

- Detailed – gathering information on ground conditions and assessing risk. This may be achieved using statutory powers of entry available to the local authority under S108 of the Environment Act 1995.
- Strategic – inspecting a local authority area to identify priorities.

See also *Duty to Inspect*.

Land affected by contamination – land where contaminants are present but not with sufficient risk to meet the Part IIA definition.

Liability group – a group of appropriate persons, these can be either Class A or Class B persons.

Normal (levels of contaminants) – the natural presence of substances at typical concentrations, or from anthropogenic sources and also considered typical, and not considered to pose an unacceptable risk.

Orphan linkage – a significant pollutant linkage where no Class A or B persons can be found.

Pollutant – see contaminant.

Pollutant linkage – see contaminant linkage.

Pollution of controlled waters (PCW) – the entry of any poisonous, noxious or polluting matter or any solid waste matter. Pollution/ significance and significant possibility are described at length in the Statutory Guidance including the four 'Categories of Harm'.

Receptor – includes human beings, relevant property, relevant ecosystems and controlled waters. These are considered at length in the Statutory Guidance.

Register – a public register required to be maintained by each enforcing authority with particulars relating to *Contaminated Land* in its area. Any particulars on an Environment Agency register should also be on the local authority register as primary *Contaminated Land* authority.

Remediation – the conventional meaning of this term relates to restoring land in some way. The meaning for the purposes of Part 2A, however, is very different. In summary it can include:

- Investigations and assessments (assessment action);
- Decontamination or otherwise breaking pollutant linkages (remedial treatment action);
- Monitoring (monitoring action).

Remediation statement – a statement prepared and published by the responsible person detailing the remediation actions which are being, have been, or are expected to be done, and when.

Risk – the probability (and likely frequency) of a defined exposure hazard, and, the seriousness of the consequences (to a Part IIA receptor).

See also unacceptable risk.

Risk summary – something which must be produced for all *Contaminated Land* to explain the contaminant linkages, the risks identified, the uncertainties apparent and any possible remediation. This must be done in a way that can be understood by the layperson.

Significant linkage – see contaminant linkage.

Special Sites – types of land to be identified as Special Sites are detailed in the *Contaminated Land* (England) Regulations 2006. These are generally where the Environment Agency or its predecessors have had, or still have, responsibility for regulation.[13]

Strategic inspection – see inspection types.

Substance – see contaminant.

Unacceptable risk – one that would give grounds for the land to be considered *Contaminated Land*.

Section 2 Notes

8 SI 2000 No27.

9 SI 2006 No1380 – Part IIA is further established in Scotland by the Contaminated Land (Scotland) Regulations (SSI 2000 No 178) as amended by SSI 2005 No 658 and the Scottish Government's Statutory Guidance: Edition 2. https://www.gov.scot/publications/environmental-protection-act-1990-part-iia-contaminated-land-statutory-guidance/

10 SI 2012 No263.

11 These are ground water within the saturated zone, inland fresh water and the sea and coastal waters within three miles (i.e., controlled waters).

12 For comparison under the Building Regulations made under the Building Act 1984 Approved Document C: *Site preparation and resistance to contaminants and moisture* defines contaminant as any substance which is or may become harmful to persons or buildings including substances which are corrosive, explosive, flammable, radioactive or toxic.

13 See for example *R (on the application of National Grid Gas plc (formerly Transco plc)) (Appellants) v. Environment Agency (Respondents) (Civil Appeal from Her Majesty's High Court of Justice) [2007] UKHL 30 on appeal from [2006] EWHC 1083 (Admin)*. This is known as the Bawtry Gasworks Case and judgement is available at https://publications.parliament.uk/pa/ld200607/ldjudgmt/jd070627/grid-1.htm and illustrating some of the complexities of the legislation.

SECTION 3: TOWN & COUNTRY PLANNING AND THE DEVELOPMENT OF CONTAMINATED LAND

The World Commission on Environment and Development, now better known as the Brundtland Commission, published its report *Our Common Future* in 1987:

> There is little doubt that your Commission's report in 1987 was the single most important document of the decade on this subject, bringing the phrase **'sustainable development'** into all our vocabularies. [21]

Sustainable development and the sustainable development goals (SDGs) have been discussed in Chapter 1, but the earlier section of this chapter has shown that modern regulation relating to land contamination has its origins in town and country planning.

The emphasis has always been on development of contaminated sites rather than enforced remediation, and despite the implementation of Part IIA of the Environment Act 1990 in the year 2000, it still is. Remediation of *land affected by contamination* is inevitably a time-consuming and expensive process. It was recognised that the only possible way this could be widely funded would be through the proceeds of development.

Again, there is much guidance on the subject, and the bulk of this merely refers the reader to further, very wordy guidance. Here a simple introduction to the planning system is provided with some key points emphasised:

Firstly, town and country planning is itself complex and it is neither possible nor desirable for EHPs to try to become experts. Similarly, the environmental protection work of the EHP is complex and it is impossible for planning professionals to become experts in the field. Nevertheless, each should have some insight into the other's world.

Planning and environmental health are two different horses from the same stable. It is essential that the planning professional and the EHP work closely together. The best outcomes arise from collaboration and co-operation. Both subject areas comprise a number of fascinating and challenging disciplines and each must complement the other. To do this effectively both must understand the restricted framework within which each must work. The most frequently heard phrase uttered from the lips of the planning professional is 'on balance'. Accurate data carefully assessed and evaluated by a person with sufficient knowledge and experience enables a well-balanced and considered judgement.

Governments continually seek to streamline the planning process to prevent delays to development, seeing that as essential for the economy (without always thinking through the longer-term or wider consequences). The best way to do this is to ensure there are sufficient well-qualified officers within local planning authorities (LPAs) to be able to consider and make difficult decisions on often highly sophisticated assessments of environmental impacts and remediation proposals.

The 'History' section of this chapter refers to the foundations of the current land-use planning system and guidance and it is worth noting that very little has changed.

How does planning work?

Planning law in the UK is comprised of Primary and Secondary legislation. The former are Acts which are approved following the passage of a Bill through parliament, whilst the latter are known as statutory instruments, usually in the form of Regulations or Orders. Together these provide the framework and detail of how the planning system should operate. In addition there are planning circulars which provide non-statutory advice and guidance on particular issues to expand on subjects referred to in legislation. They are used to explain policy and regulation more fully. Many circulars are quasi-legislative and include a direction or requirement to take specific action or provide guidance on implementation of aspects of planning policy.

Case law is an essential part of the process as this provides authoritative guidance on the interpretation of the law and also the actions of public bodies via Judicial Review (administrative law).[14] This is important in ensuring procedural correctness and clarity in interpretation and application. Many aspects of

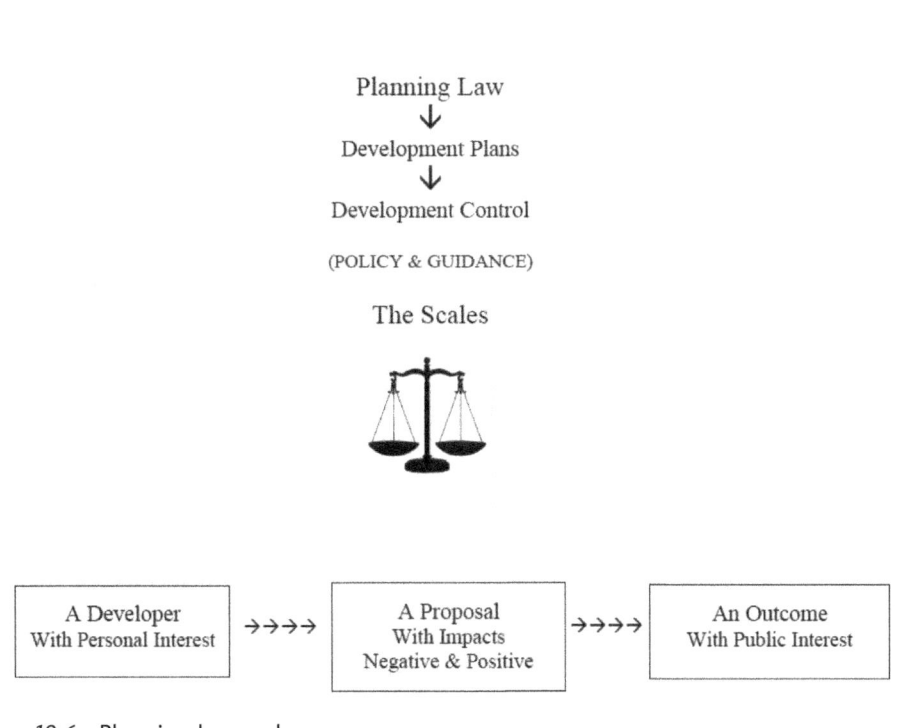

Figure 19.6 Planning law and process

case law are subsequently included in circulars and other guidance.

Difficult decisions have to be made at every step relating to:

- Procedure – If an application is not dealt with, in accordance with the law the decision can be challenged;
- Conflict Regulation – In many applications there is (sometimes considerable) conflict between the many and varied private and public interests;
- Social Regulation – Public safety, well-being and the interests of amenity;
- Environmental Regulation – Protection of biodiversity, landscape and sometimes control of noise and other forms of pollution.

There is inevitable overlap between other forms of control such as environmental permitting regimes, environmental protection and health and safety at work legislation, building control, licensing and so on. The input of EHPs is essential so priorities can be presented clearly and LPAs do not act beyond their powers (*ultra vires*). At each stage of the process transparency is required to show all the necessary information has been identified, gathered and clearly presented, to enable informed debate.

Development – All 'development' is controlled through the planning system, therefore it is important to know what it is. In the UK development is defined as:

> The carrying out of building, engineering, mining or other operations in, on, over or under land, or the making of any material change in the use of any buildings or other land.

This is a broad definition and to assist quite a lot of attention has been paid by the courts to what is, and is not, considered development. For more information see the gov.uk website.[15]

From a land contamination point of view, engineering works are considered development; therefore it can sometimes be argued that a major remediation should be dealt with as a separate application rather than ancillary to the primary development. This can be very useful for the purposes of management and enforcement. It is wise to ensure a complex programme of remediation is properly completed and verified before construction commences.

Development plans – In the UK, Town & Country Planning is described as 'Plan-led'. This is basically planning in advance, identifying land for infrastructure and major developments, areas to be protected, identifying need, etc. Plans are strategy documents that direct developers to some areas, and away from others. They are informed by policies produced at national level and aim to achieve a degree of consistency in the decision-making process.

For LPAs these are known as 'Local Development Plans' (LDPs) and include the authority-wide plan as well as any Neighbourhood Plan, which has equal status. This is often a much-neglected area in environmental health either due to lack of time, experience or both. If possible, an EHP should take an active part in the creation and review of LDPs.

Development control – Where control is required an application for planning permission must be submitted to the LPA. Then it is intended that the planning system should operate fairly and proportionately balancing private and public interests. Whilst the planning system must support economic and social progression, it must also control developments that will have a significant adverse impact.

There are two types of application, 'outline' and 'full'. Outline is the first part of a two-stage process where if the development is approved in principle, certain matters are 'reserved' for the second stage. 'Full' completes the process in one and if approved the development can commence, subject to planning conditions. It should be standard practice for the development control section of the LPA to circulate all planning applications within the EH team as there are so many impacts which overlap. Where EH is a statutory consultee they must be given 14 days to respond.

Historically, development control has been considered by successive Governments to be too restrictive, slow and costly. One of the key improvements in the process in recent years has been the early involvement of stakeholders in pre-application discussions in an attempt to resolve conflicts.

As pointed out, success can only be achieved by employing sufficient well-qualified and experienced officers to be able to consider the many and varied complex applications submitted each week. Cutting staff results in the opposite to what the Government is trying to achieve.

Policy, Guidance and Material Considerations (MCs) – In Figure 19.6 there are two words added after development control, 'Policy & Guidance'. Within the many and various administrative systems of Town & Country Planning, decision makers are only able to take into consideration matters that are classed as *material* – this is key. Lay persons often object to proposals for a long list of reasons, but if they are not *material* they have to be ignored.

Unfortunately, material considerations are not easy to define. In essence they are anything that is relevant to the development that can be controlled under planning law. We are guided on the latter in various forms of policy documents, guidance, circulars and case law (and many other sometimes idiosyncratic ways). Those persons in favour of a development seek to accumulate material considerations they deem to be relevant into the left side of the scales, whilst those against do likewise into the right side. Needless to say, the amount of weight to be afforded to any MC is a judgement in itself.

For the non-planner it is invariably policy and guidance where they must concentrate

their time and effort, and it should be noted, they seem to be changing constantly.

National policy – the National Planning Policy Framework (NPPF)

The seriousness of the situation facing the planet was spelled out by the Prince of Wales in 1992 [21] when he quoted the Royal Society and the US National Academy of Sciences:

> The future of our planet is in the balance. Sustainable development can be achieved, but only if irreversible degradation of the environment can be halted in time.

Globally the situation has not improved in the 30 years since – quite the reverse, despite the rhetoric.

The National Planning Policy Framework (NPPF) was introduced by the Government in 2012 with one key word running throughout the document – *sustainable*. We have discussed the sustainable development goals (SDGs) in Chapter 1 but one definition of sustainable relating to development is:

> Capable of being maintained at a steady level without exhausting natural resources or causing severe ecological damage.

The Minister defined it in the forward to the NPPF in 2012 as:

> Ensuring that better lives for ourselves don't mean worse lives for future generations.[16]

The basis of the philosophy is that sustainable development should, and I quote: 'go ahead without delay – a presumption in favour of sustainable development'. So the safe remediation of despoiled land and its subsequent re-use, represents a perfect example of sustainable development. Cleaning up pollution and protecting important greenfield, agricultural and biologically diverse sites in the process.

The NPPF has been revised and updated several times with the latest version completed in July 2021.[17] The NPPF is a *material consideration* in all planning decisions and should be considered the first point of reference. It is updated regularly, so readers should check on the latest version and updates.

To achieve sustainable development there are three overarching objectives as follows (précised):

- Economic – to build a strong economy by supplying the right types of land in the right places for development and infrastructure;
- Social – to support communities by providing development to satisfy need for housing, taking into consideration good design, adequate services, open space and well-being;
- Environmental – to protect and enhance the natural, built and historic environments by effective use of land, improving biodiversity, minimising waste, pollution and climate change.

Planning is very much about finding the right (or best under the circumstances) locations for the right development. Strategic plans concentrate on this process identifying land for large-scale housing and employment sites, as well as smaller sites within towns and villages.

In addition, needs can be met in other ways such as the development of brownfield sites which otherwise may not have been considered an appropriate location. To this end there is now a requirement in planning for Brownfield Registers.

Brownfield Registers are lists of previously developed land that LPAs consider to be appropriate for residential development, having regard to criteria in the Town & Country Planning (Brownfield Land Registers) Regulations 2017. LPAs can trigger a grant of permission in principle for residential development on suitable sites where they follow the required procedures.

Each authority should have completed a register by the end of 2017 and thereafter should review it annually. It should be noted that these registers are very different from the ones originally proposed in section 143 of the Environmental Protection Act 1990, in that they include only sites that are considered suitable for residential development.

Substantial weight should be given to reusing brownfield sites providing opportunities to remediate despoiled, degraded, derelict, contaminated or unstable land.

Part 15 of the NPPF considers the environment. This requires several protective matters to be taken into consideration, but only that. It is important to understand that we are told the Framework should be read as a whole and policies should be applied in a way that is proportionate. So again, policies and other material considerations, for and against, are placed in the scales and a decision is made as to how much weight is given to each. The following are a list of things taken from Part 15 that must be considered in relation to the protection of the environment:

- The geological and agricultural value of soils;
- Preventing unacceptable risk of pollution or land instability;
- Cleaning up contamination is a significant factor in favour of development;
- After clean up the land should not be capable of being determined *Contaminated Land* under Part IIA;
- All work relating to contamination should be undertaken by a competent person;
- Responsibility for safe development lies with the developer/landowner.

As mentioned, there is an inevitable overlap of environmental control with existing permitting and other regulatory regimes. Currently the final paragraph of Part 15 considers 'the acceptable use of land'. It is worth repeating this in its entirety:

> The focus of planning policies and decisions should be on whether proposed development is an acceptable use of land, rather than the control of processes or emissions (where these are subject to separate pollution control regimes). Planning decisions should assume that these regimes will operate effectively. Equally, where a planning decision has been made on a particular development, the planning issues should not be revisited through the permitting regimes operated by pollution control authorities.

Input from EHPs at this stage is crucial as there is often a mistaken belief that pollution control regimes prevent pollution. They never have and never will do, they merely limit it to what can be achieved (BAT – BATNEEC – BPEO, etc. – terms that are considered elsewhere in this book). That being the case, the judgement as to whether the proposed development is an acceptable use of land has to be made against the residual pollution which will continue after the control regime has been successfully applied and enforced. This requires expert knowledge on the part of the EHP, especially processes that are authorised by the Environment Agency.

Another common misunderstanding is that all 'nuisances' that concern people can be controlled by EHPs via enforcement action under Part III of the Environmental Protection Act 1990; they cannot. Firstly, only those nuisances deemed to be statutory nuisances under law can be considered for action. Secondly, they must amount to a statutory nuisance. Thirdly, there is a defence of best practicable means (BPM) in certain circumstances. This means that the offender does not have to stop, remove or control the action causing the nuisance, merely do their best to do so (as defined). Again this leaves a residual

impact which must be weighed in the scales at the time of judgement in the decision-making process. (For more on Statutory Nuisance see Chapter 5.)

It is self-evident that development has some impact. It is therefore the duty of the EHP to explain clearly the impact that is likely to exist, remain or be created, and affect receptors in the local or wider environment, should permission be granted.

National guidance

There was a desire to streamline a lot of the national guidance at the time the NPPF was being developed, and indeed most was repealed or deleted. Unfortunately, however, it was then replaced with even more. There are currently 57 Planning Practice Guidance notes (PPGs) listed on the Government website together with a list of 'other planning policies' and related content.[18]

Generally, these are not lengthy but do link to many thousands of pages of other guidance. In the author's experience this is not always helpful as it is often confusing, vague, and it is practically impossible to read it all.

As an example, the PPG most relevant to the development of *Contaminated Land*, *Land affected by contamination*[19] is considered.

The first line tells us that the document 'provides guiding principles on how planning can deal with land affected by contamination'. The document then links the reader to four paragraphs of the NPPF as follows: 170 (now 174); 178 (now 183); 179 (now 184); and 183 (now 188). These changes demonstrate the difficulty in updating these interrelated and interlinked documents in real time. The contents of these four paragraphs form Government Policy and are therefore considered *material*. The content of the PPGs is merely guidance, neither policy nor Statutory Guidance. For an EHP to give weight to their views it is important to always seek to associate these with either policy or the Part IIA Statutory Guidance.

Is there anything else new in the PPG? Not really but there are some points worth emphasising:

- In the LDP, seek to strategically allocate suitable land affected by contamination for development, or a sustainability appraisal;
- Applicants are advised that early engagement with environmental health is important;
- Consider the impact of development and the likelihood of mobilising otherwise stable contaminants;
- LPAs must always be clear on the role of developers and/or landowners – they alone are responsible for the safe development and subsequent occupation of the site;
- Information and desk studies, site investigations and risk assessments (as necessary) must be prepared by a 'competent person';
- These must be provided as part of a staged process before an application can be determined;
- Site investigations, field trials, etc. may require planning permission;
- As a minimum after development, land should not be capable of being determined *Contaminated Land* under Part IIA.

A very simple flow chart has been provided to aid the decision-making process in Figure 19.7.

Planning conditions

Planning conditions are an important aspect of the on-going management of development after planning permission has been granted. This can mean during the development process and, sometimes, after completion.

In determining an application the LPA has three options:

1. To refuse permission
2. To grant unconditionally
3. To grant with conditions

It is rare for permission to be granted unconditionally. Planning conditions may improve the acceptability of the development during engineering works such as remediation and construction, for example:

- Controlling days and hours of work;
- Setting limits on noise and other pollution levels;
- Specifying monitoring and reporting;
- Controlling and routing construction traffic;
- Protecting buildings, wildlife and ecology;
- Protecting against flooding.

And after completion conditions can cover:

- Controlling use type;
- Restricting hours of use;
- Setting noise levels;
- Specifying on-going monitoring and reporting;
- Requiring new (compensatory) habitats;
- Maintaining flood defences.

The correct use of planning conditions is essential and what may and may not be considered acceptable was detailed in Circular 01/85, later updated by Circular 11/95. The Government website explains that this Circular is now 'cancelled' with the exception of Appendix A (model conditions) which is retained. The latest guidance is to be found in the relevant guidance on the use of planning conditions.[20]

Historically six well-known tests have developed from case law (*Newbury DC v Secretary of State for the Environment*)[21] and as a result are known as the *Newbury Tests*. A condition must satisfy all the tests if it is to be valid.

It must be:

- Necessary;
- Relevant to planning;
- Relevant to the development to be permitted;
- Enforceable;

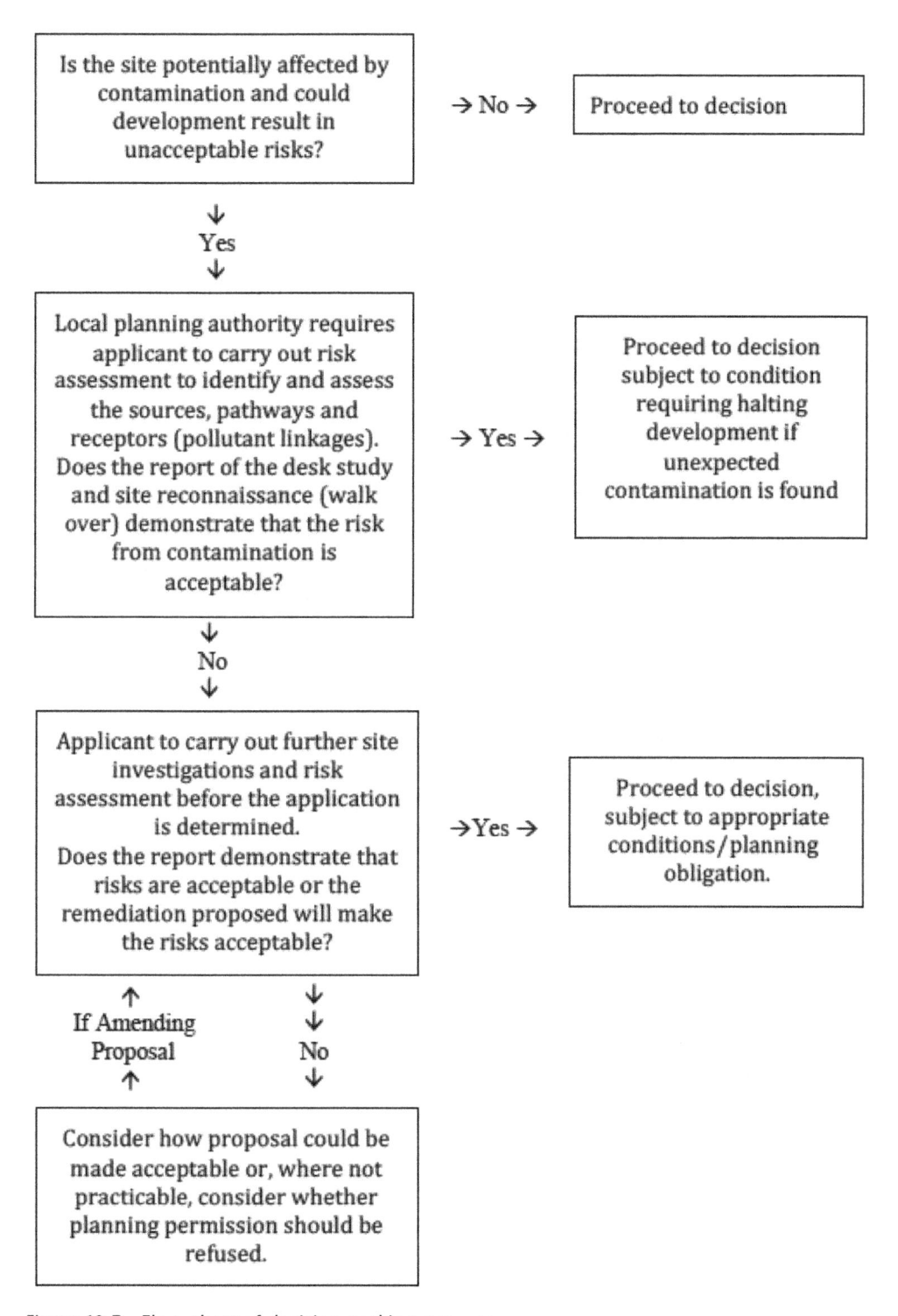

Figure 19.7 Flow chart of decision-making process

- Precise;
- Reasonable in all other respects.

A paragraph in the NPPF (currently at paragraph 56) states as follows:

> Planning conditions should be kept to a minimum and only imposed where they are necessary, relevant to planning and to the development to be permitted, enforceable, precise and reasonable in all other respects. Agreeing conditions early is beneficial to all parties involved in the process and can speed up decision-making. Conditions that are required to be discharged before development commences should be avoided, unless there is a clear justification.

Note the last sentence. There will always be strong resistance to planning conditions that must be discharged before the development commences. That is why, where a major remediation programme is required, it is advised that this be subject to a separate, stand alone, application. Sometimes these works can take several years and there can be long debate between the parties as to whether they have been completed and fully effective. Over the years the courts have considered each of the six tests in some detail and these decisions are, of course, *material*.

For the planner who is not a specialist in environmental control, and for the EHP who is not a specialist in planning practice, it is impossible for each to understand fully what sort of condition might comply with all the tests. That being the case, it must again be emphasised how important it is for the two disciplines to work closely together.

Unfortunately, as a result of lack of resources, experience or both, it is not uncommon for planning conditions to be imposed which do not comply with one, or more of the tests.

Throughout the author's career many badly written conditions, which are either unnecessary, irrelevant, vague, unenforceable and/or unreasonable have been witnessed. In one recent case, at a planning enquiry neither the local authority nor the applicant provided any expert evidence on environmental impact. As a result, the Planning Inspector granted permission with conditions which prohibited the development from proceeding. This was clearly *ultra vires* but the aggrieved party could not afford the costs of appeal so to this day the development remains void and the site detrimental to amenity of the neighbourhood.

Model conditions – As mentioned, Annex A of Circular 11/95 has been retained. In relation to the development of known or potentially contaminated sites, there are only four examples. It is advised that these be ignored as they are rely too heavily on LPA approval. For example, the first states the following:

> 56. Development shall not begin until a scheme to deal with contamination of the site has been submitted to and approved in writing by the local planning authority.

National policy (NPPF current paragraph 184) makes very clear that the responsibility for securing safe development rests with those responsible for the development of the site. Conditions like this please developers as it enables them to potentially pass all the responsibility to the LPA.

The following are a few important points to note:

- *Contaminated Land* is a very broad and complex subject and despite the developer consulting a consultancy with an experienced practitioner at Director level

---

Cambridge News headline of 13th October 2010

600 complaints as chemical factory clean-up continues

Figure 19.8 Weak planning controls and inadequate enforcement can result in serious health impacts during the remediation process and great public disquiet

when seeking quotes for site investigation work, it is often then carried out by relatively inexperienced individuals.

- High-quality desk studies are a fundamental part of the process. Environmental reports can be purchased from suppliers who provide no explanation or interpretation. These are useful as a foundation to a report but a lot more work is necessary to complete a comprehensive desk study. If an 'off the shelf report' is provided without a detailed evaluation of the data it is of no value. These are often submitted at low cost (as a loss-leader) with a recommendation for further on-site investigations, which may not be necessary.
- Where a comprehensive site investigation is necessary, a decision has to be made whether planning permission is required. This is particularly important where contaminants may be mobilised in the process.
- Note that planning officers are rarely keen on requiring planning permission for site investigations or even remediation programmes. In the author's opinion it is important to fight for a separate application where a remediation programme is complex, may take many months or even years, and risks are high.
- On-site investigations are easier to get wrong than right given the matters that have to be taken into account such as:
- How may boreholes?
- How many trial pits?
- How deep?
- Which locations?
- How many soil samples?
- How many liquid samples?
- Above or below L/D NAPLs, or an emulsion?
- Analyse for what – inorganics, organics, fibres, explosives, products?
- How many analyses are enough?
- How do you interpret the data?
- Which of the many complex algorithms should be used to quantify risk?
- Risk to whom, which or what?

The reality is that if one were to employ ten different consultants then ten different results could be provided at each stage of the process. The LPA is reliant on others to do the job properly and often they do not. It will always be difficult, and usually impossible, to 'approve in writing' with any confidence, and so it is advisable not to do so. Should things go awry, the responsibility must always lie with the developer, not the LPA or the EA. In many instances developers or their consultants or remediation engineers or contractors or subcontractors have argued as a defence for failure to address issues effectively that, 'Everything was all approved by the council and/or the Environment Agency'. This leaves the officers exposed. If nothing is approved in writing, what should be done? Firstly, for any site that gives rise for concern, pre-application discussions are paramount. The officer should explain, and provide in writing only the following information:

- The site is known to be contaminated. Please confirm that you are aware that the responsibility for safe development and secure occupancy of the site lies with the developer. Please provide details of whom this person or organisation is currently and will be post-development.
- All work relating to the contaminated site must be undertaken by 'competent persons'. Please confirm that you are aware that it is the developer's responsibility to employ suitably qualified and experienced competent persons to advise. When appointed please provide their contact details and hierarchy of responsibility.
- The LPA will have no formal relationship with any of the developer's advisors.
- Please be aware that the planning application will be determined by the LPA on the basis of the information available to it.

It is important to understand that developers rarely want to accept long-term responsibility for a complex site. It is common practice to try to get the LPA or EA to approve everything submitted to it. If that fails responsibility is passed to consultants, advisors or contractors. In many cases a new limited company, with little or no assets is created to develop a site and following development the company is dissolved.

It has always been the author's view that there should be a standard condition on all permissions where remediation is necessary requiring developers to certify that the works have been carried out correctly. Experience indicates they do not like doing this and will try to get anyone to sign it other than themselves (e.g., consultants or remediation contractors). If included as a formal condition this could not be discharged until properly completed and submitted to the LPA. This would then form part of the documentation on planning available to property conveyancers to enable them to hopefully re-assure prospective purchasers the site had been made safe. If there is no such condition imposed, the LPA are usually asked this question, and without having someone on site full time, they can never know.

It is not appropriate to provide a complete list of possible conditions here. EHPs should follow the guidance and make sure conditions are carefully drafted and comply with the *Newbury* tests.

Planning obligations

This subject area of planning is known as *planning gain*. The concept has been conceived to enable developers to provide 'things' that support a development such as schools, medical centres, highway improvements, parks, etc., or monies towards them. Positive change can be achieved through planning gain, whereas conditions tend to be restrictive. Section 106 of the Town & Country Planning Act 1990 introduced planning agreements, later to be known as planning obligations. They are still commonly referred to as 'section 106 agreements'.

These additions may allow a development to proceed that otherwise would be refused. They are directly linked to the impact caused by the proposed development and designed to make it acceptable.

Planning policy (NPPF currently paragraph 57) states that planning obligations

---

Completion Certificate:

Condition – Upon completion of the works of remediation and reclamation the following Certificate must be completed and returned to the LPA. It must be completed by the person or company developing the site, not by any third party such as advisors or contractors.

This is to CERTIFY that the scheme of decontamination and reclamation at the site(s) known as . . . . relating to planning approval . . . . was carried out between the dates of . . . . and . . . . and was completed to the specifications detailed in documents reference[1] . . . . and plan numbers[1] . . . . which were designed to afford protection from contamination[2] to all relevant receptors[2].

Signed . . . .

The developer of the site

Type here name, position and address

[1] If an extensive list please include on a separate sheet

[2] Contamination and receptor to have at least the same meaning as in Part IIA of the Environmental Protection Act 1990 plus any others considered and previously agreed to be relevant to this particular development

---

Figure 19.9: Example of Completion Certificate

must only be sought where they meet all of the following tests:

(a) necessary to make the development acceptable in planning terms;
(b) directly related to the development; and
(c) fairly and reasonably related in scale and kind to the development.[22]

Community Infrastructure Levy (CIL) – Historically there have been concerns relating to the use of planning agreements and many have been successfully challenged in the courts due to not meeting one or more of the tests. For example, there can be a temptation for local councils to seek funding for highway improvements or similar where relationship to the development is tenuous. This has resulted in the creation of the CIL which is, in effect, a charge that councils can impose to achieve infrastructure benefits. The aim is that this should be more straight forward and fairer and as such, become the principal planning gain tool.

Whereas the planning agreement is designed to focus on the area local to the site in question, the CIL goes into one large pot to be spent on infrastructure needs across the local Government area, thereby breaking the link between the development and the gain. It is, in effect, a development tax.

The Community Infrastructure Regulations 2010 were amended in 2019.[23]

Environmental Impact Assessment (EIA)[24]

EIA is a detailed process that must be followed to ensure that environmental impacts of a certain development are taken into consideration at the application stage. The Town & Country Planning (Environmental Impact Assessment) Regulations 2017,[25] revoked the previous 2011 Regulations, but there are transitional arrangements which should be considered carefully.

The decision as to whether or not a development requires an EIA is not straight forward. Those developments listed in Schedule 1 of the Regulations always require an EIA, whereas those listed in Schedule 2 may require one, depending on the scale of the development.

Screening opinions – Can be provided by the LPA to decide whether an EIA will be necessary. To do this the developer must provide sufficient information about the nature and scale of the proposal.

Scoping – May then follow to decide the requirements of an Environmental Statement (ES). This matter is complex and must be undertaken by persons sufficiently competent to decide what type and depth of detail will be necessary for the LPA to properly judge the effects of the development on the environment. A 'scoping opinion' must be provided within five weeks to allow the developer to prepare the ES as part of the planning application.

To demonstrate the difficulties this presents to LPAs, who may have very limited in-house expertise on environmental impacts, the author has experience of Judicial Review (JR) of screening opinion for a large remediation programme. The applicants (for the JR) were successful in challenging the LPAs screening opinion that an ES was not required. When confronted with the evidence, the LPA did not contest the matter in court and agreed to pay all costs associated with the case. The full EIA process was then followed. While actions like this can be considered a great success potentially protecting and enhancing the environment, they are not quite what they may seem.

Unfortunately, whilst it can be relatively straight forward to challenge procedure, it can be almost impossible to challenge the information a developer submits as part of the EIA process. The interaction of effects within and between the different compartments of the environment, and how organisms and humans are ultimately affected, can be so complex that submissions are taken on face value, regardless of their quality.

For this reason, it is important to emphasise how important it is to have well-qualified and experienced EHPs within the LPA to assist and advise decision makers.

Enforcement

Planning enforcement in relation to complex environmental issues is another difficult subject area which planning officers cannot hope to deal with without EH input.

The NPPF states at (current) paragraph 59:

> Effective enforcement is important to maintain public confidence in the planning system. Enforcement action is discretionary, and LPAs should act proportionately in responding to suspected breaches of planning control. They should consider publishing a local enforcement plan to manage enforcement proactively, in a way that is appropriate for their area. This should set out how they will monitor the implementation of planning permissions, investigate alleged cases of unauthorised development and take action where appropriate.

Every LA should have written enforcement plans/policies for both EH and planning issues. These should interrelate and make clear where environmental regulation and enforcement takes precedence over planning, and vice versa.[26]

Climate change

The importance of climate change is evident from having a dedicated chapter in this book and the topic being raised in almost every other chapter.

> The latest research shows that the world's soils hold three times as much carbon as the entire atmosphere. As part of the momentous international discussions taking place on climate change in this city at this very moment, your

Government is launching an important initiative to draw attention to this very point. They have calculated that if the quantity of carbon contained in soils could be increased by just 0.4% per year, the annual increase in carbon dioxide in the atmosphere could be halted. The same measures would, of course, also improve soil fertility and therefore our ability to feed a growing population. [22]

Scientists have been aware of the potential for climate change arising from anthropogenic emissions into the atmosphere for many years. John Sawyer was Director of research at the British Meteorological Office (1965–1976) and President of the Commission for Atmospheric Sciences at the World Meteorological Organisation (WMO) (1968–1973). In 1972 he published a study in which he accurately predicted the rate of global warming for the period between 1972 and 2000:

> The increase of 25% CO_2 expected by the end of the century therefore corresponds to an increase of 0.6°C in the world temperature − an amount somewhat greater than the climatic variation of recent centuries. [23]

In 1988 the WMO established the Intergovernmental Panel on Climate Change (IPCC),[27] now an internationally accepted authority with its work widely agreed upon by leading climate scientists and the consensus of participating Governments (currently 195).

Evidence at this time was overwhelming, therefore as part of a technical advisory committee the author wrote a report in 1990 for six local authorities titled simply, *The Greenhouse Effect*. The preface read as follows:

> Whether or not global warming is occurring now, or will occur in the future, scientific research demonstrates that human activities are adding various pollutants to the atmosphere which are detrimentally

affecting the fine balance of the gases which have given rise to and supported life on Earth to the present day. It is essential therefore that given this knowledge man takes whatever steps are necessary to protect his environment by reducing emissions of these pollutants. [24]

The aim of the report was to explain the causes and mechanisms, likely results and how local authorities could contribute to its control. One of the recommendations was that planning policy should form part of each local authority's Corporate Environmental Strategy designed to counteract the threat of global warming.

The report was dismissed, and no action was taken. Unfortunately, this seemed to be the response worldwide and instead of emissions reducing, they continued to increase. This fact, together with widespread deforestation and release of carbon dioxide and methane from the arctic tundra, accelerated the process. It is estimated that about one-third of the Earth's soil-bound carbon is in the taiga and tundra areas.

It is now nearly 50 years since John Sawyer's ground-breaking report, yet only now are we beginning to take seriously the impact of new development on climate change.

The NPPF says at (current) paragraph 20 (d):

> Strategic policies should set out an overall strategy for pattern, scale and quality of development, and make sufficient provision for:
> Conservation and enhancement of the natural, built and historic environment, including landscapes and green infrastructure, and planning measures to address climate change mitigation and adaptation.

Section 14 expands with six more paragraphs which can be summarised as follows:

- LDPs should plan for climate change (resilience to flooding, rising temperatures, etc.);

- New development should do likewise including planning for green infrastructure and measures to reduce greenhouse gas emissions;
- LPAs should encourage low carbon energy, both small and large scale sources.

Land use is not specifically mentioned but whilst re-using despoiled land is clearly a sustainable initiative, remediation can be a highly energy intensive process and building straight onto a clean piece of land next door could be argued as a 'greener' alternative when considering climate change. This demonstrates the difficulty presented to decision makers when considering what weight to give points on each side of the scale. More information and guidance is offered in the relevant Planning Guidance.[28]

The Climate Change Act 2008 requires the UK to reduce greenhouse gas emissions by 80% from 1990 levels, by 2050. The emphasis, however, is on adaptation for the inevitable changes which are to be expected as global temperatures increase.

Section 3 Notes

14 Administrative law is seen as that body of law dealing with the establishment, duties, rules and powers of government administrative agencies and is part of constitutional law.

15 https://www.gov.uk/guidance/when-is-permission-required

16 Resolution of the UN General Assembly 42/187 on the Report of the World Commission on Environment and Development says sustainable development 'implies meeting the needs of the present without compromising the ability of future generations to meet their own needs'.

17 https://www.gov.uk/government/publications/national-planning-policy-framework-2

18 https://www.gov.uk/government/collections/planning-practice-guidance

19 https://www.gov.uk/guidance/land-affected-by-contamination

20 https://www.gov.uk/guidance/use-of-planning-conditions

21 *Newbury District Council v Secretary of State for the Environment* [House of Lords] before Viscount Dilhorne, Lord Edmund-Davies, Lord Fraser

of Tullybelton, Lord Scarman and Lord Lane [1981] AC 578, [1980] 1 All ER 731, [1980] 2 WLR 379, 78 LGR 306, 40 P & CR 148, 144 JP 249, 124 Sol Jo 186, [1980] JPL 325.

22 See https://www.gov.uk/guidance/planning-obligations

23 For more detailed guidance see https://www.gov.uk/guidance/community-infrastructure-levy

24 https://www.gov.uk/guidance/environmental-impact-assessment

25 SI 2017 No571 see https://www.legislation.gov.uk/uksi/2017/571/contents/made

26 https://www.gov.uk/guidance/ensuring-effective-enforcement

27 See https://www.ipcc.ch

28 See https://www.gov.uk/guidance/climate-change

SECTION 4: GETTING THE RIGHT SORT OF HELP

'Suitably qualified' or 'competent' persons

At some point in an EHP's professional development we all finally begin to appreciate the things we will never know. This is as important as understanding the things we must know. One of the finest skills any professional can develop is the ability to judge the limits and capabilities of others, for we always need help. Our metaphorical 'little black books' should be full of the telephone numbers of reliable people who know about the important things we don't.

Land contamination is not really a subject at all; it is a collection of subjects. There is a long list of 'ologies' and branches of science which relate to land, its contamination and remediation. These can include:

- Geology and soil science/hydrogeology/hydrology;
- Toxicology (human/eco/phyto);
- Biology/ecology/botany;
- Chemistry (organic/inorganic/analytical);
- Physics;
- Building science;
- Risk.

This list does not even take into account the difficulties introduced by the social, political and legal aspects of each case. This does not mean there is no need to have some insight into all these areas and this is what this book is intended to provide as a starting point.

Because of this breadth of subjects there can be no such thing as the complete land contamination expert. Whatever qualifications a person might have and great knowledge about, for example, remediation of dense non-aqueous phase liquid (DNAPL) contaminated groundwater, they would still not be a land contamination expert. Detailed knowledge about a specific subject does not mean they have knowledge of all aspects of *Contaminated Land*. So what are we trying to achieve?

It can be argued in the context of this chapter that the aim of the responsible EHP in a local authority must be to become a Land Contamination General Practitioner (LCGP) by utilising a degree of qualified common sense. This requires an all-round knowledge of the subject gained by a combination of qualifications and experience. As has been indicated, this is highly valued by critical decision makers such as the courts, planning inspectors and committees. An astute and well-experienced LCGP is in a position to offer a balanced view taking into consideration the reports and advice of the various specialists in each field. This is crucial as often these 'experts' are not impartial; they are being paid to achieve a defined aim, usually to obtain planning permission for a difficult site.

Employing an experienced LCGP is not an option for local authorities but an unequivocal requirement to enable them to undertake a series of essential statutory functions imposed by central Government. *Land affected by contamination* has to be identified, assessed, made safe and, more often than not, developed. All of these matters are the responsibilities of local authorities.

The current (2012) Statutory Guidance [25] has several paragraphs on using external expertise. These are a selection of statements (sometimes paraphrased) which demonstrate the importance of obtaining the very best advice:

> Developing an understanding of risks in complex cases may raise issues which are beyond the expertise of any one person and may require the involvement of others to conduct a robust risk assessment. (para 3.18)

> The question of who to consult depends largely on the circumstances of the land, and the expertise and gaps in expertise of the person doing the assessment. When choosing specialist consultants, local authorities should strive as far as possible to ensure they are appropriately qualified and competent to undertake the work. Authorities, or consultants working on their behalf, may also consider seeking advice from other relevant experienced practitioners. (para 3.19)

> External experts may advise the Local Authority on regulatory decisions under the Part 2A regime, but the decisions themselves **remain the sole responsibility of the local authority** (my emphasis). (para 3.20)

> The uncertainty underlying assessments means there is unlikely to be any single "correct" conclusion on precisely what is the level of risk posed by land, and it is possible that different suitably qualified people could come to different conclusions when presented with the same information. **It is for the local authority to use its judgment to form a reasonable view**. . . . (para 3.32)

> **The local authority has the sole responsibility** for determining whether any land appears to be *Contaminated Land*. . . . It may rely on information or advice of another body such as the Environment Agency, or a suitably qualified experienced practitioner appointed for that purpose. (para 5.5)

> The Guidance in this Section does not attempt to detail technical procedures or working methods. . . . The enforcing authority may consult relevant technical documents or a suitably qualified experienced practitioner. (para 6.4)

> If the Agency is to carry out an inspection the local authority should authorise a person the Agency nominates to exercise powers of entry conferred by Section 108 of the Environment Act 1995. . . . Where this happens compliance (with the law and Guidance) and the decision as to whether the land is *Contaminated Land* **remain the sole responsibility of the Authority**. (paras 2.14 and 2.15)

Section 108 gives local authorities the power to authorise, in writing suitable persons, to investigate potentially *Contaminated Land* on their behalf. The powers which a person may be authorised to exercise include:

- To enter at any reasonable time (or in an emergency, at any time) any premises to make such examination and investigations necessary;
- To take samples, photographs, carry out tests, install monitoring equipment, etc.

It is clear, the local authority officer not only has the onerous responsibility of deciding when the balance tips towards *Contaminated Land* from *land affected by contamination*, but s/he must also recruit the right persons to advise on the many complex aspects of each case. Neither the Act nor guidance considers what may constitute a suitable person for the purposes of the investigation and assessment of land. There is no list of approved consultants or any professional organisation which oversees the training of all land contamination specialists. There is no minimum qualification and no recognised qualification. Specialists and consultants come from a range of backgrounds including:

- Environmental health and science disciplines;
- Surveyors;
- Engineers;
- Geologists;
- Soil scientists;
- Hydrologists;
- Chemists.

Over the years a number of 'bodies' have arisen offering some form of qualification/certification/accreditation/quality mark, etc. Unfortunately, it is for the authority's EHP to decide the value (or otherwise) of these when seeking outside help.

In addition, there is also the difficult issue of conflict of interest. This is a small and specialised area of activity and most active practitioners know everyone in the business and so it is essential that when an enforcing authority seeks to employ someone to provide advice on land, that person (or their company) does not have any association with potential 'appropriate persons' or developers.

To successfully employ consultants or contractors a local authority must have in its possession:

- A 'client' with responsibility and knowledge;
- A tightly drafted specification;
- A choice of consultants and selection criteria;
- Monitoring procedures and time.

The client

Who will be the client? There is nothing a consultant or contractor likes better than a client who has little idea what s/he is talking about. This is where we come back to the LCGP. Employment of a well-qualified, well-experienced person with an all-round knowledge of the subject is essential.

As an example, the author once had the need to look at a major building project which had some relatively clean imported (tipped) material on site. Consultants undertook a soil gas survey and found a few percent CO_2 in the soil. They then proposed removing all the imported material from this large site, bringing in new soil, then constructing the buildings with ventilated subfloor voids and gas-proofed floor slabs. The number of vehicle movements alone was going to paralyse this busy area for weeks. A closer look at the data revealed that even though there was a few percent CO_2, the volumes were almost

insignificant, and the risk to the occupiers of the building with conventional construction would negligible. In the end the consultants were prohibited from removing any soil and agreed only to an upgrading of the slab in accordance with Building Regulations.

The author has been speaking to and training local authority staff on land contamination issues for more than 20 years and one of the most striking matters has been the wide variety of backgrounds and basic knowledge of the personnel. A typical group may range from a PhD in volcanoes to an ONC in building maintenance. Some may have worked for several years in pollution control, some only a week. All too frequently there is an incomplete foundation knowledge on land contamination.

It is essential therefore to identify staff who have responsibility for making decisions on the identification and development of land affected by contamination, and ensure they are well trained in the basics of the subject. If this can be achieved, then they will already be one step ahead of many of the consultants likely to be retained, mainly because they are often in the same difficult position. Their staff are generally drawn from graduates from a wide range of disciplines, with their most relevant qualification being experience.

Once identified and trained, managers must be aware that that person must be authorised to act as the client when appointing external help. They must have the power to appoint (maybe together with their manager), be the main point of contact, and have the responsibility to monitor the contract.

A sure way to ensure contractors get away with inadequate work is to leave them to their own devices and fail to provide a principal contact who has the time and ability to make decisions and guide the contract along.

Specifications

All too often those producing project briefs or specifications or tender documents, leave too much open and give the contractor too much leeway. T S Eliot expresses this well when considering the 'existence' (or meaning) of a poem:

> The poem's existence is somewhere between the writer and the reader; it has a reality which is not simply the reality of what the writer is trying to 'express', or of his experience of writing it, or of the experience of the reader or of the writer as reader. Consequently the problem of what a poem 'means' is a good deal more difficult than it first appears. [26]

There should be no room for poetic licence in the interpretation of briefs or specifications. It is essential therefore that the client understands exactly what the 'project' entails and is able to 'define' it in writing in a clear and concise form to produce an unambiguous specification. It will never be possible to make any realistic comparisons during the quotation or tendering process unless the client can be sure each contractor is quoting for exactly the same work.

A comprehensive foundation of knowledge is essential to ensure the client fully understands the project context and is then able to produce a well thought-out brief or specification.

How do you write specifications? – There is some modern guidance on this subject to be found, but the most complete dates back to 1997 – CLR 12 – *A Quality Approach for Contaminated Land Consultancy* [27], which states in the introduction:

> The report focuses on the consultant's responsibilities for quality while recognising that clients also have an important role to play, particularly in the pre-commissioning phases of selecting and appointing consultants. Clients who wish to understand more clearly what they can expect from consultants may, therefore, find this a useful reference text.

It would be impractical to go into any detail here as the subject is extensive, suffice to say before commencing any project work consultants must agree and reach a clear understanding with the client on the content of the project brief. To do this the consultants need to:

- Understand the client's rationale for initiating the project;
- Be clear about the client's purpose for the output of the project;
- Be aware of the way in which the client intends to define the project brief;
- Be specific about the type of project involved;
- Provide appropriate input into the definition of the project brief;
- Review any existing information on reports relevant to defining the brief;
- Understand the requirements of any third parties such as lenders, investors or other regulatory authorities to ensure their specific requirements are met;
- Highlight any issues which may affect the implementation or output of the project;
- Respond with all the information the client needs to commission the project.

The client must try to produce a clear and unambiguous specification, paying attention to quality assurance, best value and time tables. If this cannot be achieved the contractor will always have the upper hand and resources will inevitably be wasted.

Choice of consultants and selection criteria

Local authorities may choose to employ consultants to:

- Review their *Contaminated Land* strategy;
- Undertake detailed inspection of sites/ design strategic sampling programmes and interpretation of analytical results;
- Undertake site-specific risk assessment in order to determine whether or not land should be declared contaminated under Section 78B or C;
- Provide expert witness services in court, planning enquiries or disputes with the Environment Agency;

- Design appropriate and sustainable remediation strategies for *Contaminated Land*;
- Carry out or supervise remediation of *Contaminated Land* where there is serious risk to controlled waters or imminent danger to health, either by agreement with the owner, or as work in default of a remediation notice;
- Carry out or supervise remediation of local authority-owned *Contaminated Land*.

This is quite a long list of complex and varied tasks, so after the local authority have determined what is to be done, and written a comprehensive specification, how do they employ the right person/company for the job? As emphasised, neither the Act nor the guidance considers what may constitute a 'suitable person'. The Environment Agency also is not in a position to help. It is essential therefore that the person with the client responsibility becomes aware of the companies that offer services in the field and ascertains their capability. This is best done informally.

It's good to talk – It is easy to send a simple letter to all those on the list of environmental consultants, but they are likely to have been writing to you for years telling you what they can do. A written request will only elicit another large bundle of promotional material full of impressive statements and glossy brochures, but they never tell you what they cannot do. So, it is better to speak to them. Before any more formal selection procedure is started, prepare a list of specialist companies, their contact points, and telephone them with a pre-prepared list of questions. Speak to the person most ably qualified to answer any questions, and ensure that is the case.

As with any selection process it is essential to spend time producing carefully crafted questions and selection criteria. These will of course be linked. For example, the answer to the question, what can't you do, will help you rate against the criteria of 'credibility'. Other criteria may include:

- Background and qualifications;
- Key skills;
- Recent experience;

- Quality assurance/management systems (see Chapter 8);
- Financial probity/insurance;
- Affiliations;
- Accreditations;
- Independent recommendations;
- Credibility.

A brief checklist of information requirements is included at the end of this section.

Spend time producing well thought-out selection criteria – Make sure you know who exactly will be doing the job, if they cannot or will not say, then it will be impossible to make an effective decision as this type of work is heavily dependent on the skills and experience of the individual on the case. The manager may know all the answers, but s/he will no doubt be busy elsewhere.

Some years ago, after a lengthy selection process, the author appointed a firm of consultants to advise on a thorny problem which would undoubtedly result in long and complex litigation. It was imperative therefore to employ a company with the right knowledge and experience to support us through the legal process. After interviewing the candidates, it was relatively clear who should be chosen as the consultant and who would be responsible for the contract. That person was probably the most experienced in the field at that time. When, however, the field work was carried out it was undertaken by a young graduate, who also subsequently wrote the report. The final conclusions to this report would be quoted far and wide in courts, newspapers and on the TV. When they finally arrived on my desk they could only be described as derisory. I wrote to the firm expressing my concern and suggesting an alternative form of wording. In the end they rewrote the report using my conclusions and recommendations but still took their fee.

Monitoring the contract

If the specification is unambiguous, especially the parts relating to quality assurance, then exactly what the contractor should be doing, and when, will be known.

The specification for the remediation of a large contaminated site will inevitably be complex and identify several distinct stages. A local authority officer would, however, be unlikely to undertake such a duty due to the scale of the task and the time involved. Instead, it is likely that specialists would be appointed to oversee the whole project. Then a detailed specification would be required to appoint suitable consultants with a proven track record in producing comprehensive project briefs and supervising remediation contracts. A clear statement must always be included that makes it abundantly clear why the work is being carried out and what it will ultimately achieve. It is the responsibility of the consultant to achieve this.

It is essential to emphasise to the consultant/contractor that they will be expected to deliver the contract complete and on time. Penalty clauses and staged payments may be negotiated at an early stage. No final payment should be made until the contractor has signed a statement specifying that the work has been undertaken in accordance with the specification and to any agreed limitations as to accuracy or scale. No limitations should be introduced after the start of the work.

Beware of lengthy progress reports and verification documents that put the responsibility back on the client. If several volumes of carefully thought-out statements were agreed at the outset, all that is required is unambiguous confirmation that they were achieved.

Conclusions

The complex tasks facing local authorities in dealing with any *land affected by contamination*, either by negotiation, enforcement or development, will often be daunting. To achieve a satisfactory outcome it is inevitable that some assistance will be necessary from time to time from independent sources. It is imperative therefore that:

1. Adequate resources are made available;
2. Key staff are identified at the outset and they are adequately trained for the client role;
3. Close attention is paid to specifications and project briefs;
4. Consultants/contractors are chosen carefully and time is taken to produce carefully thought-out selection criteria;
5. Contracts are effectively monitored.

Table 19.1 Checklist of information requirements

Client's information requirements	Requirements of the consultant
1 General	
1.1 Background on company capability	How long has the company been operating?
	What kind of work were they originally set up to do – is this an add on?
	Who traditionally are their clients?
1.2 Numbers and qualifications of staff	If a large company, what are the interests/sympathies of those in control? Do they consider local authorities as a serious market?
1.3 CV and availability of key staff	How many staff are available for this type of work; will they need to subcontract?
	Who will actually be doing the job, what are their qualifications and experience? Practical experience is KEY.
	Do they really understand the subject?
	Knowledge of environmental law & local Government systems an important requirement.

Table 19.1 (continued)

Client's information requirements	Requirements of the consultant
1.4 Details of QA systems including: Allocation of responsibilities Project management Technical procedures Technical review Training Assessment of external suppliers	Where appropriate, need details of quality management systems indicating whether accredited by a third party. What technical procedures to be used. Which staff responsible, which will undertake technical review. How quality of subcontractors is to be ensured.
1.5 Management of Health & Safety	Identify H&S management procedures where appropriate. Do they understand the fundamental requirements of H&S legislation?
1.6 Track record on similar projects	Ever they done similar work or is this a new departure?
1.7 Client references	Need several telephone numbers to enable rapid verification of statements made at interview. No one will ever present bad written references so conversations with referees are essential.
1.8 Financial status	May not always be necessary, but on large contracts considerable financial outlay may be required to demonstrate solvency. Bond may be required on large remediation contracts.
1.9 Details of insurance cover	Need to demonstrate insurance available for third-party liability and professional indemnity. Identify limitations/exclusions (make sure no 'pollution' exclusion clause).
1.10 Membership of professional and trade associations	May be necessary to make checks. Do corporate membership of professional organisations meet CPD requirements?
1.11 Compliance with codes of practice	Can they demonstrate knowledge of the appropriate guidance, codes of practice, etc. relevant to the job?
2 Project specific	
2.1 Technical proposal	The proposal must make it absolutely clear that work will be carried out to comply with the requirements of the specification, what the results will be, and when they will be achieved.
2.2 Project management plan/working plan	A clear timetable must be available which states what stage will be reached by when and who will be responsible to deliver.
2.2 Project management plan/working plan	A clear timetable must be available which states what stage will be reached by when and who will be responsible to deliver.
2.3 Details of subcontractors	Subcontractors will be necessary on large technical projects. Must state who they are, contact points and lines of responsibility.
2.4 Details of technical procedures	Again, the working plan must clarify all procedures and lines of responsibility.
2.5 Reporting	Reporting procedures must be made absolutely clear. It is essential not to have masses of reports landing on the desk of the client officer which puts the responsibility back on him/her. The responsibility for doing what has been agreed to the agreed standard must lie with the contractor.

(continued on next page)

Table 19.1 (continued)

Client's information requirements	Requirements of the consultant
2.6 Programme and 2.7 Financial proposal	It may be that the contractor will want to provide a guide price or include large contingency sums. The programme of work and the quotation must not be ambiguous. A lot depends on the quality of the original specification. Stage payments and timetables must be firm and with perhaps penalty clauses if they fail to deliver on time.
2.8 Conditions of engagement	Contracts need not be long and wordy, should define responsibilities of both parties, liabilities, etc. succinctly.

Source: Adapted from CLR12 [27]

SECTION 5: REMEDIATION TECHNIQUES

Lady Eve Balfour was a British organic farmer, educator, pioneer and founding figure in the organic movement and soil association.[29] She was quoted more recently as saying:

> The health of soil, plant, animal and man is one and indivisible. [28]

Remediation of *Contaminated Land* can be divided into four broad areas – engineering, biological, physical and chemical treatment. Each is a complex science in itself and it would not be appropriate to go into detail here, but it is hoped the following simple overview of the types of technique will be of value.

Some methods are designed for soil, some for groundwater. There are inevitable overlaps between each but more often than not several techniques are used together on one site.

Engineering – These methods either remove the contamination so it can be taken away for disposal (known as 'dig and dump'), or prevent the receptor gaining access to it using a physical barrier.

Before contaminated soil is disposed of it may be screened, and for example clean concrete may be re-used on site as hardcore. This can significantly reduce volumes being removed for disposal. The contaminated material may be landfilled (on or off site) or incinerated.

Landfill sites have to be licensed to take the waste, and very few sites are suitably licensed to take hazardous waste, which includes most contaminated soil (see Chapter 18).

'Dig and dump' is very much frowned upon by the authorities as it is clearly not sustainable. It often requires endless lorry movements taking contaminated material away and bringing in clean and can be a major source of pollution and nuisance.

Physical barriers can be vertical to stop lateral migration of mobile contaminants, or horizontal. Vertical barriers are mainly used to stop contaminated liquids finding their way to water courses. Horizontal barriers (caps) restrict access to the contaminated material from receptors on the surface. They may also reduce the likelihood of mobile contaminants coming up through the cap (gases, or water by capillary action) and reduce the amount of water passing through in an attempt to reduce leachate and/or pressure problems. Caps may be mineral based (rock, clays, soil, etc.), synthetic or a combination of the two.

Normal engineering works such as roads and car parks may also form a physical barrier, breaking pollutant linkages.

Biological – Treatment relies on the natural flora and fauna (fungi and bacteria) in the soil to degrade the contaminants over time. This is a slow process so it invariably has to be accelerated by creating optimum conditions for the microbes to thrive. It only works on organics and some contaminants are toxic to the bacteria.

Bio-remediation comes in several forms but all are a form of composting. Bacterial communities in the soil are a vital element in mineralising organic molecules, whether a leaf or liquid hydrocarbons such as fuel oils and lubricants. Whilst bacteria can deal with a leaf, fungi are needed to break down a tree trunk. Similarly, bacteria can deal with simple short-chain liquid hydrocarbons, but fungi are required to break down more complex long-chain and aromatic hydrocarbons oils.

The process is aerobic and requires sufficient air, moisture, warmth, nutrients (nitrogen-phosphorus-potassium based such as those in animal manures) and a neutral pH. The better the conditions, the quicker the process.

Leachate can be a major problem and systems must be in place to collect, contain, treat and properly dispose of these dangerous toxic wastes.

The main bio-remediation techniques are windrow turning, landfarming, biopiles, bioventing and monitored natural attenuation.

Physical – This treatment includes the likes of thermal methods, soil vapour extraction, soil washing, air sparging, etc. These methods mostly use physical properties to wash or drive contaminants away for disposal elsewhere. Just as a dishwasher uses high temperatures, water and detergents to remove contaminants from plates for disposal, a soil washing process does much the same.

Soil washing has been developed from the long-established mineral processing techniques from the mining industry. Chemicals tend to stick (sorb) to fine soils such as silts and clays; these can then stick to coarser particles of sand and gravel. The aim of the process is therefore to separate the various fractions, and then re-use the clean material. This way volumes requiring disposal can be reduced considerably.

There are several thermal treatment methods, but all use heat to change the physical and/or chemical properties of the contamination to make it more mobile or allow it to be burned off. Heat is applied to accelerate removal of the contamination affecting both soils and groundwater. Limited heating (warming) can also accelerate biodegradation. Techniques include hot water flushing, steam injection, radio and electrical resistive heating, and thermal desorption. The latter usually falls short of incineration and is more like cooking the soil in a large oven to drive off contaminants. In some situations, however, if refractory linings are installed in the process reactor, incineration can occur. This is obviously a very quick, but energy-intensive process with inevitable polluting potential, therefore requiring authorisation and permit.

Chemical – Such treatment is generally applied to groundwater but as the soil and groundwater are in effect one, below the water table, these processes improve both media. The treatment may be based on oxidation, reduction or hydrolysis. Soil amendment or stabilisation can use both the physical and

---

Daily Telegraph headline of 30 July 2009

'Worst child poisoning case since thalidomide'

Council's negligence blamed for babies born after steel plant was torn down

---

Figure 19.10 The author worked for four years as Expert Witness for the claimants in the Corby Steelworks case. The bulk of the environmental impact and the detrimental effect on the unborn child arose from one seriously flawed remediation technique, 'dig and dump'.

chemical properties of certain additive materials to change the ability of contaminants to become available to receptors, or leach from the soil. Contaminants can also be dissolved using solvents so they can be removed in the leachate. An engineered physical barrier can be designed to treat leachates as they pass through it (like a filter). These are known as permeable reactive barriers (PRBs).

Generally speaking, all three treatment processes can be undertaken either in situ (in the ground) or ex situ (on the surface or in a special piece of equipment or reactor). To undertake any of the treatment processes, a detailed understanding of the soil and its contaminants is necessary. For example, there will be no point in commencing a bio-remediation process if some of the contaminants are toxic to the bacteria. Similarly, it is important not to commence a chemical oxidation or reduction process if adverse reactions may result.

Treatability studies are therefore necessary, and the soils or waters must be comprehensively assessed to ensure they only contain the contaminants the process is designed to treat.

Author's postscript

Contaminated Land is a complex area for EHPs and here I have sought to provide a practical route through the intricacies of the subject and give a personal perspective from more than 25 years working in the field. An attempt has been made to highlight some of the pitfalls for the local authority EHP, but also provide positive pointers.

I have always used an informal style when writing and in training with the aim of forming a bridge between the scientist in the laboratory and the earthworks contractor on site. Historically this has been appreciated by my students, planning inspectors and the courts – most especially the courts.

The 'quotes' I have used are from learned people from all walks of life, and will hopefully help to fit care for this key part of our environment into the broader picture, that of care for the whole. Readers will note that quite a few link to papers and speeches by the present Prince of Wales. This is not just because he writes extensively on these subjects, but because I have written permission to use anything from his published work, which I am grateful for.

Finally, I will finish with one such quote, from the Indian philosopher Rabindranath Tagore. It should make you think:

> our consciousness of the world, merely as the sum total of things that exist, and as governed by laws, is imperfect. But it is perfect when our consciousness realises all things as spiritually one with it, and therefore capable of giving us joy. For us, the highest purpose of this world is not merely living in it, knowing it and making use of it, but realising our own selves in it through expansion of sympathy; not alienating ourselves from it and dominating it, but comprehending and uniting it with ourselves in perfect union [29].[30]

In my introduction I quote the native American, Chief Seattle:

> The Earth does not belong to man, man belongs to the Earth. [2]

For the sake of humanity, we all need to reunite ourselves with the environment that sustains us.

Section 5 Notes

29 See https://www.soilassociation.org/who-we-are/our-history/ and https://www.soilassociation.org/who-we-are/
30 Quote from Rabindranath Tagore, Indian poet and philosopher, the first Nobel Laureate from Asia winning the Nobel Prize for Literature in 1913.

Chapter references

[1] Roosevelt FD. (1937) *Letter to all State Governors on a Uniform Soil Conservation Law*, 26 February, and quoted by Fitzsimons M (2021) In Article in the *New Statesman*, 2 June.

https://www.newstatesman.com/spotlight/energy/2021/06/nation-destroys-its-soil-destroys-itself.

[2] Chief Seattle. (1855) From a Letter Supposedly Written to President Franklin Pierce and Attributed to Chief Seattle or Sealth of the Duwamish, but Unverified. Also Spelled Seathle, Seathl, or See-ahth, He Was a Leader of the Suquamish and Duwamish Native American Tribes.

[3] DoE Interdepartmental Committee for the Redevelopment of Contaminated Land. (1979) *ICRCL 16/78 Redevelopment of Contaminated Land: Acceptable Levels of Contaminants in Soils (consultation paper)*, Department of the Environment, London.

[4] Wilson G. (1898) *A Handbook of Hygiene and Sanitary Science*, JA Churchill, London, 8th ed.

[5] DoE. (1988) *Problems Arising from the Redevelopment of Gas Works and Similar Sites,* Environmental Resources Ltd., DOE, London. ISBN:0117520985.

[6] Kelly RT. (1979) *Site Investigation and Material Problems*, Proceedings of the Conference, Reclamation of Contaminated Land, Eastbourne, the Society of Chemical Industries, Eastbourne.

[7] HSE. (1991) *The Protection of Workers and the General Public During the Development of Contaminated Land*, Health & Safety Guidance Note HS(G) 66 (no longer current) HSE, Bootle.

[8] BSI. (2017) *Investigation of Potentially Contaminated Sites*, Code of Practice – Code of Practice BS 10175:2011+A2:2017, British Standards Institution, London. https://shop.bsigroup.com/products/investigation-of-potentially-contaminated-sites-code-of-practice-code-of-practice.

[9] MAFF. (1998) *Code of Good Agricultural Practice for the Protection of Soil*, Ministry of Agriculture Fisheries and Food, London.

[10] Batters M. (2017) *Open Letter on Soil Health*, Minette Batters President of the UK National Farmers Union, London.

[11] DoE. (1990) *Pilot Survey of Potentially Contaminated Land in Cheshire: A Methodology for Identifying Potentially Contaminated Sites*, Report prepared by Environmental Resources Ltd, DoE, London.

[12] DoE. (1994) *Framework for Contaminated Land: Outcome of the Government's Policy Review and Conclusions from the Consultation Paper "Paying for our Past"*. DoE Contaminated Land and Liabilities Division, London.

[13] Holdgate MW. (1979) – *A Perspective of Environmental Pollution*, Cambridge University Press, Cambridge.

[14] Royal Commission on Environmental Pollution (RCEP). (1984) *Tenth Report: Tackling Pollution – Experience and Prospects*, Cmd 9149, HMSO, London.

[15] Becket MJ, Simms DL. (1984) *The Development of Contaminated Land,* Proceedings of the Conference, Hazardous Waste Disposal and the Re-Use of Contaminated Land – The Society of Chemical Industries, London.

[16] Rodericks JV. (1992) *Calculated Risks – Understanding the Toxicity of Human Health Risks of Chemicals in Our Environment,* Cambridge University Press, Cambridge.

[17] British Medical Association. (1990) *Living with Risk: The BMA Guide,* Penguin, Harmondsworth.

[18] Cairney T. (Ed.) (1993) *Contaminated Land, Problems & Solutions*, Blackie Academic & Professional, Glasgow.

[19] Prince of Wales. (2009) *A Speech by HRH the Prince of Wales at the Sustainable Cities for a Better Life Seminar*, Venice, Italy. https://www.princeofwales.gov.uk/speech/speech-hrh-prince-wales-sustainable-cities-better-life-seminar-venice-italy (Accessed August 2021).

[20] Carson R. (1953) *Letter to the Editor of the Washington Post*. https://www.fws.gov/refuge/Rachel_Carson/about/rachelcarsonexcerpts2.html.

[21] Prince of Wales. (1992) *From a speech to the Brundtland Commission, the World Commission on Environment and Development*, London, 22 April. https://www.princeofwales.gov.uk/speech/speech-hrh-prince-wales-brundtland-commission-world-commission-environment-and-development (Accessed August 2021).

[22] Prince of Wales. (2015) *From a Speech Accepting the Prix Francois Rabelais a the Institut de France, Paris*, 30 November. https://www.princeofwales.gov.uk/speech/speech-hrh-prince-wales-upon-accepting-prix-francois-rabelais-institut-de-france (Accessed August 2021).

[23] Sawyer J. (1972) Man-made carbon dioxide and the "greenhouse" effect. *Nature*, 239: 23–26. https://doi.org/10.1038/239023a0.

24 Braithwaite RD. (1990) *The Greenhouse Effect*, Report for the Warwickshire Environmental Protection Council, November, Unpublished.

[25] Defra. (2012) *Environmental Protection Act 1990: Contaminated Land Statutory Guidance*, HMG Crown Copyright. https://assets.publishing.service.gov.uk/government/uploads/system/uploads/

attachment_data/file/223705/pb13735cont-land-guidance.pdf (Accessed August 2021).

[26] Eliot TS. (1933) *In Use of Poetry and the Use of Criticism – The Complete Prose of TS Eliot Critical Edition*, Volume 4, English Lion 1930–1933, 30. https://tseliot.com/prose/the-use-of-poetry-and-the-use-of-criticism (Accessed August 2021).

[27] DETR. (1997) *Contaminated Land Research (CLR) Report 12: A Quality Approach for Contaminated Land Consultancy*, Department of the Environment, Transport and the Regions, London. http://www.eugris.info/envdocs/CLR12_00.pdf (Accessed August 2021).

[28] Balfour Lady E. (1926/1996) *Quoted in the Lady Eve Balfour Memorial Lecture by HRH Prince of Wales*, 19 September. https://www.princeofwales.gov.uk/speech/speech-hrh-prince-wales-50th-anniversary-soil-association-1996-lady-eve-balfour-memorial (Accessed August 2021).

[29] Rabindranath Tagore. (1913) *Quoted in a Speech at the Unveiling of a Bust of Tagore, by HRH Prince of Wales 7th July 2011, at Gordon Square*, Bloomsbury, London. https://www.princeofwales.gov.uk/speech/speech-prince-wales-unveiling-bust-rabindranath-tagore-gordon-square-bloomsbury-london (Accessed August 2021).

Principal Contaminated Land Legislation and Guidance

Legislation

Environmental Protection Act 1990 Part 2A.
Contaminated Land (England) Regulations 2006 (UK SI 2006 No 1380) as amended by the Contaminated Land (England) (Amendment) Regulations 2012 (UK SI 2012 No 263).
Radioactive Contaminated Land (Enabling Powers) (England) Regulations 2005 as amended by Radioactive Contaminated Land (Modification of Enactments) (England) Regulations 2006 and Radioactive Contaminated Land (Enabling Powers and Modification of Enactments) (England) (Amendment) Regulations 2018 (UK SI 2018 No 429).

The Statutory Guidance

Environmental Protection Act 1990: Part 2A.
Contaminated Land Statutory Guidance – April 2012.
Environmental Protection Act 1990: Part 2A.
Radioactive Contaminated Land Statutory Guidance – June 2018.

Non-statutory guidance

Land Contamination Risk Management – October 2020 (Updated April 2021).
This replaces CLR11 (Model Procedures for Land Contamination).

British standards

BS 10175:2011 + A2:2017: Investigation of potentially contaminated sites – Code of practice.
BS 8576:2013: Guidance in investigations for ground gas. Permanent gases and volatile organic compounds (VOCs).
BS 5930:2015+A1:2020: – Code of practice for ground investigations (to assess the suitability of sites for construction operations and civil engineering works).

20

Noise and vibration

Andrew Colthurst and Steve Fisher

SECTION 1: BASIC ACOUSTICS

Introduction

This chapter is a comprehensive review of noise and acoustics as a specific topic for EHPs. It addresses the subject in different settings and so will overlap to some extent with other chapters which address the work (occupational) and domestic environments. It also addresses how noise as a matter of concern to the public, as witnessed by successive CIEH Noise Surveys, is controlled and how legislation and guidance attempts to reduce noise in the environment. As the CIEH has said, 'noise is the single largest issue of complaint made to local authorities in the UK',[1] and it is the view of the World Health Organization, that the burden of disease from noise is second in magnitude only to that from air pollution.

'Introduction' Notes

1 The CIEH has for a number of years carried out an annual noise survey that provides the most authoritative data on noise in the environment see https://www.cieh.org/news/press-releases/2021/cieh-releases-latest-noise-complaints-statistics-for-england/ and https://www.cieh.org/policy/campaigns/noise-survey/

Sound

A common definition of noise describes it as unwanted sound, the latter being the physical phenomenon and the former a subjective interpretation of sound by the listener. The definition of the word noise, as applicable to statutory nuisance, includes vibration.

Sound is a wave of consecutive increases and rarefactions of pressure within a medium, usually air in the context of environmental noise. The behaviour of sound as a wave is predictable by physical theory but often is complicated by the many potentially influential factors present in the environment that exists beyond the laboratory. To this physical complexity can be added that of the individual's response which may depend upon a range of variables including their health, situation and attitude.

In the great majority of situations, sound comprises energy at many different frequencies rather than the single or narrow band typically associated with tonal noise. The frequency content of sound is one of the main features defining its character for the listener,

DOI: 10.1201/9781003035640-20

others being the volume and the constancy or intermittency.

Noise is sometimes categorised as environmental noise, neighbour noise, neighbourhood noise and occupational noise. The Environmental Noise Directive or END (2002/49/EC) [1] defines environmental noise in the following terms:

> "'environmental noise' shall mean unwanted or harmful outdoor sound created by human activities, including noise emitted by means of transport, road traffic, rail traffic, air traffic, and from sites of industrial activity."

Neighbour noise is that which is created by people going about their lives, both inside and outside their homes. For example, it would include noise from entertainment equipment and domestic appliances as well as general airborne and impact noise from the occupation of a dwelling. Examples of external neighbour noise would be lawnmower noise, barking dogs, noisy parties and DIY activities. Neighbourhood noise includes noise from within the community such as industrial and entertainment premises, trade and business premises, construction sites and noise in the street and might be considered to extend to gatherings, such as those in the outdoor smoking areas of pubs. Occupational noise is that which is experienced in the workplace.

Sound power and sound pressure

The sound power level (L_W) is a dimensionless term. It is a logarithmic measure of the sound power against a reference level as shown in the following equation:

$$L_W = 10 \log_{10} (W / W_0) \, dB \qquad \text{[Equation 1]}$$
$$\text{where } W_0 = 10^{-12} \, W$$

The sound pressure level (L_p) also is not an absolute scale but a simple comparative scale relating two different pressures:

$$L_p = 10 \log_{10} (P / P_0)^2 \, dB \qquad \text{[Equation 2]}$$
$$= 20 \log_{10} (P / P_0) \, dB$$
$$\text{Where } P_0 = 2 \, 10^{-5} \, N / m^2$$

The relationship between sound pressure and sound power is given by the following equation which assumes that noise radiates spherically:

$$L_p = L_W - 20 \log_{10} r - 11 \qquad \text{[Equation 3]}$$
$$\text{where } r = \text{distance} (m)$$

Where the noise source is located on the ground and there is a non-absorbing ground cover between the source and receptor the relationship becomes:

$$L_p = L_W - 20 \log_{10} r - 8 \qquad \text{[Equation 4]}$$
$$\text{where } r = \text{distance} (m)$$

On the basis of Equation 4, where the specified distance is 10 m, the sound pressure level would be 28 dB less than the sound power level.

It is important when describing a sound pressure level that the distance between the measurement or prediction position and the noise source should always be specified.

Decibel

Sound is measured in decibels (dB), and that provides a means of representing a very wide numerical range of sound pressure levels in a more convenient form, albeit that the decibel scale is logarithmic rather than linear. For example, 0 dB, which sometimes is referred to as representing the threshold of hearing for a healthy young adult, equates with a pressure of 0.00002 Pa (20μPa), whilst 140 dB, often cited as the threshold of pain, represents a pressure of 200 Pa.

Logarithmic addition and subtraction of decibels is a straightforward calculation using the following basic formula:

$$L_{pTotal} = 10\log((10^{\wedge\,(L_{p1}/10)})$$
$$\pm(10^{\wedge(L_{p2}/10)})) \qquad \text{[Equation 5]}$$

For general purposes in environmental noise assessment the approximate noise level sums shown in Table 20.1 are useful to remember.

In terms of sound energy two sources of equal energy added together will increase their individual sound level by 3 dB and each subsequent doubling of the sound energy would raise the overall sound level by a further 3 dB. Summation of ten identical sources would increase the sound level by 10 dB.

Frequency

In addition to the sound pressure level it is common to describe sounds by reference to their frequency spectrum content. Frequency is described in terms of cycles per second or, more commonly, Hertz (Hz).

The frequency of audible sound for a healthy adult typically ranges between 20 Hz at the bass end of the spectrum and 20 kHz at the top end. To simplify the description of such a wide range, the frequency spectrum often is divided into octave bands or the narrower 1/3 octave bands, each of which is referred to by its centre frequency as defined by international convention. These centre frequencies are set out in Table 20.2.

Whilst a description of a noise by means of octave bands is adequate for most

environmental noise there are situations where a more specific frequency may be of interest, for example a pure tone such as a whistle or

Table 20.2 Octave and 1/3 octave band centre frequencies

Piano scale (note and Hz)	1/3 octave band centre frequency (Hz)	Octave band centre frequency (Hz)
	12.5	
	16	16
	20	
	25	31.5
28	31.5	
	40	
	50	63
	63	
	80	
	100	125
	125	
	160	
	200	250
	250	
C262	315	
	400	500
A440	500	
	630	
	800	1,000
	1,000	
	1,250	
	1,600	2,000
	2,000	
	2,500	
	3,150	4,000
4186	4,000	
	5,000	
	6,300	8,000
	8,000	
	10,000	
	12,500	16,000
	16,000	
	20,000	

Table 20.1 Examples of the logarithmic addition of decibels

Difference between the two sound levels (dB)	Change in sound levels by addition or subtraction (dB)
10	0
5	±1
0	±3

the resonant frequency of a panel. In these cases a narrow band analysis may be appropriate using measurement filters that can provide a much finer resolution of the sound spectrum, even down to a single frequency if necessary.

Knowledge of the frequency spectrum of a noise may be helpful when undertaking detailed predictions of the performance of a noise barrier, or a building façade and when specifying mitigation measures to control noise breakout by means of an acoustic enclosure or an attenuator (silencer).

Generally, there is a preponderance of low frequency energy in environmental noise. Some is due to wind noise and industrial plant but, particularly in urban situations, much of it is traffic related. Over distance, the high and medium frequency components can dissipate or be absorbed by air and ground conditions, leaving mostly low frequency noise. This often results in a modification of the frequency spectrum of noise that has been transmitted over a long distance. Within buildings the airborne sound insulation provided by a single layer wall or floor separating adjoining rooms theoretically obeys the 'mass law' which indicates that a doubling in mass produces an increase of about 6 dB in insulation and that there is an increase of the same order when the frequency is doubled. Although the difference in practice is usually reduced, the effect is for the low frequency sound to be attenuated less by the separating partition than mid and higher frequency sound, resulting in greater transmission of low frequencies from the source to the receiving room.

The diesel engines of a ship some considerable distance off-shore may be heard several miles inland as a low frequency throbbing sound because of the selective attenuation of the middle and high frequencies with increasing distance from the source. Similarly, voices in a next-door room will sound lower in pitch than they would do in the room where the conversation is in progress because of the greater loss of middle and higher frequency sound energy as it passes through the intervening partition. It is common for complaints about noise from music in neighbouring premises to focus on the bass notes or the beat of the drums and in many cases this will be due to attenuation effects.

Nevertheless, middle and high frequency noise has the capacity to disturb. This may be because the frequency content is sufficiently comprehensive for the noise to convey unwanted information, such as the words people are speaking, or because particular frequencies are clearly dominant, as in the case of a pure tone.

A-weighting

Human hearing is progressively less responsive to sound at low frequencies than it is to middle and high frequencies. Generally sound measurement equipment is designed to provide a linear response to incident sound, in other words giving equivalent significance to sound at any frequency. In order to reflect more closely the general characteristics of human hearing, the so-called A-weighting frequency spectrum has been specified in international standards and is incorporated as an electronic filter in practically all sound measurement systems. The filter results in a reduced response to low frequency noise that steadily increases until approximately 1 kHz when the curve levels off and becomes more linear from that point through the higher frequencies. This is illustrated in Figure 20.1.

Other characteristics

Other characteristics of noise that may be of significance to its perception include its intermittency, duration and level, both absolute and relative. Noise that occurs intermittently tends to attract greater attention than noise of a constant nature. Tolerance

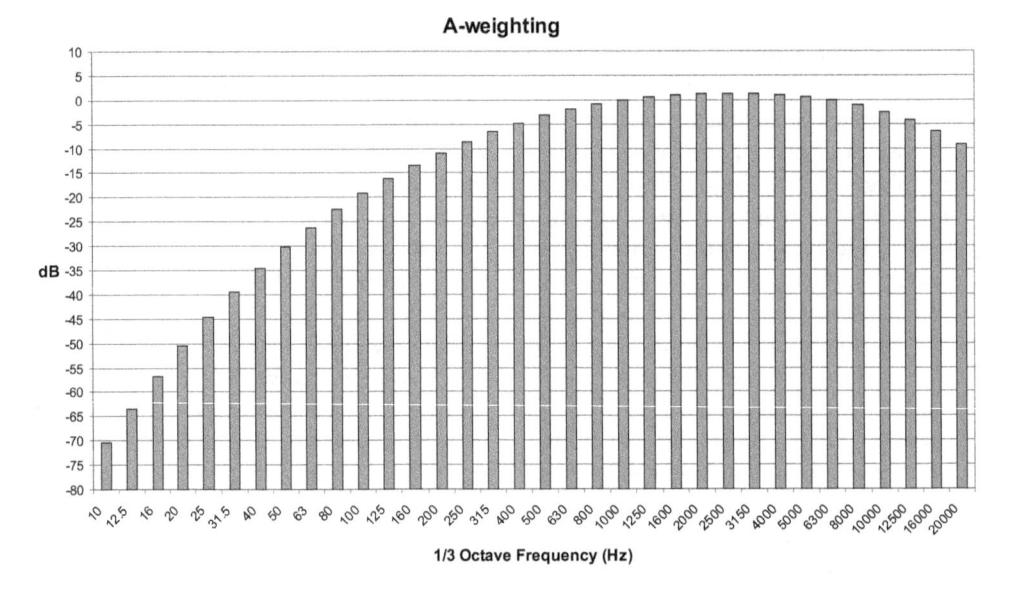

A-weighting

1/3 Octave Frequency (Hz)

Figure 20.1   A-weightings

of a noise that lasts 5 minutes usually will be greater than that of the same noise continuing for 5 hours. The effect of the level at which a noise is heard can be more subtle in that there may be an absolute noise level that will provoke a response, but that response may be more muted in a situation where the environmental noise levels are high than in the same situation where they are low relative to the specific noise being considered.

Parameters

Sound pressure level over time – In the absence of a dominant steady source, the sound level at a point, indoors or outdoors, varies continuously. For example, the variation may be over a few dB about an average value in a calm environment, or over 10 dB in a busy environment. In order to define a level to represent the sound it is usual to define an average value. The most common averaging methods are energy averaging (L_{Aeq}) and statistical averaging (L_{AN} where N is a percentage between 1 and 100).

$L_{eq,T}$ – One means of representing a time varying sound level is to sum the sound energy over the period. The $L_{eq,T}$ is the equivalent continuous sound level which, as its name suggests, is the value of the sound pressure level of a continuous, steady sound that, over the specified time period, T seconds, has the same root mean square sound pressure as the varying sound.

Representation of environmental noise invariably is by the $L_{eq,T}$ with the time period specified according to the application. For example, a daytime noise level may be expressed over a 16-hour period as an $L_{eq,16h}$, whilst a night-time noise level might be described in terms of the $L_{eq,15m}$ for an assessment which uses a method such as that contained within BS 4142:2014+A1:2019 [2].

SEL – The Sound Exposure Level (SEL) index, also sometimes referred to as the Single Event Level, and denoted L_{AE} (in dB), may be used to represent the sound energy of a discrete event, for example a train passing by. Using the SEL, which is the equivalent continuous sound level, $L_{eq,T}$ expressed as a 1-second L_{eq}, the equivalent continuous A-weighted sound pressure level can be calculated for a series of events over a given time period as follows:

$$L_{Aeq,T} = L_{AE} + 10\log_{10} N - 10\log_{10} T$$

where N = number of events during time T

[Equation 6]

Statistical parameters – By rapid sampling of the noise level over the measurement period the statistical noise level exceeded for N per cent of the total measurement time T, where N is a percentage between 1 and 100, can be calculated. For example, the $L_{A10,T}$ can be considered to be the 'average maximum' sound level and in the UK is used to describe road traffic noise. Commonly the $L_{A90,T}$ is described as representing the 'background sound level'. The $L_{A50,T}$ is the arithmetic average of the instantaneous sound levels measured during the period T.

In the example in Figure 20.2 of a variable sound level time history the $L_{A90,T}$ background sound level is 36.5 dB and the $L_{Aeq,T}$ environmental sound level is 55.2 dB.

END parameters – The END [1] was transposed into national legislation in the UK in 2006. The Directive and associated UK regulations reference two parameters – the L_{den} (day-evening-night noise indicator) as a measure of annoyance and the L_{night} as a measure of sleep disturbance, both of which are defined in Annex 1 of the END. The L_{den} is subdivided into the following three periods, with the last two receiving a weighting (+5 dB and +10 dB respectively) in the computation:

- L_{day} (day-noise indicator) means an indicator for annoyance during the day period;
- $L_{evening}$ (evening-noise indicator) means an indicator for annoyance during the evening period;
- L_{night} (night-time noise indicator) means an indicator for sleep disturbance.

The END [1] identifies the following default times for these three periods:

- Day = 07:00–19:00;
- Evening = 19:00–23:00;
- Night = 23:00–07:00.

Propagation and transmission

Distance – One of the most important factors in the propagation of sound from a source to a receiver is the distance between the two.

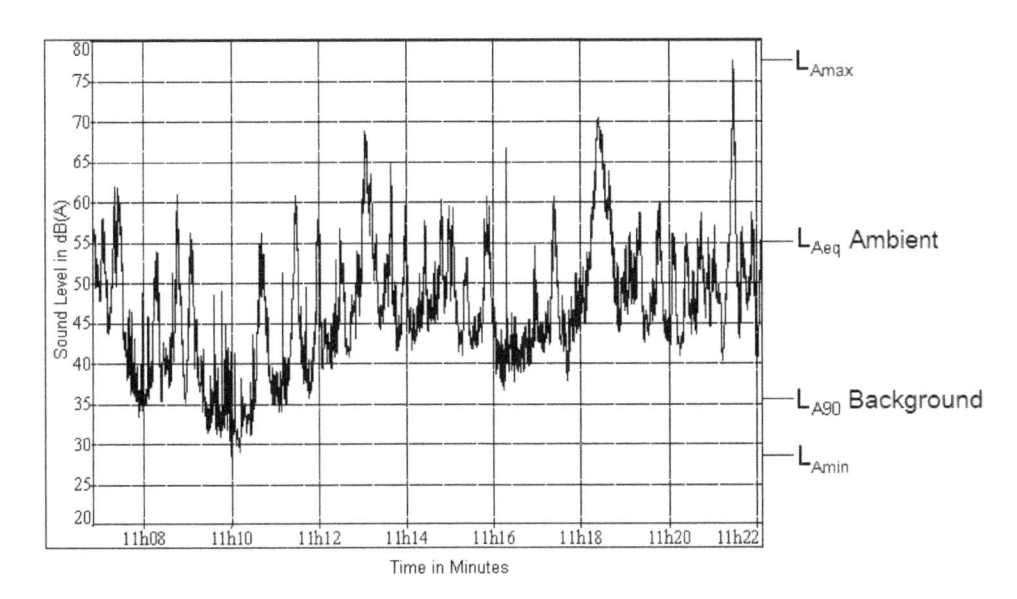

Figure 20.2 Example of noise parameters

As a general rule the rate of decay for sound from a point source such as an item of a plant or the noise of a lorry hitting a pothole would be assumed to be 6 dB per doubling of distance. For a line source such as traffic on a road or railway the equivalent figure is 3 dB per doubling of distance. These rates of attenuation of the sound may vary as a result of other factors but are commonly accepted when no more specific data are available.

Ground absorption – While there a few noise sources, such as aircraft in flight, that are normally represented as a point source radiating sound uniformly and without any major attenuating factor other than distance, the majority are relatively close to the ground. In situations where the intervening ground between source and receiver is hard and sound reflective it may be assumed that there is little or no excess attenuation resulting from absorption of sound passing over that surface. Alternatively, an acoustically absorbent landscape with vegetation may provide significant additional attenuation that should be taken into account when making predictions of sound propagation, particularly for sources very close to the ground such as road vehicles for example.

Barriers – Barriers, whether natural or human-made, can restrict the line of sight between a source and receptor, in doing so providing additional attenuation. There are a variety of equations available to determine the acoustic performance of a barrier, most of which are based on the difference in the length of the transmission path comparing the direct path between source and receptor and the path over the top of the barrier. The following conclusions can be drawn about barriers and their acoustic effects:

- The acoustic performance of the barrier is dependent on the frequency spectrum of the source – barriers are less effective at lower frequencies.
- Where the source and receiver are close to the barrier and the source is comfortably screened (often described as being in the shadow zone) 10 dB attenuation can be expected.
- The practical maximum attenuation from a barrier is 20 dB (or perhaps 25 dB in exceptional circumstances) although this is unlikely to be achieved in most situations.
- A small degree of attenuation (up to 5 dB) can be expected even if the source is just visible (often described as the source being in the illuminated zone).
- Most propagation models do not allow extra attenuation to be claimed from both absorbent ground cover and barrier. So, in certain situations (for example, at distance from a road where the intervening ground cover is acoustically absorbent) the net benefit from erecting a barrier may not be as great as expected, because the excess attenuation from the absorbent ground cover would need to be discounted.
- Beyond distances of around 200 m to 300 m between source and receiver, there can be little or no benefit from an environmental noise barrier as a consequence of the attenuation from ground absorption.
- It is essential that the barrier itself does not transmit sound. The sound reduction performance of the barrier should be 15 dB greater than the attenuation it is designed to provide in situ. In many situations a mass of 20 kg/m^2 would be sufficient, although if a superior performance is required then a mass in excess of 30 kg/m^2 might be required.
- Finally, barriers should be sufficiently long so that the transmission path around each end of the barrier does not contribute significantly to the overall noise level.

Meteorological effects – Meteorological effects may influence both the transmission of sound and the character of the sound source. Wind speed and direction, humidity, temperature, barometric pressure and precipitation are all factors that may have a bearing on the noise that is measured from a source, particularly when the separation distance from the receiver is relatively large. Generally meteorological influences are most likely to be significant when the distance between source and receiver exceeds approximately 100 m as suggested by Craven and Kerry [3] in a report prepared originally for the Department for Trade and Industry in 2001 and updated in 2007.

For external noise sources, wind gradients, temperature gradients and turbulence may have an impact on both on measurements and predictions: '*The effects are asymmetrical and, for distances of 500 m to 1000 m, typically range from increasing the level by typically 2 dB downwind to reducing it by typically 10 dB upwind*' [4].

Measurements can be particularly affected by meteorological conditions. Wind speed is potentially a cause of wind noise at the measuring microphone, even when protected with a proprietary windshield. BS 4142 [2] cites a wind speed of 5 m/s as being the effective performance limit of a microphone windshield for the purposes of that standard (see note to clause 6.3 of BS 4142). Heavy rain can also generate noise either by direct impact on the windshield/rain cover or on nearby surfaces.

Certain weather conditions may directly affect the noise source. Examples include wet road surfaces causing higher road traffic noise levels and a different frequency spectrum than would arise from the same road when dry, aircraft taking longer to get airborne and rising less quickly on take-off in hot and/or humid weather and ambient temperature determining the number of cooling fans operating on air cooled condenser units.

Measurement good practice

There are several basic objectives for any measurement, these being:

- Accuracy;
- Repeatability; and
- Robustness of the data.

In many situations the ways in which these objectives are achieved may be specified by standards or guidance but frequently there are circumstances that necessitate the exercise of judgement by the installer and/or operator of the measurement equipment in order to accommodate the practical constraints found on many sites.

Measurement location – Measurement positions established as part of a noise control requirement have to be both accessible and representative of the noise-sensitive receptor. Practical constraints, security of the measurement equipment and safety considerations often dictate the precise location but there are several factors to bear in mind when choosing a measurement site.

Almost always the measurement should be conducted where there is a direct line of sight to the sound source. This may be especially important when considering the effect on upper storeys of a building when the more convenient ground floor measurement height might be screened by an intervening obstruction such as a garden wall.

Whether the measurement is conducted in the free-field or at a façade may be relatively unimportant, but this must always be recorded so that the result can be corrected from one to the other as required. The general assumption is that a measurement made at a façade location will be 2.5 dB to 3 dB higher than if the same position was in the free-field. Façade measurements usually are made at a distance of 1 m in front of any reflective structure other than the ground, and measurements made remote from any sound reflective surface, other than the ground, are normally considered to be in the free-field. Proximity to a façade may be useful in some situations where that affords screening from some source other than the specific source being investigated.

An awareness of sound sources in the vicinity of the measurement position that may be untypical of the situation at the receptor point can be important. Positioning a microphone near trees in leaf may result in unrepresentative noise from the rustling of the leaves and/or bird song. The presence of a watercourse or even ornamental water features like a waterfall or fountain may preclude the measurement of a representative background sound level, especially at night when environmental sound levels generally are lower.

A number of standards and guidance documents include recommendations for the selection of measurement positions, such as whether these are façade or free-field and several specify the height above ground level. Commonly the ground floor is represented by a measurement height of between 1.2 m and 1.5 m and first floor by a measurement height of 4.0 m. Even greater measurement heights may be necessary in some situations.

Measurement duration – While measurement duration may be constrained by the overall timescale of the project and by budget, it nevertheless should be sufficient to provide a robust and repeatable data set covering the relevant periods of the day and the week.

For simple plant noise investigations using BS 4142 [2], for example, it may be sufficient to rely on attended measurements made over relatively short but representative measurement periods of 1 hour during the daytime or 15 minutes during the night-time during the most sensitive period of the day/week when the new plant will be operational. Proposals for a major new manufacturing facility on the other hand might well justify measurements at multiple locations over a month or more. The duration of any survey will depend on a number of factors, including development size and type. For larger developments affected by multiple sources, measurements over a number of days and nights, including a weekend, should be considered. A continuous monitoring site, or sites, might be supplemented by shorter attended measurements at satellite locations that could be related back to the results from the continuous monitoring site.

It is not unusual for measurements to be disrupted by adverse weather, equipment malfunction or other unforeseen circumstances. If possible sufficient time should be allowed to accommodate such eventualities.

Correcting for unwanted noise – In terms of measurement accuracy, the influence of other residual sources on the environmental noise being measured generally is considered to be significant if the difference between the two is less than 10 dB. BS 4142 [2] includes a calculation procedure for compensating for the effect of residual sound on the ambient sound level for differences of more than 3 dB. For differences of less than 3 dB it is assumed that the specific sound level cannot be determined by measurement alone and the advice is to determine it by a combination of measurement and calculation. This assumes that a representative measurement of the specific sound can be made at a closer location where the source sound level is more dominant. The measured level can then be corrected for distance to calculate the specific sound level at the required position.

An alternative to measuring plant in situ at a less than ideal location may be to find a similar item of a plant elsewhere that can be measured under more favourable conditions. There may also be situations where it is necessary to find a proxy measurement location to obtain a representative background or ambient sound level measurement in the absence of the specific source.

During the course of attended measurements it is sometimes desirable to exclude atypical noise events from the measurement results by use of the pause facility on the sound level meter. For unattended measurements this can be more difficult in the absence of direct knowledge of what caused any particular 'spike' in the measurement time history, but post-processing of results using an appropriate software package offers an opportunity to identify irrelevant or anomalous results and exclude them from the sound level calculations.

Reflections – Usually the reflection of sound only influences the potential accuracy of sound level measurements in the case of façade measurements, which have not been noted as such. Presenting measurement results without indicating whether they are free-field or façade risks an inappropriate and potentially significant correction being made.

Calibration – Calibration of measurement instrumentation immediately prior to and following a sound level measurement is essential to ensure accuracy within the tolerances of the equipment and the measurement method. Measurement results are likely to lack any evidential value without written or recorded calibration details that identify both the equipment used and the sound levels recorded during the calibration procedures.

Opinion varies on how regularly measurement instruments and sound level calibrators should be calibrated but as a general rule, every two years for the measurement system and annually for the calibrator provides a reasonably economical balance.

There are often two standards of calibration available, one being a check of the instrument's performance against the manufacturer's performance data (often this is all that is supplied when an instrument is purchased new), the other being a more rigorous check against relevant national performance standards, such as may be undertaken by a laboratory with the relevant United Kingdom Accreditation Service (UKAS) accreditation. For measurement results that may be presented as evidence in legal proceedings it is helpful to be able to provide copies of calibration certificates for the equipment used.

Meter settings – Although selection of the appropriate measurement range of a sound level meter for the noise being measured is rarely a consideration now due to the large dynamic range available in current instruments, the limitations of the system being deployed may still be relevant. Manufacturers' advice on the lower and upper limits of the standard measurement instruments that they supply should be borne in mind when measuring sound at particularly low or high sound pressure levels (i.e. below about 20 dB and above 120 dB). Similarly, when the measurement of sound at particularly low frequencies (i.e. below 20 Hz) is being contemplated the lower measurement limits of the measurement system should be recognised.

Initial choices in the setting up of a measurement system can be critical to the quality of the results, even with the availability of large data storage capacity and the facility to post-process data to yield various measurement parameters. For example, the choice of Fast and/or Slow meter response times may have to be made at the outset depending on the measurement instrument. Data storage rates may also be critical. Many current sound level meters allow for concurrent measurement of a number of different parameters and it is imperative to understand the influence of the sample rate and signal type when analysing the result.

SECTION 2: LAND USE PLANNING AND REGULATION

Planning and noise

There are effectively three tiers of planning policy – national, regional and local – that impact on noise management.

In England the opportunity afforded by the town and country planning system to assess the potential noise and vibration effects of proposals for new developments in advance of their implementation, at both the strategic and individual development level is the raison d'être for national guidance to local planning authorities in England that is provided within the National Planning Policy Framework (NPPF) [5].

The latest NPPF, published in July 2021, sets out the Government's planning policies for England and how these are expected to be applied. The NPPF places emphasis on the role of planning in ensuring sustainable development and on providing local communities and their representative local authorities the means to produce their own local and neighbourhood plans, which reflect their particular needs and priorities.

Three mutually dependent roles for the planning system in ensuring sustainable development are identified within the NPPF, these being economic, social and environmental. The pursuit of sustainable development is defined as seeking positive improvements in the quality of the built, natural and historic environment, as well as in people's quality of life.

The guidance on planning and noise within the NPPF is summarised in paragraph 185 of the document as follows:

'Planning policies and decisions should also ensure that new development is appropriate for its location taking into account the likely effects (including cumulative effects) of pollution on health, living conditions and the natural environment, as well as the potential sensitivity of the site or the wider area to impacts that could arise from the development. In doing so they should:

a) mitigate and reduce to a minimum potential adverse impacts resulting from noise from new development – and avoid noise giving rise to significant adverse impacts on health and the quality of life;

b) identify and protect tranquil areas which have remained relatively undisturbed by noise and are prized for their recreational and amenity value for this reason.'

The significance of the impacts referred to are explained in the Government's Noise Policy Statement for England (NPSE) [6].

The Noise Policy Statement for England (NPSE) was published in March 2010. The NPSE is the overarching statement of noise policy for England and applies to all forms of noise other than occupational noise, setting out the long-term vision of Government noise policy which is to '*Promote good health*

and a good quality of life through the effective management of noise within the context of Government policy on sustainable development'.

That vision is supported by the following objectives which are reflected in the aims for planning policies and decisions in paragraph 185 of the NPPF [5]:

'Through the effective management and control of environmental, neighbour and neighbourhood noise within the context of Government policy on sustainable development:

- avoid significant adverse impacts on health and quality of life;
- mitigate and minimise adverse impacts on health and quality of life; and
- where possible, contribute to the improvement of health and quality of life.'

The National Planning Policy Framework for England was published in July 2021 but Section 6.7 'Air Quality and Soundscape' of Planning Policy Wales, Edition 11 published in February 2021 [7] provides the majority of the country's policy on noise. A footnote on page 156 refers to the Professional Planning Guidance (ProPG) Supplementary Document 2 [8], produced by the ANC, CIEH and IoA and advises that '*ProPG has been written principally to assist with the planning process in England, but the design principles put forward in Supplementary Document 2 may also be adopted in Wales*'.

The NPPF [5] (at paragraph 187) also seeks to ensure that the grant of planning permission for a new potentially noise-sensitive development does not prejudice the continued operation of a nearby business or community facility. It confirms the principle that the applicant (or 'agent of change') for any new land use is responsible for managing its impacts. For example, the developer of new housing adjacent to a plant hire depot has a responsibility to mitigate any potential noise impact of the existing commercial activity on the proposed residential properties.

In Scotland, section 25 of the Planning (Scotland) Act 2019 [9] makes similar provision for conditional permission in cases of 'noise-sensitive development' where residents or occupiers of the development are likely to be affected by significant noise from existing activity in the vicinity of the development (a 'noise source'). Planning Policy Wales [7] also makes several references to the 'agent of change principle', primarily at paragraph 6.7.5.

The Explanatory Note to the NPSE [6] has introduced three concepts new to the assessment of noise in England as follows:

NOEL, No Observed Effect Level – is the level below which no effect can be detected and below which there is no detectable effect on health and quality of life due to noise.

LOAEL, Lowest Observed Adverse Effect Level – is the level above which adverse effects on health and quality of life can be detected.

SOAEL, Significant Observed Adverse Effect Level – is the level above which significant adverse effects on health and quality of life occur.

None of those three levels are defined numerically in the NPSE and for the SOAEL the NPSE makes it clear that the noise level is likely to vary depending upon the noise source, the receptor and the time of day/day of the week, etc. The need for more research to investigate what may represent a SOAEL for noise is acknowledged in the NPSE and the NPSE asserts that not stating specific SOAELs provides policy flexibility in the period until there is further evidence and guidance.

The NPSE concludes by explaining in a little more detail how the LOAEL and SOAEL relate to the three aims listed. It starts with the aim of avoiding significant adverse effects on health and quality of life, then addresses the situation where the noise impact falls between the LOAEL and the SOAEL when *'all reasonable steps should be taken to mitigate and minimise adverse effects on health and quality of life while also taking into account the guiding principles of sustainable development'*. The final aim envisages proactive management of noise to improve health and quality of life, again taking into account the guiding principles of Government policy on sustainable development as set out in paragraph 1.8 of the NPSE.

Planning practice guidance under the NPPF has been published by the Government as a web-based resource [10]. This includes specific guidance on noise although, like the NPPF and NPSE, the practice guidance does not provide any quantitative advice. It seeks to illustrate a range of effect levels in terms of examples of outcomes as set out in Table 20.3.

Some local authorities have developed their own interpretation of specific noise limits corresponding to the various effect levels in Table 20.3, which may differ for the type of noise source being considered. The environmental impact assessments for several large infrastructure projects undertaken since the introduction of the NPPF [5] have also sought to determine specific values for the effect levels, and examples that have been subject to public examination include the High Speed 2 (HS2) railway scheme, the Thames Tideway Tunnel sewer scheme in London and the A14 Cambridge to Huntingdon highway improvement scheme. The last project is a Highways England scheme and since the impacts and effects of this scheme were reported, Highways England has published its own guidance for assessing and reporting effects of highways noise and vibration (including LOAELs and SOAELs) in DMRB LA 111 [11], most recently updated in May 2020.

Technical advice on planning and noise in Scotland is provided in Planning Advice Note PAN 1/2011 Planning and Noise [12] published by the Scottish Government in March 2011, and an associated Technical Advice Note – Assessment of Noise [13].

The PAN itself does not assign numerical limits to noise but describes the role of the planning system with regard to the use of good design and siting of new development *'to ensure that quality of life is not unreasonably affected and that new development continues to support sustainable economic growth'*.

Table 20.3 Summary of the noise exposure hierarchy, based on the likely average response

Perception	Examples of outcomes	Increasing effect level	Action
No observed effect level			
Not present	No effect	No observed effect	No specific measures required
No observed adverse effect level			
Present and not intrusive	Noise can be heard, but does not cause any change in behaviour, attitude or other physiological response. Can slightly affect the acoustic character of the area but not such that there is a change in the quality of life.	No observed adverse effect	No specific measures required
Lowest observed adverse effect level			
Present and intrusive	Noise can be heard and causes small changes in behaviour, attitude or other physiological response (e.g. turning up volume of television; speaking more loudly; where there is no alternative ventilation, having to close windows for some of the time because of the noise). Potential for some reported sleep disturbance. Affects the acoustic character of the area such that there is a small actual or perceived change in the quality of life.	Observed adverse effect	Mitigate and reduce to a minimum
Significant observed adverse effect level			
Present and disruptive	The noise causes a material change in behaviour attitude or other physiological response (e.g. avoiding certain activities during periods of intrusion; where there is no alternative ventilation, having to keep windows closed most of the time because of the noise). Potential for sleep disturbance resulting in difficulty in getting to sleep, premature awakening and difficulty in getting back to sleep. Quality of life diminished due to change in acoustic character of the area.	Significant observed adverse effect	Avoid
Present and very disruptive	Extensive and regular changes in behaviour, attitude or other physiological response and/or an inability to mitigate effect of noise leading to psychological stress (e.g. regular sleep deprivation/awakening; loss of appetite, significant, medically definable harm, auditory and non-auditory).	Unacceptable adverse effect	Prevent

Scottish Government policy on the application of planning conditions is provided in Circular 4/1998 [14]. The PAN [12] stresses that conditions attached to a consent should meet the six tests (see later under 'Noise generating development') set out in Circular 4/1998, particularly in terms of being enforceable. There are two standard noise conditions provided in Planning Circular 4/1998: model planning

conditions addendum [15], one being for '*Minimising the effect of noise on new noise sensitive development*' and the other for '*Restricting noise from industrial or commercial development*'.

In Wales the Planning Policy Wales [7] sets out the Welsh Government's land use planning policies.

Technical Advice Note (Wales) 11 – Noise [16], published in October 1997 (see also

CL-01–15 Updates to TAN 11 Noise – Noise Action Plan (2013–18) Commitments [17]), sets out the Welsh Government's policies on noise-related planning issues. It gives guidance to local authorities in Wales on the use of their planning powers to minimise the adverse impact of noise.

On 10 February 2020 the Welsh Government published a call for evidence as part of a review of TAN 11 and the production of a new TAN 11 covering air quality and soundscape as well as noise pollution. Although the closing date for responses was 4 May 2020 no further news of progress with the review was available at the time of this chapter's publication.

TAN 11 recommends the use of four Noise Exposure Category (NEC) bands, designed to assist local planning authorities in evaluating applications for residential development in noisy areas. The definition of each NEC band depends on the noise source in question. Table 20.4 below defines the NECs, whilst Table 20.5 presents recommended ranges of noise levels for each of the NECs for dwellings exposed to noise from road, rail, air and mixed sources.

The advice in Table 20.4 mentions conditions to ensure an adequate level of protection against noise. In other parts of TAN 11 [16], reference is made to both BS 8233 [4] and World Health Organization guidance [18] and with respect to what may be considered suitable internal and external noise levels. The particular versions of both these guidance documents cited in the TAN have been superseded since it was published in 1997.

In addition to this, TAN 11 also states that during the night (23:00–07:00 hours): *'sites where individual noise events regularly exceed 82 dB L_{Amax} (S time weighting) several times in any hour should be treated as being in NEC C, regardless of the $L_{Aeq,8h}$ (except where the $L_{Aeq,8h}$ already puts the site into NEC D).'*

In Northern Ireland the Noise Policy Statement for Northern Ireland (NPSNI) was published in September 2014 [19]. The NPSNI is based on the English NPSE and consequently has many similarities.

Table 20.4 Noise exposure categories for dwellings

NEC	Planning advice
A	Noise need not be considered as a determining factor in granting planning permission, although the noise level at the high end of the category should not be regarded as desirable.
B	Noise should be taken into account when determining planning applications and, where appropriate, conditions imposed to ensure an adequate level of protection.
C	Planning permission should not normally be granted. Where it is considered that permission should be given, for example because there are no alternative quieter sites available, conditions should be imposed to ensure a commensurate level of protection against noise.
D	Planning permission should normally be refused.

Table 20.5 Recommended NECs for new dwellings near existing noise sources

Noise levels corresponding to the noise exposure categories for new dwellings $L_{Aeq,T}$ dB

Noise source	Noise exposure category			
	A	B	C	D
Road traffic				
0700–2300	<55	55–63	63–72	>72
2300–0700	<45	45–57	57–66	>66
Rail traffic				
0700–2300	<55	55–66	66–74	>74
2300–0700	<45	45–59	59–66	>66
Air traffic				
0700–2300	<57	57–66	66–72	>72
2300–0700	<48	48–57	57–66	>66
Mixed sources				
0700–2300	<55	55–63	63–72	>72
2300–0700	<45	45–57	57–66	>66

Residential accommodation

The scope of BS 8233 [4] is the provision of guidance and recommendations for the control of noise in and around buildings.

Table 20.6 Guidelines for indoor L_{Aeq} target noise levels for dwellings

Activity	Location	07:00 to 23:00	23:00 to 07:00
Resting	Living room	35 dB $L_{Aeq,16h}$	–
Dining	Dining room/area	40 dB $L_{Aeq,16h}$	–
Sleeping	Bedroom	35 dB $L_{Aeq,16h}$	30 dB $L_{Aeq,8h}$

It suggests appropriate criteria and limits for different situations, which are primarily intended to guide the design of new buildings or refurbished buildings undergoing a change of use rather than to assess the effect of changes in the external noise climate.

The Standard suggests suitable internal noise levels within different types of buildings, including residential dwellings, for steady external noise sources. For dwellings, desirable internal L_{Aeq} target noise limits for steady external noise sources are recommended, which are reproduced in Table 20.6 (without the extensive accompanying notes).

In 2017 the Professional Practice Guidance (ProPG) on Planning and Noise [20] was produced by the CIEH, IoA and ANC to provide practitioners with guidance on a recommended approach to the management of noise within the planning system in England. The scope of ProPG is limited to new residential development potentially affected primarily by existing transport sources although other sources such as industrial, commercial or entertainment premises are only specifically excluded if they are 'dominant'. Its recommended approach for new residential development reflects and extends that which is set out in Table 4 (and its notes) of BS 8233 [4] and is intended to achieve '*a multi-faceted and integrated approach to achieve good acoustic conditions, both internally and externally*'. The ProPG describes a staged approach to the assessment of whether a development site is suitable for the proposed use(s). The first stage involves an initial risk-based assessment of the proposed development site, whilst the second stage involves a detailed and systematic assessment of specific aspects informed by the outcome of the first stage. The guidance also emphasises the need for any new development to demonstrate good acoustic design.

ProPG encourages consideration of the effect of individual noise events on sleep disturbance rather than relying solely on average noise metrics such as $L_{Aeq,T}$. Appendix A to the main document includes a detailed discussion of how to deal with noise events.

With regards to noise levels in external amenity areas, BS 8233 [4] states '*it is desirable that the external noise level does not exceed 50 dB $L_{Aeq,T}$ with an upper guideline value of 55 dB $L_{Aeq,T}$ which would be acceptable in noisier environments*'. BS 8233 recognises that these guideline values are not always achievable and that a compromise might be justified in appropriate circumstances. Once again, the ProPG provides some extended guidance on the noise assessment of external amenity areas.

There are currently three World Health Organization (WHO) environmental noise guideline documents, the intention of which are to define recommended exposure levels for environmental noise in order to protect the health of the population:

- 1999 Community Noise Guidelines (CNG) [18];
- 2009 Night Noise Guidelines for Europe (NNG) [21]; and
- 2018 Environmental Noise Guidelines for the European Region (ENG) [22].

The CNG presents guideline values according to '*specific environments and appropriate effects*'. The WHO guideline noise level for '*speech intelligibility and moderate annoyance, daytime and evening*' is an internal level of noise within dwellings of 35 dB $L_{Aeq,16h}$. For '*sleep disturbance, night-time*' internal noise levels within bedrooms of 30 dB $L_{Aeq,8h}$ and 45 dB L_{Amax} are given.

The guidelines note that during the daytime period few people are highly annoyed at L_{Aeq} levels below 55 dB and few are moderately annoyed at L_{Aeq} levels below 50 dB.

These values are in general accordance with those given in BS 8233 [4]. However, whilst [4] observes that '*Regular individual noise events (for example, scheduled aircraft or passing trains) can cause sleep disturbance*' and advises that '*A guideline value may be set in terms of SEL or $L_{Amax,F}$ depending on the character and number of events per night*', it does not suggest what level might be appropriate. In the WHO Guidelines [18], the research of Vallet and Vernet [23] is referenced, which found that '*For a good sleep, it is believed that indoor sound pressure levels should not exceed approximately 45 dB L_{Amax} more than 10–15 times per night*'. Both the WHO and ProPG guidance indicate that whilst 45 dB is an appropriate limit, a significant impact is only likely where it is exceeded regularly.

In July 2015 a Parliamentary question sought information on how many people in the UK were then exposed to noise levels in excess of World Health Organization guideline levels. The Government response at that time was that it did not monitor how many people were exposed to noise levels in excess of WHO guidelines for community noise but pointed out that Defra published noise exposure figures for selected noise bands for each agglomeration in England in accordance with the END [1]. In 2022 that continues to be the position.

The guidance provided by the WHO Guidelines [18], BS 8233 [4] and ProPG [20], should all be considered in the light of the character of the area in which the proposed development is situated.

In 2009 the WHO published its 'Night Noise Guidelines for Europe' [21]. This document was the result of a working party established to '*provide scientific advice to the Member States for the development of future legislation and policy action in the area of assessment and control of night noise exposure. The working group reviewed available scientific evidence on the health effects of night noise, and derived health-based guideline values.*'

The guidelines developed by the WHO working party update and extend the earlier CNG [18]. The document is extensive but most significantly, a night noise guideline was proposed based on the scientific evidence regarding the thresholds of night-time noise indicated for the protection of the most vulnerable groups of the population (children, the chronically ill and the elderly). This is specified in terms of the $L_{night,outside}$ which is defined in the END [1]. L_{night} is defined as the 1-year L_{Aeq} (exposure to noise) over 8 hours (typically 23:00 to 07:00) outside the most exposed façade. The working group concluded that '*There is no sufficient evidence that the biological effects observed at the level below 40 dB $L_{night,outside}$ are harmful to health. However, adverse health effects are observed at [a] level above 40 dB $L_{night,outside}$*'.

Therefore, the guidelines recommend the following façade noise limits as a yearly average:

- Night noise guideline – $L_{night,outside}$ = 40 dB;
- Interim target – $L_{night,outside}$ = 55 dB.

The night noise guideline is a health-based limit value whilst the interim target is '*considered only as a feasibility-based intermediate target which can be temporarily considered by policymakers for exceptional local situations*'.

The most recent of the WHO guidelines, the ENG [22], provides recommendations for protecting human health from exposure to environmental noise originating from several specific sources: transportation (road traffic, railway and aircraft) noise, wind turbine noise and leisure noise. Leisure noise is considered only in terms of the effects on those attending leisure venues or listening to music on personal listening devices. The recommendations are rated as either 'strong' or 'conditional'. The ENG guidelines relate to external noise levels from specific sources of noise that are incident on the receptor façade, but without the contribution of the reflected sound from that façade. Consequently, the recommendations for non-specific noise in the CNG [18] remain relevant as do the recommendations for internal noise levels. Generally the CNG [18] and ENG [22] guidelines for internal and external noise levels are consistent in their accounting for the difference to be expected for a partly open window, but for aircraft noise the CNG indoor guideline levels would be associated with higher external values than those recommended by the ENG.

The ENG [22] talks predominantly in terms of the L_{Aeq}, L_{den} and L_{night} metrics that are defined over a period of time. Although the ProPG [20] was written before publication of the ENG [22] its emphasis on the potential significance of individual noise events at night remains valid in terms of WHO guidance because the NNG [21] recommendations based on metrics such as L_{Amax} and SEL have not been superseded by the ENG [22].

Looking beyond the WHO guidelines for aircraft noise, the United Nations' specialist aviation agency ICAO's main policies on land use planning and management are contained in Assembly Resolution A39–1, Appendix F [24], which urges States, where the opportunity still exists, to minimise aircraft noise problems through preventive measures, to:

'a)   locate new airports at an appropriate place, such as away from noise-sensitive areas;

b)   take the appropriate measures so that land-use planning is taken fully into account at the initial stage of any new airport or of development at an existing airport;

c)   define zones around airports associated with different noise levels taking into account population levels and growth as well as forecasts of traffic growth and establish criteria for the appropriate use of such land, taking account of ICAO guidance;

d)   enact legislation, establish guidance or other appropriate means to achieve compliance with those criteria for land use; and

e)   ensure that reader-friendly information on aircraft operations and their environmental effects is available to communities near airports.'

In the absence of any specific technical guidance on noise under the NPPF, the conclusion in paragraph 3.17 of the final version of the Government's Aviation Policy Framework (published in March 2013) [25] is possibly the most authoritative guide, in a UK planning context, to acceptable levels of aircraft noise for residential areas:

'3.17 We will continue to treat the 57 dB $L_{Aeq\ 16\ hour}$ contour as the average level of daytime aircraft noise marking the approximate onset of significant community annoyance. However, this does not mean that all people within this contour will experience significant adverse effects from aircraft noise. Nor does it mean that no-one outside of this contour will consider themselves annoyed by aircraft noise.

3.18 The Airports Commission has also recognised that there is no firm consensus on the way to measure the noise impacts of aviation and has stated that this is an issue on which it will carry out further detailed work and public engagement. We will keep our policy under review in the light of any new emerging evidence.'

The Government's Consultation Response on UK Airspace Policy [26]: A framework for balanced decisions on the design and use of airspace, published by the DfT in October 2017 reached the following conclusions:

"2.70 **The government acknowledges the evidence from recent research which shows that sensitivity to aircraft noise has increased**, with the same percentage of people reporting to be highly annoyed at a level of 54 dB LAeq 16hr as occurred at 57 dB LAeq 16hr in the past. The research also showed that some adverse effects of annoyance can be seen to occur down to 51 dB LAeq.

2.71 Taking account of this and other evidence on the link between exposure to noise from all sources and chronic health outcomes, **we will adopt the risk based approach proposed in our consultation so that airspace decisions are made in line with the latest evidence and consistent with current guidance from the World Health Organization**.

2.72 So that the potential adverse effects of an airspace change can be properly assessed, for the purpose of informing decisions on airspace design and use, **we will set a LOAEL at 51 dB LAeq 16hr for daytime, and based on feedback and further discussion with CAA we are making one minor change to the LOAEL night metric to be 45 dB LAeq 8hr rather than Lnight to be consistent with the daytime metric**. These metrics will ensure that the total adverse effects on people can be assessed and airspace options compared. They will also ensure airspace decisions are consistent with the objectives of the overall policy to avoid significant adverse impacts and minimise adverse impacts."

There is a widely shared view that representing aircraft noise annoyance using solely the L_{Aeq} and L_{den} metrics is not adequate and that additional metrics, such as the number of aircraft noise events exceeding a particular threshold level and/or L_{Amax} levels, should also be considered. Research on noise effects is complex and determining whether one metric is better than another may be affected by a great many variables, including the subjective nature of individuals' responses and a variety of non-acoustic factors.

The Government's final consultation on the policy proposals for the Aviation Strategy to 2050 and beyond was 'Aviation 2050 – The future of UK aviation – A consultation' published in December 2018.

The Independent Commission on Civil Aviation Noise (ICCAN) was established by the UK Government in January 2019 to guide the development of aviation policy. An early step was to commission a rapid evidence assessment of 'Aviation Noise and Public Health' that was published in September 2020 [27]. The research brought together previous reviews examining the existing evidence around the relationship between aviation noise and people's health. ICCAN intends to use this as the basis of efforts to improve and expand the research in this area and to determine its priorities for future research. Subsequently, in March 2021, ICCAN published its 'Report on the future of aviation noise management' [28] and a tranche of supporting documents covering eight policy areas that ICCAN identified as being fundamental to the management and understanding of aviation noise.

Non-residential noise-sensitive accommodation

It is unusual for a local planning authority to concern itself with noise conditions affecting non-residential noise-sensitive accommodation such as schools, hospitals, office or hotel developments, except in so far as there may be Building Regulation requirements. Generally, it is seen as the responsibility of the developer to ensure noise levels affecting such developments meet an acceptable standard. Nonetheless, there may be situations where the local planning authority has concerns about the suitability of a site or proposal and some relevant documents are identified here.

BREEAM (Building Research Establishment Environmental Assessment Method)[2] is a sustainability assessment that has been adopted in over 85 countries around the world and is applicable to both new build and refurbishment projects. A BREEAM assessment evaluates the procurement, design, construction and operation of a development against a range of targets based on performance benchmarks. These include assessment criteria both to ensure the building's acoustic performance, including sound insulation, meets the appropriate standards for its purpose and to assess the likelihood of noise arising from fixed installations on the new development affecting nearby noise-sensitive buildings. The rating benchmark levels facilitate comparison of an individual building's performance with other BREEAM-rated buildings and the typical sustainability performance of new UK non-domestic buildings. In the UK, most Government and public projects require a BREEAM assessment and some local authorities also require projects to undergo such an assessment. The BREEAM UK Technical Manual for New Construction Non-domestic Buildings, published in 2018, is available as an on-line document [29]. In addition to assessing commercial buildings the scope of the BREEAM International New Construction standard [30] includes residential developments.

SE 04 – Noise Pollution section of BREEAM aims to ensure that the development is designed to mitigate the impacts of noise. This includes mitigation from existing sources of noise, reducing potential noise conflicts between future site occupants, and protecting nearby noise-sensitive areas from noise sources associated with the new development. The assessment methods make reference to BS 8233 and BS 4142.

Pol 05 – Reduction of noise pollution section of BREEAM aims to reduce the likelihood of noise arising from fixed installations on the new development affecting nearby noise-sensitive buildings. In situations where there are noise-sensitive areas or buildings within an 800 m radius of the site the assessment criteria are:

'4. The noise level from the proposed site/building, as measured in the locality of the nearest or most exposed noise-sensitive development, is a difference no greater than +5 dB during the day (07:00 to 23:00) and +3 dB at night (23:00 to 07:00) compared to the background noise level.

5. Where the noise source(s) from the proposed site/building is greater than the levels described in criterion 4, measures have been installed to attenuate the noise at its source to a level where it will comply with criterion 4.'

Permitted development rights to change from an office use (Use Class B1(a)) to a residential use (Use Class C3) are subject to a number of limitations and conditions. One such condition is the need to apply to the local planning authority for prior approval, for issues such as the impact of noise from commercial premises on the intended residential occupiers (potentially permitting application of the 'agent of change' principle).

Educational facilities – Approved Document E (ADE) [31] to the Building Regulations 2010 includes the following requirements with respect to acoustic conditions in schools.

'**E4.** (1) Each room or other space in a school building shall be designed and constructed in such a way that it has the acoustic conditions and the insulation against disturbance by noise appropriate to its intended use.

(2) For the purposes of this Part – 'school' has the same meaning as in Section 4 of the Education Act 1996;[3] and 'school building' means any building forming a school or part of a school.'

In section 8 of ADE it states that '*In the Secretary of State's view the normal way of satisfying Requirement E4 will be to meet the values for sound insulation, reverberation time and internal ambient noise which are given in Building Bulletin 93*'.

Building Bulletin 93 (BB93) 'Acoustic design of schools: performance standards', provides performance standards for indoor ambient noise levels, airborne and impact sound insulation and reverberation time. The current version of BB93 was published in February 2015 by the Department for Education and the Education Funding Agency [32]. This document supersedes section 1 of

BB93 published in 2003 and should be read in conjunction with 'Acoustics of Schools: a Design Guide' [33], which contains supporting information and additional design considerations superseding sections 2 to 7 of the 2003 version of BB93.

Hospitals – The UK Department of Health has published guidance in 'Specialist Services Health Technical Memorandum 08–01-Acoustics (HTM08–01)' [34] *'for healthcare professionals to understand acoustic requirements and to help those involved in the development of healthcare facilities'*.

Whilst making the point that '*each development has special features, and these criteria may not be appropriate for all projects*' the guidance includes minimum recommended criteria for both external noise that may intrude and internal noise sources.

Offices – BS 8233 [4] includes design criteria, not only for residential accommodation (as previously noted), but also for other occupied spaces, including staff/meeting rooms, training rooms and executive and open-plan offices. BS 8233 also offers guidance on appropriate levels of sound insulation between offices and between offices and other occupied spaces to provide speech privacy.

The British Council for Offices 2019 'Guide to specification: Best practice for offices' [35] also includes internal acoustic design criteria.

Hotels – As for offices it is unusual for a local planning authority to concern itself with noise conditions affecting a new hotel development.

Most hotel chains have their own design codes for new hotels covering both external noise break-in and internal sound insulation requirements. Annex H of BS 8233 [4] includes examples of noise level criteria adopted by hotel groups.

Noise generating development

Most development has the potential to introduce new noise sources to an area even if only during the construction phase. Whilst the impact of construction work may be significant, generally it is relatively short lived. Over the longer term, sound levels from the occupation and operation of a development may be much lower than during the construction phase but could still affect the character of the 'soundscape' of an area. At its simplest this may be a contribution to the pre-existing background sound level, sometimes termed 'creeping background'. The concern about creeping background (also referred to as 'creeping ambient') is that whilst the introduction of one new sound source to an area and any resultant increase in environmental noise may be insignificant, future developments may then be assessed against this new, higher sound level resulting in the possibility of perpetual increase.

One means of guarding against creeping background is to impose conditional planning consent on individual developments limiting them to a sound emission level of 10 dB below the existing measured $L_{A90,T}$ background sound level at an appropriate control point. Clearly such a requirement can represent an onerous standard in some situations. As with any planning condition, it should meet the six tests for conditions set out in the UK Government's on-line 'Guidance – Use of planning conditions' [36], the Northern Ireland Development Management Practice Note 20 'Use of Planning Conditions' [37] and Scottish Circular 4/1998 'The use of conditions in planning permissions' [38] and should be:

- Necessary;
- Relevant to planning;
- Relevant to the development to be permitted;
- Enforceable;
- Precise; and
- Reasonable in all other respects.

A further issue to be considered is whether the limit should be set in terms of the specific sound level, or the rating sound level as described in BS 4142 [2] (see the next section). If the objective is to limit any increase in background, then the condition should be in terms of the specific sound level to avoid introducing a correction or rating penalty based on the character of the sound.

Factories and industrial premises – As noted previously, paragraph 187 of the NPPF [5] refers to the agent of change principle: *'Existing businesses and facilities should not have unreasonable restrictions placed on them as a result of development permitted after they were established. Where the operation of an existing business or community facility could have a significant adverse effect on new development (including changes of use) in its vicinity, the applicant (or 'agent of change') should be required to provide suitable mitigation before the development has been completed.'* It is important, therefore, to protect residents who live close to industrial and commercial premises since implications for the operator could be significant. Should a complaint to the local authority about a noise nuisance be investigated and upheld, a notice requiring abatement would have to be issued (see Section 5).

BS 4142 [2][4] describes methods for rating and assessing sound of an industrial and/or commercial nature, including sound from industrial and manufacturing processes and from fixed installations which comprise mechanical and electrical plant and equipment.

This Standard addresses sound from the loading and unloading of goods and materials at industrial and/or commercial premises and from mobile plant and vehicles that are an intrinsic part of the industrial and/or commercial process. The Standard is, however, not intended to be used to rate and assess sound from any of the following:

- The passage of vehicles on public roads and railways;
- Recreational activities (including all forms of motorsport);
- Music and other entertainment;
- Shooting grounds;

- Construction and demolition;
- Domestic animals;
- People; and
- Public address systems for speech.

The Standard also should not be used to determine nuisance, but can be used for the purposes of:

- Investigating complaints;
- Assessing sound from proposed, new, modified or additional source(s) of sound of an industrial and/or commercial nature; and critically
- Assessing sound at proposed new dwellings or premises used for residential purposes.

The BS 4142 [2] methodology for rating and assessing industrial and commercial sound compares the measured or predicted sound from the source in question immediately outside the dwelling, the 'specific' sound, with the 'background' sound level.

The Standard acknowledges that certain acoustic features – for example, tonality, impulsivity and intermittency – can increase the significance of impact over that expected from a basic comparison between the specific sound level and the background sound level. Where such features are present at the assessment location a correction should be applied, up to 6 dB for tonality, up to 9 dB for impulsivity, 3 dB for intermittency and another 3 dB for any other specific sound features that are readily distinctive against the residual acoustic environment. The Standard also states that where tonal and impulsive characteristics are present then *both* corrections can be taken into account (i.e. a potential maximum correction of +15 dB). The Standard is, however, silent on whether the other available corrections might also be added to obtain a potential maximum of +21 dB. In the absence of any such characteristics, the specific sound level is equivalent to the rating level.

BS 4142 [2] states that an *initial* estimate of the impact of the specific sound can be obtained by subtracting the measured background sound level from the rating level and considering the following:

'a Typically, the greater this difference, the greater the magnitude of the impact.
b A difference of around +10 dB or more is likely to be an indication of a significant adverse impact, depending on the context.
c A difference of around +5 dB is likely to be an indication of an adverse impact, depending on the context.
d The lower the rating level is relative to the measured background sound level, the less likely it is that the specific sound source will have an adverse impact or a significant adverse impact. Where the rating level does not exceed the background sound level, this is an indication of the specific sound source having a low impact, depending on the context.'

A key aspect of BS 4142 [2] is the need to consider context and in this respect the Standard advises that where the initial estimate of impact needs to be modified due to context, all pertinent factors should be taken into account, including:

1 The absolute level of sound
2 The character and level of the residual sound compared to the character and level of the specific sound
3 The sensitivity of the receptor and whether measures that secure good internal and/or outdoor acoustic conditions are present

Wind turbines – Generally the remote location of many wind farms tends to limit the size of the population exposed to noise from this source. However, this also means that wind farms typically are located in areas where environmental noise levels may be particularly low.

A methodology for the assessment and rating of noise from wind farms is set out in a Department of Trade and Industry (DTI) publication ETSU-R-97 'The Assessment and Rating of Noise from Wind Farms' [39].

Following a newspaper article published in January 2004 associating health problems with low frequency noise emissions from wind turbines at a Cornish wind farm, the DTI commissioned an independent study [40] to investigate infrasound and low frequency noise. This study concluded that there was no evidence of health effects arising from infrasound or low frequency noise emitted by wind turbine generators but found that amplitude modulation (AM) of aerodynamic noise particularly during the night hours, could result in audible internal wind farm noise levels, which might provoke an adverse reaction. AM is a regular fluctuation in the noise from a wind turbine at the blade passing frequency and may be described as a swish, whoosh, thump or whoomp.

Further research [41] commissioned by the Government was undertaken in 2007 to ascertain the prevalence of AM on UK wind farm sites, to try to gain a better understanding of the likely causes and to establish whether further research into AM was required.

In 2013 the Institute of Acoustics (IoA) published a Good Practice Guide [42] on the application of ETSU-R-97. The guide resulted from the efforts of a working group that the IoA was invited to set up by the Department of Energy and Climate Change (DECC, the successor to the DTI). The scope was limited to a consideration of the technical elements of the ETSU guidance and excluded any consideration of noise limits or AM. Six additional Supplementary Guidance Notes, published separately by the IoA in 2014, expand on some of the aspects considered in the Good Practice Guide.[5]

In 2015, DECC commissioned a study to review current evidence on the human response to AM including relevant dose-response relationships, to consider how levels of AM in a sample of noise data should be interpreted and to recommend how excessive AM might be controlled through the use of an appropriate planning condition. Phase 1 of the study was reported in October 2015 [43].

In the summer of 2016 two complementary reports were published. In the first [44], the IoA Amplitude Modulation Working Group (AMWG) identified a method for defining AM wind turbine noise and ascertaining its presence within a sample of data and for deriving an appropriate metric that, in the AMWG's view, best represented the degree of AM present. In the second [45] (Phase 2 of the aforementioned study commissioned by DECC), evidence on the effects of wind turbine AM was reviewed and the robustness of relevant research into AM was considered. Recommendations were provided on how excessive AM might be controlled through the use of a planning condition, taking into account the current policy context of wind turbine noise.

The noise issues with large commercial wind turbines generally are different from those that sometimes arise from the small 'microgeneration' wind turbines installed by householders. These latter installations can introduce airborne mechanical and aerodynamic noise which neighbours may find intrusive but may also on occasion cause structure-borne noise transmission where the device is fixed to a wall or chimney of the house.

Advice on the investigation of complaints about wind turbine noise is provided in a Defra research publication [46].

Agriculture – Agricultural noise sources tend to be seasonal but even though they may be of limited duration they can often occur at unsocial hours and throughout the week when they do occur. Auditory bird scaring devices, predominantly used in agriculture to protect crops, can be a source of nuisance to residents living close by. In the UK the National Farmers Union has published a code of practice for 'Bird deterrents and bird scarers – protecting your crops' [47] which is *'designed to minimise public aggravation whilst allowing effective crop protection'*. Agricultural machinery such as combine harvesters working late into the night may be annoying on occasion but static equipment such as irrigation pumps and grain dryers may cause more prolonged and significant noise disturbance in some situations.

Sport and leisure – The potential for conflict between noisy leisure pursuits and noise-sensitive uses, in both urban and rural areas has resulted in the development of a number of codes of practice and sporting body regulations to augment the existing statutory controls. Sporting and leisure activities remain subject to the planning and statutory nuisance control regimes. Many of the codes of practice and guidance notes that have been developed for specific sports or leisure activities are now quite old and fail to reflect developments in international and national standards and guidance. Reasons for that may be various but the increased use of environmental impact assessment, opportunities to take a more flexible approach to particular circumstances, increased sophistication in national/international standards and even the difficulty in some cases of achieving a consensus amongst stakeholders probably all play a part.

Common leisure-related noise impacts fall into two broad categories, direct and indirect. Direct impacts include clay target shoots, power boating, model boats/aircraft, car/motorbike events (including engine running during vehicle preparation, practice sessions/laps, etc.). Indirect impacts might arise from such features as access points, vehicle parking areas, mobile generator sets and other static noise sources, spectator stands, public address systems and their use for purposes other than safety announcements. Other issues may arise from the relative sensitivity of certain hours of use, days of the week, public holidays, etc. as well as the effect of event frequency and length of period between events.

Although there has been significant growth in most forms of indoor and outdoor recreational activity, much of it associated with increased prosperity and leisure time, recreational noise generally is not a significant problem in the majority of homes. In the UK National Noise Incidence Study 2000/2001 [48] recreational noise did not feature as one of the noise sources heard at any significant number of the houses that were surveyed.

Whilst many leisure activities, for example skating, bingo and fitness gyms, take place indoors and give rise to few environmental noise problems, some other indoor leisure activities, particularly music venues, are sufficiently noisy that noise breakout from the building in which they take place can be significant. However, activities that take place in the open air are often the cause of recreational noise complaints that are the most difficult to resolve. Many outdoor recreational activities have the potential to create noise disturbance but commonly complaints relate to a relatively small number of activities that include motorsport jet skis, fireworks, model aircraft (potentially including small drones), clay target shooting, shooting ranges and open-air music concerts (considered in more detail later). Illuminated all-weather sports pitches where play can continue into the evening and take place at the weekend may also be intrusive if there are noise-sensitive premises within close proximity. Sport England published a design guidance note in 2015 titled 'Artificial Grass Pitch (AGP) Acoustics – Planning Implications' [49]. It aims to promote a consistency of approach for local authority noise assessments and limits that might be set for proposals adjacent to sensitive residential areas. It highlights the importance of considering the potential for disturbance to neighbours early in the planning and design stages.

As with most sources of noise nuisance there may often be other factors contributing to the adverse reaction of complainants which may include:

- Occurrence during periods and on days of the week when residents are most likely to be at home anticipating 'quiet enjoyment' of their own recreational time;
- A perception that the participants in the sport are indulging their pleasures at the expense of residents' peace and quiet;
- A belief that it is the wealthy few inflicting the noise of their clay target shooting, motor sport, etc.; and

- The existence of technological fixes such as low noise cartridges, street legal silencers, sound attenuating marquees, etc. that are appropriate but not being used.

The main controls available to local authorities for the control of recreational noise are planning and statutory nuisance provisions.

The Town and Country Planning (General Permitted Development) (England) Order 2015 [50] permits the temporary use of any land for any purpose for not more than 28 days in total in any calendar year (although some purposes are restricted to no more than 14 days in total) and the provision on the land of any moveable structure for the purposes of that use, unless:

- It would consist of development of a kind described in Class E of this Part (temporary use of land for film-making);
- The land in question is a building or is within the curtilage of a building;
- The use of the land is for a caravan site;
- The land is, or is within, a site of special scientific interest and the use of the land is for:

  1 motor sports
  2 clay pigeon shooting
  3 any war game, or

- The use of the land is for the display of an advertisement.

The following temporary uses are restricted to no more than 14 days:

- The holding of a market, or
- Motor car and motorcycle racing including trials of speed.

Use of land in excess of the specified days would require an application for planning permission.

Technical guidance on controlling motor sport noise is scarce except for specific noise limits individual sporting bodies impose on events run under their organisation's regulations. Examples include Motorsport UK and the Auto-Cycle Union (ACU) both of which specify static tests for noise emissions from competitive motor sport vehicles which some venues may supplement with 'drive-by' noise limits to suit the particular circumstances of their location.

Two other documents providing guidance on noise from organised motor sport that are still referenced on occasion are the Noise Council's 'Code of Practice on Noise from Organised Off-Road Motor Cycle Sport', published in 1994 [51] and a 'Code of Practice for Noise for the Control of Noise from Oval Motor Racing Circuits' produced in 1997 by the National Society for Clean Air and Environmental Protection (NSCA) [52]. For example, the former is cited in the Scottish Government's current 'Assessment of noise: technical advice note' (2011) [13] and the latter in 'Neighbourhood Noise Policies and Practice for Local Authorities – a Management Guide' jointly published by Defra and the CIEH in 2006 [53]. Neither motorsport document is now readily available and in practice, although guidance in both concerning general noise mitigation measures was helpful, the current Motorsport UK and ACU advice is likely to be most relevant.

The British Ski and Wakeboard sporting body published its 'Code of Practice – Noise' in 2013 [54] which includes a method of measuring pass-by noise levels under specified conditions, for example for permits to operate on certain inland waters.

In 1982 the UK Government published its 'Code of Practice on Noise from Model Aircraft' [55] to provide guidance on minimising annoyance or disturbance caused by the sport of flying model aircraft. It includes advice on the legal controls and on operating procedures to reduce unnecessary noise.

The growth in the use of drones (small unmanned aerial systems – 'sUAS' or unmanned aerial vehicles 'UAV') both as a leisure pursuit and for commercial purposes poses the potential for noise nuisance from their characteristic 'buzzing' noise, particularly where they may be operated regularly from a particular location. An initial investigation in 2017 by NASA [56] employed psychoacoustic testing to explore differences in subjective response to noise from flyovers of drones with noise from the drive-by of road vehicles encountered in residential neighbourhoods. The results indicated that subjectively drone noise was found more annoying than noise from road vehicles. In October 2019 the UK Civil Aviation Authority (CAA) published 'The Drone and Model Aircraft Code' [57] which, whilst providing guidance on flying safely and legally, does not explicitly address noise issues. Responsibility for dealing with unmanned aircraft misuse incidents is vested in the police following a 2016 Memorandum of Understanding between the National Police Chiefs Council, the CAA, Home Office and Department for Transport.[6] The use of drones based at a commercial premise very likely will be subject to planning controls but private individuals using smaller machines (less than 250 g in weight) in domestic gardens or recreational spaces for example may only be restricted by statutory nuisance provisions or any bylaws relevant to that land.

Noise from organised clay target shooting is addressed in the CIEH publication 'Clay Target Shooting: Guidance on the Control of Noise' produced in 2003 [58]. It is not statutory guidance but is intended to promote good practice in site management, outline the method for measuring noise levels and offer guidance to enforcement officers who are investigating noise complaints. It also gives advice on practical measures site operators can take to reduce noise impacts.

If the shoot is not within 2 kilometres of noise-sensitive premises (including residential properties), and complaints are not anticipated, the guidance is that constraints on shooting times are not necessary.

The guidance says there is no fixed shooting noise level at which annoyance starts to occur but that it is less likely to occur when the mean shooting noise level (SNL) is below 55 dBA and highly likely to occur when the

SNL is over 65 dBA. The SNL is the average of the 25 highest free-field shot levels recorded in a 30 minute period. A full definition of the term and how it is calculated is detailed in Appendix 5 of the Guidance. Where planning permission is required the guidance recommends planning permission be conditioned and that it should not normally be granted if the mean SNL exceeds 55 dBA when the background noise is 45 dBA or lower.

The guidance recommends shooting is limited to a maximum cumulative duration of:

- 4 hours on Mondays to Fridays between the hours of 9:00 and 18:00;
- 3 hours between 10:00 and 18:00 on Saturdays;
- 3 hours between 10:00 and 14:00 on Sundays.

When shooting takes place more than 28 days a year further restriction on the times of operation and/or the number of days per week or weeks per year when shooting may take place may be appropriate. Advice about practical measures a site operator can take to mitigate noise from shooting sites is provided in the CIEH Guidance [58]. This might include limiting the number of shooting stands in use at any time, using low noise gun cartridges and constructing noise barriers for permanent shooting sites.

Concerts etc. – Large music events involving high powered amplification commonly are held within sporting stadia, arenas, open air sites and within lightweight buildings all over the UK. These events give pleasure to hundreds and in some cases thousands of people but have the potential to cause considerable disturbance to nearby residents.

Noise break-out from a musical event within a venue will depend upon a number of factors but of particular significance can be the placement of the stage and the orientation of the loudspeaker systems in relation to that. For large events detailed acoustic modelling to predict the noise levels at potentially noise-sensitive locations in the area may be appropriate. The modelling inputs include the parameters of the noise source, stadium/venue architecture and topographical contour data, whilst the outputs are usually the calculated noise levels at the sensitive premises and/or noise contours.

Such models take account of ground effect, natural and purpose-built barriers, spherical spreading (i.e. distance) and air absorption. It is common for the model to be constructed assuming mild downwind propagation although variations in wind velocity and direction may affect the sound propagation quite markedly during an event.

Critical elements of the noise model are likely to include:

- The location of the stage, particularly if it is limited by access and other logistical issues;
- The polar characteristics of a modern-day line array concert sound system;
- Typical outdoor concert noise level in the audience area ($L_{Aeq,T}$ 98 dB would not be unusual);
- The typical frequency characteristics of modern-day music; and
- The assumptions made regarding aspects which affect the propagation of sound such as meteorological conditions and intervening ground cover.

Low frequency noise is often a feature of rock concert music and can be a particular problem as it tends to dominate the noise heard within houses, the higher frequencies being more significantly reduced by attenuation over distance and the sound insulation characteristics of buildings. Predictions of noise from concert use should be made using the typical noise frequency spectrum characteristics of modern-day music.

For musical events such as classical or pop concerts that might be held within a stadium, the Noise Council's 'Code of Practice on Environmental Noise Control at Concerts'

[59] is still considered an appropriate source of guidance for noise, at least as far as the level at sensitive premises is concerned (see also article in Acoustics Bulletin May/June 2020 'The Pop Code: is it fit for purpose').[7] For example, Manchester City Council's 'Guidance on noise control for open air concerts and events' published in October 2018 [60] adopted the noise level guidelines in the Code of Practice [59]. The guidance is in terms of absolute noise criteria for venues with up to three concert events per calendar year, and a relative noise criterion for more frequent events.

The Code of Practice [59] is aimed specifically at '*large music events involving high powered amplification held in sporting stadia, arenas, open air sites and within lightweight buildings*'. It defines a music event as '*a concert or similar event where live or recorded music is performed by a solo or group of artists before an audience*'. For events held between 09:00 and 23:00 hours it provides target Music Noise Levels (the L_{Aeq} of the music noise measured at a particular location) that should not be exceeded at 1 m from the façade of any noise-sensitive premises (reproduced as Table 20.7).

For events that continue beyond 23:00 hours the Code recommends that the music noise should not be audible within noise-sensitive premises with their windows open for ventilation.

However, the Code is only a guide based on best practice at the time of its publication. The limits are not prescriptive and where the arrangements are satisfactory, venues may continue with limit levels which are higher or lower. Additionally, some venues are subject to low frequency noise limits in the 63 Hz and 125 Hz octave bands or using a C-weighted noise level limit, but there is currently no definitive guidance available on what might be both effective and practicable in this regard.

Generally, noise levels of about $L_{Aeq,T}$ 98 dB measured at the concert mixer position (typically about 40 m from the main loudspeakers and at a height of 10 m to 15 m) are regarded as the minimum for a concert audience to be entertained. The WHO CNG [18] recommend limiting exposure to a maximum $L_{Aeq,4h}$ of 100 dB on no more than four occasions a year and to keep the L_{Amax} below 110 dB. Although the WHO ENG talks in terms of a yearly average exposure, the recommended limits are broadly similar across the two documents.

To limit audience noise exposure, the HSE '*strongly recommends that the A-weighted equivalent continuous sound level over the duration of the event (Event L_{Aeq}) in any part of the audience area should not exceed 107 dB, and the C-weighted peak sound pressure level should not exceed 140 dB*'.[8] It also recommends that where the Event L_{Aeq} is likely to exceed 96 dB, the audience should be advised of the risk to their hearing in advance.

For illegal 'raves' (essentially defined as a gathering on land in the open air of 20

Table 20.7 The Noise Council guidelines for environmental noise control at concerts

Concert days per calendar year, per venue	Venue category	Guideline
1 to 3	Urban stadia or arenas	The MNL should not exceed 75 dB(A) over a 15 minute period
1 to 3	Other urban and rural venues	The MNL should not exceed 65 dB(A) over a 15 minute period
4 to 12	All venues	The MNL should not exceed the background noise level by more than 15 dB(A) over a 15 minute period

or more persons at which music is played through the night) there are, of course, no noise limits but police have several powers under the Criminal Justice and Public Order Act 1994 [61]. These include the removal of people attending or preparing for a rave, stopping people trying to attend and, in certain circumstances, the seizure of vehicles and sound systems being used on the site.

The Anti-social Behaviour, Crime and Policing Act 2014 [62] empowers local authorities and the police to close premises that are causing a nuisance or are associated with disorder. Premises can be closed for up to 24 hours by means of a notice served on the occupier and the owner and posted on various specified points on the premises. Furthermore, if the chief executive of the council or a senior police officer (Inspector rank and above) makes the decision, the notice can close the premises for up to 48 hours. Extensive guidance on the use of Community Protection Notices under the Act is provided by a CIEH Professional Practice Note [63].

Minerals and planning – Minerals policy in England has been provided on-line through Planning Practice Guidance (the PPG) web site since 2014, with four sections (019–022) specifically concerning noise emissions [64].

The PPG notes that there should be a noise impact assessment of proposals for mineral developments, including those for related similar processes such as aggregates recycling and disposal of construction waste, and that this should:

- Consider the main characteristics of the production process and its environs, including the location of noise-sensitive properties and sensitive environmental sites;
- Assess the existing acoustic environment around the site of the proposed operations, including background sound levels at nearby noise-sensitive properties;
- Estimate the likely future noise from the development and its impact on the neighbourhood of the proposed operations;
- Identify proposals to minimise, mitigate or remove noise emissions at source; and

- Monitor the resulting noise to check compliance with any proposed or imposed conditions.

To ensure that mineral operations are undertaken in a manner that minimises noise emissions and controls impacts at nearby noise-sensitive locations to an acceptable level, planning conditions can be imposed. Such conditions may include target noise levels, which would usually be set at the potentially affected receptor, although in some circumstances it may be appropriate to specify target levels at the site boundary or indeed some other point.

The primary noise target is that the level from mineral extraction should not exceed the background noise level by more than 10 dB(A) at noise-sensitive properties.

It is acknowledged in the PPG that such a limit may be seen as being unreasonably onerous on the mineral operators. In such circumstances, the limits should be set as near to that level as practicable during normal working hours (07:00–19:00) and should not exceed 55 dB $L_{Aeq,1h}$ (free-field). Evening (19:00–22:00) limits should not exceed the background level by more than 10 dB and should not exceed 55 dB $L_{Aeq,1h}$ (free-field) at noise-sensitive properties. Between 22:00 and 07:00 *'noise limits should be set to reduce to a minimum any adverse impacts, without imposing unreasonable burdens on the mineral operator. In any event the noise limit should not exceed 42 dB $L_{Aeq,1h}$ (free field) at a noise sensitive property.'*

Where tonal noise contributes significantly to the total site noise, the guidance notes that it may be appropriate to set specific limits for this element. Peak or impulsive noise should not be allowed to occur regularly at night.

Increased temporary daytime noise limits of up to 70 dB $L_{Aeq,1h}$ (free-field) for periods of up to 8 weeks a year may be considered where it is clear that works (such as essential site preparation and restoration work and construction of baffle mounds) will bring longer-term environmental benefits to the site and surrounding areas.

The detailed requirements for compliance monitoring (location, period, frequency, etc.) should be decided on a site-specific basis, and taking account of any monitoring required as a condition of any permit under the Environmental Permitting Regulations. Unnecessarily onerous requirements in this respect should be avoided.

Particular issues can arise from activities related to blasting – blast-induced vibration and air overpressure (airborne pressure waves with most of their energy in the sub-audible frequency spectrum below 20 Hz). The former may be significant in terms both of nuisance and building damage, the latter as a source of nuisance only. Guidance on the assessment and mitigation of vibration and air overpressure from blasting can be found in BS 5228–2:2009+A1:2014 [65] and BS 6472–2: 2008 [66].

Planning Policy Wales (PPW) [7] sets out the land use planning policies of the Welsh Government. PPW, Technical Advice Notes (TANs), Minerals TANs (MTANs) and policy clarification letters comprise national planning policy in Wales. Minerals are covered in 'Chapter 5 Productive and Enterprising Places' of the PPW. There are currently two MTANs:

- Minerals Technical Advice Note (Wales) 1: Aggregates (March 2004) [67]; and
- Minerals Technical Advice Note (Wales) 2: Coal (January 2009) [68].

The guidance on noise limits in both MTANs, whilst similar to that of the English PPG for daytime operations, is different for the evening period and for short-term operations where the limit is 3 dB lower than in the English guidance. Guidance on best practice for blasting and for noise are given in Appendices L and M respectively of MTAN 2 [68].

For Scotland, the National Planning Framework (NPF) for Scotland 3 published in 2014 [69] details the long-term planning strategy for Scotland. The work to develop NPF 4, to take the long-term strategy up to 2050, is underway. Scottish Planning Policy, published

in June 2014 and revised in December 2020 [70] includes the following advice:

'237. Local development plans should safeguard all workable mineral resources which are of economic or conservation value and ensure that these are not sterilised by other development. Plans should set out the factors that specific proposals will need to address, including:

- disturbance, disruption and noise, blasting and vibration, and potential pollution of land, air and water;
- impacts on local communities, individual houses, sensitive receptors and economic sectors important to the local economy;
- benefits to the local and national economy;
- cumulative impact with other mineral and landfill sites in the area;
- effects on natural heritage, habitats and the historic environment;
- landscape and visual impacts, including cumulative effects;
- transport impacts; and
- restoration and aftercare (including any benefits in terms of the remediation of existing areas of dereliction or instability).'

The Scottish Government has published Planning Advice Notes (PAN) including PAN 50 'Controlling the Environmental Effects of Surface Mineral Workings' and its Annex A: 'The Control of Noise at Surface Mineral Workings' [71], Annex C: 'The Control of Traffic at Surface Mineral Workings' [72] and Annex D: 'Control of Blasting at Surface Mineral Workings' [73]. Annex A gives detailed guidance on specific noise limits which are broadly similar to those for Wales.

The Strategic Planning Policy Statement (SPPS) 'Planning for Sustainable Development' published in 2015 by the Department of the Environment (Northern Ireland) [74] consolidates some 20 separate policy publications into one document and sets out strategic subject policy on important planning matters that should be addressed across Northern Ireland. The SPPS sets strategic direction for local councils to bring forward detailed operational policies within their new Local Development Plans. These Local Development Plans are in the process of being drawn up by the 11 local councils that form Northern Ireland. For minerals development, the policy approach outlined by the SPPS seeks to balance the need for minerals resources against the need to protect and conserve the environment.

Pollution prevention and control

Noise from factories and industrial premises has been discussed earlier. However, for certain industrial activities there is a separate regulatory system known as Integrated Pollution Prevention and Control (IPPC).

In England and Wales there is a common regulatory framework established by the Environmental Permitting (England and Wales) Regulations 2016 (EPR) [75]. The equivalent regimes in Scotland and Northern Ireland are different, being regulated under the Pollution Prevention and Control (Scotland) Regulations 2012 [76] and the Pollution Prevention and Control (Industrial Emissions) Regulations (Northern Ireland) 2013 (PPCR) respectively [77]. All three sets of regulations are national transpositions of the EU's Industrial Emissions Directive 2010/75/EU [78].

In Northern Ireland the IPPC system is managed by the Department of Agriculture, Environment and Rural Affairs (DAERA) and the relevant regulatory instruments are listed and briefly described in on-line guidance.[9]

The Scottish Environment Protection Agency (SEPA) is the designated regulator responsible for enforcing the PPC regime within Scotland.

In England and Wales responsibility is divided between the national regulator (the Environment Agency and Natural Resources Wales respectively) and local authorities as follows, based on the activity and the pollution risk:

- Part A(1) installations or mobile plants are regulated by the Environment Agency;
- Part A(2) and Part B installations or mobile plants are regulated by local authorities; (known as Local Authority Industrial Pollution Prevention and Control (LA-IPPC) and Local Air Pollution Prevention and Control (LAPPC) respectively) except waste operations carried out at Part B installations which are regulated by the Environment Agency;
- Waste operations or waste mobile plant carried on other than at an installation, or by Part A or Part B mobile plants, are regulated by the Environment Agency;
- Mining waste operations are regulated by the Environment Agency.

For Part A(2) and Part B processes statutory guidance from Defra is provided in the form of a two-part 'General Guidance Manual on Policy and Procedures for A2 and B Installations' that is available on-line [79]. The most recent amendments to both documents were made in March 2012 and therefore predate the latest UK EPR and PPCR regulations. Chapter 6 within Part 1 of the General Guidance Manual is very limited, comprising only a little over a page of advice on noise and vibration issues. Statutory Sector Guidance Notes[10] that set out best available techniques (BAT) for specific A(2) activities contain a little more advice on noise conditions on similar lines to the BAT conclusions for Part A(1) 'standard rules' permits (see later).

Natural Resources Wales owns and maintains much of its own guidance[11] but in some circumstances continues to use guidance produced by the Environment Agency.

There are essentially two types of permit, a 'standard rules permit' or a 'bespoke permit'. The former is a set of fixed rules for common activities whilst the latter is tailored to the particular business activities of the applicant.

For noise and vibration, a standard rules permit (listed in the on-line guidance 'Standard rules; environmental permitting' [80]) typically requires the following:

'Emissions from the activities shall be free from noise and vibration at levels likely to cause pollution outside the site, as perceived by an authorised officer of the Environment Agency, unless the operator has used appropriate measures, including, but not limited to, those specified in any approved noise and vibration management plan, to prevent or where that is not practicable, to minimise, the noise and vibration.

The operator shall:

(a) if notified by the Environment Agency that the activities are giving rise to pollution outside the site due to noise and vibration, submit to the Environment Agency for approval within the period specified, a noise and vibration management plan;

(b) implement the approved noise and vibration management plan, from the date of approval, unless otherwise agreed in writing by the Environment Agency.'

For a bespoke permit where noise and/or vibration is a significant emission, then a site-specific management plan is likely to be required as a condition.

When applying for a permit, the Environment Agency may ask for submission of a noise and vibration management plan if:

- There may be a risk of noise and vibration pollution beyond the site boundary;
- A noise impact assessment has been undertaken as part of the applicant's risk assessment.

The findings from any noise impact assessment must be considered as part of the risk assessment.

Submission of a noise and vibration management plan may also be required after a permit has been issued if noise or vibration pollution is caused beyond the site boundary, and either of these apply:

- No noise impact assessment has been undertaken;
- No noise and vibration management plan is yet in place.

The responsible organisations within each of the four countries of the UK collaborated to publish guidance in July 2021 to help holders and potential holders of permits apply for, vary and comply with their permits [81]. The Guidance covers:

- How environment agencies will assess noise from certain industrial processes;
- What the law says must be done to manage noise and vibration;
- Advice on how to manage noise – in particular how to carry out a noise impact assessment and what operators should include in a noise management plan.

In all four countries of the UK any noise impact assessment informing the development of a noise and vibration management plan must be carried out following the methodology in the Environment Agency Guidance [81].

This guidance replaces the following documents which have been withdrawn:

- Horizontal Guidance for IPPC, H3 Part 1 Regulation and Permitting [82];
- Horizontal Guidance for IPPC, H3 Part 2 Noise Assessment and Control [83];
- SEPA's Guidance on the control of noise at PPC installations [84].

In circumstances where the Environment Agency requires a noise impact assessment that uses computer modelling or spreadsheet calculations, the information listed in the on-line guidance 'Noise impact assessments involving calculations or modelling' [85] must be included.

Both the operators and regulators of IPPC processes are required to adopt relevant Best Available Techniques (BAT). BAT refers to available techniques that are best for preventing or minimising emissions and reducing impacts on the environment. Within the EU, information on BAT and emerging techniques from Member States is pooled and published as BAT Reference Documents (BREF) that typically apply to the activities of specific industry sectors. BAT conclusions are the final evaluations of Best Available Techniques and form part of every BREF. They determine the reference points used to set permit conditions for installations covered by the Industrial Emissions Directive [78].

To address the the future regime for the development of BAT within the UK following the departure from the EU 'Integrated Pollution Prevention and Control – The Developing and Setting of Best Available Techniques (BAT) Provisional Framework Outline Agreement and Concordat' was published by Defra in February 2022. It states that '*Whilst the Parties [i.e. the UK Government, Scottish Government, Welsh Government and Northern Ireland Executive] will develop and set future BATC separately from the EU, the common origin of the legislation ensures BAT will continue to be based on the same principles. The definition of BAT in UK law remains unchanged following EU exit and it forms part of our retained EU law alongside all existing BATC that were developed at the EU level (largely on a sector-by-sector basis)*'. The Framework reflects the specific circumstances in Northern Ireland that arise as a result of the Northern Ireland Protocol and remains UK wide in its scope. (See https://assets.publishing.service.gov.uk/government/uploads/system/uploads/attachment_data/file/1052051/best-available-techniques-provisional-common-framework.pdf).

Environmental impact assessment and environmental statements

The assessment of the appropriateness of a site for any proposed development can be separated into two aspects:

- The effect that noise and vibration associated with the proposed development might have on sensitive receptors – 'environmental impact assessment' or EIA; and

- The effect that existing and future sources of noise and vibration might have on the proposed development itself – 'site suitability'.

It can be useful to differentiate between the two as it is common to adopt a different assessment approach in each case. EIA requires the likelihood of significant effects to be determined, whilst the site suitability assessment is usually based on achieving appropriate target values.

The EIA Directive (85/337/EEC) [86] on the assessment of the effects of certain public and private projects on the environment has been in force since 1985. The Directive has been amended three times in 1997 (97/11/EC), 2003 (2003/35/EC) and 2009 (2009/31/EC). The initial Directive and its three amendments have been codified by Directive 2011/92/EU of December 2011, which itself was amended in 2014 by Directive 2014/52/EU.

Directive (85/337/EEC) [86] requires in Article 3 that '*The environmental impact assessment will identify, describe and assess in an appropriate manner . . . the direct and indirect effects of a project on the following factors: . . . human beings, fauna*'. In England, the Town and Country Planning (Environmental Impact Assessment) Regulations 2017 (Statutory Instrument 2017 No. 571) [87] are relevant (there are equivalent Regulations for the Devolved Administrations [88–90]) with the requirement to identify, describe and assess significant effects being set out in Regulation 4. Guidance on EIAs is also available on-line.[12]

Ideally the applicant or consultants undertaking the assessment will liaise with the local planning authority during the all-important scoping stage of the EIA process to determine the issues requiring assessment.

For projects that require an EIA the results are presented in an environmental statement (ES) that then accompanies and constitutes an integral part of the planning application.

For virtually every development for which an EIA is required some consideration of the associated noise and/or vibration impacts must be made, if only to determine that there are no significant effects. The nature of the assessment will depend on the type of the development and the prevailing relevant local, national and international guidelines.

Noise and vibration will need to be considered both in terms of temporary demolition and construction impacts and the more permanent operational issues. In some cases, decommissioning may also be a significant phase requiring assessment.

Construction phase – BS 5228:2009+A1:2014 [65, 91] '*recommends procedures for noise and vibration control in respect of construction operations and aims to assist architects, contractors and site operatives, designers, developers, engineers, local authority environmental health officers and planners*'.

For construction noise, BS 5228–1 [91] provides recommendations for basic methods of noise control where there is a need for the protection of persons living and working in the vicinity of, and those working on, construction and open sites.

The Annexes in BS 5228–1 [91] include information on:

- Relevant legislation;
- Typical noise sources and advice on mitigation;
- Sound level data for use in the prediction methods described in the standard;
- Assessing the significance of noise effects;
- The estimation of noise levels, and;
- How to implement noise monitoring.

Annex E of BS 5228–1 [91] sets out example criteria for the assessment of potential significance of construction noise impacts. Two fundamental methods of assessment could be considered, the first based on meeting an absolute standard (which itself may relate to the pre-construction ambient noise level) and the second on the magnitude of change in ambient noise that the construction noise causes.

One of these methods – the so-called ABC method – is commonly used in the UK to

Table 20.8 Example threshold of potential significant effects at dwellings

Assessment category and threshold value	Threshold value, in decibels (dB $L_{Aeq,T}$)		
	Category A [(A)]	Category B [(B)]	Category C [(C)]
Night-time (23:00–07:00)	45	50	55
Evenings and weekends [(D)]	55	60	65
Daytime Monday to Friday (07:00–19:00) Saturday morning (07:00–13:00)	65	70	75

NOTE 1   A potential significant effect is indicated if the $L_{Aeq,T}$ noise level arising from the site exceeds the threshold level for the category appropriate to the ambient noise level.

NOTE 2   If the ambient noise level exceeds the Category C threshold values given in the table (i.e. the ambient noise level is higher than the above values), then a potential significant effect is indicated if the total $L_{Aeq,T}$ noise level for the period increases by more than 3 dB due to site noise.

NOTE 3   Applied to residential receptors only.

(A)   Category A: threshold values to use when ambient noise levels (when rounded to the nearest 5 dB) are less than these values.

(B)   Category B: threshold values to use when ambient noise levels (when rounded to the nearest 5 dB) are the same as category A values.

(C)   Category C: threshold values to use when ambient noise levels (when rounded to the nearest 5 dB) are higher than category A values.

(D)   19:00–23:00 weekdays, 13:00–23:00 Saturdays and 07:00–23:00 Sundays.

determine relevant threshold categories for the assessment of construction noise, based on the existing ambient noise level. Table 20.8 and the associated notes describe the ABC method.

Independently, many local authorities have determined their own recommendations for appropriate noise and vibration limits and/or the hours and days of the week when construction noise and vibration is considered by them likely to be acceptable. A particularly comprehensive example is the 'London Good Practice Guide: Noise & Vibration Control for Demolition and Construction' published by the London Authorities Noise Action Forum in July 2016 [92]. Such an approach can assist consistent enforcement and is entirely proper to the extent that it does not purport to be a legislative requirement. It will carry most weight in any legal proceedings that may be necessary if it is adopted formally as a policy with a traceable sequence of the officers' report to committee and a record of the committee decision. Even where the local authority has a general policy in place it must remain open to the possibility that this may have to be applied flexibly to accommodate the particular circumstances.

Annex F of BS 5228–1 [91] describes a number of methods for estimating noise from construction activities. Methods are described for stationary plant and mobile plant with the starting point of each being a source sound power level or a sound pressure level at a notional distance (10 m) for each item of plant, these being selected from comprehensive lists set out in Annex C and Annex D of the Standard.

Corrections are applied for distance and screening and, with the exception of haul road calculations, for any absorbent ground cover between the construction site and receptor. An allowance is made for the percentage of the assessment period for which each item operates and then the noise levels for each plant item are combined to derive the overall L_{Aeq} level for the assessment period.

EC Directive 2000/14/AC [93] relates to the noise emission in the environment by equipment for use outdoors. In 2005 there was an Amending Directive 2005/88/EC. The EC Directive was primarily intended to reduce barriers to trade within the EC and has helped to focus efforts on the reduction of noise of construction plant at source.

The Noise Emission in the Environment by Equipment for Use Outdoors Regulations 2001 [94] implemented Directive 2000/14/EC [93]. The EU Withdrawal Act 2018 preserved the Regulations and enabled them to be amended so that they would continue to function effectively when the UK left the EU. Accordingly, the Product Safety and Metrology, etc. (Amendment etc.) (EU Exit) Regulations 2019 amended the 2001 Regulations, fixing any deficiencies that arose from the UK leaving the EU (such as references to EU institutions) and made specific provision for the GB market.

The Government guide 'Noise Emission in the Environment by Equipment for Use Outdoors Regulations 2001: As they apply to equipment being supplied in or into Great Britain after 1 January 2021' [95] was published in January 2021 to help businesses understand how the 2001 Regulations are to be applied, when placing equipment for use outdoors on the market in Great Britain post-Brexit. Whilst there is one set of UK 2001 Regulations [94], some of the provisions apply differently in Northern Ireland for as long as the Northern Ireland Protocol is in force; consequently there is different guidance for Northern Ireland [96].

There have also been developments in construction methods and plant that in many situations have introduced quieter alternatives. Examples would include vibratory compaction rather than dynamic compaction (the former utilising equipment such as vibratory rollers or pokers and the latter comprising the dropping of a weight from height), vibratory, augured and 'hush piling' techniques as alternatives to impact piling hammers and 'super silenced' plant such as air compressors and excavators.

Construction noise sources generally are regarded as temporary and, at least for some construction techniques, relatively difficult to control other than by restricting the hours of operation. While there may be opportunity for selecting quieter plant there may also be good reasons why such alternatives are unsuitable for the particular ground conditions or structure being built and the practicalities of the particular site or circumstances should always be borne in mind.

For construction vibration BS 5228–2 [65] provides, in Annex E, empirically derived formulae which relate the resulting peak particle velocity (PPV) to a number of parameters, most notably distance. Formulae are provided for the following activities:

- Vibratory compaction;
- Percussive and vibratory piling;
- Dynamic compaction;
- The vibration of stone columns; and
- Tunnel boring operations.

These formulae allow the prediction of PPV and for some of the processes provide an indicator of the probability of these figures being exceeded.

Human beings are known to be very sensitive to vibration, the threshold of perception being typically in the range of 0.14 mm/s to 0.3 mm/s PPV.

Whilst for permanent sources of vibration it generally is most appropriate to use the vibration dose value (VDV) to determine human response to vibration, in the context of demolition and construction it is most common to use PPV as this parameter is typically:

- Measured during such activity based upon concern over building damage; and
- Generated using empirically derived formulae.

BS 5228–2 [65] (Table 20.9) provides the following guidance on the effects of vibration from construction activities in terms of PPV.

Guide values for cosmetic damage to buildings are given in Table B.2 of BS 5228–2 [65] and are an order of magnitude higher than those regarded as tolerable in terms of human response.

The Considerate Constructors Scheme[13] is a scheme under which sites registered with

Table 20.9 Guidance on effects of vibration levels

Vibration level [(A), (B), (C)]	Effect
0.14 mm/s	Vibration might be just perceptible in the most sensitive situations for most vibration frequencies associated with construction. At lower frequencies, people are less sensitive to vibration.
0.3 mm/s	Vibration might be just perceptible in residential environments.
1.0 mm/s	It is likely that vibration of this level in residential environments will cause complaint, but can be tolerated if prior warning and explanation has been given to residents.
10 mm/s	Vibration is likely to be intolerable for any more than a very brief exposure to this level in most building environments.

(A) The magnitudes of the values presented apply to a measurement position that is representative of the point of entry into the recipient.

(B) A transfer function (which relates an external level to an internal level) needs to be applied if only external measurements are available.

(C) Single or infrequent occurrences of these levels do not necessarily correspond to the stated effect in every case. The values are provided to give an initial indication of potential effects, and where these values are routinely measured or expected then an assessment in accordance with BS 6472–1 or –2, and/or other available guidance, might be appropriate to determine whether the time varying exposure is likely to give rise to any degree of adverse comment.

the scheme are monitored by an experienced industry professional to assess their performance against the five parts of the Code of Considerate Practice. The five parts, Care about Appearance, Respect the Community, Protect the Environment, Secure Everyone's Safety and Value Their Workforce, each include four bullet points detailing the basic expectations of registration with the Scheme.

Operational phase – some of the more common prediction methodologies and provides some direction on the application of assessment techniques.

The UK has developed formal prediction methods for determining eligibility for insulation of residential property from both road and rail sources. The relevant regulations and associated calculation methods are:

- Road traffic: 'Calculation of Road Traffic Noise' (CRTN) 1988 [97] and the 'Noise Insulation Regulations', 1975 [98] (as amended), and
- Railways: 'Calculation of Railway Noise' (CRN) 1995 [99] and the 'Noise Insulation (Railways and Other Guided Transport Systems) Regulations' 1996 [100] (as amended).

A brief summary of the 'Noise Insulation Regulations' [98] and the 'Noise Insulation (Railways and Other Guided Transport Systems) Regulations' [100] is provided in the following section on 'Mitigation of the noise impact of transportation noise'.

Notwithstanding their primary use, the prediction methodologies CRTN [97] and CRN [99] can also be used to calculate road and rail noise for more general applications (e.g. environmental appraisal and land use planning).

The CRTN [97] sets out a five-stage process to predict road traffic noise.

Stage 1: divide the road scheme into segments.
Stage 2: for each segment determine the basic noise level (L_{A10} at 10 m) from the road taking into account the vehicle flow,

speed, proportion of heavy vehicles, road gradient and surface.
Stage 3: calculate the propagation from the segment to receptor including a correction for distance and ground cover or screening (as appropriate).
Stage 4: to obtain the noise level from segment, apply an angle of view correction and allow for reflections, whether from the façade of the receptor (if the calculation point is 1 m from a building) or from reflecting structures on the far side on the road.
Stage 5: combine the contributions from all segments to determine the overall predicted level in terms of $L_{A10,1h}$ or $L_{A10,18h}$.

Appendix A of DMRB LA 111 [11] provides some additional advice regarding the CRTN procedures.

In response to the END [1], the Government commissioned research into the conversion of the $L_{A10,T}$ levels predicted using CRTN into the parameters required for noise mapping ($L_{Aeq,T}$) [101]. In early 2006, further guidance was issued on the application of the corrections used to convert $L_{A10,18h}$ to the EU noise indices for road noise mapping [102].

The potentially rapid change of the general vehicle fleet to being powered by electricity and other non-carbon fuels has implications for road traffic noise levels, particularly in urban areas. Current research in the UK has been limited but there are clearly implications for the ability of the CRTN method to account for such a change.

Defra commissioned a study to develop a methodology to obtain new noise emission values for CNOSSOS-EU Road category 5 vehicles and provide accurate values for electric and hybrid vehicles (EV/HV). The study findings were reported in January 2020 [103]. At high speeds any benefits from electric propulsion tend to be outweighed by the dominance of rolling noise over engine and transmission noise. At constant low speed it seems that there is a noise benefit with both EV and HV vehicles, provided the HV has not switched to its internal combustion engine. At busy urban junctions where idling and accelerating internal combustion engines create much more noise than an electric vehicle more significant differences are likely to be noticeable.

As a safety measure, legislation in the UK and EU requires new vehicles to have acoustic vehicle alerting systems (AVAS) to produce a sound when they are reversing or driving below 20 km/h and in the USA below 30 km/h. This requirement is interpreted in different ways by vehicle manufacturers but has some potential to introduce novel and disturbing noises such as reversing sounders. Overall, though, it seems probable that the increasing percentage of EVs in the traffic mix will improve community response to road traffic noise exposure.

The CRN [99] sets out a six-stage process to the prediction of railway noise.

Stage 1: divide the railway into segments.

Stage 2: for each segment determine the reference noise level (SEL at 25 m) for each train type on each track taking into account the track/support structure, the number of vehicles and the segment velocity.

Stage 3: calculate the propagation from the segment to receptor including a correction for distance, ground cover or screening (as appropriate), air absorption, ballast and angle of view.

Stage 4: account for reflection effects, whether from the façade of the receptor (if the calculation point is 1 m from a building) or from reflecting structures on the far side on the road.

Stage 5: convert the SEL at the reception point to the required $L_{Aeq,T}$ by including corrections for time and the number of trains.

Stage 6: combine the contributions from all segments to determine the overall predicted level in terms of $L_{Aeq,T}$.

In 1996, a supplement to the CRN [99] was published providing noise level adjustments for Class 373 Eurostar trains [104].

In response to the END (2002/49/EC) [1], the Government commissioned two further pieces of research. One [105] investigated in detail the subject of rail and wheel roughness and its acoustic implications, whilst another [106] provided additional railway source terms for the CRN method.

In 2012 the European Commission published a document titled 'Common noise assessment methods in Europe (CNOS-SOS-EU)' [107], which is to be used by EU Member States for strategic noise mapping. In May 2015 a further EU Directive 2015/996/EC [108] (establishing common noise assessment methods according to the END (2002/49/EC) [1]) was adopted, that requires Member States to use the common assessment methods from 2019 onwards.

With the increasing use of noise prediction software, there is more opportunity for the ISO 9613-2 [109] method to be used for calculating the propagation of sound outdoors in order to predict the levels of environmental noise at a distance from a great variety of sources.

The ISO method [109] predicts the equivalent continuous sound pressure level under meteorological conditions favourable to propagation from sources of known sound emission (in other words this is a method that doesn't include any source terms and solely relates to the propagation of noise).

The ISO method [109] consists specifically of octave band algorithms (63 Hz to 8 kHz) for calculating the attenuation of sound which originates from a point sound source, or an assembly of point sources. The source(s) may be moving or stationary. Specific terms are provided in the algorithms for the following physical effects:

• Geometrical divergence (i.e. distance);
• Atmospheric absorption;
• Ground effect;
• Reflection(s) from surfaces; and
• Screening by obstacles.

Significance scales – As already noted a key objective of the EIA process is to assess whether any significant impacts exist or are likely to arise, to identify appropriate

Table 20.10 Generic significance matrix

		Receptor value/sensitivity		
		High	Medium	Low
Impact magnitude	*High*	Major adverse/beneficial	Major adverse/beneficial	Moderate adverse/ beneficial
	Medium	Major adverse/beneficial	Moderate adverse/ beneficial	Minor adverse/beneficial
	Low	Moderate adverse/ beneficial	Minor adverse/beneficial	Negligible effect
	Negligible	Minor adverse/beneficial	Negligible effect	Negligible effect

mitigation measures and ultimately to determine the significance of the residual effects. The EIA Regulations [87], [88], [89] and [90] require the assessment to cover the likely significant effects arising from impacts of a development whether:

- Permanent or temporary;
- Direct or indirect;
- Primary or secondary;
- Cumulative;
- Short, medium and long term; and
- Positive and negative.

To aid this process scales of significance are usually determined. It is common to assess the significance of effect using the environmental value (or sensitivity) of the resource or receptor and the magnitude of impact (i.e. the change in, or level of, noise), which often are set out in the form of a matrix, an example of which is set out in Table 20.10.

It is sometimes the case (for example in situations where only one category of receptor is identified) that a full matrix is not presented and that a single set of descriptors is used to describe the significance of effect. For example, in some situations, the magnitude of impact scale might be used to describe the significance of effect (Table 20.11).

Typically, moderate and major effects are deemed significant, whilst minor and negligible effects are deemed non-significant. Sometimes it is necessary only to identify

Table 20.11 Simplified significance matrix

Magnitude of impact	Significance of effect
High	Major adverse/beneficial
Medium	Moderate adverse/beneficial
Low	Minor adverse/beneficial
Negligible	Negligible effect

'significant' or 'non-significant' effects (rather than using a graded scale of significance). In this case the dashed lines within the main body of Tables 20.10 and 20.11 delineate between significant effects (above the line) and not significant effects (below the line).

Some topics have their own specific significance criteria based on relevant guidance and professional judgement. The approach to determining significance set out in DMRB LA 111 [11] is described next for construction noise, construction vibration and operational road traffic.

In DMRB LA 111 [11] construction noise is deemed to constitute a significant effect where it is determined that a major or moderate magnitude of impact will occur for a duration exceeding:

- Ten or more days or nights in any 15 consecutive days or nights;
- A total number of days exceeding 40 in any 6 consecutive months.

Table 20.12 Construction noise level – magnitude of impact

Magnitude of impact	Construction noise level
Major	Above or equal to SOAEL[1] +5dB
Moderate	Above or equal to SOAEL[1] and below SOAEL[1] +5dB
Minor	Above or equal to LOAEL[2] and below SOAEL[1]
Negligible	Below LOAEL[2]

Notes:
1 The significant observed adverse effect level or SOAEL is determined using the BS 5228–1 ABC method (section E3.2 and Table E1 of BS 5228–1, see also Table 20.8)
2 The lowest observed adverse effect level or LOAEL is the baseline noise level (dB $L_{Aeq,T}$)

Table 20.13 Construction vibration level – magnitude of impact

Magnitude of impact	Construction vibration level
Major	Above or equal to 10 mm/s PPV
Moderate	Above or equal to SOAEL[1] and below 10 mm/s PPV
Minor	Above or equal to LOAEL[2] and below SOAEL[1]
Negligible	Below LOAEL[2]

Notes:
1 The significant observed adverse effect level or SOAEL is 1.0 mm/s PPV
2 The lowest observed adverse effect level or LOAEL is 0.3 mm/s PPV

Table 20.14 Classification of magnitude of noise impacts

Magnitude of change	Change in road traffic noise, $L_{A10,18h}$ or L_{night}	
	Short term	Long term
Major	greater than or equal to 5.0	greater than or equal to 10.0
Moderate	3.0–4.9	5.0–9.9
Minor	1.0–2.9	3.0–4.9
Negligible	less than 1.0	less than 3.0

The construction noise level aligned with magnitude of impact is presented in Table 20.12.

In DMRB LA 111 [11] construction vibration is deemed to constitute a significant effect where it is determined that a major or moderate magnitude of impact will occur for a duration exceeding:

- Ten or more days or nights in any 15 consecutive days or nights;
- A total number of days exceeding 40 in any 6 consecutive months.

The construction vibration level in terms of peak particle velocity (PPV) aligned with magnitude of impact is presented in Table 20.13.

In DMRB LA 111 [11] a classification of the magnitude of impacts is given in two tables, one for an immediate short-term change in operational road traffic noise and one for a more gradual or steady-state long-term change in operational road traffic noise. Table 20.14 is a combined representation of the information in those two tables from DMRB LA 111 [11].

An initial assessment of operational noise significance is undertaken on the basis that a short-term change of negligible or minor magnitude is not significant and a short-term change of moderate or major magnitude is significant.

Where the magnitude of change in the short-term is negligible at noise-sensitive receptors, it can be concluded that the noise change will not cause changes in behaviour or response to noise and as such, will not give rise to a likely significant effect.

For noise-sensitive receptors where the magnitude of change in the short-term is minor, moderate or major a comprehensive range of contextual factors should be considered when determining whether the initial assessment of significance is retained or adjusted. The factors considered include the following:

- Whether the short-term moderate adverse change falls in the upper or lower half of the category;
- The magnitude of impact in the long term (and whether it differs from that in the short term);
- The absolute noise level;
- The location of the receptor and in particular the noise-sensitive areas;
- The acoustic context (whether the acoustic character of the area is likely to be changed by the scheme); and

- The likely perception of change by the residents (whether the scheme results in noise level changes being more acutely perceived by receptors).

Some of the contextual factors are numerical (i.e. the absolute noise level and noise level changes) and some are non-numerical (i.e. receptor setting, acoustic context and perception) and so there is inevitably an element of professional judgement and balance to be applied when determining final significance.

Where transportation sources are involved (whether as a result of a new road or railway being proposed or because existing infrastructure is affected by a proposed development) the effects of a scheme or development could be widespread. For example, a new highway might affect not only those local receptors along the immediate route corridor, but also those receptors closer to the wider transport network and indeed this might be the whole purpose of the development. Therefore, it is important for the assessment to address all the likely noise impacts. This point is illustrated by the following example in Table 20.15.

Along the route corridor dwellings within, for example, 600 m would experience various noise changes from the existing to the future situation, which are reported in the first column. It can be seen that, overall, most properties would experience an increase

Table 20.15 Number of properties affected by short-term noise changes

Change in noise level		Magnitude	Route corridor	Wider network	Combined
Increase in noise level ($L_{A10,18h}$)	0.1–0.9	Negligible	74	221	295
	1–2.0	Minor	31	35	66
	3–4.9	Moderate	40	6	46
	≥ 5.0	Major	58	35	93
	Total increase		203	297	500
No change	0	No change	0	0	0
Decrease in noise level ($L_{A10,18h}$)	0.1–0.9	Negligible	1	659	660
	1–2.0	Minor	9	458	467
	3–4.9	Moderate	70	293	363
	≥ 5.0	Major	0	24	24
	Total decrease		80	1434	1514

in noise. However, as suggested earlier, the effects of the scheme would not be confined to the route corridor. Noise changes along the wider network are reported in the second column and the combined figures in the third, with the assessment presenting a very different balance once all affected receptors have been considered.

In the section on noise generating development (Section 2) consideration was given to the assessment methodology BS 4142 [2]. This Standard determines the likely impact on the basis of the degree to which sound from the industrial/commercial source (adjusted to account for any particular characteristics) exceeds the existing background sound level.

However, this comparison only represents the initial assessment. Like the assessment approach set out in DMRB LA 111 [11] for operational road traffic, it is important that other contextual factors are considered which could modify the initial assessment. Other pertinent factors might be the absolute level of sound, the character and level of the residual sound compared to the character and level of the specific sound and the sensitivity of the receptor and whether measures that secure good internal and/or external acoustic conditions are present.

For most industrial/commercial developments the most critical aspect is likely to be the direct impact on receptors in the immediate vicinity arising from site-based plant or activities (i.e. the industrial/commercial process itself). However, the indirect effects from associated transportation links to/from the site (e.g. a rail link which carries materials or a road link enabling the work force to drive to the plant) should not be ignored and these effects might extend further away.

Aircraft noise mostly lies beyond the control of legislation enforced by local authorities except with regard to the planning controls available for landing sites and ground facilities, whether these are for an individual aircraft, a general aviation airfield or a commercial airport.

Airport noise contours do not include any contribution from ground noise which is generated airside (that part of an airport directly involved in the arrival and departure of aircraft), which also may be a source of high noise levels in areas local to the activity.

Airside ground noise is normally taken to include that from take-off to the point of becoming airborne, reverse engine thrust on landing, taxiing noise, aircraft auxiliary power units (APUs), maintenance and any ground running or engine testing facilities. The impact of these ground noise sources is cumulative with that from the aircraft noise contours and contributions from other off-airport noise sources. Birmingham Airport for example includes ground noise and noise impacts beyond the contours shown on the Strategic Noise Maps within its Noise Action Plan [110].

In October 2014 the IEMA published 'Guidelines for Environmental Noise Impact Assessment' [111].

The purpose of the guidelines is to address the key principles of noise impact assessment and they cover:

- How to scope a noise assessment;
- Issues to be considered when defining the baseline noise environment;
- Prediction of changes in noise levels as a result of implementing development proposals; and
- Definition and evaluation of the significance of the effect in noise levels (where the assessment is undertaken within an EIA).

Fundamentally there are two broad approaches that could be adopted when attempting to determine significance – an assessment against an absolute noise level or an assessment of the change or difference between the noise level attributable to a specific source and the prevailing background or ambient noise level in the absence of the specific source. It is increasingly common to consider both, rather than one or the other.

However, there is much more to consider when undertaking a noise impact assessment. The IEMA guidelines note that the assessment of noise is a complex matter with a number of factors, some of which are listed here, influencing the determination of significance:

- Averaging period;
- Time of day;
- Nature of the noise source (intermittency, etc.);
- Frequency of occurrence;
- Spectral characteristics;
- Absolute level of the noise indicator; and
- Influence of the noise indicator used.

So, for example, in the case of lightly trafficked roads the noise impact of a change in the number of vehicles in the traffic flow is dependent upon when the extra movements are likely to occur and the background sound level in the area at the time. Where existing levels of environmental noise are low, the significance of an individual vehicle passing at relatively close distance is likely to be greater than that of an individual vehicle in a high noise environment.

For the situation where a road is heavily trafficked, it is probable that individual vehicles within the total flow will have no discernible impact on noise-sensitive receiver locations. Where the flow is sufficient for the noise of individual vehicles generally to be indistinguishable, the changes in vehicle flow necessary to be readily distinguished by a typical listener are substantial.

If a quantitative scale is ultimately utilised, it is important that the noise levels are precisely specified. For example, a level of 65 dB(A) being described as significant is insufficient. What parameter is intended? Over what time period? Are there to be no exceptions? When determining these semantic scales the following points should be borne in mind:

- Is an increase of 5 dB(A) exactly balanced by a decrease of 5 dB(A)?;

- Is an increase of 5 dB(A) in road traffic noise more or less significant than an increase of 5 dB(A) in rail noise?; and
- Does an increase of 10 dB(A) from 45 dB(A) to 55 dB(A) have exactly the same effect as an increase of 10 dB(A) from 65 dB(A) to 75 dB(A)?

In summary, the approach to determining the significance of effects relies very much on reasoned argument, professional judgement, available guidelines and standards and the advice and views of appropriate stakeholders.

Site suitability – It is usual for the site suitability assessment to be based on the results of a comprehensive and robust baseline noise and, if relevant, vibration survey. If it is known that the baseline situation is likely to alter significantly in the future, then this should also be taken into account.

Survey duration depends upon a number of factors, including development size and type. The variation of noise level with time of day/night, weekday/weekend, should be considered. The appropriateness of attended, as against unattended measurements, may vary from scheme to scheme but generally continuous measurements are likely to be preferable to sample period measurements. In those situations where continuous measurements are made, the monitoring position(s) should be visited periodically to note the sources present (this is particularly important in those circumstances where there is no single dominant source).

It is generally good practice to avoid undertaking baseline measurements during school or public holidays to avoid the potential influence of atypical road traffic conditions. Assessment of the site's suitability for the proposed development will depend upon many factors including:

- For development that is noise/vibration generating, the proximity of noise/vibration sensitive receptors, both human and fauna;
- For development that is noise/vibration sensitive, the proximity to significant noise/vibration source(s);

- Scope for mitigation, for example use of screens and/or site layout; and where applicable
- Sensitivity to vibration, not only for humans and fauna but also for commercial and industrial processes (e.g. cinemas, laboratory microscopy and weighing balances).

In the section on residential accommodation (Section 2) it was identified that the ProPG [20] describes a staged approach to the assessment as to whether a development site is suitable for the proposed use(s). The first stage involves an initial risk-based assessment of the proposed development site, whilst the second stage involves a detailed and systematic assessment of specific aspects informed by the outcome of the first stage. The guidance also emphasises the need for any new development to demonstrate good acoustic design.

Section 2 also includes the target criteria from BS 8233 [4] and WHO guidelines [18] that are commonly used in the UK for residential accommodation. Should the noise or vibration source be of an industrial or commercial nature, BS 4142 [2] would be an appropriate starting point. For other noise-sensitive uses, relevant guidance documents are identified in Section 2.

Section 2 Notes

2 https://www.breeam.com/
3 1996 c.56. Section 4 was amended by Schedule 22 to the Education Act 1997 (c. 44).
4 This Standard takes great care in the use of the words 'sound' and 'noise'. Sound can be measured by a sound level meter or other measuring system. Noise is related to a human response

and is routinely described as unwanted sound, or sound that is considered undesirable or disruptive.

5 https://www.ioa.org.uk/sites/default/files/IOA%20GPG%20SGN%20No%201%20Final%20Sept%202014_0.pdf; https://www.ioa.org.uk/sites/default/files/IOA%20GPG%20SGN%20No%202%20Final%20Sept%202014_0.pdf; https://www.ioa.org.uk/sites/default/files/IOA%20GPG%20SGN%20No%203%20Final%20July%202014.pdf; https://www.ioa.org.uk/sites/default/files/IOA%20GPG%20SGN%20No%204%20Final%20July%202014.pdf; https://www.ioa.org.uk/sites/default/files/IOA%20GPG%20SGN%20No%205%20Final%20July%202014.pdf; https://www.ioa.org.uk/sites/default/files/IOA%20GPG%20SGN%20No%206%20Final%20July%202014.pdf

6 http://data.parliament.uk/DepositedPapers/Files/DEP2016-0743/160915_MOU_between_DfT_CAA_HO_Police.pdf

7 Institute of Acoustics Bulletin May/June 2020 'The Pop Code: Is It Fit for Purpose'

8 https://www.hse.gov.uk/event-safety/noise.htm#:~:text=Noise%20limits&text=However%2C%20HSE%20strongly%20recommends%20th

9 https://www.netregs.org.uk/legislation/northern-ireland-environmental-legislation/current-legislation/pollution-prevention-and-control-ppc/

10 https://www.gov.uk/government/collections/integrated-pollution-prevention-and-control-sector-guidance-notes

11 https://naturalresources.wales/permits-and-permissions/environmental-permits/environmental-permitting-regulations-guidance/?lang=en

12 https://www.gov.uk/guidance/environmental-impact-assessment; https://www.gov.scot/policies/environmental-assessment/environmental-impact-assessment-eia/; https://gov.wales/environmental-impact-assessment-guidance;https://www.daera-ni.gov.uk/articles/environmental-impact-assessment-eia

13 www.ccscheme.org.uk

SECTION 3: HEALTH EFFECTS

The health impacts of environmental noise are a growing concern among both the general public and policy-makers. At their most extreme, the potential for noise and vibration to cause physical injury is relatively well known and understood, as is the case for

industrial hearing damage and vibration white finger. Perhaps less easy to determine are the wider health impacts, particularly those due to long-term exposure and sleep disturbance. However, in recent years emerging evidence from large-scale epidemiological studies has linked the population's exposure to environmental noise with adverse health effects. Therefore, environmental noise should be considered not only as a cause of nuisance but also a concern for public health and environmental health.

Traffic noise, and air-traffic noise in particular, is an important cardiovascular risk factor and the effects of noise on health are determined partly by the extent to which the listener perceives the noise as annoying. In Government advice on reducing the causes of coronary heart disease, noise is rarely, if ever, mentioned as a significant factor but growing evidence is linking environmental noise to the development of heart conditions including arterial hypertension, stroke, heart failure and coronary artery disease.

There are currently three WHO environmental noise guideline documents which define recommended exposure levels for environmental noise in order to protect the health of the population:

- 1999 Guidelines for Community Noise (CNG) [18];
- 2009 Night Noise Guidelines for Europe (NNG) [21]; and
- 2018 Environmental Noise Guidelines for the European Region (ENG) [22].

A recent and authoritative summary of the health effects of noise, at least in the European region, is provided by the European Environment Agency (EEA) Report No 22/2019 'Environmental noise in Europe – 2020' [112] which included data for the UK in its analyses. In Chapter 3 'Health impacts of exposure to environmental noise' three 'key messages' are highlighted:

- Long-term exposure to environmental noise is estimated to cause 12,000 premature deaths and contribute to 48,000 new cases of ischaemic heart disease per year in the European territory. It is estimated that 22 million people suffer chronic high annoyance and 6.5 million people suffer chronic high sleep disturbance. As a result of aircraft noise, 12,500 schoolchildren are estimated to suffer learning impairment in school.
- Environmental noise (i.e. road, rail, aircraft and industry) features among the top environmental risks to health, with an estimated 1 million healthy years of life lost every year from health effects including annoyance, sleep disturbance and ischaemic heart disease.
- These health impacts are likely to be underestimated, with new World Health Organization evidence demonstrating effects at levels below the obligatory END [1] reporting thresholds. In addition, the END does not comprehensively cover all urban areas, roads, railways and airports across Europe.

The same report also considers the latest studies on social inequalities and vulnerability to environmental noise. In addition to the inevitable focus on the effects of high noise levels on health and well-being the EEA report highlights that one of the ENG's guiding principles is to 'reduce exposure to noise, while conserving quiet areas'. The identification and preservation of quiet or tranquil areas is promoted, not just to avoid more of the population being exposed to increased noise levels but also for the restorative effect such areas can have.

The WHO derived guidelines [18] with the aim of protecting people from the harmful effects of noise in non-industrial environments. These are based, in most cases, on a lower threshold below which the occurrence rates of any particular effect can be assumed to be negligible.

The WHO CNG [18] were followed in 2009 by publication of the WHO NNG [21] to:

'provide expertise and scientific advice to the Member States in developing future legislations in the area of night noise exposure control and surveillance, with the support of the European Commission. This guidelines document reviews the health effects of night-time noise exposure, examines exposure-effects relations, and presents guideline values of night noise exposure to prevent harmful effects of night noise in Europe. Although these guidelines are neither standards nor legally binding criteria, they are designed to offer guidance in reducing the health impacts of night noise based on expert evaluation of scientific evidence in Europe.'

Most recently, in 2018 the WHO published its ENG [22] the main purpose of which is to provide recommendations for protecting human health from exposure to environmental noise originating from various sources: transportation (road traffic, railway and aircraft) noise, wind turbine noise and leisure noise. The work represents a comprehensive review of 400 health effects studies between 1999 and 2015. The health outcomes reviewed include: sleep disturbance, annoyance, cognitive impairment, mental health and well-being, cardiovascular diseases, hearing impairment and tinnitus and adverse birth outcomes. Specific recommendations have been formulated for road traffic noise, railway noise, aircraft noise, wind turbine noise and leisure noise. Recommendations are rated as either strong or conditional.

The WHO documents provide the most comprehensive review of the health effects of noise on the population and authoritative sources of advice on the identification of specific sound levels at which the onset of various health effects are indicated.

Research more specifically targeted at the relationship between noise and health in the UK was commissioned by Defra [113][14] and considered the effects of:

- Annoyance;
- Mental health effects;
- Cardiovascular and physiological effects;
- Night-time effects, sleep disturbance;

- Cognitive effects of children; and
- Hearing impairment.

With the exception of mental health, for each of these effects the first phase of the study identified sufficient evidence to suggest a link between noise and the adverse health effect. Some of the evidence was considered potentially suitable for the estimation of a dose-response relationship. From this work the three areas considered most appropriate for more detailed investigation were cardiovascular effects, sleep disturbance and hypertension.

The research reported in [113] concluded that although there were limitations to the use of existing assessments of the dose-response relationship for myocardial infarction they could form the basis for estimating prevalence based on sound levels.

On sleep disturbance the conclusions were that, whilst this was an area of research that was well developed, with statistically robust data and dose-response relationships, there was no consensus on whether any single one of these could be used for cost benefit analysis or the monetary evaluation of adverse health effects or policy. It was also noted that no quantitative link had been established between acute or transient sleep disturbance caused by noise and any long-term adverse health effects.

The report indicated that although there is strong evidence to link noise and hypertension, hypertension effects could only be considered qualitatively when developing environmental noise management policies.

Subsequently, in 2010, an ad hoc Expert Group established by the Department of Health and, in part, funded by the Department for Environment, Food and Rural Affairs and the then Health Protection Agency, produced a report that considered the available evidence relating to the effects of environmental noise on health [114].

The Environmental Research and Consultancy Department of the UK CAA regularly publishes a six-monthly update, 'Aircraft Noise and Health Effects', providing a concise update on recent noise and health developments. CAP2257 covering published research from March 2021 to September 2021 is an example and the most recent can be accessed via the CAA web site list of environment publications [115].

A WHO report published in 2011 [116] was prepared by experts in working groups convened by the WHO Regional Office for Europe to provide technical support to policy-makers, and their advisors, on the environmental burden of disease (EBD) due to environmental noise. The report contains a summary of evidence on the relationship between environmental noise and specific health effects including cardiovascular disease, cognitive impairment, sleep disturbance and tinnitus. Annoyance was also considered.

For each of these outcomes, the EBD methodology (based on exposure-response relationship, exposure distribution, background prevalence of disease and disability weights of outcome) was applied to calculate the burden of disease in terms of disability-adjusted life-years (DALYs) (see Chapter 1 for a discussion about DALYs).

The report concluded that with conservative assumptions '*DALYs lost from environmental noise in the western European countries are 61,000 years for ischaemic heart disease, 45,000 years for cognitive impairment of children, 903,000 years for sleep disturbance, 22,000 years for tinnitus and 654,000 years for annoyance*'.

Although the extent to which the years lost across the different outcomes are additive is unclear, if all are considered together, the range of burden would be 1.0–1.6 million DALYs, which means that at least 1 million healthy life years are lost every year from traffic-related noise in the western European countries, including the EU Member States. Sleep disturbance and annoyance related to road noise constitute most of the burden of environmental noise in Western Europe.

Section 3 Note

14 Bernard Berry subsequently produced a draft technical report *Review of recent research on noise and hypertension* in October 2014 (BEL Technical Report, BEL 2013-003). https://www.research-gate.net/publication/289538274_Review_of_recent_research_on_noise_and_hypertension

SECTION 4: OCCUPATIONAL HEALTH AND SAFETY

Chapter 14 deals with occupational health and safety and as such addresses aspects of noise in the workplace.

Control of Noise at Work Regulations, 2005 [117]

Exposure to short-lived extremely high noise levels as well as long-term exposure to significantly lower levels of noise can cause physical damage to the hearing mechanism and this has been long established and accepted. It is commonly held that acoustic trauma can be caused by noise at a peak level of 140 dB (linear) or more. Generally, this will be the result of an impulse event and the mechanical damage will be instantaneous. The current UK Regulations require that the risk of hearing

damage needs to be assessed by employers when noise levels in the workplace exceed a daily or weekly personal noise exposure of 80 dB(A) (the lower exposure action value). The possibility of hearing damage is likely to depend on both the level of the sound and its duration. This less acute form of damage is cumulative and hearing loss develops gradually over years of exposure to high noise levels. It may be worsened of course by damage arising from ear infections or other causes of physiological damage as well as by the average deterioration in hearing with the ageing process.

Temporary threshold shift is a protective response of the hearing mechanism that many will have experienced at some time following short-term exposure to a high noise

Andrew Colthurst and Steve Fisher

level and may be noticeable as a reduction in hearing sensitivity on leaving a music venue for example. This effect would normally pass after a time with the hearing returning to normal. However, if the exposure is too great or over a prolonged period permanent threshold shift may result with a consequent irreversible loss of hearing acuity.

The earliest UK legislation specifically to protect the hearing of those exposed to noise in the workplace came in the form of the Noise at Work Regulations [118] made under the provisions of the Health and Safety at Work etc. Act 1974 [119]. This placed duties on employers to prevent damage to the hearing of their employees arising from exposure to excessive noise. A Guidance Note L108 (third edition) was published in 2021 [120].

The current Regulations [117] specify:

1 Exposure action values at which certain actions are required to reduce the exposure and/or to control the risk to employees; and
2 Exposure limit values above which employees must not be exposed.

The main requirements are triggered by four 'action values': daily personal noise exposures, $L_{EP,d}$, of 80 dB(A) and 85 dB(A), the lower and upper exposure action values respectively,

and 135 dB(C) and 137 dB(C), the lower and upper peak action values respectively.

There are also daily exposure and peak exposure limits of 87 dB(A) and 140 dB(C) respectively, which take into account the effect of wearing hearing protection and which the Regulations do not allow to be exceeded (see Table 20.16).

Regulation 5 places a duty upon employers to undertake a noise assessment in the workplace to ascertain whether exposures are at or above the lower action value. Such assessments are expected to identify which employees are exposed, and to provide enough information to facilitate compliance with duties under Regulations 6, 7 and 10.

Under Regulation 6, when any employee is exposed to levels at or above the upper daily exposure action value or upper peak exposure action value, the employer is required to reduce in so far as is reasonably practicable the exposure of that employee to noise, other than by the use of personal ear protection.

The provision of personal ear protection and the demarcation of hearing protection zones are covered by Regulation 7.

Regulation 9 introduces a specific duty on employers to undertake health surveillance including audiometric testing, where there is risk to health.

Under Regulation 10, the employer has a duty to each employee who is likely to be

Table 20.16 Summary of daily personal noise exposure, $L_{EP,d}$, peak action values and limit values

	Control of Noise at Work Regulations 2005	Action
Lower action value	80 dB(A)	Risk assessment
	135 dB(C)	Make hearing protection available
		Maintenance programme
		Training
Upper action value	85 dB(A)	Noise reduction at source
	137 dB(C)	Hearing Protection Zone
		Hearing protection must be used
		Health surveillance
Limit value	87 dB(A)	Immediate action to prevent exceeding limit
	140 dB(C)	Take hearing protection into account

exposed to the lower action value and above, or to the peak action value or above, to provide adequate information, instruction and training on:

- The risks to that employee's hearing that such exposure might cause;
- What steps the employee can take to minimise that risk;
- The steps that the employee has to take in order to obtain personal ear protectors; and
- The employee's obligations under the Control of Noise at Work Regulations 2005.

These Regulations are concerned with the protection of people at work and, therefore, do not deal with exposure to noise for the public.

Control of Vibration at Work Regulations 2005 [121]

These Regulations were also made under the Health and Safety at Work etc. Act 1974 [119]. A Guidance Note L140 (second edition) was published in 2019 [122].

The Regulations apply to both hand-arm vibration (HAV) and whole body vibration (WBV). The main requirements are triggered by two 'action values', one each for HAV and for WBV and two 'limit values', one each for HAV and for WBV as specified in Regulation 4.

These values and limits are expressed in terms of daily exposure over 8 hours to frequency and time weighted acceleration. For HAV the relevant figures are 2.5 m/s^2 as the exposure action value and 5.0 m/s^2 as the exposure limit value. The equivalent figures for WBV are 0.5 m/s^2 (exposure action value) and 1.15 m/s^2 (exposure limit value).

Regulation 5 imposes a duty on the employer to carry out risk assessments in the workplace.

Regulation 6 requires the employer to control or reduce the risk of exposure to vibration.

Regulation 7 requires the employer to conduct health surveillance, where the risk assessment indicates that there is potential for harm to the employee and requires the employee to cooperate with health surveillance.

Regulation 8 relates to the implementation of a suitable programme of information and training on the hazards of exposure to mechanical vibration.

SECTION 5: NOISE AND NUISANCE

Noise is one of the matters declared to be a statutory nuisance under Part III of the Environmental Protection Act 1990 [123] (see later) that applies in all countries of the United Kingdom.

As with any nuisance, the gathering of evidence is all important. For noise this is likely to include documentary evidence such as diary sheets maintained by the complainant and the contemporaneous notes made by officers witnessing the noise. In many cases it will be appropriate to support this with noise or vibration measurement data, not necessarily as a means of determining whether or not there is a nuisance, but to provide objective evidence of the level that existed at the time.

Having determined that a nuisance exists and with an abatement notice served solely on the strength of observation by an officer, it may be extremely difficult in any proceedings, be they an appeal of the notice by the recipient or a subsequent prosecution for non-compliance, to demonstrate how the noise heard or measured on a subsequent occasion might compare with that at the time the nuisance was determined. It is unlikely to be a convincing argument that officer A's perception of the noise is the same as officer B's or that an officer hearing it on one night can describe how much louder or quieter it might have been when heard again two months later for example. If any works have been carried out by the recipient of the notice, and the complainant remains unconvinced that the nuisance has been abated, without noise measurements taken at the time of the initial investigation, the extent of any improvement probably will be unquantifiable.

When measurements are made with the intention that the results be used as evidence, then it is vitally important that appropriate information is recorded. Information on precisely what equipment was used (i.e. not just the model but the serial numbers of its significant components and the calibrator) is essential for demonstrating the precision and reliability of the results by reference to the routine calibration records in case of challenge. The person undertaking the measurements and the start and finish times/dates should also be recorded in the site notes.

Details of the conditions under which the measurements were made should include at least:

- The results of the initial and final field calibration checks;
- The precise location of the microphone and whether this was free-field or façade;
- The height of the microphone above ground and if not free-field its distance from any significant reflective surface;
- Confirmation that an appropriate wind shield was used;
- The weather conditions including at least wind speed, wind direction and whether there was any precipitation; and
- The presence of other sources of noise, particularly if these are significant in relation to the noise under investigation.

Domestic and neighbour noise

Most recorded noise nuisance investigations by local authorities concern domestic and neighbour noise.[15] The range of complaints includes such matters as:

- Noisy animals;
- Vehicle repairs;
- Garden machinery;

- Disturbance arising partly or wholly from inadequate sound insulation between dwellings;
- Inconsiderate actions by occupiers (e.g. noisy television or radio);
- DIY noise;
- Mechanical noise from domestic equipment such as a washing machine or heating circulation pump; and
- Noise from misfiring audible car and intruder alarms.

These types of complaints are not always simple to deal with and gathering evidence can be difficult given that the noise often is caused at irregular and unpredictable times which may also be during 'unsocial hours' and at weekends. The relationship between neighbours may sometimes be as much of an issue as the noise itself.

In spite of the potential difficulties in investigating what may appear to be relatively trivial complaints, the distress caused by neighbour noise can be very considerable and it is important that consistent and robust procedures are followed to ensure that, within the limitations of the statutory nuisance legislation, legitimate complaints are not dismissed without an appropriate level of consideration or investigation.

For example, the stock answer to many complaints of neighbour noise is that case law[16] so far as statutory nuisance is concerned, prevents a local authority pursuing any action in cases where the primary cause of the problem is inadequate sound insulation. Whilst that is true in many situations, it does nothing to alleviate what can be a very severe problem. It should also be noted that noise is one of the hazards in the Housing Health and Safety Rating System and so in England and Wales at least, housing legislation may be available to secure resolution of the problem. Even if the details of the complaint indicate that it is a case where the floor is unmodified (and possibly unactionable as a statutory nuisance) and upgraded sound insulation is necessary to provide any significant benefit. There are other options for the EHP (an interesting illustration of why the holistic approach of the qualified EHP is important), and it can be of considerable help to a complainant to be given advice and information, perhaps in the form of a leaflet, about the methods available for remedying the noise problem and the actions available to the local authority. It is questionable whether the duty to investigate noise complaints can be adequately served by simply despatching a 'standard' letter to the person being complained about and/or diary record sheets to the complainant. It is important that potential deficiencies in the construction are considered as possibly contributing to, if not mainly responsible for the complaint.

Chapter 5 discusses the Anti-Social Behaviour Crime and Policing Act 2014 and should be read in conjunction with the next few paragraphs. In the context of noise Part 4, Chapter 1 of the Act provides for the use of Community Protection Warnings (CPW) followed by Community Protection Notices (CPN) where necessary to tackle low-level environmental issues and anti-social behaviour. Strong working relationships with the local Community Safety Partnership and police colleagues supported by regular liaison meetings can make this a particularly effective tool.

The use of CPW/Ns enables a local authority to tackle cases that statutory nuisance traditionally could not be applied to; however, the section 80 abatement notice can and should still be used where a statutory nuisance is established. Generally, the investigative processes used in nuisance investigation can still be used in establishing whether the behaviour of an individual or organisation warrants an intervention under the Anti-social Behaviour, Crime and Policing Act 2014 (ASBCP) [62]. It must be remembered that the statutory nuisance powers remain a duty whereas the CPW/N is a power that may be exercised. For example, it is not appropriate to issue a CPW/N after a section 80 notice has been issued for the same behaviour. It is permitted to issue the

CPW/N first and resort to section 80 later if Statutory Nuisance arises.

For action to be taken under the ASBCP a behaviour has to:

- Have a detrimental effect on the quality of life of those in the locality;
- Be of a persistent or continuing nature; and
- Be unreasonable.

Section 43 of the Act enables a Community Protection Notice (CPN) to be served on an individual (over 16 years of age), organisation or business whose behaviour satisfies those criteria. There is a requirement to first have issued the accused with a Community Protection Warning (CPW) before escalating to a CPN.

Periodic revisions to the Home Office statutory guidance have placed an increasing emphasis on assessing the impact of the anti-social behaviour on the victims. 'Putting victims first' stipulates that the agency considering the use of a CPW/N should first speak to members of the community to gain a proper understanding of the harm being caused to individuals and the community.

Prior to the issue of a CPW/N the officer must apply the civil evidential rules and be in possession of sufficient evidence to demonstrate 'on the balance of probabilities' that the perpetrator(s) behaved or threatened to behave in an anti-social manner. If a subsequent breach of the CPN occurs and court proceedings are being considered, then the criminal evidential requirements should be applied to demonstrate 'beyond reasonable doubt' that the perpetrator(s) have persisted with the noise-generating behaviour(s). There is no right of appeal against a CPW. If there is a failure to comply with the CPW in the specified timescale then a CPN is the next step.[17]

Where the problem is sound transmission (or other noise within the dwelling) as the result of building deficiencies and thus a hazard of noise under the Housing Health and Safety Rating System (HHSRS), then in England and Wales the local authority can take action under Part 1 of the Housing

Act 2004. It should be noted that as a matter of principle the dwelling should be capable of providing adequate protection from all potential hazards prevailing in the local external environment. This includes inter alia 'pollution including noise'.[18]

Audible intruder alarms are a source of noise that in the past have been subject to specific legislative provision, since repealed. The Code of Practice on Noise from Audible Intruder Alarms 1982 was withdrawn for England in 2014, having been superseded by more recent legislation and technical standards. The Clean Neighbourhoods and Environment Act 2005 introduced new powers for local authorities to deal with noise from audible intruder alarms in their areas. However, the guidance on minimising annoyance contained within the Code still stands and remains available on the Government web site [124].

Environmental noise

Environmental noise is taken here to include anything outside the residential environment, the main sources being industry (including agriculture), commerce, entertainment, leisure activities, road, rail and air transport. Whilst the first four of these are capable of being the cause of a statutory nuisance the last three are not by virtue of specific exemptions. However, remedial action could be available under housing legislation as in England the Decent Home Standard includes the criterion that there should be 'adequate insulation against external noise (where external noise is a problem)' as part of the requirement on reasonably modern facilities and services.[19]

One of the main differences to be considered when investigating a complaint of nuisance arising from an environmental noise source is the defence of best practicable means (BPM) (i.e. whether best practicable means have been used to control noise emanating from industrial, trade or business premises). Although this is a defence and therefore ultimately for the court to decide, at the time

of determining the existence of a statutory nuisance it would usually be prudent for the local authority to assess the likelihood that BPM might be invoked successfully. This does assume a degree of technical knowledge of the source to be controlled and the physical constraints there might be to installing mitigation. For example, whilst the application of sound attenuating high mass lagging to ductwork might seem an obvious measure to control noise breakout from the feed duct to a cyclone, if lagging the fan motor subsequently causes it to overheat and fail, or worse burst into flames, then almost certainly it would not constitute BPM. By contrast, amplified music, one of the most common sources of nuisance from commercial premises can usually be reduced simply by turning down the volume rather than adopting any sophisticated technical fix and BPM in this circumstance is much less likely to be an issue.

The advent of the Infrastructure Planning Commission in 2008 introduced new planning procedures for defined groups of large infrastructure projects but the enabling legislation, the Planning Act 2008, also removed premises consented under the new procedures from the statutory nuisance provisions of the Environmental Protection Act 1990 [123]. Subsequently, under the Localism Act 2011, the Planning Inspectorate became the Government agency responsible for operating the planning process for Nationally Significant Infrastructure Projects (NSIPs).

Noise management guide

In 2006 the CIEH and Defra jointly published a management guide on Neighbourhood Noise [53]. It was intended to assist all environmental health practitioners and other professionals involved in noise control work but in particular it aimed to encourage local authorities to:

- Review their noise policies and procedures to ensure their continued relevance and efficacy;

- To provide guidance on appropriate policies and procedures for the control of neighbourhood noise; and
- In those cases where enforcement action is necessary, encourage this in accordance with clear and consistent policies whilst retaining the flexibility to reflect local needs and circumstances appropriate to their area.

The Guide is an extensive document and provides many examples of good practice as well as references to relevant guidance, standards and legislative provisions.

Ombudsman

In situations where one of the parties to the investigation of a noise complaint, be they the complainant or the person alleged to be causing a noise nuisance, is dissatisfied with the actions taken by the local authority it is open to them to make a complaint to the Local Government Ombudsman. Those making a complaint to the Ombudsman normally are required to have exhausted the complaints procedure of the local authority about which they are complaining or to have received no reply to their complaint within a reasonable time.

The Ombudsman's role is limited to investigation of faults in the authority's handling of the matter complained of (i.e. the administrative processes and procedures followed and their application). It does not extend, for example, to questioning judgements made on matters of fact or professional opinion where those are appropriate.

Complaints to the Ombudsman regarding the handling of noise nuisance cases are not uncommon. To an extent this may be the inevitable consequence of the difference between an individual's perception of a problem that may be causing them considerable personal distress and the objective professional opinion of an investigating officer that the noise does not amount to a statutory nuisance.

In many cases the reason for complaint to the Ombudsman is the length of time that it can take to resolve the investigation, regardless of its outcome. Noise nuisance can be transient and variable making it particularly difficult to witness in some situations. Nevertheless, it is important for local authorities and officers to have clear and demonstrable strategies and procedures for the investigation of statutory nuisances and much of the CIEH/Defra publication [53] is intended to assist.

Environmental Protection Act 1990, Part III [123]

The law of statutory nuisance has evolved over the last century or more but still has its roots in the early public health legislation. Currently statutory nuisance encompasses a range of matters that has broadened to include topics such as noise and vibration where there may be no obvious or immediate potential health effects (see also Chapter 5). Section 79 of the Environmental Protection Act 1990 (as amended) declares a number of matters to be statutory nuisances including the following:

'(1)(g) noise emitted from premises so as to be prejudicial to health or a nuisance;
(1)(ga) noise that is prejudicial to health or a nuisance and is emitted from or caused by a vehicle, machinery or equipment in a street or in Scotland, road' (as introduced by the Noise and Statutory Nuisance Act 1993 [125]).

Under the provisions of the Environmental Protection Act, the local authority is required to inspect its area periodically to detect any nuisance and, where a complaint of a statutory nuisance is made by a person living within its area to take such steps as are reasonably practicable to investigate the complaint.

If the local authority is satisfied of the existence of a statutory nuisance it is obliged by section 80 to serve an abatement notice. This may require various measures including cessation of the noise, its attenuation or restriction to certain times. At its discretion the local authority can delay the service of a notice for up to seven days if it is pursuing alternative means of securing abatement of the nuisance. This discretion was introduced by provisions in the Clean Neighbourhoods and Environment Act 2005 [126]. There are attendant powers of entry, powers to seize noise-making equipment and powers to undertake works in default (where the local authority undertakes the work and recharges the owner of the property).

The Environmental Protection Act [123] provides a number of exemptions for certain matters, groups and activities. One of the most significant is the exemption of aircraft from subsection 79(1)(g).

Section 82 of the Act provides for individuals aggrieved by the existence of a statutory nuisance to complain direct to the magistrates' court. If the court is satisfied of the existence of a statutory nuisance it is obliged to issue an abatement order, which is similar to the abatement notice that would be issued by a local authority acting under section 80.

Noise Act 1996 [127]

This Act created an offence of excessive noise from domestic premises at night (23:00 to 07:00) and was introduced to give local authorities adoptive powers to investigate and, where the noise level within the complainant's dwelling exceeds the 'permitted level', to prosecute or to serve a fixed penalty notice. The powers are additional and complementary to those existing under the Environmental Protection Act [123] but are relatively little used.

The Act also serves to clarify the existing powers of seizure under statutory nuisance provisions and makes similar provision for the night-noise offence.

Control of Pollution Act 1974: Part III [128]

Much of the Control of Pollution Act 1974 (CoPA) has been superseded by the Environmental Protection Act and subsequent amendments; however, sections 60 and 61 of CoPA remain operative and give the local authority special powers for controlling noise arising from construction and demolition works, regardless of whether a statutory nuisance has been caused or is likely to be caused. Works within the scope of these provisions include repair and maintenance work and road works. These powers may be exercised either before works start or after they have started.

Contractors, or persons arranging for the works to be carried out, also have the opportunity to take the initiative and ask the local authority to make its noise control requirements known. Because there is an emphasis upon getting noise issues settled before work starts, implications exist for traditional tender and contract procedures.

Section 60 enables a local authority in whose area work is going to be carried out, or is being carried out, to serve a notice of its requirements for the control of site noise on the person who appears to the local authority to be carrying out the works. Such a notice may also be served on others appearing to the local authority to be responsible for, or to have control over, the carrying out of the works.

This notice can perform the following:

1   Specify the plant or machinery that is or is not to be used. However, before specifying any particular methods or plant or machinery a local authority has to consider the desirability, in the interests of the recipient of the notice in question, of specifying other methods or plant or machinery that will be substantially as effective in minimising noise and that will be more acceptable to the recipient.
2   Specify the hours during which the construction work can be carried out.
3   Specify the level of noise that can be emitted from the premises in question or at any specified point on those premises or that can be emitted during the specified hours.
4   Provide for any changes of circumstances. An example of such a provision might be

that if ground conditions change and do not allow the present method of working to be continued then alternative methods of working should be discussed with the local authority.

Greater detail regarding the matters that local authorities should take account of when preparing such a notice and the appeal provisions that exist are to be found in BS 5228–1 [91].

Section 61 of the Act provides a mechanism for the contractor or developer to take the initiative and approach the local authority to ascertain its noise requirements before construction work starts. If a formal application for 'prior consent' is received by the local authority it is obliged to give a decision within 28 days; failure to do so or the attachment of unnecessary or unreasonable conditions are grounds for appeal by the applicant. An application cannot be submitted in advance of any request for approval under Building Regulations.

In cases where the local authority determines that the proposals for minimising the noise from the construction activities are adequate it will issue a consent although this may be subject to conditions limiting certain aspects of the consent such as hours of use, noise levels for particular activities, etc. Provided that the applicant takes all reasonable steps to operate within the terms of the consent, even if the local authority subsequently decides to take proceedings under section 60(8), the applicant should be able to rely on the defence provided in the Act and prove that the alleged contravention amounted to the carrying out of works in accordance with a consent given under section 61.

A definition of 'best practicable means' for both the Control of Pollution Act 1974 [128] and the Environmental Protection Act 1990 [123] is provided by the following section 79(9) of the latter Act:

'(9) In this Part 'best practicable means' is to be interpreted by reference to the following provisions:

(a) 'practicable' means reasonably practicable having regard among other things to local conditions and circumstances, to the current state of technical knowledge and to the financial implications;

(b) the means to be employed include the design, installation, maintenance and manner and periods of operation of plant and machinery, and the design, construction and maintenance of buildings and structures;

(c) the test is to apply only so far as compatible with any duty imposed by law;

(d) the test is to apply only so far as compatible with safety and safe working conditions, and with the exigencies of any emergency or unforeseeable circumstances;

and, in circumstances where a code of practice under section 71 of the Control of Pollution Act 1974 (noise minimisation) is applicable, regard shall also be had to guidance given in it.'

Clean Neighbourhoods and Environment Act (Northern Ireland) 2011 [129]

In Northern Ireland the main provisions for statutory nuisance are in Part 7 of the Clean Neighbourhoods and Environment Act (Northern Ireland) 2011. The provisions are broadly similar to those in the other parts of the UK and are summarised in 'Guidance to District Councils on Part 7 (Statutory Nuisances) of the Clean Neighbourhoods and Environment Act (Northern Ireland) 2011' published by the Northern Ireland Department of the Environment [130].

Section 5 Notes

15 Annual noise statistics as published by the CIEH
16 *London Borough of Southwark and Another v. Mills* and *Baxter v. London Borough of Camden* [1999]

UKHL 40; [1999] 4 All ER 449; [1999] 3 WLR 939 (21st October, 1999) 45 EG 179

17 Detailed guidance is provided in the Home Office document *Anti-social Behaviour, Crime and Policing Act 2014: Anti-social behaviour powers Statutory guidance for frontline professionals* (revised in January 2021). (See https://assets.publishing.service.gov.uk/government/uploads/system/ uploads/attachment_data/file/956143/ASB_Statutory_Guidance.pdf)

18 HHSRS statutory Operating Guidance issued by ODPM in 2006.

19 A Decent Home: Definition and guidance for implementation issued by DCLG in June 2006 https://assets.publishing.service.gov.uk/government/uploads/system/uploads/attachment_data/file/7812/138355.pdf

SECTION 6: LICENSING

With the enactment of the Licensing Act 2003 [131] the role of local authorities in licensing was expanded. Licensing authorities are responsible for determining applications for new licences and variations of existing licences for the sale and supply of alcohol, public entertainment and late night refreshment. The Live Music Act 2012 [132] made a number of deregulatory changes to the Licensing Act and a summary of the changes is provided in paragraph 16.6 of the Home Office 'Revised Guidance issued under section 182 of the Licensing Act 2003' [133] published in 2018.

Every licensing authority is required to publish a Statement of Licensing Policy setting out how it will promote the licensing objectives in the Licensing Act 2003 [131]. Those licensing objectives are:

- Prevention of crime and disorder;
- Public safety;
- Prevention of public nuisance; and
- Protection of children from harm.

The 'prevention of public nuisance' affords local authorities a further control on the potential for noise disturbance presented by licensed premises. This is in addition to any restrictions that may have been imposed on the licensed premises at the planning stage or as a result of statutory nuisance action. Careful assessment of applications for the grant or renewal of a licence can help to achieve appropriate physical and management controls on the potential for noise from the premises to disturb residents in the vicinity.

Pubs and clubs guidance

Very often premises offering late night entertainment are a potential source of noise disturbance. This can arise from three main sources, firstly noise breakout from within the premises, secondly the noise of mechanical plant such as air handling equipment and thirdly the noise of patrons outside in the area around the venue. Generally, the first two are more readily controllable than the third over which local authorities have few sanctions.

As with any proposal for a new use of premises, the control of pubs and clubs by the appropriate use of planning powers provides the opportunity to require adequate design and physical measures to contain the noise. The imposition of a planning condition specifying the hours and days of opening is another common means of control. Sometimes it is necessary to accommodate the potentially conflicting demands of support for a viable leisure and town centre economy with the rights of residents to a standard of environment commensurate with the established character of the area. As with most planning applications it is important that each premises and its proposed use are considered on the merits of the particular case.

The IOA published a 'Good Practice Guide on the Control of Noise from Pubs and Clubs' in 2003 [134] that resulted from the deliberations of a working party comprising EHPs, acoustic consultants and, initially, members of the pub, club and entertainment industries. That guide covers the three main sources identified earlier but also includes noise from beer gardens. The one area that is specifically excluded from consideration is noise from live sporting events held at such premises.

Section 84 and Schedule 1 of the Clean Neighbourhoods Act 2005 apply to licensed premises the provisions of the Noise Act 1996 which address noise emitted from dwellings at night and the forfeiture and confiscation of equipment used to make noise unlawfully.

Inaudibility – To prevent activities within licensed premises disturbing residential occupiers in the late evening and at night some local authorities require that noise from the premises be inaudible inside adjoining or neighbouring residences between specified hours. A noise that cannot be heard is not capable of causing a disturbance.

However, there are arguments raised against the use of inaudibility requirements in connection with licensed premises and some of these are set out in a Defra report [135]. Significant issues are the variability of the hearing acuity of different individuals and the variation of the underlying background sound level in different areas at different times.

The Good Practice Guide [134] says that:

- 'for premises where entertainment takes place on a regular basis, music and associated sources should not be audible inside noise-sensitive property at any time. In the absence of the objective criteria mentioned in 2.3, what is 'regular' should be determined on a local basis to reflect local expectations and should be incorporated by local authorities in their planning and enforcement policies (see section 4); and,

- for premises where entertainment takes place less frequently, music and associated sources should not be audible inside noise-sensitive property between 23:00 and 07:00 hours. For other times, appropriate criteria need to be developed which balance the rights of those seeking and providing entertainment, with those who may be disturbed by the noise.'

The Guide goes on to qualify the meaning of audible as follows:

- 'noise may be considered not audible or inaudible when it is at a low enough level such that it is not recognisable as emanating from the source in question and it does not alter the perception of the ambient noise environment that would prevail in the absence of the source in question.'

The Guide's approach to audibility was intended to be broadly applicable to venues where entertainment takes place more than 30 times per year, not more than once in a single week and ends by 23:00 hours.

Whatever control measures are imposed on premises the six tests for conditions set out in the Government web-based resource [36] (see earlier), with modification of the second test to 'relevant to licensing', are as relevant to licensing conditions as they are to planning conditions.

SECTION 7: STRATEGIC NOISE MAPPING

Noise mapping

The Environmental Noise Directive (END, 2002/49/EC) [1] relating to the assessment and management of environmental noise was drawn up and ultimately adopted by the European Parliament and the Council of the European Union, on 25 June 2002. The Directive can be found via the link/URL in Table 20.17.

Environmental noise is defined as unwanted or harmful outdoor sound created by human activities, including noise from road, rail, airports and from industrial sites. The aim of the END [1] is to define a common approach across the European Union with the intention of avoiding, preventing or reducing, on a prioritised basis, the harmful effects including annoyance, due to exposure to environmental noise. This is to involve:

- The determination of exposure to environmental noise, through noise mapping;
- The adoption of action plans by Member States, based upon noise-mapping results, with a view to preventing and reducing environmental noise (particularly where exposure levels can induce harmful effects on human health) and to preserving environmental noise quality where it is good; and
- Ensuring that information on environmental noise and its effects is made available to the public.

The END [1] requires that Member States designate the competent authorities responsible for implementing this Directive, including:

- The making and where relevant, approving of noise maps (and action plans); and
- The collation of noise maps (and action plans).

For the first, second and third rounds of noise mapping in the UK, the responsibility for preparing the noise maps (and action plans) for all sources except airports has been taken on by the four national Governments. Major airports and non-major airports that may impact upon agglomerations are designated to undertake their own noise mapping and each publish the resultant noise action plan. The initial round of mapping took place in 2007, with the END requiring the process to be completed every five years thereafter.

The END [1] requires that Member States use common noise indicators – L_{den} and L_{night}

– to describe environmental noise levels, although the use of supplementary noise indicators is allowed. The L_{den} is an average noise level over the 24-hour period, but includes a weighting for the evening and night-time periods. The L_{night} is the average un-weighted noise level over the night-time period. In the UK the day, evening and night-time periods are determined as follows:

- Day 07:00–19:00 (12 hours);
- Evening 19:00–23:00 (4 hours); and
- Night 23:00–07:00 (8 hours).

Values of L_{den} and L_{night} can be determined either by computation or by measurement, although realistically, given the areas assessed in the UK, the END [1] requirement to provide information separately for all sources and the rolling programme set down in the END, computation (i.e. a computer-based noise model) is the only viable option.

The END [1] applies to human exposure to environmental noise in built-up areas (agglomerations) and close to major roads, railways and airports in open country. An agglomeration is defined as an area having a population in excess of 100,000 persons (250,000 for the first round of mapping) and a population density such that the Member State considers it to be an urbanised area. The noise maps include:

- Major roads with more than 3 million vehicle passages a year (6 million in 2007);
- Major railways with more than 30,000 train passages per year (60,000 in 2007); and
- Major airports (50,000 movements per year).

To assist the noise mapping process, a Good Practice Guide was produced by the European Commission [136], which can be found via the link given in Table 20.17. A second version of this Good Practice Guide was published in 2006, with Final Draft status [137].

In addition to a description of the major sources and/or agglomerations, the identification of responsible authorities and the computation/measurement methods utilised, the principal information to be sent to the Commission is an estimate of the number of people living in dwellings that are exposed to various bands of noise level in dB at 4 m above the ground on the most exposed façade. This has to be provided separately for road, rail and air traffic and (for agglomerations only) industrial sources.

A particular requirement of the Directive is that Member States have to ensure that the strategic noise maps (and action plans) are made available and disseminated to the public, and that all information is clear, comprehensible and a summary is provided setting out the most important points.

The END [1] is a form of EC legislation that is not directly applicable in the Member States but instead has to be transposed into national legislation. The Environmental Noise (England) Regulations 2006 (and subsequent amendments) apply to England [138] (see Table 20.17), although there are equivalent Regulations (and subsequent amendments) for Northern Ireland [139], Scotland [140] and Wales [141] (see also Table 20.17).

Although the UK has left the European Union as of the 1 January 2021, the English Government appears committed to continuing with the strategic noise mapping that has hitherto been undertaken in accordance with END [1]. Strategic noise maps and the associated noise action plans have to be made every five years; the most recent maps can be found via links or URLs included in Table 20.17.

In the UK, the first three rounds of noise mapping have utilised the CRTN [97] and CRN [99] methodologies for predicting road traffic noise and rail noise respectively.

Action planning

The END and associated noise maps are strategic tools intended to provide a suitable basis for the identification of areas subject to high noise levels from individual source types, or a combination of sources, and facilitate their prioritisation by considering the population exposed to noise. It is for the competent authority first to identify the noise hotspots[20] and then to investigate mitigation measures.

Based on the strategic noise maps, the appropriate competent authorities have drawn up action plans designed to manage noise and its effects for areas within agglomerations and near major roads, major railways and major airports. Each of these action plans identifies, amongst other things, mitigation measures that could be applied to manage noise.

The most recent noise action plans for the UK, following the third round of noise mapping, can be found via the links/URLs in Table 20.17. Action planning is to be repeated every five years, based on the noise maps generated the previous year.

Potential mitigation measures fall into two categories, planning policy and physical mitigation.

Planning Policy – Having identified potential noise hotspots it is important that planning policy is developed to recognise the status of these so as to avoid a situation where future developments within those areas might be approved without due consideration of the noise climate and any proposals within noise action plans for mitigating noise levels.

Physical Mitigation – There is a range of potential mitigation measures that may be appropriate to particular situations, depending on various factors such as source and receiver characteristics, local sensitivities such as visual appearance, and finally practicality and cost effectiveness. The END does not specify the target noise limits, and it is for the competent authority to determine its own criteria for the purposes of the action plans.

Section 7 Note

20 These hotspots are described as Important Areas in England, although Northern Ireland, Scotland and Wales describe them as Noise Management Areas or Noise Action Plan Priority Areas.

Table 20.17 Documents relating to noise mapping and noise action planning

---

EU

a   2002/49/EC The assessment and management of environmental noise (2002/49/EC)
    https://eur-lex.europa.eu/legal-content/en/ALL/?uri=CELEX%3A32002L0049

b   2015/996/EC Establishing common noise assessment methods according to Directive 2002/49/EC
    https://eur-lex.europa.eu/legal-content/EN/TXT/?uri=OJ:JOL_2015_168_R_0001

c   Good Practice Guide for Strategic Noise Mapping and the Production of Associated Data on Noise
    Exposure, version 1 2003
    https://www.eukn.eu/fileadmin/Lib/files/EUKN/2010/1710-strategic-noise-mapping.pdf

England

d   The Environmental Noise (England) Regulations 2006 (as amended 2010 and 2018)
    http://www.legislation.gov.uk/uksi/2006/2238/contents/made

e   Noise maps
    http://extrium.co.uk/noiseviewer.html and
    https://www.gov.uk/government/publications/strategic-noise-mapping-2019

f   Noise action plans: large urban areas, roads and railways (2019)
    https://www.gov.uk/government/publications/noise-action-plans-large-urban-areas-roads-and-
    railways-2019

Northern Ireland

g   The Environmental Noise (Northern Ireland) Regulations 2006 (as amended 2018)
    http://www.legislation.gov.uk/nisr/2006/387/contents/made

h   Round 3 Noise maps and Noise Mapping Technical Reports
    https://appsd.daera-ni.gov.uk/noisemapviewer/index.html and https://www.daera-ni.gov.uk/
    publications/round-3-noise-maps-and-noise-mapping-technical-reports

i   Noise action plans
    https://www.daera-ni.gov.uk/publications/department-regional-development-roads-environmental-
    noise-directive-round-two-noise
    https://www.daera-ni.gov.uk/industry-environmental-noise-directive-round-three-noise-action-
    plan-2019–2023
    https://www.daera-ni.gov.uk/publications/translink-ni-railways-environmental-noise-directive-
    round-two-noise-action-plan-2013
    https://www.daera-ni.gov.uk/publications/belfast-international-airport-environmental-noise-
    directive-round-two-noise-action-plan
    https://www.daera-ni.gov.uk/publications/george-best-belfast-city-airport-environmental-noise-
    directive-round-two-noise-action

Scotland

j   The Environmental Noise (Scotland) Regulations 2006 (as amended 2018)
    http://www.legislation.gov.uk/ssi/2006/465/contents/made

k   Noise maps (rounds two and three)
    https://noise.environment.gov.scot/noisemap/

l   Round three action plans (airports, agglomerations and transportation)
    https://noise.environment.gov.scot/action-planning-round-three.html

Wales

m   The Environmental Noise (Wales) Regulations 2006 (as amended 2009 and 2019)
    http://www.legislation.gov.uk/wsi/2006/2629/contents/made

n   Noise maps
    http://extrium.co.uk/walesnoiseviewer.html

o   Noise and soundscape action plan
    https://gov.wales/sites/default/files/publications/2019-04/noise-and-soundscape-action-plan.pdf

---

SECTION 8: SOUND INSULATION

Whilst generally the first objective in mitigating noise is to control it at source, there remain many situations for which sound insulation is at least part of the solution.

The distinction between 'sound insulation' and 'sound absorption' often is not understood and a few basic principles are worth bearing in mind when explaining to those not versed in acoustics the measures that are likely to be required. For example, the mistaken belief that applying egg boxes to a separating wall will improve its sound insulation is surprisingly common and the temptation for people to make their own modifications to sound insulation treatments without knowledge of how they work can compromise the best solutions.

The scope for reducing the transmission of sound from one side of a panel to the other is far greater from sound insulation than from sound absorption.

Sound absorption 'removes' sound energy, dissipating it as heat energy, and may have a small part to play in controlling noise break-out from one area to another insomuch as the overall internal reverberant sound level is reduced. However, it can only ever have a limited effect, typically no more than about 5 dB, depending on the sound absorbency of the material used, its method of application and the surface area that can be covered with it. In many situations the modifying effect that the introduction of sound absorptive surfaces to a space can have on the acoustic character of that space needs to be considered. Amongst the most critical examples are those of performance spaces such as theatres and concert halls.

Sound insulation on the other hand has the potential to all but eliminate the transfer of energy through a partition. If one imagines the effect that a heavy sound-absorbent curtain has in 'deadening' the acoustic character of an otherwise empty room, that is an illustration of sound absorption. However, if one were to stand behind the curtain and then stand outside the room behind the separating wall it is not hard to imagine the significant difference between the sound insulation provided by the lightweight curtain material and the much denser material of the wall. This is a simplistic example that ignores the other complexities of sound insulation such as flanking transmission but serves to illustrate the difference between absorption and insulation.

In general there are two approaches to airborne sound insulation. The first relies on the mass of the element, be it a wall, floor or roof and the second relies on the use of physically separated leaves of lighter material in a sealed construction. The latter will often incorporate a sound absorbent layer within the gap between panels to reduce flutter echoes within the cavity but otherwise relies on a combination of the mass of two or more leaves, differences in their resonant frequency introduced by using different materials for the layers of different thicknesses of material and the phenomenon of 'impedance mismatch' that these latter two features introduce to the structure.

Impact sound insulation requires either extra mass as for airborne sound insulation or physical isolation, commonly by including an isolating resilient material between the walking surface and the structural floor.

Building envelope

The envelope of a building may serve either to contain noise within it or to reduce noise entering the building from external noise sources such as road, rail or air traffic.

Lower levels of noise may be controlled perfectly adequately by fairly standard constructions. The acoustically weak points tend to be windows, doors and ventilation openings, although roofs may also be of relatively low sound insulation performance. As the level of noise to be contained or excluded rises so do the sound insulation performance requirements of the building envelope.

Currently there are no statutory requirements on the sound insulation performance of the external elements of a building, be they commercial or residential in use. However, at the consultation stage of the enactment of the Building Regulations 2000, proposals were included to require appropriate standards of external sound insulation for residential properties. Based on research into the performance of particular types of commonly adopted construction, general recommendations were to be provided, in a similar way to those for internal sound insulation that are detailed in Approved Document E 'Resistance to the passage of sound: 2003 Edition' (as amended) [31]. The Approved Document has been approved and issued by the Secretary of State for the purpose of providing practical guidance with respect to the Building Regulations 2010 for England and Wales. The building envelope recommendations were never implemented because it was concluded that legally that was a matter for control under planning powers and not the Building Regulations. Nevertheless, as an indication of the incremental approach to sound insulation as external noise levels rise, the recommendations of the consultation document remain very useful and are reproduced here:

Envelope constructions for external levels not exceeding 55 dB $L_{Aeq,16h}$ or 45 dB $L_{Aeq,8h}$ – at 'low noise' sites (where the external noise levels do not exceed the above levels) the internal target levels are likely to be achieved with any façade construction which complies with the other parts of the Building Regulations.

Envelope constructions for external levels not exceeding 60 dB $L_{Aeq,16h}$ or 50 dB $L_{Aeq,8h}$ – example envelope constructions are given in Table 20.18.

Envelope constructions for external levels not exceeding 65 dB $L_{Aeq,16h}$ or 60 dB $L_{Aeq,8h}$ – example envelope constructions are given in Table 20.19.

Envelope constructions for external levels not exceeding 75 dB $L_{Aeq,16h}$ or 65 dB $L_{Aeq,8h}$ – example envelope constructions are given in Table 20.20.

Envelope constructions for external levels exceeding 75 dB $L_{Aeq,16h}$ or 65 dB $L_{Aeq,8h}$ – if development is allowed at these levels, a specialist should be consulted.

The example constructions given in these tables are indicative. In Annex G of BS 8233 [4] a *'simple calculation'* and a *'more rigorous calculation'* method for the prediction of internal noise levels from an external noise source are described. It is reasonable to expect some detail to be provided with a planning application for new housing significantly affected

Table 20.18 Example envelope constructions for external noise levels not exceeding 60 dB $L_{Aeq,16h}$ or 50 dB $L_{Aeq,8h}$

Element	Example envelope construction
Wall	Solid brickwork, brick/block cavity, brick clad timber frame or timber frame with lightweight cladding.
Window	Any practical window specification, well sealed when closed.
Roof	Tiled/slated roof, 9 kg/m² plasterboard ceiling.
Ventilator	Trickle ventilators.

Table 20.19 Example envelope constructions for external noise levels not exceeding 65 dB $L_{Aeq,16h}$ or 60 dB $L_{Aeq,8h}$

Element	Example envelope construction
Wall	Solid brickwork, brick/block cavity, brick clad timber frame or timber frame with lightweight cladding.
Window	Double glazing, 10/12/6 mm, well sealed when closed.
Roof	Tiled/slated roof, 9 kg/m² plasterboard ceiling, 100 mm sound absorbing layer above the ceiling (for example mineral wool loft insulation).
Ventilator	Mechanical ventilation in bedrooms. Acoustic trickle ventilators in other (living) rooms.

Table 20.20 Example envelope constructions for external noise levels not exceeding 75 dB $L_{Aeq,16h}$ or 65 dB $L_{Aeq,8h}$

Element	Example envelope construction
Wall	Solid brickwork, brick/block cavity, brick clad timber frame.
Window	Double window 6/100/4 mm, limited to not more than 2.5 m² in area in each habitable room, well sealed when closed.
Roof	Tiled/slated roof, 20 kg/m² plasterboard ceiling, 100 mm sound absorbing layer above the ceiling (e.g. mineral wool loft insulation) and timber boarding on top of ceiling joists.
Ventilator	Mechanical ventilation throughout.

by environmental noise to demonstrate that appropriate levels can be achieved internally. However, more rigorous calculations might sensibly be left until later in the scheme's development, perhaps by imposition of a condition on the planning permission requiring submission of details of the sound insulation for approval by the local planning authority prior to commencement of the development.

On sites with higher external noise levels, achieving acceptable internal noise levels whilst at the same time ensuring adequate ventilation presents design challenges. That is particularly the case in situations where recognised standards, such as those in BS 8233 [4], cannot be achieved when open windows are the method of ventilation. Whilst some occupants might not be unduly disturbed by intrusive external noise, it is important that the design of new build properties affords occupiers acceptable internal noise levels without necessarily having to compromise when something more than background ventilation is required. For example, it is unlikely that acoustically treated trickle ventilators will provide sufficient rates of ventilation for comfortable occupation during all seasons of the year.

For many years in the United Kingdom it has been the practice to offer occupiers

whose homes are affected by public development, such as a new or altered road or railway, a compensatory sound insulation package that includes individual room ventilators. As a retrofit measure the sound attenuating mechanical ventilators provided in such cases are not universally popular but are a relatively economical means of enabling occupiers to keep windows closed whilst maintaining a reasonable level of ventilation.

For new development, the position of the noise source giving rise to high levels may be such that, in some instances, the development can be orientated to confine openable windows of noise-sensitive rooms to quieter façades. However, it is increasingly common for more sophisticated whole house ventilation strategies to be adopted, often to address a number of requirements, only one of which might be noise. The UK's Association of Noise Consultants and Institute of Acoustics has published 'The Acoustics, Ventilation and Overheating Residential Design Guide' (AVOG) [142] which provides an approach as to how the competing and technically complex aspects of thermal and acoustic comfort in new dwellings can be managed. The AVOG is intended to demonstrate good acoustic design as described in the ProPG [20], when considering internal noise level criteria, whilst also accounting for thermal and ventilation requirements.

Thermal efficiency requirements under Building Regulations are tending towards making buildings as airtight as practicable and then relying on a mechanical ventilation strategy that employs a heat recovery unit to avoid the uncontrolled thermal losses that would arise from open windows. Another reason for avoiding the need for occupiers to open windows might be because of poor air quality, for example at a façade overlooking a busy highway.

There are a variety of ventilation strategies available to developers. As with many aspects of building design it is not simply a case of appropriate specification but the care that is taken in the installation, the adequacy of instructions for its operation and, for mechanical systems, continuing routine maintenance, that together are likely to determine the overall success of the chosen method.

An increasingly common method of alternative ventilation for properties designed to have very low air leakage, is balanced supply and extract mechanical ventilation with heat recovery (MVHR). Such a system transfers heat from exhaust air to the incoming air via a passive heat exchange unit with the air being delivered from a central fan unit to the rooms in the building via ductwork. The incoming air is directed to habitable rooms including living rooms, dining rooms, bedrooms and studies, etc. Those rooms are not usually provided with extract terminals, the extraction instead being from 'wet' rooms such as the bathroom and kitchen, although it is sometimes necessary for there to be an extract terminal in a larger habitable room to avoid unacceptably high rates of air movement in the bathroom or kitchen. This flow arrangement avoids moist and or odorous air from the kitchen and bathroom being drawn into habitable rooms. The system may include a facility to boost extraction rates when the bathroom or kitchen are in use.

The provision of a 'summer bypass' facility is likely to be required to avoid an MVHR system adding to problems of overheating in summer months. Although MVHR may have a boost ventilation setting, it could still be necessary on occasion to use open windows for rapid dilution and removal of pollutants such as paint fumes, malodours, etc. or for comfort cooling.

Experience of MVHR has been mixed. Instances of poor design, installation, commissioning and/or maintenance have been identified in reviews of developments where such systems have been installed. These issues may be compounded by the inadequate instruction given to occupiers in how the systems should be used and maintained.

Change of use

The glazing and ventilation detail of the example envelope constructions described earlier in this section will be relevant to many buildings undergoing a material change of use, where the existing constructions are similar to those in the examples. The performance of alternative building envelope constructions may need to be determined individually.

Where it is necessary to retain existing windows (for example in conservation work), a secondary glazing system or other suitable alternative can be used instead of sealed units.

It should be noted that these constructions will not provide sufficient insulation in the loft, should it be used as (or converted into) a living space. Specialist advice is likely to be necessary for conversions of lofts and other spaces.

Mitigation of the noise impact of transportation noise

Statutory schemes exist for the sound insulation of residential property meeting the criteria for eligibility under the provisions of the following Regulations (as mentioned):

- 'Noise Insulation Regulations' 1975 [98] (as amended) (applicable to new or altered highways); and
- 'Noise Insulation (Railway and Other Guided Transport Systems) Regulations' 1996 [100] (as amended) (applicable to new or altered railways).

The methods for predicting the relevant noise levels at the façade of affected properties and for determining eligibility for sound insulation, are set out in Government memoranda, 'Calculation of Road Traffic Noise' [97] and 'Calculation of Railway Noise' [99].

Government legislation requires the relevant authority to insulate dwellings which meet relevant noise-related criteria as a result of noise from new or altered roads and railways. A brief summary of the relevant Regulations is given next.

A number of airports have their own individual noise insulation schemes which generally include not only the sound insulation of windows and provision of sound attenuating ventilation but also works to improve the sound insulation of roofs in appropriate cases.

Noise Insulation Regulations – The 'Noise Insulation Regulations' 1975 [98] (as amended) relate to road traffic. Three noise-related conditions have to be met:

- The relevant noise level at the façade of the affected property should not be less than the specified level of 68 dB $L_{A10,18h}$ (06:00 to 24:00 hours);
- The relevant noise level must be at least 1.0 dB(A) more than the prevailing noise level before the proposed works; and
- The anticipated noise level from the new or altered highway must make an effective contribution of at least 1.0 dB(A) to the relevant noise level.

For a new road, the relevant authority has a mandatory duty to insulate where these criteria are met whilst for an altered road (where the location, width or level of a carriageway is altered otherwise than by resurfacing) the powers are discretionary.

The 'Noise Insulation (Railways and Other Guided Transport Systems) Regulations' 1996 [100] relate to railways. As for road traffic three noise-related conditions have to be met:

- The relevant noise level at the façade of the affected property should not be less than the specified level of 68 dB $L_{Aeq,18h}$ in the day (06:00 to 24:00 hours) or 63 dB $L_{Aeq,6h}$ at night (00:00 to 06:00 hours);
- The relevant noise level must be at least 1.0 dB(A) more than the prevailing noise level before the proposed works; and
- The anticipated noise level from the new or altered railway must make an effective

contribution of at least 1.0 dB(A) to the relevant noise level.

As for road traffic, these Regulations include a mandatory duty to insulate where new or additional works are planned and a discretionary power for 'altered works'.

Internal sound insulation – In England and Wales, sound insulation between dwellings became subject to the Building Regulations in 1966, when the requirements were first grouped under the Part G – Resistance to the passage of sound. They controlled the sound insulation of separating walls and floors between new houses and flats but did not extend to flats provided by conversion of an existing dwelling.

The sound insulation requirements of the current Building Regulations 2010 are set out in considerable detail in Approved Document E (resistance to the passage of sound) [31].

Currently, in Scotland guidance on achieving the relevant standards as set in the Building (Scotland) Regulations 2004 [143] is provided in Section 5 of the Technical Handbooks [144–145].

In Northern Ireland Technical Booklets are published by the Northern Ireland Government's Department of Finance and Personnel in support of some of the technical parts of the Building Regulations (Northern Ireland) 2012 [146] (as amended). They provide construction methods that, if followed, will be deemed-to-satisfy the requirements of the Northern Ireland Building Regulations. The relevant one for sound insulation is Technical Booklet G (Resistance to the passage of sound) October 2012 [147].

The limitations introduced by case law on the pursuit of statutory nuisance action in cases of inadequate sound insulation has briefly been dealt with previously in this chapter.

SECTION 9: VIBRATION

Vibration is an oscillatory motion. The magnitude of vibration can be defined in terms of displacement (how far something moves from the equilibrium position), velocity (how fast something moves) or acceleration (the rate of change of velocity). Standards for the assessment of building damage are often given in terms of peak velocity (usually referred to as Peak Particle Velocity, or PPV), whilst human response to vibration is often described in terms of rms (root mean square) or rmq (root mean quad) acceleration. When describing vibration, it is necessary to specify whether peak values (i.e. the maximum displacement or maximum velocity) or rms/rmq values (effectively an average value) are used.

Vibration usually is quantified using the three axes, vertical, longitudinal and transverse or sometimes by a combination of the energy in all three, termed the vector sum. This is important not only to its transmission but to the sensitivity of those experiencing the vibration which tends to depend on their orientation to the vibration, something that will be different for the standing or seated receiver than for one who is lying down.

Human sensitivity to vibration

The potential effects of vibration are several but humans are susceptible to very low levels of vibration and may be disturbed or even alarmed by vibration at levels at least an order of magnitude lower than those that cause structural damage for example. As with noise the reaction may be conditioned by factors including the frequency, duration, intermittency and strength of the vibration.

Guidance on how people within buildings may react to vibration is provided by BS 6472–1:2008 [148]. There is a second part [66] that deals with the effects on inhabitants of periodic blasting.

The standard is applicable to vibration in buildings within the frequency range 0.5 Hz to 80 Hz and uses a measure called the vibration dose value (VDV). The significance of measured or predicted results can be derived from Table 20.21 which is reproduced from Table 1 of BS 6472–1:2008 [148].

Vibration may be occasional, intermittent or continuous in duration and of impulsive, variable or constant amplitude. An unexpected, impulsive vibration may be startling.

There are sometimes circumstances where the source of the vibration may be at some distance from the building within which its effects are felt. Because the human sense of vibration lacks the directivity afforded by the sense of hearing, identifying the cause of disturbance can be much more difficult. For example, complaints by occupiers of the upper floors of a modern steel frame office block that the floor was vibrating from

Table 20.21 BS 6472–1:2008 Vibration dose value ranges which might result in various probabilities of adverse comment within residential buildings

Place and time	Low probability of adverse comment m.s$^{-1.75}$ 1	Adverse comment possible m.s$^{-1.75}$	Adverse comment probable m.s$^{-1.75}$ 2
Residential buildings 16 h day	0.2 to 0.4	0.4 to 0.8	0.8 to 1.6
Residential buildings 8 h night	0.1 to 0.2	0.2 to 0.4	0.4 to 0.8

Note: For offices and workshops, multiplying factors of 2 and 4 respectively should be applied to these vibration dose value ranges for a 16 h day.
1 Below these ranges adverse comment is not expected.
2 Above these ranges adverse comment is very likely.

intermittent shocks that they feared might indicate an imminent collapse of the structure was eventually traced to a demolition site several blocks distant. The demolition contractor was cutting concrete columns and then dropping them to the excavated basement level from a height of several stories. Evidently the shock wave was being transmitted relatively efficiently to the piled foundations of the office block where the complaints were arising although nothing very significant could be detected at street level outside the complainant's building.

For road traffic vibration there are two potential sources to consider:

- Direct transmission of ground-borne vibration from impact of the vehicle tyres with road surface irregularities; and
- Airborne low frequency sound from the vehicle exciting sympathetic resonance in lighter elements of the structure of a building or its contents.

Significant levels of ground-borne vibration from even heavy-duty vehicle (HDV) traffic on public highways are rarely encountered and when they do arise, almost always a marked discontinuity in the road surface such as a pot hole or a raised (or dropped) manhole cover is the cause.

Vibration arising from the sympathetic resonance of lightweight structural elements or items within a property is more common. The rattling of loosely fitting windows and ornaments on shelves are typical manifestations of vibration induced by low frequency noise. Sometimes the low frequency sound may be sensed and described by individuals as a vibration although in fact it is unlikely that this is a direct effect, rather an association with the low frequency noise.

Low frequency airborne sound is a characteristic of the diesel engines of buses and HDVs that may be exaggerated when they are accelerating or moving off and the engine is labouring.

Damage

Minor damage, such as the development of hairline cracks for example, is usually referred to as 'cosmetic damage' but in the most severe cases significant cracking or even collapse is possible.

People experiencing vibration within their home tend to assume that it will be causing damage to the fabric of the building or even in extreme cases threatening its collapse. However, human sensitivity to vibration is such that the levels at which it becomes noticeable and even alarming are generally very much below the levels at which damage to buildings or other structures might be expected.

As a consequence it would be very unusual to be faced with the problem of having to assess potential building damage if vibration levels are limited to those at which complaints are minimised.

Measurement

The measurement of vibration can be significantly more complex than that of noise both in terms of selecting the most relevant measurement parameters and ensuring that vibration transducers are appropriately selected and mounted.

For the measurement of vibration from blasting and construction sites there are a variety of robust vibration monitors designed for use in the field with a relatively heavy steel geophone block incorporating transducers aligned on the three orthogonal axes, transverse, vertical and longitudinal. The geophone can usually be mounted in a variety of ways, for example on the surface of a firmly bedded object such as a solid stone threshold, buried in the ground, mounted on a firmly embedded ground spike or fixed by a threaded stud and nut to a surface such as a wall. The geophone can be levelled using a spirit level (sometimes built-in) and threaded pillars.

The sensitivity of the geophone needs to be selected for the range of vibration that is anticipated. A geophone that is designed primarily for measuring vibration from blasting may be insufficiently sensitive for measurements at or near the threshold of human sensitivity.

Alternative equipment might be a vibration meter, which may be a suitably specified sound level meter, used with accelerometers mounted individually on three sides of a steel block or fixed directly to the surface(s) on which the measurements are to be made. The latter method of fixing may be especially appropriate in circumstances where it is necessary to avoid 'loading' the vibrating surface by application of a heavy transducer.

The gathering of frequency data is particularly important when measuring vibration particularly when the results are required in VDV where different frequency weightings are used in the vertical and horizontal axes.

Prediction

The prediction of vibration levels at any given location is considerably more complex than the prediction of noise levels. This is largely because of the uncertainty that usually exists concerning the nature and vibration transmission characteristics of the ground or structures between the source and receiver. An additional complication often is the structural response of the building for which the predictions are being made.

For activities such as piling there is a body of empirical vibration data to be found in BS 5228–2 [65]. The document includes empirical formulae for the prediction of vibration from a number of different construction activities. It should be remembered that the predictions are of the vibration level at the ground surface. When predicting the effects inside a building, a suitable transfer function must be applied. Vibration transfer through a building is complex, and the level of amplification and attenuation of vibration can vary widely due to excitation of

particular resonances of building elements, dependent on the frequency content of the input vibration signal.

A section on the prediction of vibration from blasting using what is known as the 'scaled distance' approach is also included in BS 5228–2 [65]. Such a prediction requires the development of a scaled distance graph based on data from the measurement of vibration at a number of distances from one or more trial blasts. This method enables the vibration transmission characteristics of the intervening ground to be taken into account but its accuracy may still be influenced by factors such as variability in blasts. The method provides an indication of the vibration levels to be expected at a given distance from the blast, usually expressed as a statistical average, rather than a precise prediction.

The prediction of building response to vibration is likely to involve the use of more sophisticated techniques such as dynamic analysis computer modelling, possibly including finite element analysis of the structure, and a detailed knowledge of both the ground conditions and the building structure. Steel frame structures tend to be significantly more responsive to vibration than in situ concrete frame buildings and the height of buildings and the nature of their foundations are among the factors that may be significant.

Assessment

Guidance on the assessment of human sensitivity to vibration is available in BS 6472:2008, Parts 1 and 2 [148, 66].

As a general rule, for anything other than a temporary source of vibration such as demolition and construction work, any tangible vibration in residential property is likely to be a potential nuisance.

Guide values for the onset of damage based on the lowest vibration levels above which damage has been credibly demonstrated are set out in BS 7385–2:1993 [149]. The values are considerably higher for reinforced or framed structures, industrial and heavy commercial buildings than they are for unreinforced or light framed structures, residential or light commercial type buildings. They are also frequency dependent, being lowest for the latter categories of building at frequencies from 4 Hz up to 15 Hz. At 40 Hz and above the guide value is the same for all categories of buildings.

It is very unusual to encounter situations where damage to occupied buildings has occurred probably at least partly because occupiers are likely to have become alarmed at significantly lower levels than those normally required to result in even cosmetic damage.

SECTION 10: INTERNATIONAL REGULATION OF NOISE

In countries of the world where local legislation and/or standards on noise have not been developed to any significant extent, the default position often is either to rely on legislation and standards from other states such as Europe, the UK, the USA,[21] etc. or to use the International Finance Corporation (IFC) Environmental, Health, and Safety Guidelines (known as the 'EHS Guidelines').[22] The IFC is a member of the World Bank Group.

General EHS Guideline 1.7 Noise [150] provides basic advice on managing noise impacts from a project facility or operations where the applicable noise level guideline is exceeded at the most sensitive receptor. It is stated that noise impacts should not exceed the levels presented in Table 1.7.1 of Guideline 1.7 (reproduced as Table 20.22) or result in an increase in background level of more than 3 dB at the nearest receptor location offsite. The guideline values are for noise levels measured out of doors and are attributed to the WHO CNG [18].

General EHS Guideline 2.0 Physical Hazards [151] provides the following advice (in

section 2.3) regarding noise in the working environment:

- 'No employees should be exposed to a noise level greater than 85 dB for a duration of more than 8 hours per day without hearing protection. In addition, no unprotected ear should be exposed to a peak sound pressure level (instantaneous) of more than 140 dB(C).

- The use of hearing protection should be enforced actively when the equivalent sound level over 8 hours reaches 85 dB(A), the peak sound levels reach 140 dB(C), or the average maximum sound level reaches 110 dB(A). Hearing protective devices provided should be capable of reducing sound levels at the ear to at least 85 dB(A).

- Although hearing protection is preferred for any period of noise exposure in excess of 85 dB(A), an equivalent level of protection can be obtained, but less easily managed by limiting the duration of noise

Table 20.22 EHS Guideline 1.7 noise level guidelines

Receptor	1-hour L_{Aeq} dB(A)	
	Daytime 07:00–22:00	Night-time 22:00–07:00
Residential; institutional; educational	55	45
Industrial; commercial	70	70

Note: For acceptable indoor noise levels for residential, institutional and educational settings refer to WHO (1999) [18]

Source: International Finance Corporation (April 2007) [150]

exposure. For every 3 dB(A) increase in sound levels, the 'allowed' exposure period or duration should be reduced by 50 percent.

- Prior to the issuance of hearing protective devices as the final control mechanism, use of acoustic insulating materials, isolation of the noise source, and other engineering controls should be investigated and implemented, where feasible.
- Periodic medical hearing checks should be performed on workers exposed to high noise levels.'

The general EHS Guidelines are supported by a series of Industry Sector Guidelines that include more specific advice for particular industries [152].

Using the information and standard methodology developed by the WHO's Regional Office for Europe for its European Environment and Health Information System (ENHIS), WHO/Europe has undertaken environment and health performance reviews (EHPRs) for a number of countries that have requested it. The purpose of the EHPRs is to 'support countries in reforming and upgrading their public health systems by providing policy advice, guidance for strengthening policy-making and for planning preventive interventions, service delivery and surveillance in the field of environment and health'. These provide an overview of the regulatory controls in place or proposed by those countries at the time the particular country review was undertaken.

One area of noise management where there is a degree of international consistency relates to airports. The International Civil Aviation Organisation (ICAO) has developed policies under its 'Balanced Approach to Aircraft Noise Management'[23] which has four principal elements. Those elements are:

- Reduction at source (quieter aircraft);
- Land use planning and management;
- Noise abatement operational procedures; and
- Operating restrictions.

Noise impacts of Brexit

As at March 2021 the only change to UK legislation dealing specifically with noise that had been enacted following Britain leaving the EU on 1 January 2021 (Brexit) was the Noise Emission in the Environment by Equipment for use Outdoors Regulations 2001 [94]. These were originally introduced to transpose the provisions of the Outdoor Noise Directive 2000/14/EC [93] into UK legislation which regulates the noise emissions into the environment by outdoor equipment, such as certain types of construction plant and equipment.

Generally, it is the case that UK domestic legislation relevant to noise, be it noise nuisance, Building Regulations, Health and Safety, etc. will remain as it was prior to Brexit. Nevertheless, it is likely that over time divergence from any EU directives from which some of it was transposed will develop as the EU and UK introduce their own amendments to existing legislation and statutory instruments. Northern Ireland's special status under the Northern Ireland Protocol means that future legislation there is likely to align with EU requirements rather than any new legislation implemented in the other three nations of the UK.

Noise receives only a single incidental mention in the Environment Act 2021 that is due to come into force in parts during 2022. The Environment Act 2021 enables the Government to develop legally binding environmental targets and introduces a new independent Office for Environmental Protection (OEP) with a remit to scrutinise all Government policy to ensure the environment is at the heart of decision making. It will have the power to run its own independent investigations and enforce environmental law, including taking Government and other public bodies to court where necessary. Its provisions apply across the UK's devolved administrations but some parts apply in only one or more jurisdictions. (See https://www.legislation.gov.uk/ukpga/2021/30/contents/enacted).

The continued development of noise maps and action plans, previously undertaken to meet requirements in the END [1], looks set to continue, at least for the next round due to commence shortly.

Noise impacts of the Covid pandemic of 2020–22

The global Covid pandemic that commenced in 2020 has had some profound and potentially lasting noise impacts. National and international travel restrictions served to reduce transportation activity and with it noise levels, possibly most noticeably from aviation. Whilst that activity is likely to return gradually it may never fully recover due to the increase in home working, on-line shopping and other factors. As time progresses the introduction of new and quieter vehicles and aircraft may see even similar volumes of traffic producing lower noise levels, again tending to reduce noise levels to below pre-pandemic levels.

Conversely there seems to have been an unsurprising rise in domestic noise complaints, presumably as a consequence of people being confined to their homes and local neighbourhood to a greater extent during the pandemic. In March 2021 Police Scotland reported that for the third quarter of 2020/21 there were '*large increases in incidents of public nuisance (up from 56,936 to 123,979 or 117.8%), neighbour disputes (16,021 to 22,930 or 43.1%) and noise complaints (43,288 to 51,277 or 18.5%)* [which] *were attributed to the challenges of coronavirus*'.[24] In February 2022 the CIEH published its annual noise survey statistics for England for 2020/2125 and [153 25 (as superscript for note]. It was noted that '*Residential noise accounts for the largest proportion of noise complaints. This is the case across all regions in England.*' Compared with the last time CIEH collected data in 2019/20, the data shows a 54% increase in the number of noise complaints in 89 local authorities, which participated in both years. The survey report commented '*CIEH's noise survey captured data from three national lockdowns and constantly changing restrictions. At the time, many local authorities reported that they have received more domestic complaints during the initial lockdown period, as some people started DIY projects, whilst others juggled virtual meetings and home schooling.*

Increases in domestic noise complaints also seem to have been experienced in other parts of the world. For example, in a country where the majority of residents live in apartment buildings, the Korea Environment Corporation announced in January 2021 that a total of 42,250 apartment noise-related complaints were reported in 2020, more than double the yearly average of 20,508 and an increase of 60.9% over 2019.[26]

Research published in the journal *Science* (11 September 2020) [154] reported that lockdowns during the pandemic reduced seismic noise (a generic name for a relatively persistent vibration of the ground, due to a multitude of causes) by as much as 50% and that '*The 2020 seismic noise quiet period is the longest and most prominent global anthropogenic seismic noise reduction on record*'. This work was based on analysis of high-frequency (4 Hz to 14 Hz) seismic ambient noise (hiFSAN) data.

Section 10 Notes

21 Municipal Code Corporation (Municode) publishes legal documents for local governments in the USA https://library.municode.com

22 https://www.ifc.org/wps/wcm/connect/topics_ext_content/ifc_external_corporate_site/sustainability-at-ifc/policies-standards/ehs-guidelines

23 International Civil Aviation Organisation (ICAO) Balanced Approach to Aircraft Noise Management http://www.icao.int/environmental-protection/pages/noise.aspx

24 https://www.scotland.police.uk/what-s-happening/news/2021/march/increased-calls-to-police-scotland/

25 The CIEH has for a number of years carried out an annual noise survey that provides the most authoritative data on noise in the environment see https://www.cieh.org/media/6561/cieh-noise-survey-england-2020-21.pdf and https://www.cieh.org/policy/campaigns/noise-survey

26 https://www.koreatimes.co.kr/www/nation/2021/01/119_302696.html

Chapter references

[1] Directive 2002/49/EC of the European Parliament and of the Council of 25 June. (2002) *Relating to the Assessment and Management of Environmental Noise.* https://eur-lex.europa.eu/legal-content/en/ALL/?uri=CELEX%3A32002L0049.

[2] BSI. (2014) *Methods for Rating and Assessing Industrial and Commercial Sound,* BS 4142:2014+A1:2019, BSI, London.

[3] Craven NJ, Kerry G. (2007) *A Good Practice Guide on the Sources and Magnitude of Uncertainty Arising in the Practical Measurement of Environmental Noise,* University of Salford. http://usir.salford.ac.uk/20640/.

[4] BSI. (2014) *Guidance on Sound Insulation and Noise Reduction for Buildings,* BS 8233:2014, BSI, London.

[5] Ministry of Housing Communities & Local Government. (2021) *National Planning Policy Framework (NPPF).* https://assets.publishing.service.gov.uk/government/uploads/system/uploads/attachment_data/file/1005759/NPPF_July_2021.pdf.

[6] DEFRA. (2010) *Noise Policy Statement for England (NPSE),* March. https://assets.publishing.service.gov.uk/government/uploads/system/uploads/attachment_data/file/69533/pb13750-noise-policy.pdf.

[7] Planning Policy Wales Edition of 11 February. (2021) https://gov.wales/sites/default/files/publications/2021-02/planning-policy-wales-edition-11_0.pdf.

[8] Association of Noise Consultants (ANC), Institute of Acoustics (IoA) and Chartered Institute of Environmental Health (CIEH). (2017) *Professional Practice Guidance on Planning & Noise: New Residential Development: Supplementary Document 2 Good Acoustic Design,* May. https://www.cieh.org/media/1257/propg-document-2_good-acoustic-design.pdf.

[9] Planning (Scotland) Act. (2019) *Acts of the Scottish Parliament 2019 asp 13.* https://www.legislation.gov.uk/asp/2019/13/section/25/enacted.

[10] Ministry of Housing Communities & Local Government (2014) *Planning Practice Guidance–Noise (PPG-N),* March, updated July 2019. https://www.gov.uk/guidance/noise-2.

[11] Design Manual for Roads and Bridges, Sustainability & Environmental Appraisal. (2020) *Noise and Vibration LA 111,* Revision 2, May. https://www.standardsforhighways.co.uk/dmrb/search/cc8cfcf7-c235-4052-8d32-d5398796b364.

[12] Local Government and Communities Directorate. (2011) *Planning Advice Noise 1/2011 Planning and Noise,* March. https://www.gov.scot/publications/planning-advice-note-1-2011-planning-noise/.

[13] Environment and Forestry Directorate. (2011) *Assessment of Noise:Technical Advice Note,* March. https://www.gov.scot/publications/technical-advice-note-assessment-noise/.

[14] Local Government and Communities Directorate. (1998) *Planning Circular 4/1998 The Use Conditions in Planning Permissions,* February. https://www.gov.scot/publications/planning-circular-4-1998-use-of-conditions-in-planning-permissions/.

[15] Local Government and Communities Directorate. (1998) *Planning Circular 4/1998 Model Planning Conditions Addendum,* February. https://www.gov.scot/publications/planning-circular-4-1998-model-planning-conditions-addendum/.

[16] Planning Guidance (Wales). (1997) *Technical Advice Notes (Wales) 11, Noise,* October. https://gov.wales/sites/default/files/publications/2018-09/tan11-noise.pdf.

[17] Welsh Government. (2015) *CL-01-15 Updates to Tan 11 Noise – Noise Action Plan (2013–18) Commitments,* November. https://gov.wales/updates-technical-advice-note-tan-11-noise-cl-01-15.

[18] World Health Organization (WHO). (1999) *Guidelines for Community Noise,* WHO, Geneva. https://www.who.int/docstore/peh/noise/Comnoise-1.pdf.

[19] Department of the Environment. (2014) *Noise Policy Statement for Northern Ireland,* September. https://www.daera-ni.gov.uk/sites/default/files/publications/doe/noise-policy-statement-ni.PDF.

[20] Association of Noise Consultants (ANC), Institute of Acoustics (IoA) and Chartered Institute of Environmental Health (CIEH). (2017) *Professional Practice Guidance on Planning & Noise: New Residential Development,* May. https://www.ioa.org.uk/sites/default/files/14720%20ProPG%20Main%20Document.pdf.

[21] World Health Organization WHO. (2009) *Night Noise Guidelines for Europe.* https://www.euro.who.int/__data/assets/pdf_file/0017/43316/E92845.pdf.

[22] World Health Organization (WHO). (2018) *Environmental Noise Guidelines for the European Region.* https://www.euro.who.int/en/publications/abstracts/environmental-noise-guidelines-for-the-european-region-2018.

[23] Vallet M, Vernet I. (1991) Night aircraft noise index and sleep research results. In *Inter-Noise 91: The Cost of Noise*, Lawrence A (Ed.), Volume 1, Noise Control Foundation, Poughkeepsie, NY, 207–210.

[24] ICAO Assembly Resolution A39–1. *Consolidated Statement of Continuing ICAO Policies and Practices Related to Environmental Protection – General Provisions, Noise and Local Air Quality.* https://www.icao.int/environmental-protection/Documents/Resolution_A39_1.PDF.

[25] Secretary of State for Transport. (2013) *Aviation Policy Framework*, March. https://assets.publishing.service.gov.uk/government/uploads/system/uploads/attachment_data/file/153776/aviation-policy-framework.pdf.

[26] *Consultation Response on UK Airspace Policy: A Framework for Balanced Decisions on the Design and Use of Airspace*, DfT, October 2017. https://assets.publishing.service.gov.uk/government/uploads/system/uploads/attachment_data/file/918784/consultation-response-on-uk-airspace-policy-web.pdf

[27] ICCAN. (2020) *Aviation Noise and Public Health*, September. https://iccan.gov.uk/wp-content/uploads/2020_09_24_Aviation_Noise_and_Public_Health_ICCAN_Note.pdf.

[28] ICCAN. (2021) *ICCAN Report on the Future of Aviation Noise Management,* March. https://iccan.gov.uk/iccan-report-future-noise-management/.

[29] BRE Global Ltd. (2018) *BREEAM UK New Construction: Non-Domestic Buildings (United Kingdom)*, Technical Manual SD5078: BREEAM UK New Construction 2018 3.0. https://www.breeam.com/NC2018/content/resources/output/10_pdf/a4_pdf/print/nc_uk_a4_print_mono/nc_uk_a4_print_mono.pdf.

[30] BRE Global Ltd. (2017) *BREEAM International New Construction 2016,* Technical Manual SD233 2.0. https://tools.breeam.com/filelibrary/Technical%20Manuals/BREEAM_International_NC_2016_Technical_Manual_2.0.pdf.

[31] Ministry of Housing, Communities & Local Government. (2015) *Approved Document E Resistance to the Passage of Sound to the Building Regulations 2010. 2003 Edition Incorporating 2004, 2010, 2013 and 2015 Amendments.* https://assets.publishing.service.gov.uk/government/uploads/system/uploads/attachment_data/file/468870/ADE_LOCKED.pdf.

[32] Department for Education. (2015) *Acoustic Design of Schools: Performance Standards (BB93)*, February. https://assets.publishing.service.gov.uk/government/uploads/system/uploads/attachment_data/file/400784/BB93_February_2015.pdf.

[33] Institute of Acoustics and Association of Noise Consultants. (2015) *Acoustics of Schools: A Design Guide*, November. https://www.ioa.org.uk/sites/default/files/Acoustics%20of%20Schools%20-%20a%20design%20guide%20November%202015.pdf.

[34] Department of Health. (2013) *Health Technical Memorandum (HTM) 08–01: Acoustics.* https://www.gov.uk/government/uploads/system/uploads/attachment_data/file/144248/HTM_08-01.pdf.

[35] British Council for Offices. (2019) *Guide to Specification: Best Practice for Offices,* British Council for Offices, London.

[36] Ministry of Housing, Communities & Local Government. (2019) *Guidance – Use of Planning Conditions.* https://www.gov.uk/guidance/use-of-planning-conditions.

[37] Department of the Environment. (2015) *Development Management Practice Note 20 Use of Planning Conditions.* https://www.infrastructure-ni.gov.uk/sites/default/files/publications/infrastructure/dmpn-20-use-of-planning-conditions-v1-april-2015_0.pdf.

[38] Scottish Government, Local Government and Communities Directorate. (1998) *Planning Circular 4/1998 The Use of Conditions in Planning Permissions.* https://www.gov.scot/publications/planning-circular-4-1998-use-of-conditions-in-planning--permissions/.

[39] The Working Group on Noise from Wind Turbines. (1996) *The Assessment and Rating of Noise from Wind Farms,* ETSU-R-97, Final Report September. http://webarchive.nationalarchives.gov.uk/+/http:/www.berr.gov.uk/files/file20433.pdf.

[40] DTI. (2006) *The Measurement of Low Frequency Noise at Three UK Wind Farms: URN No: 06/1412,* DTI, London.

[41] Defra. (2007) *Research into Aerodynamic Modulation of Wind Turbine Noise: Final Report,* Defra, London (NANR233). http://usir.salford.ac.uk/id/eprint/1554/.

[42] Institute of Acoustics. (2013) *Good Practice Guide to the application of ETSU-R-97.* https://www.ioa.org.uk/sites/default/files/IOA%20Good%20Practice%20Guide%20on%20Wind%20Turbine%20Noise%20-%20May%202013.pdf.

[43] DECC. (2015) *Wind Turbine AM Review*, Phase 1 Report. https://assets.publishing.service.gov.uk/government/uploads/system/uploads/attachment_data/file/562185/Phase_1_Report_-_Wind_Turbine_AM_Review_Issue_1_291015.pdf.

[44] Institute of Acoustics. (2016) *A Method for Rating Amplitude Modulation in Wind Turbine Noise*. https://www.ioa.org.uk/sites/default/files/AMWG%20Final%20Report-09-08-2016_0.pdf.

[45] DECC. (2016) *Wind Turbine AM Review*, Phase 2 Report. https://assets.publishing.service.gov.uk/government/uploads/system/uploads/attachment_data/file/562186/Phase_2_Report_-_Wind_Turbine_AM_Review_Issue_3__FINAL_.pdf.

[46] Defra. (2011) *Wind Farm Noise Statutory Nuisance Complaint Methodology*, NANR 277, London. https://assets.publishing.service.gov.uk/government/uploads/system/uploads/attachment_data/file/69222/pb-13584-windfarm-noise-statutory-nuisance.pdf.

[47] National Farmers Union (NFU). (2012) *Bird Deterrents and Bird Scarers – Protecting Your Crop – NFU Code of Practice*. https://www.nfuonline.com/assets/4662.

[48] BRE. (2002) *The National Noise Incidence Study 2000/2001* (United Kingdom), BRE Client Report No. 206344f, BRE, London.

[49] Sport England. (2015) *Design Guidance Note Artificial Grass Pitch (AGP) Acoustics – Planning Implications*. https://sportengland-production-files.s3.eu-west-2.amazonaws.com/s3fs-public/agp-acoustics-planning-implications.pdf?eORPPBrK6irJ2FqvHWitOASeYu6U.egt.

[50] UKSI 2015 No. 596 (2015) *The Town and Country Planning (General Permitted Development) (England) Order 2015*. https://www.legislation.gov.uk/uksi/2015/596/made.

[51] The Noise Council. (1994) *Code of Practice on Noise from Organised Off Road Motor Cycle Sport*, The Noise Council, London.

[52] The National Society for Clean Air and Environmental Protection (NSCA) (1997) *Code of Practice for the Control of Noise from Oval Motor Racing Circuits*, NSCA, London.

[53] Defra, CIEH. (2006) *Neighbourhood Noise Policies and Practice for Local Authorities – a Management Guide*, Defra, CIEH, London.

[54] British Water Ski and Wakeboard. (2013) *Code of Practice – Noise*, British Water Ski and Wakeboard, London.

[55] Department of the Environment, Welsh Office, Scottish Development Department and Department of the Environment for Northern Ireland. (1982) *Code of Practice on Noise from Model Aircraft*. https://www.gov.uk/government/publications/code-of-practice-on-noise-from-model-aircraft.

[56] Christian A, Cabell R. (2017) *Initial Investigation into the Psychoacoustic Properties of Small Unmanned Aerial System Noise*, Proceedings of the 23rd AIAA/CEAS Aeroacoustics Conference, 5–9 June, The American Institute of Aeronautics and Astronautics, Denver, CO.

[57] UK Civil Aviation Authority. (2019) *The Drone and Model Aircraft Code*, October. https://register-drones.caa.co.uk/drone-code.

[58] Chartered Institute of Environmental Health. (2003) *Clay Target Shooting: Guidance on the Control of Noise*. https://www.cieh.org/media/1236/clay-target-shooting-guidance-on-the-control-of-noise.pdf.

[59] Noise Council. (1995) *Code of Practice on Environmental Noise Control at Concerts*, Noise Council, London.

[60] Manchester City Council. (2018) *Guidance on Noise Control for Open Air Concerts and Events*, October. https://www.manchester.gov.uk/downloads/download/6970/noise_control_for_open_air_venues.

[61] UK Public General Acts. (1994) *Chapter 33 Criminal Justice and Public Order Act 1994*. https://www.legislation.gov.uk/ukpga/1994/33/contents.

[62] UK Public General Acts. (2014) *Chapter 12 The Anti-social Behaviour, Crime and Policing Act 2014*. https://www.legislation.gov.uk/ukpga/2014/12/contents/enacted.

[63] Chartered Institute of Environmental Health. (2017) *Professional Practice Note – Guidance on the Use of Community Protection Notices*, Revision 3, November. https://www.cieh.org/media/1238/guidance-on-the-use-of-community-protection-notices.pdf.

[64] Ministry of Housing Communities & Local Government (March 2014) *Planning Practice Guidance – Minerals*, updated October 2014. https://www.gov.uk/guidance/minerals#Noise-emissions.

[65] BSI. (2014) *Code of Practice for Noise and Vibration Control on Construction and Open*, sites, Part 2: Vibration BS 5228–2:2009+A1:2014, BSI, London.

[66] BSI. (2008) *Guide to Evaluation of Human Exposure to Vibration in Buildings – Part 2: Blast-Induced Vibration*, BS 6472-2:2008, BSI, London.

[67] Welsh Assembly Government. (2004) *Minerals Technical Advice Note 1 Aggregates*, March.

https://gov.wales/sites/default/files/publications/2018-09/mtan1-aggregates.pdf.

[68] Welsh Government. (2009) *Minerals Technical Advice Note 2 Coal*, January. https://gov.wales/minerals-technical-advice-note-mtan-wales-2-coal.

[69] National Planning Framework for Scotland 3 published in 2014. https://www.gov.scot/binaries/content/documents/govscot/publications/advice-and-guidance/2020/12/scottish-planning-policy/documents/scottish-planning-policy/scottish-planning-policy/govscot%3Adocument/scottish-planning-policy.pdf?forceDownload=true.

[70] Scottish Planning Policy, published in June 2014 and revised in December 2020. https://www.gov.scot/publications/scottish-planning-policy/.

[71] The Scottish Office. (1996) *Planning Advice Note (PAN) 50 Controlling the Environmental Effects of Surface Mineral Workings, Annex A The Control of Noise at Surface Mineral Workings*. https://www.gov.scot/publications/planning-advice-note-pan-50-annex-controlling-environmental-effects-surface.

[72] The Scottish Office. (1998) *Planning Advice Note (PAN) 50 Controlling the Environmental Effects of Surface Mineral Workings, Annex C The Control of Traffic at Surface Mineral Workings*. https://www.gov.scot/publications/traffic-surface-mineral.

[73] The Scottish Office. (2000) *Planning Advice Note (PAN) 50 Controlling the Environmental Effects of Surface Mineral Workings, Annex D The Control of Blasting at Surface Mineral Workings*. https://www.gov.scot/publications/blasting-surface-mineral.

[74] Department of the Environment (Northern Ireland). (2015) *Strategic Planning Policy Statement (SPPS) Planning for Sustainable Development*. https://www.infrastructure-ni.gov.uk/publications/strategic-planning-policy-statement.

[75] UKSI 2016 No. 1154. (2016) *The Environmental Permitting (England and Wales) Regulations 2016*. https://www.legislation.gov.uk/uksi/2016/1154/contents/made.

[76] SSI 2012 No. 360. (2012) *The Pollution Prevention and Control (Scotland) Regulations 2012*. https://www.legislation.gov.uk/ssi/2012/360/contents/made.

[77] Northern Ireland Statutory Rules 2013 No. 160. (2013) *The Pollution Prevention and Control (Industrial Emissions) Regulations (Northern Ireland) 2013*. https://www.legislation.gov.uk/nisr/2013/160/contents/made.

[78] Directive 2010/75/EU of the European Parliament and of the Council. (2010) *Industrial Emissions (Integrated Pollution Prevention and Control)*, 24 November. https://eur-lex.europa.eu/legal-content/EN/TXT/?uri=CELEX:32010L0075.

[79] Defra. (2012) *General Guidance Manual: Policy and Procedures for A2 and B Installations Part 1 and Part 2*. https://www.gov.uk/government/publications/local-authority-pollution-control-general-guidance-manual.

[80] Environment Agency. (2020) *Standard Rules: Environmental Permitting*, November. https://www.gov.uk/government/collections/standard-rules-environmental-permitting.

[81] Environmental Agency. (2021) *Noise and Vibration Management: Environmental Permits*. https://www.gov.uk/government/publications/noise-and-vibration-management-environmental-permits/noise-and-vibration-management-environmental-permits Last updated 31 January 2022.

[82] Environmental Agency for England and Wales, Scottish Environment Protection Agency and Northern Ireland Environment and Heritage Service. (2002) *Horizontal Guidance for Noise: Part 1 – Regulation and Permitting*, IPPC H3 (part 1). https://www.sepa.org.uk/media/61299/ippc-h3-1-noise-part-1-published-september-2002.pdf.

[83] Environmental Agency for England and Wales, Scottish Environment Protection Agency and Northern Ireland Environment and Heritage Service. (2004) *Horizontal Guidance for Noise: Part 2 – Noise Assessment and Control (IPPC H3 (part 2))*. https://www.sepa.org.uk/media/61312/ippc-h3-2-noise-part-2-published-september-2002.pdf.

[84] Scottish Environment Protection Agency. (undated) *Guidance on the Control of Noise at PPC Installations*. https://www.sepa.org.uk/media/156907/ppc_noise_guidance.pdf (Accessed November 2021).

[85] Environment Agency (2018) *Guidance – Noise Impact Assessments Involving Calculations or Modelling*, October and updated November 2019. https://www.gov.uk/guidance/noise-impact-assessments-involving-calculations-or-modelling.

[86] Council Directive 85/337/EEC. (1985) *On the Assessment of the Effects of Certain Public and Private Projects on the Environment*, 27 June. https://eur-lex.europa.eu/legal-content/EN/TXT/?uri=CELEX%3A31985L0337.

[87] UKSI 2017 No. 571. (2017) *The Town and Country Planning (Environmental Impact Assessment) Regulations 2017*. https://www.

legislation.gov.uk/uksi/2017/571/contents/made.

[88] SSI 2017 No. 102. (2017) *The Town and Country Planning (Environmental Impact Assessment) (Scotland) Regulations 2017*. https://www.legislation.gov.uk/ssi/2017/102/contents/made.

[89] WSI 2017 No. 567 (W.136). (2017) *The Town and Country Planning (Environmental Impact Assessment) (Wales) Regulations 2017*. https://www.legislation.gov.uk/wsi/2017/567/contents/made.

[90] Northern Ireland Statutory Rules 2017 No. 83. (2017) *The Planning (Environmental Impact Assessment) (Northern Ireland) Regulations 2017*. https://www.legislation.gov.uk/nisr/2017/83/made.

[91] BSI. (2014) *Code of Practice for Noise and Vibration Control on Construction and Open Sites, Part 1: Noise*, BS 5228-1:2009+A1:2014, BSI, London.

[92] The London Authorities Noise Action Forum. (2016) *London Good Practice Guide: Noise & Vibration Control for Demolition and Construction*, July. https://www.cieh.org/media/1251/london-good-practice-guide-noise-vibration-control-for-demolition-and-construction.pdf.

[93] Council Directive 2000/14/EC. (2000) *On the Noise Emission in the Environment by Equipment for Use Outdoor*, 8 May. https://eur-lex.europa.eu/legal-content/EN/TXT/?uri=celex%3A32000L0014.

[94] UKSI 2001 No. 1701. (2001) *The Noise Emission in the Environment by Equipment for Use Outdoors Regulations 2001*. https://www.legislation.gov.uk/uksi/2001/1701/contents/made.

[95] Office for Products Safety & Standards. (2021) *Noise Emission in the Environment by Equipment for Use Outdoors Regulations: As They Apply to Equipment Being Supplied in or into Great Britain After 1 January 2021*, January. https://assets.publishing.service.gov.uk/government/uploads/system/uploads/attachment_data/file/951384/Guide-to-noise-emissions-regulations-2001-tp.pdf.

[96] Office for Products Safety & Standards. (2021) *Noise Emission in the Environment by Equipment for Use Outdoors Regulations: As They Apply to Equipment Being Supplied in or into Northern Ireland from 1 January 2021*, January. https://assets.publishing.service.gov.uk/government/uploads/system/uploads/attachment_data/file/950229/Guide-to-noise-emissions-regulations-2001-northern-ireland-tp.pdf.

[97] Department of Transport and Welsh Office. (1988) *Calculation of Road Traffic Noise (CRTN)*. https://www.bradford.gov.uk/Documents/Hard%20Ings%20Road%20improvement%20scheme/2b%20Compulsory%20Purchase%20Order%20and%20Side%20Road%20Order/5%20Supporting%20documents/Calculation%20of%20Road%20Traffic%20Noise%201988.pdf.

[98] UKSI 1975 No. 1763. (1975) *Noise Insulation Regulations (NIR)*. https://www.legislation.gov.uk/uksi/1975/1763/contents/made.

[99] Department of Transport. (1995) *Calculation of Railway Noise (CRN)*. https://www.thenbs.com/PublicationIndex/documents/details?Pub=DOT&DocID=251687.

[100] UKSI 1996 No. 428. (1996) *Noise Insulation (Railways and Other Guided Transport Systems) Regulations*. https://www.legislation.gov.uk/uksi/1996/428/contents/made.

[101] Abbott PG, Nelson PM. (2002) *Converting the UK Traffic Noise Index $L_{A10,18h}$ to EU Noise Indices for Noise Mapping*, PR/SE/451/02 [EPG 1/2/37]. https://moam.info/converting-the-uk-traffic-noise-index-la1018h-to-semantic-scholar_5b6c05da097c4724218b456e.html.

[102] TRL and Casella Stanger. (2006) *Method for Converting the UK Road Traffic Noise Index $L_{A10,18h}$ to the EU Noise Indices for Road Noise Mapping*. https://www.researchgate.net/publication/311964809_method_for_converting_the_uk_road_traffic_noise_index_la1018h_to_the_eu_noise_indices_for_road_noise_mapping.

[103] Erik do Graaff and Bert Peeters (M+P). (2020) *Road Traffic Noise Modelling: Development of a Methodology to Define Category 5 Vehicles*, January. http://randd.defra.gov.uk/Default.aspx?Module=More&Location=None&ProjectID=20479.

[104] Department of Transport. (1996) *Calculation of Railway Noise (Supplement 1) Procedure for the Calculation of Noise from Eurostar Trains Class 373*. https://trid.trb.org/view/471362.

[105] Hardy AEJ, Jones RRK. (2004) *Rail and Wheel Roughness – Implications for Noise Mapping Based on the Calculation of Railway Noise Procedure*. AEATR-PC&E-2003-002. http://www.xn-lrmorama-0za.ch/m5_krachmacher/pdf/railway-noise.pdf.

[106] Hardy AEJ, Jones RRK, Wright CE. (2007) *Additional Railway Noise Source Terms for 'Calculation of Railway Noise 1995'*. https://www.epd.gov.hk/eia/register/report/eiareport/eia_2332015/html/EIA/Appendices/4.%20Noise/Appendix%204.20.pdf.

[107] Institute for Health and Consumer Protection (Joint Research Council). (2012) *Common Noise Assessment Methods in Europe (CNOS-SOS-EU)*.https://op.europa.eu/en/publication-detail/-/publication/80bca144-bd3a-46fb-8beb-47e16ab603db/language-en.

[108] Commission Directive (EU) 2015/996. (2015) *Establishing Common Noise Assessment Methods According to Directive 2002/49/EC of the European Parliament and of the Council,* 19 May. https://eur-lex.europa.eu/legal-content/EN/TXT/?uri=OJ:JOL_2015_168_R_0001.

[109] ISO. (1996) *Acoustics – Attenuation of Sound During Propagation Outdoors – Part 2: General Method of Calculations.* ISO 9613–2:1996. https://www.iso.org/sites/outage/.

[110] Birmingham Airport Limited. (2019) *Birmingham Airport Noise Action Plan 2018–2023,* formally adopted by DEFRA, 11 February. https://www.birminghamairport.co.uk/media/5325/noise-action-plan-2019-2023-adopted-by-defra-110219.pdf.

[111] Institute of Environmental Management & Assessment. (2014) *Guidelines for Environmental Noise Impact Assessment,* October. http://www.iema.net/noise.

[112] European Environment Agency (EEA). (2020) *Environmental Noise in Europe – 2020,* EEA Report No. 22/2019. https://www.eea.europa.eu/publications/environmental-noise-in-europe/.

[113] Berry BF, Flindell IH. (2009) *Estimating Dose-Response Relationships Between Noise Exposure and Human Health Impacts in the UK.* Berry Environmental Ltd Report 2009–2 for Defra, London, July.

[114] HPA. (2010) *Environmental Noise and Health in the UK by the Health Protection Agency (HPA) on Behalf of an Ad Hoc Expert Group on the Effects of Environmental Noise on Health* (Chair Prof Robert Maynard). https://www.researchgate.net/profile/Andrew-Smith-36/publication/279231460_Environmental_Noise_and_Health_in_the_UK/links/559129ef08aed6ec4bf697ec/Environmental-Noise-and-Health-in-the-UK.pdf?origin=publication_detail.

[115] Civil Aviation Authority. (2021) *Aircraft Noise and Health Effects – a Six Month Update (March 2021-September 2021),* CAP2257, September. https://publicapps.caa.co.uk/docs/33/Aircraft%20Noise%20and%20Health%20Effects%20September%202021%20(CAP2257).pdf

[116] WHO. (2011) *Burden of Disease from Environmental Noise – Quantification of Healthy Life Years Lost in Europe,* WHO European Centre for Environment and Health, Regional Office for Europe, Bonn. http://www.euro.who.int/__data/assets/pdf_file/0008/136466/e94888.pdf.

[117] UKSI 2005 No. 1643. (2005) *The Control of Noise at Work Regulations.* https://www.legislation.gov.uk/uksi/2005/1643/contents/made.

[118] UKSI 1989 No. 1790. (1989) *The Noise at Work Regulations.* https://www.legislation.gov.uk/uksi/1989/1790/contents/made.

[119] UK Public General Acts 1974 Chapter 37. (1974) *Health and Safety at Work etc. Act.* https://www.legislation.gov.uk/ukpga/1974/37/contents.

[120] Health and Safety Executive. (2021) *Controlling Noise at Work: The Control of Noise at Work Regulations 2005 – Guidance on Regulations (L108)*,3rded.https://www.hse.gov.uk/pubns/priced/l108.pdf.

[121] UKSI 2005 No. 1093. (2005) *The Control of Vibration at Work Regulations.* https://www.legislation.gov.uk/uksi/2005/1093/contents/made.

[122] Health and Safety Executive. (2019) *Hand-Arm Vibration: The Control of Vibration at Work Regulations 2005 – Guidance on Regulations (L140).* https://www.hse.gov.uk/pubns/priced/l140.pdf.

[123] UK Public General Acts 1990 Chapter 43. (1990) *Environmental Protection Act.* https://www.legislation.gov.uk/ukpga/1990/43/contents.

[124] Department of the Environment, Welsh Office, Scottish Development Department and Department of the Environment for Northern Ireland. (1982) *Code of Practice on Noise from Audible Intruder Alarms 1982,* https://assets.publishing.service.gov.uk/government/uploads/system/uploads/attachment_data/file/365813/pb13762-audible-intruder-alarms.pdf.

[125] UK Public General Acts 1993 Chapter 40. (1993) *Noise and Statutory Nuisance Act.* https://www.legislation.gov.uk/ukpga/1993/40/contents.

[126] UK Public General Acts 2005 Chapter 16. (2005) *Clean Neighbourhoods and Environment Act.* https://www.legislation.gov.uk/ukpga/2005/16/contents.

[127] UK Public General Acts 1996 Chapter 37. (1996) *Noise Act.* https://www.legislation.gov.uk/ukpga/1996/37/contents.

[128] UK Public General Acts 1974 Chapter 40. (1974) *Control of Pollution Act.* https://www.legislation.gov.uk/ukpga/1974/40.

[129] Acts of the Northern Ireland Assemble 2011 Chapter 23. (2011) *Clean Neighbourhoods and Environment Act (Northern Ireland).* https://www.legislation.gov.uk/nia/2011/23/contents.

[130] Department of the Environment. (2012) *Guidance to District Councils on Part 7 (Statutory Nuisances) of the Clean Neighbourhoods and Environment Act (Northern Ireland) 2011.* https://www.daera-ni.gov.uk/sites/default/files/publications/doe/guidance-on-statutory-nuisances.pdf.

[131] UK Public General Acts 2003 Chapter 17. (2003) *Licensing Act.* https://www.legislation.gov.uk/ukpga/2003/17/contents.

[132] UK Public General Acts 2012 Chapter 2. (2012) *Live Music Act.* https://www.legislation.gov.uk/ukpga/2012/2/contents.

[133] Home Office. (2018) *Revised Guidance Issued Under Section 182 of the Licensing Act 2003,* April. https://assets.publishing.service.gov.uk/government/uploads/system/uploads/attachment_data/file/705588/Revised_guidance_issued_under_section_182_of_the_Licensing_Act_2003__April_2018_.pdf.

[134] Institute of Acoustics (IOA) (2003) *Good Practice Guide on the Control of Noise from Pubs and Clubs,* IOA, London.

[135] Defra. (2006) *Noise from Pubs and Clubs (Phase II) Final Report,* Contract NANR163, May. https://view.officeapps.live.com/op/view.aspx?src=http://randd.defra.gov.uk/Document.aspx?Document=NO01099_3733_FRP.doc.

[136] European Commission Working Group Assessment of Exposure to Noise (WG-AEN). (2003) *Good Practice Guide for Strategic Noise Mapping and the Production of Associated Data on Noise Exposure,* Version 1, December. https://www.eukn.eu/fileadmin/Lib/files/EUKN/2010/1710-strategic-noise-mapping.pdf.

[137] European Commission Working Group Assessment of Exposure to Noise (WG-AEN). (2006) *Good Practice Guide for Strategic Noise Mapping and the Production of Associated Data on Noise Exposure – Final Draft,* Version 2, January. http://sicaweb.cedex.es/docs/documentacion/Good-Practice-Guide-for-Strategic-Noise-Mapping.pdf.

[138] UKSI 2006 No. 2238. (2006) *The Environmental Noise (England) Regulations 2006.* http://www.legislation.gov.uk/uksi/2006/2238/contents/made.

[139] Northern Ireland Statutory Rules 2006 No. 387. (2006) *The Environmental Noise (Northern Ireland) Regulations 2006.* http://www.legislation.gov.uk/nisr/2006/387/contents/made.

[140] SSI 2006 No. 465. (2006) *The Environmental Noise (Scotland) Regulations 2006.* http://www.legislation.gov.uk/ssi/2006/465/contents/made.

[141] WSI 2006 No. 2629 (W.225). (2006) *The Environmental Noise (Wales) Regulations 2006.* http://www.legislation.gov.uk/wsi/2006/2629/contents/made.

[142] Association of Noise Consultants and Institute of Acoustics. (2020) *Acoustics Ventilation and Overheating – Residential Design Guide,* Version 1.1, January. https://www.association-of-noise-consultants.co.uk/wp-content/uploads/2019/12/ANC-AVO-Residential-Design-Guide-January-2020-v-1.1.pdf#:~:text=The%20Acoustics%2C%20Ventilation%20and%20Overheating%20Residential%20Design%20Guide,acoustic%20comfort%20in%20our%20work%20and%20living%20spaces.

[143] SSI 2004 No. 406. (2004) *The Building (Scotland) Regulations 2004.* https://www.legislation.gov.uk/ssi/2004/406/contents/made.

[144] Scottish Government, Local Government and Communities Directorate. (2019) *Building Standards Technical Handbook 2019: Domestic Buildings.* https://www.gov.scot/publications/building-standards-technical-handbook-2019-domestic/.

[145] Scottish Government, Local Government and Communities Directorate. (2019) *Building Standards Technical Handbook 2019: Non-Domestic Buildings.* https://www.gov.scot/publications/building-standards-technical-handbook-2019-non-domestic/.

[146] Northern Ireland Statutory Rules 2012 No. 192. (2012) *The Building Regulations (Northern Ireland) 2012.* https://www.legislation.gov.uk/nisr/2012/192/contents/made.

[147] Department of Finance and Personnel. (2012) *Building Regulations (Northern Ireland) 2012 Guidance, Resistance to the Passage of Sound. Technical Booklet G,* October. http://www.buildingcontrol-ni.com/assets/pdf/TechnicalBookletG2012.pdf.

[148] BSI. (2008) *Guide to Evaluation of Human Exposure to Vibration in Buildings – Part 1: Vibration Sources Other Than Blasting,* BS 6472-1:2008, BSI, London.

[149] BSI. (1993) *Evaluation and Measurement for Vibration in Buildings – Part 2: Guide to Damage Levels from Groundborne Vibration,* BS 7385-2:1993, BSI, London.

[150] International Finance Corporation. (2007) *General EHS Guidelines: Environmental – Noise*

Management − 1.7, April. https://www.ifc. org/wps/wcm/connect/4a4db1c5-ee97-43ba-99dd-8b120b22ea32/1-7%2BNoise. pdf?MOD=AJPERES&CVID=ls4XYBw.

[151] International Finance Corporation. (2007) *General EHS Guidelines: Occupational Health & Safety − 2.0 Occupational Health and Safety*, April. https://www. ifc.org/wps/wcm/connect/1d19c1ab-3ef8-42d4-bd6b-cb79648af3fe/2%2BOccupational%2BHealth%2Band%2BSafety. pdf?MOD=AJPERES&CVID=ls62x8l.

[152] International Finance Corporation. (2016) *Environmental, Health and Safety Guidelines − Industry Sector Guidelines*. https://

www.ifc.org/wps/wcm/connect/29f5137d-6e17-4660-b1f9-02bf561935e5/Final%2B-%2BGeneral%2BEHS%2BGuidelines. pdf?MOD=AJPERES&CVID=nPtguVM.

[153] Chartered Institute of Environmental Health. (2022) CIEH Noise Survey 2020/21 Report on findings − England , February. https:// www.cieh.org/media/6561/cieh-noise-survey-england-2020-21.pdf.

[154] Lecocq T, et al. (2020) *Global Quieting of High-Frequency Seismic Noise Due to COVID-19 Pandemic Lockdown Measures,* September. https:// science.sciencemag.org/content/369/ 6509/1338.

21

Port health

Iain Pocknell and John Ambrose

SECTION 1: PUBLIC HEALTH AT AIRPORTS

Introduction

Since the 1980s the popularity of air travel has grown year on year. According to the International Civil Aviation Organization (ICAO), globally there were 4.5 billion passengers who travelled on aircraft in 2019 [1]. The Covid pandemic has seen a major disruption to air transport and ICAO estimate that passenger numbers had dropped by 60% [2].

Over a century ago international travel was mainly by sea voyages, whereas nowadays people can fly to almost anywhere in the world in a matter of hours. However, as we have seen by the Covid pandemic, the benefits of air travel also bring increased risks, in particular the international spread of disease.

Prior to the Covid pandemic, there were other outbreaks of disease such as the outbreak of influenza (A) which originated in Mexico in 2009, the Ebola outbreak in West Africa in 2014 and the Zika virus outbreak in South America in 2016. Each of these outbreaks were declared as a 'Public Health Emergency of International Concern' (PHEIC) by the World Health Organization (WHO) [3–5]. As with the Covid pandemic,

such diseases are easily spread to other countries through international travel.

Port Health at airports is a relatively recent arm of environmental health. It comprises public health and animal health controls, with different Central Government 'Competent Authorities' overseeing the work. The work of a port health officer involves engaging with these competent authorities, as well as with a number of other airport-based organisations.

At an airport a port health officer needs to be able to respond quickly to public health events, some of which can be unusual; others may be beyond the regular responsibility of Port Health. Although not all scenarios can be covered in this chapter, it has been written to provide signposts to direct the reader to where they might need to go for help. Examples have also been included to give an illustration of how port health officers have approached certain incidents.

Port Health is a function carried out mainly by the local authority of where the airport is situated. There are some exceptions where a separate Port Health Authority (PHA) fulfils this function (e.g. Southampton PHA, Mersey PHA and London PHA). In this chapter, the term LA/PHA will be used to describe

DOI: 10.1201/9781003035640-21

these local authorities (LAs) and Port Health Authorities covering the Port Health function at airports. Where reference is made to Public Health England (PHE – UKHSA since October 2021),[1] it also refers to its counterpart organisations in the devolved areas of the United Kingdom.

International Health Regulations 2005 and related provisions

The International Health Regulations 2005 (IHR) are a set of regulations that have been accepted by the Member States making up the World Health Organization (WHO). They are published by the WHO and are concerned with controlling the international spread of disease [6]. These Regulations are not directly enforceable but require the UK Government to make provision in UK law.

The IHR 2005 changed the way the WHO dealt with the spread of disease. Previously the WHO focussed on a list of 'notifiable' diseases, but the legislative framework was incapable of adapting to new threats such as Severe Acute Respiratory Syndrome (SARS).[2] The IHR 2005 have addressed this by introducing a decision instrument, which assists countries in determining which outbreaks might constitute a Public Health Emergency of International Concern (PHEIC) that would then need reporting to the WHO.

The IHR 2005 also require countries to strengthen and maintain the capacity to respond to PHEICs. Airports that can achieve the 'core capacity' requirements for surveillance and response (as listed in Annex 1 to the regulations) are permitted to become 'designated airports'. The Department of Health has created an administrative list of designated airports in the UK, but at the time of writing, no further guidance had been issued.[3]

The Public Health (Aircraft) Regulations 1979[4] (which originally adopted the IHR 1969) were amended in 2007 so that they corresponded to the IHR 2005. This offers an interim measure for enforcement as not all the provisions of the IHR 2005 have been included in these amended regulations.

The enabling act, the Control of Disease Act 1984, was amended by the Health and Social Care Act 2012.[5] At the time of writing, however, new legislation to replace the Public Health (Aircraft) Regulations 1979 is still awaited so that all the appropriate powers required to implement the International Health Regulations 2005 will be provided. For instance, there are very limited powers in the current regulations for dealing with disinsection on aircraft and controlling vectors at airports.[6]

Public Health England which is now the UK Health Security Agency (UKSHA) are, 'Acting as the lead for the UK on the IHR including protecting the UK from international health hazards, most noticeably communicable diseases' [7]. Locally, Public Health England (PHE/UKHSA) has the operational lead concerning the implementation of the IHR 2005 at ports and port health officers will liaise most commonly through their consultant in communicable disease control (CCDC). The PHE's strategic remit and priorities were revised in response to the Covid-19 pandemic and included 'public health activity at major ports as required to respond to the outbreak' [8].

Other Government Departments also have a responsibility in issuing guidance relating to safe air travel. In response to the Covid-19 pandemic, the Department for Transport (DfT) issued guidance for operators in the aviation industry [9]. In trying to limit the spread of Covid-19, a number of approaches have been tried – to varying effect. An assessment of these different strategies is outlined in a paper prepared by the DfT and Foreign, Commonwealth and Development Office (FCDO) [10].

Public Health (Aircraft) Regulations 1979 (as amended) – cover the appointment and duties of authorised officers, including medical practitioners as well as the provision of facilities and services at airports.

The regulations give powers of entry to inspect any aircraft where:

(a) a commander has sent a message to the LA/PHA concerning suspected infectious disease on board; and

(b) there are reasonable grounds for suspecting a case of infectious disease on board.

There are also powers for medical officers and authorised officers to detain aircraft/people, disinfect/disinsect clothing, articles or require the destruction of vermin and removal of conditions on aircraft likely to convey infection, etc.

At the time of writing, the main focus of infection control was on the Covid-19 pandemic, but other outbreaks also have the potential to become a PHEIC, for example, the WHO has declared two PHEICs for Ebola Virus Disease since 2014, the last one being in 2019 [11].

It will therefore be useful to refer back to an earlier PHEIC in 2014, when the Ebola Virus Disease epidemic in Central and West Africa created fear for passengers as well as airport workers.

Many individuals sourced their information from the media and tabloid press rather than official sources. Therefore, a number of false reports of Ebola-infected individuals at Gatwick Airport were reported in the early weeks of the epidemic. These reports highlighted the urgent need for guidelines to be issued in a timely manner from a credible source (e.g. Public Health England or its successor).

Whilst the emergency services (ambulance crews) and hospitals were testing their plans concerning the control of infected individuals, it became evident that Port Health investigations had an important part to play in revealing gaps with the control of environmental contamination. During global epidemics such as this, the following areas would be useful to check:

- The provisions for cleaning and disposal of spilled bodily fluids both on aircraft and in the terminals;

- Details of disinfectants used, in particular checking for anti-viral as well as anti-bacterial properties;
- Procedures for engineers to follow concerning the removal of contaminated seats and carpets, as well as toilet removal, from aircraft;
- Removal and disposal of toilet waste from aircraft;
- The safe and proper disposal of clinical waste.

It should be noted that aircraft manufacturers prohibit some chemicals on board the aircraft due to potential corrosion of its structure. Most airlines specify which chemicals can be used and their contracted cleaning companies will hold stocks of these products.

When a notification is made, the response has to be proportionate to the risk. One of the difficulties with air travel is that the identification of an infectious disease is usually not known until several hours or maybe days after the aircraft has landed and passengers disembarked.

Notification of Suspected Infectious Disease (NOIDS) – as well as being a requirement of the Public Health (Aircraft) Regulations 1979, the Civil Aviation Authority (CAA) has also made reporting a suspected infectious disease compulsory by adopting the International Civil Aviation Organisation (ICAO) guidelines. The procedures recognise the vital role that Air Traffic Control play in communicating messages to the destination airport authorities (see paragraph 15.1, section 5, chapter 1 of CAP 493 Manual of Air Traffic Services Part 1 [12]).

There are two documents which can be sought concerning inbound aircraft:

1 'Aircraft declaration of health' (which will also indicate the nature of any disinsection that has been carried out). A model form can be found in Annex 9 of IHR 2005.

2 'Passenger locator cards' (which give the contact details of passengers should they

need to be contacted following confirmation of an infectious disease which may be of concern). A model form is also available on the WHO website [13].

Port medical response – at one time, larger airports had medical units with Port Medical Officers. However, the Consultant in Communicable Disease Control (CCDC) within Public Health England (PHE now UKHSA) now performs this function.

The LA/PHA has the legal duty to enforce the Public Health (Aircraft) Regulations 1979 (as amended) and in particular has a duty to appoint proper officers for the appropriate medical response.

The roles and responsibilities of the LA/PHA, PHE (UKHSA), together with any other stakeholders (e.g. National Health Service (NHS), NHS Ambulance Service), should be clarified at each airport. Regular meetings with all the stakeholders can assist in this process. This can take the form of 'Local Resilience Forums'. It is also good practice to formalise the roles and responsibilities in 'Port Health Plans'. There is no standard format for these plans, as they are agreed locally.

Public health and health and safety overlaps – at airports, health and safety legislation is enforced by the Health and Safety Executive (HSE) and the LA/PHA. The division of responsibility of enforcement is set out in the Health and Safety (Enforcing Authority) Regulations.[7]

On aircraft, both the HSE and the Civil Aviation Authority (CAA) enforce health and safety legislation. The division of responsibility has been set out in a Memorandum of Understanding between the HSE, HSE Northern Ireland and CAA [14].

There are often overlaps between public health issues and health and safety requirements. Where investigations are carried out concerning public health issues on aircraft, it may be prudent to liaise with the HSE and/or CAA. On occasion, joint investigations have been found to be effective. For example, an investigation about lack of notification to a LA/PHA of a pilot suffering from a suspected infectious disease also revealed other offences related to lack of notification to the Civil Aviation Authority.

Aircraft as food premises

The Food Safety (Ships and Aircraft) (England and Scotland) Order 2003[8] expanded the definition of 'premises' to include aircraft. Until that point, food safety enforcement was carried out on the supply chain to the point where the food entered the aircraft.

At one time, in-flight meals were manufactured at flight catering premises found close to airports. Some of these establishments have changed their operation from cooking meals to buying ready-made meals and assembling them onto tray sets.

Flight catering establishments should be registered as food business operators and be subject to routine interventions in accordance with the Food Standards Agency (FSA)'s Food Law Code of Practice [15].

Where the in-flight caterer is still carrying out cooking, it is likely that such premises will require approval under Retained EU Regulation 853/2004.[9] Other premises that just assemble tray sets may fall outside of the scope of approval. It should be noted that several in-flight caterers now have Primary Authority partnerships with local authorities.[10]

ITCA, the International Travel Catering Association, had published guidance, the 'Flight Catering Book' for use by the flight catering industry. However, in 2014, the association went into liquidation.

The FSA's Food Law Code of Practice requires LA/PHAs to inspect aircraft for food safety purposes. When this requirement first appeared in the Code of Practice, it was clear that coordination was going to be required between LA/PHAs, particularly as aircraft, by their very nature, are not static premises and can arrive at different airports in the UK. The Association of Port Health Authorities Airports Committee considered this issue and agreed an approach with the FSA.

In essence, UK-based airlines are required to register the airline as a food business operator with a local authority of their choice. That local authority is then responsible for carrying out the assessment of the airline's food safety management system – in effect the local authority is acting like a 'Home Authority'. A questionnaire has been designed to assist with these interventions and is available via the 'Port Health at Airports' group on the Knowledge Hub website.[11]

Initially, these interventions revealed that some airlines had very comprehensive food safety management systems whereas others were found to be minimal. For example, one airline had never carried out any temperature monitoring tests on their high-risk ready-to-eat foods stored in cool bags. Another had never tested their drinking water in the aircraft tanks. It is not always bottled water that is served to passengers: water from the tanks is often used for making up drinks of squash and hot drinks. Some aircraft have drinking water fountains in the cabin, and it is also recognised that some passengers may use the water for cleaning their teeth and this can present a problem if the tank water is of poor quality.

The Home Authority style arrangement does not preclude other local authorities from entering aircraft or carrying out inspections, for example if the local authority/PHA receives a food hygiene complaint or if routine water samples are being taken, etc. A local authority/PHA may wish to discuss issues about the food safety management system with the local authority holding that particular airline's registration form. In order to identify which local authority to contact, the 'Port Health at Airports' group on the Knowledge Hub[12] website has a list of UK-based airlines together with the local authorities that hold those airlines' respective registration forms.

Aircraft water quality

A description of the supply of water to aircraft, its use and its associated health risks can be found in the WHO Guide to Hygiene and Sanitation in Aviation [16]. LA/PHAs should be familiar with the system of water distribution at their airports.

Aircraft drinking water standards are enforced by Retained EU Directive 98/83,[13] Retained EU Regulation 178/2002[14] and Retained EU Regulation 852/2004.[15]

Powers of entry also exist in the Public Health (Aircraft) Regulations 1979 (as amended)[16] which allow authorised officers to take samples food or water for analysis or examination:

- 'With a view to the treatment of persons affected with any epidemic, endemic or infectious disease and for preventing the spread of such diseases; or
- For preventing other danger to public health.'

When these regulations were first written, these were the only powers available to port health officers to take such samples.

Sampling protocols – for water samples, protocols can be found in an Association of Port Health Authorities/PHLS Collaborative study [17] and Environment Agency procedures [18].

The following is a suggested sampling protocol used at Gatwick Airport and adapted from the previous publications. It is recommended that whatever protocol is used, it should be agreed with the LA/PHA's public examiner before taking samples.

Routinely, water can be sampled at three places in the distribution system – rising mains in the airfield, water bowsers whilst attending an aircraft and the aircraft itself. Usual tap sterilisation procedures, such as using hypochlorite or flaming taps, are inappropriate on aircraft. Hypochlorite is not permitted on aircraft due to its corrosive effect on aircraft structures.

Rising main (water supply point)

- Flush for 15 seconds;
- Disinfect the hydrant using alcohol wipes;

- Run water for a further 15 seconds to remove the disinfectant;
- Fill the 500 ml sampling bottle taking care not to contaminate the inside of the bottle or lid.

Bowsers (water carts)

- Collect sample from the filling hose;
- Flush through at least double the volume of water from the filling hose;
- Disinfect the hose outlet using alcohol wipes;
- Run water for 15 seconds to remove the disinfectant;
- Fill the 500 ml sampling bottle taking care not to contaminate the inside of the bottle or lid.

Aircraft taps/water fountains

- In the same way as the rising main.

The sample bottles should then be kept cool (2–8°C) during transportation. It is recommended that the water is examined at the laboratory within 6 hours of collection.

There may be times when water sampling is being carried out on aircraft arriving in the afternoon and evening. These samples should be kept cool (2–8°C) overnight and sent to the laboratory the following morning, to arrive at the laboratory no later than 24 hours after collection.

When taking water samples on aircraft, it is useful to also take samples from the water bowser attending the aircraft and the rising main supply point. Should the aircraft water have adverse results, the results from the bowser and rising main will show where the contamination has occurred.

The WHO Guide to Hygiene and Sanitation in Aviation[17] details the number of tanks and approximate capacity for different types of aircraft. This information enables you to ensure that sufficient samples have been taken whilst on board the aircraft.

Another consideration is whether to take samples before or after sterilising the tap head. You might want to take samples both before and after sterilising. This will help determine whether it is the tap head or the water tank that is the source of contamination.

It is useful to obtain contact information for each airline; for example, if the laboratory reports unsatisfactory results, it will be necessary to alert the airline as quickly as possible. In addition, it may be necessary to report the matter to the airline's Home or Primary Authority.

Problems have also been experienced when taking samples from aircraft registered outside the EU. For example, a foreign registered aircraft had a water sample failure due to the presence of *E. coli*. There were no direct contacts for the airline physically based at the airport. It is common for such aircraft are handled by 'station managers' which act as handling agents for the airline. The matter was therefore referred to the FSA for liaison with the 'competent authority' in the third country. It should be noted that it is relatively straightforward to deal with a water sample which can be tested in a laboratory for compliance with a stated standard. A greater challenge is assessing the airline's food safety management system in comparison with UK registered airlines.

The WHO Guide to Hygiene and Sanitation in Aviation will prove helpful in such a case as it recognises that different countries may have different standards for aircraft water. It recommends that where there is a difference, airlines should comply with the most stringent standard.

Vector control

Some of the issues discussed here and related to pest control are also considered in Chapter 17.

Disinsection of aircraft – Annex 5 of IHR 2005 requires the WHO to publish, on a regular basis, a list of areas where disinsection or other vector control measures are recommended for conveyances arriving from these areas.

At the time of writing, the list of areas had yet to be published. In its absence, the WHO publication 'International Travel and Health' [19] contains a list of countries thought to present a risk. This information is often used as a basis for recommending the routes on which airlines should be disinsecting their aircraft. However, there are limitations: it only covers malaria and yellow fever but not other mosquito-borne diseases such as dengue, dengue haemorrhagic fever, chikungunya or West Nile virus, for example. Such diseases will have different distribution patterns and therefore one would expect that the WHO's 'list of areas' when published, will take this into account.

The WHO Guide to Hygiene and Sanitation in Aviation acknowledges that this subject may be 'particularly controversial', with different countries enforcing different standards. In the UK, the Central Competent Authorities, the Department of Health and Social Care and Public Health England (now UKHSA) currently have no general policies concerning disinsection of aircraft. However, following the WHO's declaration of Zika as a PHEIC, they did issue guidance recommending disinsection for all flights return to the UK from countries with confirmed transmission of Zika [20].

There are several methods of disinsecting aircraft and these are discussed in Chapter 17. Guidelines on how to carry out disinsection together with specifications for the insecticide, have been published by the WHO in the Weekly Epidemiological Record [21]. There are also further WHO publications covering aircraft disinsection [22–24].

When monitoring airlines, the port health officer should first contact the airport operator to obtain a list of countries from where aircraft originate and to discuss the issues about which they have concern. Specific flights can then be targeted for monitoring checks. Airport operators often have live flight information on their websites and their airfield operations department can advise which numbered stand the aircraft will be allocated.

Checks at the aircraft can involve:

- Asking for the Health Part of General Declaration and checking that it has been completed fully, especially when residual disinsection has been employed;

- Collecting cans that have been used for disinsection;
- Asking about whether the aircraft holds as well as the cabin have been treated.

Further work can include:

- Comparing the number of cans of insecticide, in relation to the type and size of aircraft;
- Verifying that the active ingredients of the pesticide comply with WHO specifications;
- Following up any irregularities with the airline.

Regulation 8(2) of the Public Health (Aircraft) Regulations 1979 (as amended) provides enforcement powers for disinsection; however, the medical officer must consider this reasonably necessary for preventing the spread of infection. For example, when airlines have contacted Crawley Borough Council for advice about whether a particular aircraft route terminating at Gatwick Airport needs disinsection, a referral has been made to the CCDC at Public Health England (now UKHSA) in order to obtain their opinion about whether disinsection is necessary.

Vector control at the airport – Annex 5 of the IHR 2005 requires programmes to control vectors that may transport an infectious agent constituting a public health risk, to a minimum distance of 400 metres from those areas of point of entry facilities (i.e. airports).

The occasional case of airport malaria serves as a reminder that mosquitoes could potentially bring a disease into the UK; for example malaria was once endemic in some areas of the UK. Port Health occasionally receive complaints from concerned individuals that they may have seen a non-indigenous mosquito. Without capturing and identifying mosquitoes, LA/PHAs will not know which species are present in habitats at, or close to their airport.

Surveys include sampling standing water for mosquito larvae in ponds and drainage ditches that appear on maps. However, other standing waters, in particular where predators are absent, can include:

- Rainwater trapped within bund walls around storage tanks of oil or de-icer;
- Temporary water such as puddles;
- Any areas where tyres are stored or dumped;
- Wheelie-bins and skip storage areas;

In 2020, PHE published a National Contingency Plan for Invasive Mosquitoes [25]. The plan recognises that routine surveillance is required to confirm that invasive mosquitoes have not been imported into the UK. The plan recommends that surveillance is carried out at ports using Gravid Aedes traps and a BG Sentinel trap at Border Control Posts (BCPs) between April and November. Mosquitoes caught in the traps can be sent to PHE/UKHSA for identification. Should any turn out to be an invasive species of mosquito, a 'multi-agency Local Incident Response Group' will coordinate the response to prevent its establishment in the UK.

Section 1 Notes

1 As from October 2021 PHE was reorganised into the UK Health Security Agency (UKHSA – see https://www.gov.uk/government/organisations/

uk-health-security-agency) and the Office for Health Improvement and Disparities (OHID see https://www.gov.uk/government/organisations/office-for-health-improvement-and-disparities).

2 SARS: Severe Acute Respiratory Syndrome; caused by the SARS coronavirus, resulting in a near pandemic between the months of November 2002 and July 2003, with 8,096 known infected cases and 774 deaths worldwide.

3 Guidance, when available, can be found on the Department of Health or Public Health England websites.

4 SI 1979 No. 1434.

5 http://www.legislation.gov.uk/changes/affected/ukpga/1984/22

6 New legislation can be found on the Legislation.gov.uk website.

7 SI 1998 No. 494.

8 SI 2003 No. 1895.

9 https://www.legislation.gov.uk/all?title=853%2F2004

10 https://primary-authority.beis.gov.uk/about

11 https://khub.net/group/guest/home

12 https://khub.net/web/porthealthatairports

13 https://www.legislation.gov.uk/all?title=98%2F83

14 https://www.legislation.gov.uk/all?title=178%2F2002

15 https://www.legislation.gov.uk/all?title=852%2F2004

16 SI 1979 No. 1434, SI 2007 No. 1447 and SI 2007 No. 1603 (different legislation applies in the devolved administrations of the UK).

17 See Third Edition at https://apps.who.int/iris/bitstream/handle/10665/44164/9789241547772_eng.pdf?sequence=1&isAllowed=y

SECTION 2: PUBLIC HEALTH AT SEAPORTS

Introduction

The number of international travellers passing through UK seaports was 21.8 million in 2018; of these 2.2 million were cruise passenger numbers [26]. This has dropped significantly due to Covid-19 during 2020 and 2021.

Increasing travel and trade promote both the spread of existing diseases to new populations and regions and can also facilitate the emergence of new diseases by bringing

different populations together. Increasing levels of tourism in remote areas, the growing importation of exotic animals, consumption of novel foods and climate change all facilitate the introduction of new diseases.

International travel, migration of populations, and import and export of goods and foodstuffs make the spread of infectious diseases inevitable. Yet international trade and transport underpins the UK economy and so the avoidance of importing important infectious diseases is fundamental to future economic success.

The control of imported threats is not new and the Table 21.1 summarises historical threats and responses.

The UK until recently was relatively free from important imported infections, but public attitude in terms of acceptance of risks from infectious diseases seems likely to decline and there will be greater assurances for safety and protection. Infectious diseases are widely considered to be an enormous global health problem, are spreading faster and emerging more quickly than before and the importation of some infections into the UK is viewed by Government as the current leading threat to UK security.

Local authorities (LAs) and Port Health Authorities (PHAs) are responsible for the enforcement of infectious disease controls at seaports.

International Health Regulations 2005

The Public Health (Ships) Regulations (which originally adopted the IHR 1969) were amended in 2007 so that they corresponded to the IHR 2005. This offers an interim measure for enforcement as not all the provisions of the IHR 2005 have been included in these amended regulations.

The enabling act, the Control of Disease Act 1984, was most recently amended by the Health and Social Care Act 2012.[18] At the time of writing, however, new legislation to replace both the Public Health (Aircraft) Regulations 1979 considered previously and

Table 21.1 History of imported threats

Century	Disease	Public health response
14th	Bubonic plague	UK and European ports introduce quarantine controls on humans and goods. Measures applied inconsistently as ports and trading partner countries competed with each other for trade resulting in continuing epidemics until the 18th century.
18th	Smallpox	Protective effects of inoculation with cowpox against smallpox demonstrated and widely accepted in UK. Increased access to immunisation eradicates smallpox globally.
19th	Yellow fever, cholera, bubonic plague	International polices emerged including obligatory telegraphic notification of first cases. Measures introduced when import of disease imminent.
20th	Tuberculosis Malaria	Legislation introduced to relieve overcrowded living conditions; penicillin discovered, manufactured and supplied. Mass vaccination of UK population and animal health/food controls. Screening of immigrants at point of entry. Mass application of insecticides in British colonial Africa to destroy insect vectors.
21st	AIDS/HIV Smallpox SARS vCJD Ebola	Enhanced global surveillance, sexual health education and private partnerships to develop effective drugs. Threat of possession of smallpox virus by terrorists stimulates new and safer vaccines by the international community. Risk management. Enhanced global surveillance and international capacity developed. Government intervention along the entire 'feed to food' continuum. International response to control outbreaks.

Public Health (Ships) Regulations 1979[19] is anticipated and it is hoped that all the appropriate powers required to implement the International Health Regulations 2005 will be provided.

Public Health England (now UKHSA) are, at the time of writing, 'Acting as the lead for the UK on the IHR including protecting the UK from international health hazards, most noticeably communicable diseases' [7]. Locally, PHE has had the operational lead concerning the implementation of the IHR 2005 at ports and port health officers will liaise most commonly through their consultant in communicable disease control (CCDC).

In 2019, Covid-19 emerged and spread quickly around the world and declared a pandemic by the WHO with a global response to the disease still ongoing. Port Health Authorities played an important part in the control measures, vessels arriving were closely monitored and Maritime Declarations of Health were required to identify if crew and passengers were infected.

Infectious disease control

Port Health Authorities (PHAs) are constituted with the primary objective of preventing the introduction and dissemination of dangerous epidemic diseases through shipping activity without creating unnecessary interference with world trade.

At coastal or inland waters where there is no PHA, the relevant local authority is responsible for infectious disease control. Authorised officers at seaports have powers of entry to board ships to enforce food and public health legislation.

Food law

Retained EU Regulation 852/2004 on the hygiene of foodstuffs is implemented in the UK by the Food Hygiene (England) Regulations 2013.[20] The Food Safety (Ships and Aircraft) (England and Scotland) Order 2003[21] gives authorised officers powers of entry to enforce food law on ships and aircraft. Similar legislation has been enacted in Northern Ireland and Wales.

The general considerations and actions for achieving food safety on ships are the same as for land premises; however, ships have some unique features and constraints imposed by their physical structure, mode of operation and opportunities for sourcing food and water. The types of food hazard on ships can be different from those found in land-based food premises. As well as the 24-hour operation of ship activities, there are potential hazards associated with methods of storage of food and water on board and these may only be available when the ship is in port. This can be of particular significance on ships whose itineraries include voyages to underdeveloped countries as well as the UK.

The ship environment is a closed community; problems associated with food production on large passenger ships can have a major impact on both passengers and crew. Illness caused by the consumption of contaminated food can be extremely disruptive and on smaller commercial ships can have a significant impact on the safety of the vessel because of severe incapacitation of critical crew members. See Chapter 9 for information on ship construction relevant to sanitation and water supply.

When enforcing food law on ships, authorised officers must have regard to the Food Law Code of Practice and Practice Guidance issued by the Food Standards Agency [15] and to the Industry Guide to Good Hygiene Practice for Ship Managers and Operators [27] which contains advice on how to comply with food law.

Hygiene improvement or hygiene emergency prohibition notices may be served when serious conditions are found; however, there may be restrictions to possible improvements in the layout of food handling areas due to internationally binding conventions relating to safety and maintenance. In these circumstances, authorised officers can liaise with the Maritime Coastguard Agency to discuss appropriate actions.

In 2005, the WHO issued revised International Health Regulations (IHR) [6] extending the scope of previous regulations and imposing specific requirements on all countries. The IHR address the multiple and varied health risks that face the world today and cover existing, new (e.g. SARS, Ebola virus) and re-emerging diseases (e.g. tuberculosis) and emergencies caused by non-infectious agents (e.g. polonium). Legal sanctions are not imposed by the regulations. Instead, compliance is described as having three compelling incentives: to reduce the disruptive consequences of an outbreak, to speed its containment and to maintain good standing in the eyes of the international community and thereby avoid unnecessary trade and movement restrictions. The revised International Health Regulations 2005 (IHR) introduced a set of 'core capacity requirements' that all countries must meet in order to detect, assess, notify and report the events covered by the regulations. The Department of Health in England and the devolved administrations in Northern Ireland, Scotland and Wales are responsible for transposing the IHR into national law. In England and Wales, the Public Health (Control of Disease) Act 1984 as amended by the Health and Social Care Act 2008 and accompanying regulations, assigns powers for applying controls for health, disease and waste disposal to PHAs and LAs. Similar arrangements exist for the devolved administrations in Scotland and Northern Ireland.

Powers for applying health controls on ships are contained in the Public Health (Ships) Regulations 1979 as amended by the Public Health (Ships) (Amendment) Regulations 2007 and the Public Health (Aircraft and Ships) (Amendment) (England) Regulations 2007.[22]

These regulations provide for the notification by the ship's Master of Health of conditions on board; examination of people; inspection of incoming ships; the issuance of Ship Sanitation Control Exemption Certificate or Ship Sanitation Control Certificate (Ship Sanitation Certificates (SSCs)), introduced by the 2007 regulations; and authorise measures to be taken for preventing danger to public health.

SSCs are valid for a period of six months and are used by competent authorities (PHAs/LAs in UK) to identify and record all areas of ship-borne public health risks, together with any required control measures to be applied. All ships on international voyages are required to carry a valid SSC.

SSCs may be renewed at any port authorised to issue renewals by their national government. All authorised ports must have the capability to inspect ships, issue certificates and implement (or supervise implementation of) necessary measures for the issuance of certificates. The WHO has published the 'Handbook for the Inspection of Ships and Issuance of Ship Sanitation Certificates' [28], which provides guidance for the ship inspection and issue of SSCs.

Ship Sanitation Exemption Control Certificates are issued when there is no evidence of a public health risk found on board and the competent authority is satisfied that the ship is free of infection and contamination, including vectors and reservoirs of disease. Inspections should be carried out when the ship and holds are empty or when they contain only ballast or other material (e.g. crude oil), which makes a thorough inspection of them impossible. It may be supplemented with the Evidence Report Form (ERF) which is used to list evidence found and control measures to be performed, an example of it is in Annex 7 of the WHO Technical Handbook [29].

During the inspection to renew a SSC or a routine inspection, the ERF may be used to record the evidence found and control measures to be applied. This is attached and referenced to on the SSC, it remains a part of the SSC until it is renewed. When remedial action has been carried out, it is recorded on the ERF by a port health office at any port. The ERF was introduced to allow port health officers to record their findings during an inspection and supports the SSC. The evidence listed on the ERF must match that listed in the WHO Technical Handbook [29] and cannot be used to record defects under national or EU legislation as this is the standard that the members of the WHO have signed up to.

There may be occasions when control measures cannot be completed satisfactorily at a port (e.g. super-chlorination of water tanks and laboratory examination of re-filled tanks; see Grenfell et al. [30] on the quality of potable water). In circumstances such as these, the competent authority should make a note to that effect on the control certificate and control measures required to be applied at the subsequent port of call. Assuming the ship is allowed to depart, at the time of departure, the competent authority should notify the next known port to be visited by the ship advising all conditions found and control measures required. This is of particular importance where the public health risk may be spread internationally or may present a serious and direct danger to human health.

A ship may visit ports where neither an inspection nor control measures can be carried out. In these circumstances, and where there is no evidence of infection or contamination, the competent authority may extend the validity of an exemption certificate only for one month, allowing the ship to arrive at a port in which an inspection and any necessary control measures can be carried out.

Model certificates for SSCs outline the key physical areas of the ship to be inspected and can be found in Annex 3 of the IHR and are reproduced at Figure 21.1 [6]. The WHO has provided guidance for the inspection and issuance of Ship Sanitation Certificates [29].

World governments designate their national ports authorised to issue SSCs and send a list of those ports to the WHO which maintains up-to-date listing of the authorised ports on its website.

Under the IHR, before calling at a port, the master of a ship on an international voyage may be required to report any illness that appears to be caused by an infectious disease or other conditions on board that could present a public health risk to the PHA/LA by means of a Maritime Declaration of Health (MDH).

Model MDH forms can be found in Annex 8 of the IHR and are reproduced at Figure 21.3 [6].

In the UK, the master of a ship is required to report any suspected infectious disease on board amongst crew, passengers or animals/birds not less than 4 hours and not more than 12 hours before arrival.

If measures such as isolation/quarantine of the ship are implemented, or if the public health risk appears to be serious and/or indicates international spread of disease, the competent authority must notify the IHR National Focal Point at Public Health England (UK HSA).

Harmful contamination other than microbial contamination, for example from chemical or radio-nuclear sources can also be found on ships. In these circumstances, national and international agencies exist to deal with such emergencies and immediate assistance should be sought through LA civil emergency mechanisms.

The Department of Health and Social Care intends to update current 'port health' regulations to complete the process of implementation of IHR. Formal consultation on the new regulations which will encompass all means of international travel is anticipated in 2015. The new regulations will adopt an 'all-hazards' approach – replacing specific lists of diseases in the current regulations with more general powers to protect public health.

Illustrative examples of travel-associated infectious diseases

The rapid movement of ships from one port to another, with a likelihood of wide variation in hygiene standards and infectious disease exposure risks, often results in the introduction of communicable diseases by embarking

ANNEX 3

MODEL SHIP SANITATION CONTROL EXEMPTION CERTIFICATE/SHIP SANITATION CONTROL CERTIFICATE

Port of .... Date: ....

This Certificate records the inspection and 1) exemption from control or 2) control measures applied

Name of ship or inland navigation vessel .... Flag .... Registration/IMO No. ....

At the time of inspection the holds were unladen/laden with ...... tonnes of ...... cargo

Name and address of inspecting officer ....

Ship Sanitation Control Exemption Certificate

Areas, [systems, and services] inspected	Evidence found[1]	Sample results[2]	Documents reviewed
Galley			Medical log
Pantry			Ship's log
Stores			Other
Hold(s)/cargo			
Quarters:			
- crew			
- officers			
- passengers			
- deck			
Potable water			
Sewage			
Ballast tanks			
Solid and medical waste			
Standing water			
Engine room			
Medical facilities			
Other areas specified - see attached			
Note areas not applicable, by marking N/A.			

No evidence found. Ship/vessel is exempted from control measures.

Name and designation of issuing officer ....

Ship Sanitation Control Certificate

Control measures applied	Re-inspection date	Comments regarding conditions found

Control measures indicated were applied on the date below.

Signature and seal .... Date ....

[1] (a) Evidence of infection or contamination, including: vectors in all stages of growth; animal reservoirs for vectors; rodents or other species that could carry human disease, microbiological, chemical and other risks to human health; signs of inadequate sanitary measures. (b) Information concerning my human cases (to be included in the Maritime Declaration of Health).

[2] Results from samples taken on board. Analysis to be provided to ship's master by most expedient means and, if re-inspection is required, to the next appropriate port of call coinciding with the re-inspection date specified in this certificate.

Sanitation Control Exemption Certificates and Sanitation Control Certificates are valid for a maximum of six months, but the validity period may be extended by one month if inspection cannot be carried out at the port and there is no evidence of infection or contamination.

Figure 21.1 Model ship sanitation control exemption certificate/Ship sanitation control certificate

Areas/facilities/systems inspected[1]	Evidence found	Sample results	Documents reviewed	Control measures applied	Re-inspection date	Comments regarding conditions found
Food						
Source						
Storage						
Preparation						
Service						
Water						
Source						
Storage						
Distribution						
Waste						
Holding						
Treatment						
Disposal						
Swimming pools/spas						
Equipment						
Operation						
Medical facilities						
Equipment and medical devices						
Operation						
Medicines						
Other areas inspected						

[1] Indicate when the areas listed are not applicable by marking N/A.

Figure 21.1   (Continued)

Evidence Report Form					
This form supports the ship sanitation certificate (SSC), and provides a list of evidence found and control measures to be performed.					
When attached to the SSC, each page of this attachment needs to be signed, stamped and dated by the competent authority. If this document is used as an attachment to a pre-existing SSC, this attachment must be noted in the SSC (e.g. by using a stamp).					

Ship's name and IMO no. or registration:	Name and signature of responsible on board ship officer:
Name of issuing authority:	Actual inspection date (dd/mm/yyyy):
Date of referred SSC (dd/mm/yyyy):	SSC issued in the port of:

Indicate areas that <u>have not</u> been inspected:

☐ Quarters	☐ Galley, pantry, service area	☐ Stores	☐ Child-care facilities
☐ Medical care facilities	☐ Swimming pools/spas	☐ Solid and medical waste	☐ Engine room
☐ Potable water	☐ Sewage	☐ Ballast water	☐ Cargo holds
	☐ Other (e.g. laundry and washing machine)		

Detected health events on board	☐Yes	☐No	

Evidence code	Evidence found (brief description according to WHO checklist; draw a line under each item of evidence to ensure items are clearly separated)	Measure to be applied	Required	Recommended	Measure successfully performed (stamp and signature of re-inspecting authority)

Name of issuing inspector:	Signature of issuing inspector:	Stamp of issuing authority:	Page ____ of ____

Figure 21.2　Model evidence report form

Source: IMO, International Maritime Organization; SSC, ship sanitation certificate; WHO, World Health Organization

ANNEX 8

MODEL OF MARITIME DECLARATION OF HEALTH

To be completed and submitted to the competent authorities by the masters of ships arriving from foreign ports.

Submitted at the port of .... Date ....

Name of ship or inland navigation vessel ....………........ Registration/IMO No ....arriving from ....sailing to ....

(Nationality)(Flag of vessel) .... Master's name ....

Gross tonnage (ship) ....

Tonnage (inland navigation vessel) ....

Valid Sanitation Control Exemption/Control Certificate carried on board? Yes .... No .... Issued at .... date ....

Re-inspection required? Yes .... No ....

Has ship/vessel visited an affected area identified by the World Health Organization? Yes ..... No ....

Port and date of visit ....

List ports of call from commencement of voyage with dates of departure, or within past thirty days, whichever is shorter:

....

Upon request of the competent authority at the port of arrival, list crew members, passengers or other persons who have joined ship/vessel since international voyage began or within past thirty days, whichever is shorter, including all ports/countries visited in this period (add additional names to the attached schedule):

(1)   Name ....joined from: (1) ....(2) ....(3) ....

(2)   Name ....joined from: (1) ....(2) ....(3) ....

(3)   Name ....joined from: (1) ....(2) ....(3) ....

Number of crew members on board ....

Number of passengers on board ....

Health questions

(1)   Has any person died on board during the voyage otherwise than as a result of accident? Yes ....   No .....
       If yes, state particulars in attached schedule.   Total no. of deaths ....

(2)   Is there on board or has there been during the international voyage any case of disease which you suspect to be of an infectious nature? Yes........ No........ If yes, state particulars in attached schedule.

(3)   Has the total number of ill passengers during the voyage been greater than normal/expected? Yes ....   No .....
       How many ill persons? ....

(4)   Is there any ill person on board now? Yes .... No .... If yes, state particulars in attached schedule.

(5)   Was a medical practitioner consulted? Yes .... No ...... If yes, state particulars of medical treatment or advice provided in attached schedule.

(6)   Are you aware of any condition on board which may lead to infection or spread of disease? Yes .... No ....
       If yes, state particulars in attached schedule.

(7)   Has any sanitary measure (e.g. quarantine, isolation, disinfection or decontamination) been applied on board? Yes .... No ......
       If yes, specify type, place and date ....

(8)   Have any stowaways been found on board? Yes .... No ...... If yes, where did they join the ship (if known)? ....

(9)   Is there a sick animal or pet on board? Yes .... No ....

Note: In the absence of a surgeon, the master should regard the following symptoms as grounds for suspecting the existence of a disease of an infectious nature:

       (a)   fever, persisting for several days or accompanied by (i) prostration; (ii) decreased consciousness; (iii) glandular swelling;
              (iv) jaundice; (v) cough or shortness of breath; (vi) unusual bleeding; or (vii) paralysis.

       (b)   with or without fever: (i) any acute skin rash or eruption; (ii) severe vomiting (other than sea sickness); (iii) severe
              diarrhoea; or (iv) recurrent convulsions.

I hereby declare that the particulars and answers to the questions given in this Declaration of Health (including the schedule) are true and correct to the best of my knowledge and belief.

Signed ....

Master

Countersigned ....

Ship's Surgeon (if carried)

Date ....

Figure 21.3a   Model of Maritime Declaration of Health (two parts)

Note: [1]State: (1) whether the person recovered, is still ill or died; and (2) whether the person is still on board, was evacuated (including the name of the port or airport), or was buried at sea.

ATTACHMENT TO MODEL OF MARITIME DECLARATION OF HEALTH

Name	Class or rating	Age	Sex	Nationality	Port, date joined ship/vessel	Nature of illness	Date of onset of symptoms	Reported to a port medical officer?	Disposal of case[1]	Drugs, medicines or other treatment given to patient	Comments

Figure 21.3b

passengers and crew. In the relatively closed and crowded environment of a ship, disease may well spread to other passengers and crew; diseases may also be disseminated to the home communities of disembarking passengers and crew. A literature review by WHO identified more than 100 disease outbreaks associated with ships since 1970. This is undoubtedly an underestimate because many outbreaks are not reported and many go undetected as ships travel from port to port and from one enforcement authority to another. Such outbreaks are of concern because of their potentially serious health consequences and high costs to the industry and to international communities [31].

Outbreaks of measles, rubella, varicella, meningococcal meningitis, hepatitis A, legionellosis (see WHO [32] for more information), and respiratory and gastrointestinal illnesses are frequently reported. In recent years, norovirus and influenza outbreaks have been serious public health challenges for the international community and the shipping industry.

Gastrointestinal disease – gastroenteritis can occur in several ways; either where the person acquires a pathogenic micro-organism directly from someone else and develops illness, or where the micro-organism is carried in food or drinking water and causes infection when consumed or from environmental sources (e.g. door handles). The infections can be caused by viruses, bacteria or protozoa. Gastroenteritis can also be caused by intoxication, where a bacterium or fungus has grown on foodstuffs releasing toxins which when eaten later, causes illness.

Although passenger ships probably do not have a higher level of infectious gastroenteritis than ashore, they tend to be reported in the news more often, which may give the impression that they occur more frequently there. Ships are no different from land-based hotels or residential establishments in that both will have people becoming affected by gastroenteritis from time to time.

As with all types of gastroenteritis, good hygiene practices, both personal and food handling, together with safe food sources and drinking water integrity, are the key issues in the prevention of outbreaks. These issues are however more of a challenge on ships where bunkered water may be contaminated, potable water may become contaminated with sewage, use of sea or river water in galleys, use of unsafe food supplies and the use of poorly trained crew.

Clinical differentiation of gastroenteritis outbreaks is critical, and the presenting symptoms provide the first clues to the nature of the illness and for the early application of specific control measures to prevent spread.

Depending on the causative organism and the way people become infected, the control measures required are very different, and this is significant when trying to prevent the spread of disease.

Norovirus – is by far the most cause of outbreaks of gastroenteritis in land-based communities [33]. As indicated ships are no different from land-based hotels or residential establishments in that both will have people becoming affected by such diseases from time to time.

Norovirus is the most common pathogen implicated in gastrointestinal disease outbreaks on cruise ships. An assessment of 56 cruise ships between 2005 and 2008 showed that 95% of gastrointestinal disease was caused by norovirus. The virus can spread in food, water, person to person, and more frequently from touching contaminated surfaces; it is highly infectious and in an outbreak on a cruise ship in 1998, more than 80% of the 841 passengers were affected.

Levels of public and press interest in norovirus outbreaks on cruise liners have reached unprecedented levels in recent years in a way that is not commonly experienced shore-side.

When norovirus causes an outbreak on board ship, it is usually because someone is infected ashore and comes on board either with illness or incubating it. If they have diarrhoea or vomiting, they will excrete huge numbers of viral particles in aerosol which can be inhaled and contaminate surfaces very easily. A single case of norovirus can potentially infect a large number of people very quickly, who will go on to infect more people, producing an outbreak, which is characterised by a very rapid rise in numbers. Unless very prompt intervention is implemented, the infection will spread very easily and large protracted outbreaks (many days or weeks) with serious outcomes are then observed.

Detailed guidance for the management of norovirus can be found in the 'Guidance for the Management of Norovirus in Cruise Ships' produced by the Health Protection Agency (now UK HSA), the Association of Port Health Authorities and the Maritime and Coastguard Agency [33].

Respiratory illness – the emergence of SARS in 2003 clearly demonstrated how a previously unknown disease, causing high mortality and morbidity could spread rapidly. Fast travel and global trade facilitated transmission in the absence of relevant vaccines and drugs. Effective countermeasures were applied, but the event underlined the need for worldwide cooperation to control such contingencies. Early detection of cases and efficient international communication and coordination was an advantage in tackling the epidemic, and public health measures undertaken helped to prevent catastrophic developments. In 2009 the threat of pandemic influenza prompted governments and international bodies with responsibilities for public health protection to address preparedness plans that could mitigate the potential effects of a pandemic, and to reinforce policies, contingency plans and resources, including the alert systems and their networks.

CASE STUDY 1

MS Marco Polo

The *Marco Polo*, cruising around the UK in July 2009 had to halt its voyage 4 days into its 10-day cruise because of a norovirus outbreak. About 340 passengers and 40 crew were symptomatic and 6 were admitted to hospital. There was also an unrelated fatality. There were 769 passengers on board, giving an attack rate of 44% and 340 crew (attack rate 12%). The decision to terminate the cruise was taken by the operator in consultation with the public health authority ashore. A team of NHS doctors and nurses went on board to assist.

There had been illness on the previous cruise, from Iceland to UK, but the operator was quoted as saying 'a very small number of people on board had suffered symptoms of gastroenteritis during the cruise (from Iceland) but this is unrelated to the current outbreak'. A passenger on that trip was quoted in the national media as saying that

> despite obvious signs of a bug on the ship, there was little evidence of prevention measures – nothing was said about taking care to wash our hands etc. It was only in the last few days that someone was standing outside the restaurants and giving us the gel to use on our hands.

The operator was also quoted as saying 'port authorities gave the ship a clean bill of health when it docked in the UK'.

Several questions arise, the foremost of which has to be why the attack rate was so high and how it reached such a level so swiftly? A review of 29 norovirus outbreaks identified by the CDC Vessels Sanitation Program between 2006–2009 (ships with >500 passengers, excluding outbreaks on successive cruises) showed the mean attack rate for passengers was 7.9% (range 4.1–16.7%) and for crew 1.8% (range 0.37–4.04%). What was the reason for such incredibly rapid transmission? Was this a food-borne outbreak because of contamination by food handler(s)? Specific questions, which should be addressed by PHAs/LAs, are what policies were in place for norovirus control, were these policies implemented swiftly enough, were they implemented effectively, was there illness in food handlers, were there well-understood protocols for crew reporting gastrointestinal illness, were affected crew isolated effectively, were there deficiencies in the crew's actions, were there difficulties with passenger cooperation and was there adequate attempt to clean and decontaminate the vessel between cruises?

The statement by the operator that the infections on the two cruises were 'unrelated' is interesting. It is understood that there were 31 passengers and 2 crew with gastroenteritis on the preceding cruise from Iceland to UK. It is an important distinction as to whether the ship/operator did not understand the significance of this outbreak (they said it was 'a very small number', but the attack rate in passengers was 44%) or whether they assumed that with a change of passengers they could safely ignore it, trusting to their cleaning procedures to prevent a repetition. It is also understood that the ship did not send a MDH to PHA when it docked there between cruises. This is a contravention of national legislation and could potentially lead to legal action.

These two issues demonstrate a need for education.

If a robust reliable surveillance and inspection programme had been in place, the PHA may have been more likely to be pro-active in inspecting the vessel.

There is also a need to make available an authoritative source of information about norovirus on board ships to a wider audience.

A local councillor at the next UK port to be visited by the ship was clearly concerned, claiming that 'potentially infected passengers will be allowed to wander through the town. We have shopkeepers, businesses and people who will come into contact with cruise passengers, and they are all worried'. The consequence was that passengers were then prevented from disembarking.

Officers of PHAs/LAs are publicly accountable, which now means that they need to reassure and demonstrate their role in protecting the public, and the Freedom of Information Act and the International Health Regulations have focussed PHAs/LAs' minds on how they should respond to notifications of norovirus outbreaks and how they should be dealt with.

This outbreak also demonstrates how readily outbreaks on passenger ships will attract international media attention. This means that the actions taken (or not taken) can be subject to close scrutiny in the full glare of adverse publicity. This is always intimidating and may well impede carrying out the necessary actions. It also sets the scene for litigation by dissatisfied passengers if they feel that the ship/operator has not discharged an appropriate duty of care, either to inform them of illness on board or how diligently the on-board containment actions have been carried out.

There is an interesting postscript to this event showing how difficult it is to second-guess people's perception of risk; the operator arranged a chartered train to take passengers from Scotland to London as an alternative to staying on the ship. Only a small number chose this option — some passengers even organised a petition to be allowed to stay on board!

Conclusion

Many cruise ships have well thought out and practised procedures and there will be little disruption to travel, and good public health measures applied. Inexperience and incompetency in dealing with norovirus outbreaks can lead to heavy-handed interventions where ships can be detained unnecessarily or be forbidden to enter ports, or by inadequate responses where outbreaks are not contained and where illness spreads rapidly to large numbers of people.

The emergence of SARS is recognised as creating the momentum for the adoption of the revised International Health Regulations in 2005.

Novel H1N1 influenza (Pandemic H1N1 2009) — in autumn 2009, novel H1N1 influenza (Pandemic H1N1 2009) emerged as a global health threat. Despite the UK's reliance on international travel and trade, little consideration had been given to the practical matters involved in managing novel H1N1 influenza (Pandemic influenza) on ships and for embarking and disembarking passengers and crew.

The shipping industry lost no time in contacting PHAs and local authorities to plan for the effects of pandemic influenza on their industry. The HPA, the Passenger Shipping Association and the Association of Port Health Authorities responded by producing national and international guidance to minimise the spread of the virus, both on board and on disembarking from a ship, for the industry and enforcement officers. This 'special' guidance supplemented pre-existing guidance issued by the HPA [34].

Owing to the enormous numbers of travellers passing through and into the UK, the containment phase soon passed, and the incidence of influenza cases increased. Further guidance was necessary in order to prevent interference with travel and trade as the behaviour of the new virus became clearer (i.e. it caused a relatively mild illness, similar in severity to seasonal influenza, rather than the more highly virulent virus that was previously anticipated).

Given the worldwide distribution of distribution of Pandemic H1N1 2009 virus, no measure to prevent the introduction of the virus, either on board a large passenger ship or into a port, was likely to be completely effective. The focus of the guidance was therefore on measures to minimise the spread of the virus, both on board a ship and disembarking from a ship, as much as was practicable.

Guidance was provided for cruise ships which carried a doctor and for those ships not carrying a doctor and for ferries.

The guidance for ships carrying doctors (normally passenger ships) too account of the fact that during the pandemic, media coverage was so extensive that virtually no one was unaware of the risk of transmission of infection associated with coming into contact with other people, particularly in large groups.

It was inevitable that passenger ships would have people boarding, on a regular basis, who were or would become, infectious during the voyage. The cruise industry was advised to contact passengers who were at risk of serious complications of infection with Pandemic H1N1 2009, because of underlying medical conditions, to advise them to take this into account when choosing to cruise. It was further recommended that passengers be given guidance on precautions that they should take, either before embarkation or as a cabin health information letter.

Passengers with influenza were advised that they should not join the ship unless they agreed to remain isolated until they are no longer infectious, and the ship had sufficient capability to enable this to be carried out effectively. Screening passengers according to symptoms before joining was recommended and public health questionnaires were produced for use in identifying passengers for screening. Passengers refused boarding were given a pre-prepared advice sheet, detailing respiratory hygiene, along with a surgical face-mask to wear.

Crew with influenza-like symptoms were similarly assessed by the medical staff and isolated on board as required. On embarkation passengers and crew were reminded of the need to report respiratory illness as soon as possible and antiviral drugs were offered to those with influenza symptoms.

Prophylactic Oseltamivir (Tamiflu) was also offered to fit, healthy individuals where there were operational reasons or circumstances on board where travellers had been in close contact with someone who had influenza, on the basis that it might lessen the burden of caring for ill passengers, many of whom were older or likely to have underlying medical conditions which may make them more susceptible to complications if they become ill. Long-term prophylaxis was also considered for some essential crew (e.g. watch-keepers) to ensure safety or continue the running of the ship.

People diagnosed with influenza were requested to confine themselves to their cabins until their influenza-like symptoms have settled. Care was taken to avoid being overly draconian about the length of time a recovering person who felt well should be isolated for, particularly if they received Oseltamivir as this may have made people less likely to present to the medical service when they became ill. Passengers in isolation had their 'on board credit cards' disabled to discourage them from moving around the ship and were prevented from disembarking for shore visits by suspending their ship's shore-pass.

Infected, ill crew were grouped in isolation cabins to reduce the exposure of other crew members to infection.

Staff visiting ill people's cabins (e.g. delivering meals or normal housekeeping duties), wore surgical face-masks before entering the cabin and discarded them on leaving. Ship facilities were closed only if large numbers of crew were affected resulting in reduced staffing levels which might compromise safety.

On disembarkation at the end of a voyage, those passengers still in isolation were assessed for fitness to travel. PHAs/LAs helped to facilitate arrangements with local hotels and other accommodations. In the UK, it is not possible to impose any restriction on travel for disembarking passengers who have influenza, whether confirmed or not. However, the advice given was that where possible, it was preferable to arrange transport in a private vehicle with the minimum possible number of accompanying people, with the passenger wearing a surgical face-mask while in transit and observing respiratory hygiene measures to reduce transmission risk. If the passenger had to use public transport they were supplied with a surgical face-mask and advised to observe good respiratory hygiene as mentioned earlier but also practice social distancing where possible (keeping more than 1 metre away from other people).

As with other infectious diseases, the ship's master was required to notify cases to the PHA/LA when arriving from a foreign port. Where extra assistance from the PHA/LA was anticipated, ships were asked to give increased advance notification (24 hours).

It was important that Port Health Authorities (or local authorities), where possible, pro-actively set the expectation that, during a pandemic, influenza aboard a passenger ship is no more of a newsworthy event than influenza on shore.

Guidance for ships not carrying a doctor (normally commercial trading ships) took into account that the major issue for ships not carrying a doctor were: symptomatic crew members joining the ship; treatment of symptomatic crew; prophylaxis (and unavailability of prophylaxis) of other crew members; and ship safety when crew numbers are finite.

Although ideally, symptomatic crew who met the UK case-definition of influenza-like illness should not have been allowed to board, this was difficult in practice. Some crew had visas allowing limited travel in UK (e.g. to join a ship). Where joining crew was symptomatic, isolation was the preferred advice to avoid them infecting other crew and to follow the same advice for crew treatment and prophylaxis as for passenger ships.

For a crew member who appeared seriously unwell, it was advised that medical advice should be sought immediately with the presumption that they would be transferred ashore.

PHAs/LAs negotiated agreements with local NHS services to assist in diagnosis and treatment.

Ferry voyages may be of short duration (minutes/hours) or longer (overnight) providing cabin accommodation and meals. They differ from cruise ships in that they provide an essential public service rather than a recreational one. As such, it is not practicable to restrict passengers boarding a ferry because of influenza. The cancellation/suspension of a ferry service, particularly to an island community, because of crew incapacity is a serious issue. PHAs/LAs were advised to engage with ferry companies as part of local authority emergency planning for such an occurrence, including contingency plans and issuing public statements.

Advice was given that symptomatic passengers were seated in the most isolated accommodation area available, provided with a surgical face-mask if available and advised about cough etiquette and social distancing by pre-prepared leaflets.

On some short routes, it was suggested that where practicable, symptomatic passengers should remain in their vehicles, subject to Marine & Coastguard Agency approval.

The novel H1N1 influenza (Pandemic H1N1 2009) did not turn out to be as damaging as anticipated. However, the dependence of the UK economy on international travel and trade and the interruption that influenza may have caused mitigated for intensive planning and effort for managing influenza on board ships and at disembarkation of travellers.

The provision of well-considered, authoritative and practical advice prepared in full agreement of the shipping industry was followed by PHAs/LAs and the industry. Very little spread of influenza was observed among crew or passengers on UK voyages and disruption to international travel and trade by sea was minimal.

The guidance is regularly reviewed and continues to be a key component of capacity

planning for national and local public health responses for the international spread of infectious disease.

Section 2 Notes

18 http://www.legislation.gov.uk/changes/affected/ukpga/1984/22
19 SI 1979 No. 1435.
20 SI 2013 No. 2996.★
21 SI 2003 No. 1895.★
   ★Different legislation applies in the devolved administrations of the UK.
22 SI 1979 No. 1435; SI 2007 No. 1446; and SI 2007 No. 1603

SECTION 3: INTERNATIONAL CATERING WASTE (ICW)

Although this issue is subject to Article 22(1)(e) of the IHR 2005, the European Union had already established rules for controlling International Catering Waste (ICW). Now that the UK has left the European Union, the European legislation controlling ICW has been implemented into UK domestic legislation.

Retained EU Regulation 1069/2009[23] provides the legislative framework for the control of international catering waste (ICW) and sets out the requirements for the control, importation and disposal of ICW.

International Catering Waste includes animal product food waste, and any other material that is mixed with it (including disposable cutlery, plates, etc.) or contaminated by it, that comes from a means of transport operating internationally outside of the EU and within the EU territory. On an aircraft, such waste could be uneaten flight catering meals or a passenger's own food which they have taken on board themselves. It also includes articles that have come into contact with POAO.

Catering waste which is unloaded from all vessels, planes and vehicles is subject to waste management controls. This is to prevent any possible transmission of disease from animal by-products, such as foot and mouth disease, to animals in the UK. International catering

waste is treated as a Category 1 waste, and subject to strict controls on how it should be handled and disposed of.

Usually, Category 1 waste can only be disposed of by incineration or rendering; however an exception is made for ICW whereby it is allowed to be disposed of in approved landfill sites. Stricter rules apply for burial at landfill to prevent wildlife coming into contact with it and disseminating animal disease-causing organisms present. Guidance for waste producers and enforcement officers is available on the Gov.uk website [35].

The Animal Health Agency is responsible for overseeing the effectiveness of controls on behalf of the Department for Food, Environment and Rural Affairs (Defra). PHAs and local authorities are responsible for enforcing the regulations.

The Retained EU regulation is enforced by the Animal By-Product (Enforcement) Regulations 2013.[24] In non-unitary local authorities, the provisions for ICW fall to the County Council Trading Standards department.

Defra's Animal Health and Welfare Framework [36] requires the Animal and Plant Health Agency (APHA) to meet with the relevant local authorities on a regular basis. It is recommended that the roles of Defra and

> ## CASE STUDY 2
>
> ### Monitoring International Catering Waste
>
> At Gatwick Airport, joint waste checks are undertaken by both Crawley Borough Council and APHA. Such checks have involved picking through the waste on aircraft, in compactors and a waste reception centre in order to identify whether uneaten aircraft meals containing POAO are present. Aircraft cleaners and flight-catering staff can be a valuable source of information during checks.

the local authorities in relation to ICW are agreed and recorded for future reference.

Vessels arriving at seaports will have waste on board which will be discharged whilst the vessel is in port; this will include catering waste from the galley. Port operators will have arrangements in place for the collection of this waste, usually clearly marked bins in which the waste can be placed ready for collection for disposal at approved landfill sites.

At an airport, understanding the waste streams that operate at the port is a useful starting point. The airport operator should be able to provide details of which waste firms operate at the airport.

At one time it was widely assumed that uneaten airline meals were taken back to flight catering units where the ICW entered the Category 1 waste stream. However, this is not always the case. Checks on aircraft have shown that flight caterers may visit and empty the galleys before the aircraft cleaning crews go on board. Often ICW is present in the cabin, for example, under seats, and in seat pockets. With the flight caterers having already attended the aircraft, the waste ends up in the care of the aircraft-cleaning companies who then have responsibility for the segregation and correct disposal of the ICW.

It is likely that incorrect disposal of ICW will go undetected at an airport without some form of proactive monitoring. At many airports, APHA has taken on this role, which can involve checking aircraft, cleaning companies and waste reception facilities at the airport.

Where other contraventions are identified, APHA will refer the matter to the relevant enforcement authority depending on which legislation would be most appropriate to use. For example, a dedicated compacter skip located at an in-flight caterer's premises may be damaged or corroded allowing leachate to escape. Such an incident may be subject to a notice under the Animal By-Product (Enforcement) Regulations 2013 and therefore dealt with by the local Trading Standards officer.

For uneaten food containing POAO that is found in a passenger's baggage, the UK Border Force has the responsibility for seizing such illegally introduced POAO. Such checks take place in the Green Customs Channel in the terminal buildings.

Section 3 Notes

23 https://www.legislation.gov.uk/all?title=1069%2F2009
24 SI 2013 No. 2952 (different legislation applies in the devolved administrations of the UK, for example, SI 2014 No. 517 (W.60) in Wales and SI 2013 No. 307 in Scotland).

SECTION 4: IMPORTED FOOD, ANIMALS AND OTHER GOODS

Imported food can be divided into three areas:

- Products of animal origin (POAO);
- Non-animal origin foods (NAO);
- Organic consignments.

These three areas have different legislation and are overseen by different Central Competent Authorities. Despite this, there is some overlap between the controls. Where products of animal origin (POAO) fall outside the scope of veterinary checks under POAO legislation (e.g. certain composite products), the products will still need to satisfy the official feed and food controls regulations (OFFC) which generally cover non-animal origin foods.

Consignments subject to official controls should go through Port Health checks first, followed by HMRC controls. HMRC control the release of consignments using a database known as CHIEF (Customs Handling of Import and Export Freight). Every product has its own commodity code number (known as a Common Nomenclature code or CN code). If a CN code is entered for a product subject to restrictions, then CHIEF will not clear the consignment. However, this system relies on the honesty and accuracy of the entry made by the clearing agent. CHIEF is due to be replaced by a new system called CDS (Customs Declaration Service) [37].

CHIEF should, therefore, refer POAO consignments to the LA/PHA in order for official controls to be carried out so that a Common Health Entry Document (CHED) for POAO (CHEDP) can be issued. Customs will only release such goods with a valid CHEDP.

Certain NAO foods are also subject to statutory Port Health checks (e.g. high-risk fruit and vegetables listed in legislation, such as Retained EU Regulation 2019/1793).[25] Again, such products should not be cleared by CHIEF, but importers referred to the LA/PHA so that checks can be carried out and a Common Health Entry Document for HRFNAO (CHEDD) issued.

In 2015, a new electronic system called Automatic Licence Verification System (ALVS) [38] was rolled out by Defra. This system allows decisions made via CHEDs on the IPAFFS database to be automatically transmitted to the CHIEF database. The system benefits the trade by reducing the amount of time between Port Health clearance and Customs clearance from around two hours to just a few minutes.

EU Exit

For almost 40 years the rules and laws about food hygiene and safety were set out in European Directives, Regulations and Decisions.

The UK voted to leave the European Union in the referendum of 23 June 2016, and finally ceased to be Member State on 31 January 2020. The UK then entered a Transition Period during which time it still recognised and followed European law, including any new legislation that was issued.

The Transition Period ended at 11 pm 31 December 2020 (local time). At this point, all the existing European legislation rolled over into UK domestic law, in accordance with section 3 of the European Union (Withdrawal) Act 2018.[26] The former EU rules can be found on the Legislation.gov.uk and is now known as 'retained EU legislation'. It should be noted that any amendments to the European legislation from this date do not apply to the retained EU legislation.

At the time of writing, some food legislation had already been amended in the EU, and this has created divergence. For example, the high-risk list of products in the annexes to the UK's Retained EU Regulation 2019/1793 are now different to the EU's list in Regulation (EU) 2019/1793.[27] The similar sounding names of the regulations can cause confusion, particularly for traders who import goods both into the EU as well as GB.

The UK can also update its domestic legislation, and the retained EU legislation will eventually be amended through the parliamentary process. This means that any changes to the high-risk lists of products in the annexes to Retained EU Regulation 2019/1793 will require a statutory instrument to be made.

The situation is more complex regarding Northern Ireland. There are no land border controls between the Republic of Ireland (an EU Member State) and Northern Ireland (which is part of the UK). Instead, the Northern Ireland Protocol [39] requires Northern Ireland to follow European law, as though it was a Member State of the European Union. High-risk goods from moving from GB (i.e. England, Scotland and Wales) to Northern Ireland are now subject to controls at Border Control Posts (known as Points of Entry in Northern Ireland), although there are some easements in place. Furthermore, Northern Ireland still remains part of the UK Internal Market, but goods will have to meet the definition of a 'qualifying good' if they are to receive unfettered market access. Further information about this can be found on the Gov.uk website [40].

Official Control Regulation (OCR) – in 2019, imported food legislation underwent a great change. Over the years import legislation had developed through a number of European Directives, Regulations and Decisions, covering the imports of live animals, products of animal origin (POAO), high-risk food and feed of non-animal origin (HRFNAO), and plants and plant products. The Official Control Regulation (OCR) has consolidated this raft of legislation and provided a common framework for border checks. The OCR has been incorporated into domestic legislation as Retained EU Regulation 2017/625.[28]

While the OCR provides and overall framework of imported food controls, it includes the provision to make tertiary legislation. These are further European regulations which provide specific details about import conditions, and official controls. A list of the main EU regulations covering imported food control can be found in Table 21.2.

Table 21.2 Retained tertiary legislation made under the Official Control Regulation

2007/275 – Composite products subject to controls at a Border Control Post (BCP)[29]
2019/625 – POAO intended for human consumption[30]
2019/626 – Third country lists for POAO intended for human consumption[31]
2019/628 – Model official certificates for POAO[32]
2019/1012 – Derogation on the minimum requirements for BCPs[33]
2019/1013 – Prior notification requirements for animals and goods[34]
2019/1014 – Minimum requirements for Border Control Posts[35]
2019/1081 – Training requirements for staff performing physical checks[36]
2019/1666 – Monitoring POAO consignments to the place of destination[37]
2019/1715 – Information Management System for Official Controls (IMSOC)[38]
2019/1793 – Temporary increase of Official Controls (HRFNAO)
2019/1873 – Intensified Official Controls (IOC)[39]
2019/2007 – Lists of POAO subject to checks at a BCP[40]
2019/2074 – Rules on returned POAO[41]
2019/2098 – Animal health requirements for returned POAO[42]
2019/2122 – Personal Imports exempted from official controls[43]
2019/2124 – Rules on Transits, Transhipments and Onward Transport[44]
2019/2126 – Specific Official Controls and goods exempted from Official Controls[45]
2019/2128 – Rules on Ships Supplies and NATO or US Military Bases[46]
2019/2129 – Frequency rates for Identity and Physical Checks on POAO[47]
2019/2130 – Rules on Documentary, Identity and Physical Checks at BCPs[48]
2020–2235 – Model Official Certificates for POAO[49]

Border Control Posts (BCPs)

Imports of products of animal origin (POAO) and High-Risk Food and Feed Not of Animal Origin (HRFNAO) are required to enter GB through a Border Control Post (BCP).

Previously, BCPs were known as Border Inspection Posts, Designated Points of Entry/Import, as defined in the legislation at that time. The term Border Control Post was introduced by the Official Control Regulation and defines it as: 'a place, and the facilities belonging to it, designated by a Member State for the performance of official controls'.

Not all ports have BCPs. It is a commercial decision for the port operator whether to have a BCP and, in conjunction with the PHA/LA, can apply for designation. The designation can be for all types of products with all types of temperature (ambient, chilled and frozen goods), or it can be restricted to some types of product. The list of BCPs together with the type of designation can be found on the Gov.uk website [41].

The minimum requirements for a Border Control Post is set out in Article 64 of the OCR and Retained EU Regulation 2019/1014. The OCR provides more flexibility in how the BCP is used. In the past, Border Inspection Posts were provided for the inspection of POAO and Animal By-products only, but now the same inspection facilities can be used for HRFNAO and Plant and Plant products, provided that the risks of cross-contamination are controlled.

Products of Animal Origin (POAO)

The Trade in Animals and Related Products Regulations 2011 (TARP)[50] are the main regulations that cover veterinary checks at a Border Control Post. These regulations implement the OCR and its tertiary regulation relating to POAO.

Products subject to official controls – the Retained EU Regulation 2019/2007 and Retained EU Decision 2007/275 list the products that are subject to official controls and includes:

- Red meat and poultry;
- Fish and shellfish;
- Dairy products such as milk, butter, cheese, yoghurt;
- Honey;
- Pet food;
- Semen and embryos;
- Game trophies and blood products;
- Hay and straw;
- Composite products containing POAO such as meat, dairy products, fish.

It is not permitted to import POAO through a port into GB unless there is a suitably designated BCP for those products. Should POAO be imported to GB at a port without a BCP, it then becomes an illegal import and should be seized by the UK Border Force.

If illegally imported POAO is not detected at the port, it may still be detected at an *inland* local authority. The responsibility for seizure then falls to that local authority. The FSA's Inland Enforcement of Imported Food Controls Resource Pack on its Smarter Comms platform provides more guidance on this issue [42].

At BCPs, environmental health practitioners act as Official Fish Inspectors (OFIs) and veterinary auxiliaries. OFIs are only permitted to clear fishery products. All other POAO consignments are cleared by the Official Veterinary Surgeon (OVS). The FSA Food Law Code of Practice has been updated to reflect the OCR which bases authorisation on competence rather than listing specific qualifications [43].

Import conditions – POAO can only be imported from approved third countries. In general, products must also originate from approved premises and comply with animal and public health conditions.

In EU legislation, the term 'third country' is used to refer to countries outside of the EU. Now that the UK has left the EU, the term 'Rest of World country' (RoW) is more commonly used, particularly when making a distinction between the import conditions for RoW countries and EU Member States [44].

Import conditions are laid down in specific tertiary legislation made under the OCR. A consignment must comply with these import conditions to be permitted free circulation on the UK Internal Market.

Import legislation changes frequently, and a full list of current import legislation is maintained by Defra and FSA on their respective websites [45–46].

Generally, POAO intended for human consumption will require health certification issued by the competent authority of the exporting country containing declarations of compliance with the animal and public health criteria laid down in the import conditions. Consignments of POAO cannot be removed from the BCP until all the official controls are completed and any fees required have been paid.

Official controls – there are three main types of official control that are carried out on imported POAO: documentary, identity and physical checks.

- Documentary check;

The documentary check is the assessment of the CHEDP, animal and/or public health certificates, commercial documentation, invoices and packing lists accompanying the consignment. All consignments are subject to documentary checks.

- Identity check;

All consignments are subject to identity checks which involves a visual inspection to verify that the product, labelling including health marks on the packaging correspond to the accompanying health certificate and other documentation.

- Physical check.

Goods may also be physically checked, and may include checks on the packaging, means of transport, temperature, as well as sampling for analysis to verify compliance.

Retained EU Regulation 2019/2129 prescribes the level of checks for certain products.

Sampling of imported food and feedstuffs by PHA/LAs

Not all imported food is sampled and sent for examination or analysis. Based on previous sampling and surveillance data, current priorities for sampling imported food and feed in the UK are set out in the National Monitoring Plan on the FSA Smarter Communication platform [47] and are reviewed and updated in the light of new surveillance information produced by GB, or via intelligence from other sources, for example the International Food Safety Authorities Network (INFOSAN) [48]. See Table 21.3 for examples.

Unsatisfactory checks – products failing to satisfy import conditions may be re-exported to a country outside of the UK. However, if the consignment is deemed to be a risk to human or animal health, or where the person responsible for the consignment fails to comply with the direction to re-export, it can be sent for destruction by incineration.

Serious or repeated infringements/safeguard measures – Article 65 of the OCR requires competent authorities to take certain action where a serious or repeated infringement has been identified during veterinary checks. The FSA must be informed, and the next ten consignments should be subject to Intensified Official Controls, whereby the consignment is held at the BCP pending the results of sampling.

Serious infringements could include for example:

- Faecal contamination of meat;
- Microbiological failures;
- Excessive histamine levels in certain fish;
- Excessive contaminants such as heavy metals;
- Presence of veterinary residues;
- Any rejection on the grounds of risk to animal health.

Table 21.3 Examples of contaminants tested for at BCPs

Category of contaminant	Contaminant and food/feed product
Microbiological	*Salmonella* in fresh produce including herbs and betel leaves, bovine, ovine and poultry meat, live bivalve molluscs and gastropods *Listeria* in cooked chicken, smoked fish, dairy products *Clostridium botulinum* in vacuum packed smoked fish *E. coli* in fresh/frozen meat, in particularly looking at anti-microbial resistance.
Mycotoxins	Aflatoxins in oil seeds and corn/maize meal Moniliform, Citrinin, Cyclopiazonic acid and Sterigmatocystin in cereals and cereal products Fumonisins in cereal-based animal feed Aflatoxin B1 groundnuts for wild bird feed
Contaminants	Nitrates in lettuce, spinach and rocket
Organic contaminants	Polyaromatic hydrocarbons (PAHs) in smoked and dried fish and fish products and dietary supplements Dioxins in meat, liver, eggs, egg products, dairy products, fishmeal from marine animals Histamine in fishery products Mineral oil in vegetable oil
Inorganic contaminants	Aluminium in processed foods Mercury, lead and cadmium in fishery products and shellfish Barium in nuts Pesticides in fresh produce Hormone growth promoters in raw bovine meat Veterinary residues in sheep and porcine casings, poultry meat, aquaculture fishery products, eggs
Process contaminants	3-MCPD in soy sauce
Food contact materials	Lead and cadmium migration from ceramics Phthalate migration from gaskets in jars of sauces and pickles from the Far East
Irradiated products	Dried herbs and spices Food supplements Dehydrated Asian meals and soups
Food authenticity	Added water, meat content and foreign proteins in imported poultry products Authenticity of Basmati rice
Artificial colours	Confectionery
Unauthorised Genetically Modified (GM) organisms	Cereal products for animal feed
Melamine	High protein feed materials (e.g. maize gluten from China)

Lists of current safeguard measures for serious or repeated infringements are issued by the FSA and will be available on the IPAFFS database.

The Rapid Alert System for Food and Feed (RASFF)

RASFF [49] is an EU reporting mechanism for sharing information regarding risks associated with imported foods. Since leaving the EU, the UK only has limited access to this data. The UK can still share information about food incidents, but this is through INFOSAN which is managed by the Food and Agriculture Organization of the United Nations (FAO) and the WHO.

The FSA monitors intelligence about imported foods and sends Early Warning Notifications to LA/PHAs following an assessment of food safety signals.

Import of Products, Animal, Food and Feed System (IPAFFS)[51]

Due to the UK's departure from the EU, the UK (except Northern Ireland) no longer has access to the Trade Control and Expert System (TRACES) [50]. Instead, the UK has developed its own system known as IPAFFS [51].

Importers, or their clearing agents, are required to register onto the system and use it to notify BCPs of their consignments. Enforcement staff at BCPs use the system to enter the results of official controls.

Common Health Entry Document (CHED) for Products of Animal origin – the (CHEDP) is a document that shows that the necessary checks have been carried out on entry into the UK and the template can be found in Retained EU Regulation 2019/1715.

Importers are required to submit a declaration of the intended arrival of a product at a BCP, normally by means of Part 1 of the CHEDP, for completion by the official veterinarian or official fish inspector responsible for the BCP. On completion of all checks and the CHEDP, the document is returned to the importer or their agent. This process can be done either by hard copy or on-line. The original CHEDP must accompany the consignment to the destination.

Illegal imports – the responsibility for enforcing the controls in relation to smuggled POAO at ports is assigned to the UK Border Agency. Local authorities (at the district level) are responsible for this activity inland and

includes POAO removed from a BCP before veterinary checks have been completed. In both cases, the goods may be re-exported or, more commonly, destroyed by incineration or rendering.

Transits and transhipments – at airports, it is not unusual for transits of Products of Animal Origin to occur. A transit is where a POAO consignment arrives in GB at the Entry BCP, travels by road vehicle across GB territory to an Exit BCP before departing GB.

In this scenario, there is no intention of the freight remaining in GB and the freight has to travel with an appropriate CHEDP that demonstrates that the consignment has been cleared for transit. The freight should also be secured, usually by sealing the truck with a seal and tag – the reference number of which will be included in the CHEDP. The IPAFFS database can then be used to determine whether the exit BCP has taken control of the consignment and confirm that the consignment has left GB.

Transhipments describe consignments which arrive from a third country and are transferred from one aircraft to another aircraft or from one ship to another ship destined for another country outside of GB. The transfer from vessel to vessel happens at the same port.

It is not usual for a physical check to be carried out on goods that are being transited or transhipped, but if a consignment is found to be a danger to public or animal health, action can be taken under Regulation 21 of the TARP Regulations.

Destruction of rejected POAO – pending destruction, the POAO can be stored in a dedicated locked freezer until there is sufficient rejected product for a suitable waste collection to be made. Certain products such as whole tuna require a certain amount of defrosting prior to incineration. Rejected consignments at BCPs are classed as Category 2 Animal By-Products under Article 9 of Retained EU Regulation 1069/2009[52] (as enforced by the Animal By-Products (Enforcement) Regulations 2013).[53]

There are specific rules concerning the removal and disposal of such waste which minimises the risk to animal health (e.g. by spreading animal diseases such as foot and mouth disease), and by default, also ensures that the waste does not enter the human food chain.

During disposal, waste transfer notes are signed by the waste carrier and Certificates of destruction are produced by the approved waste plant. For Category 2 Animal By-products, the disposal has to be by incineration.

Re-exportation is the responsibility of the importer. Although this choice is sometimes given, it is rarely opted for in air-freight, due to the small and perishable nature of the consignment and the often prohibitive costs.

The choice of destruction or re-export depends on the reason for failure. Should a consignment be a hazard to human or animal health, then destruction would be the only option given. If a consignment failed due to a documentary check failure only, then re-exportation might be offered. The officer inspecting the product makes this decision at the time of inspection or may ask the advice of a specialist body, such as the FSA or APHA.

For re-exportation of a rejected consignment, the importer must advise the relevant competent authority in the country of origin. However, if the product is to be sent to another country outside the EU (i.e. not the country of origin), the importer has to arrange for the competent authority of that country to advise Port Health of its preparedness to accept the product, before the product can be released for re-export.

Although 'Transformation' is another option provided by law, it is something that rarely occurs at airport Border Control Posts, possibly due to the small size of the consignments, as well as the costs involved.

The Central Competent Authority for POAO is DEFRA's APHA (Animal and Plant Health Agency). Guidance can be found in the DEFRA BCP Manual, OVS Notes and the DEFRA BCP Compendium on the APHA website [52]. Although the FSA cover consignments of fishery products, information is still issued via the BCP Manual and OVS notes.

Illegal, Unreported and Unregulated Fishing (IUU)

IUU is the biggest global threat to fish stocks in the world and is a particular problem in countries where there is poor infrastructure and resources to undertake effective monitoring [53]. The EU accounts for 80% of the global imports of wild caught and aquaculture fishery products [54].

The EU introduced legislation in 2010 to prevent, deter and eliminate IUU fishing and applies to all imports of wild caught fish and shellfish. Aquaculture fish are exempt from the scope of these regulations. LA/PHAs enforce the controls at BCPs which involve endorsing the Catch Certificate (CC) after checks to ensure that it complies with legislation. The Legislation has now been incorporated into domestic legislation as Retained EU Regulation 1010/2009.[54]

Working with the UK's Marine Maritime Organisation, checks are made to ensure that the vessel catching the fish was not on the EU list of vessels known to be engaged in illegal fishing, the CC is valid, relates to the consignment and signed by an authorised official. Only after all these checks have been completed, will the CC be endorsed and the consignment released. In the event of non-compliance, the consignment may be seized and disposed.

High Risk Food and Feed of Non-Animal Origin (HRFNAO)

In contrast to POAO, *any* food and feed of non-animal origin (NAO) can come from *any* country in the world unless there is a restriction in place. It is this restriction that makes the product a High-Risk Food or Feed of Non-Animal Origin (HRFNAO) – the term does not have its own legal definition. There is also no requirement to notify the BCP of NAO imports, unless it is a high-risk product.

Imports of non-animal origin foods are controlled by the Official Feed and Food Controls Regulations 2009 (as amended).[55] They also cover foods which contain POAO that fall outside the official controls at BCPs, for example, certain 'composite' products. These regulations require that imported food comply with the same food safety requirements as all other food, whether or not imported.

At seaports with manifest inventory systems, PHAs are able to identify consignments of HRFNAO prior to import. At airports, this is more difficult as the manifest is still largely paper-based, and the information is rarely available to the BCP prior to import. As the checks on EU goods commence, there are also some further difficulties for certain shipments, for example roll on and roll off (RoRo) ferry traffic.

At airports, speed is of the essence. Consignments of NAO may have already left the airport for Customs clearance at an inland Customs transit shed known as an External Temporary Storage Facility (ETSF) [55]; therefore it is likely that some imported NAO on sale inland has not undergone any checks at the airport.

In the past, certain countries and types of consignments have been targeted for inspection at airports. In order to target consignments the following factors should be considered:

- The LA/PHA's own intelligence (e.g. previous history, sample results etc.);

- Results of sampling by other ports on the Risk Likelihood Dashboard;
- Intelligence from other sources (e.g. the FSA);
- Specific import conditions for certain products.

Random checks of non-animal origin foods are also carried out to check that the food complies with the food safety requirements of Retained EU Regulation 178/2002.[56]

The import conditions relating to certain HRFNAO are contained in a series of regulations made under the OCR as listed in Table 21.4.

Similar to POAO, HRFNAO has to be imported through suitably designated BCPs. Prior notification also has to be given to the LA/PHA at the BCP. 100% of consignments undergo a documentary check, which includes the submission of a Common Health Entry Document (CHEDD) on IPAFFS. Identity and physical checks are then carried out at a percentage frequency set in the regulations. On satisfactory completion of the checks, Part II of the CHEDD is then completed and given to the Food Business Operator. For inland authorities, it is the CHEDD that provides the evidence that certain products have undergone checks at the port.

The same NAO consignments can also be subject to examination by the Animal and Plant Health Agency (APHA). It is possible that both the LA/PHA and APHA could want to detain the same consignment but for different reasons. For example, in a consignment of aubergines from the Dominican

Table 21.4 Retained EU legislation containing import conditions for HRFNAO

1333/2008[57] – Controlled jelly confectionery due to a choking hazard
2011/884[58] – Unauthorised genetically modified rice from China
2016/6[59] – Controls on products affected by the accident at the Fukushima nuclear power station
2019/1793 – Temporary increase of Official Controls (HRFNAO)

Annex I – Emerging risk
Annex II – Emergency controls and compound products
Annex IIa – Prohibited products
2020/1158[60] – Controls on products affected by accident at Chernobyl nuclear power station

Republic, the LA/PHA could be checking for pesticide residues, whereas APHA Plant Health inspectors could be checking for quarantine pests. It is therefore important that both Port Health and Plant Health inspectors liaise closely.

Food Business Operators (importers) are required to pay for the checks, including sampling and analysis. The charges are not nationally set, so there is some variation between BCPs. However, all LA/PHAs need to be able to demonstrate how they have determined their charges.

Another issue of concern to Food Business Operators for products checked at airport BCPs is that the products tend to be highly perishable (e.g. fresh fruit and vegetables). Therefore, detention pending sampling results can lead to deterioration in the quality of the produce. The majority of tests on fresh fruit and vegetables is for pesticide contamination. Most BCPs are able to offer a one-day turnaround, where the products are sampled one day, and the results from the public analyst laboratory are received the following day. In order to facilitate the trade, many consignments are released for detention at the freight forwarder's External Temporary Storage Facility (ETSF) which will often be closer to the food business and also incur cheaper storage fees (as opposed to storage at an airport transit shed).

Some of the products on the high-risk list can also be imported as feedstuffs for animals. Non-animal origin feedstuffs are subject to controls by Trading Standards officers, rather than Port Health and close liaison is required between the two enforcement bodies. It should be noted that, at the time of writing, the Customs electronic processing system, CHIEF, cannot currently distinguish whether products on the high-risk list are food or feed.

The Central Competency Authority in relation to NAO is the FSA. Guidance can be found in the FSA's Food Law Practice Guidance, the FSA's Imported Food Resource Pack as well as the 'Imports' section on the FSA website.

HRFNAO exported from the EU Member States – at the time of writing there is no HRFNAO that originates from within the EU. These goods will have originated from third countries and will have been imported into an EU Member State via an appropriately designated BCP.

For certain products that require official certification (e.g. Annex II of Retained EU Regulation 2019/1793), there are some import conditions to be aware of. The regulation requires each consignment of food and feed listed in Annex II to be accompanied by an official certificate and the results of sampling and analyses performed on that consignment 'by the competent authorities of the third country of origin, or of the country where the consignment is consigned from if that country is different from the country of origin'.

Therefore, consignments of Annex II products being exported to GB will require the EU Member State to provide new official documents to accompany the import. It should also be pointed out that the templates for the official certificates for GB is now different to that of the EU, and is now available on the Gov.uk website [56].

There is no current derogation that would permit GB BCPs the evidence of checks in an EU Member State (e.g. by provision of a valid CHED-D), together with authenticated copies of the official certificate and results of analysis.

Transits of HRFNAO via EU Member States – the provisions for checks on transits is set out in Retained EU Regulation 2019/2124. Article 19 sets out the scope of the conditions for authorisation; however, this does not include HRFNAO, and such goods will not be controlled via a CHED for transit on TRACES NT.

Goods which have transited EU Members States have been controlled using the NCTS (New Computerised Transit System) which is available for all countries that form part of the Common Transit Convention. As such goods will not have been cleared for the EU

Internal Market; they are considered as 'Rest of World' imports and will require prenotification on IPAFFS, and be subjected to official controls at a suitably designated Border Control Post. This arrangement has been in place since 1st January 2021. At the time of writing, it should be noted that Dover (a busy roll-on, roll-off ferry port) has restrictions on the types of product it can handle at its BCP. Therefore, transits of any POAO and some HRFNAO cannot be brought into GB through Dover.

Organic imports

The organic checks regime covers both POAO and HRNAO which are already subject to controls already described. Its relevance for 'Environmental Health' is at the point of entry only, as Trading Standards officers enforce this legislation inland. These checks are concerned with consumer protection rather than public health, that is to prevent foods labelled as organic (and hence being sold at a higher cost) from leaving the airport without being checked.

The Organic Products Regulations 2009[61] require 6 hours advanced notification of organic consignments. These Regulations provide enforcement powers to Retained EU Regulation 834/2007 (setting out the general principles of organic production) and Retained EU Regulation 1235/2008 (which specifies the Certificate of Inspection (COI), Inspection Bodies and Third Country List).

In general, there is just a documentary check of the COI. In airfreight, certificates are unlikely to be travelling with consignments. As many of the products are perishable, it would cause difficulties if products were detained pending the arrival of the original certificate. LA/PHAs at airports are therefore permitted to sign and validate copies of COI provided that the original certificate is submitted for verification at a later date.

Guidance can be found on the DEFRA Organics Branch website [57].

Illustrative examples of imported food risks

CASE STUDY 3

Concealed paan or betel leaves

Checks were carried out at an international airport on fruit and vegetable consignments which were suspected of containing paan or betel leaves which may have been contaminated with Salmonella. As the product is likely to be chewed, without any processing, the FSA requested that Port Health should detain and sample any leaves.

It was discovered that there was often a mis-match between the commercial paperwork (e.g. invoice or packing list) and the actual contents of a consignment. Frequently paan or betel leaves were found in consignments but not listed in the commercial documentation.

Regulation 32 of the Official Feed and Food Control Regulations 2009 was used to detain the consignments pursuant to Article 65 of Retained EU Regulation 2017/625 (suspicion of non-compliance). The official controls carried out included sampling the paan leaves for microbiological examination by the Public Examiner. The regulations do not specify a time-limit but allows a consignment to be kept under official detention until it obtains the results of such official controls. For these consignments, the products were kept under detention until a laboratory report had been received.

CASE STUDY 4

Russian salmon

Seventy-two 20-tonnes containers of canned Russian salmon originating in Russia were presented as one consignment for import intended for UK 'food hampers'. The consignment of 3 and a half million cans was inspected, found to be grossly damaged and with an extensive range of canning defects.

It was discovered that the consignment had previously been entered for import into the USA. The Food and Drugs Administration had rejected the consignment for import but had allowed re-export.

Samples were taken for microbiological examination and *Clostridium* species bacteria were detected. The Authority's food examiners and a canning expert gave opinions that the food was a risk to human health.

The PHA rejected the consignment for import and sought its destruction.

The Russian owners and exporters visited the UK to examine the consignment and expressed surprise that the containers were not being kept under refrigeration. In discussion, the owners told the PHA that the cans had been pasteurised not sterilised and were required to be kept cold. There were neither instructions for temperature storage nor safe storage period present on the cans.

Disputes of ownership (and liability) followed, together with persistent and forceful representations made by both exporters and importers for the sorting and salvage of the cans based on their physical appearance.

This approach was considered to be unsatisfactory by the PHA which obtained a court judgement for the destruction of the consignment.

The fact that all 72 containers were presented as one consignment was a further risk.

Under the rules of the OCR, a percentage of a consignment must be examined. As this was one consignment, 'sorted' cans could have been placed in different containers and not been examined and the problem would have been missed with potentially widespread illness and deaths.

CASE STUDY 5

Iranian pistachios

Throughout the 1990s consignments of Iranian pistachios were sampled at UK ports at import and upon analysis were found to be heavily contaminated with aflatoxin B1, an established human carcinogen.

The import trade was largely composed of one company.

The food examiner gave the opinion that aflatoxin B1 was 'frighteningly toxic' and at the levels commonly found in imports were a serious risk to human health.

Notices to secure the destruction of a number of consignments were served by PHAs.

The importer appealed the notices and wished to re-export the consignments.

In appealing the notice, the importer disputed the scientific evidence regarding the toxicity of aflatoxin B1; described that higher levels of contamination than permitted in the

UK, were acceptable in many other countries; and asserted that there were validated techniques in Iran for sorting and decontaminating contaminated nuts. The importers wished to be allowed to re-export the nuts.

Lastly, the importer argued that if they were not permitted to re-export the nuts, then their human rights would be infringed, and they would pursue the matter in the European Court for Human Rights.

At the first court appeal hearing, the court agreed with the PHA that the consignment was dangerous; that there was no guarantee that the nuts having been allowed to leave the UK would not re-enter by some other method (e.g. intra-community trade routes), and be consumed in the UK.

The judge agreed with the PHA and ordered the destruction of the consignment.

Within one week of the court judgement the trade of Iranian nuts was switched from UK to another European country known not to actively control imported food.

Fortunately, the increasing harmonisation of EU controls for FNAO and the development of the 'high risk' list has blocked the opportunity for importers to exploit weak points in the EU food import safety control system.

Case study 6

Melons from Costa Rica

In 2009, 12 containers of melons (240 tonnes) from Costa Rica as one load were presented for import at a UK port.

Melons would not normally attract the interest of PHAs for food controls but the load was large and UK normally receives melons via intra-community trade or by air freight.

The importer contacted the PHA as he had not dealt with the supplier before and had some concerns regarding the presence of pesticides. The PHA agreed to sample the melons on the importer's behalf on the understanding that the importer would pay the cost of analysis.

As the melons were undergoing analysis, a port health officer of the PHA happened to be using the internet and came across an article on a USA ban on the import of melons from Costa Rica due to a large outbreak of Salmonellosis in the USA linked to the consumption of melons from Costa Rica.

The consignment had been detained pending the receipt of analytical results for the presence of pesticides and the PHA returned to the consignments and sampled for microbiological examination.

The results of analysis and examination confirmed the presence of illegal pesticides and Salmonella bacteria and the majority of the consignment was destroyed.

Further investigation revealed that the USA had issued the import restriction some time previously, as following the USA outbreak, many melon growing fields in Costa Rica were found to be overrun with reptiles such as iguanas which defecated over the melons, contaminating the surfaces with Salmonella.

None of this knowledge was readily available to UK PHAs.

Luck and inquisitiveness seems to have played major roles in preventing a similar food poisoning outbreak in the UK.

By using the RASFF system other member EU countries were then alerted to the potential risk from further imports. In the UK, such notifications can now be undertaken using INFOSAN.

CASE STUDY 7

Nigerian dried beans

In 2012 a routine sample was taken from dried beans loaded in a container of mixed foodstuffs from Nigeria. The container was not detained as there had been no reports of problems with this type of product. The public analyst reported the presence of the pesticide dichlorvos in excess of the MRL of 0.05 mg/kg; it is also considered to be genotoxic.

Samples were taken from the next consignments and these were detained pending results; the majority were found to contain dichlorvos above the MRL and were refused entry into the UK. All future consignments were sampled and detained at the port and it soon become established that 50% contained dichlorvos above the MRL. Other pesticides were starting to be found, again over the MRLs. At the time, dried beans were not listed in 669/2009 and the costs associated with the checks were borne by the Port Health Authority (PHA). With every rejected consignment a RASFF was submitted and checks continued for 9 months.

In 2013 the EU then took action and placed dried beans from Nigeria on the list in 669/2009 with sampling to be undertaken at a frequency of 50% at import. Now the PHA could recover the costs associated with the checks. The number of declared consignments of dried beans dropped and it was assumed that exports to the UK had stopped. However during a physical check by Customs officials of a consignment of mixed foods from Nigeria that had paperwork that did not list dried beans, sacks of dried beans were found. Customs notified the PHA who took control of the consignment and rejected the beans. Further checks by the PHA on other consignments found that the paperwork was not reliable and attempts were being made to smuggle products in that were subject to official controls.

Checks at the port increased and containers were checked for the presence of undeclared beans. If found the importer was allowed to submit a CED and official controls were carried out.

Pesticides were still being found in 50% of consignments 2 years after dried beans were put on the list in 669/2009 and this was being reported to the EU through the RASFF system. The EU then suspended imports in 2015. However, checks on containers of foodstuffs from Nigeria continue to find sacks of dried beans loaded in them. Now these are being seized and either returned to Nigeria or destroyed and costs being recovered with notifications being submitted to the RASFF system to alert other ports of the issue.

This shows that a routine sample can lead to EU controls being put in place and the value of sharing information through the RASFF system. Now that the UK has left the EU, a similar approach to identifying is being developed whereby unsatisfactory sample results will be risk-assessed by the FSA and considered for inclusion on the high-risk list.

Initial checks were borne by the PHA and not recoverable, placing pressures on its funding. Also, accompanying paperwork for consignments cannot be relied on and that good working relationship with other enforcement agencies is important for effective controls.

Live animal imports

In order to ensure that animal and human diseases are not introduced into GB via imported animals, a system of control is enforced by PHA/LAs, APHA Animal Health, the UK Border Agency and the Police.

In Northern Ireland, EU rules on the importation of live animals still apply, and further guidance can be found on the Department of Agriculture, Environment and Rural Affairs (DAERA) website [58].

There are different reasons why live animals may be imported into GB as follows:

- Live animals for human consumption (e.g. lobsters, eels, shellfish and snails);
- Commercial consignments (e.g. tropical fish, reptiles, equines and germinal products);
- Travelling pets;
- Inadvertent importation of live animals (e.g. birds, frogs, reptiles, spiders);
- Illegal imports.

Live animals for human consumption – such consignments are treated as goods under the Trade in Animals and Related Products Regulations 2011. For example, live lobsters will require to meeting the import conditions of fishery products and be subject to official controls at a Border Control Post. Guidance on importing live animals can be found on the Gov.uk website [59].

Commercial consignments of live animals and germinal products – for imports into GB, live animals need to be imported through a Border Control Post designated for live animals. The importer will need to give prenotification of at least one working day using the IPAFFS database.

Consignments of live animals will require health certificates and/or authorisations from the Animal and Plant Health Agency (APHA), depending on the type of animal or germinal product. The specific import conditions can be found in Importer Information Notes on the APHA website [60].

The rules for importing non-native animals are complex and specific guidance is available from the gov.uk website [61].

The official checks are carried out by an APHA official veterinarian (as opposed to a PHA/LA official veterinarian dealing with products of animal origin). The official controls include a check on the accompanying documentation, an identity check and may also include a physical check.

Importers are also required to check if the animal is on the list of endangered species. If so, then a permit may be needed under the Convention on International Trade in Endangered Species of Wild Fauna and Flora (CITES) [62].

Importation of pets – pets can also carry disease, and therefore there are a series of import conditions that need to be met. Owners are permitted to bring in up to five pets – more than five would be deemed as a commercial import.

Different rules apply to different countries of origin and whether they are classed as 'Part 1' or 'Part 2' listed countries, or a non-listed country. Pet passports are only accepted for pets originating from Part 1 listed countries. The pet passport lists the different treatments received by the pet. For Part 2 listed countries and non-listed countries, a health certificate would be required and includes a requirement for the pets to have had a blood-test, as well as confirmation of treatments.

Certain animals such as dogs, cats and ferrets need to be microchipped. The information in the microchip should then correspond with the accompanying paperwork. There are different rules if the animal does not arrive in the UK within five days of the owner. There are also different rules for other types of animal. Further information can be found on the Gov.uk website [63].

The Animal and Plant Health Agency have also published detailed guidance on checks to be carried out on pets by transport carriers [64]. This includes details on the types of check required, as well as guidelines on animal welfare, and what to do in the event of the death of a pet.

There are international standards for the transportation of live animals by air which are set out in the Live Animal Regulations published by the International Air Transport Association (IATA) [65]. The crates are required to be large enough for the animal to be able to stand up and turnaround and have sufficient ventilation.

Animals should be transported in a heated cargo hold with the exception of guide and assistance dogs that are permitted to travel in the aircraft cabin. However, these dogs will also need to meet all the import conditions as outlined previously.

Inadvertent importation of live animals – the Public Health (Aircraft/Ships) Regulations 1979 (as amended)[62] require the master of a ship, or commander of an aircraft, to notify the proper authority (PHA/LA), before arrival, of the presence of animals or captive birds on board, and any occurrence of mortality or sickness amongst such animals or birds. If it is not possible to provide this notification before arrival, notification must take place immediately on arrival.

All animals on board ships or aircraft must be securely confined so that they cannot escape or meet any other animal. Stowaway animals should be quarantined and can be returned to the country of origin, destroyed or given to a zoo.

Illegal imports of live animals – animals are sometimes concealed in hand-baggage on aircraft, or in vehicles entering GB. If detected, such animals are seized and placed into quarantine. Where smuggling is suspected, please seek further information from APHA's Centre for International Trade. Email: SSC.Carlisle@apha.gov.uk.

Section 4 Notes

25 Retained EU Regulation 2019/1793 https://www.legislation.gov.uk/all?title=2019%2F1793

26 European Union (Withdrawal) Act 2018 https://www.legislation.gov.uk/ukpga/2018/16/contents

27 Regulation (EU) 2019/1793 https://eur-lex.europa.eu/legal-content/EN/ALL/?uri=CELEX:32019R1793&qid=1614934516787

28 Retained EU Regulation 2017/625 https://www.legislation.gov.uk/all?title=2017%2F625&page=2

29 Retained EU Regulation 2007/275 https://www.legislation.gov.uk/all?title=2007%2F275%20

30 Retained EU Regulation 2019/625 https://www.legislation.gov.uk/all?title=2019%2F625

31 Retained EU Regulation 2019/626 https://www.legislation.gov.uk/all?title=2019%2F626

32 Retained EU Regulation 2019/628 https://www.legislation.gov.uk/all?title=2019%2F628

33 Retained EU Regulation 2019/1012 https://www.legislation.gov.uk/all?title=2019%2F1012

34 Retained EU Regulation 2019/1013 https://www.legislation.gov.uk/all?title=2019%2F1013

35 Retained EU Regulation 2019/1014 https://www.legislation.gov.uk/all?title=2019%2F1014

36 Retained EU Regulation 2019/1081 https://www.legislation.gov.uk/all?title=2019%2F1081

37 Retained EU Regulation 2019/1666 https://www.legislation.gov.uk/all?title=2019%2F1666

38 Retained EU Regulation 2019/1715 https://www.legislation.gov.uk/all?title=2019%2F1715

39 Retained EU Regulation 2019/1873 https://www.legislation.gov.uk/all?title=2019%2F1873

40 Retained EU Regulation 2019/2007 https://www.legislation.gov.uk/all?title=2019%2F2007

41 Retained EU Regulation 2019/2074 https://www.legislation.gov.uk/all?title=2019%2F2074

42 Retained EU Regulation 2019/2098 https://www.legislation.gov.uk/all?title=2019%2F2098

43 Retained EU Regulation 2019/2122 https://www.legislation.gov.uk/all?title=2019%2F2122

44 Retained EU Regulation 2019/2124 https://www.legislation.gov.uk/all?title=2019%2F2124

45 Retained EU Regulation 2019/2126 https://www.legislation.gov.uk/all?title=2019%2F2126

46 Retained EU Regulation 2019/2128 https://www.legislation.gov.uk/all?title=2019%2F2128

47 Retained EU Regulation 2019/2129 https://www.legislation.gov.uk/all?title=2019%2F2129%20

48 Retained EU Regulation 2019/2130 https://www.legislation.gov.uk/all?title=2019%2F2130

49 Retained EU Regulation 2019/2130 https://www.legislation.gov.uk/all?title=2019%2F2130

50 SI 2011 No. 1197 (different legislation applies in the devolved administrations of the UK, for example SI 2012 No. 177 in Scotland and SI 2011 No. 438 in Northern Ireland).

51 https://www.gov.uk/guidance/import-of-products-animals-food-and-feed-system

52 Retained EU Regulation 1069/2009 https://www.legislation.gov.uk/all?title=1069%2F2009

53 SI 2013 No. 2952 (different legislation applies in the devolved administrations of the UK).

54 Retained EU Regulation 1010/2009 https://www.legislation.gov.uk/all?title=1010%2F2009

55 The Official Feed and Food Controls Regulations 2009 (different legislation applies in the devolved administrations of the UK) https://www.legislation.gov.uk/primary+secondary/2009?title=official%20feed%20and%20food%20controls%20

56 Retained EU Regulation 178/2002 https://www.legislation.gov.uk/all?title=178%2F2002

57 Retained EU Regulation 1333/2008 https://www.legislation.gov.uk/all?title=1333%2F2008&sort=title

58 Retained EU Decision 2011/884 https://www.legislation.gov.uk/all?title=2011%2F884

59 Retained EU Regulation 2016/6 https://www.legislation.gov.uk/all?title=2016%2F6

60 Retained EU Regulation 2020/1158 https://www.legislation.gov.uk/all?title=2020%2F1158

61 SI 2001 No. 430 (SI 2020 No. 272 in Northern Ireland) but see also SI 2020 No. 1400, which has been made to address failures of retained EU law to operate effectively and other deficiencies arising from the withdrawal of the United Kingdom from the European Union, and to reflect the Protocol on Ireland/Northern Ireland in the EU withdrawal agreement.

62 SI 1979 No. 1434, SI 2007 No. 1447 or SI 1979 No. 1435 and SI 2007 No. 1446; and SI 2007 No. 1603 (different legislation applies in the devolved administrations of the UK).

SECTION 5: SHELLFISH

The Food Safety Act 1990 designates PHAs as food authorities that are responsible for controlling 'layings'[63] within their district which frequently abut several LA areas and which cover large areas of sea. To ensure the safe harvesting of live bivalve molluscs such as oysters, cockles, mussels, clams, razor clams and scallops from naturally occurring and farmed estuarine, river and marine stocks, PHAs and LAs are principally involved in the following matters:

Classification and monitoring of shellfish beds

Officers regularly sample each species in their area for microbiological examination to determine the presence of pathogens. Before the product can be placed on the market for human consumption, each shellfish bed must be 'classified' dependent upon the detected background level of contamination. The protocols used for the Classification of Shellfish Production Areas for England and Wales have been published on the FSA website.[64]

Shellfish bed classification

Shellfish beds are classified as A, B or C with A being the cleanest and C being the most contaminated. The classification dictates the types of treatment that the shellfish has to undergo before being placed on the market. Further information about the classification of shellfish beds and the types of treatment can be found on the FSA website [66].

Local Action Groups (LAG) have been created which are formed of a number of organisations and stakeholders such as trade bodies. PHA/LAs are responsible for the operation of the LAG which is tasked with developing a local action plan (LAP). Guidance on the operation of the Local Action Group can be found on the FSA website [67].

Section 5 Notes

63 'Laying' (as previously defined in SI 1998 No. 994) means a foreshore, bed, pond, pit, ledge, float or similar place, including a relaying area, where live shellfish are liable to be gathered, harvested or deposited.

64 Protocol for Classification of Shellfish Production Areas, England and Wales June 2019, available at https://www.food.gov.uk/sites/default/files/media/document/shellfish-classification-protocol-june-2019-en.pdf

Chapter references

[1] International Civil Aviation Organization. (2021) *The World of Air Transport in 2019.*

https://www.icao.int/annual-report-2019/Pages/the-world-of-air-transport-in-2019.aspx#:~:text=According%20to%20ICAO's%20preliminary%20compilation,a%201.7%20per%20cent%20increase (Accessed 25 August 2021).

[2] International Civil Aviation Organization. (2021) *Economic Impacts of Covid-19 on Civil Aviation.* https://www.icao.int/sustainability/Pages/Economic-Impacts-of-COVID-19.aspx (Accessed 25 August 2021).

[3] World Health Organization. (2011) *Pandemic Influenza A H1N1. Donor Report,* 1 March. https://www.who.int/csr/resources/publications/swineflu/h1n1_donor_032011.pdf (Accessed 25 August 2021).

[4] World Health Organization. (2014) *Ebola Outbreak in West Africa Declared as a Public Health Emergency of International Concern.* https://www.euro.who.int/en/health-topics/communicable-diseases/pages/news/news/2014/08/ebola-outbreak-in-west-africa-declared-a-public-health-emergency-of-international-concern (Accessed 25 August 2021).

[5] World Health Organization. (2016) *WHO Director General Summarizes the Outcome of the Emergency Committee Regarding Clusters of Microcephaly and Guillain-Barre Syndrome.* https://www.who.int/news/item/01-02-2016-who-director-general-summarizes-the-outcome-of-the-emergency-committee-regarding-clusters-of-microcephaly-and-guillain-barr%C3%A9-syndrome (Accessed 26 August 2021).

[6] World Health Organization. (2016) *International Health Regulations 2005,* Geneva, Switzerland, 3rd ed. https://www.who.int/publications/i/item/9789241580496 (Accessed 26 August 2021).

[7] Department of Health. (2014) *A Letter Outlining the Role of Public Health England,* Department of Health, London. https://assets.publishing.service.gov.uk/government/uploads/system/uploads/attachment_data/file/319708/PHE_remit_letter_pdf.pdf (Accessed 26 August 2021).

[8] Department of Health & Social Care. (2020) *A Letter Outlining the Public Health England Strategic Remit and Priorities,* Department of Health & Social Care, London. https://assets.publishing.service.gov.uk/government/uploads/system/uploads/attachment_data/file/882570/PHE_Remit_Letter_from_Jo_Churchill_to_Duncan_Selbie.pdf (Accessed 26 August 2021).

[9] Department for Transport. (2020) *Guidance: Coronavirus (Covid-19): Safer Aviation Guidance for Operators.* https://www.gov.uk/guidance/coronavirus-covid-19-safer-aviation-guidance-for-operators (Accessed 26 August 2021).

[10] Scientific Advisory Group for Emergencies, Department for Transport, and Foreign, Commonwealth & Development Office. (2021) *DfT and FDCO: International Importation, Border and Travel Measures, 21 January 2021.* https://www.gov.uk/government/publications/dft-and-fcdo-international-importation-border-and-travel-measures-21-january-2021 (Accessed 26 August 2021).

[11] World Health Organization. (2019) *Ebola Outbreak in the Democratic Republic of the Congo Declared a Public Health Emergency of International Concern.* https://www.who.int/news/item/17-07-2019-ebola-outbreak-in-the-democratic-republic-of-the-congo-declared-a-public-health-emergency-of-international-concern (Accessed 26 August 2021).

[12] Civil Aviation Authority. (2021) *CAP 493 Manual of Air Traffic Services – Part 1.* https://publicapps.caa.co.uk/docs/33/CAP493%20Edition%209%20Corrigendum%20%20(May%202021).pdf (Accessed 26 August 2021).

[13] World Health Organization. (2017) *Public Health Passenger Locator Card.* https://www.who.int/publications/m/item/public-health-passenger-locator-card (Accessed 26 August 2021).

[14] HSE, HSSENI & CAA. (2017) *Memorandum of Understanding Between the Health and Safety Executive, Health and Safety Executive Northern Ireland and the Civil Aviation Authority for Aviation Industry Enforcement Activities.* https://www.caa.co.uk/uploadedFiles/CAA/Content/Standard_Content/Our_work/About_us/Files/HSE%20CAA%20Memorandum%20of%20Understanding.pdf (Accessed 26 August 2021).

[15] Food Standards Agency. (2021) *Food and Feed Codes of Practice.* https://www.food.gov.uk/about-us/food-and-feed-codes-of-practice (Accessed 26 August 2021).

[16] World Health Organization. (2009) *Guide to Hygiene and Sanitation in Aviation,* World Health Organization, Geneva, Switzerland, 3rd ed. https://www.who.int/water_sanitation_health/hygiene/ships/guide_hygiene_sanitation_aviation_3_edition.pdf (Accessed 28 August 2021).

[17] Public Health Laboratory Service. (2003) *The Microbiological Quality of Water on Board Aircraft,* Public Health Laboratory Service, London.

[18] Environment Agency. (2010) *The Microbiology of Drinking Water (2010) Part 2 – Practices and Procedures for Sampling. Methods for the Examination of Waters and Association Materials.* http://standingcommitteeofanalysts.co.uk/Archive/The_microbiology_of_drinking_water__

part_2__practices_and_procedures_for_sampling_2010.pdf.

[19] World Health Organization. (2012) *International Travel and Health*, World Health Organization, Geneva, Switzerland.

[20] Department of Health & Social Care, Department for Transport and Public Health England. (2016) *News Story – Planes Returning from Zika Areas to Be Sprayed with Insecticide*. https://www.gov.uk/government/news/planes-returning-from-zika-areas-to-be-sprayed-with-insecticide (Accessed 28 August 2021).

[21] World Health Organization. *Weekly Epidemiological Record* (1985) 60: 45–52, 85–92, 60: 345–352; (1987) 62: 329–336; (1998) 73: 109–111; (2005) 80(21): 181–191. World Health Organization, Geneva, Switzerland.

[22] World Health Organization. (1995) *Report of the Informal Consultation on Aircraft Disinsection*, World Health Organization, Geneva, Switzerland, 6–10 November. https://www.who.int/publications/i/item/WHO.PCS.95.51.Rev (Accessed 28 August 2021).

[23] Gratz NG, Steffen R, Cocksedge W. (2000) Why aircraft disinsection? *Bulletin of the World Health Organization*, 78(8): 995–1004. https://apps.who.int/iris/handle/10665/268195 (Accessed 28 August 2021).

[24] World Health Organization. (2004) *Specification and Evaluations for Public Health Pesticides – d-Phenothrin*, World Health Organization, Geneva, Switzerland. https://www.who.int/neglected_diseases/vector_ecology/pesticide-specifications/dPhenothrin_Spec_Eval_Oct_2004.pdf?ua=1 (Accessed 28 August 2021).

[25] Public Health England. (2020) *National Contingency Plan for Invasive Mosquitoes. Detection of Incursions,* Public Health England, London. https://assets.publishing.service.gov.uk/government/uploads/system/uploads/attachment_data/file/887925/National_contingency_plan_for_invasive_mosquitoes.pdf (Accessed 28 August 2021).

[26] Department for Transport. (2021) *Sea Passenger Statistics: Data Tables (SPAS)*. https://www.gov.uk/government/statistical-data-sets/sea-passenger-statistics-spas#uk-domestic-sea-passengers (Accessed 4 October 2021).

[27] Chartered Institute of Environmental Health. (1997) *Industry Guide to Good Hygiene Practice: Catering Guide-Ships*, Chadwick House Publishing, London.

[28] World Health Organization. (2007) *Interim Technical Advice for Inspection and Issuance of Ship Sanitation Certificates*, World Health Organization, Geneva, Switzerland. https://www.who.int/ihr/travel/TechnAdvSSC.pdf (Accessed 6 October 2021).

[29] World Health Organization. (2011) *Handbook for the Inspection of Ships and Issuance of Ship Sanitation Certificates*, World Health Organization, Geneva, Switzerland. https://apps.who.int/iris/handle/10665/44594?search-result=true&query=Handbook+for+inspection+of+ships&scope=&rpp=10&sort_by=score&order=desc (Accessed 6 October 2021).

[30] Grenfell P, Little CL, Surman-Lee S, Greenwood M, Averns J, Westacott S, Lane C, Nichols G. (2008) The microbiological quality of potable water on board ships docking in the UK and channel Islands: An association of Port health authorities and health protection agency study. *Journal of Water and Health*, 6(2): 215–224, June.

[31] World Health Organization. (2007) *The World Health Report 2007 – A Safer Future Global Health Security in the 21st Century*, World Health Organization, Geneva, Switzerland. https://www.who.int/whr/2007/whr07_en.pdf (Accessed 6 October 2021).

[32] World Health Organization. (2007) *Legionella and the Prevention of Legionellosis*, Chapter 7, World Health Organization, Geneva, Switzerland. http://www.who.int/water_sanitation_health/emerging/legionella.pdf (Accessed 6 October 2021).

[33] Health Protection Agency/Maritime and Coastguard Agency/Association of Port Health Authorities. (2007) *Guidance for the Management of Norovirus Infection in Cruise Ships*, Norovirus Working Group, HPA, London. https://www.gov.uk/government/uploads/system/uploads/attachment_data/file/362998/2007_guideline_norovirus_cruiseships.pdf (Accessed 6 October 2021).

[34] Health Protection Agency. (2009) *Guidance to Shipping for Pandemic Influenza*. http://citeseerx.ist.psu.edu/viewdoc/download;jsessionid=F51E39918BDF0DFAADF20335E6D3FDD9?doi=10.1.1.182.7976&rep=rep1&type=pdf (Accessed 6 October 2021).

[35] Animal and Plant Health Agency and Department for Environment, Food & Rural Affairs. (2014) *Handling and Disposing of International Catering Waste*. https://www.gov.uk/guidance/handling-and-disposing-of-international-catering-waste (Accessed 6 October 2021).

[36] Animal and Plant Health Agency and Department for Environmental, Food and Rural Affairs. (2018) *Animal Health and Welfare Framework*. https://www.gov.uk/government/publications/animal-health-and-welfare-framework-2018/animal-health-and-welfare-framework (Accessed 6 October 2021).

[37] HM Revenue & Customs. (2014) *About the Customs Declaration Service*. https://www.gov.uk/government/collections/customs-declaration-service (Accessed 23 August 2021).

[38] HM Revenue & Customs. (2012) *Automatic Licence Verification System (ALVS) Across HM Government*. https://www.gov.uk/guidance/automatic-licence-verification-between-defra-rpa-and-hmrc (Accessed 23 August 2021).

[39] Department for Exiting the European Union. (2019) *New Protocol on Ireland/Northern Ireland and Political Declaration*. https://www.gov.uk/government/publications/new-protocol-on-irelandnorthern-ireland-and-political-declaration (Accessed 23 August 2021).

[40] HM Revenue & Customs. (2020) *Moving Qualifying Goods from Northern Ireland to the Rest of the UK*. https://www.gov.uk/guidance/moving-qualifying-goods-from-northern-ireland-to-the-rest-of-the-uk (Accessed 23 August 2021).

[41] Animal & Plant Health Agency and Department for Environment, Food & Rural Affairs. (2019) https://www.gov.uk/government/publications/uk-border-control-posts-animal-and-animal-product-imports (Accessed 23 August 2021).

[42] Food Standards Agency. (2021) *Inland Enforcement of Imported Food Controls Resource Pack*. https://smartercommunications.food.gov.uk/resource/files/2160?scrollPos=0 (Accessed 23 August 2021).

[43] Food Standards Agency. (2021) *Food Law Code of Practice*. https://www.food.gov.uk/about-us/food-and-feed-codes-of-practice (Accessed 23 August 2021).

[44] Cabinet Office. (2020) *The Border Operating Model*. https://www.gov.uk/government/publications/the-border-operating-model (Accessed 23 August 2021).

[45] Animal and Plant Health Agency. (2021) *Border Control Posts*. http://apha.defra.gov.uk/official-vets/Guidance/bip/index.htm (Accessed 23 August 2021).

[46] Food Standard Agency. (2021) *Imports and Exports*. https://www.food.gov.uk/business-guidance/imports-exports (Accessed 23 August 2021).

[47] Food Standards Agency. (2021) *National Monitoring Plan (NMP) Sampling Priorities for 2020–21*. https://smartercommunications.food.gov.uk/communications/files/5015 (Accessed 23 August 2021).

[48] Food and Agriculture Organization of the United Nations. (2021) *The International Food Safety Authorities Network (INFOSAN)*. http://www.fao.org/policy-support/mechanisms/mechanisms-details/fr/c/448741/ (Accessed 23 August 2021).

[49] European Commission. (2021) *RASFF – Food and Feed Safety Alerts*. https://ec.europa.eu/food/safety/rasff-food-and-feed-safety-alerts_en (Accessed 23 August 2021).

[50] European Commission. (2021) *TRACES – About TRACES*. https://ec.europa.eu/food/animals/traces_en (Accessed 23 August 2021).

[51] Animal and Plant Health Agency and Department for Environment, Food & Rural Affairs. (2019) *Imports of Products, Animals, Food and Feed System (IPAFFS)*. https://www.gov.uk/guidance/import-of-products-animals-food-and-feed-system (Accessed 23 August 2021).

[52] Animal and Plant Health Agency. (2021) *Border Control Posts*. http://apha.defra.gov.uk/official-vets/Guidance/bip/index.htm (Accessed 24 August 2021).

[53] Food and Agriculture Organization of the United Nations. (2021) *Illegal, Unreported and Unregulated (IUU) Fishing*. http://www.fao.org/iuu-fishing/en/ (Accessed 24 August 2021).

[54] European Market Observatory for Fisheries and Aquaculture Products. (2020) *The EU Fish Market*. https://www.eumofa.eu/documents/20178/415635/EN_The+EU+fish+market_2020.pdf (Accessed 24 August 2021).

[55] HM Revenue & Customs. (2012) *Guidance: Apply to Operate a Temporary Storage Facility*. https://www.gov.uk/guidance/temporary-storage (Accessed 24 August 2021).

[56] Department for Environment, Food & Rural Affairs and Animal and Plant Health Agency. (2020) *High Risk Food and Feed Not of Animal Origin (HRFNAO): Model Health Certificates*. https://www.gov.uk/government/publications/high-risk-food-and-feed-not-of-animal-origin-hrfnao-health-certificates (Accessed 24 August 2021).

[57] Department for Environment, Food & Rural Affairs. (2020) *Guidance: Importing and Exporting Organic Food*. https://www.gov.uk/guidance/importing-and-exporting-organic-food (Accessed 24 August 2021).

[58] Department of Agriculture, Environment and Rural Affairs. *Introduction to Importing Animals and Animal Products*. https://www.daera-ni.gov.uk/articles/introduction-importing-animals-and-animal-products (Accessed 25 September 2021).

[59] Department for Environment, Food and Rural Affairs and Animal and Plant Health Agency. (2014) *Importing Live Animal, Animal Products and High-Risk Food and Feed Not of Animal Origin from Non-EU Countries to Great Britain.*

https://www.gov.uk/guidance/importing-live-animals-or-animal-products-from-non-eu-countries (Accessed 25 September 2021).

[60] Animal and Plant Health Agency. *Imports of Live Animals and Genetic Material.* http://apha.defra.gov.uk/official-vets/Guidance/bip/iin/live-animals-gene-mat.htm (Accessed 25 September 2021).

[61] Department of Environment, Food & Rural Affairs and Animal and Plant Health Agency. (2015) *Importing Non-Native Animals.* https://www.gov.uk/guidance/importing-non-native-animals#:~:text=You%20must%20apply%20for%20a,importing%20need%20to%20be%20quarantined (Accessed 25 September 2021).

[62] Animal and Plant Health Agency and Department for Environment, Food and Rural Affairs. (2013) *Import or Export Endangered Species: Check if You Need a Cites Permit.* https://www.gov.uk/guidance/cites-imports-and-exports (Accessed 25 September 2021).

[63] Gov.uk. *Bringing Food into Great Britain.* https://www.gov.uk/bringing-food-animals-plants-into-great-britain (Accessed 25 September 2021).

[64] Animal and Plant Health Agency. (2014) *Pet Travel: Checks on Pets by Transport Carriers.* https://www.gov.uk/government/publications/pet-travel-checks-on-pets-by-transport-carriers (Accessed 25 September 2021 – though now withdrawn pending publication of new guidance).

[65] International Air Transport Association. (2021) *Live Animal Regulations (LAR).* https://www.iata.org/en/publications/store/live-animals-regulations/ (Accessed 25 September 2021).

[66] Food Standards Agency. (2021) *Shellfish Classification.* https://www.food.gov.uk/business-guidance/shellfish-classification#classification-protocol (Accessed 25 September 2021).

[67] Food Standards Agency. (2021) *Guidance for Local Action Groups (LAGs) on Handling High E.coli Results, Biotoxin Results and Pollution Events – Classification and Monitoring of Live Bivalve Molluscs.* https://www.food.gov.uk/sites/default/files/media/document/final-local-action-groups-guidance-lbm-sept2021.pdf (Accessed 25 September 2021).

Further reading and sources of information

British International Freight Association (BIFA) https://www.bifa.org/home/#.

Cabinet Office. (2007) *Security in a Global Hub: Establishing the UK's New Border Arrangements,* Cabinet Office, London.

Department of Health. (2002) *Getting Ahead of the Curve – A Strategy for Combating Infectious Diseases (Including Other Aspects of Health Protection),* Department of Health, London.

Import of Products, Animals, Food and Feed System (IPAFFS). https://www.gov.uk/guidance/import-of-products-animals-food-and-feed-system

World Health Organization. (2009) *International Health Regulations 2005 Assessment Tool for Core Capacity Requirements at Designated Airports, Ports and Ground Crossings.* http://www.who.int/ihr/ports_airports/PoE_Core_capacity_assessment_tool.pdf.

22

Environmental health in different situations

Virginia Murray, Stephen Battersby, Lt Colonel James Fawcett and Wing Commander Gary Moyes

SECTION 1: EXTREME EVENTS – THE SENDAI FRAMEWORK FOR DISASTER RISK REDUCTION 2015–2030 AND PLANNING FOR EXTREME WEATHER EVENTS AND OTHER DISASTERS

Virginia Murray, Stephen Battersby

Introduction

The following section describes the fact that disasters destroy lives and livelihoods around the world. It provides a brief summary of the history of UN-based frameworks for disaster risk reduction (DRR), a reflection on the processes leading to these frameworks, and finally focuses on Sendai Framework for Disaster Risk Reduction 2015–2030 [1].[1] It discusses some of the reasons for and importance of having a strong health focus in the Sendai Framework and the benefits of the close relationship that health, including environmental health, has with the science and technology aspects in this framework. It offers ideas on how renewing the global commitment to people's resilience, health, and well-being can be enhanced by the implementation of the Sendai Framework over the next 15 years. By using examples of extreme weather events and other disasters, the principal concepts of and the management of environmental and public health issues are summarised.

The year 2015 presented an unparalleled opportunity to align landmark UN agreements through the convergence of three global policy frameworks: the Sendai Framework for Disaster Risk Reduction 2015–2030 (March 2015), the Sustainable Development Goals (September 2015; SDGs[2] considered in Chapter 1), and the Climate Change Agreements (December 2015; COP21).[3] COP21 (the Paris Agreement) works on a five-year cycle of increasing climate actions carried out by signatory countries. Every five years, each country is expected to submit an updated national climate action plan – known as Nationally Determined Contribution, or NDC. Although each year there is a COP, in 2021 COP26 was held in Glasgow, Scotland and hosted jointly by the UK and Italy as part of the five-year review process of NDCs, and countries are obliged to set out more ambitious goals for reducing emissions. Implementation of the Paris Agreement will

DOI: 10.1201/9781003035640-22

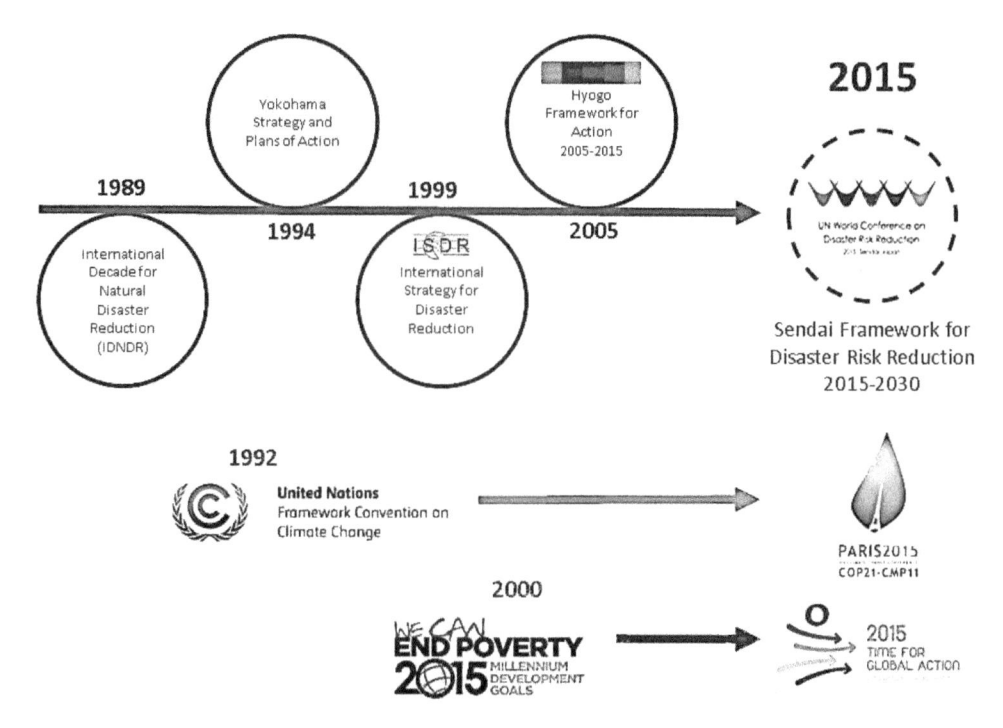

Figure 22.1 Twenty-five years of international commitments to disaster risk reduction

Source: Adapted from presentation by Andrew Maskrey, Lead Author and Head of the Risk Knowledge UNISDR [1][4]

be the key driver of international action and COP26 will review progress.

All these major global policy instruments need to align to facilitate and encourage better participation in disaster risk reduction, sustainable development, and climate-change mitigation and adaptation from the science and technology communities (Figure 22.1).

The Sendai Framework aims to achieve the following outcome: 'The substantial reduction of disaster risk and losses in lives, livelihoods and health and in the economic, physical, social, cultural and environmental assets of persons, businesses, communities and countries'. To do this the Framework seeks as its priorities for action:

- Understanding disaster risk which includes improving data collection on effects of disasters and on mapping of disasters, and encompasses disaster risk modelling, the development and evaluation of

multi-hazard early warning systems and the sharing of non-sensitive data including communications, geo-spatial data, and remote sensing of earth and climate observations. It also supports advocacy and awareness-raising activities;

- Strengthening disaster risk governance to manage disaster risk at national, regional and global levels to facilitate coordination of efforts, efficient use of limited resources and accountability which includes the development and enforcement of building codes, land use and urban planning regulations together with relevant health and safety standards with involvement of community representatives and undertaking community consultations during the development and implementation of legal frameworks to support effective disaster risk governance;
- Investing in disaster risk reduction for resilience; and

- Enhancing disaster preparedness for effective response and to support improved recovery, rehabilitation and reconstruction with the potential to 'build back better' following disasters in order to increase resilience.

The 20 years from 1998 to 2017 saw a 151% increase in direct economic losses from climate-related disasters, according to a report from the UN [2]. In that period, disaster-hit countries reported direct economic losses of US$2,908 billion of which climate-related disasters accounted for US$2,245 billion or 77% of the total. This compared with total reported losses for the period 1978–1997 of US$1,313 billion of which climate-related disasters accounted for US$895 billion or 68%.

In terms of occurrences, climate-related disasters also dominated the picture, accounting for 91% of all 7,255 major recorded events between 1998 and 2017 with floods, 43.4%, and storms, 28.2%, the two most frequently occurring disasters (see also Figure 22.3).

The greatest economic losses have been experienced by the USA, US$944.8 billion; China, US$492.2 billion; Japan, US$376.3 billion; India, US$79.5 billion; and Puerto Rico, US$71.7 billion. Storms, floods, and earthquakes placed three European countries in the top ten for economic losses: France, US$48.3 billion; Germany, US57.9 billion; and Italy, US$56.6 billion. Thailand, US$52.4 billion, and Mexico, US$46.5 billion, complete the list.

During this period 1.3 million people lost their lives and 4.4 billion people were injured, rendered homeless, displaced, or in need of emergency assistance. Some 563 earthquakes, including related tsunamis, accounted for 56% of total deaths or 747,234 lives lost.

Figure 22.2 shows that low and lower-middle income countries carried a disproportionate burden in terms of disaster deaths, experiencing 43% of all major recorded disasters in the past 20 years but the greatest proportion (68%) of fatalities.

Disasters are not natural events. They are endogenous to society and disaster risk arises when hazards interact with the physical, social, economic, and environmental vulnerabilities and exposure of populations [3]. Many of the destructive hazards are natural in origin and include earthquakes and extreme weather events resulting in floods and droughts, which has resulted in disaster risk management policy being largely event driven. Therefore, the attention of the policy community has naturally fallen on the hazards and the related physical processes that result in disasters but is now increasingly linking these with the consequences of climate change (Figure 22.3).

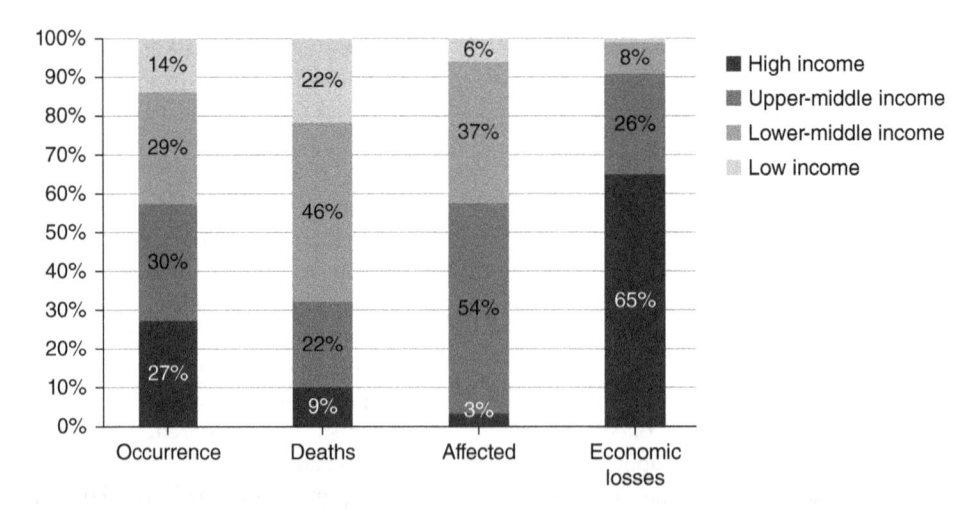

Figure 22.2   Climate-related and geophysical disasters 1998–2017 [2]

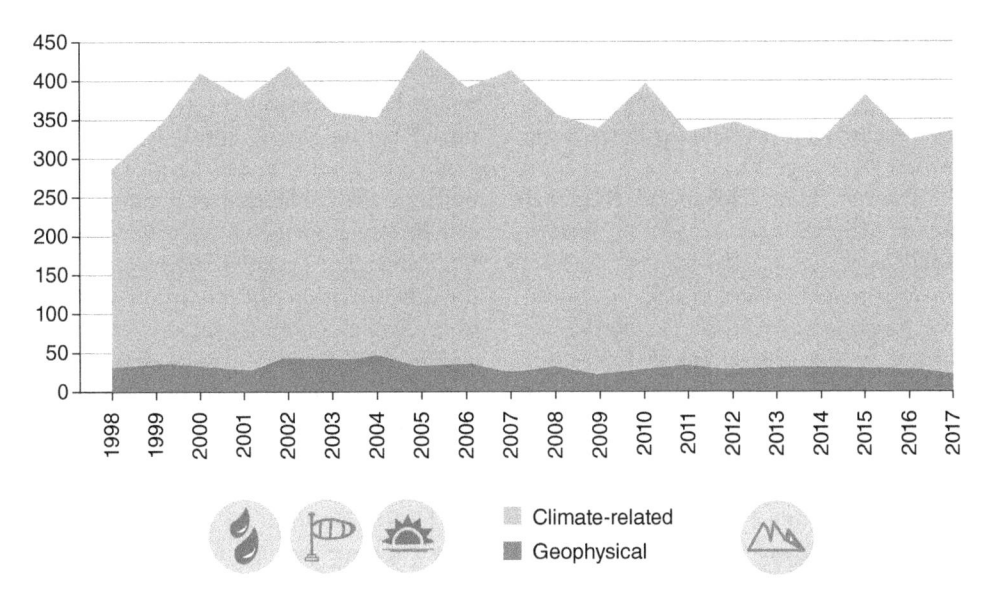

Figure 22.3   Number of disasters by major category per year 1998–2017 [2]

The Sendai Framework was born from the need to ensure DRR policy reflects the evolved understanding of the complexity of disaster risk in the 21st century. Progress in disaster risk reduction (DRR) research has shown that it is often not the hazard that determines a disaster, but the vulnerability, exposure, and ability of the population to anticipate, respond to, and recover from its effects. Implementation calls for closer collaboration among all sectors including the wider health and environmental health sectors in order to prevent, prepare for, respond to, and recover from disasters that result from the highly interdependent and evolving risks.

The following paragraphs from the Sendai Framework include actions required by public health and environmental health partners, which are agreed as priorities for WHO to act on in partnership with the then UNISDR and the UN system as well as local, national, regional, and global partners as relevant:

- In Priority 3: At National and Local Level 30(i) *Enhance the resilience of national health systems, including by integrating disaster risk management into primary, secondary and tertiary health care, especially at the local level; developing the capacity of health workers in understanding disaster risk and applying and implementing disaster risk reduction approaches in health work; and promoting and enhancing the training capacities in the field of disaster medicine; and supporting and training community health groups in disaster risk reduction approaches in health programmes, in collaboration with other sectors, as well as in the implementation of the 2005 International Health Regulations of the World Health Organization* [1] page 19;

- In Priority 3: At National and Local Level 30(j) *Strengthen the design and implementation of inclusive policies and social safety-net mechanisms, including through community involvement, integrated with livelihood enhancement programmes, and access to basic health care services, including maternal, newborn and child health, sexual and reproductive health, food security and nutrition, housing and education, towards the eradication of poverty, to find durable solutions in the post-disaster phase and to empower and assist people disproportionately affected by disasters* [1] at page 19;

- In Priority 3: At National and Local Level 30(k) *People with life threatening and chronic disease, due to their particular needs, should be included in the design of policies and plans to manage their risks before, during and after disasters, including having access to life-saving services* [1] at page 20;

- In Priority 3: At Global and Regional Level 31(e) *Enhance cooperation between health authorities and other relevant stakeholders to strengthen country capacity for disaster risk management for health, the implementation of the International Health Regulations (2005) and the building of resilient health systems* [1] at page 20;
- In Priority 4: At National and Local Level 33(c) *Promote the resilience of new and existing critical infrastructure, including water, transportation and telecommunications infrastructure, educational facilities, hospitals and other health facilities, to ensure that they remain safe, effective and operational during and after disasters in order to provide live-saving and essential services* [1] at page 21;
- In Priority 4: At National and Local Level 33(n) *Establish a mechanism of case registry and a database of mortality caused by disaster in order to improve the prevention of morbidity and mortality* [1] at page 22;
- In Priority 4: At National and Local Level 33(o) *Enhance recovery schemes to provide psychosocial support and mental health services for all people in need* [1] at page 22.

From this summary it is apparent that many of the key issues are focused on environmental and public health practice.

Public and environmental health needs in disasters

During recent decades, the world has faced a greater frequency and impact from disasters as well as a paradigm shift in the types of hazard, and the possible risks that constitute a threat to human well-being, including climate change, rapid and unmanaged urbanisation, lack of resources, poverty, and loss of biodiversity. The expansion of DRR to include risk assessments addressing vulnerability and

exposure has been compared to the widening of health activities to include prevention, which has traditionally been the preserve of public health. Public health is increasingly concerned with the total health system and not only the eradication of a particular disease affecting an individual patient [4]. The US Centers for Disease Control and Prevention link hazards to the transmission of infectious diseases, especially since water supplies and sewerage systems may be disrupted and sanitation and hygiene may be compromised by population displacement and overcrowding that led to interrupted normal public and environmental health services [5].

The five examples are briefly summarised next and these amongst many other extreme weather events and disasters have encouraged the global community to adopt a comprehensive framework for action and identify global priorities for work and practical steps that are required to achieve disaster resilience.

- The 2004 Indian Ocean earthquake and tsunami was historically exceptional in terms of its impact on lives and communities [6]. This disaster illustrated the vulnerability of multiple countries and communities to natural hazards that arise in distant locations.
- During the summer heat-waves of 2003 over 70 000 excess deaths were recorded and hospitals faced a surge in patient numbers, particularly in urban areas of several European countries [7].
- The 2011 Great East Japan Earthquake benefited from the Miyagi earthquake in 1978 when Japan was made aware of the need to strengthen older buildings to fit new building codes. Compliance with these building codes assisted in saving lives in the 2011 earthquake, which recorded a magnitude of 9.0, the strongest ever identified in Japan. Even so this earthquake and its associated tsunami resulted in 15,891 deaths, 6,152 injured, and 2,584 people reported as missing.
- In the Philippines, during Typhoon Haiyan in 2013, the major medical and public

health needs of the affected people were not injury-related, but the result of a lack of measures to prevent infectious diseases and the worsening of non-communicable diseases due to the lack of access to food, water, housing, and medicine [8].

- In Western Africa, the Ebola outbreak (2014—2015) devastated health facilities and people's trust in health care providers. The fragility of the health systems and the lack of resources to manage the isolation and treatment of patients overwhelmed the existing capacity of health care providers and local and national governments. The health disaster resulted in severe budget cuts to non-Ebola-related health services and a significant reduction in the use of health services owing to fears of cross-infection. As a result, more people are estimated to have died from childbirth, malaria, and AIDS, as well as other diseases [9].

Other than epidemics, disaster deaths are rarely due to infectious diseases, instead occurring due to a variety of causes that include blunt trauma, drowning, and air pollution, for instance, from forest fires or building collapses [5]. However, documenting extreme weather and disaster-related deaths remains complex.

Aside from physical injury and infectious diseases, disasters can leave those affected with short- and long-term mental health consequences. Significant changes can occur rapidly in people's lives when they are exposed to extreme events and disasters. These can cause great stress to people, families, and communities because of their inherent effects, such as suffering short-term fear of death and other mental health disorders [10]. Post-traumatic stress disorder (PTSD) is the most often studied manifestation of the psychosocial stress caused by disasters, but mental health impacts also include general distress, anxiety, excessive alcohol consumption, and other psychiatric disorders [11]. The multiple efforts of the health community in the policy

development process, including campaigning for safe schools and hospitals, helped to put people's mental and physical health, resilience, and well-being higher up the DRR agenda.

Those with chronic diseases could have worse outcomes and many risk dying when their medication is not available, or they lack access to health care. People with chronic diseases have ongoing medical needs that can easily be affected when health services are disrupted in disaster situations. While further understanding is required in this area, a systematic review [12] revealed that a considerable number of patients lose their medication during evacuation, many lose essential medical aids such as insulin pens, and many do not even have a record of their prescriptions with them when evacuated.

Understanding disaster risk and risk assessment

The Sendai Framework [1] recommends in paragraph 23 that

> Policies and practices for disaster risk management should be based on an understanding of disaster risk in all its dimensions of vulnerability, capacity, exposure of persons and assets, hazard characteristics and the environment. Such knowledge can be leveraged for the purpose of pre-disaster risk assessment, for prevention and mitigation and for the development and implementation of appropriate preparedness and effective response to disasters.

Understanding risk is essential for effective emergency planning and for this risk assessment is an invaluable first step. Thus, the elements of the risk management cycle include the risk assessment; risk prevention and mitigation emergency management; and recovery issues ranging from business continuity, liability, and compensation to experience feedback which are all key to the process.

In order to understand disaster risk the value of risk assessment is increasingly being used for building national risk registers. Good practice of the implementation and use of these tools benefit from close collaboration between national and local government and its many partners including public and environmental health representatives.

In the taxonomy of threats for complex risk management prepared by the Centre for Risk Studies, Judge Business School, University of Cambridge, a categorisation of disasters has been developed. These include financial shock, trade dispute, geopolitical conflict, political violence, natural catastrophe, climatic catastrophe, environmental catastrophe, technological catastrophe, disease outbreak, humanitarian crisis, external event and other unknown events [13]. Some of these are classically linked to risk assessment but this wide-ranging summary broadens the understanding of hazards which will be increasingly important in risk assessment.

Guidance on risk assessment has been provided in particular by the EC and Organisation for Economic Co-operation and Development (OECD). In 2010 the European Commission developed Risk Assessment and Mapping Guidelines for Disaster Management. The main purpose of these guidelines was to improve coherence and consistency among the risk assessments undertaken in the Member States at national level in the prevention, preparedness and planning stages and to make these risk assessments more comparable between Member States. Coherent methods for national risk assessments will support a common understanding in the EU of the risks faced by Member States and the EU and will facilitate co-operation in efforts to prevent and mitigate shared risks, such as cross-border risks [14].

OECD has developed a number of reports on risk assessments and, in 2014, with its Member States, issued a report on the recommendation on the governance of critical risks which highlights the key role of national leadership to set up a national risk management strategy and to establish the structure to foster whole of society engagement in risk

governance [15]. In this report it states that as a recommendation, countries should '*establish and promote a comprehensive, all-hazards and transboundary approach to country risk governance to serve as the foundation for enhancing national resilience and responsiveness*'. In addition, the OECD has a Summary Table of Available Tools for Risk Assessment on chemicals which address hazard and exposure assessments resources [16].

Recognising that most countries continue to have difficulties integrating risk reduction into public investment planning, urban development, spatial planning and management, and social protection, the European Commission (EC) has promoted the use of risk assessment and analysis. An EC report stated that data derived from past losses are useful for the implementation of disaster risk reduction strategies in Europe (from local to national scales) and to help understand wider disaster loss trends at global level [17]. It went on to state that to improve risk assessment and forecast methods, for which loss datasets are needed for calibrating and validating model results in particular to infer vulnerabilities, risk modelling is valuable.

By building on the available national risk assessments, the EC has prepared the first cross-sectoral overview of risks in the EU, taking into account (where possible and relevant) the future impact of climate change and the need for climate adaptation; following a consistent approach, multi-hazard national risk assessments were produced by Member States by end 2015 and followed up by assessment of national risk management capabilities and improved risk management planning [18]. The EC reported that of the 32 participating countries to the Mechanism for Civil Protection, 18 contributed to the review through their national risk assessments (NRAs) [19]. Of these contributions, nine Member States provided information on their national assessment criteria and scenario-building. It was considered that more systematic and complete information on the assessment criteria and on the risk scenarios assessed may help the Commission carry out an informed and coherent analysis of risks addressed in NRAs [20]. The UK was included at the time.

In the UK the Cabinet Office is responsible for the annual publication of the National Risk Register. Risks in the National Security Risk Assessment (NSRA) and the National Risk Register (NRR) are represented as 'reasonable worst-case scenarios'. This means that they represent the worst plausible manifestation of that particular risk (once highly unlikely variations have been discounted) to enable relevant bodies to undertake proportionate planning. They are assessed in terms of likelihood and impact and then plotted onto a matrix. Instead of plotting each individual risk onto the matrix, a number of risks have been thematically grouped, bringing together risks that share similar risk exposure and require similar capabilities to prepare, mitigate and respond. This is partly to bring similar risks together in a more usable way but is also due to the sensitivity of some of the risks assessed in the NSRA. The position of each risk category on the matrix is an average based on the positions of all the different risks that belong to that category.

As the 2020 edition of the NRR[5] says, it is the public-facing version of the National Security Risk Assessment (NSRA) – a classified cross-government and scientifically rigorous assessment of the most serious risks facing the UK or its interests overseas. The Civil Contingencies Secretariat, which is part of the Cabinet Office, is responsible for co-ordinating the production of both documents. This involves working closely with stakeholders including other UK government departments, devolved administrations, the government scientific community, intelligence and security agencies, and a range of independent experts such as industry partners and academics. The NRR provides public information on the most significant risks that could occur in the next two years and which could have a wide range of impacts on the UK. The NRR [21] also sets out what the

UK government, devolved administrations and other partners are doing about them. This public resource is for individuals and organisations wishing to be better prepared for emergencies. The 2020 NRR includes a broader range of risks including serious and organised crime, hostile state activity, antimicrobial resistance and major fires. The 2020 NRR assessments for pandemics and High Consequence Infectious Disease outbreaks do not include COVID-19 as at the time of writing, this was considered a live issue but a dedicated case study for COVID-19 is included.

The Civil Contingencies Act 2004 definition of civil emergency risks are those that could directly and significantly damage human welfare or the environment somewhere in the UK, but not events that happen overseas unless they directly affect the UK. These are taken into account in the NRR.

The definition of an emergency when assessing the expected consequences of an emergency are as follows:

- The number of fatalities that are directly attributable to the emergency;
- Illness or injury over the period following the onset of the emergency;
- Levels of social disruption to people's daily lives, from an inability to gain access to health care or schools to interruptions in supplies of essential services such as food, water, and fuel, and the need for evacuation of individuals from an area;
- Economic harm – the effect on the economy overall, rather than the cost of repairs;
- The psychological impact that emergencies may have, including widespread anxiety, loss of confidence or outrage that communities may experience.

In Figure 22.4 the vertical axis shows the impact of each risk. Level A is the lowest impact, and Level E the highest. The horizontal axis shows the annual likelihood of each risk. The likelihood range in each column, moving from left to right, is five times

Impact (of the reasonable worst case scenario using the impact indicators below)

Level E		7 25†			
Level D	34*	12 13 29			
Level C	18 28 33* 36*	14 19 21 26† 27* 38	2 3 6* 15 16 17 20		
Level B	30	24	35*	4 5 9* 10* 11* 23 32* 37	1
Level A			8* 22	31	
	< 1 in 500	1 to 5 in 500	5 to 25 in 500	25 to 125 in 500	> 125 in 500

Likelihood
(of the reasonable worst case scenario of the risk occurring in the next year)
*Risk not plotted in the 2017 NRR | †COVID-19 is not included in the risk matrix and is therefore not inclued in these risks

Malicious Attacks
1. Attacks on publicly accessible locations
2. Attacks on Infrastructure
3. Attacks on transport
4. Cyber attacks
5. Smaller scale CBRN attacks
6. Medium scale CBRN attacks
7. Larger scale CBRN attacks
8. Undermining the democratic process*

Serious and Organised Crime
9. Serious and organised crime - vulnerabilities*
10. Serious and organised crime - prosperity*
11. Serious and organised crime - commodities*

Environmental Hazards
12. Costal flooding
13. River flooding
14. Surface water flooding
15. Storms
16. Low temperatures
17. Heatwaves
18. Droughts
19. Severe space weather
20. Volcanic eruptions
21. Poor air quality
22. Earthquakes
23. Environmental disasters overseas
24. Wildfires

Human and Animal Health
25. Pandemics†
26. High consequence infections disease outbreaks†
27. Antimicroblal resistance*
28. Animal diseases

Major Accidents
29. Widespread electricity failures
30. Major transport accidents
31. System failures
32. Commercial failures*
33. Systematic financial crisis*
34. Industrials accidents - nuclear*
35. Industrials accidents - non nuclear*
36. Major fires*

Societal Risks
37. Industrial action
38. Widespread public disorder

Figure 22.4 The 2020 UK National Risk Register by overall relative impact score [21]
Source: UK Cabinet Office © Crown copyright 2020

greater than the previous column. For example, a column three risk is approximately five times more likely to occur than a column two risk. The likelihood scale is logarithmic and is reflected by the matrix boxes increasing in size, moving from the bottom left of the matrix to the top right. Risks have been plotted, using the key to the right of the matrix, and grouped under headings to reflect the chapters in the NRR.

The register demonstrates the different approach to the assessment of likelihood for hazard and threat risks; hazards and threats are shown on two separate risk matrices. The two scales are not directly comparable with one another; however, for the purposes of planning, a hazard or threat in the top right quadrant of either matrix would be given the same priority.

Early warning

The Sendai Framework recommends as one of its global targets to be implemented by 2030 and is summarised in paragraph 18 (g): '*Substantially increase the availability of and access to multi-hazard early warning systems and disaster risk information and assessments to people by 2030*' [1].

An example of early warning is the Natural Hazards Partnership (NHP) which brings together expertise from across the UK's leading public sector agencies with the aim of drawing upon scientific advice in the preparation, response and review of natural hazards. The Natural Hazards Partnership (NHP) is a collaboration of government departments and bodies from across the UK. The NHP aims to establish a forum for the exchange of natural hazards knowledge, ideas, expertise, intelligence and best practice; provide timely and consistent advice to government and emergency responders for civil emergencies and disaster response; and develop new services to assist in disaster response preparedness [22].

The partnership enables more coordinated and coherent advice for the government and the resilience community.

- Establish a forum for the exchange of knowledge, ideas, expertise, intelligence and best practice in relation to natural hazards;
- Provide a timely, common and consistent source of advice to government and emergency responders for civil contingencies and disaster response;
- Create an environment for the development of new services to assist in disaster response.

The NHP publishes a UK-wide Daily Hazard Assessment[6] providing an at-a-glance summary of a range of natural hazards including floods, geological hazards such as landslides, weather (including extreme temperatures, snow, ice, and fog) air quality, wildfire, volcanic ash, and space weather [22]. It includes the Hazard Matrix to assess 12 hazards with hazard forecasts up to five days.

Projects considering how to develop impact-based forecasting are now being developed. The World Meteorological Organisation reports scientific information is available via National Meteorological and Hydrological Services but often it does not reach responders or communities at risk in the right format and in a manner that is understandable for disaster risk reduction/management decision-makers and operational responders, as well as communities at risk in a way that can prepare and protect themselves. Interpretation of the scientific analysis and appropriate communication is key. Therefore, there is seen to be a need to strength multi-hazard early warning systems to bring out the following points:

- Definitions of hazards;
- Translation of weather into related hazards;
- The fact that non-weather hazards may have similar impacts as weather hazards hence the need for developing partnerships between hazards communities; and
- The effects of interdependency between different hazards, since interdependency could lead to similar or increased impacts. [23]

As a result, there is now an intention to potentially develop systems to support the aim to develop an impact-based forecasting initiative via the WMO Severe Weather Forecasting Demonstration Project, the Flash Flood Guidance System, and the WMO Coastal Inundation Forecasting Demonstration Project [23]. The Coastal Inundation Forecasting Demonstration Project (CIFDP) has been developing and implementing a Multi-Hazard Early Warning System (MHEWS) for coastal flooding, from both rivers and the ocean. Since its establishment in 2009, three demonstration projects have been completed – Bangladesh (2017), Caribbean (2018), Indonesia (2019) – and a fourth, Fiji, was on track for completion by the end of 2019.

Extreme weather events and other disasters

Extreme weather and other disasters can have significant impacts on population health. The magnitude of these impacts depends not only on the severity of the event (extent of flooding or intensity of a heatwave) but also on the affected communities' effective exposure, vulnerability and capacity to respond. In turn, these are determined by various underlying factors, including the physical setting and built landscape in the area of occurrence; the health and disability status of the population; and institutional and public preparedness. Although the Sendai Framework addresses an all-hazard approach, many across the world are periodically affected by various types of extreme weather events – principally heatwaves, cold weather, flooding, droughts, windstorms, and forest/wildfires – which impose heavy human costs.

The *Special Report on Managing the Risks of Extreme Events and Disasters to Advance Climate Change Adaptation* (SREX Report) published by the Intergovernmental Panel on Climate Change [24] in 2012 evaluated climate change's role in altering the characteristics of extreme events. IPCC projections for the 21st century indicate that:

- Frequency, duration and/or intensity of high temperatures and heatwaves are likely to increase;
- Heavy precipitation is likely to increase, leading to increases in local flooding in some areas;
- Droughts will intensify in central and southern Europe, the Mediterranean area and central Asia;
- Extreme coastal high-water levels are very likely to increase in height as a result of sea-level rises, worsening coastal erosion and inundation;
- Changes in heatwaves, glacial retreat and heavy precipitation will likely affect high-mountain phenomena with potential consequences such as landslides, glacial-lake outbursts and floods.

Moreover, the IPCC considers that these changes in events will have greatest impact on the sectors most directly dependent on climate, including food security and water supply. In turn, these strongly affect some of the most basic determinants of health, such as adequate nutrition and safe water.

Effective management of extreme weather and climate risks involves a portfolio of measures to reduce and transfer risk and to respond effectively to events. The SREX Report emphasises low-regret measures that could reduce the current impacts of early warning systems in Europe whilst laying the foundation for addressing projected trends. These actions include:

- Implementation of early warning systems;
- Improvements in health surveillance;
- Improvements in water supplies and drainage systems;
- Changes in land-use planning;
- Development and enforcement of building codes;
- Climate-proofing of infrastructure;
- Ecosystem management; and
- Better education and awareness.

Other good practices include the use of multi-hazard risk management approaches focusing on resilience to changing risks; targeting of vulnerable areas and groups; appropriate and timely risk communication; and better documentation of extreme events to expand the existing disaster risk management evidence base.

Extreme events and disasters address many types of events and these include heatwaves, cold weather, floods, drought, wildfires, earthquakes, tsunamis, volcanic eruptions, landslides and other mass movements, windstorms (cyclones, typhoons, and hurricanes), vectors and algal blooms, plant infections such as Oak Processionary Moth, thunderstorms and impacts such as thunderstorm asthma, impacts from space weather related to solar mass ejections, and many others. Often these events can occur in cascading ways exacerbating health and environmental impacts. Next are summaries of health impacts of heatwaves, cold weather, and flooding.

Heatwaves – are unusually hot weather. For operational purposes it can be defined as a period in which the daily maximum apparent temperature and the daily minimum temperature exceed the 90th percentile of the monthly temperature distribution for at least two days. Heatwaves can have a severe impact on health. During the summer heatwaves of 2003 over 70 000 excess deaths were recorded and hospitals faced a surge in patient numbers, particularly in urban areas of several European countries. The 2003 heatwaves were exceptionally deadly but intense heat affects Europe every year including in 2015. During the summer heatwave in Northern France in August 2003, unprecedentedly high day- and night-time temperatures for a period of three weeks resulted in 15 000 excess deaths. The vast majority of these were among older people.

Global heating is expected to cause more frequent and intense heatwaves in many regions of the world, including Europe. This trend, combined with ongoing population growth and ageing, means that heat-related deaths and illness are expected to increase in the European Region.

Heat exposure can cause severe dehydration and acute cerebrovascular accidents. It can contribute to the formation of blood clots and can further aggravate chronic pulmonary conditions, cardiac conditions, kidney disorders and psychiatric illness. Heatwaves can result in deaths from heatstroke (fatal in 10–50% of cases), dehydration and other cardiovascular and respiratory disorders. Furthermore, high temperatures worsen the harmful effects of air pollutants.

The health effects of hot weather depend on the frequency, severity, and duration of heatwaves and on the sensitivity of the individuals and communities. Those at increased risk of heat-related health problems include infants; elderly people; people who are socially isolated; patients with chronic diseases; and patients who are bed-bound, have limited mobility, or take certain medications. There is a short lag time between the occurrence of a heatwave and the onset of its health consequences – an increase in deaths may be observed within two to three days. However, heatwave-related deaths are often underreported.

For heatwave prevention, preparedness, response, and follow-up, information sheets on extreme weather, climate change, and health are of import and should consider:

- Accurate and timely heat – health warning systems.
- Heat-related health information plan such as the Public Health England Heatwave Plan for England [25].
- Measures to reduce indoor heat exposure such as provision of information on adaptation of individual behaviour (e.g. spending time in outdoor green spaces); measures to reduce indoor temperatures (e.g. low-power passive cooling through use of external shading including shutters, radiant barriers, and insulation of buildings); provision of cool rooms in care homes and other institutions; and, in the

longer term, improved urban planning, building design, and transport policies.

- Care for vulnerable population groups by identifying high-risk groups before the summer, and to plan and target interventions (advice, follow-up, and care) accordingly. Community organisations, medical practitioners and care providers play important roles in outreach to vulnerable groups.
- Preparedness of health and social care system which includes summer workforce planning and training of personnel. In the absence of specific guidance, care homes and hospitals may aim to meet the criteria for the thermal indoor environment in order to prevent heat-related illness in patients and staff. Alerting emergency departments to the onset of a heatwave enables them to prepare for increases in patient numbers. Preparedness is also essential at the individual as well as organisational level.[7]
- Long-term urban planning measures to address building design as well as energy and transport policies that will ultimately reduce heat exposure.
- Real-time health surveillance and evaluation is key as surveillance is important to detect early impacts of hot weather, with monitoring and evaluation to evaluate all phases of heat-health action plans.

Cold weather – severe cold weather is common across many countries in the European Region. This can have significant health effects as a result of low outdoor and indoor temperatures, ice, and snow. Several thousand excess deaths occur across Europe every winter, with mortality showing a greater increase in the warmer Mediterranean area than in colder northern European countries [26]. In addition, total mortality in most populations is higher in winter than in summer. Climate change (or global heating) is expected to result in a reduction of cold days but cold weather and cold wave events will still occur and their health effects need to be anticipated adequately and prevented.

Many complaints and symptoms may arise during cold weather. These include respiratory, cardiovascular, and circulatory disorders; muscle pain; and skin problems. In addition to these, cold exposure can cause injuries such as fractures and sprains from falls and accidents, as well as hypothermia, frostbite, and depression. The most common causes of cold-related deaths are coronary heart disease, cerebrovascular accidents, respiratory diseases, and hypothermia. There is a short lag time between the onset of cold weather and its health consequences – mortality rates can increase within 24 hours – but the health impacts can last for some time. The adverse health effects of cold weather are largely preventable but, given the short lag time between the onset of cold weather and its health effects, effective management is highly dependent on planning and preparedness.

Cold-related health effects can affect anyone, but some people are at greater risk. Those most vulnerable during cold spells include infants; children; teenagers; and people who are elderly, have chronic diseases, or physical or mental limitations, take certain medications or are malnourished. The economic and social status of individuals and families also plays an important role: people who are on low incomes, homeless or marginalised and suffer energy precarity are affected more severely by cold weather. Some occupations, such as agriculture, fishing, and construction, may involve greater exposure to cold. These workers must be vigilant for injuries and other health effects and wear adequate protective gear.

Certain behaviour – excess use of alcohol, some outdoor leisure activities, and inappropriate clothing – can put people at increased risk from cold exposure. Another frequent concurrent risk during cold weather is household exposure to pollution from solid fuels or burners that are not well designed for indoor use. Poisoning by carbon monoxide is a particularly severe and often fatal risk. Prolonged exposure to cold housing appears to play an important role in some countries.

In addition, snow and ice associated with cold weather can cause disruptions to health services and infrastructure and increase the risk of falls and unintentional injuries outside the home. This can lead to disturbance of the provision of routine medical and nursing care of patients with chronic diseases, and indirectly can further aggravate the health effects of cold.

For cold weather health impact prevention, preparedness, response, and follow-up, plans should consider:

- Health-related cold weather plans, policies, and strategies form the basis of good governance for public health management of extreme weather. It is important that plans are developed in advance and in coordination with other sectors such as social services, housing, energy, transport, and education – such as the UKHSA/NHS Cold weather Plan for England.[8]
- Cold weather early-warning systems developed in collaboration with meteorological services, can provide alerts which prompt agreed actions at predetermined thresholds, enabling communities and health actors to avoid or reduce the risks of emergencies – national systems already exist in many countries.
- Health system preparedness includes adequate logistics systems for transportation, telecommunications, medicines, and supplies, as well as winter workforce planning, information campaigns, and flu vaccination programmes for staff and the public, among other measures. Making hospitals and health facilities safer, prepared, and resilient protects patients, staff, and investments. Public health services can work with social service providers to identify high-risk groups and individuals in advance of cold weather.
- Multi-sector cold weather preparedness requires that a wide range of services play a fundamental role in health emergency risk management systems, and are therefore crucial for reducing exposure to cold among communities and individuals. In terms of policies and strategies, building standards for housing design, energy efficiency, and insulation can reduce indoor cold exposure in the long term. Tariffs and social-support policies to reduce the cost of energy have been applied in a variety of countries.
- An information and knowledge management strategy, specifically addressing public awareness and preparedness before a cold wave. Communities need information on the health effects of cold, how to prepare for winter, and how to protect themselves and others.
- Surge capacity means that health services should be ready to respond to likely health problems and a surge in patient numbers, and to ensure continued service delivery under any emergency, including those related to cold.
- Robust surveillance of morbidity, mortality, and health service usage during cold weather allows health service provision and public health action to be tailored to the needs of the population as a whole and to vulnerable groups in particular.
- Coordinated multi-sector cold weather responses are necessary to ensure effective preparedness, response, and follow-up to cold weather and cold waves. For instance, while health services provide additional medical support to those at risk, social services may provide additional care, warm meals, clothing, and heated shelter.

Flooding – flooding can cause loss of life and illness (including stress) as well as damage to property, infrastructure, crops, and industry, resulting in significant economic losses. It is possible to classify the several types and causes of floods by their spatial and temporal scale: heavy rainfall can cause rapid-onset flash floods, slow-onset fluvial (riverine) floods, and pluvial (surface water) floods. Storm surges and tsunami waves can cause coastal flooding. Other floods result from snowmelt, ice, or debris jams, and dam failures. Flooding is the most common natural disaster in the European Region: floods killed more than 1000 people and affected 3.4 million in Europe between 2000 and 2009. It is projected that climate change (global heating) will increase the frequency and intensity of heavy precipitation events. In the summer of 2021 at least 170 people died in Germany and Belgium as the result of flooding.[9] In the absence of any countermeasures, river flooding may affect an additional 250 000–400 000 people per year in Europe by the 2080s. Rises in sea level and storm surges, which cause coastal flooding, could affect millions more people.

Floods increase mortality and illness. Two-thirds of deaths associated with flooding result from drowning; one-third result from physical trauma, heart attacks, electrocution, carbon monoxide (CO) poisoning, and fire. Ill health associated with floods is mostly due to injuries, infections, chemical hazards and mental health problems (both acute and delayed); outbreaks of infectious diseases following floods are rare in Europe but are more common in other parts of the world. The longer-term health effects associated with flooding are less easily identified but thought to be significant. They include effects due to displacement, destruction of homes, delayed recovery, and water shortages.

Known risk factors for flood-related death and ill health are fast-flowing water; hidden hazards; water of unknown depth; driving and walking through flood water; flood-water contamination (by chemicals, sewage, and residual mud); exposure to electrical hazards during recovery and cleaning including from fallen power lines; unsafe drinking-water; food shortages and contamination; incomplete routine hygiene; and lack of access to health services.

Flooding of health facilities can result in loss of infrastructure and utilities such as water supply and electrical power at a time of increased patient admissions. Health systems can have increased difficulty in providing the routine medical and nursing care required by patients with chronic diseases such as diabetes, renal failure, cystic fibrosis, cancer, and mental illness.

Planning, prevention, and preparedness for the next flood is invaluable – this includes:

- Formulation and implementation of multi-sector emergency plans that integrate health protection as a priority and involve the health sector from inception. Clear information and communication strategies and planning for water and food supplies, evacuation and refuge areas are important topics to be addressed. As with other emergencies, the health sector should work with emergency response agencies to ensure adequate protection of health facilities and a robust health response. Plans should place emphasis on the information and knowledge management strategy, specifically addressing public awareness and preparedness.
- Flood risk forecasts are a key tool for effective health protection from flooding. Many national meteorological services already routinely prepare such forecasts.
- Identification of vulnerable groups or high-risk populations including those at risk of flooding allows more targeted communications and evacuation plans to be put in place in advance of floods.
- Hospitals can be made safe from disasters through physical reinforcement and supply-chain resilience. Mutual aid agreements between regions should be considered in advance of any flooding. New hospitals should be built to withstand local hazards and protect patients and staff and be able to provide health services in the aftermath of an emergency. Starting from the design phase, health infrastructure can be made more resilient if planning and retrofitting routinely incorporates climate change projections. Making hospitals and health facilities safer, prepared, and resilient protects patients, staff, and investments is beneficial and complies with the Sendai Framework for Disaster Risk Reduction (UNISDR 2015).

- Coordinated public health and health systems response to flooding at all levels in the public health management of floods is generally accepted, although the specifics of any response are dependent on each situation. Typical public health flood responses include search and rescue; infectious outbreak detection and control; treatment of injured, traumatised, and displaced people; and care for chronically ill individuals. For EHPs, as well as contributing to the previous issues and the emergency response, there is also the increased risk of increased surface rat infestations where sewers have surcharged. The flow of information to and from the public is, like in every public health emergency, crucial. Responses to flooding should be multi-sectoral, with collaboration between emergency services, government, utility companies (i.e. power and water), and health services.
- Ensuring availability of basic services such as clean drinking water, food, and safe shelter – needs assessments should be used to ensure fair and quick provision areas to which EHPs can make a major contribution. This includes recovery and rehabilitation of flooded houses.
- Surge capacity and continuity of operations with health services needing flexible resource allocation and mobility in order to be able to adjust their response capacity to likely health problems and surges in patient numbers. The needs of patients with chronic disease during floods can be met through pre-established continuity of operations mechanisms. Protection of hospitals and other care providers should be a priority throughout response operations.
- Robust surveillance is needed during and after flood events to identify and control infectious disease outbreaks and non-infectious health hazards; tailor health service provision to the needs of the population; and monitor vulnerable groups.
- Long-term flood risk management relies on action from multiple sectors, not the health sector alone. Structural measures include flood defences and barriers, improved flood resistance of dwellings, drainage system or dike creation, and tree planting and improved upland management to reduce run-off.[10] Key non-structural measures include land-use management policies that account for flood risk, careful urban planning, and provision of affordable flood insurance.

Conclusion

This section has focused on the significant global agreements from 2015 and developments thereafter, particularly the Sendai Framework for Disaster Risk Reduction 2015–2030, and the impact these global agreements have on environmental public health risk assessment and their application in the management of disasters using extreme weather events and other natural hazards as examples.

It covers such topics as disaster risk reduction, risk assessment, possible fates of a toxin in the body, and has identified some of the vulnerable and susceptible populations. The applications of the source–pathway–receptor concept are discussed. The value of using these concepts in chemical incident management are summarised and finally focused by considering case studies, namely an acute carbon monoxide poisoning and organophosphate exposure. In that regard although a separate chapter this is a companion section to the second section of Chapter 10.

Acknowledgements

The authors are grateful to

- Andrew Maskrey at the United Nations Office for Disaster Risk Reduction (UNISDR), for approving the use of a modified version of a figure he designed;
- UNISDR (UNDRR) for using their figures on disaster impacts 2000–2012 by economic cost, people affected and numbers killed and the number of

climate-related disasters around the world between 1980 and 2011.

- To colleagues in the UKHSA Extreme Events and Health Protection team

Section 1 Notes

1 See http://www.preventionweb.net/files/43291_sendaiframeworkfordrren.pdf

2 https://sustainabledevelopment.un.org/?page=view&nr=1064&type=13&menu=1634 The links between Sendai Framework targets and Sustainable Development Goals are set out in Fig 1 of Wright N, Fagan L, et al. (2020) Health emergency and disaster risk management: Five years into implementation of the Sendai Framework. *International Journal of Disaster Risk Science*, 11: 206–217.

3 The UN Framework Convention on Climate Change (UNFCCC) entered into force in 1994 the 21st Conference of the Parties (COP21) or the Paris Climate Conference led to a new international climate agreement and came into force in 2016 when the threshold for entry into force was achieved. The COP is the supreme decision-making body of the UNFCCC and all States that are Parties to the Convention are represented at the COP, at which they review the implementation of the Convention and any other legal instruments that the COP adopts and take decisions necessary to promote the effective implementation of the Convention, including institutional and administrative arrangements.

4 The United Nations International Strategy for Disaster Reduction (UNISDR) is now known as the United Nations Office for Disaster Risk Reduction (UNDRR) but in this section both terms are used.

5 https://assets.publishing.service.gov.uk/government/uploads/system/uploads/attachment_data/file/952959/6.6920_CO_CCS_s_National_Risk_Register_2020_11-1-21-FINAL.pdf

6 See http://www.naturalhazardspartnership.org.uk/products/dha/

7 Natural hazards negatively impact on public health. Consequently, continuous research on disaster preparedness and disaster risk reduction is vital to reduce the detrimental effect on well-being. Disaster preparedness strategies have thus far been unsuccessful in preparing individuals for the psychological stress that a natural hazard threat or impact can cause. See https://www.undrr.org/publication/psychological-preparedness-natural-hazards-improving-disaster-preparedness-policy-and and https://www.undrr.org/news/quality-preparedness-determines-quality-response

8 See UKHSA, NHS, LGA, Met Office (2021) *The Cold Weather Plan for England: Protecting Health and Reducing Harm from Cold Weather.* https://assets.publishing.service.gov.uk/government/uploads/system/uploads/attachment_data/file/1031106/UKHSA_Cold_Weather_Plan_for_England.pdf

9 https://www.abc.net.au/news/2021-07-18/europe-floods-aftermath-death-toll-germany-belgium/100302586

10 For example, 'Slowing the Flow' at Pickering North Yorkshire is a Defra pilot project exploring how natural measures can help manage flooding. It features a number of innovative, cost-effective techniques for storing more water in the landscape and reducing its rate of flow downstream – see https://ice.org.uk/knowledge-and-resources/case-studies/slowing-the-flow-at-pickering

Section 1 references

[1] UNISDR (United Nations International Strategy for Disaster Reduction). (2015) *Sendai Framework for Disaster Risk Reduction 2015–2030.* http://www.preventionweb.nct/files/43291_sendaiframeworkfordrren.pdf (Accessed August 2021).

[2] Centre for Research on the Epidemiology of Disasters (CRED). (2018) *Economic Losses, Poverty and Disasters 1998–2017*, UN Office for Disaster Risk Reduction (UNDRR), Geneva, Switzerland. https://www.unisdr.org/files/61119_credeconomiclosses.pdf (Accessed 8 August 2021).

[3] UNISDR (UN International Strategy for Disaster Reduction). (2013) *Proposed Elements for Consideration in the Post-2015 Framework for Disaster Reduction by the UN Special Representative of the Secretary – General for Disaster Risk Reduction.* http://www.preventionweb.net/files/35888_srsgelements.pdf (Accessed August 2021).

[4] Murray V, Aitsi-Selmi A, Blanchard K. (2015) The role of public health within the United Nations post-2015 framework for disaster risk reduction. *International Journal of Disaster Risk Science*, 6(1): 28–37.

[5] Lin HC, Hochberg NS. (2020) Preparing international travelers. In *Chief Centers for Disease Control and Prevention*, Brunette GW (Ed.), CDC Yellow Book 2020: Health Information for International Travel, New York. https://wwwnc.cdc.gov/travel/yellowbook/2020/

preparing-international-travelers/the-pretravel-consultation (Accessed August 2021).

[6] Rodriguez H, Wachtendorf T, Kendra J, Trainor J. (2006) A snapshot of the 2004 Indian Ocean tsunami: Societal impacts and consequences. *Disaster Prevention and Management*, 15(1): 163–177.

[7] Robine J-M, Cheung SLK, Le Roy S, Van Oyen H, Griffiths C, Michel J-P, et al. (2008) Death toll exceeded 70,000 in Europe during the Summer of 2003. *Comptes Rendus Biologies*, 331(2): 171–178.

[8] Egawa S. (2015) *The Second Report of IRIDeS Fact Finding Mission to Philippines*, 87–106. http://irides.tohoku.ac.jp/media/files/IRIDeS_Report_Haiyan_second_20150302.pdf (Accessed August 2015).

[9] Walker PG, White MT, Griffin JT, Reynolds A, Ferguson, NM, Ghani AC. (2015) Malaria morbidity and mortality in Ebola-affected countries caused by decreased healthcare capacity, and the potential effect of mitigation strategies: A modelling analysis. *Lancet Infectious Disease*. https://doi.org/10.1016/S1473-3099(15)70124-6.

[10] Williams R, Drury J. (2011) Personal and collective psychosocial resilience: Implications for children, young people and their families involved in war and disasters. In *Children, and Armed Conflict*, Cook D, Wall J, Cox P (Eds.), Palgrave Macmillan, Basingstoke and New York.

[11] Neria Y, Shultz JM. (2012) Mental health effects of hurricane sandy: Characteristics, potential aftermath, and response. *JAMA*, 308(24): 2571–2572.

[12] Ochi S, Hodgson S, Landeg O, Mayner L, Murray V. (2014) Disaster-driven evacuation and medication loss: A systematic literature review. *PLoS Currents Disasters*. https://doi.org/10.1371/currents.dis.fa417630b566a0c7dfdbf945910edd96.

[13] Coburn AW, Bowman G, Ruffle SJ, Foulser-Piggott R, Ralph D, Tuveson M. (2014) *A Taxonomy of Threats for Complex Risk Management*, Cambridge Risk Framework series; Centre for Risk Studies, University of Cambridge.

[14] European Commission (EC). (2010) *Commission Staff Working Paper Risk Assessment and Mapping Guidelines for Disaster Management*, Brussels, 21 December, SEC (2010) 1626 final. https://ec.europa.eu/echo/files/about/COMM_PDF_SEC_2010_1626_F_staff_working_document_en.pdf (Accessed August 2021).

[15] OECD (Organisation for Economic Co-operation and Development). (2014) *Recommendation of the Council on the Governance of Critical Risks*. http://www.oecd.org/mcm/C-MIN(2014)8-ENG.pdf (Accessed August 2021).

[16] OECD (Organisation for Economic Co-operation and Development). (2015) *Summary Table of Available Tools for Risk Assessment*. http://www.oecd.org/chemicalsafety/risk-assessment/summarytableofavailabletoolsforriskassessment.htm (Accessed August 2021).

[17] De Groeve T, Poljansek K, Vernaccini L. (2013) *Recording Disaster Losses: Recommendations for a European Approach*, Publications Office of the European Union, Scientific and Technical Research Reports EUR 26111. ISBN:978-92-79-32690-5. https://doi.org/10.2788/98653; http://publications.jrc.ec.europa.eu/repository/handle/111111111/29296 (Accessed August 2021).

[18] European Commission. (2014a) Communication from the commission to the European parliament, the council, the European economic and social committee and the committee of the regions. *The Post 2015 Hyogo Framework for Action: Managing Risks to Achieve Resilience*, 8 April, COM 216 final. http://ec.europa.eu/echo/files/news/post_hyogo_managing_risks_en.pdf (Accessed August 2021).

[19] European Commission. (2014b) Commission staff working document: Overview of natural and man-made disaster risks in the EU accompanying the document communication from the commission to the European parliament, the council, the European economic and social committee and the committee of the regions. *The Post 2015 Hyogo Framework for Action: Managing Risks to Achieve Resilience*, {COM(2014) 216 final}, 8 April, {SWD (2014) 133 final}, SWD (2014) 134 final. http://www.sos112.si/slo/tdocs/eu_risks_overview.pdf (Accessed August 2021).

[20] European Commission. (2014c) *Commission Staff Working Document. Overview of Natural and Man-Made Disaster Risks in the EU Accompanying the Document Communication from the Commission to the European Parliament, the Council, the European Economic and Social Committee and the Committee of the Regions The post 2015 Hyogo Framework for Action: Managing Risks to Achieve Resilience* {COM(2014) 216 final} {SWD(2014) 133 final}, 8 April, SWD(2014) 134 final. http://www.sos112.si/slo/tdocs/eu_risks_overview.pdf (Accessed August 2021).

[21] Cabinet Office. (2020) *National Risk Register*, Crown Copyright, London. https://assets.publishing.service.gov.uk/government/uploads/system/uploads/attachment_data/file/952959/6.6920_CO_CCS_s_National_Risk_Register_2020_11-1-21-FINAL.pdf (Accessed August 2021).

[22] Natural Hazards Partnership. (2020) *Who Are We?* http://www.naturalhazardspartnership.org.uk (Accessed August 2021).

[23] World Meteorological Organization (WMO). (2015) *Report of the Meeting of the Commission for Basic Systems Task Team on Impact of Multi-Hazard Prediction and Communication (CBS TT-Impact)*, Geneva, Switzerland, 16–18 February (No longer on-line). https://public.wmo.int/en/search?search_api_views_fulltext=multi-hazard%20warning%20systems; https://public.wmo.int/en and https://public.wmo.int/en (Accessed August 2021).

[24] IPCC. (2012) *Managing the Risks of Extreme Events and Disasters to Advance Climate Change Adaptation,* Field CB, Barros V, Stocker TF, Qin D, Dokken DJ, Ebi KL, Mastrandrea MD, Mach KJ, Plattner GK, Allen SK, Tignor M, Midgley PM (Eds.), A Special Report of Working Groups I and II of the Intergovernmental Panel on Climate Change, Cambridge University Press, Cambridge, New York, 582. https://www.ipcc.ch/pdf/special-reports/srex/SREX_Full_Report.pdf (Accessed August 2021).

[25] Public Health England. (2019) *Heatwave Plan for England Prepared with the Department of Health and Social Care, and NHS England, Met Office and Local Government Association.* https://assets.publishing.service.gov.uk/government/uploads/system/uploads/attachment_data/file/888668/Heatwave_plan_for_England_2020.pdf (Accessed August 2021).

[26] Fowler T, Southgate RJ, Waite T, Harrell R, Kovats S, Bone A, Doyle Y, Murray V. (2013) Excess Winter deaths in Europe: A multi-country descriptive analysis. *European Journal of Public Health.* https://doi.org/10.1093/eurpub/cku073; http://eurpub.oxfordjournals.org/content/early/2014/06/11/eurpub.cku073.short (Accessed August 2021).

SECTION 2: ENVIRONMENTAL HEALTH IN THE MILITARY CONTEXT – 'ENABLING NOT CONSTRAINING THE MISSION'

Lieutenant Colonel James Robert Fawcett MBE, RAMC
Wing Commander Gary Moyes, RAF

What is military environmental health?

The military is established with environmental health officers (EHOs) as Commissioned Officers and environmental health practitioners (EHPs) as Non-Commissioned Officers and other ranks to provide professional advice and guidance to enable the maintenance of a force that is fit for task in peace and at war. EHOs and EHPs serve in the Royal Navy (RN), the Army, and the Royal Air Force (RAF). The aim of military Environmental Health is to maintain and enhance military capability by assessing, communicating, correcting, controlling, and preventing those factors in the natural and manmade environment that potentially can adversely affect the health or survival of personnel and therefore undermine the mission.

The recognition of Environmental Health's (EH) vital role in preserving the fighting power of military forces and in providing crucial advice for humanitarian and disaster relief operations (HADRO) has, and continues to be, acknowledged by Commanders. From as far back as 1796 [1] the RN identified the importance of diet, ventilation, and cleanliness in the prevention and control of disease amongst sailors. Following the lessons learned from the Crimean War where thousands of soldiers suffered from cholera, malaria, scurvy, and cold injuries, and the Boer War with epidemics such as the typhoid outbreak at Bloemfontein in 1900, the Army established the School of Sanitation at Aldershot in 1906, which provided training to Regimental Officers and Non-Commissioned Officers for service in Sanitary Detachments which were attached to Brigades.

The role of military EH is the same today as it was historically; simply to prevent disease; maintain the health of military personnel; and

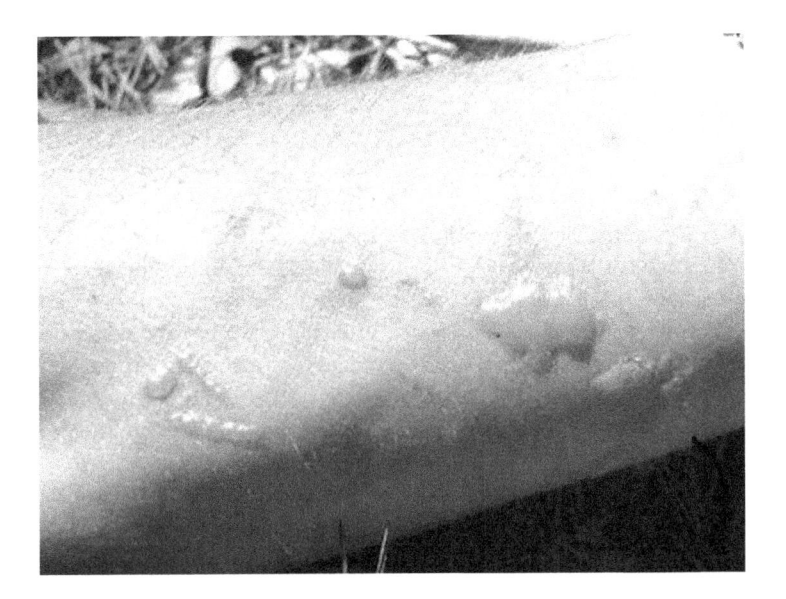

Plates 22.1 and 22.2 The Army Environmental Monitoring Team deployed to Sazan Island, Albania after Royal Marine personnel displayed symptoms of skin blistering. Eventually this was found to have been caused by contact with 'Ficus carica', Wild Fig plants.

provide specialist advice in support of HADRO and other military tasks. Specifically, EH personnel provide advice both at home and when deployed for operations and exercises on a wide range of health-related issues including:

1 Physical factors such as noise, vibration, temperature, and ventilation
2 Chemical hazards in the workplace and general environment
3 Biological hazards from parasites, bacteria, and viruses
4 Food safety and hygiene
5 The provision and maintenance of safe, potable water supplies
6 Communicable disease control (including SARS-Cov-2)
7 The control and management of pests
8 The assessment and control of environmental and industrial hazards
9 The issuing of Ship Sanitation Certificates in accordance with the International Health Regulations
10 Advising on international and national legislation and guidance in relation to the previous topic areas and assessing their impact on the military human population and the mission parameters
11 Humanitarian Assistance and Disaster Relief (sanitation, control of infectious diseases)
12 EH-related health promotion and training

The EH issues addressed by military personnel are comparable with those tackled by our civilian colleagues. The principal difference between civilian practice and military is often one of emphasis and the fact that this can require working in remote, demanding, and potentially hostile environments.

Regular RN, Army, and RAF EH personnel are currently located across the UK, Cyprus, Gibraltar, Kenya, Falkland Islands, Brunei, Belize, Canada, Iraq, Mali, and the United Arab Emirates. There are also specialist Occupational Hygiene monitoring teams located at Headquarters Field Army at Andover (Army Environmental Monitoring Team) and at RAF Henlow (Centre of Aviation Medicine), both of which rely on analytical support from the laboratories at the Institute of Naval Medicine at Gosport. These two teams support specialist occupational hygiene monitoring tasks in the UK and are operationally deployable at short notice to any location globally when the need arises and

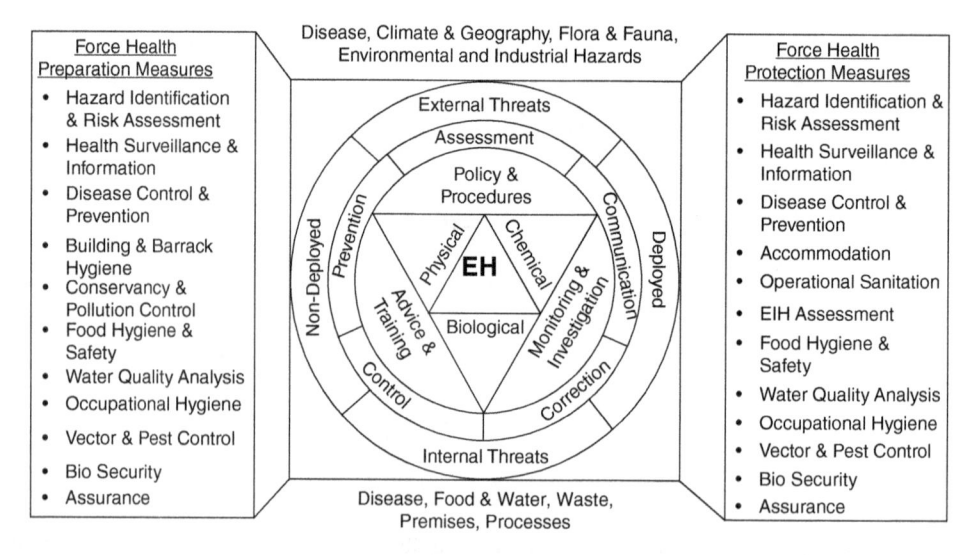

Figure 22.5 Diagram depicting the diverse subject areas military EH personnel have responsibility for, both before and during operational deployment

have recently supported operations in Sierra Leone, Estonia, Mali, and Afghanistan.

Deployed operations

Military EH personnel have been involved in every military operation in recent history, from the Falklands War in 1982 to current operations in Iraq and Mali. Deployed operations and locations have included the first Gulf War in 1991, the Balkans (Bosnia, Kosovo, Macedonia), Afghanistan, East Timor, Iraq, Mali, Nepal, UAE, Sierra Leone, and South Sudan to name but a few. The responsibility of military EH personnel on operations is primarily to provide 'force health protection' advice and assistance to the deployed military force, depending upon the nature of the operation. Clearly on HADRO deployments they will be responsible for

conducting needs assessments and prioritisation of military assistance to the affected local communities and supplementing host nation EH capability if needed; military EH was a critical element of the HADRO response in 2017 to support the British overseas territories in the Caribbean following hurricane Irma. Since March 2019, large elements of all three Services EH capabilities have been critically involved with the control and management of SARS-Cov-2 within the Ministry of Defence, overseas locations and with the provision of direct help and assistance to local authorities across England and Wales to maintain critical civil EH outputs.

The importance of EH as a key enabler and force multiplier for deployed forces is currently well recognised by the wider military community, and there is rarely an operation or overseas exercise that does not include an

Plate 22.3　In 2013 RN, Army, and RAF EH personnel deployed on Operation PATWIN, the UK Government's Humanitarian Support for the Philippines which had suffered extensive tropical storm damage. RN and Army EH personnel deployed to remote Islands by helicopter from HMS Illustrious to help in assessing needs and prioritising assistance.

EHP on the initial reconnaissance to an area, or at the very least as part of the Advance Party during deployment. EHPs are also permanent members of the major UK's spearhead formations, including the UK Commando Brigade, UK Airborne Brigade and UK Special Forces. Their role during the reconnaissance is to collate a portfolio of health threats that may pose a risk to deployed personnel. Information gathered will include local infectious disease profiles, food and water quality, sanitation, industrial pollution, waste arrangements, public health organisation and enforcement, and occasionally primary and secondary

Plates 22.4 and 22.5 Following the earthquake in Nepal on 25th April 2015, EH were among the first military responders, responsible for providing potable water, sanitation, and infectious disease advice to isolated and desperate communities

health care provision. That information is then articulated in the medical plan for the operation or exercise with relevant guidance on how to reduce the health risk from those threats to a minimum.

Routine EH activities on deployments

The nature of routine EH activities will depend on the size and distribution of the population at risk (PAR), the type of mission being undertaken, and the threats to health (natural and human-made). The work tempo will vary and will be dictated by events, such as disease outbreaks, changes in the mission posture, enemy actions, the degree of local infrastructure regeneration responsibilities, and/or suspected/actual Environmental and Industrial Hazard (EIH) episodes. In mature theatres, routine EH activities will tend to revert to largely audit, advisory, and support roles.

Plates 22.6 and 22.7 As part of the UK Government's response to the Ebola crisis EH personnel deployed on the recce to Sierra Leone in August 2014 and as part of the Advance Party in September 2015 where support was provided to military medical staff along with limited EH assistance to the wider civilian population

Re-deployment – a further peak in EH activity will tend to occur immediately prior to re-deployment of force troops and draw-down, due to involvement with, amongst other areas, site remediation and with the monitoring of biosecurity measures. Routinely, EH are one of the last specialists to withdraw from an operational area as the removal of semi-permanent infrastructure results in a return to austere living conditions.

This can be summarised diagrammatically thus (Figures 22.6 and 22.7).

Exercises

The provision of EH support to overseas exercises is identical to that delivered during support to operations. EH personnel will be involved with the reconnaissance, exercise planning, and medical and health threat assessment of exercises. EH personnel have recently deployed on exercises to Gabon, Kenya, Ghana, Botswana, Canada, Brunei, Malaysia, California, Texas, Singapore, Australia, Vietnam, Belize, Estonia, Poland, Romania, France, Slovenia, Croatia, and many other locations across the globe.

Recruitment and training

The recruitment of Environmental Health personnel differs across the three Services. The Army[11] and RAF[12] recruit and train their EHPs on an EH BSc Honours in Environmental Health degree pathway in conjunction with their academic partners Leeds Beckett University; the degree is fully accredited by the CIEH. The RN has only EHOs and recruit either qualified civilian EHPs[13] or fully qualified EHPs from the sister Services. Support to the Royal Marines is provided by RN EHOs and Army EHPs.

The School of Sanitation has undergone many changes since its inception in 1906. Today its descendant is the Department of Environmental and Occupational Health (DEOH) located at the Defence Medical Academy based at Defence Medical Services, Whittington, near Lichfield in Staffordshire. This is where the BSc Honours in Environmental Health is delivered. In addition, DEOH provides preventative medicine education to new entry Doctors, Medical Support Officers, Allied Health Professions, Medical Technicians and the Officers, and other ranks from the wider Army, Royal Marines, and RAF Regiment.

The Army environmental health reserve

Environmental health reserves have for many years been an important partner to the regular Army's environmental health capability.

Reserve EHOs and EHPs are all volunteers and are recruited from many EH employment

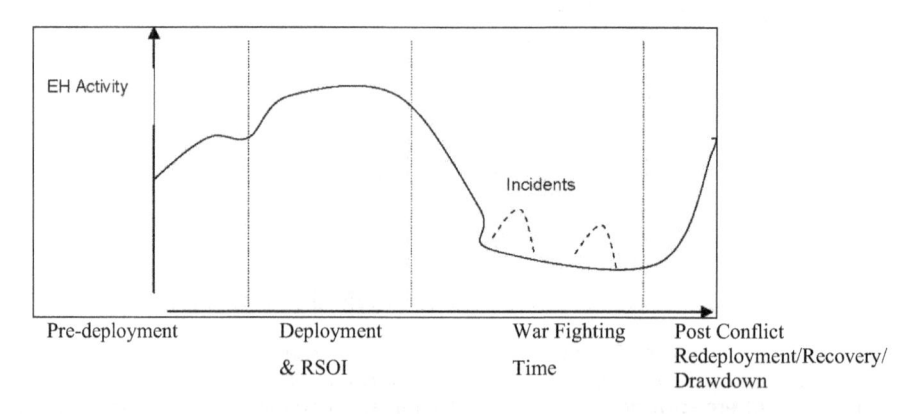

Figure 22.6 Environmental health scope and tempo

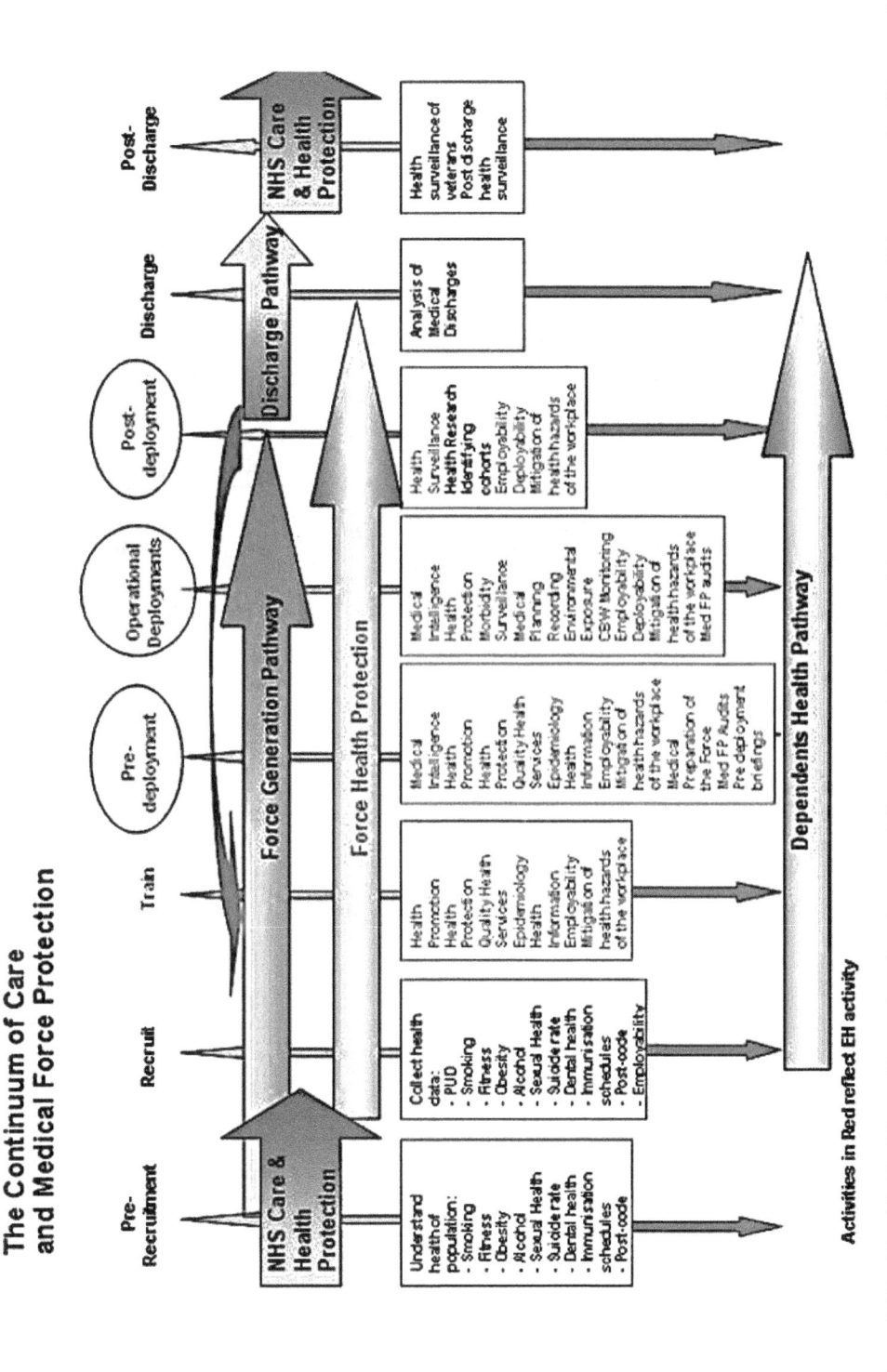

Figure 22.7 The Medical Force Protection Continuum of Care chart depicts the wide swathe of subject areas that EH personnel will be involved with to greater and lesser degrees.

Plate 22.8   2013 RN and Army EH personnel undertaking monitoring on the upper deck of HMS Illustrious during deployment in the South China Sea to advise Command in their assessment of climatic injury

groups including local government, the NHS, commercial consultancy businesses, and the self-employed. Bringing a unique set of skills to the military, they are also able to facilitate links between civilian and military organisations, professions, and practice.

Reserves provide support to regular colleagues by undertaking operational deployments and attachments, supporting exercises and training activities.

More use of reserves has been made in recent years and environmental health reserves have served in all main conflict theatres since the 1980s. Medical reserves were a mainstay of Operation HERRICK in Afghanistan, with reserve doctors, surgeons, nursing staff, and allied health professionals (which include EHOs) all providing a valuable contribution to the UKs military effort. This continued use was seen best when reserve EHOs deployed to support Joint Military Commands across

the UK during the recent SARS-Cov-2 outbreak from May 2020.

Army reserve strength and organisation

All reserve EHOs are all practising environmental health practitioners qualified to at least degree level. Army governance require all reserve officers to be a qualified member of a either the CIEH or REHIS, subscribing to their code of professional conduct and CPD schemes.

Most medical personnel are commanded under the 2nd Medical Brigade located at Queen Elizabeth Barracks, York. There are two main types of units at which the EH Reserves are held, that is 'nationally' recruited and the 'local' units. The nationally recruited units include the Operational Headquarters Support Group, 306 Hospital Support Regiment and 335 Medical Evacuation Regt. The personnel in these units provide individuals to support exercises and operations and can serve in any theatre called upon. They have a minimum 19 days per year commitment but quite a number serve for considerably more time.

The locally recruited units are the Field Hospitals and Medical Regiments based throughout the UK. They often have detachments in smaller cities and towns allowing the widest possible attraction of skilled personnel. Their role is to provide medical support, including treatment and evacuation to operations and exercises in formed units or augmentees to medical units supporting smaller operations. The structures of the nationally and locally recruited units are to undergo changes as part of the 2021 Army Future Solider programme.[14] The ability to service either nationally or locally will still be possible.

Section 2 Notes

11 https://apply.army.mod.uk/roles/army-medical-service/environmental-health-technician

12 https://www.raf.mod.uk/recruitment/roles/roles-finder/medical-and-medical-support/environmental-health-practitioner

13 https://www.royalnavy.mod.uk/careers/roles-and-specialisations/services/surface-fleet/environmental-health-officer

14 See https://assets.publishing.service.gov.uk/government/uploads/system/uploads/attachment_data/file/1037759/ADR010310-FutureSoldier Guide_30Nov.pdf

Section 2 references

[154]   Rolleston HD. (2015) James Lind pioneer of naval hygiene. *Journal of the Royal Naval Medical Service*, 1(2): 181–190. https://archive.org/stream/JRNMSVOL1Images_201507/JRNMS_VOL_1#page/n209/mode/2up (Accessed 30 July 2015).

23

A global perspective on environment and health

Andrew Mathieson

This is a chapter that includes commentaries, concepts and comments on some of the different environmental, societal and health challenges and priorities around the world that require tools and professionals with similar motivations and skills to address. This is the job of the environmental health practitioner.

Geopolitical determinants of health

It is clear from the media and academic publications the world over that geopolitical factors play an ever-increasing role determining the health and well-being of populations, identifiable groups and individuals. We need to consider the direct and indirect impact over time as well as other consequences including natural and manmade disasters, some of which have been discussed in earlier chapters such as climate change and conflict, which lead to erosion of human rights, social instability, increasing inequalities. Research associated with the ongoing SARS-CoV-2 (COVID-19) global pandemic provides evidence that there are clear interconnecting and interdependent factors involved [1, 2].

Fragile States with fragile fiscal policies (corruption), poor delegation in national administrations (overly centralised decision making), all exacerbate the existing inequalities and result in an increase in the health burdens and an ever-increasing severity of the health challenges. As we move towards 2050, there will be increased pressure on WHO, other international agencies (World Bank, UNHCR, UNICEF, etc.) and donors (both individual and nation states) to provide coordination and respond incrementally to these multi-faceted challenges. We will need to look more at influencing, indeed 'nudging' structural changes to create a more efficient and effective environment where expenditure of political and fiscal capital creates the opportunity where change can become organic. Institutional stability forms the foundation where structural changes can be scaffolded to bring about sustainable changes and provide the environment where private donors, institutions and companies once again feel confident investing in health. We are experiencing an increased recognition that many of the current models and associated evidence in terms of global health are framed from the perspective of what is commonly defined as *high-income countries*.[1] However, it is clear that the models fail to fully take notice of the dynamic nature of global challenges due to the disproportionate distribution of political and economic power around the world.

DOI: 10.1201/9781003035640-23

Evidence from the SEHA projects in Afghanistan (prior to the Taliban retaking control) demonstrated the positive impact local NGOs could have when they were allowed to deliver 'effect' free from local and national political influences. The flat decision-making structures, clear accountability (and responsibility) allowed for quick decisive decision making where necessary often to allow payment for drugs or treatments. When health professionals and policymakers have access to decision makers an important additional perspective is introduced which can improve short- and long-term changes in policy and practice, especially in countries where decisions sit at the top of a cumbersome and time-consuming autocracy. We should also review the challenges faced in Afghanistan where universal health coverage, while an ideal, is fraught with institutional, political and economic challenges [3]. Yet the large number of private clinics and hospitals offered the entrepreneurs an opportunity to socialise the idea of a low-cost insurance-backed scheme that with the support of the Ministry of Health, European Union, USAID and others, could be rolled out to provide a reinterpretation of the World Bank Group, Universal Health Coverage (UHC) model [4]. A novel insurance-backed, private UHC initiative was trialled in Kabul between 2016–2020. Its success led to forcible closure due to direct threats from local ISIS and Taliban militia. While UHC as a construct is appropriate in emerging nations, it fails to meet expectations or deliver the needs where the population already had UHC above the baseline such as pre-conflict Syria.

Syrian refugees in the surrounding nations arrived often with complex comorbidities and a need for both acute and chronic medicine. In Lebanon, Syrian refugees often drove over the border with minimal possessions and were settled in camps organised and supported by UN and NGOs. Many men quickly moved into nearby towns and cities and settled alongside the poor and vulnerable Lebanese populations (that's where the cheapest accommodation was to be found). When they had a little money, families followed. To some

Plate 23.1   NGO delivering health information in refugee camp Lebanon 2019

extent this served to hide the Syrian refugees 'in plain sight' and put significant demands on existing primary health networks. The EU provides funding to NGOs such as International Medical Corps (IMC) to support and enhance the primary health care centres where demand is greatest. This is a new and innovative model of UHC via a network of 50 primary health clinics and dispensaries rolled out across the country. The concept of the new model is to provide low-cost packages of care for families without additional payments. This new model of primary health care provides evidence that complex humanitarian problems involving middle-income countries can implement alternative solutions to achieve UHC.[2]

Climate change and security

Almost all the general and scientific population acknowledge that climate change is a real threat to humanity and our ecosystems. Increasingly, mainstream literature refers to the concept of *climate change and security* and is raised for example by McDonald [5]. However, there is no unified or agreed definition, simply a range of definitions that are becoming more closely aligned.

Added theoretical value

Climate change and security [6] differs from similar concepts such as *climate crisis* and *climate emergency*. By allowing the discourse to pivot from crisis and emergency (which can be locked into discussions around duration of the event), security allows for a more sustained discussion and indeed strategic planning by national and intergovernmental agencies. Indeed, by adding security a wider range of stakeholders can be engaged in planning. For example, inhabitants of coral islands in the Pacific, which are being inundated by an increasing number of storms, rising sea levels, less land to grow vegetables (due to increasing salt levels) and reduced access to safe drinking water, must secure new sources

of fresh water and take extreme measures to protect their island life.

This can ultimately involve abandoning one island and relocating to other islands that are often already overcrowded or migrating to larger towns which has a significant impact on urban populations.[3]

The US air force was faced with reduced access to a strategic airbase in Florida, due to storms and flooding. Instead of moving to an area with lower risk they decided to spend millions of dollars in protection measures from increasing the thickness of the runways and providing more secure buildings.[4] While this was popular politically at the time it was a poor example for the world and offers little hope for the Pacific Island States currently under threat. Concreting an island is neither pragmatic, eco-friendly nor a responsible option. It is for all nations to look at the less well-off and more vulnerable and offer sustainable, pragmatic interventions to help them overcome adversity and be allowed to live their traditional ways of life. It should be noted (for the sake of balance) that recent policy change by the Biden Administration seems to be offering new policy direction by instructing national security agencies and those responsible for foreign policy to embed 'climate change' into future missions [7].[5]

If you analyse the goals of COP26, held in Glasgow in 2021, you can see that much still have to be resolved.

1 Secure global net zero by mid-century and keep 1.5 degrees within reach
2 Adapt to protect communities and natural habitats
3 Mobilise finance
4 Work together to deliver[6]

It is disappointing that 'health', while implicit and a topic of intergovernmental and NGO discussion, did not appear in either the goals of COP26 or more than once in the 28-page 'COP26 The Glasgow Climate Pact'. Is it not time that we considered the health of both the planet and its population? Should we not place 'health' at the fore of the discussion and corresponding action?

Role of non-state actors

The traditional interpretation of security is tightly bound to the concept of national security. However, by associating the discussion of climate change with security we allow the conversation to expand and allow engagement with non-state actors such as the business sector, local government and citizen groups (not just activists which are often vilified).[7] While nation states offer pledges and promises, non-state and subnational actors are increasingly engaging with the national level to implement mitigation measures at local and community level. From the outside this may be seen as less coordinated action, but it is argued that this new approach of engaging and mobilising non-state actors offers a real opportunity for sustainable change to mitigate the impact of climate change. Australia has seen an increase in local community groups seeking funding and State/Federal support to implement strategies to mitigate risk from future natural disasters (notably bushfires and flooding) [8]. These activities include a wide range of interventions including training of volunteers to respond to local emergencies, empowering community groups, information posters/leaflets and videos, redesignating flood zones to reduce occupancy, increasing fire and flood protection to remote properties, adjusting emergency plans to absorb lessons learnt and increasing community engagement in simulation and preparedness.

We should also acknowledge the excellent work being carried out by environmental health practitioners throughout the African Region of the IFEH (most recently led by Jerry Chaka and Dr Selva Mudaly) and its affiliate African Academy (initially led by Dr Koos Engelbrecht, a gentle giant and visionary to anyone who met or worked with him). Both the African Academy and IFEH African Region are exemplars of non-state actors engaging with policymakers and politicians to help provide the evidence for policy change to help lessen the impact of climate change and to help make changes to improve health and well-being.

Regional imbalance

While the majority of the literature on climate change and security has traditionally originated from Europe and North America, academics, NGOs and governments in other regions, such as Africa, South America, Australia, Asia and the Pacific region, who are just as affected are becoming more effective at contributing to the growing body of knowledge. This increasing number of peer-reviewed articles and publications can perhaps be seen as the result of a change in government policy and associated research funding under redefining and main streaming 'climate change and security'.

Countries that are better off financially in a particular region (i.e. US, Australia, New Zealand, Japan, Korea, Malaysia in the Pacific Region) often carry out strategic missions that deliver increased technical support to remote island communities through medical and health missions (MEDCAP) and have an increased role in response to national and local emergencies to help reduce regional imbalance [9, 10]. These missions are carefully planned and often have a clear environmental health focus (WASH). However, care must be taken when designing these missions to limit the interventions to what can be supported locally without creating dependency from outside agencies. It is a mission necessity that MEDCAPs must often set aside the SPHERE standards[8] and focus on minimal intervention, maximum benefit to maximum population. This often generates criticism of MEDCAPS yet without those missions, larger policy interventions through regional support grants and strategic partnerships would not be possible. Two examples follow regarding WASH. One that is sustainable and one that was not. Answer on a postcard please!

Plate 23.3 VIP installed under WASH (designed to collapse in rough weather and be easily rebuilt)

Plate 23.2 WC installed by international NGO (no training or capacity building)

Global migration and an increasing environmental health burden

The World Migration Report 2020 [Table 1 in 11] (see Table 23.1) evidences the rise of international migrants, and associated social, health and environmental pressures, from 150 million to 272 million (2000 to 2020) and of those the number of refugees and internally displaced persons combined has increased from 35 million to nearly 67 million. While the United Arab Emirates is the country that hosts the highest proportion of international migrants, the region with the highest proportion of international migrants is Oceania.

The dates of the data estimates in the table may be different to the report publishing date (refer to the reports for more detail on dates of estimates); refer to chapter 3 of the report [11] for regional breakdowns; ★ indicates the data was not included in the report but is current for that year; ✦ as at 28 October 2019.

Consequently, Oceania (including Malaysia, Indonesia, Fiji, Australia and New Zealand) experiences a disproportionate burden in terms of access to health, housing, etc. as can be measured by the social determinants of health[9] and Sustainable Development Goals discussed in Chapter 1.[10] Yet the bulk of the world press and indeed attention focuses on populations moving north into Europe and USA. Environmental health practitioners in Oceania are well aware of the increasing demands on services and changes in the health burden associated with complex

Table 23.1 Key facts and figures from the World Migration Reports, 2000 and 2020

	2000 report	2020 report
Estimated number of international migrants	**150 million**	**272 million**
Estimated proportion of world population who are migrants	**2.8%**	**3.5%**
Estimated proportion of female international migrants	**47.5%**	**47.9%**
Estimated proportion of international migrants who are children	**16.0%**	**13.9%**
Region with the highest proportion of international migrants	**Oceania**	**Oceania**
Country with the highest proportion of international migrants	**United Arab Emirates**	**United Arab Emirates**
Number of migrant workers	**-**	**164 million**
Global international remittances (USD)	**126 billion**	**689 billion**
Number of refugees	**14 million**	**25.9 million**
Number of internally displaced persons	**21 million**	**41.3 million**
Number of stateless persons	**-**	**3.9 million**
Number of IOM Member States*	**76**	**173**
Number of IOM field offices*	**120**	**436⁺**
Sources: See IOM, 2000 and the present edition of the report for sources		

Source: [10]

Note: The dates of the data estimates in the table may be different to the report publishing date (refer to the reports for more detail on dates of estimates); refer to Chapter 3 of the report [10] for regional breakdowns; * indicates the data was not included in the report but is current for that year; *as at 28 October 2019.

migration. Population migration and rapid urbanisation cause a series of changes which can lead to increase in an environmental health burden and associated increase in premature deaths [12]. The nexus between climate change, health and migration is illustrated in Figure 23.1.

Figure 23.1 shows a conceptual framework showing human experience of environmental change as a 'driver' of migration and its potential impacts on health.

If we examine the population movement associated with the recent conflict in Syria and the pressures experienced in neighbouring countries (Southern Turkey, Iran, Iraq and Lebanon) we can see how the environmental health burden has increased. This is despite significant funding from EU [14], World Bank, WHO and other UN Agencies as well as the significant effort made by large numbers of NGOs (Red Crescent, International Medical Corps, etc.).[11] For those looking for

a more in-depth analysis they should read Regional Resilience Response Plan (3RP),[12] the associated progress reports and other publications available online. These documents evidence the coordinated international effort to provide a population level solution to a complex humanitarian situation that predates the Syrian crisis, but which is significantly impacted by that crisis and the August 2020 warehouse explosion in Beirut which compounded matters.

Syrian refugees were initially housed in temporary camps supported mainly by NGOs with funding from international donors. Over time many of the camps depopulated and either resized or closed as the occupants moved into nearby towns and cities. This internal movement often coincided with the emergency funding for food, shelter and medical supplies coming to an end and a realisation by international donors that a more sustainable funding mechanism was required.

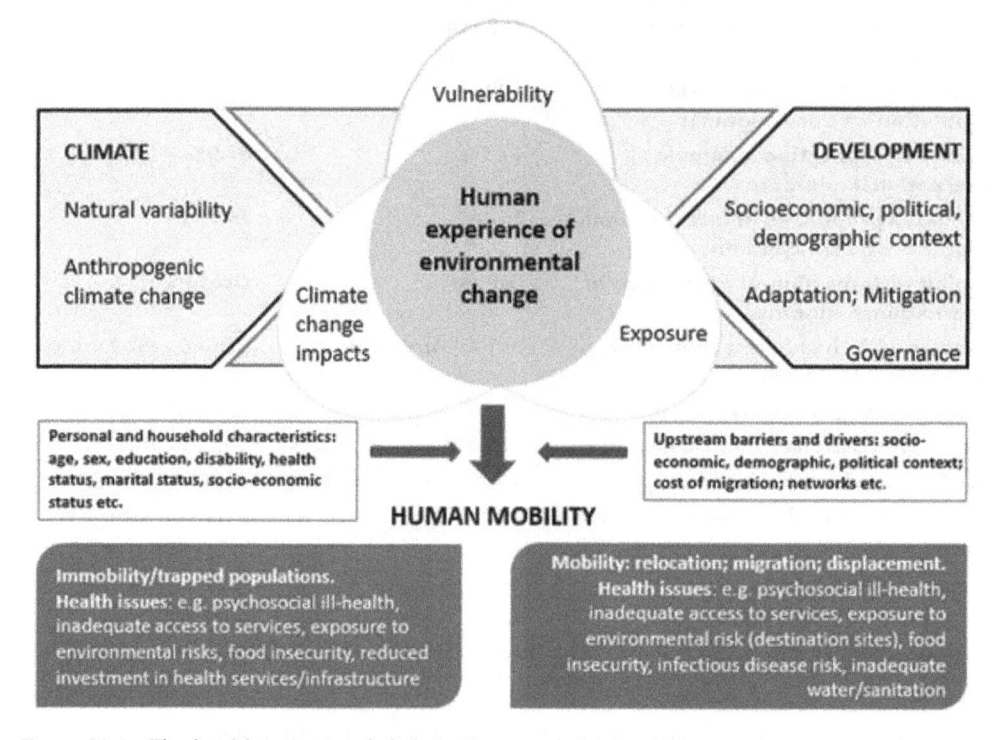

Figure 23.1 The health impacts of climate change and migration
Source: [13]

The EU Regional Trust Fund in Response to the Syrian Crisis (the 'Madad Fund')[13] was formed with funding from Directorate – General of Neighbourhood and Enlargement Negotiations (DG NEAR). Since 2018, the consortium formed by International Medical Corps UK (IMC) as lead agency, with the partners Première Urgence-Aide Médicale Internationale (PU-AMI) and the Social Promotion Foundation, has worked to improve access to quality services in primary health care, community health and mental health services for Syrian refugees and other vulnerable populations. Close coordination and consultation with the Lebanese Ministry of Public Health (MoPH) and Ministry of Social Affairs (MoSA) has been essential to the success of this action. The project is funded via MADAD and is called Reducing Economic Barriers to Accessing Health Services in Lebanon (REBAHS Lebanon) [15]. This project can be seen as a success and has

defined a new model to allow poor and vulnerable to access packages of primary health care including access to essential chronic and acute medicine at low or no cost. This new model designed and implemented by the consortium led by IMC provides a new model for middle-income countries such as Lebanon that are in crisis.

The knock on in terms of environmental health benefits are both direct and indirect as a component of the REBAHS project is to provide community workers to both Syrian refugees and vulnerable Lebanese (who are often living side by side). These community workers visit homes, ensure everyone knows what services can be accessed and coordinate the referral system to ensure follow-up visits are made, and essential services are accessed. Although primarily a public health initiative the environmental benefits are also real and measurable against both social determinants of health and Sustainable Development Goals.

Plate 23.4 Community workers from IMC visiting local Syrian and vulnerable Lebanese women to talk about health matters

Gender equity and environmental health [16]

Even a cursory glance of peer-reviewed literature demonstrates an overwhelming bias against the female gender [17]. Indeed, it is clear that females have been significantly discriminated against. The results from simple questionnaires suggest the majority of males consider females their inferior in at least one of the following: intellect, technical ability, strength, social and psychosocial aptitude. These may be the dominant 'reasons' or 'excuses' made by men for their bias; however, those reasons do not stand up to academic scrutiny [18]. The acceptance of women's increasing participation in the male-dominated labour market and the prominence of studies on the role of women in society, supported by systemic changes in both social and political norms, can trace its origins to the time of the suffragettes through two world wars, and as the age of feminism rose in the 1960s and was reenforced in the 1980s, the wave of 'active feminism' emerged. However, active feminism is not universal, and women continue to struggle for equality in many countries with the same systemic difficulties as before, and entrenched stereotypes. The 'Transforming Our World: the 2030 Agenda for Sustainable Development'[14] illustrates what is possible when governments and community groups around the world cooperate with the common aim of producing *a new global model to end poverty, promote prosperity and the well-being of all, protect the environment, fight against climate change and promote gender equality*. It is clear that to achieve goal 5 'Gender equity' we first need to embed goal 5 into the other goals, in particular those directly related to environment and health.

There is a strong argument for 'gender equity statements' to first be mandated in all research proposals as that would ensure the introduction of universal procedures for reporting of sex and gender information in study design, data analyses, results and interpretation of findings. It is clear that gender influences the health and well-being at the individual, family, community and population levels. When examined further we see differences because of gender in a variety of ways, including direct and indirect impact on the environmental, risk-taking behaviour, risks associated with occupational exposure, access to health care, access to insurance, access to transport, health-seeking behaviour and perceived bias during health care, and thus treatment outcome.

When questioned, most 'western adults' understand the term *global gender gap* [19] as most often being associated with women and women's health, employment and well-being. There are real benefits to mainstreaming gender-based analysis in terms of generating evidence and thus a clearer understanding of health as we can gender focus to the benefit of men, women and those who self-define in other ways. Health literature shows that once we categorise research by gender, and minimise associated bias, we can form distinct data sets. This would allow researchers to generate a deeper understanding of the underlying differences and similarities, get deeper insight that will lead to more innovative approaches and better solutions. Stronger evidence informs the development of more effective and focused gender policies which should meet the wider needs of society in terms of health stewardship, leadership and gender equity [20].

In many countries more women study environmental health and enter the profession. However, it is less common for women to be found in leadership positions within environmental health and this is as a result of opportunity, lack of mentoring, structural and institutional gender bias [20]. In spite of these barriers, the environmental health profession can point to an increasing number of inspirational female leaders such as Julie Barratt (President of CIEH 2021–24) and Professor Dr Susana Paixão (President of the IFEH 2020–22).

One Health . . . tackling global environmental health

A simple representation of One Health is seen in Figure 23.2 and the greater complexity is in Figure 23.3.

There is an increasing role for environmental health in managing global health threats especially in terms of zoonotic stressors and emerging communicable diseases (for example COVID-19). As populations grow, stress and damage to ecosystems increases, land and sea use changes and urban expansion and pollution all put further pressure on wild habitats and biodiversity through exploitation of natural resources and an increase in emerging and re-emerging hazards are inevitable. Thinking derived from the One Health model considers both cause and effect from the environment to the workplace and considers well-being of self. It is useful to apply a number of critical lenses and re-examine the One Health model from local, regional, national and global perspectives (the reverse of the norm); this includes local, regional national and global politics and governance of climate change and security [5]. Over the

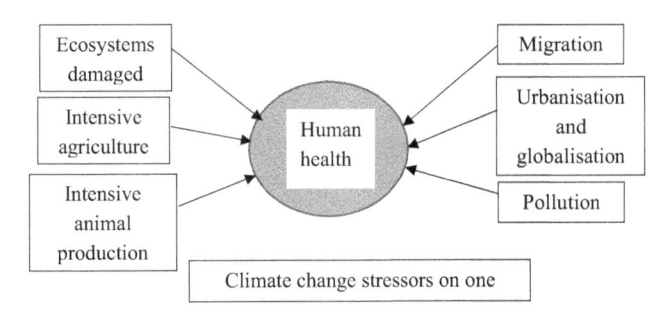

Figure 23.2   One Health . . . tackling global environmental health

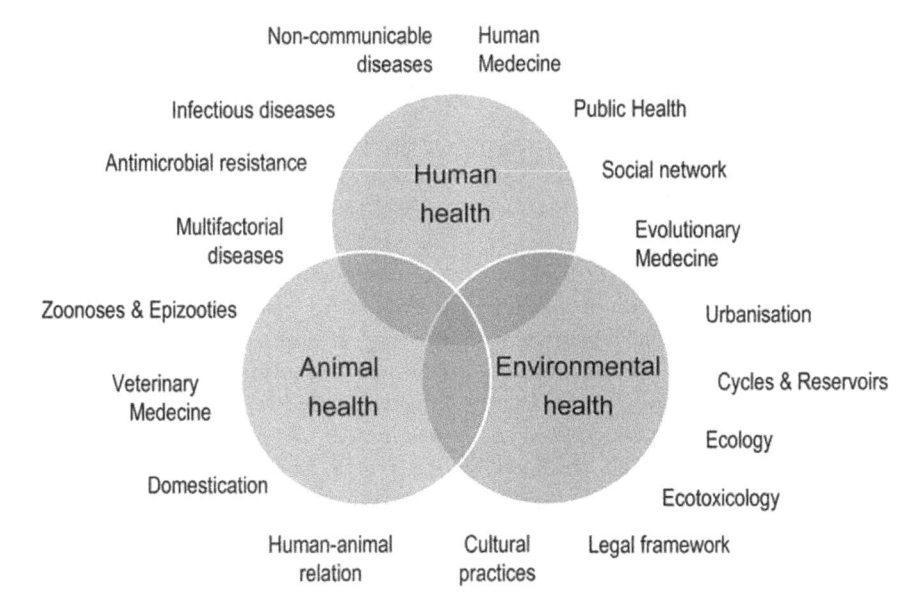

Figure 23.3   One Health
Source: [21]

last few years, WHO has identified over 100 emerging or re-emerging infectious health hazards (at least 75% zoonotic), which puts added pressure on our fragile ecosystem which in turn presents new opportunities for diseases to spread. A One Health approach enables better risk analysis and understanding on the root causes and drivers of the diseases emerging and spread and facilitates development of risk management strategies and evidence-based policy. The strength of the One Health approach is that it can be applied equally to zoonotic diseases, as well as non-communicable and communicable disease.

One Health points the way to an all-embracing systems approach that is critical to strengthen our strategies to meet Sustainable Development Goals. One Health is an integrated, unifying approach that aims to achieve optimal and sustainable health outcomes for people, animals and ecosystems. This approach has traditionally overly focused at the international and intergovernmental level and relies on national government to pursue similar policies and embed One Health priorities in domestic policy and subsequently legislation. The problems that have been identified with translating One Health into local interventions include poor media coverage, lack of socioeconomic equity, national protectionism, lack of political will, insufficient alternative strategies, fear of loss of power, weak animal health systems and inability for government to work outside of silos.

If we examine the work of EHPs (EHOs) around the world, we can see the principles of 'One Health' (if not the actual phrase) have been integrated into college/university training and professional mentoring. The work of the EHO follows the same principles as found in One Health when inspecting a 5-star hotel in Nairobi or Cape Town to inspecting street food vendors to ensure meat is fresh and handled to reduce cross-contamination and ensure food products are cooked thoroughly for immediate consumption.[15] Perhaps it is time to recalibrate 'One Health' and provide

simpler tools to weave into policy and politics both domestic and international [22].

Housing and health equity[16,17]

We need to look at the seminal work of John Snow[18] and other notable English Victorians[19] (see also Chapter 1) who first linked air, water, waste, housing, animals, environment and health. Indeed, Sustainable Development Goal 11:[20] *Make cities and human settlements inclusive, safe, resilient and sustainable*, illustrates little has changed. Despite this global priority, recent policy shifts in many nations (such as England, Australia [23], The Netherlands [24], Scotland [25]) to reduce government burden in terms of ownership and investment in social housing are likely to exacerbate these problems. As social housing shifts from government to the private sector, society will see this policy increasing the burden on acute health care service providers and increasing both the poverty and health gaps for our most disadvantaged [26]. Further the COVID-19 pandemic has resulted in significant job losses primarily in low-paid and casual sectors. This is evident in the increasing unemployment rates with the knock-on consequences of increasing economic hardship of households in social housing (whether in the public or private sector). As a result, the risk of and actual eviction, has increased during the pandemic,[21] especially amongst the most vulnerable populations [27]. It falls to environmental health organisations around the world to evidence the impact of poor housing on health and to engage with policymakers and institutions to generate more effective joined-up policies working across government departments to work towards SGD:11. Without clear, funded and deliverable national strategies that embed the core principles of SDG 11, nations will be unable to show meaningful and tangible progress by 2030.

It seems old problems need new solutions that offer both a holistic and sustainable approach. Governments need to rethink the impact of housing policy which promote the de-investing in social housing, an increase of childhood morbidity, more days of sickness, reduced learning potential and ultimately more families caught in the poverty trap. When governments reduce their investment in social housing and shifts social housing into the private market, profit trumps health. We will see increasing demand and an associated increase in costs to local health services. A truly holistic and sustainable social housing policy would realise that a balance is necessary between health and housing. By providing reasonable social housing we prevent people becoming sick, reduce the burden on health services and provide the opportunity for people to escape the poverty trap.

Environmental determinism[22,23]

Environmental determinism is a concept that rose to popularity in the late 19th century and early 20th century (sometimes replaced in the literature by Environmental Possibilism) that tried to link climate change to bad outcomes. Another definition is *also known as climatic determinism or geographical determinism, [and] relies on an approach which implies that individuals are bound to their environmental settings, especially climate*. Environmental determinism may have been an attractive concept as it seems to explain why people in warm climates are lazy (the warmth makes it less desirable/comfortable to work and hard work is not needed to stay warm), and why people in cooler environments are hardworking (a cooler environment requires more work to stay warm).

It is all too easy to fall in line with this superficial definition and fail to review the basis of the concept which is used by some to justify colonialisms. Environmental determinism gained a lot of critique and was abolished because it backed and justified the racism and imperialism that occurred in the continents of America, Africa and Asia. Consider indigenous people from many continents, they

have an especially strong connection with the land and nature. They mostly live a life in tune with the seasons and display resilience and adaption as change occurs. Any change in climate often impacts and necessitates change in local and regional harvesting, transport and storage practices, which undoubtedly has implications in terms of (local, regional and national) economy, (employment, well-being, resilience) society and culture, which all impact on human and environmental health ('One Health' concept).

Environmental determinism had a lot going for it at the time, especially for colonial powers seeking to use it as an excuse (both in retrospect and to justify past actions), to continue expanding and exploiting, not least because it offered a simple explanation: *The climate made me do it.*

In this modern age of evidence-based research we call upon governments to support the evidence of the Intergovernmental Panel on Climate Change (IPCC).[24] Many of the changes observed in the climate are unprecedented both in terms of human records and environmental samples (i.e. ice cores), and many of the changes we can detect (rising sea levels, rising temperatures, more extreme weather patterns, etc.) are part of a cascade that may take hundreds to thousands of years to reset and recover. Further research is necessary to examine the connection indigenous perspectives could bring to SDGs. Perhaps such research will provide the world with another 'aspirin', 'super food' or a more sustainable way to view both the land we live on and nature that surrounds us.

Unfortunately, climatic change will continue to force societal change and necessitate people and communities to adapt by rising to such challenges. Perhaps we can recalibrate utilising evidence and redefine the arguments and build sustainable innovative solutions in the way envisioned by COP26? It is worth socialising the term 'subsidiarity' in terms of policy and intervention. Subsidiarity allows for action to be taken at the appropriate level (without prescribing what the action is). It is a flexible concept which would seem applicable to the complex interactions of both One Health and climate change.

Effects of COVID-19 on the Sustainable Development Goals (SDGS) [28]

The world has committed to implementing the 2030 Agenda for Sustainable Development with 17 Sustainable Development Goals (SDGs) adopted by United Nations (UN) Member States in September 2015 (see also Chapter 1).[25] Even before the latest global crises caused by the pandemic of COVID-19, the world was off-track for achieving the health-related targets in the SDGs. The prolonged impact of this pandemic can only further compromise the world commitment to the 2030 Agenda for Sustainable Development. A combination of severe disruptions to essential health services due to unprecedented demands (including staff shortages, redistribution of finances and changing political arena), have all put at risk many of the 'One Health' gains we have made in routine childhood vaccination, malaria, tuberculosis, HIV, noncommunicable diseases, mental health, sexual and reproductive health, and more.

The pandemic, and corresponding response to the health threats, only serves to underline the fragile nature of international agencies (UN, WHO, WB and USAID and EU) and the real challenges faced by those agencies in trying to coordinate nations in terms of managing an effective COVID-19 response.[26] The different approaches taken by nation states to publicise, engage and often distract internal populations from the reality of the pandemic will be the subject of countless academic and political commentary for years to come. Researchers only have to look at the numbers (where accurately published)[27] to determine the impact of varying social and health policies on populations. Data is such an important component of the true picture and in this case relies on what data is available and how different people and countries

define cause of death. As seen in other epidemics such as Ebola, some people deaths were recorded as heart attack instead of Ebola. This is also likely to happen with deaths in the ongoing pandemic as some deaths will be attributed to COVID-19 while others will be recorded against the other comorbidities any one of which or combination of which may have ended their lives.

When policy advisors and lobbyists stop and reflect on how to advise politicians on future sustainable response to COVID-19, they would do well to remember the SDGs in their framing of matters that impact on health. We must integrate the SDGs into both national and international policy with greater determination, innovation and collaboration [29].

The future of environmental health working is hybrid[28],[29]

COVID-19 has accelerated changes to how we work that were already occurring before the pandemic. A growing body of literature on home, remote and office-based work is available covering topics such as in Table 23.3.

If it is agreed that the future of work is hybrid (working from home/remote working/teleworking/blend of these), steps need to be taken to redefine the workspace. If the

Table 23.2 Adapted from [30] 'WHO three strategic priorities'

1 First, to address the underlying reasons that people get sick and die: poverty; racial and gender inequities; air pollution and climate change, and other social, economic and environmental determinants of health.

2 Second, to strengthen national and global health security, including through a treaty or agreement on pandemic preparedness and response.

3 And third, to strengthen primary health care in every country, as the foundation for universal health coverage, and the first line of defence against outbreaks and diseases of all kinds.

Table 23.3 Topics for consideration when looking to introduce hybrid working

accommodation	digital transformation	work/life balance
people capability	organisational culture	training demands
work health and safety	productivity	face-to-face vs remote working

transition is to be effective, a holistic view must be taken, linked to environmental sustainability. This will require a rethinking of digital systems (hardware and coverage), managing workflows, internal and external communication, line management, team working and how colleagues and the public interact with increasingly complex databases. All this will necessitate elements of upskilling, introducing new technologies and equipment, training and coordination of personnel.

Environmental health has always had an element of independence that often involves working remotely and away from an office environment with limited direct managerial oversight. However, much of this workforce is traditionally older and experienced practitioners who are often the managers. COVID-19 has significantly reshaped the workplace and the demographics are changing to include younger people who are more technologically capable and those in a wider range of occupations [31]. Remote working, telecommuting and other innovative forms of working [32] such as home-based working and hot desking have differing impacts on various diversity groups. Younger workers experience difficulties working remotely, particularly around networking and career development [33].

These new working practices have the potential to be both positive and negative for work/family balance [34]. It can lead to work/family conflict and increased stress levels or enable employees to manage work and caring responsibilities more effectively. Teleworking could also disadvantage women due to decreased visibility to management [33].

Consider how the hybrid workplace could influence the work of environmental health practitioners. On one level the increased autonomy and freedom would be welcomed by many, yet this autonomy could be seen as a threat by managers. They have a leadership role and are to ensure that operational decisions are consistent and based on policy and best practice. Supervision and learning-on-the-job mentoring would become more challenging (as anyone who has tried to carry out supervision over teams/Zoom can testify). This author, for one, grew as a practitioner in a traditional office environment where experienced practitioners recounted their daily inspections in a discussive and informative manner that Zoom cannot replicate. As a student EHO growing up in Scotland I was privileged to work with a top team of EHOs who were keen to share knowledge and instil a solid grounding in professional practice. This author owes a debt of gratitude to unsung heroes of the profession, working for Motherwell DC in the shadow of Ravenscraig Steel Works; Jim B, Danny, Willie B, Andy A, Tom L Tom S, John S, Roddie and Dennis.

Time will tell how the hybrid workplace will develop. It is likely that different Countries will adopt different norms based on how they value and account for the elements listed in Table 23.3.

A final thought – health impacts of Australian bushfires and floods

Natural hazards (often exasperated by human influences outlined in Figure 23.2 (One Health . . . tackling global environmental health) are inherent in Australia due to its diverse climate and environment. Risks of severe floods, fires and storm are increasing both in terms of frequency and severity of consequence [35]. We need to move from complacency and indecision to taking immediate action to mitigate, minimise and reduce the residual risk – 'Natural disasters are more than just natural hazards'.

Practitioners will remember that risk is a product of the type, frequency and severity of the hazard. So, to be better prepared, we need to take the simple yet universal concepts of risk management, risk reduction/minimisation, risk communication and apply them to natural disasters. This way we can start to predict how communities will respond when they are exposed to such events. From these studies we can model, predict and plan for 'vulnerability' and then predict the ability of communities and other systems to cope with and recover from the impacts of such events (relative resilience). In the last few years Australia has experienced unprecedented bushfires and extreme floods with a massive impact on many communities.

The Australian Institute of Health and Welfare (AIHW) published the *Australian Bushfires 2019–20: Exploring the short-term health impacts in 2020* [36]. This report examines a wide range of health data sources and reports on the short-term health impacts of the bushfires including emergency department visits, prescription and purchase of asthma medicines, mental health service use and GP visits. Health professionals understand that bushfire smoke contains respirable (<2.5 PPM) and non-respirable particles (up to 10 PPM), that are toxic, in small doses, to many cells of the body [37]. Smoke exposure from bushfires can reach levels up to ten times those deemed hazardous. Short-term and extended exposure to high levels of air pollution can be associated with adverse health effects [38] (see also Chapter 16 on air quality).

The report highlights the higher use of inhalants and other asthma medications coinciding with bushfire activity or increases in air pollution from bushfire smoke [39]. Asthma-related emergency department presentations in NSW increased during the time of the bushfires. Sales of over-the-counter and prescription asthma medicines like inhalers also increased at local pharmacies. The large

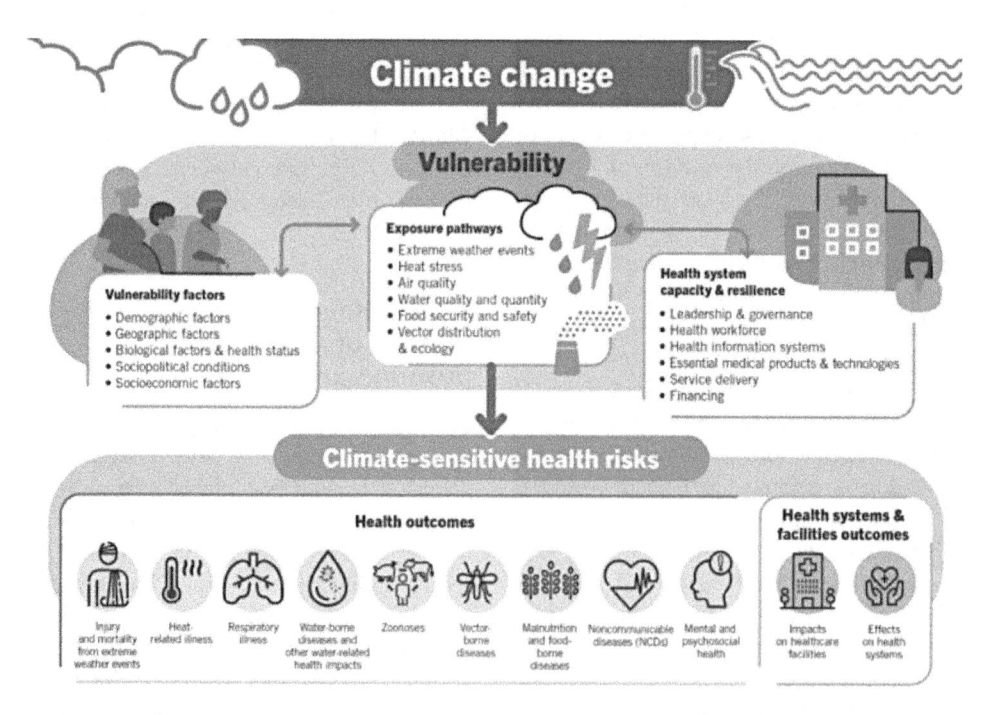

Figure 23.4 Climate-sensitive health risks, their exposure pathways and vulnerability factors (climate change impacts health both directly and indirectly, and is strongly mediated by environmental, social and public health determinants)[30]

increases in the rate of sales of inhalers during the bushfire period aligned with high levels of pollution as indicated by air quality data for south-eastern Australia. Specific Medicare Benefits Schedule mental health items for people affected by bushfire were introduced in January 2020.

Prolonged exposure to bushfire smoke over December 2019 and January 2020 is estimated to have resulted in over 400 excess deaths and over 3,000 additional hospitalisations [40, 41].

It is clear from all that has been written here and elsewhere in this book particularly Chapter 11, that climate change is having a significant and increasing impact on 'One Health'. The global health burden resulting from climate change has been highlighted here and elsewhere. Global health policy along with regional and national responses urgently needs to adapt to effectively address these hazards before events overwhelms us [42]. Work carried out by NEHA under the leadership of Dr David T. Dyjack, into workforce development serves to illustrate the actions environmental health practitioners can take to help mitigate the impact of climate change [43].

Notes

1 see https://www.nationmaster.com/country-info/groups/High-income-OECD-countries
2 For more information see https://international-medicalcorps.org/country/lebanon/
3 See 'His Pacific Island was swallowed by rising seas, so he moved to a new one' *New York Times* article by Damien Cave July 26, 2018.
4 See https://www.americansecurityproject.org/climate-energy-and-security/climate-change/climate-change-and-u-s-military-basing/
5 See https://www.defenseone.com/ideas/2021/01/pentagons-chance-get-serious-about-climate-change/171699/
6 See https://ukcop26.org/cop26-goals
7 See UN publication from 2018, Bridging the emissions gap – The role of non-state and subnational actors. https://wedocs.unep.org/bitstream/handle/20.500.11822/26093/NonState_Emissions_Gap.pdf?isAllowed=y&sequence=1
8 See https://spherestandards.org/
9 See https://www.who.int/health-topics/social-determinants-of-health#tab=tab_1)
10 See https://sdgs.un.org/goals
11 See https://aub.edu.lb.libguides.com/c.php?g=276479&p=1843038)
12 https://www.3rpsyriacrisis.org/
13 https://ec.europa.eu/trustfund-syria-region/index_en
14 https://sdgs.un.org/2030agenda
15 See Chapter 2 and Skills and interests of the environmental health professional at https://www.healthcareers.nhs.uk/explore-roles/public-health/roles-public-health/environmental-health-professional/skills-and-interests-environmental-health-professional
16 See Marmot MG. (2003) Understanding social inequalities in health. *Perspectives in Biology and Medicine*, 46(3 Suppl): S9–S23, Summer. PMID: 14563071.
17 See for example – Baker E, Beer A, Lester L, Pevalin D, Whitehead C, Bentley R. (2017) Is housing a health insult? *International Journal of Environmental Research and Public Health*, 14(6): 567. https://doi.org/10.3390/ijerph14060567
18 https://www.ph.ucla.edu/epi/snow.html#YOUTH
19 See for example https://www.bl.uk/victorian-britain/articles/health-and-hygiene-in-the-19th-century
20 See https://www.unodc.org/unodc/en/sustainable-development-goals/sdg11_-sustainable-cities-and-communities.html
21 See for example https://www.americanprogress.org/issues/poverty/reports/2020/10/30/492606/pandemic-exacerbated-housing-instability-renters-color/
22 See https://www.thoughtco.com/environmental-determinism-and-geography-1434499
23 See https://www.futurelearn.com/info/courses/remaking-nature/0/steps/16726
24 See https://www.ipcc.ch/2021/08/09/ar6-wg1-20210809-pr/
25 See United Nations-UN. Transforming our world: the 2030 Agenda for Sustainable Development. A/RES/70/1. 2015. https://www.un.org/ga/search/view_doc.asp?symbol=A/RES/70/1&Lang=E (Accessed 7 November 2021).
26 Covid-19: How to Coordinate a Global Response. While countries focused exclusively on their own needs, the virus conquered the world. https://magazine.jhsph.edu/2021/covid-19-how-coordinate-global-response (Accessed 7 November 2021).
27 See The pandemic's true death toll. Our daily estimate of excess deaths around the world. https://www.economist.com/graphic-detail/

coronavirus–excess–deaths–estimates. (Accessed 7 November 2021).

28 See What executives are saying about the future of hybrid work. https://www.mckinsey.com/business-functions/people-and-organizational-performance/our-insights/what-executives-are-saying-about-the-future-of-hybrid-work (Accessed 7 November 2021).

29 See Rewriting the future of work with hybrid workplaces. https://www.pwc.com.au/digital-pulse/report-future-of-work-hybrid-working.html (Accessed 7 November 2021).

30 This is taken and reproduced by agreement from https://www.who.int/news-room/fact-sheets/detail/climate-change-and-health

Chapter references

[1] Persaud A, Bhugra D, Valsraj K, Bhavsar V. (2021) Understanding geopolitical determinants of health. *Bulletin of the World Health Organization*, 99(2): 166–168. https://doi.org/10.2471/BLT.20.254904.

[2] Cole J, Dodds K. (2021) Unhealthy geopolitics? Bordering disease in the time of coronavirus. *Geographical Research*, 59: 169–181. https://doi.org/10.1111/1745-5871.12457.

[3] Smith SL, Shiffman J, Shawar YR, Shroff ZC. (2021) The rise and fall of global health issues: An arenas model applied to the COVID-19 pandemic shock. *Global Health*, 17(1): 33, 29 March. https://doi.org/10.1186/s12992-021-00691-7. PMID: 33781272. PMCID: PMC8006127.

[4] Dastan I, Abbasi A, Arfa C, et al. (2021) Measurement and determinants of financial protection in health in Afghanistan. *BMC Health Services Research*, 21: 650. https://doi.org/10.1186/s12913-021-06613-y.

[5] McDonald M. (2021) After the fires? Climate change and security in Australia. *Australian Journal of Political Science*, 56(1): 1–18. https://doi.org/10.1080/10361146.2020.1776680.

[6] Beilin R, Paschen J-A. (2021) Risk, resilience and response-able practice in Australia's changing bushfire landscapes. *Environment and Planning D: Society and Space*, 39(3): 514–533. https://doi.org/10.1177/0263775820976570.

[7] Baldwin A, Weiss S, Zib L, Delorit J, Chini C. (2021) Predicting lifecycle impacts of climate change on air force infrastructure. *The Military Engineer*, 113(735): 71–73. https://www.jstor.org/stable/48621916.

[8] Beilin R, Paschen J-A. (2021) Risk, resilience and response-able practice in Australia's changing bushfire landscapes. *Environment and Planning D: Society and Space*, 39(3): 514–533. https://doi.org/10.1177/02637758209 76570.

[9] Campbell K. (2021) The Australian defence force's response to the bushfire and COVID-19 crises of 2020. *United Service*, 72(1): 10–13. https://search.informit.org/doi/10.3316/informit.743977294558200.

[10] Brewster D. (2021) ADF health partnerships should be the next step in Australia's Pacific step-up. *The Strategist*. https://www.aspistrategist.org.au.

[11] UN International Organization for Migration. (2019) *World Migration Report 2020*, IOM, Geneva, Switzerland. https://publications.iom.int/system/files/pdf/wmr_2020.pdf.

[12] Anqi Lin ZZ, et al. (2021) Changes in the PM2.5-related environmental health burden caused by population migration and policy implications. *Journal of Cleaner Production*, 287. https://doi.org/10.1016/j.jclepro.2020.125051.

[13] Schwerdtle P, Bowen K, McMichael C. (2018) The health impacts of climate-related migration. *BMC Medicine*, 16(1). https://doi.org/10.1186/s12916-017-0981-7.

[14] Anholt R, Giulia Sinatti G. (2020) Under the guise of resilience: The EU approach to migration and forced displacement in Jordan and Lebanon. *Contemporary Security Policy*, 41(2): 311–335. https://doi.org/10.1080/13523260.2019.1698182.

[15] Hamadeh RS, Kdouh O, Hammoud R, et al. (2021) Working short and working long: Can primary healthcare be protected as a public good in Lebanon today? *Conflict and Health*, 15: 23. https://doi.org/10.1186/s13031-021-00359-4.

[16] Paixão S. (2021) SGDs, environmental health and gender equity. *European Journal of Public Health*, 31(Supplement_3), October. https://doi.org/10.1093/eurpub/ckab164.732.

[17] Coen S, Banister E. (Eds.) (2012) *What a Difference Sex and Gender Make: A Gender, Sex and Health Research Casebook*, Canadian Institutes of Health Research, Ottawa, Canada.

[18] Institute of Medicine (IOM). (2012) *Sex-Specific Reporting of Scientific Research: A Workshop Summary*, The National Academies Press, Washington, DC.

[19] Schwab PK, Samans R, Zahidi S, et al. (2017) *The Global Gender Gap Report 2017*, World Economic Forum, Geneva.

[20] Dhatt R, Theobald S, Buzuzi S, Ros B, Vong S, Muraya K, et al. (2017) The role of women's leadership and gender equity in leadership and health system strengthening. *Global Health, Epidemiology and Genomics*, 2: E8. https://doi.org/10.1017/gheg.2016.22.

[21] Destoumieux-Garzón D, Mavingui P, Boetsch G, et al. (2018) The one health concept: 10 years old and a long road ahead. *Frontiers in Veterinary Science*, 5: 14. https://www.frontiersin.org/article/10.3389/fvets.2018.00014; https://doi.org/10.3389/fvets.2018.00014.

[22] Dykstra MP, Baitchman EJ. (2021) A call for one health in medical education: How the COVID-19 pandemic underscores the need to integrate human, animal, and environmental health. *Academic Medicine*, 96(7): 951–953, 1 July. https://doi.org/10.1097/ACM.0000000000004072. PMID: 33769340.

[23] Baker E, Bentley R, Mason K. (2013) The mental health effects of housing tenure: Causal or compositional? *Urban Studies*, 50: 426–442.

[24] Oliver JR, Pierse N, Stefanogiannis N, Jackson C, Baker MG. (2017) Acute rheumatic fever and exposure to poor housing conditions in New Zealand: A descriptive study. *Journal of Paediatrics and Child Health*, 53: 358–364.

[25] Chantarat S, Barrett CB. (2012) Social network capital, economic mobility and poverty traps. *Journal of Economic Inequality*, 10: 299–342.

[26] Venkatapuram S. (2020) Human capabilities and pandemics. *Journal of Human Development and Capabilities*, 21(3): 280–286. https://doi.org/10.1080/19452829.2020.1786028.

[27] Benfer EA, Vlahov D, Long MY, et al. (2021) Eviction, health inequity, and the spread of COVID-19: Housing policy as a primary pandemic mitigation strategy. *Journal of Urban Health*, 98: 1–12.

[28] Shulla K, Voigt BF, Cibian S, et al. (2021) Effects of COVID-19 on the sustainable development goals (SDGs). *Discover Sustainability 2*, 15. https://doi.org/10.1007/s43621-021-00026-x.

[29] Nana Addo Dankwa AA, Solberg E. (2020) Op-ed 'Amid the Coronavirus Pandemic, the SDGs Are Even More Relevant Today Than Ever Before'. https://www.un.org/sustainabledevelopment/blog/2020/04/coronavirus-sdgs-more-relevant-than-ever-before/ (Accessed 7 November 2021).

[30] WHO. (2021) *WHO Director-General's Remarks at Session 3 – Sustainable Development at the G20 Summit*, 31 October. https://www.who.int/director-general/speeches/detail/who-director-general-s-remarks-at-session-3-sustainable-development-at-the-g20-summit--31-october-2021 (Accessed 7 November 2021).

[31] New Research Calls for Hybrid Working to Be More Sustainable. https://phys.org/news/2021-10-hybrid-sustainable.html (Accessed 7 November 2021).

[32] OECD. (2021) *Teleworking in the COVID-19 Pandemic: Trends and Prospects*. https://www.oecd.org/coronavirus/policy-responses/teleworking-in-the-covid-19-pandemic-trends-and-prospects-72a416b6/ (Accessed 7 November 2021).

[33] Williamson S, Pearce A, et al. (2021) *Future of Work Literature Review: Emerging Trends and Issues*, Report prepared for the Australian Tax Office and Department of Home Affairs, UNSW Public Service Research Group. https://apo.org.au/node/314497 (Accessed 7 November 2021).

[34] Zhang S, Moeckel R, Moreno AT, Shuai B, Gao J. (2020) A work-life conflict perspective on telework. *Transportation Research: Part A, Policy and Practice*, 141: 51–68. https://doi.org/10.1016/j.tra.2020.09.007.

[35] The Royal Commission into National Natural Disaster Arrangements. (2020) *Report*. https://naturaldisaster.royalcommission.gov.au/publications/html-report/chapter-02 (Accessed 7 November 2021).

[36] Australian Institute of Health and Welfare (AIHW). (2020) *Australian Bushfires 2019–20: Exploring the Short-Term Health Impacts*, Cat. no. PHE 276, AIHW, Canberra. https://www.aihw.gov.au/reports/environment-and-health/short-term-health-impacts-2019-20-bushfires/contents/summary (Accessed 7 November 2021).

[37] Milton LA, White AR. (2020) The potential impact of bushfire smoke on brain health. *Neurochemistry International*, 139: 104796. ISSN:0197-0186. https://doi.org/10.1016/j.neuint.2020.104796.

[38] Walter CM, Schneider-Futschik EK, Knibbs LD, Irving LB. (2020) Health impacts of bushfire smoke exposure in Australia. *Respirology*, 25: 495–501. https://doi.org/10.1111/resp.13798.

[39] Wang G, McDonald VM. (2020) Contemporary concise review 2020: Asthma. *Respirology*, 26: 804–811. https://doi.org/10.1111/resp.14099.

[40] MacIntyre CR, Nguyen PY, Trent M, Seale H, Chughtai AA, Shah S, Marks GB. (2021) Adverse health effects in people with and without preexisting respiratory conditions during bushfire smoke exposure in the 2019/2020 Australian Summer. *American Journal of Respiratory and Critical Care Medicine*, 204(3): 368–371, 1 August. https://doi.org/10.1164/rccm.202012-4471LE. PMID: 33975534.

[41] Borchers Arriagada N, Palmer AJ, Bowman DMJS, Morgan GG, Jalaludin BB, Johnston FH. (2020) Unprecedented smoke-related health burden associated with the 2019–20 bushfires in Eastern Australia. *Medical Journal of Australia*, 213: 282–283.

[42] Tong S, Ebi K. (2019) Preventing and mitigating health risks of climate change. *Environmental*

Research, 174: 9–13. ISSN:0013-9351. https://doi.org/10.1016/j.envres.2019.04.012.

[43] Welter CR, et al. (2020) Increasing environmental public health practitioner capacity to address population health challenges: Evaluation results from a workforce development project. *Journal of Environmental Health*, 82(10), June. https://www.neha.org/node/61377.

Index

Note: page numbers in *italics* indicate a figure and page numbers in **bold** indicate a table on the corresponding page. Page numbers followed by 'n' indicate a note.